The Cretaceous-Tertiary Event and Other Catastrophes in Earth History

Edited by

Graham Ryder
Lunar and Planetary Institute
3600 Bay Area Boulevard
Houston, Texas 77058-1113

David Fastovsky
Department of Geology
University of Rhode Island
Kingston, Rhode Island 02881

and

Stefan Gartner
Department of Oceanography
Texas A&M University
College Station, Texas 77843

SPECIAL PAPER

307

1996

Published by The Geological Society of America, Inc.
3300 Penrose Place, P.O. Box 9140, Boulder, Colorado 80301

Printed in U.S.A.

GSA Books Science Editor Abhijit Basu

Library of Congress Cataloging-in-Publication Data

The Cretaceous-Tertiary event and other catastrophes in earth history / edited by Graham Ryder, David Fastovsky, and Stefan Gartner.
p. cm. — (Special paper / Geological Society of America ; 307)
Includes bibliographical references and index.
ISBN 0-8137-2307-8
1. Cretaceous-Tertiary boundary. 2. Catastrophes (Geology) 3. Extinctions (Biology) I. Ryder, Graham. II. Fastovsky, David E. III. Gartner, Stefan, 1937- . IV. Series: Special papers (Geological Society of America) ; 307.
QE689.C75 1996
551.7—dc20 96-28012
CIP

Cover: Photomicrograph of a fine-grained impact breccia recovered from the Pemex Yucatan No. 6 well near the center of the Chicxulub impact crater in Yucatan. This core interval was extracted from a depth of 1,100–1,103 m below sea level. Because it contains well-preserved Upper Cretaceous foraminiferids such as the one located left of center, this rock unit had been interpreted as a Cretaceous limestone deposited on top of the impact-generated rocks and used as evidence that the Chicxulub impact did not form at the Cretaceous-Tertiary boundary. Clear indications of shock deformation, however, such as the quartz grain containing multiple sets of planar deformation features (right of center) and shards of impact-generated glass (e.g., below the bioclast) provide compelling evidence that this unit is actually part of the impact-generated rock sequence. Width of field is ~6mm. For additional information, see Sharpton et al., this volume.

10 9 8 7 6 5 4 3 2 1

Contents

Preface

Extinction is the fate of all species. In the Phanerozoic rock sequences of Earth history, there are abrupt faunal and floral transitions. Although some early workers, including Lyell and Darwin, believed that all these transitions represented gaps in the geological record, many of them are now interpreted as the record of short periods with extinction rates much greater than average: mass extinctions. Not surprisingly, they have been the subject of considerable attention and debate about both their reality and their cause or causes.

In 1980 the debate over the nature and causes of catastrophic extinction events took a dramatic turn and came under increased scrutiny. The catalyst was an extraordinary paper published in *Science* by Luis Alvarez and his colleagues at the University of California, Berkeley. That paper reported on measurements of tremendously enhanced abundances of iridium and other siderophile elements in the thin clay layer at the paleontologically identified boundary between the Cretaceous and the Tertiary divisions of geological time. The authors inferred that a large chondritic body had collided with the Earth and created a stratospheric global dust cloud that blocked sunlight and, along with other stresses, caused extinctions as food chains collapsed.

The paper ushered in a new era in which the concept of extraterrestrial bodies impacting the Earth, creating massive and sudden upheavals in the biosphere, and causing mass extinctions was at the forefront, to be investigated, scrutinized, criticized, and tested. A significant feature of this new era has been the multidisciplinary nature of the pursuit of scientific knowledge, bringing together not only traditional geological subdisciplines but specialists in climatology, cosmochemistry, impact physics, and planetary geology. The ramifications and philosophical implications are much wider than just the Cretaceous-Tertiary boundary itself, extending to the history of life on Earth in general. The term "K/T boundary," for better or worse, became commonplace in geological jargon.

In 1981, the Conference on Large Body Impacts and Terrestrial Evolution: Geological, Climatological, and Biological Implications was held at Snowbird, Utah (commonly referred to as Snowbird I). The proceedings of this conference were published as Geological Society of America Special Paper 190, *Geological Implications of Impacts of Large Asteroids and Comets on the Earth*, edited by Leon T. Silver and Peter H. Schultz. At the time, scientists were still coming to grips with the concept of a large impact causing extinctions and how an impact might have accomplished extinction, as well as beginning to accumulate evidence for it. Snowbird I participants in general agreed that the chemical traces at the Cretaceous-Tertiary boundary were probably those of an impact and that there was global evidence for it in both marine and terrestrial rock sequences. But participants believed that its consequences had not been as drastic as the three years of darkness originally proposed; the effects had not been that dire, and the physics were consistent with shorter-lived stresses. However, a sizable community rejected the reality of an impact, or the reality of a mass extinction, or both.

In 1988, a second conference, Global Catastrophes in Earth History: An Interdisciplinary Conference on Impacts, Volcanism, and Mass Mortality, was also convened at Snowbird (Snowbird II). The proceedings of this second conference, with the same name as the meeting and edited by Virgil L. Sharpton and Peter D. Ward, were published as Geological Society of America Special Paper 247. By the time of the Snowbird II conference, there had been many significant developments. One was the recognition in boundary clays of quartz containing multiple sets of planar deformation features, and another was the high-pressure mineral stishovite. These were seen as diagnostic criteria of impact by many scientists. Many boundary deposits also contained silicate melt droplets or pseudomorphs of them. Another development had been claims of periodicity in extinctions, leading to debate about the reality of the periodicity, and proposals of causes ranging from Earth generated to "death stars."

One dominant theme at the conference was whether the evidence favored an impact at the Cretaceous-Tertiary boundary or could be explained by terrestrial volcanism, a hypothesis that had several vocal adherents. Although consensus at the end of the conference was that volcanism was inconsistent with the evidence, that school of thought persists even today. (Volcanologists are not a part of the community proposing a volcanic origin.) A continuing question is that of the faunal transition itself: Just what happened and when, or how sharp is the boundary? The 1988 conference focused mostly on the Cretaceous-Tertiary boundary. No substantiated claims for identifying the necessary crater existed; the buried Manson crater in Iowa appeared to be the correct age, but there was overwhelming agreement that it was too small.

In the period following Snowbird II, two dramatic developments made it clear that another conference was desirable. The first was the recognition of unusual deposits at the Cretaceous-Tertiary boundary in the regions around the Gulf of Mexico, deposits that were postulated by some to be the result of tremendous tsunamis such as might have been caused by an impact. The second was the recognition and identification of an extremely large crater buried in the Yucatan, following an increase in attention on the Caribbean region as a probable site for the impact that was becoming widely accepted. The stratigraphic age of this crater, Chicxulub, was consistent with a Cretaceous-Tertiary boundary age. Impact melt samples retrieved from existing cores were dated as precisely boundary age using argon isotopic techniques. The wealth of research on the Chicxulub impact basin and the unusual boundary deposits in the region inspired widespread enthusiasm for a "Snowbird III" meeting. Accordingly, the Lunar and Planetary Institute and the University of Houston–Clear Lake cosponsored a conference to allow interdisciplinary discussion and critical review of the new findings, as well as to examine evidence related to other mass extinction events.

The Snowbird III conference, officially entitled New Developments Regarding the K/T Event and Other Catastrophes in Earth History, was held at the University of Houston–Clear Lake, Houston, Texas, February 9–12, 1994. The organization was aimed at, and was successful in producing, uninhibited discussion and wide participation by the 191 attendees. Examining mass extinction perspectives in general, the program of 153 contributed and invited papers, many in the form of posters, focused on Chicxulub and the Gulf of Mexico/Caribbean Cretaceous-Tertiary boundary deposits. It also included discussion of Cretaceous-Tertiary boundary deposits elsewhere; of other mass extinctions and boundaries; of the paleontological record; of killing mechanisms; and of the physics, chemistry, and geology of large impact events. Preceding the conference, approximately 50 attendees participated in a field trip (organized in conjunction with the conference) to the unusual deposits of Cretaceous-Tertiary boundary age in northeast Mexico.

At the conference there was consensus that an impact had occurred at the boundary. Proponents of a volcanic influence on boundary events were not vocal. Nonetheless, others denied that the extinctions were caused by the impact, with the argument that the extinctions were "stepped" and not of a pattern consistent with impact. Similarly, not everyone agreed that the boundary deposits in the Gulf of Mexico/Caribbean region were related to an impact, and some thought that they might not be precisely at the boundary. At issue are the precise timings and patterns of the extinctions of species.

This volume consists of 39 peer-reviewed papers appropriate to the subject matter of the conference. Most, but not all, evolved from papers presented at the conference; the editors deemed it reasonable to consider relevant papers submitted that had not been presented. The volume is broad in scope, and the review process was extensive. The papers are arranged in sections reasonably consistent with the subdivisions of the conference itself, but coverage of the topics is somewhat uneven. The review process was handled by the staff of the Lunar and Planetary Institute's Publications and Program Services department, in particular Renee Dotson, Susan Boatright, and Linda Tanner, who also kept the editors informed of problems and discrepancies and encouraged them to solve them in a timely fashion. Production of the volume was handled by the Geological Society of America.

The Scientific Organizing Committee for Snowbird III consisted of Walter Alvarez (University of California, Berkeley), David Black (Lunar and Planetary Institute), Kevin Burke (University of Houston), Robert Ginsburg (University of Miami), Lee Hunt (National Academy of Sciences), Gerta Keller (Princeton University), Luis Marín (Universidad Nacional Autónoma de México, cd. Universitaria), David Raup (University of Chicago), Virgil Sharpton (Lunar and Planetary Institute), Eugene Shoemaker (U.S. Geological Survey, Flagstaff), and Jerardo Suarez (Universidad Nacional Autónoma de México, cd. Universitaria). The Program Committee, which was responsible for reviewing and accepting abstracts and organizing them in some coherent manner, consisted of Walter Alvarez (University of California, Berkeley), Jody Bourgeois (National Science Foundation), Kevin Burke (University of Houston), Gerta Keller (Princeton University), Christian Koeberl (University of Vienna), Jose Longoria (Florida International University), Graham Ryder (Lunar and Planetary Institute), Benjamin Schuraytz (Lunar and Planetary Institute), Virgil Sharpton (Lunar and Planetary Institute), and Peter Ward (University of Washington). Logistics and administrative support for the conference were provided by the staff of the Lunar and Planetary Institute's Publications and Program Services Department, in particular Lita Holley, Sarah Enticknap, and LeBecca Simmons. The field trip agenda was scientifically arranged largely by Gerta Keller (Princeton University) and Wolfgang Stinnesbeck (Universidad Autónoma de Nuevo Leon), and the latter was responsible, along with the Lunar and Planetary Institute staff, for arranging the logistics of the field trip.

We look forward to "Snowbird IV," which will undoubtedly focus on more detailed characterization of the Chicxulub impact structure and event and how it (or a large impact in general) actually provided extinction mechanisms at the Cretaceous-Tertiary boundary. However, even at Snowbird IV there will be opponents of at least some aspects of impact-derived mechanisms. Perhaps by then we will have a more effective understanding of the character of other mass extinctions and catastrophes in the geological record as the Cretaceous-Tertiary boundary continues to stimulate research.

Graham Ryder, Lunar and Planetary Institute
David Fastovsky, University of Rhode Island
Stefan Gartner, Texas A&M University

Geological Society of America
Special Paper 307
1996

Mass extinctions: Persistent problems and new directions

David Jablonski
Department of the Geophysical Sciences, University of Chicago, 5734 South Ellis Avenue, Chicago, Illinois 60637

ABSTRACT

Many of the contentious debates on mass extinctions center of the fine temporal fabric of an extinction and the resulting implications for causal mechanisms. The controversies commonly reside in problems of resolution, correlation, and sampling, but the pitfalls to taking local or even regional data at face value are still little appreciated, resulting in massive overinterpretation of paleontological patterns. Statistical techniques are increasingly becoming available to address these problems. Linking extinction patterns to potential forcing factors is difficult because of the nature of complex biological systems: nonlinearities, thresholds, and elaborate feedbacks often rule out the reconstruction of simple cause-and-effect cascades. The same forcing factor might have different effects depending on the state of the system at the time of perturbation, and several alternative mechanisms might produce the same biotic response. Hypotheses should be framed in a way that takes this problem explicitly into account.

Patterns of selectivity are also coming under increasing scrutiny, not only as a means of eliminating hypothesized forcing factors but also as an important component of large-scale evolutionary change. Many biotic factors that enhance survivorship during times of background extinction are ineffectual during mass extinctions. Preliminary analyses for end-Cretaceous marine mollusks corroborate this contrast, showing no selectivity of genera by life position (burrowing versus exposed), bathymetric position on the continental shelf, or relative breadth of bathymetric range. Deposit-feeders have significantly lower extinction intensities than suspension-feeders, but some deposit-feeding lineages suffer losses as severe as those seen for suspension-feeders. Either different styles of deposit-feeding varied in vulnerability to the K/T event, or feeding types are not really the critical factor. Mass extinctions do not entirely disrupt large-scale evolutionary patterns, and more research is needed on continuity of major trends across extinction boundaries and postextinction rebounds.

INTRODUCTION

Our knowledge of mass extinctions has increased enormously in the decade since the first Snowbird conference on the geological implications of asteroid and comet impacts, held in 1981. Paleobiologically, new data range from improved and enlarged synoptic databases to detailed outcrop- and core-scale studies of specific events. Despite these advances, some aspects of mass extinctions have stubbornly resisted resolution. Here I will discuss some of these persistent problems and some general strategies for addressing them. I will also survey some of the questions that paleobiologists are beginning to explore in new ways, from the complexity of selectivity patterns to the nature of evolutionary recoveries.

Few workers contest that mass extinctions have punctuated the history of life or that those events were so pervasive environmentally, taxonomically, and geographically that physical forcing factors must have been involved in some way. However, consensus breaks down on the nature of those factors and on how a given perturbation (impact, volcanism, sea-level change,

Jablonski, D., 1996, Mass extinctions: Persistent problems and new directions, *in* Ryder, G., Fastovsky, D., and Gartner, S., eds., The Cretaceous-Tertiary Event and Other Catastrophes in Earth History: Boulder, Colorado, Geological Society of America Special Paper 307.

oceanic anoxic event) could actually generate the observed intensity and selectivity of biotic losses. At least two basic problems underlie these long-standing disagreements: (1) limitations to resolving the fine temporal details of taxon ranges and abundances immediately prior to and after an extinction boundary and (2) the scarcity of simple, unitary cause-and-effect relations in complex biological systems. The latter problem in turn relates to difficulties in pinpointing agents of extinction selectivity, even when such patterns are detected.

LIMITS TO RESOLUTION

The fossil record of shelly marine invertebrates is the most complete and reliable for the analysis of global diversity patterns; I will focus on this part of the record, although many of the points developed here might be generalized. The past decade of intensive work, resulting in many additions, corrections, and reinterpretations of taxonomy and stratigraphy, has only sharpened the major extinction events detected in the geologic record (see Sepkoski, 1993) (Fig. 1). With so much effort focused on the K/T event, we tend to forget that there were three extinctions of similar magnitude earlier in the Phanerozoic and that the Permo-Triassic extinction at the end of the Paleozoic dwarfed all others (see Erwin, 1993, 1994). Besides the Big Five, a number of smaller events are detectable in synoptic time-series but are not necessarily of sufficient scale geographically or taxonomically to be called mass extinctions. Some, however, such as the mid-Carboniferous extinction pulse, *do* appear to be genuine but less severe global events (Sepkoski, 1996).

Higher taxa, such as the families plotted in Figure 1, are more inclusive and thus give damped results relative to the underlying species-level patterns (Valentine, 1973; Sepkoski, 1978, 1991; Raup, 1979a). The vagaries of sampling at the species level add so much noise, however, that species extinction intensities on the global scale are usually estimated indirectly. For example, Raup's (1979b) "reverse rarefaction" technique starts with the observed family or genus loss, then uses the distribution of species within the present-day families and genera (of echinoids) to calculate the magnitude of species loss required to remove that many higher taxa assuming a random species kill. This approach carries several assumptions that require exploration. However, the convergence of estimated species losses calculated from both genus- and family-level data for each of the Big Five mass extinctions suggests that Raup's approach is fairly robust to differences in taxonomic structure, duration of the event relative to stage length, and perhaps even nonrandom species kills (Jablonski, 1995a).

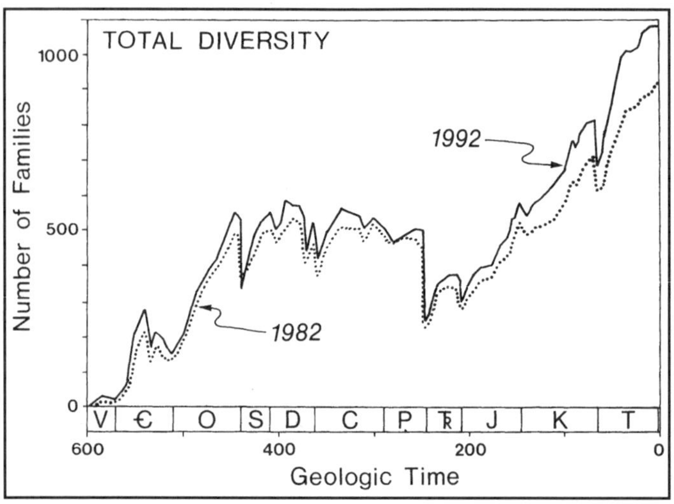

Figure 1. Comparison of Sepkoski's 1982 and 1992 diversity curves for marine families (from Sepkoski, 1993). Despite extensive changes in taxonomy and stratigraphic ranges of individual families, the overall patterns, including the mass extinction events, have remained constant (goodness-of-fit, $r^2 = 0.990$). Copyright 1993 by the Paleontological Society, and used by permission.

Local outcrops and cores are the ultimate source of the data used to analyze mass extinctions, but the pitfalls to taking local data at face value are still underappreciated. The result has sometimes been a massive overinterpretation of local and even regional paleontological patterns. At any given site, the accumulation of sediments and fossils is discontinuous, and environmental change occurs constantly at many scales. As a result, smearing of events, stepped extinction patterns, and artificially truncated ranges are inevitable on some scale in any local sequence. This is a double-edged sword, of course: as Birkelund and Hakansson (1982) pointed out at the first Snowbird meeting, stratigraphic gaps can produce artificially abrupt events just as readily as smaller breaks can produce artificially gradual or stepped extinctions (Signor and Lipps, 1982; Raup, 1989). The extinctions may have been *genuinely* abrupt or *genuinely* stepped, but the pattern must be separated from the potential biases imposed by the stratigraphic record (see also Flessa, 1990).

Also underappreciated are the processes of time-averaging and reworking, in which fossils from successive time intervals are mixed by biological and physical agents (Kidwell and Behrensmeyer, 1993). Rough estimates of the scale of these processes based on stratigraphic and taphonomic criteria (Kidwell and Bosence, 1991; Kidwell, 1993) are corroborated by radiocarbon dates for hard parts collected at the sediment surface today. For example, median ages of molluscan shells collected at the *surface* of continental shelf sediments and of nearshore sediments are 9,190 and 1,250 yr respectively (Flessa, 1993; Flessa and Kowalewski, 1994). Using the same approach, typical resolution is also 10^3 to 10^4 yr for assemblages of planktic and benthic foraminifera in deep-sea sediments (Broecker et al., 1990; Martin, 1993). Reworking and mixing can sometimes occur over vertical distances equivalent to far longer time spans, as older material is attenuated through multiple time-averaging windows (see Cutler, 1993). At a particular locality, the situation will depend on several variables, most obviously net sedimentation rate, and modern shelves may be particularly rich in relict hard parts due to Pleistocene sea-level oscillations.

However, to an order of magnitude these limits on time resolution probably hold over a wide range of conditions, at least since the mid-Mesozoic. The average scale of time-averaging may have changed through the Phanerozoic with the evolution of hard-part producers, destroyers, and bioturbators (Kidwell, 1993), but such effects have yet to be quantified. Ironically, resolution within single sections may be best in the oldest rocks, where bioturbation was least intense (Droser and Bottjer, 1993), but preservation there is generally poorest and absolute dating least precise.

Statistical protocols are available to place confidence limits on stratigraphic ranges and thereby test for artificial extinction patterns (see reviews by Marshall, 1991, and Sepkoski and Koch, 1996). These methods still have seen little use in the mass extinction literature: Most workers understandably want to take their hard-won data at face value and as virtually devoid of local overprint. In some instances they may be right, but it is very difficult to pinpoint such instances without testing patterns against an appropriate null hypothesis. To date, few statistical treatments of local or regional extinction patterns have been able to verify the stepwise or gradual patterns that sometimes emerge from a literal interpretation of the stratigraphic data (e.g., Jablonski, 1986a; Raup, 1989; Springer, 1990; Sheehan et al., 1991; Koch, 1991). The assumptions underlying most statistical approaches remain somewhat restrictive (e.g., that recovery probabilities are stochastically constant throughout a section), but several workers are reportedly close to developing more general analytical methods for putting confidence limits on stratigraphic range endpoints. These approaches will open the door to assessing apparent extinction horizons with a rigor that has been difficult to attain previously. Even now these new methods can sharpen our intuition about what we can and cannot infer from local extinction patterns.

CAUSE AND EFFECT

One of the most difficult tasks facing any biologist, and especially the paleobiologist lacking recourse to experimental manipulations, is inferring process from pattern. Biological systems are so complex, and so rich in nonlinearities, threshold effects, and elaborate feedbacks, that the thread of causation is often difficult to follow. Of particular concern to extinction analyses, a single forcing factor might have radically different effects depending on the state of the system at the time of perturbation.

Several authors have suggested that the initial physical and biological conditions prior to an external shock such as an impact may determine the style and magnitude of biotic losses (e.g., Kauffman, 1986; Jablonski, 1986a). For example, if relative sea level is high, so that the thermal inertia and other ameliorating effects of shallow seas are prominent, physical disturbances and taxonomic losses from a given perturbation might be milder than when sea level is relatively low. This suggestion has yet to be modeled rigorously, let alone tested empirically, but the probable role of initial conditions may help to explain why geologically detectable impacts such as those that formed the Late Campanian Manson crater (Izett et al., 1993), the Early Eocene Montagnais crater (Aubry et al., 1990; Jansa et al., 1990; Jansa, 1993), and the Late Pliocene tektite fields (e.g., Kyte et al., 1988; Schmidt et al., 1993) evidently had negligible biotic effects. A simpler alternative would be a threshold impactor size for biotic effects, so that the shape of the "kill curve" (Raup, 1991a) relating biotic loss to size of impactor is the overriding factor. The recalibration of the kill curve has far-reaching implications: Jansa (1993) argues, for example, that the flux of impactors large enough to cause significant extinction is far too low to sustain the 26 m.y. periodicity reported by Raup and Sepkoski (1986), Sepkoski (1990), and many other papers). It is difficult to test these two hypotheses—initial conditions versus threshold effects—given the relatively small number of events and the large number of alternative models. The problem is further complicated, of course, because the two hypotheses need not be mutually exclusive.

Not only might a single forcing factor yield different outcomes, but several alternative forcing factors might yield the same outcome. The survival of an evolutionary lineage during a mass extinction, for example, could be due to (1) its living in a habitat that was not stressed, (2) its possessing a physiology or life habit that allowed it to survive in a stressed habitat, (3) its being so widespread that its range included a locality that provided a refuge, and so on. Urgently needed is not another catalog of potential reasons for survival or extinction, but the development of protocols for testing the alternatives or, more realistically given the complexities of biotic systems, for ranking different factors. This is only partly a matter of constructing large, robust databases amenable to statistical analysis and of coming to grips with the nontrivial problem of integrating local and synoptic databases. A basic methodological change is also needed, with a shift from consistency arguments to more explicitly framed rival hypotheses. Workers must realize, for example, that with sufficient tuning the long list of potential impact-related killing mechanisms can account for virtually any conceivable extinction pattern. Therefore, while the mere fact of observed selectivity clearly is no argument against impacts, neither is it an argument uniquely in their favor. Physicists and paleobiologists alike are only now beginning to move beyond consistency arguments: Hypothesized extinction mechanisms must generate unique predictions on the timing, selectivity, or other biotic patterns of extinction if they are to advance beyond the status of plausible alternatives.

SELECTIVITY

The search for patterns of differential extinction among taxa of high rank, such as phyla and classes, has been largely unsatisfying (e.g., McKinney, 1987; Sepkoski, 1990). Major groups—for example, bivalves versus cephalopods—do seem to differ broadly in background turnover rates, and losses during mass extinction may vary with the background volatility of a given group (e.g., Sepkoski, 1984; Van Valen, 1985; McKinney, 1987; Raup and Boyajian, 1988; Valentine, 1990). However, differences among taxa of class rank are difficult to relate to biological variables, and within-group variances are large, as seen for the range of genus durations in ammonites (e.g., Ward and Signor, 1983). More coherent patterns of selectivity tend to occur mainly at lower taxonomic levels (families and genera) and/or along biogeographic and ecologic lines (see Van Valen's 1994 thoughtful review). These parameters are largely absent from synoptic databases.

Once a selectivity pattern is detected, the problem remains of inferring the selective agent from the array of alternatives. For example, many authors have argued that mass extinctions are more severe in the tropics than at higher latitudes (e.g., Stanley, 1984, 1988). The patterns are most striking in reef and related habitats, for example, the differential survivorship of shallow-water Late Devonian corals relative to deep-water genera (Sorauf and Pedder, 1986; Sorauf, 1989), and for symbiont-bearing K/T corals and rudist bivalves relative to nonsymbiotic corals and nonrudist bivalves (Rosen and Turnsek, 1989). Inferring the forcing mechanism behind this pattern is not straightforward, however, and Jablonski outlines (1995a) at least four potential mechanisms that could shape the apparent latitudinal trend in mass extinction intensities. These include low tolerances of tropical biotas to climatic excursions, fragile ecological networks in diverse tropical communities, presence of high proportions of extinction-prone endemics in the tropics, and high vulnerability to disruption of low-nutrient, low-sedimentation carbonate platforms or ramps.

The various tropical-temperate contrasts that underlie these four alternatives are poorly known and may even be erroneous, but paleontological data lend support to the last two possibilities. Regarding endemicity, few would contest that tropical provinces contain the most endemics, and many authors have indeed found that geographically restricted genera suffer more severely than do widespread ones during mass extinctions (e.g., Bretsky, 1973; Anstey, 1986; Jablonski, 1989a; Sheehan and Coorough, 1990; Robertson et al., 1990; Westrop, 1991; Erwin 1989, 1993). A similar argument might help to explain claimed latitudinal trends in survivorship of K/T planktic foraminifera, whose polar assemblages tend to be dominated by extinction-resistant cosmopolitan taxa (cf. Keller, 1993; Keller et al., 1993). More work is needed, however, to determine the generality of this approach: For example, is the contrast in biogeographic structure of tropical versus extratropical faunas sufficient to account for the quantitative pattern of extinction differentials?

The K/T patterns are complicated by the presence of what appears to be a relict, endemic fauna of mollusks in the Arctic Ocean containing perhaps a half-dozen holdover genera in the earliest Tertiary, otherwise known only from the Cretaceous (Marincovich, 1993). A similar situation may occur, although not at such high latitudes, in the earliest Silurian brachiopod fauna of the Oslo region, Norway (Baarli and Harper, 1986). Whether these short-lived survivors reflect unusual paleoceanographic conditions, geographic or habitat refugia, or generally milder extinction intensitites in polar regions is unclear.

Regarding carbonate platforms (that is, selectivity according to habitat type rather than by tropical distribution per se), Raup and Jablonski (1993) found that excluding platform-dwelling rudists from a global analysis of K/T bivalve extinction produced a remarkably homogeneous array of local extinction intensities, with tropical noncarbonate assemblages suffering no greater losses than those of high latitudes. Even with the few relicts in the Arctic Ocean, no statistically significant latitudinal trends are apparent; Antarctic extinction intensities (Zinsmeister et al., 1989) are within the statistical expectation for the number and size of these samples and involve late Maastrichtian losses of over 40%, well above background levels.

These analyses represent just a few steps toward the dissection of extinction patterns, but they represent an approach that offers promise for dealing with the complexity of biotic systems. Many other instances of extinction selectivity present comparable challenges. Differential K/T survivorship of Northern Hemisphere marsupials versus placental mammals, for example, is unlikely to derive from the reproductive mode defining these mammalian subclasses, but probably arises from some other aspect of ecology or biogeography; the milder losses reported for Southern Hemisphere marsupials (Case and Woodburne, 1988) would support this interpretation and deserves further work.

Narrow geographic ranges and tropical platform habitats aside, pinpointing selectivity during mass extinctions has been difficult. Several factors important to taxon survivorship during times of background extinction play no significant role during mass extinctions. For example, species-rich lineages tend to survive preferentially during normal times, but during mass extinctions species-richness is not a significant buffer against extinction (Jablonski, 1986b, 1989a, 1995a); the one documented exception is Permo-Triassic gastropods (Erwin, 1989).

End-Cretaceous bivalves continue to be a fertile testing ground for patterns of selectivity (Jablonski and Raup, 1995). Several authors have recorded differences in background extinction between infaunal marine organisms (e.g., bivalves like quahogs and cherrystones that burrow within the sediment) and those that live epifaunally (e.g., scallops and oysters) (Kauffman, 1978; Stanley, 1986; see also McGhee [1989] on the Late Devonian mass extinction). When we exclude the platform-dwelling rudists from our analysis (because this group appears to be a special case and, according to some authors [e.g., Kauffman, 1986], was in precipitous decline over the last few million years of the Cretaceous), we see no significant difference between K/T extinction intensities in infauna and epifauna (Table 1).

Body size is an important parameter in population biology (e.g., Peters, 1983; Calder, 1984), but for K/T bivalve genera, victims and survivors do not differ significantly in the fre-

TABLE 1. PRELIMINARY ANALYSIS OF VARIATION IN END-CRETACEOUS EXTINCTION INTENSITIES ACCORDING TO LIFE HABITS IN MARINE BIVALVE GENERA

Functional Group	Extinction Intensity (%)
I. Living position	
Epifauna (n = 99)	54 ± 10
Infauna (n = 198)	58 ± 7
II. Feeding strategy	
Suspension feeders (n = 250)	61 ± 6
Deposit feeders (n = 40)	30 ± 14
Nuculoida (n = 14)	15 ± 14
Tellinidae (n = 12)	75 ± 12
Lucinacea (n = 15)	20 ± 19
Aporrhaid gastropods (n = 25)	76 ± 17

Analyses exclude the rudists, which dwelt primarily in tropical carbonate platforms and suffered complete extinction at or near the K/T boundary; nuculoids here include superfamilies Nuculoidea and Nuculanoidea. Extinction intensities in aporrhaid gastropods from Roy, 1994. Confidence limits (95%) calculated using the method of Raup, 1991b.

quency distribution of sizes (Fig. 2). This result contrasts with several studies that conclude, on theoretical or empirical grounds, that large-bodied forms are more vulnerable to background extinction (e.g., Stanley, 1986, 1990; Pimm et al., 1988, McKinney, 1990; Brown et al., 1993; but see Martin, 1992). For the end-Cretaceous event, however, no significant survivorship differences are evident for marine bivalves, even for taxa <1 cm, which are most likely to have annual life cycles and thus be maximally vulnerable to short-term excursions in the ocean and atmospheric composition (Jablonski and Raup, 1995; see also Valentine and Jablonski, 1986).

We also find little evidence in K/T bivalves for habitat-based selectivity within a single region. Extinction intensities are statistically indistinguishable across the shelf in the Gulf and Atlantic Coastal Plain, where environmental data are most detailed (Fig. 3). Bivalve and gastropod genera that extend into the shallowest waters do no better, as a group, than those penetrating outer shelf habitats or those restricted to intermediate depth zones. Extinction intensities do not differ significantly for genera found in a single depth zone versus those occurring in two, three, or all four depth zones across the shelf (Fig. 4). Sepkoski (1987) found a similar lack of bathymetic pattern for the Paleozoic mass extinctions, although Erwin (1989) found that gastropod genera with broad environmental distributions (encompassing both basinal and platform settings) weathered the end-Permian mass extinction better than those with narrow distributions. The apparent contrast with the higher extinction intensities noted in shallow-water benthos during the Late Devonian mass extinction (e.g., Sorauf and Pedder, 1986; McGhee, 1989) may arise because the Late Devonian pattern is shaped mainly by (1) high losses in low-latitude reefal and peri-

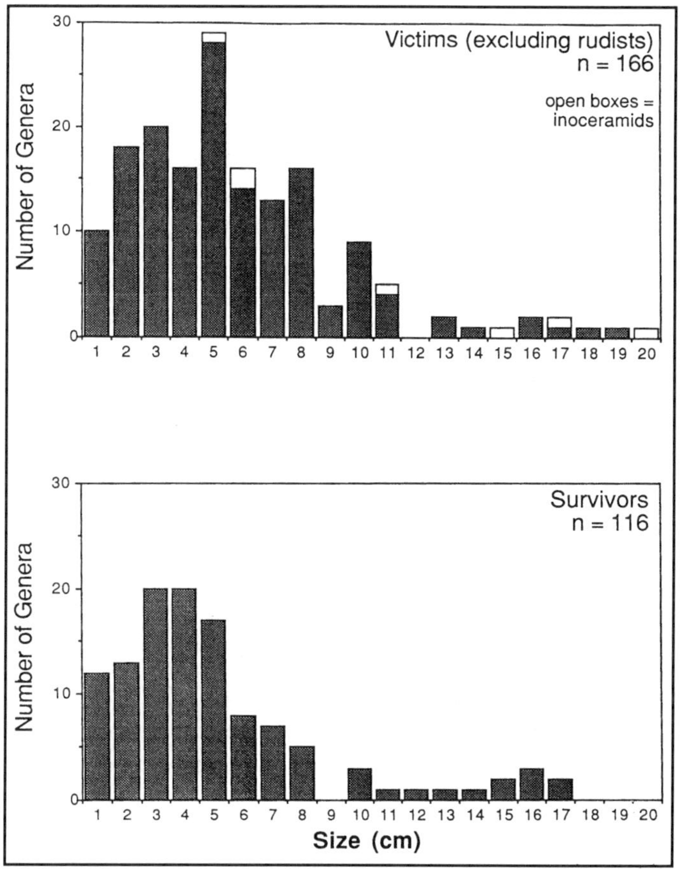

Figure 2. End-Cretaceous bivalve genera exhibit no significant differences in survivorship according to body size (p > 0.10, Kolmogorov-Smirnov test). Rudists excluded from analysis. Body size calculated (following Stanley, 1986) as the geometric mean of height and width of the largest recorded specimen of each genus.

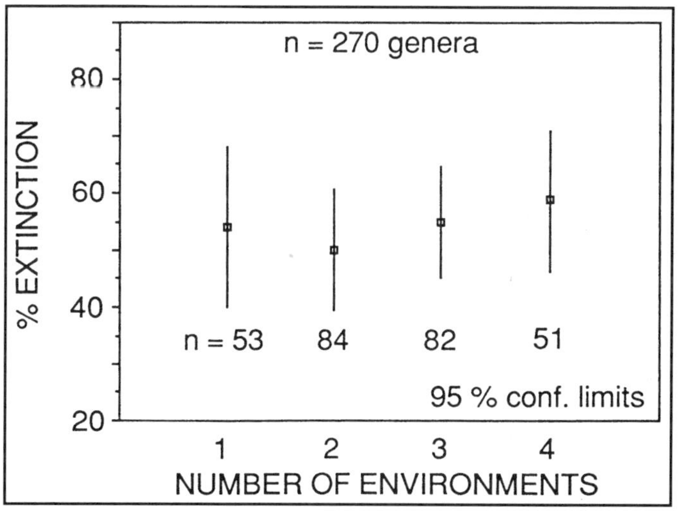

Figure 3. For end-Cretaceous bivalves and gastropods of the Gulf and Atlantic Coastal Plain, extinction intensity did not vary with position on a bathymetric gradient. See Bottjer and Jablonski (1988) for definitions and criteria for onshore-offshore environments. The 95% confidence limits are calculated using the method of Raup (1991b).

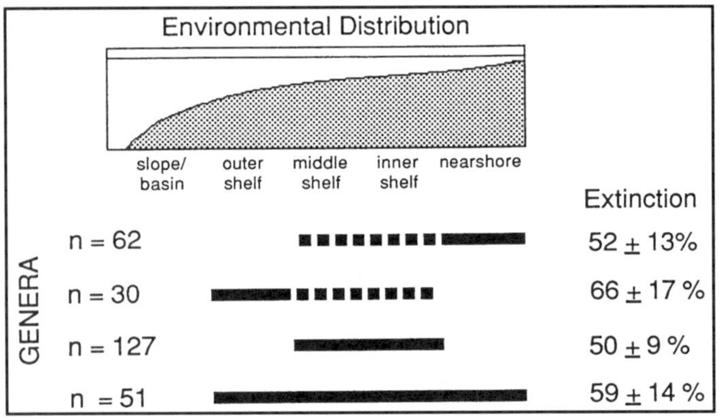

Figure 4. For end-Cretaceous bivalves and gastropods of the Gulf and Atlantic Coastal Plain, extinction intensity did not vary with breadth of distribution across the bathymetric transect shown in Figure 2. Vertical bars are 95% confidence limits calculated using the method of Raup (1991b).

reefal environments, whereas the K/T depth zones are analyzed outside the tropics to avoid confounding factors (and to exploit large and well-preserved faunas of known provenance), and (2) binary comparisons between platform versus basinal settings, whereas the K/T depth zones more finely subdivide the shelf and lack robust data from slope and deep-basin settings.

Despite the lack of selectivity recorded in the preceding analyses, K/T survivorship was not completely random but instead imposed at a different scale. The key factor was not the extent of geographic or bathymetric range within a given province but distribution among provinces. As noted above, genera found in several biogeographic regions, for example, the North American Coastal Plain, plus Europe, North Africa, or Japan, suffer significantly less extinction than those restricted to a single province. This pattern is observed for a broad array of taxa, in all of the major mass extinctions, and for a number of the lesser extinction events. The K/T and other events evidently imposed a scale of perturbation far greater than the scales that govern rate differentials during normal times. Once again, this *scale* in itself says little about the specific forcing factors involved. Neither the end-Ordovician nor the end-Permian extinction shows evidence of being impact driven but both exhibit the same kind of selectivity against geographically restricted taxa.

K/T survivorship in our global bivalve data set also varies among feeding types, but a closer examination further underscores the complexities of analyzing extinction differentials. Suspension-feeders, which collect plankton from the water column, suffer significantly greater losses (62%) than deposit-feeders, which rely on detritus and its microbial populations as a food source (34%) (Jablonski and Raup, 1995). Sheehan and Hansen (1986) and Sheehan and Fastovsky (1992) argued for similar selectivity in K/T terrestrial communities (see also McGhee, 1989, and Rhodes and Thayer, 1991, on marine invertebrates; and Levinton, 1974, for a pathfinding study).

When we subdivide K/T deposit-feeders according to evolutionary lineages, however, a more complex pattern emerges (Table 1). The Nuculoida, classic deposit-feeders in muddy sediments, show low extinction values, as do the Lucinacea, many of which have chemosymbionts that contribute to nutrition and perhaps belong in a different nutritional category altogether (Reid and Brand, 1986; Hickman, 1994). The Tellinacea show a different pattern, however. Members of this group do not fit so neatly into trophic categories: some are suspension-feeders, whereas many mainly deposit-feed but can facultatively switch to suspension-feeding depending on microhabitat and hydrodynamic conditions (e.g., Pohlo, 1969; Levinton, 1991; Kamermans, 1994; Lin and Hines, 1994) and thus should enjoy the best of both worlds. Unexpectedly, the Tellinacea show significantly higher extinction (75%) than other deposit feeders. Roy (1994) found similarly high extinction intensities in a group of detritus-feeding gastropods at the K/T boundary (76.0 ± 9%, n = 25 genera of Aporrhaidae). Numbers are painfully small here, and more work is needed on background turnover rates, but the 95% confidence limits of the Tellincea and Aporrhaidae do not overlap with those of the nuculoids and lucinaceans (Table 1).

As usual, there are several alternative explanations for the variation in extinction intensities among K/T deposit-feeders. The Nuculoida are probably taxonomically undersplit relative to the Tellinacea and Aporrhaidae—there are clearly a greater number of discrete evolutionary units in the nuculoids than is reflected in their current taxonomy. The groups also differ in how they exploit their detrital food resources, so that their respective resource pools might be buffered to different extents (see Jumars et al., 1990, who distinguish among surface deposit-feeders such as the tellinaceans, subsurface deposit-feeders such as the nuculoids, and chemautotrophic deposit-feeders such as the lucinaceans). Alternatively, the selectivity may not be on feeding strategies at all. Deposit-feeders are more tightly coupled to pelagic input than is generally recognized; in both shallow water and in the deep sea, metabolic rates, biomass, population dynamics, and even taxonomic diversity of deposit-feeders appear to track the flux of phytoplankton and other organic detritus to the seafloor (Tyler, 1988; Graf, 1989; Gooday and Turley, 1990; Lambshead and Gooday, 1990; Rex at al., 1993; Campos-Creasey et al., 1994; Kamermans, 1994; and see Levinton, 1996, regarding the K/T extinction).

To summarize on selectivity, many patterns of differential survivorship documents for background times become unimportant at the K/T boundary. The rules of the game changed during the mass extinctions: Not only were more taxa being lost, but different kinds of taxa were being lost relative to background times. Despite clear differences in survivorship among higher taxa during mass extinction events, the basis of such selectivity has been analyzed for only a few groups. Furthermore, these patterns of differential survival—as well as the absence of such patterns—are virtually neutral regarding specific killing mechanisms, but they speak volumes about the intensity of the event and its long-term evolutionary impact.

Despite their far-reaching effects, mass extinctions do not completely overturn large-scale biotic processes (although the tremendous end-Permian event came close; see Erwin, 1993, 1994; Van Valen, 1994). Many major evolutionary and ecological trends transcend even the Big Five mass extinctions. Post-Paleozoic examples include the modernization of marine communities in terms of skeletal architectures and increased burrowing and predation rates (Vermeij, 1987; see also Bambach, 1993) and on land the rise of flowering plants (Crane and Lidgard, 1990) and modern insects (Labandeira and Sepkoski, 1993). The long-term history of biotas thus involves the interplay of two kinds of dynamics: those driven mainly by background processes but able to persist during mass extinction events and those disrupted and redirected by the mass extinctions. The early Tertiary rebounds of molluscan faunas in Europe and North America follow significantly different trajectories (Jablonski, 1989b, 1995b), but both faunas are unmistakably Cenozoic in their range of adaptive types and life habits; the composition of Northern Hemisphere floras is perhaps permanently shaped by K/T extinction patterns (Wolfe, 1987) but clearly participates in the overall angiosperm radiation. To understand the role of mass extinctions in the evolutionary process, then, we need to study survivors and rebounds as well as victims and ecosystem collapse. This will be an important growth area for research in the coming decade.

ACKNOWLEDGMENTS

I thank J. John Sepkoski Jr. and Peter D. Ward for discussions and encouragement; Douglas H. Erwin, Susan M. Kidwell, and Peter M. Sheehan for valuable reviews; and David M. Raup for ongoing collaboration, some of which is described here in preliminary form. Supported by NSF Grants EAR-9005744, and EAR-9317114.

REFERENCES CITED

Anstey, R. L., 1986, Bryozoan provinces and patterns of generic evolution and extinction in the Late Ordovician of North America: Lethaia, v. 19, p. 33–51.

Aubry, M.-P., Gradstein, F. M., and Jansa, L. F., 1990, The late early Eocene Montagnais bolide: No impact on biotic diversity: Micropaleontology, v. 36, p. 164–172.

Baarli, B. G., and Harper, D. A. T., 1986, Relict Ordovician brachiopod faunas in the Lower Siluian of Asker, Oslo region, Norway: Norsk Geologisk Tidsskrift, v. 66, p. 87–98.

Bambach, R. K., 1993, Seafood through time: Changes in biomass, energetics, and productivity in the marine ecosystem: Paleobiology, v. 19, p. 372–397.

Birkelund, T., and Hakansson, E., 1982, The terminal Cretaceous extinction in Boreal shelf seas: Geological Society of America Special Paper 190, p. 373–384.

Bottjer, D. J., and Jablonski, D., 1988, Paleoenvironmental patterns in the evolution of post-Paleozoic benthic marine invertebrates: Palaios, v. 3, p. 540–560.

Bretsky, P. W., 1973, Evolutionary patterns in the Paleozoic Bivalvia: Documentation and some theoretical considerations: Geological Society of America Bulletin, v. 84, p. 2079–2096.

Broecker, W. S., Klas, M., Clark, E., Bonani, G., Ivy, S., and Wolfi, W., 1990, the influence of $CaCO_3$ dissolution on core top radiocarbon ages for deep-sea sediments: Paleoceanography, v. 6, p. 593–608.

Brown, J. H., Marquet, P. A., and Taper, M. L., 1993, Evolution of body size: Consequences of an energetic definition of fitness: American Naturalist, v. 142, p. 573–584.

Calder, W. A., 1984, Size, Function, and life history: Cambridge, Massachusetts, Harvard University Press, 431 p.

Campos-Creasey, L. S., Tyler, P. A., Gage, J. D., and John, A. W. G., 1994, Evidence for coupling the vertical flux of phytodetritus to the diet and seasonal history of the deep-sea echinoid *Echinus affinis*: Deep-Sea Research, v. 41, p. 369–388.

Case, J. A., and Woodburne, M. O., 1988, South American marsupials: A successful crossing of the Cretaceous-Tertiary boundary: Palaios, v. 1, p. 413–416.

Crane, P. R., and Lidgard, S., 1990, Angiosperm radiation and patterns of palynological diversity, *in* Taylor, P. D., and Larwood, G. P., eds., Major evolutionary radiations: Oxford, Oxford University Press, p. 377–407.

Cutler, A. H., 1993, Mathematical models of temporal mixing in the fossil record, *in* Kidwell, S. M., and Behrensmeyer, A. K., eds., Taphonomic approaches to time resolution in fossil assemblages. The Paleontological Society, Short courses in paleontology 6: Knoxville, University of Tennessee, p. 169–187.

Droser, M. L., and Bottjer, D. J., 1993, Trends and patterns of Phanerozoic ichnofabrics: Annual Review of Earth and Planetary Sciences, v. 21, p. 205–225.

Erwin, D. H., 1989, Regional paleoecology of Permian gastropod genera, southwestern United States and the end-Permian mass extinction: Palaios, v. 4, p. 424–438.

Erwin, D. H., 1993, The great Paleozoic crisis: Life and death in the Permian: New York, Columbia University Press, 327 p.

Erwin, D. H., 1994, The Permo-Triassic extinction: Nature, v. 367, p. 231–236.

Flessa, K. W., 1990, The "facts" of mass extinctions: Geological Society of America Special Paper 247, p. 1–7.

Flessa, K. W., 1993, Time-averaging and temporal resolutions in Recent marine shelly fauns, *in* Kidwell, S. M., and Behrensmeyer, A. K., eds., Taphonomic approaches to time resolution in fossil assemblages. The Paleontological Society, Short courses in paleontology 6: Knoxville, University of Tennessee, p. 9–33.

Flessa, K. W., and Kowalewski, M., 1994, Shell survival and time-averaging in nearshore and shelf environments: Estimates from the radiocarbon literature: Lethaia, v. 27, p. 153–165.

Gooday, A. J., and Turley, C. M., 1990, Responses by benthic organisms to inputs of organic material to the ocean floor: A review: Philosophical Transactions of the Royal Society of London, v. A331, p. 119–138.

Graf, G., 1989, Benthic-pelagic coupling in a deep-sea benthic community: Nature, v. 341, p. 437–439.

Hickman, C. S., 1994, The genus *Parvilucina* in the Eastern Pacific: Making evolutionary sense of a chemosymbiotic species complex: The Veliger, v. 37, p. 43–61.

Izett, G. A., Cobban, W. A., Obradovich, J. D., and Kunk, M. J., 1993, The Manson impact structure: $^{40}Ar/^{39}Ar$ age and its distal impact ejecta in the Pierre Shale in southeastern South Dakota: Science, v. 262, p. 729–732.

Jablonski, D., 1986a, Causes and consequences of mass extinctions: A comparative approach, *in* Elliott, D. K., ed., Dynamics of extinction: New York, Wiley, p. 183–229.

Jablonski, D., 1986b, Background and mass extinctions: The alternation of macroevolutionary regimes: Science, v. 231, p. 129–133.

Jablonski, D., 1989a, The biology of mass extinction: A paleontological view: Philosophical Transactions of the Royal Society of London, v. B325, p. 357–368.

Jablonski, D., 1989b, Cretaceous-Tertiary bivalves of Europe and North America: Comparison of patterns of extinction and rebound: Geological Society of America Abstracts with Programs, v. 21, p. A209.

Jablonski, D., 1995a, Extinctions in the fossil record, *in* Lawton, J. H., and May, R. M., eds., Extinction rates: New York, Oxford University Press, p. 25–44.

Jablonski, D., 1995b, The biogeography of rebounds: Comparisons among K-T bivalves: Geological Society of America Abstracts with Programs, v. 27, p. A-164.

Jablonski, D., and Raup, D. M., 1995, Selectivity of end-Cretaceous marine bivalve extinctions: Science, v. 268, p. 389–391.

Jansa, L. F., 1993, Cometary impacts into ocean: Their recognition and the threshold constraint for biological extinctions: Palaeogeography, Palaeoclimatology, Palaeoecology, v. 104, p. 271–286.

Jansa, L. F., Aubry, M.-P., and Gradstein, F. M., 1990, Comets and extinctions: Cause and effect?: Geological Society of America Special Paper 247, p. 223–232.

Jumars, P. A., Mayer, L. M., Deming, J. W., Baross, J. A., and Wheatcroft, R. A., 1990, Deep-sea deposit-feeding strategies suggested by environmental and feeding constraints: Philosophical Transactions of the Royal Society of London, v. A331, p. 85–101.

Kamermans, P., 1994, Similarity in food source and timing of feeding in deposit- and suspension-feeding bivalves: Marine Ecology Progress Series, v. 104, p. 63–75.

Kauffman, E. G., 1978, Evolutionary rates and patterns among Cretaceous Bivalvia: Philosophical Transactions of the Royal Society of London, v. B284, p. 277–304.

Kauffman, E. G., 1986, High-resolution event stratigraphy: Regional and global Cretaceous bioevents, *in* Walliser, O. H., ed., Global bioevents: Berlin, Springer-Verlag, p. 279–335.

Keller, G., 1993, The Cretaceous-Tertiary boundary transition in the Antarctic Ocean and its global implications: Marine Micropaleontology, v. 21, p. 1–45.

Keller, G., Barrera, E., Schmitz, B., and Mattson, E., 1993, Gradual mass extinction, species survivorship, and long-term environmental changes across the Cretaceous-Tertiary boundary in high latitudes: Geological Society of America Bulletin, v. 105, p. 979–997.

Kidwell, S. M., 1993, Patterns of time-averaging in the shallow marine fossil record, *in* Kidwell, S. M., and Behrensmeyer, A. K., eds., Taphonomic approaches to time resolution in fossil assemblages. The Paleontological Society, Short courses in paleontology 6: Knoxville, University of Tennessee, p. 275–300.

Kidwell, S. M., and Behrensmeyer, A. K., eds., 1993, Taphonomic approaches to time resolution in fossil assemblages. The Paleontological Society, Short courses in paleontology 6: Knoxville, University of Tennessee, 302 p.

Kidwell, S. M., and Bosence, D. W. J., 1991, Taphonomy and time-averaging of marine shelly faunas, *in* Allison, P. A., and Briggs, D. E. G., eds., Taphonomy: New York, Plenum, p. 116–209.

Koch, C. F., 1991, Species extinctions across the Cretaceous-Tertiary boundary: Observed patterns versus predicted sampling effects, stepwise or otherwise?: Historical Biology, v. 5, p. 355–361.

Kyte, F. T., Zhou, L., and Wasson, J. T., 1988, New evidence on the size and possible effects of a Late Pliocene oceanic asteroid impact: Science, v. 241, p. 63–65.

Labandeira, C. C., and Sepkoski, J. J., Jr., 1993, Insect diversity in the fossil record: Science, v. 261, p. 311–315.

Lambshead, P. J., and Gooday, A. J., 1990, The impact of seasonally deposited phytodetritus on epifaunal and shallow infaunal benthic foraminiferal populations in the bathyal northeast Atlantic: The assemblage response: Deep-Sea Research, v. 37, p. 1263–1283.

Levinton, J. S., 1974, Trophic group and evolution in bivalve molluscs: Palaeontology, v. 17, p. 579–585.

Levinton, J. S., 1996, Trophic group and the end-Cretaceous extinction: did deposit feeders have it made in the shade?: Paleobiology, v. 22, p. 104–112.

Levinton, J. S., 1991, Variable feeding behavior in three species of *Macoma* (Bivalvia: Tellinacea) as a response to water flow and sediment transport: Marine Biology, v. 110, p. 375–383.

Lin, J., and Hines, A. H., 1994, Effects of suspended food availability on the feeding mode and burial depth of the Baltic clam, *Macoma balthica*: Oikos, v. 69, p. 23–36.

Marincovich, L., Jr., 1993, Danian mollusks from the Prince Creek Formation, northern Alaska, and implications for Arctic Ocean paleogeography: Paleontological Society memoir 35 (Journal of Paleontology, v. 67, supplement to no. 5), 35 p.

Marshall, C. R., 1991, Estimation of taxonomic ranges from the fossil record, *in* Gilinsky, N. L., and Sigor, P. W., eds., Analytical paleontology. The Paleontological Society, Short courses in paleontology 4: Knoxville, University of Tennessee, p. 19–38.

Martin, R. A., 1992, Generic species richness and body mass in North American mammals: Support for the inverse relationship of body size and speciation rate: Historical Biology, v. 6, p. 73–90.

Martin, R. D., 1993, Time and taphonomy: Actualistic evidence for time-averaging of benthic foraminiferal assemblages, *in* Kidwell, S. M., and Behrensmeyer, A. K., eds., Taphonomic approaches to time resolution in fossil assemblages. The Paleontological Society, Short courses in paleontology 6: Knoxville, University of Tennessee, p. 34–56.

McGhee, G. R., 1989, The Frasnian-Famennian extinction event, *in* Donovan, S. K., ed., Mass extinctions: New York, Columbia University Press, p. 133–151.

McKinney, M. K., 1987, Taxonomic selectivity and continuous variation in mass and background extinctions of marine taxa: Nature, v. 325, p. 143–145.

McKinney, M. K., 1990, Trends in body-size evolution, *in* McNamara, K. J., ed., Evolutionary trends: Tucson, University of Arizona Press, p. 75–118.

Peters, R. H., 1983, The ecological implications of body size: Cambridge: Cambridge University Press, 329 p.

Pimm, S. L., Jones, H. L., and Diamond, J., 1988, On the risk of extinction: American Naturalist, v. 132, p. 757–785.

Pohlo, R., 1969, Confusion concerning deposit-feeding in the Tellinacea: Proceedings of the Malacological Society of London, v. 38, p. 361–364.

Raup, D. M., 1979a, Biases in the fossil record of species and genera: Bulletin of the Carnegie Museum of Natural History, v. 13, p. 85–91.

Raup, D. M., 1979b, Size of the Permo-Triassic bottleneck and its evolutionary implications: Science, v. 206, p. 217–218.

Raup, D. M., 1989, The case for extraterrestrial causes of extinction: Philosophical Transactions of the Royal Society of London, v. B325, p. 421–435.

Raup, D. M., 1991a, A kill curve for Phanerozoic marine species: Paleobiology, v. 17, p. 37–48.

Raup, D. M., 1991b, The future of analytical paleontology, *in* Gilinsky, N. L., and Signor, P. W., eds., Analytical paleontology. The Paleontological Society, Short Courses in paleontology 4: Knoxville, University of Tennessee, p. 207–216.

Raup, D. M., and Boyajian, G. E., 1988, Patterns of generic extinction in the fossil record: Paleobiology, v. 14, p. 109–125.

Raup, D. M., and Jablonski, D., 1993, Geography of end-Cretaceous marine bivalve extinctions: Science, v. 260, p. 971–973.

Raup, D. M., and Sepkoski, J. J., Jr., 1986, Periodic extinction of families and genera: Science, v. 231, p. 833–836.

Reid, R. G. B., and Brand, D. G., 1986, Sulfide-oxidizing symbiosis in lucinaceans: Implications for bivalve evolution: Veliger, v. 29, p. 3–24.

Rex, M. A., Stuart, C. T., Hessler, R. R., Allen, J. A., Sanders, H. L., and Wilson, G. D. F., 1993, Global-scale latitudinal patterns of species diversity in the deep-sea benthos: Nature, v. 365, p. 636–639.

Rhodes, M. C., and Thayer, C. W., 1991, Mass extinctions: Ecological selectivity and primary production: Geology, v. 19, p. 877–800.

Robertson, D. B. R., Brenchley, P. J., and Owen, A. W., 1990, Ecological disruption close to the Ordovician-Silurian boundary: Historical Biology, v. 5, p. 131–144.

Rosen, B. R., and Turnsek, D., 1989, Extinction patterns and biogeography of scleractinian corals across the Cretaceous/Tertiary boundary: Association of Australasian Palaeontologists Memoir 8, p. 355–370.

Roy, K., 1994, Effects of Mesozoic marine revolution on the taxonomic, morphologic and biogeographic evolution of a group: Aporrhaid gastropods during the Mesozoic: Paleobiology, v. 20, p. 274–296.

Schmidt, G., Zhou, L., and Wasson, J. T., 1993, Iridium anomaly associated with the Australasian tektite-producing impact: Masses of the impactor and of the Australasian tektites: Geochimica et Cosmochimica Acta, v. 57, p. 4851–4859.

Sepkoski, J. J., Jr., 1978, A kinetic model of Phanerozoic taxonomic diversity. I: Analysis of marine orders: Paleobiology, v. 4, p. 223–251.

Sepkoski, J. J., Jr., 1984, A kinetic model of Phanerozoic taxonomic diversity. III: Post-Paleozoic families and mass extinctions: Paleobiology, v. 10, p. 246–267.

Sepkoski, J. J., Jr., 1987, Environmental trends in extinction during the Phanerozoic: Science, v. 235, p. 64–66.

Sepkoski, J. J., Jr., 1990, The taxonomic structure of periodic extinction: Geological Society of America Special Paper 247, p. 33–44.

Sepkoski, J. J., Jr., 1991, Diversity in the Phanerozoic oceans: A partisan review, *in* Dudley, E. C., ed., The unity of evolutionary biology, Volume I: Portland, Dioscorides Press, p. 210–236.

Sepkoski, J. J., Jr., 1993, Ten years in the library: New data confirm paleontological patterns: Paleobiology, v. 19, p. 43–51.

Sepkoski, J. J., Jr., 1996, Patterns of Phanerozoic extinction: A perspective from global data bases, *in* Walliser, O. H., ed., Global bio-events and event-stratigraphy: Berlin, Springer-Verlag (in press).

Sepkoski, J. J., Jr., and Koch, C. F., 1996, Evaluating paleontologic data relating to bio-events, *in* Walliser, O. H., ed., Global bio-events and event-stratigraphy: Berlin, Springer-Verlag (in press).

Sheehan, P. M., and Coorough, P. J., 1990, Brachiopod zoogeography across the Ordovician-Silurian boundary: Geological Society of London Memoir 12, p. 181–187.

Sheehan, P. M., and Fastovsky, D. E., 1992, Major extinctions of land-dwelling vertebrates at the Cretaceous-Tertiary boundary, eastern Montana: Geology, v. 20, p. 556–560.

Sheehan, P. M., and Hansen, T. A., 1986, Detritus feeding as a buffer to extinction at the end of the Cretaceous: Geology, v. 14, p. 868–870.

Sheehan, P. M., Fastovsky, D. E., Hoffman, R. G., Berghaus, C. B., and Gabriel, D. L., 1991, Sudden extinction of the dinosaurs: Latest Cretaceous, upper Great Plains, U.S.A.: Science, v. 254, p. 835–839.

Signor, P. W., and Lipps, J. H., 1982, Sampling bias, gradual extinction patterns, and catastrophes in the fossil record: Geological Society of America Special Paper 190, p. 291–296.

Sorauf, J. E., 1989, Rugosa and the Frasnian-Famennian extinction event: A progress report: Memoirs of the Association of Australasian Palaeontologists, v. 8, p. 327–338.

Sorauf, J. E., and Pedder, A. E. H., 1986, Late Devonian rugose corals and the Frasnian-Famennian crisis: Canadian Journal of Earth Sciences, v. 23, p. 1265–1287.

Springer, M. S., 1990, The effect of random range truncations on patterns of evolution in the fossil record: Paleobiology, v. 16, p. 517–520.

Stanley, S. M., 1984, Marine mass extinctions: A dominant role for temperature, *in* Nitecki, M. H., ed., Extinctions: Chicago, University of Chicago Press, p. 69–117.

Stanley, S. M., 1986, Population size, extinction, and speciation: The fission effect in Neogene Bivalvia: Paleobiology, v. 12, p. 89–110.

Stanley, S. M., 1988, Paleozoic mass extinctions: Shared patterns suggest global cooling as a common cause: American Journal of Science, v. 288, p. 334–352.

Stanley, S. M., 1990, The general correlation between rate of speciation and rate of extinction: Fortuitous causal linkages, *in* Ross, R. M., and Allmon, W. D., eds., Causes of evolution: Chicago, University of Chicago Press, p. 103–127.

Tyler, P. A., 1988, Seasonality in the deep sea: Oceanography and Marine Biology: An Annual Review, v. 26, p. 227–258.

Valentine, J. W., 1973, Evolutionary paleoecology of the marine biosphere: Englewood Cliffs, New Jersey, Prentice-Hall, 511 p.

Valentine, J. W., 1990, The macroevolution of clade shape, *in* Ross, R. M., and Allmon, W. D., eds., Causes of evolution: Chicago, University of Chicago Press, p. 128–150.

Valentine, J. W., and Jablonski, D., 1986, Mass extinctions: Selectivity of marine larval types: Proceeding of the National Academy of Sciences, USA, v. 83, p. 6912–6914.

Van Valen, L. M., 1985, How constant is extinction?: Evolutionary Theory, v. 7, p. 93–106.

Van Valen, L. M., 1994, Concepts and the nature of selection by extinction: Is generalization possible?: *in* Glen, W., ed., Mass extinction debates: How science works in a crisis: Stanford, California, Stanford University Press, p. 200–216.

Vermeij, G. J., 1987, Evolution and escalation: Princeton, New Jersey, Princeton University Press, 527 p.

Ward, P. D., and Signor, P., III, 1983, Evolutionary tempo in Jurassic and Cretaceous ammonites: Paleobiology, v. 9, p. 183–198.

Westrop, S. R., 1991, Intercontinental variation in mass extinction patterns: Influence of biogeographic structure: Paleobiology, v. 17, p. 363–368.

Wolfe, J. A., 1987, Late Cretaceous–Cenozoic history of deciduousness and the terminal Cretaceous event: Paleobiology, v. 13, p. 215–226.

Zinsmeister, W. J., Feldmann, R. M., Woodburne, M. O., and Elliott, D. H., 1989, Latest Cretaceous/earliest Tertiary transition on Seymour Island, Antarctica: Journal of Paleontology, v. 63, p. 731–738.

Manuscript Accepted by the Society September 14, 1995

Printed in U.S. A.

Geological Society of America
Special Paper 307
1996

Impact crises and mass extinctions: A working hypothesis

Michael R. Rampino
Earth Systems Group and Department of Biology, New York University, 34 Stuyvesant Street, New York , New York 10003, and NASA, Goddard Institute for Space Studies, 2880 Broadway, New York, New York 10025
Bruce M. Haggerty
Earth Systems Group, New York University, 34 Stuyvesant St., New York , New York 10003

ABSTRACT

Based on evidence from astronomical observations, impact dynamics, and the geologic record, we explore a general hypothesis linking impacts of large asteroids and comets with mass extinctions of life. The probability of large-body impacts on the Earth derived from the flux of earth-crossing asteroids and comets and the estimated threshold impact size required to cause a global environmental disaster suggest that impacts of objects ≥ a few kilometers in diameter might be sufficient to explain the Phanerozoic record of extinction pulses. A number of extinction boundaries are known to be marked by severe environmental disturbances, including mass mortality and related extinctions (sometimes in steps), impoverished postextinction fauna and flora, and proliferation of stress-tolerant and opportunistic species, followed by gradual ecological recovery and radiation of new taxa. Abrupt negative shifts in $\delta^{13}C$ in marine sedimentary rocks at extinction boundaries suggest major biomass losses followed by low-productivity "Strangelove" oceans, and fluctuations in $\delta^{18}O$ may be interpreted as evidence of significant climatic oscillations. These biological, isotopic, and geochemical signatures seem to be consistent with the expected after effects of catastrophic impacts. Six of the extinction pulses may be associated with concurrent (in some cases multiple) impact markers (e.g., layers with high iridium, shocked minerals and/or microtektites, and large, dated impact craters). Elevated iridium levels at, or near, other extinction boundaries have characteristics suggesting a terrestrial origin, although they might be explained by collision of relatively low-Ir objects such as comets, and further work is warranted.

INTRODUCTION

Mass extinction events represent geologically brief episodes when large numbers of species (~25% to >90%) disappeared (Raup and Sepkoski, 1984; Sepkoski, 1993). Five major mass extinctions and 20 other identifiable local maxima in extinction can be recognized in the record of marine genera (% extinction) for the Phanerozoic Eon, the past ~540 m.y. (using data from Sepkoski, 1992, 1994, and personal communication, 1993) (Fig. 1). Several of these marine extinctions are known to correlate with extinctions of nonmarine organisms (e.g., tetrapods, insects, land plants) (Erwin, 1993; Benton, 1988, 1989; Labandiera and Sepkoski, 1993; Retallack, 1995), and a general correlation of marine and nonmarine extinctions is likely (Rampino, 1988; Benton, 1995), suggesting global environmental perturbations.

The extinctions, and subsequent radiations of new species, were apparently crucial events in the evolution of life, and they have provided natural markers for creating time boundaries in the geologic record. The extinction episodes are typically limited stratigraphically to a single geologic stage or substage (a few million years) but most likely took place in much shorter periods of time (or in steps) at or near the stage boundaries, which are defined largely by faunal turnover (McLaren and Goodfellow,

Rampino, M. R., and Haggerty, B. M., 1996, Impact crises and mass extinctions: A working hypothesis, *in* Ryder, G., Fastovsky, D., and Gartner, S., eds., The Cretaceous-Tertiary Event and Other Catastrophes in Earth History: Boulder, Colorado, Geological Society of America Special Paper 307.

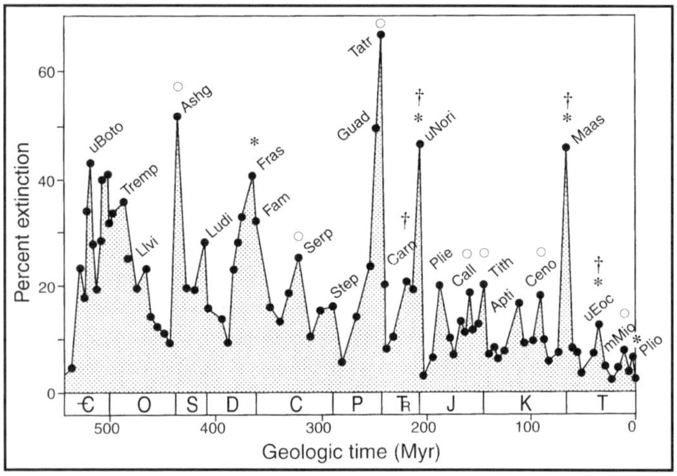

Figure 1. Percent extinction of marine genera per geologic stage (or substage) during the Phanerozoic (culled genus-level data from Sepkoski, 1992, 1994, and personal communication, 1993). The following local maxima are recognizable in Sepkoski's data (end of stage in Ma, as dated by recent geologic time scales, may not be exactly coincident with the extinctions): Pliocene (Plio) (1.6), Mid-Miocene (mMio) (11.2), Upper Eocene (uEoc) (36.6), Maastrichtian (Maas) (66.4), Cenomanian (Ceno) (92), Aptian (Apti) (113), Tithonian (Tith) (144), Callovian (Call) (163), Pliensbachian (Plie) (193), Norian (uNori) (208), Carnian (Carn) (225), Tatarian (Tatr) (245), Guadelupian (Guad) (253), Stephanian (Step) (286), Serpukhovian (Serp) (320), Famennian (Fam) (360), Frasnian (Fras) (367), Eifelian (380), Ludlovian (Ludl) (414), Ashgillian (Ashg) (438), Caradocian? (448), lower Llanvirnian (Llvi) (478), Trempeleauan (Tremp) (505), Dresbachian (515), Botomian (uBoto) (520), Proterozoic/Cambrian? (540) The dates for the Cambrian are revised after Bowring et al. (1993). Note that Cambrian extinction peaks may be anomalously high as a result of the relatively poor record of diversity. The "five major extinctions" referred to in the text are the Ashgillian, Frasnian, Tatarian, Norian, and Maastrichtian events. Symbols used indicate diagnostic stratigraphic evidence of impact (asterisks), possible stratigraphic evidence of impacts (open circles), and large, dated impact craters (daggers) (see text).

1990). Some mass extinctions are known to be associated with sudden widespread mass mortality events that decimated populations of organisms living in a broad range of environments, leading to the disappearance of many species. Such sudden mass mortalities might be expected as a result of ecological disasters of global proportions (McLaren and Goodfellow, 1990; Raup, 1991a, b, 1992; Rampino and Haggerty, 1995).

The causes of the mass extinctions remain a subject of intense debate. Although many exotic factors, ranging from supernovae to disease pandemics, have been proposed, environmental changes such as fluctuations of climate, sea level, and ocean chemistry, interacting with the global biota in perhaps quite complex ways, remain the favored ideas (e.g., Donovan, 1989; Hallam, 1989; Erwin, 1993). With the discovery of considerable evidence for the impact of a comet or asteroid at the Cretaceous-Tertiary (K/T) mass extinction boundary (65 Ma) (Alvarez et al., 1980; for a recent review, and some counterarguments, see Glen, 1994), much attention has focused on the relationship between large-body impacts and mass extinctions.

Since 1980, various groups have searched for possible evidence of impacts (iridium anomalies, shocked minerals, microtektites, etc.) at other recognized extinction boundaries. Although some correlations (or near correlations) of extinctions and possible evidence of impacts have been noted (e.g., Kyte, 1988; Orth, 1989; Orth et al., 1990), many studies of geologic boundaries have concluded that impact evidence is questionable or lacking. In many cases, elevated levels of iridium were reported, but because of nonchondritic element ratios and/or relatively small amplitude of the Ir peaks (typically ranging in hundreds of parts per trillion, rather than in parts per billion, as at the K/T boundary), the possibility of impact was rejected in favor of other explanations (Orth et al., 1990).

A contrasting approach to the problem of impacts and extinctions can be derived from astronomical observations of Earth-crossing asteroids and comets, which allow estimates of the expected mean times between collisions of objects of various sizes with the planet (e.g., Shoemaker et al., 1990; Morrison, 1992; Chapman and Morrison, 1994; Toon et al., 1994). The expected flux of impactors agrees well with estimates based on dated impact craters on the Earth (Grieve, 1991). Impact of objects greater than a few kilometers in diameter are predicted to result in far-field effects that should constitute widespread environmental disasters (Morrison, 1992; Chapman and Morrison, 1994; Toon et al., 1994). The expected flux of impactors of various sizes and the calculated environmental effects of large impacts suggest that impact events might provide a general explanation of the mass extinctions in the history of life (Alvarez, 1986; McLaren and Goodfellow, 1990; Raup, 1990, 1991a, b, 1992; Rampino and Haggerty, 1994).

Raup (1989) observed that "the subject involves so many separate scientific disciplines—from paleontology to astrophysics—that no one individual is competent to judge the merits of all the arguments and counterarguments." In light of these findings, however, we think it is useful to reexamine and attempt to synthesize recent studies of various mass extinctions and related geologic boundaries. The benefit of such a synthetic approach is that general patterns may emerge, and recognition of such patterns might allow a better assessment of the viability of a general working hypothesis linking catastrophic impacts and mass extinctions.

EVIDENCE FROM IMPACT STUDIES

A consideration of the expected time history of large impacts and their predicted effects on the biosphere generally supports an impact theory of mass extinctions. Astronomical observations of Earth-crossing asteroids and comets, and the cratering records of the inner planets, predict that in an ~100-m.y. period, the earth should be hit by several asteroids and comets larger than a few kilometers in diameter (these release ~5×10^{22} J or ~10^7 Mt TNT equivalent) and perhaps one ~10-km diameter (~5×10^{23} J, 10^8 Mt) object (Fig. 2) (Shoemaker et al., 1990; Chapman and Morrison, 1994).

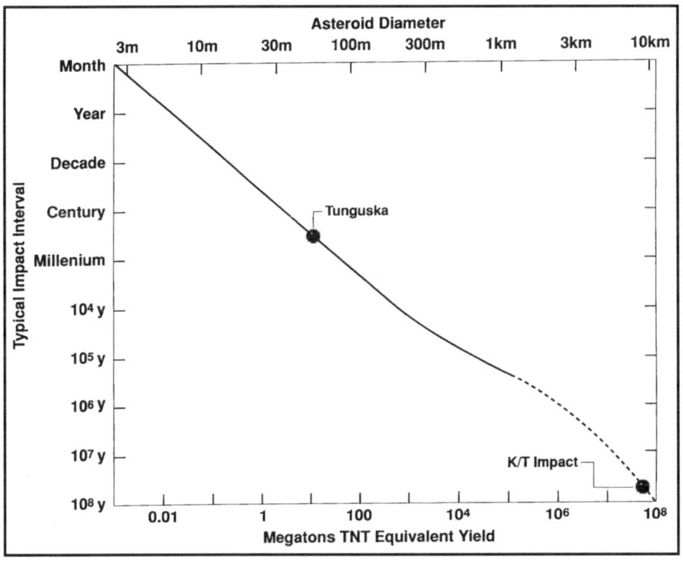

Figure 2. Frequency of impacts on the Earth estimated from the present population of comets and asteroids and evidence from impact craters. The megaton TNT equivalents of impact energy are shown (after Chapman and Morrison, 1994). Shown are the 1908 Tunguska impact event in Siberia, a once-in-several-centuries occurrence, and the K/T impact event, a once-in-about-100-m.y. event.

Many studies have concluded that large impacts of these magnitudes would result in major global environmental disasters primarily related to production of dense clouds of fine ejecta and the deposition of large amounts of heat in the atmosphere (see Morrison, 1992; Chapman and Morrison, 1994; Toon et al., 1994). An impact $\geq 10^8$ Mt is estimated to create and ballistically distribute $\geq 10^{16}$ kg of submicron-sized ejecta over the globe (Morrison, 1992; Chapman and Morrison, 1994), and this amount of fine dust is calculated to bring global atmospheric transmission down to $<10^{-2}$, which is the limit of photosynthesis (Gerstl and Zardecki, 1982). Calculations and experiments suggest that the few months of darkness that would result would have been sufficient to produce the severe extinction seen among pelagic phytoplankton (the base of the marine food chain) at the K/T boundary (Milne and McKay, 1982; Griffis and Chapman, 1988). At times of heavy atmospheric dust loading, the earth's surface receives little solar radiation, and drastic surface cooling is inevitable. Model calculations predict that land-surface temperatures could decrease by ~15°C in less than a week (Pollack et al., 1983; Covey et al., 1990).

The impact fireball and the energy emitted globally by the reentering impact ejecta are capable of igniting combustible material, resulting in widespread wildfires (Melosh et al., 1990; Toon et al., 1994). Calculations by Toon et al. (1994) suggest that global wildfires would be produced by an impact releasing somewhere between $\sim 5 \times 10^7$ and 10^8 Mt. Large amounts of soot discovered at the K/T boundary support the burning of a significant fraction of the terrestrial biomass (Anders et al., 1986; Wolbach et al., 1988), and the soot would add significantly to atmospheric opacity. Fires might also produce large volumes of CO_2, CO, CH_4, N_2O, and reactive hydrocarbons, leading to photochemical smog and possibly an increased greenhouse warming (Wolbach et al., 1988).

Atmospheric shock waves from a large impact are calculated to create large amounts of NO, possibly producing nitric acid rain with a pH of ~1 and NO_x in the stratosphere that could rapidly remove the ozone layer. Several other potentially disastrous environmental effects of large impacts have been proposed, including enhanced greenhouse effect from water vapor injected into the atmosphere by an oceanic impact (Croft, 1982; O'Keefe and Ahrens, 1982) or from CO_2 released by impact into a carbonate-rich terrane (O'Keefe and Ahrens, 1989) and the possible creation of large amounts of H_2SO_4 aerosols, and thus a marked cooling, if the target rocks contained sulfur-rich evaporite deposits (Sigurdsson et al., 1992). At present, these scenarios are less well constrained than the direct results of global dust clouds and reentry of ejecta.

With these considerations in mind, we can utilize the astronomical observations to predict a record of impact-induced mass extinctions during the Phanerozoic Eon (the last 540 m.y.). Calculations suggest that the atmospheric dust and smoke and the reentry heating effects of a $\geq 10^8$ Mt impact might cause a more severe mass extinction event than an $\sim 10^7$ Mt impact (Melosh et al., 1990; Toon et al., 1994). The expected impact history thus predicts ~5 major mass extinctions and about 25 ± 5 less severe events. This agrees well with the Phanerozoic record of extinctions that can be interpreted as showing five major and ~20 minor extinction events (Fig. 1). The result suggests that the entire record of extinction events could be explained by comet and asteroid impacts, as has been suggested by several workers (e.g., Urey, 1973; Alvarez, 1986; McLaren and Goodfellow, 1990; Raup, 1991b). This working hypothesis is testable, in principle, by examining the stratigraphic evidence for impacts at, or near, geologic boundaries that involved mass extinction.

STRATIGRAPHIC EVIDENCE OF LARGE IMPACTS

Diagnostic evidence of impacts at or near extinction boundaries

The discovery of the K/T impact layer prompted the search for impact signatures at other geological boundaries (see Kyte, 1988; Orth, 1989; Orth et al., 1990). Among the materials considered most diagnostic of impact are shocked minerals (including shocked quartz, stishovite, shocked zircons, etc.), impact glass (microtektites/tektites), microspherules with structures indicating high-temperature origin, and Ni-rich spinels (see Glen, 1994). Shocked minerals and tektite glass seem to be quite rare in the geologic record (e.g., Schmitz et al., 1994). These kinds of material have been reported in stratigraphic horizons, regionally to globally, at, or relatively close to, five recorded boundaries (Late Pliocene, Eocene-Oligocene transi-

TABLE 1. STRATIGRAPHIC EVIDENCE OF IMPACT DEBRIS AT/OR NEAR EXTINCTION BOUNDARIES

Age	Evidence	References
Pliocene (2.3 m.y.)	Microkrystites, microtektites?	Kyte, 1988; Peng, 1994 Margolis et al., 1991
Late Eocene (~36 m.y.)	Microtektites, tektites, microspherules, shocked quartz	Montanari, 1990
Cretaceous/Tertiary (65 m.y.)	Microtektites, tektites, shocked minerals, stishovite, Ni-rich spinels	Alvarez et al., 1980; Glen, 1994
Triassic/Jurassic (~205 m.y.)	Shocked quartz	Bice et al., 1992
Frasnian/Famennian (~367 m.y.)	Microtektites	Claeys et al., 1992 Claeys et al., 1994 Wang et al., 1994

tion, Cretaceous-Tertiary, Triassic-Jurassic, and Frasnian-Famennian) (Table 1).

Iridium anomalies in the geologic record

An iridium anomaly in the parts per billion to tens of parts per billion range (hundreds of times the background iridium in sediments containing mostly normal meteoroid ablation) is globally well documented at the K/T boundary. In the search for impact-related iridium anomalies, the K/T boundary has been used as the standard of reference. Criteria for impact signature have thus consisted of high (ppb) levels of iridium and chondritic or near-chondritic element ratios. Kyte (1988) argued, however, that the very high Ir concentrations at the K/T boundary anomaly should not be considered typical of stratigraphic iridium anomalies expected from impacts.

We note that the search for iridium has resulted in reports of elevated iridium levels (commonly ≥10 times background values) at or near a number of extinction boundaries (Table 2). The Ir levels are generally significantly weaker (100s of ppt) than the K/T iridium anomaly, and the anomalies are commonly associated with nonchondritic element abundance patterns (Kyte, 1988; Orth et al., 1990; Wang et al., 1991; Wang et al., 1993a). This circumstance has led to the general conclusion that the Ir peaks are probably unrelated to impact processes (Orth et al., 1990), although Kyte (1988) listed six such sedimentary horizons as "possible" impact horizons (the Cenomanian-Turonian, Callovian-Oxfordian, Early-Middle Jurassic, Permian-Triassic, Frasnian-Famennian, and Proterozoic-Cambrian).

A number of workers have suggested ways in which Ir could be concentrated by various biological and nonbiological methods (e.g., Rampino, 1982; Colodner et al., 1992). Elevated iridium levels created by scavenging of Ir from seawater would contain some fraction of dissolved extraterrestrial Ir (Kyte, 1988), however, and times of increased input of extraterrestrial Ir (vaporized or dissolved meteoritic material) should lead to enhanced precipitation of iridium. Preliminary studies of long stratigraphic intervals also suggest that elevated iridium levels may be uncommon in the geologic record away from boundaries (Kyte and Wasson, 1986; Alvarez et al., 1990).

It is also noteworthy that more diagnostic evidence of impacts has been reported associated with weak iridium anomalies at or near the Eocene-Oligocene transition (microtektites, microspherules, shocked quartz), Triassic-Jurassic (shocked quartz), and Late Devonian (microtektites) extinction boundaries. Microtektite layers are commonly associated with only background levels of Ir or mild Ir anomalies. For example, the widespread Australasian microtektites (0.76 Ma), clearly of impact origin, are associated with an iridium anomaly of only ~160 ppt (Koeberl, 1993).

In assessing the nature of Ir anomalies in the geologic record, it is important to note that impact-related Ir anomalies might show lower values than seen at the K/T boundary and/or nonchondritic element abundance patterns, for a number of reasons:

1. Meteorites differ in iridium content from $>10^3$ ppb (some irons) to $\sim10^{-2}$ ppb (eucrites and achondrites), and terrestrial impact melts range from >30 ppb to background (0.01 ppb); thus impacts of different kinds of extraterrestrial material may produce distributed iridium anomalies of various concentrations (Palme, 1982).

Comet impacts will produce significantly lower amounts of iridium than asteroid impacts. Because the rms collision velocity of comets is almost three times greater than that of asteroids (~60 km s^{-1} versus ~20 km s^{-1}), the amount of iridium produced per Joule of impact energy (or for a crater of a given size) for a comet would be only ~1/9 that of an asteroid. Taking an extreme range, the velocity of a head-on collision with a long-period comet is ~6.6 times that of collision with the slowest asteroids; thus the Ir yield per J of such a comet could be only ~1/44 that of an asteroid. Furthermore, the ice-rich composition of comets (≥50% ice?) also means that their Ir content is probably much less than that of asteroids and could lead to

TABLE 2. ELEVATED IRIDIUM REPORTED AT GEOLOGIC BOUNDARIES*

Locality	Ir (ppt)	References
	Pliocene	
Core EL13-3, southeast Pacific	As much as 5,000	Kyte, 1988
	Middle–Late Miocene	
DSDP Site 588B, South Pacific	152	See Orth, 1989
	Late Eocene	
Various localities	~4,000 330	Ganapathy, 1982; Montanari et al., 1993
	Cretaceous/Tertiary	
Worldwide	1,000s	See Glen, 1994
	Cenomanian/Turonian	
Western US, Colombia, S.A., Western Europe	560	See Orth, 1989; Orth et al., 1993
	Jurassic/Cretaceous	
Central Siberia	ave. 7,800	Zhakarov et al., 1993
	Callovian/Oxfordian	
Spain and Poland	1,000–2,400?	Brochwicz-Lewinski et al., 1985
	Middle/Late Jurassic	
Italy	3,200	Rocchia et al., 1986
	Triassic/Jurassic	
Europe	400	McLaren and Goodfellow, 1990; Olsen et al., 1990
	Permian/Triassic	
Changxing, China	8,000?	See Orth, 1989
Meishan, China	600 ± 400?	Xu et al., 1989
Meishan, China	2,000?	Xu and Yan, 1993
Nammal, Pakistan	366	Xu et al., 1989
Bolzano, Italy	230	Xu et al., 1989
San Antonio, Italy	3,000?	Brandner, 1988
Tesero, Italy	135	Oddone and Vanucci, 1988
Casera Federata, Italy	100–145	....
Butterloch, Austria	90–95	....
Carnic Alps, Austria	165 230	Attrep et al., 1991
Lalung, India	73 114	Bhandari et al., 1992
	Mississippian/Pennsylvanian	
Texas	380	See Orth, 1989
	Devonian/Carboniferous	
Oklahoma	560	See Orth, 1989
S. China	150	Chai et al., 1987
	Frasnian/Famennian	
China	230	Wang et al., 1991
Australia	300	Orth, 1989; Wang et al., 1994
Europe	75–160	See Orth, 1989
	Ordovician/Silurian	
Anticosti Island, Canada	58	See Orth, 1989
Scotland	250	See Orth, 1989
China	230	Wang et al., 1993b
	Proterozoic/Cambrian	
China	2,900	See Orth, 1989 Xu et al., 1989

*Results widely considered questionable marked by ?.

actual comet yields of Ir per J less than 1/100 that of asteroids (van den Bergh, 1994).

2. Theoretical studies suggest that some very large impacts may produce relatively weak iridium anomalies because much of the vaporized impactor is blown off of the earth in the energetic collision (Vickery and Melosh, 1990).

3. The primary carrier of the iridium—for example microspherules—may be inhomogeneously mixed in the boundary layer, a situation that might lead to large variations in measured Ir content even at the same geologic locality (Xu et al., 1989; Xu and Yan, 1993).

4. Many extinction events are represented primarily in shallow-water deposits, where bioturbation, reworking, and diagenesis of sediments could dilute Ir concentrations. Mobility of Ir and other trace metals under changing redox conditions and biological activity might significantly change the elemental ratios of trace-metal anomalies (McLaren and Goodfellow, 1990; Colodner et al., 1992; Wang et al., 1993b).

A MODEL FOR IMPACT-INDUCED EXTINCTIONS

The K/T boundary

As there is now considerable evidence for an impact at the K/T boundary, McLaren and Goodfellow (1990) argued that other impact-related extinction boundaries might be expected to show a number of features in common with the K/T. These include (1) a globally synchronous or near-synchronous mass mortality level (or levels); (2) a marked negative shift in $\delta^{13}C$ in marine carbonates, indicating a biomass loss and drop in productivity (the Strangelove ocean effect), coupled with a proliferation of opportunistic species, followed by recovery and radiation of surviving biota; (3) a marked negative shift in $\delta^{18}O$, suggesting a brief global warming; (4) a positive shift in $\delta^{34}S$, suggesting anoxic ocean bottom waters; and (5) a reduction in marine biogenic $CaCO_3$. McLaren and Goodfellow (1990) reported that published studies suggested such a common sequence at extinction horizons in the Late Triassic, Late Permian, Frasnian-Famennian, Late Ordovician, and possibly the Proterozoic-Cambrian.

There has been much debate about the nature of the fossil record at mass extinction boundaries such as the K/T, and several studies have suggested that the extinctions were gradual or stepped (e.g., Sloan et al., 1986; Keller, 1988, 1989; Keller et al., 1993; but see Huber et al., 1994). Signor and Lipps (1982) pointed out that the vagaries of preservation would cause truncations of species ranges below the true level of extinction and thus would tend to make abrupt extinctions appear gradual. Some recent tests of the effects of sampling on faunal occurrence of vertebrates and invertebrates at the K/T boundary (Sheehan et al., 1991; Raup, 1989) have concluded that sections that seem to show gradual extinction could be explained by the Signor-Lipps Effect. Recent studies suggest that bone beds and concentrations of skeletal remains should not be expected at mass-extinction boundaries (Cutler and Behrensmeyer, 1994).

Carbon-isotope shift. The K/T boundary is marked by a worldwide negative anomaly in $\delta^{13}C$ of ≥2‰ in the carbonate fine fraction (largely nannoplankton debris) and in individual planktonic foraminifers (Fig. 3) lasting for several hundred thousand years (Smit, 1990). The boundary is also commonly marked by a drastic reduction in $CaCO_3$ deposition in the deep sea, reflecting primarily the mass extinction of calcareous nannoplankton and planktonic foraminifera (Caldeira and Rampino, 1993) (Fig. 4). Biogenic pelagic $CaCO_3$ deposition was depressed for up to 1 m.y. after the boundary event in some places.

A mass mortality event would have significant effects on the cycling of carbon in the ocean. One predicted effect would be an abrupt negative shift in $\delta^{13}C$ of several per mil in marine carbonates and the organic fraction (Zachos and Arthur, 1986; Magaritz, 1989; McLaren and Goodfellow, 1990). The negative shift in $\delta^{13}C$ indicates a virtual cessation of global marine productivity, the so-called Strangelove-ocean effect, accompanied by a homogenization of the $\delta^{13}C$ content of surface and deep waters (Hsü and McKenzie, 1985, 1990; Kump, 1991). Recently, it has been suggested that a shift in $\delta^{13}C$ in surface waters might result from widespread combustion of terrestrial biomass, with transfer of isotopically light carbon to ocean surface waters (Ivany and Salawitch, 1993). Such a mass mortality event might be coincident with mass extinctions, but it should be emphasized that the severity and relative timing of the extinctions of various organisms may depend on many factors, both biological—including population size, distribution, habitat, and ecology—and geological—including the state of climate, sea level, and ocean chemistry.

Oxygen- and sulfur-isotope shifts. Although oxygen-isotope studies across the K/T boundary have given somewhat conflicting results (see Caldeira and Rampino, 1993), in a number of the most complete K/T boundary sections a marked ~2‰ negative shift in $\delta^{18}O$, beginning at the boundary and lasting at least sev-

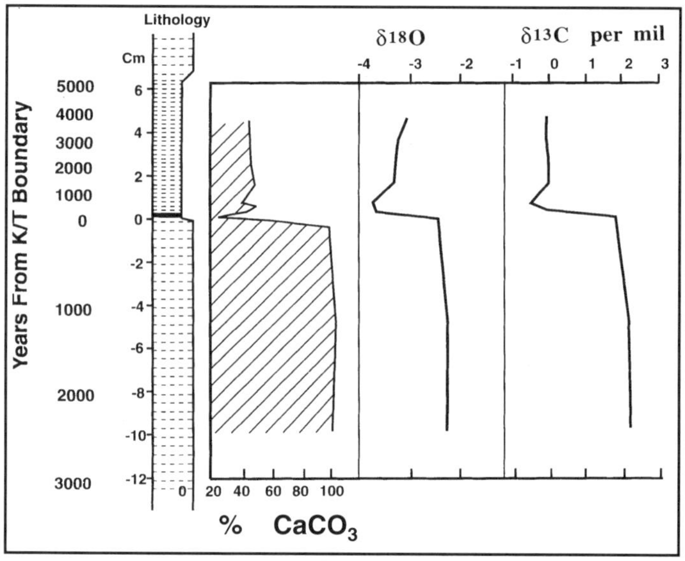

Figure 3. Detailed K/T boundary section from Agost, Spain, showing negative oxygen- and carbon-isotope shifts at the K/T boundary (as marked by an iridium anomaly) and severe reduction in pelagic biogenic calcium carbonate deposition (after Smit, 1990).

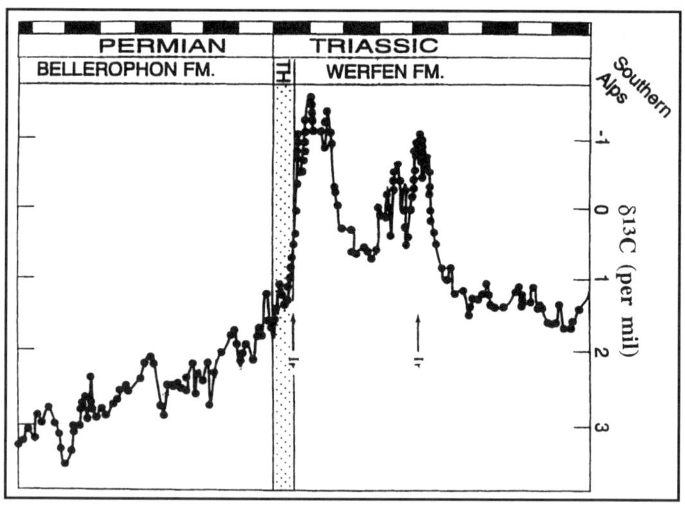

Figure 4. Carbon-isotope curve at the Permian-Triassic boundary in Core GK-1 from the Carnic Alps, Austria. Abrupt decrease in $\delta^{13}C$ of ~3 per mil at or near the boundary with partial recovery and subsequent decreases of 0.5 to >1 per mil in the earliest Triassic, over about 50 m of section. The bars on the horizontal scale represent standard time units (STUs) of ~110,000 years based on results of graphic correlation studies with other sections (Sweet et al., 1992). TH = the Tesero Horizon of the Werfen Fm (stippled). The most rapid negative shifts in $\delta^{13}C$ take place over ~4 m, possibly ≤50,000 yr. Ir measurements for the core show elevated Ir (lower anomaly = 165 ppt; upper anomaly = 230 ppt; on a carbonate-free basis) at times of negative shifts in $\delta^{13}C$ (Attrep et al., 1991).

eral thousand years, has been interpreted as representing a rapid warming of up to ~8°C (Fig. 3) (Smit, 1990).

It has also been suggested that short-term impact-induced cooling might help to trigger long intervals of decreased temperatures, such as ice ages or more severe periods within ice ages, through the positive feedback effects of increased snow and ice cover (Kyte et al., 1988). In that case, a marked positive oxygen-isotope anomaly, indicating cooling and growth of ice sheets, might occur at about the same time as some mass extinctions.

A sharp positive sulfur-isotope excursion reported at some K/T boundary sections suggests a depletion of oxygen in ocean waters in at least some areas at the time (Kajiwara and Kaiho, 1992), a condition that is consistent with organic-rich (Meyers and Simoneit, 1989) and pyrite-rich sediments in some K/T boundary sections. These conditions may have been a result, rather than a cause, of the extinctions.

Severe ecological disturbances. Plant fossils and palynological data from several sites reveal evidence for a profound ecological disturbance at the K/T boundary, with a general floral extinction most pronounced in megathermal broad-leaved evergreen vegetation and lowest in broad-leaved deciduous species. Among conifers, evergreen species were especially hard hit. These patterns suggest that plants that could enter a dormancy period were more likely to survive (Boulter et al., 1988). Land animals were also hit very hard at the end of the Cretaceous, with extinction at the family level of 83% for birds, 54% for reptiles (including all of the 24 families of dinosaurs), and ~26% for mammals (data from Benton, 1993).

Late Cretaceous palynomorph assemblages commonly show severe reductions in abundance and diversity at the boundary, followed by a thin barren interval and a sudden rise in spore abundance—the so-called fern spike. Wolfe and Upchurch (1986) recognized five vegetational-floristic phases across the K/T boundary in the western United States, from a high-diversity, broad-leaved evergreen flora in the latest Cretaceous, followed by the brief fern-dominated regime at the boundary, through recovery to a low-diversity rain forest over perhaps 1.5 m.y. in the earliest Tertiary. Increase in fungal spores at the boundary has also been observed (Bhandari et al., 1994). The increase in sporomorphs and land-derived organic matter reported in some marine K/T boundary sections (Brinkhuis and Zachariasse, 1988) also suggests a perturbation of terrestrial environments.

Plankton extinction in the oceans seems to have been abrupt and catastrophic (e.g., Smit, 1990; Pospichal, 1994; but see Keller 1988, 1989; Keller et al., 1993). The proliferation and episodic blooms of opportunistic species (e.g., acritarchs, dinoflagellates, coccoliths such as *Thoracosphaera*, etc.) and possibly bacteria (a "respiring ocean," Hollander et al., 1993) in the oceans just after the K/T boundary indicate a disturbed and generally impoverished and low-diversity marine ecosystem (Brinkhuis and Zachariasse, 1988; Alcala-Herrera et al., 1992).

EVIDENCE OF ENVIRONMENTAL PERTURBATIONS AND POSSIBLE IMPACTS AT GEOLOGICAL BOUNDARIES

Late Pliocene

Recent detailed study of ocean drill sites in the North Atlantic shows abrupt shifts in lithology and oxygen-isotopic composition in the latest Pliocene (Sancetta et al., 1992), with the occurrence of three especially severe glacial events between ~2.4 and 2.3 Ma. Sancetta et al. (1992) interpret the period as an "interruption" of the climate system, marked by a decline in primary productivity and calcite preservation, increased off-shelf transport (lowered sea level), and apparently hyperaridity on the continents. The sudden events at ~2.3 to 2.4 Ma may constitute an abrupt steplike shift from one climate regime to another (Shackleton et al., 1984; Raymo et al., 1989; Groot, 1991). A high-resolution sea-surface temperature record from the northeast Atlantic has provided additional evidence of a steplike cooling of ~9°C at about 2.4 Ma, coincident with near full-glacial $\delta^{18}O$ values, and the first major episode of ice rafting in the North Atlantic (Dowsett and Loubere, 1992).

The same Late Pliocene time interval is marked by the most recent extinction event of marine genera identified by Sepkoski (1993), including the disappearance of the characteristic molluscan fauna in the western Atlantic (Stanley and Campbell, 1985). The mixed marine, freshwater, and terrestrial organisms, including abundant aquatic birds (Jones et al., 1991; Emslie and Morgan, 1994), in spectacular bone and shell beds in Florida that are dated at between ~2.5 and 2 Ma, suggest concentration as a result of some kind of catastrophic event(s).

Kyte et al. (1981) discovered evidence (high iridium, microspherules, meteoritic debris) for the impact of a small (~0.5 km) asteroid into the southeast Pacific Ocean in the Late Pliocene (about 2.3 m.y. ago) and later suggested that it may have triggered the concurrent latest Pliocene increase in Northern Hemisphere ice volume (Kyte et al., 1988). Recently, Peng (1994) reported a large number of microtektites, dated by magnetic stratigraphy as between 2.14 and 2.30 Ma, in a North Pacific deep-sea core. The North Pacific microtektite layer is correlated with the extinction of three characteristic late Tertiary radiolarians. Peng (1994) also reported that microtektites have been found in clay between the Wuchen loess and paleosol in China, dated at ~2.4 Ma.

The Middle to Late Miocene

The Middle to Late Miocene was a time of significant global changes, including global cooling, major growth of the East Antarctic Ice Sheet, changes in deep-water circulation, and fluctuations of sea level (Flower and Kennett, 1993, 1994). A significant environmental change near the Early-Middle Miocene boundary is indicated by the initiation of Monterey Formation sedimentation in California at ~17.85 Ma, with a pronounced positive shift in $\delta^{18}O$ at ~16.3 Ma (Woodruff and Savin, 1991; Hodell et al., 1991; DePaolo and Finger, 1991; Compton et al., 1990). Sepkoski's (1992) data show elevated rates of extinction of marine organisms in the Middle Miocene (~3 times that of the Early Miocene). The Miocene faunal turnovers in mammals began at ~16 Ma in North America and ~17 Ma in Europe (Savage, 1988), and reptiles apparently show a drop in diversity in the Late Miocene (Benton, 1988).

Glassy microspherules (some containing olivine crystallites and showing impact scars) from 0.3 to 2 mm in diameter have been reported from DSDP Site 32 in the East Pacific Ocean (Vallier et al., 1977; Melson et al., 1988). The microspherules are basaltic in composition (~50% silica), and a K/Ar age determination of 16.3 ± 2 Ma suggests an Early to Middle Miocene age. Elevated levels of Ir have not been detected in these spherules (F. Kyte, personal communication, 1994). Although these microspherules and associated rock fragments have been interpreted as derived from volcanic eruptions, an impact origin cannot yet be ruled out. A scan of Upper Miocene sediments from DSDP Site 588B in the South Pacific (dated at about 11.7 Ma) detected a weak peak of Ir of 152 ppt over a background of 11 ppt, with elevated levels of Cr, Co, and Ni (see Orth, 1989).

The Eocene-Oligocene transition (Mid-Tertiary Crisis)

Marine species suffered a ~35% extinction from Middle or Late Eocene through Early Oligocene times (Sepkoski, 1992 data). This "Mid-Tertiary Crisis" interval represents the major faunal turnover in the last 65 Ma, but differs from the K/T events in that the Eocene-Oligocene (E/O) transition took place in a series of discrete steps over several m.y. (Hut et al., 1987; Berggren and Prothero, 1992; Prothero, 1994). For example, planktonic foraminifera seem to show several stepwise extinction events over 1 to 2 m.y. in the Late Eocene, each characterized by generally less than 15% species extinction but with a total turnover involving >60% of the plankton population (Keller, 1986). The lack of a clear-cut boundary has resulted in formal definition of the end of the Eocene (at ~34 Ma) by the last occurrence of Hantkeninid foraminifera, although the climatic and biological events that mark the E/O transition may have begun at ~37 Ma or even earlier (Prothero, 1994). The Eocene-Oligocene transition is also marked by mammalian extinctions in North America and elsewhere (see Berggren and Prothero, 1992; Prothero, 1994) and the "Grande Coupure," a significant extinction of European amphibians, reptiles, birds, and mammals (Hartenberger, 1987; Benton, 1988).

The Mid-Tertiary Crisis also involved several episodes of cooling and ice-volume increase, evidenced by abrupt global increases in $\delta^{18}O$ in planktonic and benthic foraminifera (Miller et al., 1991; Berggren and Prothero, 1992; Prothero, 1994). For example, Zachos et al. (1992) found the E/O transition to be marked by very rapid climatic changes, with a spike of ice-rafted sediment in the southern Indian Ocean (the onset of continental East Antarctic glaciation), and by rapid positive shifts in the carbon- and oxygen-isotope records of planktonic and benthic organisms at about 35.8 Ma. Although the positive oxygen-isotope shift was permanent, the pulse of ice rafting apparently lasted less than 100,000 yrs, perhaps as little as 10,000 yrs.

The Mid-Tertiary Crisis interval is accompanied by evidence of multiple impacts, represented by microtektite and microspherule layers (at least one with shocked quartz), and iridium peaks, although apparently none of the ejecta layers has worldwide distribution (Miller et al., 1991; Montanari et al., 1993). A major problem has been to determine the number, age, and correlation of the various impact-related layers (e.g., Miller et al., 1991; Montanari, 1990; Montanari et al., 1993). Some impact layers may correlate with faunal turnover or climatic events (e.g., Sanfilippo et al., 1985; Hut et al., 1987), whereas others apparently do not. For example, one microtektite layer may be coincident with a 68% decrease in species diversity of Gulf Coast mollusks and the extinction of six species of planktonic foraminifera in the equatorial Pacific (Hut et al., 1987).

Several iridium anomalies with values up to ~330 ppt have been reported from Upper Eocene sediments spanning about 1 m.y. (e.g., Alvarez et al., 1982; Montanari, 1990; Montanari et al., 1993). Ganapathy (1982) reported a peak of ~4,000 ppt associated with clinopyroxene microspherules in Caribbean Sea cores. A major impact structure, Popigai in Siberia (~100-km diameter), and several smaller craters (Azuara, Spain; Chesapeake, United States) were formed during the E/O transition interval. Thus, although the evidence seems to argue against a one-to-one correlation of the impact events with climatic and extinction events, the concentration of impact evidence in the mid-Tertiary is intriguing, and the relationships might be complex.

The Cenomanian-Turonian boundary

An apparent stepwise pattern of biotic change in the marine realm occurs over ~1 to 2 m.y. at the Cenomanian-Turonian (C/T) boundary (Kauffman, 1986, 1995; Jarvis et al., 1988) ($^{40}Ar/^{39}Ar$ age determinations suggest an age of 94 ± 0.4 Ma for the C/T boundary in Colorado; Collom, 1990.) Extinction rates among reptiles, mammals, and birds were also apparently high in the Cenomanian (Benton, 1988). The latest Cenomanian is marked by a general increase in organic material in marine sediments and a positive shift (with large fluctuations) in $\delta^{13}C$ in marine carbonates and organics (Kauffman, 1986). A negative oxygen-isotope shift of ≥2‰ has been reported close to the boundary and may correlate with a negative shift in $\delta^{13}C$ (Peryt and Wyrwicka, 1993).

Two closely spaced iridium peaks with values up to 560 ppt, coinciding with extinction steps (and several smaller spikes), occur widely in western North America; in Colombia, South America; and in some localities in Europe (Orth et al., 1993; Kauffman, 1995). Microspherules and shocked quartz are currently being searched for in C/T boundary sections, and grains with possible shock lamellae occur in Colorado (M. Rampino, unpublished data). Two well-dated impact structures have ages that overlap the C/T boundary: Boltysh, Russia (25-km diameter, 88 ± 3 Ma) and Steen River, Canada (25-km diameter, 95 ± 7 Ma).

The Jurassic-Cretaceous boundary

The Jurassic-Cretaceous (J/K) boundary is still poorly defined but is currently taken at the base of the Berriasian Stage (Ogg and Lowrie, 1986; Remane, 1991). Dating of the boundary is also uncertain; an average of 18 published time scales since 1959 gives 138 Ma, with a range from 129 to 144 Ma, whereas the most recent Mesozoic time scale gives 144.2 ± 2.6 Ma for the J/K boundary (Gradstein et al., 1994).

Several workers have argued that the Jurassic-Cretaceous boundary (the Tithonian–Berriasian Stage boundary) does not correspond to an important faunal change (Remane, 1991; Zhakarov et al., 1993). Sepkoski's latest (1992, and personal communication, 1993) data, however, give a 12.7% extinction of marine genera in the upper Tithonian relative to only 7.9% in the lower Tithonian and 5.2% in the Berriasian Stages. Reptiles show a marked peak in extinction at the end of the Jurassic (Benton, 1988), and several lines of dinosaurs became extinct or greatly reduced in numbers, possibly setting the stage for angiosperm radiation (Bakker, 1978). Recent work in the western United States has supported a latest Jurassic extinction of dinosaurs and radiation of new forms in the earliest Cretaceous (R. Bakker, personal communication, 1988).

A thin layer of phosphatic limestone suggested as marking the J/K boundary in Siberia has been reported to contain high levels of Ir (averaging about 8 ppb), other platinum group elements in near-chondritic proportions, and microspherules (Zhakarov et al., 1993). Recent work suggests that this layer is actually in the lower Cretaceous, at the early-middle Berriasian boundary (~142 Ma) (J. Ogg, personal communication, 1994).

The Triassic-Jurassic boundary

Marine groups show an abrupt extinction (34.1% of marine genera in the upper Norian) at the Triassic-Jurassic (Tr/Jr) boundary (Sepkoski, 1992), and the few calcareous nannoplankton then in existence exhibit an almost complete turnover at the boundary (Hallam, 1990). Negative shifts in oxygen- and carbon-isotope ratios have been noted in some European marine boundary sites (McLaren and Goodfellow, 1990).

Extinction rates for land vertebrates were also high in the latest Triassic (Benton, 1988), and the vertebrate extinction seems to have occurred within ≤700,000 years but is still poorly constrained (Olsen et al., 1990). Weems (1992) has argued that the Late Triassic vertebrate extinctions were gradual or perhaps stepwise, but this interpretation is based largely on tabulations of family diversity, with limited resolution of possible abrupt extinction events in the record. The floral transition at the end of the Triassic is quite marked in the Newark deposits of eastern North America, just prior to the eruption of the Newark Basalts (201 ± 1 Ma). The pollen and spore transition, showing a fern spore spike (similar to that seen at the K/T boundary) followed by gradual replacement of flora, is estimated to have taken place in <20,000 yrs (Olsen et al, 1990).

Iridium peaks of ~200 to 400 ppt over a background of ~30 ppt have been reported at the Tr/Jr boundary in Europe, and the occurrence of three closely spaced layers of probable shocked quartz in Europe suggests possible multiple impacts (McLaren and Goodfellow, 1990; Bice et al., 1992). The ~100 km diameter Manicouagan structure in Quebec (212 ± 2 Ma) is a suspect.

The Permian-Triassic boundary

Although the latest Paleozoic extinctions are commonly regarded as gradual in nature and/or composed of a series of extinction episodes ranging from mid-Permian to Early Triassic (Maxwell, 1992; Erwin 1993, 1994), recent compilation of fossil data suggests that the Permian extinctions occurred in two waves, at the end of the Guadalupian Stage and at the end of the final Permian Tatarian Stage, separated by ~5 m.y. (Stanley and Yang, 1994). Recent studies from a number of localities are converging on the conclusion that the end-Tatarian extinction event was severe and abrupt (see Xu, 1991; Sweet et al., 1992; Eshet et al., 1995; Retallack, 1995).

The Permian-Triassic (P/Tr) boundary (U/Pb age of 251 ± 3.4 Ma) is associated with the disappearance of up to ~96% of species in the oceans (Raup, 1979). Planktonic and nektonic groups and groups with planktonic growth stages show an abrupt crash at the boundary (Valentine, 1986; Li et al., 1991), followed by proliferation of acritarchs, and biogenic carbonate decreases in many sections. The boundary is marked by a global environmental disturbance followed by an explosion of stress-tolerant and opportunistic organisms. In the shallow oceans, the extinctions are followed by a period dominated by algal stromatolites (Schubert and Bottjer, 1992) and primitive brachiopods. On land, many major insect groups, 99% of reptile genera, and >95% of land plants disappeared (Labandiera and Sepkoski, 1993; Retallack, 1995; Eshet et al., 1995). Typical Late Permian gymnosperm palynomorphs disappear at a global stratigraphic horizon that contains only fungal spores and reworked organic matter, followed by gradual recovery of lycopods and gymnosperms (Balme, 1979; Eshet, 1990; Visscher and Brugman, 1988; Retallack, 1995; Eshet et al., 1995).

The P/Tr boundary is marked by significant shifts in carbon, oxygen, sulfur, and strontium isotope ratios (Xu et al., 1989; Magaritz et al., 1992; Magaritz et al., 1988; Holser et al., 1991). The negative carbon-isotope shift at the boundary has been described as gradual (Erwin, 1993; Holser et al., 1991; Magaritz et al., 1988). However, the data from the complete GK-1 Core from the Carnic Alps show that prior to the boundary $\delta^{13}C$ was decreasing by only ~0.02‰ m^{-1} (Fig. 4). At the boundary, the rate changed abruptly to >0.5‰ m^{-1}, and the major decrease occurs over ≤4 m of sediment, which may represent a time of ≤50,000 yrs.

The two large negative excursions in $\delta^{13}C$ seen in the GK-1 Core are marked by Ir peaks that reach 165 ppt and 230 ppt above a background of ~30 ppt Ir (Fig. 4) (Holser et al., 1991; Attrep et al., 1991). Furthermore, Ir/Al plots show that the Ir in these two peaks is the result of processes that are clearly different from those responsible for the Ir in the bulk of the section (Attrep et al., 1991). The major carbon-isotope anomaly is followed by abrupt negative shifts in $\delta^{18}O$ (Holser and Magaritz, 1992; Magaritz et al., 1988) that may indicate a drastic warming. A marked and abrupt positive shift in $\delta^{34}S$ seen at the boundary in some sections suggests the development of anoxic conditions in the oceans (Xu et al., 1989).

Two or more iridium peaks in the 100s of ppt range have been reported in the boundary zone at a number of localities (the Austrian Alps, Italy, China, and India) (Erwin, 1993; Xu et al., 1989; Holser and Magaritz, 1992; Magaritz et al., 1988; Holser et al., 1991; Bhandari et al., 1992) (Table 2). Larger Ir anomalies reported from some Chinese sections (>2 ppb at Shangsi; Xu et al., 1989) and from Italy (~4 ppb; Brandner, 1988) have not been reproduced in other laboratories (see Orth, 1989; Xu et al., 1989). One possible problem is that the carrier phase for the Ir may be discontinuously distributed in the boundary layer(s), for example, microspherules are reportedly common in Chinese sections (Xu et al., 1989). On the other hand, the boundary clay in China seems to have a volcanic signature (Zhou and Kyte, 1988).

Recently, an Ir peak of 2 ppb was again reported in China, this time in a very thin band ~10 cm above the base of the Triassic in the Meishan section, coincident with a spherule-rich layer and a negative $\delta^{13}C$ shift of 6 per mil (Xu and Yan, 1993). The ~40 km wide Araguinha impact structure in Brazil has an $^{40}Ar/^{39}Ar$ age of 247 ± 5.5 Ma (Engelhardt et al., 1992) but is probably too small to be connected (by itself) with such a major mass extinction.

The Devonian-Carboniferous boundary

Extinction patterns in the Devonian may be complex, but data for families and genera indicate three major events: in the late Givetian, at the Frasnian-Famennian boundary, and in the late Famennian (Devonian-Carboniferous boundary). The late Famennian is apparently marked by ~30% extinction of marine genera (Sepkoski, 1992, 1994). This extinction includes the apparently sudden disappearance of well-known groups of ammonoides, trilobites, placoderm, and agnathan fish and the collapse of a dominant late Famennian plant community. Extinctions among other groups may have been more gradual or stepwise, although this could be a Signor-Lipps effect. Earliest Carboniferous faunas apparently had very low diversity (Wang et al., 1993a).

A major sudden negative $\delta^{13}C$ shift has been reported in the lowermost Carboniferous in the Muhua section in China, where oxygen isotopes show fluctuations of 2 to 3‰ that might be climatically induced (Xu et al., 1985). In fact, several recognizable extinction events among brachiopods during the Devonian seem to be associated with negative $\delta^{13}C$ anomalies (Talent et al., 1993).

Iridium measurements of the Devonian-Carboniferous boundary in Europe, Canada, the United States, and China yielded values from 138 to 258 ppt above background values of 16 to 45 ppt (Orth, et al., 1990). However, the ratios of Ir to other important elements, such as Au, Ni, and Co, are not chondritic, and the Ir anomalies at the boundary are associated with the deposition of the Hangenberg black shale in Europe and similar deposits in other basins (see Orth et al., 1990). Wang et al. (1993a) suggested that the anomalies might be the result of precipitation of Ir from seawater as a result of a sudden change in paleo-redox conditions. They did not rule out the possibility that development of ocean anoxic conditions on a global scale and enhanced dissolved iridium in seawater were the result of an extraterrestrial impact or impacts. Shocked quartz or microtektites were not searched for.

Six impact craters (Grieve, 1991; Grieve and Shoemaker, 1994) have radiometric ages that overlap the estimated age of the late Givetian to Devonian-Carboniferous boundary (~375 to 360 Ma), although none of these seems large enough to be associated with a major mass extinction event. An unusually long lived impactor shower—perhaps an asteroid surge as predicted by Shoemaker et al. (1990)—is a possibility.

The Frasnian-Famennian boundary

Recent studies suggest that the Frasnian-Famennian (F/F) extinctions involved catastrophic collapse of plankton communities and the marine ecosystem (McGhee, 1988, 1989), although the extinctions of various groups may have occurred in discrete steps (McLaren and Goodfellow, 1990; Schindler, 1990). The boundary is marked by sudden impoverishment of fauna, an unusual layer rich in sponge spicules, and a brief burst of cyanobacterial and algal activity, suggesting an explosion of opportunistic species (Geldsetzer et al., 1993). Sandberg et al. (1988) suggested that the major marine extinction event was very abrupt (<20,000 years), with the demise of Frasnian reefs having taken place possibly ~1 m.y. earlier. They also noted that debris flow and chaotic deposits, possibly related to tsunami, are common at the boundary in many places (e.g., Australia, China, Europe, Nevada). Floral extinctions are indicated by a 43% drop in species diversity of spores from the Frasnian to the Famennian (Boulter et al., 1988).

A negative shift in $\delta^{13}C$ of 2 to 3‰ occurs at a number of localities (e.g., Australia, Europe, China, northwest Canada), indicating a loss of ocean productivity (McLaren and Goodfellow, 1990; Goodfellow et al., 1989; Geldsetzer et al., 1993). A large positive sulfur-isotope shift suggests reducing conditions in the oceans (Geldsetzer et al., 1987). Fluctuations of $\delta^{18}O$ of

~1 to 2‰ may indicate climatic fluctuations of up to 8°C at the boundary (Li et al., 1991; Wang et al., 1991).

Ir levels of ~75 to 160 ppt and up to ~300 ppt over a background of ~20 to 30 ppt (see Orth, 1989; Orth et al., 1990; Wang et al., 1991, 1994; Geldsetzer et al., 1993) have been reported at or near the Frasnian-Famennian boundary or within the Famennian (~1 to 1.5 m.y. after the F/F boundary) in widely separated areas (e.g., China, Australia). In China, a large carbon-isotope shift of –3.5‰ coincides with the Ir anomaly (Wang et al., 1991). High levels of other trace metals are also present (Goodfellow et al., 1989; Geldsetzer et al., 1993). The iridium peak in Australia, which occurs above the designated F/F boundary (based on conodonts), may have been formed by the trapping of oceanic Ir by the cyanophyte *Frutexites* after the mass extinction (McLaren, 1985; McLaren and Goodfellow, 1990). Although these elevated Ir levels were originally interpreted as nonimpact in origin, the subsequent discovery of glassy microtektites associated with the late Frasnian and Famennian (Wang, 1992; Claeys et al., 1992; Claeys et al., 1994) suggests a possible reinterpretation of the relatively weak Ir peaks as impact related. At least two relatively large impact craters (Charlevoix, Quebec, 54 km in diameter, 357 ± 15 Ma, and Siljan, Sweden, 55 km in diameter, 368 ± 1.1 Ma) have dates that overlap the estimated age of the F/F boundary (Grieve, 1991).

The Ordovician-Silurian boundary

The major Late Ordovician crisis is marked by extinction of both benthic and pelagic faunas at a globally recognizable event horizon at the base of the *Glyptograptus persculptus* graptolite zone (Brenchley, 1989; McLaren and Goodfellow, 1990; Wang et al., 1993b, c). Until 1985, the Ordovician-Silurian (O/S) boundary was recognized at the base of that zone, which also marked a change from shelly faunas to shaly graptolite facies (Brenchley, 1989). The recent redefinition of the Ordovician-Silurian boundary entailed the inclusion of the *G. persculptus* zone within the Ordovician at the type section in Dobbs Linn, Scotland. However, in some places (for example, China) the base of the zone is still utilized as the boundary (Wang et al., 1993c).

The Late Ordovician marine extinctions may have occurred in two main steps with some possible minor ones (Sheehan, 1988; Brenchley, 1989). In some localities, three separate episodes of extinction within <500,000 yrs are reported to have occurred in the latest Ordovician, with a pattern that suggests that trilobites, echinoderms, and plankton living in temperate latitudes disappeared in the first event, followed by brachiopods and corals in shallow tropical seas in the later events. Incomplete preservation and sampling (the Signor-Lipps Effect), however, may be contributing to the apparently stepped record. Phytoplankton and organisms with planktonic larval stages show an abrupt global extinction at or near the boundary, coincident with extinctions among benthic organisms (Colbath, 1986).

The base of the *G. persculptus* graptolite zone is also marked by a negative (3 to 4‰) carbon-isotope anomaly as well as negative oxygen-isotope anomalies and by a positive sulfur-isotope excursion (Long, 1993; McLaren and Goodfellow, 1990; Goodfellow et al., 1992; Wang et al., 1993b). Cooling and sea-level fall associated with glaciation may have been a factor in the Late Ordovician extinctions, as shallow, epicontinental seas populated by warm-water adapted species were drained (Sheehan, 1988; Wang et al., 1993b), and glaciation may have lasted only 0.5 to 1 m.y. (Brenchley et al., 1994). On the other hand, Long (1993) recently estimated maximum global glacio-eustatic sea-level changes of ±36 m (±14 × 10^6 km^3 of ice) in the Late Ordovician. The much greater fluctuations of climate and sea level during the Pleistocene glaciations, with variations of sea level of ±120 m (±50 × 10^6 km^3 of ice), caused no detectable marine extinction events (Wise and Schopf, 1981), although it must be noted that epicontinental seas were not widespread in the Pleistocene, and hence fewer organisms were at risk.

An iridium peak (up to 230 ppt) and other trace-metal anomalies have been reported at or near the boundary in mudstones or "boundary clays" in China (Wang et al., 1992) and northwestern Canada (Goodfellow et al., 1992; Wang et al., 1993c; Melchin et al., 1991), although some sections are reported to show only background amounts of Ir (Wang et al., 1993c). Goodfellow et al. (1992) reported siderophile and chalcophile trace-metal anomalies at the Ordovician-Silurian boundary in the Northern Yukon Territory, coinciding with a reduction in carbonate content, negative oxygen- and carbon-isotope excursions, and a positive sulfur-isotope shift. They suggested that these anomalies might be impact-related and that an Ir spike could have been missed by the spotty sampling. Orth et al. (1986) found only a weak Ir peak (58 ppt) in a clay layer at the recognized O/S boundary at Anticosti Island, Quebec (no shocked quartz was found), but maximum values of 250 ppt Ir were reported from the Dobbs Linn, Scotland, boundary site (Table 2).

The Proterozoic-Cambrian boundary

Rocks spanning the Proterozoic-Cambrian (Pz/C) boundary are marked by profound changes in global biota from nonskeletal organisms to those with mineralized skeletons and by changes in trace fossils (Magaritz, 1989; Myrow and Hiscott, 1993). Multiple $\delta^{13}C$ excursions span the uppermost Proterozoic to Lower Cambrian (Botomian) interval (~10 m.y.) (Hsü et al., 1985; Brasier et al., 1994). In Siberia, the Pz/C boundary zone is marked by a gradual positive shift in $\delta^{13}C$ of ~8‰, followed by a rapid negative shift of ~5‰ in the lowermost Tommotian that coincides with the appearance of the earliest brachiopods and archaeocyathids. A major negative excursion of ~3.5‰ also occurred in the Botomian, possibly coincident with a biotic crisis (Brasier et al., 1994). These fluctuations have been interpreted as indicating marked changes in oceanic productivity and biomass. Oxygen-isotope fluctuations of up to ±4‰ are reported in the same sequences, commonly with a pronounced negative shift at about the same level as the most negative carbon-isotope excursion (Xu et al., 1989; Aharon et al., 1987; Magaritz et al., 1986).

The "China C marker," separating the Tommotian and Atdabanian Series (~527 Ma), at the first appearance of trilobites, may be marked by a strong iridium anomaly (averaging ≥4,000 ppt) and an increase in Os and other trace elements correlating with major negative carbon- and oxygen-isotope anomalies (Xu et al., 1989; Fan et al., 1984). The large Lake Acraman impact structure (≥160 km in diameter) in Australia has an inferred age near the Pz/C boundary; however, its ejecta layer in late Precambrian sediments is apparently older than the China C marker (Gostin et al., 1989).

CORRELATION OF EXTINCTIONS AND EVIDENCE OF IMPACTS

Of the ~24 well-defined Phanerozoic extinction peaks (local maxima) in the genus-level data of Sepkoski (1992), six (if one includes the PZ/C boundary) seem to be associated with significant stratigraphic evidence of impacts (large Ir anomalies, shocked quartz, and/or microtektites) (Table 1). At least six more are associated with "possible" evidence of impact consisting of lower concentrations of Ir but still anomalous with respect to background values (Table 2). Several other boundaries (e.g., the Jurassic-Cretaceous, Callovian-Oxfordian, Devonian-Carboniferous) may show significant iridium enrichment, but data are scarce (Table 2). Although little work on iridium levels in sediments away from boundaries has been done, studies thus far suggest that iridium anomalies above background may be uncommon in the geologic record (Kyte and Wasson, 1986; Alvarez et al., 1990). For example, Kyte and Wasson (1986) found no iridium peaks in a slowly deposited deep-sea core over an interval of ~30 m.y. between the K/T boundary peak and smaller Ir enrichments in the Late Eocene.

A simple test of the significance of this correlation might be to consider the chances of hitting a target time series consisting of 24 extinction events (with the age error bars in this case taken as the difference in ages given by two recent time scales; Palmer, 1983; Harland et al., 1990), with a time series consisting of the six times of diagnostic stratigraphic evidence of impacts. The chance of such a correlation calculated in this way is very low (about 10^{-7}). The accidental correlation of all 12 "diagnostic" and "possible" indicators of impact with the mass extinction record is of even lower probability. Thus, given the assumption that the two time series are related, the correlation of mass-extinction events and evidence of large impacts is extremely unlikely to be an accidental occurrence.

THE KILL CURVE

Making use of fossil-range data, Raup (1990, 1991a, b, 1992) constructed a "species kill curve" that represents a best fit to the mean waiting time between extinctions of various severity. He also computed the waiting time of terrestrial impact craters of various sizes, and a possible theoretical relationship between mass extinctions and impact cratering was then derived (Fig. 5).

Using the known rate of impacts and the hypothesized relationship between impacts and mass extinctions, Raup (1990) performed Monte Carlo simulations to determine the appearance of the extinction record supposing that impacts are sufficient to explain the entire record of the past 600 m.y. The results of these simulations compared well with the actual extinction record based on several criteria, including the severity of extinction events, spacing of geologic stages defined by faunal turnover, and the general level of background extinction.

The theoretical curve can be compared with data representing the known large (>70 km) impact craters with relatively well defined ages (Grieve, 1991; Grieve and Shoemaker, 1994) that overlap (within error bars) the ages of mass-extinction boundaries (Rampino and Haggerty, 1994) (Table 3). An additional crater, the Duolun structure in North China (~100 km in diameter, with an estimated age of ~136 Ma; Wu, 1987) may correlate with the Jurassic-Cretaceous boundary event (~20% extinction of marine genera). The observed points agree with the predicted curve within the envelope of error estimated by Raup (1992). On the other hand, the association of craters of ~80- to 100-km diameter with extinctions ranging from ~25 to 60% extinction of species suggests a step or threshold in the kill curve in that range, perhaps representing the minimum impact energy (~a few times 10^7 Mt) at which the effects of thermal radiation from reentering ejecta (e.g., global wildfires) become important (Melosh et al., 1990; Toon et al., 1994).

The shape of the species kill curve suggests that for craters smaller than ~50 to 60 km in diameter there will be no associated extinction pulse that stands above the ~20 to 25% level of background species extinction (~10% genera) (see Jansa et al., 1990), and indeed there are well-dated craters in this size range

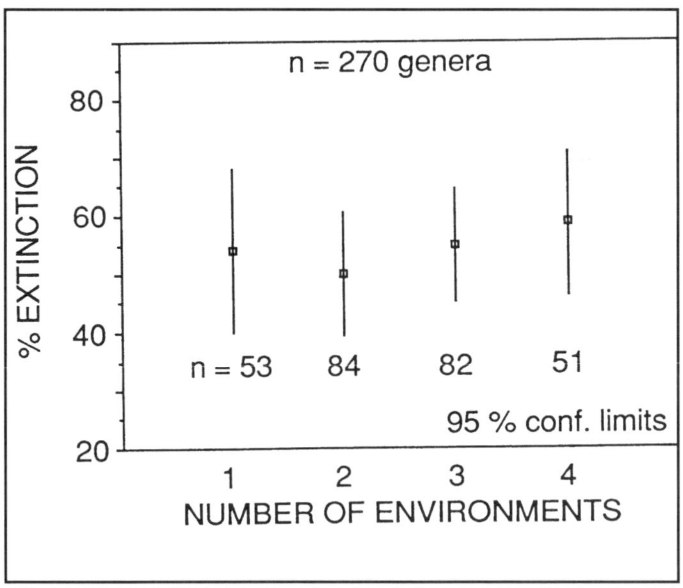

Figure 5. Kill curve for Phanerozoic marine species, with estimated error bars, plotted against estimated size of impact craters associated with extinctions of various kill magnitudes, assuming that the two are actually related (Raup, 1991a, b). Estimated sizes of the largest craters with age error bars overlapping mass extinctions (Table 3) are plotted against the species kill magnitude of those extinctions.

TABLE 3. LARGE IMPACT CRATERS* AND CORRELATIVE(?) EXTINCTIONS†

Name	~Diameter (km)	Age (Ma)	Extinction	Genera (%)	Species (%)
Puzech-Katunki	80	220 ± 10	Carnian	20.1	43
Popigai	100	35 ± 5§	Late Eocene	11.4	25.5
Manicouagan	100	212 ± 2	Late Triassic	34.1	62
Chicxulub	200?	65.2 ± 0.4	K/T	38.5	67

*Grieve, 1991.
†Sepkoski, personal communication, 1993 data.
§Latest work indicates date of 36 + 1 Ma (J. Garvin, personal communication, 1993).

(e.g., Kara, Russia, ~65 km; Montagnais, Nova Scotian Shelf, ~45 km) that apparently do not correlate with significant pulses in extinction.

EVIDENCE FOR POSSIBLE PERIODICITY IN MASS EXTINCTIONS

Periodic components

The mean occurrence rate of the 24 extinction pulses shown in Figure 1 is one every ~23 m.y. Raup and Sepkoski (1984) identified 12 extinction events at the family level by geologic stages over the last 250 m.y. and reported a statistically significant 26.4 m.y. periodicity in the extinctions time series. Later studies found a similar periodicity in extinction of genera at the substage level (Raup and Sepkoski, 1986). Periods of ~26 to 31 m.y. have been derived using various subsets of extinction events (family and genus levels), different geologic time scales, and various methods of time-series analysis. The regularity, statistical significance, and reality of the dominant underlying periodicity have been subjects of intense debate (e.g., Stigler and Wagner, 1987; and see Rampino and Haggerty, 1996, for a discussion). Rampino and Stothers (1986) reported detection of an ~29 m.y. component in the record of extinctions of nonmarine vertebrates. It is important to note that the detection of a periodic component in extinctions does not imply either that the record contains a strict periodicity or that every extinction event follows a 26-m.y. timescale; the extinction record might be a mixture of periodic and random events (Rampino and Haggerty, 1995).

Using the data of Sepkoski (1992, 1994, and personal communication, 1993), we performed Fourier analysis (treating the extinction peaks as spikelike delta functions; see Rampino and Caldeira, 1992, 1993) on the 0- to 253-Ma record and found the highest spectral peak at 26.5 m.y. (Fig. 6). One of the problems in studies of periodicity has been the shortness of the record and the reported apparent lack of evidence of a statistically significant periodicity in extinction events prior to 250 m.y. ago. Fourier analysis of an extended record of 24 extinctions going back to the Early Cambrian (~540 Ma), more than doubling the length of the series, resulted in a spectrum with the highest peak at 27.3 m.y. (Fig. 6).

As a test for significance of the 27.3-m.y. peak, we examined the record (0 to 515 Ma) with the most power at 27.3 m.y. A measure of significance was obtained by generating 1,000 pseudodata sets, each containing the same number of randomly dated pseudoevents over the same time interval. For this analysis, the probability of generating higher spectral power at 27.3 m.y. was ~2%, but in further tests the probability of generating higher spectral power at any period between 10 and 65 Ma fell below the 95% confidence level.

We also tested the robustness of the ~27-m.y. peak by performing Fourier analysis on a series of truncated extinction time series starting from 0 to 540 Ma and subtracting one extinction at a time back to 253 Ma (Fig. 6). The robustness of the ~27 m.y. periodicity in the extinction time series is attested to by the fact that a stable peak between 26.5 and 27.3 m.y. remains the dominant feature of the Fourier spectrum in the truncated extinction data sets.

A proposed astronomical explanation for the possible periodicity in mass extinctions

If a general connection between impacts and mass extinctions exists, then the periodic component may be related to a similar period in galactic dynamics. In this model, increased flux of long-period comets (comet showers) occurs during the periodic passage of the Solar System through the central plane of the Milky Way Galaxy (~26 to 36 m.y. depending on galactic models) (Rampino and Stothers, 1984a; Clube and Napier, 1984). Models suggested for time modulation of the flux of Oort cloud comets include gravitational perturbations of the Oort comet cloud by adiabatically varying galactic tides (Matese et al., 1995) and/or by near encounters with massive objects concentrated in the plane (Rampino and Stothers, 1984a, b), although components of these models remain problematic (e.g., Weissman, 1986, and personal communication, 1994). Alternative models for producing periodic comet showers involving a companion star to the sun (Whitmire and Jackson, 1984; Davis et al., 1984) or a tenth planet (Whitmire and Matese, 1985) have also been suggested. Showers of impactors might explain the stepped nature of some extinction events (Hut et al., 1987).

Does the terrestrial record of impact craters contain a similar periodic component? Time-series analyses of terrestrial impact craters have provided some evidence of a possible 28- to 32-m.y. periodicity in impacts (Rampino and Stothers, 1984a, b; Alvarez and Muller, 1984; Rampino and Stothers, 1986; Yabushita 1991, 1992) and suggest that the impact record is a mixture of periodic and random events (Trefil and Raup, 1987; Fogg, 1989). Several studies have concluded that, considering the problems in dating, the observed differences in the periodicity detected in impact craters and mass extinctions are to be expected (Stothers, 1989; Fogg, 1989). The ~30-m.y. period is detected when an extended set of ~40 dated impact craters over the past 250 m.y. is used, but the signal can become much weaker when smaller subsets of the best-dated craters are analyzed. These studies are plagued by small number statistics, and other workers have not detected a significant ~30-m.y. component in subsets of craters (see Grieve, 1991; Grieve and Shoemaker, 1994). Notwithstanding the problems of detecting a significant periodicity, the impact-crater record has been interpreted as suggesting pulses at ~0, 35, 65, and 95 Ma (Hut et al., 1987), close to suggested times of galactic plane crossings. Increased fluxes of cosmic dust (evidenced by ^{3}He peaks), possibly from comet showers, occur at ~65 and 35 Ma and in the last few m.y. (Farley, 1995).

CONCLUSIONS

The numbers and dynamics of earth-crossing asteroids and comets predict that impactors greater than a few kilometers in diameter should collide with the earth on average every few tens of millions of years. Large impacts, releasing $>10^7$ Mt TNT equivalent of explosive energy, are calculated to produce global-scale disasters involving possible loss of photosynthesis, "impact winter," and other effects, with impact events releasing $\sim 10^8$ Mt TNT and larger producing additional severe effects such as a global energy pulse and wildfires from reentering ejecta. These environmental effects have been predicted to lead to mass mortality and subsequent extinctions and might also destabilize the climatic system, leading to longer-term environmental changes.

Extinction events, which may occur in steps, are commonly associated with negative shifts in carbon isotopes in marine sediments, suggesting abrupt mass mortality; with abrupt shifts in oxygen isotopes, indicating sudden climatic changes, and sometimes with positive sulfur-isotope anomalies, suggesting lowered oxygenation of ocean waters. In an increasing number of cases, the extinction boundaries are found to be

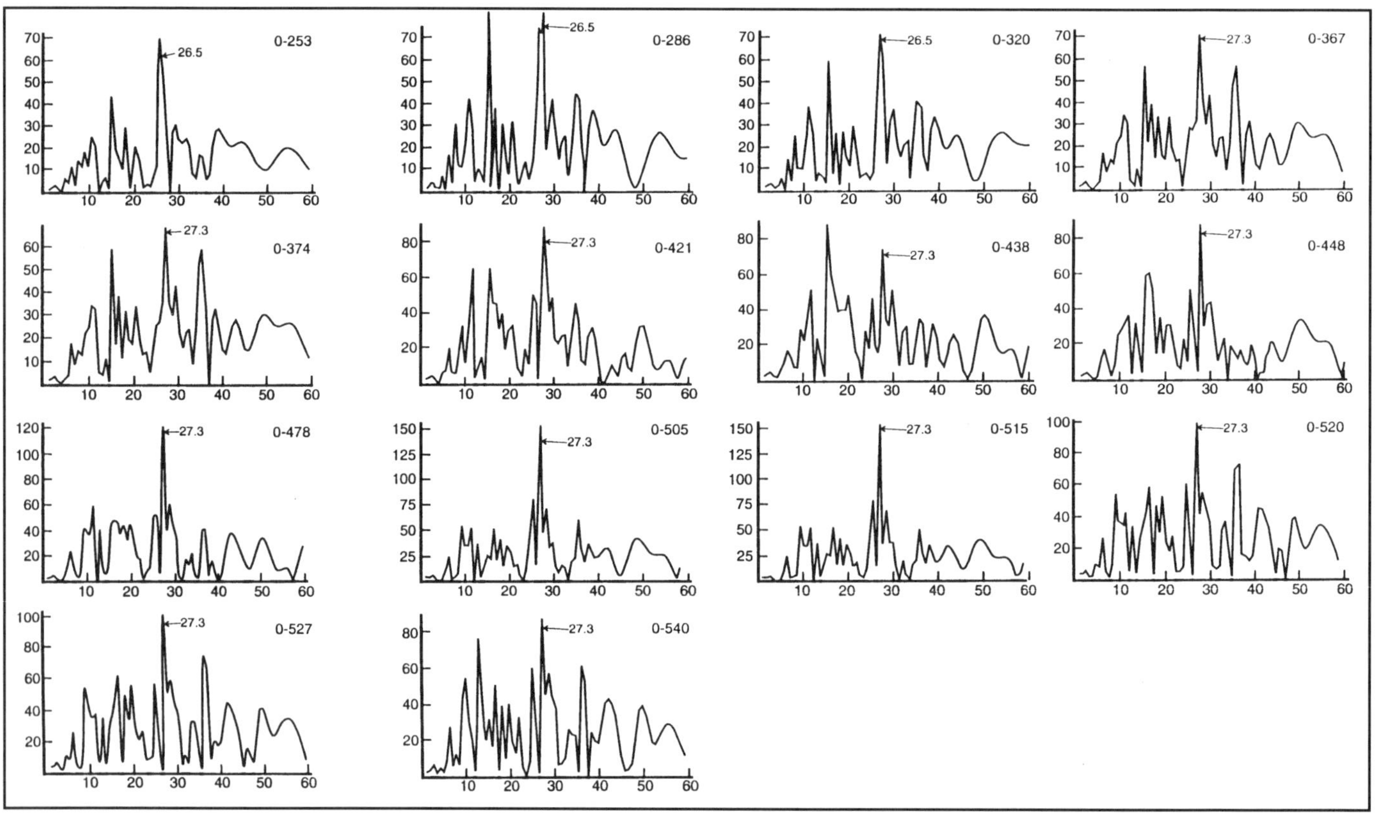

Figure 6. Fourier power spectrum for extinction events from Figure 1 (0 to 540 Ma), computed as described in Rampino and Caldeira (1993). X-axes = period in m.y.; y-axes = spectral power. The time series was truncated, one extinction event at a time, from 540 Ma to 253 Ma. The highest peak remained in the range 26.5 to 27.3 m.y. in all but one of the truncations, where the 27.3 m.y. peak was the second highest. The results for the 0- to 253-Ma sequence gives 26.5 m.y. period, similar to that obtained in previous studies (Raup and Sepkoski, 1984, 1986; Rampino and Caldeira, 1992, 1993).

marked by die-off of marine plankton, decimation of terrestrial plant communities, impoverishment of postextinction ecosystems, and proliferation of opportunistic survivor species, with gradual recovery.

Six pulses of marine extinction in the last 540 m.y. seem to be associated with evidence of impacts—layers containing high siderophile trace-element anomalies (especially iridium) and/or shocked minerals, tektites, and microtektites and large impact craters with age error bars overlapping the age of the extinction. Some other boundaries are associated with elevated iridium concentrations above background that might be related to impacts of Ir-depleted objects, possibly comets. We believe that although uncertainties remain in many aspects of the problem and much more work needs to be done, the principle of parsimony suggests that a general relationship between large impacts and mass extinctions is a reasonable working hypothesis.

ACKNOWLEDGMENTS

We thank D. M. Raup and P. M. Sheehan for critical reviews and W. T. Holser, E. G. Kauffman, C. J. Orth, J. J. Sepkoski Jr., E. M. Shoemaker, J. Smit, R. B. Stothers, and S. Yabushita for data, discussions, and criticism. M.R.R. was partially supported by NASA Grant NAGW-1697.

REFERENCES CITED

Aharon, P., Schidlowski, M., and Singh, I. B., 1987, Chronostratigraphic markers in the end-Precambrian carbon isotope record of the Lesser Himalaya: Nature, v. 327, p. 699–702.

Alcala-Herrera, J. A., Grossman, E. L., and Gartner, S., 1992, Nannofossil diversity and equitability and fine-fraction $\delta^{13}C$ across the Cretaceous/Tertiary boundary at Walvis Ridge Leg 74, South Atlantic: Marine Micropaleontology, v. 20, p. 77–88.

Alvarez, L. W., Alvarez, W., Asaro, F., and Michel, H. V., 1980, Extraterrestrial cause of the Cretaceous/Tertiary extinction: Experimental results and theoretical interpretation: Science, v. 208, p. 1095–1108.

Alvarez, W., 1986, Toward a theory of impact crises: Eos (Transactions, American Geophysical Union), v. 67, p. 649–658.

Alvarez, W., and Muller, R. A., 1984, Evidence from crater ages for periodic impacts on the Earth: Nature, v. 308, p. 718–720.

Alvarez, W., Alvarez, L. W., Asaro, F., and Michel, H. V., 1982, Iridium anomaly approximately synchronous with terminal Eocene extinctions: Science, v. 216, p. 886–888.

Alvarez, W., Asaro, F., and Montanari, A., 1990, Iridium profile for 10 million years across the Cretaceous-Tertiary boundary at Gubbio (Italy): Science, v. 250, p. 1700–1702.

Anders, E., Wolbach, W. S., and Lewis, R. S., 1986, Cretaceous extinctions and wildfires: Science, v. 234, p. 261–264.

Attrep, M., Jr., Orth, C. J., and Quintana, L. R., 1991, The Permian-Triassic of the Gartnerkofel-1 Core (Carnic Alps, Austria): Geochemistry of common and trace elements II—INAA and RNAA, *in* Holser, W. T., and Schönlaub, H. P., eds., The Permian-Triassic boundary in the Carnic Alps of Austria (Gartnerkofel Region): Abhandlungen der Geologischen Bundesanstalt, v. 45, p. 123–137.

Bakker, R. T., 1978, Dinosaur feeding behaviour and the origin of flowering plants: Nature, v. 274, p. 661–663.

Balme, B. E., 1979, Palynology of Permian-Triassic boundary beds of Kap Stosch, east Greenland: Meddelelser Om Grønland, v. 200, p. 1–35.

Benton, M. J., 1988, Mass extinctions in the fossil record of reptiles: Paraphyly, patchiness, and periodicity (?), *in* Larwood, G. P., ed., Extinction and survival in the fossil record: Oxford, Oxford University Press, p. 269–294.

Benton, M. J., 1989, Mass extinctions among tetrapods and the quality of the fossil record: Philosophical Transactions of the Royal Society of London, v. B325, p. 369–386.

Benton, M. J., ed., 1993, The fossil record 2: London, Chapman and Hall, 845 p.

Benton, M. J., 1995, Diversification and extinction in the history of life: Science, v. 268, p. 52–58.

Berggren, W. A., and Prothero, D. R., 1992, Eocene-Oligocene climatic and biotic evolution: An overview, *in* Prothero, D. R., and Berggren, W. A., eds., Eocene-Oligocene climatic and biotic evolution: Princeton, Princeton University Press, p. 1–28.

Bhandari, N., Shukla, P. N., and Azmi, R. J., 1992, Positive europium anomaly at the Permo-Triassic boundary, Spiti, India: Geophysical Research Letters, v. 19, p. 1531–1534.

Bhandari, N., Gupta, M., Pandey, J., and Shukla, P. N., 1994, Chemical profiles in K/T boundary section of Meghalaya, India: Cometary, asteroidal or volcanic: Chemical Geology, v. 113, p. 45–60.

Bice, D. M., Newton, C. R., McCauley, S., Reiners, P. W., and McRoberts, C. A., 1992, Shocked quartz at the Triassic-Jurassic boundary in Italy: Science, v. 259, p. 443–446.

Boulter, M. C., Spicer, R. A., and Thomas., B. A., 1988, Patterns of plant extinction from some palaeobotanical evidence, *in* Larwood, G. P., ed., Evolution and survival in the fossil record: Oxford, Clarendon, p. 1–36.

Bowring, S. A., Grotzinger, J. P., Isachsen, C. E., Knoll, A. H., Pelechaty, S. M., and Kolosov, P., 1993, Calibrating rates of Early Cambrian evolution: Science, v. 261, p. 1293–1298.

Brandner, R., 1988, The Permian-Triassic boundary section in the Dolomites (Southern Alps, Italy), San Antonio section: Ber. Geologischen Bundes-anstalt, v. 15, p. 49–56.

Brasier, M. D., Corfield, R. M., Derry, L. A., Rozanov, A. Yu., and Zhuralev, A. Yu., 1994, Multiple $\delta^{13}C$ excursions spanning the Cambrian explosion to the Botomian crisis in Siberia: Geology, v. 22, p. 455–458.

Brenchley, P. J., 1989, The late Ordovician extinction, *in* Donovan, S. K., ed., Mass extinctions, processes and evidence: New York, Columbia University Press, p. 104–132.

Brenchley, P. J., Marshall, J. D., Carden, G. A. F., Robertson, D. B. R., Long, D. G. F., Meidla, T., Hints, L., and Anderson, T. F., 1994, Bathymetric and isotopic evidence for a short-lived Late Ordovician glaciation in a greenhouse period: Geology, v. 22, p. 295–298.

Brinkhuis, H., and Zachariasse, W. J., 1988, Dinoflagellate cysts, sea level changes and planktonic foraminifers across the Cretaceous-Tertiary boundary at El Haria, northwest Tunisia: Marine Micropaleontology, v. 13, p. 153–191.

Brochwicz-Lewinski, W., Gasiewicz, A., Melendez, G., Siqueiros, L., Suffczynski, S., Szatkowski, K., Tarkowski, R., and Zbik, M., 1985, A possible Middle/Upper Jurassic boundary event: International Geological Congress, 27th, Moscow, USSR, Abstracts.

Caldeira, K., and Rampino, M. R., 1993, Aftermath of the end-Cretaceous mass extinction: Possible stabilization of the carbon cycle and climate: Paleoceanography, v. 8, p. 515–525.

Chai, Z.-F., Mao, X.-Y., Ma, S.-L., Bai, S.-L., Orth, C. J., Zhou, Y.-Q., and Ma, J.-G., 1987, Geochemical anomaly of the Devonian-Carboniferous boundary at Huangmao, Guangxi, China: International Geological Correlation Project 199, Rare Events in Geology, Abstracts, p. 29.

Chapman, C. R., and Morrison, D., 1994, Impacts on the Earth by asteroids and comets: Assessing the hazard: Nature, v. 367, p. 33–40.

Claeys, P., Casier, J-G., and Margolis, S. V., 1992, Microtektites and mass extinctions: Evidence for a Late Devonian asteroid impact: Science, v. 257, p. 1102–1104.

Claeys, P., Kyte, F. T., and Casier, J.-G., 1994, Frasnian-Famennian boundary: Mass extinctions, anoxic oceans, microtektite layers, but not much iridium? *in* New developments regarding the KT event and other catastrophes in Earth history: Houston, Texas, Lunar and Planetary Institute Contribution 825, p. 22–24.

Clube, S. V. M., and Napier, W. M., 1984, Comet capture from molecular clouds: A dynamical constraint on star and planet formation: Monthly Notices of the Royal Astronomical Society, v. 208, p. 575–588.

Colbath, G. K., 1986, Abrupt terminal Ordovician extinction in phytoplankton associations, southern Appalachians: Geology, v. 14, p. 943–946.

Collom, C. J., 1990, The taxonomic analysis of mass extinction intervals: An approach to problems of resolution as shown by Cretaceous ammonite genera (global) and species (western interior of the United States), *in* Kauffman, E. G., and Walliser, O. H., eds., Extinction events in Earth history: Berlin, Springer-Verlag, p. 265–276.

Colodner, D. C., Boyle, E. A., Edmond, J. M. N., and Thomson, J., 1992, Post-depositional mobility of platinum, iridium and rhenium in marine sediments: Nature, v. 358, p. 402–404.

Compton, J. S., Snyder, S. W., and Hodell, D. A., 1990, Phosphogenesis and weathering of shelf sediments from the southeastern United States: Implications for Miocene $\delta^{13}C$ excursions and global cooling: Geology, v. 18, p. 1227–1230.

Covey, C., Ghan, S. J., Walton, J. J., and Weissman, P. R., 1990, Global environmental effects of impact-generated aerosols: Results from a general circulation model: Geological Society of America Special Paper 247, p. 263–270.

Croft, S. K., 1982, A first-order estimate of shock heating and vaporization in oceanic impacts: Geological Society of America Special Paper 190, p. 143–151.

Cutler, A. H., and Behrensmeyer, A. K., 1994, Bone beds at the boundary: Are they a realistic expectation? *in* New developments regarding the K/T event and other catastrophes in Earth history: Houston, Texas, Lunar and Planetary Institute Contribution 825, p. 28.

Davis, M., Hut, P., and Muller, R. A., 1984, Extinction of species by periodic comet showers: Nature, v. 308, p. 715–717.

DePaolo, D. J., and Finger, K. L., 1991, High-resolution strontium-isotope stratigraphy and biostratigraphy of the Miocene Monterey Formation, central California: Geological Society of America Bulletin, v. 103, p. 112–124.

Donovan, S. K., 1989, Introduction, *in* Donovan, S. K., ed., Mass extinctions, processes and evidence: New York, Columbia University Press, p. xi–xiv.

Dowsett, H. J., and Loubere, P., 1992, High resolution late Pliocene sea-surface temperature record from the Northeast Atlantic Ocean: Marine Micropaleontology, v. 20, p. 91–105.

Emslie, S. D., and Morgan, G. S., 1994, A catastrophic death assemblage and paleoclimatic implications of Pliocene seabirds of Florida: Science, v. 264, p. 684–685.

Engelhardt, W. v., Matthai, S. K., and Walzebuck, J., 1992, Araguainha impact crater, Brazil. I: The interior part of the uplift: Meteoritics, v. 27, p. 442–457.

Erwin, D. H., 1993, The great Paleozoic crisis: New York, Columbia University Press, 327 p.

Erwin, D. H., 1994, The Permo-Triassic extinction: Nature, v. 367, p. 231–236.

Eshet, Y., 1990, The palynostratigraphy of the Permian-Triassic boundary in Israel: Two approaches to biostratigraphy: Israel Journal of the Earth Sciences, v. 39, p. 1–15.

Eshet, Y., Rampino, M. R., and Visscher, H., 1995, Fungal event and palynological record of ecological crisis and recovery across the Permian/Triassic boundary: Geology, v. 23, p. 967–970.

Fan, D., Yang, R., and Huang, Z., 1984, The Lower Cambrian black shale series and the iridium anomaly in South China, *in* Academia Sinica, Developments in Geoscience, Contribution to 27th Geological Congress, Moscow: Beijing, Science Press, p. 215–224.

Farley, K. A., 1995, Cenozoic variations in the flux of interplanetary dust recorded by ^{3}He in a deep-sea sediment: Nature, v. 376, p. 153–156.

Flower, B. P., and Kennett, J. P., 1993, Middle Miocene ocean-climate transition: High-resolution oxygen and carbon isotopic records from Deep Sea Drilling Project Site 588A, southwest Pacific: Paleoceanography, v. 8, p. 811–843.

Flower, B. P., and Kennett, J. P., 1994, The middle Miocene climatic transition: East Antarctic ice sheet development, deep ocean circulation and global carbon cycling: Palaeogeography, Palaeoclimatology, Palaeoecology, v. 108, p. 537–555.

Fogg, M. J., 1989, The relevance of the background impact flux to cyclic impact/mass extinction hypothesis: Icarus, v. 79, p. 382–395.

Ganapathy, R., 1982, Evidence for a major meteorite impact on the Earth 34 million years ago: Implications for Eocene extinctions: Science, v. 216, p. 885–886.

Geldsetzer, H. H. J., Goodfellow, W. D., McLaren, D. J., and Orchard, M. J., 1987, Sulfur-isotope anomaly associated with the Frasnian-Famennian extinction, Medicine Lake, Alberta, Canada: Geology, v. 15, p. 393–396.

Geldsetzer, H. H. J., Goodfellow, W. D., and McLaren, D. J., 1993, The Frasnian-Famennian extinction event in a stable cratonic shelf setting: Trout River, Northwest Territories, Canada: Palaeogeography, Palaeoclimatology, Palaeoecology, v. 104, p. 81–95.

Gerstl, S. A., and Zardecki, A., 1982, Reduction of photosynthetically active radiation under extreme stratospheric aerosol loads: Geological Society of America Special Paper 190, p. 201–210.

Glen, W., ed., 1994, The mass-extinction debates: How science works in a crisis: Stanford, Stanford University Press, 370 p.

Goodfellow, W. D., Geldsetzer, H. H. J., McLaren, D. J., Orchard, M. J., and Klapper, G., 1989, Geochemical and isotopic anomalies associated with the Frasnian-Famennian extinction: Historical Biology, v. 2, p. 51–72.

Goodfellow, W. D., Nowlan, G. S., McCracken, A. D., Lenz, A. C., and Gregoire, D. C., 1992, Geochemical anomalies near the Ordovician-Silurian boundary, Northern Yukon Territory, Canada: Historical Biology, v. 6, p. 1–23.

Gostin, V. A., Keays, R. R., and Wallace, M. W., 1989, Iridium anomaly from the Acraman impact ejecta horizon: Impacts can produce sedimentary iridium peaks: Nature, v. 340, p. 542–544.

Gradstein, F. M., Agterberg, F. P., Ogg, J. G., Hardenbol, J., van Veen, P., Thierry, J., and Huang, Z., 1994, A Mesozoic time scale: Journal of Geophysical Research, v. 99, p. 24,051–24,074.

Grieve, R. A. F., 1991, Terrestrial impact: The record in the rocks: Meteoritics, v. 26, p. 175–194.

Grieve, R. A. F., and Shoemaker, E. M., 1994, The record of past impacts on Earth, *in* Gehrels, T., ed., Hazards due to asteroids and comets: Tucson, University of Arizona Press, p. 417–462.

Griffis, K., and Chapman, D. J., 1988, Survival of phytoplankton under prolonged darkness: Implications for the Cretaceous-Tertiary boundary darkness hypothesis: Palaeogeography, Palaeoclimatology, Palaeoecology, v. 67, p. 305–314.

Groot, J. J., 1991, Palynological evidence for Late Miocene, Pliocene, and Early Pleistocene climatic changes in the Middle U.S. Atlantic coastal plain: Quaternary Science Reviews, v. 10, p. 147–162.

Hallam, A., 1989, The case for sea-level change as a dominant causal factor in mass extinction of marine invertebrates: Philosophical Transactions of the Royal Society of London, v. B325, p. 437–455.

Hallam, A., 1990, The end-Triassic mass extinction event: Geological Society of America Special Paper 247, p. 577–583.

Harland, W. B., Armstrong, R. L., Cox, A. V., Craig, L. E., Smith, A. G., and Smith, D. G., 1990, A geologic time scale: Cambridge, Cambridge University Press, 263 p.

Hartenberger, J.-L., 1987, Modalités des extinctions et apparitions chez les Mammiferes du Paléogene d'Europe: Mémoires de la Société Géologique de France 150, p. 133–143.

Hodell, D. A., Mueller, P. A., and Garrido, J. R., 1991, Variations in the strontium isotopic composition of seawater during the Neogene: Geology, v. 19, p. 24–27.

Hollander, D. J., McKenzie, J. A., and Hsü, K. J., 1993, Carbon isotope evidence for unusual plankton blooms and fluctuations of surface water CO_2 in "Strangelove Ocean" after terminal Cretaceous event: Palaeogeography, Palaeoclimatology, Palaeoecology, v. 104, p. 229–237.

Holser, W. T., and Magaritz, M., 1992, Cretaceous/Tertiary and Permian/Triassic boundary events compared: Geochimica et Cosmochimica Acta, v. 56, p. 3297–3309.

Holser, W. T., Schönlaub, H. P., Boeckelmann, K., and Magaritz, M., 1991, The Permian-Triassic of the Gartnerkofel-1 Core (Carnic Alps, Austria): Synthesis and conclusions, *in* Holser, W. T., and Schönlaub, H. P., eds., The Permian-Triassic boundary in the Carnic Alps of Austria (Gartnerkofel region): Abhandlung der Geologischen Bundesanstalt, v. 45, p. 213–232.

Hsü, K. J., and McKenzie, J. A., 1985, A "Strangelove" ocean in earliest Tertiary, *in* Sundquist, E. T., and Broecker, W. S., eds., Carbon cycle and atmospheric CO_2, natural variations Archean to Present: Washington, D.C., American Geophysical Union, Geophysical Monograph 32, p. 487–492.

Hsü, K. J., and McKenzie, J. A., 1990, Carbon-isotope anomalies at era boundaries; Global catastrophes and their ultimate cause: Geological Society of America Special Paper 247, p. 61–70.

Hsü, K. J., Oberhansli, H., Gao, J. Y., Shu, S., Haihong, C., and Krahenbuhl, U., 1985, "Strangelove ocean" before the Cambrian explosion: Nature, v. 316, p. 809–811.

Huber, B. T., Liu, C., Olsson, R. K., and Berggren, W. A., 1994, The Cretaceous-Tertiary boundary transition in the Antarctic Ocean and its global implications: Comment: Marine Micropaleontology, v. 24, p. 91–99.

Hut, P., Alvarez, W., Elder, W. P., Hansen, T., Kauffman, E. G., Keller, G., Shoemaker, E. M., and Weissman, P. R., 1987, Comet showers as a cause of mass extinctions: Nature, v. 329, p. 118–126.

Ivany, L. C., and Salawitch, R. J., 1993, Carbon isotope evidence for biomass burning at the K-T boundary: Geology, v. 21, p. 487–490.

Jansa, L. F., Aubry, M-P., and Gradstein F. M., 1990, Comets and extinctions; cause and effect? Geological Society of America Special Paper 247, p. 223–232.

Jarvis, I., Carson, G. A., Cooper, K., Hart, M. B., Horne, D., Leary, P. N., Rosenfeld, A., and Tocher, B. A., 1988, Chalk microfossil assemblages and the Cenomanian-Turonian (late Cretaceous) oceanic anoxic event: Cretaceous Research, v. 9, p. 3–103.

Jones, D. S., MacFadden, B. J., Webb, S. O., Mueller, P. A., Hodell, D. A., and Cronin, T. M., 1991, Integrated geochronology of a classic Pliocene fossil site in Florida: Linking marine and terrestrial biochronologies: Journal of Geology, v. 99, p. 637–648.

Kajiwara, Y., and Kaiho, K., 1992, Oceanic anoxia at the Cretaceous/Tertiary boundary supported by the sulfur isotope record: Palaeogeography, Palaeoclimatology, Palaeoecology, v. 99, p. 151–162.

Kauffman, E. G., 1986, High-resolution event stratigraphy: Regional and global Cretaceous bioevents, *in* Walliser, O. H., ed., Global bio-events: Berlin, Springer-Verlag, p. 279–335.

Kauffman, E. G., 1995, Global change leading to biodiversity crisis in a greenhouse world: The Cenomanian-Turonian (Cretaceous) mass extinction, *in* Effects of past global change on life: Washington, National Academy Press, p. 47–72.

Keller, G., 1986, Stepwise mass extinctions and impact events: Late Eocene to early Oligocene: Marine Micropaleontology, v. 10, p. 267–293.

Keller, G., 1988, Extinction, survivorship and evolution of planktic foraminifera across the Cretaceous/Tertiary boundary at El Kef, Tunisia: Marine Micropaleontology, v. 13, p. 239–263.

Keller, G., 1989, Extended Cretaceous/Tertiary boundary extinctions and delayed population change in planktonic foraminifera from Brazos River, Texas: Paleoceanography, v. 4, p. 287–332.

Keller, G., Barrera, E., Schmitz, B., and Mattson, E., 1993, Gradual mass extinction, species survivorship, and long-term environmental changes across the Cretaceous-Tertiary boundary in high latitudes: Geological Society of America Bulletin, v. 105, p. 979–997.

Koeberl, C., 1993, Extraterrestrial component associated with Australasian microtektites in a core from ODP Site 758B: Earth and Planetary Science Letters, v. 119, p. 453–458.

Kump, L. R., 1991, Interpreting carbon-isotope excursions: Strangelove oceans: Geology, v. 19, p. 299–302.

Kyte, F. T., 1988, The extraterrestrial component in marine sediments: Description and interpretation: Paleoceanography, v. 3, p. 235–247.

Kyte, F. T., and Wasson, J. T., 1986, Accretion rate of extraterrestrial matter: Iridium in marine sediments deposited 33–67 Ma ago: Science, v. 232, p. 1225–1229.

Kyte, F. T., Zhou, Z., and Wasson, J. T., 1981, High noble metal concentrations in a late Pliocene sediment: Nature, v. 292, p. 417–420.

Kyte, F. T., Zhou, Z., and Wasson, J. T., 1988, New evidence on the size and possible effects of a Late Pliocene asteroid impact: Science, v. 241, p. 63–65.

Labandiera, C. C., and Sepkoski, J. J., Jr., 1993, Insect diversity in the fossil record: Science, v. 261, p. 310–315.

Li, Z., Zhan, L., Yao, J., and Zhou, Y., 1991, On the Permian-Triassic events in South China—Probe into the end Permian abrupt extinction and its possible causes: Saito Ho-on Kai Special Publication 3, p. 371–385.

Long, D., 1993, Oxygen and carbon isotopes and event stratigraphy near the Ordovician-Silurian boundary, Anticosti Island Quebec: Palaeogeography, Palaeoclimatology, Palaeoecology, v. 104, p. 49–59.

Magaritz, M., 1989, ^{13}C Minima follow extinction events: A clue to faunal radiation: Geology, v. 17, p. 337–340.

Magaritz, M., Holser, W. T., and Kirschvink, J. L., 1986, Carbon-isotope events across the Precambrian/Cambrian boundary on the Siberian Platform: Nature, v. 320, p. 258–259.

Magaritz, M., Bar, R., Baud, A., and Holser, W. T., 1988, The carbon isotope shift at the Permian/Triassic boundary in the southern Alps is gradual: Nature, v. 331, p. 337–339.

Magaritz, M., Krishnamurthy, R. V., and Holser, W. T., 1992, Parallel trends in organic and inorganic carbon isotopes across the Permian/Triassic boundary: American Journal of Science, v. 292, p. 727–739.

Margolis, S. V., Claeys, P., and Kyte, F. T., 1991, Microtektites, microkrystites, and spinels from a Late Pliocene asteroid impact in the Southern Ocean: Science, v. 251, p. 1594–1597.

Matese, J. J., Whitman, P. G., Innanen, K. A., and Valtonen, M. J., 1995, Periodic modulation of the Oort Cloud comet flux by the adiabatically changing galactic tide: Icarus, v. 116, p. 255–268.

Maxwell, W. D., 1992, Permian and Early Triassic extinction of non-marine tetrapods: Palaeontology, v. 35, p. 571–583.

McGhee, G. R., Jr., 1988, The Late Devonian extinction event: Evidence for abrupt ecosystem collapse: Paleobiology, v. 14, p. 250–257.

McGhee, G. R., Jr., 1989, The Frasnian-Famennian extinction event, *in* Donovan, S. K., ed., Mass extinctions, processes and evidence: New York: Columbia University Press, p. 133–151.

McLaren, D. J., 1985, Mass extinction and iridium anomaly in the Upper Devonian of Western Australia: A commentary: Geology, v. 13, p. 170–172.

McLaren, D. J., and Goodfellow, W. D., 1990, Geological and biological consequences of giant impacts: Annual Review of Earth and Planetary Science, v. 18, p. 123–171.

Melchin, M. J., MacCracken, A. D., and Oliff, F. D., 1991, The Ordovician-Silurian boundary on Cornwallis and Truro islands, Arctic Canada: Preliminary data: Canadian Journal of the Earth Sciences, v. 28, p. 1854–1862.

Melosh, H. J., Schneider, N. M., Zahnle, K. J., and Latham, D., 1990, Ignition of global wildfires at the Cretaceous/Tertiary boundary: Nature, v. 343, p. 251–254.

Melson, W. G., O'Hearn, T., and Fredriksson, K., 1988, Composition and origin of basaltic glass spherules in pelagic clay from the eastern Pacific: Marine Geology, v. 83, p. 253–271.

Meyers, P. A., and Simoneit, B. R. T., 1989, Global comparisons of organic matter in sediments across the Cretaceous/Tertiary boundary: Organic Geochemistry, v. 16, p. 641–648.

Miller, K. G., Berggren, W. A., Zhang, J., and Palmer-Julson, A. A., 1991, Biostratigraphy and isotope stratigraphy of Upper Eocene microtektites at Site 612: How many impacts? Palaios, v. 6, p. 17–38.

Milne, D. H., and McKay, C. P., 1982, Response of marine plankton communities to a global atmospheric darkening: Geological Society of America Special Paper 190, p. 297–304.

Montanari, A., 1990, Geochronology of the terminal Eocene impacts; An update: Geological Society of America Special Paper 247, p. 607–616.

Montanari, A., Asaro, F., Michel, H. V., and Kennett, J. P., 1993, Iridium anomalies of Late Eocene age at Massignano (Italy), and ODP Site 689B (Maud Rise, Antarctica): Palaios, v. 8, p. 420–438.

Morrison, D., 1992, The spaceguard survey: Report of the NASA international near-earth object detection workshop: Pasadena, California, NASA, Jet Propulsion Laboratory, 52 p.

Myrow, P. M., and Hiscott, R. N., 1993, Depositional history and sequence stratigraphy of the Precambrian-Cambrian boundary stratotype section, Chapel Island Formation, southeast Newfoundland: Palaeogeography, Palaeoclimatology, Palaeoecology, v. 104, p. 13–35.

Oddone, W., and Vannucci, R., 1988, PGE and REE geochemistry at the B-W boundary in the Carnian and Dolomite Alps (Italy): Memorie della Societá Geologica Italiana, v. 34, p. 121–128.

Ogg, J. G., and Lowrie, W., 1986, Magnetostratigraphy of the Jurassic/Cretaceous boundary: Geology, v. 14, p. 547–550.

O'Keefe, J. D., and Ahrens, T. J., 1982, Impact mechanics of the Cretaceous-Tertiary impact bolide: Nature, v. 298, p. 123–127.

O'Keefe, J. D., and Ahrens, T. J., 1989, Impact production of CO_2 by the Cretaceous/Tertiary extinction bolide and the resultant heating of the Earth: Nature, v. 338, p. 247–249.

Olsen, P. E., Fowell, S. J., and Cornet, B., 1990, The Triassic/Jurassic boundary in continental rocks of eastern North America; A progress report: Geological Society of America Special Paper 247, p. 585–593.

Orth, C. J., 1989, Geochemistry of the bio-event horizons, *in* Donovan, S. K., ed., Mass extinctions: Processes and evidence: New York, Columbia University Press, p. 37–72.

Orth, C. J., Gilmore, J. S., Quintana, L. R., and Sheehan, P. M., 1986, Terminal Ordovician extinction: Geochemical analysis of the Ordovician-Silurian boundary, Anticosti Island, Quebec: Geology, v. 14, p. 433–436.

Orth, C. J., Attrep, M., Jr. and Quintana, L. R., 1990, Iridium abundance patterns across bioevent horizons in the fossil record: Geological Society of America Special Paper 247, p. 45–59.

Orth, C. J., Attrep, M., Jr., Quintana, L. R., Elder, W. P., Kauffman, E. G., Diner, R., and Villamil, T., 1993, Elemental abundance anomalies in the late Cenomanian extinction interval: a search for the source(s): Earth and Planetary Science Letters, v. 117, p. 189–204.

Palme, H., 1982, Identification of projectiles of large terrestrial impact craters and some implications for the interpretation of Ir-rich Cretaceous/Tertiary boundary layer: Geological Society of America Special Paper 190, p. 223–233.

Palmer, A. R., 1983, The Decade of North American Geology 1983 geologic time scale: Geology, v. 11, p. 503–504.

Peng, H., 1994, An extraterrestrial event at the Tertiary-Quaternary boundary: *in* New developments regarding the KT event and other catastrophes in Earth history: Houston, Texas, Lunar and Planetary Institute Contribution 825, p. 88–89.

Peryt, D., and Wyrwicka, K., 1993, The Cenomanian/Turonian boundary in Central Poland: Palaeogeography, Palaeoclimatology, Palaeoecology, v. 104, p. 185–197.

Pollack, J. B., Toon, O. B., Ackerman, T. P., Mckay, C. P., and Turco, R. P., 1983, Environmental effects of an impact-generated dust cloud: Implications for the Cretaceous-Tertiary extinctions: Science, v. 219, p. 287–290.

Pospichal, J. J., 1994, Calcareous nannofossils at the K/T boundary, El Kef: No evidence for stepwise, gradual, or sequential extinctions: Geology, v. 22, p. 99–102.

Prothero, D. R., 1994, The Late Eocene–Oligocene extinctions: Annual Review of Earth and Planetary Science, v. 22, p. 145–165.

Rampino, M. R., 1982, A non-catastrophist explanation for the iridium anomaly at the Cretaceous/Tertiary boundary: Geological Society of America Special Paper 190, p. 455–460.

Rampino, M. R., 1988, Are marine and nonmarine extinctions correlated?: Eos (Transactions, American Geophysical Union), v. 69, p. 889, 895.

Rampino, M. R., and Caldeira, K., 1992, Episodes of terrestrial geologic activity during the past 260 million years: A quantitative approach: Celestial Mechanics and Dynamical Astronomy, v. 54, p. 143–159.

Rampino, M. R., and Caldeira, K., 1993, Major episodes of geologic change: Correlations, time structure and possible causes: Earth and Planetary Science Letters, v. 114, p. 215–227.

Rampino, M. R., and Haggerty, B. M., 1994, Extraterrestrial impacts and mass extinctions of life, *in* Gehrels, T., ed., Hazards due to asteroids and comets: Tucson, University of Arizona Press, p. 827–857.

Rampino, M. R., and Haggerty, B. M., 1995, Mass extinctions and periodicity: Science, v. 269, p. 617–619.

Rampino, M. R., and Haggerty, B. M., 1996, The "Shiva hypothesis": Impacts, mass extinctions, and the Galaxy: Earth, Moon, and Planets v. 72, p. 441–460.

Rampino, M. R., and Stothers, R. B., 1984a, Terrestrial mass extinctions, cometary impacts and the Sun's motion perpendicular to the galactic plane: Nature, v. 308, p. 709–712.

Rampino, M. R., and Stothers, R. B., 1984b, Geological rhythms and cometary impacts: Science, v. 226, p. 1427–1430.

Rampino, M. R., and Stothers, R. B., 1986, Geologic periodicities and the Galaxy, *in* Smoluchowski, R., Bahcall, J. N., and Matthews, M. S., eds., The galaxy and the solar system: Tucson, University of Arizona Press, p. 241–259.

Raup, D. M., 1979, Size of the Permo-Triassic bottleneck and its evolutionary implications: Science, v. 206, p. 217–218.

Raup, D. M., 1989, The case for extraterrestrial causes of extinction: Philosophical Transactions of the Royal Society of London, v. B325, p. 421–435.

Raup, D. M., 1990, Impact as a general cause of extinction; a feasibility test: Geological Society of America Special Paper 247, p. 27–32.

Raup, D. M., 1991a, A kill curve for Phanerozoic marine species: Paleobiology, v. 17, p. 37–48.

Raup, D. M., 1991b, Extinction: Bad genes or bad luck?: New York, W. W. Norton, 210 p.

Raup, D. M., 1992, Large-body impact and extinction in the Phanerozoic: Paleobiology, v. 18, p. 80–88.

Raup, D. M., and Sepkoski, J. J., Jr., 1984, Periodicity of extinctions in the geologic past: Proceedings of the National Academy of Sciences USA, v. 81, p. 801–805.

Raup, D. M., and Sepkoski, J. J., Jr., 1986, Periodic extinctions of families and genera: Science, v. 231, p. 833–836.

Raymo, M. E., et al., 1989, Late Pliocene variation in Northern Hemisphere ice sheets and North Atlantic deep water circulation: Paleoceanography, v. 4, p. 413–446.

Remane, J., 1991, The Jurassic-Cretaceous boundary: Problems of definition and procedure: Cretaceous Research, v. 12, p. 447–453.

Retallack, G. J., 1995, Permian-Triassic life crisis on land: Science, v. 267, p. 77–80.

Rocchia, R., Boclet, D., Bonte, P., Castellarin, A., and Jéhanno, C., 1986, An iridium anomaly in the Middle-Lower Jurassic of the Venetian region, northern Italy: Journal of Geophysical Research, v. 91, p. E259–E262.

Sancetta, C., Heusser, L., and Hall, M. A., 1992, Late Pliocene climate in the Southeast Atlantic: Preliminary results from a multi-disciplinary study of DSDP Site 532: Marine Micropaleontology, v. 20, p. 59–75.

Sandberg, C. A., Ziegler, W., Dreesen, R., and Butler, J. L., 1988, Part 3: Late Frasnian mass extinction: Conodont event stratigraphy, global changes, and possible causes: Courier Forschungsinstitute Senckenberg, v. 102, p. 263–307.

Sanfilippo, A., Riedel, W. R., Glass, B. P., and Kyte, F. T., 1985, Late Eocene microtektites and radiolarian extinctions on Barbados: Nature, v. 314,

p. 613–615.

Savage, R. J. G., 1988, Extinction and the fossil mammal record, *in* Larwood, G. P., ed., Extinction and survival in the fossil record: Oxford, Clarendon, p. 319–334.

Schindler, E., 1990, Die Kelwasser-Krise (hohe Frasne-Stufe, Ober-Devon): Gottinger Arbeiten zur Geologie und Palaontologie 46, 115 p.

Schmitz, B., Jeppsson, L., and Ekvall, J., 1994, A search for shocked quartz grains and impact ejecta in early Silurian sediments on Gotland, Sweden: Geological Magazine, v. 131, p. 361–367.

Schubert, J. K., and Bottjer, D. J., 1992, Early Triassic stromatolites as post–mass extinction disaster forms: Geology, v. 20, p. 883–886.

Sepkoski, J. J., Jr., 1992, A compendium of fossil marine animal families (second edition): Milwaukee Public Museum Contributions to Biology and Geology 83, 156 p.

Sepkoski, J. J., Jr., 1993, Ten years in the library: New data confirm paleontological patterns: Paleobiology, v. 19, p. 43–51.

Sepkoski, J. J., Jr., 1994, Extinction and the fossil record: Geotimes, v. 39, p. 15–17.

Shackleton, N. J., Backman, J., and 15 others, 1984, Oxygen isotope calibration of the onset of ice-rafting and history of glaciation in the North Atlantic region: Nature, v. 307, p. 620–623.

Sheehan, P. M., 1988, Late Ordovician events and the terminal Ordovician extinction: New Mexico Bureau of Mines and Mineral Resources Memoir 44, p. 405–415.

Sheehan, P. M., Fastovsky, D. E., Hoffmann, R. G., Berghaus, C. B., and Gabriel, D. L., 1991, Sudden extinction of the dinosaurs: Latest Cretaceous, Upper Great Plains, U.S.A.: Science, v. 254, p. 835–839.

Shoemaker, E. M., Wolfe, R. F., and Shoemaker, C. S., 1990, Asteroid and comet flux in the neighborhood of Earth: Geological Society of America Special Paper 247, p. 155–170.

Signor, P. W. III, and Lipps, J. H., 1982, Sampling bias, gradual extinction patterns and catastrophes in the fossil record: Geological Society of America Special Paper 190, p. 291–296.

Sigurdsson, H., D'Hondt, S., and Carey, S., 1992, The impact of the Cretaceous/Tertiary bolide on evaporite terrane and generation of major sulfuric acid aerosol: Earth and Planetary Science Letters, v. 109, p. 543–559.

Sloan, R. E., Rigby, J. K., Van Valen, L. H., and Gabriel, D., 1986, Gradual dinosaur extinction and simultaneous ungulate radiation in the Hell Creek Formation: Science, v. 232, p. 629–632.

Smit, J., 1990, Meteorite impact, extinctions and the Cretaceous-Tertiary boundary: Geologie en Mijnbouw, v. 69, p. 187–204.

Stanley, S. M., and Campbell, L. D., 1985, Neogene mass extinction of western Atlantic molluscs: Nature, v. 293, p. 457–459.

Stanley, S. M., and Yang, X., 1994, Two extinction events in the Late Permian: Science, v. 266, p. 1340–1344.

Stigler, S. M., and Wagner, M. J., 1987, A substantial bias in non-parametric tests for periodicity in geophysical data: Science, v. 238, p. 940–945.

Stothers, R. B., 1989, Structure and dating errors in the geologic time scale and periodicity of mass extinctions: Geophysical Research Letters, v. 16, p. 119–122.

Sweet, W. C., Yang, Z., Dickins, J. M., and Yin, H., 1992, Permo-Triassic events in the eastern Tethys—an overview, *in* Sweet, W. C., Yang, Z., Dickins, J. M., and Yin, H., eds., Permo-Triassic events in the eastern Tethys: Cambridge, Cambridge University Press, p. 1–8.

Talent, J. A., Mawson, R., Andrew, A. S., Hamilton, P. J., and Whitford, D. J., 1993, Middle Palaeozoic extinction events: Faunal and isotopic data, *in* Carr, P. F., ed., Centre for Isotope Studies research report 1991–92: North Ryde, New South Wales, CSIRO Mineral Research Laboratories, p. 98–106.

Toon, O. B., Zahnle, K., Turco, R. P., and Covey, C., 1994, Environmental perturbations caused by asteroid impacts *in* Gehrels, T., ed., Hazards due to asteroids and comets: Tucson, University of Arizona Press, p. 791–826.

Trefil, J. S., and Raup, D. M., 1987, Numerical simulations and the problem of periodicity in the cratering record: Earth and Planetary Science Letters, v. 82, p. 159–164.

Urey, H. C., 1973, Cometary collisions and geological periods: Nature, v. 242, p. 32–33.

Valentine, J. W., 1986, The Permian-Triassic extinction event and invertebrate developmental modes: Bulletin of Marine Science, v. 39, p. 607–615.

Vallier, T. L., Bohrer, D., Moreland, G., and McKee, E. H., 1977, Origin of basalt microlapilli in lower Miocene pelagic sediment, northeastern Pacific Ocean: Geological Society of America Bulletin, v. 88, p. 787–796.

van den Bergh, S., 1994, Astronomical catastrophes in earth history: Publications of the Astronomical Society of the Pacific, v. 106, p. 689–695.

Vickery, A. M., and Melosh, H. J., 1990, Atmospheric erosion and impactor retention in large impacts, with application to mass extinctions: Geological Society of America Special Paper 247, p. 289–300.

Visscher, H., and Brugman, W. A., 1988, The Permian-Triassic boundary in the Southern Alps: A palynological approach: Memorie della Societa Geologica Italiana, v. 34, p. 121–128.

Wang, K., 1992, Glassy microspherules (microtektites) from an Upper Devonian limestone: Science, v. 256, p. 1547–1550.

Wang, K., Orth, C. J., Attrep, M., Jr., Chatterton, B. D. E., Hou, H., and Geldsetzer, H. H. J., 1991, Geochemical evidence for a catastrophic biotic event at the Frasnian/Famennian boundary in south China: Geology, v. 19, p. 776–779.

Wang, K., Chatterton, B. D. E., Attrep, M., Jr., and Orth, C. J., 1992, Iridium abundance maxima in the latest Ordovician mass extinction horizon, Yangtze Basin, China: Terrestrial or extraterrestrial?: Geology, v. 20, p. 39–42.

Wang, K., Attrep, M., Jr., and Orth, C. J., 1993a, Global iridium anomaly, mass extinction, and redox change at the Devonian-Carboniferous boundary: Geology, v. 21, p. 1071–1074.

Wang, K., Chatterton, B. D. E., Attrep, M., Jr., and Orth, C. J., 1993b, Late Ordovician mass extinction in the Selwyn Basin, northwestern Canada: Geochemical, sedimentological, and paleontological evidence: Canadian Journal of the Earth Sciences, v. 30, p. 1870–1880.

Wang, K., Orth, C. J., Attrep, M., Jr., Chatterton, B. D. E., Wang, X., and Li, J-j., 1993c, The great latest Ordovician extinction on the South China Plate: Chemostratigraphic studies of the Ordovician-Silurian boundary interval on the Yangtze Platform: Palaeogeography, Palaeoclimatology, Palaeoecology, v. 104, p. 61–79.

Wang, K., Geldsetzer, H. H. J., and Chatterton, B. D. E., 1994, A Late Devonian extraterrestrial impact and extinction in eastern Gondwana: Geochemical, sedimentological, and faunal evidence, *in* Dressler, B. O., Grieve, R. A. F., and Sharpton, V. L., eds., Large meteorite impacts and planetary evolution: Geological Society of America Special Paper 293, p. 111–120.

Weems, R. E., 1992, The "terminal Triassic catastrophic extinction event" in perspective: A review of Carboniferous through Early Jurassic terrestrial vertebrate extinction patterns: Palaeogeography, Palaeoclimatology, Palaeoecology, v. 94, p. 1–29.

Weissman, P. R., 1986, The Oort Cloud and the Galaxy: Dynamical interactions, *in* Smoluchowski, R., Bahcall, J. N., and Matthews, M. S., eds., The Galaxy and the Solar System: Tucson, University of Arizona Press, p. 204–237.

Whitmire, D. P., and Jackson, A. A., 1984, Are periodic mass extinctions driven by a distant solar companion?: Nature, v. 308, p. 713–715.

Whitmire, D. P., and Matese, J. J., 1985, Periodic comet showers and Planet X: Nature, v. 313, p. 36–38.

Wise, K. P., and Schopf, T. J. M., 1981, Was marine faunal diversity in the Pleistocene affected by changes in sea level?: Paleobiology, v. 7, p. 394–399.

Wolbach, W. S., Gilmour, I., Anders, E., Orth, C. J., and Brooks, R. R., 1988, Global fire at the Cretaceous-Tertiary boundary: Nature, v. 334, p. 670–673.

Wolfe, J. A., and Upchurch, G. R., 1986, Vegetation, climatic and floral changes at the Cretaceous-Tertiary boundary: Nature, v. 324, p. 148–152.

Woodruff, F., and Savin, S. M., 1991, Mid-Miocene isotope stratigraphy in the deep sea: High resolution correlations, paleoclimatic cycles and sediment preservation: Paleoceanography, v. 6, p. 755–806.

Wu, S., 1987, The Duolun impact crater, China: International Geological Correlation Project 199 Meeting, Beijing, Abstracts.

Xu, D., and Yan, Z., 1993, Carbon isotope and iridium event markers near the Permian/Triassic boundary in the Meishan section, Zhejiang Province, China: Palaeogeography, Palaeoclimatology, Palaeoecology, v. 104, p. 171–176.

Xu, D., Ma, S., Chai, Z., Mao, X., Sun, Y., Zhang, Q., and Yang, Z., 1985, Significance of a $\delta^{13}C$ anomaly near the Devonian-Carboniferous boundary at the Muhua Section, South China: Nature, v. 314, p. 154–156.

Xu, D., Yan, Z., et al., 1989, Astrogeological events in China: New York, Van Nostrand Reinhold, 264 p.

Xu, G., 1991, Stratigraphical time-correlation and mass extinction event near Permian-Triassic boundary in South China: Journal of China University of Geosciences, v. 2, p. 36–46.

Yabushita, S., 1991, A statistical test for periodicity hypothesis in the crater formation rate: Monthly Notices of the Royal Astronomical Society, v. 250, p. 481–485.

Yabushita, S., 1992, Periodicity in the crater formation rate and implications for astronomical modeling, *in* Clube, S. V. M., Yabushita, S., and Henrard, J., eds., Dynamics and evolution of minor bodies with galactic and geological implications: Dordrecht, Kluwer, p. 161–168.

Zachos, J. C., and Arthur, M. A., 1986, Paleoceanography of the Cretaceous/Tertiary boundary event: Inferences from stable isotopic and other data: Paleoceanography, v. 1, p. 5–26.

Zachos, J. C., Breza, J. R., and Wise, S. W., 1992, Early Oligocene ice-sheet expansion on Antarctica: Stable isotope and sedimentological evidence from Kerguelen Plateau, southern Indian Ocean: Geology, v. 20, p. 569–573.

Zhakarov, V. A., Lapukhov, A. S., and Shenfil, O. V., 1993, Iridium anomaly at the Jurassic-Cretaceous boundary in northern Siberia: Russian Journal of Geology and Geophysics, v. 34, p. 83–90.

Zhou, L., and Kyte, F. T., 1988, The Permian-Triassic boundary event: A geochemical study of three Chinese sections: Earth and Planetary Science Letters, v. 90, p. 411–421.

Manuscript Accepted by the Society September 14, 1995

Printed in U.S.A.

Geological Society of America
Special Paper 307
1996

The unique significance and origin of the Cretaceous-Tertiary boundary: Historical context and burdens of proof

Graham Ryder
Lunar and Planetary Institute, 3600 Bay Area Boulevard, Houston, Texas 77058-1113

ABSTRACT

The abruptness and intensity of the Cretaceous-Tertiary boundary have been deemphasized by some authors over recent years, mainly by those skeptical of an impact origin for the boundary. However, it was recognized at the birth of stratigraphy as both abrupt and of major importance. It was used to define the change from the Mesozoic to the Cenozoic; the boundary has become continually more precisely defined and its global sequences more correlatable. It is now unique in being an event boundary marked by an iridium-bearing layer of global extent, rather than an arbitrary boundary in a sequence of little change. The Permian-Triassic boundary, in contrast, is arbitrary and the transition is not yet proven to be abrupt, the extinctions that define it perhaps having taken place in pulses over several millions of years. Some of those who have denied the importance (and in some cases even the existence) of an impact in the Cretaceous-Tertiary extinctions have placed burdens of proof on the impact hypothesis that they do not place on strictly terrestrial mechanisms. Terrestrial mechanisms have always been unsatisfactory (or at least unconvincing for global, massive, multienvironment faunal change) and are now even more so. Some authors have required of the impact hypothesis attributes that are not inherent in it, including particular patterns of extinction selectivity and timing.

INTRODUCTION

Considerable scientific and public interest has attended the debate over the last 15 years about the nature and causes of the faunal transition from the Cretaceous Period to the Paleogene Period, more commonly known as the Cretaceous-Tertiary boundary. The debate is noteworthy for its multidisciplinary participation, which has led to great advances as well as some difficulties and confusions. Immediately prior to the impact hypothesis of Alvarez et al. (1980), the boundary was generally accepted as an extremely sharp one (e.g., Bramlette, 1965; Alvarez et al., 1980; Russell, 1979). Some authors, particularly (but certainly not restricted to) some paleontologists from both invertebrate and vertebrate disciplines, have recently deemphasized the abruptness of the transition and its importance, particularly in relation to other mass extinctions but also in absolute terms. This has been in parallel with attempts to deny any particular significance to any impact in the transition, whatever the reality or characteristics of an impact. In part this denial reflects the continuing influence of uniformitarianism in its erroneous guise of requiring slow, gradual change as well as the multiple meanings of "uniformitarianism" conflated by Lyell (1830, 1832, 1833) that facilitate unwitting circular arguments. One of those meanings of Lyell's was *uniform rates of change.*

This chapter, an amplification of Ryder (1994), has two main goals. The first is to briefly review the historical context and recognition of the Cretaceous-Tertiary boundary and to claim that the boundary truly represents the most clear example of a rapid mass extinction event. The boundary is historically recognized as critical or major and has become more precisely defineable with time since its original recognition. Its correlation with an iridium (Ir)-depositing or Ir-concentrating event is now undoubted, and the Ir is an important component of the definition of the boundary stratotype at El Kef, Tunisia. The

Ryder, G., 1996, The unique significance and origin of the Cretaceous-Tertiary boundary: Historical context and burdens of proof, *in* Ryder, G., Fastovsky, D., and Gartner, S., eds., The Cretaceous-Tertiary Event and Other Catastrophes in Earth History: Boulder, Colorado, Geological Society of America Special Paper 307.

boundary is unique in being an essentially nonarbitrary event boundary. Other mass extinction transitions, including the end-Permian, may ultimately prove to be precisely definable and nonarbitrary, but at present they are not.

The second goal is to indicate that some opponents of the impact-generated mechanisms for extinction have required of it burdens of proof that they do not require of their own nonimpact mechanisms. This even applies to some workers who agree that the Ir-bearing layer was a consequence of an impact. They require of the impact hypothesis attributes that are not inherent in it; that is, they have set up inappropriate straw men in their attempts at counterrevolution. For instance, it has been common to claim that all extinctions are instantaneous following an impact (e.g., Hallam, 1989). A few workers continue to deny that an impact occurred, attributing to igneous processes effects that have not been observed in action, have not been deduced from the geological record, and are not inherently plausible. However, this extreme position will be almost ignored in this chapter.

RECOGNITION AND HISTORICAL DEFINITION OF THE K/T AND OTHER BOUNDARIES

Disregarding the historical context of the concept of a Cretaceous-Tertiary boundary, some authors have recently deemphasized or questioned both the abruptness of the transition and its importance. It has been claimed that interest in the boundary reflects comparatively unimportant characteristics such as our childlike interest in the extinction of dinosaurs, our anthropocentric view that is biased in favor of the rise of mammals, the controversy (and hence journalistically favored polarization) stemming from the impact hypothesis, and the mere closeness of the boundary to the present. Thus Stanley (1987) in his review states that ". . . these factors have led us to overrate the terminal Cretaceous crisis." Erwin (1993) listed six "really important events in the history of life on Earth" but added that he included the Cretaceous-Tertiary events only to forestall a "lot of caterwauling from a few mammals." Even Stephen Jay Gould (in Raup, 1991), who not only accepts but advocates the impact as the trigger of extinctions and accepts the particular significance of the extinctions, nonetheless suggested that it is "particularly dear to our hearts because it wiped out the dinosaurs and gave mammals a chance." Even though insiders realize that some such statements may have been made tongue-in-cheek, the statements diffuse and confuse the significance of the boundary and therefore the significance of studying it.

The Cretaceous-Tertiary boundary has been recognized as both major in importance and abruptness since the early decades of the nineteenth century, well before dinosaurs or iridium became geologically useful concepts. It was the first major boundary recognized. That it is most critically the Mesozoic-Cenozoic boundary is somewhat obscured by its more popular name of Cretaceous-Tertiary. Notwithstanding the unclear definitions that attend what is meant by a mass extinction, I contend that the Cretaceous-Tertiary boundary extinctions represent the most rapid or catastrophic and widespread example of a mass extinction. In comparison, the massive Permian-Triassic extinctions apparently took place over a period of several millions of years and cannot, at least yet, even be postulated as a (single) event.

The importance of the boundary and the faunal change

The recognition of a major and abrupt change that is the Cretaceous-Tertiary boundary took place (under catastrophic thinking, admittedly) at about 1810 at the very beginning of fossil-based stratigraphy. Cuvier had previously established the fact of extinction. The studies in the Paris Basin by Cuvier and Brongniart (1811) defined the sharp transition. They found no transition between the chalk and the overlying "plastic clay," which had distinctive fossils. They recognized that there was probably a separation in time between the two, based on depositional, not paleontological, evidence. In 1813, d'Halloy distinguished "cretaceous" from "mastozootic" (= Tertiary) as the upper two of four stratigraphic divisions of rocks, although precise limits at a type locality were not indicated (published as d'Halloy, 1822). The name "mastozootic" demonstrates that the faunal change was as important as the lithological change in making the distinctions. No Permian-Triassic boundary was then recognized; it was subsumed in the second group from the (now) upper Paleozoic through Jurassic (yet d'Halloy was more than familiar with the rocks that Murchison later included in the Permian). The recognition of a sharp boundary corresponding with the modern Cretaceous-Tertiary boundary preceded the concepts of mass extinctions (in a modern sense), dinosaurs, micropaleontology, iridium, thin sections, isotopes, and radioactivity and preceded the theory of natural selection. The term "Dinosaur" was invented by Richard Owen in 1841 with the recognition of nine genera at that time.

Phillips (1841, 1860) recognized major and significant divisions of stratigraphy, labeling them as Paleozoic, Mesozoic, and Cenozoic. He adopted the name Paleozoic from Adam Sedgwick, who had coined it to combine his Cambrian with Roderick Murchison's Silurian. Phillips followed d'Halloy's distinction of Permian rocks from Triassic rocks (around 1830) and recognized the critical significance of the Mesozoic-Cenozoic change, at the Cretaceous-Tertiary boundary. This corresponded exactly with d'Halloys division. Murchison's Permian (Murchison et al., 1845) originally included rocks now recognized as Triassic, so the Permian-Triassic boundary in the rocks of Europe was not obvious to everyone at around 1840. Apparently d'Halloy's description in 1834 was closer to our modern definition of Permian than was Murchison's (Erwin, 1993). Under the then recently established uniformitarian principles (in all their conflated forms), passages or transitions were expected. Boundaries were invented for convenience of systematic arrangement and in consequence were expected to be arbitrarily assigned. This was (and remains) the case for the Permian-Triassic boundary.

The expected passage beds for the Mesozoic-Cenozoic boundary were never found. That they did not appear seems to have been an embarrassment for Lyell and perhaps even more so later for Darwin. Lyell had subdivided the Cenozoic on the basis of molluscs yet noted that the faunal change from uppermost Cretaceous to lowermost (then) Eocene was equivalent to the entire Tertiary faunal change (Lyell, 1833, Chapter 23). Lyell accepted, on the authority of Deshayes, that the marked discordance extended "to all other departments of the animal kingdom, and to fossil plants." Although the relevant strata in Britain are indeed missing (though at worst only a few million years worth, not sixty), Lyell was familiar with European strata, where more complete successions occur. Even in England a near-complete succession occurs in Dorset. Lyell thought that the "missing" sequences (which under his uniformitarian concepts corresponded with missing time) might be found somewhere—that the hiatus resulted from shifting basins of deposition. Although Darwin was also familiar with this faunal change, his only mention of it concerns the abrupt extinction of the ammonites. Darwin's (1859, Chapter 11) explanation for the "wonderfully sudden" extinction of the ammonites, and other groups, is the "probable wide interval of time between our consecutive formations."

These original distinctions of the Mesozoic from the Cenozoic were founded in macrofossil sequences and included the presence of belemnites, ammonites, and rudists in the Cretaceous, quite distinct from the marine macrofauna of the overlying Tertiary. It was recognized that mammals were fairly common in Tertiary rocks (hence "mastozoic") but rare to absent in Mesozoic ones, and with the recognition of the Dinosauria it was realized that these animals dominated the large Mesozoic faunas but were completely absent from the Cenozoic. Such distinctions applied to the uppermost Cretaceous (the Maastrichtian stage, defined in 1849 by Dumont) and the lowermost Tertiary (the Danian stage defined in 1847 by Desor). Although the latter stage was originally assigned to the Mesozoic, further work showed that the Danian did not correlate with Cretaceous elsewhere. As paleontology and biostratigraphy progressed, it became clear that the distinctions shown by the macrofaunas were paralleled by distinctions in the microfaunas, and these microfaunas were better able to show that the distinction of Cretaceous from Tertiary was both abrupt and global in extent. Microfossils provide a much better statistical database from which to make conclusions than do macrofossils.

A review of the perspective just prior to the interjection of the Ir data into the debate shows that many and varied paleontologists were cognizant of the reality of the importance of the boundary. Colbert (1965) took pains to show that the "age of reptiles" was not the same as the Mesozoic; the reptiles he described included those of the Permian. He referred to the extinctions at the end of the Cretaceous as The Great Extinction. While noting that there were others, he claimed that none of these had the finality of the several parallel extinctions that marked the end of the Mesozoic world. Bramlette (1965) claimed that the extinction of some large populations thriving towards the end of Mesozoic time was more demonstrably abrupt than that at the end of the Paleozoic. He emphasized invertebrate extinction, whereas Colbert (1965) emphasized vertebrate extinction. The invertebrates were most obviously and strikingly represented by marine planktonic forms and those forms dependent on plankton, such as ammonites and belemnites.

Nonetheless, the recognition of an abrupt faunal break in the record does not automatically lead to the recognition of an abrupt faunal break in reality. The immense faunal break, even where it occurs without significant lithological change, has been used to infer the presence of a sedimentological and time break even with no other supporting evidence. For example, for northeast Mexico sections, Hay (1960) claimed that the faunal change from the Mendez Formation (Maastrichtian) to the Vasquez Formation (Danian) spoke unequivocably for a hiatus of some extent, even though there is "little field evidence for an erosional break." The withdrawal of the sea, erosion, and subsequent rise of the sea is couched in uniformitarian terms, and arguments that the Mendez is more indurated than the Vasquez, that its microfossils are more recrystallized, and that there is an apparent weathering zone at the top of the Mendez do not seem to be borne out by more modern studies. Instead, these arguments are a clear demonstration of the power of theory to induce particular observations, in this case according to then-accepted philosophies of uniformitarianism that included (rightly or wrongly) the concept of slow, gradual change. A similar argument was made by Rainwater (1960) for the Cretaceous-Tertiary rocks in the Gulf Coastal Plain: "An important unconformity separates the two, as evidenced by a great change in faunas." He concluded that this probably represented considerable time.

Overall, it is safe to infer that even prior to the impact hypothesis of Alvarez et al. (1980), many paleontologists and biostratigraphers recognized that the Cretaceous-Tertiary boundary was both an abrupt and important transition. They nevertheless had a strong propensity to impose upon the boundary some cause that was mundane, noncatastrophic, and uniformitarian (in the then-accepted meaning) (e.g. Newell, 1962, 1963).

The precision and correlation of the Cretaceous-Tertiary and other boundaries

There appears to have been little difficulty in pinning down the Cretaceous-Tertiary boundary in most marine sequences since the mid-1800s. Most specialists concur that the boundary can be identified with relative ease and precision in local sections (Russell, 1979). Nonmarine sequences have been a greater problem. A question as to whether the Danian in its type area should be included with the Cretaceous was resolved negatively when the Danian was shown to correspond with a facies of rocks already defined as post-Cretaceous. Inclusion of the Danian within the Mesozoic appears to have been more influenced by its chalk facies at the type localities than by its fauna; both microfaunas and macrofaunas of the Danian show affinities with Tertiary life-forms. Although there were claims based on molluscan affinities for including the Danian within the Mesozoic, Rosenkrantz (1960) convincingly showed that lamellibranch, gastropod, and cephalopod assemblages in Denmark were much more like overlying Tertiary than underlying Cretaceous assemblages. Arbitrary assignment of the Danian to the Cretaceous would merely make the rapid transition recognized be within the Mesozoic rather than at the end; this may be what has happened to the near-end Ordovician mass extinction.

The Cretaceous-Tertiary boundary has become continually more tightly defined and recognizable, not less, and particularly so with the introduction of the study of microfossils. The recognition of a global iridium-rich layer (whatever origin is invoked) at the paleontologically defined boundary has led to the layer's use as a marker bed, much as bentonite beds have been used in more local stratigraphies. The new boundary stratotype at El Kef includes the Ir-bearing layer as an important component. It has been possible to precisely correlate some marine and nonmarine records because of the Ir-bearing layer. Some terrestrial disruptions recorded by pollen and other floral attributes took place geologically simultaneously with marine foraminiferal changes. Both ammonite (Ward et al., 1991) and dinosaur (Sheehan et al., 1991; Sheehan and Fastovsky, 1992) fragments, slightly or not at all reworked, persist to the boundary, and dinosaur footprints have been found less than 37 cm below the boundary (Pillmore et al., 1994). Rare fragments of reworked dinosaur fragments in floodplains of the lowermost Tertiary are irrelevant.

Some arguments against a rapid turnover are merely permissive rather than demonstrative, for example, that of MacLeod and Keller (1993) that many sections contain a hiatus and that faunal turnover is thus gradual. The conclusion from such evidence is that faunal change might be gradual, not that it is.

The precision and correlation at the Cretaceous-Tertiary boundary are not possible for the Permian-Triassic boundary, for which correlation of marine and nonmarine sequences has proven elusive and equivocal even among marine sequences. That part of the problem may be the incompleteness of the record as a result of regression does not eliminate the reality. Indeed, the placing of the Permian-Triassic boundary has changed with time and author and remains contentious (Erwin, 1993). It is arbitrarily defined on the appearance of various genera, e.g., *Otoceras*. The oldest definitions were partly lithological. Even the modern differences in placing in fact represent stratigraphic differences corresponding with several million years. The complexity of the Permian-Triassic transition in the record and its underlying causes have been eloquently described and discussed by Erwin (1993, 1994a, b). The transition is indeed marked by the passage beds expected by Murchison and others, with mixed fauna (not reworked) of Permian and of Triassic aspect at continuous sequences. The apparent contrast of the brachiopod-rich faunas of the Permian with the mollusc-rich faunas of the Triassic has been accentuated by the general absence of the Late Permian marine record in the regions with which the early stratigraphers were familiar. There was a massive transition from the Permian to the Triassic, but because the transition took place over several million years, unlike the Cretaceous-Tertiary transition, any boundary is arbitrary. There is no single horizon corresponding with the Ir-bearing layer of the Cretaceous-Tertiary boundary. The extinctions perhaps took place in several pulses (Ward, 1994; Erwin,1993). Stanley and Yang (1994) have presented evidence that the extinctions took place in two main pulses, one at the end of the Guadalupian and the other at the end of the Tatarian. The ultimate end-Permian pulse is somewhat reduced in intensity from previous suggestions. The resolution available for this final pulse is not yet as distinct as that of the Cretaceous-Tertiary; the total Changxinian (end-Tatarian) is probably about 2 to 3 million yr (Harland et al., 1989), although Stanley and Yang (1994) *presume* that a shift in C-isotopes over the final 100,000 years relates to the extinctions.

None of the other major boundaries in the stratigraphic record correspond with mass extinctions, except perhaps at the Triassic-Jurassic transition. The boundaries are instead arbitrarily defined, in accordance with the expectations of stratigraphers in the middle part of the last century. Thus the Silurian-Devonian boundary definition has long been subject to debate as to its placing and correlation (Holland, 1965; Earp, 1967). Its eventual definition places it at a horizon at which there was no sudden or dramatic change in life. The Triassic-Jurassic transition and mass extinction, which includes the extinction of several groups of reptiles, amphibians, and ammonites, does not appear to have been abrupt, and the boundary is arbitrarily defined.

The other mass extinctions, such as those of the Upper Ordovician and the Frasnian-Fammenian in the Devonian, do not correspond with major named boundaries. They were recognized as mass extinctions much later than was the Cretaceous-Tertiary boundary. Indeed, the Ordovician itself is a stratigraphic latecomer that is largely arbitrary, with only recently defined boundaries. The Upper Ordovician extinction was completed prior to the end of the Ordovician (Sheehan and Coorough, 1994) and appears to have tracked a glacial epoch. The extinction interval appears to have been 2 to 3 m.y. long and to have had several pulses, although Brenchley et al. (1994) have inferred a shorter duration for the glaciation.

The Frasnian-Fammenian extinction has long been recognized as abrupt and global and indeed so abrupt that an impact-induced catastrophic extinction was suggested by McLaren (1970, 1983). Many groups, including trilobites, corals, stromatoporoids, and brachiopods, were severely affected. However, this boundary has not been historically clearly defined. Harland et al. (1989) placed the boundary at the top of the *Palmatolepis triangularis* zone, whereas the International Subcommission on Devonian Stratigraphy (in Cowie, et al., 1989) defined the boundary at the base of that zone, corresponding with perhaps 1 m.y. (or more) earlier. This placement closely corresponds with McLaren's (1983) extinction horizon and is marked by a drastic reduction in marine biomass, ecosystem diversity, and extinction (McLaren, 1970, 1983; Claeys et al., 1994). It is not known whether this extinction is gradual over several conodont zones or instantaneous at the base of the *P. triangularis* zone. Events marking the boundary may have been spread over 2 m.y. (Claeys et al., 1994). No global layer equivalent to the Ir-bearing layer of the Cretaceous-Tertiary boundary is available for correlation, although glassy spherule layers are locally present at different horizons.

The Cretaceous-Tertiary boundary extinction is not just a poor relative of the Permian-Triassic, or any other, mass extinction. It was quite different, was far more abrupt and clear (i.e., the rate was much faster), and affected many more ecological niches, even if the total number of orders, families, or genera that became extinct was smaller than in some other mass extinctions. The Cretaceous-Tertiary transition and the Permian-Triassic transition are not really comparable in characteristics or in causes: There are mass extinctions and there are mass extinctions. The Cretaceous-Tertiary boundary is an event boundary; that is, something happened. This is in contrast with the previous stratigraphic aim of placing boundaries where abrupt and drastic changes in lithology or fossil content did not occur (McLaren, 1970); that is, no major geological events took place. Such rapid transitions were more or less assumed to indicate an hiatus. This makes the Cretaceous-Tertiary boundary unique in the stratigraphic record as currently defined and understood: a boundary at which something happened. Ultimately other such boundaries might be defined, possibly including the Frasnian-Fammenian. There is no doubt that the nature of the Cretaceous-Tertiary boundary has been under great scrutiny over the last decade and a half, more so than other boundaries. Nonetheless, this scrutiny cannot be taken to mean that other mass extinctions will necessarily eventually become so contracted in duration; a major reason for the impact hypothesis and the scrutiny of the Cretaceous-Tertiary boundary derives from its very abruptness.

STRAW MEN, COUNTERREVOLUTIONS, AND BURDENS OF PROOF

Many of those who are at best skeptical and commonly opponents of the impact-extinction hypothesis for the Cretaceous-Tertiary boundary changes have set up straw men and imputed to that hypothesis attributes neither inherent nor generally included in it by impact proponents. Such imputations are that nearly all of the uppermost Maastrichtian extinctions are impact caused (e.g., Keller et al., 1993), that virtually everything was instantaneously wiped out worldwide (e.g., Hallam, 1989), and that extraterrestrial catastrophic causes would produce more random extinction than terrestrial gradualistic causes (e.g., MacLeod, 1994). Some workers have demanded exact burdens of proof of "impactors" but have not demanded such proofs of themselves; for instance, "impactors" alone are criticized for not explaining the survival of crocodiles and turtles (although some impact scenarios have suggested survival mechanisms with detritus-based food chains).

Proponents of terrestrial causes, such as sea-level changes, have usually failed to portray the mechanisms much beyond speculation, and such mechanisms are currently little advanced and are even less consistent with current constraints than they were prior to the publication of Alvarez et al. (1980). Although many mechanisms, such as habitat fragmentation as a result of sea-level changes, might explain particular extinctions, they cannot yet be held capable of causing global extinctions over many different types of environment. Similar criticism can be made of those invoking a volcanic or mantle hypothesis (e.g., McCartney et al., 1990; Officer et al., 1992). Obviously there is a requirement to understand what is inherent in or predictable with an impact-caused extinction or any other type of extinction hypothesis.

Reality of a boundary impact event

The presence of an iridium-rich layer, shocked quartz, and the identification of perhaps the largest Phanerozoic impact crater—with a stratigraphic and radiometric age indistinguishable from that of the boundary elsewhere—have convinced most workers that there was a boundary impact event. There remain considerable differences in opinion of the significance, if any, of that event for extinctions. An important historical point is that the abruptness of extinctions led to the analysis for iridium at a particular horizon as a potential determinant of sedimentation rate. That measurement led to the inference of an impact at the time represented by that horizon. This in turn led to the hypothesis that the impact was responsible for the extinctions, from the inference that this coincidence was causal (and extinctions obviously cannot be held responsible for impacts). The impact hypothesis is not in itself dependent on a specific mechanism(s), although such mechanisms were reasonably suggested, including rapid and dramatic climate changes from dust clouds and other atmospheric effects. Many counterrevolutionary papers (i.e., those that deny an impact cause) over the last decade, however, give the impression that it is those who invoke an impact who have required a particular paleontological significance, for instance that impact proponents *claim* abrupt extinction. Yet it was never the case that an impact was inferred and that then there was a search for associated extinctions. It is an ironic reversal that some paleontologists chose to reduce the significance of the boundary after the impact was inferred. Rather than evaluate the record in the light of an impact, they chose to construct inappropriate straw men.

Impact and extinction timing patterns

Some paleontologists have made an effort to show that (stepped) extinctions were taking place in the Late Cretaceous up to the boundary; they follow with the specious conclusion that the impact had no significant effect (e.g., Keller, 1989; Keller et al., 1993). Exactly what "stepped extinction" means has not been defined, but it seems to imply some relationship among the steps rather than just some random, preboundary extinctions. Clearly some explanation is required for why planktonic foraminifera (for instance) should go extinct in steps in some clear order. Identification of relevant "stepped extinction" requires exceptionally fine stratigraphic resolution and needs to take full account of the Signor-Lipps effect. The steps also need to be shown to be globally correlative. The impact hypothesis, however, in no way requires that *all* or even most of the upper Maastrichtian extinctions resulted from the impact, and unless the "stepped extinctions" are causally related to each other they have no bearing on the reality of impact extinctions.

The impact hypothesis does demand that those extinctions that essentially define the change from Mesozoic to Cenozoic took place fairly rapidly as a result of (and therefore not before) the impact. This demand is fulfilled in the drastic microfaunal changes, the floral changes, and the extinction of ammonites, belemnites, dinosaurs, and probably marine reptiles that took place within analytical stratigraphic uncertainty of the boundary (e.g., Coccioni and Galeotti, 1994; Sheehan et al., 1991; Sheehan and Fastovsky, 1992; Ward et al., 1991). That inoceramids did not survive as far as the boundary, or that some foraminifera went extinct earlier, is (or may be) correct but does not deny the later catastrophic extinction. There were obviously continuing extinctions within the upper Cretaceous. Some of those extinctions historically linked with the Cretaceous-Tertiary boundary, including inoceramids, can no longer be used to define it.

The persistence of genera, or even species, through the boundary, even if they went extinct shortly after, does not deny the reality of impact-caused exinctions either. Even though there were almost certainly considerable geologically immediate effects of an impact, longer-term effects seem unavoidable. Thus Hallam's (1989) statement that exact contemporaneity of extinctions is required by the impact hypothesis is at best misleading. There is an unnecessary polarization of concepts: that either extinctions are not related to an impact or that all extinctions are related to an impact. In reality, some extinctions could precede and be unrelated to an impact, others could be an almost direct or near-term result of impact effects such as atmospheric heating or darkness, and others could be slower responses, perhaps over a few years to many tens of thousands of years, to a dramatically changed environment that is still in biotic instability. Keller (1989) has suggested that an extended period of extinctions that she infers over the Cretaceous-Tertiary boundary is incompatible with the impact hypothesis for extinctions. Keller et al. (1993) similarly state that impact proponents have long argued that a large bolide impact caused the extinction of nearly the entire Cretaceous biota and then use their conclusion that there were gradual extinctions at high latitudes to deny the general significance of an impact. Whether or not there was an extended period of extinctions, such a test imputes to the impact hypothesis an attribute that it does not have and that impact proponents have not (usually) required of it. Catastrophic collapse does not necessarily mean instantaneous and total collapse.

Impact and extinction selectivity patterns

Those skeptical of impact-related extinction have turned to demonstrating the case for selective extinctions, such as particular groups of animals or latitude-dependent effects, and concluding that these are more compatible with a nonimpact scenario than with an impact one (e.g., Keller et al., 1993; MacLeod and Keller, 1993; Archibald, 1994). Such a conclusion depends upon a knowledge of extinction mechanisms in both gradual and abrupt situations, the very things researchers are currently attempting to understand. Ecosystems today are complex and incompletely understood, and that of the terminal Cretaceous is far more removed from observation and analysis. Perhaps an analog would be to investigate the remains of a ruined house of cards and attempt to determine which cards were first removed and how. Thus Archibald's (1994) claim that the pattern of extinction of vertebrates is incompatible with an impact hypothesis relies on a claim of understanding of the relevant ecosystems that is currently not possible. It ultimately may have merit, once the relevant impact environmental effects are better known. It is presently inconsistent and inappropriate to claim that impact is incompatible with crocodilian survival yet claim that sharks would be decimated by sea-level regression and habitat fragmentation (Archibald, 1994, Table 1). Similarly, the claim by Clemens and Nelms (1993) that an impact cannot explain the extinction of some high-latitude Dinosauria has no firm basis; it is certainly not a crucial experiment. A few years ago few researchers would have thought that dinosaurs could have lived at such high latitudes at all; their discovery emphasizes how much we have to learn about how various dinosaur groups lived.

To suggest that gradual climatic changes or sea-level changes caused abrupt extinctions is to divorce the direct mechanism from the indirect one, with the same problems of interpretation and association as the impact scenario. Few total killing mechanisms (or alternatively reproduction-inhibiting mechanisms) have yet been certainly established for any hypothesis. Many suggested mechanisms are not demonstrated but highly speculative, and even such dramatic climate changes as the cooling of the last Ice Ages did not result in significant, and certainly not a mass, extinction. Decimation of the large vertebrate species had a human, not climatic, cause. Keller et al. (1993) claimed that changes created by an impact at the end of the Cretaceous would *not* have affected tropical more than high-latitude environments. This is an unfounded claim, not only because that impact took place in the tropics. It is certainly an undemonstrated claim.

CONCLUSIONS: THE UNIQUE BOUNDARY AT THE MESOZOIC-CENOZOIC TRANSITION

The abruptness and importance of the Cretaceous-Tertiary boundary have been recognized from the birth of stratigraphy, and the fundamental change from Mesozoic to Cenozoic has come to be continually more tightly constrained over 180 years. The significance of the boundary is not merely a product of dinosaur-fascination and the fantasies of those who postulate that an impact took place at the end of the Cretaceous Period. Rather, the significance of the boundary is a product of its unique sharpness and the important faunal changes. The Cretaceous-Tertiary boundary is a much more abrupt boundary than that of the Paleozoic-Mesozoic or of any other mass extinction or stratigraphic boundary as currently described. The Cretaceous-Tertiary boundary is unique in being essentially nonarbitrary and generally easily, precisely, and indisputably identifiable in at least marine sections. It is also unique in being a definite event boundary.

An impact hypothesis provides the most consistent and coherent understanding of the origin of the Cretaceous-Tertiary boundary, relating features as distinct as planktonic foraminiferal extinctions in Denmark, dinosaur extinction in Montana, tremendous loss of productivity at the boundary, iridium-bearing layers in the Raton Basin, shocked quartz in New Zealand, rapidly deposited spherule beds in a deep-water sequence in northeast Mexico, and a large buried crater in the Yucatan. Terrestrial-origin hypotheses such as sea-level changes as favored by Newell (1962, 1963) or volcanic activity (McCartney et al., 1990; Officer et al., 1992; Courtillot, 1990) do not have such unifying attributes without claiming coincidences and even when claiming speculations (such as shocked quartz and iridium to be from volcanic sources) as truths. Those who are skeptical of the impact hypothesis should not demand more of it than is inherent in it; too many straw men have led to confusion on what are the relevant issues and tests.

ACKNOWLEDGMENTS

This chapter is LPI Contribution 862. Reviews by Peter Sheehan, David Archibald, and an anonymous reviewer and comments by David Fastovsky led to considerable improvements (I hope) of the chapter. None of them necessarily agrees with any individual or combined polemics of this contribution. The Lunar and Planetary Institute is operated by the Universities Space Research Association under contract NASW-4066 with the National Aeronautics and Space Administration.

REFERENCES CITED

Alvarez, L. W., Alvarez, W., Asaro, F., and Michel, H. V., 1980, Extraterrestrial cause for the Cretaceous-Tertiary extinction: Science, v. 208, p. 1095–1108.

Archibald, J. D., 1994, Testing KT extinction hypotheses using the vertebrate fossil record, *in* New developments regarding the KT event and other catastrophes in Earth history: Houston, Texas, Lunar and Planetary Institute Contribution 825, p. 6–7.

Bramlette, M. N., 1965, Massive extinctions in biota at the end of Mesozoic time: Science, v. 148, p. 1696–1699.

Brenchley, P. J., Marshall, J. D., Carden, G. A. F., Robertson, D. B. R., Long, D. G. F., Meidla, T., Hints, L., and Anderson, T. F., 1994, Bathymetric and isotopic evidence for a short-lived Late Ordovician glaciation in a greenhouse period: Geology, v. 22, p. 295–298.

Claeys, P., Kyte, F. T., and Casier, J.-G., 1994, Frasnian-Fammenian boundary: Mass extinctions, anoxic oceans, microtektite layers, but not much iridium? *in* New developments regarding the KT event and other catastrophes in Earth history: Houston, Texas, Lunar and Planetary Institute Contribution 825, p. 22–24.

Clemens, W. A., and Nelms, L. G., 1993, Paleoecological implications of Alaskan terrestrial vertebrate fauna in latest Cretaceous time at high paleolatitudes: Geology, v. 21, p. 503–506.

Coccioni, R., and Galeotti, S., 1994, K-T boundary extinction: Geologically instantaneous or gradual event? Evidence from deep-sea benthic foraminifera: Geology, v. 22, p. 779–782.

Colbert, E. H., 1965, The age of reptiles: New York, W. W. Norton, 228 p.

Courtillot, V., 1990, A volcanic eruption: Scientific American, v. 263, p. 85–92.

Cowie, J. W., Ziegler W., and Remane, J., 1989, Stratigraphic Commission accelerates progress, 1984 to 1989: Episodes, v. 12, p. 79–83.

Cuvier, G., and Brogniart, A., 1811, Essai sur la geographie mineralogique des environs de Paris (Stratigraphy of the Paris Basin): Memoires de la Classe des Sciences Mathematiques et Physiques de l'Institut Imperial de France 1810, p. 1–278.

Darwin, C., 1859, The origin of species (1st ed.): London, John Murray, 490 p.

d'Halloy, J. J. O., 1822, The systematic classification of geologic formations: Annales des Mines, v. 7, p. 353–376.

Earp, J. R., 1967, The Siluro-Devonian boundary: Geological Magazine, v. 104, p. 401–403.

Erwin, D. H., 1993, The great Paleozoic crisis: Life and death in the Permian: New York, Columbia University Press, 327 p.

Erwin, D. H., 1994a, The Permo-Triassic extinction: Nature, v. 3647, p. 231–236.

Erwin, D. H., 1994b, The end-Permian mass extinction, A complex, multicausal extinction, *in* New developments regarding the KT event and other catastrophes in Earth history: Houston, Texas, Lunar and Planetary Institute Contribution 825, p. 33–34.

Hallam, A., 1989, Great geological controversies: New York, Oxford Science Publications, Oxford University Press, 244 p.

Harland, W. B., Armstrong, R. L., Cox, A., Craig, L. E., Smith, A. G., and Smith, D. G., 1989, A geologic time scale 1989: Cambridge, Cambridge University Press, 263 p.

Hay, W. M., 1960, The Cretaceous-Tertiary boundary in the Tampico embayment, Mexico: Proceedings, International Geological Congress, 21st, Copenhagen, pt. 5, p. 70–77.

Holland, C. H., 1965, The Siluro-Devonian boundary: Geological Magazine, v. 102, p. 213–221.

Keller, G., 1989, Extended period of extinctions across the Cretaceous/Tertiary boundary in planktonic foraminifera of continental shelf sections: Implications for impact and volcanism theories: Geological Society of America Bulletin, v. 101, p. 1408–1419.

Keller, G., Barrera, E., Schmitz, B., and Mattson, E., 1993, Gradual mass extinction, species survivorship, and long-term environmental changes across the Cretaceous-Tertiary boundary in high latitudes: Geological Society of America Bulletin, v. 105, p. 979–997.

Lyell, C., 1830, Principles of geology, Volume 1: London, John Murray, 511 p.

Lyell, C., 1832, Principles of geology, Volume 2: London, John Murray, 330 p.

Lyell, C., 1833, Principles of geology, Volume 3: London, John Murray, 398 p., and appendices.

MacLeod, N., 1994, An evaluation of criteria that may be used to identify species surviving a mass extinction, *in* New developments regarding the KT event and other catastrophes in Earth history: Houston, Texas, Lunar and Planetary Institute Contribution 825, p. 75–77.

MacLeod, N., and Keller, G., 1993, Hiatus distributions and mass extinctions at the Cretaceous/Tertiary boundary: Geology, v. 19, p. 497–501.

McCartney, K., Huffman, A. R., and Tredoux, M., 1990, A paradigm for endogenous causation of mass extinctions, *in* Sharpton, V. L., and

Ward, P. D., eds., Global catastrophes in Earth history; an interdisciplinary conference on impacts, volcanism, and mass mortality: Geological Society of America Special Paper 247, p. 125–138.

McLaren, D. J., 1970, Time, life, and boundaries: Journal of Paleontology, v. 44, p. 801–815.

McLaren, D. J., 1983, Bolides and biostratigraphy: Geological Society of America Bulletin, v. 94, p. 313–324.

Murchison, R. I., De Verneuil, E., and von Keyserling, A., 1845, The geology of Russia in Europe and the Ural Mountains: London, John Murray.

Newell, N. D., 1962, Paleontological gaps and geochronology: Journal of Paleontology, v. 36, p. 592–610.

Newell, N. D., 1963, Crises in the history of life: Scientific American, v. 208, p. 77–93.

Officer, C. B., Drake, C. L., Pindell, J. L., and Meyerhoff, A. A., 1992, Cretaceous-Tertiary events and the Caribbean caper: GSA Today, v. 2, p. 69–70, 73–75.

Phillips, J., 1841, Paleozoic fossils of Cornwall, Devon, and West Somerset: Great Britain Geological Memoir, 160 p.

Phillips, J., 1860, Life on Earth. Its origin and succession: Cambridge, Macmillan.

Pillmore, C. L., Lockley, M. G., Fleming, R. F., and Johnson, K. R., 1994, Footprints in the rocks—New evidence from the Raton Basin that dinosaurs flourished on land until the terminal Cretaceous impact event, *in* New developments regarding the KT event and other catastrophes in Earth history: Houston, Texas, Lunar and Planetary Institute Contribution 825, p. 89–90.

Rainwater, E. H., 1960, Paleocene of the Gulf Coastal Plain of the United States of America: Proceedings, International Geological Congress, 21st, Copenhagen, pt. 5, p. 97–116.

Raup, D., 1991, Extinction: Bad genes or bad luck?: New York, W.W. Norton, 210 p.

Rosenkrantz, A., 1960, Danian mollusca from Denmark: Proceedings, International Geological Congress, 21st, Copenhagen, pt. 5, p. 193–198.

Russell, D. A., 1979, The Cretaceous-Tertiary boundary problem: Episodes, v. 1979, p. 21–24.

Ryder, G., 1994, KT boundary: Historical context, counter-revolutions, and strawmen, *in* New developments regarding the KT event and other catastrophes in Earth history: Houston, Texas, Lunar and Planetary Institute Contribution 825, p. 101–103.

Sheehan, P. M., and Coorough, P. J., 1994, Were all extinction events caused by impacts?, *in* New developments regarding the KT event and other catastrophes in Earth history: Houston, Texas, Lunar and Planetary Institute Contribution 825, p. 110–112.

Sheehan, P. M., and Fastovsky, D. E., 1992, Major extinctions of land-dwelling vertebrates at the Cretaceous-Tertiary boundary, eastern Montana: Geology, v. 20, p. 556–560.

Sheehan, P. M., Fastovsky, D. E., Hoffman, R.G., Berghaus, C. B., and Gabriel, D. L., 1991, Sudden extinction of the dinosaurs: Latest Cretaceous, Upper Great Plains, U.S.A.: Science, v. 254, p. 835–839.

Stanley, S. M., 1987, Extinction: New York, Scientific American Library, Freeman, 242 p.

Stanley, S. M., and Yang, X., 1994, A double mass extinction at the end of the Paleozoic era: Science, v. 266, p. 1340–1344.

Ward, P. D., 1994, Dissecting the KT extinction: Components and comparisons with the Permo-Triassic and "modern" mass extinctions, *in* New developments regarding the KT event and other catastrophes in Earth history: Houston, Texas, Lunar and Planetary Institute Contribution 825, p. 127.

Ward, P. D., Kennedy, W. J., MacLeod, K. G., and Mount, J. F., 1991, Ammonite and inoceramid bivalve extinction patterns in Cretaceous/Tertiary boundary sections of the Biscay region (southwestern France, northern Spain): Geology, v. 19, p. 1181–1184.

Manuscript Accepted by the Society September 14, 1995

Printed in U.S.A.

Geological Society of America
Special Paper 307
1996

Observations on the mass-extinction debates

William Glen
U.S. Geological Survey, MS 930, 345 Middlefield Road, Menlo Park, California 94025, and Stanford University Press, Stanford, California, 94305-2235

ABSTRACT

During a decade-long, real-time, historical study of the impact/volcanism/mass-extinction debates, a diversity of primary historical data, including tape-recorded interviews and opinion surveys on debated issues, has been acquired from hundreds of scientists representing the many disciplines engaged. I argue that the current extinction debates have impugned uniformitarianism, and will have further-reaching conceptual and philosophical effects than even those of the plate tectonics revolution, which strengthened, rather than detracted from, uniformitarianism. At the advent of the Alvarez–Berkeley group impact hypothesis of Cretaceous-Tertiary (K/T) boundary mass-extinction in 1980, the plate tectonics paradigm, which perfectly accommodated uniformitarianism, had already come to form the basis for well-received, endogenous mass-extinction and faunal-change hypotheses. For that reason especially did uniformitarianism contribute to a poor reception of the impact hypothesis, which flew in the face of uniformitarianism. Impact theory's reception was further chilled because it was based on evidence (iridium) unfamiliar to the community charged with its appraisal, its mechanism of extinction cause was improbable in terms of canonical knowledge, and it was authored by aliens to paleobiology (the discipline mainly charged with its assessment). Irrespective of the causal hypothesis chosen, that choice became the strongest predictor of how the chooser would select and apply standards in all aspects of the debates. Correlation also appeared—although less strong—between disciplinary specialty and choice of hypothesis and between disciplinary specialty and the assessment of evidence. Most paleontologists were straightaway against impact, and most paleontologists who later accepted boundary impact(s) continued to deny impact(s) as the main extinction cause. Many paleontologists regarded the duration, severity, and other aspects of the K/T mass extinction in terms of the fate of their own fossil groups. Vertebrate paleontologists were almost unanimously opposed to impact-as-extinction-cause; in contrast, micropaleontologists—especially those treating planktonic calcareous forms—were most often favorably inclined. Almost all specialists intimately familiar with impacting and earth-crossing bodies were sympathetic to all of impact theory. All volcanologists I interviewed rejected the volcanist hypothesis. Authors and published supporters of endogenous alternatives to the impact hypothesis, in all cases examined save one, were opposed to the impact hypothesis. However poorly informed, scientists seldom failed to subscribe to one of the extinction hypotheses. The mind-sets (gestalts) of opposed polemicists seemingly precluded agreement on any of the important debated issues. Impactors (advocates of impact-as-extinction-cause) and volcanists (proponents of volcanism-as-extinction-cause) commonly used different standards of appraisal and weighted the same evidence differently. Collabo-

Glen, W., 1996, Observations on the mass-extinction debates, *in* Ryder, G., Fastovsky, D., and Gartner, S., eds., The Cretaceous-Tertiary Event and Other Catastrophes in Earth History: Boulder, Colorado, Geological Society of America Special Paper 307.

rative teams—with members often from distant, isolated disciplines but almost always supporters of the same mass-extinction hypothesis—formed overnight, with many members redirecting their careers in order to participate. Impact's opponents sought to demonstrate a lack of the ubiquitous, globally uniform effects claimed for an impact. It is shown here that a perceived or claimed anomaly (stair-step extinctions), in terms of one theory (single-impact), can grow into a standard of appraisal that can force the revision of the challenged theory (single impact) into an accommodating theory (multiple impacts). The search for impact sites and the remapping and dating of great flood basalt bodies are being actively pursued. The successful application of new methods, techniques, and instruments across a diverse and growing research front has facilitated the formulation of increasingly sophisticated models of the effects of impacts, mantle plumes, and other processes on sublithospheric, crustal, and biospheric processes, but other important likely effects of impacting and explosive volcanism are yet to be modeled. The intensity, fast pace, and disciplinary scope of this continuing upheaval are providing extraordinary views of the workings of science.

INTRODUCTION

This is an interim report of a continuing longitudinal examination, begun in 1983, of the history of the science, behavior of the scientists, and the philosophy underpinning the mass-extinction debates. Such a study of a major, broadly interdisciplinary, multitheoretical upheaval, *as it unfolds*, is unique and aimed at providing data, which have not been otherwise obtainable, about how science works in a crisis. After completing *The Road to Jaramillo* in 1982, which was a detailed, internalist, intellectual history of several research programs in earth science that intersected in 1966 to trigger the plate tectonics revolution, I realized that the inadequacies that remained at its conclusion could only have been remedied by further data that might have been obtained *while continental drift was being debated.* I had begun my study of the drift debates after they were resolved. I thus sought a current debate in its incipient stage as my next case for study.

The current debates began slowly in 1979 with the advent of the Alvarez group supernova hypothesis, but after the group replaced the supernova hypothesis with the meteorite-impact hypothesis in 1980, and significant, predicted, supporting evidence for impact mounted seemingly overnight, the mass-extinction debates accelerated into a compelling case for study by 1983. It was also in that year that the volcanism hypothesis—now being argued—was readvanced by Charles Officer and Charles Drake as an alternative to the impact hypothesis. At that time I put aside other less-promising studies of ongoing conflicts in earth science in order to study the mass-extinction debates.

In 1984, the emergence of the closely related hypotheses of periodic mass extinctions (Raup and Sepkoski) and periodic bolide impacting (Rampino and Stothers; Alvarez and Muller) greatly increased the scope and audience of the debates. At that point the conflict accelerated into an upheaval of seemingly unprecedented interdisciplinary proportions that has spawned a reticulum of conflicted theories and corollaries over the past fifteen years, with most of the major issues still unresolved.

Some of the larger conclusions presented here, and others, are discussed further in *The Mass-Extinction Debates: How Science Works in a Crisis* (Stanford University Press, May 1994). A book is now under way that will detail the evidence supporting the larger ideas presented here, and others. The most complete single list of publications (almost 2,000) treating the scientific components of the debated issues can be found in Shaw, 1994.

CONTEXT FOR THE RECEPTION OF THE ALVAREZ–BERKELEY GROUP HYPOTHESIS

The mass-extinction hypothesis of the Alvarez group was only one of many offered up over the course of more than two centuries, but it was based on a form of evidence that was uniquely compelling—and it could be tested practically, economically, and almost immediately (the endogenous "arctic spillover" hypothesis of Stefan Gartner and J. Keany [1978a, b] was testable but virtually prohibitive in cost). The litany of untestable hypotheses or those that were simply ignored because they flouted orthodoxy—including some of meteorite impact—reaches back to Pieerer de Maupertuis, who suggested in 1742 that comets had repeatedly struck the Earth and snuffed out life en masse. A number of eighteenth and nineteenth century authors reformulated his hypothesis with little real change, including the French astronomer Laplace in 1797. The theory of M. W. De Laubenfels (1956) may be the most notable of the virtually untestable impact theories advanced in this century; he invoked the modern Siberian, Tunguska bolide (exploding meteorite) event as a model to explain the K/T boundary extinction. René Gallant's (1964) little-known, important, precocious book (correctly urged on me by Hugh Torrens in 1986; Torrens, 1994) on the litho- and biospheric effects of great meteorite impacts proposed that disturbances of both the Earth's rotation and physical environment would have produced the mass extinctions that punctuate Earth's history. Gallant believed, as do current "impactors" who hold that meteorite impact caused the K/T boundary extinction(s), that the survivors of global catastrophes

". . . would be not the 'fittest,' but the "lucky' "—a view that has since been profoundly expanded in an engaging and popular book by Dave Raup (1991). Harold C. Urey (1973)—who had earlier published on the origin of tektites and cometary collisions (1957, 1963)—discussed a wide range of environmental effects that followed from an impact and surmised it "even probable that [such a collision] destroyed the dinosaurs and initiated the Tertiary division of geologic time." He wondered ". . . *if tektites might not be found at some other boundaries between the Eocene, Paleocene and Cretaceous periods?*" [italics added]. (The canonically framed, geologic mind-set that militated against mounting such a search at that date is discussed in Glen, 1994a).

Near the close of Digby McLaren's Presidential Address to the Paleontological Society in 1970 we find the first impact theory of mass extinction (370 million years ago) based on a careful analysis of the fossil record. However, it too was not practically testable, and strong anticatastrophic sentiment and the community's unfamiliarity with bolide impacts relegated the theory to virtual oblivion—a fate common to all extraterrestrial mass-extinction hypotheses prior to that of the Alvarez–Berkeley group.

Exogenous causal hypotheses of earthly biotic change, such as solar flares, supernovae, and meteorite impacts, were simply not seriously entertained—nor was that of a volcanic cause (Vogt, 1972). In contrast, hypotheses of gradual, endogenous causes, such as climate and sea-level change, and derivative effects such as a phytoplankton collapse either triggering extinctions at higher trophic levels (Bramlette, 1965), or changing atmospheric carbon dioxide and oxygen levels (Tappan, 1968) were generally systematically addressed, however variously received. The terminal Cretaceous carbon-cycle perturbation theory of Dewey McLean (1978)—a precursor to his volcanism hypothesis of 1981, which he advanced as the first alternative to the Alvarez group hypothesis—called neither for catastrophe nor geologic instantaneity. McLean's theory attracted much interest, and McLean was encouraged to pursue his novel idea by several distinguished earth scientists, including paleontologists (McLean, oral communication, 1986, 1994).

Most influential in conditioning thinking about mass extinctions—just as geology's new paradigm of plate tectonics was being fit to Earth history during the 1970s—were several extensive articles, especially those by James Valentine and Eldridge Moores. They invoked plate theory as the causal mechanism behind faunal changes, and strongly linked multiple aspects of the fossil record to plate movements (Valentine and Moores, 1970, 1972; Schopf, 1974; Simberloff, 1974; etc.). Valentine and his supporters showed how the splitting and assembly of the continents explained changing taxonomic diversity, including massive reductions over a several-million-year interval at the times of major time scale boundaries. And with only a small conceptual leap, important parts of the mass-extinction record thus seemed to make sense if one accepted such a lengthy time frame and plate movements to generate such faunal-change boundaries. Valentine and Moores (1970, 1972), and later Karl Flessa and John Imbrie (1973), showed how continental collisions would reduce preferred living space, produce more severe seasonal climates, and open biotas to new forms of competition, and James Hays and Walter Pitman (1973) used plate movements to alter sea level and deteriorate climates. All such essays invoked a plate-driven forcing function to change climates, alter ocean currents, modify continental shelf area, or increase competition, and all, except one, accorded perfectly with the then vague, widely accepted notion that the major extinction boundaries had been brought into being within the time frame for mountain building. That time interval for orogeny, however poorly constrained, was authenticated for decades in textbooks at a few million years.

The warm reception of the plate-driven extinction hypothesis was in large measure due to its accommodation of uniformity. The hypothesis was warmed from still another source: The panoply of environmental changes attributed to plate movements was already very familiar to the paleontologic community. Norman Newell (1967, 1972, etc.) an influential figure, and others had earlier explicated environmental changes as the effects of well-documented sea-level changes—eustatic changes were simply a fact of geologic life, by then a seemingly trusted and problem-free component to include within any new theory.

Only Stefan Gartner and J. Keany in their testable "Arctic spillover model" (1978a, b; Gartner and McGuirk, 1979)—which employed a plate-tectonics mechanism to open an isolated and brackish Arctic into the North Atlantic—surmised that many fossil groups had expired in the terminal Cretaceous with geologic *instantaneity* (Thierstein and Berger [1978] immediately extended such a model to several different mass-extinction horizons). The Gartner and Keany hypothesis of extinction, which predicted that drilling the Arctic Ocean floor—an extremely costly test—would reveal fresh-water, youngest Cretaceous sediments, was virtually ignored, even by Gartner's close colleagues (Gartner, oral communication, 1994). When the hypothesis was addressed, it was generally rejected. It appears not to have been denied because of untestability—which appears the case for all other pre-Alvarez hypotheses to varying degrees if the "testability" mind-set of the involved disciplines is considered—but seemingly because of Gartner's claim of "instantaneous" extinction. Gartner meant "within no more than a couple of hundred years." Thus, although Gartner and Keany embraced the uniformity of a plate tectonics causal mechanism in their hypothesis, they simultaneously invoked an effect of instantaneous, global, catastrophic character that flew in the face of uniformity. That put them beyond the pale (recall that uniformity of that date allowed only local catastrophes; note also that Gartner and Keany were focused on the severely altered K/T boundary coccolith record [Gartner and Keany, 1979], a fact that accords with a surmise to which we shortly turn, holding that the fate of a given taxonomic group at the boundary conditioned the theory choice of its students).

Plate theory, ascendant as geology's paradigm by the early 1970s, and incorporated into generally well-received, related hypotheses of faunal change that sought to explain mass extinctions, thus appears a significant force in helping to shape resistance to the impact hypothesis. Plate theory gave us a radical, new picture of planetary mechanics that showed the rate and direction of movement of pieces of the Earth's shell (but did not explain plate or sublithospheric dynamics). That revolutionary picture of plate kinematics not only accommodated the gradualist views of Hutton, Lyell, and Darwin in the first half of the nineteenth century, but even furthered these views by demonstrating a suitably slow, Lyellian pace for tectonic processes resulting from plate motions that averaged only a few centimeters per year. In contrast to plate theory, the current upheaval surrounding the impact hypothesis has led to a reassessment of the role of catastrophe and its time frame, altered the definition of uniformity in the minds of many, and forced a reexamination of Earth's history with a truly new mind-set. That emerging new gestalt is already reshaping assessments of the evolutionary record, man's origins, his place on the planet, and life's fit to the cosmic scheme (see Gould, 1994).

I have argued that the impact/volcanism debates will have further-reaching conceptual and philosophical effects than even those of the plate tectonics revolution (Glen, 1988; Gould, 1991, 1994; Glen, 1994D). It might be said that since the victory of uniformitarianism over catastrophism in the last century—after which, talk of catastrophic extinction was virtually disallowed—paleontology has lacked consensus about an overarching conceptual scheme by which to explain mass extinction. Even the plate tectonics–driven solutions—however well received—left many with lingering doubts about their power to stress the great diversity of taxa that were extirpated on a global scale in a variety of both marine and continental environments.

Karl Flessa spoke to that point in 1979, just prior to the advent of the impact hypothesis, in his article on extinctions in the Encyclopedia of Paleontology. He appears to have expressed a widespread paleontologic view, if not the orthodox one at that date, that held for a multiplicity of endogenous causal factors when he remarked that "it is improbable that any single cause is responsible for all the major extinctions of the past 600 m yr. Nor is it likely that any one specific environmental change is responsible for all the extinctions that occurred during an episode of biotic crisis." The persistence of Flessa's view can be read a decade later when the impact hypothesis had been argued successfully to many geologists, but only few paleontologists were moved to embrace it. In an overview of the then current mass-extinction debates that generally reflected the views of most paleontologists—and still does today—Antoni Hoffman (1989) concluded that "Perhaps then mass extinctions are not the biotic consequences of any single phenomenon of one or another sort, but rather rare incidences of more than one major change in the physical environment accidentally clumped together within relatively short intervals of geological time, say 2–4 million years in duration."

The term *mass extinction*, and close scrutiny of the idea of mass extinction, does not appear until the middle of this century. Since then, we find a host of largely uniformity-friendly hypotheses that range from the well-received and provocative to the fatuous trying to explain the massive taxonomic changes that mark the major boundaries of the geologic timescale. All went untested, wanting in practicality, methodology, or the likely preservation of supporting evidence. In 1980, sensu stricto, a true paradigm for mass kills appears still absent.

Given the significance of extinction to studies of evolution and the fossil record, one might marvel at how little attention was historically afforded the question of the nature of extinction or mass-extinction cause. Although interest in the question rose somewhat in the 1970s, only with the advent of the impact hypothesis did research programs mount feverishly to fuel this modern upheaval in science.

THE ARCHITECTURE OF AND EVIDENCE FOR THE ALVAREZ–BERKELEY GROUP HYPOTHESIS

The Alvarez–Berkeley group consisted of the Nobel laureate physicist, Luis Alvarez; his son, Walter Alvarez, a geologist; and nuclear chemists Frank Asaro and Helen Michel. They published their hypothesis of mass extinction by meteorite impact in 1980. It was based on the discovery of an unprecedented, highly anomalous concentration of the platinum group element iridium in a distinctive, pencil-thick clay layer, 65 million years old, near the northern Italian town of Gubbio. The layer which represented the K/T boundary had been deposited between limestones that were paleontologically and magnetostratigraphically well defined. The K/T boundary marked the time of the second-greatest and best-known extinction of life in Earth history. It included the entire dinosaur stock, many other terrestrial animal groups, and about 75% of all oceanic species. Some even surmise that a count of the taxa made extinct at such a boundary may be an understated proxy for environmental stress or total biomass loss, since only a few survivors of a populous taxon will keep it off the extinction roll (McLaren and Goodfellow, 1990).

Because iridium is vanishingly scarce in the Earth's crust, its anomalous concentration was attributed to an extraterrestrial source. In 1979, the Alvarez group first postulated a supernova as the cause of the iridium anomaly, but after evidence against a supernova was found in the same year (Alvarez et al., 1979a, b), the hypothesis was revised in 1980 and an iridium-rich, single meteorite replaced the supernova. The single-meteorite–single-impact hypothesis was further altered a few years later as evidence for multiple, lesser extinction horizons (spread over one to two million years) near the K/T boundary was found that led to a revised hypothesis of multiple impacts in order to accommodate the multiple extinctions.

The single-impact meteorite hypothesis of 1980 was founded partly on two tacit assumptions, retrodicted two cataclysmic events, and prompted by historically unprecedented evidence (siderophile element anomalies in meteoritic ratios), made predictions guiding experiments that would test the impact component of the hypothesis. The first assumption of the hypothesis—however weakly suppositive—held that a mass extinction at the end of the Cretaceous period occurred with a magnitude and circum-mundane effect that made it distinguishable from "normal" or "background" extinctions. The idea of such a mass extinction was accepted by the great majority of paleontologists, but because a very small group, including some highly accomplished members such as J. Wyatt Durham and John C. Briggs (Briggs, 1994; see Glen, 1994b, c), did not accept the concept of mass extinction at any horizon, the existence of a K/T boundary mass extinction, in small part, was considered an assumption.

The second assumption of the hypothesis, which was more strongly suppositive than the first, was that the mass extinction was instantaneous ("geologic instantaneity"—as read from the historical literature—ranges from days to hundreds or even thousands of years). Most paleontologists believe that extinctions are not instantaneous in the geologic sense, but instead took hundreds of thousands or millions of years; that is largely an assumption, since there is no definitive temporal frame for constraining extinction intervals.

The first of the two major retrodictions brought a 10-km meteorite to strike the Earth an instant before the mass extinction at the K/T boundary. The second principal retrodiction grew from the first—the impact supposedly altered the biosphere violently and thus caused the mass extinction. Overnight, the compelling impact hypothesis triggered research programs across several different disciplines. They were organized to test the prediction that platinum-group element anomalies and, implicitly, companion forms of impact evidence distributed by impact-generated, cataclysmically energetic, transport modes—including debris fractions blown into ballistic trajectories—would be discovered in, or stratigraphically proximal to, the K/T boundary layer around the globe.

Such a prediction was unprecedented because no empirical or abstract geologic reference frame existed that could guide prediction of the generation, global dispersal, and preservation of bolide impact products. The "impactors" (subscribers to the whole of the impact hypothesis; Glen, 1990) claimed that evidence such as iridium anomalies, shocked quartz, and tektites was "diagnostic of impact." However, those "diagnostics"—the last two of which were widely held—were all opened to question by critics. Because of long geologic tradition, few in the geological community—except those in specialties such as planetary and lunar geology and impacting studies—believed that impacting was an important agent in shaping anything other than early Earth history. Presentations in historical geology courses—see any textbook of the last century save that of Seyfert and Sirkin, 1979—did not include impacting. Meteorites were never thought to have influenced the fossil record, almost all of which is confined to rocks formed during the last 550 or so million years of our planet's 4.55-billion-year history.

In keeping with the prediction of the impact hypothesis, during the 1980s more than 100 localities displaying iridium anomalies within the K/T boundary layer were found around the world in rocks representing both marine and terrestrial depositional environments. And other kinds of evidence—some strong, some tenuous—were found that appeared to have been wrought by an impact.

Two influential forms of evidence appeared within a half-decade of impact theory's rise that went far in drawing further attention to the question of impact: quartz grains bearing several intersecting sets of shock lamellae (known now from almost three dozen localities) and spherules that were interpreted by some as altered glass tektites (which are formed from impact-melted rock droplets), now known from more than 60 localities.

Arguments that the spherules represent impact-generated tektites have been buttressed in the last few years by the discovery around the Caribbean basin of spherules with a clay rind enclosing unaltered glass cores of tektitelike glass, which are now generally regarded as impact products (Izett et al., 1990; Sigurdsson et al., 1991); that is because their chemistry compares closely with rocks at the postulated impact site at the tip of the Yucatan Peninsula, and their oxygen-isotope ratios do not match those of volcanic products (Blum and Chamberlain, 1992). Especially was the oxygen-isotope evidence a severe blow to the cause of the "volcanists" (those who embrace massive volcanism as the K/T extinction cause), because throughout most of the debates they claimed that the spherules were volcanic and specifically denied an impact origin.

A new form of evidence for impact was reported in 1991 by David Carlisle and Dennis Braman, who found microscopic diamonds (3 to 5 nanometers) in a ratio to the iridium of 1.22:1 in a K/T boundary bed of continental origin in Alberta, Canada. Such a ratio is close to that found in type C2 chondritic meteorites (elemental ratios previously had indicated a meteorite but not the specific type). In the following year Carlisle (1992) reported that the carbon-isotope ratio in the diamonds approximated that of interstellar dust and did not accord with a terrestrial origin. Carlisle and Braman advanced still further support for impact in lectures in 1993 (published in 1995): their diamond-bearing bed also held 51 amino acids, including 18 known only from carbonaceous chondritic meteorites. Gilmour et al. (1992) also found diamonds in the K/T boundary layer at two different localities in the United States, but unlike Carlisle (1992), who believed that the diamonds are remnants of the bolide, Gilmour et al. thought that the diamonds were probably derived from the shock-induced transformation of carbon compounds in the target rock by the impact. That surmise was based on the carbon and nitrogen isotopic analysis of the diamond. Carlisle (1995, and oral communication, July 1994) responds to Gilmour et al. by noting that "Marine and lacustrine deposits would not have either amino acids or extra-terrestrial diamonds because they could not settle out from seawater or lakewater. The impact-generated diamonds are much larger and therefore can settle out; they are also much scarcer than the bolide-component, extra-terrestrial diamonds; thus, in marine and lacustrine deposits, impact-generated diamonds predominate, whereas in certain kinds of terrestrial deposits diamonds brought to Earth by the bolide predominate."

Whether one believes that the diamonds found at the K/T boundary were contained in the bolide, or generated on Earth by impact pressures, both possibilities require a bolide impact—thus diamonds within the K/T boundary layer constitute evidence for impact.

Impact theory's critics alluded to the absence of the postulated 200-km-wide, K/T impact crater for the first few years of the debates, but they never emphasized that point because most earthly craters have been removed or hidden by erosion, sedimentation, volcanism, or the subduction of tectonic plates—Gene Shoemaker, who had spent decades trying to focus geology's eyes on Earth's impacting history, made that clear to those who listened (Shoemaker, 1977). The more recent evidence of Carlisle and Braman, Blum and Chamberlain, and Izett and others by the summer of 1992, when added to earlier evidence, attracted new sympathizers to the impact cause. But many subscribers to impact, mainly among the paleontologists, remained unconvinced that impact(s) was the exclusive or main cause of the extinction(s) at or proximal to the K/T boundary.

The search for the K/T impact site(s) focused on the tip of the Yucatan Peninsula in Mexico by about late 1990 (Penfield and Camargo, 1981; Hildebrand and Penfield, 1990; Kring and Boynton, 1992). The site contained the Chicxulub impact structure, and although its age was then only approximated, its crustal makeup matched the enigmatic mineralogy of the K/T boundary layer(s) that contained debris of both continental and oceanic character (Kyte and Smit, 1986). The crater was then isotopically dated at precisely K/T boundary age (65 Ma; Swisher et al. 1992), and elevated iridium levels were found in purported impact melt rock within the crater (Sharpton et al., 1992). Its 200-km diameter spanned continental rocks that matched the age of the precisely dated, unaltered glass tektite cores from the K/T layer around the Caribbean basin.

During 1991 and 1992, a swarm of reports from different Caribbean sites analyzed the purported products of impact at Chicxulub by composition, age, and structural context. Many were convinced of an impact by the new evidence—even some long-standing impact opponents—*but only few of the paleontologists who converted to impact(s) regarded impact(s) as the main cause of the K/T boundary extinction(s).*

THE EARLY FATE OF THE IMPACT HYPOTHESIS

The impact hypothesis was generally not well received initially. Such might have been expected for a number of reasons: The geological community charged with its appraisal was generally unfamiliar with the phenomenon of impacting; the instantaneous, worldwide effects of a great impact seemed to contravene uniformitarianism (geology's philosophical mainstay; Gould, 1965, 1989; Goodman, 1967; Shea, 1982; Marvin, 1990; Glen 1994c; Hsü, 1994). The iridium anomaly (a mysteriously unfamiliar element to most geologists) was a "black box," product of nuclear chemistry, and it had been discovered and presented by disciplinary strangers speaking a barely intelligible language. Westrum (1978) spoke of anomalies in science and surmised that scientific communities protect their internal processes from external interference—this appeared to be the case in these debates.

Earth science—not particularly "hard" and distinctly historical in several of its subdisciplines—had a history of suffering badly in its conflicts with the "magisterial authority" (S. Toulmin's term) of the "hard" or "restricted sciences," that is, those sciences with powerful laws that bestow great predictive and explanatory power (Pantin, 1968). Lord Kelvin, in the late nineteenth century, impugned uniformity with the second law of thermodynamics and thus disallowed enough time in which to develop life within the canonical geologic timeframe. Sir Harold Jeffreys, in the first half of this century, used physical laws to preclude granitic ships from sailing through basaltic seas, and by dropping a rock and finding its magnetism thus disturbed, surmised that rocks could not possibly be used to reconstruct the ancient magnetic fields of the earth. Émile Argand's interpretation of massive overthrust sheets in alpine structure in 1923 was also met with disdain by those wielding the scepter of physical laws. Still other authority figures from "restricted sciences" pontificated on issues that appeared beyond the canons and tools of Earth science.

Geologists were thus infused with robust skepticism about the conclusions of disciplinary aliens. Luis Alvarez, the Nobelist in physics, and nuclear chemists Frank Asaro and Helen Michel appear indeed to have suffered a reception that was conditioned in part by their identification with those unhappy aspects of Earth sciences history. The presence of the geologist Walter Alvarez as a senior author in that group seems only to have barely warmed the chilled reception the hypothesis encountered, especially among those most unfamiliar with its main evidence—the siderophile elements.

Motivated by the Alvarez group hypothesis, and after two failed proposals to win support for what most regarded as a needle-in-the-haystack search for shocked quartz in the K/T boundary layer, Bruce Bohor, through the use of his own resources, made the astounding discovery of shocked quartz at the predicted K/T horizon (Bohor et al., 1984). Unlike the "black-box" iridium of "alien" nuclear chemists—which was not well understood by most geologists and seemed somehow inapplicable or dismissible to others—Bohor's evidence, tools, and methods were generally comprehensible to many geologists; he attracted a swarm of the formerly inattentive and fence straddlers to the impact hypothesis.

Although the impact hypothesis had two major discrete theses (impact, and impact-as-extinction-cause), interview respondents and journalists more often than not spoke of the hypothesis as a single unified entity, without differentiating the impact from its role as mass-extinction cause.

Interviews with scientists in this study indicated that their freely given opinions were typically not well informed—a phenomenon that largely accorded with my impressions in earlier historical studies. Commonly encountered were unambiguous responses to both general and specific questions by scientists who were poorly informed (unfamiliar with the most important papers on the issues in question). Ignorance of older, fundamental information was sometimes evident, but more often they had not read much of the burgeoning new literature; the tide of publications in the mass-extinction debates kept most from keeping abreast. Nevertheless, few respondents, however poorly informed they might have been, abstained from giving opinions.

It was apparent in this study—as it was in studies of continental drift history in the 1970s (Glen, 1975, 1982)—that *scientists very commonly claimed to have been better informed than they actually had been about debated issues that had eventuated in ascendant, important ideas.* Most interviewees in the continental drift study claimed to have kept informed of the continental drift debates over the decades prior to the advent, in 1966, of the proof that confirmed drift (Hallam, 1973; Glen, 1982). Few earth scientists were broadly informed of the details of the arguments over drift before the ascendance of the plate tectonics paradigm. Drift was seldom mentioned in university curricula, and when broached, it was typically through ridicule.

The same behaviors are exhibited in the mass-extinction upheaval: Scientists commonly claim to have known more than they actually did about issues that were later resolved to become historically important. That common tendency appears to heighten with the stature or notoriety that the issue attained.

Few of the earth scientists I questioned at random prior to the appearances, in 1984, of the important, audience-bolstering hypotheses of periodic mass extinctions (Raup and Sepkoski) and periodic impacts (Rampino and Stothers; Alvarez and Muller) were well informed about either the Alvarez group impact hypothesis or the McLean volcanist hypothesis (1981). Interviews over the following years provided a very different picture: The incremental resculpting of memory over the long term to accommodate the evolving needs of mental life was manifest; it is now a well known phenomenon with important implications for historians (Spence, 1982; Rubin, 1986; Glen, 1994c, 1996). How such reconstruction of memory among members of a collaborative team led to the production of an apocryphalized, collective account of a signal discovery in the mass-extinction debates will be detailed in my forthcoming book. I had encountered such troublesome aspects of data collection that accrued with passing time in earlier studies (Glen, 1982), and my concerns about them and the possibility of remedies residing in real-time studies of ongoing scientific debate were voiced by others, including Lemke and his collaborators in their valuable essays (Lemke et al., 1980; Nitecki et al., 1978) on the drift controversy. Hoffman and Nitecki (1985) then did a brief, real-time opinion survey among scientists regarding the mass-extinction debates; their results, where applicable, are largely in accord with some presented here.

HOW PRIOR INTELLECTUAL COMMITMENT AND DISCIPLINE CONDITIONED THEORY RECEPTION

At the appearance of the impact hypothesis all, save one, of the authors or published sympathizers of endogenous alternatives to the impact hypothesis were disposed against impact. They included the authors or advocates of a great variety of causal mechanisms: sea-level changes, climatic changes, plate tectonics movements, and volcanically induced environmental stress among them.

It was generally the case that disciplinary identity was a strong predictor of how one would assess theory; some disciplines predisposed their members unanimously, or almost so, to one hypothesis choice. A few disciplines, however, conditioned such choices only slightly.

The vast majority of paleontologists eschewed both the impact and the impact-as-extinction-cause postulates of the Alvarez group hypothesis for about a half-decade following its advent; in the subsequent half-decade several further forms of evidence were found that persuaded many of an impact. However, most of those paleontologists who later came to embrace impact did not then, and still do not, accept impact as the primary cause of the K/T mass extinction.

Only a few paleontologists—a handful of magisters (the most distinguished and influential) among them—agreed to a connection between impact and mass extinction early on and remain so disposed. As further evidence for impact grew, the small minority of subscribers to impact-as-extinction-cause among the paleontologists increased only somewhat—an increase far less than was found in other subspecialty communities of Earth science.

Very few paleontologists rejected the idea that mass extinctions actually occurred in Earth history. Those who denied the phenomenon of mass extinction thought that the very fragmentary character of the stratigraphic and fossil records taken together with extraordinary intersections of a number of poorly understood factors had been misconstrued as massive taxonomic change (mass extinction) within a brief time span (Newell, 1982; Briggs, 1990, 1994; Glen, 1994b, c).

Members of the communities of meteoritics, cosmo- and geochemistry, planetary geology, impacting studies, and allied fields that disposed their members to familiarity with the cometary and asteroidal flux of the solar system were much more likely than others to embrace the whole of impact theory. Expectedly, those who had treated the dynamics, effects, or products of impacting generally were most receptive to the impact hypothesis.

Most volcanologists (volcanic rock specialists) did not favor the explosive volcanist hypothesis of mass extinction (of Charles Officer and Charles Drake [1983, 1985]—who are not volcanologists); however, volcanologists did not embrace the alternative impact hypothesis in its entirety, or in part, in unusual numbers. Volcanologists said they were opposed to the explosive volcanist hypothesis especially because they did not believe that volcanoes could produce pressures high enough to generate shocked quartz with multiple sets of intersecting planar lamellae (which the orthodox held as diagnostic of a meteorite impact). No volcanologists interviewed gave credence to the postulate that quartz bearing shock-diagnostic features could be made by volcanic explosion—they thought the idea untenable in terms of their discipline's canons.

How and why the volcanologists as a group rejected the volcanist hypothesis clearly demonstrated that discipline conditions the reception of theory and the assessment of evidence. The important role of disciplinary specialty also emerged repeatedly in this study for other areas of science.

Paleontology was among the fields in which there was clear evidence of the influence of subdiscipline on the reception of novel ideas and hypothesis choice. The nature, and particularly the time frame, of the K/T mass extinction(s) was perceived by many paleontologists as a function of the fossil taxa in which they specialized. Those who treated groups that suffered greatly at the extinction boundary (e.g., marine calcareous planktonic microfossils) generally regarded the environmental stress at the K/T extinction boundary as more severe, the time interval for extinction as briefer, and cause(s) of the extinction as more stressful than paleontologists who dealt with groups that were significantly less disturbed (e.g., marine, deep-water, benthos). Specialists in the strongly affected taxa were more likely than the average paleontologist to embrace the impact hypothesis in its entirety. There were, however, a significant number of paleontologists for whom the influence of fossil group or subject specialty (disciplinary gestalt) was not recognizable in the choosing process.

Vertebrate paleontologists, in greater proportion than specialists of any other subdiscipline within or beyond paleontology, objected to the impact hypothesis (both the impact and impact-as-extinction-cause postulates). Vertebrate fossils, virtually confined to stream, lake, and other continental basins, are extremely rare as compared to fossils entombed in marine rocks that constitute most of the Earth's sedimentary shell. The vertebrate fossil record inclines its scholars to a mind-set that differs from most other paleontologic specialties. Several paleontologists mentioned that important factor, noting that vertebrate paleontology (VP) was a subculture different from paleontology as a whole. The same, however, might be said of other paleontologic subspecialties, notably micropaleontology.

Unlike vertebrate paleontology, which is focused on the biology of its fossils, micropaleontology is instead concerned largely with reconstructing ancient environments, locating stratigraphic position, and dating rocks. Micropaleontologists have served mainly in providing services to geology that derive largely from the usefulness of many planktonic microfossils (Lipps, 1981; Glen 1994c). Micropaleontology started in the study of foraminifera that were easily recovered from drill cores and thus of great value in oil exploration and development. Most micropaleontologists, employed in the oil industry, have not treated biologic problems as have most other paleontologic subspecialists. The record of microfossils is singular—generally readable in greater detail than that of other fossil groups and strongly reflecting taxonomic changes that have been interpreted as "knife-blade sharp" boundaries. Such obtained particularly for the calcareous, planktonic microfossils that suffered greatly right at the iridium-marked K/T boundary. The mind-set or gestalt of micropaleontologists is thus clearly different from that of specialists in most other fossil groups—it was a mindset perhaps closer to that of geologists than other paleontologists. We might thus expect micropaleontologists to be somewhat differently oriented than other kinds of paleontologists in the reception of ideas impinging on their microscopic, abundant, and generally most readily available objects of study.

THE POLEMIC MODE

How the participants in these debates argued seems to reflect the history of their disciplines and the kinds of evidence from which they built their briefs. Although the impactors had formulated a radical and catastrophic hypothesis, they argued mainly from orthodox standards and canonical knowledge. They held that their new empirical evidence lent itself to experimentation that could test clear predictions facilitated by components of the impact hypothesis.

The prediction that impact products were distributed globally by impact-induced, uniquely energetic transport, including a ballistically transported fraction, was central to the impact hypothesis. Implicit in that prediction was the expectation—or at least the hope—that careful search would show features in common at the K/T boundary, such as similar stratigraphy and marker beds, concentrations of iridium, and so on.

Impact's detractors focused on precisely the opposite: the wide range of differences in the nature of the boundary interval around the world. They argued that the differences that could be found from site to site meant that the extinction cause was neither circum-mundane nor geologically instantaneous. The impactors riposted that local differences in sedimentary processes and sources and diagenetic changes expectedly produce unique stratigraphic intervals at distant localities—that such differences were likely to alter or mask the effects of an impact at globally distributed K/T boundary sites. The impactors anticipated that the signatures of an impact would become increasingly legible through the turbid overlay of local differences in sedimentation, diagenetic changes, and the host of other factors that varied with distance to impact target; as more was learned about a new subject and methods and instruments evolved, finer discrimination would become possible. Such a hope for the impactors seemed warranted in part because the global effects of a massive impact lay essentially outside the realm of both stratigraphic theory and practice on Earth.

The anti-impactors tabulated the variations in distribution of the platinum-group elements, spherules, mineralogy of the boundary layer(s) and other forms of evidence claimed by the impactors and countered that earthly agencies had produced the purported impact signatures. The anti-impactors called on volcanoes, locally differing and anomalous sedimentation conditions, including the influence of organisms, and other possible means for concentrating chemicals or minerals in order to explain what was found at the numerous K/T boundary localities. The primary aim of impact's opponents was to demonstrate that an impact had not produced ubiquitous, globally uniform effects and to simultaneously provide alternative cause(s) for what was found.

The anti-impactors opened to question a number of suppositions that hid in established principles and methods of stratigraphy and related disciplines; they thus prompted much fruitful research at unprecedented levels of resolution. Such refined work required the development of new methods, techniques, and instruments; the scrutiny—with a new eye—of classical stratigraphic boundary sections; the search for new field localities; and detailed reexamination of specimens and samples that lay in museums, archives, and laboratories.

The Alvarez group built an automated coincidence spectrometer—designed by Luis Alvarez—that enabled them to measure the siderophile content of large numbers of rock samples collected at very short stratigraphic intervals through K/T boundary sections from many localities. It processed a sample in 10 minutes as compared to the older methods requiring hours or days. The effort to design and build that machine had been prompted by one of the earliest criticisms of the impact hypothesis: Since no significant stratigraphic interval has ever before been examined elementally at the level of parts per trillion to demonstrate iridium's rarity within the stratigraphic record, how can we be sure the K/T iridium anomaly is as rare as claimed? That question subsided as several research groups collected data during the 1980s that showed clearly that iridium anomalies were rare in the stratigraphic record. The paper of Frank Kyte and John Wasson (1986) of U.C.L.A. was particularly important: It demonstrated, through a deep-sea core representing tens of millions of years, an iridium anomaly present only at the K/T boundary. Anti-impactors, of course, and others pointed out that since sedimentation rates in the deep sea are exceedingly slow, such a core was not likely to have captured the many lesser signals in the depositional record that are more likely to be found in strata deposited rapidly in basins on the continents and the shallow continental shelf. Frank Asaro, the nuclear chemist, has noted that an iridium signal or spike shows up more clearly above background in rapidly deposited sediments than in sediments deposited slowly (oral communication, October 18, 1994). The objections of the volcanists to the impact hypothesis led to still other productive, novel lines of research, but the early, main task of the volcanists was clear: show how to produce an iridium anomaly around the globe at the K/T boundary in the absence of a meteorite impact.

By 1983, it was demonstrated that airborne particles blown from Kilauea volcano in Hawaii were enriched in iridium 17,000 times the value found in Hawaiian basalt (Zoller et al., 1983). The volcanists held that although the iridium in the erupting Kilauean basalt rock was at least two orders of magnitude less than that of the K/T boundary layer, it was highly concentrated in gases and sublimates. Olmez, Finnegan, and Zoller (1986) later emphasized that iridium had not been found in many studies of volcanic gases—only the 1983 paper of Zoller provided an exception. That seemed to detract from the force of the volcanists' empiric volcanic iridium evidence. However, little was known about the origin, transport, and chemical differentiation of magmas in volcanism; those uncertainties allowed the volcanists to continue to underpin their argument with the modern volcanic iridium data.

Toutain and Meyer furthered the volcanist cause in 1989 when they found that sublimates from Piton de la Fournaise volcano in the Indian Ocean were anomalously rich in iridium; they suggested that such hot-spot volcanoes could be related to the events at the K/T boundary. Iridium from active volcanos still seems to be the most potent of the various lines of volcanist evidence. Piton de la Fournaise volcano not only spews iridium, it sits astride the hot spot in the Indian Ocean over which the Indian tectonic plate passed, apparently as the million-cubic-kilometer Deccan basalt sequence was erupted.

The bulk of the Deccan volume was believed to have accrued in just a few strong pulses within a less than one-million-year interval that includes the K/T boundary (Baksi, 1990; Courtillot, 1990; Baksi and Farrar, 1991). Although environmental-effects questions about such a seemingly horrific volcanic scenario are still not well defined, such volcanism was an attractive alternative to comet showers to explain stepped extinctions and the "smeared" iridium occurrence stratigraphically proximal to the main K/T iridium signal dated at 65 Ma (Izett et al., 1991; Swisher et al., 1992). However, Venkatesan et al. (1993) dated a 2.5-km-thick composite section in the western margin of the Deccan basalt province showing a range in age of 67 to 62.5 Ma, which is at least 3 m.y. longer than earlier estimates for the duration of the entire Deccan eruption interval. The volcanists also faced the new challenge of the elemental analyses of Bhandari et al. (1993), who concluded that the Deccan volcanic contribution of iridium and other trace elements near the K/T boundary in India "is inadequate to explain the Iridium inventory in the marine K/T boundary layer." A workshop on the Deccan Traps held in April 1994 (Singh and Gupta, 1994) concluded that "the mass extinction at the K/T boundary could be a combined effect of Deccan volcanism and bolide impact." It was also reported at the workshop that "interpretation of gravity anomalies in the Bombay region indicates the presence of a possible magma-conduit structure at a shallow Moho boundary, which was suggested to have been caused by a large bolide impact." That appears a cheering note—however tenuous—to those who have invoked bolides as the cause of flood basalts (Rampino, 1987; Alt et al., 1988; etc.)

By 1989, more than 2,000 professional publications had appeared addressing various aspects of the complex upheaval

(Glen, 1990). The tide of new information was daunting; citation indexes grew increasingly tardy (and still are) as the debates entrained additional subdisciplines and the literature continued to burgeon.

The difficult question of what properly constitutes the stratigraphic record of the K/T boundary mass extinction was seldom addressed early on in the debates. Prior to the widespread recognition in the mid 1980s of what appeared to be extinction steps proximal to the great iridium-marked extinction horizon, most outside of paleontology discussed the boundary as a single horizon at which the great number of taxa disappeared. Some focused on the idea of a "knife-blade sharp" stratigraphic change that is marked by the K/T boundary layer(s) of variable thickness (limited to a few centimeters until 1991) holding the iridium signature, shocked quartz, and other forms of stratigraphically confined evidence for catastrophe. The question of the time interval represented by the events marking the profound biotic and lithologic transformation from one era to the next lies at the center of the debates.

It is the question of that boundary's temporal frame that the Alvarez group has said drove them to the search that exploded in their epochal discovery of iridium (Alvarez et al., 1980; L. Alvarez, tape-recorded, March 2, 1984). Whether the K/T boundary is stratigraphically sharp or very thin, or tens to hundreds of centimeters of strata representing hundreds of thousands or even one to two million years is still being argued.

Mainly in marine stratigraphic sections the boundary is represented by a single event marker bed (from pencil thick up to a half meter) that contains iridium, shocked quartz, and other evidence held to be diagnostic of impact. However, in continental K/T sections in the western United States, two discrete microstratigraphic units were recognized by Izett and Pillmore in 1985. That triggered argumentation among several groups (Hildebrand and Boynton, 1990; Bohor et al., 1987a, b; Bohor, 1988; Izett, 1990) about the nature and origin of the units. The arguments grew; at present the question as to whether the origin and timing of these two discrete units reflect one or more impacts is unresolved (Shoemaker and Izett, 1992).

A longer boundary interval is suggested by a disputed series of extinction steps of variable magnitude—but all much weaker than that of the main K/T boundary horizon—that are in some cases described as stratigraphically coincident with stable chemical isotope spikes (Kauffman, 1984, 1988; Keller, 1989; Mount et al., 1986; etc.). Also, at some localities the iridium and other purported evidence for catastrophe have been claimed to occur through a stratigraphic section of a few meters super- and subjacent to the main iridium spike and horizon of greatest faunal change (Asaro et al., 1988; Crockett et al., 1988). The "shoulders" of the iridium spike, especially, triggered debates in which the volcanists called on iridium infusions over the long term, but the impactors oppositely held that movement of elements after burial could produce such a "smeared" distribution. Continuing argumentation is fueled by the lack of understanding of how various elements move about within sediments and rocks, especially platinum group species such as iridium (Colodner et al., 1992). The attention that the pattern of stepped extinctions at the K/T boundary received was heightened by extinction steps reported at other horizons (Glen, 1994b).

Only after the debates were under way for several years did the pattern of extinction steps, above and below the K/T terminal boundary layer(s), receive much attention and become a main line of argument for the anti-impactors—mainly the volcanists. This and other issues were seldom made clear by scientists in debates, and media representatives were often remiss; however, as evidence for possible multiple impacts mounted, even Richard Kerr of *Science*, who closely followed the arguments, noted in 1988 that growing evidence appeared to be incompatible with the single-impact hypothesis.

There is an extreme range in opinion regarding the character of mass extinctions, from those who deny their existence entirely (Briggs, 1994), to those who describe them as representing biotic change over an interval approximating that of orogenies (a few million to tens of millions of years), to those who attribute them to instantaneous catastrophic causes (effected within days to years). Here, as in other geological conflicts, argumentation defied resolution for want of a precise temporal frame.

The concept of stepped extinctions seemed to match the punctuated, repetitional coughing of volcanoes and militated against the hypothesis of a single great killing episode. Thus the volcanists sought to expand the reception of the stepped-extinction idea. After several years of confrontational arguments, compelled by the evidence for stepped extinctions, many impactors, including the Alvarez group, abandoned their defense of the single impact hypothesis. The lesser extinction horizons, some with stable isotope signatures, appeared to detract from the single impact idea, and so too did the "iridium shoulders" above and below the main iridium spike (Asaro et al., 1988; Crockett et al., 1988). The impactors accommodated stepped extinctions by expanding their hypothesis to include multiple impacts produced by comet showers over a period of one to two or more million years (Hut et al., 1985, 1987). That revised theory was not accepted by all other impactors. Among the adherents to single impact were notable geochemists who continue to attribute the evidence at the K/T boundary to a single impact or to multiple instantaneous impacts (F. Kyte, oral communication, 1994).

As the impactors and volcanists expanded the debates, the old, uniformity-friendly hypotheses such as climate and sea-level changes, plate movements, and others, which were by then regarded—either alone or in concert with others—as only weakly explanatory of mass extinctions, were displaced from center stage.

CHOOSING STANDARDS, APPRAISING EVIDENCE, AND REWRITING BRIEFS

The debates demonstrate that both theoretical camps invoked different standards by which they evaluated evidence and also gave different weight to the same evidence. The combatants in the shocked-quartz controversy are a good example (Bohor et al., 1987a; Carter and Officer, 1989; Officer and Drake, 1989; Grieve, 1990). The volcanists explicitly said that shocked quartz with multiple sets of intersecting planar lamellae—long considered a canonical standard of appraisal for impact origin—could instead also be generated by volcanic explosion. That claim of the volcanists—although never substantiated to the satisfaction of the specialists charged with its assessment—appeared to impugn what had been a theretofore seemingly unassailable diagnostic or standard.

Standards are seemingly inextricable parts of the systems of knowledge from which they grow and, once emplaced, play key roles in the future growth of the system. Standards (which may include principles, methods, techniques, rules, gauges, widely cited data, etc.) are considered basic, conventional, and normal, and perhaps most important, standards are considered official—sanctioned by the institutions charged with authenticating and regulating the discipline in which they reside. Standards seem almost immune to change in quiet times, but during times of conflict or upheaval they may change almost as rapidly as other parts of knowledge systems. Standards were challenged and impugned in these debates as crisis mounted and were at times revised, displaced, and discarded. Whether standards are a more useful measure of rates and kinds of conceptual change in science than other parts of knowledge systems is questionable.

How a hypothesis-challenging anomaly evolved into a powerful standard of appraisal that eventually drove the reformulation of the impact hypothesis can be found in the stepped-extinctions episode. From the outset, the concept of extinction steps was opposed by the majority of the impactors; they argued from several different lines of evidence: the Signor-Lipps effect (Signor and Lipps, 1982), which explains why extinctions may appear more gradual than they are; inadequate sampling; methodological errors in gathering evidence; and so on. However, the concept of stepped extinctions (Kauffman, 1984, 1988; Keller, 1989) grew stronger as more evidence accrued: Osmium isotope ratios (Luck and Turekian, 1983); spikes of stable isotopes at lesser extinction horizons (Mount et al., 1986); enigmatic "shoulders" on the main iridium spike bearing lesser iridium spikes and shocked quartz (Asaro et al., 1988; Crockett et al., 1988); and indications that there were extinction steps present at horizons in addition to the K/T.

Two iridium peaks were found subjacent to the 92-million-year-old (upper Cenomanian) extinction boundary in which the lower peak coincided with the first in a series of stepwise extinctions and the upper peak matched the third of the extinction steps (Orth et al., 1987, 1988). Sandro Montanari's (1990) valuable integration of the scattered, confusing complex of data on the Eocene/Oligocene stratigraphic interval showed heightened change in sea life over a few million years that included several impact events marked by tektites and microtektites. Even though the latter cases did not include a "mass extinction," they did seem to fit models for the time intervals ascribed to comet showers and hinted at a series of larger impacting bodies capable of causing extinction steps.

I argue that a standard of appraisal evolved from anomalous data and grew into an increasingly serious problem for the single-impact hypothesis. That problem defied solution by the community that had embraced the single-impact hypothesis, and a localized state of normlessness (lack of norms or standards, or anomie) resulted (Merton, 1951; Ben-David, 1971); over the next few years, within that intractable problem state, increasing numbers of workers came to regard extinction steps as a normative assumption that gained stature as further supporting data accrued.

What began life as an anomaly grew into a standard of appraisal that assumed the central role in challenging the single-impact hypothesis of mass extinction and forced the formulation of a new hypothesis of multiple impacts (Hut et al., 1985, 1987). The new multiple-impacts hypothesis—with impacts spread over one to three million years—made sense of the multiple extinction horizons. Thus, a new theoretical framework that accommodated stepped extinctions as a standard of appraisal was born.

A classification such as the one presented here, with discrete stages in the evolutionary sequence from anomaly to standard, is necessarily an artifice. It freezes stages within a dynamic continuum that is really free of sharp boundaries and must be further qualified with a corollary: *The rate of evolution from an anomaly into a hypothesis-impugning standard varied with the discipline in which the evolution took place.* The anomaly evolved into a standard of appraisal at different rates in different disciplinary communities. For example, the paleontologist/impactors were early in moving toward revising single-impact theory, whereas the nuclear and geochemist/impactors were very late. It has earlier been noted that several geochemists still deny the evidence for stepped extinctions and tenaciously hold to either the single-impact or the near-simultaneous, multiple-impacts hypothesis.

Earlier hypotheses of multiple impacts had been published; however, they had appeared before the strong supporting evidence for the issue of stepped extinctions had come to the fore. A theory of near-instantaneous, multiple impacts by the geochemist Frank Kyte and others came out in 1980, and another more extensive hypothesis by the astrophysicists Victor Clube and Bill Napier, published in 1984, had multiple impacts occurring over sufficient time to accommodate the time frame for the gradual extinction of the dinosaurs as argued by Bill Clemens (1982) and several others.

Some impactors who had played important roles in supporting the single-impact hypothesis (or multiple-near-instantaneous impacts) eschewed the stepped-extinction evidence and, unlike most of the other impactors, denied the comet-shower-impacts hypothesis. The stratigraphic/elemental evidence that those tenacious multiple-near-instantaneous impactors had contributed by their own efforts (Kyte and Wasson, 1986, etc.) and the fact that they were the authors of the first near-instantaneous, multiple-impacts hypothesis (Kyte et al., 1980) appear to have disposed them to eschew the long-term, multiple-impacts hypothesis. (They still do.) Their behavior accords with the conclusion drawn here earlier that scientists who, at the advent of the impact hypothesis, were the published authors or published supporters of endogenous alternatives to impact, in all cases examined save one, rejected impact.

MIND-SET, COLLABORATION, AND VANGUARDS

Advocates of the competing extinction hypotheses almost always screened, reconfigured, or reordered evidence at some cognitive level to reconcile it with their chosen hypothesis. Combatants also sometimes argued from obsolete data or omitted adversarial evidence. One may simply pick any of the many review articles written over the last several years by the protagonists of any camp to find that no review article written by a scientist member, save one, was balanced in space or words in presenting the arguments of both the volcanists and impactors. Some articles were almost free of citations of the strongest opposing arguments. But that appears to be the way that science has always worked. Sciences of the mind are yet to explain how the embrace of a hypothesis manifestly alters cognition and perhaps other forms of intellectual activity, but even casual observation reveals that it does so profoundly.

I thus read Sutherland's review (1988) of the impact/volcanism debates—that he sent me in 1990—with much surprise. The review is a balanced treatment of the evidence and issues at the center of the heated dialectic. His case appears singular in supporting my generalization that theory choice strongly and clearly conditioned the assessment of ideas and evidence in these debates—just as it so clearly did in the case of continental drift. Sutherland briefly presented his own hypothesis in that paper, which held that volcanism and impacting both, coincidentally and independently, caused the K/T extinctions. Sutherland does not believe that impacts triggered the volcanism. His impartial presentation of both the impactor and volcanist hypotheses is required for his own hypothesis, which encompasses both volcanism and impacting. Sutherland equally validates those processes, each at the center of competing hypotheses, because his mind-set or gestalt is a bipartite one. It is a gestalt that permits him to present a very balanced overview of both impacting and volcanism because those hypotheses are subsumed as components of his own hypothesis. *Because he was unique in his own bipartite-theory-engendered gestalt, the generalizations earlier argued here regarding the effects of single theory choice do not apply to him.*

Different forms of evidence in the debates had remarkably different persuasive power in different communities. The iridium evidence was powerfully convincing for those who initially embraced the impact hypothesis; however, it was incomprehensible to many others unfamiliar with the nuclear chemical evidence of siderophile elements. The discovery of shocked minerals at the K/T boundary extinction horizon by Bruce Bohor (Bohor et al., 1984), four years after the advent of the impact hypothesis, was crucial in convincing many uncommitted geologists to accept the impact idea of the Alvarez hypothesis (the mass extinction effects are another matter). The shocked-quartz data—unlike the exotic iridium of disciplinary aliens—were familiar to certain subdisciplinary communities within Earth science, were proffered by one of geology's practitioners, and were written in their professional language in the foremost American science journal.

The data of different disciplines varied as to whether or how they could be comprehended, ordered, or analyzed by students in distant disciplines. Especially different were the rates at which new information was assimilated to the extent that the discipline and its institutions took up a revised intellectual posture (recall the stepped-extinctions case). Bauer (1992) was direct in his conclusion that different branches of science are very dissimilar in many important respects. For example, paleontology is different from nuclear chemistry or optical physics. Historical subjects like paleontology do not lend themselves to predictive generalizations as powerful, or experimentation as exact, as that of the "hard" sciences; neither does paleontology inculcate a mind-set and pattern of expectations that is similar to that of physical science. Such disciplinary differences foster wholly different gestalts (Glen, 1994b).

Those who wavered as the flow of data alternated to favor one or the other hypothesis were very few and not present at all in certain specialties. Most held firmly to the hypothesis they chose at the outset of the debates. I encountered neither a planetary geologist, meteoriticist, or impacting-studies specialist who abandoned the impact-as-extinction-cause component of the hypothesis, nor a vertebrate paleontologist who converted to embrace it.

The historical, nonrepeatable character of the stratigraphic record defies uniformity in its classification. Geologic maps of the same region by different workers always show significant differences between them. Such differences partly arise from "play" in the system of field mapping and stratigraphy that are practiced in no small part with tools of tacit knowledge. Stratigraphic units that are laterally persistent in their properties over regional areas are uncommon, and vanishingly rare intercontinentally.

Stratigraphic science passed through a tortuous early history because of the difficulties of recognizing lateral changes of every sort within rock units. Such lateral variations (facies) may be reflected in fossil content, lithology, or even time value.

Because most sedimentary beds are unreliable as time markers, rare, geologically instantaneous, air-fall ash beds from explosive volcanism are extremely valuable. They frequently act as guides in correlation because of their wide distribution and mineralogic fingerprints. However, even ash beds—with only rare exception—are limited in usefulness to only parts of single continents. Magnetostratigraphy also provides important global correlation markers, but it too is limited: It must almost always be guided by biostratigraphy and radiometric dating.

Although the complex of techniques, methods, and principles of correlation facilitated progress over the past decade, much uncertainty remained in efforts to correlate marine with continental rocks of K/T boundary age, even when both were richly fossiliferous. Historically, almost always such attempts at correlation had eluded stratigraphic science. That is so because seldom does the same stratigraphic section contain time-diagnostic index fossils of both the sea and continents in a spatial and structural context that permits their correlation. During the first half of the debates many claimed that it was impossible to correlate K/T boundary extinctions on land with those in the sea. But in 1986, Saito and others recognized a marked change in the spores and pollen (continental) right at the base of the K/T boundary clay as determined by planktonic foraminifera (marine) in Hokkaido. Few knew of, or seemed to pay attention to, that profoundly important Japanese stratigraphic section, even though its importance was reemphasized by Anthony Hallam in 1988.

The tortuous history of stratigraphic science makes it clear that *never before had it been claimed—as did the impactors—that a global stratigraphic marker horizon containing components of fingerprint value had been formed instantly, in the geologic sense, through cataclysmically energetic, impact-driven, Earth-encompassing transport processes.*

The reception of impact theory was conditioned by geology's uncertainties and the notion of "play" in deciphering the stratigraphic record; the nonrepetitive and inexact nature of a largely historical science; and a history of failed or only partially successful search for precise correlation tools. What then—to those charged with evaluating it—was the likelihood of the existence of a globally correlative rock bed holding exotic materials that had come into being through processes alien to the history of their science. Even planetary geologists who classified and mapped impact-ejecta layers on the Moon had never demonstrated such a layer to be circum-planetary (Wilhelms, 1987).

The collaborative teams observed in this study were commonly made up of specialists from unrelated or distantly related fields who redirected their research and careers. *The collaborators in these debates, with almost no exceptions, shared a commitment to the same hypothesis of mass extinction cause.* The teams were more diversely comprised than any alluded to by Kuhn (1970) and others—who have emphasized the impediments to communication between those imprinted with the paradigms of different disciplines—but were easily recognizable by their special conferences, distribution of draft manuscripts, and close networking in both communication and citations. Because collaborators were from distant fields, they were not likely to possess in common the information, methodological rules, or cognitive aims and values that would be expected of team members working in concert. However, a shared belief in the same hypothesis of mass-extinction cause and the need to resolve problems of the broadest significance to several branches of science (the height of the stakes were clear to them) engendered intellectual rapport, accommodation, and the formation of commensurable views among scientists from vastly different fields. That surprising phenomenon buttressed my surmise from earlier studies that *subscription to the same hypothesis—especially a strongly debated one—conditions the cognition of facts, methods, standards of appraisal, and much more, such as to foster commensurable views even among those from unrelated disciplines.*

As growing complexity increasingly restricted understanding of mathematical modeling to very small groups within certain subject fields, the crop of potential referees shrank. Some of those unable to understand modeling became children of faith, and others simply eschewed models as things unworthy of their trust. Mathematical modeling—much in keeping with other new methods that required an inordinate personal investment to master—has generally had a cool reception early on. An example of such appeared in seeking referees for both Herbert Shaw's (1987) article that pioneered the application of nonlinear-dynamics analysis to the earth sciences, and his monumental book (1994) prompted by the impact debates, that presents a boldly encompassing new theory that reinterprets the effects of impacting of the Earth through the lens of chaos theory. Very few in 1986 in the earth sciences were able to use nonlinear analysis; at that time the conceptual bridge between conventional mathematics and chaos theory had not yet been built. Shaw's career digression to master those exotic new methods and the field of celestial mechanics (equally alien to almost all earth scientists), however heuristically and predictively promising, marked him as a curious object among his peers. Support is hard to obtain for those who advance far beyond orthodoxy. While arguing that meteoritic impacting of the planets is not a random process, Shaw performs an astounding synthesis of uniquely broad interdisciplinary scope. His book is the only extant detailed presentation of virtually all of the conceptual components entrained in the mass-extinction debates—he cites two thousand references.

ACKNOWLEDGMENTS

This study has benefited from the technical support of the U.S. Geological Survey, Menlo Park, California; The Center for the History of Physics of the American Institute of Physics, New York City; and The Tremont Research Institute, San Francisco, California. The National Science Foundation provided funding through grants SES-8420419, SES-8619545, DIR-8821587, and SBR-9311216.

REFERENCES CITED

Alt, D., Sears, J. M., and Hyndman, D. W., Nov., 1988, Terrestrial maria: The origins of large basalt plateaus, hotspot tracks and spreading ridges: Journal of Geology, v. 96, p. 647–662.

Alvarez, L. W., Alvarez, W., Asaro, F., and Michel, H. V., 1980, Extraterrestrial cause for the Cretaceous-Tertiary extinction: Science, v. 208, p. 1095–1108.

Alvarez, W., and Muller, R. A., 1984, Evidence from crater ages for periodic impacts on the Earth: Nature, v. 308, p. 718–720.

Alvarez, W., Alvarez, L. W., Asaro, F., and Michel, H. V., 1979a, Anomalous iridium levels at the Cretaceous/Tertiary boundary at Gubbio, Italy: Negative results of tests for a supernova origin, *in* Christensen, W. K., and Birkelund, T. B., eds., Proceedings, Cretaceous-Tertiary Boundary Events Symposium, 2nd, University of Copenhagen, p. 69.

Alvarez, W., Alvarez, L., Asaro, F., and Michel, H. V., 1979b, Anomalous iridium levels at the Cretaceous/Tertiary boundary at Gubbio, Italy: Negative results of tests for a supernova origin: Geological Society of America Abstracts with Programs, v. 11, p. 378.

Alvarez, W., Alvarez, L.W., Asaro, F., and Michel, H. V., 1979c, Experimental evidence in support of an extraterrestrial trigger for the Cretaceous-Tertiary extinctions [abs.]: Eos (Transactions, American Geophysical Union), v. 60, p. 734.

Asaro, F., Michel, H. V., Alvarez, L. W., Alvarez, W. and Montanari, A., April 19, 1988, Impacts and multiple iridium anomalies: Eos (Transactions, American Geophysical Union), v. 69, p. 301–302.

Baksi, A. K., 4 Dec. 1990, Timing and duration of Mesozoic-Tertiary flood-basalt volcanism: Eos (Transactions, American Geophysical Union), v. 71, p. 1835–1836.

Baksi, A. K., and Farrar, E., May 1991, ^{40}Argon/^{39}Argon dating of the Siberian Traps, USSR: Evaluation of the ages of the two major extinction events relative to episodes of flood-basalt volcanism in the USSR and the Deccan Traps, India: Geology, v. 19, p. 461–464.

Bauer, H. H., 1992, Scientific literacy and the myth of the scientific method: Champaign, University of Illinois Press, 180 p.

Ben-David, J., 1971, The scientist's role in society: New York, Prentice-Hall, 207 p.

Bhandari, N., Gupta, M., and Shukla, P. N., 1993, Deccan volcanic contribution of Iridium and other trace elements near the K/T boundary in India: Chemical Geology, v. 103, p. 129–139.

Blum, J. D., and Chamberlain, C. P., 21 Aug., 1992, Oxygen isotope constraints on the origin of impact glasses from the Cretaceous-Tertiary boundary: Science, v. 257, p. 1104–1107.

Bohor, B. F., 1988, Shock-induced microdeformation in quartz and other mineralogical indications of an impact event at the Cretaceous-Tertiary boundary: Tectonophysics, v. 171, p. 360–361.

Bohor, B. F., Foord, E. E., Modreski, P. J., and Triplehorn, D. M., 1984, Mineralogic evidence for an impact event at the Cretaceous-Tertiary boundary: Science, v. 224, p. 867–869.

Bohor, B.F., Modreski, P. J., and Foord, E.E., 8 May 1987a, Shocked quartz in the Cretaceous- Tertiary boundary clays: Evidence for a global distribution: Science, v. 236, p. 705–708.

Bohor, B. F., Triplehorn, D. M., Nichols, D. J., and Millard, H. T., Jr., 1987b, Dinosaurs, spherules and the "magic" layer: A new K-T boundary clay site in Wyoming: Geology, v. 15, p. 896–899.

Bramlette, M. N., 1965, Massive extinctions in biota at the end of Mesozoic time: Science, v. 148, p. 1696–1699.

Briggs, J. C., 1990, Global extinctions, recoveries and evolutionary consequences: Chicago, University of Chicago Press, Evolutionary Monographs, v. 13, p. 1–47.

Briggs, J. C., 1994, Mass extinctions: fact or fallacy?, *in* Glen, W., ed., The mass extinction debates: How science works in a crisis: Stanford, California, Stanford University Press. p. 230–236.

Carlisle, D. B., 1992, Diamonds at the K/T boundary: Nature, v. 357, p. 119–120.

Carlisle, D. B., 1995, Dinosaurs, diamonds and things from outer space: The great extinctions: Stanford, California, Stanford University Press, 241 p.

Carlisle, D. B., and Braman, D. R., 22 Aug. 1991, Nanometre-size diamonds in the Cretaceous/Tertiary boundary clay of Alberta: Nature, v. 352, p. 708–709.

Carter, N. L., and Officer, C. B., 1989, on Microscopic lamellar deformation features in quartz: discriminative characteristics of shock-generated varieties: Comment and reply: Geology, v. 17, p. 477–480.

Clemens, W. A., 1982, Patterns of survival of the terrestrial biota during the Cretaceous/Tertiary transition, *in* Silver, L. T., and Schultz, P. H., eds., Geological implications of impacts of large asteroids and comets on the Earth: Geological Society of America Special Paper 190, p. 407–413.

Clube, S. V. M., and Napier. W. M., 1984, The microstructure of terrestrial catastrophism: Monthly Notices of the Royal Astronomical Society, v. 211, p. 953–968.

Colodner, D. C., Boyle, E. A., Edmond J. M., and Thomson, J., 30 July 1992, Post-depositional mobility of platinum, iridium and rhenium in marine sediments: Nature, v. 358, p. 402–404.

Crockett, J. H., Officer, C. B., Wezel, F. C., and Johnson., G. D., Jan. 1988, Distribution of noble metals across the Cretaceous/Tertiary boundary at Gubbio, Italy: Iridium variations as a constraint on the duration and nature of Cretaceous/Tertiary boundary events: Geology, v. 16, p. 77–80.

Courtillot, V. E., Oct. 1990, A volcanic eruption: Scientific American, v. 263, p. 85–92.

De Laubenfels, M. W., January 1956, Dinosaur extinction: One more hypothesis: Journal of Paleontology, v. 30, p. 207–212.

Flessa, K. W., 1979, Extinction, *in* Fairbridge, R. W., and Jablonski, D., eds., The encyclopedia of paleontology: Stroudsberg, Pennsylvania, Douden, Hutchinson and Ross (New York, distributed by Academic Press), p. 300–305.

Flessa, K. W., and Imbrie, J., 1973, Evolutionary pulsations: Evidence from Phanerozoic diversity patterns, *in* Tarling, D. H., and Runcorn, S. K., eds., Implications of continental drift to the earth sciences: London, Academic Press, p. 247–285.

Gallant, R., 1964, Bombarded Earth: London, John Baker, 256 p.

Gartner, S., and Keany, J., 1978a, The terminal Cretaceous event: An oceanographic solution for a geological puzzle: Geological Society of America Abstracts with Programs, v. 10, p. 406.

Gartner, S., and Keany, J., Dec., 1978b, The terminal Cretaceous event: A geologic problem with an oceanographic solution: Geology, v. 6, p. 708–712.

Gartner, S., and Keany, J., 1979, The coccolith succession across the Cretaceous/Tertiary boundary in the subsurface of the North Sea (Ekofisk), *in* Christensen, W. K., and Birkelund, T. B., eds., Proceedings, Cretaceous-Tertiary Boundary Events Symposium, 2nd: University of Copenhagen, p. 103–105.

Gartner, S., and McGuirk, J. P., Dec., 1979, Terminal Cretaceous extinction scenario for a catastrophe: Science, v. 206, p. 1272–1276.

Gilmour, I., Russell, S. S., Arden, J. W., Lee, M. R., Franchi, I. A., and Pillinger, C. T., 4 Dec. 1992, Terrestrial carbon and nitrogen isotopic ratios from Cretaceous-Tertiary boundary nanodiamonds: Science, v. 258, p. 1624–1626.

Glen, W., 1975, Continental drift and plate tectonics: Columbus, Ohio, Charles E. Merrill Publishing, 188 p.

Glen, W., 1982, The road to Jaramillo: Critical years of the revolution in earth science: Stanford, California, Stanford University Press, 459 p.

Glen, W., 1988, Styles of conflict, past and present: Lecture, American Association for the Advancement of Science, 69th Annual Meeting, Pacific Division, Oregon State University, June 18, 1988.

Glen, W., July–Aug., 1990, What killed the dinosaurs?: American Scientist, v. 78, p. 354–370.

Glen, W., ed., 1994a, The mass extinction debates: How science works in a crisis: Stanford, California, Stanford University Press, 371 p.

Glen, W., 1994b, What the impact/volcanism/mass-extinction debates are about, *in* Glen, W., ed., The mass extinction debates: How science works in a crisis: Stanford, California, Stanford University Press, p. 7–38.

Glen, W., 1994c, How science works in the mass-extinction debates, *in* Glen, W., ed., The mass-extinction debates: How science works in a crisis: Stanford, California, Stanford University Press, p. 39–91.

Glen, W., 1994d, How different disciplines have responded to the Alvarez–Berkeley group hypothesis: Geological Society of America Abstracts with Programs, p. A-282.

Glen, W., 1996, Myth, muse, and mind: Problems in the rational reconstruction of science history: presented at a working conference (28–30 April, 1994, Stanford University) on Interviews in Writing the History of Recent Science: Cambridge, Harvard University Press (in press).

Goodman, N., 1967, Uniformity and simplicity, *in* Albritton, C. C., ed., Uniformity and simplicity: Geological Society of America Special

Paper 89, p. 93–99.

Gould, S. J., 1965, Is uniformitarianism necessary?: American Journal of Science, v. 263, p. 223–228.

Gould, S. J., Oct. 1989, An asteroid to die for: Discover, p. 60–65.

Gould, S. J., 1991, Introduction, *in* Raup, D. M., Extinction: Bad genes or bad luck?: New York, W.W. Norton, p. xiii–xvii.

Gould, S. J., Oct. 1994, Jove's thunderbolts: Natural History, p. 7–12.

Grieve, R.A., 1990, Shocked minerals and the K/T controversy: Eos (Transactions, American Geophysical Union), v. 71, p. 1792.

Hallam, A., 1973, A revolution in the earth sciences: From continental drift to plate tectonics: Oxford, Oxford University Press, 127 p.

Hallam, A., 1988, A compound scenario for the end-Cretaceous mass extinctions, *in* Lamolda, M., Kauffman, E. G., and Walliser, O. H., eds., Paleontology and evolution: Extinction events: Revista Española de Paleontología, Special Issue, p. 7–20.

Hays, J. D., and Pitman, W. C., 1973, Lithosphere plate motion, sea level changes and climatic and ecological consequences: Nature, v. 246, p. 18–22.

Hildebrand, A. R., and Boynton, W. V., 18 May 1990, Proximal Cretaceous/Tertiary boundary impact deposits in the Caribbean: Science, v. 248, p. 843–847.

Hildebrand, A.R., and Penfield, G.T., 4 Dec. 1990, A buried 180 km-diameter probable impact crater on the Yucatan Peninsula, Mexico: Eos (Transactions, American Geophysical Union), v. 71, p. 1425.

Hoffman, A., 1989, Arguments on evolution: A paleontologist's perspective: Oxford, Oxford University Press, 274 p.

Hoffman, A., and Nitecki, M. H., Dec. 1985, Reception of the asteroid hypothesis of terminal Cretaceous extinctions: Geology, v. 13, p. 884–887.

Hsü, K. J., 1994, Uniformitarianism vs. catastrophism, *in* Glen, W., ed., The mass-extinction debates: How science works in a crisis: Stanford, California, Stanford University Press. p. 217–229.

Hut, P., Alvarez, W., Elder, W., Kauffman, E. G., Hansen, T. A., Keller, G., Shoemaker, E. M., and Weissman, P., 1985, Comet showers as possible causes of stepwise mass extinctions[abs.]: Eos (Transactions, American Geophysical Union), v. 66, p. 813.

Hut, P., Alvarez, W., Elder, W., Hansen, T. A., Kauffman, E. G., Keller, G., Shoemaker, E. M., and Weissman, P., 10 Sept. 1987, Comet showers as a cause of mass extinctions: Nature, v. 329, p. 118–126.

Izett, G.A., 1990, The Cretaceous/Tertiary boundary interval, Raton Basin, Colorado and New Mexico, and its contents of shock-metamorphosed minerals; Evidence relevant to the K-T boundary impact extinction theory: Geological Society of America Special Paper 249, 100 p.

Izett, G. A., and Pillmore, C. L., 1985, Abrupt appearance of shocked quartz at the Cretaceous-Tertiary boundary, Raton Basin, Colorado and New Mexico: Geological Society of America Abstracts with Programs, v. 17, p. 617.

Izett, G. A., Maurrasse, F. J.-M. R., Lichte, F. E., Meeker, G. P., and Bates, R., Nov. 1990a, Tektites in Cretaceous-Tertiary boundary rocks on Haiti: U.S. Geological Survey Open File Report 90-365, 31 p.

Izett, G. A., Dalrymple, G. B., and Snee, L. W., 1991, ^{40}Ar-^{39}Ar age of Cretaceous-Tertiary boundary tektites from Haiti: Science, v. 252, p. 1539, 1543.

Kauffman, E. G., 1984, The fabric of Cretaceous marine extinctions, *in* Berggren, W. A., and Van Couvering, J., eds., Catastrophes in earth history—The new uniformitarianism: Princeton, New Jersey, Princeton University Press, p. 151–246.

Kauffman, E. G., 1988, Stepwise mass extinctions, *in* Lamolda, M., Kauffman, E. G., and Walliser, O. H., eds., Paleontology and evolution: Extinction events: Revista Española de Paleontología. Special Issue, p. 58–71.

Keller, G., June 1989, Extended Cretaceous/Tertiary boundary extinctions and delayed population change in planktonic foraminifera from Brazos River, Texas: Paleoceanography, v. 4, p. 287–332.

Kerr, R. A., 12 Feb. 1988, Was there a prelude to the dinosaur's demise: Science, v. 239, p. 729–730.

Kring, D. A., and Boynton, W. V., 9 July 1992, Petrogenesis of an augite-bearing melt rock in the Chicxulub structure and its relationship to K/T impact spherules in Haiti: Nature, v. 358, p. 141–143.

Kuhn, T. S., 1970, The structure of scientific revolutions (second edition, enlarged): Chicago, Illinois, University of Chicago Press, p. 1–210.

Kyte, F. T., and Smit, J., 1986, Regional variations in spinel compositions: An important key to the Cretaceous-Tertiary event: Geology, v. 14, p. 485–487.

Kyte, F. T., and Wasson, J. T., 6 June 1986, Accretion rate of extra- terrestrial matter: Iridium deposited 33 to 67 million years ago: Science, v. 232, p. 1225–1229.

Kyte, F. T., Zhou, Z., and Wasson, J. T., 18 Dec. 1980, Siderophile-enriched sediments from the Cretaceous-Tertiary boundary: Nature, v. 288, p. 651–656.

Lemke, J. L., Nitecki, M. H., and Pullman, H., Sept. 1980, Studies of the acceptance of plate tectonics, *in* Oceanography: The past: Proceedings, Third International Congress on the History of Oceanography: New York, Springer-Verlag, p. 614–621.

Lipps, J. H., 1981, What, if anything, is micropaleontology?: Paleobiology, v. 7, p. 167–199.

Luck, J. M., and Turekian, K. K., 11 Nov. 1983, Osmium-187/Osmium-186 in manganese nodules and the Cretaceous-Tertiary boundary: Science, v. 222, p. 613.

Marvin, U. B., 1990, Impact and its revolutionary implications for geology, *in* Sharpton, V. L., and Ward, P. D., eds., Global catastrophes in Earth history: Geological Society of America Special Paper 247, p. 87–91.

Maupertuis, P. L. M. de, 1742, Lettre sur la cométe [issued anonymously in Paris].

McLaren, D. J., 1970, Time, life and boundaries: Journal of Paleontology, v. 44, p. 801–815.

McLaren, D. J., and Goodfellow, W. D., 1990, Geological and biological consequences of giant impacts: Annual Review of Earth and Planetary Science, v. 18, p. 123–171.

McLean, D. M., Aug. 1978, A terminal Mesozoic "greenhouse": Lessons from the past: Science, v. 201, p. 401–406.

McLean, D. M., 1981, Cretaceous-Tertiary extinctions and possible terrestrial and extraterrestrial causes, *in* Proceedings, Workshop held in Ottawa, Canada, 19–20 May, 1981 by the K-TEC group: Syllogeous No. 39, Ottawa, National Museums of Canada, p. 143–144.

Merton, R. K., 1951, Social structure and anomie, *in* Merton, R. K., ed., Social theory and social structure (second edition): New York, Free Press. p. 131–194.

Montanari, A., 1990, Geochronology of the terminal Eocene impacts; An update, *in* Sharpton, V. L., and Ward, P. D., eds., Global catastrophes in Earth history: Geological Society of America Special Paper 247, p. 607–616.

Mount, J. F., Margolis, S. V., Showers, W., Ward, P., and Doehne, E., 1986, Carbon and oxygen isotope stratigraphy of the Upper Maastrichtian, Zumaya, Spain: A record of oceanographic and biologic changes at the end of the Cretaceous Period: Palaios, v. 1, p. 87–92.

Newell, N. D., 1967, Revolutions in the history of life: Geological Society of America Special Paper 89, p. 62–91.

Newell, N. D., 1972, The evolution of reefs: Scientific American, v. 226, p. 54–65.

Newell, N. D., 1982, Mass extinctions-Illusions or realities, *in* Silver, L. T., and Schultz, P. H., eds., Geological implications of impacts of large asteroids and comets on the Earth: Geological Society of America Special Paper 190, p. 257–263.

Nitecki, M. H., Lemke, J. L., Pullman, H. W., and Johnson, M. E., 1978, Acceptance of plate tectonic theory by geologists: Geology, v. 6, p. 661–664.

Officer, C. B., and Drake, C. L., 25 March 1983, The Cretaceous-Tertiary transition: Science, v. 219, p. 1383–1390.

Officer, C. B., and Drake, C. L., 1985, Terminal Cretaceous environmental events: Science, v. 227, p. 1161–1167.

Officer, C. B., and Drake, C. L., June 20, 1989, Cretaceous/Tertiary extinctions: Eos (Transactions, American Geophysical Union), v. 70, p. 25.

Olmez, I., Finnegan, D. L., and Zoller, W. H., 1986, Iridium emissions from

Kilauea Volcano: Journal of Geophysical Research, v. 91, p. 653–663.

Orth, C. J., Attrep, M., Jr., Mao, X. Y., Kauffman, E. G., and Diner, R., October, 1987, Iridium abundance peaks at upper Cenomanian stepwise extinction boundaries: Geological Society of America Abstracts with Programs, v. 19, p. 796.

Orth, C. J., Attrep, M., Jr., Mao, X. Y., Kauffman, E. G., Diner, R., and Elder, W. P., April 1988, Iridium abundance maxima in the Upper Cenomanian extinction interval: Geophysical Research Letters, v. 15, p. 346–349.

Pantin, C. F. A., 1968, The relations among the sciences: New York, Cambridge University Press, 206 p.

Penfield, G. T., and Camargo Z., A., 1981, Definition of a major igneous zone in the central Yucatan platform with aeromagnetics and gravity [abs.]: Society of Exploration Geophysics, 51st Annual Meeting, Technical Program 37.

Rampino, M. R., 11 June, 1987, Impact cratering and flood basalt volcanism: Nature, v. 327, p. 468.

Rampino, M. R., and Stothers, R. B., 1984, Geological rhythms and cometary impacts: Science, v. 226, p. 1427.

Raup, D. M., 1991, Extinction: Bad genes or bad luck?: New York: W.W. Norton, 210 p.

Raup, D. M., and Sepkoski, J. J., 1984, Periodicity of extinctions in the geologic past: Proceedings of the National Academy of Sciences, v. 81, p. 801–805.

Rubin, D. C., ed., 1986, Autobiographical memory: New York, Cambridge University Press, 298 p.

Saito, T., Yamanoi, T., and Kaiho, K., 1986, End Cretaceous devastation of terrestrial flora in the boreal Far East: Nature, v. 323, p. 253–255.

Schopf, T. J. M., 1974, Permo-Triassic extinctions: Relation to sea-floor spreading: Journal of Geology, v. 82, p. 129–143.

Seyfert, C. K., and Sirkin, L. A., 1979, Earth history and plate tectonics (second edition): New York, Harper and Row, 600 p.

Sharpton, V. L., Dalrymple, G. B., Marin, L. E., Ryder, G., Schuraytz, B. C., and Urrutia-Fucugauchi, J., 29 Oct. 1992, New links between the Chicxulub impact structure and the Cretaceous/Tertiary boundary: Nature, v. 359, p. 819–821.

Shaw, H. R., 1987. The periodic structure of the natural record and nonlinear dynamics: EOS (Transactions of the American Geophysical Union) v.68, p. 1651–1664.

Shaw, H. R., 1994, Craters, cosmos and chronicles: A new theory of the Earth: Stanford, California, Stanford University Press, 688 p.

Shea, J. H., 1982, Twelve fallacies of uniformitarianism: Geology, v. 10, p. 455–460.

Shoemaker, E. M., 1977, Astronomically observable crater-forming projectiles, *in* Roddy, D. J., Pepin, R. O., and Merrill, R. B., eds., Impact and explosion cratering: Planetary and terrestrial implications: New York, Pergamon Press, p. 617–628.

Shoemaker, E. M., and Izett, G. A., 1992, K/T boundary stratigraphy: Evidence for multiple impacts and possible comet stream: Papers presented at international conference on large meteorite impacts and planetary evolution, 31 Aug–2 Sept., 1992, Sudbury, Ontario, Canada: Houston, Texas, Lunar and Planetary Science Contribution 790, p. 66–68.

Signor, P. W, III, and Lipps, J. H., 1982, Sampling bias, gradual extinction patterns and catastrophes in the fossil record, *in* Silver, L. T., and Schultz, P. H., eds., Geological implications of impacts of large asteroids and comets on the Earth: Geological Society of America Special Paper 190, p. 291–296.

Sigurdsson, H., D'Hondt, S., Arthur, M. A., Bralower, T. J., Zachos, J. C., Van Fossen, M., and Channell, J. E. T., 1991, Glass from the Cretaceous-Tertiary boundary in Haiti: Nature, v. 349, p. 482–487.

Simberloff, D. S., 1974, Permo-Triassic extinctions: Effects of area on biotic equilibrium: Journal of Geology, v. 82, p. 267–274.

Singh, R. N., and Gupta, K. R., 1994, Workshop yields new insight into volcanism at Deccan Traps, India: Eos (Transactions, American Geophysical Union), v. 75, p. 356.

Spence, D. P., 1982, Narrative truth and historical truth: Meaning and interpretation in psychoanalysis: New York, W.W. Norton, 320 p.

Sutherland, F. L., 1988, Demise of the dinosaurs and other denizens—by cosmic clout, volcanic vapours or other means: Royal Society of New South Wales, Journal and Proceedings, v. 121, p. 123-164

Swisher, C. C., III, and 11 others, 1992, Coeval $^{40}Ar/^{39}Ar$ Ages of 65.0 million years ago from Chicxulub crater melt rock and Cretaceous-Tertiary boundary tektites: Science, v. 257, p. 954–958.

Tappan, H., 1968, Primary production, isotopes, extinctions and the atmosphere: Paleogeography, Paleoclimatology, Paleoecology, v. 4, p. 187–210.

Thierstein, H. R., and Berger, W. H., 1978, Injection events in ocean history: Nature, v. 276, p. 461–466.

Torrens, H. S., 1994, "No impact": René Gallant (1906–1985)'s book *Bombarded Earth* (1964): Geological Society of America Abstracts with Programs, v. 26, p. A-281.

Toutain, J.-P., and Meyer, G., 1989, Iridium-bearing sublimates at a hot-spot volcano (Piton de la Fournaise, Indian Ocean): Geophysical Research Letters, v. 16, p. 1391–1394.

Urey, H. C., 1957, Origin of tektites: Nature, v. 179, p. 556–557.

Urey, H. C., 1963, Cometary collisions and tektites: Nature, v. 197, p. 228–230.

Urey, H. C., 1973, Cometary collisions and geological periods: Nature, v. 242, p. 32–33.

Valentine, J. W., and Moores, E. M., 1970, Plate tectonics regulation of faunal diversity and sea level: A model: Nature, v. 228, p. 657–659.

Valentine, J. W., and Moores, E. M., 1972, Global tectonics and the fossil record: Journal of Geology, v. 80, p. 167–184.

Venkatesan, T. R., Pande, K., and Gopalan, K., 1993, Did Deccan volcanism predate the Cretaceous/Tertiary transition?: Earth and Planetary Science Letters, v. 119, p. 181–189.

Vogt, P. R., 1972, Evidence for global synchronism in mantle plum convection, and possible significance for geology: Nature, v. 240, p. 338–342.

Westrum, R., 1978, Science and social intelligence about anomalies: The case of the meteorites: Social Studies of Science, v. 8, p. 461–493.

Wilhelms, D. E., 1987, The geological history of the Moon: U.S. Geological Survey Professional Paper 1348, 302 p.

Zoller, N. H., Parrington, J. R., and Phelan Kotra, J. M., 1983, Iridium enrichment in airborne particles from Kilauea volcano: Science, v. 222, p. 1118.

MANUSCRIPT ACCEPTED BY THE SOCIETY SEPTEMBER 14, 1995

Printed in U.S.A.

Geological Society of America
Special Paper 307
1996

A model of the Chicxulub impact basin based on evaluation of geophysical data, well logs, and drill core samples

Virgil L. Sharpton
Lunar and Planetary Institute, 3600 Bay Area Boulevard, Houston, Texas 77058-1113
Luis E. Marín
Instituto de Geofísica, Universidad Nacional Autónoma de México, Ciudad Universitaria, Mexico City, CP 04510, Mexico
John L. Carney
Amoco Production Company, 501 WestLake Park Boulevard, Houston, Texas 77253
Scott Lee, Graham Ryder, and Benjamin C. Schuraytz
Lunar and Planetary Institute, 3600 Bay Area Boulevard, Houston, Texas 77058-1113
Paul Sikora
Amoco Production Company, 501 WestLake Park Boulevard, Houston, Texas 77253
Paul D. Spudis
Lunar and Planetary Institute, 3600 Bay Area Boulevard, Houston, Texas 77058-1113

ABSTRACT

Abundant evidence now shows that the buried Chicxulub structure in northern Yucatan, Mexico, is indeed the intensely sought-after source of the ejecta found worldwide at the Cretaceous-Tertiary (K/T) boundary. In addition to large-scale concentric patterns in gravity and magnetic data over the structure, recent analyses of drill-core samples reveal a lithological assemblage similar to that observed at other terrestrial craters. This assemblage comprises suevite breccias, ejecta deposit breccias (Bunte Breccia equivalents), fine-grained impact melt rocks, and melt-matrix breccias. All these impact-produced lithologies contain diagnostic evidence of shock metamorphism, including planar deformation features in quartz, feldspar, and zircons; diaplectic glasses of quartz and feldspar; and fused mineral melts and whole-rock melts. These features indicate a range of peak pressures from <8 GPa to >60 GPa. In addition, elevated concentrations of Ir, Re, and Os, in meteoritic relative proportions, have been detected in some melt-rock samples from the center of the structure. Isotopic analyses, magnetization of melt-rock samples, and local stratigraphic constraints identify this crater as the source of K/T boundary deposits.

Although the size and precise morphology of this buried basin are not yet established conclusively, the evidence gathered to date from geophysical data, well logs, and drill-core samples indicates that the Chicxulub basin may be ~300 km in diameter. Chicxulub's circular gravity depression appears to be nearly 300 km in diameter; at other terrestrial craters the edge of the gravity anomaly roughly coincides with the topographic basin rim. Superimposed on this broad depression are local concentric anomalies consistent with the structure of a multiring impact basin. Depths to the K/T boundary in well logs also are consistent with an ~300-km basin but are not as expected of a crater less than 200 km in diameter. Thus, steep gravity gradients located between 75 to 105 km from center may outline a central region of deep deformation associated with transient crater collapse rather than the final basin rim as previously assumed. Similar steep, inward-facing gradients delimit deep deformation at other terrestrial craters. Well data indicate that the silicate basement within ~160 km of Chicxulub's cen-

Sharpton, V. L., Marín, L. E., Carney, J. L., Lee, S., Ryder, G., Schuraytz, B. C., Sikora, P., and Spudis, P. D., 1996, A model of the Chicxulub impact basin based on evaluation of geophysical data, well logs, and drill core samples, *in* Ryder, G., Fastovsky, D., and Gartner, S., eds., The Cretaceous-Tertiary Event and Other Catastrophes in Earth History: Boulder, Colorado, Geological Society of America Special Paper 307.

ter is ~1 km deeper than in wells farther out. This is consistent with late-stage collapse of the crater flanks to yield a final basin ~300 km in diameter but is hard to reconcile with a much smaller crater. Finally, the ejecta thicknesses exceed 600 m in regions where they previously were estimated to be less than about 90 m. Such observations provide an intriguing—albeit not airtight—case for an ~300-km-diameter multiring impact basin.

BACKGROUND

The buried Cretaceous-Tertiary (K/T) source crater in northernmost Yucatan (Plate 1) was first recognized as a zone of concentric anomalies in gravity surveys circa 1948 (Cornejo-Toledo and Hernandez-Osuna, 1950), prompting an exploratory drilling program by Petróleos Mexicanos (Pemex) that began in the early 1950s. By then, the feature was known as the Chicxulub structure after the small Mayan village of Chicxulub near the structure's center, the place where the first exploratory well was drilled (Antonio Camargo, personal communication, 1994). The Pemex drilling program comprised eight deep holes, with intermittent core recovery, and was completed in the mid-1970s. This effort revealed unusual occurrences of Upper Cretaceous crystalline rocks and breccias—thought at the time to indicate a large volcanic field—amid the typical Mesozoic carbonate platform sequence of this region (Lopez-Ramos, 1983).

Penfield and Camargo (1981) first reported Chicxulub's unusual features and proposed that they could signal a large buried impact structure. Even though popular reports at that time (Byars, 1981; Sky and Telescope, 1982) suggested this could be the elusive site of the K/T impact event proposed by Alvarez et al. (1980), Chicxulub went virtually unnoticed for almost a decade. In the meantime, impact enthusiasts explored various scenarios to mitigate the absence of a recognized K/T crater large enough to account for the environmental crisis predicted by the Alvarez hypothesis. By the second Snowbird meeting in 1988 (Sharpton and Ward, 1990), two potential solutions to this problem had gained broad acceptance: (1) a large K/T impact site on the ocean floor would be difficult to recognize and could have been destroyed during 65 million years of plate recycling; (2) many smaller impact events, occurring either simultaneously or in a narrow time window, could initiate the K/T biospheric crisis without the need for a 200-km-class crater. Both these ideas were convenient and both were seriously considered even though there was very little, if any, robust evidence in their support.

During the last few years, however, as analyses of K/T boundary materials yielded clear evidence of a single (Bohor, 1990) continental impact site (Bohor, 1990; Sharpton et al., 1990a, b; Izett, 1990; Izett et al., 1990), in the vicinity of the Gulf of Mexico (Bourgeois et al., 1988; Hildebrand and Boynton, 1990), attention turned finally toward the Yucatan Peninsula of Mexico (Sharpton et al., 1990b; Hildebrand and Boynton, 1990). Even then, however, the rediscovery of the Chicxulub structure was not a clever act of scientific detective work, as it is sometimes portrayed. As late as 1991, the group conducting the most ambitious search for the K/T crater was still combing ocean floor in the Colombian Basin off the coast of Nicaragua. It took the intervention of Carlos Byars, the reporter for the *Houston Chronicle* who had interviewed Penfield and Camargo in 1981, to put the planetary community on the track to Chicxulub almost a decade later.

Armed with this information, it was only a matter of time until Chicxulub became the center of intense scientific scrutiny; drill core samples were soon located, and a number of laboratories began to uncover more compelling evidence that the Chicxulub structure is a buried impact crater and to strengthen its temporal link to the KT extinction event (Hildebrand et al., 1991; Pope et al., 1991, 1993; Kring and Boynton, 1992; Sharpton et al., 1992, 1993, 1994a, b; Swisher et al., 1992; Blum et al., 1993; Krogh et al., 1993; Koeberl et al., 1994; Schuraytz and Sharpton, 1993; Schuraytz et al., 1994). Focus is now shifting from proving the origin and age of the Chicxulub impact basin to gaining a better understanding of the crater as well as the geological and environmental consequences of the impact event that produced it.

Here we assemble and assess the geophysical and geological characteristics of the Chicxulub basin as understood to date, including the general stratigraphic sequence of impact lithologies and target rocks obtained from drill-core and well-log data. We combine these constraints with those derived from terrestrial and planetary crater studies to construct a general model of the Chicxulub structure as an ~300-km-diameter impact basin and discuss its implications.

THE GEOPHYSICAL EXPRESSION OF THE CHICXULUB STRUCTURE

Morphology and structure of impact-generated landforms

Exploration of the inner solar system during the latter half of the twentieth century has shown that large-body impact played a major role in shaping ancient planetary landscapes such as those preserved on Mercury and the Moon. It is there, on the ancient surfaces of these small airless planets, that the best examples of large impact basins are observed (e.g., Spudis, 1993). The largest basins exhibit multiple concentric rings and can exceed 2,000 km in diameter. In contrast, Earth's active surface processes quickly destroy its impact record; only ~140 impact craters have been identified so far and many are small, bowl-shaped, or simple craters. Larger complex structures, greater than about 4 km in diameter on Earth, are rarer, older, and more modified; the truly large impact basins, those greater than ~50 km in diameter, are extensively modified and have retained few, if any, clues to their original surface configuration. But such erosional and tectonic activity, combined with extensive drilling and geophysical surveys, also affords the only information on crater substructure from which to understand the third dimension of large impact craters—anywhere in the solar system.

The smallest pristine craters on any planet are simple bowl-shaped depressions with raised rims; these features are produced by the downward displacement and the excavation flowfield initiated by a system of impact-driven compression and rarefaction waves (see Melosh, 1989, for a complete treatment of the physics of impact). But larger-magnitude impacts create transient conditions within the target that initiate large-scale, gravity-driven modification of this bowl-shaped crater form during the final stages of the impact event. Two processes take place immediately and almost simultaneously: The floor of the initial or "transient crater" rebounds upward, producing a central structure that, when exposed, takes the form of a central peak composed of the originally deep target rocks comprising the crater subfloor; and the crater margins collapse inward along either a series of normal faults (producing terraces) or along a single scarplike fault. This late-stage modification results in a final basin that is broader and shallower than the transient crater and is characterized (ideally) by a central peak, broad flat floor deposits, and a terraced rim zone. (N.B.: The "transient crater" is a theoretical construct that aids conceptualization of the cratering process; however, given the scale and complexity of planetary cratering events, there could be considerable overlap in time between excavation and modification, i.e., central rebound may begin prior to the termination of excavation and outward flow at the distal parts of the excavation cavity; consequently a true transient crater may not fully form before modification destroys it.) The modified basin is filled with a mixture of melt rock and highly shocked rock debris, both originating near the impact site, as well as a considerable amount of weakly shocked or unshocked debris, much of which entered the basin during slumping and enlargement. For consecutively larger impact events, the central peak is replaced by clusters of peaks, and finally by a peak ring.

At the very largest diameters, impact basins are characterized by two or more concentric rings in addition to—and outside of—the central peak ring. The canonical multiring basin is the Orientale basin on the Moon (Fig. 1). Multiring basins have been studied intensely over the course of 30 years of planetary investigations (e.g., Baldwin, 1963; Schultz, 1976; Head, 1977; Melosh and McKinnon, 1978; Pike and Spudis, 1987; Melosh, 1989; Spudis, 1993), yet even now they are not well understood. The surficial expression of basin rings, particularly the distal ones, can be quite subtle, especially when the basin has been modified by eons of subsequent geological activity, as most have. There is even debate over whether or not multiring basins exist at all on some terrestrial planets (see discussion in Chapter IX, Melosh, 1989), and one faction maintains that a planetary asthenosphere is necessary for multiring basin formation (Melosh and McKinnon, 1978; Melosh, 1989). This lack of consensus has led to a generally hesitant approach to interpreting even the largest impact features on Earth as the modified remains of multiring basins.

The recent Magellan mission provided the first high-resolution unencumbered views of the surface of Venus. Recognition that multiring basins occur on that planet (Schaber et al., 1992; Alexopoulos and McKinnon, 1994), so different from the Moon and Mercury, yet so similar to Earth in terms of its surface gravity, atmospheric conditions, and target strength, has strengthened the proposition that multiring-basin morphology is an ultimate consequence of increased impact energy and not related to some planetary property such as the presence of an asthenosphere. The diameter at which the transition from peak-ring basins to multiring basins takes place appears to follow the same inverse dependence on planetary gravity as the transition from simple to complex craters and the transition from central-peak to peak-ring basins. All impact basins greater than ~140 km on Venus exhibit the attributes of multiring basins; this and the basic similarities between Venus and Earth suggest that impact features of this size on Earth should correspond to multiring basins as well. Three terrestrial craters identified to date exceed this size: the Sudbury structure in Ontario, the Vredefort structure in South Africa, and the Chicxulub structure.

Gravity data over Chicxulub

Penfield and Camargo (1981) originally recognized two concentric zones in gravity anomaly and aeromagnetic data over the Chicxulub structure: an inner zone ~60 km in diameter, characterized by a local gravity high and high-frequency magnetic anomalies approaching 1,000 nT, and an outer gravity trough with low amplitude (5 to 20 nT) magnetic anomalies. Hildebrand et al. (1991), primarily on the basis of these patterns proposed that the Chicxulub crater was a double-ring or peak-ring basin with a rim diameter of ~180 km. More recently, however, the gravity data over the northern Yucatan have been reprocessed (Sharpton et al., 1993), and additional concentric patterns outside the original two have been identified (Plate 1 and Fig. 2), providing geophysical evidence that the Chicxulub basin is ~300 km in diameter and may be a multiring impact basin.

The gridded Bouguer gravity anomaly map of the Yucatan region is shown in Plate 1B. Gravity anomalies in the vicinity of the Chicxulub basin (Plate 1C) range from –16.4 mgal (10^{-5} m s^{-2}) to +53.6 mgal. The Chicxulub basin is expressed as a broad, nearly circular region in which gravity values are 20 to 30 mgal lower than regional values. A distinct 15- to 20-mgal high occupies the geometric center which we place at 21.3° N latitude and 89.6° W longitude. Figure 2 shows gravity profiles through the center of the structure for each 10° of azimuth and clearly reveals multiple rings expressed as local maxima in the gravity anomaly data. Like topographic rings at large multiring basins on other planets, these rings are spaced systematically such that distances between successive rings are approximately multiples of $\sqrt{2}$ (Pike and Spudis, 1987; Spudis, 1993; Sharpton et al., 1993). Major topographic rings characteristic of multiring impact basins include (1) the innermost or central peak ring, thought to represent the topographic expression of the central uplift (Grieve et al., 1981; Melosh, 1989, p. 163–168),

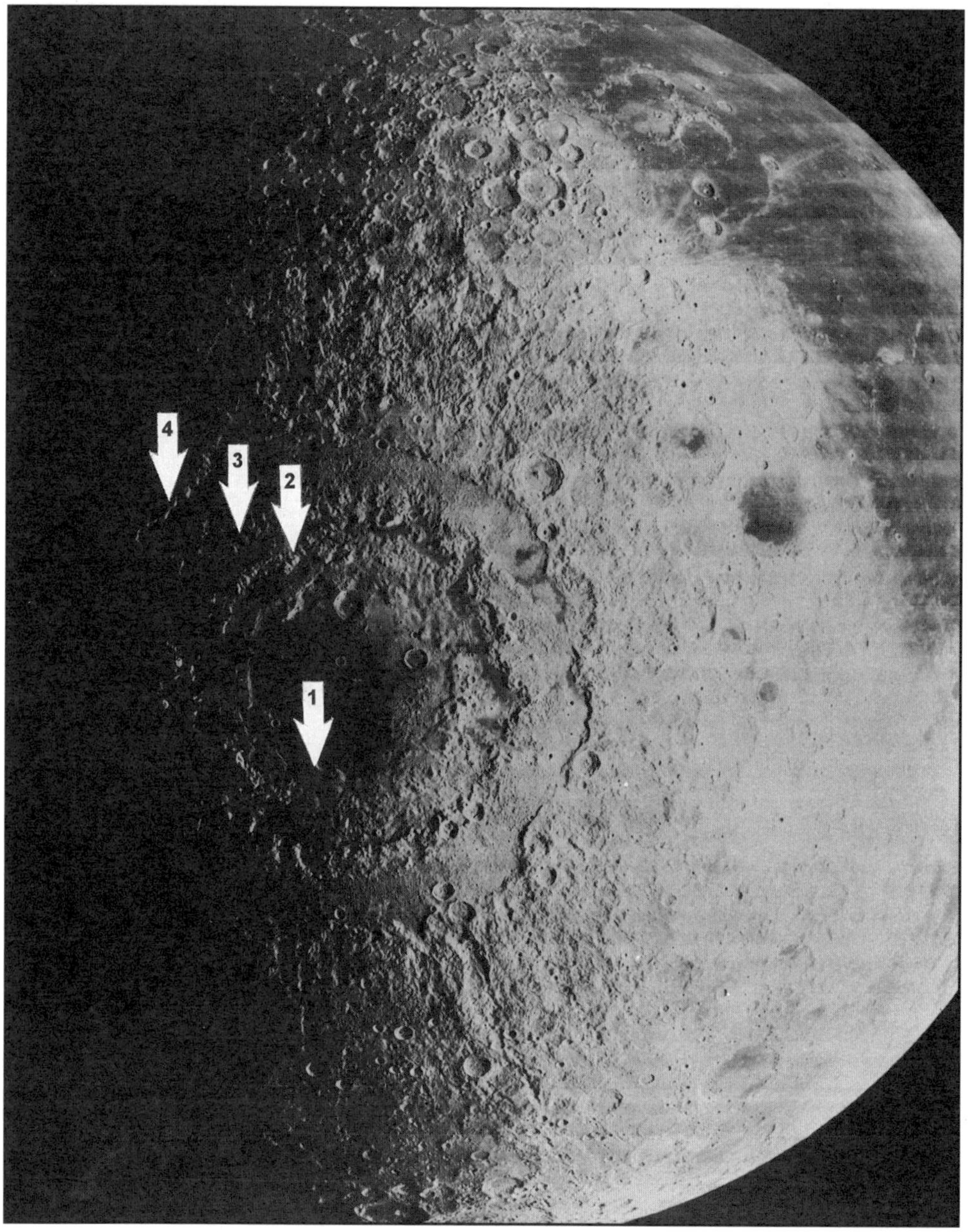

Figure 1. The 930-km Orientale Multiring Basin on the western limb of the Moon as seen from the Lunar Orbiter IV spacecraft. North is toward the top of the page. Numbered arrows indicate the four main basin rings (e.g., Howard et al., 1974; Wilhelms, 1987, Chapter 4, Spudis 1993). 1 is the central peak ring; 2 and 3 are the Inner and Outer Rook Mountain rings, respectively; and Ring 4 is the Cordilleran Scarp.

(2) one or more intermediate rings marking the margins of annular shelves stepping down toward the center, and (3) the topographic basin rim (Spudis, 1993, p. 7, Fig. 1.2).

The central peak ring. The first gravity ring at the Chicxulub structure has a crest diameter of ~100 km and probably corresponds to the central peak ring, that is, equivalent to the innermost ring at the Orientale Basin on the Moon (Ring 1; Fig. 1). In their 170-km-diameter model of the Chicxulub crater, Pilkington et al. (1994) place the central peak ring at the annular low between the central gravity high and this first ring of local gravity maxima. They attribute their ring of local negative anomalies to a thick wedge of low-density breccia that they claim makes up the central peak ring. In their model, this breccia wedge floats on a thick pool of impact melt rock. Such a peak ring—physically decoupled from, and unrelated to, the central uplift—is at odds with all observations of central structures of terrestrial craters (e.g., Howard et al., 1974; Grieve et al., 1981) and evaluation of planetary crater records (Spudis, 1993). From

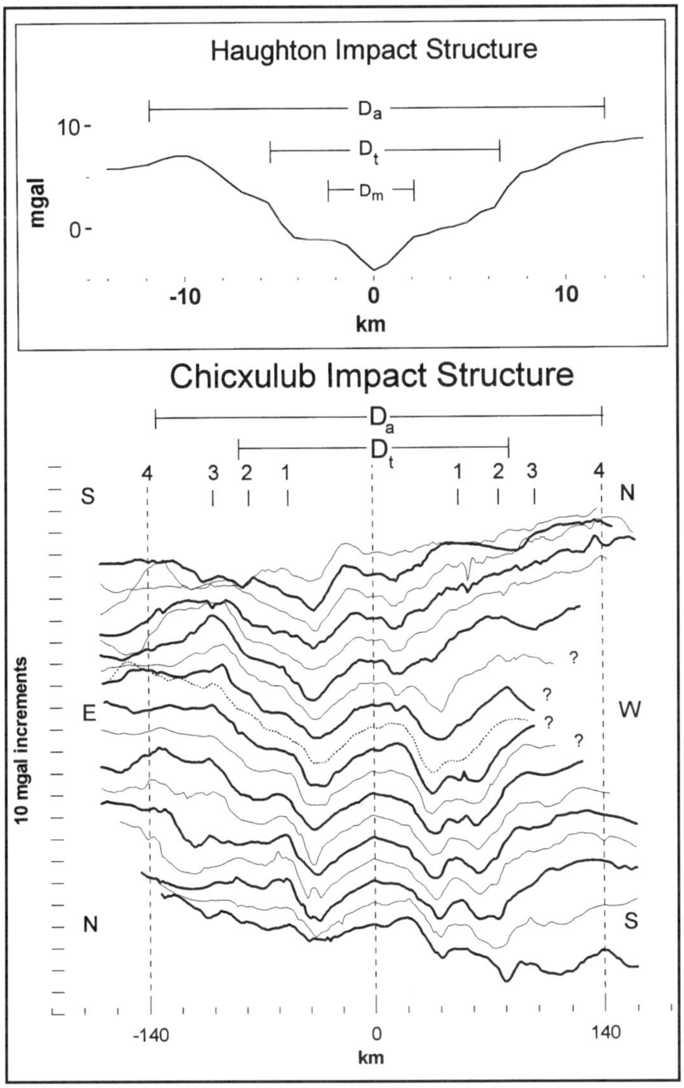

Figure 2. Radial gravity profiles taken across the Chicxulub basin at 10° intervals of azimuth. D_t and D_a refer to the transient crater diameter and final or apparent crater diameter resulting from this analysis. Numbers denote average position of gravity rings shown in Plate 1. Profiles are offset vertically by 10 mgal; the annotated vertical axis shows the 0 mgal value for each profile, beginning from the bottom with N to S. Dashed profile is due E to W. The value of the center point of each gravity profile is 10.4 mgal. Inset: The Haughton impact structure also shows a marked steepening in gravity gradient at the edge of the transient crater. Note also that the zone of intense magnetization is considerably smaller than D_t.

studies of such terrestrial craters as the 100-km-diameter Popigai structure (Masaitis et al., 1980; Masaitis, 1990) and the 24-km-diameter Ries crater (e.g., Pohl et al., 1977) it is well known that the central peak ring is the upper surface of deep crustal massifs that comprise the central uplift. Images of well preserved ring basins on other planets demonstrate that this ring often, but not always, appears as a topographic ring, cropping out through the melt-rock and breccia deposits.

The Sudbury and Vredefort impact structures are both well-studied ancient impact basins interpreted to be between ~180 and 300 km in diameter, and although their outer reaches have experienced considerable erosion and deformation, they retain important structural and lithologic information on the deep central portions of large impact basins. Recent reconstructions of the Sudbury basin (e.g., Grieve et al., 1991; Roest and Pilkington, 1994) show that the location of the peak ring corresponds to a ring of gravity maxima located near the edge of the preserved sequence (corresponding to the inner portion of the original Sudbury impact basin). The Vredefort structure is heavily eroded but has a well-expressed concentric gravity pattern similar to that over the central one-third of Chicxulub but approximately 60% as large (e.g., Antoine et al., 1990; Stepto, 1990). As described by Nicolaysen (1990), the so-called collar strata seem most likely to correspond to the structural and lithological expression of the peak ring: These rocks are highly deformed and locally overturned, and they separate the deep rocks of the crystalline core from the modestly deformed rim synclinorium of the shallower Transvaal strata. A distinct ring of gravity highs, with a diameter of 60 to 70 km, is located over this sequence. Consequently at both Sudbury and Vredefort, the central peak ring is expressed as a gravity high, not a gravity low.

Intermediate rings: Delimiting the zone of deep deformation at Chicxulub. Gravity values increase abruptly and significantly between ~70 and ~105 km from the center (Plate 1 and Fig. 2); this region exhibits the steepest gravity gradients associated with the Chicxulub structure. Its distal edge defines a prominent gravity ring (Ring 3), which Hildebrand et al. (1991) and Pilkington et al. (1994) interpret as the basin rim. Such steep inward-facing gradients define a sharp boundary to the major mass deficiency in the basin center. But because the crater topography has been completely filled with platform sediments, there should only be a very small density contrast between the crater fill and the surrounding upper crustal rocks of the crater walls. Therefore it seems unlikely that the steepest gradients observed in the structure would be associated with the rim zone, a region that, in other terrestrial impact structures, is characterized by shallow inward slumping (e.g., Pilkington and Grieve, 1992).

Although the largest terrestrial basins are either heavily modified or are associated with strong regional variations that hinder precise extraction of the gravity anomalies associated with the outer portions of the basin, assessment of some smaller terrestrial craters can provide help in linking gravity gradients with basin structure. Both the 24-km-diameter Ries crater (Jung et al., 1969; Pohl et al., 1977) and the 24-km-diameter Haughton crater (Pohl et al., 1988) are well-preserved craters with distinct gravity signatures characterized by a broad depression, the edge of which marks the location of the topographic rim (Fig. 2). In both cases, there is a sharp inward steepening in gradient near the edge of the central zone of excavation and deep deformation as determined by independent techniques (e.g., for Haughton crater, see Scott and Hajnal, 1988, and Bishoff and Oskierski, 1988). Furthermore, recent gravity and topography data for the lunar Orientale basin by the Clementine space probe (Spudis et al., 1994; Zuber et al., 1994) show that the steepest Bouguer anomalies are located over and within the Outer Rook Ring (Ring 3; Fig. 1) and not associated with the Cordilleran Ring (Ring 4; Fig. 1), the topographic rim of this multiring basin. We believe, therefore, that Rings 2 and 3 at Chicxulub would correspond to the Inner and Outer Rook Mountain Rings at Orientale (Rings 2 and 3; Fig. 1) and that the steep gradients observed between these two rings at Chicxulub could mark the outward extent of excavation and deep deformation associated with the collapsed "transient" crater.

This measurement should not be taken as a precise estimate of the "true" transient crater diameter (the diameter of the crater as measured from the preimpact surface) as used by modelers to define scaling relationships (e.g., Melosh, 1989; Chapter 7). During collapse, the transient crater walls are severely deformed, and the zone of intense deformation is broadened somewhat. This leads to an enlargement of the gradient ring associated with the boundary between the transient crater and the more distal slumped outer basin floor. We suspect that our estimate based on gravity data would be somewhat larger than the transient crater's "rim-crest" diameter; therefore, it would overestimate the apparent diameter by >25% (Melosh, 1989; p. 112). Consequently, the most appropriate estimate of the "true" transient crater diameter resulting from analysis of the gravity data would be ~105 to 160 km.

The topographic basin rim. The outermost ring at Chicxulub is a broad, discontinuous gravity feature whose crest is located at a diameter of ~260 to 300 km. This ring is most conspicuously expressed to the south of the basin center as it crosses and interrupts the southward-trending troughlike gravity low in Plate 1; B and C, and toward the east and north of the crater center. Sharpton et al. (1993, 1994a) have argued that this ring marks the modified topographic rim of the Chicxulub multiring basin. This ring is located at the edge of the broad circular gravity low associated with the Chicxulub basin; the anomaly edge corresponds approximately to the topographic rim at other terrestrial craters. We emphasize, however, that although the edge of the gravity depression is associated with the rim-crest region at many other terrestrial craters (e.g., Pilkington and Grieve, 1992), gravity data alone cannot pinpoint the rim location, particularly if, as is the case for Chicxulub, the basin is completely filled with sediments that are similar in density to the upper target sequence. The outermost ring is probably not an exterior ring (i.e., located outside the topographic rim), as observed at some large multiring basins on other planets, because it has a detectable gravity expression and because of its association with the edge of the basin's circular gravity low. Consequently, this ring corresponds to the Cordilleran Scarp at Orientale (Ring 4; Fig. 1). As discussed in a later section, this interpretation is supported by well data.

Magnetic data

The magnetic anomaly map of the Chicxulub structure, compiled from aeromagnetic surveys, is shown in Plate 1; D. An ~80-km-diameter zone of high-frequency dipolar magnetic anomalies with amplitudes of up to 1,000 nT occupies the center of the structure (as determined by concentric gravity anomaly patterns). This strongly magnetic zone is flanked by a region of lower amplitude (5 to 20 nT) high-frequency magnetic anomalies extending out to a radial distance of ~100 km. The magnetic signature of the Chicxulub structure is distinct from the regional magnetic characteristics of the attenuated continental crust of the Yucatan Peninsula. The central zone of high-amplitude magnetic anomalies is located within the first gravity ring and appears to be offset westward from the center of the gravity anomalies.

Pilkington et al. (1994) suggest that these magnetic anomalies reflect a thick sheet of impact melt rocks covering the crater floor, with the strongly magnetic central zone due to a deep central melt pool analogous, they state, with the craters Dellen and Mien. If other terrestrial craters similar to Chicxulub actually demonstrated a clear association between magnetic patterns and melt-rock distribution, then magnetic patterns would provide a reasonable constraint to crater size, as they propose. But terrestrial craters display highly variable magnetic characteristics (e.g., Pilkington and Grieve, 1992), and it is not the case that magnetic anomalies unambiguously delineate melt sheets in general. Both Mien (5-km diameter) and Dellen (15-km diameter) are impact structures on purely crystalline terrains, conducive to the formation of coherent melt sheets. On the other hand, the Chicxulub basin formed by impact into volatile-rich target rocks, and other craters into such lithological sequences rarely contain appreciable coherent melt sheets (e.g., Kieffer and Simonds, 1980). Haughton crater, for instance, has no coherent melt sheet but shows a central zone of high magnetization (Pohl et al., 1988) that is considerably smaller than the crater floor defined on the basis of surface mapping and seismic studies (Fig. 2).

Vredefort experienced such extensive erosion that, if a melt sheet was formed, it has subsequently been removed by erosion and only dikes of "bronzite granophyre" (French and Nielsen, 1990) remain as evidence of wholesale melting during impact. Nonetheless, Vredefort has magnetic anomalies that are comparable in amplitude and frequency to those observed over Chicxulub (Antoine et al., 1990; Hart et al., 1995). The magnetic highs are inboard of the uplifted, highly deformed collar strata (Antoine et al., 1990) that, as we discussed previously, seem to correspond with the central peak ring. The cause of the magnetization appears to be impact-induced thermal metamorphism of the deformed deep crustal rocks composing the central uplift (Hart et al., 1995). We therefore suggest that the principal source of the magnetization at Chicxulub could also be the highly deformed and uplifted deep crystalline basement inside the peak ring. Core samples of Chicxulub's melt rocks are strongly magnetic (Sharpton et al., 1992; Urrutia-Fucugauchi et al., 1994), and melt rock deposits may contribute to its magnetic signature. However, it is not clear whether or not there is a coherent, laterally extensive sheet of impact melt rocks within the Chicxulub basin or, if it does exist, what its dimensions are.

GENERAL STRATIGRAPHIC FRAMEWORK FROM WELL DATA

The eight wells drilled into and around the Chicxulub structure provide a stratigraphic framework that is consistent with a large impact structure formed at the K/T boundary. For core-sample locations, thc convention for identifying wells is as described in the caption of Plate 1. Core intervals are preceded by an N (for *núcleo*, Spanish for core), so a complete description of the fourteenth core interval recovered from the Yucatan No. 6 well would be Y6-N14. All depths are referenced to sea level. The stratigraphic correlations derived from drill core samples and well logs are summarized in Fig. 3.

Pre-impact stratigraphy, target characteristics, and basin shape

The rocks involved in the Chicxulub impact event are silicate basement of continental affinity overlain by a shallow-water platform sequence consisting of Lower Cretaceous (or older) evaporites and younger limestones and dolomites. The wells on the flanks of the structure (Yucatan-1 and Yucatan-4) indicate between 300 and 800 m of evaporite in the platform section; however, the evaporites seem to thicken toward the center of the structure. It seems likely that outwardly directed compression during the early stages of the impact event (Croft, 1981a, b) could have structurally thickened the sequence at radial distances equivalent to Ticul-1 or even Yucatan-2, but scarcity of well data and uncertainties in the size of the structure allow other explanations as well. In any event, it seems unlikely that there is as much anhydrite as previously estimated by Lopez-Ramos (1983) who, relying upon reports based on analysis of well cuttings, mistook the anhydrite-dolomite breccia in the Upper Cretaceous section (Fig. 3) for interbedded anhydrite and dolomite layers. As we discuss below, we now recognize that this unit is debris ejected from the Chicxulub basin and deposited onto the preexisting Upper Cretaceous rocks around the impact site.

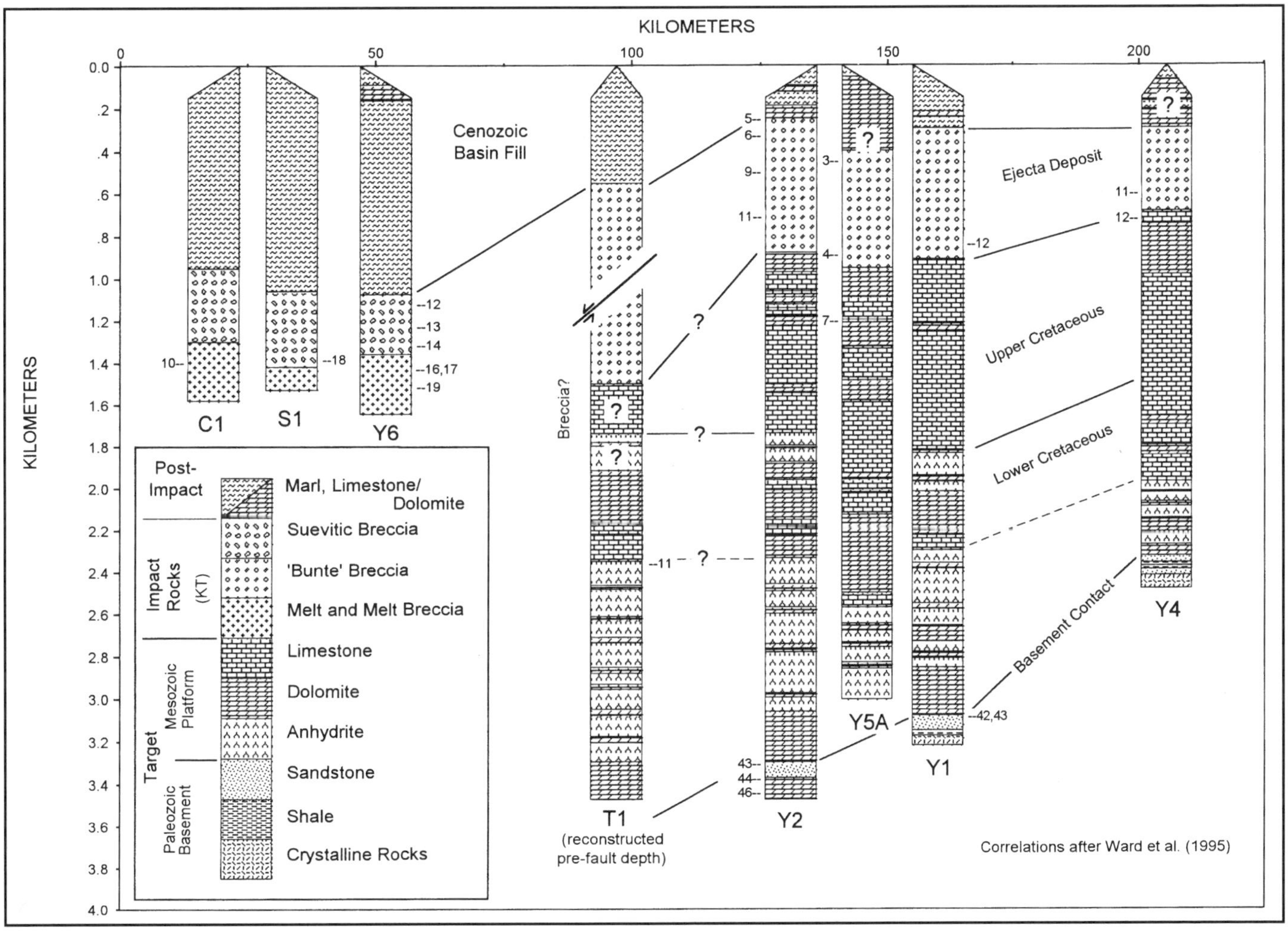

Figure 3. Stratigraphic correlations across the Chicxulub basin based on well logs and sample analysis. Wells are arranged in distance from the basin center. The samples we have studied are identified by numbers. Other lithological assignments are based on visual inspection of cores, updating previous published and unpublished reports. The stratigraphic thickness of the breccia unit in T1 reflects the pre-fault conditions we infer from analysis of well-log data and drilling reports. True (present) depths can be calculated by removing 200 m from the depths of all units below the fault shown. Major stratigraphic correlations are based on the work of Ward et al., 1995.

Based on well data alone, the Chicxulub basin extends ~135 to 150 km from center to the vicinity of Yucatan-2, Yucatan-5A, and Yucatan-1, where the K/T boundary is nearest to the present surface (Fig. 3) . This distance corresponds to the location of the fourth and outermost ring of gravity anomalies (Plate 1 and Fig. 2). Filling the basin are Tertiary limestones and marls ranging in thickness from slightly over 1 km near the center to less than 300 m at the basin rim. The total thickness of platform rocks at the time of the impact is constrained by depth to basement outside the structure (removing the thickness of the Tertiary section) and is ~2 km in Y4 as well as in several other wells located east and south of the structure between 200 and 300 km from the center, such as Quintana Roo 1, Tower Hill 1 and 2, and the Basil Jones 1 wells (Lopez-Ramos, 1983).

In wells located within ~160 km of the center, that is, Ticul-1, Yucatan-1, Yucatan-5A, and Yucatan-2, the silicate basement appears to be 750 to 1,000 m deeper than in more distant wells (Fig. 3). This is consistent with faulting and sagging associated with downward collapse and widening of the basin that occurs during the final modification stage of the impact process (Croft, 1981 b). The deepening of the basement contact in wells within the Chicxulub basin appears to be too abrupt to be explained solely by a regional basement tilt as previously proposed (e.g., Viniegra, 1981): Yucatan-5A is less than 70 km west of Yucatan-4, yet reached total depth of 3,003 m still in platform rocks whereas the basement contact at Yucatan-4 was reached at ~2350 m—at least 650 m higher. Outcrop patterns in northern Yucatan (Plate 1; A) indicate a modest northward-trending epirogenic rise located slightly west of the peninsula's center, suggesting that Cenozoic units on the eastern side of the basin tilt gently to the east, not west. There are also no indications of significant westward regional tilting of the pre-Cenozoic units in offshore seismic profiles located ~40 km north of the basin center (Camargo Z. and Suárez R., 1994). Nor are there indications of basement contacts as deep as those around the Chicxulub basin in any of eight onshore wells drilled between 160 and 450 km of the center of Chicxulub (Lopez-Ramos, 1983).

The Upper Cretaceous–Lower Cretaceous contact is also substantially deeper in wells inside the basin (Fig. 3), consistent with downward (and inward) displacement of the pre-Cenozoic crustal column during collapse of the transient crater (e.g., Melosh, 1989). We believe, therefore, that these vertical offsets, which could exceed 1 km, reflect structural displacements induced during the collapse (modification) stage of this impact event. These observations support the interpretation that the outermost gravity ring, located ~140 km from the center of Chicxulub, marks the topographic basin rim (Sharpton et al., 1993).

Impact lithologies

Foremost in support of the impact origin of the Chicxulub structure is a lithological sequence similar to that observed at several other large well-studied terrestrial impact structures such as the Ries crater (von Engelhardt, 1972; Hörz et al., 1983), Haughton crater (Redeker and Stöffler, 1988), and the Popigai structure (Masaitis et al., 1980; Masaitis, 1990). Elements of this characteristic rock assemblage are also observed at other more eroded impact structures, such as the Sudbury structure. The sequence consists of suevitic breccias, weakly shocked crater fill and ejecta deposits equivalent to the Bunte Breccia at the Ries crater, melt rocks, and melt-matrix breccias.

The Chicxulub suevite breccia. Three Pemex exploratory wells are located near the center of the concentric gravity anomalies over the Chicxulub structure (Plate 1) and all intercepted breccias and crystalline silicate melt rocks at depths of ~1 km below sea level (e.g., Lopez-Ramos, 1983). Core samples from the Y6 and S1 include suevite (Plate 2; A-C), a polymict, clastic-matrix breccia containing impact-generated glass and melt-rock clasts analogous to impact-produced breccias observed at several other terrestrial craters such as Ries (e.g., Stöffler et al., 1977), Wanapitei (Dressler, 1982; Grieve and Ber, 1994), and Haughton (Redeker and Stöffler, 1988). Suevite characteristically contains deformed clasts of melt rock and glass. Like these other crater suevite units, the suevite deposits within Chicxulub are predominantly composed of highly shocked rocks from the lower parts of the target stratigraphy and contain abundant glass and deformed melt clasts (Sharpton et al., 1992).

The Chicxulub suevite recovered from the Pemex wells seems to be graded, ranging from sand-sized clasts near the top (Y6-N13; 1,100 to 1,103 m) to pebble-sized clasts at Y6-N14 (1,208 to 1,211 m) and S1-N18 (1,365 to 1,368 m) (Plate 2; A-C). This grading indicates that the breccia sequence most likely represents fall-back ejecta that was either sorted during transport through the atmosphere or ocean or subsequently reworked. At the Ries crater, sorted suevite is also observed but only near the top of the suevite sequence (Stöffler, 1977; Newsom et al., 1990), suggesting the possibility that the Chicxulub suevite deposits may extend much deeper than the current drill cores reach.

Pilkington et al. (1994) envisage the peak ring of the Chicxulub impact crater as a thick wedge of this upper breccia floating on a thick pool of impact melt rock. In addition to the problems with this interpretation discussed earlier, the evidence that these upper breccias are normally graded, and thus reworked, provides further evidence that they are unrelated to a primary element of crater morphology like the central peak ring.

Constraints on biostratigraphy and paleoenvironment. The carbonate unit recovered from Y6-N12 (1,000 to 1,003 m) immediately above the suevite contains a well-preserved fossil assemblage consisting mainly of abundant planktic foraminifera of the genera *Globigerina* and *Globorotalia*. The following age-significant species were identified (Marín et al., 1994): *Globorotalia pseudobulloides*, *Globorotalia trinidadensis*, and *Globigerina triloculinoides*. Each of these forms has the base of its stratigraphic range in the lower Danian; consequently, the carbonates containing these fossils are Paleocene and not Upper Cretaceous as previously suspected (Lopez-Ramos, 1983). The Y6-N13 suevite sample (Plate 2; A) contains no forams younger than latest Maastrichtian, consistent with a K/T age for the Chicxulub-forming impact event.

The fossil assemblage within the suevite samples Y6-N13 and N14 (Plate 2; A, B, F, G) includes orbitoid foraminiferids, mollusk fragments, and algal fragments. This shallow-water assemblage indicates that water depths prior to the impact event were at most a few tens of meters. By the time the Y6-N12 interval was deposited (Lower Danian), the water depth in this portion of the basin (Plate 1; A) was considerably deeper, based on the abundant occurrences of planktonic forams in general. Frequent occurrences of *Globorotalias* suggest an outer neritic environment (approaching 200-m water depths) or beyond. A single occurrence of *Nuttallides* implies an upper bathyal environment. Based on our evaluation of the stratigraphic correlations from well data (Fig. 3), relief across the basin during Lower Danian was ~700 to 800 m.

Characteristics of the silicate basement as revealed in suevite samples. Centimeter-sized fragments of fine- to coarse-grained silicate basement are common in Y6-N14 and S1-N18 breccias and indicate a medium- to high-grade metamorphic terrain. The predominant basement rock is pink, medium- to coarse-grained granitic gneiss (Plate 2; H-L) containing 10 to 20% quartz, 60 to 80% alkali feldspar, and 10 to 20% low-calcium plagioclase. The results of an electron microprobe survey of feldspar grains from Y6-N14 are summarized on the ternary diagram shown in Figure 4. Minerals such as biotite, pyroxene, or hornblende are rare in the granitic clasts, most likely because of their greater susceptibility to postshock alteration and their lower melting temperature. Sphene, apatite, and zircon exist as trace minerals. Grain boundary relationships within these fragments reveal substantial shearing and recrystallization before impact. In addition, ~20% of the phyllosilicate clasts in Y6-N14 contain ≤1-mm-wide irregular domains of relict glasses (Plate 2; R) (Sharpton et al., 1992) that are similar in major element chemistry to feldspars in the granitic clasts; however, microprobe analyses of these glasses consistently yield low totals (~91 wt%), suggesting that they have incurred secondary hydration associated with diagenesis. In conjunction with the unmelted granitic clasts observed in Y6-N19 melt-rock samples, these mineral glasses demonstrate that granitic gneiss is a major constituent of Chicxulub melts. Other basement clasts include quartz-mica schist, metaquartzite, and rare occurrences of mylonite. These rock types are similar, in terms of their mineralogy, structural fabric, and texture, to the shocked lithic clasts distributed within the K/T boundary (Sharpton et al., 1990a, b; Izett, 1990).

Evidence of shock metamorphism. The suevite unit intercepted by wells Yucatan-6 and Sacapuc-1 (Plate 1 and Fig. 3) contains clasts of silicate basement and platform rocks showing abundant and unequivocal evidence of shock metamorphism including: (1) planar deformation features in quartz, feldspar (Sharpton et al., 1992), and zircons (Krogh et al., 1993) (Plate 2; F-L), indicating dynamic pressures up to 23 GPa; (2) shock mosaicism in quartz and feldspar grains; (3) diaplectic glasses (i.e., glasses produced from mineral clasts

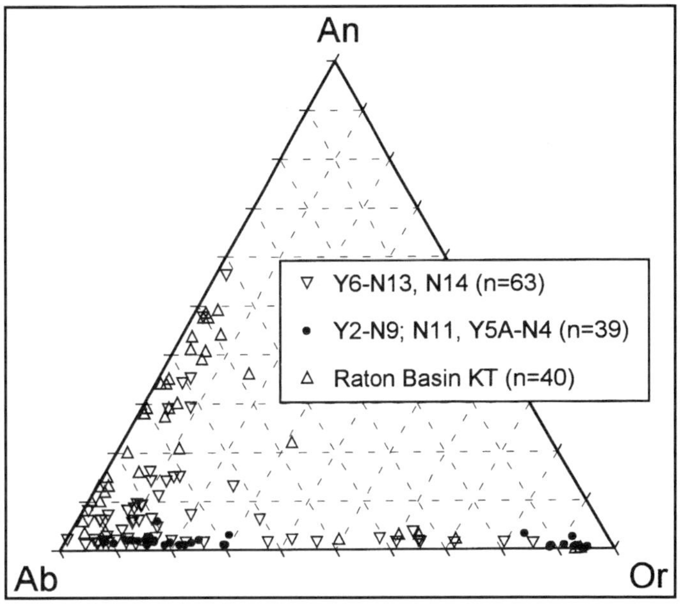

Figure 4. Feldspar cation compositions plotted on Anorthite-Albite-Orthoclase (An-Ab-Or) ($CaAl_2Si_2O_8$-$NaAlSi_3O_8$-$KAlSi_3O_8$) ternary diagram. This diagram shows the similarity between shocked feldspar grains extracted from the K/T boundary layer in Raton, New Mexico, and those from the central suevite (Y6 samples) and the distal Bunte Breccia–type deposits of Yucatan-2 and Yucatan-5A.

by the destruction of crystal order during shock compression but without fusion) from quartz and feldspar (maskelynite), which record peak pressures in the range of 30–45 Gpa; (4) fused mineral glasses, particularly of alkali feldspar, indicating shock pressures above 45 Gpa (Plate 2; R); and (5) whole rock melts (>60 GPa). Examples of these features found in various Chicxulub samples are shown in Plate 2 (F-R).

Melt rocks and melt matrix breccias. Below the deposits of suevite (Fig. 3) are occurrences of impact melt and melt-matrix breccia (Plate 2; D, E, Q) showing ubiquitous evidence of chemical and thermal disequilibrium and superheating as is characteristic of impact melt rocks. Melt-rock textures of the various core intervals are distinctly different (Plate 2), most notably in the size and abundance of undigested clasts, variations in color, grain size, and porosity of the matrix and in evidence of alteration. Brief descriptions of these melt rocks follow. For more information see Schuraytz et al. (1994).

The C1-N10 sample (Plate 2; E, S) is distinct from Y6 samples with respect to both the virtual absence of unmelted clasts and the coarser grain size of the matrix. The matrix is dominated by an intersertal arrangement of subhedral to euhedral pyroxene (augite) up to 0.7 mm in length and plagioclase (andesine to labradorite) showing a range of crystal morphologies and ranging up to 1 mm in length. The mesostasis of C1-N10 includes alkali feldspar and plagioclase ranging from oligoclase to pure albite.

Approximately 35% of Y6-N17 is undigested silicate basement clasts. These clasts show a bimodal size distribution dominated by single mineral fragments and polycrystalline domains of highly deformed, recrystallized quartz and feldspar ranging from <1 mm up to ~4 mm in length. Subhedral augitic pyroxene prisms (10 to 70 μm long) form coronas surrounding quartz clasts and pervade the interiors of more highly disrupted granitic domains, with aggregates up to 1 mm. The primary constituents of the matrix are subhedral to euhedral microphenocrysts of augite and plagioclase ranging from 5 to 15 μm in length. These are set in a porous, cryptocrystalline mesostasis containing albite, quartz and potassium feldspar. Anhydrite and gypsum constitute ~8% of total rock volume in the thin sections we have studied, mostly as veins and cavity fillings, but also as clasts.

Samples of Y6-N19 (Plate 2; D) reveal a melt matrix breccia containing 2- to 11-cm-wide, angular-to-subrounded melt clasts of varied textures. The dominant melt clast type is very similar to the surrounding matrix, and in some cases the boundary between them is difficult to discern. This material is also essentially similar to Y6-N17, consisting of 5- to 15-μm-long, subhedral-to-euhedral pyroxene and plagioclase in a cryptocrystalline quartzofeldspathic mesostasis showing variable porosity. In contrast to Y6-N17, however, undigested silicate basement clasts in Y6-N19 are typically larger and show clear evidence of shock metamorphism (Plate 2; D).

Evidence of shock metamorphism in the Chicxulub melt rocks.. Undigested basement clasts retain indications of shock metamorphism (Plate 2; M) but are often overprinted with subsequent thermal effects such as recrystallization and annealing.

A recrystallization effect commonly associated with shock-deformed quartz clasts observed in the Chicxulub melt rocks is the occurrence of "ballen structures" characterized by distinctive perliticlike fracture patterns that define individual optical domains (Plate 2; N). Similar features have been observed at the Ries crater (Engelhardt, 1972), Lappäjavi crater (Carstens, 1975; Bischoff and Stöffler, 1981), and West Clearwater Lake structure (McIntyre, 1968; Phinney et al., 1978). These ballen structures seem to document the strain associated with devitrification of either an impact-fused quartz glass known as lechatelierite (Carstens, 1975), or diaplectic (shock-vitrified without fusion) quartz glass (Bishoff and Stöffler, 1981). Stöffler and Langenhorst (1994) have shown that diaplectic glass displays a higher degree of long-range order than normal fused glasses quenched from a melt (e.g., silicic volcanic glasses) and effectively retains a "memory" of its previously crystalline state. Consequently, they note, diaplectic glass reverts to α-quartz + cristobalite during annealing, instead of just cristobalite as is the case with fused silica glass. Bischoff and Stöffler (1981) suggested that ballen structures record a range of peak shock pressures; those in which the ballens have different optical orientations, such as observed in the Chicxulub melt rocks, record shock pressures between 45 GPa and 55 GPa.

Target affinity and meteoritic contamination. Except for anomalous Ir enrichments in several specimens attributed to nonuniform dissemination of the projectile (Sharpton et al., 1992; Schuraytz and Sharpton, 1994), our analyses suggest that the melts were derived exclusively from continental crust and platform sediment target lithologies, with no evidence of a significant mantle or oceanic crustal signature. These results are supported by Sr, Nd, O, and Os-Re isotopic studies on the C1-N10 samples with regard to both the continental affinity of the target rocks (Blum et al., 1993) and the heterogeneous distribution of up to 3% meteoritic contamination (Sharpton et al., 1992; Schuraytz and Sharpton, 1994; Koeberl et al., 1994). Considering current constraints on excavation depth (15 to 25 km) of the Chicxulub impact event (Sharpton et al., 1994b) and the potential lithologic diversity within this volume, the observed chemical variability (andesitic to dacitic) is rather small and in keeping with the gross compositional homogeneity of melt rocks from other terrestrial impact structures, such as Manicouagan (Grieve and Floran, 1978) and West Clearwater (Simonds et al., 1978). However, given that Chicxulub melt compositions were derived from very small samples compared to those available from other well-exposed melt sheets, some heterogeneity might be expected. Furthermore, because currently available Chicxulub specimens represent an inordinately small sampling of the upper ~100 m of known melt rock, the limited compositional range should be regarded as tentative, as should comparisons with smaller structures where the upper part of the melt sequence has been eroded. Finally, all samples from Chicxulub have been modified to some degree by hydrothermal alteration (Schuraytz et al., 1994).

Are these rocks from a coherent melt sheet? Several workers have interpreted these melt rocks as the Chicxulub melt sheet (Hildebrand et al., 1991; Pilkington et al., 1994), however, the current constraints on lateral and vertical continuity of these melt zones are very poor and it seems premature to interpret these deposits as necessarily representative of a coherent melt sheet. Isolated melt zones are observed in the Sudbury structure associated with the fall back sequence (Onaping Formation); similar "stratiform melt bodies," several tens to a few hundred meters thick have been documented within the suevite breccias from the 100-km-diameter Popigai structure (Masaitis et al., 1980; Masaitis, 1990) and the 45-km-diameter Montagnais (Jansa et al., 1989, 1990). Furthermore, some large impact basins—Montagnais, Popigai, and Kara (Masaitis et al., 1980), for instance—show no evidence of the thick, laterally continuous melt sheets observed at others such as Manicouagan and Mistastin. Observations that suggest the melt rocks from Chicxulub represent local pods of melt, possibly within a thick suevite unit, rather than a thick coherent melt sheet include (1) the typically fine grain size of the melt matrix, (2) the clast-rich nature of the melt, including the presence of xenocrysts of anhydrite and carbonate rocks, (3) the presence of diaplectic glasses, (4) the chemical variability in the melt rock assemblage (Table 1 in Schuraytz et al., 1994), and (5) the nonuniform distribution of siderophile elements within the melt rocks. Consequently, a coherent melt sheet at Chicxulub, if it exists at all, may reside at considerably greater depth than any drilling to date has reached.

The Chicxulub multiring basin and the K/T boundary. Melt rocks within the Chicxulub crater have experienced varying levels of hydrothermal alteration and albitization (Sharpton et al., 1992; Schuraytz and Sharpton, 1993; Schuraytz et al., 1994); however ^{40}Ar-^{39}Ar determinations on relatively pristine melt-rock samples from the center of the basin indicate a crystallization age of 65.0 ± 0.4 Ma (Sharpton et al., 1992) or 64.98 ± 0.05 Ma (Swisher et al., 1992) which is at or very near the K/T boundary. Evaluation of the magnetization of these and other samples shows that they cooled during an episode of reversed geomagnetic polarity, consistent with a K/T boundary age (Sharpton et al., 1992; Urrutia-Fucugauchi et al., 1994).

Analysis of the Rb-Sr, O, and Nd-Sm isotopic systems confirms a chemical link between the Chicxulub melt rocks and the impact glasses contained in the K/T boundary unit at Beloc, Haiti (Blum et al., 1993); the Beloc tektite-bearing unit appears to be equivalent to the lower, kaolinite-rich member of the K/T boundary couplet in the Western Interior. Krogh et al. (1993) recently determined U-Pb ages of zircon xenocrysts from Chicxulub breccia sample Y6-N13 to be ~545 Ma, with a small percentage showing a 418-Ma age. The population of 545-Ma zircons matches the crystallization age of zircons extracted from the upper member (magic layer, fireball layer, etc.) of the K/T doublet in western North America, and both ~545-Ma and 418-Ma zircons have been detected in the tektite-rich K/T boundary layer at Beloc (Krogh et al., 1993; Kamo and Krogh, 1995). Unmelted breccia clasts, representing the silicate basement involved in the Chicxulub basin-forming event, are medium- to high-grade continental crust (Plate 2; H-L) similar to the lithic clasts observed within K/T boundary sediments from around the world (Bohor et al., 1984, Bohor, 1990; Izett, 1990; Sharpton et al., 1990a, b; 1992; 1994a). For instance, Figure 4 shows the match between feldspar compositions of shocked basement clasts within Chicxulub suevites and ejecta deposits with shocked feldspars from the upper member of Raton Basin K/T boundary layer. These observations provide compelling evidence that the Chicxulub impact event is responsible for both the shocked minerals observed in the upper member of the K/T boundary sequence and the glass-rich lower member. Consequently all evidence to date points to a singular, extremely energetic, and deadly impact event at the K/T boundary.

Ejecta and allogenic impact breccia deposits (Bunte Breccia equivalents). The central portion of the Chicxulub basin, showing high-amplitude magnetic and gravity anomalies and deeply buried occurrences of suevite breccias and melt rocks, is flanked by a broad annular trough, extending outward to ~150 km from center, where the pre-Tertiary sequence is depressed by up to 1 km. Filling this trough and covering outlying areas is a breccia sequence consisting of centimeter-scale clasts of dolomite, limestone, and anhydrite in a clayey-micritic matrix. This unit was completely overlooked by Lopez-Ramos (1983), who compiled his lithostratigraphic data for the Chicxulub region primarily from analyses of drill cuttings. Marshall (1974) was the first to document this "clastic anhydrite-dolomite lithofacies" and Hildebrand et al. (1991) first reported evidence of shock-deformed quartz clasts within sample Y2-N6 near the top of this sequence. However, Hildebrand et al. (1991) did not interpret the entire breccia sequence as Chicxulub ejecta. Instead they envisaged only a "90-m thick ejecta horizon," with the Y2-N6 sample representing the base of this unit. The ejecta thickness at the Yucatan-2 well site has been a fundamental line of evidence used for the 170-km estimate of Chicxulub's diameter (Hildebrand et al., 1991; Hildebrand, 1992, 1993; Kring, 1993; Pilkington et al., 1994). We now have compelling evidence, however, that the whole anhydrite-dolomite–bearing breccia unit is the continuous ejecta deposit at Chicxulub as independently proposed by Pope et al. (1993) and Sharpton et al. (1994a, b).

We have found trace quantities (<1% volume) of shocked but unmelted silicate basement clasts and melt-rock clasts with small xenocrysts of shocked quartz and feldspar (Plate 2; O-Q) in all our samples throughout the anhydrite-dolomite breccia shown in Figure 3. Shocked feldspar crystals within larger melt clasts have compositions similar to those found within the Chicxulub suevite breccias and the K/T boundary layer (Figure 4). This evidence demonstrates that the anhydrite-dolomite breccia unit is the Chicxulub crater's continuous ejecta deposit and is analogous to the well-studied Bunte Breccia at the Ries Crater (Hörz and Banholzer, 1980; Hörz et al., 1983). It is likely that the proportion of silicate basement apparent in these breccia samples has been diminished by the effects of chemical alteration and secondary mineralization working on the highly vulnerable shocked and partially melted basement clasts. At the Yucatan-2 well site (where this unit previously was estimated to be ≤90 m thick), this unit is >600 m thick, and the top of the unit is located ~250 m below sea level.

Pilkington et al. (1994) noted that published reports do not show an Upper Cretaceous breccia unit occurring at the Ticul-1 well (~95 km from center) and claim that this provides strong evidence that Chicxulub cannot be a 300-km-class impact basin. However, with the exception of the work conducted at the University of New Orleans (Marshall, 1974; Weidie et al., 1978; Ward et al., 1985; Weidie, 1985), this breccia was not recognized in *any* wells; Lopez-Ramos (1983) shows a sequence of interbedded anhydrite and carbonate units in the uppermost Cretaceous section in Yucatan-1, Yucatan-2, Yucatan-4, and Yucatan-5A at stratigraphic levels where breccia was subsequently reported by Marshall (1974) and Weidie et al. (1978). This circumstance was primarily due to the fact that the University of New Orleans (UNO) group focused on core studies, whereas previous reports were based predominantly on analyses of drill cuttings. Ticul-1 core samples were not available to the UNO group, so it is not surprising that the breccia was not recognized there.

Several lines of evidence indicate that the Upper Cretaceous section at Ticul-1 is indeed predominantly, if not totally, composed of anhydrite-dolomite impact breccia. Well reports show the K/T boundary at Ticul-1 to be located at 550 m, and the Upper Cretaceous section down to at least 1,200 m consists of interbedded evaporites, marls, limestone, and dolomite with signs of reworking. These well-cutting descriptions are similar to the descriptions of the Upper Cretaceous unit at Yucatan-2, Yucatan-1, Yucatan-4, and Yucatan-5A that is now recognized to be impact breccia. Furthermore, the only primary (bedded) anhydrites recovered in drill cores in northern Yucatan are located in the Lower Cretaceous section. Therefore, occurrences of Upper Cretaceous anhydrite at Ticul-1, where the section is located at greater depth than at Yucatan-2, Yucatan-5A, and Yucatan-1, most likely signify breccia containing Lower Cretaceous anhydrite excavated during impact.

Ward et al. (1995) have conducted detailed stratigraphic analysis of the wells in the vicinity of Chicxulub based on well logs, biostratigraphy, and lithostratigraphy of existing samples. They find the pre-Cenozoic section in the Ticul-1 well to be 200 to 300 m shallower than similar strata at the Yucatan-2 well. We attribute this to the intersection of the Ticul-1 core with the northward-dipping normal fault system shown in Plate 1 (Ticul Fault). This is consistent with the present topography across the scarp, which constrains the vertical offset on the fault to be in excess of ~100 m. Ward et al. (1995) show this fault intersecting the core at ~1-km depth, still within the zone where Pemex well reports show evidence of impact breccias. Consequently, more than 200 m of the original breccia thickness have been removed by normal faulting, and we infer a conservative lower bound of 900 m for the ejecta thickness at Ticul-1. Well logs and coring efforts provide some indications that it might be considerably thicker.

No coring was undertaken at Ticul-1 in the section above 1 km. However, 10 1-m core intervals were attempted in the Cretaceous section between depths of 1,150.8 and 2,137.7 m, but recovery was minimal, averaging less than 5%. The recovered fragments from these core tests were primarily anhydrite clasts that ranged in size between 2 and 10 cm. This poor recovery indicates a highly friable anhydrite-bearing material at these depths, possibly a reworked allogenic breccia deposit or shattered target rocks. Visual inspection of existing core samples taken from greater depths and dip-meter data show these rocks to be anhydrite blocks with considerable variation in the attitudes of bedding; these blocks are pervaded by thin veins of breccia consisting of anhydrite and dolomite clasts in a dolomitic mudstone matrix (e.g., T1-N11; 2,177 to 2,182 m; 100% recovery). This sequence could represent the original evaporite-rich platform target disrupted by ejecta emplacement and other impact processes.

Because there is no evidence of Upper Cretaceous bedded evaporites, anhydrite clasts may provide a lower constraint on the proportion of "primary ejecta" mixed with local material in this deposit. At Yucatan-2, anhydrite clasts (Plate 2; T), averaging ~0.5 mm to ~3 cm in length, account for 10 to 20% of the total volume in samples we have analyzed. If, as the well logs suggest, bedded anhydrite constitutes no more than about one-third of the total pre-Tertiary platform sequence, then the breccia at Yucatan-2 is at least 30 to 60% primary ejecta. Based on analyses of ejecta deposits at the Ries crater, such an enrichment is expected only very near the final basin rim of large impact structures (Hörz et al., 1983).

A STRUCTURAL MODEL OF THE CHICXULUB IMPACT BASIN

A structural model of the Chicxulub basin based on the evidence presented above is shown in Figure 5. Although there are still many unknown and poorly constrained aspects to this buried basin, we suggest that this model is more consistent with current knowledge of this structure and the cratering process than is the model proposed by Hildebrand et al. (1991) and Pilkington et al. (1994). The model of Pope et al. (1993) is in good agreement with our own interpretation of the location of the topographic rim. In that model, the center of the cenote ring (Plate 1A) defines the basin center (located east of the center based on gravity anomalies) and that offset results in the slightly lower basin size estimate of 240-km diameter rather than the ~260- to 300-km diameter we derive from our gravity analysis. However, given the lack of well data and gravity stations in the offshore western part of the basin as well as uncertainties associated with the cenote ring, either estimate is permitted. Evaluation and refinement of these models are possible through acquisition of high-quality seismic data across the Chicxulub basin, additional gravity data for the western portion of the basin, and a program of deep and shallow drilling within and around the basin.

Analysis of the ejecta deposits surrounding the Ries crater indicates that approximately 30% of the Bunte Breccia represents "primary" material derived from within the excavation cavity, the rest being local materials incorporated into the breccia during its energetic emplacement (Hörz et al., 1983). Accepting this estimate for Chicxulub and assuming the average thickness of the Chicxulub ejecta is 400 m, there would be more than 20,000 km^3 of primary ejecta located between 135 and 300 km from center around the Chicxulub structure. Because this breccia unit is derived predominantly from the 2 to 2.5-km-thick sequence of Mesozoic platform rocks, the excavation cavity would have to be at least 110 km in diameter just to account for the breccia deposited within that range. Additional ejected materials deposited inside or beyond this zone as well as the platform rocks that were melted or vaporized would increase this estimate, as would any additional ejecta that has been removed by erosion.

The conservative lower constraint on transient crater "true" size (>110 km) presented herein, therefore, is in reasonable agreement with the 130 ± 25-km-diameter estimate based on our analysis of gravity data. The Schmidt-Holsapple crater scaling relationship, derived primarily from theoretical and laboratory-scale constraints (Schmidt and Housen, 1987), indicates that such a crater could be generated by a 15-km-diameter comet (ρ = 1,100 kg m^3) striking the Yucatan Platform at 57 km s^{-1} (corresponding to the mean impact velocity of a long-period comet; Weissman, 1982). This object's size is somewhat

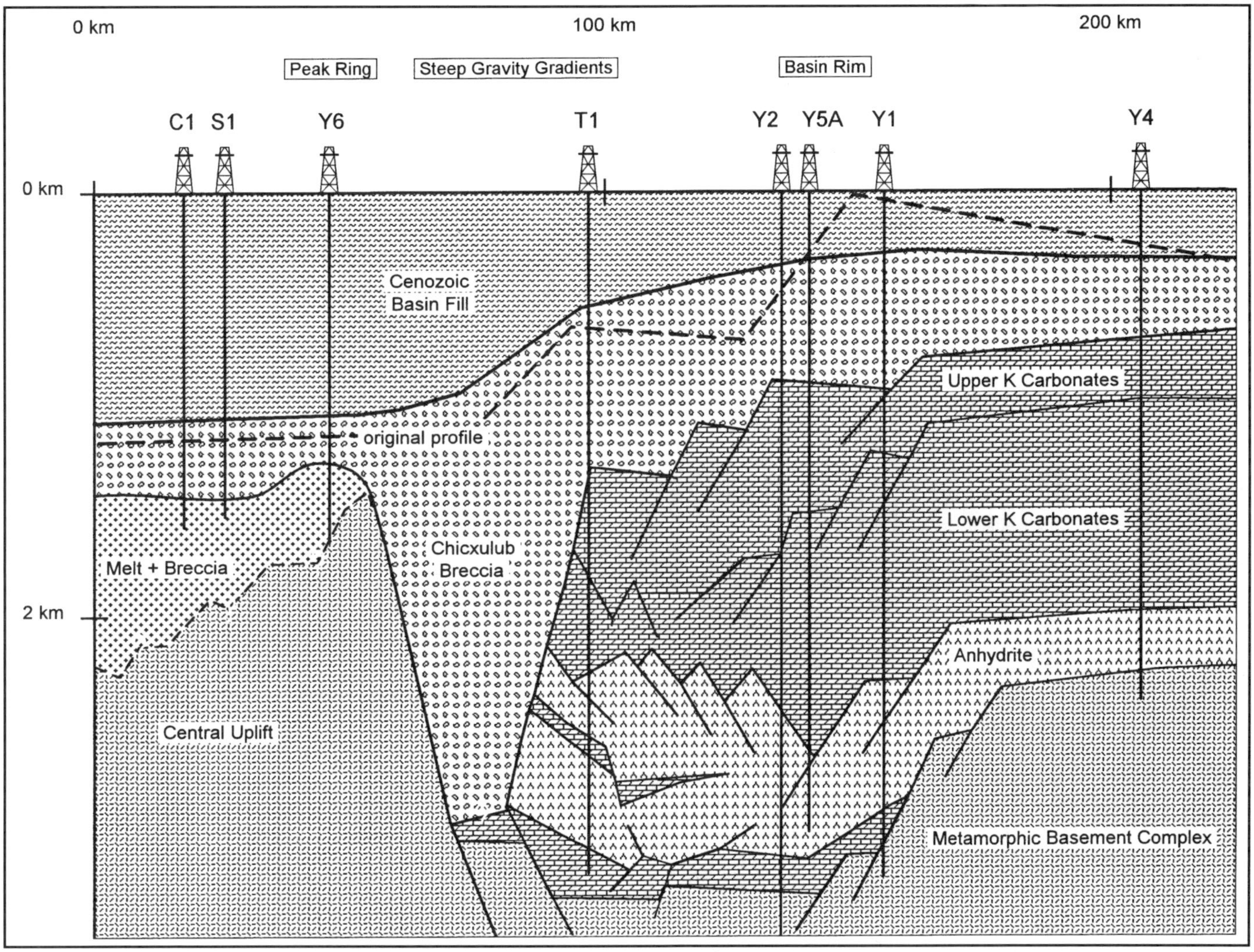

Figure 5. Schematic model of the Chicxulub impact basin that we derive from the analysis in this chapter. This simplified cross section shows the generalized present-day configuration of the crater consistent with geophysical data and well information. Erosion at the time of impact could rearrange the upper crater units significantly and reduce crater topography. Original profile shown as dashed line is reconstructed from the generalized topographic profile of the 270-km Mead Basin on Venus (Fig. 6).

larger than the canonical 10-km-diameter chondritic asteroid proposed by Alvarez et al. (1980) from their estimate of the amount of iridium distributed worldwide at the K/T boundary. However, recent analysis of the population of potential Earth-striking objects indicates that craters greater than 100 km in diameter are approximately five times more likely to have been produced by comets than by asteroids (Shoemaker et al., 1990). If the object were a comet, chondritic material might represent only a small proportion (<10% to 50%) of its total mass, the rest being ices, carbon, and other volatile components.

A 15-km-diameter comet containing 30 wt% chondritic material could account for the estimated iridium distributed worldwide at the K/T boundary (~2.5×10^{11} kg Ir) if all Ir were partitioned into the boundary layer. But, as we have shown, some projectile remains in the melt rocks deposited within the Chicxulub impact basin, and particularly at cometary impact velocities, there is a reasonable likelihood that some was expelled from the Earth (Vickery and Melosh, 1990). Given the above, we suspect that the K/T event was generated by an ~15-km-diameter long-period comet. However, because the integrated iridium mass within the K/T boundary only provides a lower constraint on the size of the impacting body, larger, more energetic objects cannot be ruled out.

We suspect that the best planetary analog to Chicxulub is the 270 km Mead basin on Venus (Fig. 6); altimetry from the Magellan mission indicates that the rim of this feature rises 300 to 400 m above the surrounding plains (Fig. 5). Because the sea covering the Yucatan platform would most likely have been less than 50 m deep prior to impact, the rim deposits and much of the ejecta blanket must have stood above sea level and exposed to erosion, even if the crater is significantly smaller than we estimate. Consequently, the crater shape has been substantially modified from that of a pristine impact basin. Whatever remnants of the original outer crater walls survived the water-laden impact event itself would be reworked quickly in the tidal environment of the Yucatan platform. The preserved portions of the rim and ejecta units therefore represent minimum thicknesses. This suggests that ejecta deposits at the rim, that is, near Yucatan-2 or Yucatan-5A, may have been ~1 km thick prior to erosion and redistribution.

At the Ries crater, which exhibits the best preservation state of a complex crater's ejecta deposits, the weakly shocked, shallow-sourced Bunte Breccia is overlain by suevitic ejecta containing considerably more deep-seated basement clasts and exhibiting a higher degree of shock deformation. Similar breccia units have not been reported from around the Chicxulub crater, possibly because they have been eroded from the high-standing rim zone. We would expect to find these deposits, however, just inside the topographic rim (equivalent to the "megablock zone" at Ries) where such suevitic ejecta would be deposited below sea level and most likely to be preserved.

Well data indicate a smaller proportion of the anhydrite within the Mesozoic platform sequence involved in the Chicxulub basin–forming event than previously reported (Lopez-Ramos, 1983). Consequently, the contribution of vaporized sulfate-bearing lithologies to the killing potential of this impact event requires reevaluation, particularly if the impact crater is as small as 170 to 180 km in diameter (Hildebrand et al., 1991; Pilkington et al., 1994). We suggest that vaporization of carbonates in addition to sulfates played a role in the environmental collapse. An early transient period of global cooling prompted by sulfur aerosols could be followed by a longer-lasting period of global warming resulting from enhanced concentrations of CO_2 in the atmosphere. The extreme temperature excursions produced by such a dipolar temperature shift could provoke enhanced stresses in the biosphere, giving rise to the K/T extinctions.

Figure 6. East-looking Magellan radar image of the Mead impact basin, the largest exposed impact structure on Venus (270-km diameter). This multiring basin is located at 12.5°N; 57.0°E. Mead may be the best planetary analog of the buried Chicxulub basin. The rim to floor depth of this feature is ~1,100 m. The outer ring is the topographic basin rim and corresponds to Ring 4 in our Chicxulub gravity model. The rim stands 300 to 400 m above the surrounding plains. Between the rim and the next prominent ring is a broad flat shelf that on average is ~350 m below the surrounding plains (~700 m below the maximum rim height). The inner scarplike ring (corresponding to Ring 3 in our model) marks a sharp drop in elevation of ~350 m to the crater center. The two innermost rings are covered by impact deposits. North is toward the top of the page.

Plate 1. **A**, Surface geology, ring locations from gravity data, and wells in the vicinity of the Chicxulub multiring impact basin. The three wells that penetrated impact melt rocks and breccias beneath the carbonate cover rocks are C1 (Chicxulub 1), S1 (Sacapuc 1), and Y6 (Yucatan 6). Other well sites shown are Yucatan 1 (Y1), Yucatan 2 (Y2), Yucatan 5A (Y5A), and Ticul 1 (T1). The Yucatan 4 (Y4) well site is located off the map, ~65 km east of Y5A. Carbonate units at the surface are Q (Quaternary; <2 Ma), Tu (Upper Tertiary; ~1.6 to 38 Ma), To (Oligocene; ~25 to 38 Ma), Te (Eocene; ~38 to ~55 Ma), and Tpal (Paleocene; ~55 to 66.0 Ma). Hatchured lines represent the Ticul fault system. Dashed lines indicate trends of ringlike zones of cenotes or sinkholes. **B,** Relief-shaded gravity anomaly data for the Yucatan region between 18° to 23°N latitude and 92° to 88°W longitude. Colors are ordered spectrally with violet and deep blue representing lowest gravity values and red representing the highest. Linear elements present in offshore data are shiptrack artifacts. Data were shaded using a low illumination angle from the north to emphasize broad low-amplitude anomalies. All data have been Bouguer corrected using a model slab density $\rho = 2{,}200$ kg m^{-3}. **C,** Relief-shaded gravity anomaly data for the Chicxulub region shown in A and D. Color coding is the same as in the regional gravity map B. Gravity anomalies in this region range from –16.4 mgal to 53.6 mgal. **D,** Aeromagnetic data for the region shown in A and C. Blues and greens represent negative anomalies; reds and yellows are positive. Data were acquired from a flight elevation of 500 m. Anomaly amplitudes exceed 900 nT near the center of the crater.

Plate 1A

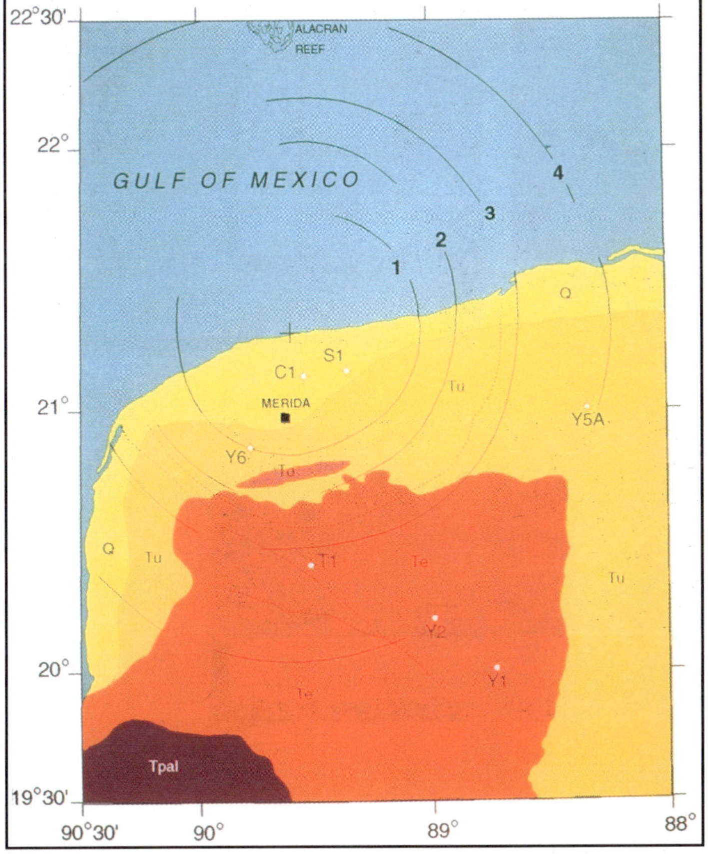

Plate 1B

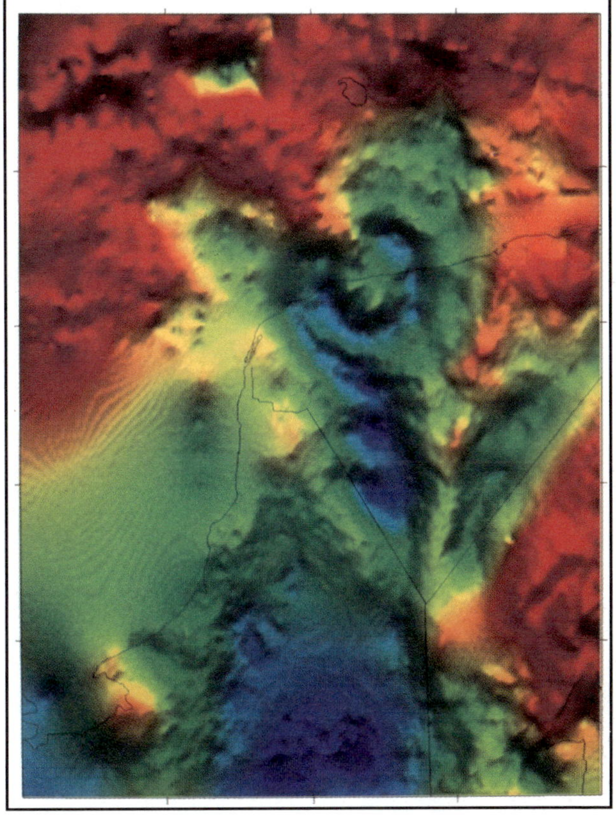

Plate 1C

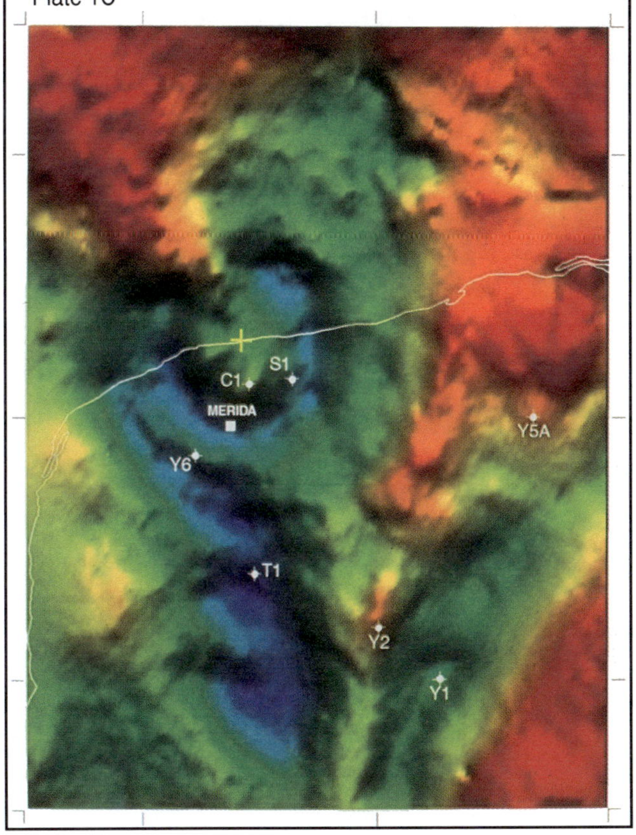

Plate 1D

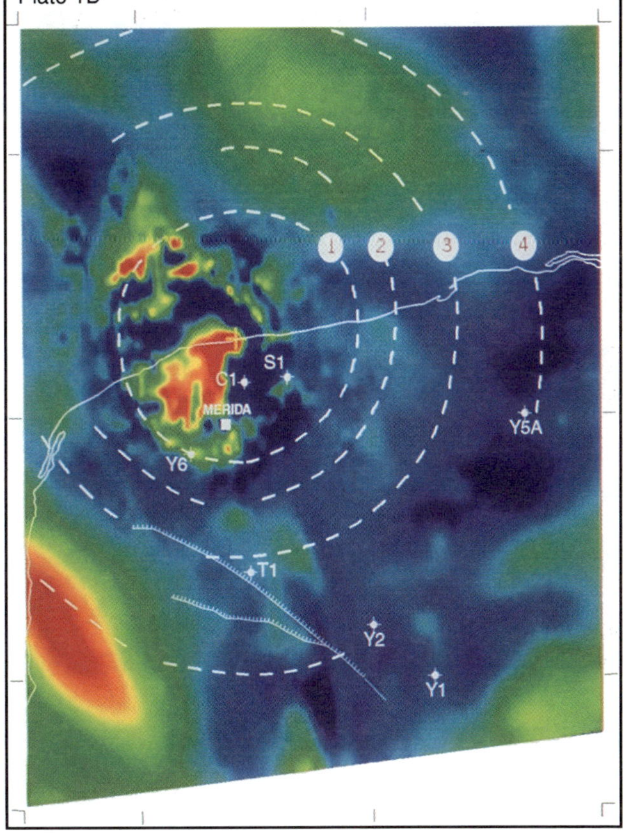

Plate 2 (caption on page 72)

K

L

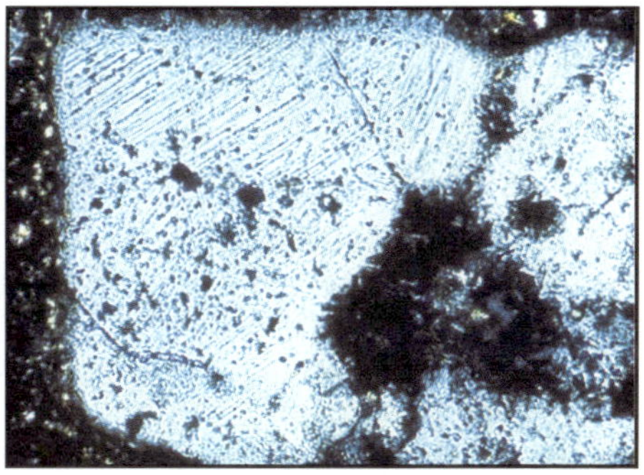

M

N

O

P

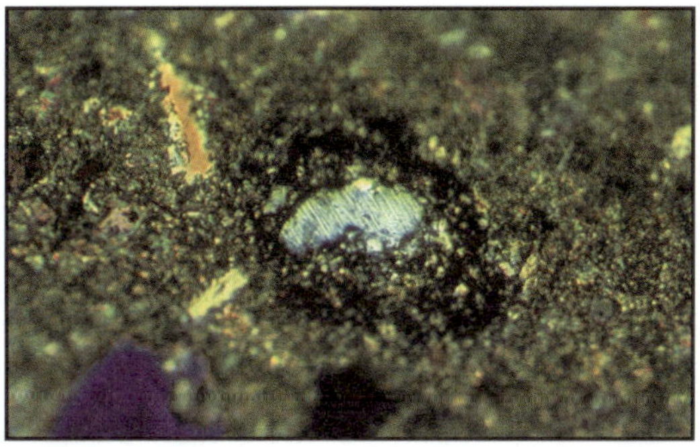

Q

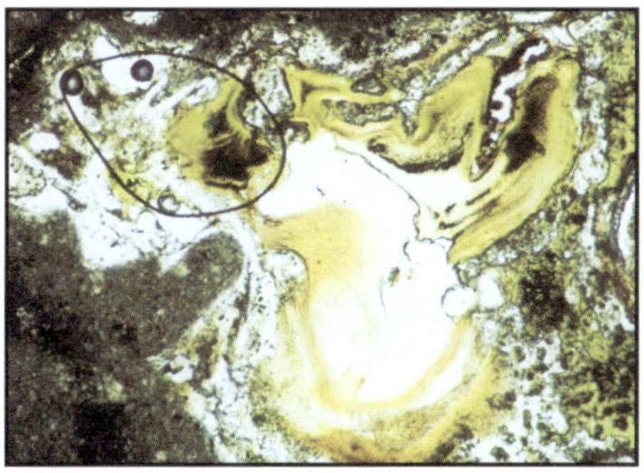

R

S

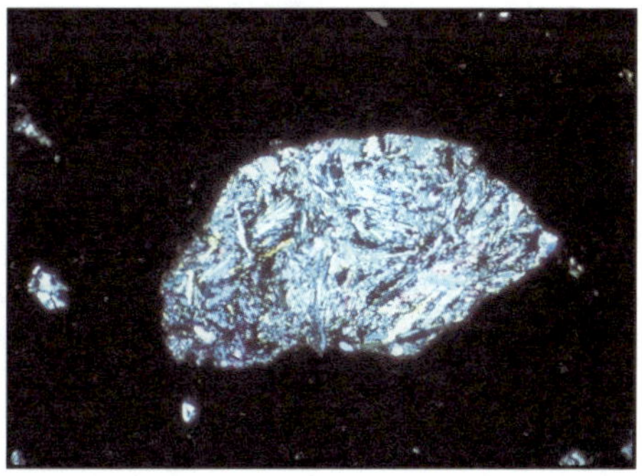

T

Plate 2 (on previous page spread). Photographs and photomicrographs of drill core samples from within the Chicxulub impact basin. **A,** Saw-cut surface of Y6-N13 suevite breccia. **B,** Saw-cut surface of Y6-N14 suevite breccia. **C,** Saw-cut surface of S1-N18 suevite breccia. **D,** Saw-cut surface of Y6-N19 melt matrix breccia. **E,** Saw-cut surface of C1-N10 melt rock. **F and G,** Cross-polarized photomicrographs of Y6-N11 showing shock-deformed quartz clasts with multiple sets of planar deformation features, glass shards, and bioclasts. Width of field is ~4 mm. **H–L,** Cross-polarized photomicrograph of shock-deformed basement clasts from samples Y6-N11, Y6-N14, and S1-N11, showing granitic mineral assemblage and, in some, a primary metamorphic foliation. Width of field is ~2 mm. **K and L,** Cross-polarized photomicrograph of shock-deformed alkali feldspar clasts from Y6-N14. **M,** Cross-polarized photomicrograph of shock-deformed quartz grain from Y6-N19 melt rock showing corona of augite. Width of field is ~1 mm **N,** Cross-polarized photomicrograph of quartz clast with augite corona from Y6-N19 showing ballen structures. These perliticfracturelike features result from recrystallization of diaplectic quartz glass. Width of field is ~2 mm. **O,** Cross-polarized photomicrograph of a shock-deformed quartz grain within Y5A-N4 sample of anhydrite-dolomite breccia (Bunte Breccia equivalent). Width of field is ~0.05 mm. **P,** Cross-polarized photomicrograph of a shock-deformed quartz grain from within Y2-N11 sample of anhydrite-dolomite breccia. Width of field is ~0.2 mm. **Q,** Cross-polarized photomicrograph of anhydrite-dolomite breccia sample Y1-N12 showing a shock-deformed quartz grain enveloped within a small pod of partially altered melt rock. Width of field is ~0.2 mm. **R,** Plane-polarized photomicrograph of mineral glass of alkali feldspar composition from Y6-N14 sample. Width of field is ~1.0 mm. **S,** Cross-polarized photomicrograph shows typical igneous texture of C1-N10 with microphenocrysts of plagioclase and augite in a potassium feldspar-rich mesostasis. Width of field is ~9 mm. **T,** Cross-polarized photomicrograph of a clast of anhydrite from Y2-N9 breccia sample. Width of field is ~9 mm.

ACKNOWLEDGMENTS

We thank Antonio Camargo, and Juan Manual Quesada (Pemex), and Gerardo Suárez (Universidad Nacional Autónoma de México) for their assistance and encouragement in this research project. Parts of this research were made possible through research grants provided by the National Aeronautics and Space Administration's Venus Data Analysis Program (to VLS) and Planetary Geology and Geophysics Program (to VLS) and by the National Science Foundation's Continental Dynamics Program. Bevan French and David Fastovsky are gratefully acknowledged for their detailed and constructive reviews, as are Bill Ward, Robbie Herrick, and Gene Perry for their helpful input. Sue Sharpton assisted in manuscript preparation, and Steve Hokanson and Debra Rueb assisted in preparation of artwork. The Lunar and Planetary Institute is operated by Universities Space Research Association under Contract NASW-4574 with NASA. This is Lunar and Planetary Institute Contribution Number 873.

REFERENCES CITED

Alexopoulos, J. S., and McKinnon, W. B., 1994, Large impact craters and basins on Venus, with implications for ring mechanics on the terrestrial planets, *in* Dressler, B. O., Grieve, R. A. F., and Sharpton, V. L., eds., Large meteorite impacts and planetary evolution: Geological Society of America Special Paper 293, p. 29–50.

Alvarez, L. W., Alvarez, W., Asaro, F., and Michel, H. V., 1980, Extraterrestrial cause for the Cretaceous-Tertiary extinction: Science, v. 208, p. 1095–1108.

Antoine, L. A. G., Nicolaysen, L. O., and Niccol, S. L., 1990, Processed and enhanced gravity and magnetic images over the Vredefort structure and their interpretation: Tectonophysics, v. 171, p. 63–74.

Baldwin, R. B., 1963, The measure of the Moon: Chicago, University of Chicago Press, 488 p.

Bishoff, L., and Oskierski, W., 1988, The surface structure of the Haughton impact crater, Devon Island, Canada: Meteoritics, v. 23, p. 209–220.

Bishoff, A., and Stöffler, D., 1981, Thermal metamorphism of feldspar clasts in impact melt rocks from Lappäjarvi crater, Finland [abs]: Lunar and Planetary Science Conference, 12th, Abstracts of Papers: Houston, Texas, Lunar and Planetary Institute, p. 77–79.

Blum, J. D., Chamberlain, C. P., Koeberl, C., Marín, L. E., Schuraytz, B. C., and Sharpton, V. L., 1993, Isotopic comparison of K/T boundary impact glass with melt rock from the Chicxulub and Manson impact structures: Nature, v. 364, p. 325–327.

Bohor, B. F., 1990, Shocked quartz and more: Impact signatures in Cretaceous/Tertiary boundary clays, *in* Sharpton, V. L., and Ward, P. D., eds., Global catastrophes in Earth history: An interdisciplinary conference on impacts, volcanism, and mass mortality: Geological Society of America Special Paper 247, p. 335–342.

Bohor, B. F., Foord, E. E., Modreski, P. J., and Triplehorn, D. M., 1984, Mineralogic evidence for an impact event at the Cretaceous-Tertiary boundary: Science, v. 224, p. 867–869.

Bourgeois, J., Hansen, T. A., Wiberg, P. L., and Kauffman, E. G., 1988, A tsunami deposit at the Cretaceous-Tertiary boundary in Texas: Science, v. 241, p. 567–570.

Byars, C., 1981, Mexican site may be link to dinosaurs disappearance: Houston Chronicle, December 13, p. 1, continued on p. 18.

Camargo, Z., A., and Suárez, R., G., 1994, Evidencia sismica del crater de impacto de Chicxulub [Seismic evidence of the Chicxulub Impact Crater]: Boletín de la Asociación Mexicana de Geofisicos de Exploracion, v. 34, p. 1–28 (in Spanish).

Carstens, H., 1975, Thermal history of impact melt rocks in the Fennoscandian shield: Contributions to Mineralogy and Petrology, v. 50, p. 145–155.

Cornejo-Toledo, A., and Hernandez-Osuna, A., 1950, Las anomalias gravimetricas en la cuenca salina del istmo, planicie costera de Tabasco, Campeche y Pensula de Yucatan: Boletín de la Asociación Mexicana de Geólogos Petroleros, v. 2, p. 453–460.

Croft, S. K., 1981a, The excavation stage of basin formation: A qualitative model, *in* Schultz, P. H., and Merrill, R. B., eds., Multi-ring basins: New York, Pergamon Press, p. 207–225.

Croft, S. K., 1981b, The modification stage of basin formation: Conditions of ring formation, *in* Schultz, P. H., and Merrill, R. B., eds., Multi-ring basins: New York, Pergamon Press, p. 227–257.

Dressler, B. O., 1982, Geology of the Wanapitei Lake area, District of Sudbury: Ontario Geological Survey report, v. 213, 131 p.

Engelhardt, W. v., 1972, Shock produced rock glasses from the Ries crater: Contributions to Mineralogy and Petrology, v. 36, p. 265–292.

French, B. M., and Nielsen, R. L., 1990, Vredefort Bronzite Granophyre:

Chemical evidence for origin as a meteorite impact melt: Tectonophysics, v. 171, p. 119–138.

Grieve, R. A. F., and Ber, T. J., 1994, Shocked lithologies at the Wanapitei impact structure, Ontario, Canada: Meteoritics, v. 29, p. 621–631.

Grieve, R. A. F., and Floran, R. J., 1978, Manicouagan impact melt, Quebec. 2: Chemical interrelations with basement and formational processes: Journal of Geophysical Research, v. 83, p. 2761–2771.

Grieve, R. A. F., Robertson, P. B., and Dence, M. R., 1981, Constraints on the formation of ring impact structures based on terrestrial data, *in* Schultz, P. H., and Merrill, R. B., eds., Multi-ring basins: Proceedings of Lunar and Planetary Science, Volume 12, Part A: New York, Pergamon, p. 37–57.

Grieve, R. A. F., Stöffler, D., and Deutsch, A., 1991, The Sudbury Structure: Controversial or misunderstood?: Journal of Geological Research, v. 96, p. 22, 753, 764.

Hart, R. J., Hargraves, R. B., Andreoli, M. A. G., Tredoux, M., and Doucouré, C. M., 1995, Magnetic anomaly near the center of the Vredefort structure: Implications for impact related magnetic signatures: Geology, v. 23, p. 277–280.

Head, J. W., 1977, Origin of outer rings in lunar multi-ringed basins: Evidence from morphology and ring spacing, *in* Roddy, D. J., Pepin, R. O., and Merrill, R. B., eds., Impact and explosion Cratering: New York, Pergamon Press, p. 563–573.

Hildebrand, A. R., 1992, Geochemistry and stratigraphy of the Cretaceous/Tertiary boundary impact ejecta [Ph.D. thesis]: Tucson, University of Arizona, 358 p.

Hildebrand, A. R., 1993, The Cretaceous/Tertiary boundary impact (or the dinosaurs didn't have a chance): Journal of the Royal Astronomical Society of Canada, v. 87, p. 77–118.

Hildebrand, A. R., and Boynton, W. W., 1990, Proximal Cretaceous-Tertiary boundary deposits in the Caribbean: Science, v. 248, p. 843–847.

Hildebrand, A. R., Penfield, G. T., Kring, D. A., Pilkington, M., Camargo, Z. A., Jacobsen, S., and Boynton, W. V., 1991, A possible Cretaceous-Tertiary boundary impact crater on the Yucatán peninsula, Mexico: Geology, v. 19, p. 867–871.

Hörz, F., and Banholzer, G. S., 1980, Deep-seated target materials in the continuous deposits of the Ries Crater, Germany, *in* Papike, J. J., and Merrill, R. B., eds., Proceedings of the Conference on Lunar Highlands Crust: New York, Pergamon Press, p. 211–231.

Hörz, F., Ostertag, R., and Rainey, D. A., 1983, Bunte Breccia of the Ries: Continuous deposits of large impact craters: Reviews of Geophysics and Space Physics, v. 21, p. 1667–1725.

Howard, K. A., Wilhelms, D. E., and Scott, D. H., 1974, Lunar basin formation and highland stratigraphy: Reviews of Geophysics and Space Physics, v. 12, p. 309–327.

Izett, G. A., 1990, The Cretaceous/Tertiary boundary interval, Raton Basin, Colorado and New Mexico, and its content of shock-metamorphosed minerals: Evidence relevant to the K-T boundary impact-extinction theory: Geological Society of America Special Paper 249, 100 p.

Izett, G. A., Maurrasse, F. J.-M. R., Lichte, F. E., Meeker, G. P., and Bates, R., 1990, Tektites in Cretaceous-Tertiary boundary rocks on Haiti: U.S. Geological Survey Open-File Report OF-90-635, 31 p.

Jansa, L. F., Pe-Piper, G., Robertson, P. B., and Friedenreich, O., 1989, Montagnais: A submarine impact structure on the Scotian Shelf, eastern Canada: Geological Society of America Bulletin, v. 101, p. 450–463.

Jansa, L. F., Aubry, M.-P., and Gradstein, F. M., 1990, Comets and extinctions; Cause and effect? *in* Sharpton, V. L., and Ward, P. D., eds., Global catastrophes in Earth history: An interdisciplinary conference on impacts, volcanism, and mass mortality: Geological Society of America Special Paper 247, p. 223–232.

Jung, K., Shaaf, H., and Kahle, H. G., 1969, Ergebnisse gravimetrischer Messungen im Ries [Results of gravimetric measurements in the Ries]: Geologica Bavarica, v. 61, p. 337–342 (in German).

Kamo, S. L., and Krogh, T. E., 1995, Chicxulub crater source for shocked zircon crystals from the Cretaceous-Tertiary boundary layer, Saskatchewan: Evidence from new U-Pb data: Geology, v. 23, p. 281–284.

Kieffer, S. W., and Simonds, C. H., 1980, The role of volatiles and lithology in the impact cratering process: Reviews of Geophysics and Space Physics, v. 18, p. 143–181.

Koeberl, C., Sharpton, V. L., Schuraytz, B. C., Shirey, S. B., Blum, J. D., and Marín, L. E., 1994, Evidence for a meteoritic component in impact melt rock from the Chicxulub structure: Geochimica et Cosmochimica Acta, v. 58, p. 1679–1684.

Kring, D. A., 1993, The Chicxulub impact event and possible causes of KT boundary extinctions, *in* Boaz, D., and Dornan, M., eds., Proceedings, Annual Symposium of Fossils of Arizona, 1st: Mesa, Arizona, Mesa Southwest Museum and Southwest Paleontological Society, p. 63–79.

Kring, D. A. and Boynton, W. V., 1992, Petrogenesis of an augite-bearing melt rock in the Chicxulub structure and its relationship to K/T impact spherules in Haiti: Nature, v. 358, p. 141–144.

Krogh, T. E., Kamo, S. L., Sharpton, V. L., Marín, L. E., and Hildebrand, A. R., 1993, U-Pb ages of single shocked zircons linking distal K/T ejecta to the Chicxulub crater: Nature, v. 366, p. 731–734.

Lopez-Ramos, E., 1983, Geologia de Mexico (third edition): Mexico City, Universidad Nacional Autónoma de México, p. 269–301 (in Spanish).

Marín, L. E., Sharpton, V. L., Urrutia-Fucugauchi, J., Sikora, P., and Carney, C., 1994, The "Upper Cretaceous Unit" in the Chicxulub multiring basin: New age based on planktic foraminiferal assemblage [abs.], *in* New developments regarding the KT event and other catastrophes in Earth history: Houston, Texas, Lunar and Planetary Institute Contribution 825, p. 77.

Marshall, R. G., 1974, Petrology of subsurface Mesozoic rocks of the Yucatan Platform, Mexico [M.S. thesis]: New Orleans, Louisiana, University of New Orleans, 97 p.

Masaitis, V. L., 1990, Impactites from Popigai Crater, *in* Dressler, B. O., Grieve, R. A. F., and Sharpton, V. L., eds., Large meteorite impacts and planetary evolution: Geological Society of America Special Paper 293, p. 153–162.

Masaitis, V. L., and five others, 1980, The geology of astroblems: Leningrad, Nedra, 231 p. (in Russian).

McIntyre, D. B., 1968, Impact metamorphism at Clearwater Lake, Quebec, *in* French, B. M., and Short, N. M., eds., Shock metamorphism of natural materials: Baltimore, Maryland, Mono Book, p. 363–366.

Melosh, H. J., 1989, Impact cratering: A geological process: New York, Oxford University Press, 245 p.

Melosh, H. J., and McKinnon, W. B., 1978, The mechanics of ringed basin formation: Geophysical Research Letters, v. 5, p. 985–988.

Newsom, H. E., Graup, G., Iseri, D. A., Geissman, J. W., and Keil, K., 1990, The formation of the Ries Crater, West Germany: Evidence of atmospheric interactions during a large cratering event, *in* Sharpton, V. L., and Ward, P. D., eds., Global catastrophes in Earth history: An interdisciplinary conference on impacts, volcanism, and mass mortality: Geological Society of America Special Paper 247, p. 195–206.

Nicolaysen, L. O., 1990, the Vredefort structure: An introduction and a guide to recent literature: Tectonophysics, v. 171, p. 1–6.

Penfield, G. T., and Camargo Z., A., 1981, Definition of a major igneous zone in the central Yucatán platform with aeromagnetics and gravity, *in* Technical Program, Abstracts and Bibliographies, Society of Exploration Geophysicists, 51st Annual Meeting, Tulsa, Oklahoma, Society of Exploration Geophysicists, p. 37.

Phinney, W. C., Simonds, C. H., Cochran, A., and McGee, P. E., 1978, West Clearwater Quebec, impact structure: Part II: Petrology: Proceedings, Lunar and Planetary Science Conference, 9th, Volume 2: New York, Pergamon Press, p. 2659–2693.

Pike, R. J., and Spudis, P. D., 1987, Basin-ring spacing on the Moon, Mercury, and Mars: Earth, Moon, and Planets, v. 39, p. 129–194.

Pilkington, M., and Grieve, R. A. F., 1992, The geophysical signature of terrestrial impact craters: Reviews of Geophysics, v. 30, p. 161–181.

Pilkington, M., Hildebrand, A. R., and Ortiz-Aleman, C., 1994, Gravity and magnetic field modeling and structure of the Chicxulub Crater, Mexico:

Journal of Geophysical Research, v. 99, p. 13147–13162.

Pohl, J., Stöffler, D., Gall, H., and Ernstson, K., 1977, The Ries impact crater, *in* Roddy, D. J., Pepin, R. O., and Merrill, R. B., eds., Impact and explosion cratering: New York, Pergamon Press, p. 343–404.

Pohl, J., Eckstaller, A., and Robertson, P. B., 1988, Gravity and magnetic investigations in the Haughton impact structure, Devon Island, Canada: Meteoritics, v. 23, p. 235–238.

Pope, K. O., Ocampo, A. C., and Duller, C. E., 1991, Mexican site for the K/T Crater?: Nature, v. 351, p. 105.

Pope, K. O., Ocampo, A. C., and Duller, C. E., 1993, Surficial geology of the Chicxulub impact crater Yucatán, México: Earth, Moon and Planets, v. 63, p. 93–104.

Redeker, H.-J., and Stöffler, D., 1988, The allochthonous polymict breccia layer of the Haughton impact crater, Devon Island, Canada: Meteoritics, v. 23, p. 185–196.

Roest, W., and Pilkington, M., 1994, Restoring post-impact deformation at Sudbury: A circular argument: Geophysical Research Letters, v. 21, p. 959–962.

Schaber, G. G., and nine others, 1992, Geology and distribution of impact craters on Venus: What are they telling us?: Journal of Geophysical Research, v. 97, p. 13257–13301.

Schmidt, R. M., and Housen, K. R., 1987, Some recent advances in scaling of impact and explosion cratering: International Journal of Impact Engineering, v. 5, p. 543–560.

Schultz, P. H., 1976, Moon morphology: Austin, University of Texas Press, 926 p.

Schuraytz, B. C., and Sharpton, V. L., 1993, Chicxulub—K/T melt complexities: Nature, v. 362, p. 503–504.

Schuraytz, B. C., and Sharpton, V. L., 1994, Siderophile-element distribution in Chicxulub melt rocks: Forensic chemistry on the KT smoking gun, *in* New developments regarding the KT event and other catastrophes in Earth history [abs.]: Houston, Texas, Lunar and Planetary Institute Contribution 825, p. 106–108.

Schuraytz, B. C., Sharpton, V. L., and Marín, L. E., 1994, Petrology of impact-melt rocks at the Chicxulub multiring basin, Yucatán, Mexico: Geology, v. 22, p. 868–872.

Scott, D., and Hajnal A., 1988, Seismic signature of the Haughton structure: Meteoritics, v. 23, p. 239–247.

Sharpton, V. L., and Ward, P. D., eds., 1990, Global catastrophes in Earth history: An interdisciplinary conference on impacts, volcanism, and mass mortality: Geological Society of America Special Paper 247, 631 p.

Sharpton, V. L., Schuraytz, B. C., Burke, K., Murali, A. V., and Ryder, G., 1990a, Detritus in K/T boundary clays of western North America: Evidence against a single oceanic impact, *in* Sharpton, V. L., and Ward, P. D., eds., Global catastrophes in Earth history: An interdisciplinary conference on impacts, volcanism, and mass mortality: Geological Society of America Special Paper 247, p. 349–359.

Sharpton, V. L., Schuraytz, B. C., and Jones, J., 1990b, Arguments favoring a single continental impact at the KT boundary: Meteoritics, v. 25, p. 408–409.

Sharpton, V. L., and five others, 1992, New links between the Chicxulub impact structure and the Cretaceous/Tertiary boundary, Nature, v. 359, p. 819–821.

Sharpton, V. L., and 9 others, 1993, Chicxulub multi-ring impact basin: Size and other characteristics derived from gravity analysis: Science, v. 261, p. 1564–1567.

Sharpton, V. L., Marín, L. E., and Schuraytz, B. C., 1994a, The Chicxulub multiring impact basin: Evaluation of geophysical data, well logs, and drill core samples [abs.], *in* New developments regarding the KT event and other catastrophes in Earth history: Houston, Texas, Lunar and Planetary Institute Contribution 825, p. 108–110.

Sharpton, V. L., Marín, L. E., and Schuraytz, B. C., 1994b, Constraints on excavation and mixing during the Chicxulub impact event [abs.], *in* 25th Lunar and Planetary Science Conference: Houston, Texas, Lunar and Planetary Institute, p. 1255–1256.

Shoemaker, E. M., Wolfe, R. F., and Shoemaker, E. S., 1990, Asteroid and comet flux in the neighborhood of Earth, *in* Sharpton, V. L., and Ward, P. D., eds., Global catastrophes in Earth history: An interdisciplinary conference on impacts, volcanism, and mass mortality: Geological Society of America Special Paper 247, p. 155–170.

Simonds, C. H., Phinney, W. C., McGee, P. E., and Cochran, A., 1978, West Clearwater, Quebec impact structure, Part I: Field geology, structure and bulk chemistry: Proceedings, Lunar and Planetary Science Conference, 9th, Volume 2: New York, Pergamon Press, p. 2633–2658.

Sky and Telescope, 1982, Possible Yucatan impact basin: Sky and Telescope, v. 63, p. 249–250.

Spudis, P. D., 1993, The geology of multi-ring impact basins: Cambridge, Cambridge University Press, 263 p.

Spudis, P. D., Reisse, R. A., and Gillis J. J., 1994, Ancient multiring basins on the Moon revealed by Clementine laser altimetry: Science, v. 266, p. 1848–1851.

Stepto, D., 1990, The geology and gravity field in the central core of the Vredefort structure: Tectonophysics, v. 171, p. 75–103.

Stöffler, D., 1977, Research drilling Nördlingen 1973: Polymict breccias, crater basement, and cratering model of the Ries impact structure: Geologica Bavarica, v. 75, p. 443–458.

Stöffler, D., and Langenhorst, F., 1994, Shock metamorphism of quartz in nature and experiment. I. Basic observation and theory: Meteoritics, v. 29, p. 155–181.

Stöffler, D., Ewald, U., Ostertag, R., and Reimold, W. U., 1977, Ries deep drilling. I: Composition and texture of polymict breccias: Geologica Bavarica, v. 75, p. 163–189.

Swisher, C. C., III, and 11 others, 1992, Coeval $^{40}Ar/^{39}Ar$ ages of 65.0 million years ago from Chicxulub melt rock and Cretaceous-Tertiary boundary tektites: Science, v. 257, p. 954–958.

Urrutia-Fucugauchi, J., Marín, L., and Sharpton, V. L., 1994, Reverse polarity magnetized melt rocks from the Cretaceous/Tertiary Chicxulub structure, Yucatan peninsula, Mexico: Tectonophysics, v. 237, p. 105–112.

Vickery, A. M., and Melosh, H. J., 1990, Atmospheric erosion and impactor retention in large impacts, with application to mass extinctions, *in* Sharpton, V. L., and Ward, P. D., eds., Global catastrophes in Earth history: An interdisciplinary conference on impacts, volcanism, and mass mortality: Geological Society of America Special Paper 247, p. 289–300.

Viniegra O., F., 1981, Great carbonate bank of Yucatan, southern Mexico: Journal of Petroleum Geology v. 3, p. 247–278.

Ward, W. C., Weidie, A.E., and Back, W., 1985, Geology and hydrogeology of the Yucatan and Quaternary geology of northeastern Yucatan Peninsula: New Orleans, New Orleans Geological Society Publications, 160 p.

Ward, W. C., Keller, G., Stinnesbeck, W, and Adatte, T., 1995, Yucatan subsurface stratigraphy: Implications and constraints for the Chicxulub impact: Geology, v. 23, 873–876.

Weidie, A. E., 1985, Geology of Yucatan Platform, *in* Ward, W. C., Weidie, A. E., and Back, W., eds, Geology and hydrogeology of the Yucatan and Quaternary geology of northeastern Yucatan Peninsula: New Orleans, New Orleans Geological Society Publications, p 1–19.

Weidie, A. E., Ward, W. C, and Marshall, R. M., 1978, Geology of Yucatan Platform, *in* Ward, W. C., and Weidie, A. E., eds., Geology and hydrogeology of northeastern Yucatan: New Orleans, New Orleans Geological Society Publications, p. 3–29.

Weissman, P. R., 1988, Terrestrial impact rates for long and short-period comets, *in* Silver, L. T., and Schultz, P. H., eds., Geological implications of impacts of large asteroids and comets on the Earth: Geological Society of America Special Paper 190, p. 15–24.

Wilhelms, D. E., 1987, The geologic history of the Moon: U.S. Geological Survey Professional Paper 1348, 328 p.

Zuber, M. T., Smith, D. E., Lemoine, F. G., and Neumann, G. A., 1994, The shape and internal structure of the Moon from the Clementine Mission: Science, v. 266, p. 1839–1843.

Manuscript Accepted by the Society September 14, 1995

Printed in U.S.A.

Geological Society of America
Special Paper 307
1996

Ejecta blanket deposits of the Chicxulub crater from Albion Island, Belize

Adriana C. Ocampo
Jet Propulsion Laboratory, California Institute of Technology, MS183-601, 4800 Oak Grove Drive, Pasadena, California 91109
Kevin O. Pope
Geo Eco Arc Research, 2222 Foothill Boulevard, Suite E-272, La Cañada, California 91011
Alfred G. Fischer
Department of Earth Sciences, University of Southern California, University Park, Los Angeles, California 90089

ABSTRACT

An extraordinary deposit disconformably overlies Late Cretaceous dolomite on Albion Island in northern Belize. The basal unit is a 1-m-thick bed of clay-rich dolomite and calcite silt containing abundant 1- to 20-mm-diameter spheroids of dolomite and spheroidal to angular clasts of clay containing relict vesicles. Both the dolomite and clay spheroids are flattened near the basal contact. This spheroid bed is overlain by 15 m of diamictite that contains millimeter- to meter-sized clasts (maximum 7.5 m) floating in a matrix of dolomite silt. The clasts are mostly dolomite but include limestones with Early to Middle Cretaceous rudists and foraminifera. Clasts are commonly fractured and range from well-rounded pebbles and cobbles to highly angular flakes. Polished facets and multiple sets of striations on single clasts indicate severe abrasion. The diamictite bed also contains abundant clay clasts with relict vesicles as well as mud balls, pockets of calcite pisoids, and authigenic quartz. Metamorphic rocks and detrital quartz are restricted to rare sand-size chips found in insoluble residues.

We propose that the Albion Island deposits are the most proximal, exposed section of ejecta from the Chicxulub crater yet found. The Albion spheroid bed is similar to spherule beds described from elsewhere in the Gulf Coast and Caribbean region, but the dolomite spheroids are thus far unique and may be the tektite-equivalent of vaporized carbonates. The only correlate for the Albion diamictite is the coarse ejecta unit drilled on the rim of the Chicxulub crater. The Albion diamictite shares many characteristics with the Bunte breccia of the Ries crater. We propose that the spheroid bed was deposited by the vapor plume and the diamictite by ballistic sedimentation coupled with debris flow processes.

INTRODUCTION

An iridium-rich ejecta deposit at the Cretaceous/Tertiary (K/T) boundary lead Alvarez et al. (1980) to propose that the impact of a comet or asteroid 65 m.y. ago caused the mass extinction recorded at this geological boundary. Subsequent studies throughout the world have identified impact ejecta deposits at the K/T boundary. The recent recognition that the proposed K/T impact formed the Chicxulub crater in the northern Yucatán Peninsula (Hildebrand et al., 1991; Pope et al., 1991) has focused research on the Gulf of Mexico–Caribbean region. It is in this region that the ejecta deposits are thickest and K/T tektites are largest, and it is here that impact-wave deposits have been inferred (e.g., Smit and Romein, 1985;

Ocampo, A. C., Pope, K. O., and Fischer, A. G., 1996, Ejecta blanket deposits of the Chicxulub crater from Albion Island, Belize, *in* Ryder, G., Fastovsky, D., and Gartner, S., eds., The Cretaceous-Tertiary Event and Other Catastrophes in Earth History: Boulder, Colorado, Geological Society of America Special Paper 307.

Bourgeois et al., 1988; Hildebrand and Boynton 1990; Izett, 1991; Maurrasse and Sen, 1991; Alvarez et al., 1992; Smit et al., 1992; Pitakpaivan et al., 1994).

In this chapter we report findings from a K/T boundary section located in a quarry on Albion Island, Belize, which at only 364 km from the center of the Chicxulub Crater (Fig. 1) is the most proximal ejecta exposure yet known. The Albion Island section provides an example of deposits intermediate between the crater rim (known only from drilling) and more distal ejecta, slumps, and impact-wave deposits.

REGIONAL GEOLOGY AND STRATIGRAPHY OF ALBION ISLAND

The geology of northern Belize is not well known because there is a paucity of good outcrops in this tropical, low-relief region and because weathering, solution collapse, and recrystallization of the carbonates and evaporites have destroyed much of the original textures, bedding, and fossils. Quarrying activity on Albion Island provides one of the best bedrock exposures in this region. The island was formed by a bifurcation in the Hondo River around a resistant carbonate block with local relief of about 60 m. The provisional geologic map of Belize (Fig. 2) identifies all of the carbonate rocks in this area as belonging to the Miocene to Pleistocene Orange Walk Group (Cornec, 1986), based primarily upon the work of Flores (1952). Uplifted blocks of Upper Cretaceous carbonates of the Barton Creek Formation are mapped 30 to 50 km south of Albion Island, along a series of northeast-southwest trending faults of the Río Hondo fault zone. Albion Island is located along one of these faults (Fig. 2). As we shall demonstrate, the island is also one of these uplifted blocks of Upper Cretaceous carbonate.

The regional stratigraphy (Fig. 3) has been established in a series of exploratory oil wells based upon the analysis of foraminifera and correlation of lithological units (Geology and Petroleum Office, 1986). The Remate well records place the K/T boundary at a depth of 375 m within dolomites equivalent in age to Barton Creek Formation (Fig. 3). The K/T boundary rises to a depth of about 190 m in the San Pablo 1 well, which is located 32 km southwest of Remate and 14 km northeast of Albion Island. The well at Blue Creek, 36 km southwest of Albion Island, encountered Cretaceous Barton Creek Formation at the surface, which contradicts the provisional geologic map (Fig. 2). The nature of the K/T boundary in these wells is uncertain, and some sections may not be complete. Micropaleontological studies of the Remate well suggest that the earliest Paleocene may be absent (Ramanathan, 1985), although no lithologic break is found. The Río Hondo fault zone appears to be a series of en

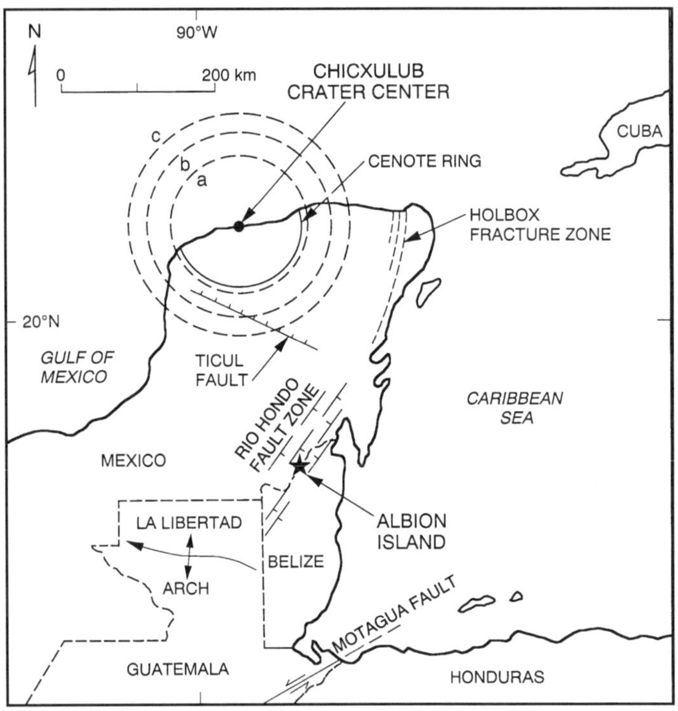

Figure 1. Map showing the location of Albion Island, Belize, with respect to the buried Chicxulub impact crater and other geological structures of the region. Dashed circles represent proposed crater diameters: (a) 180 km (Hildebrand et al., 1991; Pilkington et al., 1994), (b) 240 km (Pope et al., 1993), and (c) 300 km (Sharpton et al., 1993).

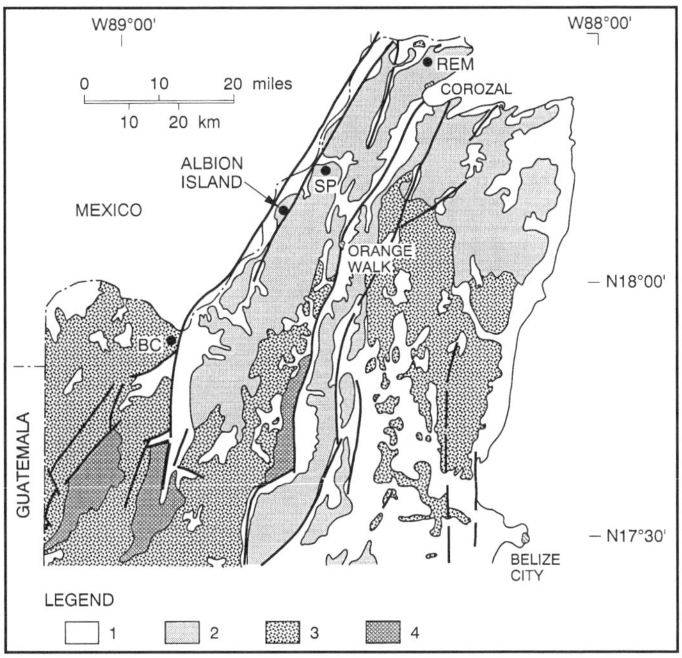

Figure 2. Geological map of northern Belize after Cornec (1986). Black dots mark exploratory oil wells and the Albion Island quarry shown in the cross section in Figure 3: REM—Remate; SP—San Pablo; BC—Blue Creek. Bold lines are faults. Key: (1) Quaternary carbonates and alluvium; (2) Miocene-Pleistocene carbonates and evaporites of the Orange Walk group; (3) Paleocene-Eocene carbonates of the Doubloon Bank group; (4) Upper Cretaceous carbonates and evaporites of the Barton Creek Formation.

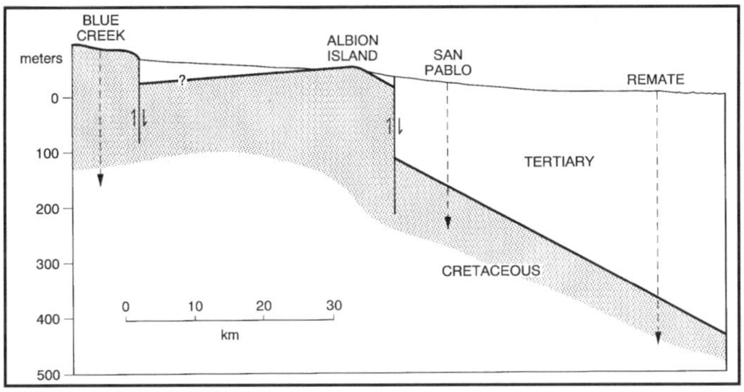

Figure 3. Geological cross section of northern Belize showing the location of the K/T boundary (bold line). Location of exploratory oil wells Remate, San Pablo, and Blue Creek is shown in Figure 2.

echelon faults with progressive uplift to the south and west; however, when this uplift occurred is unknown. Albion Island is the northernmost block with any significant relief.

The quarry on Albion Island exposes a 42-m-thick section (Figs. 4 and 5). A sharp disconformity separates the hard crystalline dolomites being quarried (Barton Creek Formation) from overlying clay-rich spheroid and coarse diamictite beds, here termed the Albion Formation. We propose that this disconformity marks the K/T boundary.

BARTON CREEK DOLOMITE

The basal portion of the Albion quarry section is composed of massive and thinly bedded, medium-crystalline dolomites, pale gray buff in color, such as is common through much of the Cretaceous of the region. Dark-gray laminations centimeters to millimeters thick are common, especially near the top. Thin sections indicate that the dark colors are due to organic matter and are associated with benthic foraminifera that are too poorly preserved for identification. The dolomites have been heavily recrystallized. Solution features including minor collapse breccias and rare clay layers are present. These rocks are typical of shallow and restricted carbonate platform deposits of the region.

We separated the dolomite from calcite in a foraminifera-rich dolomite sample from just below the proposed K/T boundary with dilute HCl for strontium isotope analysis. We conclude that the $^{87}Sr/^{86}Sr$ ratios reflect seawater compositions at the time of crystallization because the Sr and Rb concentrations are within the typical range for seawater (97.61 and 118.30 ppm for the Sr residue and leach and 0.2248 and 0.3490 ppm for the Rb residue and leach). The $^{87}Sr/^{86}Sr$ ratios from dolomite residue and calcite leach are 0.70796 and 0.70786 respectively, which suggests a terminal Cretaceous age, based upon comparisons with dated Sr isotope ratios (Fig. 6).

The dolomites dip slightly away from the center of the quarry and are much fractured. Minor faults with displacement of up to 1 m are spaced at about 5- to 10-m intervals. These faults do not cut the overlying strata but create a wavy contact

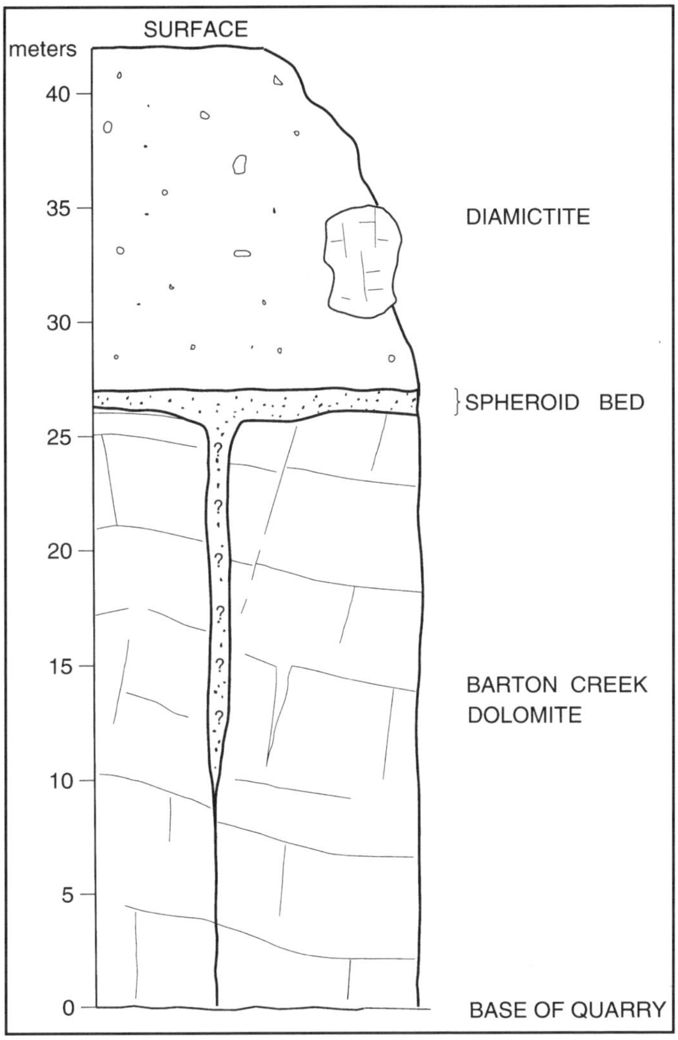

Figure 4. Geological cross section from the Albion Island quarry showing the three principal stratigraphic units: the Barton Creek dolomite and the Albion spheroid bed and diamictite.

between the dolomite and spheroid bed (Fig. 5). The deformation of the dolomites thus occurred prior to the deposition of the spheroid bed, which filled in depressions. The spheroids also occur in fractures up to 1 m wide to a depth of about 13 m (Fig. 4 and 5), and some fractures may have been open at the time of spheroid bed deposition.

THE SPHEROID BED

The spheroid bed is an informal name given to the 90- to 150-cm-thick member that lies between the Barton Creek dolomite and the overlying coarse Albion diamictite. The spheroid bed is heterogeneous, and although it contains rounded and angular clasts, its distinctive feature is the abundance of millimeter- to centimeter-size spheroidal bodies of dolomite and clay that we refer to as spheroids (Figs. 7 through 10). These spheroids are matrix supported in a weakly consolidated, clay-rich, fine-grained (10 to 40 μm) dolomite and calcite silt matrix.

Figure 5. Albion Island quarry. Note the disconformity between the Barton Creek dolomite and overlying Albion Formation that extends around the quarry. Also note the abundant vertical fractures in the dolomite and the wavy appearance of the contact between the Barton Creek dolomite and spheroid bed.

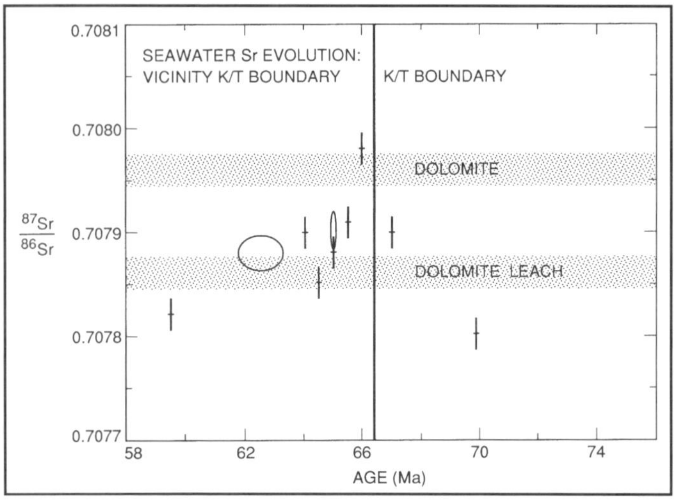

Figure 6. Results of strontium isotope analyses of the Barton Creek dolomite. Horizontal stippled bands give the $^{87}Sr/^{86}Sr$ values (1 sigma) for the whole dolomite and the dolomite leach (calcite removed with weak HCl). Sample is from a foraminifera-rich sample collected 1 m below the base of the spheroid bed. Bars (Hess et al., 1986) and ellipses (DePaolo and Ingram, 1985) give the global trend in $^{87}Sr/^{86}Sr$ values about the K/T boundary (vertical line), set by convention to 66.5 Ma.

Figure 7. Spheroid bed outcrop. Note large dolomite spheroid (white) near center of photograph. Dark clasts are clay. Hammer handle width is 2 cm.

Detailed examination of four spheroid bed profiles indicates that as many as four layers are present. They are differentiated by varying amounts of clay and calcite and by mottled colors ranging from orange to pink to white. Boundaries between these layers are gradational. Field observations indicate that the percentages of clay and spheroids are variable but never exceed 50% of the bed (by volume). Clay-rich layers contain sheer planes with slickensides.

Dolomite spheroids with a modal diameter of 2 to 4 mm compose about 20 to 30% (by volume) of the bed (Fig. 8). Shapes range from spherical, to oblate spheroids, to triaxial ellipsoids, but include lumpy and subangular forms. Oblate shapes are more common near the base, where elongate forms lie roughly parallel to the basal surface (Fig. 8) and the spheroids appear to be squashed. The spheroids are finer grained than the matrix (Fig. 9). Large dolomite spheroids, 1 to 2 cm in diameter, compose about 1% of the bed. Some of these show concentric alternations of fine and coarse grain sizes or a coarser rind. The

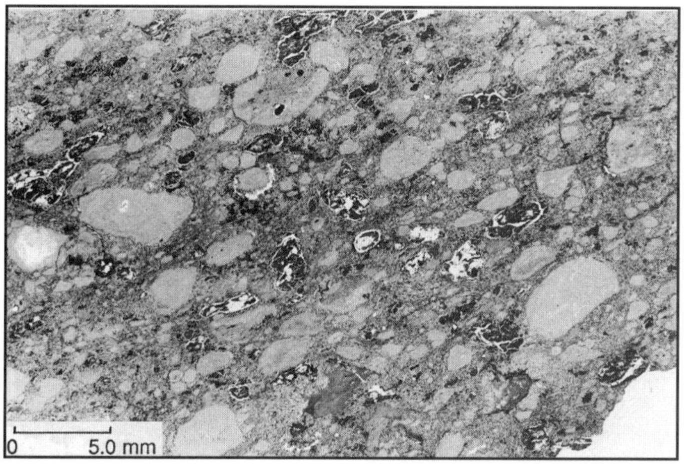

Figure 8. Thin section (natural light) of spheroid bed near base. Light-gray angular to spheroidal clasts are dolomite. Dark clasts with voids (white) are clay. Note alignment of clasts parallel to bedding (angled lower left to upper right in thin section) and apparent flattening of spheroids (oblate shapes).

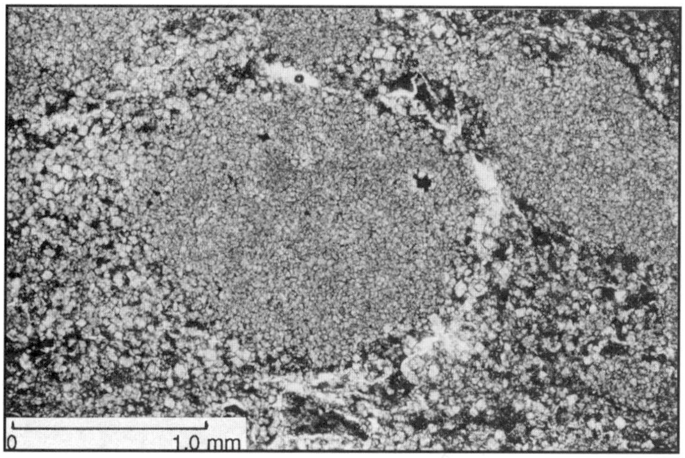

Figure 9. Thin section (natural light) of dolomite spheroid. Note that the spheroid is finer grained than the silt matrix.

upper portion of the spheroid bed contains about 1% dark-gray rounded to highly angular, fine-grained, 1- to 3-mm-diameter dolomite clasts, some of which are plastically deformed. These are absent from the lower portion of the bed.

The clay clasts also occur as spheroids (Fig. 10), but they are more commonly angular or lenticular bodies. Clay clasts are green, orange, and red in outcrop, with the variations in color often being a product of alteration (many green clasts have red or orange rims). The basal layer in all four profiles examined contains abundant, shiny, translucent green, flattened oblate, clay clasts. These clasts, like the dolomite spheroids, also appear to be squashed. At the basal contact, spheroids surround lumps of the underlying dolomite and appear "welded" to the dolomite surface.

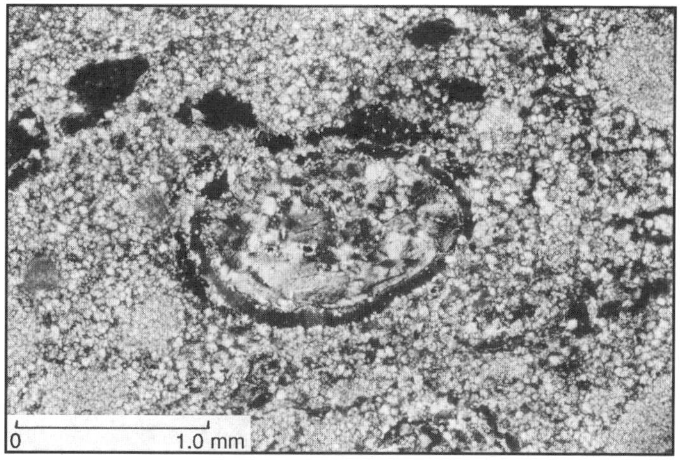

Figure 10. Thin section (crossed nicols) of clay spheroid from spheroid bed. Note void (black) around the spheroid, which is typical. Dark patches in center are nearly isotropic and transparent under natural light and may be partially devitrified glass.

THE DIAMICTITE

The upper member of the Albion Formation is a coarse diamictite containing rounded to highly angular carbonate clasts that range in size from less than 1 mm to about 7.5 m. The minimum thickness of this unit is 15 m, but since it extends to the surface we do not know the original thickness. The entire unit is matrix supported by weakly consolidated (except at the base) micritic dolomite silt in the 10-to 60-μm range. There is no clear indication of sorting, stratification, or grading. The basal contact is characterized by an abrupt occurrence of large clasts within a lithified matrix. Although the abrupt change in grain size and lithification gives the boundary a sharp appearance, in several locations there appears to be little change in the matrix composition across the boundary, and dolomite spheroids are present in the diamictite.

Clasts are commonly fractured, and those greater than 30 cm are always fractured (Fig. 11). The largest clast (7.5 m) found in the quarry has a grooved facet (Fig. 12). Many smaller clasts were found with polish and striations (Fig. 13), the latter in some cases in multiple directions on the same clast. We found one 3-m clast with a 7-cm-thick coating of dolomite mud (Fig. 11). Many of the smaller dolomite clasts, including those studied in thin section, also contain dolomite coatings or rinds. A cluster of rounded lumps of fine-grained dolomite ranging in size from 15 to 60 cm was also found (Fig. 14). These lumps have a few dolomite clast inclusions and are very similar to the mud balls reported from alluvial fan mudflow (debris flow) deposits (e.g., Bull, 1964). In one location we found a 2 m × 1 m exposure of a pocket of radially fibrous calcite spheres (pisoids) (Fig. 15). Most of these pisoids are about 1 cm in diameter, but a few are smaller and form amalgamated masses.

Thin sections (Fig. 16) of the fine fraction of the diamictite demonstrate that many of the observations made at the outcrop

Figure 11. Edge of 3-m-diameter dolomite clast. Note extensive fracturing and 7-cm-thick mud coating. Albion diamictite. Hammer 30 cm long.

Figure 12. Right half of 7.5-m-diameter dolomite clast (outlined in black for clarity) showing faceted and grooved surface on upper right of clast. Albion diamictite.

also occur at the microscopic level. Millimeter-sized clasts are also poorly sorted, rounded to angular, and matrix supported and include breccia clasts and dolomite clasts with rinds of coarser dolomite (Figs. 16 and 17). There is commonly a void between clasts and the surrounding matrix (Fig. 16), which in some cases retains gypsum filling. Some clasts contain dark, presumably carbonaceous, cores that are rounded and sharply bounded and may be replacement features, although similar dark dolomites occur as individual spheroids (Figs. 16 and 17). Thin-section analysis of one of the mud balls noted above shows that it is composed of fine-grained euhedral dolomite and an unidentified silicate, possibly a member of the palygorskite group, along with sparse authigenic quartz.

All of the medium to large (greater than 15 cm) clasts we examined in the field are a dolomite that is not notably different from the underlying dolomite in the quarry. Thin sections, however, reveal minor components of clearly exotic carbonate clasts that have sedimentary textures or fossil preservation markedly different from the underlying Barton Creek dolomite (Fig. 18). One such clast of limestone contains rudist (*Sellaea* sp.) fragments of Barremian to Albian age (Erle Kauffman, personal communication, 1994). Another contains foraminifera (*Nezzatinella piccardi*) of Albian to early Campanian age (Richard Boettcher, personal communication, 1993). These limestones must have come from more than 1 km below the Cretaceous surface of the Yucatán platform and are exposed nowhere near this region. Similar limestones are reported from the deep drilling near the Chicxulub crater (Marshall, 1974; Weidie, 1985), and we interpret these clasts as primary ejecta from the Chicxulub crater.

Authigenic quartz crystals up to 4 cm long and commonly intergrown with carbonate are found in the diamictite. Acid leach residues of the diamictite matrix yielded an authigenic quartz grain with serpentine inclusions (Fig. 19) and two granitic clasts (Fig. 20). Other quartz clasts contain Class 1 deformation features (Robertson et al., 1968), such as irregular planes of inclusions and undulatory extinction, but we observed no definitive evidence of shock metamorphism.

The most abundant clast type that we interpret as exotic is

Figure 13. Striated dolomite clasts. Albion diamictite.

Figure 14. Dolomite mud balls. Albion diamictite. Hammer 30 cm long.

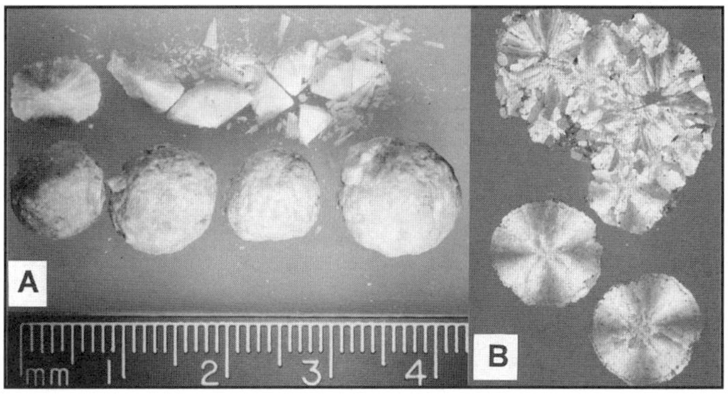

Figure 15. Calcite pisoids. A, whole and fractured pisoids showing fibrous radial structure. B, thin section (crossed nicols) showing radial structure of single pisoids and amalgamated clusters. Albion diamictite. The origin of the pisoids is uncertain, but they may be the product of hydrothermal activity (see Addendum).

clay. Clay clasts compose 1 to 10% of the diamictite by volume as measured in thin section. A 10% clay abundance by weight was measured from the acid leach of a kilogram sample. Nontronite, a common alteration product of glass, was identified with X-ray diffraction. Thin sections show the presence of a variety of clay minerals, of which smectites appear most abundant. We interpret these clay clasts as altered glass. Many of the clay clasts have vesicular structure, and a few retain the optical properties of glass (Figs. 21, 22, and 23). Although exotic clasts are critical to our interpretation of the diamictite unit, they may comprise no more than 10% of the unit, depending upon the source of the large population of dolomite clasts.

The weakly consolidated dolomite silt matrix in the diamictite and spheroid bed presents a special problem. The grains are individual rhombs: some have sharp corners; others are rounded. Such dolomite silts have been reported from the deep-sea floor and occur locally on carbonate platforms as solution residues of dolomitic limestones (Fischer, 1953), but no other large-scale occurrence is known to us.

COMPARISONS AND CORRELATIONS

We propose that the disconformity between the Barton Creek dolomite and spheroid bed represents the K/T boundary. Our placement of the boundary is based upon our interpretation of the Albion spheroid bed and diamictite as Chicxulub impact ejecta, since no Tertiary sediments are exposed at the quarry. We correlate the spheroid bed with the spherule beds reported from K/T sections elsewhere in the Gulf Coast and Caribbean region (e.g., Hildebrand and Boynton, 1990; Izett, 1991; Maurrasse and Sen, 1991; Smit et al., 1992, 1994; Pitakpaivan et al., 1994). We chose the name spheroid bed to emphasize that it is not identical to the previously described spherule beds. We correlate the Albion diamictite with the coarse detrital unit interpreted as the "continuous deposit" of ejecta and scoured surface debris found on the rim of the Chicxulub crater (Pope et al., 1993).

The closest comparison for the Albion Island spheroid bed is the spherule bed described from Mimbral (Smit et al., 1992). The Mimbral bed varies from 0 to 100 cm thick and is composed of 20% to more than 90% calcite and chlorite-smectite spherules with relict vesicles, along with subangular to subrounded limestone clasts similar in size to the spherules. Some of these clasts are plastically deformed. Most of the spherules at Mimbral range from 1 to 5 mm. Even though all of these characteristics have parallels in the Albion Island spheroid bed, there is a notable absence of dolomite at Mimbral. The Mimbral bed apparently does not contain the large centimeter-sized spheroids found at Albion Island; however, centimeter-sized spherule fragments are reported from Haiti (Hildebrand and Boynton, 1990; Hildebrand, 1993). The fabric of the Albion spheroid bed is similar to the imbricated fabric reported for the melt ejecta layer in K/T boundary deposits in the western interior of North America (Pollastro and Bohor, 1993), although the Albion fabric appears to be mostly the result of diagenic flattening.

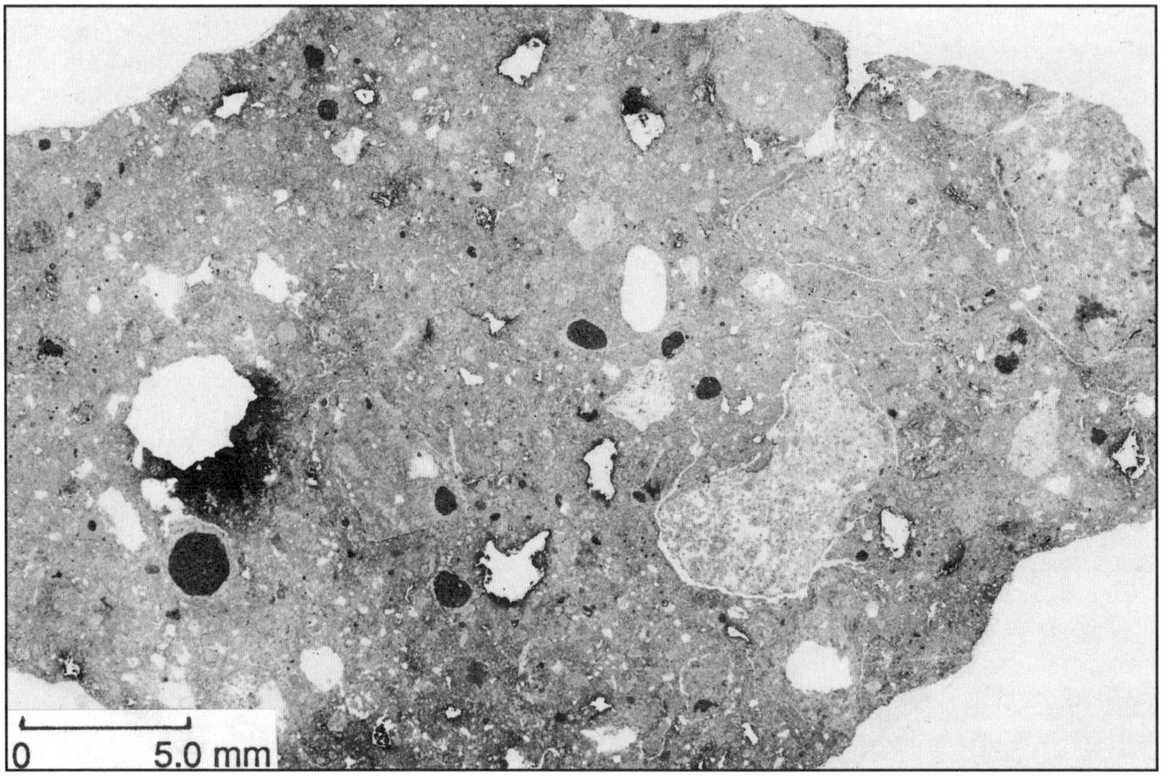

Figure 16. Thin section (natural light) of diamictite sample. Note variety of clasts, several with voids (thin white bands) around clasts, and dark dolomite spheroids (also shown in Fig. 17). These dark spheroids also occur within clasts and may be secondary replacements.

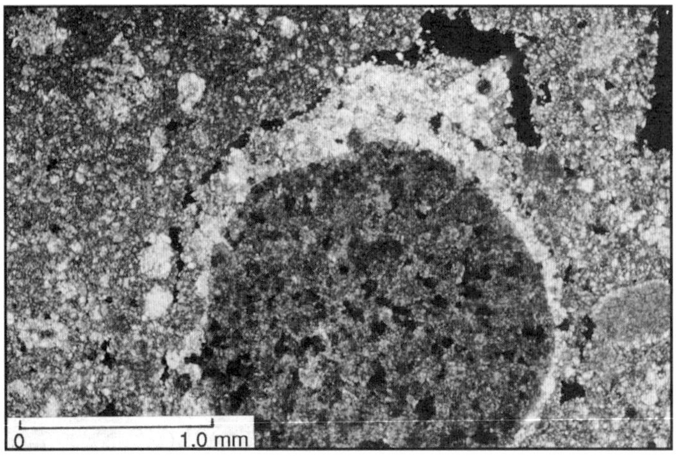

Figure 17. Thin section (crossed nicols) of dark, carbonaceous (?) dolomite spheroid with dolomite rind. Albion diamictite.

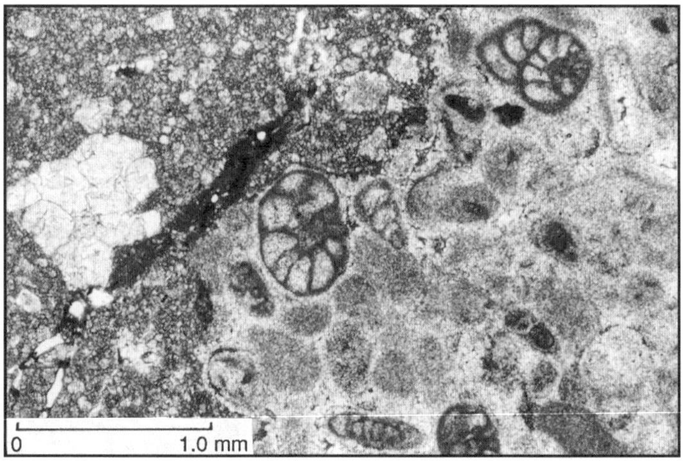

Figure 18. Thin section (natural light) of fossiliferous limestone clast. Albion diamictite.

The clay spherules from the Gulf Coast and Caribbean sites are interpreted as altered tektites (Hildebrand and Boynton, 1990; Izett, 1991; Maurrasse and Sen, 1991; Smit et al., 1992; Pitakpaivan et al., 1994), but spherical clay clasts are much more common at those sites than at Albion Island. We interpret the clay clasts with relict vesicles and rare devitrified glass shards from the Albion Island spheroid bed and diamictite as impact glass, but their shapes compare better with the "Flädle" in suevite glass from Ries crater (e.g., Hörz, 1965) than with tektites. Similar observations have been made for clay clasts in well Y-2 from the upper portion of the ejecta drilled on the rim of the Chicxulub crater (Hildebrand et al., 1991). There is, however, a

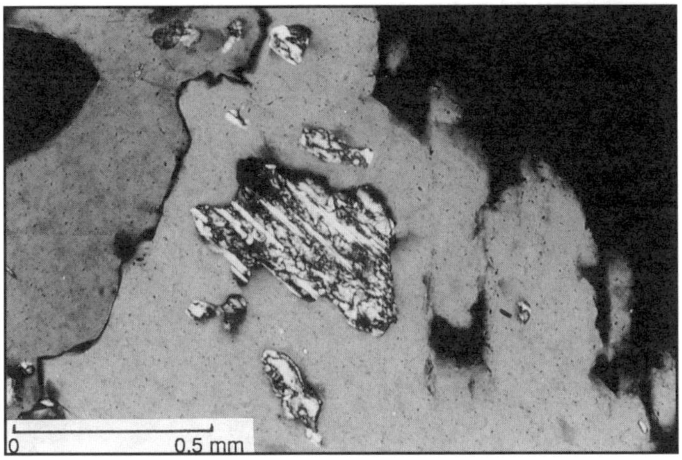

Figure 19. Thin section (crossed nicols) of authigenic quartz grain with serpentine inclusions. Albion diamictite.

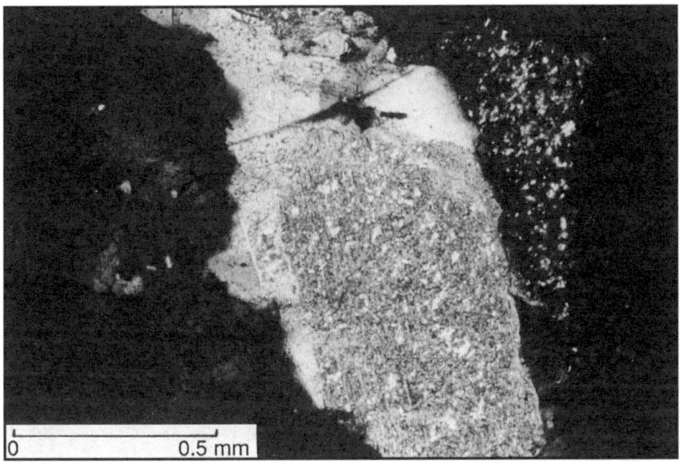

Figure 20. Thin section (crossed nicols) of granitic schist clast. Albion diamictite.

discrepancy in size: The smallest "Flädle" from Ries correspond to the larger Albion examples. Kieffer and Simonds (1980) suggest that impacts into carbonate targets produce a rapid expansion of gases that disperses large amounts of melt with the ejecta. The abundant clay clasts in the Albion deposits may represent such a process. The fact that classic tektite forms are rare at Albion Island may reflect the much shorter trajectory of the Albion glass, given its proximity to Chicxulub.

The dolomite spheroids, which make up most of the spheroids at Albion Island, may include clasts rounded by abrasion, but others have concentric structures that imply accretionary growth. These spheroids with concentric zones of fine and coarse dolomite resemble the accretionary lapilli from Ries, which have similar concentric structures that may have formed in a water-rich vapor plume (Graup, 1981). We suggest that some if not most of these spheroids are the tektite equivalent of shock-heated carbonates and sulfates, present in the upper Chicxulub target rocks. Such heating would drive off CO_2, SO_3, and SO_2, producing large amounts of quicklime (CaO, and possibly MgO) (Brett, 1992; Sigurdsson et al., 1992; Pope et al., 1994; Ivanov et al., this volume). This quicklime may have formed spheroids in the vapor plume. The dolomite spheroids may have precipitated directly from the plume through the reverse reaction between the oxides and the CO_2 produced. Such a scenario has been proposed for carbonates found in the Haughton impact crater in Canada (Martinez et al., 1994). Similarly, the dolomite silt in the Albion deposits may have formed by nucleation of dolomite in the cooling vapor plume.

An alternative explanation is that the dolomite spheroids formed after deposition through diagenesis of spheroids composed of the highly unstable Ca and Mg oxides. The flattening of the dolomite spheroids, most evident at the base of the bed, suggests that the transition from oxide to the now hard, crystalline carbonate may have succeeded emplacement. The dolomite spheroids with rinds may be carbonate clasts that were calcined on the surface but not completely degassed. CO_2 depletion has

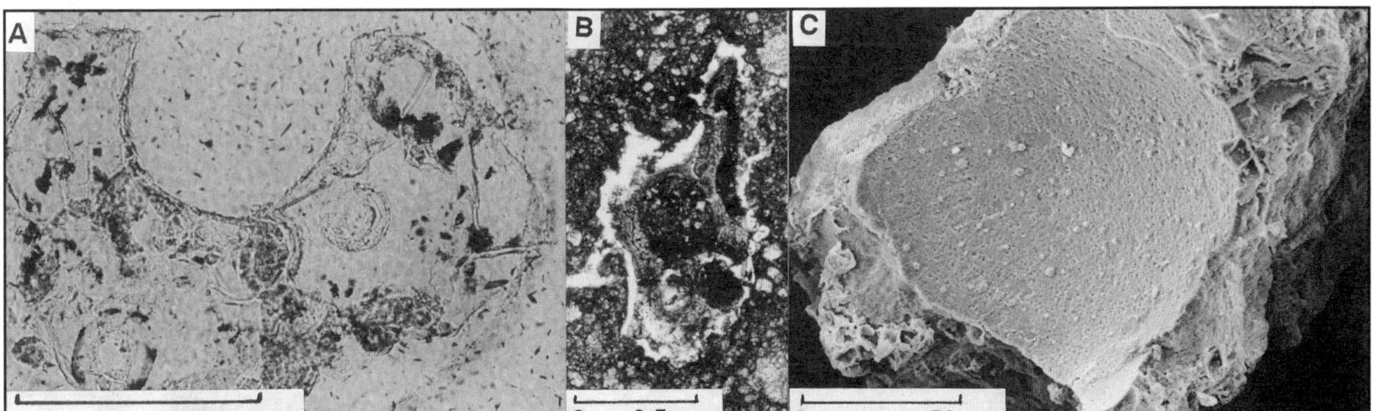

Figure 21. A, thin section (natural light) of devitrified, vesicular glass shard. B, thin section (natural light) of dolomite-filled vesicle in clay clast. C, electronmicrograph of a clay clast vesicle interior. Albion diamictite.

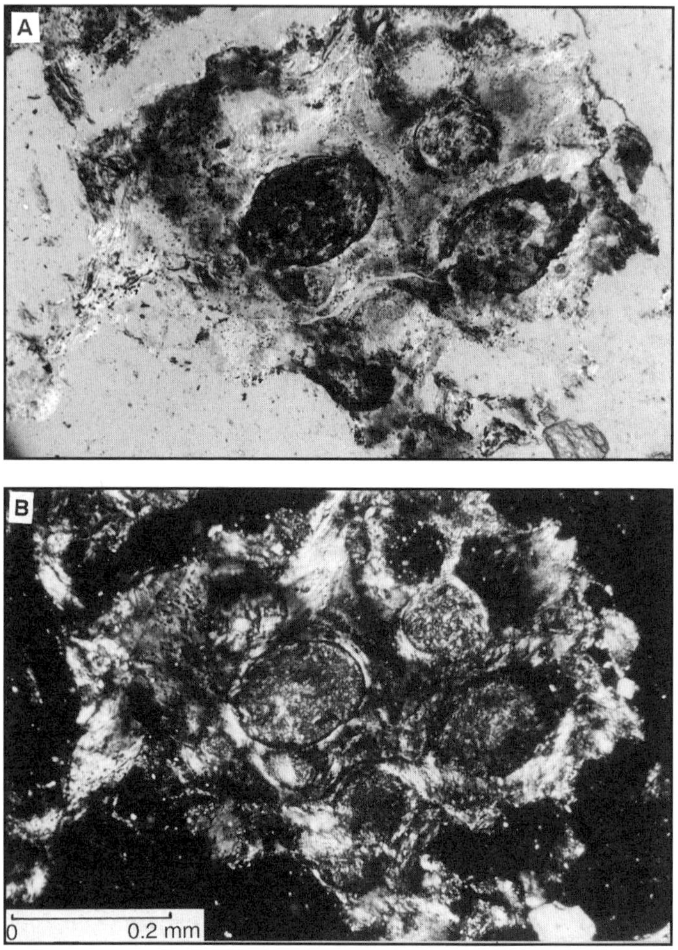

Figure 22. Thin section of vesicular shard: A, natural light; B, crossed nicols. Note that some portions of the shard are isotropic (translucent regions in A are completely dark in B) and may be mostly glass. Bright areas in A are clay, which also fills most of the vesicles. Albion diamictite.

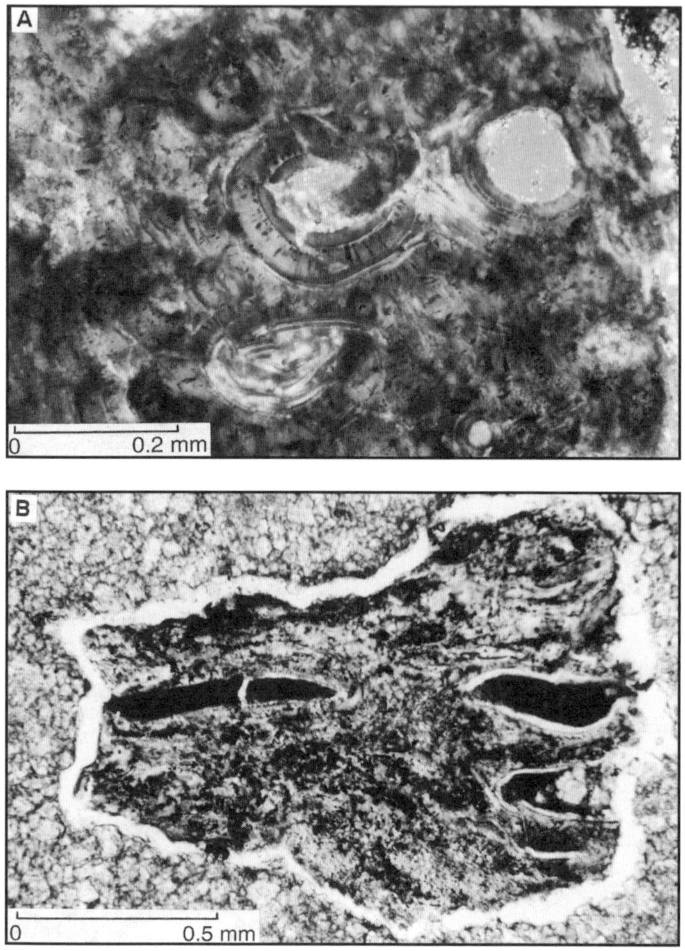

Figure 23. Thin sections of altered vesicular shards in natural light. Note squashed vesicles in A and B and concentric and radial growths of clay minerals partially filling vesicles in B. Albion diamictite.

been proposed for the origin of reaction rims around limestone clasts at Ries crater (Kieffer and Simonds, 1980). These rinds, like the quicklime spheroids, would have altered back to carbonate.

The Albion diamictite is petrographically similar to the approximately 600-m-thick coarse detrital unit recorded in petroleum exploration drilling on the rim of the Chicxulub crater, originally described as the detrital anhydrite facies (Marshall, 1974; Weidie, 1985). Thin sections from the Chicxulub and Albion deposits (Fig. 24) show very similar texture and composition; for example, both contain rounded and angular clasts floating in a dolomite silt matrix. Major differences at Chicxulub are the abundance of anhydrite, which is nearly absent at Albion Island, and the paucity of clay clasts, which are abundant at Albion Island. Previously (Pope et al., 1993), we interpreted the Chicxulub detrital deposit as the "continuous deposit" resulting from ballistic sedimentation as described by Oberbeck (1975). This interpretation was tentative, however, since the Chicxulub deposits are known only from drilling with limited sampling and macroscale features are unknown.

The Albion diamictite provides a more complete picture of the Chicxulub ejecta blanket, exposing a facies comparable to the well-studied Bunte breccia of the Ries crater (Hörz et al., 1977; Chao et al., 1978; Hörz and Banholzer, 1980; Hörz et al., 1983). Important characteristics of the Bunte breccia apparently shared by the Albion diamictite are as follows: (1) they are extremely poorly sorted (clast sizes range over eight orders of magnitude for Bunte and five orders for Albion); (2) they are totally chaotic with no stratification (indicative of turbulent flow); (3) the distal sections contain greater than 90% local material (most of the large Albion clasts may be local, but this is not confirmed); (4) breccias within breccias; and (5) there are clasts with striations, polish, and possible hinge fractures. Although highly shocked clasts (greater than 10 GPa) occur at Ries, they are rare: less than 0.1% by weight. None have been found in the Albion diamictite.

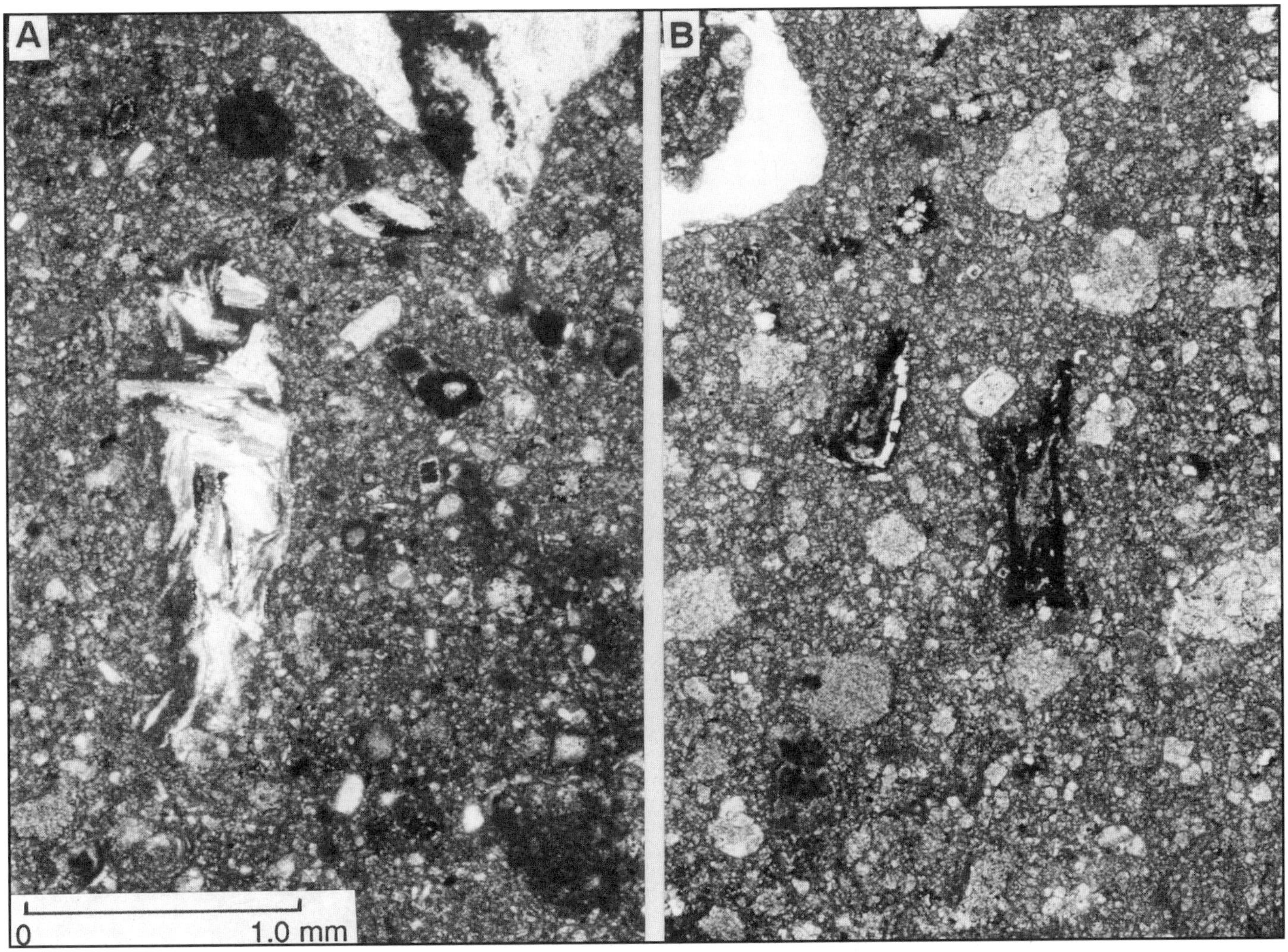

Figure 24. A, thin section (crossed nicols) of ejecta blanket sample from Yucatan 2 well, located about 130 km from Chicxulub crater center. Depth 700 m (sample Y2 N11). B, thin section (natural light) of Albion diamictite sample. Note similarities in clast sizes and morphologies. Both samples have dolomite silt matrix. Large clast near the center of A is anhydrite and in B is clay, illustrating the main compositional differences between the two deposits.

All of the above characteristics are either consistent with or indicative of ballistic sedimentation processes, as argued by Hörz and his colleagues (Hörz et al., 1977; Hörz and Banholzer, 1980; Hörz et al., 1983). Nevertheless, the preservation of the spheroid bed beneath the diamictite indicates that emplacement of this member at Albion Island did not involve secondary impacts or turbulent flow with any erosive power. This subject is discussed in more detail in the next section.

DEPOSITIONAL PROCESSES AND THE K/T EVENT

We propose the following sequence of events following the Chicxulub impact that are recorded in the sediments on Albion Island. We begin with a carbonate platform containing extensive interbedded evaporites. By analogy with other platforms, such as the Bahama-Florida platform, we may assume that this one oscillated between shallow marine, hypersaline evaporitic, and emergent conditions leading to freshwater incursions and that much of the carbonate was converted to dolomite relatively soon after deposition. The Albion Island location was surrounded by shallow seas receiving very little clastic sediment, hundreds of kilometers from any land mass with significant relief (Bateson and Hall, 1977). The first event was the air blast from the leading edge of the expanding vapor plume that may have stripped away surficial sediments, soil, or water, exposing foraminifera-bearing lagoonal carbonates that were probably already dolomitized. This event probably began about 10 sec after impact, judging from the proposed greater than 30 km/sec velocities for the initial blast proposed by Melosh (1989). The velocity of the expanding plume can surpass that of the impactor because gas expansion adds an additional velocity component. Although the bulk of plume traveled upward, spreading out on top of the atmosphere, the initial expansion may have been quasi-equidirectional until atmospheric drag slowed the lateral blast (Melosh, 1989; Vickery and Melosh, 1990).

The next event was the arrival of the seismic shock wave from the impact. This may be recorded by the fractures in the Barton Creek dolomite. Fracturing would have happened in less than a minute after impact and may have been more or less contemporaneous with the onset of spheroid deposition. The spheroid bed was probably deposited in a few minutes, as the main body of the vapor cloud passed over the site with an average velocity estimated at about 10 km/sec (Melosh, 1989). Expanding vapors would accelerate rapidly and quickly "outrun" the ballistic ejecta. As the superheated cloud passed through the early stages of the ejecta curtain it may have entrained the smaller clasts (Vickery, 1986), perhaps abrading and calcining the outer portions of the carbonate clasts to produce the large spheroids with rinds.

The last event recorded on Albion Island is the deposition of the Albion diamictite. We interpret the diamictite as having originated through a combination of ballistic and debris flow sedimentation. The presence at Albion Island of carbonate clasts from deep in the Yucatán platform indicates contribution of debris by ballistic sedimentation. Nevertheless, the preservation of the underlying spheroid bed argues against any secondary impact scouring at Albion Island but does not preclude the possibility that secondary impacts occurred nearby. The diamictite is probably a debris flow deposit that, initiated by ballistic sedimentation, continued to flow great distances, partly because of the abundance of dolomite mud and water in the matrix. The Albion diamictite contains large boulders floating in a mud matrix, mud balls, mud coatings on clasts, and minimal erosion at its base, all common characteristics of debris flow deposits. Rampino (1994) recently summarized the debris flow–like character of ballistic ejecta deposits. Hörz and Banholzer (1980) argue that turbulent flow can extend as much as a crater radius beyond ballistic termination. The Albion diamictite may be similar to the fluidized ejecta blankets on Mars.

Albion Island, 364 km from the center of Chicxulub, lies near the theoretical limit (Melosh, 1989) of continuous deposition and significant secondary cratering for either the 240-km or 300-km-diameter proposed for Chicxulub by Pope et al. (1993) and Sharpton et al. (1993) respectively. The island lies 100 to 200 km beyond this limit for the 170- to 180-km-diameter crater proposed by Hildebrand and colleagues (Hildebrand et al., 1991; Pilkington et al., 1994). Although the location of the Albion diamictite with respect to Chicxulub is consistent with all three crater diameters noted above, rather extensive fluidized flow must be invoked for the smaller diameter. Given the debris flow character of the diamictite, such long-distance flow is clearly reasonable.

CONCLUSIONS

Seismic disruption of basal sediments and spherule deposition occurred throughout the Gulf of Mexico region (Smit et al.,1994) and appear to be the hallmarks of early-stage, proximal events following the Chicxulub impact. Distal ejecta are characterized by two distinct stratigraphic units. The lower unit, called the "ejecta" or "melt ejecta" layer, contains spherules, altered glass shards, and other impact debris; the upper unit, called the "fireball" layer, contains abundant shocked quartz and is rich in siderophile elements, most notably iridium (e.g., Hildebrand and Boynton 1990; Hildebrand 1993; Pollastro and Bohor, 1993). Although the origin of this two-part stratigraphy is controversial, arguments have been presented that the lower bed contains ejecta of limited distribution that was wholly or partially ballistically emplaced and that the upper bed contains globally distributed material from the vapor plume originally deposited on top of the atmosphere (Hildebrand and Boynton, 1990; Hildebrand, 1993; Pollastro and Bohor, 1993).

The spheroid bed on Albion Island may be the proximal equivalent of the lower "ejecta" layer. Nevertheless, we suggest that this spheroid bed contains ejecta entrained by the laterally expanding vapor plume, and possibly material precipitated from the rapidly cooling plume, but little material deposited along ballistic trajectories. The spheroids and associated debris traveled faster than, and were detached from, the main ejecta curtain, thus explaining their stratigraphic position beneath the bulk of the ballistic ejecta (Albion diamictite), which arrived later. Whether or not this scenario applies in toto to the widely distributed "ejecta" layer is uncertain; however, given that the Chicxulub impact into a nearly 3-km-thick sequence of evaporites and carbonates released over a trillion tons of CO_2, SO_3, SO_2, and water vapor (Pope et al., 1994; Ivanov et al., this volume), we propose that the importance of volatiles in the formation and transport of the ejecta has been greatly underestimated.

The Albion diamictite is the distal portion of the true ejecta blanket as defined by Melosh (1989, p. 89). The ejecta blanket is the zone of thick, continuous deposition that surrounds most craters and is distinct from more distal ballistic ejecta deposits. The Albion Island site may be near the limit of the ejecta blanket distribution (reconnaissance of the region in 1994 and 1995 failed to locate any other examples). The Albion diamictite is an extension of the thick detrital unit drilled on the rim of the Chicxulub crater, but it is genetically distinct from other K/T boundary breccias and coarse deposits reported from the Chicxulub vicinity, which are probably slump deposits (e.g. Alvarez et al., 1992; Hildebrand et al., 1994; Montanari et al., 1994).

The globally distributed "fireball" layer that typically caps K/T boundary deposits is not present at Albion Island, where the sequence is truncated by the surface. Evidence of other late-stage impact processes, such as those related to impact waves, are similarly absent. Nevertheless, the Albion Island section is thus far unique in its record of the early stages of the Chicxulub impact event. Future work at the Albion Island quarry will no doubt produce many more insights into the final seconds of the Cretaceous and the process of large impacts on Earth.

ADDENDUM

Ongoing studies at the Albion Island quarry have added important new information, which we briefly summarize below.

The Late Cretaceous age of the dolomite underlying the Albion spheroid bed has been confirmed by fossils. A fossiliferous limestone lens within the dolomite yielded nerineid gastropods showing infolded wall patterns characteristic of the latest (Campanian-Maastrichtian) members of the family (Jan Smit, personal communication, 1995). The same lens contains fossil crabs of a new species, *Carcineretes planetarius,* a genus that is also restricted to the Upper Cretaceous, probably Maastrichtian (personal communication, Francisco Vega Vera and Rodney Feldmann, 1995).

Two recent observations suggest that Albion Island was subaerial at the time of the Chicxulub impact: (1) caliche deposits are found in several locations just below the spheroid bed, capping the Barton Creek dolomite, and (2) the apex of an anticlinal fold in the Barton Creek dolomite was eroded slightly before deposition of the ejecta, indicating that uplift occurred prior to the impact.

It has also become clear that the Albion diamictite contains evidence of hydrothermal activity. Hydrothermal alteration appears concentrated in one area of the quarry a few hundred square meters in size, where authigenic quartz is found in vugs and in veins within clasts. Spheroidal intergrowths of fibrous quartz, calcite, and fibrous palygorskite occur in pockets in this zone and are probably the product of circulating hot fluids. The calcite pisoids described above may also be of hydrothermal origin. Brecciated and recemented clasts are associated with the authigenic quartz, possibly the product of steam explosions. Authigenic quartz, possibly also of hydrothermal origin, is present in the underlying Barton Creek dolomite, which suggests that the hydrothermal activity may not be restricted to the ejecta deposits. The timing of the hydrothermal activity remains uncertain. Conceivably hot fluids circulated in the ejecta as it cooled, or the seismic shock of the impact may have triggered local hydrothermal activity, but equally well the hydrothermal overprint may be much younger and unrelated to the impact.

ACKNOWLEDGMENTS

This work was funded by the Exobiology Program of the National Aeronautics and Space Administration and by The Planetary Society. We thank Erle Kauffman, Richard Boettcher, Jan Smit, Francisco Vega Vera, and Rodney Feldmann for identifying fossils; Virgil Sharpton for providing the Chicxulub core sample; and Alan Hildebrand and Friedrich Hörz for valuable comments on an earlier draft of this paper. We thank Joy Crisp for providing the facilities for the petrographic analyses. We gratefully acknowledge the support provided by the Geology and Petroleum Office, Ministry of Natural Resources, Government of Belize, and the hospitality of the Albion Stone Company.

REFERENCES CITED

Alvarez, L. W., Alvarez, W., Asaro, F., and Michel, H. V., 1980, Extraterrestrial cause for the Cretaceous-Tertiary extinction: Science, v. 208, p. 1095–1108.

Alvarez, W., Smit, J., Lowrie, W. Asaro, F., Margolis, S. V., Claeys, P., Kastner, M., and Hildebrand, A., 1992, Proximal impact deposits at the Cretaceous-Tertiary boundary in the Gulf of Mexico: A restudy of DSDP Leg 77 Sites 536 and 540: Geology, v. 20, p. 697–700.

Bateson, J. H., and Hall, I. H. S., 1977, The geology of the Maya Mountains, Belize: Institute of Geological Sciences Overseas Memoir 3: London, Her Majesty's Stationery Office, 43 p.

Bourgeois, J., Hansen, T. A., Wiberg, P. L., and Kauffman, E. G., 1988, A tsunami deposit at the Cretaceous-Tertiary boundary in Texas: Science, v. 241, p. 567–570.

Brett, R., 1992, The Cretaceous-Tertiary extinction: A lethal mechanism involving anhydrite target rocks: Geochimica et Cosmochimica Acta, v. 56, p. 3603–3606.

Bull, W. B., 1964, Alluvial fans and near-surface subsidence in western Fresno County, California: U.S. Geological Survey Professional Paper 437-A, 70 p.

Chao, E. C. T., Hüttner, R., and Schmidt-Kaler, H., 1978, Principal exposures of the Ries Meteorite Crater in southern Germany: Munich, Bayerisches Geologisches Landesamt, 84 p.

Cornec, J. H., 1986, Provisional geologic map of Belize: Belmopan, Geology and Petroleum Office, Ministry of Natural Resources, 1:250,000.

DePaolo, D. J., and Ingram, B. L., 1985, High-resolution stratigraphy with strontium isotopes: Science, v. 227, p. 938–941.

Fischer, A. G., 1953, Petrology of Eocene limestones in and around the Citrus–Levy County area, Florida: Florida Geological Survey Reports of Investigations, v. 9, pt. 2, p. 41–70.

Flores, G., 1952, Geology of northern British Honduras: American Association of Petroleum Geologists Bulletin, v. 36, p. 404–409.

Geology and Petroleum Office, 1986, Petroleum prospect and potential of concession block-1 in Belize, Central America: Belmopan, Geology and Petroleum Office, Ministry of Natural Resources, 256 p.

Graup, G., 1981, Terrestrial chondrules, glass spherules and accretionary lapilli from the suevite, Ries crater, Germany: Earth and Planetary Science Letters, v. 45, p. 407–418.

Hess, J., Bender, M., and Schilling, J.-G., 1986, Seawater $^{87}Sr/^{86}Sr$ evolution from Cretaceous to present: Science, v. 231, p. 979–984.

Hildebrand, A. R., 1993, The Cretaceous/Tertiary boundary impact (or the dinosaurs didn't have a chance): Journal of the Royal Astronomical Society of Canada, v. 87, p. 77–118.

Hildebrand, A. R., and Boynton, W. V., 1990, Proximal Cretaceous-Tertiary boundary impact deposits in the Caribbean: Science, v. 248, p. 843–847.

Hildebrand, A. R., Penfield, G. T., Kring, D. A., Pilkington, M., Camargo Z., A., Jacobsen, S. B., and Boynton, W. V., 1991, Chicxulub crater: A possible Cretaceous/Tertiary boundary impact crater on the Yucatan Peninsula, Mexico: Geology, v. 19, p. 867–871.

Hildebrand, A. R., and 14 others, 1994, The Chicxulub crater and its relation to KT boundary ejecta and impact-wave deposits; *in* New developments regarding the KT event and other catastrophes in Earth history: Houston, Texas, Lunar and Planetary Institute Contribution 825, p. 49–50.

Hörz, F., 1965, Untersuchungen an Riesgläsern: Beitäge zur Mineralogie und

Petrographie, v. 11, p. 621–661.

Hörz, F., and Banholzer, G. S., Jr., 1980, Deep seated target material in the continuous deposits of the Ries Crater, Germany, *in* Papike, J. J., and Merrill, R. B., eds., Proceedings, Conference on the Lunar Highlands Crust: New York, Pergamon Press, p. 211–231.

Hörz, F., Gall, H., Hüttner, R., and Oberbeck, V. R., 1977, Shallow drilling in the "Bunte Breccia" impact deposits, Ries Crater, Germany, *in* Roddy, D. J., Pepin, R. O., and Merrill, R. B., eds., Impact and explosion cratering: New York, Pergamon Press, p. 425–448.

Hörz, F., Ostertag, R., and Rainey, D. A., 1983, Bunte Breccia of the Ries: Continuous deposits of large impact craters: Reviews of Geophysics and Space Physics, v. 21., p. 1667–1725.

Izett, G. A., 1991, Cretaceous-Tertiary boundary rocks on Haiti and their bearing on the Alvarez impact extinction hypothesis: Journal of Geophysical Research, v. 96. p. 20879–20905.

Kieffer, S. W., and Simonds, C. H., 1980, The role of volatiles and lithology in the impact cratering process: Reviews of Geophysics and Space Physics, v. 18, p. 143–181.

Marshall, R. H., 1974, Petrology of subsurface Mesozoic rocks of the Yucatan Platform, Mexico [Masters Thesis]: New Orleans, Louisiana, University of New Orleans, 97 p.

Martinez, I., Agrinier, P., Scharer, U., and Javoy, M., 1994, A SEM-ATEM and stable isotope study of carbonates from the Haughton impact crater, Canada: Earth and Planetary Science Letters, v. 121, p. 559–574.

Maurrasse, F.J.M.R., and Sen, G., 1991, Impacts, tsunamis, and the Haitian Cretaceous-Tertiary boundary layer: Science, v. 252, p. 1690–1693.

Melosh, H. J., 1989, Impact cratering: A geologic process: New York, Oxford University Press, 245 p.

Montanari, A., Claeys, P., Asaro, F., Bermudez, J., and Smit, J., 1994, Preliminary stratigraphy and iridium and other geochemical anomalies across the KT boundary in the Bochil section (Chiapas, southeastern Mexico), *in* New developments regarding the KT event and other catastrophes in Earth history: Houston, Texas, Lunar and Planetary Institute Contribution 825, p. 84–85.

Oberbeck, V. R., 1975, The role of ballistic erosion and sedimentation in lunar stratigraphy: Review of Geophysics and Space Physics, v. 13, p. 337–362.

Pilkington, M., Hildebrand, A. R., and Ortiz-Aleman, C., 1994, Gravity and magnetic field modeling and structure of the Chicxulub crater, Mexico: Journal of Geophysical Research, v. 99, p. 13147–13162.

Pitakpaivan, K., Byerly, G. R., and Hazel, J. E., 1994, Pseudomorphs of impact spherules from a Cretaceous-Tertiary boundary section at Shell Creek, Alabama: Earth and Planetary Science Letters, v. 124, p. 49–56.

Pollastro, R. M., and Bohor, R. F., 1993, Origin and clay-mineral genesis of the Cretaceous/Tertiary boundary unit, western interior of North America: Clays and Clay Minerals, v. 41, p. 7–25.

Pope, K. O., Ocampo, A. C., and Duller, C. E., 1991, Mexican site for K/T impact crater?: Nature, v. 351, p. 105.

Pope, K. O., Ocampo, A. C., and Duller, C. E., 1993, Surficial geology of the Chicxulub impact crater, Yucatan, Mexico: Earth, Moon, and Planets, v. 63, p. 93–104.

Pope, K. O., Baines, K. H., Ocampo, A. C., and Ivanov, B. A., 1994, Impact winter and the Cretaceous/Tertiary mass extinctions: Results of a Chicxulub asteroid impact model: Earth and Planetary Science Letters, v. 128, p. 719–725.

Ramanathan, Rm., 1985, Foraminiferal micropaleontology report on the Remate Well no. 1, onshore, Belize: Belmopan, Geology and Petroleum Office, Ministry of Natural Resources, 15 p.

Rampino, M. R., 1994, Tillites, diamictites, and ballistic ejecta of large impacts: Journal of Geology, v. 102, p. 439–456.

Robertson, P. B., Dence, M. R., and Vos, M. A., 1968, Deformation in rock-forming minerals from Canadian craters: *in* French, B. M., and Short, N. M., eds., Shock metamorphism of natural materials: Baltimore, Maryland, Mono Book Corporation, p. 433–452.

Sharpton, V. L., and 9 others, 1993, Chicxulub multiring impact basin: Size and other characteristics derived from gravity analysis: Science, v. 261, p. 1564–1567.

Sigurdsson, H., D'Hondt, S., and Carey, S., 1992, The impact of the Cretaceous/Tertiary bolide on evaporite terrain and the generation of major sulfuric acid aerosol: Earth and Planetary Science Letters, v. 109, p. 543–559.

Smit, J., and Romein, A. J. T., 1985, A sequence of events across the Cretaceous-Tertiary boundary: Earth and Planetary Science Letters, v. 74, p. 155–170.

Smit, J., and eight others, 1992, Tektite-bearing, deep-water clastic unit at the Cretaceous-Tertiary boundary in northwestern Mexico: Geology, v. 20, p. 99–103.

Smit, J., Roep, Th. B., Alvarez, W., Montanari, S., and Claeys, P., 1994, Stratigraphy and sedimentology of KT clastic beds in the Moscow Landing (Alabama) outcrop: Evidence for impact-related earthquakes and tsunamis, *in* New developments regarding the KT event and other catastrophes in Earth history: Houston, Lunar and Planetary Institute Contribution 825, p. 119–120.

Vickery, A. M., 1986, Effect of an impact-generated gas cloud on the acceleration of solid ejecta: Journal of Geophysical Research, v. 91, p. 14139–14160.

Vickery, A. M., and Melosh, H. J., 1990, Atmospheric erosion and impactor retention in large impacts, with application to mass extinctions, *in* Sharpton, V. L., and Ward, P. D., eds., Global catastrophes in Earth history: Geological Society of America Special Paper 247, p. 289–300.

Weidie, A. E., 1985, Geology of Yucatan platform, *in* Ward, W. C., Weidie, A. E., and Back, W., eds., Geology and hydrology of the Yucatan and Quaternary geology of northeastern Yucatan Peninsula: New Orleans, New Orleans Geological Society, p. 1–19.

NOTE ADDED IN PROOF

Further study has shown that the radially fibrous calcite "pisoids" are more widely distributed in the ejecta. Blocks of carbonate composed of interlocked spheroidal to asymmetrical radial bodies are incompatible with accretionary pisolite growth and suggest quench crystallization from carbonate melt. Continuing field studies have identified additional coarse ejecta deposits 100 km farther south in central Belize.

Manuscript Accepted by the Society September 14, 1995

Geological Society of America
Special Paper 307
1996

Implications of magneto-mineralogic characteristics of the Manson and Chicxulub impact rocks

Maureen B. Steiner
Department of Geology and Geophysics, University of Wyoming, Laramie, Wyoming 82071

ABSTRACT

The rocks within the 75-Ma Manson impact structure have two unusual characteristics of remanent magnetization. First, the normal-polarity characteristic magnetization has an approximately antipodal reversed polarity overprint. The reversed remanence is held only in specific stability ranges, namely low coercivities (≥7.5 mT), low unblocking temperatures (80 to 260 °C), and unblocking temperatures above the magnetite Curie point (580 to 680 °C). The other peculiar feature is the fact that the characteristic remanent magnetization has two distinct maximum unblocking temperatures: between 300 to 400 °C and between 500 to 540 °C.

These unusual features led to examination of rocks from another impact structure for comparison; samples were examined from the Chicxulub structure, of similar age (65 Ma) and relatively nearby. Remarkably, the Chicxulub samples show much the same magnetization characteristics. Chicxulub samples exhibit a pronounced remanence loss at low unblocking temperatures, 350 to 400 °C, in addition to unblocking temperatures of 540 to 580 °C. An opposite polarity, antipodal remanence also is removed from the characteristic magnetization and at the same low unblocking temperatures as displayed by the Manson samples.

Saturation magnetization, thermomagnetic, and reflected-light characteristics of the Chicxulub and Manson impact rocks indicate dominantly a magnetite magnetization carrier. Pyrrhotite is almost wholly absent in the rocks of both impact structures, while a hemo-ilmenite, formed from the oxidation of magnetite, is present in the Manson rocks. The presence of a low-temperature opposite polarity remanence with the same unblocking temperatures in the rocks of two impact structures suggests a mineralogical self-reversal origin, perhaps a hemo-ilmenite with a composition within the self-reversal range. The high-temperature opposite polarity remanence in the Manson structure (stable to 680 °C) is carried by hematite. The prominent 300 to 400 °C remanence loss in both structures only can be carried by magnetite.

Both the low-temperature remanence loss and the opposite polarity high-temperature remanence are postulated to have had an origin in the hydrothermal activity commonly associated with large impact craters. Hydrothermal circulation prolongs cooling, and magnetization processes, beyond the simple cooling from the heat generated by impact mechanics. In the case of the Manson structure, cooling persisted for up to 1 m.y. after impact, and the impact occurred within 0.7 to 1.5 m.y. of a reversal of geomagnetic field polarity. Chemically active hydrothermal fluids may have created a chemical remanence in hematite, and the possible overlap of the duration of fluid flow with the change of field polarity would have allowed some remanence to be generated in the succeeding

Steiner, M. B., 1996, Implications of magneto-mineralogic characteristics of the Manson and Chicxulub impact rocks, *in* Ryder, G., Fastovsky, D., and Gartner, S., eds., The Cretaceous-Tertiary Event and Other Catastrophes in Earth History: Boulder, Colorado, Geological Society of America Special Paper 307.

polarity interval. The magnetization loss at 300 to 400 °C is hypothesized to have been a viscous magnetization in magnetite, resulting from the extended duration of high temperatures accompanying the hydrothermal circulation. Potentially, studies of this type of remanence may provide information regarding thermal conditions or duration of hydrothermal regimes.

INTRODUCTION

The Manson impact structure of northwestern Iowa (Fig. 1) is one of the largest (35 km diameter) and best-preserved impact structures in the United States. As a result of evidence indicating that it could have been a Cretaceous-Tertiary (K/T) impact, an extensive program of investigation was conducted. The Manson impact appeared to be related to the K/T boundary because the boundary deposits of western North America contain shocked minerals of comparatively large grain size, and grains appeared to be largest in the vicinity of the Manson impact structure (Bohor and Izett, 1986; Izett, 1990). Furthermore, the basement mineralogy of the Manson impact structure was of the appropriate range of compositions (Hartung et al., 1990) to have been the source of the shocked minerals found at the K/T boundary. Therefore, breccia materials from 40-year-old cores drilled in the structure were dated (Kunk et al., 1989); the resulting age of 65.8 ± 0.4 Ma suggested that the Manson impact had occurred at the K/T boundary.

Because of this result, a new coring program was initiated, and 12 holes were cored into the structure (Anderson et al., 1993; Roddy and Shoemaker, 1994). Radiometric dating and paleomagnetic investigation were undertaken on the new core material, and both showed that the Manson impact structure was not of the same age as the K/T boundary. Authigenic sanidine $^{40}Ar/^{39}Ar$ analyses indicate that the Manson structure is about 75 m.y. old (Dalyrmple and Izett, 1994; Izett et al., 1993a; ages recomputed according to an age of 520.4 Ma for the MMhb-1 laboratory standard). Paleomagnetic study shows that all brecciated materials filling the Manson structure have a characteristic magnetization of normal polarity (Steiner and Shoemaker, 1993a, 1994). It is well established that the K/T boundary occurred within a reversed polarity interval, Chron C29r (e.g., Thierstein, 1982; Alvarez and Asaro, 1990); therefore, both paleomagnetic and geochronologic studies demonstrated that the Manson impact did not occur at the K/T boundary.

Peculiar features in the paleomagnetic and rock-magnetic signatures of the Manson impact rocks led to a desire to inves-

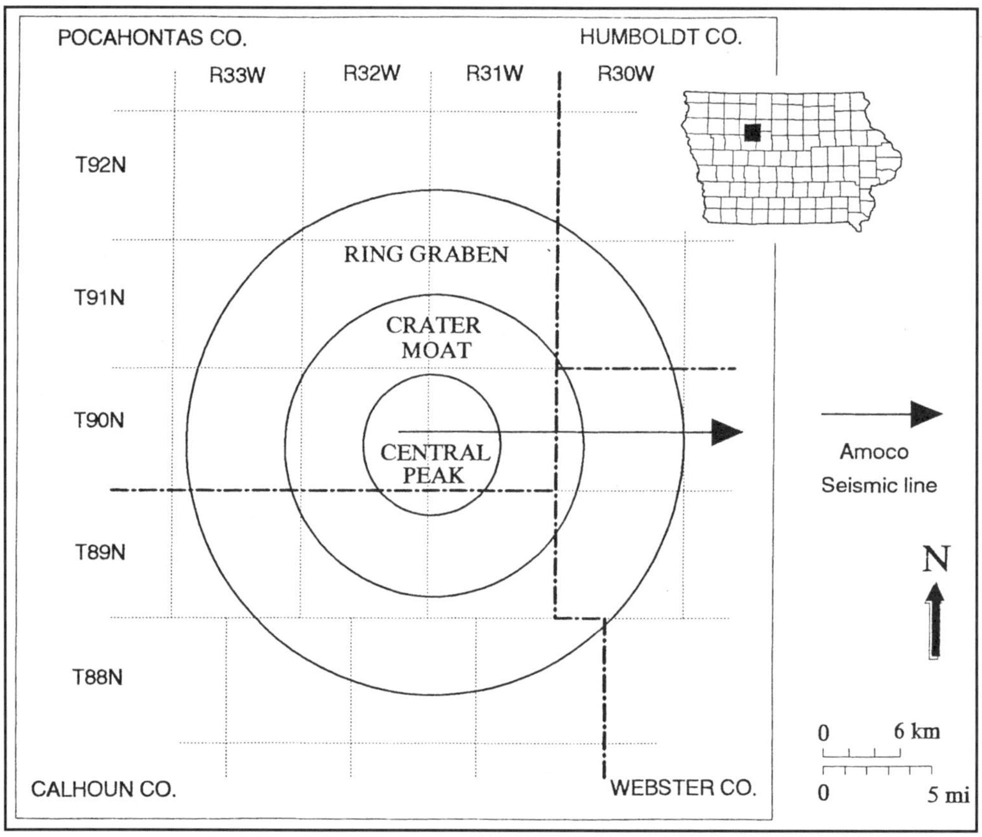

Figure 1. Location of the Manson impact structure in Iowa (from Hartung et al., 1990).

tigate samples from another impact structure for comparison. The relatively nearby K/T boundary Chicxulub impact structure was formed in a basement mineralogy similar to that of the Manson structure, as evidenced by the similarity in composition of shocked mineral grains from the K/T boundary layer and the Manson basement mineralogy (Izett, 1990; Hartung et al., 1990) and similarities of some Yucatan basement exposures (Steiner and Walker, 1996) to the clast compositions at Manson. Furthermore, Chicxulub and Manson have similar ages, having both impacted in the Late Cretaceous, although the Chicxulub structure is 10 m.y. younger than the Manson structure. Finally, both impacts occurred during the period of relatively frequent geomagnetic field reversals following the Cretaceous long normal polarity superchron.

Access to Chicxulub samples was extremely limited, and as a result, only three samples were obtained for study. However, the remarkable similarity of the magnetic characteristics between Chicxulub and Manson rocks suggests that despite the small number, the Chicxulub samples may be representative of that structure.

All samples were measured with an SCT cryogenic magnetometer that has a sensitivity limit of >5 × 10^{-6} A/m. The magnetometer and demagnetization equipment are housed within a magnetically shielded room in which the field is 150 nT. The thermal demagnetization furnace has a cooling region with a field of less than 10 nT.

MAGNETIZATION CHARACTERISTICS OF THE MANSON IMPACT ROCKS

The basic paleomagnetic data for the Manson impact rocks are to be reported in a future publication on the results of the new Manson coring program. Those results are summarized here for purposes of comparing Manson and Chicxulub rocks.

The Manson structure is filled by a number of superposed impact-generated breccias. The magnetic characteristics of the three main breccias were studied in one of the newly cored holes, M1 (Steiner and Shoemaker, 1993a, 1994). In the M1 core, the highest breccia is composed of entirely sedimentary materials. The immediately underlying breccia contains clasts of igneous and metamorphic rocks in a very dark, dense, finely crystalline matrix; both matrix and clasts are partially melted (discontinuous melt sheet of Izett et al., 1993b). The lowest breccia also has igneous and metamorphic clasts, but these are contained in a granular or "sandy" matrix of either light gray or green colors. The clasts, matrix, and the complete range of clast compositions of all three breccias were studied. Samples also were studied from the granular-matrix breccia in one of the 40-year-old cores (core 2A; Hoppin and Dryden, 1958).

Remanent magnetization

Initial magnetization intensities of the melt-matrix breccia were 1 A/m (matrix) and 0.5 to 1.0 A/m (clasts). The granular-matrix breccia had intensities of 0.5 A/m (matrix) and 0.01 to 1.0 A/m (clast); the sedimentary breccia matrix intensity was 1 × 10^{-3} A/m. Demagnetization (both alternating field [AF] and thermal) on pairs of samples of clasts and of matrix gave the same characteristic directions in all cases.

AF demagnetization of the melt- and granular-matrix breccias generally resulted in simple univectorial decay of the remanent magnetization (Fig. 2a). One-half of the remanence typically was lost at 18 mT in the melt matrix and at 36 mT in the granular matrix. Clast magnetizations in both breccias generally were two component but showed a univectorially decaying magnetization above 10 mT, which is considered the characteristic magnetization. AF demagnetization of the sedimentary breccia matrix yielded very similar characteristics.

Thermal demagnetization insignificantly changed the magnetization directions of matrix and clasts from their initial directions. Demagnetization of both matrix and clasts of the melt- and granular-matrix breccias yielded a linear decay of remanence between the temperatures of 260 to 300 °C and the magnetite Curie temperature (about 580 °C; Figs. 2b–2d). By 570 to 580 °C, ≥98% of the initial intensity of magnetization is lost in all samples, and large amounts of remanence commonly are lost between 500 and 580 °C (observable in Fig. 4).

The melt-matrix breccia contains material melted by the Manson impact and therefore is most likely to contain the signature of the geomagnetic field at the time of impact. The characteristic remanent direction of the melt matrix as well as the other two breccias is a normal polarity magnetization and is identical between clasts and matrix. This magnetization is exhibited up to the magnetite Curie temperature and also more weakly above that temperature.

In addition to the characteristic normal-polarity magnetization in these impact breccias, a reversed-polarity magnetization is present. Reversed magnetization was observed in all three breccias, including the sedimentary breccia, and in both clasts and matrix. The direction of the reversed magnetization generally is poorly defined, because its stability ranges largely overlap those of the characteristic normal-polarity magnetization. It is mostly observed as part of a mixed two-component magnetization in which the reversed, negative inclination component dom inates. Remanent directions at successive demagnetization steps and of the vectors removed between steps indicate that the two superposed vectors were the characteristic normal direction and a reversed direction. In a few samples, unblocking temperatures are not completely overlapping, and in these samples, the reversed component appears to be approximately antipodal to the normal characteristic magnetization (Fig. 3).

Reversed-polarity magnetization was observed in three types of data. A reversed polarity vector was removed at low temperatures, between 80 and 260 °C. Its presence was most obvious in the melt and granular breccias, both in matrix and clasts, but it is visible in the sedimentary breccia as well. Intensities in a number of samples were observed to rise through this interval (Fig. 3a–c), and the removed vector directions are clearly of reversed polarity (Fig. 3d and e).

A second observation of reversed magnetization was in the

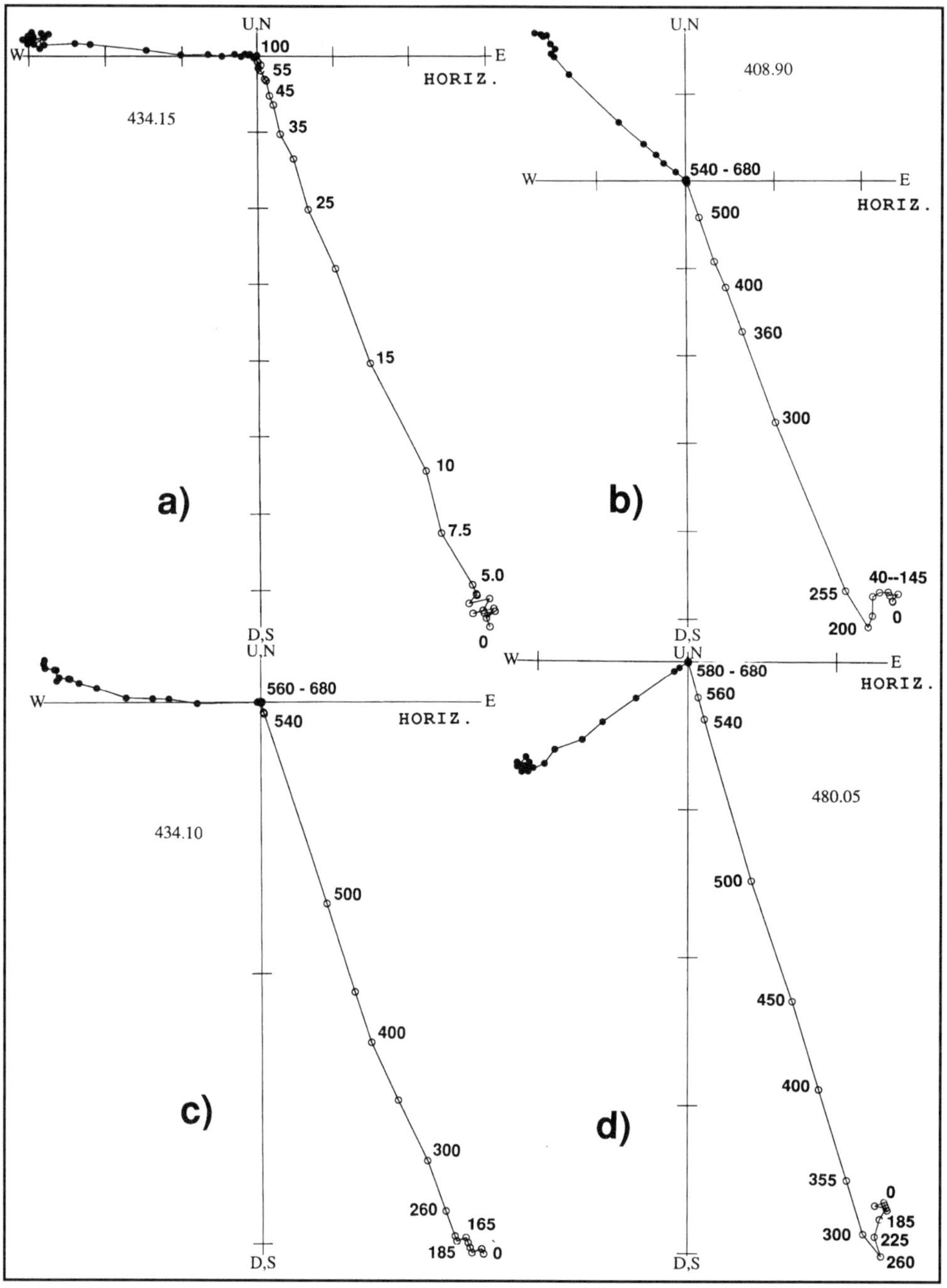

Figure 2. Examples of typical responses to demagnetization of the melt-matrix samples from the Manson impact structure. a, Alternating fields; numbers indicate values of alternating field in mT. b through d, Thermal demagnetization; numbers indicate degrees Centigrade. Scale divisions: a and b, 1×10^{-1} A/m; c and d, 5×10^{-1} A/m. These orthogonal axes plots show inclination plotted in the plane of the declination. The horizontal (vertical) components of the magnetization are shown as solid (open) circles. Inclination may be read directly from the vertical component as the angle between the data point and the horizontal axis. Abbreviations: N, north; S, south; U, up; D, down.

Figure 3. Examples of the reversed polarity magnetization in Manson samples. a, Magnetization loss with increasing temperature in two melt-matrix and one clast sample, respectively. The remanent intensity at each temperature step (J) is normalized to the original intensity (JO). b, Vectors removed at low temperatures (140 to 260 °C) in the melt-matrix sample shown in a form a great circle, approximately including the characteristic normal polarity direction. c, Reversed magnetization observed as subtracted vectors within a clast sample. The nearly antipodal reversed magnetization is removed at both low (55 and 150 °C) and high (610, 620 °C) temperatures. d, Reversed remanence observed in the remanent magnetization at temperatures above the magnetite Curie point. e, Removed vectors during AF demagnetization show reversed remanence (open circles) at fields between 0.7 to 4.5 mT. Equal area stereographic projections; open (solid) symbols indicate negative (positive) inclinations, hence reversed (normal) polarity.

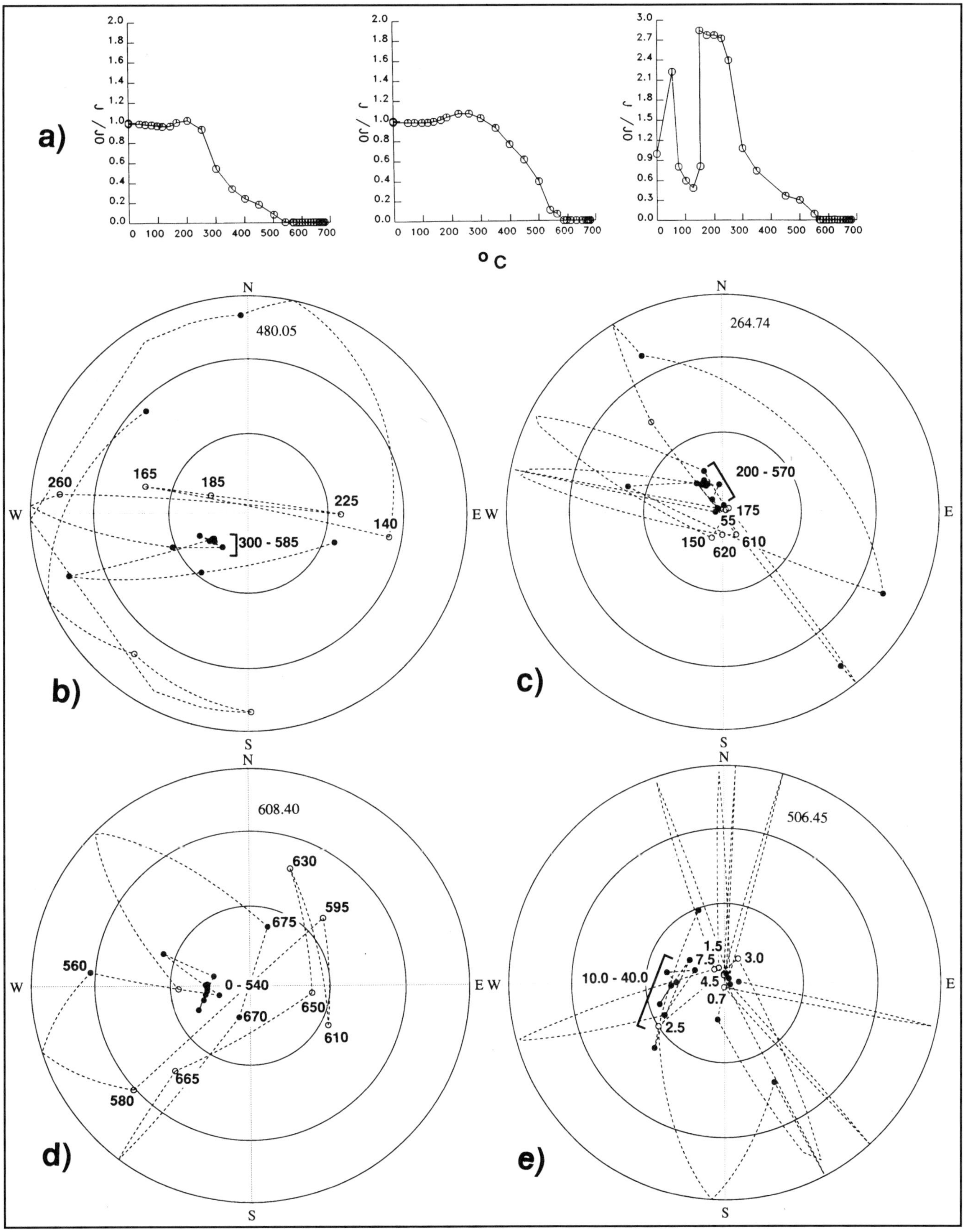
a)
J / J0
°C
N
S
W
E
480.05
165
185
260
225
140
300 - 585
b)
264.74
200 - 570
175
55
150
620
610
c)
608.40
630
595
675
560
0 - 540
670
650
610
665
580
d)
506.45
1.5
3.0
7.5
10.0 - 40.0
4.5
0.7
2.5
e)

remanent magnetization at and above the magnetite Curie point. Remanence directions above the magnetite Curie point were less well defined than those below, in spite of the fact that intensities were about three orders of magnitude above the noise level of the magnetometer. Negative inclination directions are extremely common in most samples above 580 °C, both in the remanent magnetization (Fig. 3f) and in the vectors removed between the heating steps (Fig. 3e). About 50% of the samples display negative inclinations by 610 to 630 °C (with intensities = 10^{-2} to 10^{-3} A/m).

The third observation of reversed polarity magnetization was in the removed vectors during very low field, 0.10 to 0.75 mT, AF demagnetization. More than 50% of the samples of clasts and matrix in both the melt and granular breccias displayed the removal of a reversed polarity magnetization (Fig. 3g), although the behavior was most pronounced in the melt breccia.

The mean inclinations of the Manson breccias are essentially similar to the expected Cretaceous inclination at this site (64 ± 4°; from Gordon and Van der Voo, 1995). The sedimentary breccia mean inclination (66 ± 6°) is identical to that of the Cretaceous, whereas both the underlying melt-matrix and the granular-matrix breccias have a steeper inclination, 78 ± 4°. That the sedimentary breccia magnetization mimics the Cretaceous field may represent a postdepositional remanence; such a remanence is consistent with a magnetization by sedimentary processes, the mechanism by which this breccia was formed. The steeper inclination of the melt-matrix breccia probably represents a secular variation direction of the Cretaceous geomagnetic field, a concept supported by cooling models from other impact structures.

Magnetic mineralogy

Manson samples display significant losses of remanent magnetization at two temperatures: 300 to 375 °C, and >500 °C (Fig. 4). Of the samples from the melt- and granular-matrix breccias, 75% show either a large loss at 300 to 375 °C or dual, near-equal, losses at the two temperatures; clasts and matrix show this characteristic equally. Figure 4 shows samples representative of the entire core, from the top of the melt breccia to the base of the core (which terminated in the granular-matrix breccia). As the figure shows, this pronounced approximately 300 to 375 °C unblocking temperature persists through both breccias, to the base of the cored section, and the unblocking temperature appears to increase in the lower part of the sequence. The sedimentary breccia also shows prominent magnetization loss beginning at 300 °C; some samples display distinct peak losses at 300 and 500 °C, wheareas others show large losses between 300 and 500 °C without any sharp delineation between those temperatures.

The significant loss of magnetization between 500 and 580 °C in most samples suggests a low-titanium titanomagnetite carrier. Thermomagnetic measurements were performed to determine the magnetization carriers and especially the carrier of the magnetization with 300 to 375 °C unblocking temperatures. Measurements were made on a representative suite of 12 clast and matrix samples from the melt and granular breccias. Samples were chosen to include a number with well-displayed low unblocking temperature or the reversed polarity remanences.

Thermomagnetic measurements were made in vacuums of 1×10^{-6} Torr and fields of 0.30 to 0.45 T. All samples show loss of all but a few percent of magnetization by 580 °C (Fig. 5). All samples show a single pronounced drop in magnetization between 550 and 575 °C. No difference exists between the melt-matrix and the granular-matrix breccias, nor between clasts and matrix. A percent or two of the magnetization persists to 680 °C, indicating the presence of hematite. None of the thermomagnetic curves show a loss of magnetization at 300 to 375 °C, in dramatic contrast to the display of large unblocking of remanent magnetization in that temperature range by thermal demagnetization (Fig. 5).

Isothermal remanence acquisition (IRM) experiments were performed on samples from all three breccias. Clasts and matrix are again the same in their characteristics of magnetization acquisition. Near-saturation of magnetization is achieved in all samples in magnetizing fields of about 0.15 T (Fig. 6). Saturation at these fields suggests magnetite as the dominant magnetic mineral in all rocks. An additional 1 to 4% of magnetization is gained in fields between 0.15 and 0.5 T, indicating the presence of another mineral whose magnetization requires higher fields for saturation (most likely hematite).

The opaque minerals of the melt- and granular-matrix breccia were examined petrographically in reflected light; use of oil immersion and magnifications up to 2,000× allowed observation of grains ≤0.5 microns. Fifteen polished thin-sections of clasts and matrix were examined, with the emphasis on those samples that showed either significant remanence loss at 300 to 375 °C or significant reversed polarity magnetization.

The main minerals observed were magnetite and ilmenite. A number of samples showed some of the magnetite to be highly oxidized, corresponding to the C6-C7 oxidation stage of Haggerty (1976). Samples of this oxidation state contain rutile, pseudobrookite, and hemo-ilmenite (Haggerty, 1976). The magnetite in these samples commonly was observed in two grain sizes: very large grains (30 to 500 microns) and relatively small grains, (≤0.5 to 15 microns). Probable hematite was observed as a pigmentary red staining of silicates with occasional rare grain grown to a fine specularite grain size (≤0.25 micron). Occasionally hematite was observed replacing a few magnetite grains within large clusters of such grains. In a diabase sample, hematite also was observed in a high-temperature oxidation-exolution association with ilmenite. Pyrrhotite, potentially expected because of the 300 °C unblocking temperatures, was observed in only one of the samples and only as a rare component.

CHICXULUB MAGNETIZATION

Three core pieces of melt-matrix material from the Chicxulub melt-matrix breccia were examined. All pieces were from Chicxulub core Yucatan-6. One was from level N-17 at 1295.5 to

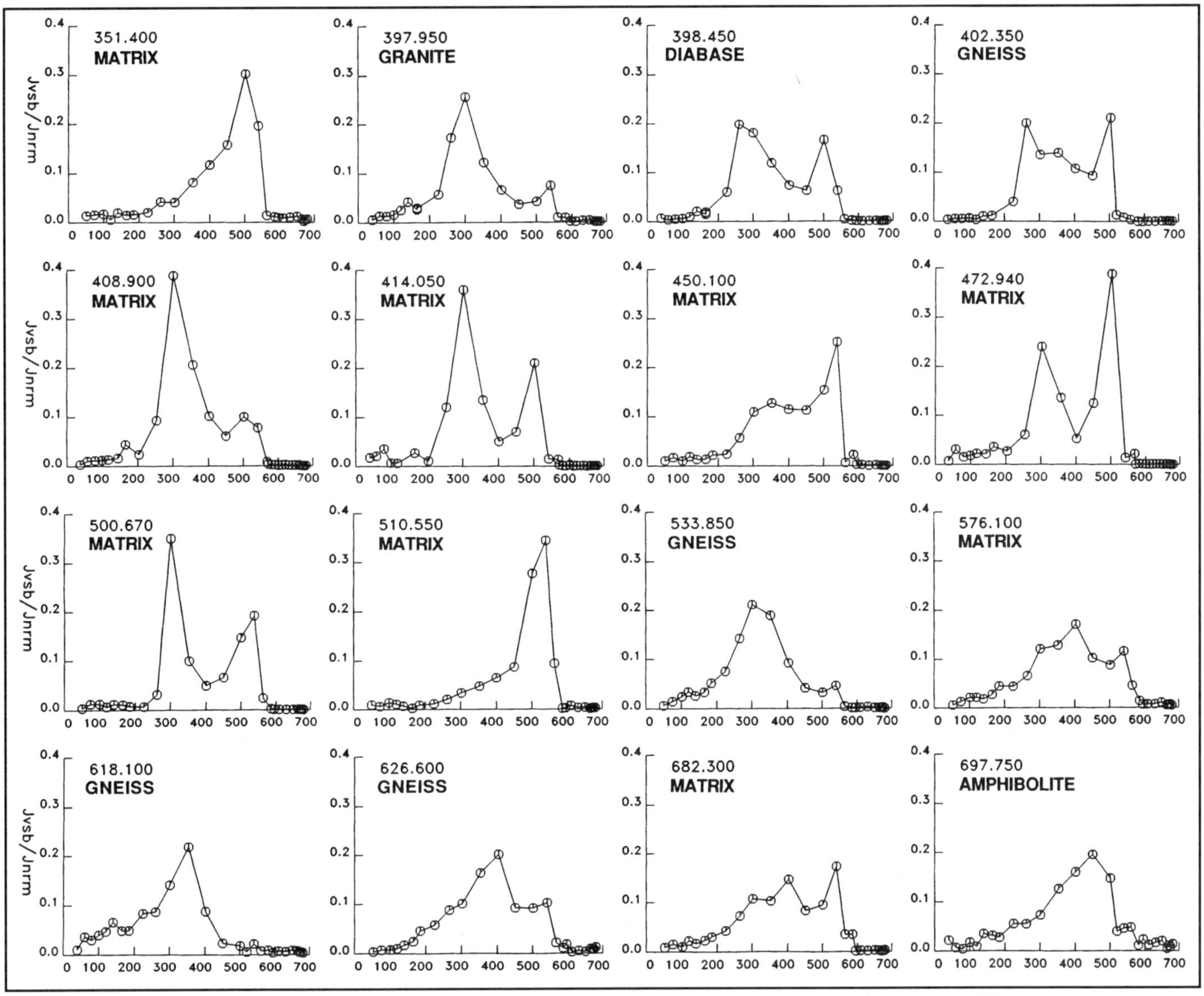

Figure 4. Incremental remanence loss during thermal demagnetization, showing the pronounced loss at 300 °C. The intensity lost between two successive temperature steps (Jvsb) is plotted as a function of the total initial intensity for that sample (Jnrm). Data shown are from representative melt- and granular-matrix breccias and clasts from the M1 core. Samples are displayed in sequence from the top of the melt breccia (upper left) down to the base of the granular breccia (lower right).

1299.0 m; a second was from level N-19 (listed as "parte 6"); the third piece had incomplete location information and was from either level N-17 or N-19. Despite the depth uncertainty, this piece was particularly important because it was a fully oriented core segment, the only one of the three studied. The oriented piece had both the red and black lines of the international convention used to designate the up-direction of the core; the most common version of the convention, black on the right and red on the left, was used to determine the orientation of this piece. Another of the segments (N-17) was a quarter-round piece of core that had a lengthwise red line down one edge of the segment but no adjacent black line on the contiguous exterior surface, representing about one-fourth of the circumference. It thus was assumed that the black line, normally marked closely adjacent to the red line, had been cut off in sawing, and using that assumption, the orientation of this piece was inferred. The third core piece had no orientation markings at all.

Two samples were cut from the fully oriented core piece, and one each from the other two pieces. Extensive AF demagnetization measurements and IRM acquisition measurements on Chicxulub materials already had been made by Urrutia-Fucugauchi and colleagues (Sharpton et al., 1992; Urrutia-Fucugauchi et al., 1993). This study concentrated on thermal demagnetization, thermomagnetic measurements, and reflected light microscopy of these samples; AF demagnetization of one sample was performed for comparison.

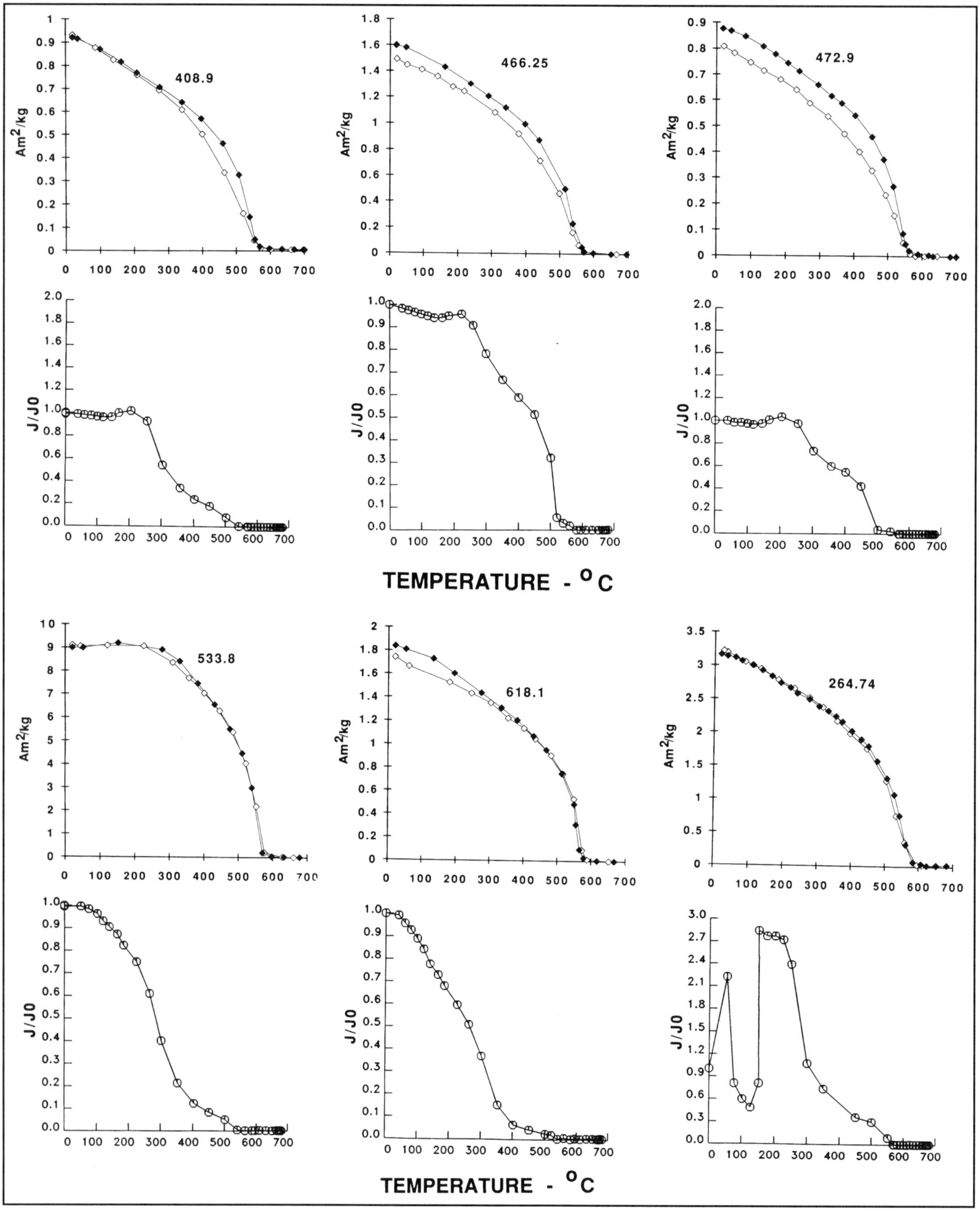
408.9
466.25
472.9
Am2/kg
J/J0
TEMPERATURE - oC
533.8
618.1
264.74
Am2/kg
J/J0
TEMPERATURE - oC

Figure 5. Thermomagnetic curves and intensity loss during thermal demagnetization for Manson samples. The top three thermomagnetic and remanence loss curves are from samples of the melt-matrix; the curves for the lower three samples are from clasts from the melt- and granular-matrix breccias. Temperature is in degrees Centigrade; solid (open) symbols indicate heating (cooling).

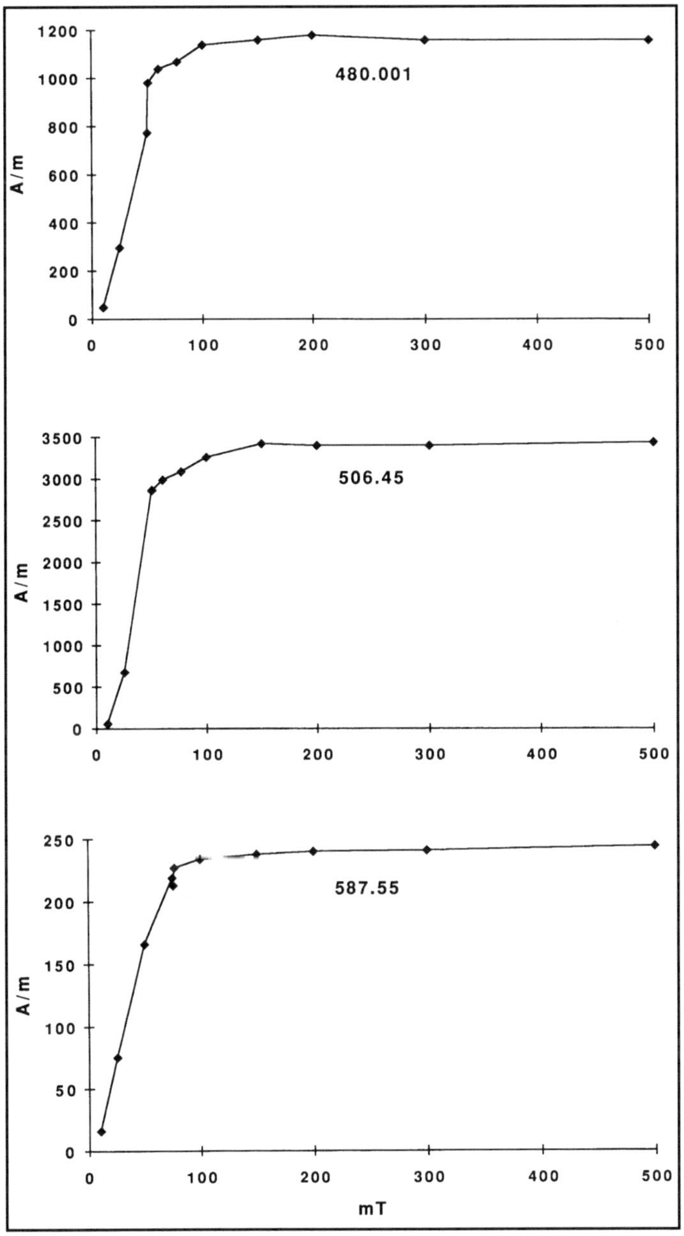

Figure 6. Isothermal Remanent Magnetization (IRM) acquisition in Manson samples. Typical responses are shown for a, matrix, melt-matrix breccia; b, clast (granular-matrix breccia); c, matrix, granular-matrix breccia.

Magnetization

Initial magnetization intensities were 0.3 to 0.6 A/m. AF demagnetization on one specimen of the fully oriented core piece (Fig. 7a) showed behavior very similar to that obtained by Urrutia-Fucugauchi et al. (1993; J. Urrutia-Fucugauchi, personal communication, 1993). Magnetization decayed univectorially above 7.5 mT, with half of the remanence removed by 27 mT. The directions of the magnetization vectors removed above 7.5 mT were the same as the remanent direction. The remanent magnetization of this sample has reversed polarity.

Thermal demagnetization behavior was markedly similar to that of the Manson samples. All three Chicxulub samples displayed a univectorial decay of magnetization above 260 to 300 °C (Figs. 7b through d). This magnetization, considered the characteristic remanence, is displayed from 300 °C up to 580 °C in one sample, and in two samples the characteristic direction persists until 640 °C. Thereafter, remanent directions are scattered.

Below 300 °C, all three samples displayed an increase in remanent intensities (observable in Fig. 10). Vectors removed in the temperature range of 60 to 225 °C were almost exclusively of opposite declinations and inclinations to the magnetization displayed between 300 and 580 °C. The directions of the magnetization removed from one of the samples at two temperatures (150 and 200 °C) were fully antipodal to the characteristic magnetization direction (Fig. 8a); inclinations and declinations both are 180° apart. The two other samples exhibited somewhat more overlapping stability magnetization components, but still each displayed antipodal or near-antipodal removed magnetization directions within that temperature range. One sample displayed one antipodal and one near-antipodal removed direction in that range (Fig. 8b), and the third sample also showed one removed vector with a direction nearly antipodal to the characteristic direction within that temperature range.

Both AF and thermal demagnetization of specimens of the fully oriented Chicxulub sample yielded a reversed polarity direction (Fig. 7a and 7c), and the same characteristic direction was obtained by each method. Orientation of the sample possessing only the red orientation line (N-17), in which the position of the missing black line seemingly could be reasonably assumed, resulted in a positive inclination, that is, normal polarity magnetization, for that sample. Discussions with geologists in both Pemex Oil Company (which drilled the core) and in Shell Oil Company led to the conclusion that occasionally the red-black line convention is inconsistently applied, that is, that it sometimes is reversed. This information and the consistency of the fully oriented sample with oriented samples measured by Urrutia-Fucugauchi (Sharpton et al., 1992; Urrutia-Fucugauchi et al., 1993; Urrutia-Fucugauchi, personal communication, 1993) suggest that the most reasonable conclusion is that the samples of this study all contain reversed polarity remanence. Therefore the other two Chicxulub samples also are shown in Figure 7 as exhibiting reversed polarity magnetization.

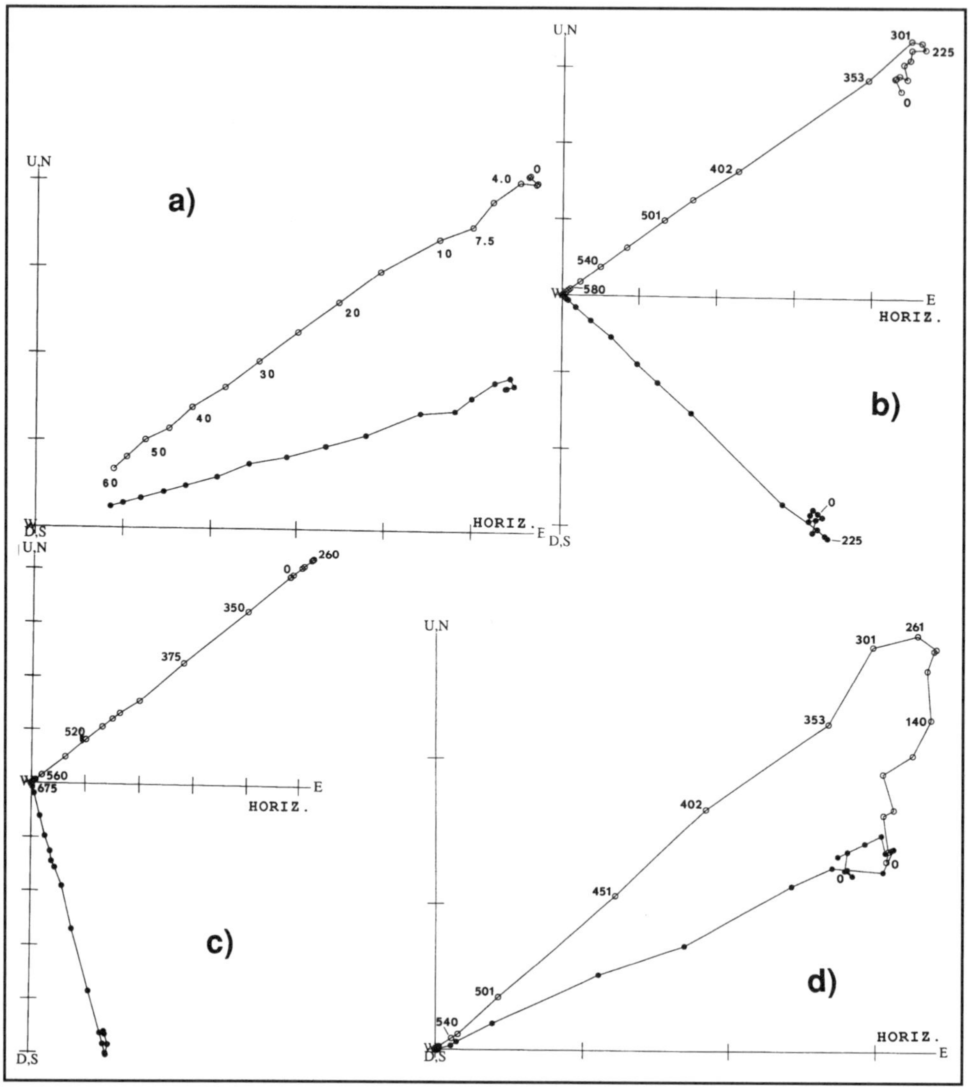

Figure 7. Chicxulub demagnetization responses. a, Alternating field demagnetization (sample N-17/19; scale divisions = 5×10^{-2} A/m. Thermal demagnetization: b, sample N-17, scale divisions = 1×10^{-1} A/m; c, sample N-17/19, scale divisions = 1×10^{-1} A/m; d, sample N-19, scale divisions = 5×10^{-2} A/m. Symbols and plotting convention as in Figure 2.

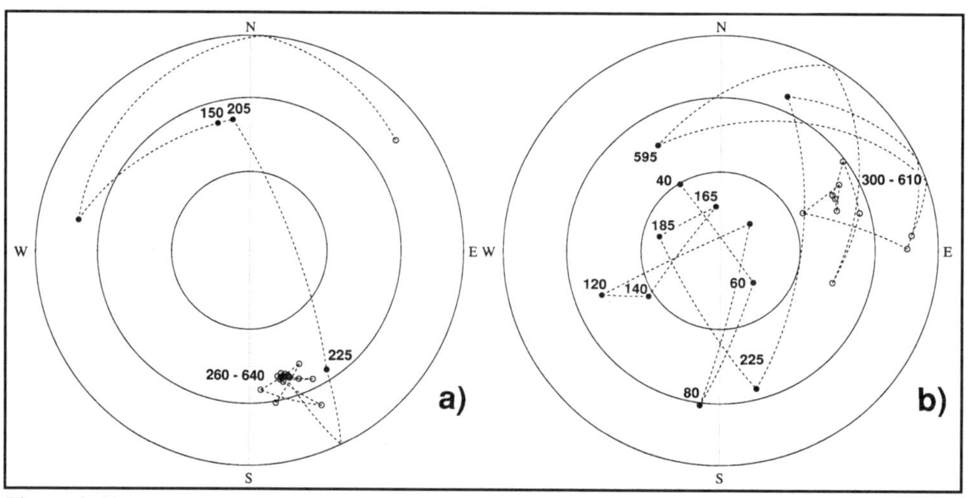

Figure 8. Vectors removed during thermal demagnetization. a, Sample N-17/19; b, sample N-19; plotting conventions as in Figure 3.

Principal component analysis of the directions of the three thermally demagnetized samples gave inclinations of –36.9° (N-17), –41.7° (N-19), and –39.0° (N-17/19); the inclination of the AF demagnetized sample is –36.8° (N-17/19). The Yucatan Peninsula and Chicxulub structure had achieved its present position relative to North America by Cretaceous time (e.g., Pindell, 1985); therefore, the Late Cretaceous paleodirection can be calculated from the North American reference pole. The inclination of the geomagnetic field at the Chicxulub site in the Late Cretaceous was –43.1 (from Gordon and Van der Voo, 1995), and therefore the inclinations of the three Chicxulub samples appear to be those of the Late Cretaceous field.

The origin of the low-temperature opposite-polarity remanence removed from the Chicxulub samples is less straightforward than the situation of the Manson samples because it is a normal-polarity remanence. A normal-polarity magnetization imprinted in recent geomagnetic fields could mimic a Late Cretaceous–early Tertiary normal-polarity magnetization. Drilling-induced magnetization can be excluded as the origin, because generally the normal-polarity inclinations are far from vertical (Fig. 8; Table 1). Those inclinations also are distinct from the inclination of the presently observed field at Chicxulub (51.7°).

However, at the Chicxulub site the theoretical axial dipole inclination (37.9°) and the Cretaceous inclination (43.1°) are very similar, and therefore the low-temperature normal-polarity remanence is similar both to a recent Tertiary magnetization and to the early Tertiary normal-polarity interval postdating the Chicxulub impact (Chron C28n). Although these directions are similar in inclination, the declination of the Late Cretaceous field at this site was 339.7°, separated by 20° from that of the axial dipole late Tertiary field (0.0°). The declinations of the antipodal to near-antipodal normal-polarity low-temperature remanence are within 4 to 7° of those of the characteristic reversed remanence in all three samples. These essentially antipodal declinations suggest that the low-temperature normal-polarity remanence is not a recent magnetization acquisition and also suggest that the normal and reversed magnetizations are of similar age.

Magnetic mineralogy

Remarkably, the Chicxulub samples show the same pronounced mid-temperature loss of remanent magnetization during thermal demagnetization as do the Manson samples. The Chicxulub samples show significant unblocking of remanent magnetization between 350 and 400 °C (Fig. 9), in addition to significant remanence loss above 500 °C. The magnetic mineralogy of these samples was examined for an explanation.

Thermomagnetic measurements reveal high Curie temperatures for all three of the Chicxulub samples (Fig. 10): 525 °C (N-17), 520 °C (N-19), and 555 °C (N-17/19). The thermomagnetic curves are very slightly irreversible, indicating some minor oxidation during heating (Fig. 10). The thermomagnetic curves are shown adjacent to the remanence loss during thermal demagnetization to illustrate the fact that there is no magnetization loss corresponding to the pronounced remanence loss at low temperatures; the thermomagnetic curves indicate only a magnetite Curie point. IRM experiments performed by Urrutia-Fucugauchi and colleagues (1993; 1994) also indicate primarily magnetite as the main magnetic mineral (magnetization saturation in fields of ≤0.2 T).

TABLE 1. THERMAL DEMAGNETIZATION DIRECTIONS OF CHICXULUB SAMPLES

Sample	N-17		N-17/19		N-19	
Low-T, normal polarity	318.3	20.7	349.0	38.6	249.4	41.2
Char., reversed polarity	131.5	–36.9	163.6	–39.0	65.6	–41.7

The three Chicxulub samples, which are all melt-matrix material, were studied petrographically; reflected-light examination indicated abundant finely crystalline magnetite, commonly around <0.25 microns to 5 microns. Some ilmenite is present as large composite grains. A red pigmentary staining is present, visible mainly at high power (≥50×). Very little iron sulfide was observed; a very few grains per slide of pyrite were evident, but no pyrrhotite was observed. Hemo-ilmenite also was not observed; the sample exhibiting the clearest indication of dual-polarity magnetization (Figure 8a) was examined in particularly great detail for hemo-ilmenite and pyrrhotite series minerals.

DISCUSSION

The Chicxulub impact rocks display characteristics of remanent magnetization that are largely identical to those of the Manson structure. Even though the small number of Chicxulub samples studied was potentially unrepresentative, the similarities of magnetic mineralogy, unblocking temperatures, and dual polarity remanence between Chicxulub and Manson rocks suggests that these samples probably are reasonably representative of the rocks of the Chicxulub structure. Further support is drawn from the fact that the magnetization characteristics among the Manson samples is so similar that a random selection from that population would likely provide a representative characterization of the Manson remanent magnetization. The two most striking similarities between the Manson and Chicxulub samples are the large remanence loss at low unblocking temperatures (Figs. 4 and 9), and the presence of dual polarity remanences.

Mineralogy and 300 to 400 °C unblocking temperatures

Magnetite is the dominant magnetic mineral in both Manson and Chicxulub rocks. In both cases, Curie points are 525 to 560 °C, suggesting an essentially titanium-free magnetite. The persistence of remanence to 680 °C suggests that hematite is also a remanence carrier. Isothermal remanent magnetization experiments show >96% saturation of magnetization in fields of about 0.15 to 0.20 T in rocks of both structures, again indicative of magnetite. Manson samples displayed a few percent further gain of magnetization in fields up to 0.5 T, which is consistent with the presence of minor hematite. Optical examination of grains down to <0.5 micron confirms a dominance of magnetite and shows minor amounts of hematite in both structures. The petrographic abundance of magnetite and ilmenite in Manson and Chicxulub rocks and the presence of rutile and pseudobrookite in the Manson samples agree with the observations of El Gorsey (1968; also Chao, 1968) of the opaque minerals common in impact rocks.

All observations suggest that magnetite is the dominant magnetic mineral, in agreement with observations from numerous other impact structures (see Pohl, 1971). Although hematite also is present, the 10-to-1 difference in spontaneous magneti-

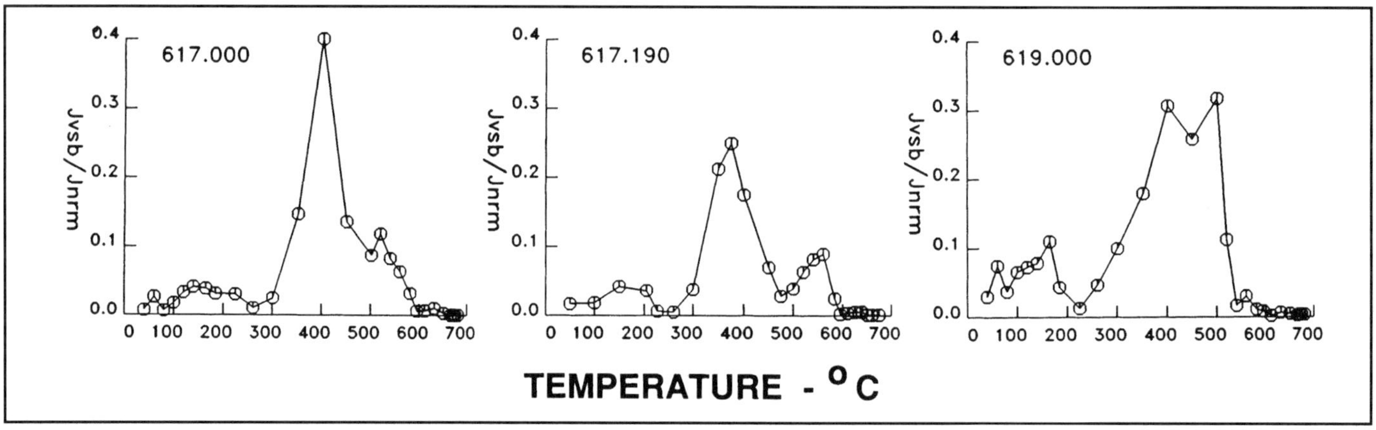

Figure 9. Intensity loss during thermal demagnetization of the three Chicxulub samples. Plots use the same convention as Figure 6.

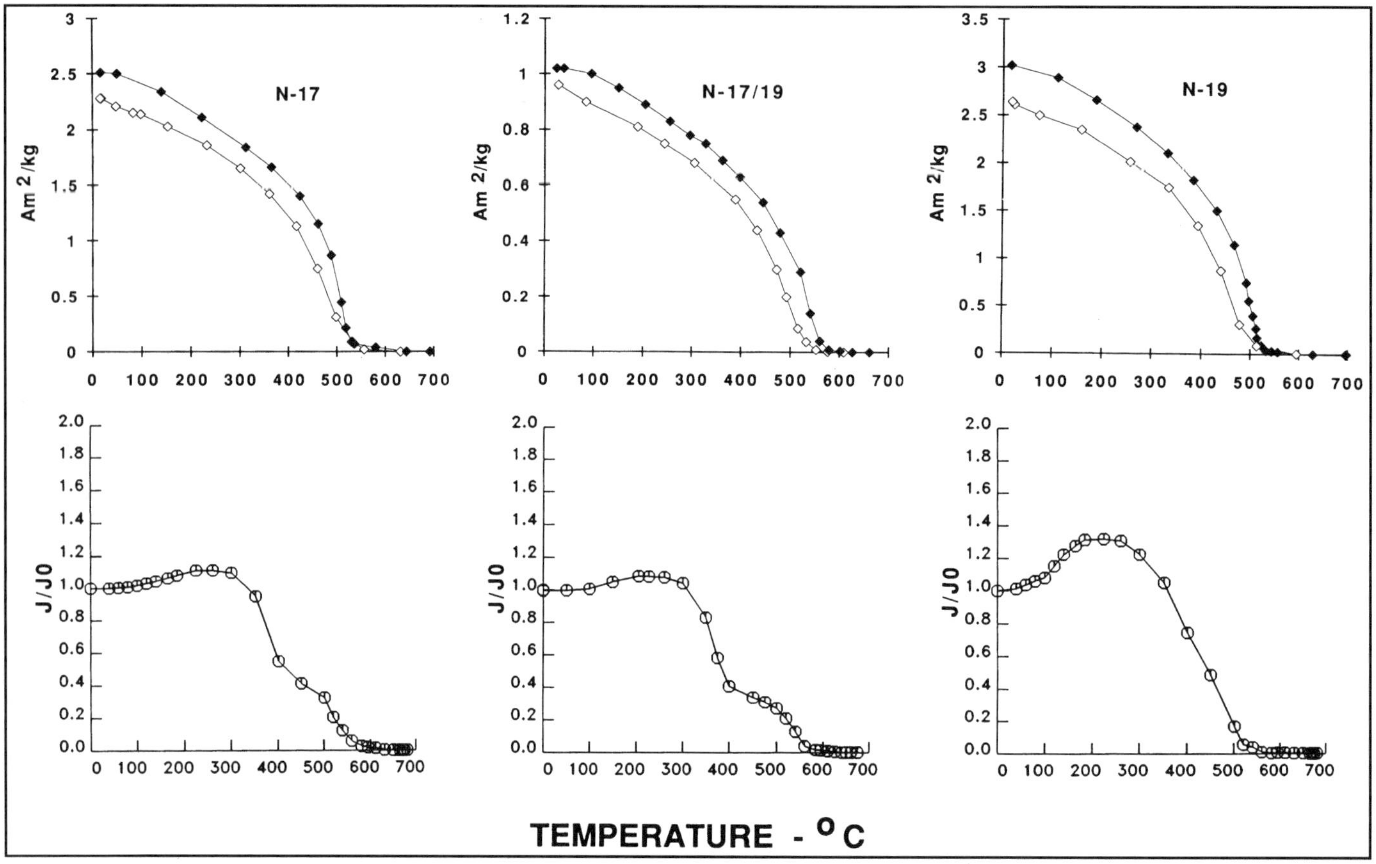

Figure 10. Thermomagnetic curves and intensity loss during thermal demagnetization for Chicxulub samples. Temperature is in degrees Centigrade; solid (open) symbols indicate heating (cooling).

zations and the relatively small amount observed optically indicate that hematite's contribution to the remanent magnetization is relatively small.

Pyrrhotite can be a carrier of magnetic remanence. Although pyrrhotite has been observed in impact rocks, with the exception of the Finnish Lappajarvi structure (Personen et al., 1992), pyrrhotite normally is a minor component (El Gorsey, 1968; Pohl, 1971). Pyrrhotite was expected in the Manson samples because of the unblocking temperatures around 300 °C. But both Chicxulub and Manson samples that prominently display such low unblocking temperatures do not exhibit a Curie point in that range (Figs. 5 and 10). Moreover, with the exception of a single Manson sample, pyrrhotite was not optically observed in the rocks from either structure and particularly was conspicuously absent in samples that showed the nearly complete remanence loss at the 300 to 400 °C unblocking temperatures (e.g., samples 397, 533, and 697 in Fig. 4 or sample 618 in Fig. 5).

Even had pyrrhotite been present, it could not explain remanence unblocking above about 320 °C, its Curie temperature; the low unblocking temperatures in Chicxulub and Manson samples range between 300 and 400 °C. Certain compositions of hemo-ilmenite have Neel temperatures in this range, but the corresponding saturation magnetization is too low to produce the observed remanence losses, which are comparable in magnitude to magnetite. Maghemite inverts to hematite in this temperature range, but no indication of maghemite is observed in the thermomagnetic curves, nor was maghemite common petrographically. Therefore, mineralogy seems incapable of explaining the 300 to 400 °C unblocking temperatures in Manson and Chicxulub impact rocks, and another explanation must be sought.

Recently, the ability of magnetite to acquire a viscous magnetization (VRM) at elevated temperatures has been investigated (Kelso and Banerjee, 1993). Although not shown by the natural samples of their experiments, the synthetic samples demonstrated a marked increase in VRM between 200 and 400 °C (Kelso and Banerjee, 1993). A cursory reexamination of the samples of this study, which showed the most pronounced 300 to 400 °C remanence losses, indicates that these samples frequently have very large grains (100 to 300 microns) of magnetite. Therefore, an elevated temperature VRM in magnetite may be the explanation of the prominent 300 to 400 °C unblocking temperatures in Manson and Chicxulub rocks. In impact structures, there is a substantial probability of an extended duration of elevated temperatures as a result of hydrothermal circulation that commonly follows impact; this will be discussed in the next section.

Dual-polarity magnetization

Both Manson and Chicxulub rocks show the phenomenon of dual-polarity magnetization. The presence of opposite-polarity magnetization could originate in several ways: mineralogic self-reversal of magnetization, a magnetization process that spanned a reversal of the geomagnetic field, or remagnetization at some time long after the impact event.

Self-reversal is known to occur within certain restricted compositional ranges of both pyrrhotite and hemo-ilmenite. Optical and thermomagnetic measurements exclude pyrrhotite. However, high degrees of oxidation of magnetite were observed, such that some of the original magnetite is now largely composed of pseudobrookite, rutile, and hemo-ilmenite. It is possible that the hemo-ilmenite of this alteration may have a composition within the self-reversing range. The fact that two impact structures of vastly different crater sizes, 35 and 200 to 300 km, both show the presence of a weak opposite-polarity magnetization in the same unblocking temperature range (80 to 260 °C) is highly suggestive of a mineralogical origin. It therefore is speculated that such a self-reversing hemo-ilmenite may account for the low unblocking temperature reversed-polarity magnetization. The fact that an opposite-polarity remanence also is exhibited in very low coercivities in some of the Manson rocks is consistent with the properties of some self-reversing hemo-ilmenite (e.g., Nord and Lawson, 1992). Obviously additional studies (microprobe, X-ray, and so on) are needed to confirm the presence of a self-reversing hemo-ilmenite, but because Chicxulub and Manson both display the phenomenon, it must be related to impact crater dynamics.

In Manson rocks, however, opposite-polarity magnetization is also observed at high unblocking temperatures (580 to 680 °C). The remanence remaining at temperatures above the magnetite Curie point has much weaker intensities than the normal-polarity remanence held up to the magnetite Curie point and persists to 680 °C. It therefore probably is held in hematite. Both normal and reversed-polarity remanences are observed 580 °C, although reversed-polarity magnetization is more common; the two magnetizations generally display considerable overlap in unblocking temperatures. At high temperature, the Chicxulub samples behave similarly to the Manson samples, but are too few in number to warrant conclusions about their high-temperature remanence.

The >580 °C reversed-polarity remanence cannot be explained by self-reversal, because hematite with such high unblocking and Neel temperatures lies outside the compositional range of self-reversal in the hematite-ilmenite series. Secondary magnetization long after the impact event is a possibility but seems unlikely because of the quiescent history of the region of the Manson impact following the impact. No tectonic activity has affected the Manson region since the impact; subsequent to impact, the area has undergone only erosion and glaciation in the Pleistocene, culminating in the deposition of about 30 m of glacial till (Hartung et al., 1990). For these reasons, and because of the presence of the 300 to 400 °C unblocking temperatures, magnetization over an extended time duration, which included a reversal of the geomagnetic field, was considered.

Impact near the time of a change of geomagnetic field polarity is probable because the Manson impact occurred in the Late Cretaceous, a time of frequent reversals of the geomagnetic field. The 75-Ma age (Dalyrmple and Izett, 1994) and the normal polarity of the melt rocks suggests that the impact occurred very late in Chron C33n (see Gradstein et al., 1994; Cande and Kent, 1992; 1995). The radiometric age and uncertainty and the variation among time scales suggest that the impact occurred within 0.7 to 1.5 m.y. of the polarity change to succeeding Chron C32r. Because the Late Cretaceous magnetic polarity time scale has not yet been definitively tied to radiometric ages, Izett et al. (1993a) suggested that the normal polarity of the impact rocks conceivably also could reflect the short (≥100 k.y.) normal polarity event within Chron C32r (see Cande and Kent, 1992; Gradstein et al., 1994). The short duration of this interval would imply impact even closer to a polarity reversal. But at present, the 75-Ma age seems most compatible with impact in Chron C33n, especially by comparison with the most recent revision of the time scale (Cande and Kent, 1995).

Magnetization of a duration extending across a polarity reversal could well have been caused by hydrothermal circulation activity that commonly follows large impacts (e.g., Newsom, 1980; Koeberl et al., 1989; McCarville and Crossey, 1996). This activity easily could extend the duration of the magnetizing processes in impact rocks beyond simple cooling from the mechanical events of impact. The hydrothermal activity both chemically alters portions of the impacted rocks and precipitates new minerals (see Newsom, 1980; Komor et al., 1988; Koeberl et al., 1989; Masaitis and Naumov, 1993; McCarville and Crossey, 1996).

Detailed mineralogical examination of the Manson impact rocks has demonstrated that a high-temperature hydrothermal system operated after the impact (Crossey and McCarville, 1994; McCarville and Crossey, 1994, 1996). Diagenetic minerals formed by that activity include garnet, indicating by comparison to modern geothermal systems, that fluid temperatures were 275 to 360 °C (Crossey and McCarville, 1994; McCarville and Crossey, 1994, 1996). Further, thermal modeling of the Manson structure led Crossey and McCarville (1994) to estimate that the time required to cool to about 50 to 75 °C was more than 1 m.y. A duration of elevated temperature for a significant portion of 1 m.y., the high fluid temperatures, and impact during a time of frequent changes of geomagnetic polarity make it quite likely that the Manson rocks could have been magnetized in two polarities of the geomagnetic field.

The other unusual aspect of the impact magnetizations, the 300 to 400 °C unblocking temperatures, striking because of its occurrence in both impact structures, also seems to demand a mechanistic origin related to impact processes. The presence of 300 to 400 °C unblocking temperatures is ubiquitous in all three of the impact breccias examined at Manson, including, to a lesser degree, the sedimentary breccia at the top of the crater-fill sequence. Furthermore, more detailed examination of the extensive data set of the Manson structure shows that the temperature of the lower-temperature unblocking increases with depth in that structure (Fig. 11). Figure 11 displays all samples from the melt- and granular-matrix breccias from the M1 core. Both clasts and matrix show the phenomenon, and it does not differ between clasts and matrix (Figs. 11 and 4).

As Figure 11 shows, nearly all samples exhibit a 500 to 540 °C unblocking temperature and fully three-fourths of all samples exhibit 300 to 400 °C unblocking temperatures. But the unblocking temperature increases down-core from an average of 300 °C to about 375 °C. The change appears to occur abruptly, although a continuous line also could be fit to the data. An abrupt change appears to occur at about 168-m (550-ft) depth; this level is slightly below the base of a transitional zone (146 to 162 m or 480 to 530 ft), in which lithologies of the melt-matrix and granular-matrix breccias are intermixed.

Essentially, Figure 11 indicates that the melt-matrix breccia has lower low-temperature unblocking temperatures than the granular-matrix breccia. A likely explanation for such a rela-

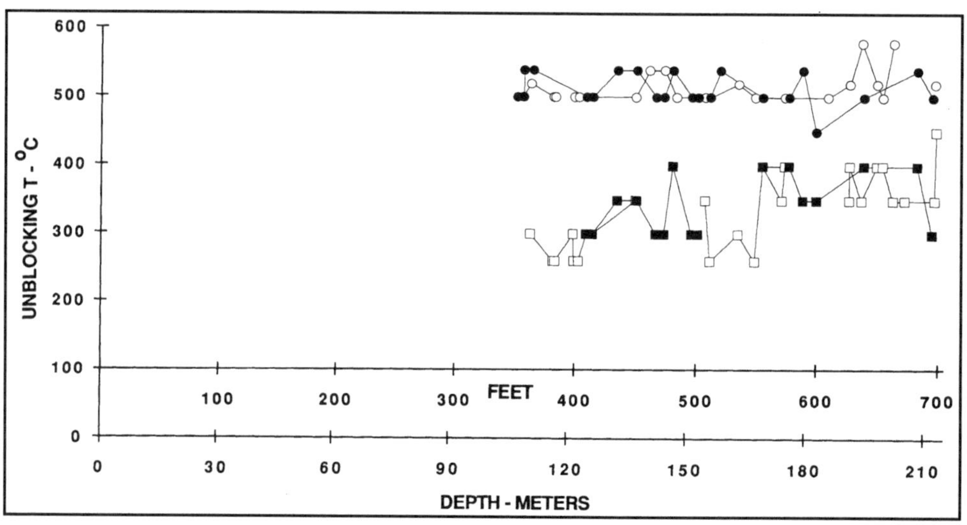

Figure 11. The unblocking temperatures of major remanence loss(es), shown for all samples from Manson core M1. Matrix and clast samples are distinguished by solid and open symbols respectively.

tionship is that these unblocking temperatures are related to the hydrothermal activity. The depth relationship could be explained in several ways. During hydrothermal activity, the granular-matrix breccia may have been subjected to higher temperatures than the melt-matrix breccia because of its greater depth, or the relative differences in porosities of the breccia matrixes could cause a difference in the amount of hydrothermal fluids that circulated in each, thus causing a difference in the temperatures reached within the two breccias. Another, or an additional, explanation may be that the base of the far more impermeable melt-matrix breccia may have acted as a barrier to fluid circulation, thus limiting the temperatures reached in the melt-matrix breccia.

The presence of these anomalous and quite similar low unblocking temperatures in both the Chicxulub and Manson structures appears to be a strong argument for an origin by an impact-related mechanism, and postimpact hydrothermal circulation is a likely candidate. It cannot be coincidence that two large impact structures both show such unblocking anomalies and that neither shows any mineralogical origin; hence, the origin must be in a process common to both. Although the mechanism that generated the remanent magnetization unblocked at 300 to 400 °C presently is not known, the elevated temperature VRM discussed earlier is a very reasonable possibility. If temperatures, particularly at Manson, remained high for 0.5 to 1.0 m.y., a thermoviscous remanence in magnetite seems inescapable.

Hydrothermal activity appears to provide the most logical vehicle for creating both the narrowly blocked, low-temperature (300 to 400 °C) magnetization at Manson and Chicxulub and the high-temperature opposite-polarity remanence at Manson. The hypothesis of hydrothermal activity as the origin for these unusual remanence features suggests the following scenario. Initial acquisition of magnetization occurred with the relatively rapid cooling of the melt rock and associated breccias. Hydrothermal circulation began soon after impact, while the geomagnetic field still maintained the same polarity as at impact, and caused additional acquisition of the characteristic polarity magnetization with 300 to 400 °C unblocking temperatures. That the geomagnetic field had not changed polarity is indicated by the fact that the remanence removed at 300 to 400 °C is identical in direction, in both structures, to that lost at ≥500 °C. Subsequently, the field polarity changed and a remanence in the opposite polarity was acquired in the Manson structure. The polarity change was presumably during a late stage of hydrothermal activity in order for the oxygen fugacity to have become more oxidizing, allowing hematite to be generated and imprinting a weak, opposite-polarity remanence observed above 580 °C.

CONCLUSIONS

The rocks filling the Manson and Chicxulub impact structures show nearly identical remanent magnetization characteristics, and the remanence of both contains several of the same anomalous features. Rocks of both structures show low unblocking temperatures in magnetite, between 300 to 400 °C, in addition to unblocking temperatures at ≥500 °C. Both structures also display a low-stability magnetization of opposite, approximately antipodal, polarity to the characteristic magnetization. Additionally, the Manson structure displays opposite-polarity remanence at unblocking temperatures above the magnetite Curie point.

No evidence was found in the rocks of either structure of a magnetic carrier that would, by virtue of Curie or Neel temperatures or inversion, lose its remanence between 300 and 400 °C. The principal magnetic mineral is magnetite; therefore, the remanent magnetization held in 300 to 400 °C unblocking temperatures is unexplained by mineralogy. The low–unblocking temperature opposite-polarity magnetization may be explicable by mineralogy, as highly oxidized magnetite, probably transformed to hemo-ilmenite, pseudobrookite, and rutile, was observed in some of the Manson rocks. However, the high-temperature opposite-polarity remanence presence is unexplained by mineralogy.

Hydrothermal activity is the likely mechanism that gave rise to both the prominent low–unblocking temperatures and the high–unblocking temperature opposite-polarity remanence. Recent studies have shown quite high temperatures associated with the hydrothermal activity at Manson, and thermal modeling suggested that cooling may have extended up to 1 m.y. after impact. The geomagnetic field reversal frequency in the Late Cretaceous was such that the Manson impact occurred near the time of a polarity reversal (between 0.7 and 1.5 m.y.). Hydrothermal circulation of up to 1 m.y. duration may have caused magnetization to extend into the succeeding, postimpact polarity interval of the geomagnetic field. High temperatures for such a significant time postimpact would probably create a thermo-viscous magnetization in magnetite, hence the prominent 300 to 400 °C unblocking temperatures. More detailed experiments with these rocks may reveal additional constraints on the temperature and duration of the hydrothermal activity.

ACKNOWLEDGMENTS

I am grateful to the National Science Foundation for support of this study (NSF grant EAR-9209112). I gratefully acknowledge Walter Alvarez for providing the samples of Chicxulub rocks. Discussions with Laura Crossey, Peter Shive, Ron Frost, David Dunlop, Ozden Ozdmir, Ken Hoffman, and Subir Banerjee have improved the concepts advanced in this paper, and I deeply appreciate their input.

REFERENCES CITED

Alvarez, W., and Asaro, F., 1990, An extraterrestrial impact: Scientific American, v. 263, p. 78–84.

Anderson, R. R., Witzke, B. J., and Hartung, J. B., 1993, Descriptions and preliminary interpretations of cores recovered from the Manson Impact Structure (Iowa): Lunar and Planetary Science Conference, 24th, p. 35–36.

Bohor, B. F., and Izett, G. A., 1986, Worldwide size distribution of shocked quartz at the K/T boundary: Evidence for a North American impact site:

Lunar and Planetary Science Conference, 18th, Pt. 1, p. 68–69.

Cande, S. C., and Kent, D. V., 1992, A new geomagnetic polarity time scale for the Late Cretaceous and Cenozoic: Journal of Geophysical Research, v. 97, p. 13917–13951.

Cande, S. C., and Kent, D. V., 1995, Revised calibration of the geomagnetic polarity timescale for the Late Cretaceous and Cenozoic: Journal of Geophysical Research, v. 100, p. 6093–6095.

Chao, E. T. C., 1968, Pressure and temperature histories of impact metamorphosed rocks—based on petrographic observations, *in* French, B. M., and Short, N. M., eds., Shock metamorphism of natural materials: Baltimore, Maryland, Mono Book Corporation, p. 135–158.

Crossey, L. J., and McCarville, P., 1994, Post-impact hydrothermal systems: Manson Impact Structure, *in* Proceedings, International Symposium on the Observation of the Continental Crust Through Drilling, 7th, Santa Fe, New Mexico, p. 211–214.

Crossey, L. J., Kudo, A., and McCarville, P., 1994a, Post-impact fluid flow in complex craters: An example from the Manson impact structure: Geological Society of America Abstracts with Programs, v. 26, p. A-403.

Crossley, L. J., Kudo, A. M., and McCarville, P., 1994b, Post-impact hydrothermal systems: Manson Impact Structure, Iowa: Lunar and Planetary Science Conference, 25th, p. 299–300.

Dalyrmple, B., and Izett, G., 1994, ^{40}Ar/^{39}Ar age of the Manson impact structure, Iowa: Geological Society of America Abstracts with Programs, v. 26, p. A-337.

El Gorsey, A., 1968, The oqaque minerals in impactite glasses, *in* French, B. M., and Short, N. M., eds., Shock metamorphism of natural materials: Baltimore, Maryland, Mono Book Corporation, p. 531–553.

Gordon, R., and Van der Voo, R., 1995, Mean paleomagnetic poles for the major continents and the Pacific Plate, *in* Ahrens, T. J., ed., Gobal earth physics: A handbook of physical constants: AGU Reference Shelf 1, Washington, D.C., American Geophysical Union, p.–228.

Gradstein, F. M., Agterberg, F. P., Ogg, J. G., Hardenbol, J., van Veen, P., Thierry, J., and Huang, Z., 1994, A Mesozoic timescale: Journal of Geophysical Research, v. 99, p. 24051–24074.

Haggerty, S. E., 1976, Oxidation of opaque mineral oxides in basalts, *in* Rumble, D., III, ed., Oxide minerals, short course notes, Volume 3: Washington, D.C., Mineralogical Society of America, p. Hg1-100.

Hartung, J. B., Kunk, M. J., and Anderson, R. R., 1990, Geology, geophysics, and geochronology of the Manson impact structure: Geological Society of America Special Paper 247, p. 207–221.

Hoppin, R. A., and Dryden, J. E., 1958, An unusual occurrence of Pre-Cambrian crystalline rocks beneath glacial drift near Manson, Iowa: Journal of Geology, v. 66, p. 694–699.

Izett, G. A., 1990, The Cretaceous-Tertiary (K/T) boundary interval Raton basin, Colorado and New Mexico, and its content of minerals—Evidence relevant to the Cretaceous-Tertiary impact-extinction theory: Geological Society of America Special Paper 249, 100 p.

Izett, G. A., Cobban, W. A., Obradovich, J. D., and Kunk, M. J., 1993a, The Manson Impact Structure: ^{40}Ar/^{39}Ar age and its distal impact eject in the Pierre Shale in southeastern South Dakota: Science, v. 262, p. 729–732.

Izett, G. A., Reynolds, R. L., Rosenbaum, J. G., and Nishi, J. M., 1993b, A discontinuous melt sheet in the Manson Impact Structure [abs.]: Lunar and Planetary Science Conference, 24th, p. 705–706.

Kelso, P. R., and Banerjee, S. K., 1993, Elevated temperature viscous remanent magnetization of natural and synthetic multidomain magnetite: Earth and Planetary Science Letters, v. 122, p. 43–56.

Koeberl, C., Fredriksson, K., Gotzinger, M., and Reimold, W. U., 1989, Anomalous quartz from the Roter Kamm impact crater, Namibia: Evidence for post-impact hydrothermal activity?: Geochimica et Cosmochimica Acta, v. 53, p. 2113–2118.

Komor, S. C., Valley, J. W., Brown, P. E., and Collini, B., 1988, Fluid inclusions in granite from the Siljan Ring impact structure and surrounding regions, *in* Boden, A., and Eriksson, K. G., eds., Deep drilling in crystalline bedrock, Volume 1: Berlin, Springer-Verlag, p. 180–208.

Kunk, M. J., Izett, G. A., Haugerud, R. A., and Sutter, J. F., 1989, ^{40}Ar/^{39}Ar dating of the Manson Impact Structure: A Cretaceous-Tertiary boundary crater candidate: Science, v. 244, p. 1565–1568.

Masaitis, V. L., and Naumov, M. V., 1993, Puchezh-Katunki impact crater: Preliminary model of hydrothermal circulation system: Meteoritics, v. 28, p. 390–391.

McCarville, P., and Crossey, L. J., 1994, Post-impact alteration of the Manson impact structure; Manson, Iowa: Geological Society of America Abstracts with Programs, v. 26, p. A-397.

McCarville, P., and Crossley, L. J., 1996, Post-impact alteration of the Manson Impact Structure, *in* Koeberl, C., and Anderson, R., eds., The Manson Impact Structure: Geological Society of America Special Paper 302 p. 347–376.

Newsom, H. E., 1980, Hydrothermal alteration of impact melt sheets with implications for Mars: Icarus, v. 44, p. 207–216.

Nord, G. L., and Lawson, C. A., 1992, Magnetic properties of ilmenite$_{70}$-hematite$_{30}$: Effect of transformation-induced twin boundaries: Journal of Geophysical Research, v. 97, p. 10897–10910.

Personen, L. J., Marcos, N., and Pipping, F., 1992, Paleomagnetism of the Lappajarvi impact structure: Tectonophysics, v. 216, p. 123–142.

Pindell, J., 1985, Alleghenian reconstruction and subsequent evolution of the Gulf of Mexico, Bahamas and proto-Caribbean: Tectonics, v. 4, p. 1–39.

Pohl, J., 1971, On the origin of magnetization of impact breccias on earth: Zeitschrift für Geophysik, v. 37, p. 549–555.

Roddy, D. J., and Shoemaker, E. M., 1994, Manson Impact Structure research program: Summary through February 1994, *in* Proceedings, International Symposium on the Observation of the Continental Crust Through Drilling, 7th: Santa Fe, New Mexico, p. 1–5.

Sharpton, V. L., Dalrymple, G. B., Marin, L. E., Ryder, G., Schuraytz, B. C., and Urrutia-Fucugauchi, J., 1992, New links between the Chicxulub impact structure and the Cretaceous-Tertiary boundary: Nature, v. 359, p. 819–821.

Steiner, M. B., and Shoemaker, E. M., 1993a, Two-polarity magnetization in the Manson impact breccia: Lunar and Planetary Science Conference, 24th, p. 1347–1348.

Steiner, M. B., and Shoemaker, E. M., 1993b, The Late Cretaceous Manson Impact Structure: Eos (Transactions, American Geophysical Union), v. 74, p. 386.

Steiner, M. B., and Shoemaker, E. M., 1994, The Late Cretaceous Manson Impact Structure: Eos, (Transactions, American Geophysical Union), v. 75, p. 123.

Steiner, M. B., and Walker, J. D., 1996, Late Silurian plutons on the Yucatan Peninsula: Journal of Geophysical Research (in press).

Thierstein, H. R., 1982, Terminal Cretaceous plankton extinctions: A critical assessment, *in* Silver, L. T., and Schultz, P. H., eds., Geological implications of impacts of large asteroids and comets on the Earth: Geological Society of America Special Paper 190, p. 385–399.

Urrutia-Fucugauchi, J., Marin, L., Sharpton, V. L., and Quezada, J. M., 1993, Reverse polarity magnetized melt rocks from the Chicxulub impact structure, Yucatan Peninsula, Mexico: Lunar and Planetary Science Conference, 24th, p. 1455.

Urrutia-Fucugauchi, J., Marin, L., and Sharpton, V. L., 1994, Reverse-polarity magnetized melt rocks from the KT Chicxulub structure, Yucatan Peninsula, Mexico, *in* New developments regarding the KT event and other catastrophes in Earth history: Houston, Texas, Lunar and Planetary Science Institute Contribution 825, p. 126–127.

MANUSCRIPT ACCEPTED BY THE SOCIETY SEPTEMBER 14, 1995

Printed in U.S.A.

Geological Society of America
Special Paper 307
1996

Mega-impact melt petrology (Chicxulub, Sudbury, and the Moon): Effects of scale and other factors on potential for fractional crystallization and development of cumulates

Paul H. Warren
Institute of Geophysics, University of California, Los Angeles, California 90095-1567; and Mineralogical Institute, Graduate School of Science, University of Tokyo, Tokyo 113, Japan
Philippe Claeys*
Department of Geology and Geophysics, University of California, Berkeley, California 94720
Esteban Cedillo-Pardo
Instituto Mexicano del Petróleo, Eje Central Lazaro Cardenas 152, 07739 Mexico DF, Mexico

ABSTRACT

Compared to the granophyre that forms the upper one-third of the Sudbury Igneous Complex, available melt products from the Chicxulub Basin are consistently more fine grained, and their compositions are more andesitic (less differentiated from bulk continental crust). Available lunar impact melt products are also fine grained and little differentiated among themselves. The origin of the Sudbury Igneous Complex is controversial, but assuming it formed mainly as impact melt, the size of the Complex and the extent of vertical layering within it were probably strongly influenced by the high melting/displacement ratio engendered because the impact occurred in a region previously warmed by the Penokean Orogeny. Potential for impact melt to undergo fractional crystallization is probably correlated with size of the impact structure, but the most important variable, the melting/displacement (*m*/*d*) ratio, is not a simple function of structure diameter. Planetary *g* and preimpact *T* of the melting zone both strongly influence *m*/*d*, which in turn influences (1) the proportion of suspended clastic debris in the melt, (2) the average *T* of these clasts (which in any event are much colder than, and thus act to chill, the melt), and (3) the proportion of the melt that is disseminated by ejection from the transient crater. The magnitude of these shifts can only be crudely calibrated, due to uncertainties regarding the shape of the melting zone, the size/shape of the ejection zone in relation to the transient crater, and the manner in which modification (collapse) of the transient crater mingles clastic debris with the impact melt. But clearly, (1) and (2) significantly increase, while (3) significantly decreases, as *m*/*d* varies from the low values (≤0.2) associated with simple craters to the extremely high values (0.45–0.6) associated with Chicxulub, Sudbury, and the largest lunar basins. In lunar impacts, differentiation of the impact melt is hampered by systematically lower *m*/*d* compared to terrestrial craters of similar size and by the comparatively high (solid crust like) density of typical impact melts. However, in a few of the largest impacts, most of the unejected melt forms in the upper mantle, where it probably remains while crystallizing into a series of mafic cumulates essentially similar to the preimpact upper mantle. On the earliest Moon (the "magma

*Present address: Institut für Mineralogie, Museum für Naturkunde, Invalidenstrasse 43, D-10015 Berlin, Germany.

Warren, P. H., Claeys, P., and Cedillo-Pardo, E., 1996, Mega-impact melt petrology (Chicxulub, Sudbury, and the Moon): Effects of scale and other factors on potential for fractional crystallization and development of cumulates, *in* Ryder, G., Fastovsky, D., and Gartner, S., eds., The Cretaceous-Tertiary Event and Other Catastrophes in Earth History: Boulder, Colorado, Geological Society of America Special Paper 307.

ocean" era and the first few hundred million years thereafter) the background temperature of upper mantle was so hot that episodes of large-scale impact melting must have been practically indistinguishable from pulses of enhanced endogenous magmatism. If the putative Procellarum basin formed with a transient crater diameter of ~1,450 km, most of the impact melt–derived rocks in the western nearside megaregolith should be a distinctively mantle-rich (roughly 50%) variety formed and ejected in this one impact.

INTRODUCTION

Interest in the Chicxulub (Yucatán) structure usually focuses on the catastrophic global-ecological effects associated with formation of a Phanerozoic impact basin (Penfield and Camargo Z., 1981; Swisher et al., 1992), but Chicxulub is also inherently interesting as one of the two or three largest impact structures on Earth. Sharpton et al. (1993) identify four rings, the widest of which has diameter $D \sim 300$ km; suggesting that the most prominent ring, with $D \sim 170$ km, is close to the rim of the transient crater. In contrast, Hildebrand et al. (1995) advocate a structural D of ~170 km and a transient crater D of ~90 km. For purposes of discussion, we will take the structural "rim" diameter of the basin to be ~240 km, which conveniently coincides with the average of several recent estimates for the Sudbury (Ontario) basin (Deutsch and Grieve, 1994; Spray and Thompson, 1995). The only possibly larger impact basin recognizable on Earth is the deeply eroded Vredefort (South Africa) structure, for which Therriault et al. (1993) estimate D was originally 190 to 300 km. It should be admitted that some skepticism persists toward the impact hypothesis for all three of these structures (e.g., Meyerhoff et al., 1994).

One of the most important effects of a large impact is the sudden conversion of nearly all of the impactor's kinetic energy into heat, mostly in a roughly hemispherical region extending deep below the epicenter of the impact (Melosh, 1989). Vast volumes of impact melt may be produced, and at least in principle these melts may undergo fractional crystallization (i.e., form cumulates) to produce a layered complex. Understanding the fate of melts generated by basin-scale impacts is crucial to interpretation of early crustal genesis. The ancient lunar crust is scarred by numerous multiring impact basins: The compilation of Wilhelms (1987) includes 17 definite basins (plus 14 "probable and possible" ones) with apparent diameter ≥500 km. It has been suggested (e.g., Alfvén and Arrhenius, 1976; Wetherill, 1981) that impact melting was primarily responsible for differentiating the Moon's large (64-km thick) crust and that many highland rocks customarily interpreted as products of endogenous lunar magmatism (e.g., Warren and Wasson, 1977) may actually be impact melt products (Grieve et al., 1991). Distinguishing between the roles of impact-generated magmatism versus endogenous, mantle-derived magmatism is one of the most fundamental problems of lunar science, with profound implications regarding thermal and geochemical evolution of the Moon and indirectly of its "sister planet" Earth.

A debate in the mid-1970s (e.g., Warner et al., 1974; Wood, 1975) led to a general consensus that even large-scale impact melts undergo little differentiation. This conclusion was based primarily on (1) observations that terrestrial impact melt products tend to be chemically homogeneous (Phinney and Simonds, 1977; Grieve and Floran, 1978) and (2) an inference that thorough mixing with, and digestion of, cooler clastic debris causes impact melts to cool too rapidly for significant differentiation (Simonds et al., 1976; Onorato et al., 1978). Also supporting this interpretation, the relatively few lunar rocks with preserved coarse cumulate textures tend to be vastly more compositionally diverse than fine-grained, siderophile-rich lunar rocks of obvious impact melt derivation (e.g., Warren and Wasson, 1977; Norman, 1994b). Some skepticism persists (Delano and Ringwood, 1978), but for most of the past two decades it has been widely assumed that even basin-scale impact melts do not undergo significant igneous differentiation (e.g., Simonds et al., 1976; Spudis and Ryder, 1981; Lindstrom et al., 1990).

Since the 1970s, much has been learned about the theory of impact processes (e.g., Melosh, 1989) and about the petrology of lunar and terrestrial impact craters. Some of the most significant new observations pertain to the Sudbury Igneous Complex (SIC), a sill-like body of grossly layered, medium- to coarse-grained, roughly noritic igneous rocks near the center of the Sudbury structure. The SIC is famous for the rich Ni-Cu sulfide deposits associated with the "Sublayer" igneous rocks and Footwall Breccia that form its basal contact. The SIC was formerly interpreted, even by geologists recognizing the impact origin of the overall structure, as a product of mainly endogenous terrestrial heating, that is, essentially a "normal" layered intrusion, formed as an indirect consequence of the impact. However, estimates of the structure's size have recently increased (Deutsch and Grieve, 1994; Spray and Thompson, 1995) to the point at which the implied volume of impact melt is nearly 10 times the estimated (present-day, partly eroded) mass of the SIC (Grieve and Cintala, 1992). Also, isotopic evidence (such as highly positive ε_{Nd}, highly negative ε_{Sr}) suggests that the bulk of the material of the SIC is of crustal, not mantle, derivation (Faggart et al., 1985; Naldrett et al., 1986; Deutsch et al., 1989; Walker et al, 1991). Thus, the SIC has recently been interpreted as a product of igneous differentiation (i.e., fractional crystallization) of a large mass of impact melt, with negligible addition of endogenously derived magma (Grieve et al., 1991; Avermann et al., 1992). This interpretation is still very controversial (e.g., Chai and Eckstrand, 1993; Norman, 1994b).

Lunar samples with clear indications of impact melt origin

(e.g., high siderophile concentrations) are common, and although the vast majority have broadly noritic to anorthositic-noritic bulk compositions, at several sites close to individual large basins distinctive secondary compositional traits prevail, and such groups of samples are generally viewed as characteristic of the individual nearby basins (e.g., Lindstrom et al., 1990; McKinley et al., 1984). The best-sampled lunar basin is probably Serenitatis (structural $D \sim 740$ to 880 km), the margin of which was sampled by Apollo 17; sample linkages with the Imbrium (~1160 km) and Nectaris (~860 km) basins are widely invoked but not so firmly constrained (Wilhelms, 1987). Like the "melt sheets" of terrestrial craters smaller than Chicxulub and Sudbury (Phinney and Simonds, 1977), lunar impact melt-derived samples of apparent basin provenance tend to be rather uniform in composition, but Grieve et al. (1991) and to a lesser degree Spudis (1992) have recently questioned the assumption that basin-scale lunar impact melts undergo no significant igneous differentiation.

In this chapter, we endeavor to apply constraints from modern cratering theory (e.g., Melosh, 1989; Chen et al., 1994) in conjunction with petrologic observations, including some new data for Chicxulub melt rocks, to assess effects of scale as well as other factors on potential for fractional crystallization and cumulate development by large-scale impact melts.

PETROLOGIC COMPARISONS: SUDBURY, THE MOON, CHICXULUB

Sudbury

The Sudbury Igneous Complex features pronounced large-scale (vertical) compositional layering and is occasionally likened to classic layered intrusions of the Skaergaard-Stillwater variety. As reviewed by Naldrett and Hewins (1984), SIC rocks range from a mafic norite with pyroxenes that have Mg/(Mg + Fe) as high as 78 mol% to a thick capping granophyre, estimated (Collins, 1934) to make up ~63% of the total Complex and consisting of ~90% quartz + feldspar with only 10% mafic minerals, including pyroxenes with Mg/(Mg + Fe) as low as 25 mol%. The granophyre layer is enriched in K and incompatible elements by a factor of ~2.8 and depleted in CaO by a factor of ~3.6, compared to the norite layer (Collins, 1934; Kuo and Crocket, 1979; Chai and Eckstrand, 1993). However, in several respects the SIC is relatively undifferentiated. The SIC rocks are comparatively fine grained (grains seldom more than 2 mm in longest dimension), they completely lack graded or fine-scale igneous layering, and even contacts between the large-scale layers are strictly gradational (Naldrett and Hewins, 1984). An intrusion emplaced in mainly a single stage, as impact melt probably tends to be emplaced, cannot be expected to engender layering as complex as that found in typical episodically intruded igneous complexes. Nonetheless, considering the stark compositional disparity between the norite and granophyre layers, Chai and Eckstrand (1993) infer from remarkably uniform trace element contents within each of these layers that the SIC formed by intrusion of two separate magmas. These magmas partly intermixed, but in both cases internally differentiated (formed cumulates) only to a very slight extent.

Textures of the noritic and gabbroic SIC rocks were interpreted by Naldrett and coworkers (e.g., Naldrett and Hewins, 1984) as indicative of cumulus origin. However, Naldrett and Hewins (1984) did not specify any basis for this inference. The term "hypidiomorphic granular," applied by these authors to most of the noritic and gabbroic rocks, is seldom associated with cumulates. None of the examples of SIC "main mass" textures shown by Naldrett and Hewins (1984) manifestly features a cumulus framework as required to meet Irvine's (1982) carefully formulated definition of a cumulate (cf. Wadsworth, 1985). Undoubted cumulus crystals (Wager and Brown, 1967) are generally coarser than these examples (these show 1.5 mm maximum dimension and typically have aspect ratios of 3:1). Warren and Wasson (1977) stipulated grain size ≥3 mm in their listing of possible indications that a lunar rock is of "pristine" non–impact melt origin. The "main mass" SIC textures shown by Naldrett and Hewins (1984) would have been interpreted by Warren and Wasson (1977) or Warren (1990) as probable impact melt products.

The scarcity of perceptible clasts in most SIC layers (all except the uppermost granophyre—cf. the section "Interpretation of impact 'melt rock' textures: Clast Darwinism") does not guarantee that these layers were once completely molten. In addition to the trace element evidence of Chai and Eckstrand (1993), Sr isotopic data can be interpreted to suggest that the SIC was never even grossly homogeneous (Naldrett et al., 1986; Deutsch et al., 1990). It should be admitted, however, that Gibbins and McNutt (1975) found relatively little $^{87}Sr/^{86}Sr$ (initial) diversity among SIC rocks, apart from a strong norite/ granophyre disparity, and suggested that Sr isotopes in the upper part of the SIC were pervasively altered by metamorphism. Naldrett et al. (1986) also reported a large spread in ε_{Nd}, roughly correlating with their Sr-initial diversity, but a set of possibly more precise Nd measurements by Faggart et al. (1985), unfortunately without accompanying Sr measurements, indicates rather uniform ε_{Nd}.

Above the SIC, extending the full length and width of the Sudbury Basin, is a complex layer of breccias known as the Onaping Formation, with an estimated thickness of 1.6 to 1.8 km (Muir and Peredery, 1984; Avermann and Brockmeyer, 1992). Grieve et al. (1991) interpret the upper 60 to 70% of the Onaping as a welded mass of back-fallen ejecta mixed with subordinate, heterogeneously distributed impact melt, but they interpret the lower 30 to 40% (i.e., the ~300-m-thick Basal Member and apparently the "melt body" parts of the Gray Member) as part of the main Sudbury "melt sheet." Avermann and Brockmeyer (1992) likewise interpret the Basal Member (and related "isolated bodies in the Gray and Black Members") as the "upper zone of the Sudbury impact melt system," even though they describe the lower half of the Basal Member as crystalline melt breccia consisting of "~80–90%" clasts.

Chicxulub

We have studied seven thin sections of impact melt–derived rocks from Chicxulub: one from core C1, position N9, near the geographical center of the structure and a short distance (≤4 m) above the C1/N10 sample of Sharpton et al. (1992) and Schuraytz et al. (1994); and three each from positions N17 and N19 of core Y6, the same (to within a few m) as the original positions of samples described previously by Hildebrand et al. (1991), Kring and Boynton (1992), Sharpton et al. (1992), and Schuraytz et al. (1994). Our samples came from ~1,390 (C1/N9), ~1,297 (Y6/N17), and ~1,378 (Y6/N19) m below sea level.

The matrix portions of all six Y6 thin sections are aphanitic and are dominated by grains <10 μm across (Fig. 1), albeit leucocratic, texturally diverse but generally much coarser grained lithic clasts are incongruously sprinkled throughout (~20 vol%). The extraordinarily wide range of feldspar compositions in the Y6/N19 samples (Fig. 2a) tends to confirm the textural evidence for origin as a clast-rich impact magma (cf. Cedillo-Pardo et al., 1992). Pyroxene compositions show considerable variation between matrix portions of two nearby Y6/N19 samples. Aside from En-Fs-Wo variations (Fig. 2b), the Y6/N19-10 pyroxenes are remarkably rich in Al_2O_3: 7/20 analyses >4.0 wt%, maximum = 5.6%, whereas the Y6/N19-11 pyroxenes all have ≤1.42 wt%.

Making up roughly half of the Y6/N19-11 thin section is a large fragment of a granophyric or myrmekitic rock (Fig. 1e) of extraordinary mineralogy: The only major minerals are quartz + feldspar (compositionally diverse: Fig. 2), and ~8% anhedral, locally vermicular, anhydrite. Superficially (i.e., mainly quartz + feldspar, intergrown in a granophyric texture) this fragment resembles the granophyre layer that constitutes the upper half of the Sudbury Igneous Complex (Naldrett and Hewins, 1984). The fragment is not genetically analogous, however, because it contains at least two grains of quartz with multiple sets of planar elements indicative of shock metamorphism. Thus, at least the quartz (and presumably also the intergrown feldspar) of this clast was part of the basement at the time of the impact. The origin of the anhydrite is an interesting second-order question. Its largely vermicular habit tends to suggest that it (or more likely melt or vapor that precipitated it) was injected into a porously brecciated granitic protolith during the early stages of cooling of the impact melt. Volcanic eruptions occasionally produce anhydrite, in cases in which the parent magma features unusually high sulfur content and probably also high oxygen fugacity (Carroll and Webster, 1994).

Among our samples, the coarsest matrix by far is that of C1/N9, with grains mostly ~0.3 mm in maximum dimension. Texturally and mineralogically, this sample strongly resembles C1/N10 (Schuraytz et al., 1994). Only six clasts are discernible, all very fine grained (~20 μm), totaling ~2 vol% of the rock (Fig. 1a). A minor proportion (1%) of the matrix consists of isolated uncommonly coarse (up to 1.6 mm) feldspar and pyroxene grains similar to those described as phenocrysts (up to 1 mm) in C1/N10 by Schuraytz et al. (1994). We retain this term for C1/N9 grains coarser than twice the prevalent 0.3 mm size, including three of the most diopside-rich pyroxenes analyzed (Fig. 2b). However, in C1/N9 the-grain size distribution (apart from the matrix/clast contrast) is not markedly bimodal, and we suspect the coarser grains are strongly altered relicts from a suite of disintegrated target rocks.

C1/N9 and C1/N10 are the coarsest Chicxulub impact melt products yet sampled. Most other impact melt-like rocks (from elsewhere in the C1 and Y6 cores and from the S1 core) are described as "glass" or "microcrystalline" (Hildebrand et al., 1991; Quezada Muñeton et al., 1992; Kring and Boynton, 1992) or as "fine- to medium-grained coherent crystalline" (Sharpton et al., 1992). Sharpton et al. (1992) described Y6/N19 as having a "medium- to coarse-grained melt rock matrix," but a more detailed recent description by three of the same authors (Schuraytz et al., 1994) indicates that their sample from Y6/N19, like the two we studied, consists mainly of an aphanitic matrix with grains ~10 μm across. The textural contrast between C1/N9 (and C1/N10) and the other Chicxulub samples follows an important trend among all impact melt products: Matrix grain size tends to anticorrelate with abundance of clasts (Simonds et al., 1976). This trend arises because abundant clasts have a chilling effect on the melt.

Compositionally, the sampled Chicxulub impact melt products are similar to the bulk continental crust, and to the overall composition of the Sudbury Igneous Complex (Table 1). But they are consistently SiO_2 poor, MgO rich, and K_2O poor in comparison to the *upper* (granophyre) portion of the SIC (SiO_2 ~ 68.4wt%) and even in comparison to the scattered impact-melted components of the Onaping Formation (e.g., the relatively large "melt bodies," average SiO_2 ~ 69.0 wt%), based on numerous analyses by Muir and Peredery (1984) and Chai and Eckstrand (1993). If the Chicxulub impact magma initially formed with a composition similar to bulk SIC (which is very similar to bulk continental crust), the Chicxulub samples show little effect of the fractional crystallization that putatively produced the vertical layering of the SIC.

All known samples of Chicxulub impact melt products are fine grained and clast rich compared to the main mass of the SIC. Only 150 m from the edge of the SIC, average grain size (maximum dimension) is ~0.75 mm (Naldrett and Hewins, 1984), and SIC rocks almost uniformly lack perceptible clasts. Rocks with textures roughly similar to the Chicxulub samples are fairly common in the Onaping Formation, however, as thin "veinlets" or subrounded fragments "a few m" across and as larger, fluidal textured "melt bodies" with typical dimensions of the order 100 m (Muir and Peredery, 1984). Possibly the available Chicxulub impact melt products are all from an Onaping analog, overlying a SIC-like large mass of relatively coarse grained rocks.

However, the Chicxulub samples are fine grained even compared to analogous Onaping materials. Onaping rocks are either suevitic breccias, with vastly higher fragmental compo-

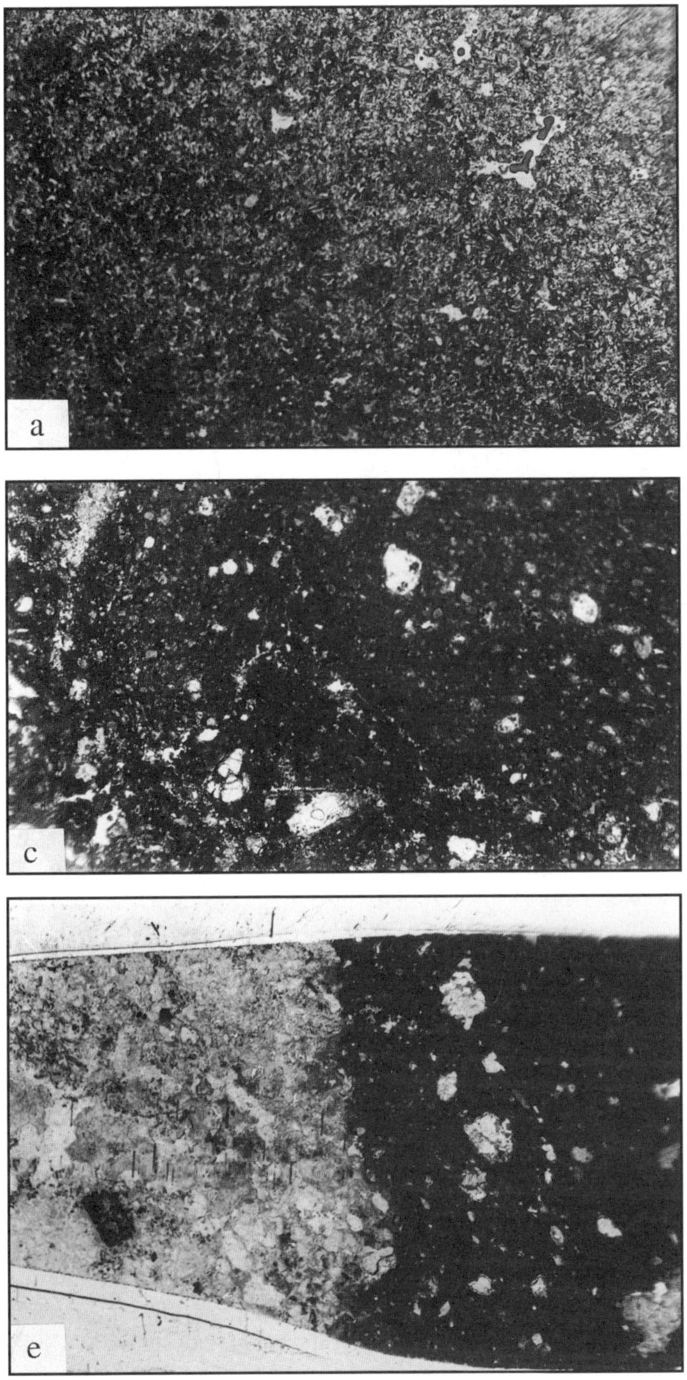

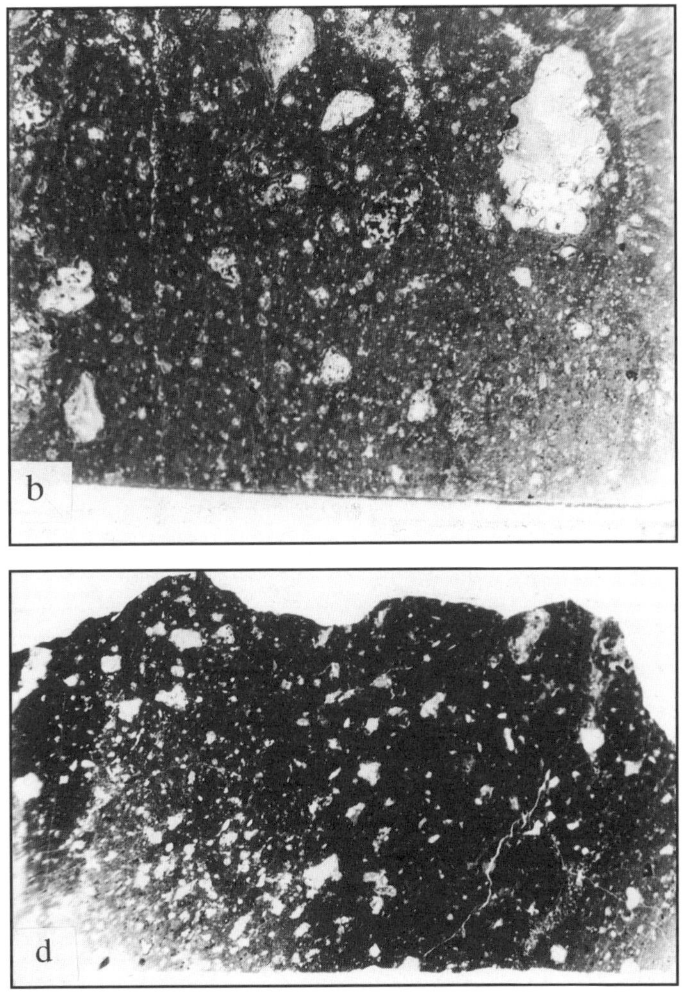

Figure 1. Chicxulub thin sections, viewed in transmitted light, all at similar scale (each view is ~27 mm wide): (a) C1/N9, (b) Y6/N17-2, (c) Y6/N17-9, (d) Y6/N19-10, (e) Y6/N19-11. Left half of (e) shows part of the anhydrite-rich granophyre.

nents (as opposed to melt-derived matrix components) than available Chicxulub impact melt products, or "melt body" materials. The melt bodies typically have aphanitic chilled margins, but their cores are coarse grained compared to the Chicxulub samples: generally 0.2 to 0.5 mm but with quartz up to 4 mm across (Muir and Peredery, 1984). The finer grain size of the Chicxulub samples is probably not a consequence of systematically smaller dimensions of individual Chicxulub "melt bodies," because according to Lopez Ramos (1975) the Chicxulub cores C1, S1, and Y6 all contain layers of pure "extrusive andesite" over 200 m thick. The Y6/N17 samples came from near the middle of a 60-m-thick continuous "extrusive andesite" layer, and the Y6/N19 samples came from ~12 m below the top of a continuous layer ~220 m thick (Meyerhoff et al., 1994). Possibly the Chicxulub impact melt was chilled with unusual efficiency because (as seems likely, at least near the top of the melt complex) the suspended clasts originally included large proportions of carbonate and sulfate, which tended to undergo heat-consuming vaporization (cf. the section "Interpretation of impact 'melt rock' textures: Clast Darwinism"). Another potentially crucial Chicxulub/Sudbury contrast will be discussed in the section "Near-Penokean Sudbury versus stable Chicxulub."

VOLUME OF IMPACT MELT AS A FUNCTION OF CRATER/BASIN SIZE

Large-scale fractional crystallization (formation of cumulates) is only possible where cooling is sufficiently slow to

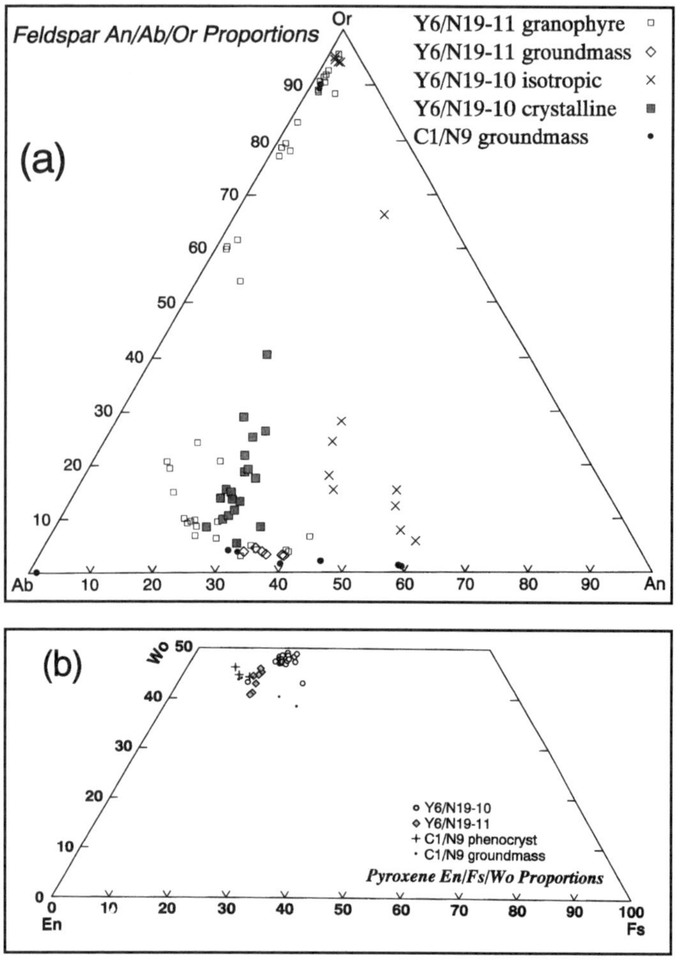

Figure 2. Phase-compositional (electron microprobe) data for Chicxulub impact melt products: (a) An-Ab-Or proportions in feldspars and feldspathic (isotropic) glasses and (b) En-Fs-Wo proportions in pyroxenes. Data for Y6/N19-11 "granitic" feldspars refer to a large clast of anhydrite granophyre that apparently existed (possibly sans anhydrite) before the impact (see text).

allow diffusive processes (such as chemical diffusion and thermal convection) within the magma to deplete the overall melt in the constituents of successive layers of crystals forming (or at least accumulating) only along the fringes of the magma. In general, cooling rate correlates inversely, and potential for convection correlates positively, with the total volume of a magma body. Thus, the total volume of melt formed in an impact is among the most obvious factors in determining the potential for fractional crystallization by the impact melt.

In this chapter, we estimate impact melt volume as a function of crater size by averaging results from two different semiempirical methods. One method is based on equation. 7.10.2 of Melosh (1989):

$$\text{melting/displacement (mass ratio)} = 1.6 \times 10^{-7} (g\,D_{tc})^{0.83}\, v_i^{0.33} \quad (1)$$

where g is acceleration of gravity and v_i is impactor velocity. Following Melosh (1989), we equate the "displacement" volume with the volume of the transient crater. The transient crater is a zone that is mobilized in horizontally outward (including downward) directions during the early stages of crater formation. Subsequent peripheral collapse (and, in large impacts, central uplift) modifies the transient crater into a shallower but wider final crater. Substituting from an equation on p. 119 of Melosh (1989), assuming that depth/D for the transient crater = 0.3, and neglecting possible density contrast between the unmelted displaced matter and the melt, equation (1) implies that the impact melt volume is:

$$V_{melt,1} = 6 \times 10^{-9}\, \pi\, g^{0.83}\, D_{tc}^{2.17}\, v_i^{0.33} \quad (2)$$

The second method is a similar power-law relationship proposed by Grieve and Cintala (1992):

$$V_{melt,2} = c\, D_{tc}^{3.85} \quad (3)$$

where c is a constant related to the impactor density and velocity (for a chondritic impactor at 15 to 25 km s^{-1}, c = 0.000621 to 0.000715) and D_{tc} is the diameter of the transient crater. For nonterrestrial application, results from equation (3) are multiplied times $g^{0.83}$, by analogy to equation (2). For large craters such as those considered in this chapter, equation (2) gives ~30% higher melt volumes than equation (3).

Estimating the transient crater diameter D_{tc} based on the observed "final," "structural" or "crater rim" diameter is problematical, especially for large multiring basins. In this chapter, we approximate the translation by averaging results from the semiempirical models of Chyba (1991) and Croft (1985). These models imply that for Chicxulub, Sudbury, and all the lunar multiring basins that will be mentioned, D_{tc} is uniformly 0.45 to 0.54 × the final D. Differences between results from the Chyba (1991) and Croft (1985) models are almost negligible. Nonetheless, for large basins these models are highly uncertain because both are predicated upon the same reasonable but unproven assumption: that the outermost prominent ring on each multiring basin is analogous to the solitary rim that defines diameter for a smaller crater.

Ring formation may be a fundamentally different process from the collapse that transforms a transient crater into a nonringed crater (Melosh, 1989, 1995). The abundance of large (nominal D ~ 1,000 km) multiring basins on the Moon (and with decreasingly robust statistics also on Mercury and Mars) is three times greater than would be expected based on any simple extrapolation from abundances of smaller peak-ring craters (Strom and Neukum, 1988). One interpretation for this surfeit would be that impactors (asteroids and/or comets) with the right combination of size and velocity to form ~1000-km basins are more common than would be extrapolated from abundances of less powerful impactors. Alternatively, however, the surfeit of ~1000-km basins may be an indication that the outermost prominent ring on a multiring basin is generally far more distant from the rim of the transient crater than would be extrapolated from D_{tc}/D_{rim} relationships among smaller craters. If so,

TABLE 1. COMPOSITIONS OF MEGA-IMPACT MELT PRODUCTS AND SIMILAR(?) MATERIALS, AND CORRESPONDING MODEL MELT DENSITIES AND VISCOSITIES*

	SiO_2	TiO_2	Al_2O_3	FeO	Fe_2O_3	MnO	MgO	CaO	Na_2O	K_2O	P_2O_5	H_2O	Sum	ρ (kg/m³)	μ (kg/m/s)
Chicxulub impact melt rocks (modeled at *T* = 1150° C)															
N10 - Sharpton	64.4	0.5	14.9	4.6	....	0.1	2.8	5.5	4.3	2.7	....	....	99.8		
N17 - lit. avg.	62.1	0.5	13.9	4.8	....	0.1	3.0	9.6	4.6	2.2	0.09	....	100.8		
N19mtrx - Sharpton	61.2	0.4	14.7	3.9	....	0.1	2.7	9.3	2.7	2.9	....	....	97.9		
Average†	62.6	0.5	14.5	4.4†	.......†	0.1	2.8	8.1	3.9†	2.6	0.09	.......†	99.5	2450	670
Sudbury Igneous Complex (modeled at *T* = 1150° C)															
Sudbury, ??initial§	59.4	0.8	15.7	5.8	1.7	0.08	3.9	5.6	3.0	2.2	0.24	1.3	99.7	2530	360
Collins, 1934 average	63.1	0.7	14.7	5.2	1.5	0.09	2.9	4.0	3.2	2.9	0.23	1.2	99.8	2440	720
Earth's continental crust (modeled at *T* = 1150° C)															
Literature average	60.5	0.7	16.4	3.3	3.1	0.14	2.9	5.7	3.4	2.4	0.2	1.2	100.0	2500	270
Chilled margins (possible parent melts) of classic layered intrusions (modeled at *T* = 1230° C)															
Muskox	50.7	1.1	13.6	9.1	1.2	0.18	9.7	11.2	1.8	0.6	0.10	0.59	99.8	2690	13
Skaergaard	48.1	1.2	17.2	8.4	1.3	0.16	8.6	11.4	2.4	0.25	0.10	1.06	100.2	2690	11
Stillwater	50.7	0.5	17.6	9.9	0.3	0.15	7.7	10.5	1.9	0.24	0.09	0.48	99.9	2680	29
Lunar impact melt rocks of possible basin origin (modeled at *T* = 1280° C)															
A15 average (Imbrium)	47.3	1.5	17.8	8.9	0.0	0.12	11.9	11.0	0.6	0.33	0.5	0.00	99.9	2720	8
A16G2 (Nectaris)	45.2	0.8	21.9	8.2	0.0	0.08	10.2	13.2	0.5	0.17	0.19	0.00	100.3	2720	11
A17 average (Serenitatis)	46.2	1.5	18.0	9.2	0.0	0.12	12.3	11.2	0.7	0.26	0.28	0.00	99.6	2730	6

*Densities (P = 1 bar) calculated in the manner of Bottinga and Weill, 1970, using updated molar volume constraints from Lange and Carmichael, 1990.

Viscosities (P = 1 bar) are averages of results from methods of Shaw, 1972; Persikov et al., 1987; and Bottinga and Weill, 1972 (for hydrous melts, Bottinga and Weill, 1972, style results are extrapolations based on ratios of hydrous/anhydrous results by other methods).

Temperatures used for calculations are based on liquidus estimates using "MELTS" program of M. S. Ghiorso (see text).

Chicxulub compositions are from Sharpton et al., 1992, and Hildebrand et al., 1991. Average continental crust composition is based on two estimates in Condie, 1982, and one from Taylor, 1982. Layered intrusion data are from Carmichael et al., 1974. Average compositions for lunar impact melt rocks are based on data of Christian et al., 1976; Lindstrom et al., 1990; McKinley et al., 1984; and Rhodes et al., 1974.

†For Chicxulub, density and viscosity estimates assume 3.5 wt% Na_2O, 3.5 wt% FeO, 1.0 wt% Fe_2O_3, and 1.0 wt% H_2O (Na_2O contents of Chicxulub samples may be high due to secondary hydrous alteration).

§Possible initial composition of Sudbury Igneous Complex, assuming a 2:1 mix of norite and granophyre instead of canonical 37:63 mix.

then any model that regards the outermost prominent ring as a crater rim will systematically overestimate the size of the transient crater and thus the volume of impact melt. Despite these caveats, we adopt as a nominal working model the customary view that the outermost prominent ring of a multiring basin defines a structural diameter analogous to the rim diameter of a smaller crater.

An additional area of uncertainty is the impact velocity; a range of 12 to 70 km/s, encompassing nearly all conceivable major impacts, would shift the implied melting/displacement ratio (i.e., the estimated V_{melt}) by a factor of 0.9 to 1.5. The velocities adopted for modeling purposes are the average velocities for Earth-asteroidal and Moon-asteroidal impacts, as estimated by Chyba (1991).

Figure 3 shows—for Chicxulub, Sudbury, various lunar basins, and a few moderately large terrestrial craters—V_{melt} calculated from equations (1) and (3), expressed as thickness of an idealized melt sheet. The idealized melt sheet contains 100% of the impact melt, aggregated into a volume of shallow-cylinder shape, with the cylinder's diameter equal to the average of (1) D_{tc}, and (2) half the "final" nominal basin D (models (1) and (2) are nearly identical, for most of the basins considered). This choice for diameter of the melt sheet is based on rough analogy to the Grieve et al. (1991) interpretation of the Sudbury structure and to interpretations from drill-hole and gravity results for the Chicxulub region (Sharpton et al., 1993; Hildebrand et al., 1995), which suggest that the main mass of impact melt was limited to a region roughly 100 to 120 km in diameter, that is, 0.6 to 1.3 times the (variously inferred) Chicxulub transient crater D.

Figure 3 also shows two estimates for the depth to the base of the zone of impact melting. The shape for the initial distribution of the melt formed in a major impact is unfortunately not well constrained. Grieve and Cintala (1992) assume a roughly hemispherical shape for the zone of melting, but Tonks and Melosh (1993) give equal consideration to a model in which the melt region is a nearly complete sphere truncated on one side by the planet surface, saying "reality lies somewhere between these two extremes." Note that the depth to base of melting zone is generally many times greater than the thickness of the melt sheet because the melt sheet's diameter, in these models, is far greater than the diameter of even a hemispherical melting zone. Realistically, the original disposition of the impact melt is more diffuse, forming largely as veins and pockets amid surviving solids; compare the impact melts in the deeply eroded Vredefort structure (French and Nielsen, 1990) and realize, as discussed in a later section, not all of the melt can be expected to aggregate into the melt sheet. Thus, the ratio (depth to base of melting zone)/(thickness of idealized melt sheet) implied by Figure 3 should be regarded as a lower limit.

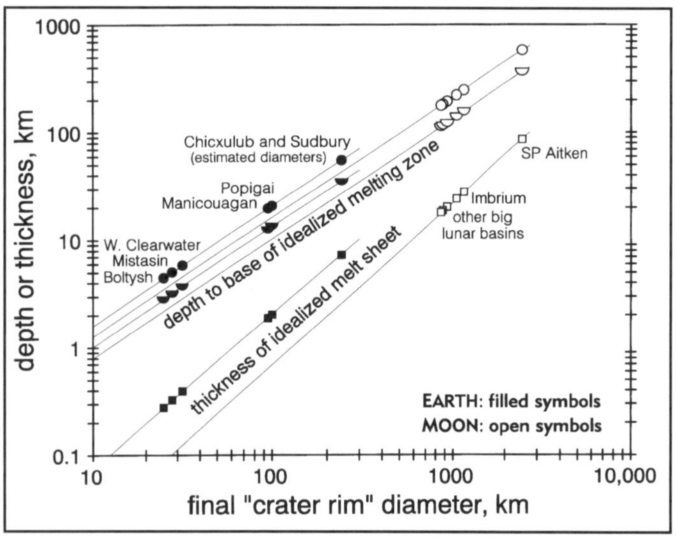

Figure 3. Model predictions for impact melt volume as a function of transient crater size, estimated from equations 2 and 3. Also shown are translations from total melt volume into maximum depth of melting (assuming hemispherical or spherical shape for the melting zone) and into idealized thickness of a "melt sheet" (see text).

Comparison with observed craters (Grieve and Cintala, 1992) suggests that the slope of the D_{tc} versus V_{melt} relationship (i.e., the relative increase in V_{melt} as a function of D_{tc}) is modeled fairly well, at least up to transient crater D of the order 60 km (final crater D of the order 100 km, terrestrial g), by both equations (2) and (3). However, we suspect that the impact melt volumes predicted by (2) and (3) may be systematically high. For craters with estimated D_{tc} in the range 4 to 60 km, V_{melt} predicted by equation (3) is generally about two times higher than estimated actual volume of impact melt (Grieve and Cintala, 1992). Most of these "observed volumes of impact melt rocks" are highly uncertain, and in some cases Grieve and Cintala (1992) regard them as "minimum" estimates. Note, however, that such estimates customarily regard any and all "melt rock" as former melt, whereas equation (1) suggests that for craters of this size range the "melt rocks" are initially only roughly half molten. For Sudbury, the idealized melt sheet thickness (Fig. 3) is three times the average observed thickness of the SIC: 2.5 km, although the maximum observed thickness is ~4 km (Naldrett and Hewins, 1984).

IMPORTANCE OF THE MELTING/ DISPLACEMENT RATIO

Equation (1) implies that the melting/displacement ratio in major impacts varies enormously as a function of crater size as well as planetary g (Fig. 4a). In this section, we examine the implications of these variations for crystallization behavior of impact melts.

Accelerated cooling of magma during digestion of suspended clastic debris

The common expression "impact melt sheet" can be seriously misleading. An aggregated "melt sheet" is in general not truly a melt but an impact-generated magma (melt/crystal sus-

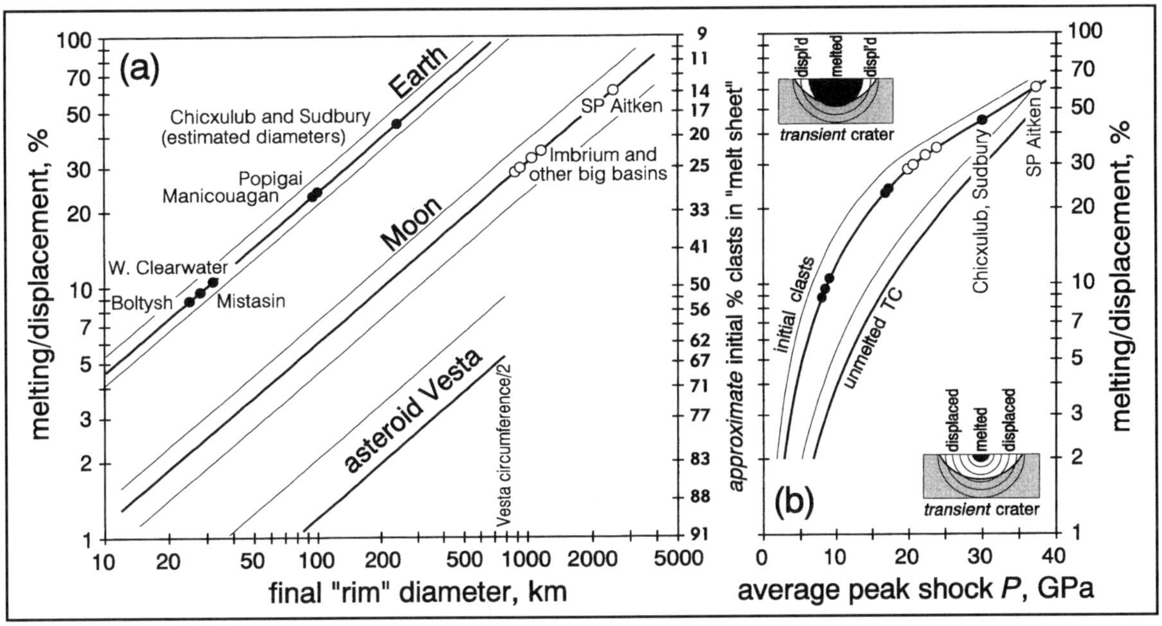

Figure 4. (a) Melting/displacement ratio calculated from equation 1, shown as a function of crater size and impact velocity. For Earth and Moon, three curves shown correspond to Chyba's (1991) impact velocity estimates for collision with an average short-period comet, an average asteroid, and an extremely slow asteroid. For asteroid Vesta, two curves correspond to impact velocity of 5 to 26 km/s. Also shown, at right, are translations (very approximate!) into initial proportion of clasts in the "melt sheet" (see text). (b) Average peak-shock pressure of clastic debris suspended in the impact melt and of the entire unmelted portion of the transient crater, calculated by combination of equation 6 with the geometrical model shown in the inset cartoons (concentric half circles represent contours of peak shock P), and further assumptions described in the text. The lighter curves show results assuming a value of 2.2 for the exponent in equation 6, to illustrate the effect of the higher values favored by some shock-attenuation models (Melosh, 1989).

pension), with large proportions of suspended solids that survive the impact and are swept along into (or buoyantly ascend into or fall back down into) the melt as it aggregates to form the sheet. An important distinction between such an impact-generated magma and a typical endogenously produced phenocryst-bearing magma is that the impact magma's suspended solids are for the most part much cooler than the impact melt. Until these suspended solids either settle out or dissolve into the melt, they accelerate the cooling of the melt, and thus effectively dampen the potential for fractional crystallization.

Petrologic studies of impact melt–derived rocks at a variety of moderately large terrestrial impact craters indicate that recognizable clasts of diverse provenance tend to be evenly distributed throughout the sheet, presumably due to turbulent mixing during aggregation of the melt (Phinney and Simonds, 1977). Only at Manicouagan does the sheet feature a relatively heterogeneous clast distribution (Grieve and Floran, 1978). Presumably, the melt temperature is also largely homogenized through this turbulent mixing. However, studies of shock metamorphic effects, applied to terrestrial as well as lunar impact melt breccias, indicate that the suspended solids are originally remarkably cool. In the estimation of Simonds et al. (1976), 50 to 100°C would be representative for the clastic matter in "large terrestrial impacts"; in this context, "large" presumably refers to craters the size of Manicouagan ($D \sim 95$ km). As will be discussed in a later section, average initial clast T is probably not uniformly so low but is a function of the size of the crater and the preimpact regional geology. In any case, the initial clasts are much cooler than the initial melt.

The melt's potential for fractional crystallization will be diminished initially through heat loss in warming the clasts, but even after the clasts are warmed to near-liquidus T, melting or assimilating them may consume significant latent heat. The initial clasts will generally be significantly removed in composition from equilibrium with the melt. As Bowen (1928) emphasized, assimilative reactions between disequilibrium solids and silicate melts are always either endothermic or, if exothermic, cause an increase in the solid/melt ratio. As an extremely simple example of this effect, consider an impact magma that initially contains a large proportion of quartz as suspended clasts, but in which fractional crystallization of the melt would require deposition of a thick sequence of gabbroic or noritic cumulates before onset of quartz crystallization (cf. the SIC, where a thick layer of norite formed before and below the granophyre). If a large proportion of the initial quartz settles among the gabbroic early crystals, the ultimate vertical layering, in terms of SiO_2, is considerably dampened. Avoidance of this dampening effect requires that rapid cooling and/or crystallization of the melt supply latent heat for melting the quartz. At roughly 1150°C, converting, say, 20% of the initial magma from quartz to melt would require latent heat equivalent to cooling by 70 degrees. This is only an extremely simple example, but the same basic effect operates if initial mafic silicate or feldspar crystals have to be "made over" into more refractory solid solution compositions. Of course, once significant crystallization of equilibrium solids commences, latent heat plays a different role, slowing the cooling of the magma.

If the impact magma has a low viscosity and a favorable density contrast between suspended solids and melt, the initial suspended solids might conceivably settle out before greatly cooling the melt. However, thermal equilibration between the clasts (most of which are expected to be small because of pervasive brecciation) and the melt probably takes place within a time scale of the order 10^2 to 10^3 seconds (Onorato et al., 1978). Application of Stokes Law (e.g., equation 6-229 of Turcotte and Schubert, 1982) to even the most favorable crater considered in Figure 3 (Boltysh, assuming melt viscosity of 400 kg m^{-1} s^{-1} and clast-melt density contrast of 400 kg m^{-3}) indicates that only clasts >0.1 m in diameter would settle more than 10% of the idealized melt sheet thickness in 10^3 seconds. Note, too, that if cool clasts become concentrated, either by settling or some effect of turbulence, into the lower half of a sill-like magmatic system (presumably cooling mainly through its roof), their cooling effect will tend to forestall thermal convection.

Simonds et al. (1976) estimated that the Serenitatis and Manicouagan impact magmas both initially consisted of ~30% cool clasts. Using this for a very rough calibration, the right side of Figure 4a shows the melting/displacement scale translated into initial clast contents, modeled by simply assuming that the impact magma represents all of the melt mixed with suspended solids equivalent to 10% of the displaced (transient crater) mass:

$$\phi = 1 - \frac{m/d}{m/d + 0.1} \tag{4}$$

where ϕ is the (initial) fraction of suspended solids and m/d is the melting/displacement (mass) ratio. This method for estimating ϕ as a function of crater size (and g) is obviously very approximate, but in view of the great relative variation in m/d, and thus probably also ϕ, among the craters considered, even a crude estimate is worthwhile.

Figure 5 shows the potential cooling effect of the suspended solids, modeled purely in terms of specific heat consumed in warming the initial suspended solids (i.e., ignoring likely further cooling due to latent heat in melting or "makeover" of the solids). In this model, the diminution of melt temperature ΔT_m is related to the rise in temperature of the crystals ΔT_x by:

$$\Delta T_m = \Delta T_x \left(\frac{\phi C_{p,x}}{f C_{p,m}} \right) \tag{5}$$

where f is the (initial) fraction of melt ($f + \phi \equiv 1$), and $C_{p,x}$ and $C_{p,m}$ are the heat capacities of the crystals and melt, respectively. We assume that $C_{p,m} / C_{p,x} \approx 1.1$ (Stebbins et al., 1984; Newton, 1987). Figure 5 indicates that a wide range of scenarios is possible, extending in principle from negligible cooling, in impacts with very high melting/displacement ratios or very hot initial suspended solids, to rapid quenching of the melt, in cases in which the suspended solids are abundant and cool.

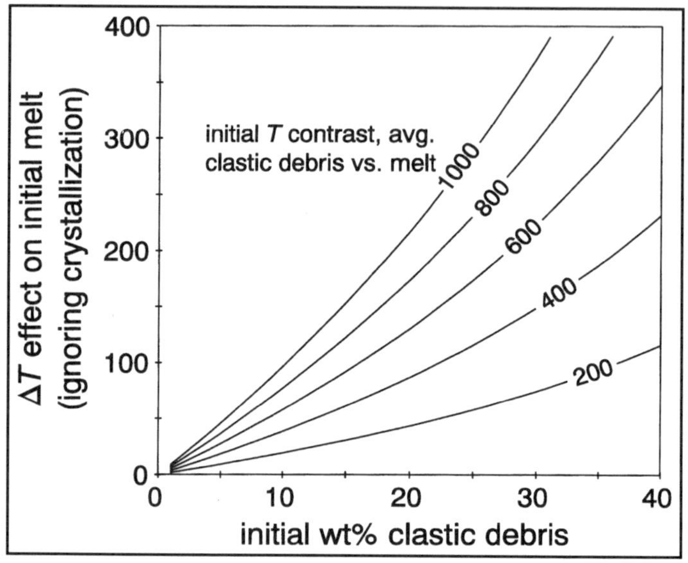

Figure 5. Potential cooling effect of solid debris suspended in an impact magma, modeled in a highly simplified (i.e., probably conservative; see text) fashion, based on equation 5.

Average initial temperature of suspended clastic debris

In principle, the average initial temperature of the clasts can be estimated based on cratering models. Passage of an impact shock wave through a rock induces a rise in postshock T that is roughly in direct proportion to the maximum shock pressure. Unfortunately, even for solid silicates, experimental constraints (e.g., Raikes and Ahrens, 1979; Langenhorst, 1994) establish this P-T relationship only to within a factor of ~1.4, and material properties (most notably porosity) can also have a major influence. The greatest uncertainty, however, concerns the provenance of the clasts in relation to the geometry of the transient crater and the melting zone.

The geometrical relationship between the zone of melting (assumed hemispherical to spherical) and the transient crater (parabolic) is clearly sensitive to the melting/displacement ratio (Figure 4b). Because peak shock P falls sharply with distance from the zone of melting (in which the average peak shock P is just high enough to induce general melting, i.e., roughly 60 GPa: Melosh, 1989), only clasts derived from close to the zone of melting will be greatly heated in advance of the diffusive heating that occurs at the expense of the melt.

In general, attenuation of peak shock P as a function of radial distance r is expected to follow $P \propto r^{-n}$, where $n \approx 2$ (e.g., Melosh, 1989; Chen et al., 1994) (P decays more rapidly in a central zone, less than roughly three projectile radii from the point of an impact, but this zone is virtually all melted or vaporized, or else ejected from the crater, in cratering events as large as those considered here). A P-decay model recommended by Melosh (1989) is:

$$P(r) = \rho_0 [C + Su_0(r_0/r)^{1.87}] u_0 (r_0/r)^{1.87} \quad (6)$$

where ρ_0 is the preimpact target density, C and S are equation-of-state constants appropriate to the target material, r_0 is a reference radius ~ the radius of the impactor r_i, and u_0 is the initial particle velocity (= 0.5 × the impact velocity) at $r = r_0$. Based on Melosh's (1989) Table AII.2, we assume C = 3,580 m s^{-1} and S (which is nondimensional) = 1.35. Modeling the zone of melting as a hemisphere with radius r_m, we determine r_i/r_m using Melosh's (1989) equation 7.10.1a and r_{tc}/r_m as a function of the melting/displacement ratio (assuming that depth/D for the transient crater = 0.3 and neglecting possible density contrast between the unmelted displaced matter and the melt). We take r_0/r_i to be 1.2, which gives, for impact velocity of the order 12 km/s, $P \approx 59$ GPa at $r = r_m$. Exact choices for most of these parameters are not important, because the goal is simply to set up a model in which P, attenuating according to equation (6), is roughly 60 GPa at $r = r_m$. Given this model for distribution of peak shock P around the point of impact, we then can calculate (using models dividing the volume of a cube with side length = D_{tc} into 10^6 cells) peak shock P within the unmelted portion of the transient crater as a function of the melting/displacement ratio.

As in equation (4), we assume that only a volume equivalent to 10% of the volume of the transient crater contributes clastic debris to the "melt sheet." For modeling purposes, we assume that this clastic material is exclusively derived from the uppermost available (unmelted) portion of the transient crater. A random sampling of the unmelted portion of transient crater would clearly be inappropriate because: (1) rarefaction effects dampen peak shock P for all locations close to the surface (Chen et al., 1994), (2) much of the near-surface component of the transient crater that is not melted is ejected to points beyond the perimeter of the sheet, (3) collapse processes during modification of the transient crater add materials originally beyond the transient crater rim to the mix of clasts in the sheet, and (4) in reality the initial melting is not so well concentrated into a hemisphere (or sphere), and initial proximity to veins and pockets of melt probably enhances the fraction of uncommonly hot solids that become clasts in the sheet. Note that (1) and (3) tend to reduce the expected average clast peak shock P (and thus T), whereas (2) and (4) have the opposite effect. We assume that (1) + (3) more than offset (2) + (4). Our model does not directly address all these complexities, but to roughly compensate for their probable net effect, we model the clastic material as exclusively derived from the uppermost unmelted portion of the transient crater, leading to lower average clast peak shock P_{ac} than would be predicted by averaging over the entire unmelted portion of the transient crater. Results from this procedure, assuming a hemispherical melting zone, are shown in Figure 4b.

The absolute values of these predicted pressures are obviously uncertain. Analogous models assuming a spherical shape for the melting zone result in higher P_{ac} by a factor of about 1.5. However, several aspects of Figure 4b are probably significant. The "unmelted TC" curve should represent a conservative upper limit on P_{ac} for all but the highest melting/displacement (m/d) impacts. We expect significant variation in P_{ac}, if only as a function of the m/d ratio. For m/d in the range of Chicxulub/Sudbury–sized terrestrial basins or the South Pole Aitken lunar basin, we expect P_{ac} (and thus, in general, the average initial clast ΔT) to be roughly half the level associated with general melting of the target. For the many lunar basins with D in the 800 to 1,200-km range, we expect P_{ac} and average initial clast ΔT to be well below half the level associated with general target melting. Even the largest of impact melts on an asteroid should have clasts that are on average negligibly warmed from their preimpact T. In any impact that induces significant melting, finite proportions of clasts will form at both extremes of initial T: from virtually unwarmed to the solidus T. Nonetheless, the average initial clast T (i.e., P_{ac}), and consequently the statistical variance in initial clast T, must be strongly correlated with m/d.

Interpretation of impact "melt rock" textures: Clast Darwinism

A low proportion of perceptibly surviving clasts in a large impact "melt sheet," such as (according to many interpretations) the Sudbury Igneous Complex, does not necessarily imply that the proportion of clasts in the newly formed sheet was equally low or that portions of the sheet without perceptible clasts were once completely molten. Simonds et al. (1976) observed that degree of clast digestion in impact melt systems is correlated with initial clast/melt ratio. We now know that initial clast/melt ratio is in general a function of the melting/displacement ratio, that is, a function of (primarily) crater size and g (Figure 4), and that clasts in a large impact melt system probably feature high statistical variance in their initial T. In a sufficiently large (high m/d) impact, only a minority of the coarsest and coolest clasts will survive in perceptible form.

Clasts are probably initially coolest near the top of the sheet, where their provenance is relatively shallow (i.e., cooler before the impact and partially protected from shock by rarefaction effects) and the clast(cool)/melt(hot) ratio is relatively high (as near any margin of the system). Clasts are also more likely to survive near the top of the system because cooling is generally fastest in that direction. Avermann and Brockmeyer (1992) describe the uppermost granophyre of the SIC as a "low fragment content (~5%) impact melt" in "largely gradational" (cf. Muir and Pederery, 1984) contact with a mostly unmelted (80 to 90%) breccia layer above. In both of these layers, the clasts are exclusively metasediments of relatively shallow provenance, and yet "many fragments are corroded or even disintegrated." The rate at which within the granophyre perceptible clast content grades downward to zero was not discussed by Avermann and Brockmeyer (1992), but the zero point, and the initial clast content in general, probably cannot be determined by petrographic observation. Even where no portion of a rock from a large impact melt system consists of clasts still in perceptible form, a significant portion may consist of former clasts that have completely disintegrated into individual crystals or unrecognizable small clumps of crystals, yet have never completely melted. And still more may consist of material that became locally 100% molten too late (as a result of the chilling and stagnating effects of unmelted clastic matter nearby) to ever be truly assimilated into the main mass of melt.

Proportion of melt ejected from the transient crater

As reviewed by Melosh (1989), cratering models and experiments indicate that a region extending to a depth of ≈ one-third the depth of the transient crater is ejected beyond the transient crater rim, that is, to regions generally too distal to contribute to the main melt sheet. The exact shape of the ejected zone is uncertain, but in cross section it probably resembles the dual-asymmetric-parabolic shape shown in Figure 6, where depth/diameter is exactly 0.3 for the parabolic transient crater and 0.1 for the ejection zone, and the detailed shape of the ejection zone is modeled after Melosh's (1989) Fig. 5.13. Assuming this geometry, we can calculate the proportion of melt ejected from the transient crater as a function of melting/displacement and various assumptions regarding the shape of the melting zone. In principle, these proportions could be calculated by application of Pappus's Rule concerning volumes of solids of revolution. In practice, we found it easier to calculate the volume proportions using the same approach as for Fig. 2b, that is, plugging equations for the boundaries of the ejection, melting, and transient crater volumes into a computer program that divides the overall volume into 5×10^6 cells. These calculations ignore the relatively minor proportion (roughly one-tenth of the melt volume) of impact vapor, which probably forms mainly within the volume treated as ejected melt in the calculations. The total ejecta (melted + unmelted) is uniformly 42 vol% of the transient crater.

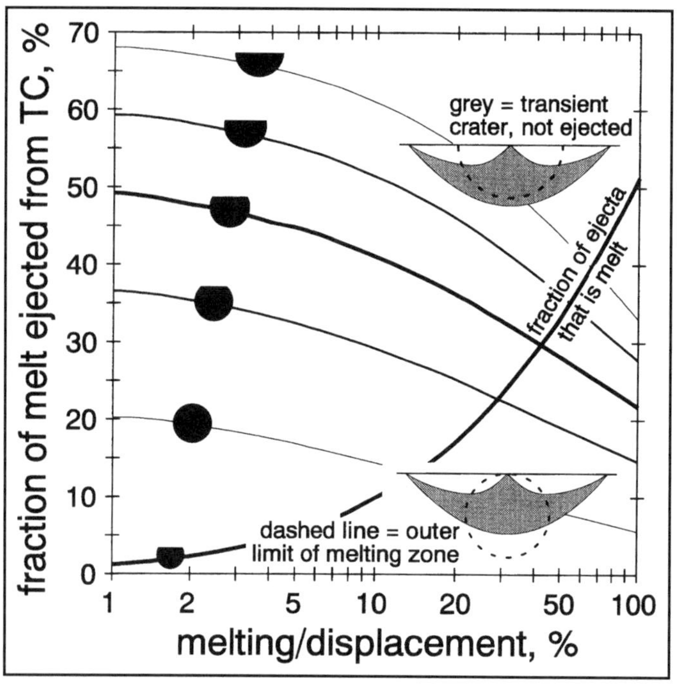

Figure 6. Fraction of melt ejected from the transient crater calculated as a function of melting/displacement and shape of the melting zone (indicated by black symbols). Inset cartoons show shapes assumed for transient crater and ejection zone (consistent among all models shown). Also shown, for an intermediate-shaped melting zone, is fraction of the ejecta that is melt.

As discussed above, the shape of the melting zone is probably intermediate between spherical and hemispherical. Figure 6 shows results assuming shapes where a sphere is truncated at 0%, 12.5%, 25%, 37.5%, and 50% of its diameter. For the 50%-truncated (i.e., hemispherical) model, and even for the 25%-truncated model, low-moderate melting/displacement ratios (cf. Fig. 4a) result in large proportions of the total impact melt being ejected from the transient crater. However, even the hemispherical model results in most of the melt remaining in (or below) the transient crater if melting/displacement is greater than about 30%. Figure 6 also shows, for the 25%-truncated model, the fraction of the ejecta that is melt. Note that this fraction increases steeply, constituting significant proportions of the total ejecta, after melting/displacement exceeds roughly 20%.

INFLUENCES OF REGIONAL GEOLOGY: MAINLY COMPOSITIONAL EFFECTS

Melt/solid density relationships

The impact melt that forms within the unejected portion of the transient crater, or in large impacts partly below the transient crater, will not aggregate into a compact near-surface "sheet" with 100% efficiency. Large proportions of adjacent unmelted material mix with the melt during the processes of rebound, collapse, and so on, that affect regions just outside the melting zone. Shortly after the impact, buoyancy forces cause the melt to migrate apart from the solids, but the melt also starts to crystallize as it is cooled by the solids. In general, the aggregation efficiency will be sensitive to the density contrast between the impact melt and the unmelted materials that initially surround it.

A melt's density can be precisely estimated as a function of composition by summing the partial molar volumes of the oxide components (Bottinga and Weill, 1970). Variations related to temperature and to even moderate ranges of pressure are well constrained (Lange, 1994). Compared to the Chicxulub and Sudbury melts or to average continental crust, lunar impact melts are consistently richer in dense FeO, MgO, and CaO and poorer in the low-density components SiO_2, alkalis, and H_2O (Table 1). As a result, even though lunar melts form at higher *T*, they are denser (Fig. 7) by roughly 300 kg m^{-3}. The Chicxulub melt was probably rich in CO_2 (roughly 1,420 kg m^{-3}, but not included in these calculations) and in sulfur species, for which molar densities are unfortunately unknown (Lange, 1994). However, we find a strong ($r = 0.97$) correlation between the low-T, low-P densities of solid oxides and Lange's (1994) corresponding 1,400°C partial molar densities, and this correlation extrapolates to ~2,000 kg m^{-3} for SO_4^{2-} (assumed, by analogy with CO_2 and CO_3^{2-}, to have similar density to SO_3). Thus, carbon and sulfur probably both tended to make the Chicxulub melt's density even lower than suggested by Figure 7.

The density of the upper crust in the Sudbury region is estimated from gravity data to be ≈2730 kg m^{-3} (McGrath and Broome, 1994), which—assuming porosity is typical of the upper ~10 km of continental crust, at roughly 3 to 6% (Meissner, 1986)—implies a pore-free density of roughly 2,810 kg m^{-3}. This is probably a conservatively low value for the pore-free density of any large region of continental crust. It is higher than the density of typically andesitic impact melts (Figure 7) by roughly 270 to 380 kg m^{-3} (i.e., roughly 13%—depending on assumptions regarding the poorly constrained contents of low-density oxides in the melts). In contrast, a conservatively high estimate for the average density of the lunar highland crust (2,950 kg m^{-3}, Haines and Metzger, 1980) is only 150 to 200 kg m^{-3} (~6%) denser than a typical lunar impact melt (Fig. 7; in this case concentrations of the low-density oxides are negligibly low).

Density relationships will also be affected by the target region: pre-impact porosity as well as porosity created by brecciation during the impact. Terrestrial gravity and crystallization from pore fluids eliminate most porosity within a few kilometers of the surface. In a typical sequence of fine-grained carbonate sediment, porosity falls exponentially from 50% near the surface to 4% at 5 km depth (Moore, 1989). Extrapolation of this trend suggests that average porosity of the upper 30 km is roughly 3%. Thus, porosity probably diminished the average bulk density of the outer 30 km of the target by ~2% at Chicxulub, and by even less at Sudbury. The lunar crust is pervasively brecciated, with vacuous (0 kg m^{-3}) porosity. Even excluding the most porous breccia type (regolith breccias, possibly abundant only in the outer few tens of meters of the crust), rock porosities are typically 4 to 17% and average roughly 11% (Warren and Rasmussen, 1987). Most of this porosity is probably squeezed out in the outer 2 to 3 km, but seismic data suggest significant porosity persists down to 20 to 25 km (Taylor, 1982). Pressure at the base of the lunar crust (~64 km) is about the same as at 12 km in the Earth.

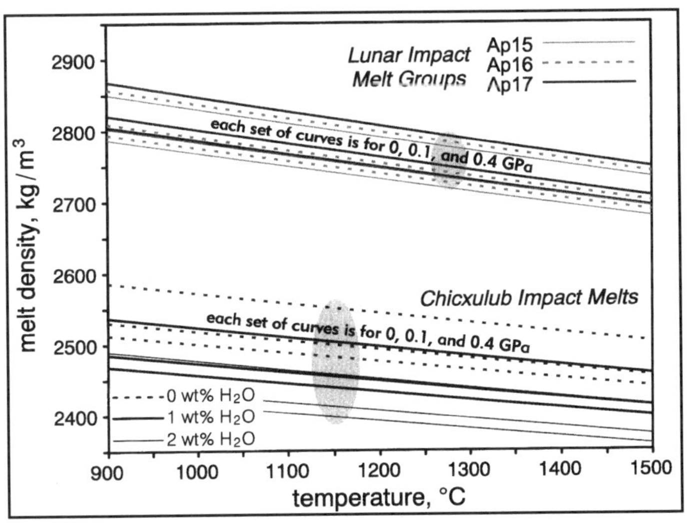

Figure 7. Densities of melts with compositions of average Chicxulub impact melt rock and three different lunar impact melt types (Table 1), calculated from partial molar volumes of oxides (Lange, 1994) for a range of pressures (the pressure gradient in the outer Moon is 0.005 GPa/km). Regarding liquidus T for these melts, see Table 1.

As a by-product of shock P, large-scale average porosity induced by brecciation during an impact is probably proportional to $(r_0/r)^n$, where by analogy to equation (6) n is roughly 2. Unless the melting/displacement ratio is extremely high (say ≥100%), the solids within a short distance around and above the impact melt will inevitably include a large proportion that have acquired new porosity of the order 10% or higher (based on the typical porosities of lunar breccias). If the density of the impact melt is not at least 10% lower than the pre-impact density of the target crust, the melt will have neutral or even negative buoyancy in relation to a large component of the surrounding rocks. Thus, Figure 7 implies that at Sudbury the melt was probably buoyant relative to all but a small fraction of the most porously-brecciated rocks and that at Chicxulub (where pre-impact porosity was significant) the melt buoyancy was marginal in relation to a slightly higher fraction of the surrounding rocks. But in lunar impacts, melt buoyancy is typically marginal or negative relative to even moderately brecciated (<<10% new porosity) varieties of crustal rock.

After the melt "sheet" aggregates, melt/solid density relationships might still be important, because at least some mechanisms of cumulate formation are enhanced by large melt/solid density contrast (Wager and Brown, 1967). At Sudbury, the major liquidus phases of the SIC were pyroxene, plagioclase, and possibly hypersthene in the lower (norite) zone and plagioclase, K-feldspar, and quartz in the upper (granophyre) zone. Major liquidus phases of the Chicxulub impact magma were probably similar (although the pyroxene and plagioclase were probably more Ca rich). All these phases are at least 150 kg m^{-3} denser than any plausible SIC or Chicxulub melt. In contrast, one of the main liquidus phases of lunar impact melt, plagioclase, would have density similar to (actually ~70 kg m^{-3} less dense than) the typical initial melt.

Viscosity and potential for convective stirring

The efficiency with which a large magma body differentiates may be limited by the rate at which fluid (i.e., mainly convective) motions supply fresh melt to crystal/melt interfaces. The tendency for a magma to convect is governed by the

Rayleigh number Ra, which is proportional to thickness^3, g^1, and μ^{-1}, where μ is the viscosity (Turcotte and Schubert, 1982). Viscosity of a silicate melt can be estimated as a function of composition. Table 1 shows viscosity estimates based on averaging results from three different but similar methods (Bottinga and Weill, 1972; Shaw, 1972; Persikov et al., 1987). These methods all give similar results, but for the terrestrial melts uncertainties regarding H_2O (especially for Chicxulub, where the H_2O content is only estimated based on analogy to Sudbury) translate into order-of-magnitude (or worse) uncertainties in the estimated viscosities. However, it seems safe to conclude that compared to classic terrestrial layered intrusions (1) lunar impact melts have at least comparably low viscosity and (2) unless Sudbury/Chicxulub–type impact melts have much higher H_2O than estimated for the SIC by Collins (1934), they probably (mainly as a result of their high SiO_2 contents) have unfavorably higher viscosity.

Suspended crystals, such as marginally buoyant plagioclase in a lunar impact melt, may significantly enhance viscosity. Figure 11 of Marsh (1981) suggests that in typical lunar impact melts (46 wt% SiO_2), 20% suspended crystals would enhance μ by roughly nine times.

Impact into a volatile-rich terrain

As noted by Kieffer and Simonds (1980), high contents of volatile materials within a target terrain can lead to two significant effects. Latent heat of vaporization may significantly enhance the chilling effect of volatile-rich clasts suspended in an impact melt (e.g., at 1,100°C the heat content of 1 kg of half-melted silicate is matched by 0.3 kg of steam or 0.5 kg of CO_2 gas). Also, explosive expansion of the newly formed gases may enhance dispersal of the impact melt to points far from the transient crater (cf. Grieve and Cintala, 1992). However, only vaporization and expansion starting from deep within the melt zone will have a major influence on the fraction of melt blown out of the transient crater (most of the upper melt zone is probably ejected, anyway; see Fig. 6). Major enhancement of dispersal from a melt zone many tens of kilometers deep, as the melt zones at Chicxulub and Sudbury putatively were (Fig. 3), would require a thickness of volatile-rich sediments greater than roughly 10 km. The Chicxulub region features a relatively thick pile of carbonates and anhydrites, but these sediments give way to basement granite and schist at depths of typically 2.1 to 3.2 km (Lopez Ramos, 1975). Even if sediments were piled >10 km thick, H_2O would probably not contribute significantly to melt dispersal from a Chicxulub/Sudbury sized–transient crater. Sediment porosity (i.e., free water content) typically reduces to 4% at 5-km depth and 0.3% at 10-km depth (Moore, 1989). Moreover, solubility of H_2O in silicate melts is roughly proportional to $P^{0.5}$, so that by 5-km depth the solubility of H_2O in andesitic melt is already ~5 wt% (Holloway and Blanck, 1994). Solubilities of CO_2 and (at least under reducing conditions) SO_2 are not so sensitive to P (Holloway and Blanck, 1994; Carroll and Webster, 1994).

INFLUENCES OF REGIONAL GEOLOGY: MAINLY THERMAL EFFECTS

Near-Penokean Sudbury versus stable Chicxulub

If the transient craters of the Sudbury and Chicxulub events really were 120 or more km in diameter (Deutsch and Grieve, 1994; Sharpton et al., 1993), melting extended so deep that the regional geothermal gradients strongly influenced the melting/displacement ratio, the average depth provenance of the melt feeding the central "sheet," and the average initial clast T. In the nominal (Fig. 3) model of 240-km basin genesis, assuming a hemispherical shape, one-third of the melt originally forms shallower than 8 km, two-thirds forms shallower than 17 km, and the deepest melting occurs at 35 km; a spherical shape shifts these implications to 21, 34, and 55 km, respectively. Much of the shallowest-forming melt (~45%, for m/d ~45% and hemispherical melting) gets ejected to positions beyond the transient crater rim (Fig. 6) where it can scarcely aggregate into the main mass of impact magma. Thus, median depth of origin for the main aggregation of impact magma is 20 to 30 km, for melting zone shape ranging from hemispherical to spherical.

The Yucatán was a stable, cool carbonate platform when the Chicxulub event occurred. But the Sudbury impact hit a region already warmed by the Penokean Orogeny, which produced folding and granitic magmatism along a roughly east-west axis centered around the upper peninsula of Michigan (Brocoum and Dalziel, 1974; Clifford, 1990). The timing of the Penokean Orogeny has recently been clarified by 37 precise U-Pb-zircon ages reported for Proterozoic granitic intrusives and gneisses from northern Wisconsin (as close as 500 km from the SIC) by Sims et al. (1989). Of these 37 ages, 29 (78%) are clustered in the range 1831 to 1871 Ma and 23 (62%) are in the range 1835 to 1862 Ma. The timing could hardly coincide more closely with the Sudbury event: 1850 ± 1 Ma, based on the same technique (Krogh et al., 1984). The Sudbury event presumably enhanced (and redirected) Penokean magmatism, but it probably was not the primary cause, because the orogenic activity commenced well before 1850 Ma (Sims et al., 1989) and was concentrated hundreds of kilometers west of Sudbury (Clifford, 1990). Davidson et al. (1992) have even argued that Penokean magmatism did not extend into southern Ontario, but Card (1992) has critiqued their interpretation.

During orogeny, the regional near-surface dT/dP is enhanced to as much as three times a normal continental geotherm, that is, to roughly 30 to 40 K/km instead of roughly 10 to 15 K/km (Turner, 1968) (a further slight enhancement to the Sudbury geotherm could be expected from the ~1.44 × higher rate of radiogenic heat production at 1850 Ma in comparison to the 65-Ma age of Chicxulub). An orogenic geotherm could easily bring most of a Sudbury-sized impact melting zone (more than half deeper than 20 to 30 km, based on the nominal models) to a state of incipient melting—before the impact. The latent heat of silicate rock melting is only ~20% of the heat required to melt starting from a near-surface T of the order 0°C (Fig. 8). Assuming the orogenic geotherm has not already induced pre-impact melting but has warmed the melting zone (average T) to, say, 800°C instead of the 300°C associated with a more normally cool geotherm, the same heat input that yields x km^3 of melt for the cool geotherm will yield 1.8x km^3 for the orogenic geotherm.

The principal effects of a higher melting/displacement ratio have been described in the section "Importance of the Melting/Displacement Ratio." In addition, increasing m/d, for constant crater/basin size, increases average depth provenance of the melt. If impact melt volume is truly as large for a 240-km basin as estimated here (Fig. 3), then considering the further enhancement associated with an orogenic terrain, it seems inevitable that mantle-derived magma contributed to the SIC—although presumably almost exclusively to its lower, norite portion—as suggested by Chai and Eckstrand (1993).

Recent interpretations of the Sudbury Igneous Complex as nearly pure impact melt may have gone too far. Some field studies (e.g., Cowan and Schwerdtner, 1992) suggest that the emplacement of the SIC was a prolonged process. Chai and Eckstrand (1993) find difficulty modeling the distribution of incompatible elements within the SIC under a single-magma hypothesis. In terms of the isotopic constraints, showing that the SIC is not mostly mantle derived is hardly the same as showing that no mantle component whatsoever is present. Even the former conclusion is only suggested, not required, by the ε_{Sr} and ε_{Nd} evidence. Hawkesworth et al. (1984) noted that the voluminous Karoo basaltic flows and dikes of southern Africa have ε_{Sr} and ε_{Nd} remarkably similar to the range for the SIC, and yet the Karoo magmas appear from nonisotopic geochemical evidence not to have assimilated large proportions of crust (Cox, 1983). Norman (1994a) interprets the Os isotopic evidence as suggesting endogenous derivation for most of the SIC. Even if formed by "pure" impact melting, assuming the melting zone was as deep as implied by the transient crater diameter estimates (120 to 140 km) of Deutsch and Grieve (1994), the SIC should be regarded as somewhat anomalous. Without prior warming of the deep crust by the Penokean Orogeny, both the extent of regional melting and the extent of fractional crystallization of the impact magma would have been substantially reduced.

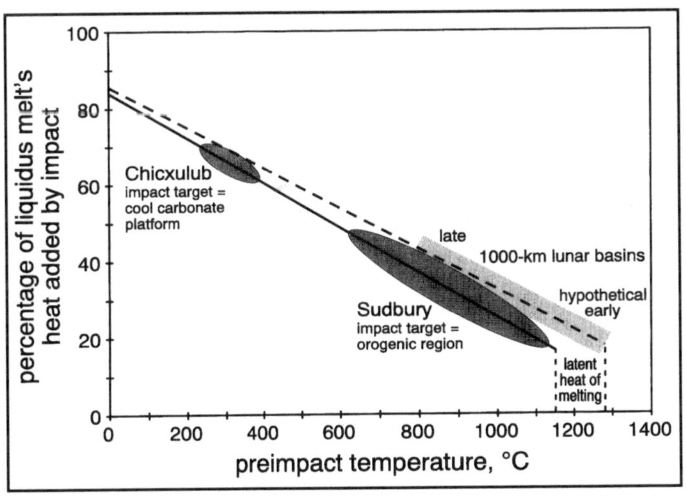

Figure 8. In large-scale impacts where the target crust has been prewarmed by orogeny, as at Sudbury, or by primordial global heating, as on the early Moon, impact melt may contain mainly endogenous (non-impact-derived) heat. Specific heat is assumed to be 1200 J kg^{-1} K^{-1} and latent heat (modeled as a function of liquidus T) is assumed to be 3.35 × 10^5 J kg^{-1} for terrestrial impact melting and 4 × 10^5 J kg^{-1} for lunar impact melting.

The ancient Moon

In terrestrial multiring basin studies (e.g., Sharpton et al., 1993; Deutsch and Grieve, 1994; and nearly all of our own discussion to this point), the transient crater diameter is generally assumed to be roughly half of the D of the outermost prominent ring; that is, a Croft (1985) (or Chyba, 1991) type model is applied, assuming the outermost prominent ring is analogous to the rim of a smaller crater. Extending this approach to the larger lunar multiring basins, such as South Pole Aitken (D ~ 2,500 km, implying D_{tc} ~ 1,170 km) and the dubious "Procellarum" basin (D ~ 3,200 km, implying D_{tc} ~ 1,450 km) (regarding Procellarum, see Zuber et al., 1994), results in some remarkable implications.

In impacts with D_{tc} more than roughly five times the thickness of the crust, the crustal portion of the transient crater is largely ejected. For lunar impacts with D_{tc} greater than about 300 km, the melt that escapes ejection (and thus is positioned to aggregate into a sheetlike central mass) becomes increasingly rich in matter of mantle derivation. The mantle fraction can be calculated as a function of D_{tc} and the assumed geometries of the ejection and melting zones (Fig. 9). Unless the melting zone is almost perfectly hemispherical in shape, these calculations indicate that in any lunar basin with D_{tc} >> 400 km (i.e., final D >> 720 km), the main mass of melt is >50% of mantle origin; in the cases of Imbrium (D_{tc} ~ 600 km) and South Pole Aitken, most likely about 85% and 99%, respectively. Of course, this initial melt will quickly become contaminated with crustal (but mostly unmelted) debris through collapse of the walls of the transient crater. Still, a central melt aggregation of preponderantly mantle derivation is fundamentally different from one of mainly crustal provenance.

Petrologic evidence suggests that the earliest lunar crust formed by flotation of plagioclase-rich cumulates over a Moon-wide "ocean" of extraordinarily FeO-rich magma, with molar Mg/(Mg + Fe) ~ 0.4 (Warren, 1990). At least initially, therefore, the uppermost lunar mantle was probably extremely FeO rich, by terrestrial standards. Models of mare basalt genesis also indicate a generally FeO-rich (and locally also TiO_2-rich) upper mantle (e.g., Taylor, 1982). An impact melt large enough to be preponderantly composed of such material would almost cer-

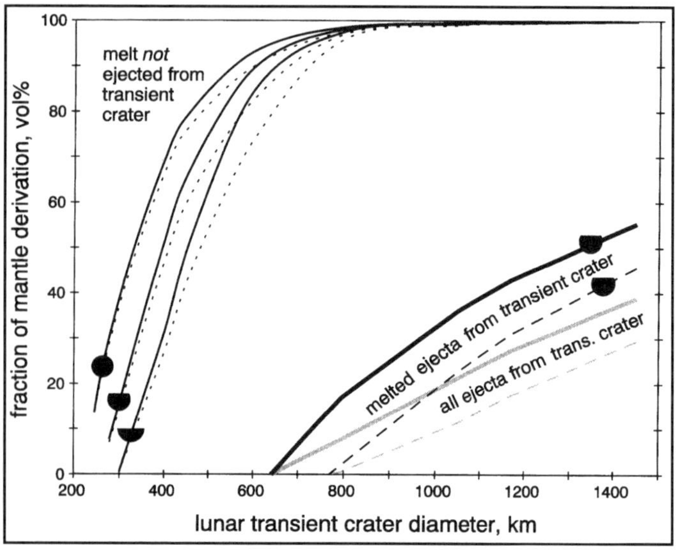

Figure 9. Mantle vs. crustal provenance of materials involved in lunar multiring basin formation, calculated analogously to Figure 6, with melting/displacement estimated from equation (1). Solid curves assume depth of maximum excavation = 0.1 D_{tc} (as in Fig. 6); dashed lines assume 0.083 D_{tc}. Crust/mantle boundary is assumed uniform at the global average depth of 64 km (Zuber et al., 1994). Models (Croft, 1985; Chyba, 1991) suggest that lunar transient craters in the 400 to 1450 km diameter range collapse into final structures with diameters of 720 to 3,200 km.

tainly be denser than the anorthositic lunar crust yet strongly buoyant relative to unmelted mantle of the same composition. Such an impact melt would probably tend to spread into a broad, thin zone of melt sandwiched between lowermost crust and uppermost unmelted mantle. Eventually, by precipitation of mafic cumulates, it would solidify into a new zone of lower crust plus (preponderantly) uppermost mantle. Assuming that crust/mantle differentiation preceded this cycle, little net differentiation would probably be achieved. Most of the crust in the center of the basin would consist of unmelted rubble emplaced by lateral collapse of the transient crater. The net effect would be a region of thinned crust, probably with enrichments of incompatible elements concentrated from a broad region of the upper mantle toward where the latest crystallization of the melt occurs, directly under the thin-crusted, yet abnormally hot, basin (which thus acquires enhanced potential for subsequent basaltic volcanism).

A more important question concerns impact melt differentiation when crust/mantle differentiation was still far from complete. Melts associated with lunar basins with $D << 1{,}000$ km would have little tendency to differentiate, because of their low melting/displacement ratios and unfavorable melt/crust density ratios. Most known basins with $D \sim 1{,}000$ km are thought to have formed roughly 600 m.y. after the main (4.5 to 4.4 Ga) phase of lunar crust formation (Ryder, 1990), but at least in principle many older basins may have been obliterated by ongoing bombardment. Individually, each such impact would make a modest contribution to the early melting that drove crust/mantle differentiation. For example, the total melt formed in a South Pole Aitken–sized event is nominally (assuming $D_{tc} = 1{,}170$ km) 0.3 vol% of the lunar mantle. However, all the next 10 smaller basins combined (Imbrium included) generated only 0.54 times as much melt as South Pole Aitken alone.

The cumulative effect of many giant impacts, closely spaced in time, would begin to take on many of the characteristics that are normally associated with endogenous (nonimpact) melting. By all accounts, the Moon's crust formed so quickly after origin of the Solar System (e.g., Taylor, 1982) that the early crust, moonwide, must have been simultaneously at or very near the solidus. It was also probably insulated by a several-kilometers-thick layer of vacuous-porous impact rubble (megaregolith). Under these circumstances, cooling would have been remarkably slow. For example, models by Warren et al. (1991) suggest that T did not fall below 950°C at a depth of 50 km until 300 to 500 m.y. after crystallization from the magma ocean. If the average preimpact temperature of the melted zone is greater than ~600 to 700°C (depending upon liquidus T, and thus composition), then more than 50% of the liquidus melt's heat is endogenous to the planet rather than added by the impact (Fig. 8).

Together, Figures 8 and 9 suggest that any 1,000-km-scale impact basins formed much earlier than 4.0 Ga would have produced central magmas so dominated by mantle material, and by endogenous heat, that the normally obvious distinction between impact melt and endogenous magma becomes fuzzy and even potentially misleading. On the earliest Moon, whatever new melts were forming by impacts were merely enlarging an already global magma ocean. As the magma ocean era gave way to the era of Mg-suite intrusives (4.5 to 4.2 Ga; e.g., Warren et al., 1991), big impacts may have increasingly influenced the siting and timing of magmatism, but the heat that drove the magmatism was primarily a holdover from the magmasphere. As the Moon cooled, big impacts kept occurring and possibly even increased in frequency at ~ 3.9 Ga (Ryder, 1990), but by then the era of volumetrically significant crustal genesis was long past.

To the degree that melt plays a key role in formation of basin rings (Spray and Thompson, 1995; cf. Melosh, 1995), even the final size of a multiring structure may be largely a function of the T (and other properties) of the target terrain before the impact. Both the average preimpact T (on a still-warm early Moon) and the bulk composition (Fig. 9) of the future melt zone are profoundly different for events with $D_{tc} >> 300$ km. Strom and Neukum's (1988) surfeit of "craters" with nominal $D \sim 800$ km may be an artifact of effectively overestimating D_{tc} for multiring basins (by regarding the D of the outermost prominent ring as analogous to the final D of smaller craters and peak-ring basins). The "nominal" models discussed in this chapter are subject to the same caveat.

In the South Pole Aitken event, assuming D_{tc} was the "nominal" 1,170 km, mantle material composed a significant proportion of the total ejecta (roughly 30 vol%) and an even higher proportion (roughly 45 vol%) of the melted component of the ejecta (Fig. 9). Melt (of combined mantle + crust derivation) also represents a large fraction, roughly one-third, of the total ejecta from such a high melting/displacement (*m/d*) impact (Fig. 6). Thus, the megaregolith in regions close to the basin rim should contain high proportions of distinctively mantle-rich impact melt. Ponds of such material might be tentatively detected through analysis of Clementine data, but a rigorous test of this hypothesis would require petrologic study of samples (e.g., impact melt glasses in a regolith sample) from the South Pole Aitken region. The mantle component should be even more prominent in impact melt from the Procellarum region, assuming that the proposed Procellarum basin truly had D_{tc} ~ 1,450 km. Roughly 40 vol% of the melt formed in such a high *m/d* impact is probably ejected (Fig. 6), and the total volume of melt ("nominal" D_{tc} model) is 2.3 times the volume formed in all of the next 10 smaller basins, including South Pole Aitken, combined. Thus, most of the impact melt splashed into the western nearside megaregolith should be a distinctively mantle-rich (roughly 50%—Fig. 9) variety formed and ejected in this one impact. Impact melt–derived samples from the western Apollo sites should be reexamined with this possibility in mind, but it seems more likely that either the "nominal" multiring basin D_{tc} model, or the Procellarum impact hypothesis itself, is incorrect.

CONCLUSIONS

Compared to the granophyre that forms the upper one-third of the Sudbury Igneous Complex, available melt products from the Chicxulub Basin are consistently more fine grained, and their compositions are more andesitic (i.e., less differentiated from bulk continental crust). The origin of these differences is unclear, but the simplest explanation would be that the Sudbury impact featured significantly higher melting/displacement, in part (if not wholly) because the Sudbury region had been previously warmed by the Penokean Orogeny. Possibly the available Chicxulub impact melt samples are more analogous to the Sudbury Onaping Formation impact breccia layer (in which case the deeper main mass of Chicxulub impact magma has not yet been sampled), but the Chicxulub samples are melt rich (clast poor) compared to most of the Onaping and SiO_2 poor compared to the more melt rich parts of the Onaping. If the Chicxulub impact magma initially formed with a composition similar to the bulk SIC (which is very similar to bulk continental crust), the Chicxulub samples show little effect of the fractional crystallization that putatively produced the gross vertical layering of the SIC.

Impact melt crystallization behavior is probably correlated with the size of the impact structure, but the most important variable, the melting/displacement (*m/d*) ratio, is not a simple function of structure diameter. Planetary *g* and preimpact *T* of the melting zone both strongly influence *m/d*, which in turn influences (1) the proportion of suspended clastic debris in the melt, (2) the average *T* of these clasts (which in any event are much colder than, and thus act to chill, the melt), and (3) the proportion of the melt that is disseminated by ejection from the transient crater. Our proposed calibrations for these shifts are very approximate, owing to uncertainties regarding the shape of the melting zone, the size/shape of the ejection zone in relation to the transient crater, and the manner in which modification (collapse) of the transient crater mingles clastic debris with the impact melt. Nonetheless, the effects are clearly important, and through them (primarily (1)), *m/d* strongly influences the potential for an impact melt to undergo fractional crystallization.

In most lunar impacts, differentiation of the impact melt is hampered by systematically lower *m/d* compared to terrestrial craters of similar size, and by the comparatively high density of typical impact melts. However, in a few of the largest impacts, most of the unejected melt forms in the upper mantle, where it probably remains while crystallizing into a series of mafic cumulates essentially similar to the preimpact upper mantle. On the earliest Moon (the "magma ocean" era and the first few hundred million years thereafter), *T* in the upper mantle was so high that episodes of large-scale impact melting must have been practically indistinguishable from pulses of enhanced endogenous magmatism. If the putative Procellarum basin truly formed with a transient crater diameter of ~1,450 km, most of the impact melt–derived rocks in the western nearside megaregolith should be a distinctively mantle-rich (roughly 50%) variety formed and ejected in this one impact.

We urge international support for deep drilling near the center of the Chicxulub structure and that the South Pole Aitken region be given highest priority in future lunar exploration. Improved constraints on the impact melts within these two as yet poorly studied basins would clarify the potentially crucial role of impact melting in early planetary and lunar differentiation.

ACKNOWLEDGMENTS

We thank Brian Stewart and J. Manual Grajales-Nishimura for helpful discussion, John Christie for petrographic advice, and M. Norman and I. Ridley for constructive reviews of a poorly organized initial manuscript. This research was supported by NASA grants NAGW-3808 (to P. H. Warren, at Los Angeles) and NAGW-3008 (to W. Alvarez, at Berkeley).

REFERENCES CITED

Alfvén, H., and Arrhenius, G., 1976, Evolution of the Solar System: Washington, D.C., NASA, 599 p.

Ashwal, L. D., 1993, Anorthosites: Berlin, Springer-Verlag, 422 p.

Avermann, M., and Brockmeyer, P., 1992, The Onaping Formation of the Sudbury Structure (Canada): An example of allochthonous impact breccias: Tectonophysics, v. 216, p. 227–234.

Avermann, M., and eight others, 1992, Sudbury Project (University of Münster—Ontario Geological Survey): (1) Summary of results—An updated impact model (abstract), *in* Papers Presented to the International Conference on Large Meteorite Impacts and Planetary Evolution: Houston, Lunar and Planetary Institute Contribution 790, p. 5–6.

Bottinga, Y., and Weill, D. F., 1970, Densities of liquid silicate systems calculated from partial molar volumes of oxide components: American Journal of Science, v. 269, p. 169–182.

Bottinga, Y., and Weill, D. F., 1972, The viscosity of magmatic silicate liquids: A model for calculation: American Journal of Science, v. 272, p. 438–475.

Bowen, N. L., 1928, The evolution of the igneous rocks: Princeton, Princeton University Press, 334 p.

Brocoum, S. J., and Dalziel, I. W. D., 1974, The Sudbury Basin, the Southern Province, the Grenville Front, and the Penokean Orogeny: Geological Society of America Bulletin, v. 85, p. 1571–1580.

Card, K. D., 1992, Circa 1.75 Ga ages for plutonic rocks from the Southern Province and adjacent Grenville Province: What is the expression of the Penokean Orogeny?: Discussion: Radiogenic Age and Isotopic Studies: Report 6, Geological Survey of Canada, Paper 92-2, p. 227–228.

Carmichael, I. S. E., Turner, F. J., and Verhoogen, J., 1974, Igneous petrology: New York, McGraw-Hill, 739 p.

Carroll, M. R., and Webster, J. D., 1994, Solubilities of sulfur, noble gases, nitrogen, chlorine, and fluorine in magmas, *in* Carroll, M. R., and Holloway, J. R., eds., Volatiles in magmas: Reviews in Mineralogy, v. 30, p. 231–279.

Cedillo-Pardo, E., Grajales N., J. M., Margolis, S. V., and Alvarez, W., 1992, Petrography and chemistry of Chicxulub melt rocks, Yucatan Peninsula, Mexico: Geological Society of America Abstracts with Programs, v. 24, p. A333.

Chai, G., and Eckstrand, O. R., 1993, Origin of the Sudbury Igneous Complex, Ontario—differentiate of two separate magmas: Geological Survey of Canada, Current Research, Part E, Paper 93-1E, p. 219-230.

Chen, G., Tyburczy, J. A., and Ahrens, T. J., 1994, Shock-induced devolatilization of calcium sulfate and implications for K-T extinctions: Earth and Planetary Science Letters, v. 128, p. 615–628.

Christian, R. P., Berman, S., Dwornik, E. J., Rose, H. J., Jr., and Schnepfe, M. M., 1976, Composition of some Apollo 14, 15, and 16 lunar breccias and two Apollo 15 fines [abs.]: Lunar and Planetary Science, v. 7, p. 138–140.

Chyba, C. F., 1991, Terrestrial mantle siderophiles and the lunar impact record: Icarus, v. 92, p. 217–233.

Clifford, P. M., 1990, Mid-Proterozoic deformational and intrusive events along the Grenville Front in the Sudbury-Killarney Area, Ontario, and their implications, *in* Gower, C. F., Rivers, T., and Ryan, B., eds., Mid-Proterozoic Laurentia-Baltica: Ottawa, Geological Society of Canada Special Paper 38, p. 335–350.

Collins, W. H., 1934, The life-history of the Sudbury Nickel Irruptive (1): Petrogenesis: Transactions of the Royal Society of Canada, v. 28, p. 123–177.

Condie, K. C., 1982, Plate tectonics and crustal evolution (second edition): New York, Pergamon, 310 p.

Cowan, E. J., and Schwerdtner, W. M., 1992, Intrusive origin of the Sudbury Igneous complex: Structural and sedimentological evidence (abstract), *in* Papers Presented to the International Conference on Large Meteorite Impacts and Planetary Evolution: Houston, Lunar and Planetary Institute (Contribution no. 790), p. 17–18.

Cox, K. G., 1983, The Karoo province of southern Africa: Origin of trace element enrichment patterns, *in* Hawkesworth, C. J., and Norry, M. J., eds., Continental basalts and mantle xenoliths: Nantwich, England, Shiva, p. 139–157.

Croft, S. K., 1985, The scaling of complex craters: Proceedings, Lunar and Planetary Science Conference, 15th, p. C828–C842.

Davidson, A., van Breemen, O., and Sullivan, R. W., 1992, Circa 1.75 Ga ages for plutonic rocks from the Southern Province and adjacent Grenville Province: What is the expression of the Penokean orogeny? *in* Radiogenic Age and Isotopic Studies: Report 6, Geological Survey of Canada Paper 92-2, p. 107–118.

Delano, J. W., and Ringwood, A. E., 1978, Siderophile elements in the lunar highlands: Nature of the indigenous component and implications for the origin of the Moon: Proceedings, Lunar and Planetary Science Conference, 9th, p. 1087–1102.

Deutsch, A., and Grieve, R. A. F., 1994, The Sudbury Structure: Constraints on its genesis from Lithoprobe results: Geophysical Research Letters, v. 21, p. 963–966.

Deutsch, A., Lakomy, R., and Buhl, D., 1989, Strontium- and neodymium-isotopic characteristics of a heterolithic breccia in the basement of the Sudbury impact structure: Earth and Planetary Science Letters, v. 93, p. 359–370.

Deutsch, A., Brockmeyer, P., and Buhl, D., 1990, Sudbury again: New and old isotope data [abs.]: Lunar and Planetary Science, v. 21, p. 282–283.

Faggart, B. E., Jr., Basu, A. R., and Tatsumoto, M., 1985, Origin of the Sudbury complex by meteoritic impact: Neodymium isotopic evidence: Science, v. 230, p. 436–439.

French, B. M., and Nielsen, R. L., 1990, Vredefort Bronzite Granophyre: Chemical evidence for origin as a meteorite impact melt: Tectonophysics, v. 171, p. 119–138.

Gibbins, W. A., and McNutt, R. H., 1975, The age of the Sudbury Nickel Irruptive and the Murray granite: Canadian Journal of Earth Sciences, v. 12, p. 1970–1989.

Grieve, R. A. F., and Cintala, M. J., 1992, An analysis of differential impact melt-crater scaling and implications for the terrestrial record: Meteoritics, v. 27, p. 526–538.

Grieve, R. A. F., and Floran, R. J., 1978, Manicouagan impact melt, Quebec. 2: Chemical interactions with basement and formational processes: Journal of Geophysical Research, v. 83, p. 2761–2771.

Grieve, R. A. F., Stöffler, D., and Deutsch, A., 1991, The Sudbury Structure: Controversial or misunderstood?: Journal of Geophysical Research, v. 96, p. 22753–22764.

Haines, E. L., and Metzger, A. E., 1980, Lunar highland crustal models based on iron concentrations: Isostasy and center-of-mass displacement: Proceedings, Lunar and Planetary Science Conference, 11th, p. 689–718.

Hawkesworth, C. J., Marsh, J S., Duncan, A. R., Erlank, A. J., and Norry, M. J., 1984, The role of continental lithosphere in the generation of the Karoo volcanic rocks: Evidence from combined Nd- and Sr-isotope studies: Special Publications of the Geological Society of South Africa, v. 13, p. 341–354.

Hildebrand, A. R., Penfield, G. T., Kring, D. A., Pilkington, M., Camargo Z., A., Jacobsen, S. B., and Boynton, W. V., 1991, Chicxulub crater: A possible Cretaceous/Tertiary boundary impact crater on the Yucatán Peninsula, Mexico: Geology, v. 19, p. 867–871.

Hildebrand, A. R., Pilkington, M., Connors M., Ortiz-Aleman, C., and Chavez, R. E., 1995, Chicxulub crater size and structure as revealed by horizontal Bouger gravity gradients and cenote distribution [abs.]: Lunar and Planetary Science, v. 26, p. 603–604.

Holloway, J. R., and Blanck, J. G., 1994, Application of experimental results to C-O-H species in natural melts, *in* Carroll, M. R., and Holloway, J. R., eds., Volatiles in magmas: Reviews in Mineralogy, v. 30, p. 187–230.

Irvine, T. N., 1982, Terminology for layered intrusions: Journal of Petrology, v. 23, p. 127–162.

Kieffer, S. W., and Simonds, C. H., 1980, The role of volatiles and lithology in the impact cratering process: Reviews of Geophysics and Space Physics, v. 18, p. 143–181.

Kring, D. A., and Boynton, W. V., 1992, Petrogenesis of an augite-bearing melt rock in the Chicxulub structure and its relationship to K/T impact spherules in Haiti: Nature, v. 358, p. 141–144.

Krogh, T. E., Davis, D. W., and Corfu, F., 1984, Precise U-Pb zircon and baddelyite ages for the Sudbury area, *in* Pye, E. G., Naldrett, A. J., and Giblin, P. E., eds., The geology and ore deposits of the Sudbury Structure: Toronto, Ontario Geological Survey Special Volume 1, p. 431–446.

Kuo, H. Y., and Crocket, J. H., 1979, Rare earth elements in the Sudbury Nickel Irruptive: Comparison with layered gabbros and implications for Nickel Irruptive genesis: Economic Geology, v. 74, p. 590–605.

Lange, R. L., 1994, The effect of H_2O, CO_2 and F on the density and viscosity of silicate liquids, *in* Carroll, M. R., and Holloway, J. R., eds., Volatiles in magmas: Washington, D.C., Mineralogical Society of America, p. 331–369.

Langenhorst, F., 1994, Shock experiments on pre-heated a- and b-quartz. II. X-ray and TEM investigations: Earth and Planetary Science Letters, v. 128, p. 683–698.

Lindstrom, M. M., Marvin, U. B., Holmberg, B. B., and Mittlefehldt, D. W., 1990, Apollo 15 KREEP-poor impact melts: Proceedings, Lunar and Planetary Science Conference, 20th, p. 77–90.

Lopez Ramos, E., 1975, Geological summary of the Yucatan Peninsula, *in* Nairn, A. E. M., and Stehli, F. G., eds., The ocean basins and margins: Volume 3: The Gulf of Mexico and the Caribbean: New York, Plenum, p. 257–282.

Marsh, B. D., 1981, On the crystallinity, probability of occurrence, and rheology of lava and magma: Contributions to Mineralogy and Petrology, v. 78, p. 85–98.

McGrath, P. H., and Broome, H. J., 1994, A gravity model for the Sudbury Structure along the Lithoprobe seismic line: Geophysical Research Letters, v. 21, p. 955–958.

McKinley, J. P., Taylor, G. J., Keil, K., Ma, M. -S., and Schmitt, R. A., 1984, Apollo 16: Impact melt sheets, contrasting nature of the Cayley Plains and Descartes Mountains, and geologic history: Proceedings, Lunar and Planetary Science Conference, 14th, p. B513–B524.

Meissner, R., 1986, The continental crust: A geophysical approach: Orlando, Academic, 426 p.

Melosh, H. J., 1989, Impact cratering: A geologic process: New York, Oxford University Press, 245 p.

Melosh, H. J., 1995, Under the ringed basins: Nature, v. 373, p. 104–105.

Meyerhoff, A. A., Lyons, J. B., and Officer, C. B., 1994, Chicxulub Structure: A volcanic sequence of Late Cretaceous age: Geology, v. 22, p. 3–4.

Moore, C. H., 1989, Carbonate diagenesis and porosity: Amsterdam, Elsevier, 338 p.

Muir, T. L., and Peredery, W. V., 1984, The Onaping Formation, *in* Pye, E. G., Naldrett, A. J., and Giblin, P. E., eds., The geology and ore deposits of the Sudbury Structure: Toronto, Ontario Geological Survey Special Volume 1, p. 139–199.

Naldrett, A. J., and Hewins, R. H., 1984, The Main Mass of the Sudbury Igneous Complex, *in* Pye, E. G., Naldrett, A. J., and Giblin, P. E., eds., The geology and ore deposits of the Sudbury Structure: Toronto, Ontario Geological Survey Special Volume 1, p. 235–251.

Naldrett, A. J., Rao, B. V., and Evensen, N. M., 1986, Contamination at Sudbury and its role in ore formation, *in* Gallagher, M. J., Ixer, R. A., Neary, C. R., and Prichard, H. M., eds., Metallogeny of basic and ultrabasic rocks: London, Institute of Mining Metallurgy, p. 75–91.

Newton, R. C., 1987, Thermodynamic analysis of phase equilibria in simple mineral systems, *in* Carmichael, I. S. E., and Eugster, H. P., eds., Thermodynamic Modeling of Geological Materials: Minerals, fluids and melts: Reviews in Mineralogy, v. 17, p. 1–33.

Norman, M. D., 1994a, Sudbury Igneous Complex: Evidence favoring endogenous magma rather than impact melt (abstract): Lunar and Planetary Science, v. 25, p. 1007–1008.

Norman, M. D., 1994b, Sudbury Igneous Complex: Impact melt or endogenous magma? Implications for lunar crustal evolution, *in* Dressler, B. O., Grieve, R. A. F., and Sharpton, V. L., eds., Large meteorite impacts and planetary evolution: Boulder, Colorado, Geological Society of America Special Paper 293, p. 331–341.

Onorato, P. I. K., Uhlmann, D. R., and Simonds, C. H., 1978, The thermal history of the Manicouagan impact melt sheet, Quebec: Journal of Geophysical Research, v. 83, p. 2789–2798.

Penfield, G. T., and Camargo Z., A., 1981, Definition of a major igneous zone in the central Yucatan platform with aeromagnetics and gravity [abs.]: 51st Annual International Meeting, Society of Exploration Geophysicists, v. 51, p. 38–39.

Persikov, E. S., Bukhtiyarov, P. G., and Kalinicheva, T. V., 1987, Effects of composition, temperature, and pressure on magma fluidity: Geokhimiya, v. 3, p. 483–498.

Phinney, W. C., and Simonds, C. H., 1977, Dynamical implications of the petrology and distribution of impact melt rocks, *in* Roddy, D. J., Pepin, R. O., and Merrill, R. B., eds., Impact and explosion cratering: New York, Pergamon, p. 771–790.

Quezada Muñeton, J. M., Marín, L. E., Sharpton, V. L., Ryder, G., and Schuraytz, B. C., 1992, The Chicxulub impact structure: Shock deformation and target composition [abs.]: Lunar and Planetary Science, v. 23, p. 1121–1122.

Raikes, S. A., and Ahrens, T. J., 1979, Post-shock temperatures in minerals: Geophysical Journal of the Royal Astronomical Society, v. 58, p. 717–747.

Rhodes, J. M., Rodgers, K. V., Shih, C., Bansal, B. M., Nyquist, L. E., Wiesmann, H., and Hubbard, N. J., 1974, The relationships between geology and soil chemistry at the Apollo 17 landing site: Proceedings, Lunar Science Conference, 5th, p. 1097–1117.

Ryder, G., 1990, Lunar samples, lunar accretion and the early bombardment of the Moon: Eos, v. 71, p. 313–323.

Schuraytz, B. C., Sharpton, V. L., and Marín, L. E., 1994, Petrology of impact-melt rocks at the Chicxulub multiring basin, Yucatán, Mexico: Geology, v. 22, p. 868–872.

Sharpton, V. L., Dalrymple, G. B., Marín, L. E., Schuraytz, B. C., and Urrutia-Fucugauchi, J., 1992, New links between the Chicxulub impact structure and the Cretaceous/Tertiary boundary: Nature, v. 359, p. 819–821.

Sharpton, V. L., and 9 others, 1993, Chicxulub multiring impact basin: Size and other characteristics derived from gravity analysis: Science, v. 261, p. 1564–1567.

Shaw, H. R., 1972, Viscosities of magmatic silicate liquids: An empirical method of prediction: American Journal of Science, v. 272, p. 870–893.

Simonds, C. H., Warner, J. L., Phinney, W. C., and McGee, P. E., 1976, Thermal model for impact breccia lithification: Manicouagan and the Moon: Proceedings, Lunar Science Conference, 7th, p. 2509–2528.

Sims, P. K., Van Schmus, W. R., Schulz, K. J., and Peterman, Z. E., 1989, Tectono-stratigraphic evolution of the Early Proterozoic Wisconsin magmatic terranes of the Penokean Orogen: Canadian Journal of Earth Sciences, v. 26, p. 2145–2158.

Spray, J. G., and Thompson, L. M., 1995, Friction melt distribution in a multiring impact basin: Nature, v. 373, p. 130–132.

Spudis, P. D., 1992, Basaltic impact melts in the Apollo collections: How many impacts and which events are recorded? [abs.], *in* Ryder, G., Schmitt, H. H., and Spudis, P. D., eds., Workshop on the geology of the Apollo 17 landing site: Houston, Lunar and Planetary Institute Technical Report Number 92-09, p. 55–57.

Spudis, P. D., and Ryder, G., 1981, Apollo 17 impact melts and their relation to the Serenitatis basin, *in* Schultz, P. H., and Merrill, R. B., eds., Multiring basins: New York, Pergamon, p. 133–148.

Stebbins, J. F., Carmichael, I. S. E., and Moret, L. K., 1984, Heat capacities and entropies of silicate liquids and glasses: Contributions to Mineralogy and Petrology, v. 86, p. 131–148.

Strom, R. G., and Neukum, G., 1988, The cratering record on Mercury and the origin of impacting objects, *in* Vilas, F., Chapman, C. R., and Matthews, M. S., eds., Mercury: Tuscon, University of Arizona Press, p. 336–373.

Swisher, C. C. III, and 11 others, 1992, Coeval ^{40}Ar-^{39}Ar ages of 65.0 million years ago from Chicxulub crater melt rock and Cretaceous-Tertiary boundary tektites: Science, v. 257, p. 954–958.

Taylor, S. R., 1982, Planetary science: A lunar perspective: Houston, Lunar and Planetary Institute, 512 p.

Therriault, A. M., Reid, A. M., and Reimold, W. U., 1993, Original size of the Vredefort structure, South Africa (abstract): Lunar and Planetary Science, v. 24, p. 1419–1420.

Tonks, W. B., and Melosh, H. J., 1993, Core formation by giant impacts: Conditions for impact melt region formation [abs.]: Lunar and Planetary Science, v. 24, p. 1435–1436.

Turcotte, D. L., and Schubert, G., 1982, Geodynamics: Applications of continuum physics to geological problems: New York, Wiley, 450 p.

Turner, F. G., 1968, Metamorphic petrology: New York, McGraw-Hill, 403 p.

Wadsworth, W. J., 1985, Terminology of postcumulus processes and products in the Rhum layered intrusion: Geological Magazine, v. 122, p. 549–554.

Wager, L. R., and Brown, G. M., 1967, Layered igneous rocks: San Francisco, Freeman, 588 p.

Walker, R. J., Morgan, J. W., Naldrett, A. J., Li, C., and Fassett, J. D., 1991, Re-Os isotope systematics of Ni-Cu sulfide ores, Sudbury Igneous Complex, Ontario: Evidence for a major crustal component: Earth and Planetary Science Letters, v. 105, p. 416–429.

Warner, J. L., Simonds, C. H., and Phinney, W. C., 1974, Impact-induced fractionation in the olunar highlands: Proceedings, Lunar Science Conference, 5th, p. 379–398.

Warren, P. H., 1990, Lunar anorthosites and the magma-ocean plagioclase-flotation hypothesis: Importance of FeO enrichment in the parent magma: American Mineralogist, v. 75, p. 46–58.

Warren, p. H., and Rasmussen, K. L., 1987, Megaregolith insulation, internal temperatures, and bulk uranium content of the Moon: Journal of Geophysical Research, v. 92, p. 3453-3465.

Warren, P. H., Haack, H., and Rasmussen, K. L., 1991, Megaregolith insulation and the duration of cooling to isotopic closure within differentiated asteroids and the Moon: Journal of Geophysical Research, v. 96, p. 5909–5923.

Wetherill, G. W., 1981, Nature and origin of basin-forming projectiles, *in* Schultz, P. H., and Merrill, R. B., eds., Multi-ring basins: New York, Pergamon, p. 1–18.

Wilhelms, D. E., 1987, The geologic history of the Moon: U.S. Geological Survey Professional Paper 1348, 302 p.

Wood, J. A., 1975, Lunar petrogenesis in a well-stirred magma ocean: Proceedings, Lunar Science Conference, 6th, p. 1087–1102.

Zuber, M. T., Smith, D. E., Lemoine, F. G., and Neumann, G. A., 1994, The shape and internal structure of the Moon from the Clementine mission: Science, v. 266, p. 1839–1843.

Manuscript Accepted by the Society September 14, 1995

Printed in U.S.A.

Geological Society of America
Special Paper 307
1996

Degassing of sedimentary rocks due to Chicxulub impact: Hydrocode and physical simulations

B. A. Ivanov
Institute for Dynamics of Geospheres, Moscow, Russian Federation
D. D. Badukov and O. I. Yakovlev
Vernadsky Institute, Moscow, Russian Federation
M. V. Gerasimov
Space Research Institute, Moscow, Russian Federation
Yu. P. Dikov
Institute of Ore Deposits, Petrography, Mineralogy, and Analytical Chemistry, Moscow, Russian Federation
K. O. Pope
Geo Eco Arc Research, 2222 Foothill Boulevard, Suite E-272, La Cañada, California 91011
A. C. Ocampo
Jet Propulsion Laboratory, California Institute of Technology, MS183-601, 4800 Oak Grove Drive, Pasadena California, 91109

ABSTRACT

The first phase of our work on the degassing of sedimentary rocks due to the Chixulub impact event includes both a two-dimensional (2D) hydrocode simulation of the impact and an experimental study of the geochemistry of shocked sedimentary rocks evaporated by an energetic laser pulse. The hydrocode simulation provides estimates of the total amount of sediments shocked above 100 GPa in the range from 200 to 1,000 km^3 depending upon crater diameter and other impact parameters. The experimental study contains the possible geochemical evolution of highly shocked anhydrite and its reactions with quartz and water and gives estimates of the percentage of sulfur injected into the atmosphere. Combining (1) volume estimates of highly shocked sediments, (2) geological data on the anhydrite content in sediments, and (3) the experimental estimates of the percentage of degassed sulfur from the highly shocked anhydrite, we estimate the total amount of sulfur delivered to the atmosphere to be in the range from 10^{16} to 10^{17} g. The result is important for studies of local and global effects of the Chicxulub impact on the terrestrial environment at the end of the Cretaceous.

INTRODUCTION

An important aspect of the geology of the Chicxulub crater is the presence of a 2- to 3-km-thick layer of volatile-rich sediments overlying the crystalline basement (Fig. 1). Impact vaporization of sulfates (anhydrite and gypsum) and carbonates (limestone and dolomite) in the sediments released large volumes of SO_2 and SO_3 and CO_2 into the atmosphere, which may have caused global climate changes responsible for the Cretaceous/Tertiary mass extinction event (Brett, 1992; Sigurdsson et al., 1992; Pope et al., 1994). The release of sulfur gases can cause global cooling through the formation of highly reflective sulfuric acid clouds. Here the magnitude of cooling is dependent upon the amount as well as the oxidation state (SO_2 and SO_3) of sulfur degassed (Pope et al., 1994). The impact release of CO_2 can cause global warming as a result of the well-known greenhouse effect (O'Keefe and Ahrens, 1989). However, our preliminary studies (Pope et al., 1994) suggest that the potential for

Ivanov, B. A., Badukov, D. D., Yakovlev, O. I., Gerasimov, M. V., Dikov, Yu. P., Pope, K. O., and Ocampo, A. C., 1996, Degassing of sedimentary rocks due to Chicxulub impact: Hydrocode and physical simulations, *in* Ryder, G., Fastovsky, D., and Gartner, S., eds., The Cretaceous-Tertiary Event and Other Catastrophes in Earth History: Boulder, Colorado, Geological Society of America Special Paper 307.

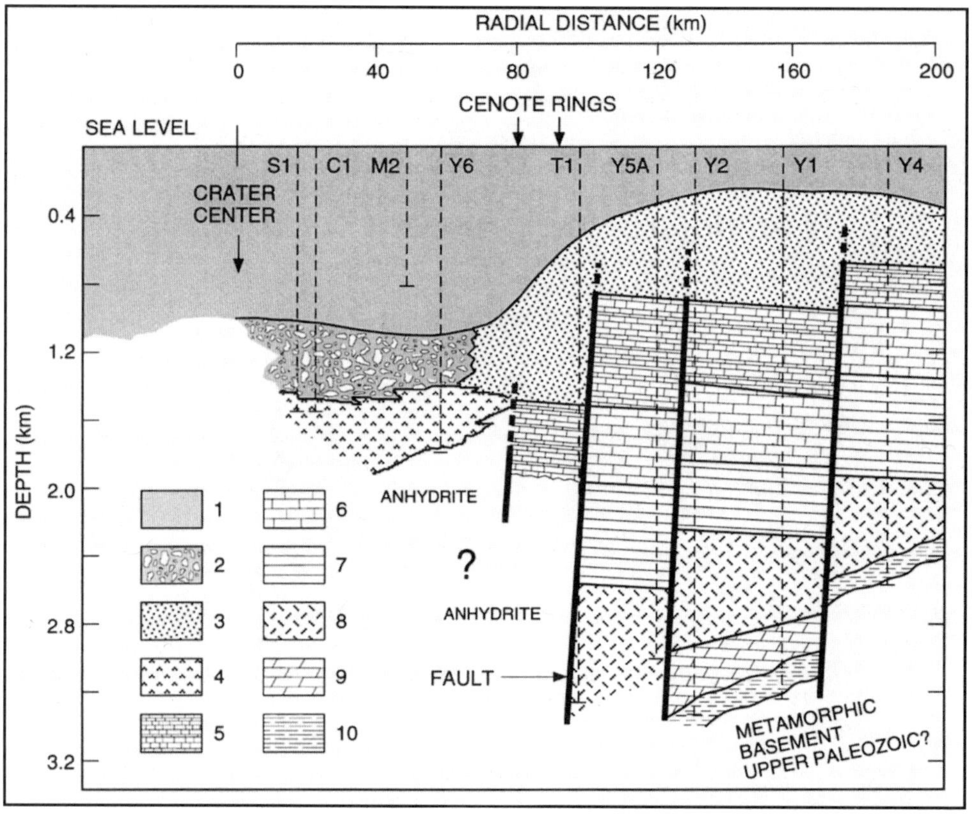

Figure 1. Subsurface geology of the Chicxulub structure from Pope et al. (1993), modified with data from Sharpton et al. (this volume). Data from exploratory oil wells drilled by Petroleos Mexicanos (PEMEX) are plotted as a function of their radial distance from the center of the crater inferred from the center of the Cenote Ring (a semicircle of sink holes that apparently demarcates the floor of the crater; see Pope et al., 1991, 1993). Vertical exaggeration is 50 times. All faults are inferred from offset of beds except the one recorded in drill logs of the Ticul 1 well. Faults inside of Y2 are interpreted as ring faults; others may not be related to the crater. Key for PEMEX wells: Y1—Yucatan 1, Y2—Yucatan 2, Y4—Yucatan 4, Y5A—Yucatan 5A, Y6—Yucatan 6, S1—Sacapuc 1, C1—Chicxulub 1, M2—Merida 2, and T1—Ticul 1. Key for rock units (interpretations in parentheses): 1—Tertiary marine sediments. 2—Bentonitic breccia (impact breccia and suevite). 3—Conglomerate and breccia composed of angular to rounded pebble-sized detritial clasts of anhydrite, dolomite, and limestone in a dolomitic micrite matrix (ejecta blanket deposits, including possible crater rim collapse material in T1). 4—Crystalline melt rocks and glass (impact melt sheet). 5—Upper Cretaceous mollusk, pelletoid micrite, and laminated fossiliferous dolomite. 6—Lower Cretaceous pelleted rudist biomicrite, biosparite, milinoid pelmicrite with anhydrite interlaminae. 7—Lower Cretaceous laminated dolomite. 8—Lower Cretaceous anhydrite. 9—Lower Cretaceous milinoid dolomite and quartz-silty dolomite. 10—Jurassic (?) arenite, shale, anhydrite, and dolomite. The bottom 10 m of Y6 and the lower half of T1 contain anhydrite, but the stratigraphy between these two wells is not well constrained.

cooling due to vaporization of sulfates far exceeds the potential for warming due to the vaporization of carbonates. Therefore, in this chapter, we focus primarily on the vaporization of sulfates. The vaporization of carbonates was investigated by Tyburczy and Ahrens (1986) and by Lange and Ahrens (1986).

Our study is in two parts: The first is a numerical simulation of the Chicxulub impact, and the second is a theoretical and experimental study of high-temperature vaporization of sulfates that includes simulations of shock heating with an intensive laser pulse. In the case of Chicxulub, the sedimentary layer is relatively thin in comparison with the diameter of the projectile. Such a geometry does not permit accurate estimates with previously derived relationships for vaporized mass of the target material (O'Keefe and Ahrens, 1977). We provide a computer simulation of an asteroid impact into a two-layer target to derive volume estimates of highly shocked sediments.

The possibility of generating long-term climatic effects depends on the model applied. For example, global warming (O'Keefe and Ahrens, 1989) or global cooling (cf. Pope et al., 1994). The scale of global and local effects and their nature depends critically on basic volumetric estimates of products delivered to the atmosphere by an impact and on the kind of models one uses to evaluate the biospheric and climatic change. Here we concentrate on the issue of volumetric estimates—

specifically, on estimates of the volume of highly shocked sediments evaporated during the asteroid impact. In the laser experiments, we vaporized several milligrams of sulfate under controlled atmospheric conditions. A plume of expanding vapors is produced. Some plume constituents are condensed with a special cold trap. Their chemical analyses provide some insights into the vaporization process as well as the chemical reactions among plume constituents. We conclude with a synthesis of our results and estimates of the amount of sulfur degassed by Chicxulub impact.

The diameter of the Chicxulub basin is not known yet with certainty. The current estimates include specific values of 180 km (Hildebrand et al., 1994), 240 km (Pope et al., 1993), and 300 km (Sharpton et al., 1993). The problem is that the crater is buried by younger sediments, and the same details of its structure, revealed by drilling and geophysical studies, are differently interpreted by researchers. So, now we definitely know that Chicxulub is an impact crater (or basin), but we need to determine the proper impact energy and outline the range of probable parameters of the projectile. We use diameters of 180 km and 300 km to constrain all the range of possible projectile parameters. The exact scaling for such an undefined crater diameter may not now be the critical problem, as other uncertainties may prove more important. Nevertheless, when the crater diameter becomes better known, all models will need to be iterated with a proper scaling law.

A silicate asteroid with a diameter of 10 to 20 km striking the Earth at an impact velocity of 20 km/s (the most probable for Earth-crossing asteroids) would create a crater like Chicxulub (Roddy et al., 1987). Variations in the asteroid's size, composition and impact velocity give an intrinsic uncertainty for all the estimates of a single event. An oblique impact may result in additional complex phenomena (Schultz and Gault, 1990). All these details need to be studied. Here we make initial (in some sense, simple) estimates of the volumes of highly shocked sediments created by a vertical impact with enough kinetic energy to form the Chicxulub structure. This initial step is necessary to lay the groundwork for further modeling.

NUMERICAL SIMULATION

We use the concept of "the late-stage equivalence" (Dienes and Walsh, 1970) to vary the projectile parameters while maintaining the same cratering efficiency. According to this concept, for the same target all high-velocity projectiles would create the same crater if the characteristic lengths (diameters), L, and impact velocities, v, of the projectiles satisfy the condition:

$$Lv^{\mu} = \text{constant}$$

The sense of the late-stage equivalence is simple: Once the crater is much larger than the projectile, the far cratering flow field is insensitive to the specific parameters of an impact. One can create the same size crater with an impact of a large, slow projectile or a small, fast one. The simplest approach here is the energetic scaling: All projectiles with the same kinetic energy would create craters of the same dimension. The kinetic energy is proportional to the projectile mass, m, multiplied by the square of the impact velocity: $KE = 0.5\ m\ v^2$. The mass of a projectile is proportional to δL^3, so KE is proportional to $L^3 v^2$. To keep the constant KE, one needs to keep constant the value of $L^3 v^2$, or—just the same—the value of $L v^{2/3}$, so for the energy scaling the value $\mu = 2/3 = 0.67$.

The simple energetic scaling does not work for real impacts, because part of the impact energy is used to heat the target and projectile material. The larger the impact velocity, the higher the irreversible heat loss, so faster projectiles are less effective for cratering. This phenomenon results in a value of μ less than $2/3$. Centrifuge experiments by Schmidt and his colleagues, as well as their theoretical treatments (cf. Schmidt and Housen, 1987), confirm this condition and show that for nonporous targets the value of the exponent, μ, is in the range of about 0.55 to 0.58.

Applying this concept, we used projectile size and velocity combinations of 10 km at 20 km/s, 7 km at 38 km/s, and 14 km at 11 km/s, all of which would produce the supposed crater with a diameter of 180 km. For the larger crater we used a 20-km projectile at 20 km/s and a 14-km projectile at 38 km/s. All projectiles were assumed to be cylinders with diameter equal to height. The "slow," 11 km/s projectile for the larger crater was not computed because of the large projectile diameter, which conflicted with spatial limitations of computational grid.

HYDROCODE

We simulate the vertical impact of an asteroid into a target with two layers. Such a geometry allows us to use the relatively simple SALE (Simplified Arbitrary Lagrangian Eulerian) code (Amsden et al., 1980), which solves hydrodynamic equations for a grid of four-corner cells. The simplest projectile geometry for this type of a grid is a cylindrical one. Initially, the lower boundary of the cylindrical projectile is placed on the top of the target surface.

The first step in the hydrocode simulation of the Chicxulub event is to calculate the shock-wave propagation in the target with a 2- to 3-km-thick upper layer, which is thin in comparison to the projectile diameter (10 to 20 km). To resolve the plane shock wave, high-order codes typically need four or so rows of cells. If a shock wave enters a thin soft layer from a relatively more competent basement, one needs to have approximately four rows to resolve the shock-wave entrance from the crystalline basement plus four (or so) rows to resolve the shock-wave interaction with a free surface. The problem is complicated when the shock-wave front has a complex nonplanar geometry. With the code we used, one needs to have at least 10 rows of cells describing the sedimentary layer to resolve multiple shock fronts with an affordable artificial viscosity. Thus, the cell dimension is limited to the range of a few hundreds meters. In this case, we need to cover the projectile with several tens of cells. The need for a fine resolution in the projectile volume indicates that the shape of the projectile may play an important role in the region close to it. This effect of the projectile shape seems to be a natural extension of the present work and is currently under study.

In our calculations, 250- × 250-m cells have been used to construct a 250 × 200 grid. The 3-km sedimentary layer here is divided into 12 rows. The cylindrical projectiles are 7 to 20 km in diameter, and they are represented in the beginning of the calculation as rectangles of 40 (radius) × 80 (height) cells for the largest (20 km) projectile, and one of 28 × 56 cells for the smallest (7 km) projectile.

The SALE code allows us to calculate the problem in the Lagrangian mode, when the grid is embedded in the material, as well in the Eulerian mode, when the material moves through a fixed grid. A most important aspect of our model is the use of a mixed regime of calculations, when part of the grid is a Lagrangian one while the other part is an Eulerian one. We used this feature of the code to hold the Lagrangian (impenetrable) boundary between sedimentary layer and the crystalline basement. The mixed Eulerian/Lagrangian regime has been used in the most critical part of the projectile's penetrating into the target. This allowed the grid to slowly diffuse through the media, maintaining a reasonable geometry of the computational grid.

To simplify the calculations, we assumed that the projectile consists of the same material as the sedimentary layer. The atmosphere has been modeled as a gas, based on parameters of vaporized sediments (see below). This simplification allows us to use a single material boundary in the calculations. The price for this simplifications is decreased accuracy in the description of the atmospheric shock wave and some artificial mixing of "atmosphere" and "projectile." The energy transferred to the atmosphere before the impact was estimated analytically. To keep the energy balance, all this energy was placed in the atmosphere around the projectile with a spatial distribution similar to the strong cylindrical explosion in the atmosphere. The energy density corresponds to the atmospheric density at a given altitude.

EQUATION OF STATE

We ran several variants of the code, varying the equation of state of target layers. The Tillotson equation of state (EOS) for rocks (Allen, 1967; see also Melosh, 1989) has been used with a minor modification in the expansion range (Ivanov, 1994). The Tillotson EOS is a reasonable first approximation for sedimentary and crystalline rocks.

According to the Tillotson EOS, in the compression region where density, ρ, is greater than the initial density, ρ_0, pressure, p, is a function of the density and specific internal energy, E:

$$p = A\mu + B\mu^2 + \rho E\left(a + \frac{b}{\frac{E}{E_0\eta^2} + 1}\right) \qquad (1)$$

where $\eta = \rho/\rho_0 - 1$ and $\mu = \eta - 1$.

TABLE 1. PARAMETERS FOR THE TILLOTSON EQUATION OF STATE

Material	ρ_0 (kg/m^3)	A, GPa	B, GPa	a	b	E_0 (MJ/kg)
Limestone	2,700	40	67	0.5	0.6	10
Wet tuff	1,970	10	6	0.5	1.3	11
Aluminum	2,700	75.2	65	0.5	1.63	5.0
Granite, lpp*	2,630	60	40	0.5	1.5	5.0
Granite, hpp*	3,944	230	119	0.5	0.189	40

*lpp is the low-pressure phase; hpp is the high-pressure phase. The energy of the lpp to hpp phase transition was assumed to equal to 0.12 MJ/kg. Parameters for granite were obtained by fitting to the experimental Huginiot adiabat.

To model dynamic properties of the sedimentary layer, we used the constants for wet tuff and limestone and the aluminum and granite constants for the crystalline basement cited by Melosh (1989). Parameters A, B, E_0, a, and b are listed in Table 1 for substances we studied. The reason for using aluminum to simulate crystalline rocks is that aluminum is close to typical rocks in density (2,700 kg/m^3) and bulk sound velocity (around 5 km/s). In contrast to typical granitelike rocks, aluminum has no solid-phase transition, which is good for testing the code. The comparison of our results for granite and aluminum as a basement demonstrates that the difference in their Hugonix curves has no critical effect on the estimate of the volume of highly shocked sediments.

In the range of densities, $\rho < \rho_0$ we used a smoothed transition to the ideal gas equation with an effective heat capacity ratio equal to a + 1. Details of the modified EOS for the expanded state have been described elsewhere (Ivanov, 1994). As for all materials, we used a = 0.5 (see Table 1); the undisturbed atmosphere in calculations is described as an ideal gas with a temperature of 300 K.

The target material and the atmosphere are placed into the normal terrestrial gravity field at isothermal conditions. Thus, we have initially the nonuniform isothermal atmosphere of an ideal gas and a proper lithostatic pressure beneath the ground surface. The pressure at the ground surface was equal to 0.1 Mpa (1 bar).

PRELIMINARY RESULTS OF NUMERICAL SIMULATION

Figures 2 and 3 illustrate typical output of numerical calculations. Figure 3 shows contours of the artificial viscosity, which is the parameter proportional to the rate of compression. The artificial viscosity is the best parameter to represent a shock wave's position at a given moment of time. The principal peculiarity of the rock wave pattern in the Chicxulub case is the higher shock-wave velocity in the more competent crystalline basement. As the result, the shock wave in the basement overtakes the shock wave in the sedimentary layer. The shock wave

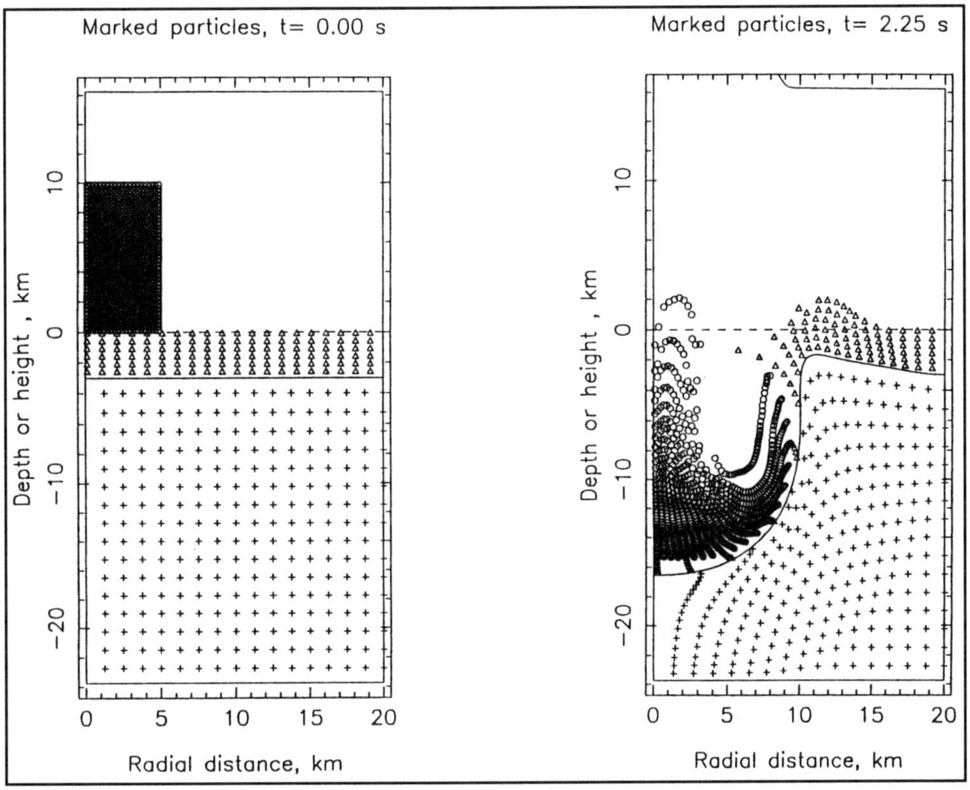

Figure 2. Computational geometry at the beginning of the calculations (left) and 2.25 s after impact (right) for a projectile with a height and diameter of 10 km and an impact velocity of 20 km/s. The left vertical axis is the axis of a cylindrical symmetry. Circles show the positions of marked particles, placed on the projectile. Triangles show position of marked particles, placed initially on the sedimentary layer (every second row and fourth column of computational cells). The crystalline basement material is marked with crosses (every fifth row and column). At t = 0 (left), circles on the projectile overlap one another, making total shading of the projectile. At t = 2.25 s (right), the projectile is totally decelerated and begins to expand upward near the center line; the sedimentary rocks under the projectile are squeezed into a thin layer at the surface of the transient cavity in the basement. The motion of the basement material was computed in a pure Lagrangian regime.

in the sediments is controlled by the refraction of the shock wave from the crystalline basement at radial distances larger than approximately 10 km.

The maximum shock pressure is recorded during the calculation for each marked particle. Values of these maximum pressures can be overlain on the initial target/projectile geometry after the calculation, thus providing a suitable means for calculating volumes of variously shocked materials (Fig. 4). For example, one can readily see the highly shocked sedimentary material originally situated just beneath the projectile as the projectile "footprint" material. The maximum shock pressures outside of the projectile footprint rapidly decay with radial distance. The complex shape of the maximum pressure contours is due to shock-wave refraction from the basement to sediments and to shock reflection from the free surface. A good discussion of the shock-wave interference at the free surface is given by Melosh (1989).

We examined the space-time history of several marked particles initially embedded in the sedimentary layer to outline differences in motion between footprint and out-of-footprint material (Fig. 5). The footprint material moves down and will be ejected at a later state of cratering. The out-of-footprint material is ejected in the first 2 to 3 sec after impact. This material is shocked to lower pressures than is the footprint material. More advance calculations are needed to show how this pattern would change for the oblique impact.

The principal result of the simulation described here is an estimate of sedimentary rock volume compressed above a given shock pressure. Figure 6 and Table 2 summarize computational results in the form of cumulative volumes of sediments, shocked to pressures higher than a given value for a range of pressures from 10 to 100 Gpa. The volume of sediments shocked above 100 Gpa varies from about 200 to 1,000 km^3 for the range of impact parameters.

It is important to note that more than half of this volume is the "plug" of sediments pushed down by the projectile. This material will be released to the atmosphere relatively late and will be mixed with projectile material and maybe with material

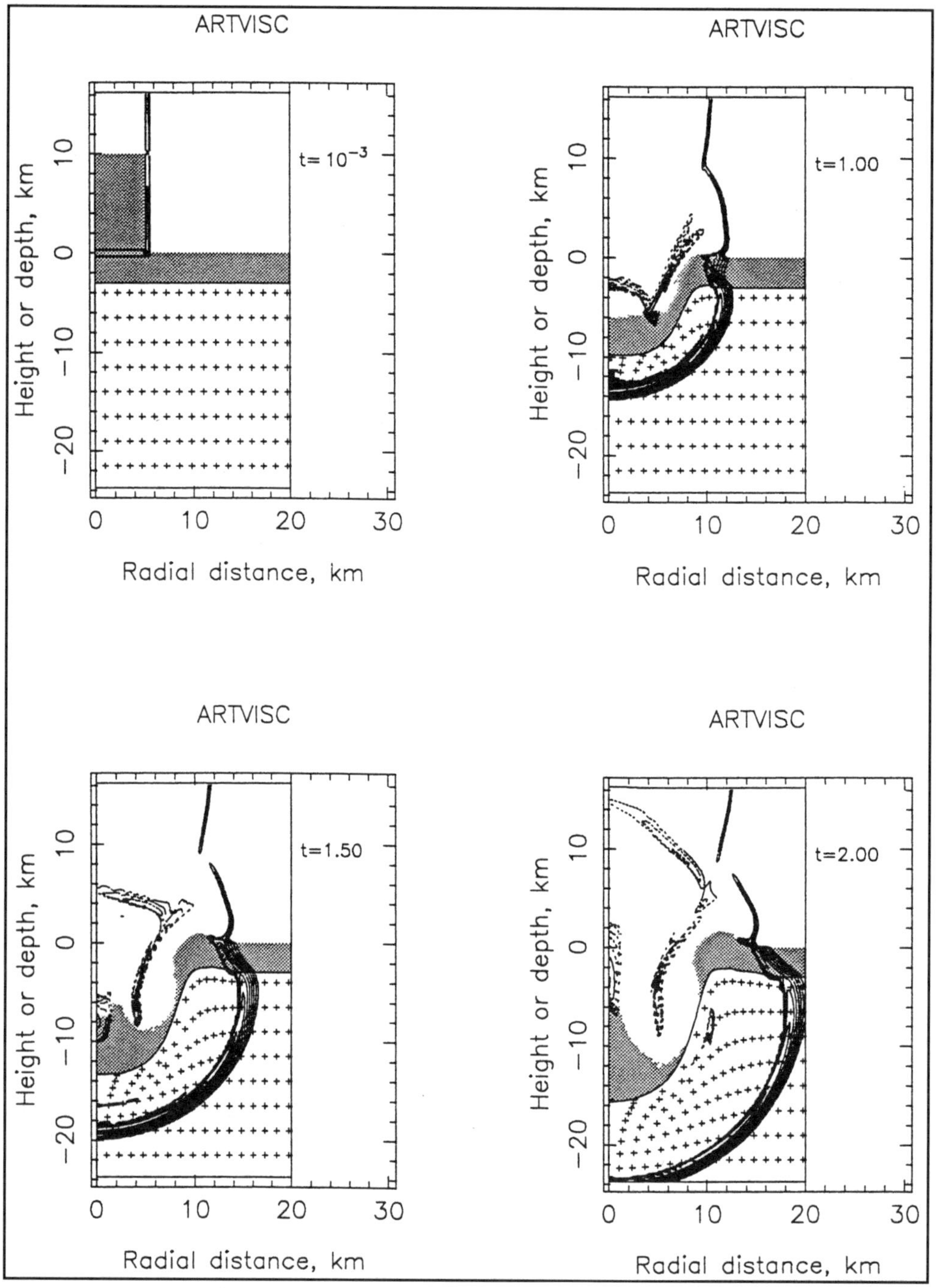

Figure 3. Artificial viscosity (ARTVISC) field evolution during first 2.0 s after impact (the same projectile parameters as in Fig. 2). Time t in seconds is designated on each plot. Artificial viscosity contours show zones of the highest rate of material compression. Therefore high concentrations of isolines represent the geometry of the shock-wave fronts. Note the oblique shock front in sediments after 1.0 s: Shock waves in the basement overrun waves in sediments, and the main compression of the sediments is due to shock waves refracted from the crystalline basement. Crosses mark crystalline rocks. Gray shading corresponds to the sedimentary rocks with a density above 1,000 kg/m^3.

derived from the crystalline basement. The results for out-of-footprint volumes are shown in Figure 7. Note that the difference between results for the 180-km and 300-km craters is smaller than for total volumes (Fig. 6), but the volume of highly shocked (above 100 GPa) out-of-footprint material is still between 40 and 200 km^3 for all variants except the slowest (11 km/s) case. Our current results are restricted by the simplification of setting the EOS of the projectile to that of the sedimentary layer. Actual asteroids may be more competent, which would increase the initial impact pressure. The initial impact

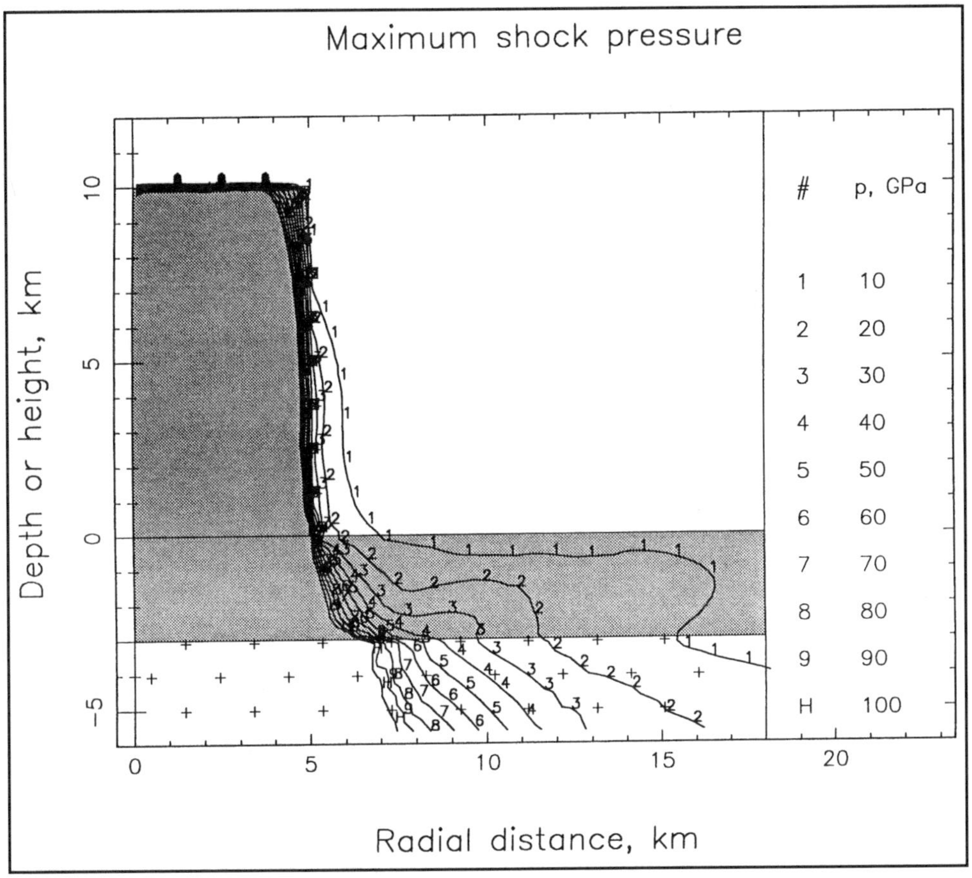

Figure 4. An example of the maximum isobar plot (the same projectile parameters as in Fig. 2). To construct such a plot, the value of the maximum shock pressure experience has been recorded for each marked particle during the calculation. These calculated values have been superimposed on the initial target/projectile geometry (t = 0) to show how much the material will be compressed during the impact event. Maximum shock pressure isobars marked with numbers (1–9) and one letter (H); corresponding shock pressure levels (in GPa) are shown at the legend to the right. Note that most of the highly shocked sediments are just beneath the projectile. This volume of sediments (so-called projectile footprint material) will move downward initially, being compressed between the projectile and crystalline material at the surface of the growing transient cavity (see Fig. 2). The material outside the projectile footprint will be ejected relatively early into the atmosphere.

pressure would decrease for oblique impacts. The balance of these factors will be the subject of future studies.

HIGH-TEMPERATURE VAPORIZATION OF SULFATES

We need to discuss properties of real rocks and minerals before we can apply the results of the numerical calculations to natural processes. For numerical simulations, we used the Tillotson EOS that gives a reliable description of shock compression but fails to predict with the same accuracy the state of materials after release. To partially close this gap, we need to use additional information about the thermodynamical and geochemical behavior of rocks and rock-forming minerals.

In this section, we present (1) estimates of shock pressures needed to induce anhydrite decomposition after release and (2) experimental results on the geochemistry of sulfates vaporized with a laser pulse.

Estimates of shock pressures required to vaporize sulfates

An estimate of the amount of volatiles released by an impact into sulfate-rich rocks can be obtained using Hugoniot data and thermodynamic parameters from the following reactions of decomposition of anhydrite and gypsum:

$$CaSO_4 = CaO + SO_3 \quad (2)$$

$$CaSO_4 \cdot 2H_2O = CaO + SO_3 + 2H_2O \quad (3)$$

Outgassing of sulfates can also occur by their reactions with silicates. For simplicity we consider the reaction with quartz only:

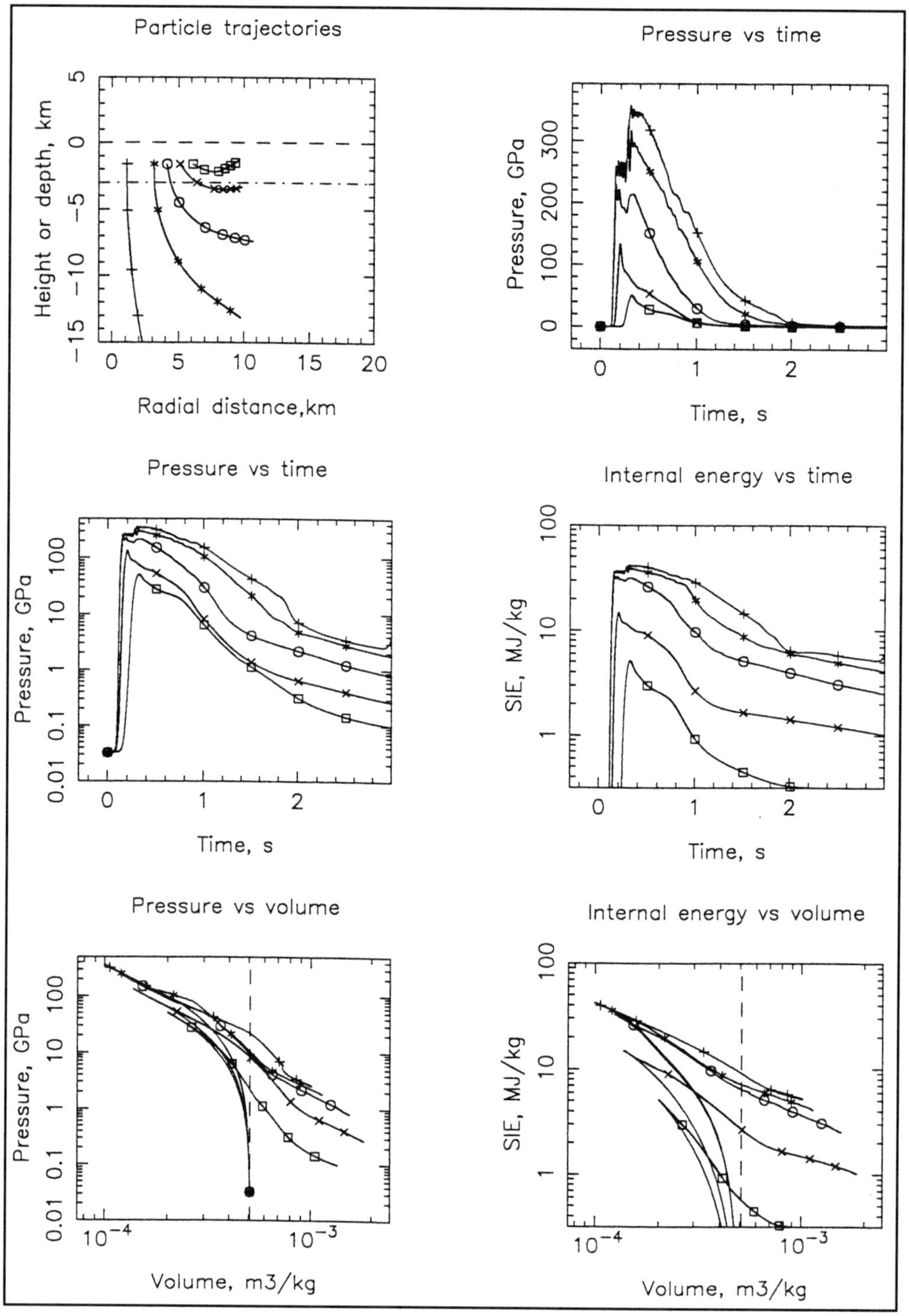

Figure 5. Trajectories and thermodynamic parameters for five marked particles inside the sedimentary layer, originally placed at a depth of 2 km. Different particles are distinguished by unique symbols (cross, asterisk, circle, x, and square) at all plots for each 0.5 s. Plots show (left to right, top to bottom): particle trajectories, pressure versus time (linear pressure scale), pressure versus time (logarithmic pressure scale), specific internal energy versus time, pressure versus specific volume, and specific internal energy (SIE) versus specific volume. The vertical dashed lines in the last two plots designate the initial specific volume of the sedimentary rocks. The first three particles are inside the projectile footprint and experience shock pressures above 100 GPa, but would be ejected later. The last two particles are outside the footprint. They are shocked up to 130 (x) and to 50 GPa approximately and would be ejected a few seconds after impact. Note that the thermodynamic release path for the particles marked with an x is close to the conditions of the laser experiment described in this chapter: The release isentrope passes through the initial specific volume with a specific internal energy of 2 to 3 MJ/kg, which roughly corresponds to the heated spot temperature of 2,000K that was estimated for the conditions in the laser experiment.

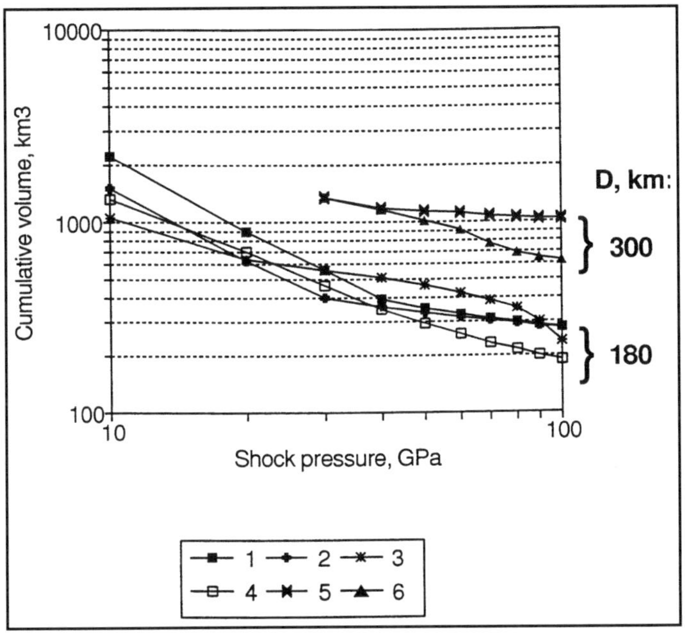

Figure 6. Total (footprint plus out-of-footprint) cumulative volume (km^3) of sediments, compressed to shock pressure larger than a given value in GPa. The lower group of curves corresponds to the crater diameter of 180 km. The upper two curves are for the crater diameter of 300 km. Numbers on the legend correspond to calculated variants (see Table 2 for the description). The lack of data below 30 GPa for the 300-km-crater case is due to shock wave interaction with the boundaries of the calculation grid.

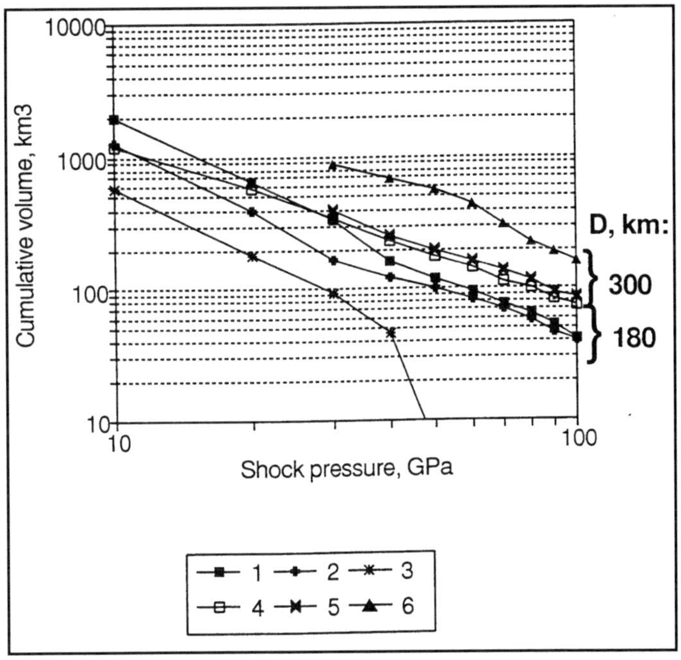

Figure 7. Cumulative volume (km^3) of out-of-footprint sediments shocked above a given pressure (in GPa). See explanation for the legend in Figure 6.

TABLE 2. VOLUMETRIC ESTIMATES OF HIGHLY SHOCKED SEDIMENTS

Projectile Diameter (km)	Impact Velocity (km/s)	Type of EOS*	Volume (km^3) of Sediments† Compressed Above 100 GPa	30 GPa
		180-KM CRATER		
10	20	1	280 (42)	560 (330)
10	20	2	280 (40)	400 (160)
14	11	2	236 (-)	560 (94)
7	38	2	190 (74)	460 (350)
		300-KM CRATER		
20	20	1	1,050 (110)	2,050 (1,100)
20	20	2	1,020 (80)	1,200 (250)
14	38	2	629 (160)	1,340 (880)

*#1 type of EOS is the Tillotson EOS with parameters for limestone (as sediments) and aluminum (as crystalline rocks). #2 type of EOS is the Tillotson EOS for wet tuff (as sediments) and granite (as crystalline rocks). The granite EOS also takes into account the quartz/stishovite-like phase transition.

†The value in brackets indicates the volume of shock sediments outside the projectile footprint.

$$CaSO_4 + SiO_2 = CaSiO_3 + SO_3 \tag{4}$$

Let us assume that all processes in the system are in equilibrium, and the ending state of the system does not depend on possible paths leading to this state. Hence, we can suggest that reactions start after the decompression to normal atmospheric pressure of 0.1 MPa (1 bar). Starting temperatures of the outgassing reactions should depend on the ending pressure of SO_3. To calculate the temperatures we assume that kinetic rates of the reactions and the cooling rate are much faster than the rate of mixing of SO_3 with ambient atmosphere.

The high temperature needed to start the decomposition is achieved as a result of a growth of internal energy behind the shock wave. The matter is heated at the shock front, and during the release the material temperature decreases. The gain of the specific internal energy, ΔE after the shock compression-decompression cycle can be written as (Zel'dovich and Raizer, 1967; Tyburczy and Ahrens, 1986):

$$\Delta E = 0.5P(V_0 - V_H) - \int_{V_{00}}^{V_H} P(V)dV \tag{5}$$

where V_0 is the initial specific volume, V_H is the specific volume under the shock pressure p, V_{00} is the specific volume after release, and $p(V)$ is a path of adiabatic release from the Hugoniot state p and V_H. The release path $p(V)$ can be calculated with the proper equation of state. Here we employed Mie-Gruneisen EOS of anhydrite, gypsum, and an equimolar mixture of anhydrite with quartz. This EOS differs from those used in our numerical simulations, but it is more suitable for thermodynamic estimates. The release curves from shock states $p - V_H$ were calculated according to an expression for isentropes (Baum et al., 1975, p. 640). To keep calculations simple, we do not take into account any possible phase changes in anhydrite and gypsum that take place by their shock compression (Simakov et al., 1974). Rough estimates of the phase transition energy and some assumptions about the shape of release curves with phase transitions suggest a maximum possible relative error of 10% in ΔE. The values are comparable with an accuracy of ΔE computations.

In the case of partial decomposition of anhydrite, the gain of the internal energy is spent in (1) heating of the undecomposed part of the matter and reaction products, (2) passage of the reaction, and (3) the mechanical work of the outflowing gas against an ambient pressure. The heating continues up to the temperature of incipient decomposition T_c, which can be obtained from thermochemical data. Hence, for the case of partial outgassing, we can write:

$$\Delta E = (1 - x)E^A + xE^R + x\Delta H^{T_C} + W \tag{6}$$

where x is a mole fraction of produced SO_3 and/or CaO produced from one mole of the reacted material, E^A and E^R are the energies spent on heating of anhydrite and the reaction products, ΔH^{T_c} is enthalpy of the reaction in the decompressed material by T_c, and W is a mechanical work of expanded SO_3 at constant ambient pressure p. Other values in Eq. 6 are:

$$E^A = \int_{298}^{T_c} C_p^A dT$$

and

$$W = pV^{SO_3}$$

where $C_p{}^A$ is the specific heat of the minerals or the anhydrite-quartz mixture at constant pressure and V^{SO_3} is the modular volume of SO_3 at ambient pressure p of 0.1 Mpa (1 bar) and temperature T_c. The value of ΔH^{T_c} can be calculated with values of enthalpy of the reaction at standard parameters, equal ΔH^{298}, and the difference of specific heats of the decomposed (reacted) material and products of the reaction $\Delta C_p{}^A$, using the relationship

$$\Delta H^{T_c} = \Delta H^{298} + \int_{298}^{T_c} \Delta C_p^A dT \tag{7}$$

It should be noted that for the reactions of decomposition of anhydrite, it is necessary to add an enthalpy of melting to the right part of Eq. 7, because the temperature of decomposition of anhydrite is higher than its melting point. On this base, the mole fraction of SO_3 gas generated by partial decomposition of anhydrite, gypsum, and quartz-anhydrite mixture can be written as

$$x = \frac{\Delta E - \int_{298}^{T_c} C_p^A dT}{\Delta H^{T_c} + pV^{SO_3}} \tag{8}$$

Parameters used in calculations are listed in Table 3. The calculated degree of outgassing of minerals is shown at Figure 8.

According to the calculations described above, pure anhydrite begins to decompose with SO_3 outgassing at the release from the shock loading to 80 Gpa. Complete decomposition may be reached after a shock pressure of 180 Gpa, and 50%

TABLE 3. PARAMETERS USED BY CALCULATIONS OF THE DEGREE OF SHOCK-INDUCED SO_3 LOSS

	Gypsum	Anhydrite	Anhydrite-quartz Mixture
Hugoniot data*			
C_0, km/s*	2.45†	3.25†	2.55§
S*	1.80†	1.72†	1.73§
Pressure range, GPa	>10	>15	>38
Initial specific volume V_0, m³/kg	4,310 • 10⁻⁴	3,357 • 10⁻⁴	3,484 • 10⁻⁴
Gruneisen parameter	$0.9(V/V_0)$**	$1.6(V/V_0)$**	$1.1(V/V_0)$**
Temperature of the reaction T_c,K	1,165	1,845	900**
ΔH^{298}, kJ/kg	1,047‡	2,939‡	1,616‡
ΔH^{Tc},kJ/gm	591	2,801	1,608

*The Hugoniot data are presented in the form $U_s = C_0 + su_p$, where U_s and u_p are shock-wave velocity and particle velocity respectively.
†Simakov et al., 1974.
§Calculated.
**Assumed.
‡Karpov et al., 1971, and Robie et al., 1978.

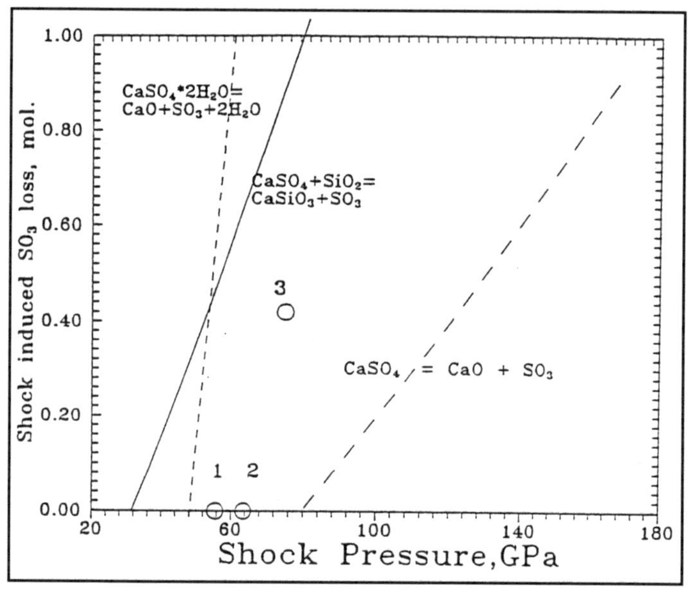

Figure 8. Calculated degree of shock-induced loss of SO_3 as a function of shock pressure for gypsum, anhydrite, and an equimolar anhydrite + quartz mixture. Circles present the experimental data from the shock recovery experiments with gypsum (1), anhydrite (2), and the anhydrite + quartz mixture (3) (Badjukov et al., 1995).

decomposition is achieved at 130 Gpa. These high shock pressures may be decreased by reactions with SiO_2 and water. Provided that thermodynamic equilibrium conditions are achieved (which may be difficult to reach in a dynamic process), SO_3 outgassing begins at the release from 30 to 40 Gpa shock loading and is completed at 60 to 80 Gpa.

It should be noted that the above-mentioned reactions critically depend on the degree of mixing. This question needs to be studied separately. However, the results of numerical simulations described in the previous section show that the sedimentary layer under the projectile (footprint material) is compressed and deformed into a thin layer between the projectile and basement material. The material here experiences a high shear rate, which is the proper condition for turbulent mixing of highly shocked (and partially vaporized) projectile, crystalline basement, and various sedimentary strata. A shear flow with a high strain rate is also calculated for the out-of-footprint zone, which also may create favorable conditions for mixing. The code we used is not advanced enough to simulate such a process.

Preliminary data from shock recovery experiments with anhydrite, gypsum, and anhydrite-quartz mixtures (see the progress report by Badjukov et al., 1995) support our estimates. Full discussion of these data is beyond the scope of this chapter and will be published elsewhere. Here, we summarize the main (but tentative) results. Recovery experiments show no decomposition of anhydrite shocked at 63 Gpa. Gypsum loaded to 56 Gpa is transformed to anyhdrite and demonstrates incipient decomposition, which can be the result of a reaction of gypsum with minor silicate component of the target. Shock loading to 75 Gpa of an anhydrite-quartz mixture induced a reaction between the components. A 0.46 mole part of anhydrite takes place in the reaction, which is lower than the calculated value. This discrepancy may result from kinetic constraints due to the high cooling rate of specimens. We also need to mention that the reaction's kinetic presents problems for any predictions of geochemical transformations during actual impact processes.

Reliable values of shock pressures required for partial or complete anhydrite decomposition are still not available. For equilibrium conditions the estimates of shock pressures needed for an effective (say, 50%) decomposition after release vary from 50 to 150 Gpa, depending on how far from equilibrium the natural processes go and how different reactive materials may be mixed. In this chapter we select the reference value of 100 Gpa to calculate the amount of degassed sulfur. This reference value seems to give the correct order-of-magnitude estimates. But accuracy below a factor of 3 or so cannot be achievable from the results of our current investigations.

A significant role of the sulfate-silicate interaction may arise from the high ability of a newly formed CaO to react with SO_3 in the case of impact-induced outgassing of pure gypsum and anhydrite. This reverse reaction can take place in cooling ejecta plumes and may decrease the degree of outgassing. As an analog of ahydrite decomposition, we refer to the study of shocked carbonates from the Haughton crater by Martinez et al. (1994). They suggested that there was a back-reaction of released CO_2 with shock-produced oxides. Moreover, the investigation of polycrystalline calcite shocked in a closed container (Ivanova et al., 1994) demonstrates that the material loaded up to 150 GPa and higher does not show the presence of the CaO phase formed by decomposition, although there are some textural traces of the processes of decomposition and new formation of calcite. On the other hand, the completeness of the reverse reaction at open conditions can be because of (1) expansion of the gas in a plume, (2) reactions with other components, like H_2O and CO_2, and (3) decay of SO_3 to SO_2 and O_2. Some outlook at the problem may be done using experiments with laser evaporation.

Pulse laser heating experiments

The chemistry of a fireball produced by impact is rather complex, taking into consideration a wide range of temperature and pressure, condensation of vaporized rocks, and, as a consequence, the heterogeneous character of chemical reactions. In such a case an experimental approach to study the impact chemistry seems to be helpful. Here we present some results of experiments on high-temperature evaporation of gypsum ($CaSO_4 \cdot 2H_2O$) and anhydrite, ($CaSO_4$).

Partial vaporization during an impact occurs at the stage of expansion of shocked matter at which the pressure of vapor phase can exceed the ambient pressure and be able to disrupt condensed matter to form vapor bubbles. Physically this point correspond to the intersection of a release adiabate with melt-vapor equilibrium curve (Fig. 9).

We use a pulsed Nd-glass laser with density of luminosity at the specimen surface adjusted to produce partial or close-to-complete vaporization. The objective of simulations with the laser technique is to rapidly bring a small amount of sample to the vapor state (without shock compression) and to provide adiabatic expansion of a vapor cloud.

Analysis indicates that vaporization begins as an equilibrium process at the beginning (Gerasimov et al., 1985), but the final composition of the vapor phase is determined by quenching when a characteristic time of chemical reactions exceeds the characteristic time of cloud expansion. So, the question of correlation between laser simulation products and real impact products is the question of correlation of quenching products in the first and the second cases. Analysis also indicates a weak dependence of chemical composition of expanded vapor cloud on temperature (Gerasimov et al., 1985).

The detailed description of the laser experiment can be found elsewhere (Gerasimov et al., 1985, 1994). Briefly the scheme of experimental setup is the following. Samples are vaporized in a 500-cm^3 hermetic cell filled in one series of experiments with air and in another one with helium at 1 atm. About 10 mg of sample is vaporized by the laser pulse with ~600 J luminous energy output at ~10^7 W/cm^2 density of luminosity. Simple comparison of the maximum temperature in

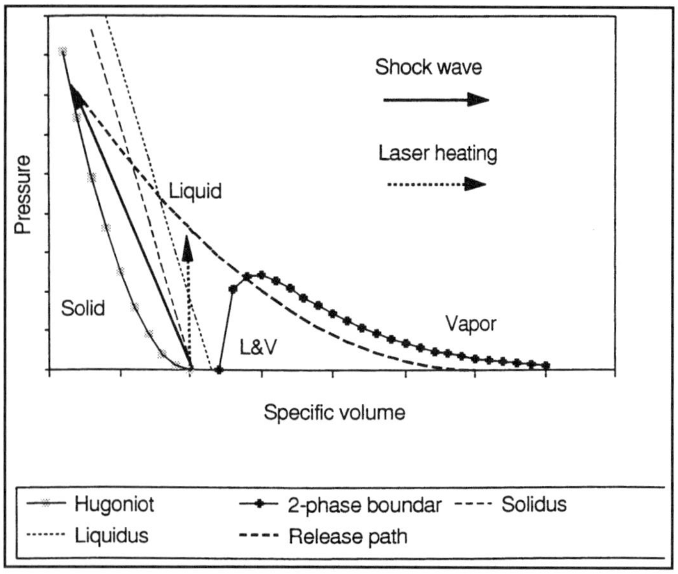

Figure 9. Pressure-volume diagram showing Hugoniot and a release adiabat intersecting the two-phase equilibrium curve close to the critical point. Shock wave compresses the matter and increases its entropy. The matter descends during the release along the release path and arrives at the two-phase region, where it begins to be vaporized (partially or completely). The objective of the pulse laser experiments is to heat matter without compression and to put it onto the same release adiabat without shock compression.

experiments and the specific internal energy at release adiabats indicates that experiments reproduce the expansion of the matter shocked approximately to 100 to 120 GPa. The hot and dense initial state of evaporated cloud with a following adiabatic expansion inside a cell could to a certain extent simulate the expansion of vapor cloud during impact processes.

At a distance of ~5 cm from the place of vaporization a witness plate of Ni foil was mounted to collect products of condensation that are produced inside an expanding and cooling vapor cloud. Analysis of initial samples and of condensed films is performed using X-ray photoelectron spectroscopy (XPS). The accuracy of XPS analyses is within 5 relative %. Condensed films were etched layer by layer (with a step of 200 to 400 A) by a beam of argon ions. For every layer, XPS analysis was performed, providing the information of the cross-section of the film. XPS analyses give both chemical elemental composition of the condensate and the distribution of elements between different phases.

We used samples of gypsum and anhydrite to investigate the role of water in high-temperature chemistry of vapors. Experiments in air and in inert helium help to study the possible role of atmospheric oxygen. The initial chemical composition of anhydrite was Ca—16.8, S—17.9, O—65.3 atom%; and for gypsum it was Ca—9.6, S—9.9; O—53.1, H—27.4 atom%. The concentration of hydrogen was calculated based on the value of oxygen which is bound in water O(-H). The deficit of water compared with the gypsum formula indicates that about 28% of it was dehydrated to anhydrite.

Mean chemical composition of condensed films produced in air and in helium is presented for gypsum and for anhydrite in Table 4. We can distinguish with XPS analysis the following phases: calcium sulfide (column 1), calcium in the form of pure oxide CaO (column 2), calcium bound with sulfur (column 3), sulfur in sulfide form (column 4), sulfur in the form of sulfite (S^{4+}, column 5), sulfur in the form of sulfate (S^{6+}, column 6), sulfur absorbed as SO_3 molecules (column 7), oxygen bound to calcium (column 8), and oxygen bound to sulfur (column 9). No distinct hydroxides were detected.

The presence of sulfur in the condensate shows that not all evaporated sulfur is released to the atmosphere. Some part of it is trapped during condensation. A comparison of chemical compositions of the condensed film and of the initial sample gives an estimate of the amount of sulfur-bearing gases that evolved into the atmosphere. If we admit that calcium is almost totally precipitated in the condensate, then the ratio of Ca to S in the initial sample and in the condensed film gives the information about the amount of sulfur that is degassed into the atmosphere. Such a comparison shows that in experiments in air only about 30% of sulfur was degassed from vaporized anhydrite (without account of adsorbed SO_3), and for gypsum this value was larger, rising up to 55%. Experiments in helium demonstrate 60 to 70% of sulfur degassed.

The amount of calcium bound to sulfur (column 3) is approximately the same as the amount of sulfate and sulfite sulfur (columns 6 and 5) in the condensate. This observation

TABLE 4. MEAN CHEMICAL COMPOSITION (ATOM. %) AND CHEMICAL STATE OF ELEMENTS IN CONDENSED FILMS OBTAINED DURING LASER PULSE VAPORIZATION OF GYPSUM (GYP) AND OF ANHYDRITE (ANH) IN AIR AND IN HELIUM

Elements	Ca			S				O	
Chemical State	Ca-(S)	Ca-(O-Ca)	Ca-(O-S)	S^{2+}	S^{4+}	S^{6+}	SO_3 Abs.	O-(Ca)	O-(S)
Column	1	2	3	4	5	6	7	8	9
GYP (He)	2.4	23.8	8.6	2.4	4.4	4.0	0.04	24.1	30.4
ANH (He)	....	18.5	10.4	....	2.7	7.6	0.9	18.2	41.8
ANH (air)	....	7.5	9.7	....	2.9	8.7	2.8	8.1	60.4
GYP (air)	....	15.8	8.8	....	2.1	7.7	1.3	16.5	47.7

shows that anhydrite and calcium sulfite are the main sulfur-bearing forms in condensates of vaporized anhydrite and gypsum. Outer layers of condensed films, which are formed mainly by precipitation from a cooled vapor cloud after the end of the laser pulse, have a ratio of S/Ca notably higher (S/Ca ~ 0.8 to 1) than the inner parts of films (S/Ca ~ 0.15 to 0.5), which are precipitated at the early stage of plume expansion when vapors were hotter (Fig. 10). This observation indicates that the efficiency of trapping of sulfur-bearing gases strongly depends on temperature and is higher at lower temperatures. Trapping of sulfur at low temperatures seems to be almost complete and is restricted only by availability of S or Ca atoms.

A generally accepted scheme of anhydrite decomposition during impact can be presented by reaction (Tyburczy and Ahrens, 1993):

$$2CaSO_4 = 2CaO + 2SO_2 + O_2$$

We do not measure the composition of gases in the "fireball" in our experiments, but we can estimate the possible chemical composition based on the chemical composition of the condensate. About 20% of sulfur in anhydrite and gypsum condensed films in experimental runs in air is presented in the form of adsorbed SO_3. Concentration of adsorbed SO_3 and the ratio of sulfate to sulfite phase is higher for experiments in air in comparison with runs in inert helium atmosphere. This observation

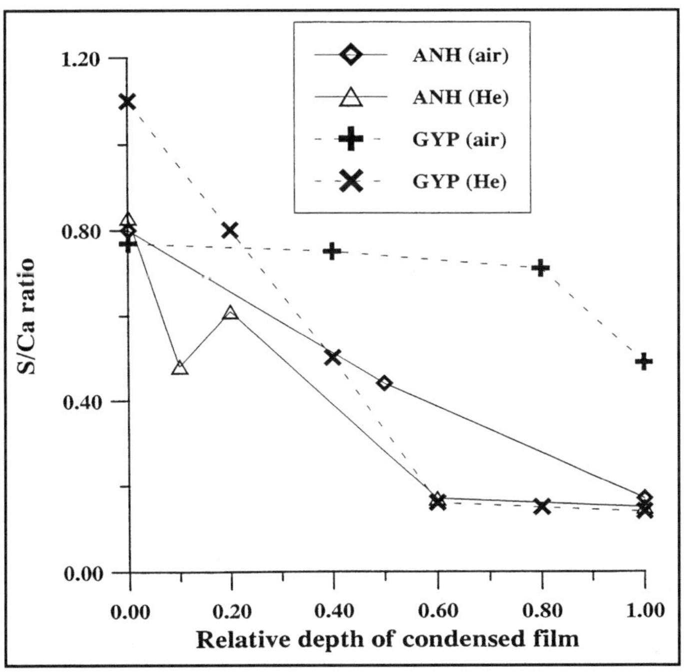

Figure 10. S/Ca ratio profiles across condensed films for laser pulse vaporization of anhydrite (ANH) and gypsum (GYP) in the experimental cell filled with air and helium. Outer layers of condensed films (small relative depths), which are formed later in time by precipitation from a cooled vapor cloud, have S/Ca ratio notably higher than the deeper parts (at a high relative depth) of the film, which are precipitated from the early hot vapors.

strongly indicates the important role of oxidation of sulfur, probably at the expense of atmospheric oxygen:

$$SO_2 + {}^1/_2O_2 = SO_3$$

The resultant chemical reaction of simple decomposition of anhydrite seems to give a rather good description of chemical process in the cloud:

$$CaSO_4 = CaO + SO_3$$

The absence of hydrated phases and a relative deficit of sulfur in the condensate from gypsum in comparison with anhydrite can result from the formation of sulfuric acid (H_2SO_4). A possible path of gypsum decomposition can be presented by the reaction:

$$CaSO_4 \cdot 2H_2O = CaO + H_2SO_4 + H_2O$$

In any case the presence of noticeable quantities of SO_3 in the products of vaporization makes rather realistic the possibility of direct mass production of sulfuric acid during the fireball expansion by reaction of SO_3 with evaporated water from the ocean. The presence of water in the system strongly influences redox conditions. Vaporization of gypsum in inert atmosphere resulted in a relative reduction of the products up to production of sulfide phase, maybe as a result of reduction by hydrogen.

In connection with the Chicxulub impact event, our studies show the possibility of the following effects. Volatilization of sulfur during the impact did not result in its total degassing as SO_2 into the atmosphere. A noticeable portion of sulfur is trapped by forming anhydrite and sulfite phases. Concentrations of trapped sulfur in experiments can be considered as low limit values, since quenching in a real impact occurs at lower temperatures in comparison with laser experiments, creating better conditions for trapping. It is necessary to note here that in a real, chemically more complex situation calcium can be used, for example, to build silicates, and as a consequence it may be removed from the calcium-sulfur system. This effect would provide less effective trapping of sulfur by the loss of calcium. A large portion of sulfur gases could be represented by SO_3. Oxygen of the atmosphere may also be used for oxidation of sulfur to SO_3, giving in turn local depletion in oxygen concentration that may also be an effect of the impact that is important for the biosphere. Involving in the impact process water from the ocean or from dehydration of target gypsum possibly results in direct mass production of sulfuric acid.

DISCUSSION AND CONCLUSIONS

Two parts of the presented study are still poorly coupled: (1) Existing equations of state for numerical simulations do not adequately describe the geochemical processes needed for physical modeling; (2) Laser evaporation experiments have relatively short time scales and a restricted range of initial conditions. Nevertheless we draw some tentative conclusions, with the understanding that final answers require more research.

We can estimate the maximum mass of sulfur that may be delivered to the atmosphere in the form of sulfur-bearing gases based on estimates for volumes of sediments shocked to the pressure required to decompose anhydrite. The estimated anhydrite content in the Chicxulub sedimentary layer varies from ~23%, if one counts only the 700-m-layer of anhydrite at the base of a 3-km sedimentary sheet, to 60%, if one adds to the basal layer thin anhydrite layers above (Lopes Ramos, 1979). As anhydrite contains about 25 mass% of sulfur, the average sulfur content in the sediments is about 6 to 15 mass%. At the average density of 2,500 kg/m^3, every 100 km^3 of sediments (2.5×10^{17} g) contains approximately from 1.5×10^{16} g to 3.8×10^{16} g of sulfur. If all the sulfur from sediments shocked above 100 GPa (200 to 1000 km^3; see Table 2) were degassed, the total amount of sulfur delivered to the atmosphere would be between 3.0×10^{16} g and 3.8×10^{17} g. This large uncertainty here may be decreased by better constraints on the crater dimensions; however, uncertainties related to the obliqueness of the impact will also need to be resolved.

The aforementioned maximum estimates may be increased if scaling will confirm that a larger projectile is needed to explain a 300-km crater (Holsapple, 1994). Nevertheless other factors tend to decrease the maximum estimate, for example, the reverse reaction in the cooling ejecta plume. The completeness of the reverse reaction can be restricted because of fast plume expansion. Laboratory experiments give some constraints for the process. Currently we have experimental estimates for anhydrite and gypsum at only one initial temperature. For these tested conditions, 30% to 60% of sulfur may be degassed. These percentages, if valid, decrease the estimates given above by a factor of 2 to 3. We need to expand the experimental range of initial conditions and rates of expansion to better understand the dynamic geochemistry of anhydrite at high rates of heating and release. This work is now in progress.

From the point of view of cratering mechanics, an important problem is the manner in which early shocked and released sediments are delivered to the atmosphere. Most of the highly shocked sediments (footprint material) would travel down to the transient cavity before being released to the atmosphere. During this travel, sediments may be mixed with vapors of the projectile and crystalline material. Material from different sedimentary layers also may be mixed. The geochemistry of these processes is a challenge for future investigations.

The remaining part of the highly shocked sediments (out-of-footprint material) may be involved in the early high-speed ejecta. This material seems to be mostly a mixture of different types of sediments. A large amount of water vapors also would have a prominent effect at the geochemical state of impact products. Clearly the study of the impact production of gases and their transport to the atmosphere and stratosphere has several consequent stages that require thorough investigation.

ACKNOWLEDGMENTS

All authors appreciate the support of the Exobiology Program of the National Aeronautical and Space Administration. One of authors (BAI) appreciates the partial financial support from the Russian Fundamental Research Foundation (Project 94-05-16329). We thank M. Boslough and M. Cintala for very useful reviews.

REFERENCES CITED

Allen, R. T., 1967, Equation of state of rocks and minerals: Internal report to DASA under contract DA 49-146-XZ-462: San Diego, California, General Dynamics Reports GA MD-7834, 23 p.

Amsden, A. A., Ruppel, H. M., and Hirt, C. W., 1980, SALE: A simplified ALE computer program for fluid at all speeds: Los Alamos, New Mexico, Los Alamos Scientific Laboratory Report LA-8095, 100 p.

Badjukov, D. D., Dikov, Yu. P., Petrova, T. L., and Pershin, S. V., 1995, Shock behavior of calcite, anhydrite, and gypsum [abs.]: Lunar and Planetary Science Conference 26; Houston, Texas, Lunar and Planetary Institute, p. 63–64.

Baum, F. A., Orlenko, L. P., Stanyukovich, K. P., Tshcelyshev, V. P., and Shechter, B. I., 1975, Physics of explosion: Moscow, Nauka Press, 704 p. (in Russian).

Brett, R., 1992, The Cretaceous-Tertiary extinction: A lethal mechanism involving anhydrite target rocks: Geochimica et Cosmochimica Acta, v. 56, p. 3603–3606.

Dienes, J. K., and Walsh, J. M., 1970, Theory of impact: Some general principles and the method of Eulerian codes; *in* Kinslow, R., ed., High-velocity impact phenomena: New York and London, Academic Press, p. 46–104.

Gerasimov, M. V., Mukhin, L. M., Dikov, Yu. P., and Rekarsky, V. I., 1985, Mechanisms of the early differentiation of the Earth: Vestnik Academii Nauk SSSR, no. 9, p. 10–25 (in Russian).

Gerasimov, M. V., Dikov, Yu. P., Yakovlev, O. I., and Wlotzka, F., 1994, Trapping of water vapor from the atmosphere by condensate during pulse vaporization: Geokhimia, no. 4, p. 597–608 (in Russian).

Hildebrand, A. R., Penfield, G. T., Kring, D. A., Pilington, M., Camgargo Z., A., Jacobsen, S. B., and Boynton, W. V., 1994, Chicxulub crater: A possible Cretaceous/Tertiary boundary impact crater on the Yucatan Peninsula, Mexico: Geology, v. 19, p. 867–871.

Holsapple, K. A., 1994, Recipes for impact cratering [abs.]: Lunar and Planetary Science Conference 25; Houston, Texas, Lunar and Planetary Institute, p. 559–560.

Ivanov, B. A., 1994, Geomechanical models of impact cratering: Puchezh-Katunki structure, *in* Dressler, B. O., Grieve, R. A. F., and Sharpton, V. L., eds., Large meteorite impacts and planetary evolution: Geological Society of America Special Paper 293, p. 81–91.

Ivanova, L. A., Litvinov, B. V., Kozlov, E. A., Zhugin, Yu. N., Abakshin, E. V., Uszapovskaya, Z. F., Nartava, N. V., and Yarovoy, P. N., 1994, Some results of high-pressure affecting on carbonate: Doklady Rossiiskoi Akademii Nauk, v. 336, p. 795–798.

Karpov, I. K., Kiselev, A. I., and Letnikov, F. A., 1971, The chemical thermodynamics in petrology and geochemistry: Irkutsk, USSR Academy of Sciences, 210 p. (in Russian).

Lange, M. A., and Ahrens, T. J., 1986, Shock-induced CO_2 loss from CaCO: Implications for early planetary atmospheres: Earth and Planetary Science Letters, v. 77, p. 409–418.

Lopes Ramos, E., 1979, Geologia de Mexico, Volume 3: Mexico City, Universidad Nacional Autonoma de Mexico, 445 p.

Martinez, I., Agrinier, P., Scharer, U., and Javoy, M., 1994, A SEM-ATEM and stable isotopy study of carbonates from the Haughton impact crater,

Canada: Earth and Planetary Science Letters, v. 121, p. 559–574.

Melosh, H. J., 1989, Impact cratering: A geologic process: New York, Oxford University Press, and Oxford, Clarendon Press, 245 p.

O'Keefe, J. D., and Ahrens, T. J., 1977, Impact-induced energy partitioning, melting, and vaporization on terrestrial planets, *in* Proceedings, Lunar and Science Conference, 8th: New York, Pergamon Press, p. 3357–3374.

O'Keefe, J. D., and Ahrens, T. J., 1989, Impact production of CO_2 by the Cretaceous/Tertiary extinction bolide and the resultant heating of the Earth: Nature, v. 338, p. 247–249.

Pope, K. O., Ocampo, A. C., and Duller, C. E., 1991, Mexican site for K/T impact crater?: Nature, v. 351, p. 105.

Pope, K. O., Ocampo, A. C., and Duller, C. E., 1993, Surficial geology of the Chicxulub impact crater, Yucatan, Mexico: Earth, Moon, and Planets, v. 63, p. 93–104.

Pope, K. O., Baines, K. H., Ocampo, A. C., and Ivanov, B. A., 1994, Impact winter and the Cretaceous/Tertiary extinctions: Results of a Chicxulub asteroid impact model: Earth and Planetary Science Letters, v. 128, p. 719–725.

Robie, R. A., Hemingway, B. S., and Fisher, J. R., 1978, Thermodynamic properties of minerals and related substances at 298.15 K and 1 bar (10^5 Pascal's) pressure and at higher temperatures: U.S. Geological Survey Bulletin 1452, 456 p.

Roddy, D. J., Schuster, S. H., rosenblatt, M., Grant, L. B., Hassig, P. J., and Kreyenhagen, K. N., 1987, Computer simulations of large asteroid impacts into oceanic and continental sites—preliminary results on atmospheric, cratering and ejecta dynamics: International Journal of Impact Engineering, v. 5, p. 525–541.

Schmidt, R. M., and Housen, K. R., 1987, Some recent advances in the scaling of impact and explosion cratering: International Journal of Impact Engineering, v. 5, p. 543–560.

Schultz, P. H., and Gault, D. E., 1990, Prolonged global catastrophes from oblique impacts, *in* Sharpton, V. L., and Ward, P. D., eds., Global catastrophes in Earth history: Geological Society of America Special Paper 247, p. 239–262.

Sharpton, V. L., and nine others, 1993, Chicxulub multiring basin: Size and other characteristics derived from gravity analysis: Science, v. 261, p. 1564–1567.

Sigurdsson, H., D'Hondt, S., and Carey, S., 1992, The impact of the Cretaceous/Tertiary bolide on evaporite terrain and generation of major sulfuric acid aerosol: Earth and Planetary Science Letters, v. 109, p. 543–559.

Simakov, G. V., Pavlovsky, M. N., Kalashnikov, N. G., and Trunin, R. F., 1974, Shock compressibility of twelve minerals: Izvestia for Physics of Solid Earth, no. 8, p. 488–492.

Tyburczy, J. A., and Ahrens, T. J., 1986, Dynamic compression and volatile release of carbonates: Journal of Geophysical Research, v. 91, p. 4730–4744.

Tyburczy, J. A., and Ahrens, T. J., 1993, Impact-induced devolatilization of $CaSO_4$ anhydrite and implications for K/T extinctions: Preliminary results [abs.]: Lunar and Planetary Science Conference 24: Houston, Texas, Lunar and Planetary Institute, p. 1449–1450.

Zel'dovich, Ya. B., and Raizer, Yu. P., 1967, The physics of shock waves and high-temperature hydrodynamic phenomena, Volumes 1 and 2: New York, Academic Press, 460 p.

Manuscript Accepted by the Society September 14, 1995

Printed in U.S.A.

Geological Society of America
Special Paper 307
1996

Trajectories of ballistic ejecta from the Chicxulub Crater

Walter Alvarez
Department of Geology and Geophysics, University of California, Berkeley, California 94720

ABSTRACT

Ejecta from the Chicxulub Crater includes partially or completely altered glassy tektites and microtektites, shocked quartz, microkrystites, and anomalous iridium. Debris from large craters should include both ejecta-curtain and fireball material, but it is not yet clear which mechanism dispersed each of the ejecta types from Chicxulub. Ejecta launched beyond the atmosphere will follow ballistic trajectories corresponding to partial elliptical orbits ending at the point of reimpact. This chapter shows that the Earth's rotation affects ballistic trajectories in two ways: (1) The rotational velocity is added to the launch velocity of eastbound ejecta and subtracted from that of westbound ejecta. As a result, fast eastbound ejecta reaches higher altitudes than fast westbound ejecta. (2) The Earth rotates beneath ejecta in flight, displacing the site of reimpact westward along parallels of latitude. For ejecta-curtain material launched with an elevation angle close to 45° there is a forbidden zone beyond about 100° east of an equatorial impact that cannot be reached by eastbound ejecta; however, ejecta-curtain debris might not be launched with sufficient velocity for this effect to be important. Fireball ejecta can probably be launched at velocities approaching escape velocity and at elevation angles lower than 45°, and thus can reach the area forbidden to ejecta-curtain material. Full study of the ballistics of the KT boundary ejecta strewnfields remains to be undertaken, but the approach taken here should eventually contribute to understanding the distribution of Chicxulub ejecta.

INTRODUCTION

The development of the impact theory for the Cretaceous-Tertiary (KT) boundary extinction has previously been reviewed (Alvarez and Asaro, 1990; Smit, 1990). The initial discovery of anomalous iridium at the KT boundary (Alvarez et al., 1980) was interpreted as evidence for a large-body impact. This was supported by the discovery at the boundary of spherules interpreted as originating from droplets of impact melt (Smit and Klaver, 1981; Montanari et al., 1983) and then by the discovery at this horizon of quartz grains showing the lamellae produced by shock (Bohor et al., 1984). There was no good candidate for the predicted impact crater until 1991, when Hildebrand et al. (1991) proposed that the Chicxulub Structure in the northern part of the Yucatán, previously interpreted as an impact crater by Penfield and Camargo (1981), was the KT impact site. This identification has been supported by studies of the petrography, geochemistry (Blum et al., 1993; Koeberl, 1993), and radiometric age (Swisher et al., 1992) of melt rocks from the Chicxulub subsurface and by the discovery of tektite-bearing clastic deposits of probable tsunami origin around the Gulf of Mexico (Bourgeois et al., 1988; Maurrasse and Sen, 1991; Alvarez et al., 1992b; Smit et al., 1992b).

The five types of material interpreted as impact ejecta that have been found at the KT boundary, either locally or worldwide, are discussed below. Now that Yucatán has been identified as the impact site, it is possible to investigate the mechanisms and paths by which these types of ejecta material reached their sites of deposition. Ballistic transport outside the atmosphere is essential in these emplacement mechanisms because the atmosphere is incapable of supporting sand-size grains during transport to sites thousands of kilometers away. Oberbeck (1975) investigated ballistic erosion and sedimentation in the ejecta blankets surrounding lunar craters, and Argyle (1989) examined

Alvarez, W., 1996, Trajectories of ballistic ejecta from the Chicxulub Crater, *in* Ryder, G., Fastovsky, D., and Gartner, S., eds., The Cretaceous-Tertiary Event and Other Catastrophes in Earth History: Boulder, Colorado, Geological Society of America Special Paper 307.

the global pattern of ejecta from a large terrestrial crater. With the wider distribution of the KT boundary ejecta deposits it appears that a new area of investigation is emerging, which might be called "global ballistic sedimentology" and might eventually be applied to other large impact events.

In this chapter the interactions between ejecta and atmosphere are largely ignored, because the purpose of the chapter is to understand the dynamics of ejecta during ballistic flight. Atmospheric drag and wind transport of the re-entering ejecta during settling will slightly distort the depositional patterns calculated here, but these 10- to 100-km effects are small compared to 1,000- to 10,000-km ballistic trajectories. Of more concern are interactions during launch, which raise the question of whether ejecta can escape the atmosphere with sufficient velocity for long-range ballistic travel. If ejecta are slowed by friction immediately after launch, as is surely the case with very small impacts, ballistic flight cannot be achieved, and the calculations presented here are irrelevant. In large impacts, however, the expanding hot cloud or "fireball" of vapor derived from impactor and target rock encounters the density gradient of the atmosphere and actually accelerates the ejecta upward to speeds approaching escape velocity (Jones and Kodis, 1982). Thus it seems likely that as the vapor density is thinned by expansion, entrained ejecta particles will be transferred into long-range ballistic trajectories beyond the atmosphere. Even a fireball several hundred kilometers in diameter would be small compared to the full trajectory, which can thus be considered to first approximation to begin at the impact site. Numerical simulations of the KT impact (Roddy et al., 1987) indicate that ejecta-curtain material can also rise tens of kilometers, thus effectively passing through the atmosphere and achieving ballistic flight, although perhaps not at the velocities necessary to travel long distances. The present chapter should thus be understood as a study of global ballistics, which may or may not be applicable to the KT impact depending on the nature of the impact mechanism, which is not yet understood. On the other hand, if ballistic dynamics are found to explain patterns of ejecta distribution, this may provide information useful for understanding the impact event.

OBSERVED CHICXULUB EJECTA

Five types of material have been interpreted as ejecta from the Chicxulub Crater. They are listed below, not in the order of their discovery but in the approximate order in which they would have been deposited.

Ejecta blanket

Large fragments of target rock, which would have traveled slowly and not very far, are to be expected as part of the continuous ejecta blanket surrounding the crater out to a few crater radii. These expected breccias appear to be present in five wells just outside the perimeter of the Chicxulub Crater and have been compared to the Bunte Breccia of the Ries Crater (Sharpton et al., 1994). They have also been found to outcrop in a quarry in Belize (Ocampo et al., this volume).

Proximal microtektites

Tektites and microtektites are droplets of impact glass, characterized by very low water content and by the absence of crystals. Spherules, largely replaced by alteration products, especially clay and calcite, have been found in Haiti (Hildebrand and Boynton, 1990), eastern Mexico (Alvarez et al., 1992a; Smit et al., 1992b), and the southern United States (Pitakpaivan and Hazel, 1992). These are evidently altered tektites, and in a few localities unaltered cores of tektitelike glass have been found in the spherules (Izett, 1991; Maurrasse and Sen, 1991; Sigurdsson et al., 1991; Smit et al., 1992a, b). The altered tektites typically contain abundant small bubbles, a characteristic that aids in their identification in the field (Alvarez et al., 1992a; Smit et al., 1992a).

Shocked quartz

Shocked quartz grains are abundant in a layer a few millimeters thick resting immediately on top of the KT boundary clay in the western interior of North America, from New Mexico to Alberta (Bohor, 1990; Izett, 1990). Shocked quartz at the KT boundary was considered to be very rare and fine grained elsewhere, until its recognition in a number of Pacific deep-sea cores (Kyte et al., 1994; Bostwick and Kyte, this volume).

Distal microkrystites

Spherules found in the KT boundary layer far from Chicxulub lack bubbles, have not as yet yielded preserved glass, and commonly show a feathery, dendritic, or radial-fiber internal texture (Smit and Klaver, 1981; Montanari et al., 1983; Smit et al., 1992a). The primary mineralogy is almost everywhere replaced by such alteration products as glauconite and K-feldspar, but in one case (DSDP site 577a on the Shatsky Rise in the Pacific) primary clinopyroxene with dendritic habit is preserved (Smit, 1990).

Iridium

Anomalous iridium concentrations have been measured at the KT boundary in over 100 sites around the world. The Ir must occur in extremely fine particles, and the carrier phase or phases have not yet been identified. The column densities of Ir vary over two orders of magnitude in no obvious pattern and for no obvious reason. In sites around the Gulf of Mexico, where the KT event is represented by a few meters of ballistic and tsunami deposits, the Ir anomaly occurs at the very top of the sequence, also indicating that it resides in very fine particles that took a long time to settle.

DISPERSAL VIA EJECTA CURTAIN AND FIREBALL

Hildebrand and Boynton (1988) and Bohor (1990) subdivide the double KT boundary layer in the western interior of North America into an "ejecta layer" or "melt-ejecta layer" below and a "fireball" layer above, containing the shocked quartz. Smit et al. (1992a) inferred that the KT tektites represent the melted part of the ejecta plume and the microkrystites represent the more energetic, vaporized part. These interpretations can be evaluated on the basis of (1) current understanding of ejecta deposits from large craters and (2) calculation of ballistic trajectories for ejecta material.

Ejecta from large impacts is thought to be dispersed by two different mechanisms (Melosh, 1988). In the first mechanism, short-range debris travels as part of an ejecta curtain, a coherent front of particles each on its own ballistic trajectory. Ejecta curtains are regularly observed in hypervelocity impact experiments in the laboratory (Schultz and Gault, 1982). The velocities, times, and positions of launch are such that at any moment most of the ejecta lies on a downward-pointing conical surface centered on the impact site, and the cone expands with time. The launch angle of ejecta-curtain materials is commonly thought to be close to 45° (Melosh, 1988, p. 80, 92), and that angle is assumed throughout this chapter. There are indications, however, that this is not strictly true (Oberbeck, 1975), and variable launch angle in ejecta curtain material will need to be considered if the approach used here is carried farther.

In the second mechanism, rock vapor produced in a large impact will expand as a fireball or vapor plume (Melosh, 1988, p. 68–71, 211–213). The fireball accelerates upward as it encounters decreasing atmospheric pressure and blows out of the top of the atmosphere, launching its entrained debris into ballistic trajectories (Jones and Kodis, 1982). This process has (fortunately) never been observed, and numerical modelers have not fully explored it, but velocities in the expanding fireball probably approach or even exceed escape velocity (Melosh, 1988, p. 69), and elevation angles at launch should not be restricted to 45°.

The difference in behavior between the ejecta curtain and the fireball is evidently due to the frequency of interactions between particles. The solid fragments and liquid droplets in the ejecta curtain have limited contact with each other, and each one therefore follows a near-ballistic trajectory. In the expanding vapor of the fireball, simple ballistic trajectories are not possible because of thermal interactions between gas molecules. The presence or absence of vapor therefore distinguishes the two main mechanisms of ejecta dispersal.

Melting and vaporization take place in impact ejecta during release from irreversible shock compression, which deposits its energy in the shocked material. Decompression melting occurs in a wide range of rock types after shocking to pressures of 35 to 100 GPa. (For comparison, the static pressure at the base of the continental crust is about 1 GPa.) Vaporization occurs during release from shock pressures of a few hundred GPa in silicate rocks, but in limestone CO_2 begins to be generated after shock to only tens of GPa (Kieffer and Simonds, 1980; Boslough et al., 1982).

BALLISTIC DYNAMICS

Debris ejected from the young craters Copernicus and Tycho blankets the lunar surface out to one or two crater radii and extends for long distances along rays that follow great circles. This pattern indicates that ejecta was concentrated at certain azimuths but at varying velocities and/or elevation angles. The straight rays are consistent with the slow rotation rate of the Moon (360°/28 days).

Until the recognition of the worldwide distribution of Chicxulub impact ejecta at the KT boundary, the few known terrestrial impact-ejecta strewnfields were local, or regional at most. Terrestrial ejecta no longer lie exposed at the surface, so ray patterns cannot be detected. For these reasons there is apparently no thorough treatment of ballistic ejecta dynamics on the Earth, although this is partly subsumed in the general equations of Dobrovolskis (1981). In Melosh's monograph (1988, p. 87) approximations are given for calculating range and time of flight for short-range ejecta, along with a more complicated range equation for more distal ejecta. Understanding the worldwide KT strewnfield requires a full treatment based on a spherical, rotating Earth.

Obviously there are no atmospheric effects on lunar ejecta. On Earth the atmosphere strongly affects the ejecta from small impacts. For the ca. 200-km-diameter Chicxulub Crater, even ejecta traveling only a fraction of a crater diameter will spend most of its time outside most of the atmosphere, which has a scale height of about 10 km; for calculating the trajectories of far-traveling ejecta the atmosphere can be ignored.

Calculating trajectories

The trajectories followed by ejecta particles are determined by forces applied (1) during their time of launch, (2) during the ballistic phase outside the atmosphere, and (3) during atmospheric reentry. (1) An impact-ejecta particle is accelerated either as the ground rises to form the ejecta curtain or during the expansion of the fireball. (2) Once clear of the atmosphere, the particle goes into a ballistic condition with no further propulsive forces acting on it and affected only by the Earth's gravity. In addition, when viewed in the reference frame of the Earth's surface, ballistic ejecta are affected by Coriolis pseudoforces resulting from the rotation of the planet. (3) Upon reentering the atmosphere, the particles undergo frictional deceleration; smaller particles may drift in the wind before reaching the ground or the sea.

The apparent acceleration of an ejecta particle during its ballistic phase is given in vector notation by

$$\mathbf{a}' = \mathbf{g} - 2\omega \times \mathbf{v}'$$

where $\mathbf{a}'$ is the apparent acceleration in the rotating-Earth framework, $\mathbf{g}$ is the gravitational acceleration, ω is the Earth's angular rotation vector, and $\mathbf{v}'$ is the particle's velocity in Earth coordinates. The general equation for apparent acceleration in the rotating-Earth reference frame (Fowles, 1986, p. 126–129) includes terms for axipetal acceleration of a particle constrained to stay on the surface and for accelerations in the Earth's rotation rate and orbital velocity, but these are not applicable to the case of impact ejecta. Coriolis accelerations are generally viewed as acting in the plane of the Earth's surface, but that is only because the usual particle motions of winds and ocean currents are parallel to the surface. In the more general case, Coriolis acceleration can act in any direction. Additional complicated accelerations act while the particle is in the expanding fireball and again when it returns to the atmosphere.

One way to calculate a ballistic trajectory would be to integrate the gravitational and pseudoforce accelerations throughout the time of flight of the particle. This is complicated and, fortunately, unnecessary. It is more convenient to treat the particle, while it is outside the atmosphere, as a satellite orbiting the Earth. In this condition it feels the gravitational attraction of the Earth as if the Earth's mass were concentrated at a point at the center of the planet, and the particle therefore follows an elliptical orbit with the center of the Earth at one focus, by Kepler's first law. Of course this orbit is not completed, for the particle encounters the surface at its point of reimpact (Fig. 1). The partial ellipse of the trajectory begins at the point of "primary impact," where an incoming object strikes the Earth, and ends at the point of "secondary impact" or "reimpact."

The partial ellipse of the ballistic trajectory lies in a plane fixed in the inertial reference frame of the Earth as the Earth orbits the Sun, and the curvature of the Earth's orbit during the brief time the ejecta is aloft can be neglected for these purposes. The tidal pull of the Moon causes the ellipse to be slightly nonplanar, but this effect also is negligible except for ejecta on extremely high trajectories. Instead of calculating Coriolis accelerations, the effect of the Earth's rotation can be treated by determining how much the Earth rotates beneath the ballistic particle before it reimpacts (Fig. 2). Clearly this affects the longitude of reimpact but not the latitude. This approach greatly simplifies the calculation of ballistic trajectories.

It is worth considering the appropriate conditions for the two approaches. The Coriolis formulation is used in studies of atmospheric and oceanic circulation. In these situations a parcel of fluid is part of a continuum and is constantly propelled by pressure-gradient forces. The force of gravity that acts on the parcel is balanced by the supporting force of whatever fluid or solid layer it is traveling over, and geophysical fluid dynamicists treat this balance in terms of a "platform of dynamic equilibrium" (Stommel and Moore, 1989). This is very different from the ejecta situation in which the particle coasts after an initial impulse and is unsupported. A further difference is that atmospheric and oceanic currents are deflected by the Earth's rotation over periods of days to months, whereas ejecta is deflected only during flight times of minutes to hours.

The Coriolis formulation will be useful in calculating the flow of a base surge developed when the fireball and its entrained ejecta collapse back on the impact site and begin to spread outward on top of the atmosphere or along the Earth's surface.

The computational procedure used here is based largely on the dimensionless treatment of Dobrovolskis (1981), with the appropriate conversions to Earth values (1 Earth mass ≡ 1 = 5.9733×10^{24} kg; 1 Earth radius ≡ 1 = 6,371 km; 1 dimension-

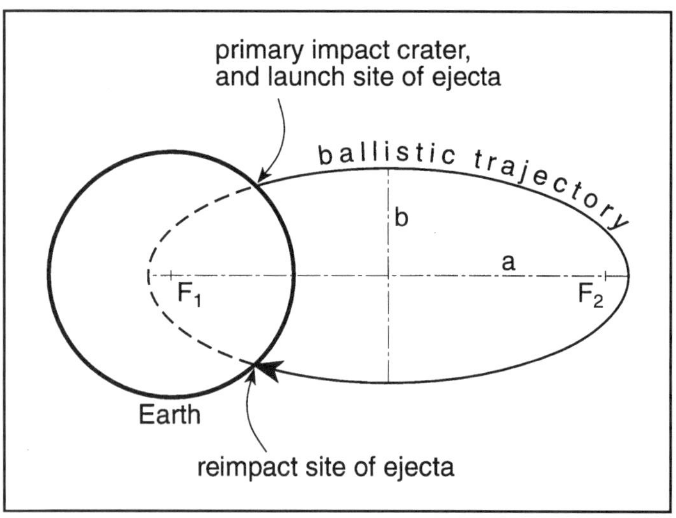

Figure 1. Ballistic trajectory as a partial elliptical orbit. The semi-major axis, a, and semi-minor axis, b, describe the ellipse. One focus, F_1, lies at the center of the Earth.

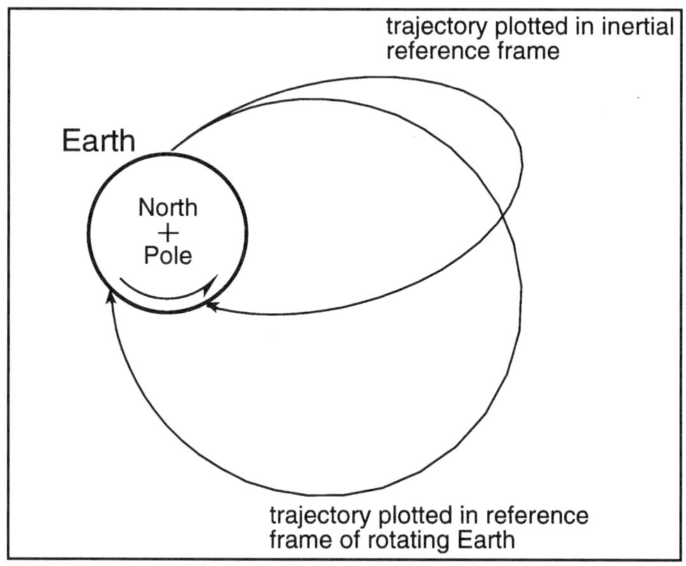

Figure 2. The reimpact point of an ejecta particle that spends hours aloft is displaced westward because the Earth rotates beneath it. When plotted in an inertial reference frame, the trajectory is an ellipse. When plotted in the reference frame of the rotating Earth, the trajectory is distorted in a complicated way. This diagram shows ejecta launched due west from an equatorial impact, with an elevation angle of 45° and velocity of 10 km/sec, plotted in the two coordinate systems.

less time unit = 806.29 sec, which is the value needed to make g, the Earth's surface gravity, and G, the gravitational constant, both = 1). Dobrovolskis's treatment finds reimpact sites but does not allow determination of position in the orbit as a function of time; this has been done following the treatment of Dubyago (1961).

Effect of the Earth's rotation

The Earth's rotation affects ballistic trajectories in two ways. First, at the time of launch the eastward rotational velocity of the Earth is added to the initial launch velocity in Earth coordinates, with the vector sum giving the take-off velocity in inertial coordinates. Second, as the ejecta particle follows its planar elliptical orbit in inertial coordinates, the Earth turns beneath it.

The first effect is the addition of the Earth's eastward rotational velocity at the time of launch. The Earth's eastward velocity at the equator is 40,030 km/day = 0.463 km/sec = 0.059 in dimensionless Earth units (assuming a spherical Earth with radius = 6,371 km). The eastward component added to the launch velocity of ejecta falls off as the sine of the colatitude of the impact site and as the cosine of the elevation angle of the launch. NASA takes advantage of this effect by launching spacecraft eastward from its low-latitude facility at Cape Canaveral. This component has little effect on the semi-major axis, a, of the elliptical orbit until the initial velocity (in Earth coordinates) reaches about 9 km/sec, but the effect on a is dramatic above about 10 km/sec (Fig. 3).

The second effect is due to the rotation of the Earth beneath the orbiting ejecta. Low-velocity ejecta reimpact so quickly that the Earth does not rotate very far during its time of flight, but high-velocity particles may stay aloft for hours, so this effect becomes important. The reimpact site is displaced westward along a parallel of equal latitude, relative to where it would be on an unrotating Earth.

The combined result of these two effects can be seen clearly in the case of ejecta launched eastward and westward from an equatorial impact. Figure 4 shows the case of ejecta launched at 45° from vertical (as in the ejecta curtain), for several ejection velocities, viewed in the frame of a rotating Earth. This is the relevant reference frame for determining where ejecta will be found in stratigraphic context. For low ejection velocities (6 and 8 km/sec) the Earth's rotation has little effect, because the added velocity does not appreciably increase the semimajor axis, and the ejecta reimpacts before the Earth rotates very far. At higher velocities the effects become noticeable. Eastbound trajectories have larger semimajor axes and therefore reach greater altitude than westbound ones. In addition, there are dramatic effects because of the Earth's rotation. In the case of eastbound ejecta, trajectories for velocities up to 9 km/sec have rounded crests and the reimpact sites are progressively farther downrange. However, at 9.5 km/sec the crest is pointed and at 10 km/sec it loops backward; the reimpact sites in these cases are deflected back

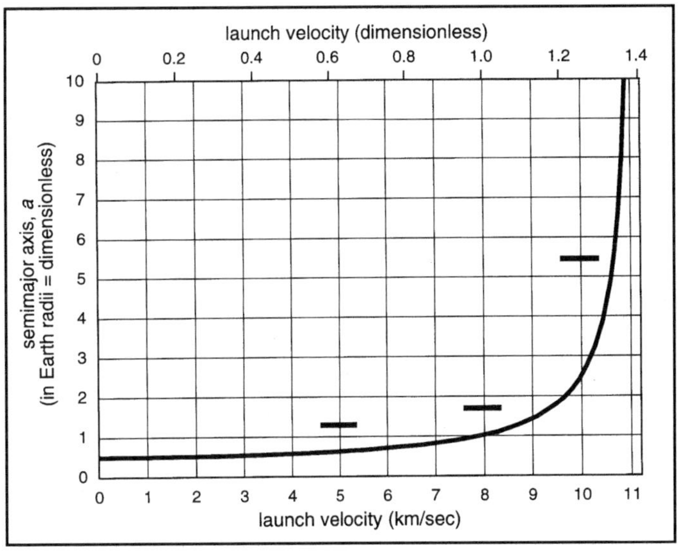

Figure 3. The semi-major axis, a, of the orbit followed by an ejecta particle depends only on launch velocity. At low velocities a is very slightly greater than 0.5 Earth radii, that is, the particle does not get far above the surface. With increasing launch velocity, a increases very slowly, until around 9 to 10 km/sec, when the increase in a becomes steep. Escape velocity ($a = \infty$) is at 11.2 km/sec = dimensionless $\sqrt{2}$. Dark bars centered on 5, 8, and 10 km/sec show plus or minus the eastward velocity of the equator (0.463 km/sec). This value or some fraction of it (depending on launch azimuth and elevation angle and on latitude) is added to the velocity of eastbound ejecta or subtracted from the velocity of westbound ejecta. As a result, eastbound orbits have greater values of a than equivalent westbound orbits. Because of the steep rise of the curve beyond about 10 km/sec, this effect is important for fast ejecta.

uprange. The reason for these patterns is that for apogees far from Earth, the forward rotation rate (slowest at apogee, by Kepler's second law) is about the same as the rotation rate of the Earth (for 9.5 km/sec) or even slower (for 10 km/sec).

Reimpact pattern of ejecta-curtain material

It is clear from Figure 4 that because of the Earth's rotation there is a forbidden zone for 45°-elevation-angle ejecta, beyond about 100° east of an equatorial impact, which cannot be reached directly by eastbound ejecta, although it can be reached by westbound ejecta traveling more than 180° or by eastbound ejecta on very high orbits that reimpact far behind the launch site. Figure 5a shows the map pattern of reimpact for ejecta from an equatorial impact site, launched in all directions, with a constant 45° elevation angle, for a variety of launch velocities. At 6 and 8 km/sec, the locus of reimpact is nearly circular, but at higher velocities the reimpact locus is distorted westward because of the Earth's rotation. The greatest eastward range occurs at about 9 km/sec, and the forbidden zone lies to the east of this limit. Figure 5b shows an analogous plot for an impact at the 65-Ma Chicxulub paleolatitude of 26.6° N.

The paths followed by ejecta particles in reaching their

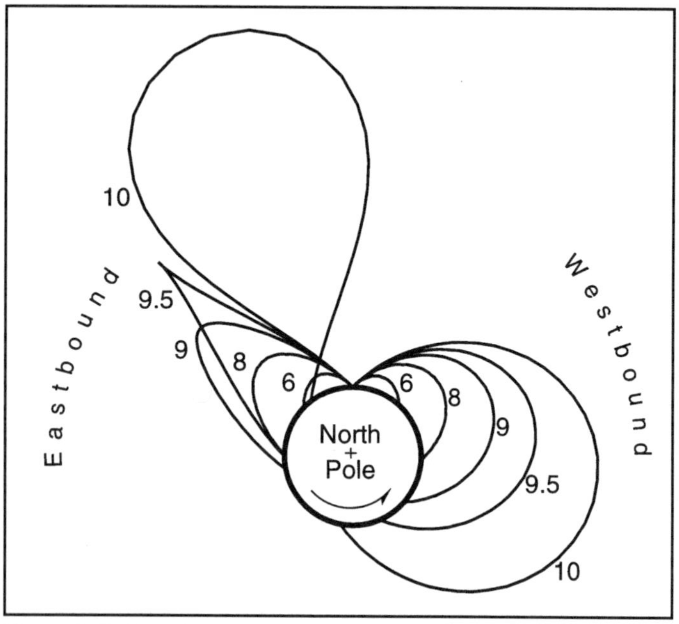

Figure 4. Trajectories of eastbound and westbound ejecta, launched at 45° elevation angle (as in the ejecta curtain) from an equatorial impact, at several velocities (labeled in km/sec). View looks down on Earth from above the north pole, and trajectories are plotted in the reference frame of the rotating Earth. (If plotted in inertial coordinates, all trajectories would be elliptical.) Higher velocity trajectories are deflected to the west because of the Earth's rotation, and eastbound trajectories go higher than westbound ones, because the Earth's rotational velocity is added to eastbound velocities and subtracted from westbound velocities.

reimpact sites are shown in Figure 6, as the traces of the point on the Earth's surface directly beneath the particle. These maps should be examined with reference to Figure 4, noting first that the particle is at several Earth radii at apogee, and second that the cusp for 9.5 km/sec and the backward loop for 10 km/sec are traced out by the particles departing at N 60° E and S 60° E (and of course at N 90° E, although this is not visible because the map path on this projection lies on a straight line).

These trajectories and reimpact patterns with their cusps, backward loops, and forbidden zones clarify the effect of the Earth's rotation on ballistic ejecta dynamics. However, they may not be directly relevant to the Chicxulub ejecta, for the following reason. These trajectories are all for particles launched at a 45° elevation angle, which is characteristic of the ejecta curtain. However, the solid and liquid ejecta of the ejecta curtain may not be launched with velocities sufficient to stay aloft long enough to develop cusps, loops, or forbidden zones. Ejecta launched at higher velocities has sufficient internal energy to vaporize and thus becomes part of the fireball where elevation angles are not restricted to 45°.

Reimpact pattern of fireball material

Figure 7 shows trajectories of eastbound ejecta from equatorial impacts for a variety of velocities and for elevation angles ranging from 15 to 75°. It is clear that low elevation angles permit eastbound ejecta to enter the zone forbidden to ejecta departing at 45° from horizontal. Argyle (1989) used a computer

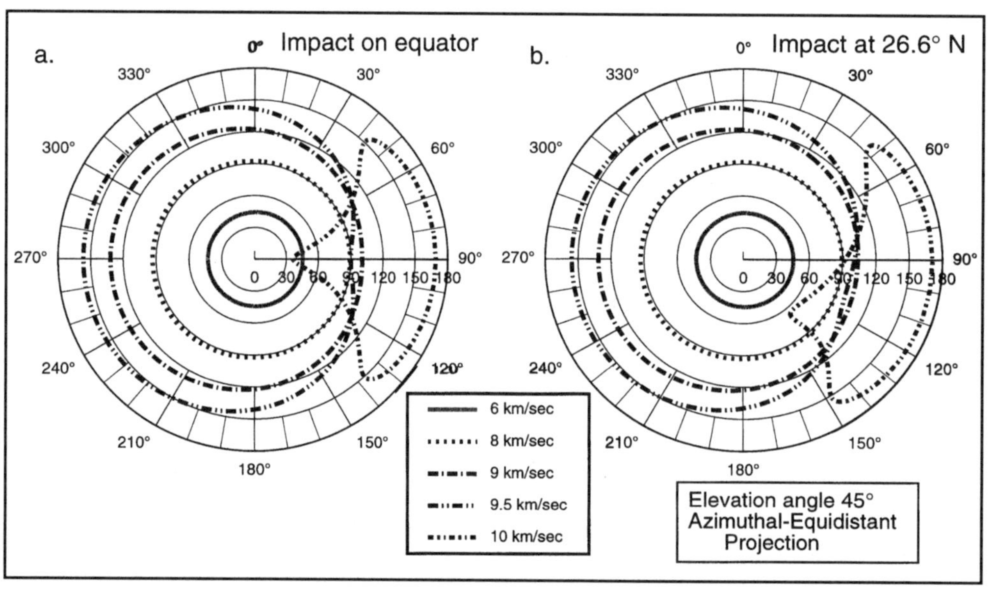

Figure 5. Reimpact loci for ejecta launched at 45° elevation angle and a range of velocities from an equatorial impact (a) and an impact at the 65-Ma Chicxulub paleolatitude of 26.6° N (b). The irregular area enclosing 120° and 150° E is the forbidden zone, unreachable by 45°-elevation ejecta with velocities ≤10 km/sec. Some care is needed in interpreting these azimuthal-equidistant projections, centered at the impact site, because of the serious distortion beyond 90° from the center, with the outer ring representing the point antipodal to the impact.

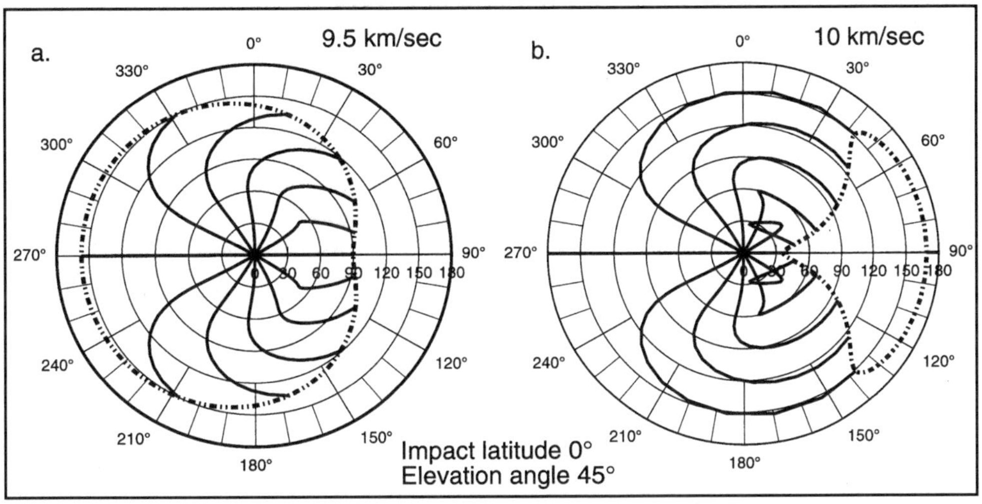

Figure 6. Plots corresponding to Figure 5, showing the track of the sub-particle point on the Earth's surface, for take-off azimuths separated by 30°, with elevation angles of 45° and velocities of 9.5 km/sec (a) and 10 km/sec (b). Cusps in a and loops in b correspond to those features in Figure 4.

simulation to explore the global pattern of ejecta thickness that would result from a fireball that bursts through the atmosphere, expands with hemispheric symmetry, and distributes particles worldwide on ballistic trajectories. The pattern shows a large concentration around the impact site and a smaller concentration near the antipodal point, but displaced westward by the Earth's rotation.

Finally, since these patterns of ejecta distribution are most relevant to the Chicxulub impact, Figures 8A, B, and C show the patterns for low-angle, moderate-angle, and steep-angle launch from Chicxulub, on a continental reconstruction for Cretaceous-Tertiary boundary time. Ejecta launched at a low

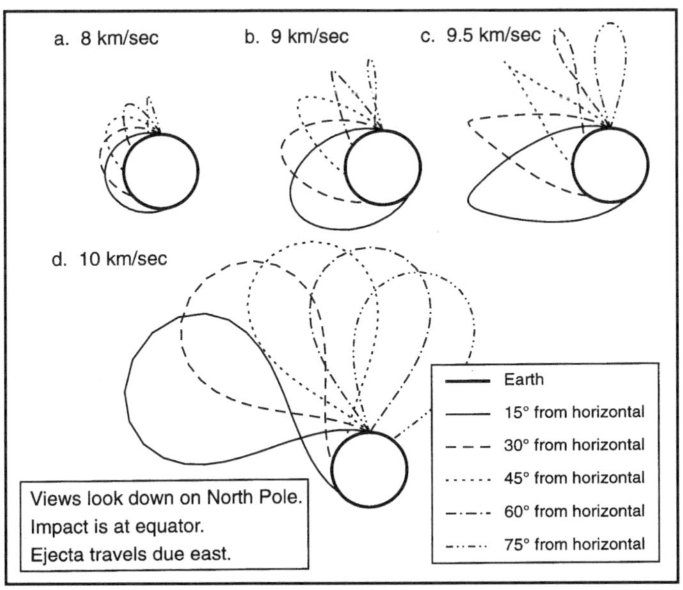

Figure 7. For eastbound ejecta from an equatorial impact, elevation angles less than 45° allow the particles to enter the zone forbidden to 45° ejecta.

elevation angle (20°) can reach any point on the globe (Fig. 8A), provided its velocity is at least 8.5 km/sec. Since low-angle ejecta does not stay aloft for long, the impact loci of ejecta are not seriously distorted until they are close to the antipodal point. Ejecta launched at 8.5 km/sec lands on a locus doubled over into a pretzel shape, and the locus for 10 km/sec develops a serious distortion.

Ejecta launched at a moderate angle (45°) displays a small forbidden zone around India (Fig. 8B). This zone can be reached by ejecta launched westward at >10 km/sec but is inaccessible to eastbound ejecta of any launch velocity. If ejecta-curtain material can reach velocities around 9 to 10 km/sec with launch angles around 45°, it may display this pattern. Sites in Madagascar, India, and neighboring oceans will be critical for testing this possibility.

Steep ejecta (70°) displays a very large zone forbidden to eastbound ejecta, covering all of the old world continents (Fig. 8C). This map clearly reveals the uprange impacts of ejecta launched eastward at a steep angle and high velocity.

DISCUSSION

This chapter has shown that the ballistic trajectories available to impact-ejecta particles on a rotating Earth may be complicated and nonintuitive, especially for high-velocity ejecta. In the case of the KT boundary ejecta from Chicxulub, it is tempting to attribute the glassy tektites and microtektites to ejecta-curtain dispersal and the microkrystites to fireball dispersal. Yet this leaves the dispersal mechanism of the shocked quartz in the western interior of North America unclear. It will be critical to know the maximum launch velocity at which ejecta derived from different target rocks can avoid vaporization, for this will control the boundary between ejecta curtain and fireball and will determine the maximum range of ejecta-curtain debris. The for-

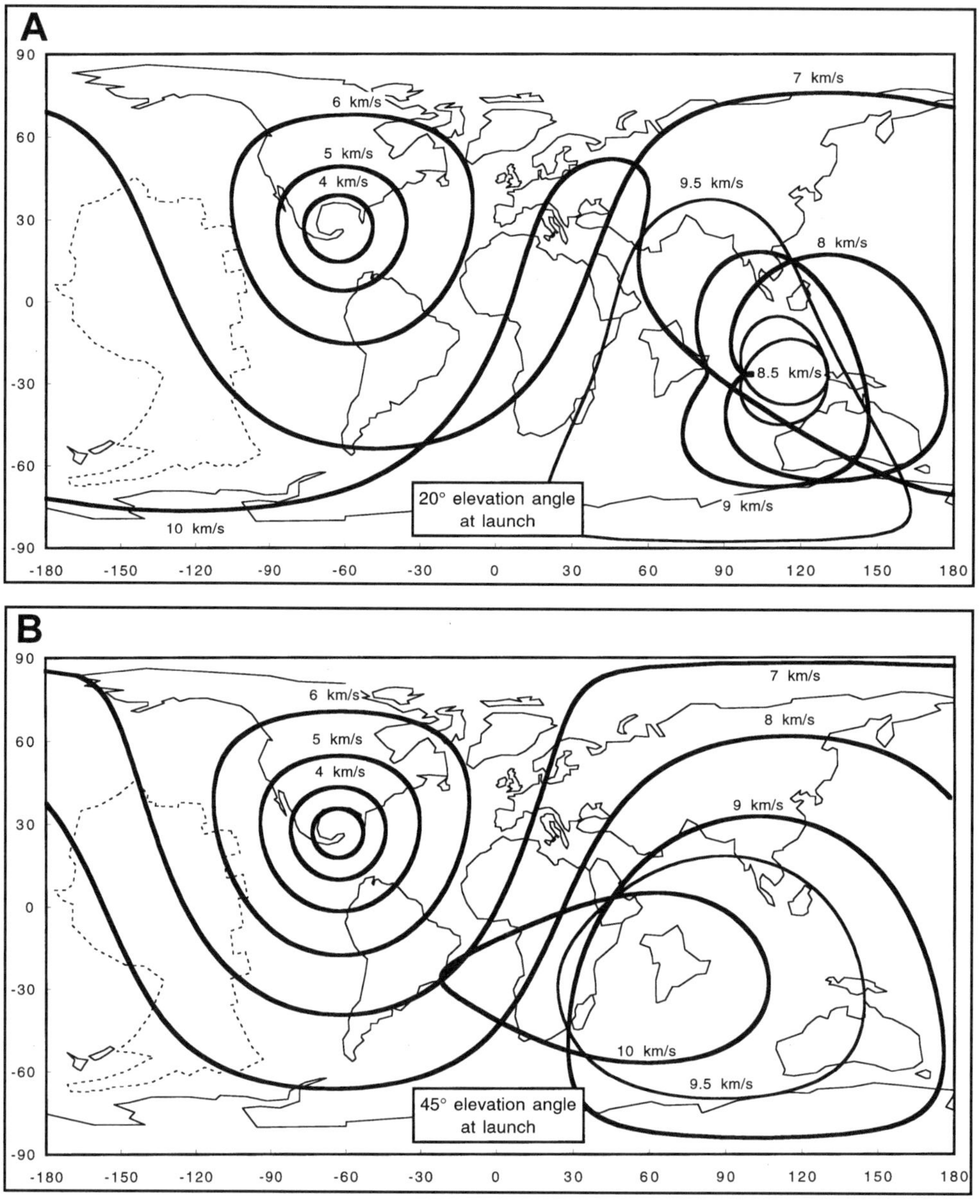

Figure 8. Maps of the impact loci of Chicxulub ejecta plotted on a Cretaceous-Tertiary boundary continental reconstruction (continental positions were calculated from the 66-Ma rotation parameters [65-Ma parameters not listed] of the University of Chicago Paleogeographic Atlas Project, David Rowley, personal communication, 1994). Dashed line is the limit of Pacific Ocean crust still preserved, shown in its 66-Ma position. A, 20° launch angle. B, 45° launch angle. C, 70° launch angle.

bidden zone, unreachable by eastbound ejecta with 45° elevation angles, provides a testable prediction, but unfortunately there are few well-studied sites in the KT-boundary forbidden zone (taking account of subsequent plate motion). The possibility that Chicxulub was a low-angle impact (Alvarez et al., 1993; Schultz, 1993) would strongly affect the pattern of proximal ejecta. The further development of global ballistic sedimentology will require field study of ejecta horizons worldwide, petrographic study of ejecta materials, better understanding of the patterns of launch directions and velocities based on computer simulations of very large impacts (Roddy et al., 1987), and calculation of the ballistic trajectories followed by the ejecta.

ACKNOWLEDGMENTS

This study was supported by NASA Grant NAGW-3008 and NSF Grant EAR-91-05297. It benefited from ongoing discussions with Philippe Claeys and Daniel B. Karner, and I appreciate the careful comments of Mark Cintala, an anonymous reviewer, and editor Graham Ryder.

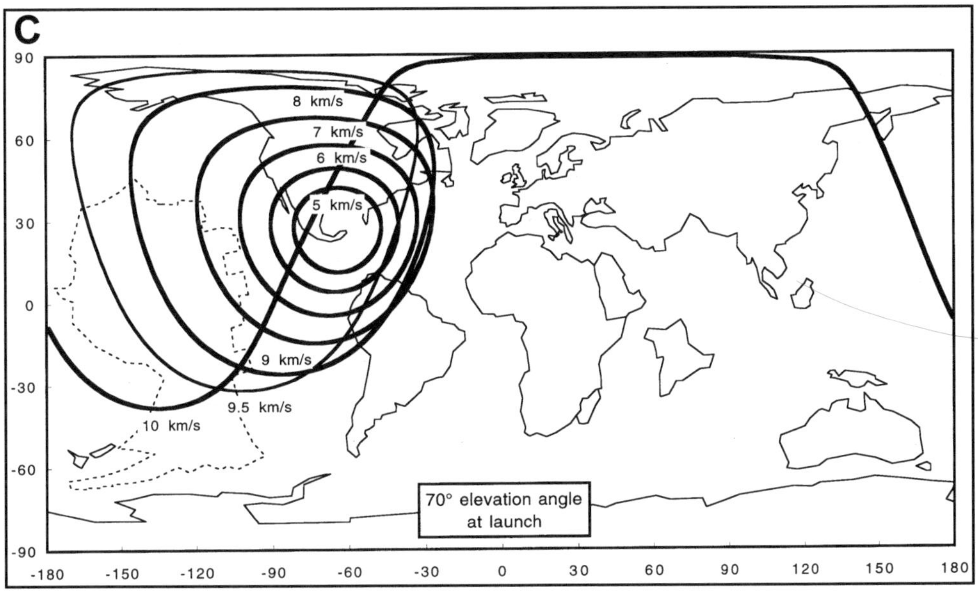

REFERENCES CITED

Alvarez, W., and Asaro, F., 1990, What caused the mass extinction? An extraterrestrial impact: Scientific American, v. 263, p. 78–84.

Alvarez, L. W., Alvarez, W., Asaro, F., and Michel, H. V., 1980, Extraterrestrial cause for the Cretaceous-Tertiary extinction: Science, v. 208, p. 1095–1108.

Alvarez, W., Grajales N., J. M., Martinez S., R., Romero M., P. R., Ruiz L., E., Guzman R., M. J., Zambrano A., M., Smit, J., Swinburne, N. H. M., and Margolis, S. V., 1992a, The Cretaceous–Tertiary boundary impact–tsunami deposit in NE Mexico: Geological Society of America Abstracts with Programs, v. 24, p. A331.

Alvarez, W., Smit, J., Lowrie, W., Asaro, F., Margolis, S. V., Claeys, P., Kastner, M., and Hildebrand, A. R., 1992b, Proximal impact deposits at the Cretaceous-Tertiary boundary in the Gulf of Mexico: A restudy of DSDP Leg 77 Sites 536 and 540: Geology, v. 20, p. 697–700.

Alvarez, W., Claeys, P., and Burns, E., 1993, A candidate KT boundary impact crater in Alabama: Eos (Transactions, American Geophysical Union), v. 74, p. 387.

Argyle, E., 1989, The global fallout signature of the K-T bolide impact: Icarus, v. 77, p. 220–222.

Blum, J. D., Chamberlain, C. P., Hingston, M. P., Koeberl, C., Marin, L. E., Schuraytz, B. C., and Sharpton, V. L., 1993, Isotopic comparison of K/T boundary impact glasses with melt rock from the Chicxulub and Manson impact structures: Nature, v. 364, p. 325–327.

Bohor, B. F., 1990, Shocked quartz and more; Impact signatures in Cretaceous/Tertiary boundary clays, *in* Sharpton, V. L., and Ward, P. D., eds., Global catastrophes in Earth history: Geological Society of America Special Paper 247, p. 335–342.

Bohor, B. F., Foord, E. E., Modreski, P. J., and Triplehorn, D. M., 1984, Mineralogic evidence for an impact event at the Cretaceous-Tertiary boundary: Science, v. 224, p. 867–869.

Boslough, M. B., Ahrens, T. J., Vizgirda, J., Becker, R. H., and Epstein, S., 1982, Shock-induced devolatilization of calcite: Earth and Planetary Science Letters, v. 61, p. 166–170.

Bourgeois, J., Hansen, T. A., Wiberg, P. L., and Kauffman, E. G., 1988, A tsunami deposit at the Cretaceous-Tertiary boundary in Texas: Science, v. 241, p. 567–570.

Dobrovolskis, A., 1981, Ejecta patterns diagnostic of planetary rotations: Icarus, v. 47, p. 203–219.

Dubyago, A. D., 1961, The determination of orbits: New York, Macmillan, 434 p.

Fowles, G. R., 1986, Analytical mechanics: Philadelphia, Saunders College Publishing, 342 p.

Hildebrand, A. R., and Boynton, W. V., 1988, Provenance of the K/T boundary layers, *in* Global catastrophes in Earth history; an interdisciplinary conference on impacts, volcanism, and mass mortality (Oct. 20–23, 1988, Snowbird, Utah, USA): Houston, Texas, Lunar and Planetary Institute Contribution 673, p. 78–79.

Hildebrand, A. R., and Boynton, W. V., 1990, Proximal Cretaceous-Tertiary boundary impact deposits in the Caribbean: Science, v. 248, p. 843–846.

Hildebrand, A. R., Penfield, G. T., Kring, D. A., Pilkington, M., Camargo Z., A., Jacobsen, S. B., and Boynton, W. V., 1991, Chicxulub Crater—A possible Cretaceous Tertiary boundary impact crater on the Yucatan Peninsula, Mexico: Geology, v. 19, p. 867–871.

Izett, G. A., 1990, The Cretaceous/Tertiary boundary interval, Raton Basin, Colorado and New Mexico, and its content of shock-metamorphosed minerals; evidence relevant to the K/T boundary impact-extinction theory: Geological Society of America Special Paper 249, 100 p.

Izett, G. A., 1991, Tektites in Cretaceous-Tertiary boundary rocks on Haiti and their bearing on the Alvarez impact extinction hypothesis: Journal of Geophysical Research, v. 96, p. 20879–20905.

Jones, E. M., and Kodis, J. W., 1982, Atmospheric effects of large-body impacts: The first few minutes: Geological Society of America Special Paper 190, p. 175–186.

Kieffer, S. W., and Simonds, C. H., 1980, The role of volatiles and lithology in the impact cratering process: Reviews of Geophysics and Space Physics, v. 18, p. 143–181.

Koeberl, C., 1993, Chicxulub Crater, Yucatán: Tektites, impact glasses, and the geochemistry of target rocks and breccias: Geology, v. 21, p. 211–214.

Kyte, F. T., Bostwick, J. A., and Zhou, L., 1994, The KT boundary on the Pacific plate, *in* New developments regarding the KT event and other catastrophes in Earth history: Houston, Texas, Lunar and Planetary Institute Contribution 825, p. 64–65.

Maurrasse, F. J.-M. R. and Sen, G., 1991, Impacts, tsunamis, and the Haitian Cretaceous-Tertiary boundary layer: Science, v. 252, p. 1690–1693.

Melosh, H. J., 1988, Impact cratering: A geologic process: Tucson, University of Arizona Press, 272 p.

Montanari, A., Hay, R. L., Alvarez, W., Asaro, F., Michel, H. V., Alvarez, L. W., and Smit, J., 1983, Spheroids at the Cretaceous-Tertiary boundary are altered impact droplets of basaltic composition: Geology, v. 11, p. 668–671.

Oberbeck, V. R., 1975, The role of ballistic erosion and sedimentation in Lunar stratigraphy: Reviews of Geophysics and Space Physics, v. 13, p. 337–362.

Penfield, G. T., and Camargo Z., A., 1981, Definition of a major igneous zone in the central Yucatán platform with aeromagnetics and gravity: Society of Exploration Geophysicists Technical Program, Abstracts, and Biographies, v. 51, p. 37.

Pitakpaivan, K., and Hazel, J. E., 1992, Pseudomorphs of microtektites at the Cretaceous/Tertiary boundary in Alabama and Mississippi: Geological Society of America Abstracts with Programs, v. 24, p. A361.

Roddy, D. J., Schuster, S. H., Rosenblatt, M., Grant, L. B., Hassig, P. J., and Kreyenhagen, K. N., 1987, Computer simulations of large asteroid impacts into oceanic and continental sites; preliminary results on atmospheric, cratering, and ejecta dynamics: International Journal of Impact Engineering, v. 5, p. 525–541.

Schultz, P. H., 1993, Could Chicxulub be the result of an oblique impact?: Eos (Transactions American Geophysical Union), v. 74, p. 388.

Schultz, P. H., and Gault, D. E., 1982, Impact ejecta dynamics in an atmosphere: Experimental results and extrapolations: Geological Society of America Special Paper 190, p. 153–174.

Sharpton, V. L., Marin, L. E., and Schuraytz, B. C., 1994, The Chicxulub multiring impact basin: Evaluation of geophysical data, well logs, and drill core samples, *in* New developments regarding the KT event and other catastrophes in Earth history: Houston, Texas, Lunar and Planetary Institute Contribution 825, p. 108–110.

Sigurdsson, H., D'Hondt, S., Arthur, M. A., Bralower, T. J., Zachos, J. C., Van Fossen, M., and Channell, J. E. T., 1991, Glass from the Cretaceous-Tertiary boundary in Haiti: Nature, v. 349, p. 482–487.

Smit, J., 1990, Meteorite impact, extinctions and the Cretaceous-Tertiary boundary: Geologie en Mijnbouw, v. 69, p. 187–204.

Smit, J., and Klaver, G., 1981, Sanidine spherules at the Cretaceous-Tertiary boundary indicate a large impact event: Nature, v. 292, p. 47–49.

Smit, J., Alvarez, W., Montanari, A., Swinburne, N., Van Kempen, T. M., Klaver, G. T., and Lustenhouwer, W. J., 1992a, "Tektites" and microkrystites at the Cretaceous Tertiary boundary: Two strewnfields, one crater?: Proceedings of Lunar and Planetary Science, v. 22, p. 87–100.

Smit, J., Montanari, A., Swinburne, N. H. M., Alvarez, W., Hildebrand, A. R., Margolis, S. V., Claeys, P., Lowrie, W., and Asaro, F., 1992b, Tektite-bearing, deep-water clastic unit at the Cretaceous-Tertiary boundary in northeastern Mexico: Geology, v. 20, p. 99–103.

Stommel, H. M., and Moore, D. W., 1989, An introduction to the Coriolis force: New York, Columbia University Press, 297 p.

Swisher, C. C., Grajales-Nishimura, J. M., Montanari, A., Margolis, S. V., Claeys, P., Alvarez, W., Renne, P., Cedillo Pardo, E., Maurrasse, F. J.-M. R., Curtis, G. H., Smit, J., and McWilliams, M. O., 1992, Coeval $^{40}Ar/^{39}Ar$ ages of 65.0 million years ago from Chicxulub Crater melt rock and Cretaceous-Tertiary boundary tektites: Science, v. 257, p. 954–958.

Manuscript Accepted by the Society September 14, 1995

Geological Society of America
Special Paper 307
1996

Coarse-grained, clastic sandstone complex at the K/T boundary around the Gulf of Mexico: Deposition by tsunami waves induced by the Chicxulub impact?

J. Smit and Th. B. Roep
Institute of Earth Sciences, Vrije Universiteit, de Boelelaan 1085, 1081 HV Amsterdam, Netherlands
W. Alvarez
Department of Geology and Geophysics, University of California, Berkeley, California 94720
A. Montanari
Observatorio Geologico, Coldigoco, Apiro di Frontale, Umbria, Italy
P. Claeys*
Department of Geology and Geophysics, University of California, Berkeley, California 94720
J. M. Grajales-Nishimura and J. Bermudez
Instituto Mexicano del Petroleo, Apartado Postal 14-805, 07730 Mexico D.F., Mexico

ABSTRACT

The K/T boundary in marine deposits in and around the Gulf of Mexico at a distance of <800 km from the Chicxulub impact structure is marked by a series of unusual sandstone beds in almost any outcrop. These sandstone beds have a complex architecture, varying with depth of deposition. At neritic to upper-bathyal depths (100 to 500 m) the sandstone beds can consistently be subdivided into three (sometimes four) successive lithological units. From bottom to top, the units decrease in grain size and change in composition and texture. The first unit (I) is characterized by millimeter-sized particles: spherules and other splash-form particles with internal filled vesicles and limestone clasts; both are interpreted as altered impact ejecta. Those ejecta, mixed with clay/marl rip-up clasts from underlying formations, washed-out foraminifers, and various material scavenged from the seafloor, fill shallow channel-like depressions. The second unit (II) consists of a sequence of several lenticular sandstone layers made up of a mixture of foraminifers, bioclasts, plant remains, and terrigenous material. The sandstones may be massive, graded, or parallel- or current-ripple laminated. Some rare spherules and limestone clasts infrequently occur at the base of each layer. The Unit II sands are deposited in very shallow, stacked channels or erosional depression infillings and display a variety of sedimentary features containing evidence for repeated up-section changes in current direction and current strength. The third unit (III) consists of strings of fine sand ripples alternating with thin silt layers. Only Unit III and the overlying unit (IV) contain anomalous iridium and Ni-rich spinels, concentrated in the silt layers. Unit IV is a graded silt-mudstone, representing a quiet depositional phase of fine material, settling—over at least a few days—together with the fine iridium-rich dust.

The coarser basal units in particular show internal unconformable stratigraphic relationships, but all the units and beds are amalgamated on top of each other, with-

*Present address: Institut für Mineralogie, Museum für Naturkunde, Invalidenstrasse 43, D-10015 Berlin, Germany.

Smit, J., Roep, Th. B., Alvarez, W., Montanari, A., Claeys, P., Grajales-Nishimura, J. M., and Bermudez, J., 1996, Coarse-grained, clastic sandstone complex at the K/T boundary around the Gulf of Mexico: Deposition by tsunami waves induced by the Chicxulub impact?, *in* Ryder, G., Fastovsky, D., and Gartner, S., eds., The Cretaceous-Tertiary Event and Other Catastrophes in Earth History: Boulder, Colorado, Geological Society of America Special Paper 307.

out interruption by layers of normal, background sediment. There is no unequivocal evidence (such as intercalated burrowed layers) that the deposition of the entire sandstone complex took more than a few days.

(Re)deposition by a series of large, waning tsunami waves is the most likely explanation for the texture of the K/T sandstone beds, although locally gravity flows may have assisted in the transport of sands from nearshore areas. In shallow shelf-seas (Alabama, Texas) earliest Danian sea-level fluctuations have modified or (partially) removed the K/T sandstone complex. In the shallowest marine K/T site studied, the Parras Basin in northeastern Mexico, no sign of the K/T sandstone complex has been found.

INTRODUCTION

The K/T boundary in the Gulf of Mexico region is often marked by a graded unit of sandstone beds ranging in thickness from 5 cm to 9 m (Fig. 1) (Ross and Scotese, 1988; Smit et al., 1992b; Alvarez et al., 1992b). The sandstone complex is intercalated in a more or less continuous relatively shallow-water fine-grained clay or chalk sequence in Alabama and Texas and in a pelagic marl sequence in eastern Mexico, Haiti, and DSDP holes 536/540 (Alvarez et al., 1992a). We refer in this chapter to these sandstone beds as the K/T sandstone complex.

Elsewhere in the world—preserved only in the most continuous sections (El Kef, Tunisia; Agost, Caravaca, and Zumaya, Spain; Stevns Klint, Denmark; Woodside Creek, New Zealand)—the K/T boundary is marked by a few millimeters-thick layer of (red-stained) clay (Smit, 1990). This clay layer is characterized by anomalous enrichment of siderophile elements (iridium) (Alvarez et al., 1980), shocked minerals (Bohor, 1990), spherules that resemble microkrystites (Smit et al., 1992a), and spherules with Ni-rich spinels (Kyte and Smit, 1986; Montanari et al., 1983). The thin clay layer is therefore interpreted by many as an altered distal impact ejecta—or fallout—layer. The Ir-rich clay is immediately overlain by a few centimeters-thick detrital marl or clay layer, impoverished in microfossils in comparison with marls from just below the ejecta layer. The detrital clay probably reflects effects of low-productivity, "strangelove" ocean conditions (Hsü and McKenzie, 1990) following the mass mortality and extinction at the K/T boundary.

The discovery of the 180- to 300-km-diameter Chicxulub crater in Mexico—the best candidate for the K/T impact site (Hildebrand et al., 1991)—close to all the outcrops with sandstone units prompts the question of the origin of the sandstone complex: Is it really related to the K/T boundary and does it show signs of (Chicxulub) impact-related processes?

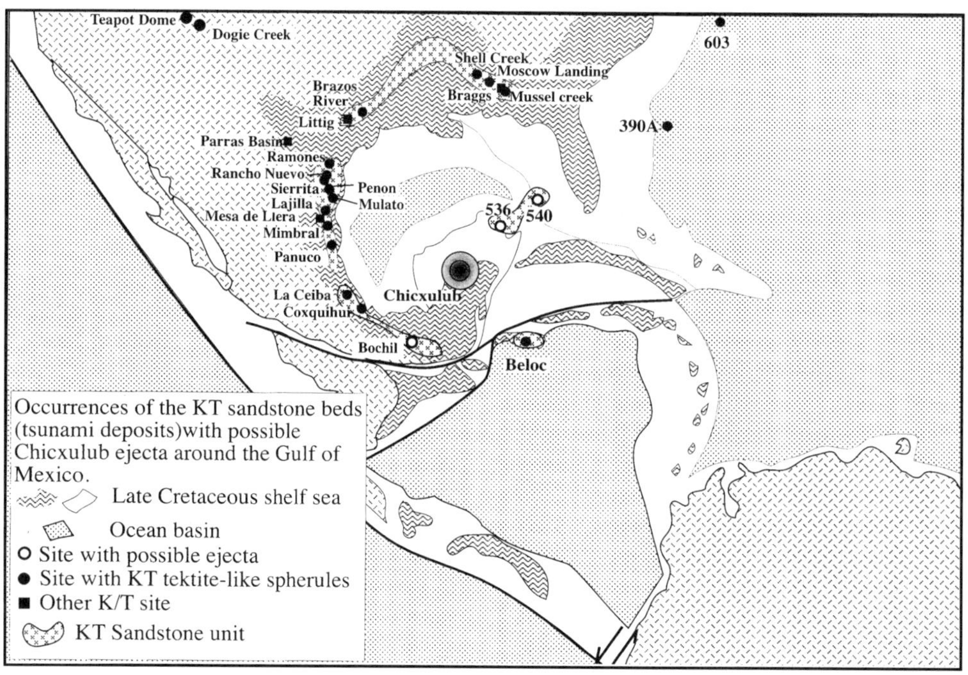

Figure 1. Map of the Gulf of Mexico, about 65 m.y. ago (Ross and Scotese, 1988), showing the Chicxulub impact and the location of outcrops with K/T sandstone complex deposits and sites where spherules (altered ejecta) have been found. 603, 390A, 536, and 540 are Deep Sea Drilling project sites.

The timing of the deposition of the sandstone complex is controversial: Does the sandstone complex occur precisely at the K/T boundary as proposed by Bourgeois et al. (1988), Smit et al. (1992b), and Smit and Romein (1985)? Or does the unit predate the K/T boundary (Jiang and Gartner, 1986; Keller et al., 1993, 1994b; Stinnesbeck et al., 1993, 1994b)? The question has also been posed of whether the beds of the sandstone complex were deposited in one, pulsating event or during several events spread over several thousands of years (Keller et al., 1994b; Stinnesbeck et al., 1993).

The K/T sandstone complex is sedimentologically complex, differing in architecture and composition from place to place. Therefore, many different interpretations have been offered for its origin. The beds have been interpreted as nonmarine sandstones (Kellum, 1937; Muir, 1936), shallow-water deposits (Morgan, 1931), low-stand channel infills (Mancini and Tew, 1993; Savrda, 1991), storm deposits (Hansen et al., 1987), gravity flows (turbidite and debris flows) (Bohor and Betterton, 1993; Stinnesbeck et al., 1993), and tsunami deposits (Bourgeois et al., 1988; Smit et al., 1992b; Smit and Romein, 1985).

The oldest reports on the sandstone deposits at or near the K/T boundary in Mexico between the Upper Cretaceous Mendez and Paleocene Velasco formations date back to the 1930s (Kellum, 1937; Muir, 1936). They interpreted the sandsone unit as shallow-water or nonmarine deposits marking a hiatus and an unconformity. Muir (1936) described a "ripple marked" sandstone and/or a bentonite bed between the Mendez and Velasco (Tamesi) Formations. He reported the contact between the Mendez and the Velasco (Tamesi) Formations in the Arroyo de Mimbral (Membral) outcrop as follows: "The base of the Velasco fm consists of mixed material, including inclusions from the Mendez, filling irregularities at the unconformity between the two formations. Next comes a sandstone, variable in thickness and local in extent." The worm track and plant remains in the sandstones were seen as evidence for a nonmarine origin. Morgan (1931) noted an abrupt change in (planktic) faunal composition between the Mendez and Velasco shales and suspected an unconformity because a "bentonitic conglomerate" occurs at the base of the Velasco Formation in the Tampico area. Hay (1960) described an angular unconformity between the Mendez and Velasco Formations at Mimbral and in addition inferred a major regression at the K/T boundary.

In Texas, the conglomeratic Littig member of the Paleocene Kincaid Formation often unconformably overlies the Cretaceous Corsicana Formation where it marks a considerable hiatus (Jiang and Gartner, 1986; Hansen et al., 1987). However, in the Brazos River outcrops, 1 to 4.5 m of the Kincaid Formation is preserved below the Littig member, containing a graded sandstone unit at its base (here called the K/T sandstone complex). Ganapathy (1980) mentioned that the sandstone unit occurred just below the K/T iridium anomaly (and consequently assigned it a Late Cretaceous age) at Brazos River but did not further interpret it. Smit and Romein (1985) interpreted the Brazos River sandstone unit as a K/T boundary turbidite or tsunami-generated deposit, indicating a nearby impact. Hansen et al. (1987) lithologically subdivided the K/T sandstone unit in detail and mentioned a possible tsunami origin but favored a tempestite interpretation, later supported by Montgomery et al. (1992).

Bourgeois et al. (1988) elaborated on the tsunami interpretation and calculated high current velocities at an estimated water depth of 50 to 100 m, where even hurricane storm waves cannot transport coarse material. Hansen et al. (1987) and Bourgeois et al. (1988) suspected that the 25-cm-thick micritic or silty limestone and mudstone layers overlying the highest sandstone layers could represent silt and clay sediment that settled from the water column following the initiating—storm or tsunami—event. To test this idea, we performed a detailed analysis of the grain-size distribution in this interval (see below). Considerable confusion exists as to where the K/T boundary should be positioned in the Brazos River sections. For example, Hansen et al. (1987), Montgomery et al. (1992), and Smit and Romein (1985) favor a K/T boundary at the base of the sandstone complex, whereas Jiang and Gartner (1986) and Keller (1989a) favor a K/T boundary position at the top, or above the sandstone unit. The positioning of the K/T boundary is further discussed below.

In Alabama, the sandstone complex occurs as discontinuous irregular bedded sandstone bodies, known as the basal Clayton sands (see, e.g., Habib et al., 1992; Mancini and Tew, 1993; Mancini et al., 1989). Those sandstone bodies were interpreted—applying sequence stratigraphic methods—as transgressive infilling of low-stand channels (Mancini and Tew, 1993) or low-stand incised valley fills (Savrda, 1991). An origin by tsunami waves was shortly discussed but rejected (Savrda, 1991, 1993) Another approach was taken by Pitakpaivan et al. (1994) and Smit et al. (1994b), who reported pseudomorphs of impact spherules from the very base of the Clayton sands at Shell Creek and Moscow Landing, Alabama, suggesting a close temporal relation with the Chicxulub impact.

Recently, the K/T clastic complex has been studied in the classical Borrega Canyon–Arroyo de Mimbral section (Bohor, 1994; Keller et al., 1994b; Smit et al., 1992b, Stinnesbeck et al., 1994a) and in many other K/T outcrops in eastern Mexico (Alvarez et al., 1992b), Chiapas (Montanari et al., 1994), Guatemala (Hildebrand et al., 1994), and DSDP sites 540 and 536 (Alvarez et al., 1992a). Below we will report our observations on some of these sites.

METHODS

In this chapter, we present sedimentologic and stratigraphic data of surface outcrops in Alabama, Texas, Mexico, and Haiti. Foraminiferal faunas were studied in thin sections and washed residues, obtained using standard laboratory techniques. Thin sections were petrographically analyzed to evaluate variations in composition and texture of the different sublayers of the K/T sandstone unit, with the aim of establishing the (bio)stratigraphic and sedimentologic relationships of the pre-, syn-, and post-K/T units.

The grain-size distribution of the insoluble residue of

55 samples across the K/T sandstone complex in the Brazos River section was determined with a Fritsch A-22 laser particle-sizer. The samples were dissolved in HCl and oxidized with H_2O_2 to remove carbonate, organic material and some authigenic phases, such as pyrite. The remaining residue was treated in $Na_4(P_2O_7)$ to prevent coagulation of the clay particles. The laser diffraction patterns were translated to a grain-size distribution according to the Fraunhofer model (Hess and Gatzemeier, 1991). From a few key profiles the iridium content was determined by coincidence neutron activation methods (F. Asaro, personal communication, 1991; Mimbral and Brazos River).

THE K/T BOUNDARY SEQUENCE IN THE GULF OF MEXICO AREA

Although the details may differ locally, there appears to be a consistent depositional sequence in all outcrops of the K/T sandstone complex (Fig. 2). The K/T sandstone complex can be subdivided into four lithologic units. The lower three were defined earlier (Smit et al., 1992b) and can be recognized in most outcrops. However, in some outcrops (e.g., Brazos River) a fourth unit occurs on top of the sequence.

1. The basal Unit I consists of poorly sorted, coarse-grained sediments, usually pebbly sandstones, filling irregular scours and channels in the topmost Cretaceous formations. Unit I is principally characterized by and locally rich in a peculiar type of spherules. The most common forms are spherules and droplets with an internal bubbly texture, interpreted as altered impact ejecta. The various sublayers of Unit I are laminated and contain variable mixtures of spherules, small limestone clasts, rip-up clasts and boulders from underlying formations, planktic and benthic foraminifers, and—rarely—phosphatic and glauconitic debris. Size grading within Unit I is often not apparent. Lithic grains occur but are less abundant than in the overlying units. The most common texture is large-scale, channel-fill cross-bedding. Some larger channels display accretion-type lateral infilling. Burrowing has not been observed within the basal Unit I or extending from it at the bottom.

2. Unit II usually displays a stack of shallow lenses (or channels?) filling in shallow erosional depressions or blankets of fining-upward medium- to fine-grained sandstones, displaying a wide variety of sedimentary structures, such as parallel lamination associated with primary current-lineation, two- and three-dimensional—lunate and linguid—ripples, in-drift and in-phase climbing ripples of variable dimensions, and rare antidune-like ripple structures. In the case where Unit I is absent and Unit II is directly in contact with the underlying fine-grained sediments, flute- and groove-casts occur at the base. Current directions inferred from such sedimentary structures indicate variable—often 180° different—current directions in successive sublayers. The sandstones are highly variable in composition but usually form a mixture of foraminifers and lithic grains, with minor plant debris. Armored mudballs frequently occur at the base, with an armor of the above-mentioned bubbly spherules and lime clasts.

3. Unit III consists of thinning- and fining-upward, small-scale cross-bedded, fine sandstone layers alternating with thin layers of silt/mud draping over the sand layers.

In particular the thin silt layers contain anomalous iridium concentrations (Smit et al., 1992b) and Ni-rich spinels (Robin et al., 1994). The top ripple layers of Unit III are burrowed but mostly at the upper surface. The topmost rippled sandstone layer is often severely bioturbated. The ichnofabric of the trace fossils is usually tiered. Different types of burrows occur at different levels.

4. Unit IV overlies Unit III, sometimes gradually, sometimes more abruptly. In thin section the very base of Unit IV

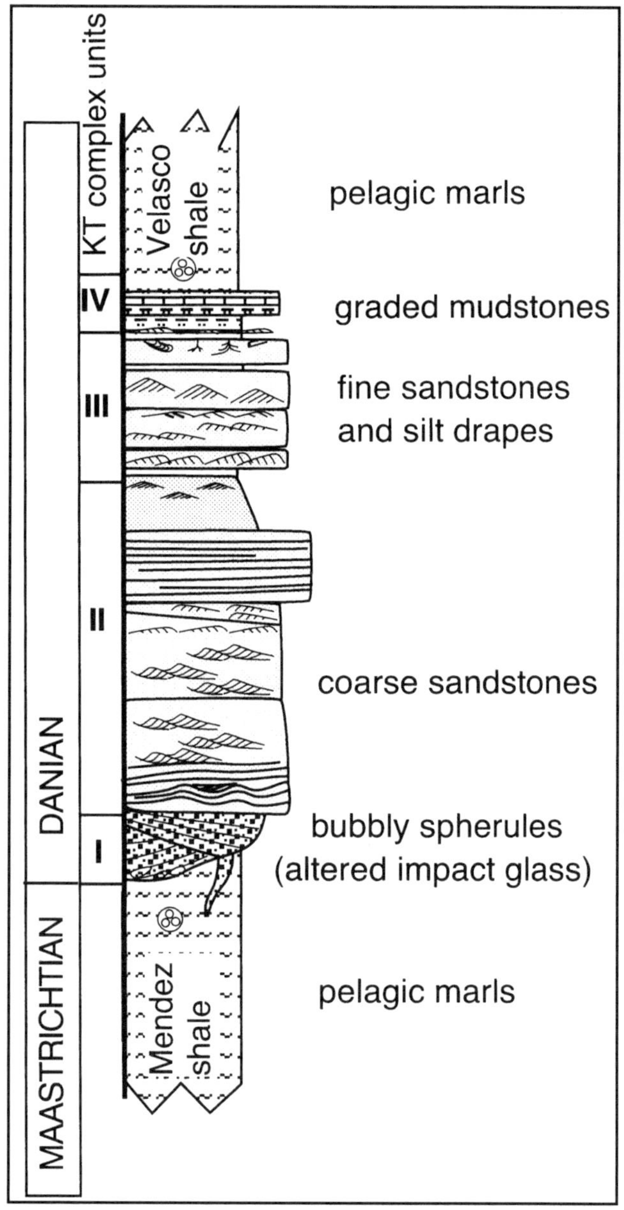

Figure 2. Lithologic column of the K/T sandstone complex (La Lajilla, northeastern Mexico), showing the main lithological subdivisions (Units I through IV), and the position of the K/T boundary.

still shows alternations of millimeters-thin fine sand and silt layers, passing upward into siltstones. The most conspicuous lithologic feature of Unit IV is a 5- to 10-cm-thick calcareous siltstone or mudstone layer (limestone unit F of Hansen et al., 1987), which can be traced over most K/T outcrops in northeastern Mexico and at Brazos River.

THE STRATIGRAPHIC POSITION OF THE K/T BOUNDARY IN THE GULF COAST AREA

The placement of the K/T boundary in the Brazos River sections in Texas and in sections in eastern Mexico containing the K/T sandstone complex is controversial, as mentioned in the introduction (Smit et al., 1994a; Stinnesbeck et al., 1994b). Some place the K/T boundary above the complex (Beeson et al., 1994; Jiang and Gartner, 1986; Keller et al., 1993, 1994b; Stinnesbeck et al., 1993, 1994b) and some at the base of the clastic beds (Bourgeois et al., 1988; Hansen et al., 1987; Montgomery et al., 1992; Olsson and Liu, 1993; Smit et al., 1992b; Smit and Romein, 1985). Formally, the Global Stratotype Section and Point (GSSP) of the K/T boundary is defined (Cowie et al., 1989) in the El Kef section in Tunisia. At El Kef the K/T boundary is defined at the base of the boundary clay, directly above marls with an abundant and diverse Cretaceous planktic flora and fauna. The boundary clay at El Kef includes at its base a few-millimeters-thick—often rusty—clay layer, anomalously enriched in iridium. The layer also contains crystalline spherules (altered microkrystites; Smit et al., 1992a), Ni-rich spinels, and shocked minerals and is regarded by many as a distal ejecta—or fallout—layer. In other words, the K/T boundary in its type section is defined directly below a thin layer with impact ejecta. Given the hypothesis that this layer is globally produced by one near-synchronous impact event, it would make most sense to chronostratigraphically place the K/T boundary at the beginning of the event that produced the ejecta layer. Consequently, *all* deposits with the ejecta from the K/T impact event are Paleocene in age.

When locating the K/T boundary in sections other than the type locality, one can choose among several tools to make the correlation to the K/T boundary stratotype section at El Kef. Biostratigraphically, the correlation can best be achieved by using the mass-extinction–mass-mortality horizon, although that level may often be blurred by reworking. Using the first appearance datum (FAD) of the new Paleocene planktic species such as *Globigerina minutula* (=*Globoconusa conusa* sensu [Keller, 1988]) is more ambiguous. At El Kef, the FAD of *G. minutula* is 23 cm, and the FAD of *Globigerina eugubina*—a widely used biomarker for the basal Paleocene—is about 25 cm above the K/T boundary at El Kef. According to work currently underway by H. Vonhof and J. Smit, this means that the FADs of these species are at least 5 k.y. above the K/T boundary in the Kef section, using the average sedimentation rate of 40 m/m.y. of the Upper Cretaceous and Paleocene at El Kef. Another way to locate the K/T boundary is by means of the worldwide dispersed ejecta, including iridium-rich dust, shocked minerals, and spherules produced by the presumed K/T impact event. The K/T boundary should be placed at the base of all these ejecta. We intend to demonstrate in this chapter that the sandstone complex at the K/T boundary postdates the planktic mass mortality and was deposited in one complex but coherent event and that the ejecta found in the base of the sandstone units (altered impact glass, shocked minerals) and at the top of the sandstone complex (iridium anomaly), although separated by a few meters, are easiest to explain as products of one and the same impact event. If this option is followed, the K/T boundary in the Gulf Coast outcrops should be placed at the base of the ejecta-bearing clastic deposits. When using a single tool such as the iridium anomaly, one could favor the K/T boundary being drawn just below the Ir anomaly, well above the base of the sandstone unit. However, the iridium anomaly in the Gulf Coast outcrops is dispersed and usually consists of multiple peaks. Because the Ir carrier is very fine grained (a particulate "carrier" for the iridium has yet to be found), it is not surprising that Ir settled more slowly through the atmosphere and water column than the coarse ejecta and that both occur in separate levels in the K/T sandstone unit. The mass-extinction horizon in the Gulf Coast outcrops, that is, the level where the abundance of Cretaceous planktic foraminifers drops by an order of magnitude, locates the K/T boundary again at the base of the clastic (ejecta) deposits: The last (hemi)pelagic sediments with abundant Cretaceous planktic faunas invariably occur below the K/T sandstone unit. The FAD of Paleocene foraminifers may be used as proxy for the K/T boundary as well. The FAD of Paleocene foraminifers in the Brazos River sections (Keller, 1989a; Beeson et al., 1994) is about 25 cm above the highest rippled sandstone layer. However, as said above, it should be kept in mind that the FAD of unequivocal Paleocene foraminifers is several thousands of years above the type K/T boundary in the Kef section and probably also in the Brazos River sections.

SECTIONS CONTAINING K/T SANDSTONE COMPLEX IN THE U.S. GULF COAST AND MEXICO

Moscow Landing, Alabama

The easternmost outcrop studied containing the K/T sandstone complex is at Moscow Landing, Alabama. The K/T sandstone complex occurs here as a series of discontinuous lenticular sandstone bodies, known as the Clayton basal sands of the Clayton Formation, overlying the fossiliferous upper Maastrichtian Prairie Bluff Chalk with an erosional contact. The Prairie Bluff Chalk is a gray sandy fossiliferous chalk, containing about 1 m below the K/T boundary a conspicuous phosphatic macrofossil lag bed, interpreted as a condensed section (Mancini et al., 1989). Bedding planes dip slightly to the south (2°), but as a result of repeated offsets by small normal faults, the K/T boundary remains exposed for more than 1 km along the east bank of the Tombigbee River.

The basal Clayton sands are usually interpreted in sequence stratigraphic terms (Habib et al., 1992; Mancini and Tew, 1993;

Mancini et al., 1989) as early Danian low-stand channel-fill deposits, overlain by a transgressive surface with a *Thalassinoides* burrow-fabric and transgressive deposits, respectively. However, the situation is more complex, as illustrated in Figures 3 and 4. Over most of the outcrop, the Danian transgressive surface overlies paraconformably the Prairie Bluff Chalk, but the same surface also overlies the coarse-grained basal Clayton sand bodies (Fig. 3). Within the 200- to 400-m-wide tilted blocks, bounded by the small normal faults mentioned above, the bedding planes of the Prairie Bluff Chalk are not disturbed and remain mostly parallel. However, near the faults, bedding is chaotic, displaying a mixture of soft and semisoft sediment deformation. In some rotated, about 5-m-wide, domains, the bedding planes are tilted vertically and are truncated by the Danian transgressive surface (Fig. 4). Some of these faults also offset the basal Clayton sands but do not offset the early Danian transgressive surface. Drag along these faults plastically deforms—even overturns—the basalmost layers of the Clayton basal sands. In some of the faults Clayton sands have been injected. The next coarse sand deposit of a younger part of the Clayton Formation again truncates the tilted bedding planes of the basalmost Clayton layers, showing that movements along the fault plane occurred during deposition of the Clayton basal sands.

The lowermost infillings of <3-m-wide and <1-m-thick "pockets" of the basal Clayton sands consist of a poorly sorted conglomerate containing Cretaceous macrofossils and Prairie Bluff chalks clasts, in a matrix of coarse sands. Some of the chalk clasts are plastically deformed, in particular near the base. Characteristic components of the sand matrix are green, sparry calcite-filled spheroids and droplets, 2 to 3 mm in diameter, with an external lining of clay minerals (Fig. 5A). The spheroids contain internal cavities, also filled with sparry calcite.

Identical spheroids were found in the nearby outcrops of Shell Creek and Lynn Creek, and those were interpreted as altered splashform tektites, by analogy with spheroids from Beloc and Mimbral filled with bubble-cavity–rich glass (Pitakpaivan et al., 1994).

The basal conglomerate has the appearance of a mass-flow and is found only at those places where the Prairie Bluff Chalk is slumped and deformed. The bottom of the "pockets" follows the slump structures of the deformed Prairie Bluff Chalk. The conglomerate is overlain and truncated by 10- to 50-cm-thick lenticular layers of very poorly sorted, poorly graded, parallel-laminated coarse sandstones and pebbly sandstones, rich in Cretaceous macrofossils and chalk clasts. The laminated textures are typical for high current strength (upper flow-regime) and rapid sedimentation. The sand matrix dominates, but the typical bubbly spheroids of the underlying conglomerate matrix were not found. As mentioned above, the lowermost layers were drag-tilted near the faults and truncated by the overlying sand layers. The geometry of the sand bodies filled with conglomerates and sandstones is asymmetrical. They appear as triangular wedges in cross section, with their thickest part at the southern end, near the faults (Fig. 4). The pockets with spheroid-rich conglomerate occur only at the southern—thickest—end of the wedges. At none of the bedding surfaces or in the sandstone layers were borings or *Thalassinoides*-type burrows observed other than those penetrating down from the Danian transgressive surface.

A few symmetrical channels filled with graded conglomerate and nonlaminated coarse sand—with chalk boulders as large as 90 cm—are also paraconformably overlain by the Danian transgressive surface. The infilling layers are not deformed in contrast to the triangual wedges and are thus presumably younger. Those channels are incised into the underlying chaotically bedded part of Prairie Bluff Chalk and presumably also

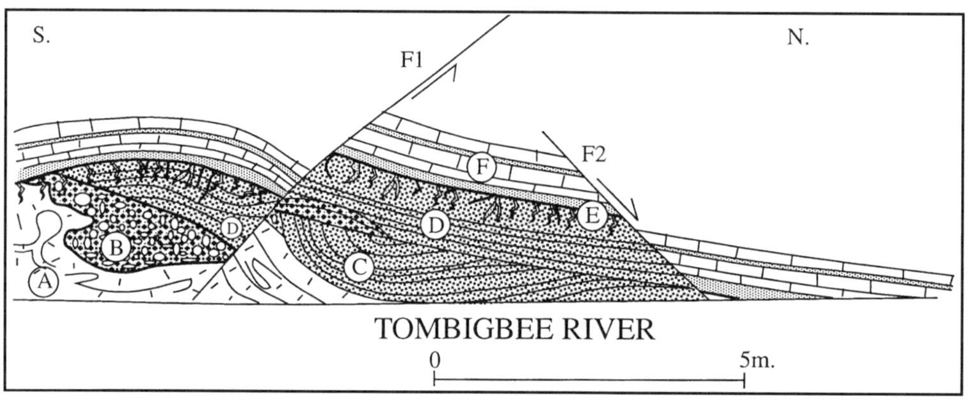

Figure 3. Moscow Landing, Alabama. Line drawing (after a photograph) of one of the spherule-(altered tektites–) bearing K/T complex channels. A, Deformed chalks of the Prairie Bluff Formation. B, Spherule-bearing conglomerate "pocket," or deformed channel. C, Coarse-grained pebbly sandstone–filled channel, upturned near fault F1. D, Subsequent coarse laminated pebbly sandstone and coarse sandstone layer, truncating the upturned end of channel C. E, Transgressive surface truncating B–D channels as well as a symmetrical channel (not shown) of the preceding low-stand incision. Borings penetrate into the upturned end of channel D. F, Clayton Formation graded sandstones, limestones, and phosphatic lag deposits.

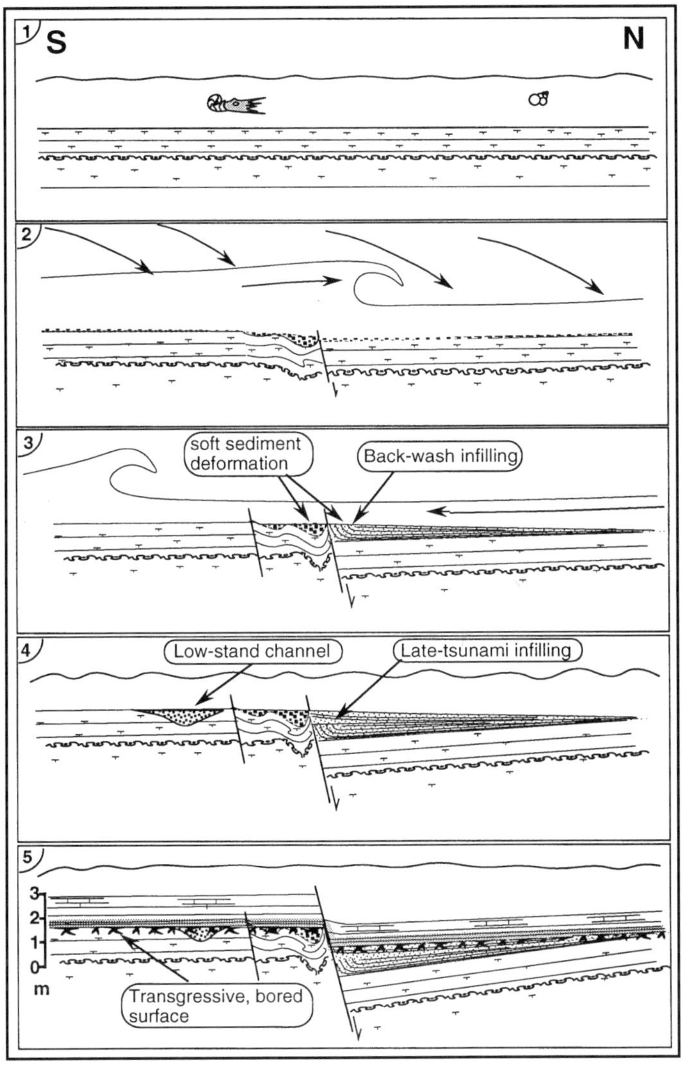

Figure 4. Cartoon of the sequence of events at the K/T boundary in the Moscow Landing, Alabama, outcrop. Only locally, at the downthrown side of the faulted blocks, has an almost complete sequence of the K/T clastic beds been preserved. 1, Situation just before the K/T boundary. The Prairie Bluff chalks are undisturbed and parallel bedded and contain a phosphatic lag bed, interpreted as maximum flooding surface. 2, Arrival of large waves, possibly preceded by earthquakes, causes slumping and faulting. The ejecta from the Chicxulub impact arrive at about the same time and are reworked into conglomerate pockets loaded into the slumped parts. 3, Continued faulting causes deformation of the edge of the sandstone channel infill. Some channels appear to originate as a result of the faulting and are infilled with coarse terrigenous (backwash?) sands. 4, Triangular wedges caused by continued faulting are further filled with pebbly sandstones, and the early Danian sea-level fall–caused low-stand channels are filled with basal Danian sediments. 5, The low-stand phase and subsequent transgression erodes part of the low-stand channel, the tsunami deposits (Units II–IV), and the top of the Cretaceous and leaves a bored hard ground. Continued faulting causes offset of the transgressive surface and the subsequent Clayton Formation deposits.

into the basalmost Clayton sand wedges, although this could not be clearly observed in the outcrop. Habib and Talvirska (1994) demonstrated the presence of earliest Danian dinoflagellate cysts in one of the symmetrical channels. We, however, did not observe any Danian foraminifers or dinoflagellate cysts in the underlying triangular sand wedges.

The trangressive surface overlies the Prairie Bluff Chalk, the basalmost wedges, and the symmetrical channels. The *Thalassinoides*-type burrows extending down from the Danian transgressive surface penetrate a few decimeters into the top of the Prairie Bluff Chalk as well as into the top of both types of channels. The burrows penetrate parallel to the bedding planes where the basal Clayton sands were tilted vertically, showing that the transgressive surface postdates deformation of the basal Clayton sand bodies. The burrow-fillings contain *Morozovella pseudobulloides* (Mancini et al., 1989), demonstrating a hiatus of at least 50 k.y. (Groot et al., 1989) between the sandstone channels underlying the transgressive surface and the marlstones and limestones of the upper part of the Clayton Formation overlying the transgressive surface. The Clayton marlstone contains at the base a 10-cm-thick lag bed with quartz grains, phosphatic pebbles, and reworked Cretaceous fossils (Mancini et al., 1989). This layer is continuous over the entire outcrop but thickens over the slumped parts of the Prairie Bluff Chalk, filling up shallow depressions. The next overlying layers are two continuous 15-cm-thick limestone beds separated by a 10-cm-thick sandy marl. The thickness of these layers is independent of the underlying type of lithology. Mancini et al. (1989) reported a similar phosphatic lag bed in the Braggs and Mussel Creek sections, overlying a transgressive surface on top of the Prairie Bluff Chalk (Braggs) or a transgressive surface with associated *Thalassinoides* burrows (Savrda, 1993) on both the Prairie Bluff Chalk and basal Clayton sands (Mussel Creek). The basal Clayton sands at Mussel Creek fill more or less symmetrical channels, similar to the symmetrical channel at Moscow Landing described above, and are reported to contain Danian microfossils (Mancini et al., 1989) and several tiers of *Thalassinoides* burrows (Savrda 1993). Bubbly spheroids were not reported, nor have we found those. Thus sediments comparable to those in the triangular wedges at Moscow Landing seem to be lacking in these outcrops.

Laterally, from north to south along the outcrop, a consistent sequence is observed across several of the K/T sandstone wedges (Fig. 3). The upthrown, not deformed part of the Prairie Bluff Chalk ends with a series of normal faults. In between the faults the intensity of deformation increases until the chalk layers are completely slumped. The basal mass-flow conglomerates were deposited against, and partially upon the slumped chalk, partially upon the nondeformed downthrown end of the next Prairie Bluff Chalk block. The lenticular pebbly sandstone layers were deposited against and partially upon and farther to the north of the basal conglomerates, gradually pinching out to the north.

In general we agree with the interpretation of Mancini et al.

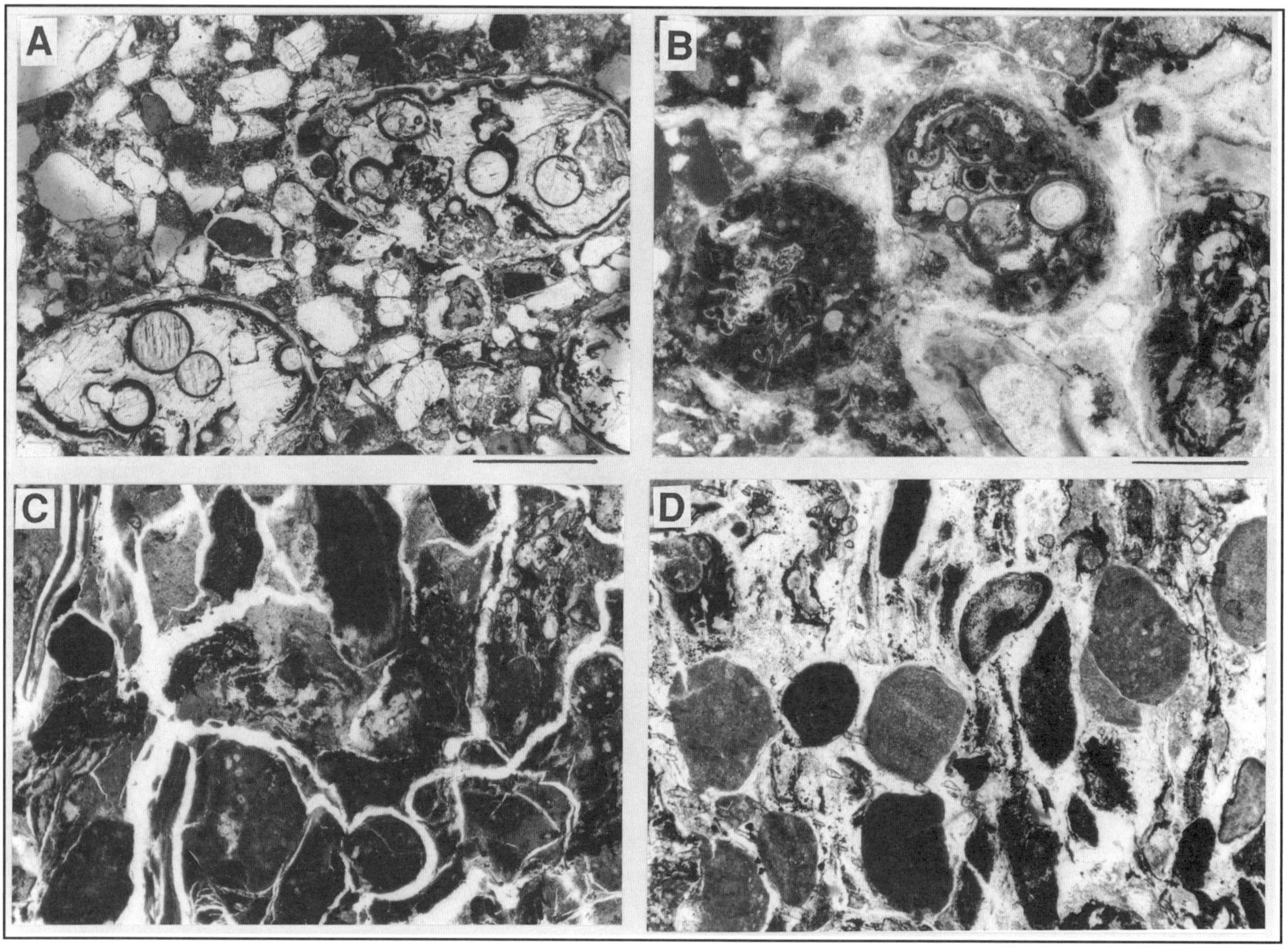

Figure 5 (on this and facing page). Micrographs of thin sections of the bubbly spherules and limestone fragments, interpreted as ejecta. A, Bubbly calcitic spherule (Moscow Landing). B, Bubbly spherule (Darting Minnows Creek, near Brazos River, Texas). C, Remnant of a bubbly spherule from Brazos 1 section. D, Limestone fragment (Darting Minnows Creek). E, F, Bubbly spherules (Mimbral, Mexico). G, H, Concentric banded limestone fragment, with some bubble? cavities (Mimbral, Mexico). (Bar = 0.5 mm.)

(1989), Mancini and Tew (1993), and Savrda (1993) that most of the sequence at Moscow Landing and all of the sequences found at Mussel Creek and Braggs can be explained in a sequence stratigraphic scheme. However, we interpret the Moscow Landing basal Clayton sands as two different types of deposits: a basal sequence consisting of several nonburrowed, rapidly deposited, partially mass-flow–type deposits (K/T sandstone Units I and II), followed by low-stand ravinement valleys filled with Danian low-stand deposits. The interpretion of the Moscow Landing K/T sequence is best illustrated with the cartoon of Figure 4. We assume that faulting and associated slumping of the Prairie Bluff Chalk started just before and remained active during deposition of the Clayton basal sands. Because the thickness and texture of the Prairie Bluff Chalk layers are not influenced by these faults, the faults were presumably not active in the Upper Cretaceous. The major faults were later rejuvenated, because the major tilting phase—offsetting the K/T boundary as described above—took place after formation of the Danian transgressive surface and deposition of the Clayton Formation. During and immediately following the initial phase of faulting and slumping the mass-flow conglomerates with green bubbly spheroids and the laminated pebbly sandstone layers were rapidly deposited. Some sandy material (without bubbly spheroids) is even injected into open faults. The above-mentioned faulting phase and subsequent slumping and deposition of these mass-flow units may be explained by seismic shaking resulting from the Chicxulub impact event, followed by deposition of the spheroid-rich ejecta. The ejecta are mixed with rip-up chalk clast (K/T sandstone Unit I) and redeposited by strong currents, believed to be induced by passage of large tsunami

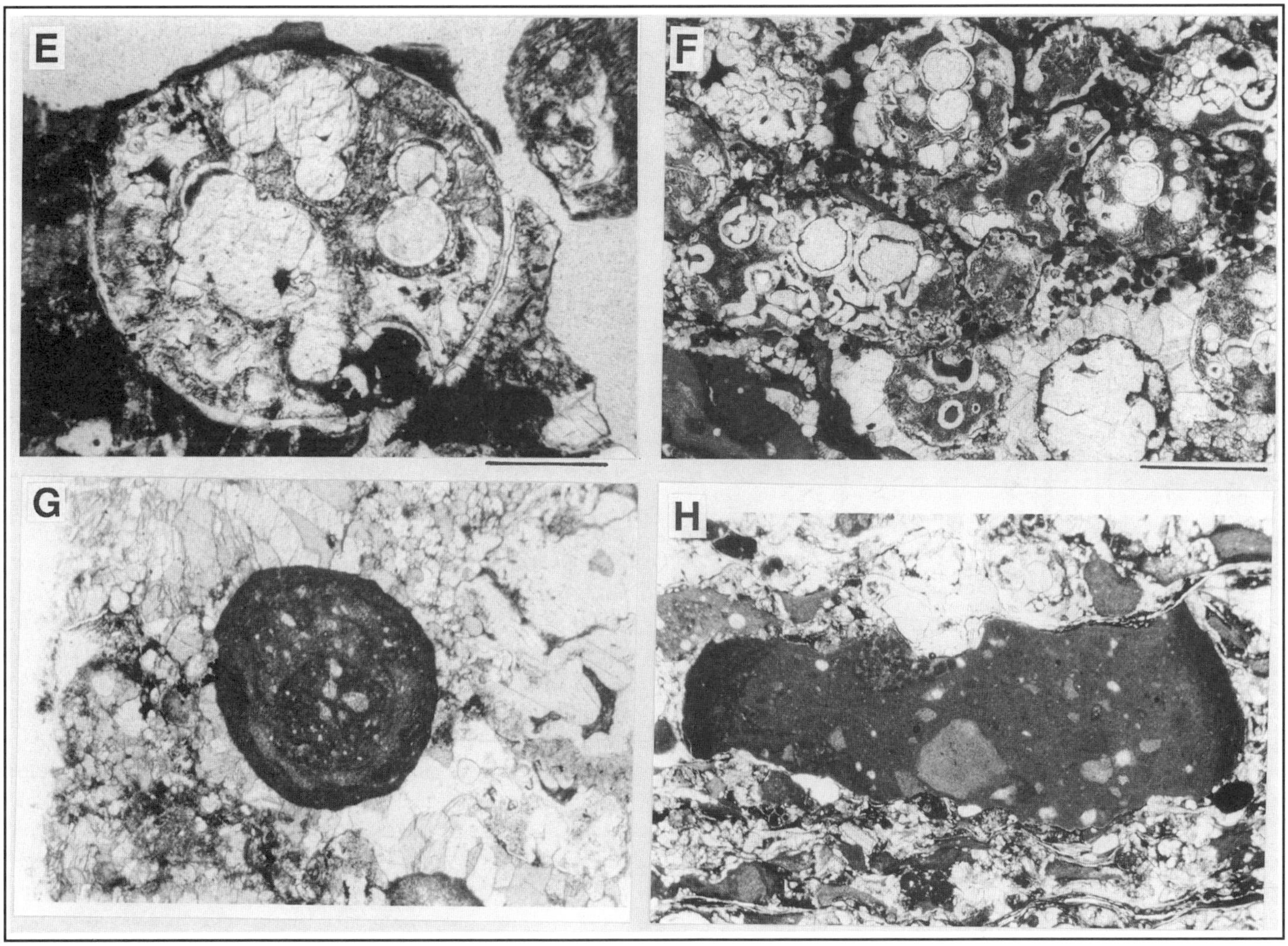

waves. Those deposits are followed by deposition of mass-flows and coarse sandstones with coarse (?near-shore) terrigenous material (K/T sandstone Unit II). Any later fine-grained phases (K/T sandstone Units III and IV), as observed in other K/T sections around the Gulf of Mexico that may also have been associated with tsunami waves, were removed during low-stand erosion, preceding the burrowed Danian transgressive surface.

The next phase is infilling of incised lowstand channels, with early, but not earliest, Danian material.

Brazos River, Texas

The many studies of the Brazos River sections have led to publication of numerous papers that convey a variety of interpretations. The controversy about the placement of the K/T boundary in the Brazos sections has already been mentioned. At low-water level of the Brazos River, the K/T sandstone complex can be traced over a considerable distance in the river bed. In the Brazos 1 section the lithologic sequence across the K/T sandstone complex has been subdivided into units A through J (Hansen et al., 1987), a subdivision that we will follow herein (Fig. 6). The lower unit (unit A, Corsicana Fm, >15 m thick as exposed in the river bed) consists of fossiliferous upper Maastrichtian mudstones deposited at mid- to outer-shelf depths (75 to 200 m; Bourgeois et al., 1988). Occasional shell stringers in the Corsicana Formation about 10 m below the K/T boundary and the low-diversity planktic foraminiferal fauna suggest a water depth closer to the lower estimate, near storm wave base (50 to 75 m). Unit B/C (0 to 45 cm thick) is a laminated, poorly lithified, graded skeletal—shell hash—sandstone, rich in phosphate, glauconite, shell debris, and small clay clasts, resting with a scoured contact on unit A. At several places the basal part of unit B/C contains boulders of black claystone and micritic limestone.

Recently, we have found abundant green spheroids in the basal coarse-grained sandstone (unit B/C) in the Darting Minnows Creek outcrop, about 3 km southwest of the Brazos 1 outcrop (Bourgeois et al., 1988). The spheroids are dominantly composed of dull olive-green clay minerals. As at Moscow Landing, they contain numerous internal cavities filled with calcite (Fig. 5B). Those spheroids were earlier probably mis-

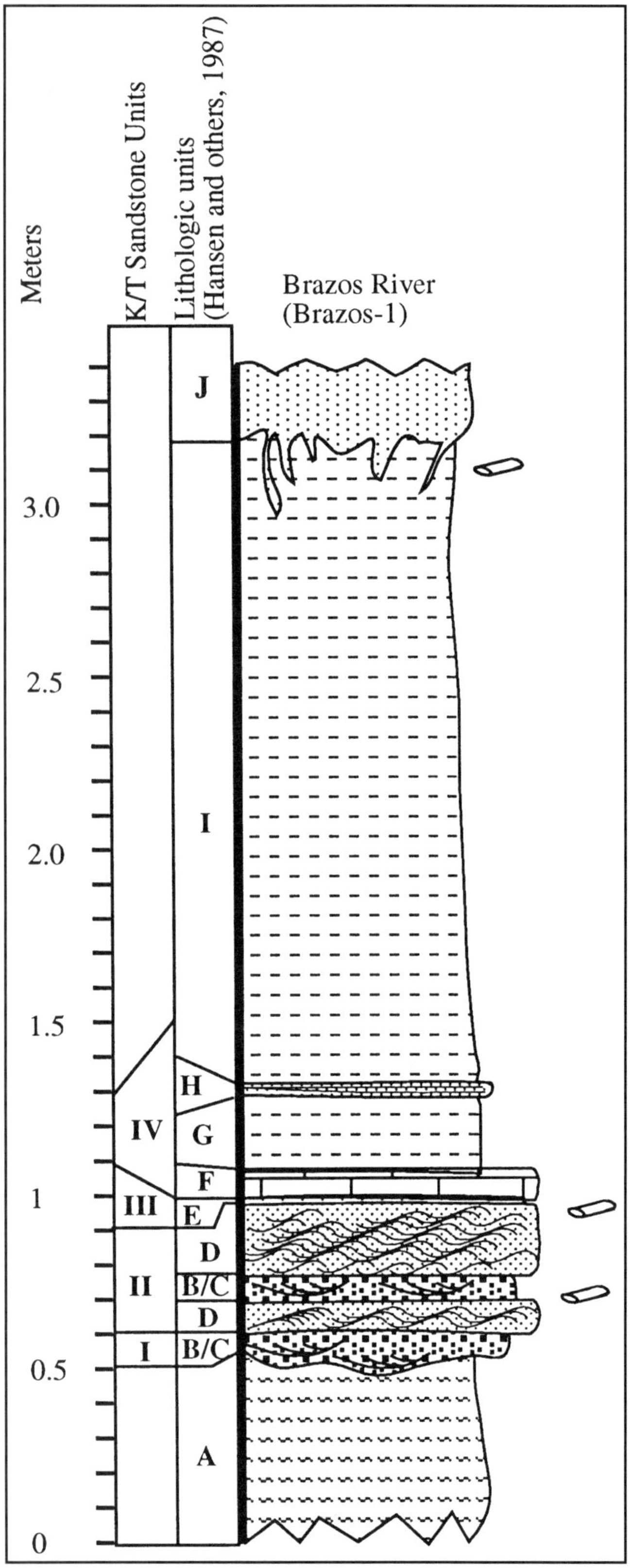

Figure 6. Lithologic column of the Brazos River section, showing the subdivision A–J of Hansen et al. (1987) (see text) and the subdivision into K/T sandstone complex units I–IV.

taken for glauconite grains. However, glauconite grains—also present in the shell hash—can be easily distinguished by their bright green colors, irregular outline, and absence of spherical cavities. Reexamination of earlier thin sections of the coarse "shell hash" of the Brazos 1 and 3 and Cottonmouth Creek (CMC) outcrops (Hansen et al., 1987) reveals that spheroids occur frequently also in those outcrops, in the lower B/C units (Fig. 6). However, the spheroids are strongly corroded and hardly recognizable. Often only an aggregate of the calcite-filled cavities remains (Fig. 5C). Another characteristic component of the basal sandstone is abundant, rounded carbonate fragments, 1 to 5 mm in diameter (Fig. 5D). These fragments often display irregular, concentric banding and contain rare, internal spherical cavities. The fragments are coarsely crystalline in the interior and have a <0.1-mm-thick, darker, often-corroded rim of micro-sparry calcite. An $^{87}Sr/^{86}Sr$ ratio of 0.707564 (K. Beets, personal communication, 1992) is consistent with a Campanian age of the carbonate (MacDougall, 1988). Carbonate fragments of the same size and with an identical texture were found in all other Gulf Coast basal K/T sandstone beds, including DSDP site 540 and Beloc, Haiti. The fragments are easily identifiable in thin section (Fig. 5D, G, H) in the basal sands of all Brazos River sections and appear to have better survived diagenetic alteration than the bubbly spheroids.

A salient lenticular layer of well-lithified, current-rippled medium- to fine-grained sandstone (unit D, 0 to 35 cm thick) quite sharply overlies the basal sands (unit B/C). The ripple structures are invariably asymmetrical and include steeply climbing ripples, climbing ripples-in-drift, and linguoid ripples. These ripple types change vertically and laterally into each other. Oscillation-type ripple structures, hummocky cross-stratification, or other symmetrical or asymmetrical wave ripple structures have not been observed, although it is often hard to distinguish the ripple structures observed in a section perpendicular to the current direction from wave ripple structures.

Current directions measured in unit D of the Brazos 1 and 3 outcrops are dominantly toward N100–110E but shift to N180 near the top of the layer. The basal sands and rippled sandstone (units B through D) repeat up to three times stratigraphically in the outcrops in the Brazos river bed, Brazos 3, CMC, and Darting Minnows Creek (DMC) (Hansen et al., 1987; Bourgeois et al., 1988). The second sequence (units B through D) is more continuous, and the second current-rippled sandstone (unit D) is often the thickest. The Brazos 1 section contains only one unit B through D sequence, but it could not be observed whether the first or second (units B through D) sequence outcrops in the Brazos 1 section.

Conspicuous, 1-cm-diameter, tubelike, branching *(Ophiomorpha?)* burrows occur on the top surface of both the first and the second rippled sandstone (unit D). These burrows were not observed in the interior of the rippled sandstone. It is not clear whether these burrows penetrated from above and spread out at these particular surfaces after deposition of the entire sandstone sequence or were produced shortly after deposition of the rippled sandstone layers.

The rippled sandstone (unit D) is overlain—in ascending order—by a thin silt layer (unit E, 1 cm thick), a weathering-resistant silty limestone (unit F, 8 cm thick), and a silt/mudstone (unit G, 20 cm thick) (Fig. 6). The first (rare) Paleocene biota appear at the top of the mudstone (unit G) (Jiang and Gartner, 1986; Hansen et al., 1987). A single laminated fine sand layer (unit H, 2 to 3 cm thick) follows, overlain by fossiliferous mudstones, containing extremely rare Paleocene planktic foraminifers in the upper part (unit I, 2.5 m thick) capped by a burrowed transgressive surface. Large *Ophiomorpha*-type burrows associated with the transgression extend downward for 0.5 m into the mudstones of unit I. The Littig Member of the Kincaid Formation, a sandy, glauconitic, phosphatic lag bed containing reworked Cretaceous fossils (unit J, >30 cm thick), overlies the transgressive surface.

It has been suggested by some authors that the silty limestone and mudstone (units F, G) represent either the grading settling "tail" of the tsunami disturbance (Bourgeois, 1991; Bourgeois et al., 1988) or low-energy sedimentation after a storm (Hansen et al., 1987). Alternatively, it has been suggested that no size grading is apparent in those units and that they represent a homogeneous background sedimentation, not related to the deposition of the sandstone units (Beeson et al., 1994; Keller, 1989b; Keller et al., 1993; Keller et al., 1994a; Stinnesbeck, et al., 1993; Stinnesbeck et al., 1994a).

To verify these hypotheses, the grain-size distribution of the HCl-insoluble and H_2O_2-treated residue of 60 samples across the K/T boundary sequence of the Brazos 1 outcrop (units A through I) was analyzed. HCl-insoluble residues were used to eliminate biases that could be caused by such factors as the difference in size between Cretaceous and Paleocene foraminifers. Samples from the Corsicana mudstones, the basal skeletal sandstone, and the rippled sandstone (units A through D) were taken from the outcrop in the Brazos river bed, and the remaining samples (units E through I) came from the Brazos 1 outcrop in the west bank of the river, about 20 m to the northwest. The silty limestone and overlying mudstone (units F, G) above the last sandstone ripples were sampled in detail (on average every 0.8 cm) (Fig. 6).

The grain-size distributions are presented as histograms (Fig. 7). Percentages of the 60 histograms, arranged in ascending stratigraphic order, were contoured at 1% intervals (Fig. 8). The detailed sampled interval of units E through G is presented in Figure 9.

Several trends can be observed in the grain-size distributions that are not easily visible in outcrop. The Corsicana mudstones (unit A) can be subdivided into a slightly siltier lower part (0.5 to 10 m below the K/T boundary) and a more clay-rich upper part (0 to 0.5 m below the K/T boundary). The rippled sandstone (unit D) is fining upward, consistent with inferred settling from suspension-rich currents decreasing in strength. Above unit D no fine sand has been stirred up anymore, with the exception of one 1- to 2-cm-thick very fine sand layer (unit H). The silt, silty limestone, and mudstone (units E through G) above the rippled sandstone (Figs. 7 through 9) show a bimodal size distribution, with peaks in the silt (Phi 5.75 to 6.5) and clay (Phi 9) size-fraction (Fig. 9). The silt fraction gradually decreases upward while the clay fraction (Phi 9) gradually increases, until near the top of unit G the clay fraction dominates. In the silty limestone (unit F) the modal grain size of the silt fraction shifts gradually from Phi 5.75 to Phi 6.5 (19 to 11μ, Fig. 9). The graded grain size distribution through units E through G strongly suggests that a suspended cloud of silt and clay settled out during low-energy conditions to form those units, directly on top of the K/T sandstone units, contradicting the suggestion (Beeson et al., 1994, Keller et al., 1993) that these silt-mudstones are part of normal background sedimentation. The grain-size distributions of the unit A mudstones are comparable to those of units E through G, suggesting

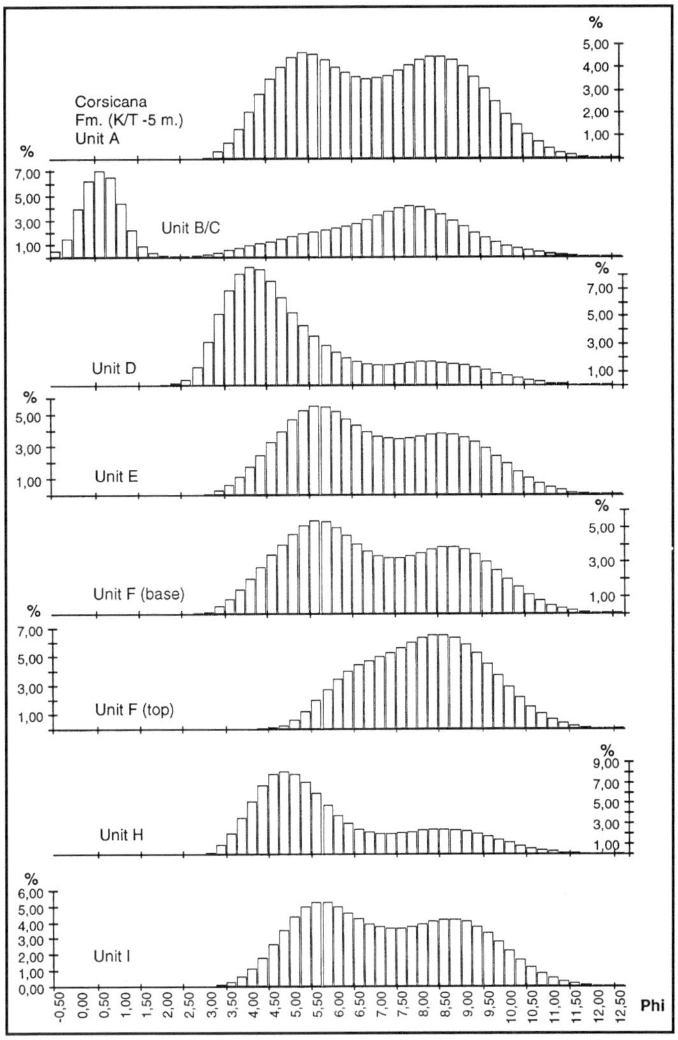

Figure 7. Selected histograms of the grain-size distribution of the different lithological units (see Fig. 6). Grain size in Phi units.

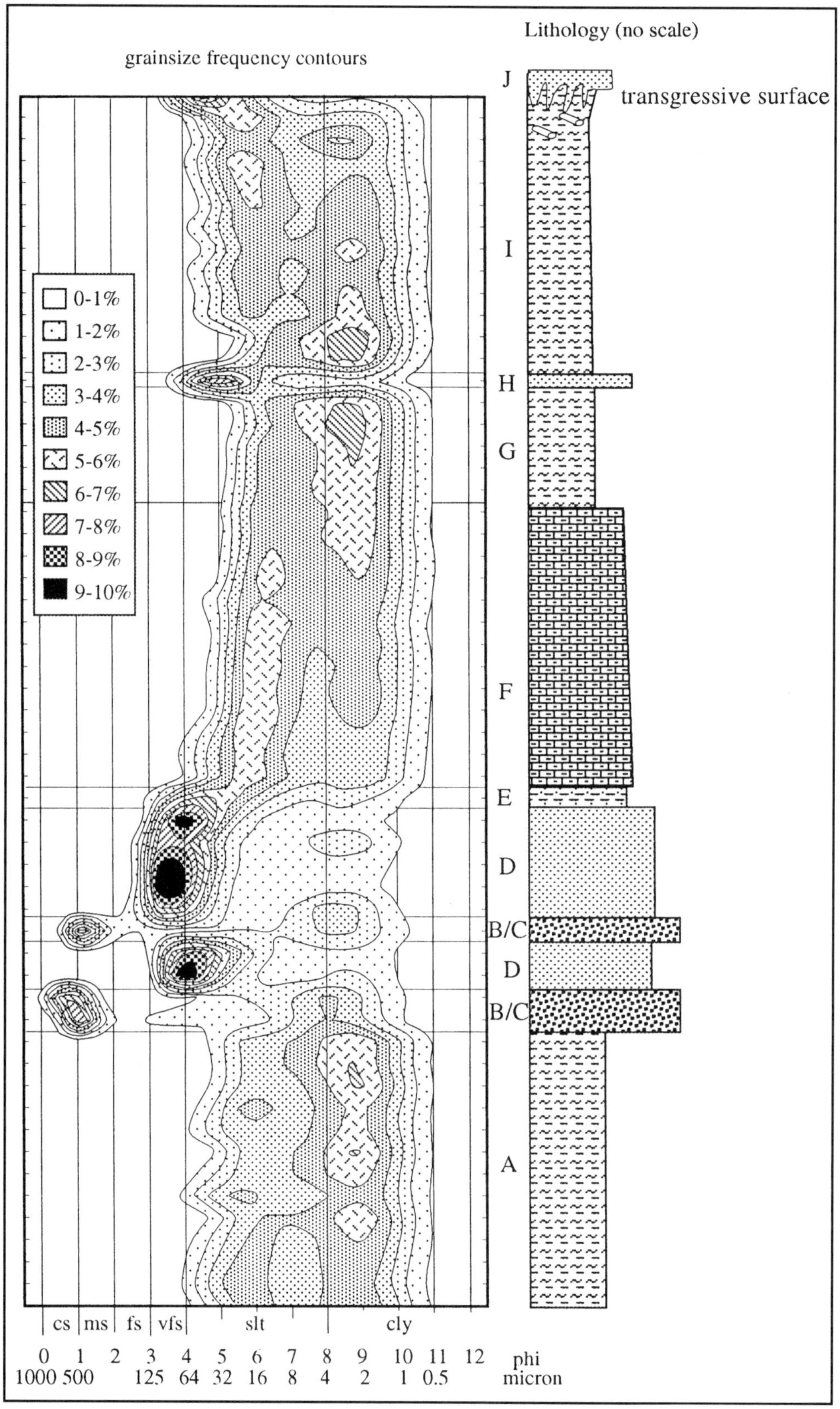

Figure 8. Contour plot of grain-size distribution in 60 samples across the K/T boundary in the Brazos 1 section. The samples are plotted in stratigraphic order but note that the true sample spacing is different. Therefore, the lithologic column represents lithologic expression of the samples in the correct stratigraphic order but has no vertical scale. The contours are at 1% spacing, and scores are computed for each 1/4 Phi class. A–J, lithologic units of Hansen et al. (1987). Abbreviations: cs = coarse sand, ms = medium sand, fs = fine sand, vfs = very fine sand, slt = silt, cly = clay.

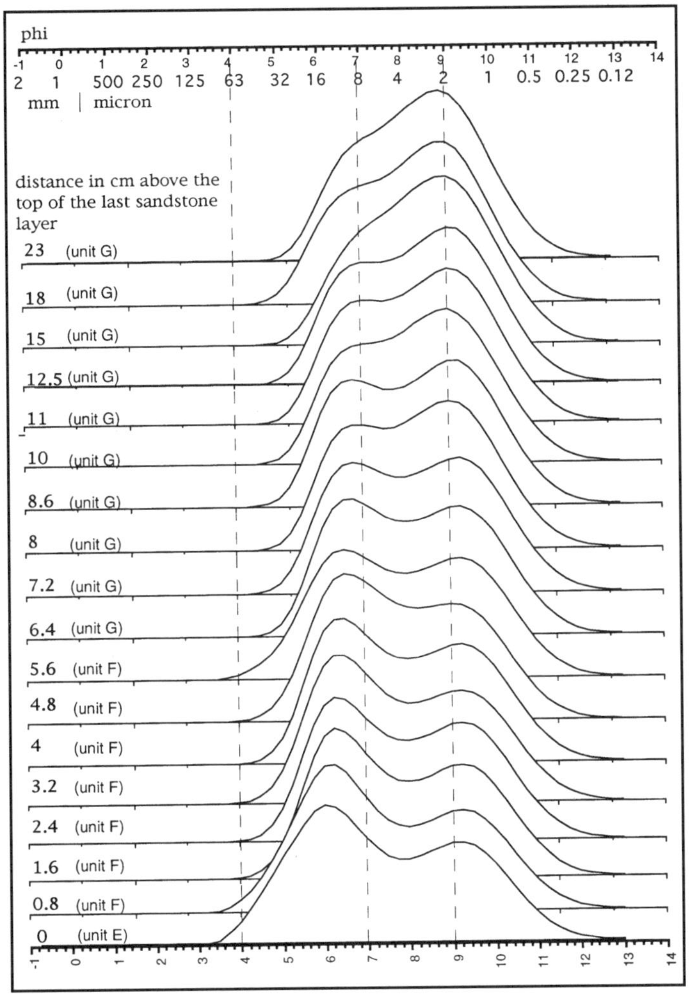

Figure 9. Brazos 1 outcrop. Detail of Units E–G (K/T sandstone complex Unit IV). Grain-size histograms are in stratigraphic order, showing the gradual shift of a bimodal distribution, with a dominating silt component, to a distribution with a dominant clay fraction.

that resuspended Corsicana Formation mudstones could have deposited units E through G. All iridium peaks found in the Brazos 1 section (Asaro et al., 1982), occur in this graded interval (units E through G, Fig. 10).

The thin, very fine sandstone (unit H) layer at 25 cm above the rippled sandstone indicates an increase of energy, with or without connection with the K/T sandstone beds. Near the top of the mudstones of unit I (Fig. 8) grain size is coarsening upward toward the transgressive surface. The settling time for the graded units E through G can be estimated using Stokes law. In Figure 10 the percentile scores of a few selected grain-size classes are plotted against stratigraphic thickness. The grain sizes with diameter >63μ (Phi 4) have settled within 1.5 cm from the base of unit E. Grain size 32μ (Phi 5) decreases to a minimum at 13 cm above the base of unit E. Assuming a homogeneous silt/clay suspension cloud of 50-m thickness (Table 1)—the minimum estimated water depth—the first 13 cm of units E through G could have been deposited in about 20 hs. The settling time of the finer tail (Phi >6) of the graded units E through G is harder to estimate. However, the fine suspended particles (Phi 8 to 9.5, 4 to 1.4μ) apparently continued to settle, as the clay fraction at the top of unit G and base of unit I (Fig. 8) dominates the size distribution. If this interval is still part of the settling phase, the fraction of 4 to 16μ might have settled between 1 week and a few months. Figure 10 shows that all of the iridium peaks occur within this graded interval.

Arroyo de Mimbral, Mexico

The K/T sandstone complex crops out in the south bank of the Arroyo de Mimbral, about 4.5 km east of the electric power lines (Fig. 11). Thirty-five m of well-bedded upper Maastrichtian (*Abathomphalus mayaroensis* Zone) Mendez Formation marls are exposed below the K/T sandstone unit. Bedding planes in the last meter below the K/T sandstone complex are barely visible.

The K/T sandstone complex in the eastern, best-exposed, part of the outcrop is depicted in Figure 11, showing the meter markings (m.mk.) that were painted on the outcrop. Unit I is deposited in four separate shallow depressions or scours along the outcrop, with a maximum width of 20 m and a depth of 1.2 m at m.mk. 28. A 2-cm-thick bentonite bed that occurs between 10 and 50 cm below the base of the depressions in the Mendez shales was never breached but runs parallel to the base of the depressions. This shows that the depth of scouring was less than 1.2 m, and that the depressions were accentuated by loading. The depressions are channel-like, and the axis of the best exposed, middle channel (m.mk. 18–38) strikes more or less north-south, as determined with the orientation of the basal scours at m.mk. 24 and 36–38. This channel is laterally infilled from the west. The lateral accreting layers are inclined to as much as 15° with respect to the overlying Unit II layers and are truncated by the overlying Unit II sandstones at m.mk. 32–38, giving the (false) impression of an angular unconformity (Hay, 1960). The different layers of the sideward shifting channel fill are composed of a variable mixture of spherules, droplets, and other splash forms (some with a glass interior; Fig. 5E, F), small subrounded microsparitic limestone clasts (Fig. 5G, H), foraminifers, terrigenous grains, and flat Mendez rip-up clasts containing Maastrichtian foraminifers. Texture and sorting differ from layer to layer. Some layers have the character of a poorly sorted debris-flow, with clasts supported in a sand matrix. Others are slightly better sorted, displaying low-angle to very low angle large-scale cross-bedding, probably due to channel-migration. Some layers are a well-sorted packstone of foraminifers, spherules, and a few detrital quartz grains. Such layers are well cemented and weather out. In some layers rip-up clasts of the Mendez dominate. These rip-up clasts are strongly deformed and commonly fused together. The boundary between the Mendez clasts is often outlined by thin stringers of spherules. The fused clasts seemingly

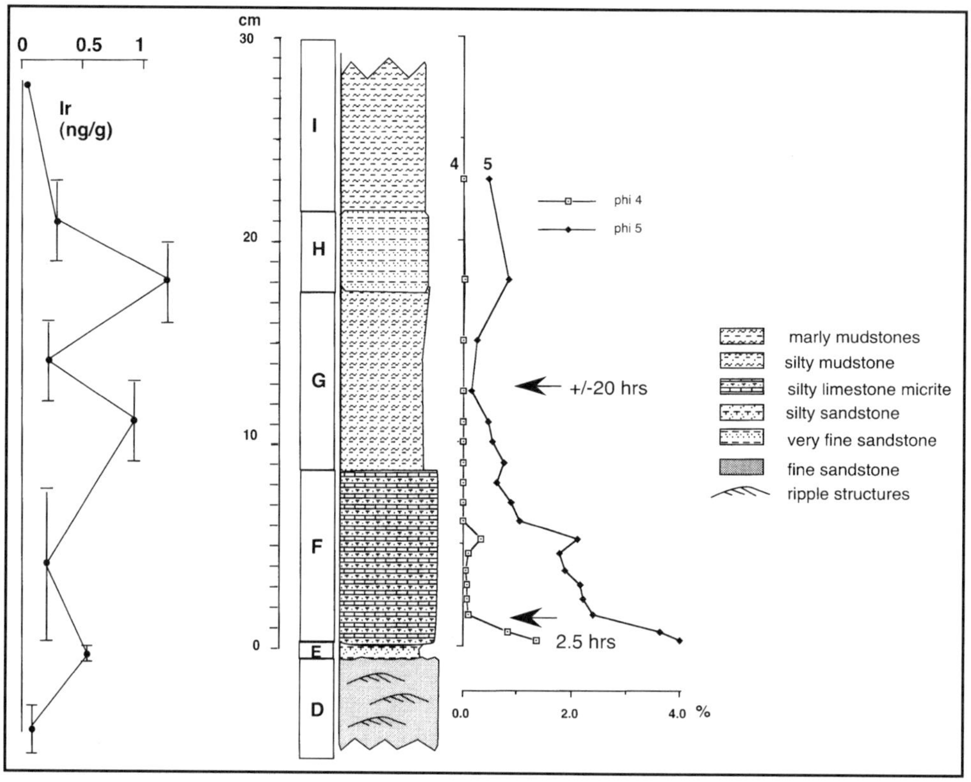

Figure 10. Profiles of percentile score of two grain-size (Phi) classes from the histograms of Figures 7 through 9, in true stratigraphic thickness. Ir concentrations in ng/g (nanogram/gram) are from Asaro et al. (1982). The duration in hours is taken from Table 1 and estimates the settling time of silt of Phi 4 and 5 from a 50-m-thick water column, after the deposition of the last sandstone (unit D). The clay fraction would settle after weeks (Table 1). The Ir carrier is apparently fine grained, because it settles down together with the finest clay fraction, several days to weeks after the impact.

TABLE 1. SETTLING TIME OF FINE SEDIMENT SUSPENDED IN THE WATER COLUMN COMPUTED FOR DIFFERENT WATER-COLUMN THICKNESS

Size Class (phi)	Micron	Settling Velocity (m/h)	Water-column Height in Meters (time - hours) 2	5	10	25	50	500
3	125.00	57.60	0.03	0.09	0.17	0.43	0.87	8.68
4	64.00	19.08	0.10	0.26	0.52	1.31	2.62	26.21
4.5	40.00	3.60	0.56	1.39	2.78	6.94	13.89	138.89
5	32.00	2.52	0.79	1.98	3.97	9.92	19.84	198.41
6	16.00	0.61	3.27	8.17	16.34	40.85	81.70	816.99
7	8.00	0.18	11.11	27.78	55.56	138.89	277.78	2,777.78

form "continuous" layers and can give the impression of normal hemipelagic sedimentation alternating with coarse-grained sand layers. This impression has led to estimates of the duration of sedimentation of the K/T sandstone complex of tens of thousands of years (Lyons and Officer, 1992).

The Unit I channels are progressively deformed toward the edges of the channels (m.mk. 34–38 and 44–48). The dip of the lateral accreting layers increases, and near m.mk. 38 and 45 those layers are even overturned. The underlying Mendez shale follows a similar deformation pattern. The disappearance of bedding planes in the Mendez shales just below the sandstone units could be due to soft-sediment deformation of the Mendez marls. The above-mentioned bentonite bed enclosed in the Mendez shale is folded at several places, showing that the shales were deformed. Near m.mk. 36–46 the Mendez shales are strongly deformed and contain frequent slickensides, without obvious preferred orientations. We infer that the Mendez shales at time of deposition of the K/T sandstone complex were still extremely soft and could be easily deformed. As a result of uneven loading of the rapidly accumulating incipient channel fills, the soft Mendez oozes were pushed away from the center of the channels into the large flamelike or diapirlike structures

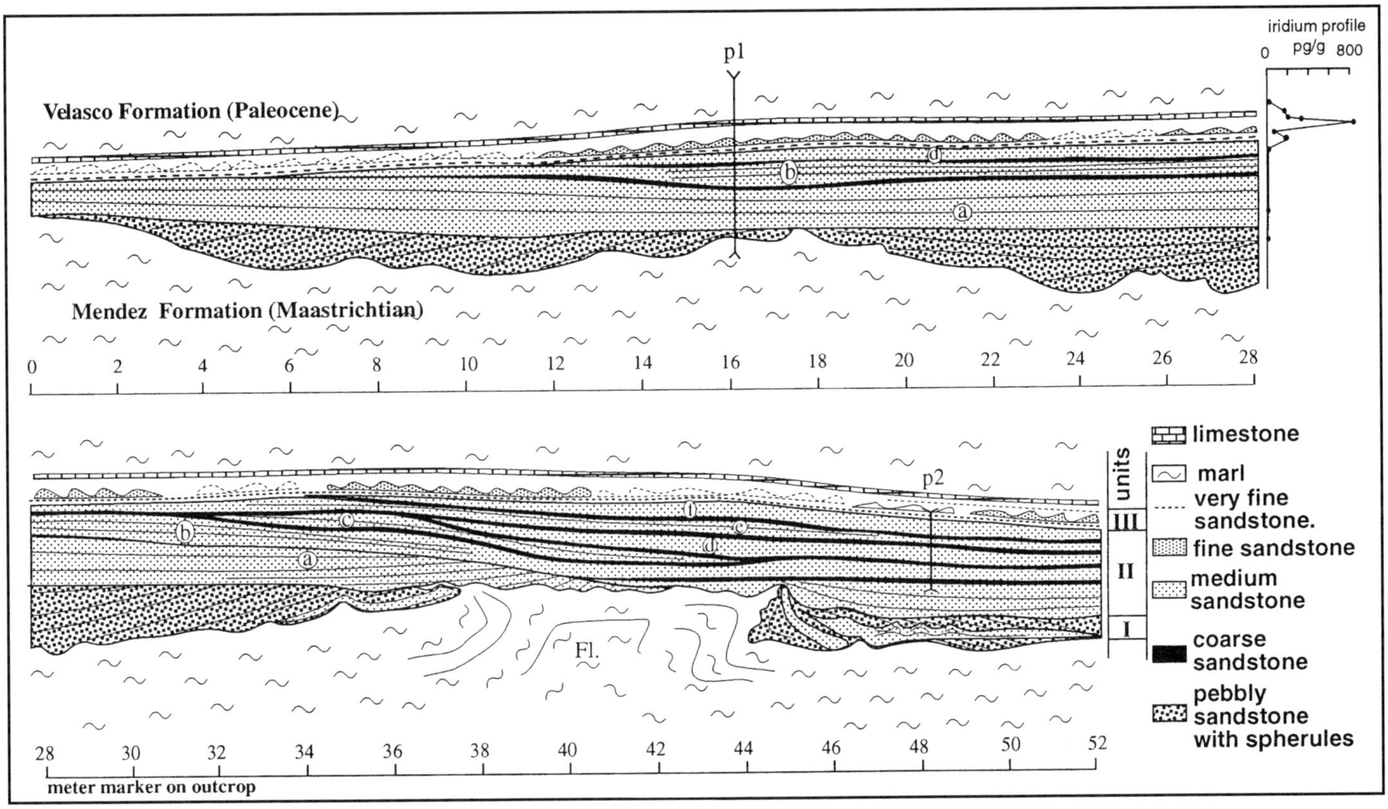

Figure 11. Arroyo de Mimbral, Mexico. Simplified cross section through the eastern outcrop of the K/T sandstone complex. Meter markings (see text) are painted on the outcrop. Ir abundances (pg/g = picogram/gram) (Table 3; Smit et al., 1992b) are measured at the section p1. The petrographic thin-section analysis (Table 2) is at p1 and p2. F1 = Mendez shale "diapir" or large, deformed flame structures between two channels. A–F, different lenticular parallel-laminated layers of Unit II, each with a scoured, coarse base. Note that near the shale "diapir" F1, the edges of Units I and II are upturned and that each successive channel or layer is less deformed and either "onlaps" on or truncates the steeper, previously deposited one.

observed between the channels (m.mk. 38–44). The sand layers near the depression rims (m.mk. 38 and 44) were dragged along and were oversteepened or even overturned by this diapiric movement. One can infer that the diapiric movements continued during the infilling of the channels, because successive layers onlap on more steeply inclined previously deposited layers (m.mk. 44–48). A spectacular example of such large flames or "diapirism" can also be observed at the Rancho Nuevo, Mexico, site (Alvarez et al., 1992b).

Current velocities involved in deposition of Unit I are apparently variable and mostly high, based on the coarse grain size. Unit I lacks adequate current-direction indications. The direction of lateral accretion shows infill from the west in one channel near m.mk. 28–38 and from the east in the next channel near m.mk. 44–48.

Unit II is made up of at least six fining-upward, stacked, thinning-upward lensing calcareous sandstone bodies (Fig. 11, a through f). Often the base of a higher sandstone body scours into the top of the lower ones and truncates the bedding. The latter is most obvious between m.mk. 36–42, because the angle of dip of the laminations of the lower (Fig. 11, a) sandstone body has been increased in a way similar to that of Unit I near the Mendez "diapir." Unit II differs petrographically and texturally from Unit I. The sandstone is pervasively parallel laminated and is well sorted, except in the axial, basal parts of the basal (Fig. 11, a) sandstone body. In Unit II detrital terrigenous grains are increasingly more common. In some laminae the sand-fraction is dominated by foraminifers identical to those in the Mendez shales; other laminae are rich in detrital quartz grains (Fig. 12A, B). Minor amounts of biotite crystals and charcoal fragments (Kruge et al., 1994) occur as well. Spherule remains similar to those found in Unit I also occur in the coarse basal part of several Unit II (Fig. 11b, c) sand bodies. Other components of Unit II, in particular in the basal part of sandstone body (Fig. 11a) near m.mk. 22–26, include rare siliceous sponges, plant debris, and rounded clasts of Mendez shale, some of which are 80 cm in size (m.mk. 29).

The axial part of the basal Unit II channels contains multiple layers rich in plant debris and pieces of wood. Some of the wood fragments have borings and were probably water satu-

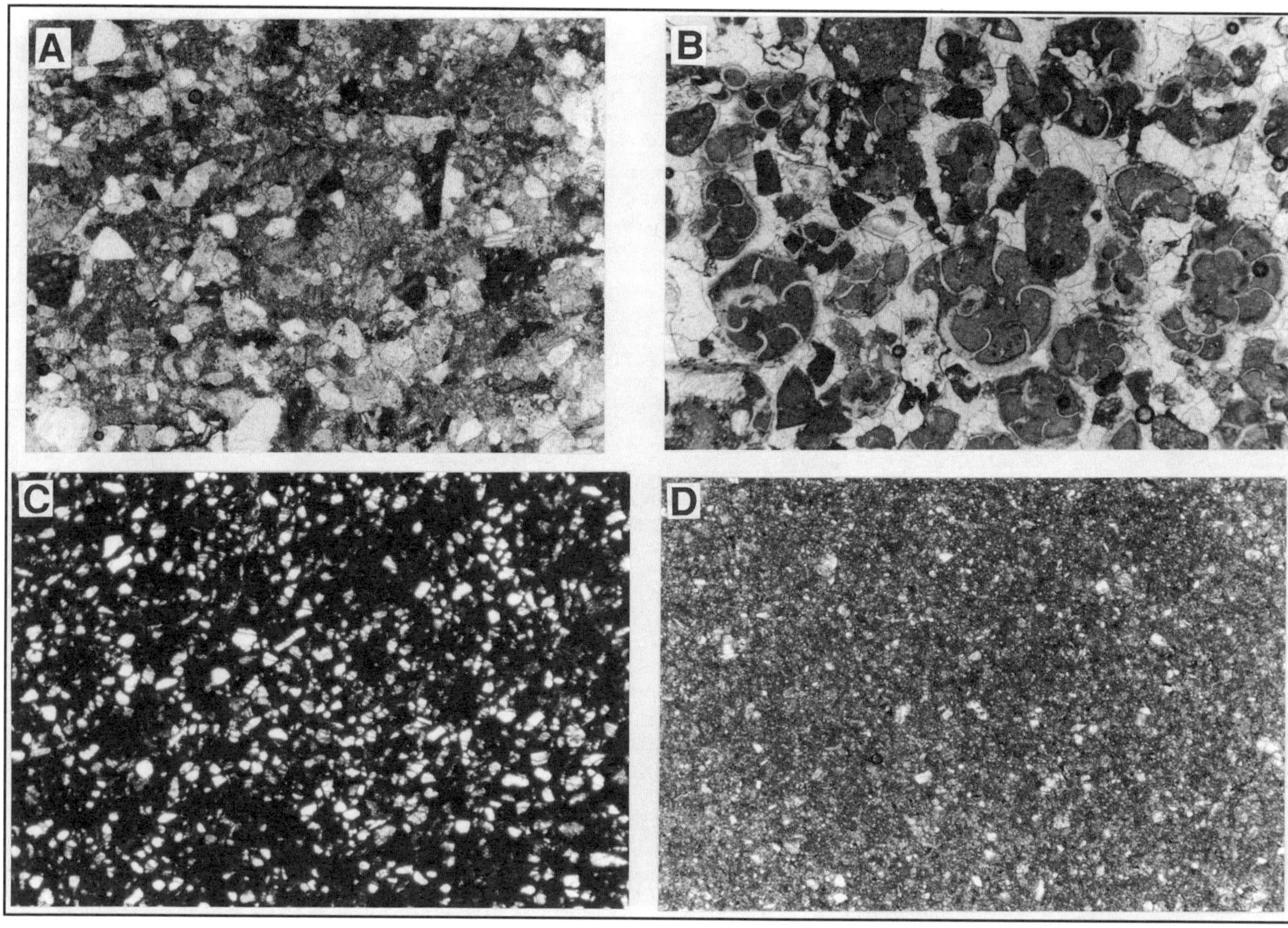

Figure 12 (on this and facing page). Thin-section micrographs of the Mimbral, Mexico, section. A, Unit II (Sample M5) terrigenous sandstone with "dirty" matrix. B, Foraminiferal grainstone with sparry calcite matrix, Unit II (sample M18). C, Fine sandstone flaser from top of Unit III (see Fig. 13). The matrix is entirely composed of iron oxides, falsely resembling the impact fallout layer, or "fireball" layer (Stinnesbeck et al., 1993) of distal K/T sections. D, Micrite (cf. Fig. 13, level 300 cm). E, Basalmost Velasco shale, poor in Cretaceous planktic foraminifers. F, Mendez shale, 5 to 10 cm below the K/T sandstone Unit I, rich in Cretaceous planktic foraminifers.

rated before they were transported. They are not the remains of fresh plants. Small charcoal fragments occur in increasing numbers in the higher sandstone bodies (Kruge et al., 1994).

Unit II overlies Unit I and the Mendez diapirs with a sharp contact at most places and truncates the layers of Unit I. However, in the middle of the Unit I channels, the transition from Unit I to II is rapid but gradual and nonerosive (between m.mk. 22–30) over a thickness of about 15 cm. The uppermost layers in Unit I show changes leading up to Unit II. In the transition zone detrital quartz content increases, and the Mendez shale rip-up clasts are progressively more rounded and often armored with a peppering of impact spherules. The rip-up clasts preserve internal layering and apparently were more resistant to deformation than the Mendez clasts within Unit I. We assume that these "armored mudballs" are eroded from deeper, more-compacted Mendez beds. Sedimentary structures in Unit II are dominated by semiparallel laminae associated with primary current lineations on the bedding-plane surfaces, having different orientations at different levels (N75°E–N180°E). A single 2-cm-thick layer with low-angle cross-bedding occurs at m.mk. 12 (current direction N120°E ± 15°). Flute-casts were observed near m.mk. 38–42 where Unit II directly overlies the Mendez marls. These flute-casts show a N350°E ± 10° current direction.

Unit III conformably overlies Unit II, where the first thin (l cm thick) silt layer indicates the beginning of a series of alternating silt and thin current-rippled fine sandstone layers (Fig. 13). The first silt layer is also the first layer enriched in Ir. The rippled layers consist of asymmetric, unidirectional current ripples. Symmetric ripples or ripples with bidirectional internal

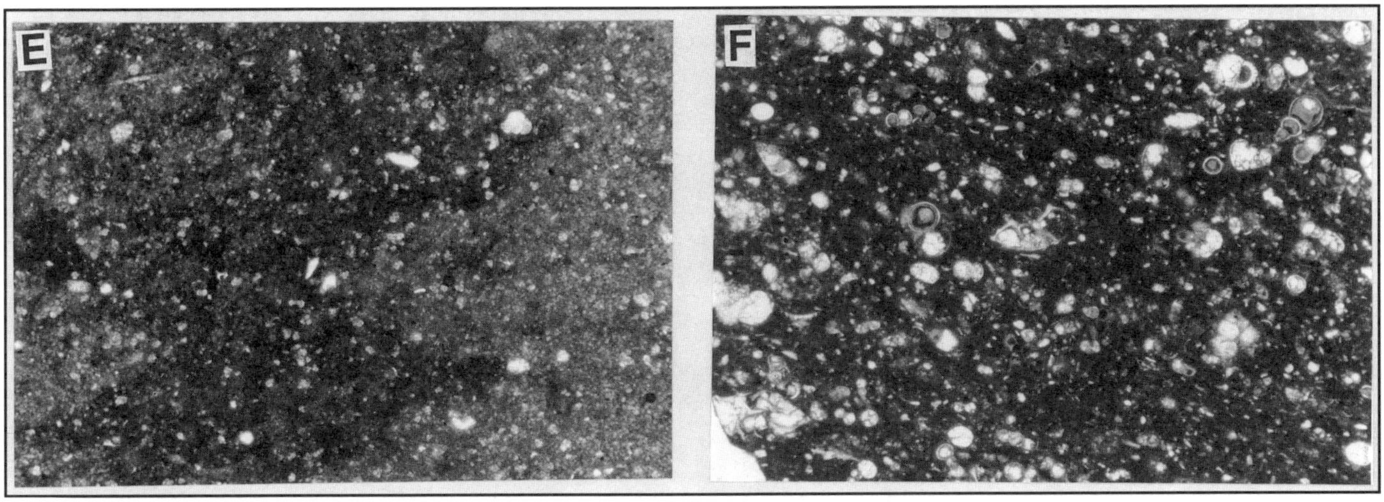

structure indicative of oscillation (wave) ripples were not observed. Some ripples bear a resemblance to hummocky cross-stratification structures on the rock-face cut at right angles to the current direction (Lyons and Officer, 1992), but at a rock-face cut parallel to the current direction they show their unidirectional cross-bedded character. The top surface of the highest rippled layer is bioturbated. Some *chondrites* burrows also occur at the top of the next-to-last ripple and in the silt layer in between. We have not observed any traces of burrows in any of the lower ripples or ripple surfaces of Unit III or in Units I and II.

The highest ripples are overlain by a 4-cm-thick silt layer, containing a 2- to 5-mm-thick flaserlike fine sandstone layer that is most enriched in Ir (Fig. 13; Table 2). This layer can be traced over the entire outcrop, and at the westernmost outcrops of the K/T sandstone complex (outcrop Mimbral 2 of Stinnesbeck et al. [1994]) the thin flaserlike sandstone layer is red stained by iron-oxides. Stinnesbeck et al. (1994a) and Hildebrand et al. (1991) termed this layer as "red K/T layer" and "fireball layer," respectively, but in thin section (Fig. 12C) it is clearly a sandstone with goethite matrix, not compatible with an interpretation as a K/T red ejecta or fireball layer. The 4-cm silt layer is overlain by a 7-cm-thick micritic limestone layer, still silty and graded (Fig. 12D; Fig. 13). This layer appears correlative with the graded silty limestone layer (unit F of Hansen et al., 1987) at Brazos River. The first small Danian foraminifers (*G. eugubina*?) appear about 12 cm above the silty limestone layer in shales of the Velasco Formation that are somewhat darker than the Mendez shales. The first fine-grained sediments above the rippled layers and all of the subsequent Velasco shales are highly impoverished in Cretaceous planktic foraminifers (Fig. 12E) in comparison with the uppermost Mendez shales underlying the clastic beds (Fig. 12F). Those few present are probably reworked.

Petrography of the Mimbral sandstone sequence

Twenty-nine thin sections were analyzed from samples taken every 10 cm in Units II and III in a vertical section at

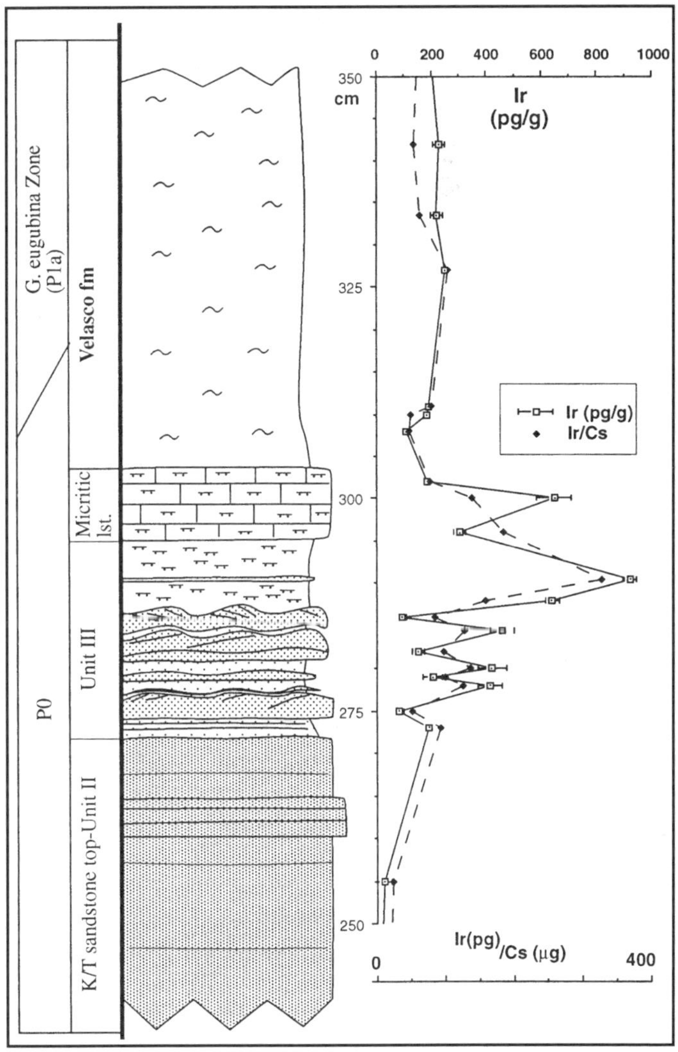

Figure 13. Mimbral, northeastern Mexico. Detail of the top of the K/T sandstone complex near meter mark 16 (cf. Fig. 11). Lithological column traced from a photograph. Ir and Cs data from Table 3 (F. Asaro, personal communication, 1991). Ir is normalized to Cs to eliminate dilution effects by carbonate.

TABLE 2. GRAIN-SIZE TRENDS IN THIN SECTIONS FROM UNIT II OF THE MIMBRAL K/T SANDSTONE COMPLEX

	General						Limeclasts				Terrig. Detritus	
	Muddy Matrix				Sparry Calcite Cement							
Samples	Crs sand/fine gravel	Crs-med sand	Med-fine sand	Fine sand	Terr. sand	Foram grainst.	Crs	Med	Med-fine	No lst grains	Med	Med-fine v fine
M29						X		X			X	
M28						X		X			X	
M27					X				X		X	
M26					X				X		X	
M25						X		X			X	
M24					X				X		X	
M23						X		X			X	
M22						X		X			X	
M21					X				X		X	
M20						X		X			X	
M19					X				X		X	
M18						X		X			X	
M17						X			X		X	
M16	X						X					
M15			X					X				X
M14				X						X		X
M13				X					X			X
M12				X						X		X
M11				X					X			X
M10			X						X			X
M9			X						X			X
M8			X						X			X
M7			X					X				X
M6			X					X			X	
M5		X						X				X
M4			X					X				X
M3			X					X				X
M2		X					X					X
M1	X						X				X	

TABLE 3. Ir, Fe, AND Cs ABUNDANCES IN THE MIMBRAL SECTION

Position (m)	Ir (ppt)	Fe (%)	Cst (ppm)
–0.14	13 ± 4	2.90	3.55
+0.13	18 ± 5	4.96	0.30
+0.15	4 ± 9	12.95	0.73
+0.78	16 ± 4	1.09	0.98
+1.48	7 ± 4	1.41	0.36
+1.83	8 ± 9/7	1.19	0.54
+1.98	25 ± 32/20	1.01	1.06
+2.01	192 ± 12	1.72	2.04
+2.03	83 ± 15	1.22	1.58
+2.06	414 ± 44	2.34	3.26
+2.07	206 ± 36	1.72	2.04
+2.08	423 ± 50	2.28	3.05
+2.1	153 ± 24	1.29	1.57
+2.125	457 ± 44	2.14	3.52
+2.14	92 ± 18	0.79	1.07
+2.16	642 ± 24	2.16	4.05
+2.185	921 ± 23	1.61	2.80
+2.24	306 ± 21	1.34	1.64
+2.28	650 ± 63	2.22	4.67
+2.3	187 ± 9	1.86	2.40
+2.36	109 ± 22	1.34	2.26
+2.38	184 ± 7	0.67	3.68
+2.39	191 ± 8	1.17	2.36
+2.55	250 ± 12	1.58	2.40
+2.615	218 ± 20	2.62	3.46
+2.7	228 ± 23	2.80	4.17
+2.84	22 ± 4	0.82	1.16
+3.39	79 ± 8	2.82	4.17
+4.49	41 ± 6	2.57	3.98

m.mk. 16, supplemented at m.mk. 48 for the part of Unit II missing there. The most common composition is a foraminiferal grainstone with abundant lime-clasts and with minor amounts of terrigenous grains, alternating with layers rich in terrigenous detritus. A basic twofold subdivision could be made of Unit II sandstone (Table 3). The basal part has essentially a "dirty muddy matrix" (Fig. 12A); the upper part is a clean, better-sorted sandstone with a sparry calcite matrix (Fig. 12B). Petrographically, three different groups of sediment can be distinguished. The first group—the lower part of Unit II (M2 through 15)—has a "dirty" aspect due to smeared, lime-mud matrix. The sand is a polymict mixture of medium-fine to fine-grained terrigenous sand, with foraminifers. Lime-mud grains occur, roughly fining up from coarse grained at the bottom to medium and fine grained at the top. The second group—Samples M16 through 29 (=upper part of Unit II, starting with a small erosive channel with spherules)—have a cleaner and better-sorted aspect, with a sparry calcitic cement. Medium-grained calcareous sandstones with medium-grained terrigenous grains alternate with foraminiferal grainstones. The foraminifers have an infilling of greenish clay. The third group—samples M30 through 42 (Unit III)—shows a crude grading from laminated and cross-bedded silty and very fine sandstones to siltstones. The topmost samples contain burrows.

La Lajilla, Mexico

At La Lajilla (Alvarez et al., 1992b) the K/T sandstone complex crops out at both sides of the overflow channel of the Lajilla reservoir dam. The upper Mendez shales are rich in Upper Maastrichtian foraminifers (Fig. 14A). The well-developed planktic fauna and high planktic/benthic foraminifer ratio suggest deposition at upper bathyal (±500 m) water depth. Bedding is not visible in the top 0.5 m of the Mendez Formation, possibly as a consequence of soft sediment deformation, similar to that in the Mimbral outcrop. The Lajilla K/T sandstone complex can also be subdivided into three units (Fig. 15). Unit I fills shallow (<50 cm) scours in the top Mendez. Unit I sands with K/T spherules are locally "injected" into the top Mendez beds. Unit I displays a crude low-angle cross-bedding with alternating layers of pebbly sandstone and sandstone, dominantly composed of a mixture of spherules (Fig. 14B) and Cretaceous foraminifers. Internal truncations occur, but measurement of current directions was not possible. Unit II sharply overlies Unit I and has locally truncated Unit I. The beds of Unit II are well lithified and are more continuous than at Mimbral; individual layers (e.g., the parallel laminated bed at 60 cm) could be traced over 150 m, and a strongly eroded outcrop at the other side of the dam (300 m southwest) shows the same lithostratigraphic subdivisions. Unlike Mimbral, Unit II displays a wide variety of (current) ripple structures that allow a large number of paleocurrent measurements (Figs. 15 through 18). Where possible, the direction of migration of the ripples was measured on bedding planes, and otherwise a three-dimensional view of the ripples was obtained, using vertical sections at a perpendicular angle to each other (Fig. 16). The ripple structures observed are all compatible with origin as current-ripples. Ripples indicative of oscillation (waves) were not found. The ripple types observed vary from straight crested to linguoid, possibly lunate, ripples-in-drift to climbing ripples, with transitions in between. The basal 10 cm of Unit II displays ripple structures—associated with parallel-laminated beds—somewhat reminiscent of standing-wave– (antidune?-) type ripples (Fig. 17). A prominent, well-lithified bed, similar to the Mimbral Unit II beds, occurs at 55 to 62 cm. This bed is parallel laminated, and in thin section it shows an alternation of foraminifer-rich and darker lithic grain-rich laminae. Unit III begins at 78 cm, where the first thin silt layer appears. Unit III further consists of fine-sand layers with ripple structures alternating with (draping) thin silt layers. The top of Unit III (105 to 107 cm) displays a very fine scale flaserlike alternation of silts and millimeter-sized ripples (Fig. 19). Iridium concentrations are high in this interval (0.8 ng/g, R. Rocchia, personal communication, 1994). The top layers of Unit III are burrowed (Fig. 15). Unit III is capped by a silt layer 4 cm thick, overlain by a lithified silty micritic limestone layer (Fig. 14C), again similar to the one at Brazos River and Mimbral. This silty limestone layer is followed by a rhythmically bedded series of gray hemipelagic shales of the Velasco Formation, containing basal Danian foraminifers. The silty limestone and the first few centimeters of the Velasco Formation contain high iridium concentrations (R. Rocchia, personal communication, 1994). Although the silty limestone layer is greenish gray as is the Mendez Formation, the limestone layer is greatly depleted in Cretaceous foraminifers compared with the Mendez (compare Figs. 14A, B). Danian foraminifers have not been observed in the silty limestone, but they appear in the Velasco shales just above it. Therefore, we infer that the silty limestone layer is not Maastrichtian in age, as believed by some (Keller et al., 1994), but basal Danian (P0 Zone) in age.

The paleocurrent directions measured are shown in Figures 15 and 18. The current directions are bidirectional when integrated over the entire thickness of the K/T sandstone unit. The dominant directions are away and toward the Chicxulub crater, but that may be a coincidence. Within individual layers the current directions appear stable and unidirectional (Fig. 15), as they do when measured at different places in the same layer. Those opposite current directions demonstrate that the individual sandstone layers are deposited by currents repeatedly changing direction in almost 180°. The bidirectional currents are somewhat reminiscent of tidal-cycle currents, but such currents are difficult to explain in a semienclosed-basin environment and at a water depth of 500 m.

The sedimentology of Unit II at first glance makes one think of turbidity currents with repetitive Bouma sequence (A)B-C alternations. The opposing current directions then could be explained by reflected turbidites (Best and Bridge, 1992). The general setting of these opposing cross-beds of Unit II in the large-scale energy decreasing-up sequence I through III and in the smaller-scale energy decreasing-up sequence of Unit II is thought to be too small and gradual compared to energy differences between a supposed initiating turbidity current and a reflected one. Also Unit III is quite different from a turbidite setting, regarding the preservation of the ripple forms sharply overlain by a mud drape, contrary to the more gradual transition from Bouma C to D in a true turbidite.

A better explanation seems to be an origin by landward-breaking wave surges alternating with the seaward return flow of a series of large tsunami waves. Gravity flows may have helped to transport clastic material from the coast, ~70 km to the west, into the deeper basin but are not directly involved in the deposition of the K/T sandstone unit. The general thinning- and fining-upward trend shows the decreasing energy of tsunami waves, and in the final stages current velocity sufficiently decreases in between individual wave arrivals to allow the settling of the fine silt and clay of Unit III.

El Peñon, Mexico

El Peñon is located on a hilltop where the "flagstones" covering the dam of the Porvenir reservoir were quarried, mainly from Units II and III of the clastic complex. These units are therefore exposed over several acres, allowing investigation of the sedimentary structures preserved on the surface of the sandstone layers (Figs. 20, 21).

As at Mimbral and Lajilla, current measurements demonstrate periodic changes in current direction, consistent with the previously mentioned hypothesis of deposition by breaking tsunami waves.

El Peñon may be important for the correct interpretation of the many burrow structures that occur on and in the top of the K/T sandstone complex in almost every outcrop of the Gulf Coast. Most burrows (*Zoophycos, Chondrites*) occur in the topmost sand layers only, and all those seem to penetrate from above *after* deposition of the K/T sandstone unit. The burrow fabric show a distinct tiering; different types of burrows occur on different bedding planes. This tiering leads to alternation of more intensively burrowed beds and hardly burrowed beds, a

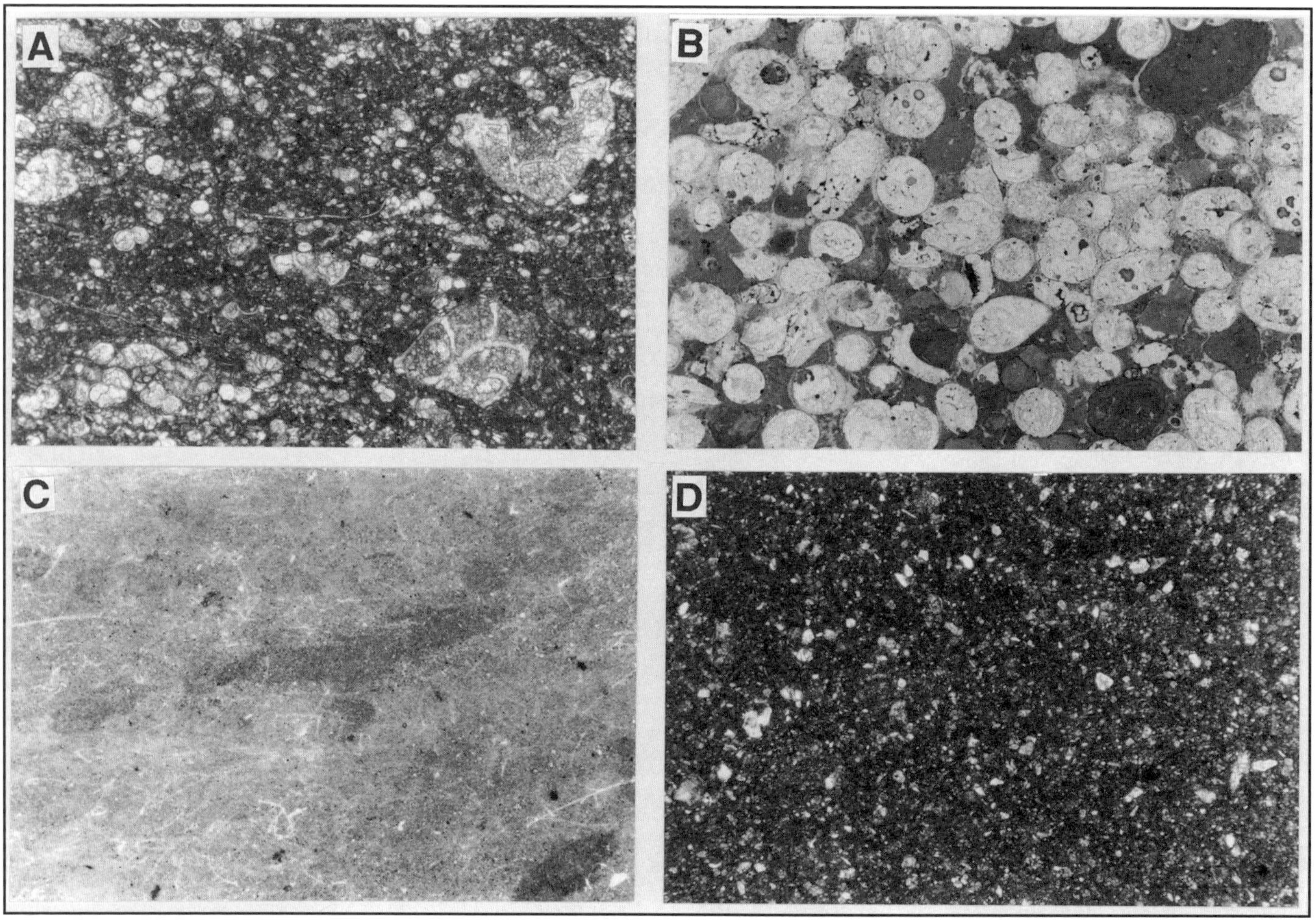

Figure 14 (on this and facing page). Thin-section micrographs. A, La Lajilla, upper Maastrichtian Mendez shale. B, La Lajilla, bubbly calcite spherules and limestone fragments (Unit I). C, Basalmost Velasco shale. D, Silty limestone (cf. Fig. 15, 112 cm) poor in foraminifers. E, Coxquihui, Mexico. Calcitic, bubbly spherule. F, La Ceiba, Mexico. Bubbly spherule composed of clay minerals with tar filling. Matrix is rich in coarse quartz grains and orbitoid fragments. G, H, Beloc, Haiti. G, Lowermost Paleocene chalk, poor in foraminifers. H, Uppermost Cretaceous chalk, rich in planktic foraminifers.

situation that might easily be mistaken for repeated sediment-depositional events followed by longer periods of burrowing. An *Ophiomorpha*-type burrow penetrates up to 50 cm below the top of the clastic complex (Fig. 21). These burrows consist of 1-cm-diameter, up to 3 m long, frequently branching, usually straight horizontal tubes, which follow the thin silt layers in between Unit III rippled sandstone layers. The horizontal tubes radiate from a central bundle of vertical burrows, observed at El Peñon to penetrate through several sandstone layers. The burrowing shrimplike animal is apparently capable of penetrating 50 cm down through sand layers before spreading out. Thus as far as we can establish, these burrows also *postdate* the deposition of the K/T sandstone unit. We assume that similar *Ophiomorpha*-type burrows at the same position in the Brazos River K/T clastic layer originate the same way, although there we did not observe the vertical central tubes.

The basal 2.75 m of Unit II at El Peñon are characterized by massive bedding, without any clear sedimentary structures except some water-escape structures and faint parallel laminations. A similar lack of sedimenary structures is also common in Unit II of nearby K/T outcrops Las Bruselas, Porvenir, Rancho Nuevo, La Sierrita, Loma las Rusias, and Los Ramones (Alvarez et al., 1992b). Apparently this massive bedding is shared by K/T outcrops in the northeastern Mexico area but is absent farther south. The water-escape structures indicate rapid deposition of massive sand or leading to massive sand by obliterating the earlier sedimentary structures.

Mesa de Llera, Mexico

A few hundred meters east of the microwave tower on the Mesa de Llera, next to the road between Ciudad Victoria and

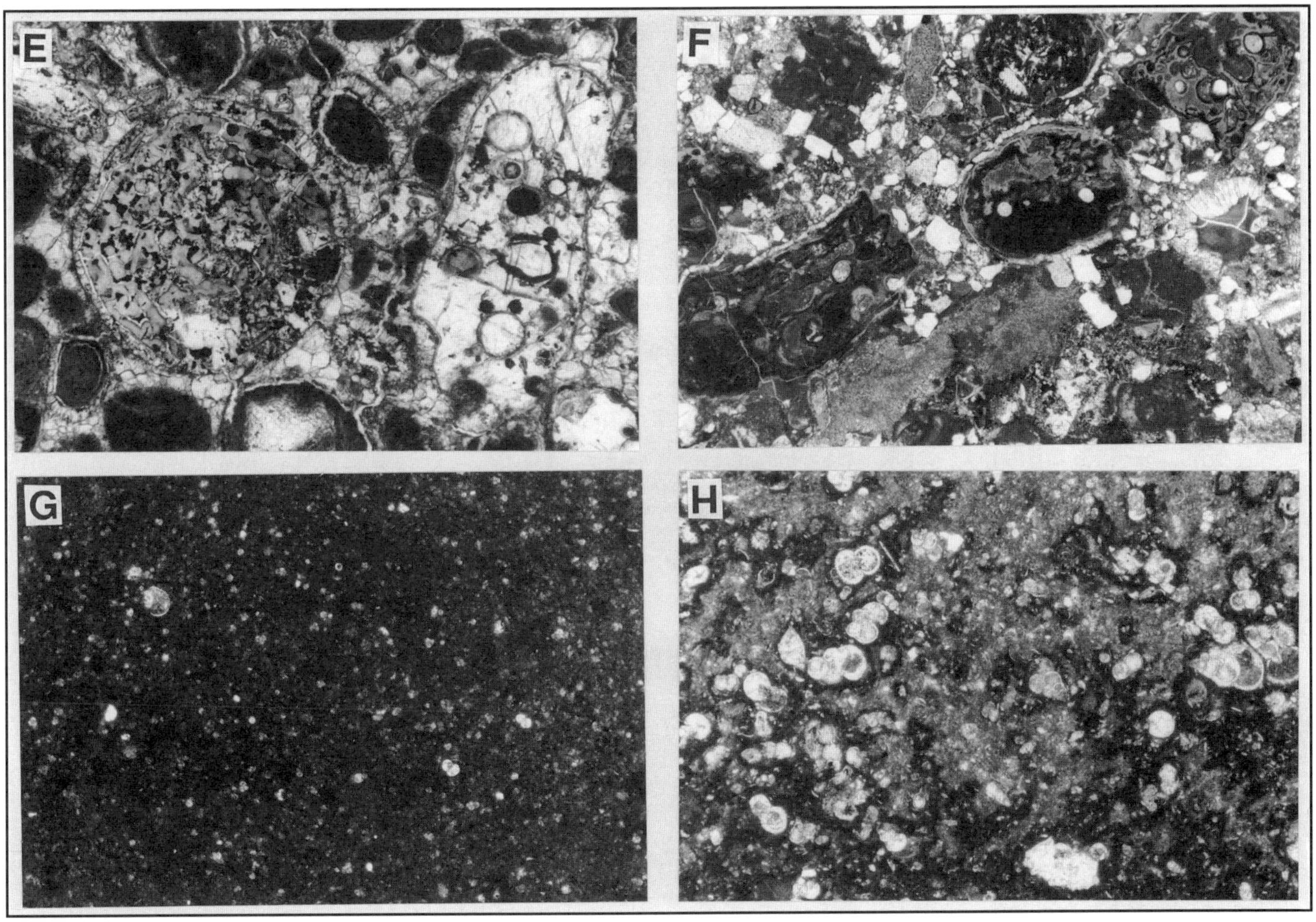

Ciudad Valles, the K/T sandstone complex crops out in several small clearings in the mesquite chapparal bushes, below cappings of Eocene basalts (23° 21.8′N, 98°59.9′W). The K/T sandstone complex is only 3 cm thick and consists of bioturbated ripples of very fine sand of Unit III. Unit I and II are absent and also could not be traced in nearby outcrops. The K/T sandstone complex overlies indurated Mendez marls of *A. mayaroensis* Zone age and is directly overlain by a 5-cm-thick silty limestone layer. This limestone layer is overlain by *G. eugubina* Zone age marls of the Velasco Formation. Although the Velasco and Mendez marls crop out at many places along the Mesa de Llera, evidence for thick K/T sandstone layers has not yet been found.

La Ceiba, Mexico

About 7 km south of La Ceiba (20°19.8′N, 97°41.0′W), along the road from La Ceiba to Tlaxcalaltongo, the K/T sandstone complex crops out over several tens of meters and is also exposed 200 m farther to the north in a dirt road. The sandstone complex changes in thickness from 1.2 m in the roadside outcrop to 0.25 m along the dirt road (Fig. 22). The sequence consists of a channelized lower Unit I downcutting in foraminifera-rich Mendez marls of *A. mayaroensis* Zone age. Unit I contains bubbly spherules altered to green clay minerals (smectite?), frequently filled with tar, and abundant (>50%) large detrital quartz grains and mica-crystals (Fig. 14E). Unit I also contains bioclasts of shallow-water origin (Orbitoid foraminifers, bryozoans). The thickness of Unit I varies from 1 to 25 cm. Unit II, as in other sections in northeastern Mexico, is a graded sandstone displaying climbing ripples indicating bidirectional currents. Unit III consists of fine sand ripples alternating with silt layers, overlain by Velasco shales containing a P0 Zone and *G. eugubina* Zone fauna. Only the top sandstone ripples contain burrow traces.

Coxquihui, Mexico

Just 30 km southeast of La Ceiba, a badly weathered K/T sandstone complex crops out in the roadside, just east of the village of Coxquihui. The K/T sandstone complex is 88 cm thick and overlies 90 cm of exposed red marls of *A. mayaroensis* Zone age. The K/T sandstone complex is poorly graded and faintly layered. Two graded layers, 50 and 38 cm thick respectively, are separated by a layer in which a few cobbles of whitish Upper Cretaceous micritic limestone are intercalated in a spherule-rich

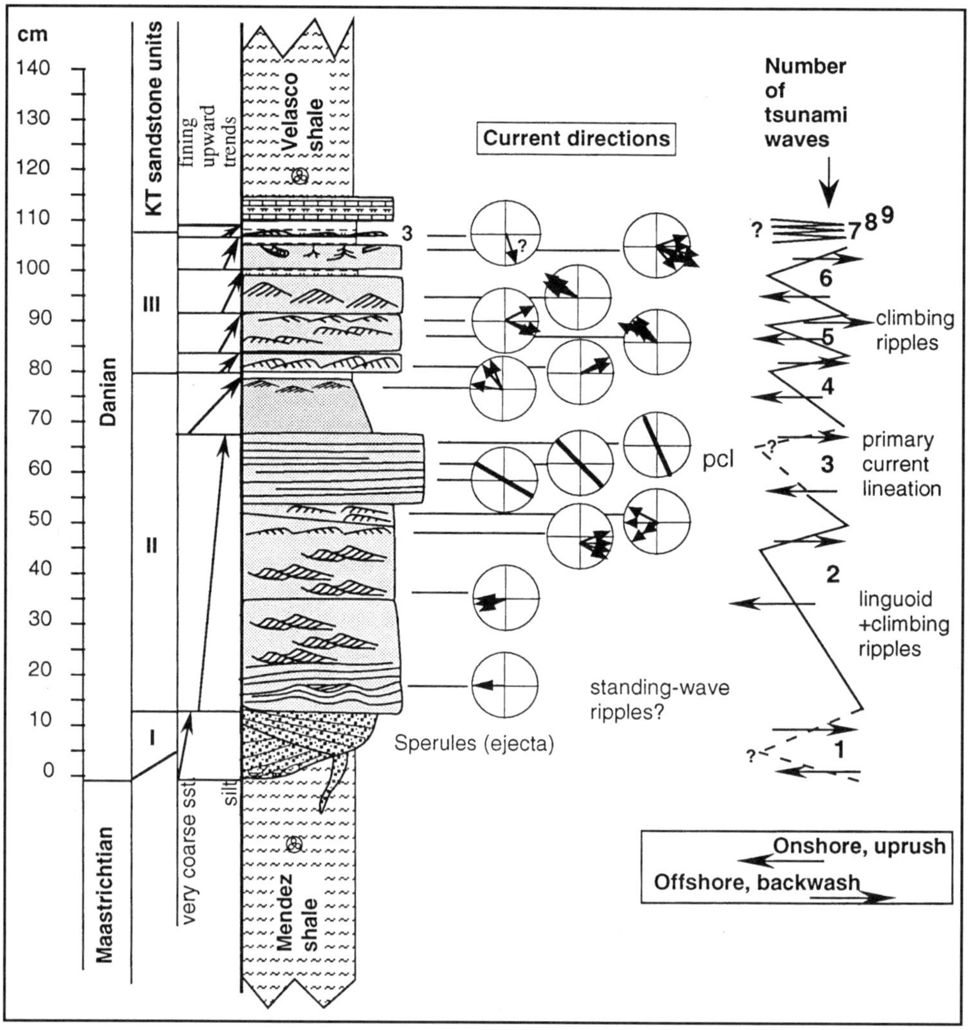

Figure 15. La Lajilla, Mexico. Stratigraphic column through the K/T sandstone complex at La Lajilla, Mexico. 40 km east of Ciudad Victoria. Measured paleocurrent directions are plotted next to the stratigraphic level. At right the total number of current reversals is indicated, inferred to indicate the number of passages of tsunami waves. Pcl = primary current lineation.

matrix. In contrast to the other K/T clastic beds, no Units II and III could be distinguished; almost the entire bed consists of bubbly spherules. The spherules are composed of sparry calcite (Fig. 14F); unfortunately, we have found no relict glass in a sample of 800 g dissolved in HCl. Red pelagic sediments directly overlying the K/T sandstone complex are of *G. eugubina* Zone age. In contrast to the other K/T outcrops in eastern Mexico, the K/T sandstone complex is intercalated in red marls that may indicate deposition in deep (>600 m) water, below the oxygen minimum zone. Although we have not found preserved glass, the K/T sandstone complex is reminiscent of the Beloc outcrop on Haiti, in that it is essentially composed of Unit I.

OTHER OUTCROPS IN THE GULF OF MEXICO

Other outcrops of the K/T sandstone complex are known from the literature, and we have also studied them. We report here some of our observations, in particular those that have a bearing on interpretations in the literature.

Beloc, Haiti

The K/T boundary sandstone complex crops out along the road from Carrefour Dufort to Jacmel, within rhythmically bedded pelagic chalks of the Beloc Formation (Maurasse et al., 1979). The K/T sandstone complex is exposed at several places along the road, because the road descends almost at the same angle as the local dip of the layers. The Beloc Formation is gently folded but not severely deformed, although many minor faults occur. However, package-sliding and slumping also occur at many horizons, and the intensity of synsedimentary deformation differs from place to place. None of the K/T sandstone outcrops has escaped synsedimentary deformation, and each

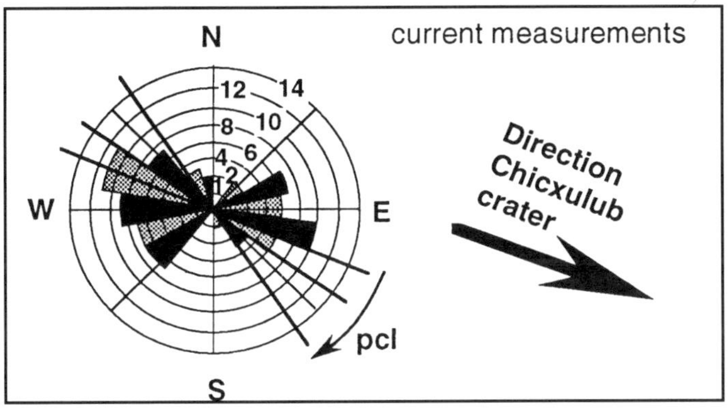

Figure 16. La Lajilla, Mexico. Rose-diagram of all paleocurrent direction measurements of Units II and III. The dominant measured directions are toward and away from the Chicxulub crater.

Figure 17. Photograph of Unit II standing wave(?) ripples 5 to 10 cm above the top of Unit I (visible at the bottom, resting on Mendez shales).

outcrop therefore has a different lithological sequence and expression. The K/T sandstone complex at Beloc at most places essentially has only Unit I preserved, a layer characterized by bubbly spherules and other splash forms and limeclasts similar to the clasts observed in Unit I in Gulf Coast outcrops. The spherules have a glass interior preserved only in the two southernmost outcrops, which appear least weathered (outcrops B and M; Jéhanno et al., 1992). The northern two outcrops (outcrops A and H; Jéhanno et al., 1992) have the thickest Unit I spherule layers (15 to 30 cm) deposited in shallow channels showing crude mega–cross-bedding. The spherule layer is size graded. The size grading continues into a fine sand-silty calcarenitic layer of about 20 cm thickness, which shows low-angle cross-bedding. Near the top of the limestone layer several 1- to 2-mm-thick lenses of fine sandstone occur that are locally stained by iron oxides and are rich in iridium and Ni-rich spinels (Jéhanno et al., 1992). In view of the continuous grading

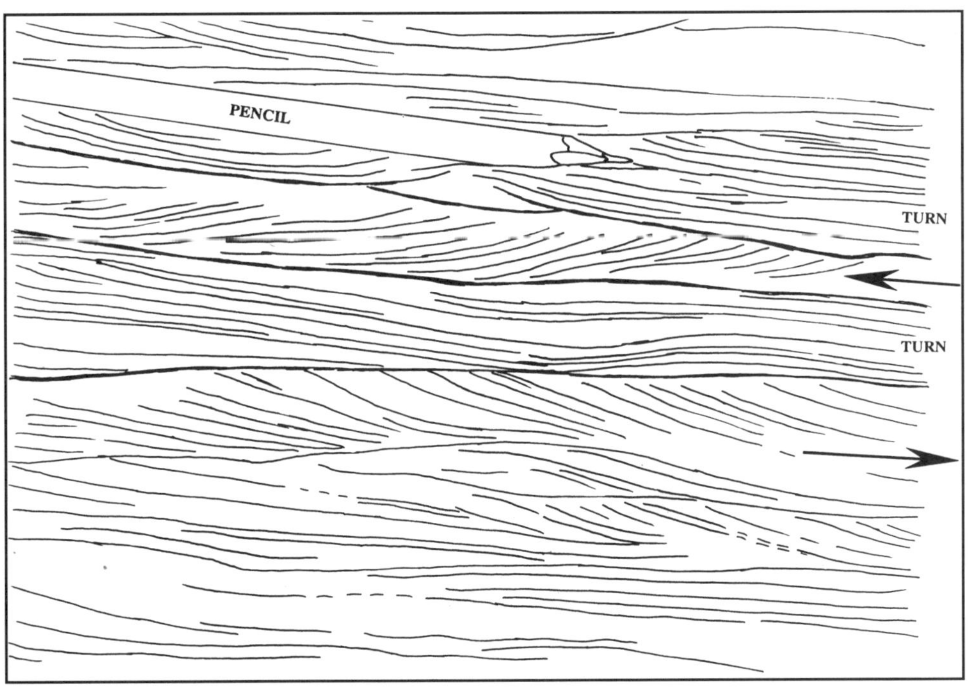

Figure 18. La Lajilla, Mexico. Drawing after a photograph showing two changes in migration of climbing ripples in Unit II, between 35 and 49 cm (cf. Fig. 15). Pencil for scale. Near the bottom current direction is approximately N270°E, just below the pencil approximately N90°E, and at the top again approximately N270°E.

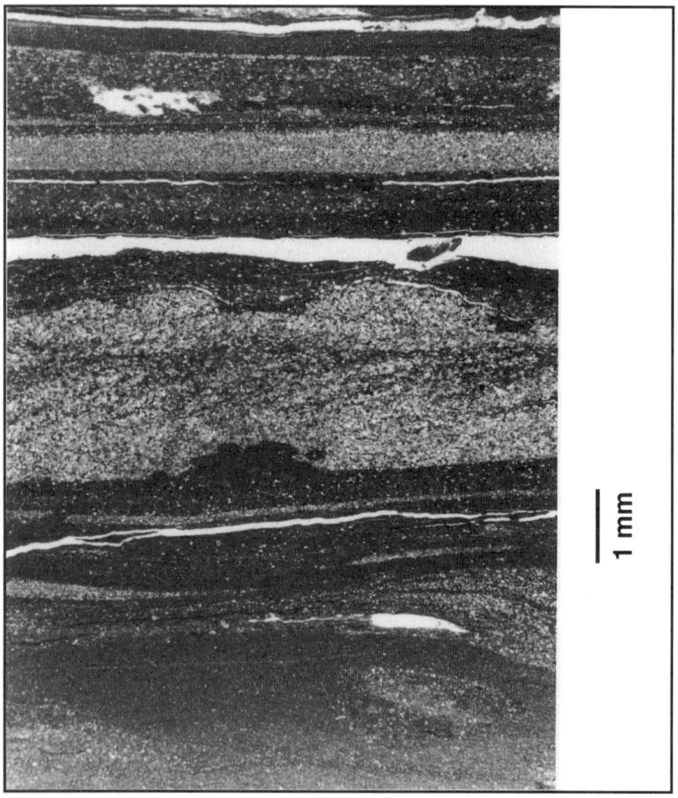

Figure 19. La Lajilla, Mexico. Thin section of topmost part of unit (III) at cm 108 (Fig. 15). At least three 0.2- to 2-mm thick (cross-bedded) fine sandstone flasers alternate with fine mud and silt, showing evidence for changes in current strength.

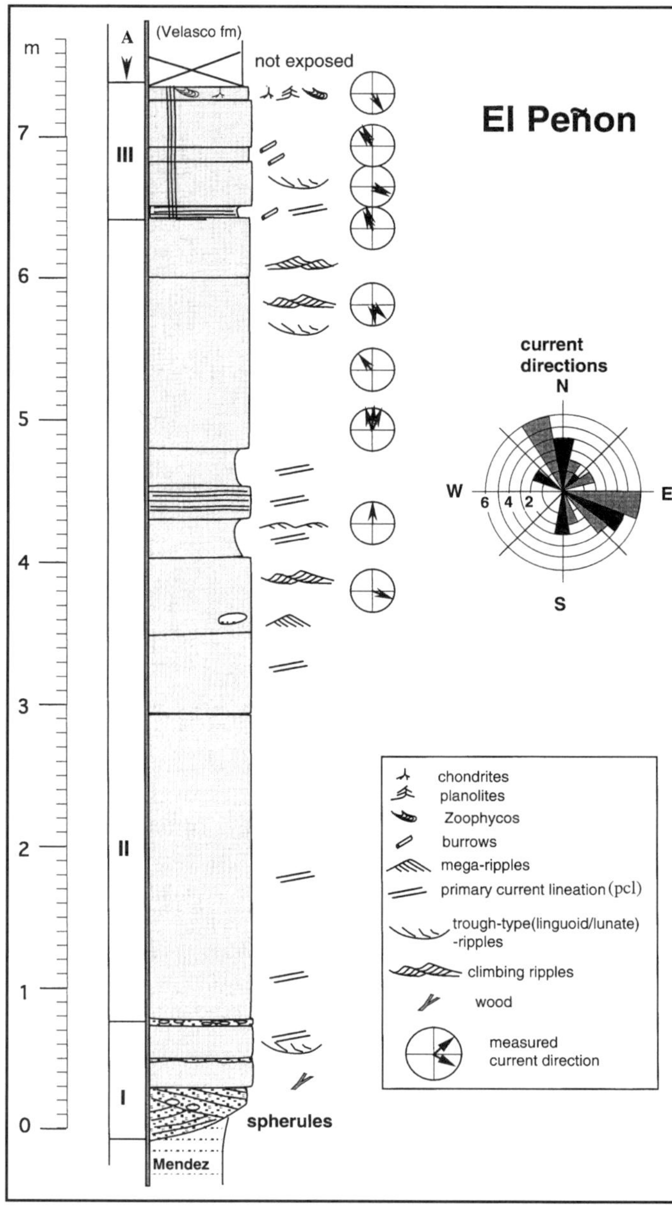

Figure 20. El Peñon, northeastern Mexico. Stratigraphic column through the K/T clastic complex. Paleocurrent directions of individual levels are plotted on the right-hand side and combined in the rose-diagram. The top and contact with overlying Velasco shale are not exposed here.

and pervasive cross-bedding we believe that all these units were rapidly deposited in one depositional sequence, not interrupted by periods of slow background sedimentation.

Jéhanno et al. (1992) also have described the lithology of these sections. Sigurdsson et al. (1991), Izett (1991), and Koeberl (1992) discussed the chemistry of the relict glass from the interior of the K/T spherules and concluded that the glass is of impact origin. Jéhanno et al. (1992) prefer a volcanic origin of glass based on two arguments: the high Fe^{3+}/Fe^{2+} ratio in the glass and the stratigraphic position of the K/T sandstone unit, supposedly below the K/T boundary. Although the high Fe^{3+}/Fe^{2+} is not yet explained, Jéhanno et al. (1992) did not discuss the extremely low volatile content of the glass (Koeberl, 1992), which is not consistent with volcanic origin. We have performed a few simple crushing experiments of glass fragments with relatively large (0.5 mm) fresh bubble cavities inside the glass. Ten glass fragments were crushed immersed in glycerine under a microscope. When the glass was crushed, the glycerine almost completely filled the bubble cavities in all fragments. This means that the gas pressure inside the bubble cavities is extremely low, consistent with a tektite glass origin.

Jéhanno et al. (1992) positioned the K/T boundary at the thin iron-oxide-stained, Ir-rich sand laminae and thus believed the K/T sandstone complex to be below the K/T boundary. The chalks directly below the K/T sandstone complex are rich in Maastrichtian planktic foraminifers (Fig. 14H), and the cross-bedded limestone and following sediments above the spherule layer are extremely poor in Cretaceous foraminifers (Fig. 14G). This interval is moreover rich in calcispheres *(Pithonella)* reworked from shallower-water sediments. These are thus part of the K/T sandstone unit and not of Cretaceous age because the Cretaceous fossils are clearly reworked.

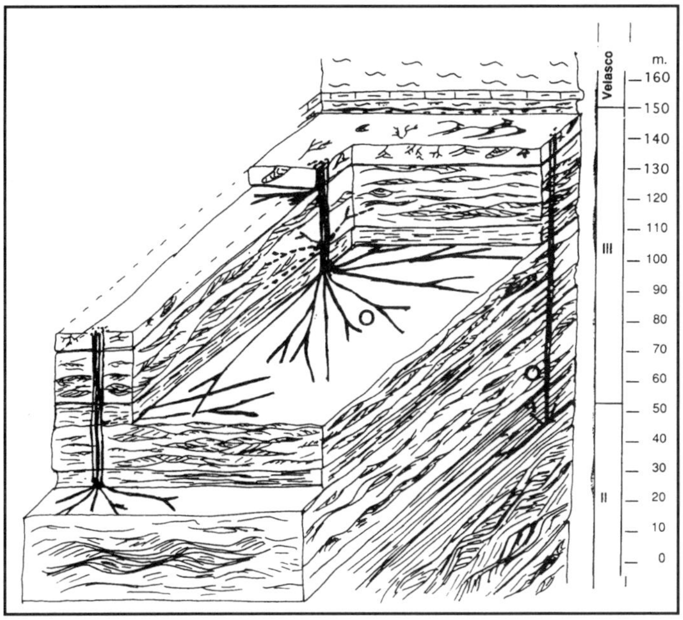

Figure 21. El Peñon, northeastern Mexico. Reconstruction of the topmost rippled sand and silt layers of Unit III, showing the tiered deep and shallow burrow types. The star-shaped, 5-m diameter *ophiomorpha*-type burrow structures follow the interbedded silt layers, up to a depth of 50 cm below the top of the K/T sandstone complex. In the middle of the "star" they are connected to a number of vertical tubes, observed to penetrate from above through at least 20 cm of sandstone above the star-structure and here interpreted to reach all the way to the surface, although that has not yet been observed.

Bochil, Chiapas, Mexico

The K/T sandstone complex crops out along a dirt road to the PEMEX well Soyalo-1, 9.4 km from the turnoff from the road between Bochil and Tuxtla Gutierrez (Montanari et al., 1994). The K/T sandstone complex occurs within a sequence (from bottom to top) of >80 m Campanian–Early Maastrichtian platform limestones with rudists, 30 m of late Maastrichtian pelagic marls containing several turbidites and debris flows, a >50-m-thick debris flow, and >15 m pelagic Paleocene marls. The Maastrichtian debris flows contain, among others, limestone cobbles, orbitoid foraminifers, and rudists. The K/T sandstone complex is underlain by a graded, >50-m-thick polymict mass-flow containing platform limestone blocks as large as 10 m. Those blocks consist of rudist-limestones, orbitoid limestones, and miliolid (lagoonal) limestones of the same composition as the cobbles in the Maastrichtian mass-flows. These components are thus locally derived and unlikely to be coarse ejecta from the Chicxulub impact. The matrix of the mass-flow contains loose orbitoids (Fig. 23A) and rudists.

The mass-flow grades into a 105-cm-thick, laminated and crudely cross-bedded sandstone with lime and bioclasts (orbitoids, rudist fragments, miliolids [*Murciella* sp.]; Fig. 23B). Remains of bubbly spherules have not yet been identified in the sandstone. The sandstones are overlain by a 5-cm-thick marl

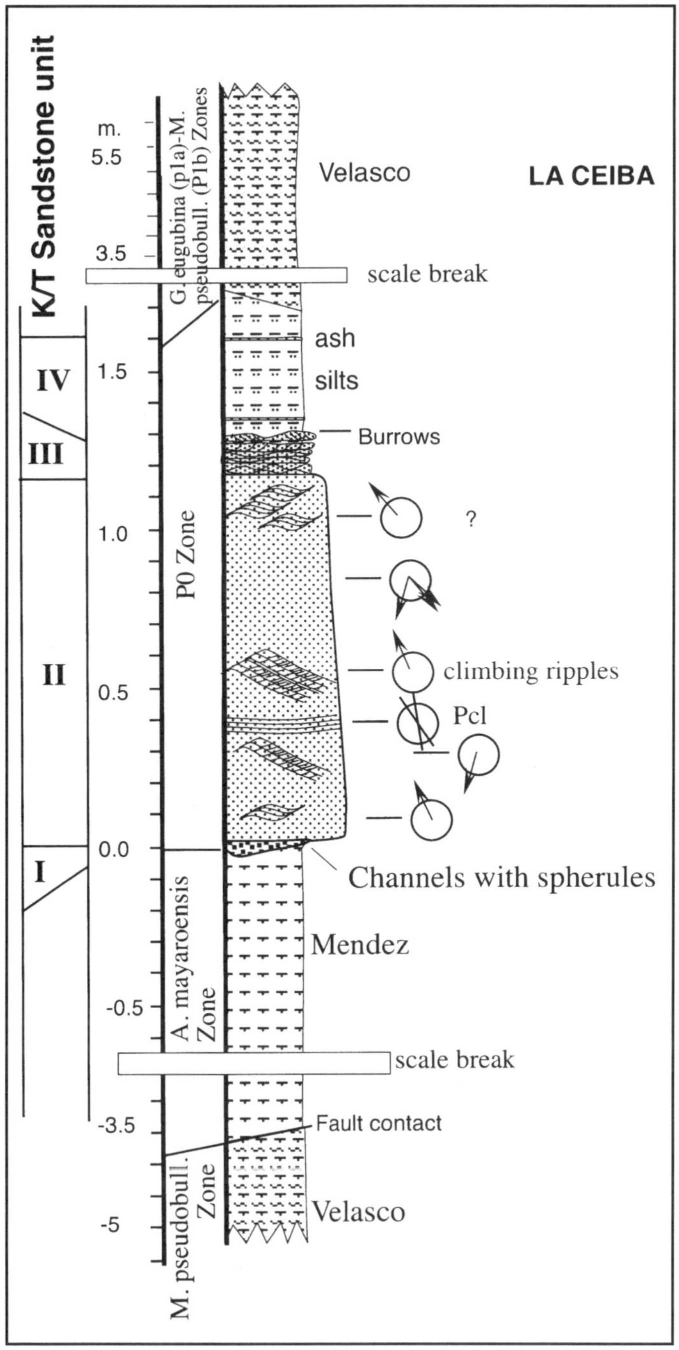

Figure 22. Stratigraphic column of the K/T sandstone complex at La Ceiba, east-central Mexico, with measured current directions. Pcl = primary current lineation. Unit II contains dominantly climbing ripple structures.

enriched in iridium, containing fine sand laminae locally stained by iron oxides (Montanari et al., 1994). Some poorly defined burrows occur in the marl. The marl is overlain by a 10-cm-thick silty limestone, similar to other outcrops in the Gulf Coast, followed by marls containing *G. eugubina.* The 50-m-thick mass-flow underlying the K/T sandstone complex may actually

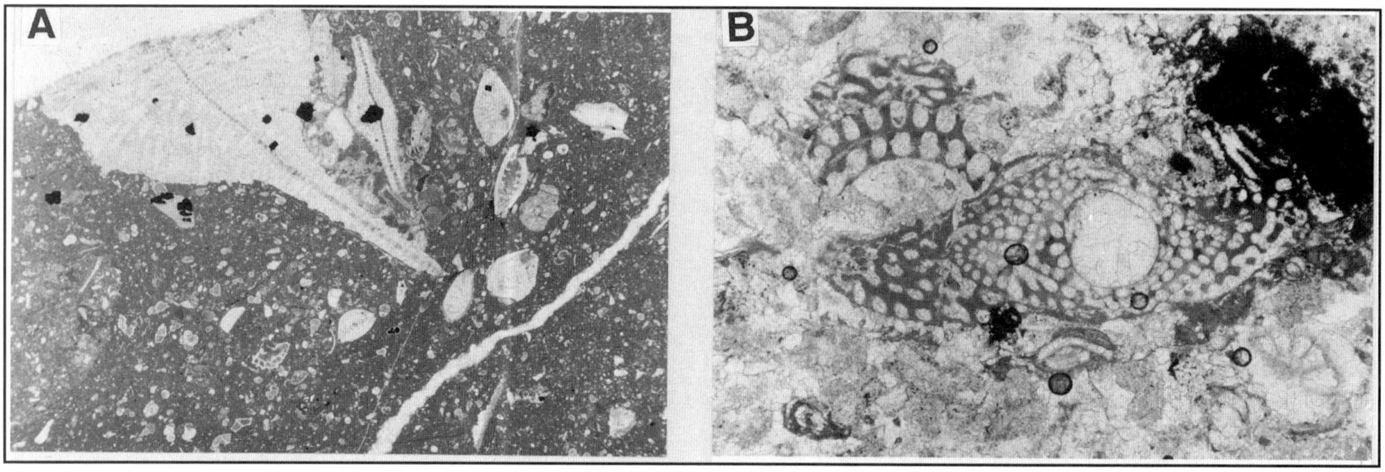

Figure 23. Bochil, Chiapas, Mexico. Thin sections of mass-flow components. A, Loose orbitoid foraminifer, from base of 50-m-thick mass-flow, underlying the K/T sandstone complex. B, Loose, *Murciella* sp., an upper Maastrichtian miliolid (lagoonal) foraminifer. These *Murciella* appear at the top of the K/T sandstone, as in nearby outcrops in Guatemala (Hildebrand et al., 1994), and may have been flushed from lagoons on the nearby carbonate platform.

belong to the depositional sequence of the K/T sandstone because it appears to be grading into the K/T sandstone. However, mass-flows that are similar but thinner, containing the same components, occur already in the Maastrichtian.

Apparently the Bochil area was at a slope setting at K/T boundary time, near a carbonate ramp, where thick mass-flows could have been triggered by the seismic shaking of the Chicxulub impact.

Los Ramones, Mexico

A small K/T sandstone complex outcrop occurs in the east bank of the Rio Pesqueria (Alvarez et al., 1992b). Unit I is almost absent and contains very few bubbly spherules at the base, mostly as armoring of the many clay pebbles. On some of the mud-clasts oriented current-crescents occur upstream of the spherules of the armoring. Unit II is massive, contains many stringers of clay pebbles, and fills a single channel (max. 440 cm thick, axis strikes N140°E) that is loaded into the Maastrichtian Mendez Formation. A Unit III is absent, but the flat top of the sandstone complex is marked by several 30-cm-large crescent scours (direction N140°E), with superimposed current ripples (N230°E) running at an oblique angle to the channel axis.

Overlying sediments are not exposed for over 5 m; the first sediments exposed are Velasco shales, containing *M. pseudobulloides*.

Rancho Nuevo, Mexico

Two adjacent channels, ~7 m wide (strike of channel axis N130°E), are exposed on the east bank of the Rio San Juan (Alvarez et al., 1992b). Unit I is 0 to 15 cm thick, channelized, almost entirely composed of bubbly spherules. Unit II is massive, maximum 6.5 m thick in the center of the channels, reduced to zero between the two channels. The massive sandstones display faint water-escape structures. Between the two channels (5 m apart) the Mendez shale is pushed up in large, deformed "flame-structures" or diapirs. At the base of Unit II frequent flute-casts occur, showing a current direction parallel to the channel axis (N110°–140°E), but some deviating directions were also measured (N80°E–N260°E). At the base imprints of siliceous sponges occur. Unit III is poorly exposed, but several (30 to 40 cm thick) fine sandstone layers with climbing ripples occur on top of Unit II. The transition from Unit III sandstone to the Paleocene marls is not exposed over 30 cm; then follow Paleocene Velasco marls containing *G. eugubina.*

Parras basin, northeastern Mexico

The K/T boundary has been described by Kauffman and Hansen (1987), who described a chaotic sedimentary breccia/conglomerate as a giant storm or tsunami bed within the dominant siliciclastic offshore to shoreface Maastrichtian-Danian siliciclastic sequence of the Parras basin. The K/T boundary was drawn between the sudden extinction of *Exogyra* and the appearance of a Paleocene mollusc *(Cucullea)* in a prominent middle shoreface sandstone bar, with abundant hummocky cross-stratification. We have investigated the outcrop, but we have not found any evidence for a chaotic breccia in the K/T interval. We found exfoliated sandstone boulders resembling a boulder bed. This site is clearly the shallowest, more or less continuous marine sequence across the K/T boundary, regarding the highly fossiliferous sequence of the wave-rippled offshore sandstone bars across the K/T boundary.

DISCUSSION

The sandstone layers occurring at the K/T boundary around the Gulf of Mexico are a complex deposit, size graded overall, with a distinct sequence of lithologies. We interpret this sequence to be the result of tsunami-wave activity initiated by the Chicxulub impact event on northern Yucatan. The K/T sandstone can be divided into four subunits, each one representing a different phase in the impact-tsunami scenario. We first discuss the tsunami-wave hypothesis, then alternatives given by other authors.

The basal Unit I invariably contains spherules and droplets, 0.5 to 10 mm in size, that we interpret as remains of molten ejecta—impact glass or (micro) tektites—because at Mimbral and Beloc some spherules contain a relict core of impact glass (Smit et al., 1992b). The tektites are invariably altered, but the same basic morphology is found in all K/T localities, including Beloc, and therefore can be easily recognized in the field and under a microscope. More than 99% of the tektites contain abundant bubble cavities, which are diagenetically infilled with clay minerals, or calcite. The bubble-cavity infillings may weather out, leading to interpretation as separate spherules, and the illusion of "accretionary lapilli," whereas they are actually bubble-cavity infillings enclosed in a spherule of clay minerals (Lyons and Officer, 1992). We also interpret the abundant rounded limestone fragments as ejecta, because invariably they are associated with the tektites and are highly similar in different places, whereas other detritus differs in composition from place to place, depending on the local source area. The ejecta are not primary deposits but are reworked into scours or channels gauged in the seafloor and mixed with local components, because the deposits are invariably cross-bedded.

Locally, we believe there is evidence for strong seismic activity, at least preceding the arrival of the first tsunami waves. At Moscow Landing, the ejecta deposits are associated with faults and soft-sediment deformation. The faults were active just before the deposition of the ejecta, because these deposits occur only near the faults and slumped areas. At Moscow Landing, La Lajilla, and El Peñon, material with ejecta is injected into fault planes below the K/T sandstone unit, indicating continued faulting during deposition of the K/T sandstone unit. At Bochil and in Guatamala (Hildebrand et al., 1994) the K/T sandstone complex is preceded by thick mass-flow conglomerates, probably triggered by seismic activity and moving downslope just before the arrival of the first tsunami waves, possibly before the arrival of the first ejecta. At Rancho Nuevo, Los Ramones, La Lajilla, and Mimbral, the Unit I and II channels are deeply loaded into the soft Mendez marls. Seismites—single, thick mass-flows of turbidites triggered by an earthquake—have some of the same characteristics as the K/T sandstone complex (Kleverlaan, 1989). It is reported that the seafloor has been deformed as a direct result of the earthquake and the loading by the seismite.

Unit II usually displays a set of well-sorted lenticular sandstones, with a wide variety of sedimentary structures indicating currents that change direction several times up-section, often by 180°. The material of Unit II is a mixture of locally derived material and material from nearshore areas (plant debris, terrigenous grains, shallow-water foraminifers) and differs from place to place, although some of the Unit I ejecta still occur at the base of some layers. The sedimentary structures—strong parallel lamination, climbing ripples, water-escape structures—show that these sandstones were deposited rapidly by strong currents varying in strength and direction. In none of the localities investigated by us did we observe that the lenticular sandstone layers are interrupted by layers of normal hemipelagic deposition, or layers with burrows, that would represent long periods of time between deposition of the individual sandstone layers.

We interpret the repeated change in current directions to be a result of deposition by up-surge and back-surge of several large tsunami waves, best observed in the La Lajilla section. In some localities (Beloc, Coxquihui), Unit II does not occur. Since the bulk of the material is derived from nearshore material (shallow-water bioclasts, coarse terrigenous grains, and plant debris) and winnowed foraminifers, we infer that these localities are either too far from the shore or are at a water depth below tsunami wave-base.

Unit III is essentially a continuation of Unit II, but represents the phase in which tsunami waves decrease in strength to such extent that fine sediment is able to settle in between ripple layers of fine sand deposited during wave surges. These mud-drapes are the first sediments to be enriched in iridium. Clearly, the deposition of iridium-rich phases is decoupled from the deposition of the coarse ejecta. Assuming a one-impact scenario, we infer that the iridium carrier particles are much finer grained and settle much more slowly than the coarse ejecta. Burrows occur in and at the top of the last sandstone ripples. Rare chondrites also occur on the surface of some of the lower sand ripples and the silt in between but not inside those lower sand ripples. Because the burrow-fabric is tiered, we conclude that the top surface was colonized after, not during, deposition of the sand ripples.

Unit IV represents the phase when currents induced by tsunami-waves are no longer capable of transporting fine sand at the seafloor, and silt and clay are able to settle on the seafloor. Regarding the large amount of winnowed foraminifers in the K/T sandstone unit, it is expected that most of the fine sediment has been washed out and remained suspended in the water-column during passage of the tsunami-waves. The mudstone interval is size graded, and we therefore infer that this graded interval represents an interval of days to weeks after the initiating tsunami event. The burrows present in this interval (which is not bioturbated) are difficult to interpret. These burrows could be due to a colonizing episode of weeks to months but could also be due to organisms trying to burrow back again in the seafloor sediment after being "uprooted" by the tsunami currents.

The K/T sandstone complex as a whole forms one continuous overall size-graded sequence not interrupted by sediments or burrowed surfaces, indicating long periods of time in between the different sublayers.

This scenario has been challenged by several authors, who believe that the K/T sandstone complex can be interpreted differently. Stinnesbeck et al. (1993) conclude that the sandstone complex predates the K/T boundary and belongs to a normal turbidite system derived from the deltaic sediments of the Difunta group. Savrda (1993) interprets the sandstone complex as a normal low-stand sediment, because at Mussel Creek in Alabama bedding surfaces truncate burrow structures. Bohor (1994) interprets the K/T sandstone complex as an impact-triggered debris flow and turbidite.

At a first glance the K/T sandstone complex has many features in common with a turbidite fan, such as erosive base and channels, introduction of shallow marine benthos, plant debris and sands in a deeper marine environment, flute casts, loadings, plane beds with primary current lineation, and climbing ripples. However, comparing the K/T complex with generalized turbidite-fan models of Mutti and Normark (1987) and Walker (1978), it is clear that the K/T complex is not part of a distal outer fan or of the fan lobes. In the K/T complex none of the following occur: normal turbidites with Bouma sequences, separated by mud; thickening-up turbidite sequences (due to prograding lobes) and thinning-up compensation sequences (infillings between lobes); thicker mud intercalations between lobes (due to channel and lobe switch). The work of Mutti and Normark (1987) and Mutti and Ricci Lucchi (1972; their Fig. 16) enables us to quickly compare features, which they hold for characteristic of channels, lobes, and the transition zone in between.

Our K/T complex has the following in common with "channels of turbidite fans":

- erosional channels, in which infilling layers truncate against the scour (rarely occurring at Mimbral);
- depositional channels, in which infilling layers converge to the channel edge (at Mimbral in Units I and II and at Rancho Nuevo in Unit II);
- clast supported conglomerates (rare, except at Brazos River, Bochil, and Moscow Landing);
- thin-bedded overbank deposits (Unit III could be interpreted as such, but no wedging away from the channels is observed and instead opposing current directions or at Los Ramones even a winnowed channel top with current ripples moving at right angles to the channel); and
- outsize mud clasts and armored mud clasts.

Not observed in the K/T sandstone complex were the following:

- massive, graded or inversely graded infillings of channels (instead, Unit I generally shows internally laminated fills, sometimes as at Mimbral with lateral accretion, indicating shifting channels); and
- chaotic units pointing to sliding (instead vertical loading is seen, notably at Mimbral, Los Ramones, and Rancho Nuevo).

Our K/T complex unit has the following in common with "transition zone of turbidite fans":

- "zones of roughness" or levels with irregular scour, indicating bypass (such levels probably have been encountered at the base of the Unit 2 channel at Ramones and Rancho Nuevo, but that needs more study);
- low-amplitude megaripples (transition zone megaripples, according to Mutti and Ricci Lucchi [1972] are clast-supported parallel and cross-bedded units indicative of bypass); and
- stack of shallow channels (at Mimbral, Mulato, and Darting Minnows Creek; in our case, however, never with massive or graded fill and never with mud-drapes over a scoured surface).

Our K/T complex has nothing in common with "lobes of turbidite fans" except for the even parallel-bedded pattern (only at Lajilla) and repeating parts of the Bouma sequence B and C.

Concluding, we may say that our sandstone complex in some aspects resembles the channels and transition parts of a turbidite fan. The scarcity of mud-supported conglomerates, absence of chaotic masses, presence of bedded channel infillings, and indication of lateral migration of some of the channels of Unit I at Mimbral and Unit II at Mulato point to a low-gradient slope.

Having said this, we reject the hypothesis of the K/T sandstone complex as having been deposited as a turbidite-fan on the basis of the following arguments:

1. In all outcrops studied so far, the eventlike character is indicated by a general fining-up and thinning-up of our sandstone unit. Fining-up is clearly indicated by the conglomeratic and spherule rich Unit I, overlain by the sandy Unit II and the current-rippled flaser/mud alternation of Unit III and grading of Unit IV. Despite repetition of beds and sedimentary structures with similar features, pointing to a strongly pulsating current, the vertical order is invariably such that a higher unit indicates progressive waning of the current. This can be deduced from upward thinning of similar beds or channels or from the size and steepening of climbing rippled intervals.

2. The presence of a relatively thin and lensing sandstone complex over a vast area of at least 2,000 km between Brazos River, Texas, and La Ceiba (Bochil) in Mexico with a consistent superposition of spherule-rich channels, sandstone layers, and a current-rippled sandstone/mud alternation. Over this vast area, the unit varies between only a few centimeters to 10 m, generally between 1 and 3 m. In turbidite fans such thicknesses occur only in the distal parts of the fan.

3. Common reversals of currents as revealed by the orientation of ripple cross-bedding and flutes and without the current sense from the orientation of channel axes and primary

current lineation. Such reversals also may be seen in turbidites reflected or deflected by fault scarps or in narrow basins (Pickering and Hiscott, 1985) but can hardly be expected to occur in the ancient Gulf of Mexico in turbidites over such a vast area.

4. The uniqueness of the sandstone complex within long tracts of Cretaceous marl and Paleocene marls without turbidites (or only very distal ones such as at Rancho Nuevo) between exactly the latest Cretaceous and earliest Paleocene marls and characterized by spherules at the base and high Ir concentration at the top of the sandstone unit.

5. The absence of longer time intervals between the beds as may be deduced from the lack of interbedded mud (except for Unit 3 at the top) and intercalated bioturbated levels. The only instance where we found two burrow levels separated by a scarcely burrowed sandstone bed was in the top of the K/T sandstone complex at El Peñon. However, we found that the horizontal star-shaped tube systems are connected to vertical tubes in the center, penetrating through several-decimeters-thick sandstone layers above the star-structure. Although we have not observed these vertical tubes to penetrate to the surface of the sandstone complex, there is no evidence to the contrary.

6. Channel cut and channel fill occurred directly after each other, as is indicated by only shallow downcutting in the order of 0.5 m, followed by meters-deep loading with upward-decreasing intensity of plastic deformations. Such fills were observed at Mimbral in Unit I and partially in Unit II and at Rancho Nuevo and Ramones in Unit II. Injection dikes with spherules in the Mendez Formation likewise point to connected cut-and-fill events at Lajilla and Moscow Landing. Rapid sedimentation is further indicated by water-escape structures and sand-in-sand loading at Rancho Nuevo, Ramones, and El Peñon.

7. The occurrence of mud clasts sparsely armored with spherules at the base of Unit II. These mud lumps just before sedimentation of Unit II sands were struck by the current, leaving minute, oriented current-crescents upstream of the spherules. These unique features were seen at Ramones, La Sierrita (Alvarez et al., 1992b), and Rancho Nuevo; they indicate a short time lapse between Units I and II. At Sierrita such an armored mud clast was picked up by the current and deposited in Unit II in a clear traction carpet (Lowe, 1982; Mutti and Ricci Lucchi, 1972) in the sandstone.

In conclusion, the general fining-up and thinning-up of the beds; the unique occurrence at the K/T boundary; the (impact-derived) spherules at the base of the unit, shelf-derived sand in the middle, and (cosmic) Ir concentration at the top; the relatively thin deposit but widespread occurrence; and rapid deposition by opposing and deviating currents cannot be explained by deposition in a turbidite fan or series of fans. On the contrary, these arguments fit well an impact scenario with associated tsunami deposits at the K/T boundary.

However, especially in the shallow-water areas of Alabama and Texas, other noncatastrophic factors such as eustatic sea-level changes have contributed locally to the final architecture of the sandstone units. An early, but not earliest, Danian sea-level lowstand may explain many features of the K/T boundary in the Gulf coastal plain and elsewhere. It is common knowledge that the K/T boundary at shallow-marine settings contains a hiatus due to erosion. Figure 24 shows the possible interactions of early Danian eustatic sea-level changes in the Gulf Coast sections (after Mancini and Tew, 1993). The onset of the sea-level drop is hard to estimate, but some biostratigraphic data are available. The burrows of the transgressive surface, in Brazos River and at Moscow Landing, are of *M. pseudobulloides* age. The low-stand phase has locally eroded down into earliest Danian, K/T sandstone complex, or Late Maastrichtian sediments, as shown at Moscow Landing. Some remnants of the low-stand (ravinement valleys, Habib et al., 1992; Habib and Talvirska, 1994) are still preserved in Moscow Landing and Mussel Creek but should not be confused with the K/T sandstone complex. These low-stand infills contain earliest Danian dinoflagellates. Also, as seen in the grain-size analysis of Brazos River (Fig. 7), there is a coarsening-upward trend in the top of the sequence, where the first *G. eugubina* appear and where Jiang and Gartner (1986) first found the basal Danian *Cruco-placolithus primus.*

Some have expressed doubts that the clastic beds are related to the Chicxulub impact, or have a bearing on the mass extinction of, among others, planktic foraminifers (Stinnesbeck et al., 1993, Keller et al., 1993; Jéhanno et al., 1992). Keller et al. (1994b) presented relative abundance plots of planktic foraminifers in the Mimbral section, suggesting the continuation of "thriving" planktic populations above the clastic beds. However, comparison of thin sections from pelagic marls just below (Figs. 12F, 14A, H) and above the clastic beds (Figs. 12E, 14C, G) clearly show that the Cretaceous planktic populations were decimated just above the clastic beds. The few Cretaceous specimens present are for the largest part most likely reworked, because they occur in cross-bedded sediments. But even if some of these were survivors, then their ecological significance in the planktic populations above the clastic beds is negligible.

CONCLUSIONS

The set of sandstone beds as found in intermediate water depths (50 to 500 m) in Gulf Coast outcrops from Alabama to Chiapas, Mexico—designated here as the K/T sandstone complex—is best explained as a deposit of large tsunami waves, caused by the impact of a large extraterrestrial body at Chicxulub, Yucatan, Mexico.

The resulting sequence of sedimentary events is as follows:

1. Seismic shaking of the Chicxulub impact locally caused faulting and slumping (Moscow Landing, La Lajilla, Mimbral) and triggered thick mass-flows (Bochil, Mexico, Guatemala).

2. During the earthquakes, or shortly afterward, coarse ejecta (tektites, limestone clasts) fell down around the Gulf of Mexico.

3. These ejecta were immediately reworked with local (rip-up) material into channel-like deposits by currents caused by

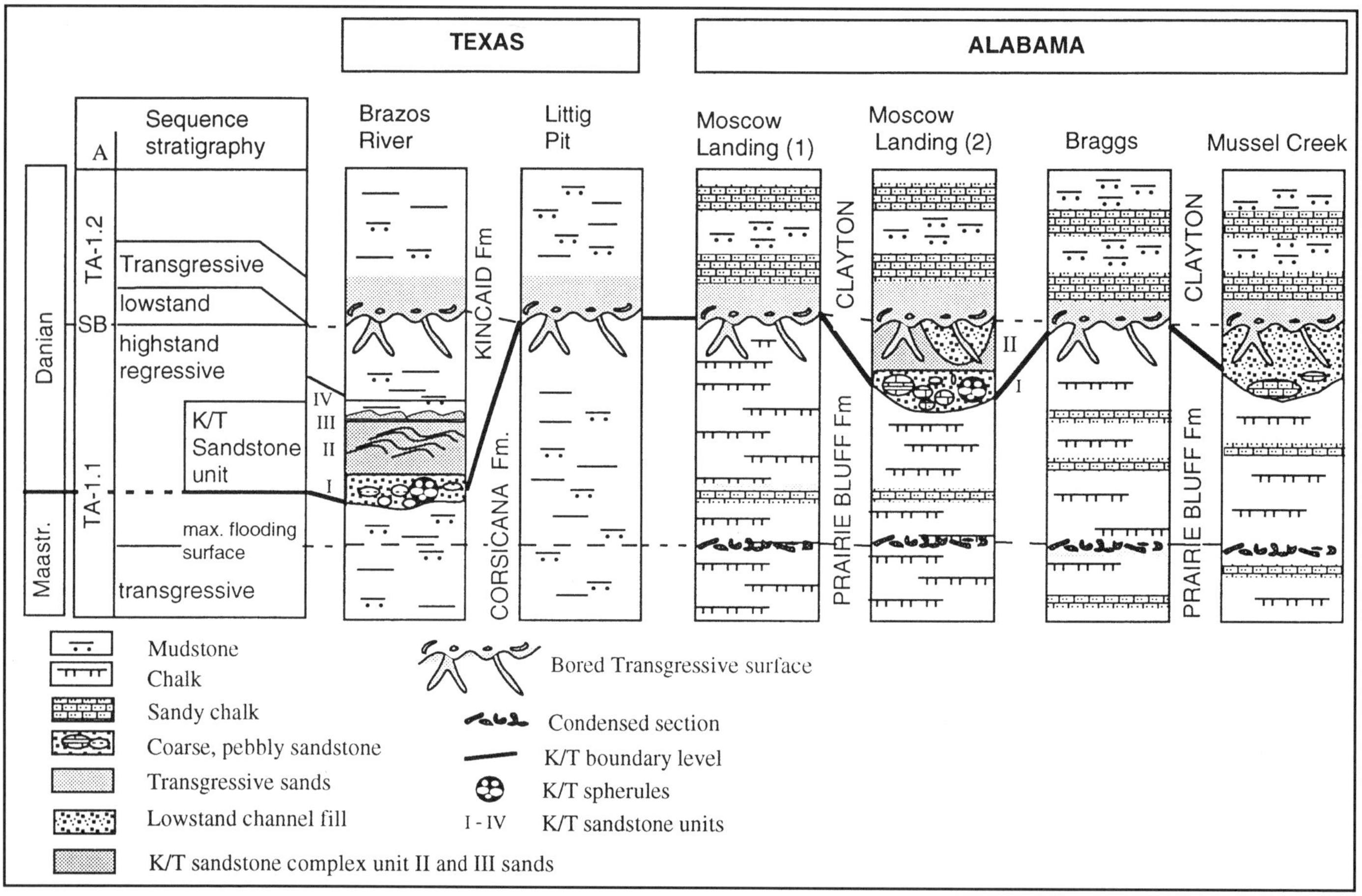

Figure 24. K/T boundary sections in Alabama and Texas, interpreting the K/T sandstone complex in a sequence stratigraphic context (after Mancini and Tew, 1993). We infer that the K/T sandstone complex occurs in a regressive sequence in all the outcrops, between a phosphatic lag-bed indicating a high stand and incised low-stand valleys and a transgressive surface. The channelized sands at Mussel Creek probably belong to incised channels from the early Danian sea-level fall but are not part of the tsunami deposits (Savrda, 1993). The conspicuous bored hard ground can be traced all over the U.S. Gulf coastal plain and is basal Danian (*M. pseudobulloides* Zone) in age. A similar hard ground occurs in the shallow water K/T sequences in Holland (Curffs quarry) and Denmark (Stevns Klint) at about the same biostratigraphic level. The K/T sea-level low stand was formerly considered to be K/T or latest Maastrichtian in age (Mancini et al., 1993). Only at the Brazos River has the sea-level low stand not removed (part of) the K/T boundary tsunami deposits.

the passage of large tsunami waves or were moved downslope by mass-flows (Beloc, Coxquihui). These channels loaded into the (?seismic-softened) seafloor sediments (Unit I).

4. Subsequent tsunami waves, possibly assisted by mass-flows, transported terrigenous sand, plant debris, and shallow-water fossils from nearshore areas into deeper basins, where they deposited a sequence of thick lenticular sandstone beds, showing evidence for currents from opposite directions, consistent with the water movements—upsurge and return-flow—caused by large waves (Unit II).

5. The final waning tsunami waves only transported fine sand in the next deposits (Unit III), but fine mud was now allowed to settle that contained the first enrichments of iridium.

6. After the deposition of the last fine sandstone flaserlike layers, the suspended cloud of fine material settled out between a few hours and days, to form a graded mudstone deposit. The bulk of the iridium was deposited in this graded interval. Apparently, this extraterrestrial material was fine grained, because it arrived more slowly than the coarse tektite ejecta. Based on the settling time of the fine muds in the Brazos River and the duration of several passages of the tsunami waves and their travel times across the Gulf of Mexico (1 to 6 hs), we estimate that the iridium settled between a few days and a week.

7. The first background sediments, deposited on top of the K/T sandstone complex, are greatly depleted in planktic foraminifers. This shows that the mass extinction coincided closely with the deposition of the K/T sandstone complex and thus most likely was caused by the consequences of the Chicxulub impact.

Placement of the K/T boundary sandstone complex in the top Cretaceous, below the K/T boundary (Stinnesbeck et al., this volume), is based on the use of the wrong criteria for correlation to the K/T boundary type-locality in El Kef, Tunisia.

ACKNOWLEDGMENTS

Reviews by J. Bourgeois and E. Clifton greatly improved the manuscript. Funding was provided by a grant from the Royal Dutch Academy of Sciences (J. Smit), and NWO, the Dutch Science Foundation (Th. B. Roep, J. Smit).

REFERENCES CITED

Alvarez, L. W., Alvarez, W., Asaro, F., and Michel, H. V., 1980, Extraterrestrial cause for the Cretaceous-Tertiary extinction: Science, v. 208, p. 1095–1108.

Alvarez, W., Smit, J., Lowrie, W., Asaro, F., Margolis, S. V., Claeys, P., Kastner, M., and Hildebrand, A. R., 1992a, Proximal impact deposits at the Cretaceous-Tertiary boundary in the Gulf of Mexico: A restudy of DSDP Leg 77 Sites 536 and 540: Geology, v. 20, p. 697–700.

Alvarez, W., and 8 others, 1992b, The Cretaceous-Tertiary boundary impact–tsunami deposit in NE Mexico: Geological Society of America Annual Meeting, p. A331.

Asaro, F., Michel, H. V., Alvarez, L. W., Maddocks, R. F., and Bunch, T., 1982, Iridium and other geochemical profiles near the Cretaceous-Tertiary boundary in a Brazos river section in Texas, *in* Maddocks, R. F., ed., Texas ostracoda, International Symposium on Ostracoda, 8th, Guidebook of Excursion and Related Papers: Houston, University of Houston, p. 238–241.

Beeson, D., Gartner, S., Keller, G., MacLeod, N., Medus, J., Rocchia, R., and Robin, E., 1994, The K/T boundary along the Brazos River, Falls County, Texas: Multidisciplinary stratigraphy and depositional environment [abs.], *in* New developments regarding the K/T event and other catastrophes in Earth history: Houston, Texas, Lunar and Planetary Institute Contribution 825, p. 9–10.

Best, J., and Bridge, J., 1992, The morphology and dynamics of low-amplitude bedwaves upon upper stage plane beds and the preservation of planar laminae: Sedimentology, v. 39, p. 737–752.

Bohor, B. F., 1990, Shocked quartz and more; Impact signatures in Cretaceous/Tertiary boundary clays, *in* Sharpton, V. L., and Ward, P. D., eds., Global catastrophes in Earth history; An interdisciplinary conference on impacts, volcanism, and mass mortality: Geological Society of America Special Paper 247, p. 335–342.

Bohor, B. F., 1994, K/T clastic deposits, northeastern Mexico: Single-pulse debris-flow/turbidite units associated with impact, *in* Keller, G., Stinnesbeck, W., Adatte, T., MacLeod, N., and Lowe, D. R., eds., Field guide to Cretaceous-Tertiary boundary sections in northeastern Mexico: Houston, Texas, Lunar and Planetary Institute Contribution 827, p. 102–106.

Bohor, B. F., and Betterton, W. J., 1993, Arroyo el Mimbral, Mexico, K/T unit: Origin as debrisflow/turbidite, not a tsunami deposit: Proceedings, Lunar and Planetary Science Conference, v. 23, p. 143–144.

Bourgeois, J., 1991, Extended period of extinctions across the Cretaceous/Tertiary boundary of continental shelf sections: Implications for impact and volcanism theories: Discussion and reply: Geological Society of America Bulletin, v. 103, p. 434–436.

Bourgeois, J., Hansen, T. A., Wiberg, P. L., and Kauffman, E. G., 1988, A tsunami deposit at the Cretaceous-Tertiary boundary in Texas: Science, v. 241, p. 567–570.

Cowie, J. W., Zieger, W., and Remane, J., 1989, Stratigraphic Commission accelerates progress, 1984–1989: Episodes, v. 112, p. 79–83.

Ganapathy, R., 1980, A major meteorite impact on the Earth 65 million years ago: Evidence from the Cretaceous-Tertiary boundary clay: Science, v. 209, p. 921–923.

Groot, J. J., de Jonge, R. B. G., Langereis, C. G., ten Kate, W. G., and Smit, J., 1989, Magnetostratigraphy of the Cretaceous-Tertiary boundary at Agost, Spain: Earth and Planetary Science Letters, v. 94, p. 385–397.

Habib, D., and Talvirska, I., 1994, Dinoflagellate stratigraphy and diversity in the early Danian: Geological Society of America Abstracts with Programs, v. 24, p. A358.

Habib, D., Moshkovotz, S., and Kramer, C., 1992, Dinoflagellate and calcareous nannofossil response to sealevel change in Cretaceous-Tertiary boundary sections: Geology, v. 20, p. 165–168.

Hansen, T., Farrand, R. B., Montgomery, H. A., Billman, H. G., and Blechschmidt, G., 1987, Sedimentology and extinction patterns across the Cretaceous-Tertiary boundary interval in East Texas: Cretaceous Research, v. 8, p. 229–252.

Hay, W. W., 1960, The Cretaceous-Tertiary boundary in the Tampico Embayment, Mexico: Proceedings, International Geological Congress, v. 5, p. 70–77.

Hess, W. F., and Gatzemeier, M. B., 1991, Vergleichende Untersuchung von ausgewahlten Beugungs Spectrometern: Chemische Ingenieur Technologie, v. 63, p. 378–381.

Hildebrand, A. R., Penfield, G. T., Kring, D. A., Pilkington, M., Camargo, Z. A., Jacobsen, S. B., and Boynton, W. V., 1991, Chicxulub crater: A possible Cretaceous/Tertiary boundary impact crater on the Yucatan peninsula, Mexico: Geology, v. 19, p. 867–871.

Hildebrand, A. R., and 13 others, 1994, The Chicxulub crater and its relation to the K/T boundary ejecta and impact-wave deposits, *in* New developments regarding the K/T event and other catastrophes in Earth history: Houston, Texas, Lunar and Planetary Institute Contribution 825, p. 49–50.

Hsü, K. J., and McKenzie, J. A., 1990, Carbon isotope anomalies at era boundaries; global catastrophes and their ultimate cause, *in* Sharpton, V. L., and Ward, P. D., eds., Global catastrophes in Earth history; An interdisciplinary conference on impacts, volcanism, and mass mortality: Geological Society of America Special Paper 247, p. 61–70.

Izett, G. A., 1991, Tektites in Cretaceous-Tertiary boundary rocks on Haiti and their bearing on the Alvarez impact extinction hypothesis: Journal of Geophysical Research, v. 96, p. 20879–20905.

Jéhanno, C., Boclet, D., Froget, L., Lambert, B., Robin, E., Rocchia, R., and Turpin, L., 1992, The Cretaceous-Tertiary boundary at Beloc, Haiti: No evidence for an impact in the Caribbean area: Earth and Planetary Science Letters, v. 109, p. 229–241.

Jiang, M. J., and Gartner, S., 1986, Calcareous nannofossil succession across the Cretaceous/Tertiary boundary in east-central Texas: Micropaleontology, v. 32, p. 232–255.

Kauffman, E. G., and Hansen, T. A., 1987, The Cretaceous-Tertiary boundary in northern Mexico: Conference on Palaeontology and Evolution: Extinction Events, Abstracts: Leioa, Spain, II Jornadas de Paleontologia, p. 181.

Keller, G., 1988, Extinction, survivorship and evolution of planktic Foraminifera across the Cretaceous/Tertiary boundary at El Kef, Tunisia: Marine Micropaleontology, v. 13, p. 239–263.

Keller, G., 1989a, Extended Cretaceous/Tertiary boundary extinctions and delayed population change in planktonic foraminifers from Brazos River, Texas: Paleoceanography, v. 4, p. 287–332.

Keller, G., 1989b, Extended period of extinctions across the Cretaceous/Tertiary boundary in planktonic foraminifera of continental shelf sections: Implications for impact and volcanism theories: Geological Society of America Bulletin, v. 101, p. 1408–1419.

Keller, G., MacLeod, N., Lyons, J. B., and Officer, C. B., 1993, Is there evidence for Cretaceous-Tertiary boundary-age deep-water deposits in the Caribbean and Gulf of Mexico?: Geology, v. 21, p. 776–780.

Keller, G., Stinnesbeck, W., Adatte, T., Lopez-Oliva, J. G., and MacLeod, N., 1994a, The K/T boundary clastic deposit in northeastern Mexico as product of non-catastrophic geologic processes? *in* Keller, G., Stinnesbeck, W., Adatte, T., MacLeod, N., and Lowe, D. R., eds., Field guide to Cretaceous-Tertiary boundary sections in northeastern Mexico: Houston, Texas, Lunar and Planetary Institute Contribution 827, p. 65–94.

Keller, G., Stinnesbeck, W., and Lopez-Oliva, J. G., 1994b, Age, deposition and biotic effects of the Cretaceous/Tertiary boundary event at Mimbral,

NE Mexico: Palaios, v. 9, p. 144–157.

Kellum, L. B., 1937, The geology and biology of the San Carlos mountains, Tamaulipas, Mexico: Ann Arbor, University of Michigan Press, 200 p.

Kleverlaan, K., 1989, Tabernas fan-complex: A study of a Tortonian fan complex in a Neogene basin, Tabernas, Province of Almeria, SE Spain [Ph.D. thesis]: Amsterdam, Vrije Universiteit, 97 p.

Koeberl, C., 1992, Water content of glasses from the K/T boundary, Haiti: An indication of impact origin: Geochimica et Cosmochimica Acta, v. 56, p. 4329–4332.

Kruge, M. A., Stankiewicz, B. A., Krelling, J. C., Montanari, A., and Bensley, D. F., 1994, Fossil charcoal in Cretaceous-Tertiary strata: Evidence for catastrophic firestorm and megawave: Geochimica et Cosmochimica Acta, v. 58, p. 1393–1397.

Kyte, F. T., and Smit, J., 1986, Regional variations in spinel compositions: An important key to the Cretaceous/Tertiary event: Geology, v. 14, p. 485–487.

Lowe, D. R., 1982, Sediment gravity. II: Depositional models with special reference to the deposits of high-density turbidity currents: Journal of Sedimentary Petrology, v. 5, p. 279–297.

Lyons, J. B., and Officer, C. B., 1992, Mineralogy and petrology of the Haiti Cretaceous/Tertiary section: Earth and Planetary Science Letters, v. 109, p. 205–224.

MacDougall, J. D., 1988, Seawater strontium isotopes, acid rain, and the Cretaceous-Tertiary boundary: Science, v. 239, p. 485–489.

Mancini, E. A., and Tew, B. H., 1993, Eustasy versus subsidence: Lower Paleocene depositional sequences from southern Alabama, eastern Gulf Coastal plain: Geological Society of America Bulletin, v. 105, p. 3–17.

Mancini, E. A., Tew, B. H., and Smith, C. C., 1989, Cretaceous-Tertiary contact, Mississippi and Alabama: Journal of Foraminiferal Research, v. 19, p. 93–104.

Maurasse, F. J.-M. R., Pierre-Louis, F., and Rigaud, J. J.-G., 1979, Upper Cretaceous to Lower Paleocene pelagic calcareous deposits in the southern Peninsula of Haiti: Their bearing on the problem of the Cretaceous-Tertiary boundary: Transactions, Latin American Geological Congress, 4th, p. 328–338.

Montanari, A., Hay, R. L., Alvarez, W., Asaro, F., Michel, H. V., Alvarez, L. W., and Smit, J., 1983, Spheroids at the Cretaceous-Tertiary boundary are altered impact droplets of basaltic composition: Geology, v. 11, p. 668–671.

Montanari, A., Claeys, P., Asaro, F., Bermudez, J., and Smit, J., 1994, Preliminary stratigraphy and iridium and other geochemical anomalies across the K/T boundary in the Bochil section (Chiapas, southeastern Mexico), *in* New developments regarding the K/T event and other catastrophes in Earth history: Houston, Texas, Lunar and Planetary Institute Contribution 825, p. 84–85.

Montgomery, H., Pessagno, E., Soegaard, K., Smith, C., Munoz, I., and Pessagno, J., 1992, Misconceptions concerning the Cretaceous/Tertiary boundary at the Brazos River, Falls County, Texas: Earth and Planetary Science Letters, v. 109, p. 593–600.

Morgan, H. J., Jr., 1931, The Velasco-Mendez contact in the vicinity of the Ebano Field, Mexico: Journal of Paleontology, v. 5, p. 42–47.

Muir, J. M., 1936, Geology of the Tampico region, Mexico: Tulsa, American Association of Petroleum Geologists, 300 p.

Mutti, E., and Normark, W. R., 1987, Comparing examples of modern and ancient turbidite systems: Problems and concepts, *in* Legget, J. K., Zuffa, G., Graham, G., and Trotman, A., eds., Marine clastic sedimentology: Concepts and case studies: London, Graham and Trotman, p. 1–38.

Mutti, E., and Ricci Lucchi, F., 1972, Le torbiditi dell' Appennino Settentrionale: Introduzione all' analisi di facies: Memoria Società Geologica Italiana, v. 11, p. 161–199.

Olsson, R. K., and Liu, C., 1993, Controversies on the placement of Cretaceous-Paleogene boundary and the K/P mass extinction of planktonic foraminifera: Palaios, v. 8, p. 127–139.

Pickering, K. T., and Hiscott, R. N., 1985, Contained (reflected) turbidity currents from the Middle Ordovician Cloridorme Formation, Quebec, Canada: An alternative to the anti-dune hypothesis: Sedimentology, v. 32, p. 373–394.

Pitakpaivan, K., Byerly, G. R., and Hazel, J. E., 1994, Pseudomorphs of impact spherules from a Cretaceous Tertiary boundary section at Shell Creek, Alabama: Earth and Planetary Science Letters, v. 124, p. 49–56.

Robin, E., Rocchia, R., Lyons, J. B., and Officer, C. B., 1994, Deposition of channel deposits near the Cretaceous-Tertiary boundary in northeastern Mexico: Catastrophic or "normal" sedimentary deposits?: Reply; Is there evidence for Cretaceous-Tertiary boundary age deep-water deposits in the Caribbean and Gulf of Mexico?: Reply: Geology, v. 22, p. 958.

Ross, M. I., and Scotese, C. R., 1988, A hierarchical tectonic model of the Gulf of Mexico and Caribbean region: Tectonophysics, v. 155, p. 139–168.

Savrda, C. E., 1991, Ichnology in sequence stratigraphic studies: An example from the Lower Paleocene of Alabama: Palaios, v. 6, p. 39–53.

Savrda, C. E., 1993, Ichnosedimentologic evidence for a noncatastrophic origin of Cretaceous-Tertiary boundary sands in Alabama: Geology, v. 21, p. 1075–1078.

Sigurdsson, H., D'Hondt, S., Arthur, M. A., Bralower, T. J., Zachos, J. C., van Fossen, M., and Channell, J. E. T., 1991, Glass from the Cretaceous-Tertiary boundary in Haiti: Nature, v. 349, p. 482–487.

Smit, J., 1990, Meteorite impact, extinctions and the Cretaceous-Tertiary boundary: Geologie en Mijnbouw, v. 69, p. 187–204.

Smit, J., and Romein, A. J. T., 1985, A sequence of events across the Cretaceous-Tertiary boundary: Earth and Planetary Science Letters, v. 74, p. 155–170.

Smit, J., Alvarez, W., Montanari, A., Swinburne, N., Kempen, T. M. v., Klaver, G. T., and Lustenhouwer, W. J., 1992a, "Tektites" and microkrystites at the Cretaceous-Tertiary boundary: Two strewnfields, one crater?: Proceedings, Lunar and Planetary Science Conference, v. 22, p. 87–100.

Smit, J., and 8 others, 1992b, Tektite-bearing, deep-water clastic unit at the Cretaceous-Tertiary boundary in northeastern Mexico: Geology, v. 20, p. 99–103.

Smit, J., Roep, T. B., Alvarez, W., Claeys, P., and Montanari, A., 1994a, Deposition of channel deposits near the Cretaceous-Tertiary boundary in northeastern Mexico: Catastrophic or "normal" sedimentary deposits?: Comment; Is there evidence for Cretaceous-Tertiary boundary age deep-water deposits in the Caribbean and Gulf of Mexico?: Reply: Geology, v. 22, p. 953–959.

Smit, J., Roep, T. B., Alvarez, W., Montanari, A., and Claeys, P., 1994b, Stratigraphy and sedimentology of K/T clastic beds in the Moscow Landing (Alabama) outcrops: Evidence for impact related earthquakes and tsunamis, *in* New developments regarding the K/T event and other catastrophes in Earth history: Houston, Texas, Lunar and Planetary Institute Contribution 825, p. 119–120.

Stinnesbeck, W., and 10 others, 1993, Deposition of channel deposits near the Cretaceous-Tertiary boundary in northeastern Mexico: Catastrophic or "normal" sedimentary deposits?: Geology, v. 21, p. 797–800.

Stinnesbeck, W., Keller, G., and Adatte, T., 1994a, K/T boundary deposits in northeastern Mexico: Bolide impact or sea-level lowstand? [abs.], *in* New developments regarding the K/T event and other catastrophes in Earth history: Houston, Texas, Lunar and Planetary Institute Contribution 825, p. 121–123.

Stinnesbeck, W., Keller, G., Adatte, T., and MacLeod, N., 1994b, Deposition of channel deposits near the Cretaceous-Tertiary boundary in northeastern Mexico: Catastrophic or "normal" sedimentary deposits?: Reply; Is there evidence for Cretaceous-Tertiary boundary age deep-water deposits in the Caribbean and Gulf of Mexico?: Reply: Geology, v. 22, p. 955–956.

Walker, R. G., 1978, Deep-water sandstone facies and ancient submarine fans: Models for exploration for stratigraphic traps: American Association of Petroleum Geologists Bulletin, v. 62, p. 932–966.

MANUSCRIPT ACCEPTED BY THE SOCIETY SEPTEMBER 14, 1995

Geological Society of America
Special Paper 307
1996

A sediment gravity flow hypothesis for siliciclastic units at the K/T boundary, northeastern Mexico

Bruce F. Bohor
U.S. Geological Survey, Federal Center, Box 25046, MS 972, Denver, Colorado 80225

ABSTRACT

Coarse siliciclastic beds bracketed by deep-water marls occur at several Cretaceous/Tertiary (K/T) boundary sites in northeastern Mexico. These siliciclastic units are composed of impact spherules and sands deposited by sediment gravity flows, which may have been mobilized on a shallow shelf by a tsunami-like wave or seismic disturbances due to the K/T impact event. Several reasons are presented for these units being impact related and redeposited within a short time span after the event. Other arguments are presented in opposition to a primary depositional origin by tsunami-like impact megawaves on the sea floor. All of the textural evidence in these clastic units supports their primary deposition by sediment gravity currents, and not by waves. The siliciclastic unit at Arroyo el Mimbral is shown to closely resemble the classic idealized Bouma turbidite sequence, except for the basal spherule-bearing bed that shows evidence of emplacement from a fluidized flow. The fluidized and turbidity flows that deposited these units may have originated at multiple incised canyons along a linear source area (shallow shelf) bordering the Sierra Madre mountains. The spherules and sands comprising these units most probably were swept off of the shoreface and shelf by backwash from a tsunami-like wave generated by an asteroid impact on the Yucatán peninsula and deposited at basinal depths as turbidites.

INTRODUCTION

Distal ejecta layers from the Cretaceous/Tertiary (K/T) impact event have been found at sites in marine rocks worldwide ever since the discovery by Alvarez et al. (1980) of elevated Ir values in a thin clay layer bracketed by limestones precisely at the K/T boundary near Gubbio, Italy. However, it has been only recently that proximal ejecta deposits, which are significantly thicker and coarser than known distal ejecta layers, have been discovered enclosed within marine rocks. The reinterpretation of ~50-cm-thick spherule-bearing beds on Haiti as proximal impact ejecta (Hildebrand and Boynton, 1990) contributed to the eventual identification of the putative K/T impact crater on the Yucatán peninsula (Hildebrand et al., 1991).

The Haitian K/T boundary ejecta deposit represents a primary airfall deposit that was modified by local slumping and sliding in a deep oceanic environment (Jéhanno et al., 1992). These localized redepositional processes did not obliterate the overall primary aspect of the Haitian deposits, however. Other proximal K/T boundary deposits were identified in cores by Alvarez et al. (1992) in the Gulf of Mexico, but the depositional mode of these isolated samples was difficult to interpret.

Smit et al. (1992) relocated a clastic-bearing K/T boundary site at Arroyo el Mimbral in northeastern Mexico (State of Tamaulipas) that had been described originally by Muir (1936). The Mimbral outcrop displays the cross section of a mounded siliciclastic deposit, filling and overspreading one or more low-relief channels scoured into upper Maastrichtian Mendez Formation deep-water marls. At the La Lajilla outcrop northwest of Arroyo el Mimbral, this siliciclastic unit is somewhat thinner and not as well developed, but most of the sedimentary units described from Mimbral can be found at La Lajilla as well (see diagrams in Stinnesbeck et al., 1993, their Fig. 2; Keller et al., 1994b, their Figs. 10 and 32). In comparison to Mimbral, both

Bohor, B. F., 1996, A sediment gravity flow hypothesis for siliciclastic units at the K/T boundary, northeastern Mexico, *in* Ryder, G., Fastovsky, D., and Gartner, S., eds., The Cretaceous-Tertiary Event and Other Catastrophes in Earth History: Boulder, Colorado, Geological Society of America Special Paper 307.

the spherule bed and lower parallel-laminated division of the sandstone unit are drastically diminished in thickness at La Lajilla. Nine additional sites containing siliciclastic units enclosed in marine calcareous rocks at the K/T boundary were reported by Alvarez et al. (1992). These occur along a semilinear trend that extends 300 km north-northwest from Arroyo el Mimbral (Fig. 1). In addition, other sites to the south-southeast of Arroyo el Mimbral have been discovered recently (Smit et al., 1994a). The clastic sequences at almost all of these sites consist of two compositionally distinct units: a basal spherule-bearing unit and an overlying siliceous sandstone unit.

Probably more has been written about sediment density flow deposits in the past 35 years than on any other subject in sedimentology. Because of the vastly different geological situations possible for the generation and deposition of these flows, it is not unexpected that various authors have expressed differing opinions about the exact hydraulic mechanisms involved and what to call them. The classification "sedimentary gravity flows," which encompasses turbidity currents, fluidized sediment flows, grain flows, and debris flows (see Middleton and Hampton, 1973), will be modified to "sediment gravity flows" and used in this chapter to describe the generalized mechanism that transported the materials forming the siliciclastic units from a shallow-water environment out into the Mendez basin.

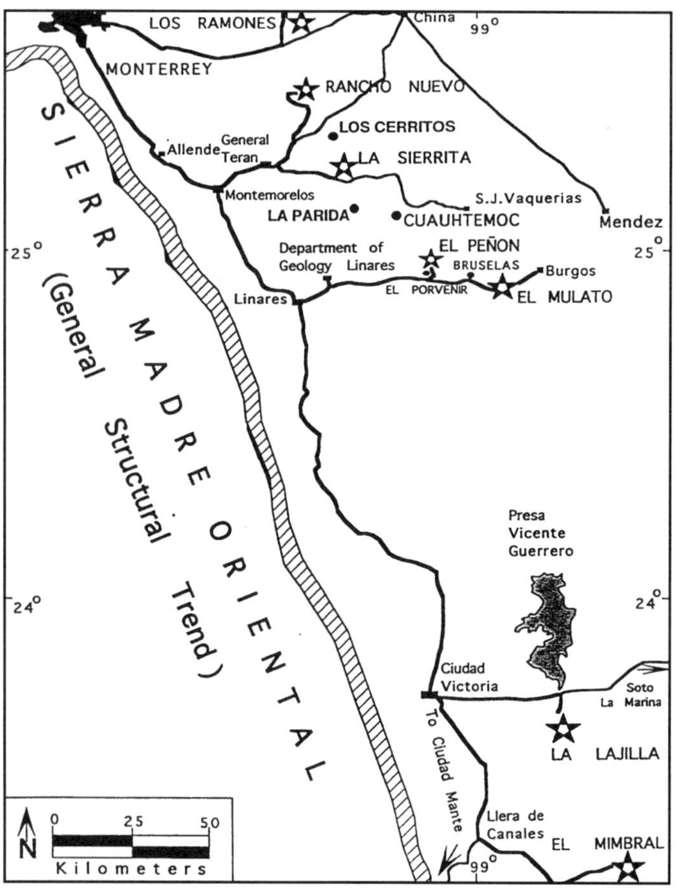

Figure 1. Location map of sites containing siliciclastic units at the K/T boundary in northeastern Mexico. From Keller et al. (1994b).

Bohor and Betterton (1993) and Bohor (1994a, b) investigated the Arroyo el Mimbral and La Lajilla sites and interpreted the lower spherule-bearing unit as a debris-flow deposit (herein designated as a fluidized-flow deposit) and the upper sandstone unit as a turbidite. In their scenario, the mobilization of source materials for both units into sediment gravity flows and deposition from turbidity and other density currents was initiated by disturbances associated with the K/T impact event. Other workers, however, have interpreted these siliciclastic beds quite differently. Smit et al. (1992, 1994a, b) considered these clastic units to have been the result of primary wave deposition by tsunamis (megawaves) generated by the K/T impact. On the other hand, Stinnesbeck et al. (1993) and Keller et al. (1994a) interpreted these same deposits as incised-valley fillings that accumulated over a long time period and were generally unrelated to the K/T impact.

Obviously, these three markedly different interpretations cannot all be correct, and they reflect widely divergent theories about what transpired at the end of the Mesozoic Era. This chapter is an evaluation of these three competing theories based on textural, compositional, and sedimentological features of clastic units at the K/T boundary in northeastern Mexico. Evidence will be presented showing that a sediment gravity flow origin fits the field and laboratory data better than other hypotheses. No radical change of the deep-water sedimentational paradigm is needed in order to incorporate this theory into an impact-related scenario.

DESCRIPTION OF THE CLASTIC UNITS

Nearly a dozen outcrops at the K/T boundary in northeastern Mexico display channelized, coarse siliciclastic deposits bracketed by fine-grained, deep-water marls (Keller et al., 1994b). Close examination of these Mexican clastic deposits reveals two compositionally distinct units: a calcite-cemented bed containing poorly sorted and graded clayey spherules, overlain by a thick, normally graded, medium- to fine-grained, laminated to rippled, calcite-cemented, siliceous sandstone unit. These two siliciclastic units at the Mimbral site are illustrated by Smit et al. (1992, 1994a, b), Stinnesbeck et al. (1993), and Keller et al. (1994a, b). Some of the other known K/T boundary sites with siliciclastic units in northeastern Mexico are described in detail in Keller et al. (1994b). All of these authors designate the basal spherule-bearing bed as Unit 1 and subdivide the overlying sandstone into Units 2 and 3.

Spherule bed

The lithology and composition of the lower spherule-bearing bed is quite different from that of the overlying sandstone bed. The spherule bed (up to 1 m in thickness near the center of the channel at Mimbral) is poorly sorted and inversely graded at its base, with some slight horizontal stratification in its upper portion. The base of the spherule bed is erosionally scoured into the underlying marl of the Mendez Formation, forming a low-relief submarine channel profile (Fig. 2). In-situ rip-up clasts of Mendez marl at and near the base of this bed indicate erosion concurrent with the bed's emplacement (Figs. 3 and 4).

The bed itself is composed of clayey spherules generally 1–3 mm in diameter, set in a matrix of clay, carbonate minerals, and microfossils. Clasts of Mendez marl up to 10 cm in length are present at various levels within this bed and are generally aligned parallel to its base (Fig. 3). White anhedral clasts (1–8 mm in diameter) of recrystallized carbonate rock occur both as scattered grains and somewhat concentrated into dilute lenses within the lower portion of the bed (Fig. 4). These altered limestone clasts may have been derived from platform carbonate target rocks of Campanian age (Smit et al., 1994a), ballistically transported from the impact site on the Yucatán peninsula. Some of the diagenetically altered clayey spherules enclose rare relict glass fragments and complete scalloped glass cores (Fig. 5) that are similar in composition to the relict impact glass cores in the Haitian spherule beds (Table 1). Both Smit et al. (1992) and Keller et al. (1994b, their Fig. 22) report finding rare shocked quartz grains with multiple sets of lamellae in this bed. Upper Maastrichtian benthic foraminifers derived from sediment deposited in shallow neritic shelf environments have been iden-

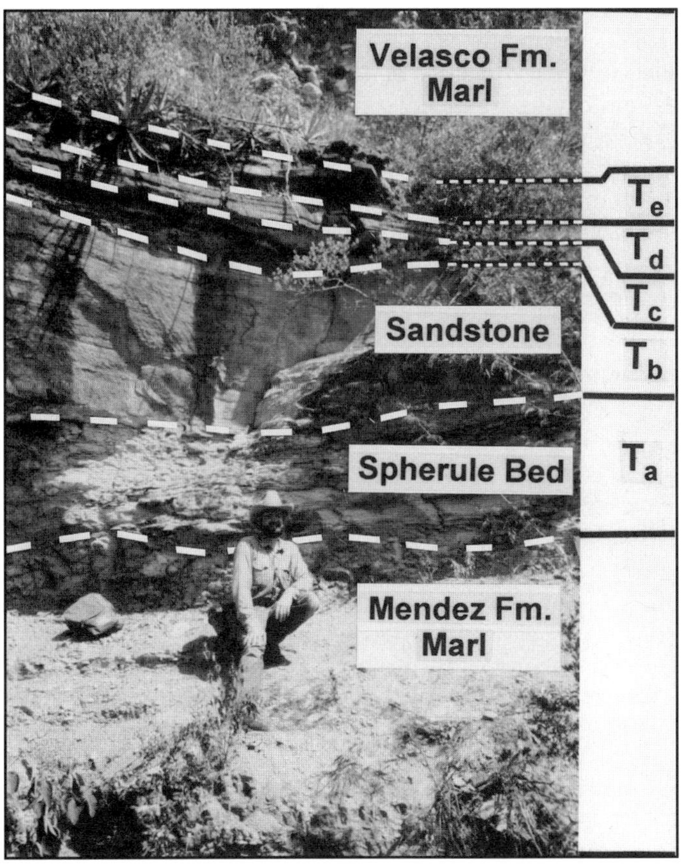

Figure 2. Photo of siliciclastic unit at Arroyo el Mimbral, near mid-channel. White dashes delineate the various subunits designated as idealized Bouma turbidite divisions (T_a–T_e).

Figure 3. Polished section of spherule bed at La Lajilla, cut normal to channel axis. Note piece of Mendez Formation marl being pried up from base and suspended marl clasts (arrows).

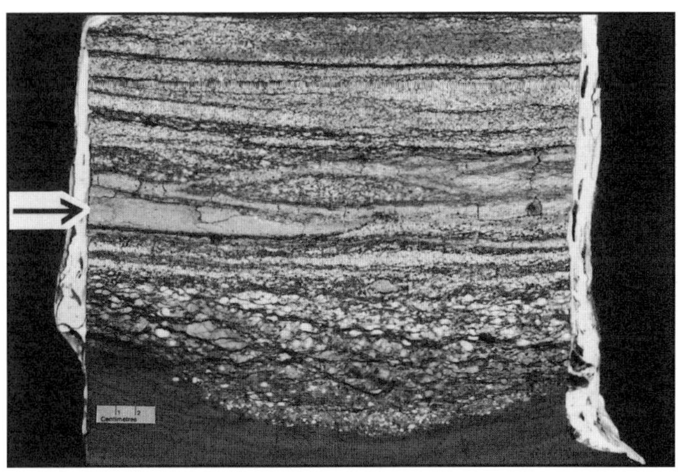

Figure 4. Polished section of spherule bed at La Lajilla, cut parallel to channel axis; current flowed left to right. Note scouring of base into Mendez Formation marl and inverse grading of basal portion. White lens (arrow) near middle of photo is almost wholly composed of fossil tests.

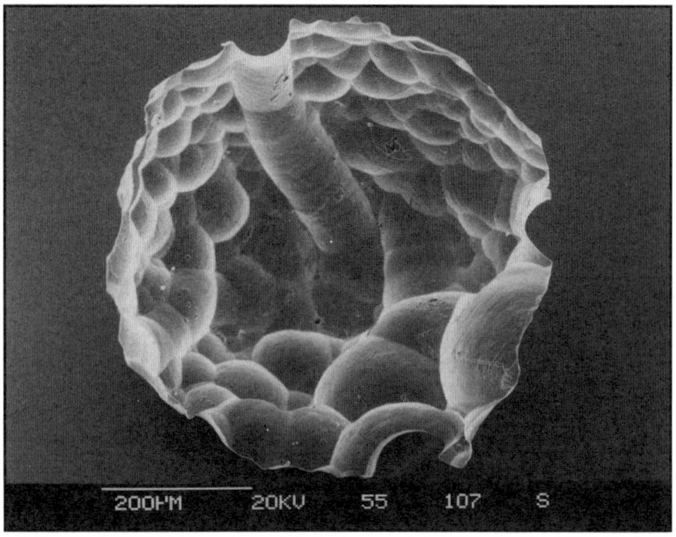

Figure 5. Scanning electron microscope (SEM) photo of relict, scalloped, black-glass spherule core from the spherule bed, Arroyo el Mimbral.

tified in the spherule bed (Keller et al., 1994a; Stinnesbeck et al., 1993).

Polished blocks of the spherule bed at La Lajilla (Figs. 3 and 4) show some apparent stratification or lensing, especially in the upper portion of this unit, but no cross-stratification. Closer examination of this stratification, however, reveals that some of it is due to strung-out clay streaks. This clay appears to be the diagenetic alteration product of glass spherules and fragments (e.g., Bohor and Glass, 1995) and contributed to this pseudo-stratification after being plastically deformed and somewhat mobilized during compaction. Other aspects of the primary stratification seem to be due to local concentrations of recrystallized carbonate clasts and lenses of foram tests (light-colored thin lens near middle of bed, Figs. 3 and 4). The spherule bed exhibits grading, locally inverse at the bottom and normal thereafter (Fig. 4). The upper surface of the bed is somewhat wavy where it is in contact with the overlying sandstone.

Sandstone unit

The overlying quartz-rich, poorly graded, medium-grained sandstone body (~3 m in maximum thickness at Mimbral) is plane-parallel laminated in its lower portion directly above the spherule bed (Fig. 6). This thick parallel-laminated subunit is succeeded upward by a ripple-drifted, cross-laminated to wavy-bedded fine sand bed. A second parallel-laminated fine-sand to silty bed forms a thin subunit above the rippled bed and is capped by a thin interval of alternating silt and clay layers (all beds are labeled with Bouma idealized turbidite sudivisions in Fig. 6). In the lithostratigraphic columns of Mimbral and La Lajilla presented by Smit et al. (1992, 1994a) and Keller et al. (1994b, their Figs. 10 and 32), this upper parallel-laminated zone is not drawn or described as such, even though it is clearly shown photographically in Figs. 35 and 39 of the latter authors' field guidebook paper. The siliciclastic sequence ends with the resumption of normal hemipelagic marl sedimentation, similar to that forming the overlying Velasco Formation (Paleocene).

Smit et al. (1992) identified the sandstone unit as a calcarenite, but in reality it is better described as a quartzose sandstone cemented by calcite. This sandstone unit also contains shallow-water foraminifera similar in age and type to those in the underlying spherule bed (Keller et al., 1994a). The laminated portion of the sandstone unit slightly overlaps the channel scoured out and then filled by the spherule unit (Fig. 7). Both Smit et al. (1992) and Stinnesbeck et al. (1993) found organic fragments (including wood and leaves) and marl clasts near the base of this thick, parallel-laminated sandstone bed; faint sole markings also were seen on the basal surface of this bed. One

TABLE 1. COMPARISON OF MIMBRAL AND HAITI RELICT GLASS ANALYSES

	Mimbral, Mexico, Relict Glasses					Beloc, Haiti, Glasses	
	Dark Brown	Yellow	Dark Green	Light Green	K-rich*	Dark† Brown	Yellow§
SiO_2	60.43	50.38	60.68	60.82	66.20	63.09	49.06
Al_2O_3	15.75	13.78	15.41	14.88	18.73	15.21	13.19
FeO	6.42	5.15	3.63	4.35	5.67	5.44	5.14
MgO	3.79	5.64	1.16	2.46	2.64	2.74	4.09
CaO	7.21	23.66	3.76	5.21	0.84	7.26	24.54
K_2O	1.27	0.61	1.11	0.91	3.68	1.59	0.63
Na_2O	2.65	2.15	1.46	0.99	0.84	3.53	2.10
TiO_2	0.78	0.64	0.83	0.70	0.02	0.67	0.64
MnO	....	....	....	....	0.00	0.14	0.16
Total	98.34	101.2	88.10	90.38	98.83	99.76	99.56

Note: Table data taken from Bohor et al., 1993.
*Smit et al., 1992.
†Sigurdsson et al., 1991.
§Maurrasse and Sen, 1991.

Figure 6. Clastic unit at La Lajilla, with suggested equivalent Bouma sequence notations. Scale at base spans spherule bed (T_a), overlain by parallel-laminated sandstone with undulating upper surface (T_b). In the overlying cross-stratified zone (T_c), wavy and distorted bedding grades upward to climbing ripples and is overlain by another parallel-laminated zone (T_d). Ir- and spinel-bearing fireball layer occurs at base of the capping hemipelagic fallout zone (T_e).

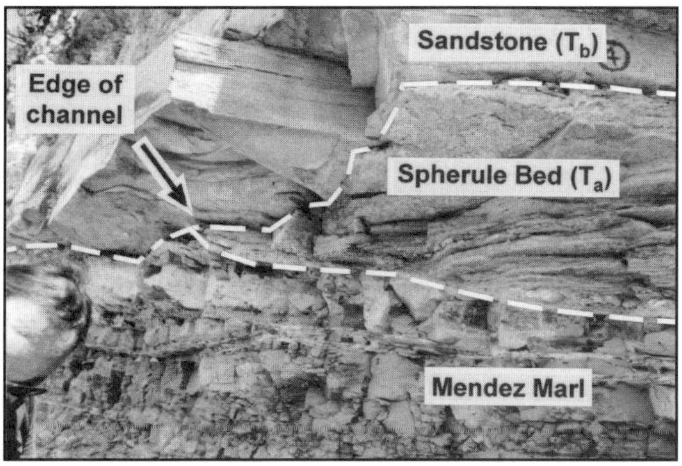

Figure 7. Edge of basal channel deposit filled with spherules at Arroyo el Mimbral. Note overlap of channel edge by overlying parallel-laminated sandstone. Partial head of person at far lower left shows scale.

or two thin stringers or lenses of spherules also may occur near the base of the sandstone bed at some outcrops, but individual spherules are not randomly dispersed within the sandstone unit.

Other investigators have divided the sandstone unit into two separate bodies. Smit et al. (1992, 1994b) placed the division at the base of the "ripple beds," on their contact with the thick, parallel-laminated bed. Stinnesbeck et al. (1993) and Keller et al. (1994a) placed a boundary between their units at about the same place, supposedly marking a disconformity between the underlying thick, parallel-laminated sandstone (their Unit 2) and the wavy cross-bedded and upper parallel-laminated beds (their Unit 3). All of the previous authors also found an Ir anomaly (~0.8–0.9 ppb) in a clay-rich zone at the top of or just above their Unit 3. We have found rare magnesioferrite spinel crystals (Fig. 8) at the position of the Ir peak in the La Lajilla section, but none from a similar position at Mimbral. We made a brief search for shocked quartz grains in the uppermost Ir-bearing layers at both sections but were unsuccessful in finding any with clearly expressed shock features. Not only are these layers massively contaminated with ordinary detrital quartz sand, but the grains are also heavily coated with secondary quartz overgrowth; both conditions make the identification of shocked grains very difficult.

DISCUSSION OF ORIGIN OF THE SILICICLASTIC UNITS

Incised valley–fill, non-impact-related hypothesis

Stinnesbeck et al. (1993) rejected any influence of impact waves and postulated that these siliciclastic units were derived from distant shallow shelf environments and deposited in incised valleys at relatively shallow depths over an extended period of time prior to the K/T boundary event. These "valleys" were proposed to have been incised in the shelf, slope, and basin by a major eustatic sea-level low stand near the end of the Maastrichtian. No modes of transport or deposition were proposed by these authors for these clastic units. Later, Keller et al. (1994a) followed much the same scenario but specified that the clastic units that infilled these incised valleys were emplaced by normal marine sedimentary processes, including gravity flows, turbidity currents, and debris flows. Both of these groups of authors proposed that the deposition of these clastic units took place over thousands of years in shallow water. They based this interpretation on biostratigraphy, apparent disconformities between units, apparent bioturbation within units, settling rates, and other sedimentological features such as grading, climbing

Figure 8. SEM photo of magnesioferrite spinel octahedron from fireball layer at top of T_d (Unit 3) at La Lajilla.

ripples, cross-bedding, and symmetrical ripples at the top of the sandstone unit (interpreted as wave ripples). Most importantly, these authors deny any catastrophic trigger for these clastic deposits and claim a volcanic rather than impact origin for the relict glass in the spherule bed.

The shallow-water part of these authors' scenario can be easily criticized from outcrop data. The sedimentological features of grading, climbing ripples, cross-bedded ripple sets, and parallel lamination also are characteristic of deep-water turbidites (Kuenen, 1964). The shallow-water benthic forams that occur within both the spherule bed and the sandstone unit are not found in the enclosing deep-water marls, indicating that the clastic deposit is allocthonous (resedimented) neritic material brought into basinal depths by sediment gravity currents (e.g, turbidity currents). The apparent disconformities between units are nonerosional, reflecting flow-regime changes and not diastems. The only evidence of bioturbation occurs at the very top of the siliciclastic unit (e.g., at El Peñon), and it only commenced following the deposition of the uppermost layers (Smit et al., 1994a); bioturbation of the uppermost layers is also characteristically present in basinal turbidites (Kuenen, 1964). The hummocky cross-stratification described by Keller et al. (1994a) as wave ripples at the top of the sandstone unit has been interpreted as current ripples by Smit et al. (1994b). The sketch of an idealized valley-filling deposit in Keller et al. (1994a, their Fig. 95b) conforms neither to outcrop characteristics nor to actual cross sections shown in Keller et al. (1994b) and Smit et al. (1992). Instead, the currents that deposited the spherule beds also scoured shallow channels in basinal marls, and these are overtopped by sandstone bodies that extend laterally somewhat beyond these scour-channel margins (Fig. 7). Multiple channels or branches may be present at the Mimbral outcrop. Most of the compression of underlying marl beds is due to differential compaction and loading by the clastic unit. No evidence exists for pre-existing surface morphology, such as deeply incised valleys, in these basinal marls.

Stinnesbeck et al. (1993) and Keller et al. (1994a) rejected an impact origin for the spherules in the spherule bed, considering them to be volcanic instead. However, the relict glass in the Mimbral spherule bed has been correlated compositionally with glass from Haitian spherule beds (Table 1). Both of these ejecta glasses have been dated and give the same radiometric age as melt-glass from within the K/T impact crater in Yucatán (Swisher et al., 1992). Also, shocked quartz grains with multiple sets of lamellae have been identified from the spherule bed at Mimbral (Smit et al., 1992). Thus, on the basis of these strongly indicative impact signatures, there is little reason to doubt the impact origin of most of the material in the spherule bed.

Tsunami deposition hypothesis

Smit et al. (1992) stated that the basal part (the spherule bed) of the clastic unit at Mimbral was deposited in place on the sea floor by an impact-induced tsunami or megawave operating at depths they estimated as greater than 400 m (but may have been as much as 1,000 m). These authors admit that the K/T bolide could not have directly generated such an enormous wave by impacting a shallow carbonate shelf in the area of northern Yucatán but proposed instead that several other impact-associated mechanisms theoretically could have done so. The K/T tsunami concept was proposed (Bourgeois et al., 1988; Hildebrand and Boynton, 1988) before a crater was identified, when the possibility still existed that the bolide might have impacted into oceanic crust in a deep ocean basin. This oceanic-impact theory was based on erroneous geochemical interpretations of the marine K/T boundary layer (Shaw and Wasserburg, 1982; Hildebrand and Boynton, 1987, 1988, 1990) that mistook the basaltic signature of chondritic bolide components in the fireball layer as indicating a deep-ocean impact target site. This interpretation was not supported by the mineralogy of this layer (Bohor, 1990) and has now been further invalidated by the discovery of the putative impact crater on the northern end of the Yucatán peninsula. The impact site was a shallow carbonate platform at the end of the Maastrichtian, underlain at depth by crystalline rocks of continental crust (Lopez Ramos, 1981).

Smit et al. (1992) proposed that the sandstone unit (overlying the basal spherule unit) at Mimbral resulted from backwash off the coast and shallow shelf due to an impact megawave, but they do not say how these sediments were transported into their present basinal location. They considered the possibility of a turbidity current origin for the sandstone body at Mimbral, but rejected it in part because (1) its occurrence was apparently unique in the locally exposed Mendez-Velasco strata; (2) they did not find any Bouma sequences in it; (3) a strong contrast in lithologies existed between locally derived intraclasts and the ejecta products of the spherule bed, and the far-transported sands and wood fragments with borings of the sandstone unit; (4) there were indications of alternating directions of infill in channel-like features; and (5) they found successive laminated beds that are slightly lenticular and overlapping without basal channeling. These five points do not disprove a turbidite (sediment gravity flow) origin, however, for the following reasons.

1. The siliciclastic units found at the K/T boundary apparently are not unique in the upper Maastrichtian. Keller et al. (1994a, p. 94) report that clastic bodies are commonly found elsewhere within the Mendez Formation, for example, at Linares, Mexico.

2. The siliciclastic unit at Mimbral is composed almost entirely of clearly recognizable classic Bouma turbidite sequences. Comparison of the Mimbral section (Fig. 9a) with an idealized Bouma sequence (Fig. 9b) shows that the subdivisions of the sandstone unit at Mimbral correspond directly with the classic turbidite sequence T_b-T_e in the correct stratigraphic order. The normal basal subdivision of the idealized Bouma sequence (T_a of Fig. 9b) is modified somewhat in these K/T sequences because of compositional and textural differences between its spherical ejecta components and the coarse sand that would normally be present here in a typical sand-rich tur-

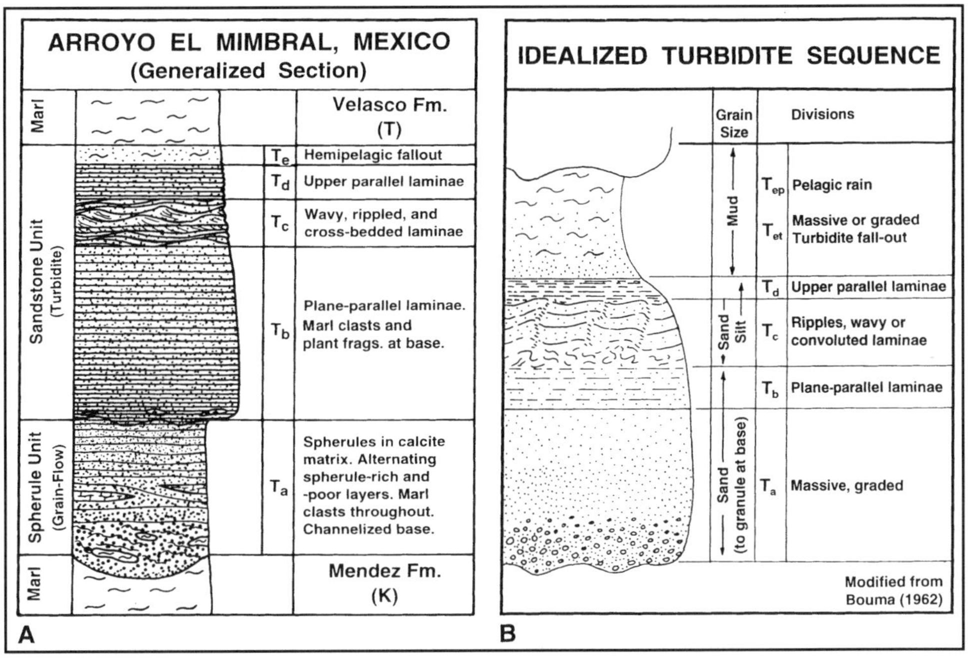

Figure 9. A, Diagram of stratigraphic section through siliciclastic unit at Arroyo el Mimbral. Unit is subdivided and labeled with corresponding idealized Bouma turbidite divisions. Thickness of unit is about 3 m. B, Idealized complete T_{a-e} Bouma turbidite sequence, showing hemiplagic subdivision as the T_e unit (from Howell and Normark, 1982). Compare with sections through the siliciclastic unit at Arroyo el Mimbral shown in A.

bidite. Summaries of turbidite characteristics are presented by Kuenen (1964), Sanders (1965), and Walker and Mutti (1973).

3. Lithological contrast is strong between components of the spherule bed and the sandstone unit because of the unique circumstances of emplacement of the spherules on a shallow-water shelf or upper-slope environment by ballistic impact ejection. The deep-water Mendez marl intraclasts of the spherule bed are rip-up clasts formed by scouring, but this bed also contains shallow-water benthic forams indicative of transport from a distant shallow-shelf environment (Stinnesbeck et al., 1993; Keller et al., 1994a). The overlying sandstone unit also contains similar benthic forams as well as wood and other organic debris that indicate that it was redeposited after erosion from a shallower-water site. In comparison, the clastic sediments of turbidites also usually originate in relatively shallow water and are transported into deeper basins by sediment gravity (turbidity) currents (see Sanders, 1965, p. 193).

4. It is not clear what channel-like features Smit et al. (1992) are referring to when they use alternation of infilling directions as an objection to a turbidite origin. The spherule bed was deposited in shallow channels that were simultaneously being eroded by the passage of dense sediment gravity flows. As in any channel, deposits here would be expected to show some degree of oblique bedding, especially near channel margins. Figures 3 and 4 show polished blocks of the spherule bed at La Lajilla at mid-channel, cut both normal and parallel to the channel axis, and no evidence of cross-bedding is evident in either of these blocks. On the other hand, unidirectional impact waves could not be expected to form deposits with alternating directions of infill either. Furthermore, this wave could not cut out a narrow channel parallel to its direction of propagation and simultaneously fill it with spherules.

If these alternating directions refer to cross-laminations in the sandstone unit, then a different situation applies. At La Lajilla, Smit et al. (1994b) purported to have measured current directions on cross-laminated layers that were oriented almost 180 degrees from each other (their Fig. 101) and interpreted this as indicative of the back-and-forth passage of up to nine different tsunami megawaves or seiches. Almost all of the exposure faces at this site are perpendicular to the apparent channel axis, however, not parallel to it. If care is not taken to measure cross-laminae parallel to the channel axis, such measurements oblique or normal to this implied direction of flow often can provide misleading information (Walker, 1970). Many authors have cautioned against the use of cross-laminae for directional purposes in flysch deposits (Enos, 1969) and have recommended that sole markings and grain imbrication measurements be used instead (e.g., Sestini and Pranzini, 1965). Some authors have even gone so far as to refuse to use cross-laminae for determining current directions because of their general unreliability for this purpose (e.g., Macdonald and Tanner, 1983).

Antidunes were reported at La Lajilla by Smit et al. (1994b, their Fig. 101), and current directions measured by them from these features were generally almost opposite to their other primary directional indicators. However, Skipper (1971) reported up-current inclinations measured on antidune ripples from turbidite upper flow regimes; in his study, individual sets of laminae were opposed at different levels, producing a "herringbone" pattern.

5. Lenticularity and slight overlapping of laminated units without apparent basal channeling has little to do with the rejection of these units as turbidites, because these features often can exist in channelized turbidite architecture as well.

As has been shown above, most of the objections to a turbidite origin for these clastic units presented by Smit et al. (1992, 1994b) cannot be justified. On the other hand, the rationale for tsunami (megawave) primary deposition of these clastic beds as promulgated by these authors is extremely vulnerable to criticism on the basis of hydraulic theory. Smit et al. (1994b) emphasized that all of the internal features of these clastic units at Mimbral and La Lajilla are due to *currents*, without exception. I agree with this assessment and further stipulate that these currents were caused by gravity (turbidity) flows that deposited the sandstone units and formed the antidunes, parallel laminae, cross-stratified laminae, climbing ripples (ripple-drift), wavy bedding, and so on. However, Smit et al. (1992, 1994b) further stated that impact-generated tsunamis or megawaves *directly* deposited the basal spherule unit and caused the current-derived internal features of the sandstone unit by the repetitive passage of reversing seiche megawaves. This is extremely unlikely for these types of waves, because neither individual waves nor wave trains at sea can deposit anything and are capable only of rearranging what is already on the bottom. When waves move into shallower water, the circular orbits of the water particles at the surface become progressively more elliptical with increasing depth. On a shoaling shelf, this elliptical motion is replaced by a straight-line to-and-fro motion of the water (Johnson and Baldwin, 1986). Thus, waves can move grains back and forth on the sea floor only in shallow water and can create strong return currents (rips) only after breaking upon a shoreface. Figure 10, taken from Harvey (1976), illustrates this orbital motion of water during the passage of a wave in both deep and shallow water.

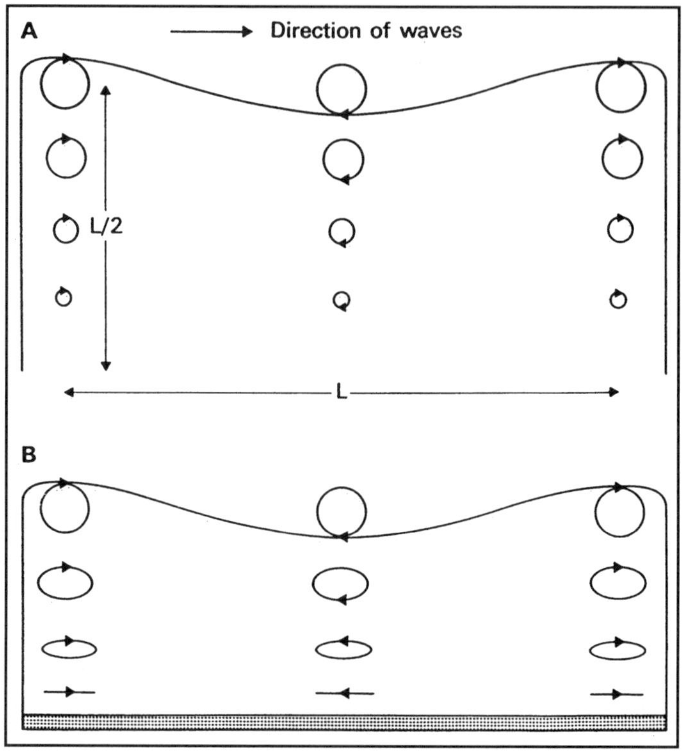

Figure 10. Diagram showing water particle motion in deep sea, A, and in shallower water, B. Note that water (and any suspended sediment) moves in circular orbits upon the passage of a wave in the deep sea (from Harvey, 1976).

It is clear that all of the internal sedimentological features of these siliciclastic units are formed by currents, and microfossil evidence indicates that materials comprising both the spherule-bearing and sandstone units originated in shallower water. Logically then, neither of these clastic units could have been directly deposited *in situ* by tsunami-like impact waves. By way of example, a rare literature reference to a deep-water deposit ascribed to tsunami waves (from the caldera collapse of Santorini volcano in the Mediterranean) described it as composed of single, graded, but otherwise featureless units called homogenites (Kastens and Cita, 1981). These homogenites were partially composed of abyssal muds liquified and stirred up by pressure oscillations from the passage of the tsunami wave and partially from debris-flow/turbidites triggered on nearby shelf slopes by this wave. Note that this featureless deposit in no way (except for the presence of grading) resembles the K/T channelized siliciclastic deposits in northeastern Mexico that display almost all the sequences of an idealized Bouma turbidite (e.g., Middleton and Hampton, 1973).

Sediment gravity flow hypothesis

Field evidence clearly supports a sediment gravity flow genesis for these siliciclastic deposits and indicates that these flows were closely related to the K/T impact event, probably being triggered by it. However, insufficient outcrops exist to demonstrate conclusively how these resedimented deposits are tied to a definite source area, so that a only a plausible scenario of how they may have been mobilized and transported out into the Mendez basin can be presented here. Analyses of sedimentary features in both the spherule bed and the overlying sandstone unit will be marshalled to support the sediment gravity flow origin of these deposits and to suggest an impact-related, multiple-source scenario for their initiation and subsequent transport.

Spherule bed. The siliciclastic deposit at Arroyo el Mimbral (Fig. 9a) closely resembles an idealized Bouma turbidite sequence (Fig. 9b), except possibly for the basal portion (T_a). This latter subdivision is composed of coarse impact spherules that were probably transported and deposited by some sort of fluidized flow mechanism, rather than in turbulent suspension. The proximal gravity (debris-type) flow initiating transport of the spherules may have changed into a fluidized flow farther downslope and out into the basin (e.g., Howell and Normark, 1982). This fluidized-flow mode probably was accompanied by a turbidity current that suspended shelf sands and silts above and slightly behind it in a less-dense turbid suspension, as had been shown to occur experimentally by Hampton (1972).

The presence of relatively undeformed marl clasts within the spherule bed suggests that this unit was emplaced by a non-turbulent, laminar-flow, cohesionless current such as a fluidized flow, rather than by a turbidity current. The reversed grading at the bottom of the spherule unit, the relative lack of primary laminations, and the presence of marl rip-up clasts (see Kano and Takeuchi, 1989) throughout the bed are characteristic of these types of gravity-flow deposits (e.g., Hiscott, 1994, who described them as debris flows). This single-pulse scenario could account both for the exotic gravity-flow characteristics of the spherule bed and for the idealized turbidite sequence of the sandstone unit above it.

The basal spherule-bearing bed at Mimbral is composed of altered impact melt ejecta that was originally deposited ballistically onto a shallow shelf, according to ecological indications from included forams. Spherule deposition may have been restricted to the shelf and shoreface locales, implying a possible raylike distribution of melt ejecta from the impact crater, as was suggested by Pollastro and Bohor (1993) for more distal ejecta in the Western Interior. Although the spherules were originally deposited on top of sands on the shelf and shoreface, they now occupy a basal position in these basinal clastic units. Two different mechanisms may explain this reversal of stratigraphic position. The first involves spherule morphology and mass. These large (1–3 mm) spherules were solid glass with hydration rims (see Bohor and Glass, 1995) shortly after deposition, so they may have rapidly migrated to the bottom of a turbulent flow unit because of size and density differences between them and the comobilized shelf sands and silts. A second mechanism involves the sequence of erosion or mobilization. As the impact-generated wave receded, strong ebb (rip) currents generated by the backwash would have swept recently deposited spherules off the shelf and shoreface and into the mouths of canyons on the shelf. This erosion of the last-deposited spherules first and the sweeping of them into the canyons prior to erosion of the underlying shelf sands could have caused a reversal of initial stratigraphic positions when these sediments were finally redeposited by sediment gravity currents in the basin.

Sandstone unit. The thick, parallel-laminated quartzose sandstone unit (T_b) rests disconformably on top of the spherule bed, separated from it by a slightly undulating surface. This apparent disconformity between the spherule bed and the sandstone unit may be caused by differences in flow regimes and sediment densities, and its development has been noted in other turbidites (Walker, 1967). Standing waves at the top of subdivision T_a or "churning" by interfacial waves (Middleton and Hampton, 1973) may be responsible for this undulating contact between these two types of current flow deposits (laminar and turbulent). Figure 11 (from Walker, 1967) shows this undulating contact between these units in a turbidite. Thus, this contact probably represents a paraconformity and not an erosional disconformity. Similarly, the contact between Units 2 and 3 of Stinnesbeck et al. (1993) and Keller et al. (1994a) also is not a disconformity, and instead reflects the change from an upper flow regime to a lower flow regime within the turbidity current.

The entire sandstone unit is normally graded and fines upward to alternating silts and clays at the top. Shelf sands in turbulent suspension followed along the same shallow courses scoured out by the spherule-bearing fluidized-flow unit and slightly overlapped both sides of the channels created by the latter (Fig. 7). Except for grain size, timing, and scale of the units involved, this model can be compared with that of chaotic silty debris-flow deposits and sandy turbidites that were transported as channelized flows for more than 600 km across the floor of the Gulf of Mexico during the Pleistocene. Figure 12, taken from Nelson et al. (1992), shows a generalized diagram of these channelized flow units. Comparison of the spherule bed with these Pleistocene debris flows is only presented here to show that the fluidized mode of transport can reach far out into a basin, is usually channelized, and includes suspended, oriented mud clasts. Due to the unique nature of the K/T event that instantly loaded a substantial thickness (probably ~1 m) of glass spherules onto prexisting shelf sands and silts, these clastic units

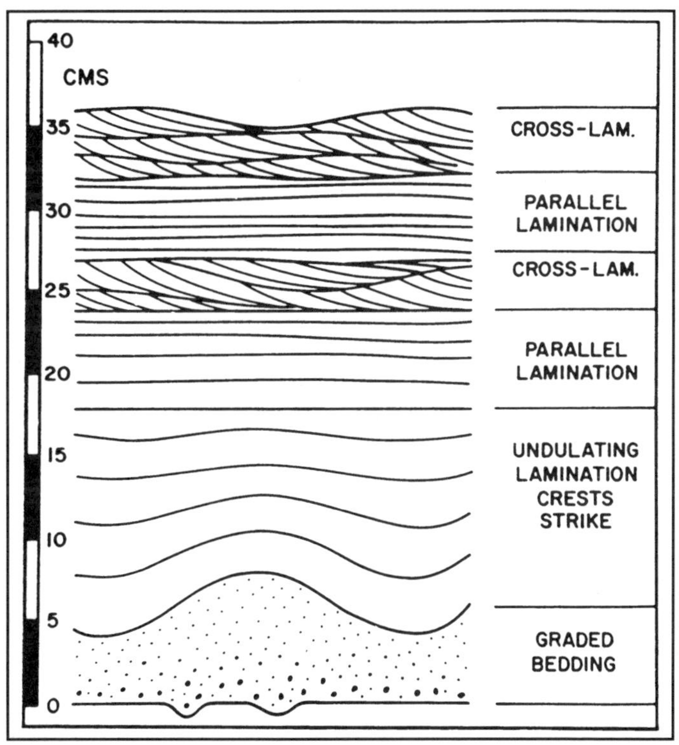

Figure 11. Diagram modified from Walker (1967) showing standing wave–form development at contact between T_a and T_b divisions (base of parallel-laminated unit). Uppermost unit in turbidite pictured is a repeat of cross-laminated division (T_c); this repetition is not characteristic of idealized Bouma sequence.

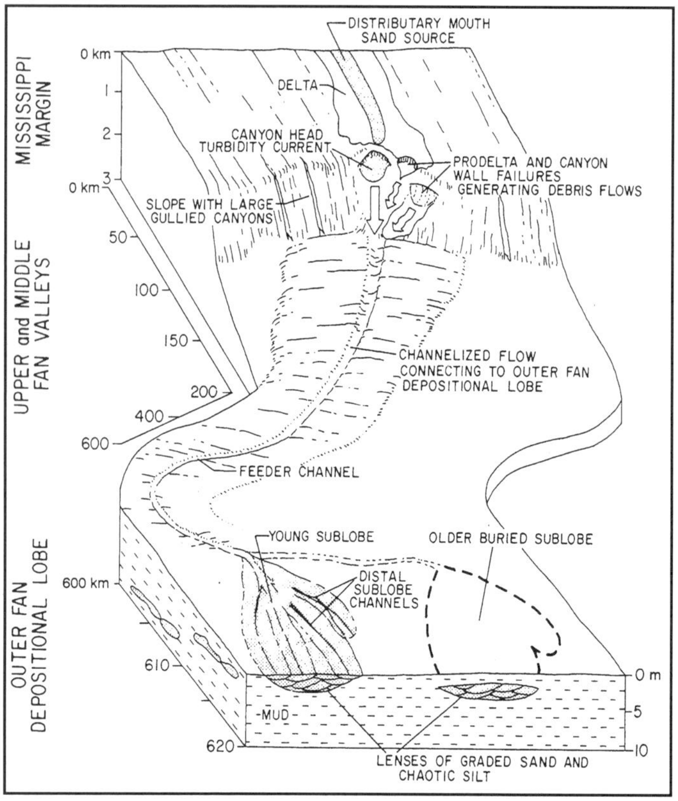

Figure 12. Conceptual diagram from Nelson et al. (1992) depicting sublobe stratigraphy and lithology of chaotic silt debris flows and graded sand turbidites of Mississippi outer fan during the Pleistocene.

may not closely resemble *any* other known deposits resulting from gravity flow of normal shelf-sediment accumulations. Most turbidites described in the literature were deposited in a submarine fan environment as broad, repetitive sheets, and not as single pulses of channelized sediment gravity flows.

Fireball layer. Coincident with the uppermost layers of alternating silt and clay at the top of the sandstone unit (e.g., Walker and Mutti, 1973) is a clay-rich layer that contains an Ir anomaly and Ni-rich magnesioferrite spinel crystals; the latter signatures identify it as the K/T fireball layer (Bohor, 1990), diluted with the fine tailings of the turbidity flow and normal hemipelagic sedimentation. Thus, these clastic flow units were mobilized from shallow water and emplaced in the basin after the impact had finished distributing the spherules of the melt-ejecta layer ballistically via an ejecta curtain and before the air-fall fireball layer had finished settling through both the atmosphere and a thick water column. This scenario appears to conform with the dual-dispersion-mode model for the K/T impact event (Bohor and Betterton, 1992; Pollastro and Bohor, 1993) and suggests that these clastic units were deposited in a matter of days after impact. Jéhanno et al. (1992) calculated that the settling of most components of the fireball layer (0.2–0.05 mm particles) at the deep-water Haiti K/T boundary sites took from a few days to more than a week in ~2,000 m of water. In the ~500 m of water estimated for the depth of the Mendez basin, these fireball components would have settled to the bottom over a period from one to several days.

Emplacement scenario for sediment gravity flow deposits

The trend of channel axes of the 10 siliciclastic deposits at the K/T boundary in northeastern Mexico is north-northwest–south-southeast and parallels the axis of the Mendez basin. During the Late Cretaceous, a shallow shelf bordered the rising Sierra Madre mountain range, more than 75 km west-southwest of where these turbidites are located (Stinnesbeck et al., 1993; Keller et al., 1994b). The paleoslope of the Mendez basin strikes toward the south-southeast (Keller et al., 1994b). It is postulated that when the linear-sourced sediment gravity flows debouched from canyons in the shelf/slope, they turned and flowed down the axis of the basin along the strike of the paleoslope. Examples of this change from lateral to longitudinal flow direction are well known from the literature. Walker (1970) cites several references to this reorientation of flow when the axial trough is reached. Hsü et al. (1980) and Haughton (1994) provide models and examples of this longitudinal transport. A Cretaceous example is reported from the Chugach terrane in the Gulf of Alaska (Nilsen, 1985), where linear-sourced lateral gravity flows became longitudinally flowing turbidity currents out in the forearc trough. A similar system probably existed in northeastern Mexico at the end of the Cretaceous, and multiple linear slump sources (e.g., Fig. 17 in Reading and Richards, 1994) may have been activated on the sand-rich shelf and slope edge by impact-related disturbances. Primary current directional indicators predominantly reflect a longitudinal direction of transport (Smit et al., 1994b), with a few directional anomalies from lower-flow-regime cross-laminated beds that may be due to the influence of bottom currents upon the waning phase of the turbidity flows. The multiple linear-source mechanism could explain the thickness variations observed among the dozen or so known sites. If these siliciclastic units were being fed from a single source at the northwest end of the Mendez basin as suggested by Keller et al. (1994a), then they should progressively thin and change to more distal facies to the southeast; this does not seem to be the case with respect to these outcrops. Furthermore, not all of the outcrops of these siliciclastic units show the almost complete sequence of idealized turbidite units displayed in the Mimbral outcrop; divisions T_a and T_b are sometimes missing or modified in outcrops of sections northwest of Mimbral. These thickness and stratigraphic variations seem to negate the concept of a single source for these deposits. Thus, this scenario provides a predictive model for the location and characterization of future outcrops as well as constraints on impact-generated, deep-water sedimentation throughout the Gulf of Mexico.

Significance of the controversy over origin

Smit et al. (1992, 1994a, b) and Bohor and Betterton (1993) and Bohor (1994a, b) agree that these siliciclastic units are directly related to the K/T impact event, as confirmed by the impact signatures in the spherule bed and the fireball layer. However, these two groups of authors disagree about how these clastic units got to their present locations in a deep-water basin. The Smit group(s) invoked an impact megawave to perform the actual deposition of the spherule bed on the sea floor and to create a series of oscillating waves (seiches) that deposited the sandstone unit. Bohor and coworkers rejected the notion of an impact wave large enough to reach basinal depths as well as the ability of such waves to deposit anything on the deep-sea floor. Instead, their scenario invoked sediment gravity flows as the means of transport of impact spherules and shelf sands from shallow-water environments and their redeposition as channelized turbidites within deep-water marls of the Mendez Formation. The significance of this difference extends beyond the interpretation of these few outcrops and will define the paradigm for the K/T event. Trying to force an undocumented megawave mechanism upon these deep-sea siliciclastic units, when they can be more readily explained as the product of normal, well-documented sedimentary processes, is not scientifically sound procedure. Both the internal features and their stratigraphic sequence in the siliciclastic units are consistent with those of the classic idealized Bouma turbidite, with the possible exception of the basal spherule unit. However, even this unit could be reconciled as part of a turbidite sequence (e.g., the S_3 subdivision of Lowe, 1980).

SUMMARY AND CONCLUSIONS

Based on field and laboratory studies, the siliciclastic deposits in northeastern Mexico are interpreted as sediment gravity flows (mainly turbidites whose basal units may have been deposited from fluidized flows) that were probably emplaced in a single-pulse mode. The basal spherule units are composed of altered impact-glass droplets ejected from the putative K/T impact crater on the Yucatán peninsula. The impact nature of the spherule bed was previously confirmed by identification of shocked quartz and relict impact glass.

The almost instantaneous loading of a thick deposit of coarse glass spherules into shallow water constitutes a unique geologic event. An impact-modulated scenario for the mobilization and transport of these materials postulates that a shallow sandy shelf was located west of the present sites in the Mendez basin, bordering the Sierra Madre Oriental mountains. Thick, unstable deposits of impact spherules may have been mobilized along multiple linear sources on the shelf, forming sediment gravity flows by backwash from impact-generated waves or by slumps caused by seismic events associated with impact. These sediment gravity currents would have flowed down canyons previouly incised into the shelf/slope; the dense flows may have helped generate their own superjacent turbidity currents from comobilized unconsolidated shelf sands. This process resulted in the deposition of siliciclastic units, consisting of basal beds of glass spherules emplaced by laminar fluidized flow overlain by shelf sands deposited from turbidity currents, in distal shallow channels among deep-water basinal marls. Channel scouring by the basal fluidized-flow unit is exemplified by the presence of rip-up clasts of underlying Mendez marl within this bed, which is deposited in single or multiple typical shallow channels displaying clearly defined edges. The present orientation of these channelized deposits may be due to longitudinal flow of the turbidity currents upon reaching the axis of the Mendez basin.

Both the spherule unit and the sandstone unit contain benthic foraminifers of late Maastrichtian age derived from a shallow neritic environment, whereas benthic taxa are rare in the underlying, deep-water Mendez marl. This strongly suggests that the entire siliciclastic deposit is allochthonous, having been transported from shallow water and resedimented in its present basinal setting via sediment gravity flows.

The final fine-grained phase of turbidite deposition and recommencement of hemipelagic rain coincided with the arrival of the last portion of the fireball layer at the sea floor; components identifying this airfall fireball layer include an Ir anomaly and magnesioferrite spinel crystals. The presence of a capping fireball layer limits the time of formation of this clastic unit to a period of several days or so following impact, in accordance with the dual-dispersion-mode model of a single, large, K/T impact event.

No evidence exists to suggest that these siliciclastic units were deposited in shallow water in preincised valleys over a period of thousands of years. Based on impact signatures, the spherule bed is clearly composed of melt-ejecta materials derived from the K/T impact event, and an Ir-bearing fireball layer capping the siliciclastic units, along with a lack of interunit bioturbation, constrains their deposition to a very short time period measured in days.

The tsunami theory of deposition fails to explain these units on the basis of both hydraulics and impact location. Waves cannot form such deposits in the deep sea and, moreover, no truly gigantic wave could have been generated from the known impact site location on a very shallow carbonate shelf in Yucatán. It is difficult to ignore the close feature-for-feature correspondence between the sandstone units of these siliciclastic deposits and an idealized Bouma turbidite sequence. The possibility that any deep-sea wave-formed deposit would so closely resemble a channelized, current-deposited turbidite in both the type of internal features and their exact sequence seems extraordinarily remote.

ACKNOWLEDGMENTS

Thanks are due to Walter Alvarez for providing outcrop locations for Arroyo el Mimbral and La Lajilla, and to Homer Montgomery, Emile Pessagno, and James Carter from the University of Texas at Dallas for accompanying the author and Bill Betterton on a field trip to these sites; Bill Betterton provided technical support in the field and in the lab. Reviews by Joan Bourgeois and Robert Dott provided valuable suggestions for improving this manuscript. Partial support for this research was provided by NASA Grant T-5715P.

REFERENCES CITED

Alvarez, L. W., Alvarez, W., Asaro, F., and Michel, H. V., 1980, Extraterrestrial cause of the Cretaceous-Tertiary extinction: Science, v. 208, p. 1095–1108.

Alvarez, W., Smit, J., Lowrie, W., Asaro, F., Margolis, S. V., Claeys, P., Kastner, M., and Hildebrand, A., 1992, The Cretaceous-Tertiary boundary impact tsunami deposit in NE Mexico [abs.]: Geological Society of America Abstracts with Programs, v. 24, p. A331.

Bohor, B. F., 1990, Shocked quartz and more: Impact signatures in Cretaceous/Tertiary boundary clays, *in* Sharpton, V. L., and Ward, P. D., eds., Global catastrophes in Earth history: Geological Society of America Special Paper 247, p. 335–342.

Bohor, B. F., 1994a, KT clastic deposits, northeastern Mexico:Single-pulse debris flow/turbidite units associated with impact, *in* Keller, G., Stinnesbeck, W., Adatte, T., MacLeod, N., and Lowe, D. R., eds., Field guide to Cretaceous-Tertiary boundary sections in northeastern Mexico: Houston, Texas, Lunar and Planetary Institute Contribution 827, p. 102–106.

Bohor, B. F., 1994b, Debris flow/turbidite clastic units at the KT boundary, northeastern Mexico [abs.], *in* Papers presented to the Conference on new developments regarding the KT event and other catastrophes in Earth history: Houston, Texas, Lunar and Planetary Institute, p. 13–14.

Bohor, B. F., and Betterton, W. J., 1992, Ejection and dispersal mechanisms of the K/T impact [abs.]: Lunar and Planetary Science Conference XXIII, Houston, Texas, Lunar and Planetary Institute, p. 135–136.

Bohor, B. F., and Betterton, W. J., 1993, Arroyo el Mimbral, Mexico, K/T unit: Origin as debris-flow/turbidite, not a tsunami deposit [abs.]: Lunar and Planetary Science Conference XXIV, Houston, Texas, Lunar and Planetary Institute, p. 143–144.

Bohor, B. F., and Glass, B. P., 1995. Origin and diagenesis of the K/T impact spherules—From Haiti to Wyoming and beyond: Meteoritics, v. 30, p. 182–198.

Bohor, B. F., Glass, B. P., and Betterton, W. J., 1993, Spherules from Haiti, Wyoming and Ivory Coast: Origin and diagenesis [abs.]: Lunar and Planetary Science Conference XXIV, Houston, Texas, Lunar and Planetary Institute, p. 145–146.

Bouma, A. H., 1962, Sedimentology of some flysch deposits: Amsterdam, Elsevier Publishing, 168 p.

Bourgeois, J., Hansen, T. A., Wilberg, P. L., and Kauffman, E. G., 1988, A tsunami deposit at the Cretaceous-Tertiary boundary in Texas: Science, v. 247, p. 567–570.

Enos, P., 1969, Anatomy of a flysch: Journal of Sedimentary Petrology, v. 39, p. 680–723.

Hampton, M. A., 1972, The role of subaqueous debris flow in generating turbidity currents: Journal of Sedimentary Petrology, v. 42, p. 775–793.

Harvey, J. G., 1976, Atmosphere and ocean: Our fluid environments: Sussex, England, Artemis Press, 143 p.

Haughton, P. D. W., 1994, Deposits of deflected and ponded turbidity currents, Sorbas basin, southeast Spain: Journal of Sedimentary Research, v. A64, p. 233–246.

Hildebrand, A. R., and Boynton, W. V., 1987, The K/T impact excavated oceanic mantle: Evidence from REE abundances [abs.]: Lunar and Planetary Science Conference XVIII, Houston, Texas, Lunar and Planetary Institute, p. 427.

Hildebrand, A. R., and Boynton, W. V., 1988, Provenance of the K/T boundary layers [abs.], *in* Global catastrophes in Earth history: An interdisiplinary conference on impacts, volcanism, and mass mortality: Houston, Texas, Lunar and Planetary Institute Contribution 673, p. 78–79.

Hildebrand, A. R., and Boynton, W. V., 1990, Proximal Cretaceous-Tertiary boundary impact deposits in the Caribbean: Science, v. 248, p. 843–847.

Hildebrand, A. R., Penfield, G. T., Kring, D. A., Pilkington, M., Camargo Z., A., Jacobsen, S. B., and Boynton, W. V., 1991, Chicxulub Crater: A possible Cretaceous/Tertiary boundary impact crater on the Yucatán Peninsula, Mexico: Geology, v. 19, p. 867–871.

Hiscott, R. N., 1994. Traction-carpet stratification in turbidites—fact or fiction?: Journal of Sedimentary Research, v. A64, p. 204–208.

Howell, D. G., and Normark, W. R., 1982, Sedimentology of submarine fans, *in* Scholle, P. A., and Spearing, D., eds., Sandstone depositional environments: Tulsa, Oklahoma, American Association of Petroleum Geologists, p. 365–404.

Hsü, K. J., Kelts, K., and Valentine, J. W., 1980, Resedimented facies in Ventura Basin, California, and model of longitudinal transport of turbidity currents: American Association of Petroleum Geologists Bulletin, v. 64, p. 1034–1051.

Jéhanno, C., Boclet, D., Froget, L., Lambert, B., Robin, E., Rocchia, R., and Turpin, L., 1992, The Cretaceous-Tertiary boundary at Beloc, Haiti: No evidence for an impact in the Caribbean area: Earth and Planetary Science Letters, v. 109, p. 229–241.

Johnson, H. D., and Baldwin, C. T., 1986, Shallow siliciclastic seas, *in* Reading, H. G., ed., Sedimentary environments and facies: Oxford, Blackwell Scientific, p. 229–282.

Kano, K., and Takeuchi, K., 1989, Origin of mudstone clasts in turbidites of the Miocene Ushikiri Formation, Shimane Peninsula, Southwest Japan: Sedimentary Geology, v. 62, p. 79–87.

Kastens, K. A., and Cita, M. B., 1981, Tsunami-induced sediment transport in the abyssal Mediterranean Sea: Geological Society of America Bulletin, v. 92, p. 845–857.

Keller, G., Stinnesbeck, W., Adatte, T., Lopez-Oliva, J. G., and MacLeod, N., 1994a, The KT boundary clastic deposits in northeastern Mexico as product of noncatastrophic geologic processes? *in* Keller, G., Stinnesbeck, W., Adatte, T., MacLeod, N., and Lower, D. R., eds., Field guide to Cretaceous-Tertiary boundary sections in northeastern Mexico: Houston, Texas, Lunar and Planetary Institute Contribution 827, p. 65–94.

Keller, G., Stinnesbeck, W., Adatte, T., MacLeod, N., and Lowe, D. R., eds., 1994b, Field guide to Cretaceous-Tertiary boundary sections in northeastern Mexico: Houston, Texas, Lunar and Planetary Institute Contribution 827, 110 p.

Kuenen, Ph. H., 1964, Deep-sea sands and ancient turbidites, *in* Bouma, A. H., and Brouwer, A., eds., Turbidites: Amsterdam, Elsevier, Developments in Sedimentology, v. 3, p. 3–33.

Lopez Ramos, E., 1981, Geologia de Mexico. Volume 1: Mexico City, Universidad Nacional Autónoma de Mexico, 446 p.

Lowe, D. R., 1980, Sediment gravity flows. II: Depositional models with special reference to the deposits of high-density turbidity currents: Journal of Sedimentary Petrology, v. 52, p. 279–298.

Macdonald, D. I. M., and Tanner, P. W. G., 1983, Sediment dispersal patterns in part of a deformed Mesozoic back-arc basin on South Georgia, South Atlantic: Journal of Sedimentary Petrology, v. 53, p. 83–104.

Maurrasse, F. J., and Sen, G., 1991, Impacts, tsunamis, and the Haitian Cretaceous-Tertiary boundary layer: Science, v. 252, p. 1690–1693.

Middleton, G. V., and Hampton, M. A., 1973, Sediment gravity flows: Mechanics of flow and deposition, *in* Turbidites and deep water sedimentation: Short Course Lecture Notes, Part 1, Anaheim, California: Los Angeles, California, Society of Economic Paleontologists and Mineralogists Pacific Section, p. 1–38.

Muir, J. M., 1936, Geology of the Tampico region, Mexico: Tulsa, Oklahoma, American Association of Petroleum Geologists, 280 p.

Nelson, C. H., Twichell, D. C., Schwab, W. C., Lee, H. J., and Kenyon, N. H., 1992, Upper Pleistocene turbidite sand beds and chaotic silt beds in the channelized, distal, outer-fan lobes of the Mississippi fan: Geology, v. 20, p. 693–696.

Nilsen, T. H., 1985, Chugach turbidite system, Alaska, *in* Bouma, A. H., Normark, W. R., and Barnes, N. E., eds., Submarine fans and related turbidite systems: New York, Springer-Verlag, p. 185–198.

Pollastro, R. M., and Bohor, B. F., 1993, Origin and clay-mineral genesis of the Cretaceous/Tertiary boundary unit, Western Interior of North America:

Clays and Clay Minerals, v. 41, p. 7–25.

Reading, H. C., and Richards, M., 1994, Turbidite systems in deep-water basin margins classified by grain size and feeder system: American Association of Petroleum Geologists Bulletin, v. 78, p. 792–822.

Sanders, J. E., 1965, Primary sedimentary structures formed by turbidity currents and related sedimentation mechanisms, *in* Middleton, G. E., ed., Primary sedimentary structures and their dynamic interpretation: Society of Economic Paleontologists and Mineralogists Special Publication 12, p. 192–219.

Sestini, G., and Pranzini, G., 1965, Correlation of sedimentary fabric and sole marks as current indicators in turbidites: Journal of Sedimentary Petrology, v. 35, p. 10–108.

Shaw, H. F., and Wasserburg, G. J., 1982, Age and provenance of the target materials for tektites and possible impactites as inferred from Sm-Nd and Rb-Sr systematics: Earth and Planetary Science Letters, v. 60, p. 155–177.

Sigurdsson, H., D'Hondt, S., Arthur, M. A., Bralower, T. J., Zachos, J. C., Van Fossen, M., and Channell, J. E. T., 1991, Glass from the Cretaceous-Tertiary boundary in Haiti: Nature, v. 349, p. 482–487.

Skipper, K., 1971, Antidune cross-stratification in a turbidite sequence, Cloridorme Formation, Gaspé, Quebec: Sedimentology, v. 17, p. 51–68.

Smit, J., and 8 others, 1992, Tektite-bearing, deep-water clastic unit at the Cretaceous-Tertiary boundary in northeastern Mexico: Geology, v. 20, p. 99–103.

Smit, J., Roep, Th. B., Alvarez, W., Claeys, P., Montanari, A., and Grajales, M., 1994a, Impact-tsunami–generated clastic beds at the KT boundary of the Gulf coastal plain: A synthesis of old and new outcrops [abs.], *in* Papers of the international conference on new developments regarding the KT event and other catastrophes in Earth history: Houston, Texas, Lunar and Planetary Institute Contribution 825, p. 117–119.

Smit, J., Montanari, A., and Alvarez, W., 1994b, Tsunami-generated beds at the KT boundary in northeastern Mexico: *in* Keller, G., Stinnesbeck, W., Adatte, T., MacLeod, N., and Lower, D. R., eds., Field guide to Cretaceous-Tertiary boundary sections in northeastern Mexico: Houston, Texas, Lunar and Planetary Institute Contribution 827, p. 95–101.

Stinnesbeck, W., and 10 others, 1993, Deposition of channel deposits near the Cretaceous-Tertiary boundary in northeastern Mexico: Catastrophic or "normal" sedimentary deposits?: Geology, v. 21, p. 797–800.

Swisher, C. C., III, and 11 others, 1992, Coeval $^{40}Ar/^{39}Ar$ ages of 65.0 million years ago from Chicxulub crater melt rock and Cretaceous-Tertiary boundary tektites: Science, v. 257, p. 954–958.

Walker, R. G., 1967, Upper flow regime bed forms in turbidites of the Hatch Formation, Devonian of New York State: Journal of Sedimentary Petrology, v. 37, p. 1052–1058.

Walker, R. G., 1970, Review of the geometry and facies organization of turbidites and turbidite-bearing basins, *in* Lajoie, J., ed., Geological Society of Canada Special Paper 7, p. 217–252.

Walker, R. G., and Mutti, E., 1973, Turbidite facies and facies associations, *in* Middleton, G. V., and Bouma, A. H. (cochairmen), Turbidites and deep sea sedimentation: Los Angeles, California, Society of Economic Paleontologists and Mineralogists, Pacific Coast Section, p. 119–157.

Manuscript Accepted by the Society September 14, 1995

Printed in U.S.A.

Geological Society of America
Special Paper 307
1996

K/T boundary coarse-grained siliciclastic deposits in northeastern Mexico and northeastern Brazil: Evidence for mega-tsunami or sea-level changes?

Wolfgang Stinnesbeck
Facultad de Ciencias de la Tierra, Universidad Autónoma de Nuevo León, Apartado Postal 104, 67700 Nuevo León, Mexico
Gerta Keller
Department of Geological and Geophysical Sciences, Princeton University, Princeton, New Jersey 08544

ABSTRACT

Coarse-grained siliciclastic and limestone breccia deposits have been described from over a dozen K/T boundary sections in northeastern Mexico and variously interpreted as the result of an impact-generated mega-tsunami or sea-level low stand. In addition, a K/T boundary sequence in northeastern Brazil has been reinterpreted as the result of two bolide impacts. In both regions, the clastic deposits are presumed to have originated as a result of a bolide impact on Yucatan. We review the age, biostratigraphy, lithology, and depositional environment of these deposits in order to evaluate the time of deposition and their stratigraphic relationship to the K/T boundary event. Our observations indicate that the near–K/T boundary siliciclastic deposits of northeastern Mexico cannot be explained by deposition resulting from a single catastrophic event. They appear to have been deposited by normal sedimentary processes over an extended time period spanning thousands of years. This is indicated by the presence of multiple bioturbation horizons, discrete layers of normal hemipelagic sedimentation within the clastic deposits, and mineralogically and sedimentologically discrete horizons correlatable over 300 km. Such correlations do not support chaotic single-event deposition.

Clastic deposition occurred sometime during the last 170 to 200 k.y. of the Maastrichtian but ended sometime before the K/T boundary. This is indicated by the presence of the index species *Plummerita hantkeninoides*. These deposits stratigraphically correlate with the limeclast breccia of the Poty section in northeastern Brazil, which also predates the K/T boundary. Benthic foraminifers indicate that deposition of the limestone breccia at Poty and siliciclastic sediments in northeastern Mexico coincided with a eustatic sea-level low stand that lowered sea level by 70 to 100 m. The coincidence of clastic deposition with the latest Maastrichtian sea-level low stand suggests that they are causally related. However, since clastic deposition is not necessarily related to sea-level regressions, and the glass-bearing deposits in northeastern Mexico are unique, a pre–K/T boundary event (bolide impact or volcanism) cannot be excluded.

INTRODUCTION

Coarse-grained siliciclastic and limestone breccia deposits of Cretaceous-Tertiary boundary or near–K/T boundary age are present in sedimentary sequences in the Gulf states of the United States, the Gulf of Mexico, northeastern and southern Mexico, Guatemala, the Caribbean area, and northeastern Brazil (Fig. 1). They have generally been interpreted as sea-level low-stand deposits (Donovan et al., 1988; Baum and Vail, 1988; Mancini et al., 1989; Keller, 1989; Savrda, 1991, 1993; Habib et al., 1992; Mosh-

Stinnesbeck, W., and Keller, G., 1996, K/T boundary coarse-grained siliciclastic deposits in northeastern Mexico and northeastern Brazil: Evidence for mega-tsunami or sea-level changes?, *in* Ryder, G., Fastovsky, D., and Gartner, S., eds., The Cretaceous-Tertiary Event and Other Catastrophes in Earth History: Boulder, Colorado, Geological Society of America Special Paper 307.

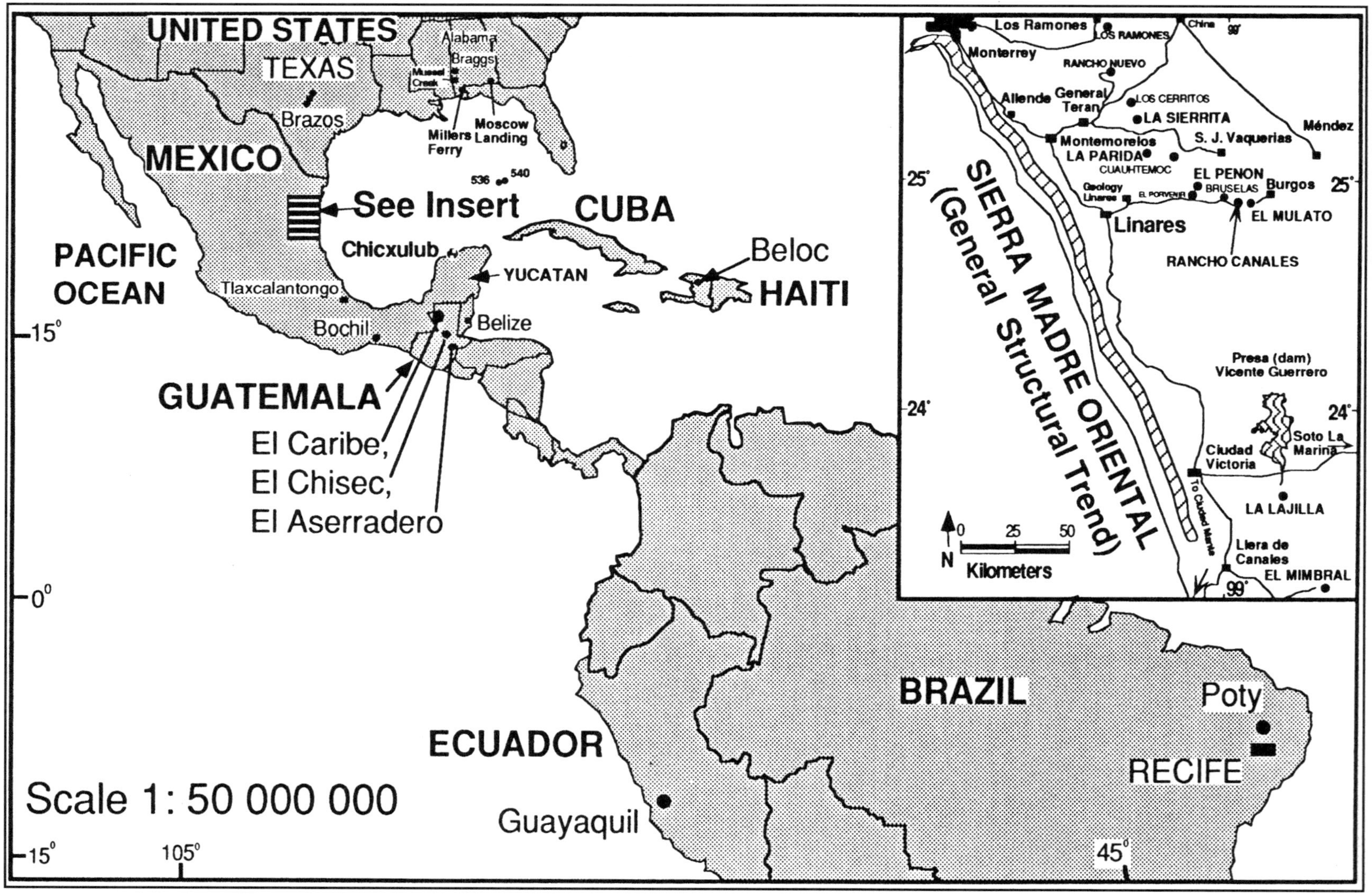

Figure 1. Location of Cretaceous-Tertiary boundary sections with near-K/T clastic deposits from Texas to Brazil. Insert shows area of location of northeastern Mexico sections.

kovitz and Habib, 1993; Stinnesbeck et al., 1993a; Keller et al., 1993a, 1994a). This interpretation has been challenged with the recent suggestion that the circular subsurface structure at Chicxulub in northern Yucatan represents the crater of the K/T boundary bolide impact (Hildebrand et al., 1991; Pope et al., 1991; Swisher et al., 1992; Sharpton et al., 1992, 1993). In this scenario, these deposits are interpreted as impact-generated mega-tsunami deposits (Hildebrand and Boynton, 1990; Hildebrand et al., 1994; Smit et al., 1992, 1994a, b; Alvarez et al., 1992; Bohor and Betterton, 1993; Montanari et al., 1994; Habib, 1994; Albertão et al., 1994; Kousoukos, 1995), as earlier suggested by Bourgeois et al. (1988) for coarse siliciclastic event beds along the Brazos River, Texas.

It is unknown, however, what mega-tsunami deposits should look like: whether clastic deposition should be chaotic or stratified, locally channelized or sheetlike, or even if such deposits would differ substantially from submarine gravity flows, turbidites, or sea-level low-stand deposits. Nevertheless, if these clastic sediments represent single-event deposition over a period of hours to days, they must be coeval and have resulted from the same catastrophic event. If a bolide impact on Yucatan triggered deposition, then clastic sediments should also contain impact products such as iridium, Ni-rich spinels, shocked quartz grains, and glassy microtektites. In this chapter we discuss the age, stratigraphy, lithology, and depositional environment of clastic deposits from northeastern Mexico and northeastern Brazil, based on our recent field and laboratory observations. The biostratigraphy is based on the planktic foraminiferal zonation of Keller (1988a) as further refined and updated by Canudo et al. (1991) and Keller (1993). Pardo et al. (1996) have added a new biozone, *Plummerita hantkeninoides,* to characterize the topmost 170 to 200 k.y. of the Maastrichtian (upper part of C29R below the K/T boundary).

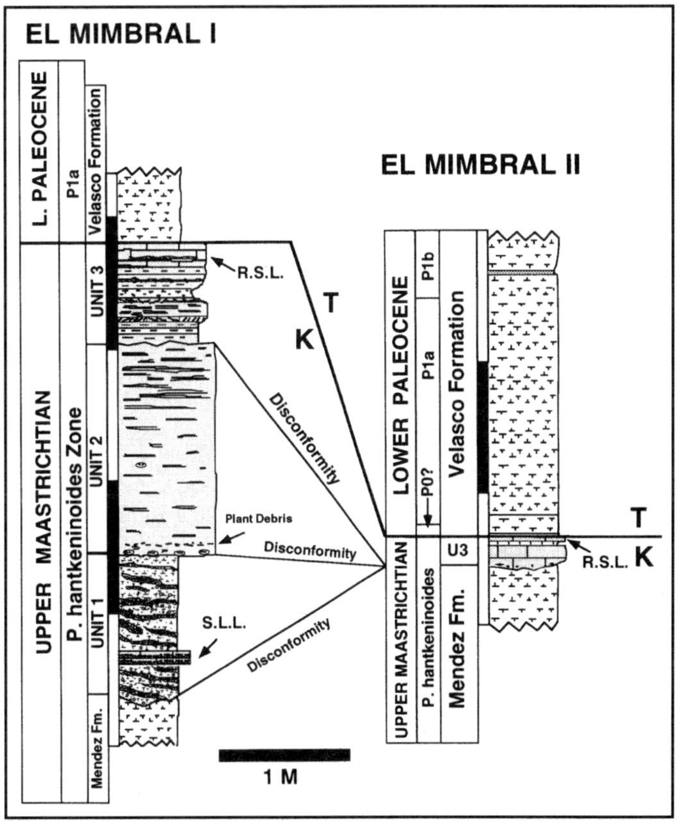

Figure 2. Lithostratigraphy and biostratigraphy across two K/T boundary transects: El Mimbral I across the middle of the channelized siliciclastic deposit and El Mimbral II across the edge where the siliciclastic deposit is only 20 cm thick. S.L.L. refers to the sandy limestone layer within unit 1. R.S.L. refers to the rippled sandy limestone layer that caps unit 3.

NORTHEASTERN MEXICO

In northeastern and east-central Mexico, coarse-grained siliciclastic deposits that occur regionally in K/T boundary sections over a distance of 600 km (Fig. 1) have been described by a number of investigators (Smit et al., 1992, 1994a, b; Longoria and Gamper, 1992; Stinnesbeck et al., 1993a, 1994a, 1996; Bohor and Betterton, 1993; Longoria and Grajales-Nishimura, 1993; Keller et al., 1994a, b; Ekdale and Stinnesbeck, 1994; Adatte et al., this volume; Lopez-Oliva and Keller, this volume). The clastic deposits consist of a complex sequence of sandstones, silts, and minor shale and horizons rich in calcareous spherules at the base as shown in Figure 2 for the El Mimbral section. These coarse-grained sediments discontinuously truncate fine-grained marls of the Maastrichtian Mendez Formation and underlie shaly marls of the Tertiary Velasco Formation. Planktic foraminiferal assemblages indicative of the *Plummerita hantkeninoides* zone of the latest Maastrichtian (Pardo et al., 1996) are present in the 50 cm to 100 cm of the Mendez Formation below the clastic deposits (Figs. 2 and 3). *Plummerita hantkeninoides* is also present within the clastic sediments and within a thin Maastrichtian-age Mendez-type marl layer above (shown for Lajilla in Fig. 3; Keller et al., 1994a, b; Lopez-Oliva and Keller, this volume). In four sections (Mimbral, Lajilla, Mulato, and Parida) Tertiary shales directly overlie the clastic deposits, whereas in other sections (e.g., Ramones, Sierrita, Peñon, Fig. 4) Tertiary sediments are not exposed (Keller et al., 1994a). In three of the four sections (Lajilla, Mulato, and Parida, Figs. 1, 4), a 5- to 10-cm-thick marl layer of latest Maastrichtian age (*P. hantkeninoides* Zone) overlies the clastic deposits and underlies Tertiary sediments as shown for Lajilla in Figure 3 (Keller et al., 1994a; Lopez-Oliva and Keller, this volume). Smit et al. (1994b) suggested that this Maastrichtian marl layer represents settling through the water column after the tsunami event. However, there is no grain-size grading or size sorting of foraminifera (Adatte et al., this volume). The presence of these marl layers, with their faunas of abundant planktic foraminifers, thus indicates that normal hemipelagic sedimentation similar to the Mendez Formation resumed after deposition of the siliciclastic sediments and that the clastic deposits are stratigraphically below, and hence predate, the K/T boundary event.

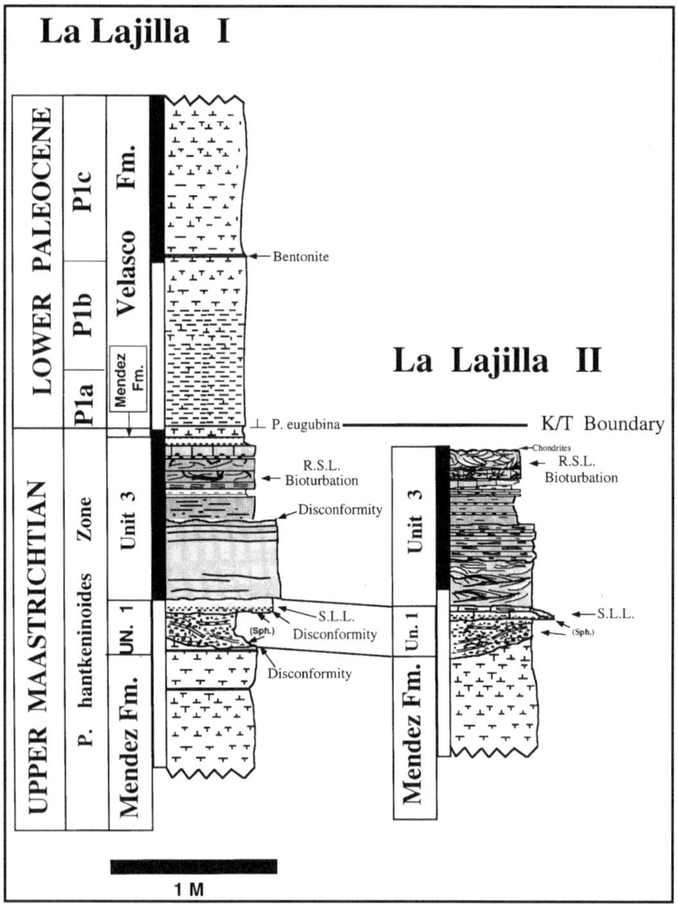

Figure 3. Lithostratigraphy and biostratigraphy across two transects at La Lajilla. Note that at Lajilla I a thin layer of Mendez Formation marl caps the clastic deposit and indicates that clastic deposition predates the K/T boundary. S.L.L. refers to sandy limestone layer in unit 1, which at Lajilla is at the top of unit 1 because of erosion. R.S.L. refers to the rippled sandy limestone layer that caps unit 3.

The earliest Danian Zone P0 is generally not present in northeastern Mexico, with the possible exception of El Mimbral, where a 3- to 4-cm-thick clay layer, barren of planktic foraminifera, caps the clastic deposit (Keller et al., 1994a, b). Immediately above this clay layer at Mimbral, and directly overlying the Maastrichtian-age marl layer at La Lajilla, La Parida, and El Mulato, six to eleven planktic foraminiferal species, all characteristic of Zone P1a, simultaneously appear (Lopez-Oliva and Keller, this volume). This indicates that at least Zone P0 and the lower part of Zone P1a are missing. It is likely that the uppermost Maastrichtian sediment is also missing, though we lack biostratigraphic resolution to evaluate this possibility. A short hiatus at the K/T boundary is present in K/T boundary sections worldwide (Keller, 1989, 1993; Canudo et al., 1991; Keller and Benjamini, 1991; MacLeod and Keller, 1991a, b; Keller et al., 1993a).

Siliciclastic deposits of northeastern Mexico can be subdivided into three units that are defined by lithology, sedimentology, and mineralogy (Stinnesbeck et al., 1993a, 1996; Keller et al., 1994a; Adatte et al., this volume). They can be correlated over a distance of more than 300 km (Figs. 1 and 4). The basal unit 1 is characterized by abundant and poorly cemented spherules, foraminiferal tests, gravel-sized intraclasts of marl from the underlying Mendez Formation, extraclasts of limestone containing Turonian-age foraminifers, and minor quartz. The spherules have been interpreted to be altered glassy microtektites (Smit et al., 1992, 1994a; Bohor and Betterton, 1993). Spherules range from 1 to 5 mm in diameter. They are generally infilled with blocky calcite or greenish microcrystalline phyllosilicates, including glauconite and smectite. Many large spherules contain smaller spherules. Some spherules enclose foraminifers, sediment grains, or rutile crystals. Others are enclosed by a thin organic coating that suggests a biogenic origin as either algal resting cysts (Hansen et al., 1986) or oncolites. We found no glassy spherules, although rare glass fragments are present. These glass fragments are similar in chemistry and low water contents to glass from the Beloc section in Haiti (Smit et al., 1992, 1994b; Stinnesbeck et al., 1993a; Bohor and Betterton, 1993). The glass from Beloc has been interpreted as being of impact origin by Izett (1991a, b), Blum and Chamberlain (1992), Sigurdsson et al. (1991), Koeberl and Sigurdsson (1992), and Koeberl (1994). A volcanic origin for these glasses has been suggested by Jehanno et al. (1992), Lyons and Officer (1992), and Robin et al. (1994).

Within the unit 1 spherule layer is a 5- to 25-cm-thick, well-cemented, sandy limestone layer (Fig. 2). In some sections (e.g., Lajilla, Fig. 3), this sandy limestone layer tops unit 1 because the spherule-rich sediments above are eroded. There are no spherules within the sandy limestone layer, except at the top and bottom surfaces. Convolute bedding is observed at the top, which is draped by the overlying spherule-rich sediments at Mimbral (Fig. 2). Microlayering of finer- and coarser-grained sediments can be observed in thin sections within the sandy limestone layer (W. Ward, 1994, written commun.). This sandy limestone layer is widespread over 300 km (Mimbral, Lajilla, Peñon, Rancho Canales, Fig. 1) and is not a local diagenetic feature as suggested by Smit et al. (1992, 1994a).

Unit 2 disconformably overlies the spherule-rich unit 1 and consists of massive, and sometimes weakly laminated, sandstone. Microconglomeratic to gravel-sized intra- and extraclasts are commonly present at the base, and at Mimbral discrete layers of plant debris are present (Fig. 2). At Peñon, the base of unit 2 contains spherule-filled burrows that are truncated by the overlying sand layers. This indicates that burrowing of the sediment occurred during deposition of unit 2 and that deposition was interrupted by erosion. Thus, unit 2, like unit 1, indicates multievent deposition over a longer time period than a few days (Ekdale and Stinnesbeck, 1994).

An erosional unconformity marks the contact between unit 2 and the overlying unit 3 (Figs. 2, 3). Unit 3 consists of interlayered sand, shale, and silt beds with diverse sedimentological features, such as horizontal laminations, ripple marks, small-scale cross-bedding, flaser bedding, and convolute laminations. These fine-grained layers contain typical Maastrichtian planktic foraminiferal assemblages. Transported shallow-water species, which are commonly present in sandstones, are absent in these layers. This suggests periods of normal hemipelagic sedimentation alternating with periods of more rapid deposition and terrigenous influx. Two distinct layers enriched in zeolites are present within this unit (Adatte et al., this volume). They are correlatable over a distance of 300 km and suggest episodes of influx of volcanoclastic matter during deposition. Unit 3 is topped by one or more resistant sandy limestone layers, which are always bioturbated, most commonly by *Chondrites, Thalassinoides, Ophiomorpha, Planolites*, and *Zoophycos* (Stinnesbeck et al., 1993a; Keller et al., 1994a; Ekdale and Stinnesbeck, 1994). Populations of *Chondrites* and *Ophiomorpha* were truncated repeatedly by overlying sand layers in unit 3. The repeated truncations indicate successive colonization episodes of the accreting substrate and hence deposition over a time period of at least months (Ekdale and Stinnesbeck, 1994).

Sediment transport of the three clastic units was from deltaic or inner neritic environments into deeper waters. This direction is indicated by plant debris including wood and leaves (in abundance near the base of unit 2 at Mimbral), common glauconite, and the presence of abundant benthic foraminifers characteristic of shallower neritic environments. Assemblages include *Bolivinoides incrassata gigantea, Praebulimina reussi, Anomalinoides acuta, A. praeacuta, Cibicidoides succedens*, and *Globorotalites michelianus*, which are rare both in the underlying Mendez marls and overlying Velasco shales (Keller et al., 1994a, b). Although many of these species may have been transported, absence of deeper-water taxa suggests that deposition occurred in a shallower environment than either below or above.

Sediment transport from shallower into deeper water probably occurred via submarine canyons. This is suggested by the distribution of clastic sediments in channelized deposits. But the presence of Turonian-age foraminifers (e.g., *Marginotruncana sigali, M. pseudolinneiana*) and clasts within units 1, 2, and 3 also indicates deeply incised canyons into the Turonian Agua Nueva Formation (Stinnesbeck et al., 1993a; Keller et al., 1994a, b). No Turonian-age sediments were exposed in the region at K/T boundary time. The Agua Nueva Formation was then buried under several hundred meters of sediment.

Clastic deposition thus appears to have occurred over an extended time period and by means of multiple events that included periods of colonization by invertebrates alternating with periods of erosion and deposition. In addition, biostratigraphy indicates that clastic deposition began during the last 170 to 200 k.y. of the Maastrichtian and ended sometime before the K/T boundary event. Benthic foraminifers indicate that deposition occurred during a low sea level. Based on these observations, we suggest that the coarse-grained siliciclastic sediments were deposited during the latest Maastrichtian sea-level low stand and the Mendez marls below and above were deposited during high sea levels, as shown in Figure 5. In this scenario, deposition of the entire clastic deposit occurred during the sea-level low stand. The unconformity at the base of the clastic deposit represents a type 1 sequence boundary associated with a falling sea level. Deltaic sediments are exposed at this time, incised valleys and canyons are formed, and deltaic sediments from near-shore areas are transported into deeper waters, depositing unit 1. Continued lowering of the sea level results in erosion and bypass of inner shelf sediments and deposition of massive sand lenses in submarine canyons and fans, thus depositing unit 2. Deposition of unit 3 occurred during a stable low sea level. At this time erosion of shelf sediments decreased, resulting in deposition of distal turbidites (coarse-grained layers) alternating with normal hemipelagic sedimentation (fine grained layers). A rising sea level during the latest Maastrichtian deposited the hemipelagic marls observed above the clastic deposits at Lajilla, Mulato, and Parida.

NORTHEASTERN BRAZIL: THE POTY QUARRY

The Poty quarry section north of Recife in northeastern Brazil (Fig. 1) contains one of the most complete transitions across the K/T boundary in South America (Mabesoone et al., 1968; Ashraf and Stinnesbeck, 1988; Stinnesbeck, 1989; Stinnesbeck et al., 1993b; Albertão et al., 1994; Stinnesbeck and Keller, 1995, 1996; Koutsoukos, 1995). Maastrichtian limestones and marls of the Gramame Formation contain the late Maastrichtian coccolithophorid index species *Micula murus* as well as typical latest Maastrichtian planktic foraminiferal assemblages, including *Plummerita hantkeninoides* (Fig. 6). Sediment deposition occurred in a marine shelf area (water depth of about 150 m). This type of deposition is indicated by the presence of pelagic elements such as ammonites, radiolaria, coccolithophorids, and shallow-water benthic and surface-water–dwelling planktic foraminifera (e.g., rugoglobigerinids, hedbergellids, guembelitrids, heterohelicids, racemiguembelinids) but the near absence of most deeper-dwelling globotruncanids. The presence of echinoids (*Hemiaster*) and annelids (*Hamulus*) also indicates shallow marine shelf conditions. Environmental conditions changed about 70 cm below the K/T boundary, coincident with deposition of a 20-cm-thick limestone breccia (unit C, Fig. 6). At this time, water depth was reduced from an outer-middle to shallow middle-inner neritic (50 to 100 m) environment, as indicated by abundant *Discorbinella burlingtonensis, Lenticulina midwayensis*, and a few *Quinqueloculina*. Thus, shallowing of 50 to 100 m is indicated at this time. The presence of vertebrate bones and teeth, phosphatic lumps, phosphatized foraminifers, glauconite, terrigenous quartz, and serpulids in the breccia and at the base of the overlying limestone (units C and D) also suggests condensed sedimentation in a shallow water environment. Climate cooled at this time, as indicated by the disappearance of all tropical and subtropical palynomorph taxa within and also above the breccia (Ashraf and Stinnesbeck, 1988; Stinnesbeck and Keller, 1996). Upsection from the breccia layer, detrital limestones contain abundant phosphatized lumps, including many phosphatic spherules, and quartz (unit D). The megafauna consists

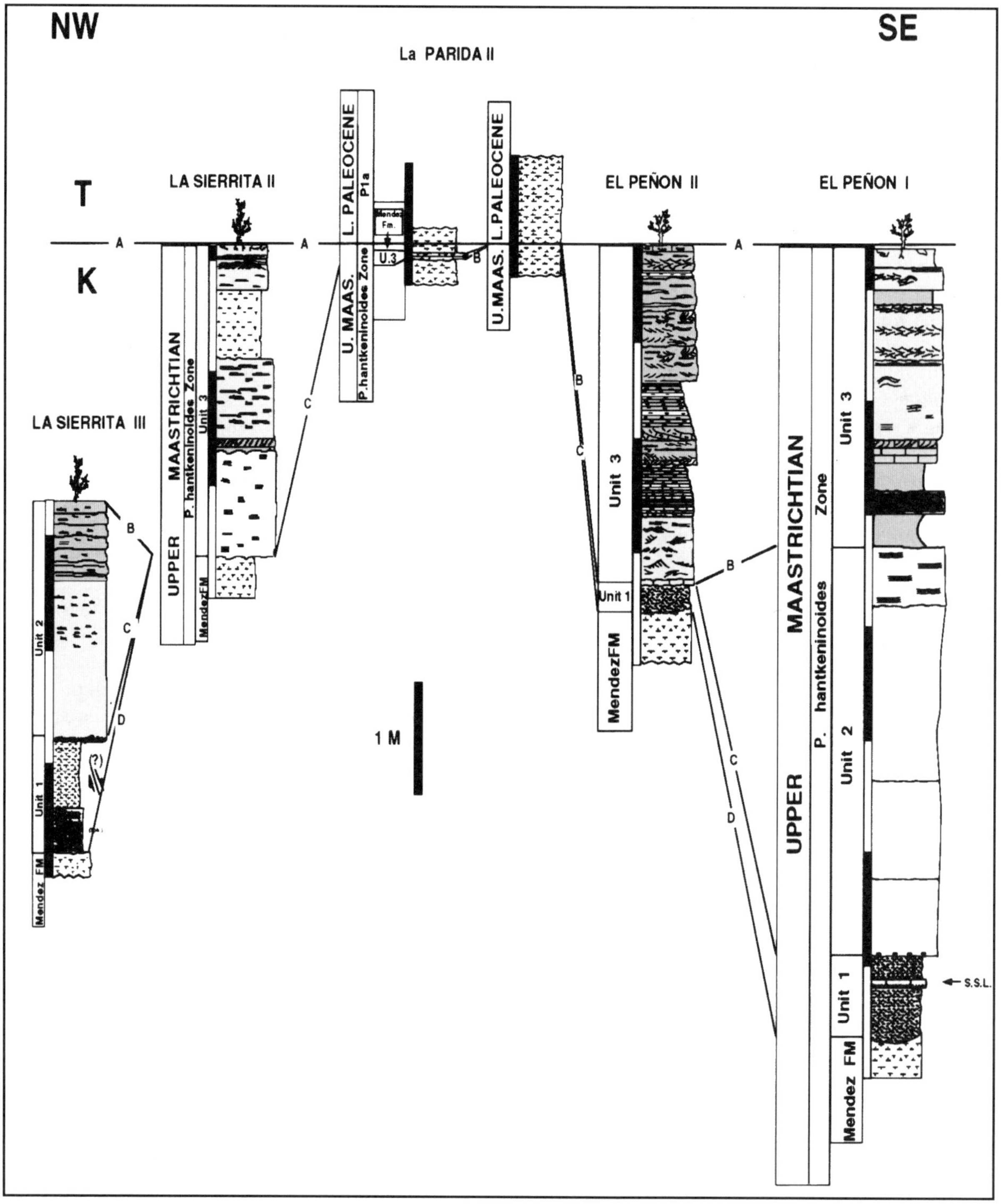

Figure 4 (on this and facing page). Lithostratigraphic correlation of K-T boundary sections in northeastern Mexico. Note the correlatability of the units of the clastic deposits. Note also that in various sections unit 2 is missing. We interpret these sections to have been located away from the central parts of the channels.

of serpulids, occasional oysters, pteriid bivalves, and snails. Abundant and more diverse benthic foraminiferal assemblages (e.g., common *Alabamina midwayensis, Discorbinella burlingtonensis* and few *Lenticulina midwayensis, Cibicidoides, Nonionella, Gaudryina, Spiroplectammina* and *Dentalina*) as well as increasing presence of deeper-dwelling planktic foraminifers (e.g., globotruncanids and pseudotextularids) indicate deepening and a sea-level transgression. By K/T boundary time, a middle neritic water depth was established. The transgression appears to have continued across the K/T boundary, which is

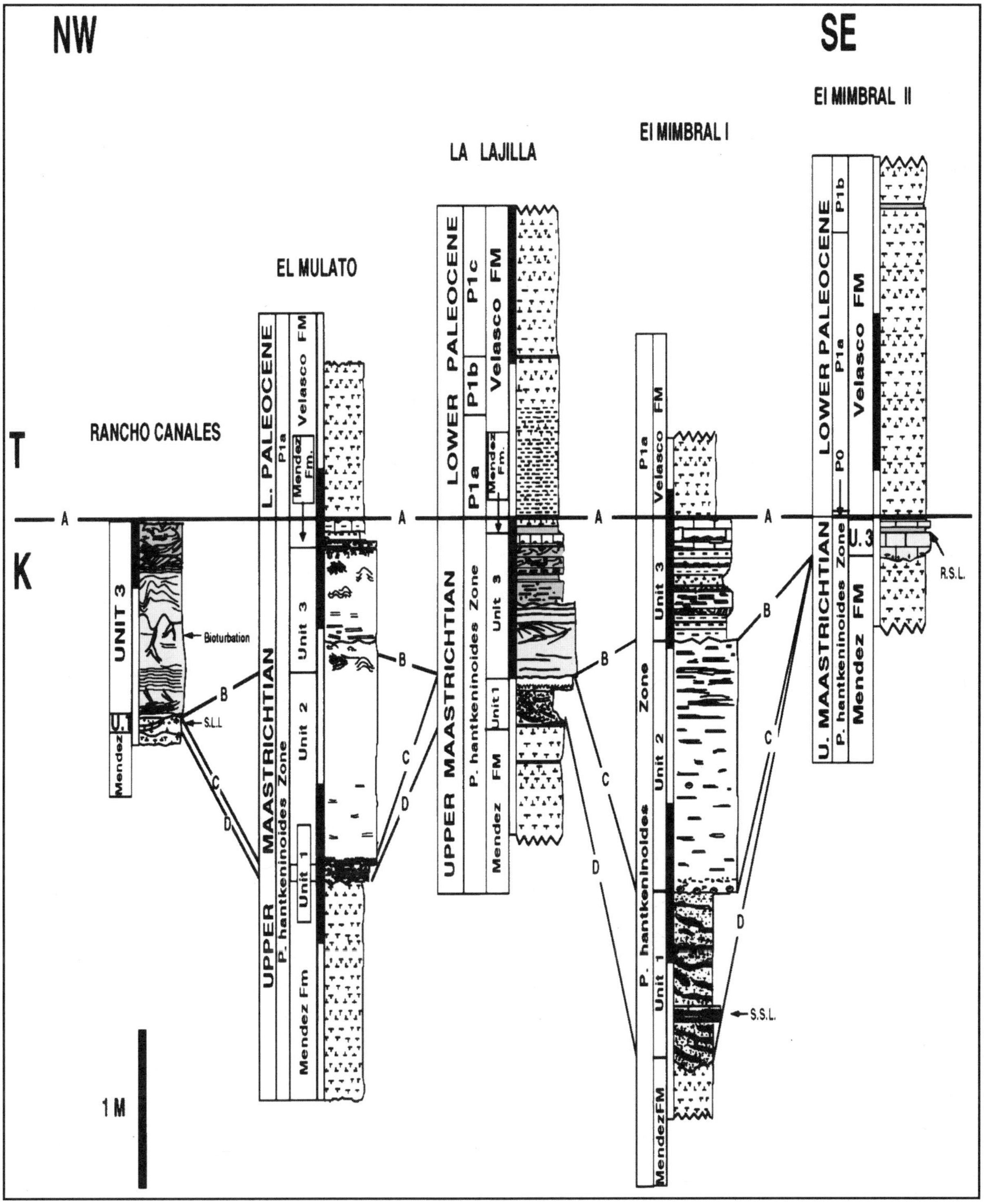

marked by a thin clay layer that contains an iridium anomaly (Fig. 6). The first appearance of Tertiary species, including *Parvularugoglobigerina eugubina*, occurs above the clay layer and suggests a condensed interval or short hiatus.

Recently, Albertão et al. (1994) and Koutsoukos (1995) suggested that the K/T boundary is at the base of the breccia layer (unit C), which they interpret to represent a Yucatan impact-generated tsunami deposit. The iridium anomaly, they suggest, represents a second Danian bolide impact. Their interpretation is based on the putative presence of rare Danian species (e.g., *Guembelitria irregularis, Woodringina claytonensis, W. hornerstownensis, Eoglobigerina simplicissima*, and *E. fringa*) within and above the breccia layer and spherules that they considered to be microtektites (see Koutsoukos, 1995, plate 1). We were

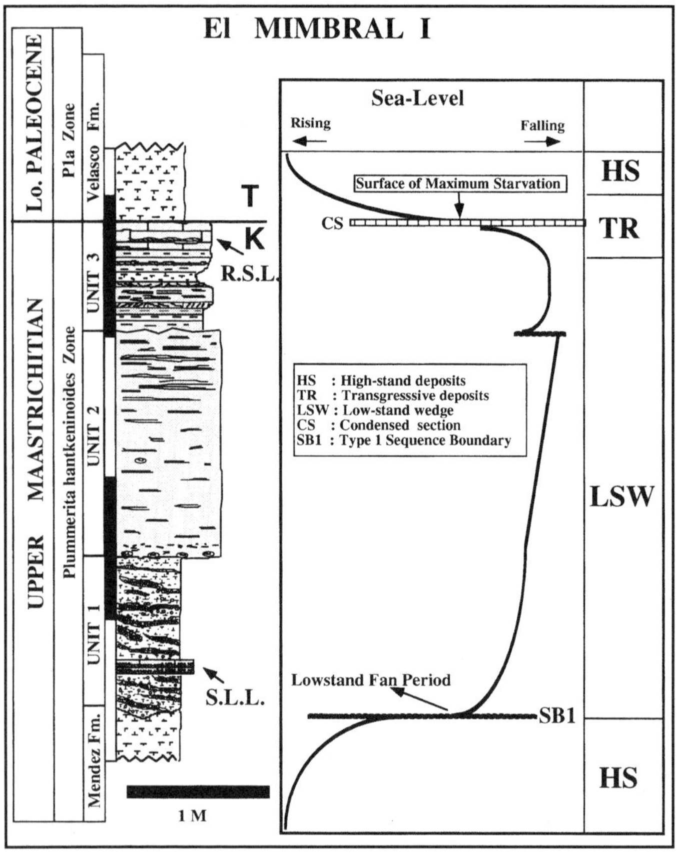

Figure 5. Lithostratigraphy and biostratigraphy of the El Mimbral I transect with interpreted history of sea-level changes and sediment deposition. In this model, clastic deposition occurred during the late Maastrichtian sea-level low stand.

unable to confirm the presence of Tertiary species below the iridium anomaly. *Guembelitria irregularis*, which Koutsoukos (1995) considers a Tertiary species, first appears in the late Maastrichtian (Keller, 1988a; Keller et al., 1993b). Koutsoukos's figure 4, which he identified as *W. claytonensis*, is identical to *G. irregularis* in his figure 1 in both numbers of chambers and chamber arrangements as well as apertures. Figure 7, identified as *P. eugubina*, is nondiagnostic and shows a rim that is not characteristic of *P. eugebina*. Figure 8 (*E. fringa*) and figures 9, 10, and 11, identified as *E. eobulloides* and *Parasubbotina* cf. *pseudobulloides*, are poor illustrations and nondiagnostical to these species; for example, *P. pseudobulloides* has five chambers rather than the four shown in the illustration. The illustration of *E. eobulloides* suggests compressed rather than round chambers. Koutsoukos's figure 12 of *Praemurica taurica* is morphologically identical to *H. monmouthensis*, a Cretaceous species. Contrary to Koutsoukos's claim that cancellate surface-wall texture does not appear before the Danian (p. 378), the significance of wall textures is still little understood and no such claim can be made at this time. In our examined section we failed to find any Danian species below the iridium anomaly. If Danian species are present in Koutsoukos's section, it is possible that they may be contaminants or that in our section K/T erosion was less severe, resulting in Maastrichtian sediments preserved above the breccia. As evidence for a second Danian impact, Albertão et al. (1994) and Koutsoukos (1995) cite the iridium anomaly and glassy microtektites. We have found many spherules that could be mistaken for microtektites, but they are all phosphatic and dissolve in acid. Moreover, we observed these phosphatic spherules throughout the section from below the breccia to above the iridium anomaly (Stinnesbeck and Keller, 1995). Our stratigraphic investigation indicates that the Ir anomaly marks the K/T boundary event. The breccia layer, which predates the K/T boundary, is stratigraphically within the *P. hantkeninoides* Zone and hence coeval with siliciclastic deposits in northeastern Mexico.

Impact-generated tsunami deposition for the breccia layer and overlying detrital limestones is also challenged by the occurrence of well-preserved and unbroken, very fragile, thin-shelled bivalves (e.g., *Pteria*) that are not likely to have been transported far. The presence of abundant serpulid colonized limestone clasts within the breccia (prior to the deposition of the overlying detrital limestone) indicates an extended time of nondeposition. Moreover, both benthic and planktic foraminifera reflect distinct shallowing within the breccia layer, followed by deepening above (Fig. 6). These faunal assemblages suggest environmental changes accompanied by normal long-term sediment deposition.

Our observations indicate that the K/T boundary transition at Poty, like the clastic deposits in northeastern Mexico, was deposited during a time of oceanic instability. Within this environment the limestone breccia 70 cm below the K/T boundary could mark a regional truncation of the underlying Gramame Formation (Fig. 6). This sequence boundary appears to be coeval with the latest Maastrichtian sea-level low stand near the base of the *P. hantkeninoides* Zone. The breccia could represent deposition in an incised valley. Upsection, lithology and faunas indicate gradually rising seas.

LATE MAASTRICHTIAN: GLOBAL SEA-LEVEL LOW STAND

Based on seismic reflection lines, Haq et al. (1987) noted that a major eustatic sea-level low stand occurred about 500 k.y. before the K/T boundary. A sea-level low stand below the K/T boundary has been confirmed based on benthic and planktic foraminifers, spores and pollen, dinoflagellates, and macrofossils from numerous K/T boundary sequences across latitudes. This sea-level low stand and subsequent transgression across the K/T boundary clay are well documented in marine sections from Denmark (Nye Klov, Stevns Klint; Schmitz at al., 1992; Keller et al., 1993b), Spain (Agost and Caravaca; Keller, 1992; Pardo et al., 1996), Tunisia (El Kef; Keller, 1988a, b, 1992), Texas (Hansen et al., 1993; Keller, 1992), Mexico (Keller et al., 1994b), Brazil (Stinnesbeck, 1989; Stinnesbeck and Keller, 1995, 1996) and Seymour Island, Antarctica (Askin, 1992;

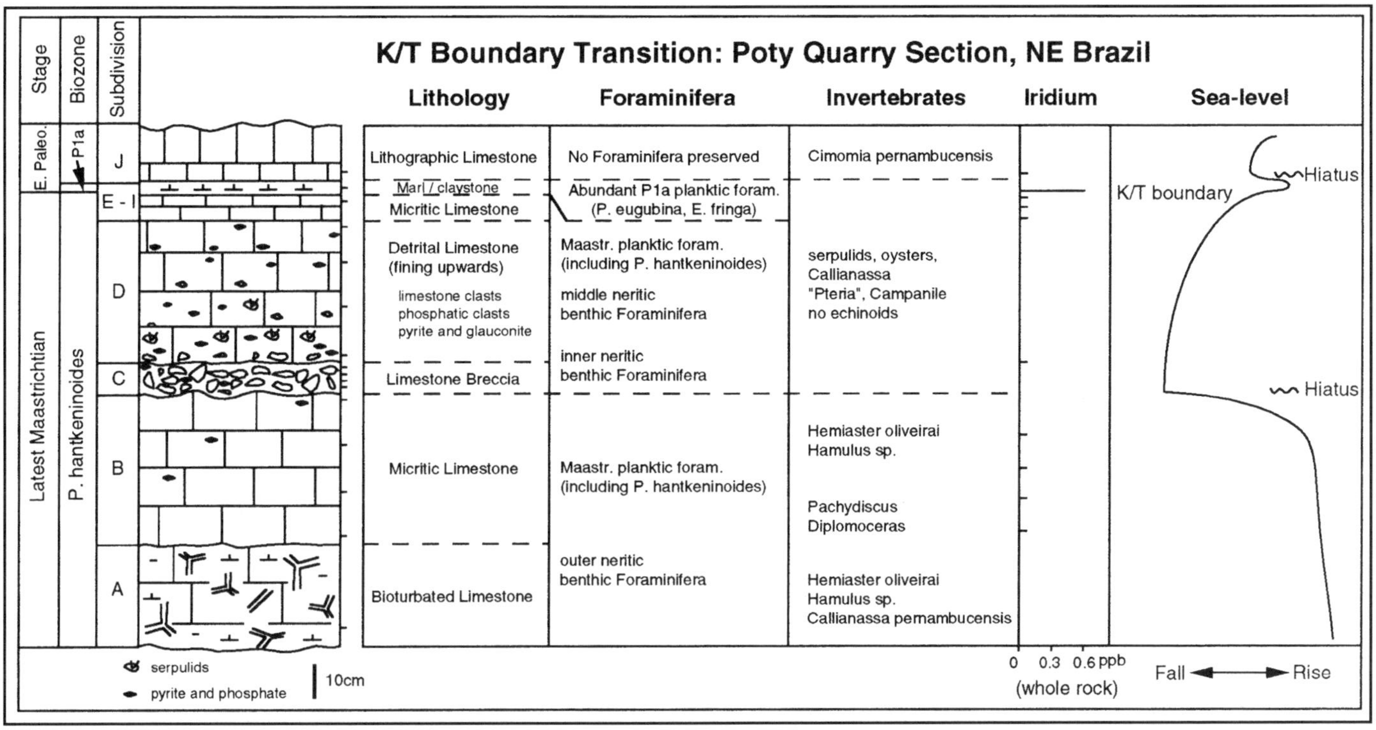

Figure 6. The Cretaceous-Tertiary boundary transition at the Poty Quarry in northeastern Brazil, showing characteristic lithology, foraminifera and invertebrates, Ir anomaly, and sea-level changes. Lithologic subdivisions A through J and Ir data from Albertão et al. (1994). Note that the K/T boundary based on the first appearance of Tertiary planktic foraminifera coincides with the Ir anomaly. Sea-level low stands and short hiatuses mark the base of the limestone breccia and Zone P1a.

Askin and Jacobson, 1996). Sea level lowered between 70 to 100 m at this time (Schmitz et al., 1992; Pardo et al., 1996). In many deep-sea sections, the K/T boundary transition is very condensed or contains a hiatus (MacLeod and Keller, 1991a, b; Keller et al., 1993a). As a result, pre-K/T, as well as post-K/T records that are marked by sea-level low stands, may be missing, as has been shown for the Caribbean area by Keller et al. (1993a). In continental shelf sections, such as Stevns Klint and Nye Klov, the late Maastrichtian sea-level low stand coincides with a hiatus (Schmitz et al., 1992).

It has been difficult to determine the precise age of this latest Maastrichtian sea-level low stand because of a lack of biostratigraphic resolution as well as paleomagnetic control. Two developments have recently made it possible to determine both the timing of the sea-level regression and the subsequent transgression. These are good paleomagnetic control at Agost in Spain (Groot et al., 1989), one of the most complete K/T boundary transitions, and the discovery of the short-ranging index species *P. hantkeninoides* to mark the latest Maastrichtian (Pardo et al., 1996). As a result, Pardo et al. (1996) have analyzed the late Maastrichtian and early Tertiary depositional environments at Agost and El Kef in order to determine the magnitude and tempo of sea-level fluctuations. Their results are shown in Figure 7.

The latest Maastrichtian sea-level low stand occurred near the base of the *P. hantkeninoides* Zone and within the lower part of Chron 29R below the K/T boundary. Based on paleomagnetic stratigraphy, sediment accumulation during the interval of Chron 29R below the K/T boundary averaged 2 cm/1,000 yr (Pardo et al., 1996). The base of the *P. hantkeninoides* Zone (first appearance of index species) occurs between 170 to 200 k.y. below the K/T boundary. Benthic foraminiferal data indicate that sea-level lowering began at least 50 k.y. before the *P. hantkeninoides* Zone, and sea level remained low for about 100 k.y. The stable isotope record indicates maximum global cooling and possibly Antarctic glaciation coinciding with this sea-level low stand (Barrera, 1994). Benthic foraminifers indicate a deepening environment and the onset of low oxygen conditions beginning 100 k.y. before the K/T boundary and accelerating during the last 50 k.y. (Pardo et al., 1996). This sea-level rise is accompanied by rapid global warming in the stable isotope record (Stott and Kennett, 1990; Barrera, 1994) and the incursion of low-latitude warm-water species into high latitudes (Schmitz et al., 1992; Keller et al., 1993b). The sea-level rise continues across the K/T boundary and temporarily reaches a maximum high stand during deposition of the boundary clay (Zone P0, Fig. 7). During the early Danian, the sea level is generally rising but is interrupted by two short-term low stands at the P0/P1a and P1a/P1b Zone boundaries. These sea-level low stands are associated with

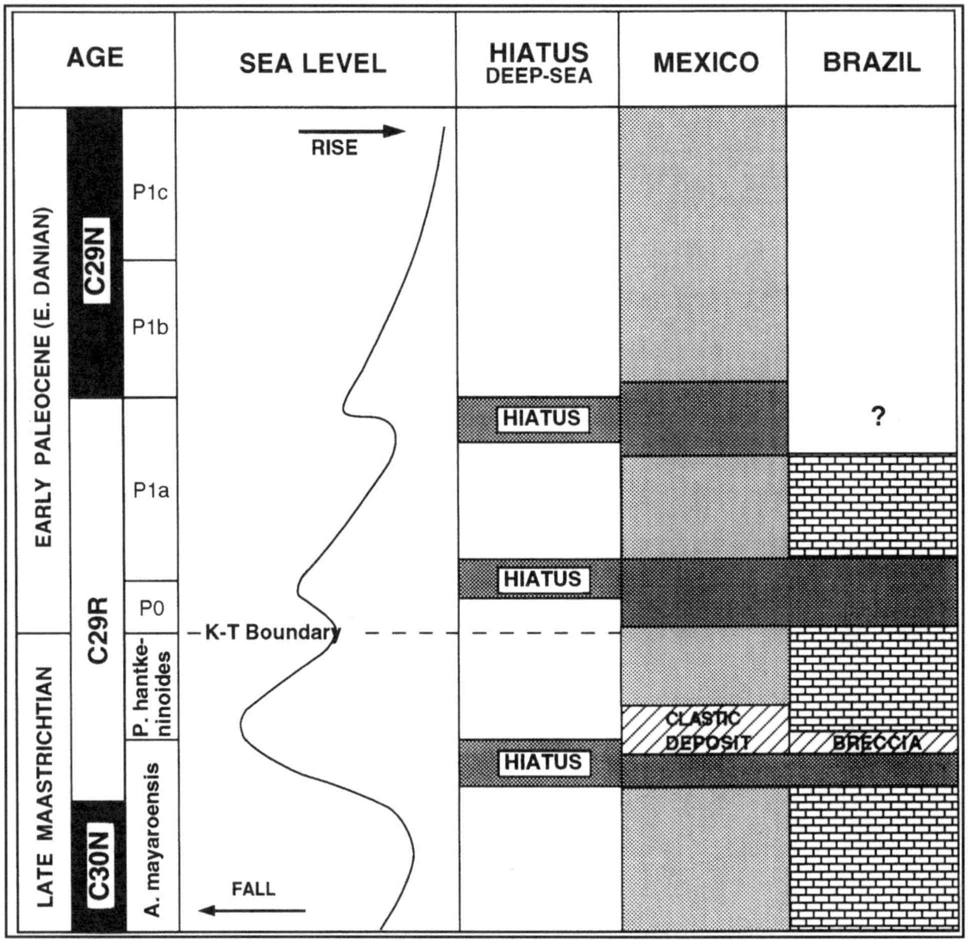

Figure 7. Sea-level changes across the K/T transition based on benthic foraminiferal analysis at Agost as well as K/T boundary sections worldwide. Paleomagnetic time scale from Agost by Groot et al. (1989). Age and correlation of Mexican and Brazilian sections to the Agost section based on biostratigraphy; interpretation of sea-level changes based on microfossils and macrofossils. Note that clastic and limestone breccia deposition in Mexico and Brazil coincided with the latest Maastrichtian sea-level regression and low stand. Normal hemipelagic sedimentation resumed during the sea-level rise at the end of the Maastrichtian. Hiatuses coincided with sea-level low stands.

short hiatuses in many sections worldwide (MacLeod and Keller, 1991a, b; Keller et al., 1993a).

The sea-level curve shown in Figure 7, based on the Agost section, is corroborated in sections across latitudes and strongly indicates eustatic trans-K/T sea-level changes, as noted earlier. The sections in Mexico and Brazil also reflect these fluctuating sea levels. Based on high-resolution biostratigraphy and benthic and planktic faunal assemblages, these sections can be correlated to the sea-level curve in Figure 7. The clastic sediments were deposited during the sea-level low stand within the *P. hantkeninoides* Zone, and clastic deposition ended below the K/T boundary. In both Mexico and Brazil (as well as in Spain, Tunisia, and Denmark), the K/T boundary is well marked by the first Tertiary planktic foraminifers and an iridium anomaly. In Mexican sections, short hiatuses are present at the P0/P1a and P1a/P1b boundaries in most sections examined. In the Brazil section, the P0/P1a hiatus is also present; the section was not studied above Zone P1a.

SUMMARY AND CONCLUSIONS

We examined the K/T boundary coarse-grained siliciclastic deposits of northeastern Mexico and limestone breccia of northeastern Brazil in order to evaluate the timing and nature of these deposits and their relationship to global sea-level changes and/or the proposed K/T boundary impact on Yucatan. Our observations indicate that deposition occurred during the last 170 to 200 k.y. of the Maastrichtian during a sea-level low stand, as indicated by the presence of the index species *P. hantkeninoides* in sediment layers below, within, and above. Clastic deposition ended below the K/T boundary event, as indicated by the return to normal hemipelagic sedimentation during the latest Maastrichtian. We observed no evidence of chaotic single-event deposition by an impact-generated megatsunami. Deposition appears to have occurred over an extended time period spanning thousands of years. For clastic deposits in northeastern Mexico, this is indicated by the occurrence of multiple bioturbation horizons, and discrete horizons of normal hemipelagic sedimentation, discrete horizons enriched in zeolites and by the presence of mineralogically and sedimentologically discrete horizons. For the limestone breccia deposit and overlying detrital limestone of the Poty section in northeastern Brazil, deposition over an extended time period is indicated by gradually changing benthic micro- and macrofossil assemblages with no chaotic mixing of the faunal elements of different depositional environments. The tsunami interpretation fails to account for these gradual faunal changes and the mineralogically and lithologically discrete and correlatable sediment layers, whereas the presence of normal hemipelagic sediments of latest Maastrichtian age above the clastic deposits invalidates a K/T boundary age. We conclude that the clastic deposits in Mexico and Brazil predate the K/T boundary event and hence are not causally related to it. We demonstrate that these clastic deposits coincide with a eustatic sea-level low stand and could have been deposited as a result of it. However, since sea-level low stands do not necessarily result in coarse-grained clastic or limestone breccia deposits and since the glass present at the base of the clastic deposits in northeastern Mexico is unique, a pre–K/T boundary event (bolide impact or volcanism) cannot be excluded.

ACKNOWLEDGMENTS

We gratefully acknowledge support from NSF grants INT-9314080, OCE-9021338, the Petroleum Research Fund of the American Chemical Society PRF No. 26780-AC8, and Mexico Conacyt grant No. L120-36-36. Our investigations have benefited from numerous discussions with colleagues, including Thierry Adatte, Bill Ward, Don Lowe, Ferran Colombo, and Tony Ekdale. We thank the reviewers—B. H. Wilkinson, R. Buffler, and S. Gartner—for many helpful suggestions and critical comments.

REFERENCES CITED

Albertão, G. A., Koutsoukos, E. A. M., Regali, M. P. S., Attrep, M., Jr., and Martins, P. P., Jr., 1994, The Cretaceous-Tertiary boundary in southern low-latitude regions: Preliminary study in Pernambuco, northeastern Brazil: Terra Nova, v. 6, p. 366–375.

Alvarez, W., Smit, J., Lowrie, W., Asaro, F., Margolis, S. V., Claeys, P., Kastner, M., and Hildebrand, A. R., 1992, Proximal impact deposits at the Cretaceous-Tertiary boundary in the Gulf of Mexico: A restudy of DSDP Leg 77 Sites 536 and 540: Geology, v. 20, p. 697–700.

Ashraf, A. R., and Stinnesbeck, W., 1988, Pollen und Sporen an der Kreide-Tertiärgrenze im Staate Pernambuco, Nordost-Brasilien: Palaeontographica, Abt. B, v. 208, p. 39–51.

Askin, R. A., 1992, Preliminary palynology and stratigraphic interpretations from a new Cretaceous/Tertiary section from Seymour Island: Antarctic Journal, U.S. v. 25, p. 42–44.

Askin, R. A., and Jacobson, S. R., 1996, Palynological change across the Cretaceous-Tertiary boundary on Seymour Island, Antarctica: Environmental and depositional factors, *in* MacLeod, N. and Keller, G., eds., The Cretaceous-Tertiary boundary mass extinction: Biotic and environmental events: New York, W.W. Norton (in press).

Barrera, E., 1994, Global environmental changes preceding the Cretaceous-Tertiary boundary: Early-late Maastrichtian transition: Geology, v. 22, p. 877–880.

Baum, G. R., and Vail, P. R., 1988, Sequence stratigraphic concepts applied to Paleogene outcrops, Gulf and Atlantic Basins, *in* Wilgus, C. K., et al., eds., Sea-level changes—An integrated approach: Society of Economic Paleontologists and Mineralogists Special Publication 42, p. 309–328.

Blum, J. D., and Chamberlain, C. P., 1992, Oxygen isotope constraints on the origin of impact glasses from the Cretaceous-Tertiary boundary: Science, v. 257, p. 1104–1107.

Bohor, B. F., and Betterton, W. J., 1993, Arroyo El Mimbral, Mexico, K/T unit: Origin as debris flow/turbidite, not a tsunami deposit [abs.]: Lunar and Planetary Science, v. 24, p. 143–144.

Bourgeois, I., Hansen, T. A., Wiberg, P. L., and Kauffman, E. G., 1988, A tsunami deposit at the Cretaceous-Tertiary boundary in Texas: Science, v. 241, p. 567–570.

Canudo, I., Keller, G., and Molina, E., 1991, K/T boundary extinction pattern and faunal turnover at Agost and Caravaca, SE Spain: Marine Micropaleontology, v. 17, p. 319–341.

Donovan, A. D., Baum, G. R., Blechschmidt, G. L., Loutit, L. S., Pflum, C. E., and Vail, P. R., 1988, Sequence stratigraphic setting of the Cretaceous-Tertiary boundary in central Alabama, *in* Wilgus, C. K., et al., eds., Sea-level changes—An integrated approach: Society of Economic Paleontologists and Mineralogists Special Publication 42, p. 299–307.

Ekdale, A. A., and Stinnesbeck, W., 1994, Sedimentological significance of trace fossils in the KT "Mimbral Beds" of northeastern Mexico: Geological Society of America, Abstracts with Programs, v. 26, p. A395.

Groot, J. J., de Jonge, R. B. G., Langereis, C. G., ten Kate, W. G. H. Z., and Smit, J., 1989, Magnetostratigraphy of the Cretaceous-Tertiary boundary at Agost (Spain): Earth and Planetary Science Letters, v. 94, p. 385–397.

Habib, D., 1994, Biostratigraphic evidence of the K/T boundary in the eastern Gulf Coastal Plain, north of the Chicxulub crater, *in* New Developments regarding the K/T event and other catastrophes in Earth history: Houston, Texas, Lunar and Planetary Institute Contribution 825, p. 45.

Habib, D., Moshkovitz, S., and Kramer, C., 1992, Dinoflagellate and calcareous nanofossil response to sea-level change in Cretaceous-Tertiary sections: Geology, v. 20, p. 165–168.

Hansen, H. J., Gwozdz, R., Bromley, R. G., Rasmussen, K. L., Vogensen, E. W., and Pedersen, K. R., 1986, Cretaceous-Tertiary boundary spherules from Denmark, New Zealand and Spain: Geological Society of Denmark Bulletin, v. 35, p. 75–82.

Hansen, T. A., Upshaw, B., Kauffman, E. G., and Gose, W., 1993, Patterns of molluscan extinction and recovery across the Cretaceous-Tertiary boundary in east Texas: Report on new outcrops: Cretaceous Research, v. 14, p. 685–706.

Haq, B. U., Hardenbol, J., and Vail, P., 1987, Chronology of fluctuating sea levels since the Triassic: Science, v. 235, p.1156–1166.

Hildebrand, A. R., and Boynton, W. V., 1990, Proximal Cretaceous/Tertiary boundary impact deposits in the Caribbean: Science, v. 248, p. 843–847.

Hildebrand, A. R., Penfield, G. T., Kring, D. A., Pilkington, M., Camargo, Z. A., Jacobson, S. B., and Boynton, W. V., 1991, Chicxulub crater: A possible Cretaceous/Tertiary impact crater on the Yucatan Peninsula: Geology, v. 19, p. 867–869.

Hildebrand, A. R., and 14 others, 1994, The Chicxulub crater and its relation to the KT boundary ejecta and impact wave deposits, *in* New developments regarding the K/T event and other catastrophes in Earth history: Houston, Texas, Lunar and Planetary Institute Contribution 825, p. 49.

Izett, G. A., 1991a, K-T boundary tektites from near Beloc, Haiti: Lunar and Planetary Science Conference, 22nd, Houston, Texas, Abstracts, v. 22, p. 625–626.

Izett, G. A., 1991b, Tektites in Cretaceous-Tertiary boundary rocks on Haiti, and their bearing on the Alvarez impact extinction hypothesis: Journal of Geophysical Research, v. 96, p.20879–20905.

Jéhanno, C., Boclet, D., Froget, L., Lambert, B., Robin, E., Rocchia, R., and Turpin, L., 1992, The Cretaceous-Tertiary boundary at Beloc, Haiti: No evidence for an impact in the Caribbean area: Earth and Planetary Science Letters, v. 109, p. 229–241.

Keller, G., 1988a, Extinction, survivorship and evolution of planktic foraminifera across the Cretaceous-Tertiary boundary at El Kef, Tunisia: Marine Micropaleontology, v. 13, p. 239–263.

Keller, G., 1988b, Biotic turnover in benthic foraminifera across the Cretaceous-Tertiary boundary at El Kef, Tunisia: Paleogeography, Paleoclimatology, Paleoecology, v. 66, p. 153–171.

Keller, G., 1989, Extended period of extinctions across the Cretaceous/Tertiary boundary in planktic foraminifera of continental shelf sections: implications for impact and volcanism theories: Geological Society of America Bulletin, v. 101, p. 1408–1419.

Keller, G., 1992, Paleoecologic response of Tethyan benthic foraminifera to the Cretaceous/Tertiary boundary transition, *in* Takayanagi, Y., and Saito, T., eds., Studies in benthic Foraminifera, BENTHOS '90, Sendai, 1990: Tokyo, Tokyo University Press, p. 77–91.

Keller, G., 1993, The Cretaceous-Tertiary boundary transition in the Antarctic Ocean and its global implications: Marine Micropaleontology, v. 21, p. 1–45.

Keller, G., and Benjamini, C., 1991, Paleoenvironment of the eastern Tethys in the early Paleocene: Palaios, v. 6, p. 439–464.

Keller, G., MacLeod, N., Lyons, J. B., and Officer, C. B., 1993a, Is there evidence for Cretaceous/Tertiary boundary age deep water deposits in the Caribbean and Gulf of Mexico?: Geology, v. 21, p. 776–780.

Keller, G., Barrera, W., Schmitz, B., and Mateson, E., 1993b, Gradual mass extinction, species survivorship and long-term changes across the Cretaceous-Tertiary boundary in high latitudes: Geological Society of America Bulletin, v. 35, p. 979–997.

Keller, G., Stinnesbeck, W., Adatte, T., MacLeod, N., and Lowe, D., 1994a, Field Guide to Cretaceous/Tertiary boundary sections in northeastern Mexico: Houston, Texas, Lunar and Planetary Institute Contribution 827, 110 p.

Keller, G., Stinnesbeck, W., and Lopez-Oliva, J. G., 1994b, Age, deposition and biotic effects of the Cretaceous/Tertiary boundary event at the Arroyo El Mimbral, NE Mexico: Palaios, v. 9, p. 144–157.

Koeberl, C., 1994, Deposition of channel deposits near the Cretaceous/Tertiary boundary in northeastern Mexico: Catastrophic or "normal" sedimentary deposits?: Comment: Geology, v. 22, p. 957.

Koeberl, C., and Sigurdsson, H., 1992, Geochemistry of impact glasses from the K/T boundary in Haiti: Relation to smectites and new types of glass: Geochimica et Cosmochimica Acta, v. 56, p. 2113–2129.

Koutsoukos, E. A. M., 1995, The Cretaceous-Tertiary boundary in southern low-latitude regions: Preliminary study in Pernambuco: Reply: Terra Nova, v. 7, p. 378–382.

Longoria, J. F., and Gamper, M. A., 1992, Planktonic foraminiferal biochronology across the K/T boundary from the Gulf coastal plain of Mexico: Implications for timing the extraterrestrial bolide impact in Yucatan: Boletin Associacion Mexicana de Petroleros Geologos, v. 42, p. 19–40.

Longoria, J. F., and Grajales-Nishimura, J. M., 1993, Field trip guide to selected K/T boundary localities in Tamaulipas and Nuevo Leon, northeast Mexico: Revista de la Sociedad Mexicana de Paleontologia, v. 6, p. 73–85.

Lyons, J. B., and Officer, C. B., 1992, Mineralogy and petrology of the Haiti Cretaceous/Tertiary section: Earth and Planetary Science Letters, v. 109, p. 205–224.

Mabesoone, J. M., Tinoco, I. M., and Coutinho, P.I ., 1968, The Mesozoic-Tertiary boundary in northeastern Brazil: Paleogeography, Paleoclimatology, Paleoecology, v. 4, p. 161–185.

MacLeod, N., and Keller, G., 1991a, Hiatus distribution and mass extinctions at the Cretaceous/Tertiary boundary: Geology, v. 19, p. 497–501.

MacLeod, N., and Keller, G., 1991b, How complete are Cretaceous/Tertiary boundary sections? A chronostratigrahic estimate based on graphic correlation: Geological Society of America Bulletin, v. 103, p. 1439–1457.

Mancini, E. A., Tew, B., and Smith, C. C., 1989, Cretaceous-Tertiary contact, Mississippi and Alabama: Journal of Foraminiferal Research, v. 19, p. 93–104.

Montanari, A., Claeys, P., Asaro, F., Bermudez, J., and Smit, J., 1994, Preliminary stratigraphy and iridium and other geochemical anomalies across the K/T boundary in the Bochil section (Chiapas, southeastern Mexico), *in* New developments regarding the K/T event and other catastrophes in Earth history: Houston, Texas, Lunar and Planetary Institute Contribution 825, p. 84.

Moshkovitz, S., and Habib, D., 1993, Calcareous nannofossil and dinoflagellate stratigraphy of the Cretaceous/Tertiary boundary, Alabama and Georgia: Micropaleontology, v. 39, p. 167–191.

Pardo, A., Ortiz, N., and Keller, G., 1996, Latest Maastrichtian and K/T boundary foraminiferal turnover and environmental changes at Agost, Spain, *in* MacLeod, N., and Keller, G., eds., The Cretaceous/Tertiary boundary mass extinction: Biotic and environmental events: New York, W.W. Norton (in press).

Pope, K. O., Ocampo, A. C., and Duller, C. E., 1991, Mexican site for K/T impact crater: Nature, v. 351, p. 105–108.

Robin, E., Rocchia, R., Lyons, J. B., and Officer, C. B., 1994, Deposition of channel deposits near the Cretaceous/Tertiary boundary in northeastern Mexico: Catastrophic or "normal" sedimentary deposits?: Reply: Geology, v. 22, p. 958.

Savrda, C. E., 1991, Ichnology in sequence stratigraphic studies: An example from the lower Paleocene of Alabama: Palaios, v. 6, p. 39–53.

Savrda, C. E., 1993, Ichnosedimentologic evidence for a noncatastrophic origin of Cretaceous/Tertiary boundary sands in Alabama: Geology, v. 21, p. 1075–1078.

Schmitz, B., Keller, G., and Stenvall, O., 1992, Stable isotope and foraminiferal changes across the Cretaceous/Tertiary boundary at Stevns Klint, Denmark: Arguments for long-term oceanic instability before and after bolide impact: Paleogeography, Paleoclimatology, Paleoecology, v. 96, p. 233–260.

Sharpton, V. L., Dalrymple, G. B., Marin, L. E., Ryder, G. Schuraytz, B. C., and Urrutia- Fucugauchi, J., 1992, New links between the Chicxulub impact structure and the Cretaceous-Tertiary boundary: Science, v. 359, p. 819–821.

Sharpton, V. L., and 9 others, 1993, Chicxulub multi-ring impact basin: Size and other characteristics derived from gravity analysis: Science, v. 261, p. 1564–1567.

Sigurdsson, H., D'Hondt, S., Arthus, M. A., Bralower, T. J., Zachos, J. C., Fossen, M., and Channell, J. E. T., 1991, Glass from the Cretaceous/Tertiary boundary in Haiti: Nature, v. 349, p. 482–486.

Smit, J., and 8 others, 1992, Tektite bearing deep-water clastic unit at the Cretaceous-Tertiary boundary in northeastern Mexico: Geology, v. 20, p. 99–103.

Smit, J., Roep, T. B., Alvarez, W., Claeys, P., and Montanari, A., 1994a, Deposition of channel deposits near the Cretaceous/Tertiary boundary in northeastern Mexico: Catastrophic or "normal" sedimentary deposits? and Is there evidence for Cretaceous-Tertiary boundary age deep water deposits in the Caribbean and Gulf of Mexico: Comment: Geology, v. 22, p. 953–954.

Smit, J., Roep, T. B., Alvarez, W., Montanari, A., and Claeys, P., 1994b, Stratigraphy and sedimentology of K/T clastic beds in the Moscow Landing (Alabama) outcrop: Evidence for impact related earthquakes and tsunamis, *in* New developments regarding the K/T event and other catastrophes in Earth history: Houston, Texas, Lunar and Planetary Institute Contribution 825, p. 119.

Stinnesbeck, W., 1989, Fauna y microflora en el límite Cretácico-Terciario en el estado de Pernambuco, Noreste de Brasil: Contribuciones a los Simposios sobre el Cretácico de América Latina, *in* Spalletti, L. A., ed., Pte. A:

Eventos y Registro Sedimentario: Buenos Aires, n.p., p. 215–230.

Stinnesbeck, W., and Keller, G., 1995, The Cretaceous-Tertiary boundary in southern low-latitude regions: Preliminary study in Pernambuco: Comments: Terra Nova, v. 7, p. 375–378.

Stinnesbeck, W. and Keller, G., 1996, Environmental changes across the Cretaceous-Tertiary boundary in northeastern Brazil, *in* MacLeod, N. and Keller, G., eds., The Cretaceous-Tertiary boundary mass extinction: Biotic and environmental events: New York, W.W. Norton (in press).

Stinnesbeck, W., and 10 others, 1993a, Deposition of channel deposits near the Cretaceous/Tertiary boundary in northeastern Mexico: Catastrophic or "normal" sedimentary deposits? Geology, v. 21, p. 797–800.

Stinnesbeck, W., Ashraf, A. R., and v. Salis Perch-Nielsen, K., 1993b, Estudos paleontologicos no límite Cretácico-Terciario no Estado de Pernambuco: Recife, Universidade Federal de Pernambuco, Departamento de Geologia, Estudos Pesquisas, v. 10, p. 141–156.

Stinnesbeck, W., Keller, G., Adatte, T. and MacLeod, N., 1994a, Deposition of channel deposits near the Cretaceous/Tertiary boundary in northeastern Mexico: Catastrophic or "normal" sedimentary deposits? and Is there evidence for Cretaceous-Tertiary boundary age deep water deposits in the Caribbean and Gulf of Mexico?: Reply: Geology, v. 22, p. 955–956.

Stinnesbeck, W., Keller, G. and Adatte, T., 1994b, K/T boundary sections in southern Mexico (Chiapas): Implications for the proposed Chicxulub impact site, *in* New developments regarding the K/T event and other catastrophes in earth history: Houston, Texas, Lunar and Planetary Institute Contribution 825, p. 120–121.

Stinnesbeck, W., Keller, G., Adatte, T., Lopez-Oliva, J. G., and MacLeod, N., 1996, Cretaceous-Tertiary boundary clastic deposits in NE Mexico: Bolide impact or sea-level lowstand?, *in* MacLeod, N. and Keller, G., eds., The Cretaceous-Tertiary boundary mass extinction: Biotic and environmental events, W.W. Norton, New York (in press).

Stott, L. D., and Kennett, J. P., 1990, The paleoceonographic and paleoclimatic signature of the Cretaceous/Paleogene boundary in the Antarctic: Proceedings of the Ocean Drilling Program, Scientific Results, v. 113, p. 829–848.

Swisher, C. C., and 11 others, 1992, Coeval $^{40}Ar/^{39}Ar$ ages of 65 million years ago from Chicxulub crater melt rock and Cretaceous-Tertiary boundary tektites: Science, v. 257, p. 954–958.

Manuscript Accepted by the Society September 14, 1995

Geological Society of America
Special Paper 307
1996

Lithostratigraphic and mineralogic correlations of near K/T boundary clastic sediments in northeastern Mexico: Implications for origin and nature of deposition

Thierry Adatte
Institut de Géologie, 11 rue Emile Argand, 2007 Neuchâtel, Switzerland
Wolfgang Stinnesbeck
Facultad de Ciencias de la Tierra, Universidad Autónoma de Nuevo León, Ap. Postal 104, 67700 Linares, N.L., Mexico
Gerta Keller
Department of Geological and Geophysical Sciences, Princeton University, Princeton, New Jersey 08544

ABSTRACT

We interpret the near K/T boundary clastic deposits of northeastern Mexico as deposited over an extended time period, during the last 170 to 200 k.y. of the Maastrichtian and by normal sedimentary processes that include gravity flows and turbidity currents, rather than impact-generated tsunami waves. This deposition scenario is indicated by multiple horizons of bioturbation within the top unit 3 and near the base of the middle unit 2; the presence of thin layers enriched in fine clay-minerals and planktic foraminifera, suggesting hemipelagic sedimentation within unit 3; the presence of a marl layer of Maastrichtian age above the clastic deposit; the occurrence of distinct layers enriched in zeolites in unit 3; and the presence of lithologically, sedimentologically, and mineralogically distinct units and subunits that are correlatable over more than 300 km. Such correlations do not support a chaotic deposition as predicted for an impact-generated tsunami event. We interpret the clastic beds of northeastern Mexico as having accumulated during the major eustatic sea-level lowstand near the end of the Maastrichtian. In this scenario, the unconformity at the base of the clastic deposit represents a type 1 sequence boundary, where deltaic sediments were eroded and transported into deeper waters, depositing the spherule-rich layer of unit 1. Continued sea-level lowering resulted in erosion and bypass of shelf sediments and the deposition of the sandstone of unit 2. Subsequently, stabilization of the sea-level lowstand resulted in episodes of decreased erosion and sediment transport alternating with normal hemipelagic sedimentation, thus depositing the sand and the silt and shale layers of unit 3. The sea-level rise during the last 50 to 100 k.y. of the Maastrichtian resulted in the normal hemipelagic sedimentation observed in the pre–K/T boundary marl layer above the clastic deposit.

INTRODUCTION

The controversy over the nature of the Cretaceous-Tertiary transition and the causes of the associated mass extinction was altered fundamentally in 1980 with the discovery of the now-famous iridium anomaly at the K/T boundary at Gubbio, Italy (Alvarez et al., 1980). The discovery of similar anomalies elsewhere and the proposition that these anomalies and the K/T extinctions resulted from the impact of a large extraterrestrial bolide have spurred over a decade of unparalleled research on

Adatte, T., Stinnesbeck, W., and Keller, G., 1996, Lithostratigraphic and mineralogic correlations of near K/T boundary clastic sediments in northeastern Mexico: Implications for origin and nature of deposition, *in* Ryder, G., Fastovsky, D., and Gartner, S., eds., The Cretaceous-Tertiary Event and Other Catastrophes in Earth History: Boulder, Colorado, Geological Society of America Special Paper 307.

the physical and biological events at and near the K/T boundary. Within a short time, the controversy resolved itself into two contrasting schools of thought: (1) the K/T event reflects the catastrophic effects of a large (10 km) bolide colliding with the Earth, and (2) the K/T mass extinction was the culmination of long-term changes in the Earth's biota reflecting major changes in the global climate, sea-level fluctuations, and massive volcanism, which may have been accelerated by a bolide impact at the K/T boundary.

With the discovery of the buried Chicxulub structure on the north coast of Yucatan, Mexico, as a possible site for the putative K/T impact, outcrops of uppermost Cretaceous and lowermost Tertiary strata around the Gulf of Mexico assumed major importance as possible sources of direct evidence regarding the nature of the K/T boundary event. Among the first such sites investigated, well before the discovery of the Chicxulub structure, were the K/T sections along the Brazos River in southern Texas, which include a coarse-grained clastic unit at or near the K/T boundary. This unit has been variously interpreted as reflecting a high-energy impact-generated tsunami (Bourgeois et al., 1988), or a sea-level lowstand deposit unrelated to an impact (Jiang and Gartner, 1986; Donovan et al., 1988; Keller, 1989). Subsequent investigations of K/T sections at Beloc, Haiti, revealed the presence of glassy spherules with compositions suggesting the melting of carbonate rocks, consistent with the presence of thick Cretaceous carbonates at the Chicxulub site. Other sections in southern Mexico and Guatemala have been interpreted to show both conformable and unconformable relationships across the K/T boundary (Keller et al., 1994a; Keller and Stinnesbeck, 1995; Stinnesbeck et al., 1994a; Montanari et al., 1994).

Recently, over a dozen outcrops have been found in northeastern Mexico spanning a distance of more than 300 km. These outcrops are structurally uncomplicated and represent comformable sequences across the Cretaceous-Tertiary transition. To date, only preliminary descriptions and interpretations of these outcrops have been published (Longoria and Gamper, 1992; Longoria and Grajales, 1994; Smit et al., 1992, 1994; Stinnesbeck et al., 1993, 1994a; Bohor and Betterton, 1993; Keller et al., 1994a, b), and many aspects of the basic geology of the outcrops remain either unresolved or controversial. An important component of these sections is a thick (up to 12 m), sedimentologically complicated, coarse-grained clastic member that is stratigraphically positioned near the K/T boundary. This clastic member has been variously interpreted as an impact-generated tsunami deposit (Smit et al., 1992, 1994), a single turbidite deposit (Bohor and Betterton, 1993), or a series of gravity flows induced by a sea-level lowstand (Stinnesbeck et al., 1993, 1994b, 1996; Keller et al., 1994a). Here, we report on the lithology and mineralogy of this clastic member in order to derive clues as to the origin, nature, and tempo of deposition. Figure 1 shows the location of sections examined. Geographic distribution of these localities indicates a north-northwest–south-south-east–trending area that is parallel to and 40 to 80 km east of the front range of today's Sierra Madre Oriental. The southernmost section is El Mimbral, located about 80 km southeast of Ciudad Victoria, the capital of the state of Tamaulipas, Mexico. The northernmost section is Los Ramones, 40 km east of Monterrey, the capital of the state of Nuevo Leon, Mexico.

METHODS

Whole-rock and insoluble-residue analyses have been carried out for all the samples by XRD 2000 diffractometer at the Geological Institute of the University of Neuchâtel. The samples were prepared following the procedure of Kübler (1987). Two sample preparation methods have been applied. Random powder of the bulk sample is used for characterization of the whole rock mineralogy. Nearly 20 g of each rock sample was ground with a "jaw" crusher to obtain small chips (1 to 5 mm) of rock. Approximately 5 g were dried at a temperature of 110 °C and then ground again to a homogenous powder with particle sizes <40 μm. Eight hundred mg of this powder were pressed (20 bars) in a powder holder covered with a blotting paper and finally analyzed by X-ray diffractometer (XRD). Whole rock compositions were determined by X-ray diffractometer (SCINTAG XRD 2000 Diffractometer) based on methods by Ferrero (1966) and Kübler (1983). The method for the semiquantitative analysis of the bulk rock mineralogy (obtained by XRD patterns of random powder samples) used external standards (Table 1).

Clay-mineral analyses were based on methods by Kübler (1987). Ground chips were mixed with deionized water (pH 7 to 8) and agitated. The carbonate fraction was removed with the addition of HCl 10% (1.25 N) at room temperature for 20 minutes or more until all the carbonate was dissolved. Ultrasonic disaggregation was accomplished during 3-minute intervals. The insoluble residue was washed and centrifuged (5 to 6 times) until a neutral suspension was obtained (pH 7 to 8). Separation of different grain size fractions (<2 μm and 2 to 16 μm) was obtained by the timed settling method based on Stokes law. The selected fraction was then pipetted onto a glass plate and air-dried at room temperature. XRD analysis of oriented clay samples was made after air drying at room temperature and ethylene-glycol–solvated conditions. The intensities of selected XRD peaks characterizing each clay mineral present in the size fraction (e.g., chlorite, mica, and mixed layers) were measured for a semiquantitative estimate of the proportion of clay minerals present in the size fractions <2 μm and 2 to 16 μm. Therefore, clay minerals are given in relative percent abundance without correction factors (Table 2); content in swelling (% smectite) is estimated by using the method of Moore and Reynolds (1989).

LITHOLOGY AND MINERALOGY

In northeastern Mexico, near–K/T boundary clastic deposits are discontinuously present within a thick sequence of fine-grained, marly sedimentary rocks of the Mendez (Upper Cretaceous) and Velasco (Paleocene) Formations. This clastic

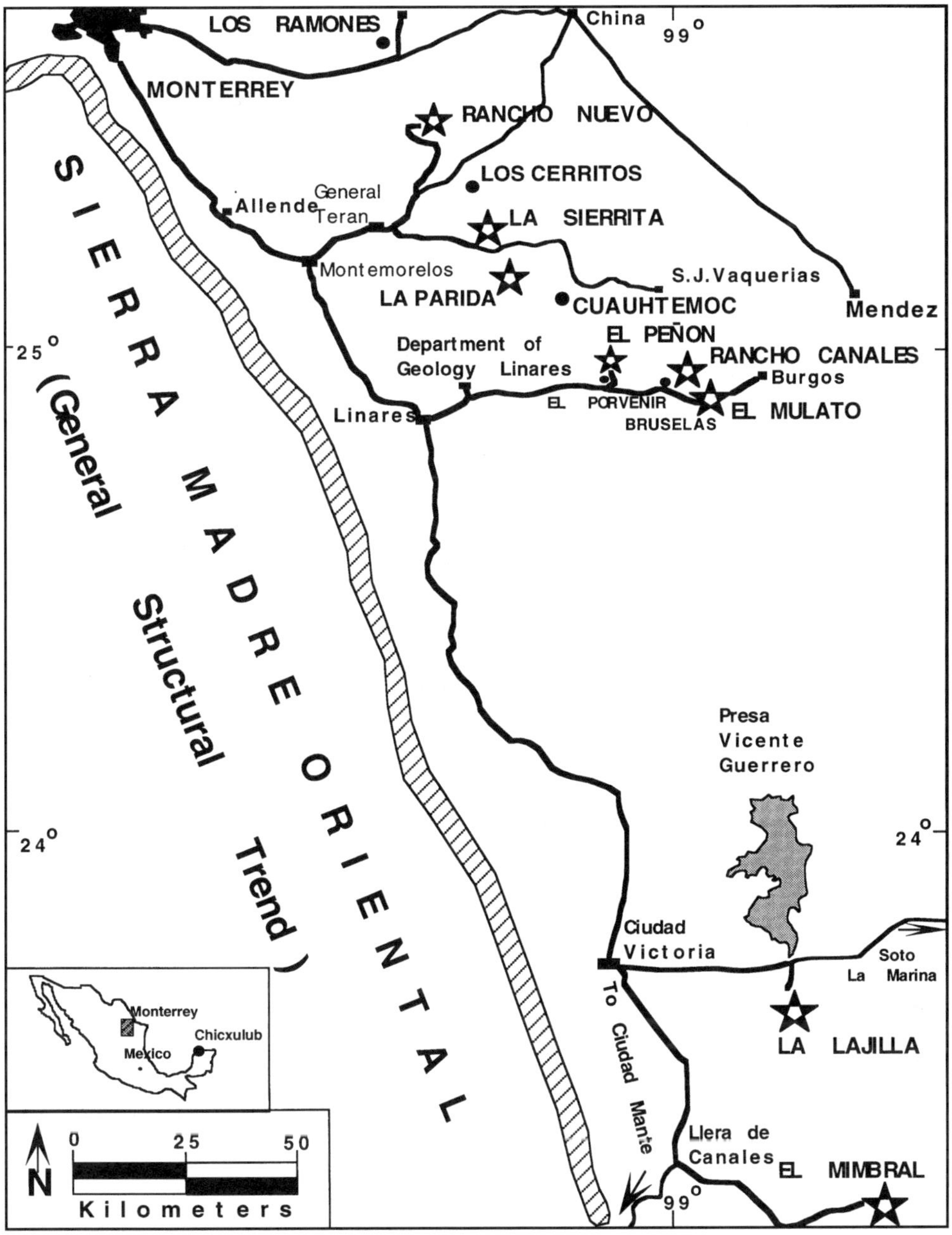

Figure 1. Location map of Cretaceous/Tertiary boundary outcrops with near–K/T boundary clastic deposits in northeastern Mexico.

member can be subdivided into three distinct units based on their lithology, sedimentology, and mineralogy: a basal spherule-rich unit 1, a middle unit 2 of horizontally laminated sandstone, and an upper unit 3 consisting of intercalated sand and silt and shales layers (Fig. 2). The thickness of each of the three units and subunits varies considerably, and the entire clastic member even disappears in some sections (e.g., La Parida). Various deposits range from 0 to 1 m for the spherule-rich unit 1, which therefore does not blanket the region; 0 to 4 m for the laminated sandstone of unit 2; and 0 to 3 m for the interlayered sand, shale, and siltstone beds of unit 3.

The three units of the clastic member are lithologically and mineralogically correlatable more than 300 km (Figs. 1, 2). In all outcrops, the clastic deposit disconformably overlies the gray marls of the Maastrichian Mendez Formation. Mineralogical analyses indicate a marly lithology for the Mendez Formation, with an average composition of 48% calcite, 30% phyllosilicates, 15% quartz and 8% plagioclase. Clay-mineral contents, such as mica (illite), chlorite, and chlorite-smectite and illite-smectite mixed-layers, are variable. This is especially the case with chlorite, which varies in the Mendez Formation from 60% (<2 μm size fraction) in the Lajilla I and Mulato sections to less

TABLE 1. AVERAGE WHOLE ROCK COMPOSITION FOR THE MARLS OF THE MENDEZ FORMATION, THREE UNITS AND SUBUNITS OF THE CLASTIC DEPOSIT, THE THIN LATEST MAASTRICHTIAN MARL LAYER ABOVE THE CLASTIC DEPOSIT AND THE EARLY TERTIARY SHALES OF THE VELASCO FORMATION

UNIT	Calcite (%)	Quartz (%)	Phyllosilicates (%)	Plagioclase (%)
Velasco Formation	35	21	36	8
Maastrichtian Thin Marl Layer	53	9	32	6
UNIT 3 : R.S.L.	36	27	22	15
UNIT 3 : Silt-Shale layers	31	22	36	11
UNIT 3 : Sandstone layers	39	28	16	17
UNIT 2 : Laminated Sandstone	34	31	19	16
UNIT 1 : Spherules rich layers	64	10	20	6
UNIT 1 : S.S.L.	43	24	23	10
UNIT 1 : Marl Layers	38	13	41	8
Mendez Formation	48	15	30	7

TABLE 2. AVERAGE PERCENT CLAY MINERAL COMPOSITION (EL PEñON, EL MULATO, AND LA LAJILLA SECTIONS, $<2\mu$ SIZE-FRACTION) FOR MARLS OF THE MENDEZ FORMATION, THREE UNITS AND SUBUNITS OF THE CLASTIC DEPOSIT, THE LATEST MAASTRICHTIAN MARL LAYER ABOVE THE CLASTIC DEPOSIT AND THE EARLY TERTIARY SHALES OF THE VELASCO FORMATION

UNIT	Zeolite (%)	Chlorite (%)	Mica (%)	Illite-Smectite (%)	Chlorite-smectite (%)
Velasco Formation	0.5	47	35	11	7
Maastrichtian Thin Marl Layer	2	32	46	15	5
UNIT 3 : R.S.L.	3	39	35	16	7
UNIT 3 : Silt-Shale layers	0	30	18	41	11
UNIT 3 : Sandstone layers	2	41	42	12	3
UNIT 2 : Massive Sandstone	0	37	43	17	0
UNIT 1 : Spherules rich layers	0	40	19	22	18
UNIT 1 : S.S.L.	0	39	37	15	9
UNIT 1 : Marl Layers	0	19	18	54	9
Mendez Formation	0	40	35	25	1

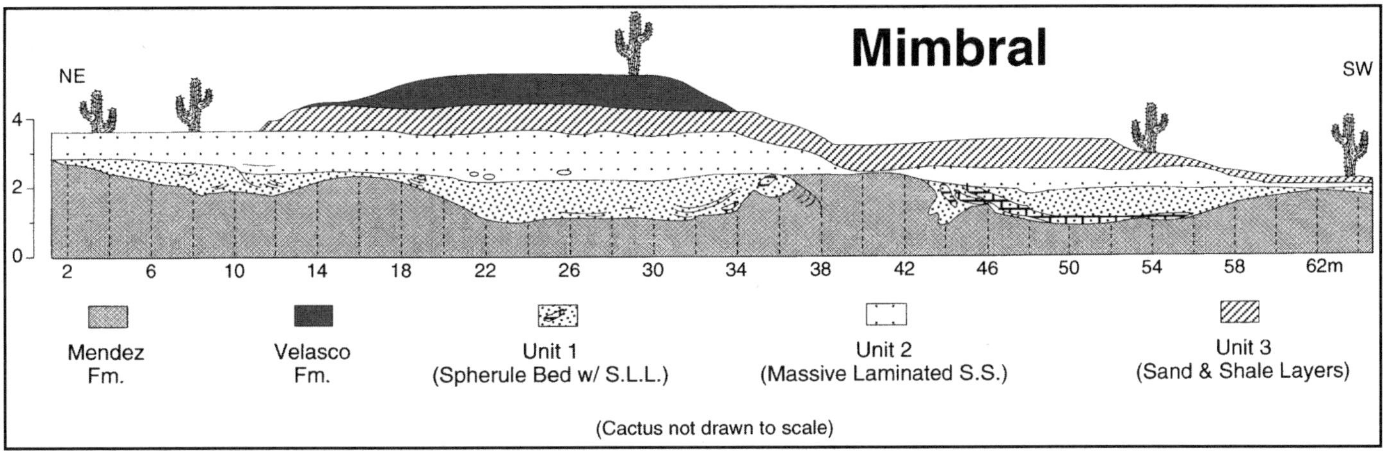

Figure 2. Sketch of El Mimbral K/T boundary outcrop showing units 1, 2, and 3. Note that units thin out laterally toward the edges of the outcrop. S.L.L. = sandy limestone layer.

than 20% in the La Sierrita and Peñon sections (Figs. 3, 4). These mineralogically different sediment layers below the unconformity suggest erosion to variable depths within the Mendez marls.

Unit 1: Spherule-rich layer

Unit 1 disconformably overlies an irregular erosional surface of the Mendez Formation and consists of repeated intercalations of soft-weathered spherule-rich layers and Mendez marls. Unit 1 has been observed at El Mimbral, La Lajilla, El Mulato, Rancho Canales, El Peñon, and La Sierrita (Figs. 1, 5) and ranges in thickness from 0 to 100 cm. It is either massive or crudely stratified with marl intercalations and locally shows large-scale trough cross-stratification. In addition to abundant spherules, unit 1 contains common foraminiferal tests including transported shallow-water benthic foraminifera (Stinnesbeck et al., 1993, 1996; Keller et al., 1994b), clasts of marl from the underlying Mendez Formation, clasts containing Turonian age foraminifera, minor quartz, and rare glass shards. The spherules range from 1 to 5 mm in diameter. Many spherules are filled with blocky calcite; some contain foraminiferal tests, limestone clasts, and rutile crystals (Stinnesbeck et al., 1993, 1994a; Keller et al., 1994a). Large spherules are often composites containing smaller spherules, as described from volcanic deposits. Many spherules with blocky calcite infillings are enclosed in an organic shell characterized by a significant content of organic carbon (0.3 to 0.5%), suggesting a biogenic origin as either algal resting cysts (Hansen et al., 1986) or oncolites. Spherules thus appear to be of multiple origins (e.g., oncolites, algal resting cysts, and volcanic) and transported from shallow carbonate shelf depositional settings (Stinnesbeck et al., 1993; Keller et al., 1994a). Other workers have interpreted the spherules as originally hollow, glassy impact-produced microtektites that have been subsequently infilled and replaced by calcite or phyllosilicate minerals (Smit et al., 1992; Bohor and Betterton, 1993). Rare glass fragments and very rare shocked quartz grains are present in this unit.

A 10- to 20-cm-thick layer of a well-cemented packstone with fine-scale laminations, termed sandy limestone layer (S.L.L.) (Stinnesbeck et al., 1993; Keller et al., 1994a) occurs within or at the top of unit 1 at Mimbral, Mulato, Peñon, Rancho Canales, and Lajilla. This S.L.L. shows only few graded spherules at the top and base in contact with the spherule-rich sediments. The top of this S.L.L. is sometimes convolute, with the overlying sediments draped across the convolute bedding surface. At Lajilla and Peñon II, the convolute bedding surface marks an erosional disconformity and the upper part of the spherule-rich layer is missing (Figs. 4, 5).

Whole-rock and clay-mineral compositions of the densely cemented spherule-rich layer of unit 1 are very similar in all outcrops and show two trends. The spherule-rich intercalations are primarily composed of calcite (64%), quartz (10%), phyllosilicates (20%), and plagioclase (6%). In contrast, marl intercalations are composed of abundant phyllosilicates (>40%) and less calcite (38%, Table 1, Fig. 6). These proportions are similar to those determined in the underlying Mendez Formation. The difference in whole rock and clay-mineral compositions suggests variable amounts of detrital influx between deposition of the sandy limestone layer and the spherule-rich layers above and below.

Unit 2: Laminated sandstone

Unit 2 consists of a horizontally laminated sandstone with mud clasts of Maastrichtian and Turonian age marls at its base, reflecting an erosional disconformity. Wood fragments and leaves are present in distinct layers near the base of unit 2 without significant coal content. Sparse coal fragments associated with dispersed organic matter are present throughout unit 2 in all sections. Sand beds grade upward to slightly finer sands containing mica and a clay-rich matrix. Both lower and upper

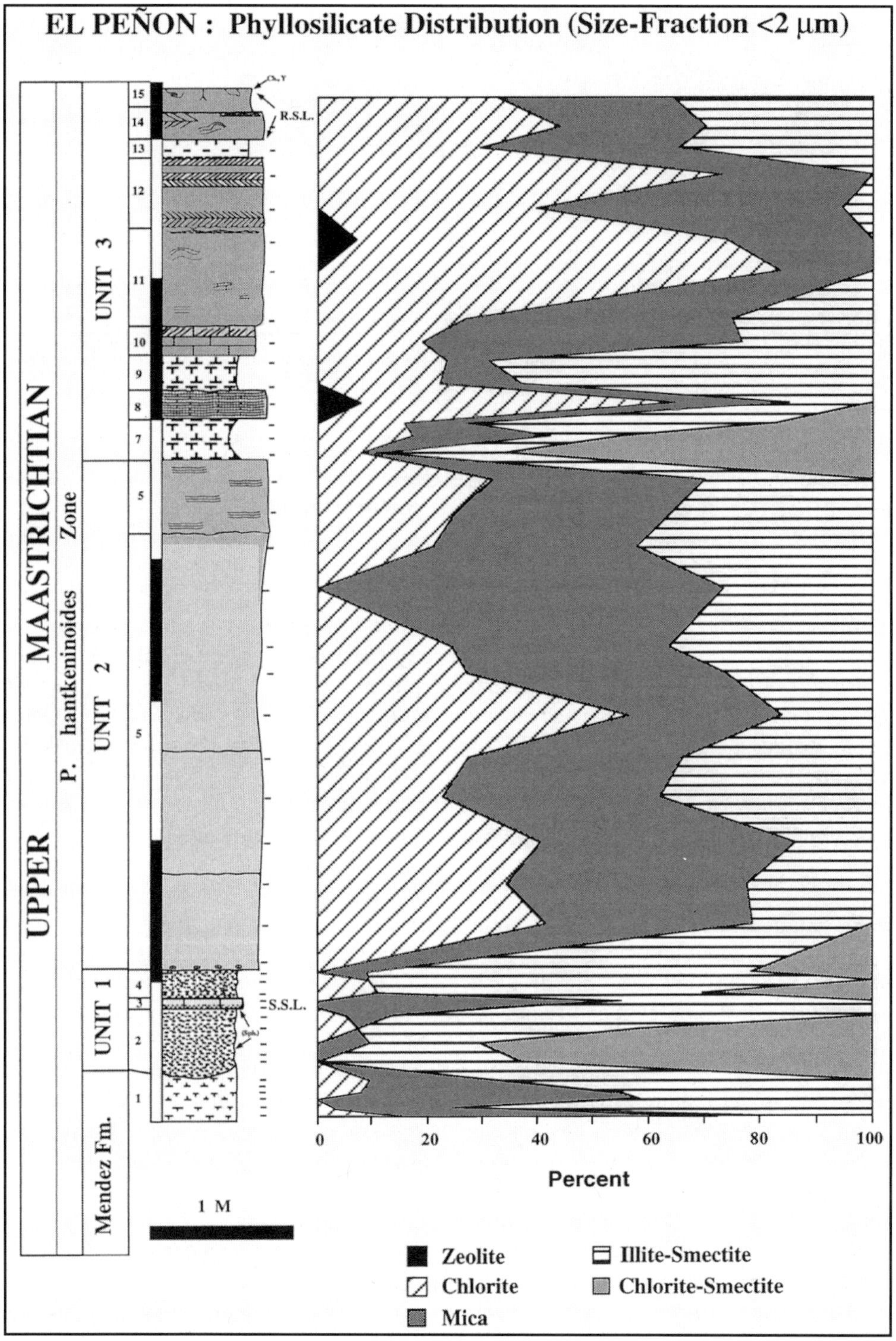

Figure 3. El Peñon I, phyllosilicate distribution from the Mendez Formation through the clastic deposit (<2 μm size fraction). Note the two layers enriched in zeolites in unit 3 and mixed layers enriched in illite-smectite and chlorite-smectite in unit 1 and in the shale layers of unit 3. S.L.L. = sandy limestone layer; R.S.L. = rippled sandy limestone.

contacts of unit 2 are disconformable. The sandstone is of variable thickness, ranging from 0 to 4 m. This unit is not present at La Sierrita II, La Parida, El Peñon II, La Lajilla, and El Mimbral II (Fig. 2). Recently, Ekdale and Stinnesbeck (1994) observed bioturbation near the base of unit 2 at El Peñon, with burrows truncated and infilled with spherule-rich sediments. This suggests discontinuous sedimentation within unit 2.

Whole-rock and clay-mineral contents of unit 2 are more regular than in unit 1 and characterized by low calcite (<40%) and phyllosilicates (19%) but higher quartz (31%) and plagioclase (albite >16%; Table 1, Fig. 6). These mineralogical data indicate a significant increase in detrital influx and more rapid deposition relative to the underlying unit 1 as well as to the overlying unit 3.

Unit 3: Interlayered sand-silt beds

Unit 3 consists of interlayered sand, shale, and siltstone beds topped by a rippled sandy limestone (R.S.L.) layer. Characteristic sedimentological features include horizontal laminations, ripple marks, flaser bedding, and convolute lamination. Some of the ripples are symmetrical and some show divergent dipping cross laminations. Smit et al. (1994) also report opposite orientation of sandstone cross laminae, suggesting bidirectional current regimes. Thin layers with normal hemipelagic sedimentation and distinct layers containing zeolites were detected within unit 3 at El Mimbral, La Lajilla, El Mulato, El Peñon, La Sierrita, and La Parida (Adatte et al., 1994). Rare plant debris may be present. Where unit 3 directly overlies unit 1 or the Mendez Formation (e.g., Peñon II, La Lajilla I and II, La Sierrita III, mud clasts are commonly found near the base, marking a disconformity. The R.S.L. layer that tops unit 3 in all outcrops (except Los Ramones) may consist of one or several layers and is marked by extensive bioturbation, most commonly *Chondrites, Zoophycos, Planolites, Thalassinoides*, and *Ophiomorpha* burrowing networks (Keller et al., 1994a, Ekdale and Stinnesbeck, 1994). Within unit 3 and the R.S.L., Ekdale and Stinnesbeck (1994) observed several horizons of trace fossils truncated by overlying sediments (e.g., Rancho Canales, Figs. 1, 5) that indicate repeated colonization and erosion and hence deposition over an extended time period. The R.S.L. layer is widely distributed and can be found where the underlying part of unit 3 and units 2 and 1 are absent (e.g., El Mimbral II, La Parida; Fig. 6).

Whole-rock and clay-mineral compositions of unit 3 are highly variable (Figs. 3 through 6). Two distinct clay-mineral associations can be identified. The R.S.L. and sand layers are characterized by high chlorite (39 to 41%) and mica (35 to 42%) and suggest increased detrital influx and probably more rapid deposition, similar to unit 2. In contrast, the shale layers are enriched in finer chlorite-smectite (11%) and illite-smectite irregular mixed layers (up to 70%, <2 μm size fraction; Table 2, Figs. 4, 5). These shale-mineral associations are similar to those of the Mendez marls (Table 2) and indicate periods of normal hemipelagic sedimentation during deposition of unit 3.

Two distinct layers rich in zeolites (clinoptilolite-heulandite) are recognized near the base and top of unit 3 (Table 2; Figs. 3 through 7) in all sections examined (Adatte et al., 1994). Additional zeolite-enriched layers associated with smectite are observed in unit 1, on top of the clastic deposit at El Mimbral and in the Maastrichtian age marly layer that overlies the clastic deposits at La Lajilla, El Mulato, and La Parida.

POSITION OF THE K/T BOUNDARY AND AGE OF THE CLASTIC DEPOSIT

The horizontal correlation line A in Figure 5 marks the K/T boundary at the top of the R.S.L. layer of unit 3 at Mimbral and above the thin Maastrichtian marl layer overlying unit 3 at Lajilla, Mulato, and Parida. The R.S.L. generally contains two or more distinct sediment layers burrowed by *Chondrites, Zoophycos, Thalassinoides*, and *Ophiomorpha*. The overlying marl layer contains typical *Plummerita hantkeninoides* Zone foraminiferal assemblages, indicating the presence of latest Maastrichtian sediments. This marl layer contains no size sorting, grain-size grading, or transported shallow-water benthic foraminifera that would suggest settling from the water column after deposition of the R.S.L. Whole-rock and clay-mineral analyses indicate that these sediments are enriched in fine phyllosilicates (>2 μm) with illite-smectite and chlorite-smectite mixed layers most abundant. No exotic minerals have been observed.

Biostratigraphic control suggests that the clastic sediment was deposited some time during the last 170 to 200 k.y. of the Maastrichtian. This age is derived from the range of the new latest Maastrichtian planktic foraminiferal markers species *P. hantkeninoides*, which indicates the last 200 k.y. of paleomagnetic chron 29R below the K/T boundary (Pardo et al., 1996). In all sections examined, except Mulato, this marker species is present either above, within, or below the clastic deposits but always below the K/T boundary (López-Oliva and Keller, this volume). This marl layer overlying the clastic deposit thus suggests that normal hemipelagic sedimentation resumed prior to the K/T boundary.

LITHOSTRATIGRAPHIC CORRELATIONS

The near K/T–boundary clastic deposits of northeastern Mexico consist primarily of transported terrigenous and shallow-water neritic debris from multiple sources (e.g., terrestrial, shallow neritic, neritic, upper bathyal) and ages (e.g., Maastrichtian, Campanian, Turonian). Nevertheless, the depositional sequence of this varied transported debris is surprisingly constant. It allows recognition of three well-defined units and subunits (e.g., S.L.L. of unit 1, R.S.L. of unit 3) based on lithologic and sedimentologic characteristics that can be traced and correlated over a distance of more than 300 km.

Figure 5 shows the lithostratigraphic correlation of eleven

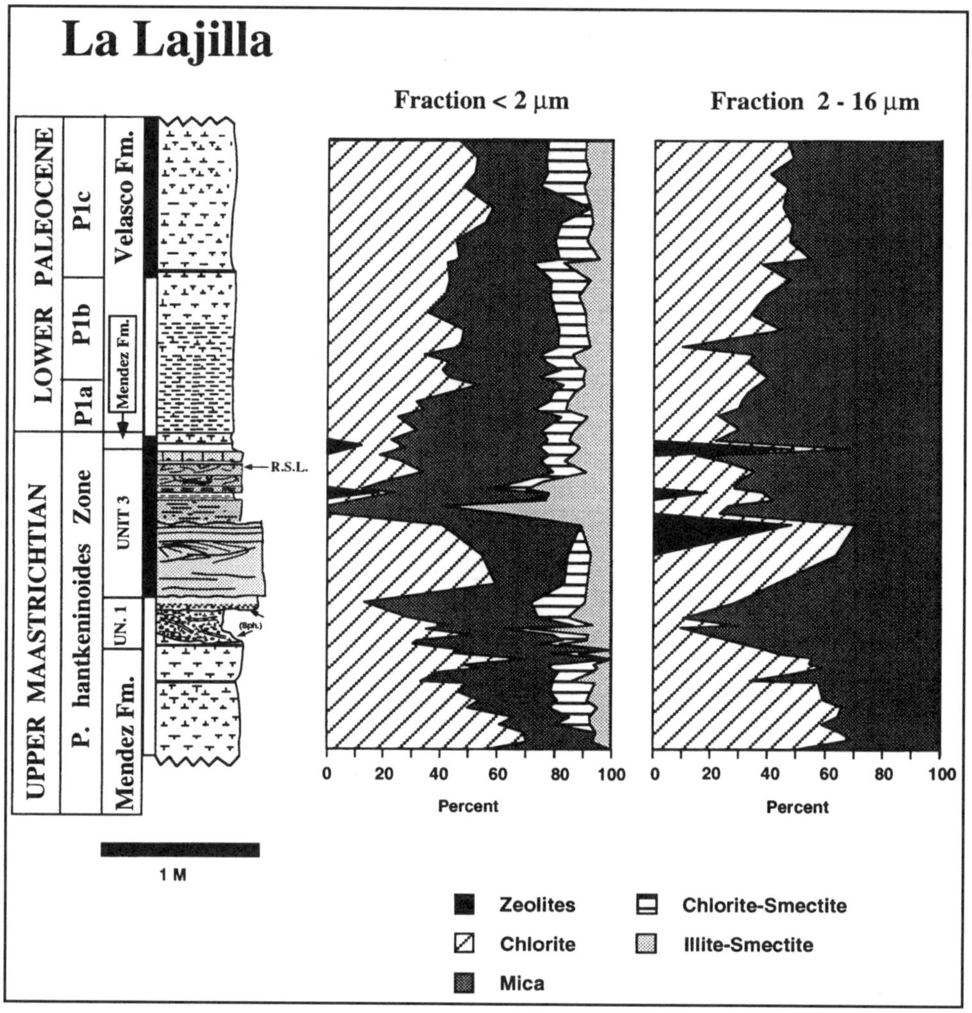

Figure 4 (on this and facing page). La Lajilla and El Mulato, phyllosilicate distribution from the Mendez Formation through the clastic deposit (<2 µm and 2- to 16 µm size fractions). Note the layers enriched in zeolites in unit 3 and mixed layers enriched in chlorite-smectite and illite-smectite in unit 1 (La Lajilla) and in the shale layers of unit 3. R.S.L. = rippled sandy limestone.

outcrops in seven localities. Horizontal correlation line A marks the K/T boundary at the top of the R.S.L. layer of unit 3 at Mimbral and above the thin Maastrichtian marl layer overlying unit 3 at Lajilla, Mulato, and Parida.

Correlation line B marks the base of unit 3 (alternating sand, silt, and shale layers) and top of unit 2 (laminated sandstone). The Ramones section (Fig. 1) is not figured because units 2 and 3 cannot be distinguished as a result of the shallower-water depositional environment. However, the presence of *P. hantkeninoides* in these sediments indicates that deposition was coeval.

Correlation line C marks the base of the laminated sandstone of unit 2 and the top of the spherule-rich layer of unit 1. Unit 2 laterally disappears first and is absent at Mimbral II, La Lajilla II, El Peñon II, and La Parida.

Correlation line D marks the top of the Mendez Formation and the base of the spherule-rich unit 1. Within unit 1 there is a distinct subunit consisting of the resistant 10- to 20-cm-thick S.L.L. that is present at El Mimbral, El Peñon, Rancho Canales, and La Lajilla. The presence of this S.L.L. is thus correlatable over a distance of more than 200 km.

MINERALOGICAL CORRELATIONS

Correlations based on whole rock composition of five localities are shown in Figure 6. Whole-rock and clay-mineral analyses of five near K/T–boundary clastic deposits in northeastern Mexico indicate three distinct and correlatable units and subunits comparable to those based on lithology and sedimentology. The sections examined include La Sierrita, El Peñon, El Mulato, La Lajilla, and El Mimbral.

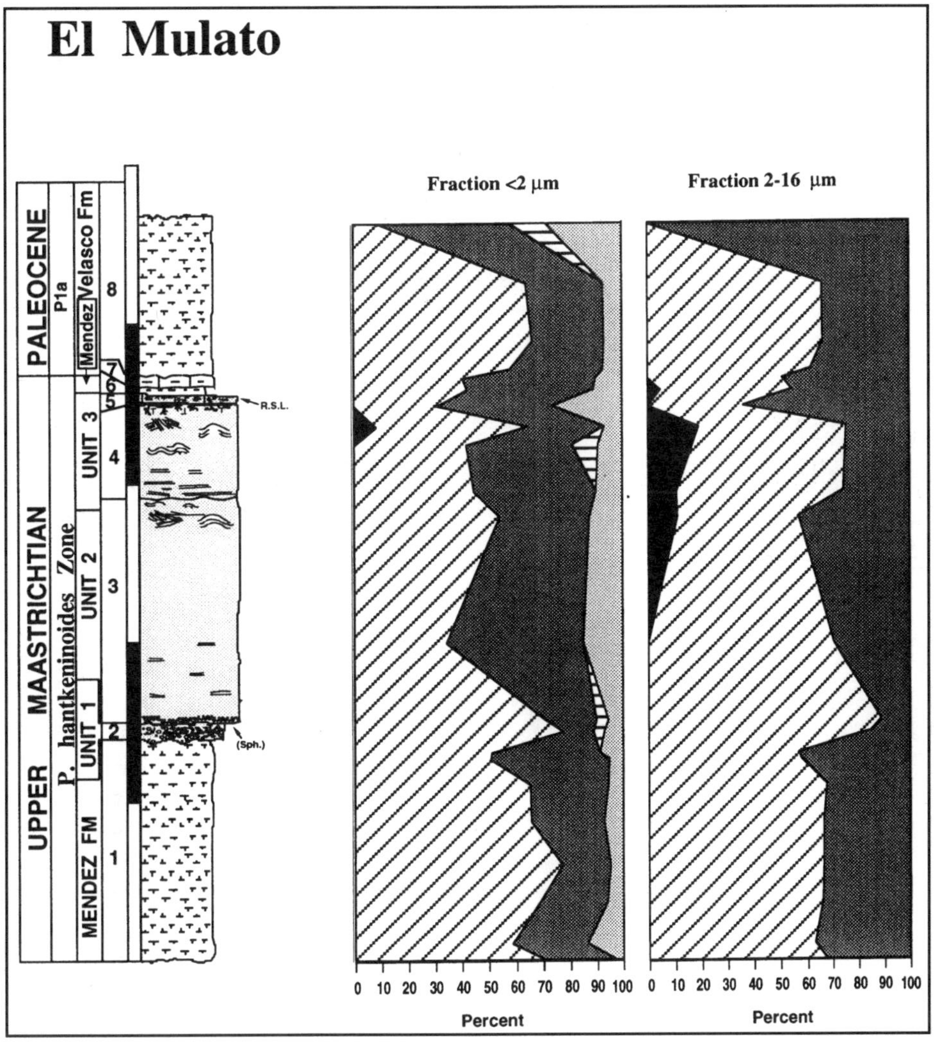

The Mendez Formation is characterized by 50% calcite, 30% phyllosilicates, 15 to 20% quartz, and 5 to 10% plagioclase, indicating a marly sediment composition (Table 1). Whole-rock composition of unit 1 marks repeated intercalations of spherule-rich layers with increased calcite (up to 60%), decreased phyllosilicates, quartz, and plagioclase, and intercalations of Mendez marls with the same composition as described above. The thick S.L.L. within unit 1 differs from these sediments by showing lower calcite but higher quartz, plagioclase, chlorite, and mica (illite). This suggests distinctly different detrital influxes during deposition of the S.L.L. and the loosely cemented spherule-rich sediments above and below.

The laminated sandstone of unit 2 is characterized by low calcite (20 to 40%) and phyllosilicates (10 to 15%) and high quartz (up to 40%) and plagioclase (albite 15 to 25%; Table 1; Fig. 7). These mineralogical data are correlatable in all outcrops and indicate increased detrital influx relative to units 1 and 3 and probably more rapid deposition. Whole-rock composition of the sand, silt, and shale layers of unit 3 is highly variable in calcite (but generally higher than in unit 2), phyllosilicates, quartz, and plagioclase. The sand and silt layers with higher quartz, plagioclase, and calcite reflect increased detrital influx and probably more rapid deposition, similar to unit 2. The shale layers are similar in composition to the underlying Mendez Formation and reflect periods of normal hemipelagic sedimentation within unit 3. The thin marl layer of latest Maastrichian age that overlies unit 3 at Mulato, Parida, and Lajilla shows a marly composition comparable to the Mendez Formation, suggesting that normal hemipelagic sedimentation resumed after deposition of the clastic member and prior to the K/T boundary event.

Correlation of zeolite-enriched layers in eight sections over a distance of more than 200 km, from El Mimbral to La Sierrita, is shown in Fig. 7. Two distinct zeolite-enriched layers (clinoptilolite-heulandite) have been recognized near the base and near the top of unit 3, below the rippled sandy limestone at El Mimbral, La Lajilla, El Mulato, La Parida, and La Sierrita.

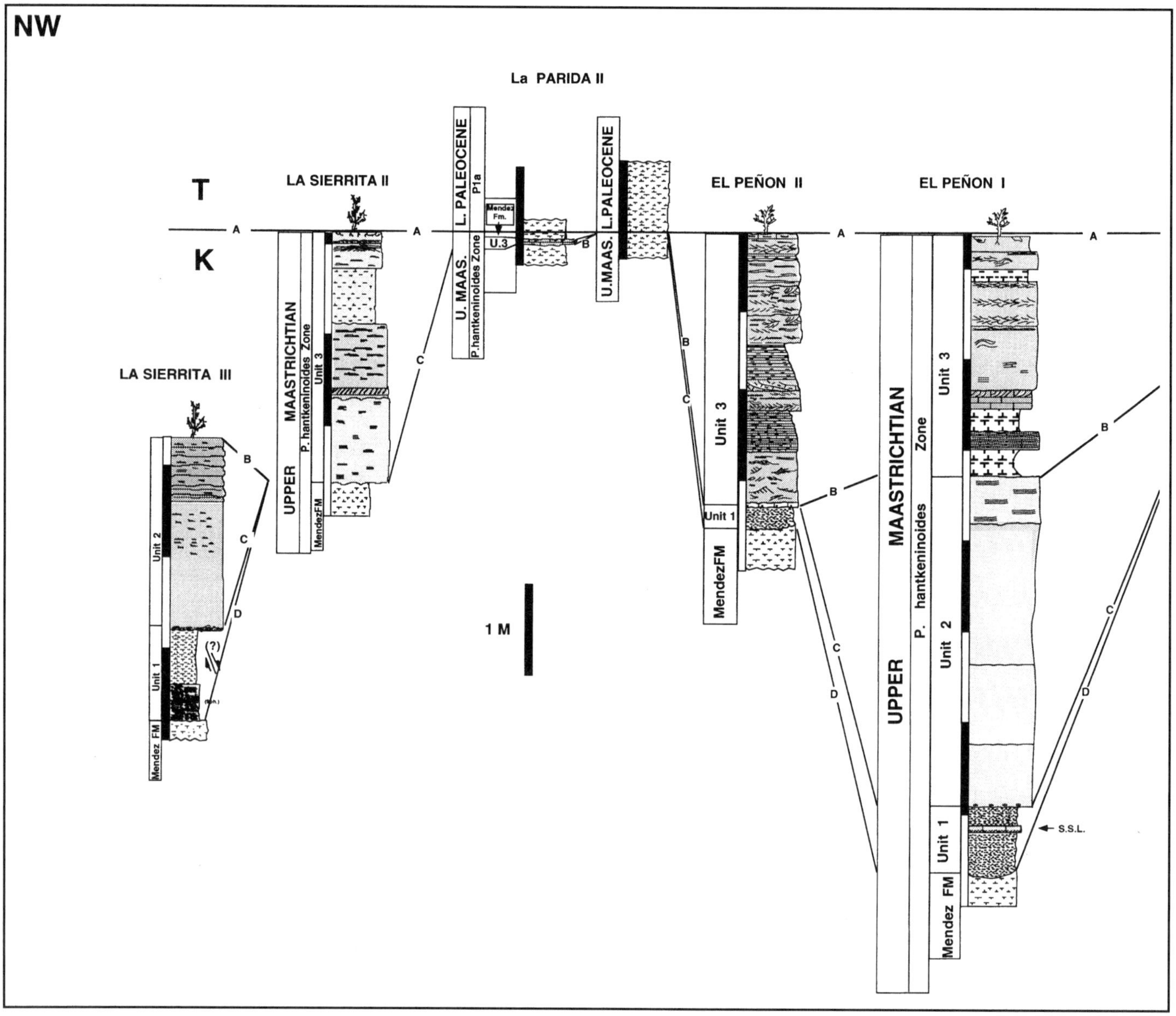

Figure 5 (on this and facing page). Biostratigraphic and lithostratigraphic correlation of ten K-T boundary outcrops in northeastern Mexico that contain near–K/T clastic deposits. S.L.L. = sandy limestone layer; R.S.L. = rippled sandy limestone.

Significant enrichment in zeolites is also observed in the thin marl layer of Maastrichtian age on top of unit 3 at La Lajilla, La Parida, and El Mulato outcrops and in the basal Tertiary clay layer at El Mimbral II.

IMPLICATIONS FOR ORIGIN AND NATURE OF DEPOSITION

The clastic deposit ranges from a few centimeters at La Parida to 7 m at El Peñon and is characterized by three lithologically and mineralogically distinct correlatable units and several subunits. Each of these units and subunits represents differing flow and depositional regimes with varying detrital influx and rates of sediment deposition. These units, subunits, and even individual zeolite-enriched layers are correlatable on a regional scale and suggest a nonchaotic deposition.

The only unusual component of the clastic deposit is the spherule-rich layer of unit 1, which contains rare glass and shocked quartz. Within unit 1 is a subunit consisting of a 20-cm-thick, well-cemented microlaminated sandy limestone layer (S.L.L.) with few graded spherules at the base and top. This S.L.L. is distinctly different from the spherule-rich sedi-

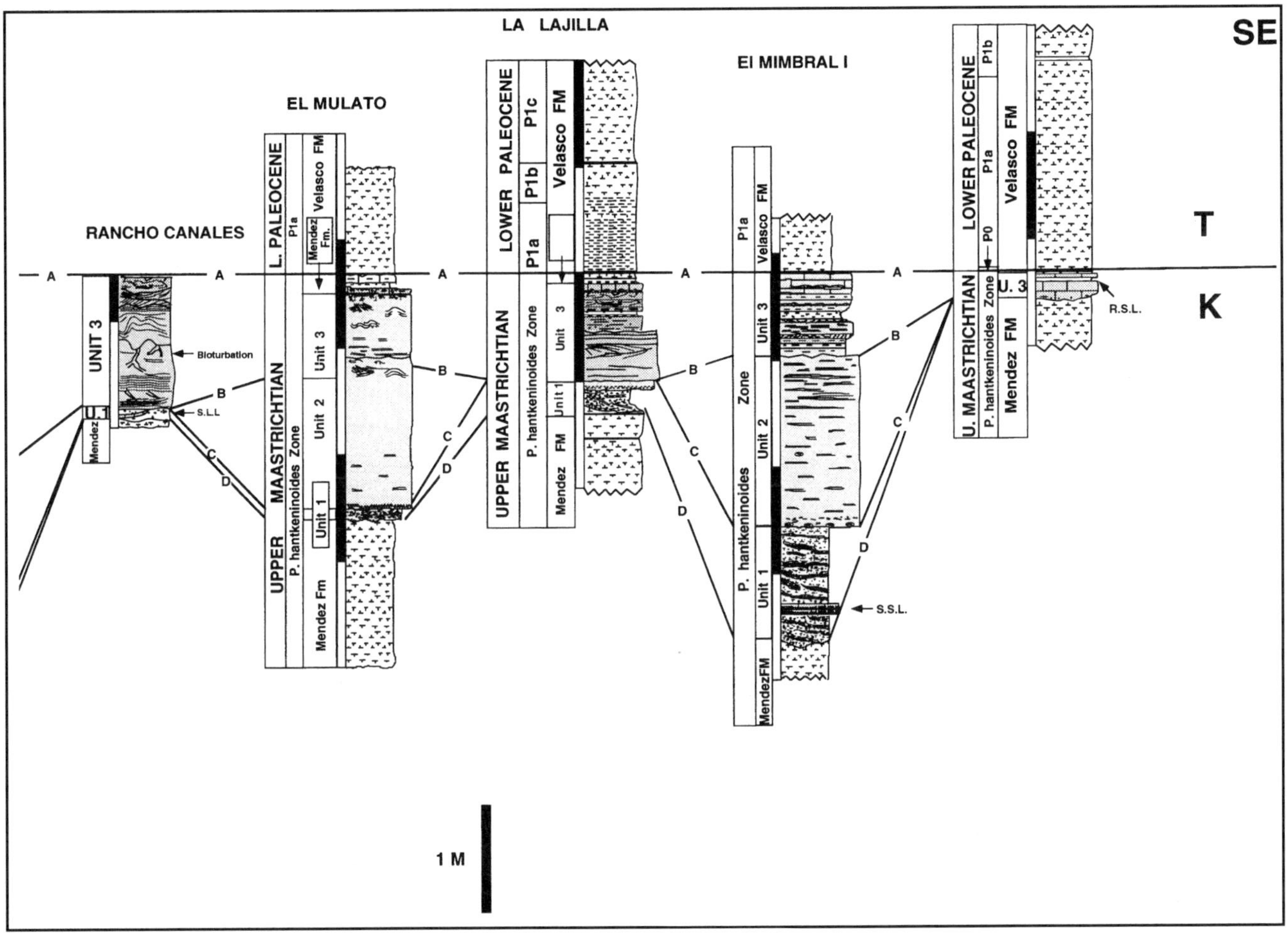

ment and generally lower in calcite and higher in quartz, plagioclase, chlorite, and mica (Tables 1, 2). This difference implies variable detrital influx and perhaps rates of deposition between the spherule rich layers and the S.L.L. Moreover, this indicates that the S.L.L. was at least semilithified prior to deposition of the spherule-rich sediments above it and also prior to deposition of the sandstone of unit 2. Thus, deposition of unit 1 occurred over a long time and not in a matter of hours as suggested by Smit et al. (1992) or Bohor and Betterton (1993). The composition of the spherule-rich layers also suggests a sediment source primarily from shallow neritic environments, as reflected by the presence of glauconite, transported shallow neritic benthic foraminifera, clasts with Turonian age foraminifers, spherules enclosing foraminiferal tests or clasts, oncolites, and spherical infilled organic tests that may be algal resting cysts. Since no Turonian age sediments (Agua Nueva Formation) were exposed in the region during K/T boundary time, these clasts were probably eroded from deeply cut canyon walls and transported. Some larger spherules contain several smaller calcite-infilled spherules that may originally have been bubble-rich glass spherules, and some spherules contain rutile crystals, also suggesting a volcanic origin (Stinnesbeck et al., 1993; Keller et al., 1994a). Rare glass shards are also present with average compositions similar to the glassy spherules found at Beloc, Haiti (Smit et al., 1992; Bohor and Betterton, 1993, Stinnesbeck et al., 1993). The origin of these glass shards is still in dispute, with some arguing for an impact origin (Izett et al., 1990; Sigurdsson et al., 1991; Blum and Chamberlain, 1992; Koeberl and Sigurdsson, 1992; Koeberl, 1994) and others arguing for a volcanic origin (Jéhanno et al., 1992; Lyons and Officer, 1992; Robin et al., 1992). In either case, the event that produced the glass appears to have predated the K/T boundary bolide impact.

The horizontally laminated sandstone of unit 2 represents increased detrital influx and more rapid deposition than unit 1. The only unusual feature is the presence of discrete layers of leaf and wood debris at the base of unit 2 at El Mimbral, aligned in a north-northwest–south-southeast direction, which suggests a major detrital influx from the coastal area and possibly reflects a collapse of the northwest delta in the Monterrey area (Fig. 1). No similar plant debris layers have been observed in any other outcrops. Very small fragments of plant debris can be found throughout the sandstone and occasionally in sandy layers of unit 3. These are not unique or restricted to K/T boundary

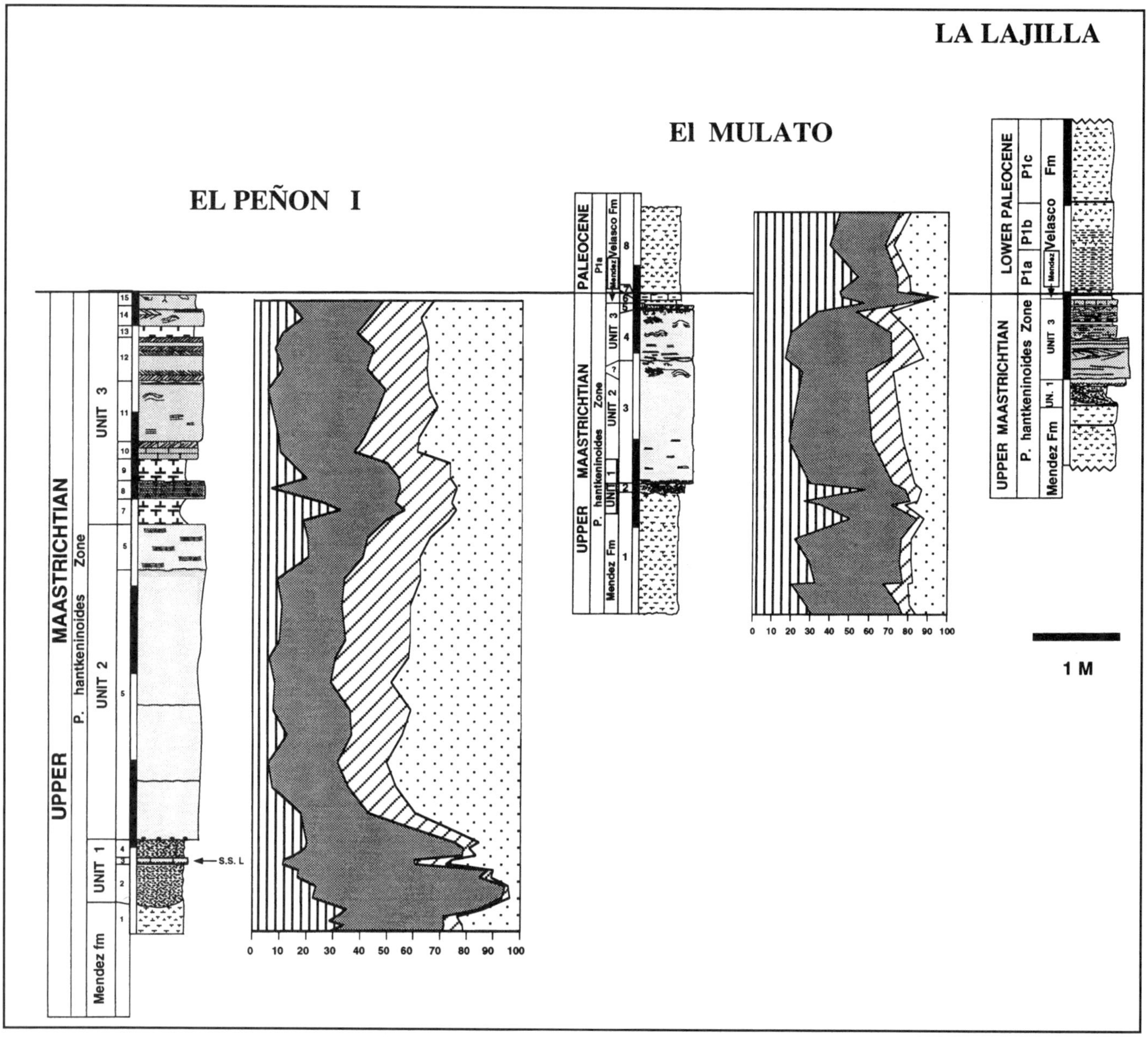

Figure 6 (on this and facing page) Mineralogical correlation of five K/T outcrops in northeastern Mexico, based on whole-rock compositions. S.L.L. = sandy limestone layer; R.S.L. = rippled sandy limestone.

deposit but are commonly found in any predominantly detrital sediments. Land plant debris is very commonly carried during floods or gravity flows. Bioturbation at the base of unit 2 suggests that deposition occured not as a single event but as multi-events (Ekdale and Stinnesbeck, 1994).

The sand and silt and shale layers of unit 3 and their mineralogic variability indicate high detrital influx and rapid deposition (sand and silt layers) alternating with normal hemipelagic deposition (shale layers). Deposition of the rippled sandy limestone at the top of the clastic deposit occurred slowly enough to permit a thriving benthic community to exist (Ekdale and Stinnesbeck, 1994). Discrete layers enriched in zeolites also indicate volcanoclastic influx. Deposition of unit 3 probably occurred over a longer time period than either unit 1 and unit 2 and under more variable environmental conditions. These different layers enriched in zeolites are correlatable from section to section over a distance of more than 300 km. An in situ diagenesis origin of these zeolites is unlikely because of the geographic distribution and excellent correlatability of the layers in different lithologies, such as sands, silts, shales, and marls (Fig. 7). We conclude that

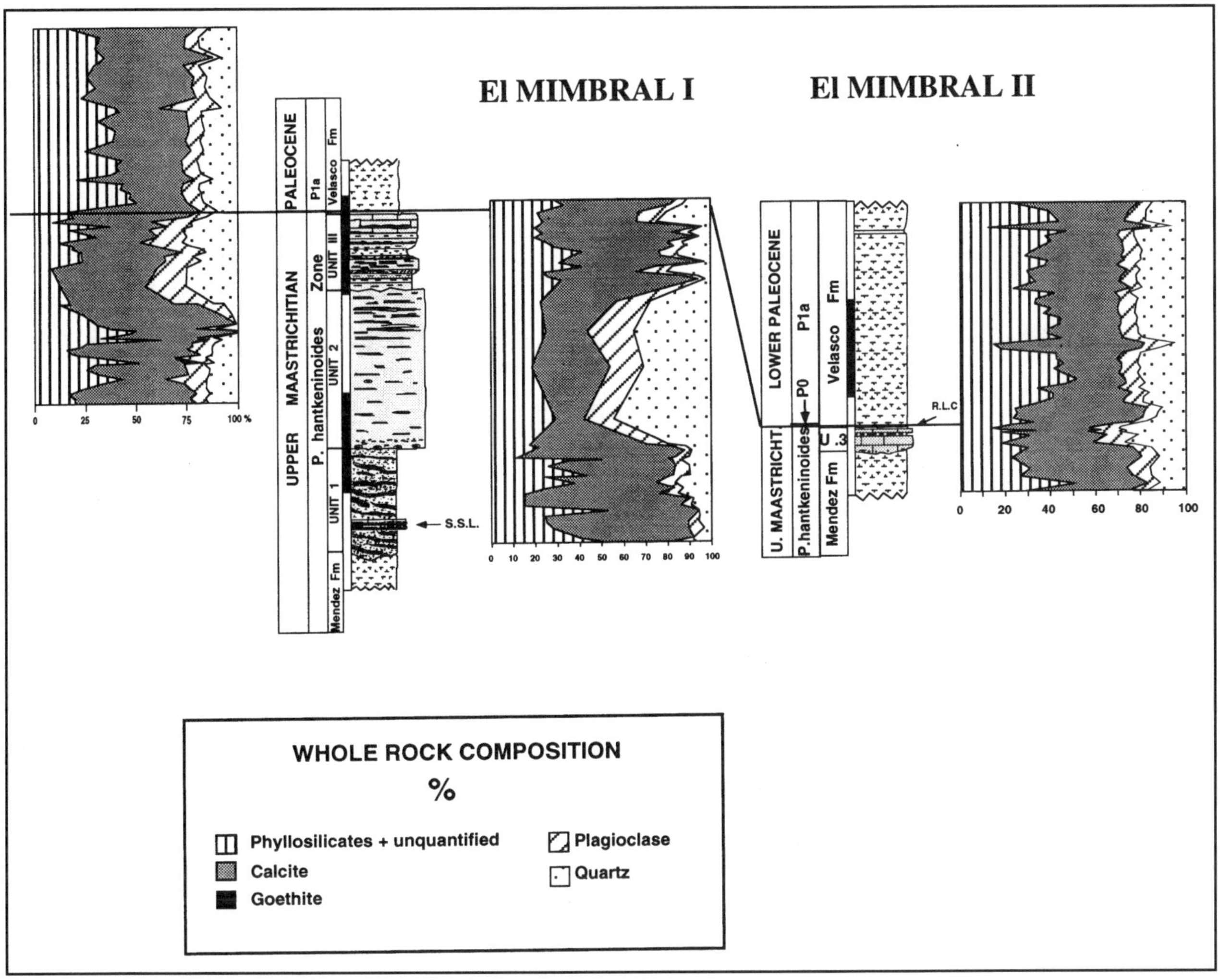

these layers are detrital in origin and indicate discrete periods of volcanoclastic influx. Whole-rock and clay-mineral analyses indicate that the shale layers and the overlying marl of Maastrichtian age are enriched in phyllosilicates with illite-smectite and chlorite-smectite mixed layers most abundant. The small size (<2 μm) of these clay minerals precludes settling through the water column within a few hours after a tsunami wave This layer implies that normal hemipelagic sedimentation resumed in the uppermost Maastrichtian after deposition of the clastic sediments (see López-Oliva and Keller, this volume). Absence of this marl layer at Mimbral and Peñon appears to be due to a short K/T boundary hiatus.

CONCLUSIONS

Our lithologic, mineralogic, and biostratigraphic analysis of K/T sections in northeastern Mexico suggests that the clastic sediments represent a series of depositional events that varied from rapid to normal hemipelagic sedimentation. Deposition of the spherule-rich layers of unit 1 was relatively rapid, with a source primarily from shallow neritic environments, transported into deeper waters. Unit 2 represents erosion and rapid redeposition of massive sand lenses. Unit 3 represents variable, but decreased, detrital influx restricted to discrete layers that alternate with periods of normal hemipelagic sedimentation.

We propose that these varied detrital influxes are related to the latest Maastrichtian eustatic sea-level lowstand (Haq et al., 1987; Donovan et al., 1988; MacLeod and Keller, 1991a, b; Keller and Stinnesbeck, 1996) that occured about 200 to 300 k.y. before the K/T boundary (Pardo et al., 1996). In this scenario, the unconformity at the base of the clastic deposit represents a type 1 sequence boundary associated with a falling sea level. Deltaic sediments were exposed, eroded, and

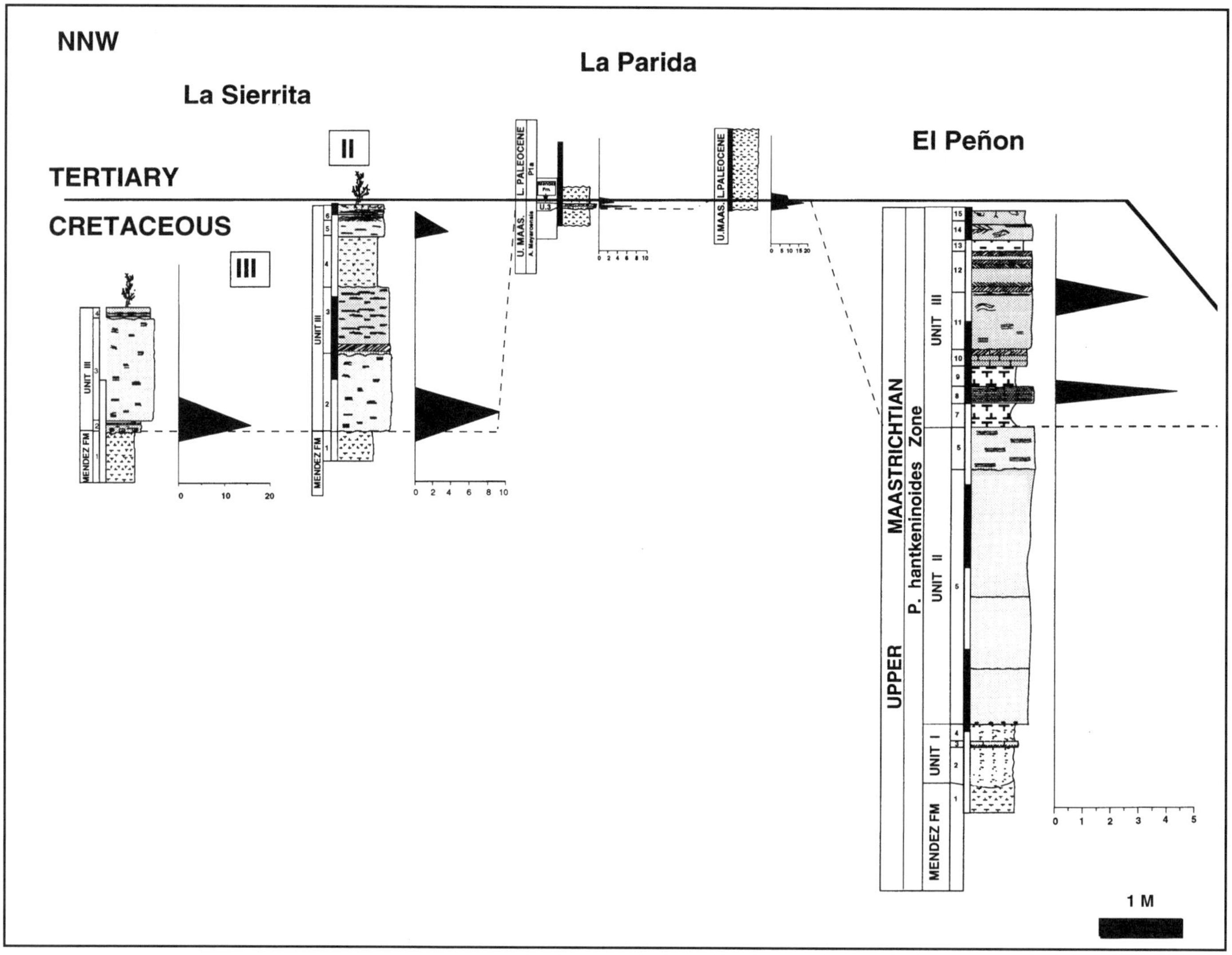

Figure 7 (on this and facing page). Correlation of zeolite-enriched layers in eight K/T boundary outcrops in northeastern Mexico.

sporadically transported into deeper waters, depositing the spherule-rich sediments of unit 1. Continued lowering of the sea level resulted in erosion and bypass, first of inner-shelf (plant debris layer) followed by outer-shelf sediments and deposition of massive or weakly laminated sand lenses (unit 2). Unit 3 was probably deposited during the subsequent stable low sea level when erosion of shelf sediments decreased, resulting in deposition of fine-grained sand and silt layers alternating with normal hemipelagic sedimentation (shale layers). Sea level began to rise prior to the K/T boundary, as suggested by deposition of Maastrichtian age marls above the clastic deposit at Parida, Mulato, and Lajilla. A rising sea level beginning during the latest Maastrichtian has also been recognized in sections from Denmark (Schmitz et al., 1992; Keller, 1989), Brazil (Stinnesbeck and Keller, 1996), and Seymour Island (Askin and Jacobson, 1996).

Lithology and whole-rock and clay-mineral compositions thus indicate the presence of distinct and correlatable units spanning more than 300 km. From these multicorrelations we conclude that deposition of units 1, 2, and 3 of the clastic deposit occurred over an extended time interval with periods of normal hemipelagic sedimentation alternating with times of increased detrital influx and more rapid sedimentation. These variable depositional environments seem to be related to the late Maastrichtian sea-level lowstand and may also have been influenced by regional tectonic and volcanic activity.

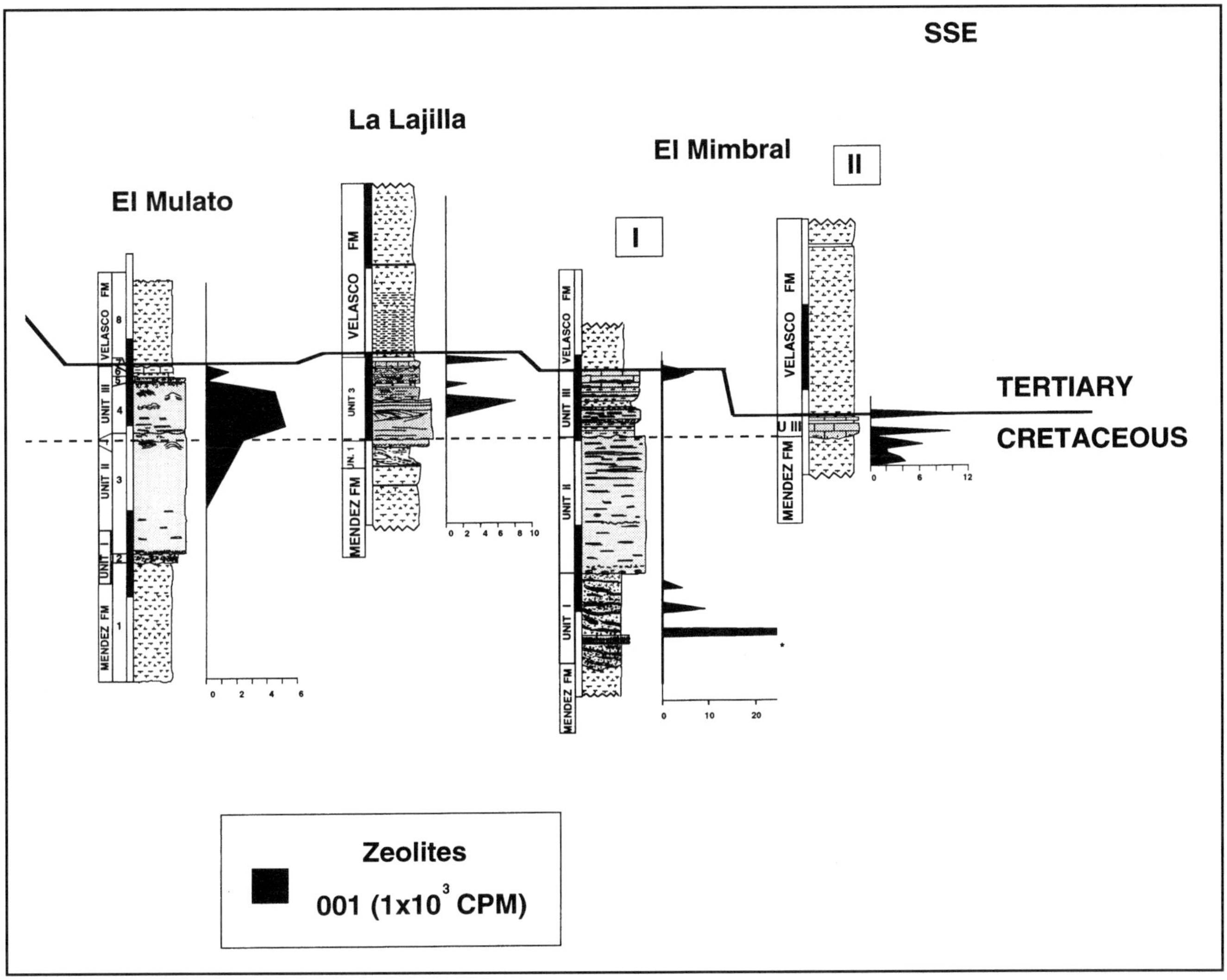

ACKNOWLEDGMENTS

Fieldwork and laboratory research were supported by grants from the National Geographic Society (no. 4620-91), NSF OCE-9021338, EAR-9115044, ACS-PRF no. 26780-AC8, CONACYT No. L120-36-36, and the Swiss National Fund no. 8220-028367. Technical support was provided by Laboratoire de Géochimie et Pétrographie de l'Institut de Géologie de Neuchâtel, Switzerland, and by B. Kübler.

REFERENCES CITED

Adatte, T., Stinnesbeck, W., and Keller, G., 1994, mineralogical correlations of near-K/T boundary deposits in northeastern Mexico: Evidence for long term deposition and volcanoclastic influence, *in* New developments regarding the K/T event and other catastrophes in Earth history: Houston, Texas, Lunar and Planetary Institute Contribution 825, p. 1–2.

Alvarez, L. W., Alvarez, W., Asaro, F., and Michel, H., 1980, Extraterrestrial cause for the Cretaceous-Tertiary Extinction: Science, v. 208, p. 1095–1108.

Askin, R. A., and Jacobson, S.R., 1996, Palynological change across the Cretaceous-Tertiary boundary on Seymour Island, Antarctica: Environmental and depositional factors, *in* MacLeod, N., and Keller, G., eds., The Cretaceous/Tertiary mass extinction: biotic and environmental events: New York, W.W. Norton (in press).

Blum, J. D., and Chamberlain, C. P., 1992, Oxygen isotope constraints on the origin of impact glasses from the Cretaceous-Tertiary boundary: Science, v. 257, p. 1104–1107.

Bohor, B. F., and Betterton, W. J., 1993, Arroyo el Mimbral, Mexico, K/T unit: Origin as debris flow/turbidite, not a tsunami deposit: Houston, Texas, Lunar and Planetary Sciences Conference, v. 24, p. 143–144.

Bourgeois, J., Hansen, T. A., Wiberg, P. L., and Kauffman, E. G., 1988, A tsunami deposit at the Cretaceous-Tertiary boundary in Texas: Science, v. 241, p. 567–570.

Donovan, A. D., Baum, G. R., Blechschmidt, G. L., Loutit, T. S., Pflum, C. E., and Vail, P. R., 1988, Sequence stratigraphic setting of the Cretaceous-Tertiary boundary in Central Alabama, *in* Wilgus, C. K., Hastings, B. S., Kendall, C. G. St. C., et al., eds., Sea-level changes: An integrated approach: Society of Economic Paleontologists and Mineralogists Special Publication 42, p. 300–307.

Ekdale, A. A., and Stinnesbeck, W., 1994, Sedimentologic significance of trace fossils in the KT "Mimbral Beds" of northeastern Mexico: Geological Society of America Abstracts with Programs, v. 26, p. A395.

Ferrero, J., 1966, Nouvelle méthode empirique pour le dosage des minéraux par diffraction R.X.: Bordeaux, France, Rapport Compagnie Française des Pétroles, unpublished report.

Hansen, H. J., Gwozdz, R., Bromley, R. G., Rasmussen, K. L., Vogensen, E. W., and Pedersen, K. R., 1986, Cretaceous-Tertiary boundary spherules from Denmark, New Zealand, and Spain: Geological Society of Denmark Bulletin, v. 35, p. 75–82.

Haq, B. U., Hardenbol, J., and Vail, P. R., 1987, Chronology of fluctuating sea levels since the Triassic: Science, v. 235, p. 1156–1166.

Izett, G. A., Maurrasse, F. J. M. R., Lichte, F. E., Meeker, G. P., and Bates, R., 1990, Tektites in Cretaceous-Tertiary boundary rocks on Haiti: U.S. Geological Survey Open-File Report 90-635, p. 1–31.

Jéhanno, C., Boclet, D., Froget, L., Lambert, B., Robin, E., Rocchia, R., and Turpin, L., 1992, The Cretaceous-Tertiary boundary at Beloc, Haiti: No evidence for an impact in the Caribbean Area: Earth and Planetary Science Letters, v. 109, p. 229–241.

Jiang, M. J., and Gartner, S., 1986, Calcareous nannofossil succession across the Cretaceous/Tertiary boundary in east-central Texas: Micropaleontology, v. 32, p. 232–255.

Keller, G., 1989, Extended period of extinctions across the Cretaceous/Tertiary boundary in planktic foraminifera of continental shelf sections: Implications for impact and volcanism theories: Geological Society of America Bulletin, v. 101, p. 1408–1419.

Keller, G., and Stinnesbeck, W., 1996, Sea-level changes, clastic deposits and megatsunami across the Cretaceous-Tertiary boundary, *in* MacLeod, N., and Keller, G., eds., The Cretaceous/Tertiary mass extinction: Biotic and environmental events: New York, W.W. Norton (in press).

Keller, G., Stinnesbeck, W., Adatte, T., MacLeod, N., and Lowe, D. R., 1994a, Field guide to Cretaceous-Tertiary boundary in northeastern Mexico: Houston, Texas, Lunar and Planetary Institute Contribution 827, 110 p.

Keller, G., Stinnesbeck, W., and López-Oliva, J. G., 1994b, Age, deposition and biotic effects of the Cretaceous/Tertiary boundary event at Mimbral, NE Mexico: Palaios, v. 9, p. 144–147.

Koeberl, C., 1994, Stinnesbeck et al., 1993 and Keller et al., 1993: Comment: Geology, v. 22, p. 957.

Koeberl, C., and Sigurdsson, H., 1992. Geochemistry of impact glasses from the K/T boundary in Haiti: Relation to smectites and a new type of glass: Geochimica et Cosmochimica Acta,v. 56, p. 2113–2129.

Kübler, B., 1983, Dosage quantitatif des minéraux majeurs des roches sédimentaires par diffraction X: Neuchâtel, Suisse, Cahiers Institut Géologie, série ADX, Volume 1, 12 p.

Kübler, B., 1987, Cristallinité de l'illite, méthodes normalisées de préparations, méthodes normalisées de mesures: Neuchâtel, Suisse, Cahiers Institut Géologie, série ADX, Volume 1, 13 p.

Longoria, J. F., and Gamper, M. A., 1992, Planktonic foraminiferal biochronology across the K-T boundary from the Gulf coastal plain of Mexico: Implications for timing the extraterrestrial bolide impact in Yucatan: Boletín de la Asociación Mexicana de Geólogos Petroleros, v. 42, p. 19–40.

Longoria, J. F., and Grajales, J.M., 1994, Field trip to selected K/T boundary localities in Tamaulipas and Nuevo Leon, northeast Mexico: Revista Sociedad Mexicana de paleontologia, v. 6, Mem. 2, p. 73–85.

Lyons, J. B., and Officer, C. B., 1992, Mineralogy and petrology of the Haiti Cretaceous/Tertiary section: Earth and Planetary Science Letters, v. 109, p. 205–224.

MacLeod, N., and Keller, G., 1991a, Hiatus distribution and mass extinctions at the Cretaceous/Tertiary boundary: Geology, v. 19, p. 497–501.

MacLeod, N., and Keller, G., 1991b, How complete are Cretaceous/Tertiary boundary sections? A chronostratigraphic estimate based on graphic correlation: Geological Society of America Bulletin, v. 113, p. 1439–1457.

Montanari, A., Claeys, P., Asaro, F., Bermudez, J., and Smit, J., 1994, Preliminary stratigraphy and iridium and other geochemical anomalies across the K-T boundary in the Bochil Section (Chiapas, SE Mexico), *in* New developments regarding the K/T event and other catastrophes in Earth history: Houston, Texas, Lunar and Planetary Institute Contribution 825, p. 84–85.

Moore, D., and Reynolds, R., 1989, X-Ray diffraction and the identification and analysis of clay-minerals: Oxford, Oxford University Press, 332 p.

Pardo, A., Ortiz, N. and Keller, G., 1996, Latest Maastrichtian and K/T boundary foraminiferal turnover and environmental changes at Agost, Spain, *in* MacLeod, N., and Keller, G., eds., The Cretaceous/Tertiary boundary mass extinction: Biotic and environmental events: New York, W.W. Norton (in press).

Robin, E., Bonté, Ph., Froget, L., Jéhanno, C., and Rocchia, R., 1992, Formation of spinels in cosmic objects during atmospheric entry: A clue to the Cretaceous-Tertiary boundary event: Earth and Planetary Science Letters, v. 108, p. 181–190.

Schmitz, B., Keller, G., and Stenwall, O., 1992, Stable isotope and foraminiferal changes across the Cretaceous/Tertiary boundary at Stevns Klint, Denmark: Arguments for long-term oceanic instability before and after bolide impact: Paleogeography, Paleoclimatology, Paleoecology, v. 96 , p. 233–260.

Sigurdsson, H., D'Hondt, S., Arthur, M. A., Bralower, T. J., Zachos, J. C., Van Fossen, M., and Channell, J. E. T., 1991, Glass from the Cretaceous-Tertiary boundary in Haiti: Nature, v. 349, p. 482–487.

Smit, J., and eight others, 1992, Tektite-bearing deep-water clastic unit at the Cretaceous-Tertiary boundary in northeastern Mexico: Geology, v. 20, p. 99–103.

Smit, J., Roep ,T. B., Alvarez, W., Claeys, Ph., and Montanari, A., 1994, Stinnesbeck et al., 1993: Comment: Geology, v. 22, p. 953–954.

Stinnesbeck, W., and 10 others, 1993, Deposition of channel deposits near the Cretaceous-Tertiary boundary in northeastern Mexico: Catastrophic or "normal" sedimentary deposits?: Geology, v. 21, p. 797–800.

Stinnesbeck, W., Keller, G., and Adatte, T., 1994a, K/T boundary sections in southern Mexico (Chiapas): Implications for the proposed Chicxulub impact site, *in* New developments regarding the K/T event and other catastrophes in Earth history: Houston, Texas, Lunar and Planetary Institute Contribution 825, p. 1–2.

Stinnesbeck, W., Keller, G., Adatte, T. and MacLeod, N., 1994b, Depositions of channel deposits near the Cretaceous-Teritary boundary in northeastern Mexico: Catastrophic or "normal" sedimentary deposits?; and Is there evidence for Cretaceous-Tertiary boundary-age deep-water deposits in the Caribbean and Gulf of Mexico?: Reply: Geology, v. 22, p. 955–956.

Stinnesbeck, W., Keller, G., Adatte, T., López-Oliva, G. J., and MacLeod, N., 1996, Cretaceous-Tertiary boundary clastic deposits in NE Mexico: Impact-tsunami or sea-level lowstand?, *in* MacLeod, N., and Keller, G., eds., The Cretaceous-Tertiary boundary mass extinction: Biotic and environmental events: New York, W.W. Norton (in press).

Manuscript Accepted by the Society September 14, 1995

Geological Society of America
Special Paper 307
1996

Age and stratigraphy of near-K/T boundary siliciclastic deposits in northeastern Mexico

J. G. López-Oliva and G. Keller
Department of Geological and Geophysical Sciences, Princeton University, Princeton, New Jersey 08544

ABSTRACT

Biostratigraphic analysis of four Cretaceous/Tertiary (K/T) boundary sections in northeastern Mexico (Mimbral, Lajilla, Mulato, Parida) reveals that all, except Mimbral, contain a thin (5- to 10-cm) layer of Maastrichtian marl above a siliciclastic deposit that was considered to be an impact generated tsunami-deposit. This marl layer contains typical late Maastrichtian planktic foraminiferal assemblages with no apparent size sorting, grain-size gradation, or reworked shallow-water benthic foraminifera, suggesting an interval of normal hemipelagic sedimentation between deposition of the siliciclastic sediments and the K/T boundary. Siliciclastic sediments were deposited prior to the K/T boundary and within the last 170 to 200 k.y. of the Maastrichtian, as indicated by the presence of the latest Maastrichtian index species *Plummerita hantkeninoides* and *Micula prinsii* either below, within, or above this deposit but always below the K/T boundary. Deposition of the siliciclastic sediments occurred not as a single event but as a series of events, as indicated by the presence of bioturbation in several layers within unit 3 and at the base of unit 2 of this deposit. Sedimentologic, biostratigraphic, and trace fossil data all point to multievent deposition prior to the K/T boundary, with depositional events separated by scouring and erosion as well as alternating episodes of terrigenous influx and hemipelagic sedimentation.

The spherule-rich sediments of unit 1, with its rare glass fragments of composition similar to that of the glass spherules from Beloc, Haiti, represent an unusual feature of the siliciclastic deposit. Whatever the origin of these glasses, the event that produced them, whether impact or volcanism, preceded the K/T boundary.

INTRODUCTION

The discovery of the circular Chicxulub structure in northern Yucatán and its interpretation as the long-sought Cretaceous/Tertiary (K/T) boundary bolide impact crater (Hildebrand et al., 1991; Pope et al., 1991; Sharpton et al., 1992, 1993) has led Smit et al. (1992, 1994a, b) to reinterpret the near-K/T boundary siliciclastic sediments of northeastern Mexico as impact-generated tsunami deposits. In contrast, Stinnesbeck et al. (1993, 1994), Stinnesbeck and Keller (this volume), and Keller et al. (1994a, b) interpreted the same deposits as a series of channelized gravity flows deposited over an extended time period during the latest Maastrichtian sea-level low stand.

The siliciclastic deposits consist of three lithologically and sedimentologically distinct units characterized by a basal spherule-rich unit 1, overlain by a massive sandstone of unit 2, and topped by alternating sand-silt layers of unit 3. In the Smit et al. (1994b, p. 118) scenario, unit 1 is "deposited by the first tsunami waves, which reworked impact ejecta which had just arrived on ballistic trajectories from Chicxulub, along with rip-up clasts and deposited them in discrete channels." The middle unit 2 "represents backwash debris of continental-margin origin

López-Oliva, J. G., and Keller, G., 1996, Age and stratigraphy of near-K/T boundary siliciclastic deposits in northeastern Mexico, *in* Ryder, G., Fastovsky, D., and Gartner, S., eds., The Cretaceous-Tertiary Event and Other Catastrophes in Earth History: Boulder, Colorado, Geological Society of America Special Paper 307.

(sand, plant remains) shaken loose as tsunami waves washed up on the coast of Mexico." And the top of unit 3 "represents several passes of a seiche capable of transporting fine sand on the floor of the deep Gulf of Mexico before the waves finally damped out." Smit et al. (1994a, fig. 1, p. 120) estimate that the first tsunami wave arrived 2 hrs after the bolide impact. This requires airborne impact material to fall from the sky and settle through several hundreds of meters of water in less than 2 hrs, then to be reworked by the tsunami wave. It also assumes that the sediment source of unit 1 is primarily impact fallout material (microtektites, shocked quartz, Ni-rich spinels, iridium) and that a siliciclastic sediment source existed southeast in the direction of Chicxulub (Smit et al., 1994a). Neither assumption is substantiated. Shocked-quartz grains are very rare in unit 1. Ni-rich spinels, iridium, or confirmed microtektites have not been observed in units 1, 2, or 3. The only potential impact materials present in unit 1 are very rare shocked-quartz grains and glass shards, the latter of chemical composition similar to that of glass spherules in the Beloc section of Haiti (Smit et al., 1992; Stinnesbeck et al., 1993; Bohor and Betterton, 1993; Keller et al., 1994a). Whether the Haiti glass spherules are of impact or volcanic origin is still in dispute, as recently discussed by Koeberl (1994) and Robin et al. (1994 and references therein).

Regardless of the mechanism of deposition and whether it was long term or short term, the stratigraphic position of this siliciclastic member at the K/T boundary at El Mimbral would suggest a cause-and-effect relationship, if confirmed in other K/T outcrops. It is therefore critical to determine the precise stratigraphic position of the siliciclastic deposit relative to the K/T boundary in outcrops throughout northeastern Mexico. If the siliciclastic deposit is directly overlain by lower Tertiary sediments as at El Mimbral and no hiatus is present, then its deposition seems closely related to a K/T boundary impact event. However, if normal Maastrichtian marls overlie the siliciclastic member, then deposition most likely preceded the K/T boundary event and there is no cause-and-effect relationship.

We have examined the stratigraphic position of siliciclastic deposits in eight outcrops spanning over 300 km in northeastern Mexico (Fig. 1; Keller et al., 1994a). In four of the outcrops, lower Tertiary sediments are present (El Mimbral I and II at the center and edge of the channelized deposit respectively, La Lajilla, El Mulato, and La Parida), and in the remainder, Tertiary sediments have been eroded. In three of the four sections (e.g., La Lajilla, El Mulato, La Parida) a thin layer of Maastrichtian marl overlies the siliciclastic deposit. In this chapter, we examine the age and biostratigraphy of Maastrichtian and lower Tertiary (Danian) sediments and the stratigraphic position of the siliciclastic deposit and discuss its implications for depositional scenarios.

CHARACTERISTICS OF SILICICLASTIC DEPOSITS

The four K/T boundary outcrops examined are similar in lithology and stratigraphy, except that the thickness of the siliciclastic member varies between 10 cm at La Parida to nearly 4 m at El Mimbral (Fig. 2). Thick siliciclastic deposits, such as at El Mimbral and El Mulato (but also at El Peñon and La Sierrita; Keller et al., 1994a), are subdivided into three lithologically distinct units that can be correlated over 300 km in a northwest-to-southeast trending direction, parallel to and 40 to 80 km east of the Sierra Madre Oriental (Fig. 1). During K/T boundary time, this topographic high provided a source of terrigenous siliciclastic sediments to the west. Characteristics of the three units of the siliciclastic deposit are here discussed based on the sequence at El Mimbral I and illustrated in Figure 2.

Unit 1: Spherule-rich layer

The base of the siliciclastic deposit disconformably overlies an undulating erosional surface of the Maastrichtian marl of the Mendez Formation in all sections examined. The basal unit 1 consists of a soft, weathered, spherule-rich and crudely stratified sediment that contains rip-up clasts of the underlying Mendez Formation and older sediments. Some of these rip-up clasts contain Turonian-age foraminifera (*Marginotruncana pseudolinneiana* and *M. sigali*), as also observed in clasts at the base of unit 2. Since Turonian-age sediments should have been deeply buried at K/T boundary time, except where exposed on the flanks of deeply incised submarine canyons, they indicate that sediment transport and deposition occurred via deep canyons rather than by broad surface reworking of recently deposited sediments by tsunami waves. The spherules range from 1 to 5 mm in diameter and are generally infilled by blocky calcite, glauconite, smectite, and greenish microcrystalline phyllosilicates (Stinnesbeck et al., 1993, 1994; Stinnesbeck and Keller, this volume). Large spherules generally have several small spherules inside; some contain foraminifers or rock fragments, suggesting they formed as oolites or oncolites. Many spherules are enclosed by a thin organic wall, indicating that they are calcite-infilled algal resting cysts. The shallow-water origins of many spherules, coupled with transported outer neritic to upper-slope benthic foraminifera (*Bolivina incrassata gigantea*, *Praebulimina reussi*, *Anomalina acuta*, *A. praeacuta*, *Cibicidoides succedens*), suggests downslope transport from shallow neritic environments (Keller et al., 1994b).

The spherule layer at El Mimbral I (also at El Peñon I, La Lajilla, El Mulato; Keller et al., 1994a) contains a 20- to 25-cm thick, well-cemented sandy limestone layer (S.L.L. of Fig. 2). In thin sections, microlayering of fine- and coarser-grained sediments can be observed (W. Ward, written communication, 1994). The S.L.L. contains few graded spherules, usually concentrated at the base and top. The upper surface of the S.L.L. is characterized by convolute bedding and draped by the overlying spherule-rich sediment suggesting at least semilithification of the S.L.L. prior to deposition of the overlying sediment. In some sections, the tops of the spherule-rich sediments of unit 1 are eroded, indicating a hiatus (e.g., at La Lajilla, El Peñon II, and El Mulato; Keller et al., 1994a). These sedimentologic vari-

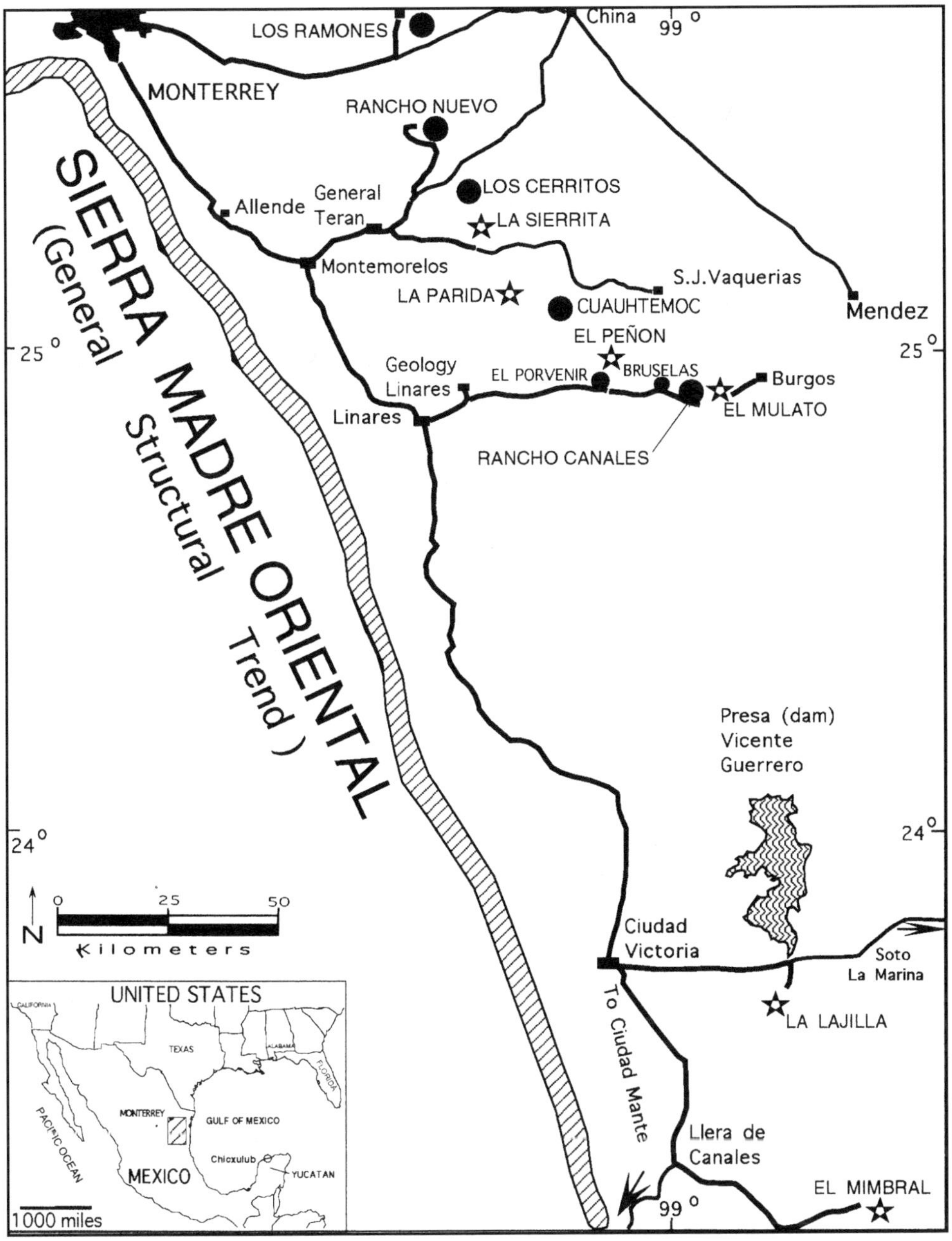

Figure 1. Location map of northeastern Mexico K/T boundary sections studied (shown by dots) and sections discussed in this chapter (shown by stars).

ations suggest that unit 1 was not deposited in a single event but represents alternating episodes of deposition and scouring.

Unit 2: Massive sandstone

Unit 2 disconformably overlies unit 1 with an undulating erosional basal contact and rip-up clasts and lenses of spherule-rich sediments in the basal part. Sediments of unit 2 consist of a massive laminated sandstone with mud clasts at its base, some of which contain Turonian-age planktic foraminifers. At El Peñon I, burrows infilled with spherule-rich sediment of unit 1 were observed at the base of unit 2 (Ekdale and Stinnesbeck, 1994, 1996). Some of these burrows are truncated, suggesting scouring before deposition of the overlying sand layers. Thus unit 2 was not deposited in a single event but contains episodes of colonization by invertebrates alternating with episodes of scouring and sediment deposition. At El Mimbral I, discrete layers of large plant debris are present near the base of unit 2, but such plant debris was not observed in any of the other outcrops. In some outcrops, unit 2 is absent, as at La Lajilla and at

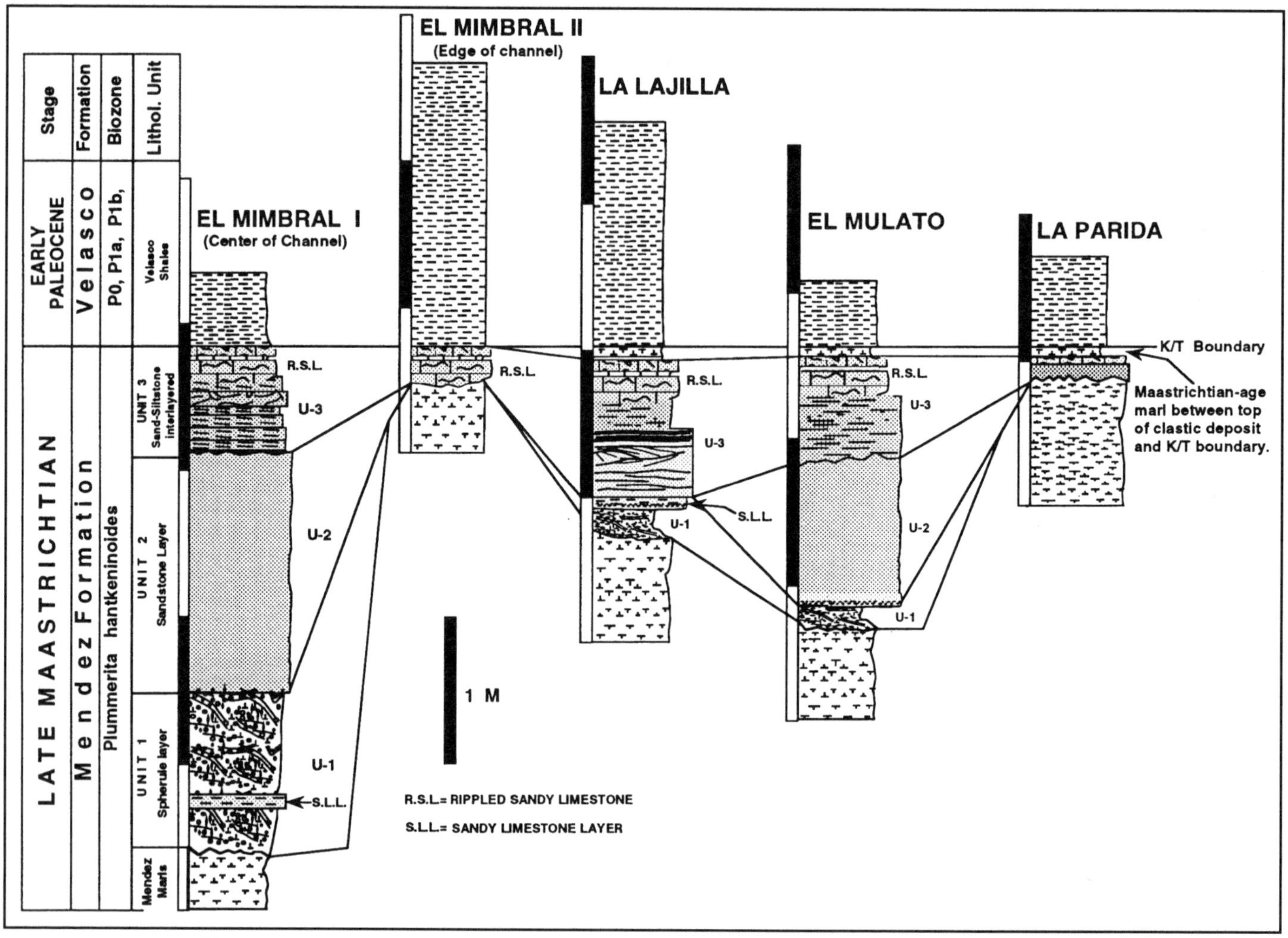

Figure 2. Lithologic and biostratigraphic correlation of K/T boundary sections at Mimbral, Lajilla, Mulato, and Parida. Note that the three units (U-1, U-2, U-3) of the siliciclastic deposits can be correlated over 250 to 300 km. Outcrops with thick siliciclastic sediments generally represent channel-fill deposits that thin toward channel edges, as shown for Mimbral II. At La Parida the siliciclastic deposit is thin and laterally disappears. S.L.L. = sandy limestone layer, R.S.L. = rippled sandy limestone.

the channel edge of the El Mimbral II outcrop (Fig. 2). Unit 2 is also missing at El Peñon II and La Sierrita (Keller et al., 1994a, b). It appears that unit 2 was primarily deposited in the central parts of the channels and thins or disappears toward channel edges (Keller et al., 1994a, b). Foraminifera are rare in unit 2. Deposition of unit 2 appears to have been more rapid than either units 1 or 3.

Unit 3: Interlayered sand-silt beds

Unit 3 consists of alternating layers of sand and silt topped by a rippled sandy limestone (R.S.L., Fig. 2). Horizontal laminations, ripple marks, small-scale cross-bedding, flaser bedding, and convolute laminations are commonly present. The fine-grained silty layers contain typical late Maastrichtian *A. mayaroensis* Zone planktic foraminifera similar to those in the Mendez Formation Reworked neritic benthic foraminifera are generally absent. This suggests that short periods of normal hemipelagic sedimentation were interrupted by episodes of terrigenous transport during deposition of unit 3. Adatte et al. (1994a, this volume) documented two fine-grained layers enriched in zeolites in all four sections examined, indicating periods of increased volcanic influx.

The R.S.L. layer at the top of unit 3 at El Mimbral I and II consists of a 25-cm-thick layer bioturbated by *Chondrites*, *Thalassinoides*, and *Zoophycos* (Keller et al., 1994a, b). Burrows are infilled with Maastrichtian sediments. This R.S.L. marks the top of all coarse siliciclastic deposits even in outcrops where units 1, 2, and the lower part of unit 3 are missing, such as at the channel edge at El Mimbral II and at La Parida (Fig. 2). In outcrops at La Lajilla, El Peñon, and La Sierrita the R.S.L. consists of two to three discrete layers separated by thin silty interlayers. Each of these layers is strongly bioturbated, and burrows are infilled with Maastrichtian sediments. Ekdale and Stinnesbeck (1994, 1996) found evidence of successive colonization episodes within unit 3, which indicates that organisms lived within the accreting substrate during deposition. Thus, sedimentologic, mineralogic, and trace fossils indicate that unit 3, like units 2 and 1, was also deposited episodically rather than as a single event.

BIOSTRATIGRAPHY

The biostratigraphy of the northeastern Mexico sections is based on planktic foraminifera and the K/T biozonation of Keller (1988, 1993), with a new *Plummerita hantkeninoides* Zone added (Pardo et al., 1996), as illustrated in Figure 3 along with other commonly used zonal schemes (Smit, 1982; Berggren and Miller, 1988; Longoria and Gamper, 1992). The various biozonations differ largely in the degree of biostratigraphic resolution, with Keller's zonation providing the highest resolution. First and last appearance datums (FAD's and LAD's) of zonal index taxa are marked in bold type in Figure 3. FAD's of other Danian species are shown in their approximate sequential order. The order of first appearances of Danian species is based on a composite database of FAD's that by graphic correlation integrates the most complete K/T boundary sections (e.g., El Kef stratotype, Agost, Caravaca, Nye Klov, Brazos River, Millers Ferry, Mimbral, ODP Site 738, and DSDP Sites 528, 577). If several of these FAD's co-occur at the same stratigraphic horizon, a hiatus or condensed interval is likely to be present. This is certainly the case where a large number of species first appear at the same stratigraphic horizon and are also numerically abundant there.

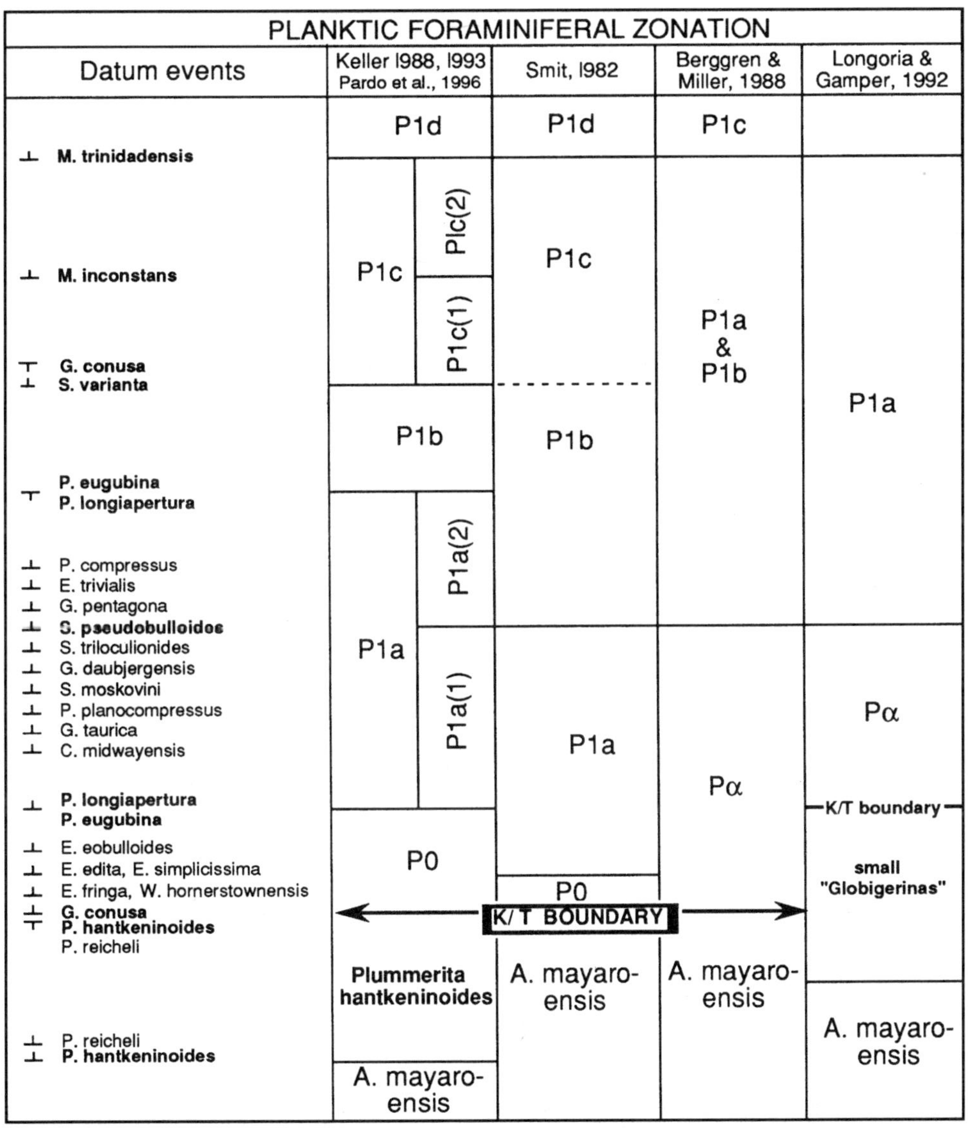

Figure 3. Correlation of commonly used planktic foraminiferal zonal schemes for the K/T boundary transition. Note the new latest Maastrichtian *Plummerita hantkeninoides* Zone that marks the last 170 to 200 k.y. of the Maastrichtian.

Upper Maastrichtian

In conventional zonal schemes, the upper Maastrichtian is characterized by the *Abathomphalus mayaroensis* Zone, which extends from the FAD of *A. mayaroensis* to the K/T boundary and includes a time span of over 1 m.y. Such long biozones provide only the most general stratigraphic control and have no power of resolving whether all or only part of the biozone is present. In evaluating the continuity of sedimentation just below the K/T boundary, the *A. mayaroensis* Zone is of limited use. In order to determine the extinction pattern prior to the K/T boundary or to evaluate the time missing at the erosional disconformity at the base of the siliciclastic deposits, a higher-resolution biostratigraphy is necessary. Pardo et al. (1996) have proposed the new *Plummerita hantkeninoides* biozone to mark the top of the Maastrichtian. This biozone is defined by the total range of *P. hantkeninoides* (or the related species *P. reicheli*), which at the El Kef stratotype spans the top 6 m of the Maastrichtian. At the Agost section where paleomagnetic data are available, *P. hantkeninoides* spans the top 3.5 m of the Maastrichtian, and its first appearance is within the lower part of C29R below the K/T boundary and within the lower part of the calcareous nannoplankton *Micula prinsii* Zone. Based on the paleomagnetic data at Agost, the *P. hantkeninoides* Zone spans the top 170 to 200 k.y. of the Maastrichtian (Pardo et al., 1996).

Evidence for Maastrichtian age of siliciclastic deposit

The age of the siliciclastic deposits can now be constrained based on the new index species *Plummerita hantkeninoides*. At El Mimbral, Lajilla, Mulato, and Parida, an erosional disconformity marks the base of the siliciclastic deposit. Erosion, however, seems to have been limited largely to local channelized areas, as suggested by the presence of *P. hantkeninoides* below the siliciclastic sediments. For instance, at El Mimbral II, this species is present within the top 50 cm of Mendez marls that directly underlie unit 3 of the siliciclastic deposit (Fig. 2; Keller et al., 1994b). At La Lajilla, *P. hantkeninoides* is present in the top 50 cm of Mendez marls below the siliciclastic deposit as well as within the 5- to 10-cm-thick Mendez marl overlying it (Fig. 4). The same is observed at La Parida in the 5- to 10-cm-thick overlying marl layer. Only at El Mulato was *P. hantkeninioides* not observed, possibly as a result of poor preservation or small sample size. Among all other sections examined that lack sediments above the siliciclastic deposits (e.g., Peñon, Sierrita, Rancho Nuevo, Rancho Canales, Ramones, Los Cerritos, Bruselas), *P. hantkeninoides* is present within as well as below the siliciclastic deposits (J. G. López-Oliva, 1996). This situation suggests that erosion at the unconformity between the Mendez Formation and the overlying siliciclastic deposit was limited and that deposition of the spherule-rich unit 1 may have started near the first appearance of *P. hantkeninoides*,and prior to the last 100 to 150 k.y. of the Maastrichtian.

The presence of the 5- to 10-cm-thick Maastrichtian age marl or marly limestone layer above the siliciclastic deposit at La Lajilla, Mulato, and Parida (though not at El Mimbral) indicates that normal hemipelagic sedimentation resumed prior to the K/T boundary. It has been suggested that this marl or marly limestone layer represents settling from the water column after siliciclastic deposition. We believe this to be unlikely because there is no size grading of planktic foraminifera, and both large and small species are present in normal proportions; and because laser counter grain-size analyses by Adatte et al. (1994 a, b, this volume) show no size grading of sediments as would be expected if sediments settled from the water column.

Furthermore, Macias Pérez (1988) reported 100 cm of Maastrichtian Mendez marl of the *M. prinsii* Zone overlying the siliciclastic deposit at the Mirador section near La Ceiba in east-central Mexico. Similarly, in the Bochil section in southern Mexico, Montanari et al. (1994) report 100 cm of stratified shale, silt, and clay of late Maastrichtian age overlying the siliciclastic deposit. Thus, in five of six outcrops, Maastrichtian marls overlie the siliciclastic sediments, indicating that El Mimbral is the exception rather than the norm. We conclude that current biostratigraphic and grain-size data strongly indicate that the siliciclastic sediments were emplaced during the latest Maastrichtian and prior to the K/T boundary event.

The time interval during which the entire siliciclastic deposit (units 1, 2, and 3) accumulated cannot be determined precisely because of the predominance of transported sediments in some layers as well as numerous disconformities. Nevertheless, typical upper Maastrichtian assemblages devoid of transported neritic benthic foraminifera are present in the silty interlayers of unit 3. These layers, along with two zeolite-enriched layers (Adatte et al., this volume), suggest that deposition during unit 3 alternated between periods of normal hemipelagic sedimentation and increased siliciclastic transport, as also observed in whole-rock and mineralogical analyses (Adatte et al., this volume). This evidence, along with evidence of repeated colonization by invertebrates within unit 3 as well as near the base of unit 2 (Ekdale and Stinnesbeck, 1994, 1996), indicates that deposition of units 1, 2, and 3 occurred over a long time period and negates the tsunami model proposed by Smit et al. (1992, 1994 a, b). Current evidence from sections in northeastern and southern Mexico suggests that siliciclastic deposition ended several tens of thousands of years before the K/T boundary and at a time of a sea-level transgression (Pardo et al., 1996; Keller and Stinnesbeck, 1996).

K/T Boundary placement

In order to evaluate the depositional sequence of the northeastern Mexico sections, uniform global criteria must be used to identify the K/T boundary. These criteria have been defined based on the global stratotype section and point (GSSP) at El Kef, Tunisia: (1) A lithological break from chalk or marl deposition of the Maastrichtian to a thin layer of dark organic-rich and $CaCO_3$-poor clay known as the boundary clay; (2) 2 to 3 mm

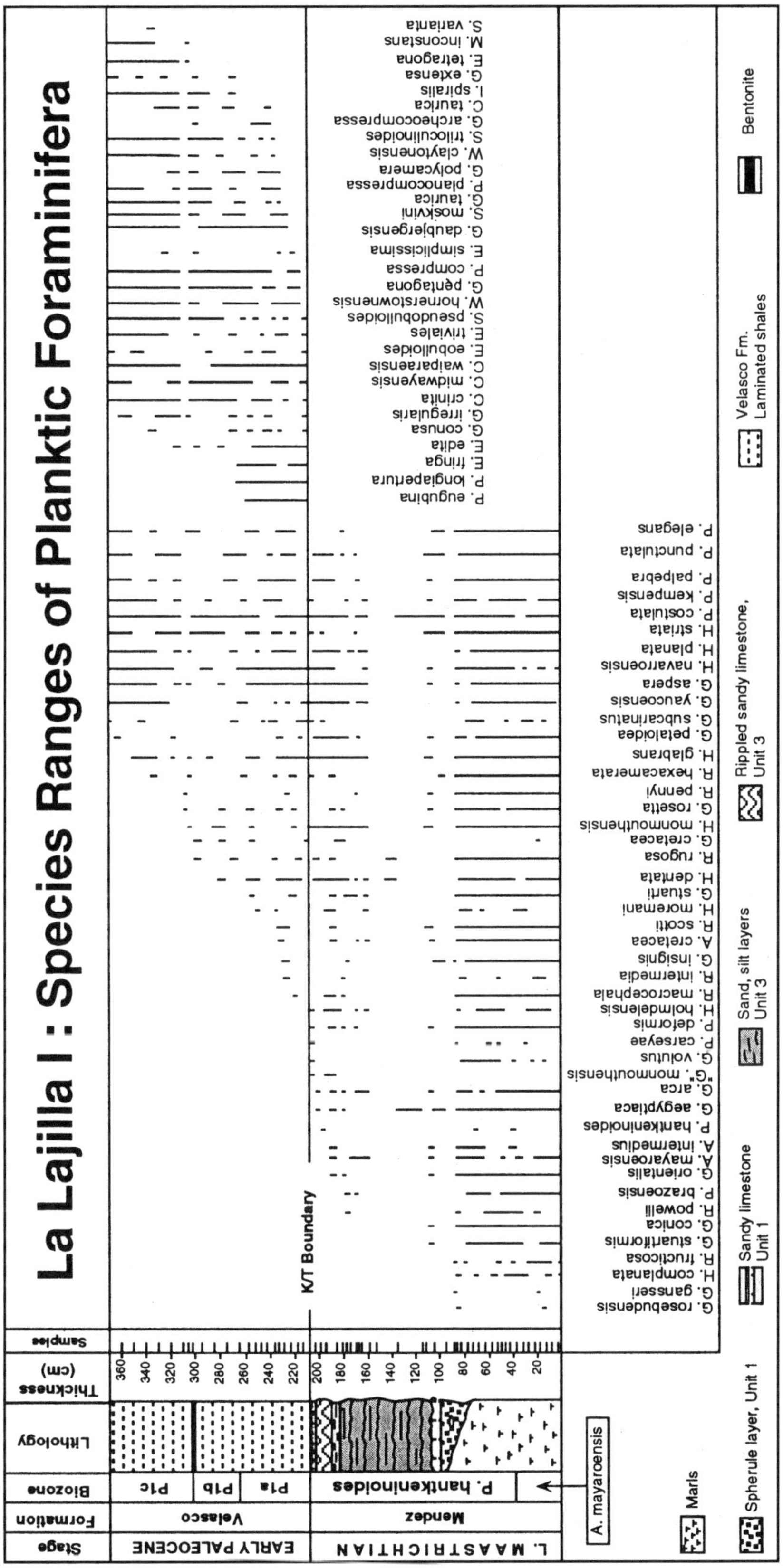

Figure 4. Planktic foraminiferal species ranges across the K/T boundary transition at La Lajilla. Note the Maastrichtian assemblage present in the thin marl layer above the siliciclastic deposit and below the K/T boundary. Note also the simultaneous appearance of 12 early Danian species above this marl layer, which indicates a hiatus with Zone P0 and the lower part of Zone P1a missing.

oxidized red layer at the base of the boundary clay; (3) maximum iridium concentrations generally found in the red layer and boundary clay, although they may trail several tens of centimeters above the boundary clay; (4) Ni-rich spinels generally concentrated in the red layer; (5) shocked mineral grains (very rare at El Kef and in marine sequences in general; Schmitz, 1988); (6) a negative $\delta^{13}C$ shift in marine plankton that is restricted to low and middle latitudes; (7) the first appearance of Tertiary planktic foraminifera at the base or within a few centimeters of the boundary clay, red layer, Ir anomaly, and Ni-rich spinels; and (8) disappearance of tropical taxa at or below criteria 1 through 7.

In the stratotype section at El Kef all of these criteria are met, and most of them are present in all of the best and most complete K/T boundary sections worldwide (e.g., Agost, Caravaca, Sopelana, Nye Klov, Brazos River, ODP Site 738). Coincidence of these lithological, geochemical, mineralogical, and paleontological criteria is unique in the geological record and virtually ensures that the placement of the K/T boundary is uniform and coeval in marine sequences across latitudes.

El Mimbral. At Mimbral, Smit et al. (1992) and Keller et al. (1994a, b) observed that in the central part of the channel (at meter mark 28), Zone P1a directly overlies the siliciclastic deposit and no clay layer is present. Keller et al. (1994b) have shown that because six Danian species that normally evolve sequentially in Zone P0 and P1a(1) (*Eoglobigerina fringa, E. edita, Parvularugoglobigerina longiapertura, P. eugubina, Globoconusa daubjergensis,* and *Woodringina hornerstownensis*) simultaneously appear in basal Danian sediments, a short hiatus or interval of nondeposition for Zones P0 and P1a(1) is indicated. This is also suggested by the high abundance of *P. eugubina, P. longiapertura,* and *G. daubjergensis* at their initial appearance, since evolutionary first appearances are generally characterized by few specimens (Keller et al., 1994b). The absence of a marl or marly limestone layer overlying the siliciclastic deposit at El Mimbral thus appears to be due to a hiatus.

At the channel edge (Mimbral II), where the siliciclastic deposit is represented only by the 20-cm-thick R.S.L. that characterizes the top of unit 3, the K/T transition is more complete and relative abundances of the same taxa are significantly lower (Keller et al., 1994b). In this transect, a 4-cm-thick clay layer with a 2- to 4-mm red oxidized layer at its base lies between the simultaneous first appearance of the seven Danian species and the top of unit 3 of the siliciclastic deposit. Thus, a short hiatus is also present at El Mimbral II. This clay layer, although devoid of planktic foraminifera, may represent the K/T boundary clay. Rare shocked-quartz grains were found in the red oxidized layer, but no Ni-rich spinels, glass, or microspherules other than pyrite framboids (Stinnesbeck et al., 1993; Keller et al., 1994a). Ir concentrations reach a maximum of 6.0 ng/g in the 4-cm clay layer and 8.0 ng/g at 8- to 10-cm above the red layer near the base of Zone P1a. Low values of Ir between 0.15 and 0.25 ng/g are present in the topmost 10 cm of the rippled sandy limestone of unit 3 (Stinnesbeck et al., 1993; Keller et al., 1994a). In the center of the channel where the basal Tertiary sediments are missing, Smit et al. (1992) report maximum Ir concentrations of 0.8 ng/g in Zone P1a directly above the top of unit 3 and low values between 0.1 and 0.4 ng/g in the topmost 10 cm of the rippled sandy limestone of unit 3. Smit et al. (1992, 1994a, b) argue that the low values of Ir in the topmost 10 cm of unit 3 of the siliciclastic deposit provide evidence that the siliciclastic unit was deposited by a tsunami wave and the Ir began to settle near the end of this event. However, the presence of low values of Ir in a porous sediment beneath relatively high concentrations of Ir values cannot be taken as evidence for primary deposition because of remobilization of rare earth elements as recently discussed by Sawlowicz (1993). Based on our analysis, we conclude that the Ir anomaly is not related to deposition of the siliciclastic sediments but is associated with deposition of K/T boundary–age sediments above.

La Lajilla. The La Lajilla section differs from the Mimbral section in that a thin marl layer (5 to 10 cm) of Maastrichtian age overlies the top of the siliciclastic deposit and Tertiary Velasco shales continue above it. No clay layer is present. Nearly all of the Maastrichtian species present in the Mendez marl below the siliciclastic deposit are also present in the marl layer above it (Tables 1, 2, and 3, Figs. 4 and 5), including the latest Maastrichtian marker species *P. hantkeninoides*. The presence of this marker species below the siliciclastic deposit indicates that deposition of this deposit occurred within the last 200 k.y. of the Maastrichtian. The presence of the thin marl layer above the siliciclastic deposit that includes small and large Maastrichtian species, without size sorting, and the lack of upward fining of grain size that would indicate suspension settling from the water column (Adatte et al., 1994a, b this volume), suggest that siliciclastic deposition also predates the K/T boundary. These data suggest that normal Maastrichtian hemipelagic sedimentation resumed after deposition of the three siliciclastic units.

The first Tertiary sediments above the thin Maastrichtian-age marl layer that tops the siliciclastic deposit at La Lajilla are characterized by the simultaneous first appearance of twelve Tertiary species (*Parvularugoglobigerina eugubina, P. longiapertura, Eoglobigerina fringa, E. edita, E. eobulloides, E. trivialis, Globoconusa conusa, Guembelitria irregularis, Chiloguembelina crinita, C. midwayensis, C. waiparaensis, and Subbotina pseudobulloides*; Table 3, Fig. 4). In K/T complete sections, these taxa evolve sequentially in Zones P0 and P1a (Fig. 3). The simultaneous first appearance of these species strongly indicates that Zone P0 and part of Zone P1a are missing. An additional indicator of a hiatus is the high relative abundance of these Danian species at the time of their first appearance (Fig. 5) because evolving species are generally rare in their initial appearance. It is likely that erosion at this hiatus also removed the topmost Maastrichtian sediments.

El Mulato. The El Mulato section is very similar to the La Lajilla section in that a thin Maastrichtian-age marl layer overlies the top of the siliciclastic deposit and Tertiary Velasco shales continue above it. No boundary clay layer is present.

TABLE 1. RELATIVE ABUNDANCE (%) OF PLANKTIC FORAMINIFERA AT LA LAJILLA, MEXICO. SAMPLES 1.1 TO 3.7 MAASTRICHTIAN MENDEZ FM. MARLS BELOW CLASTIC DEPOSIT

Note sample 1.1. is at the base of the section and sample 3.7 is just below base of clastic deposit.

	Abathomphalus mayaroensis Zone								*Plummerita hantkeninoides Zone*									
Sample No.	1.1	1.2	1.3	1.4	1.5	1.6	1.7	1.8	1.9	2.1	3.1	3.2	3.3	3.4	3.5	3.6	3.6B	3.7
Thickness in cm.	3	7	13	17	23	27	33	40	43	45	47	53	57	63	67	73	74	77
Abathomphalus intermedius	-	-	-	-	-	-	-	-	0.9	2.3	-	-	-	-	0.2	0.7	1.9	1.1
A. mayaroensis	1.3	1.1	1.1	0.3	1.6	0.6	-	0.9	0.3	-	0.7	-	-	-	2.0	0.2	0.8	1.1
Archeoglobigerina cretacea	-	0.3	0.3	0.3	4.5	3.2	6.3	4.0	3.5	7.2	1.2	2.6	1.8	0.5	2.8	2.5	5.4	3.2
Gansserina gansseri	-	-	-	-	1.6	-	-	-	-	-	-	-	-	-	-	-	-	-
Globigerinelloides aspera	0.8	1.7	0.6	1.7	1.3	1.0	2.5	1.5	0.3	2.3	1.7	1.1	1.8	0.9	2.0	5.4	1.6	-
G. rosebudensis	-	-	-	0.2	-	-	-	-	-	-	-	-	-	-	-	-	-	-
G. subcarinatus	0.3	-	-	1.0	-	1.0	0.5	0.7	-	0.4	-	-	-	-	3.3	2.7	0.4	0.4
G. volutus	-	-	-	0.5	-	0.3	-	-	0.6	-	-	0.5	0.8	1.4	0.9	1.1	0.8	0.7
G. yaucoensis	-	2.0	0.6	5.1	2.6	1.3	2.3	0.4	1.7	2.7	4.2	2.9	1.6	1.4	11.1	2.0	1.6	-
"G" monmouthensis	-	-	-	-	-	-	-	-	-	-	-	-	-	-	-	-	-	-
Globotruncanella petaloidea	1.9	4.2	3.9	2.9	1.0	1.0	2.5	1.5	1.2	1.9	4.0	2.5	2.4	0.5	4.8	5.8	-	2.1
Guembelitria cretacea	-	-	-	-	-	-	-	-	-	-	-	-	-	-	-	-	-	-
Hedbergella holmdelensis	-	-	0.8	1.3	1.0	1.0	0.5	-	-	-	1.0	0.6	0.5	1.4	0.4	1.1	-	-
H. monmouthensis	1.9	3.4	2.8	2.7	0.3	1.6	2.3	3.1	0.6	3.0	6.4	3.4	1.0	1.4	1.1	1.6	1.6	0.4
Globotruncana aegyptiaca	1.6	0.3	1.1	1.5	3.8	1.0	1.1	3.5	2.3	1.9	1.2	2.6	0.5	1.4	0.7	0.7	1.2	1.1
G. arca	2.1	4.5	5.3	1.5	5.4	5.4	4.5	1.3	0.3	2.7	1.0	2.2	2.9	-	0.7	-	0.8	0.4
G. duwi	-	-	-	-	-	-	-	-	-	-	-	-	1.0	-	0.2	-	-	-
G. insignis	5.0	4.5	3.9	1.0	1.6	0.3	2.0	2.4	2.0	2.7	0.7	1.4	2.4	1.9	2.2	1.1	1.6	1.1
G. orientalis	2.9	2.5	2.2	2.4	6.4	16.6	8.8	2.2	5.8	2.3	2.0	1.9	3.1	2.8	3.3	7.4	10.5	7.4
G. rosetta	5.6	1.7	2.5	1.0	3.5	3.5	2.5	5.9	13.4	5.7		0.9	3.4	3.3	6.3	2.9	3.5	9.9
Globotruncanita angulata	-	-	-	-	-	-	-	-	-	-	-	-	-	-	-	-	-	0.4
G conica	0.8	0.8	0.3	0.5	1.0	2.9	0.7	1.5	2.0	0.4	0.7	0.2	0.5	0.9	0.2	0.7	1.9	1.8
G. stuarti	4.2	3.4	2.8	4.4	3.5	8.3	4.7	6.2	9.9	3.8	2.5	3.2	2.4	1.4	2.0	1.8	2.3	13.1
G. stuartiformis	0.3	0.3	0.6	0.3	1.0	-	-	0.7	1.2	0.8	0.5	0.6	0.8	0.5	1.1	0.7	0.4	1.1
Heterohelix complanata	0.3	-	1.4	0.3	-	-	0.2	0.2	1.2	1.1	0.7	-	1.3	3.3	-	0.2	-	-
H. dentata	1.6	1.4	0.6	3.5	1.0	2.9	1.8	-	1.2	3.8	6.9	0.8	1.6	4.2	3.0	1.3	6.2	0.7
H. globulosa	0.3	4.2	0.6	1.5	1.9	2.9	2.9	7.0	1.2	3.0	1.2	1.7	1.8	3.8	2.2	4.3	3.9	3.2
H. navarroensis	0.5	4.7	-	2.4	1.3	-	1.4	-	2.0	2.7	4.2	2.2	0.8	2.8	1.7	2.2	1.6	0.4
H. moremani	-	0.0	-	-	-	-	0.2	0.7	0.6	-	-	-	-	-	0.2	-	1.2	1.4
H. planata	1.6	0.6	1.1	1.0	2.2	1.3	1.8	0.9	1.2	4.9	3.2	5.1	3.7	1.9	3.7	2.0	1.6	
H. striata	0.5	4.2	2.2	3.0	0.3	2.2	6.5	2.6	1.2	3.0	5.7	2.5	2.1	4.7	0.4	2.9	5.1	1.4
Pseudoguembelina costulata	18.0	19.3	14.3	9.6	10.9	3.5	13.3	-	7.3	6.8	11.9	13.9	11.5	7.1	13.0	12.5	23.0	1.8
P. kempensis	0.5	0.6	1.4	0.3	1.6	-	0.2	4.2	0.3	0.4	3.0	0.3	1.6	0.5	-	6.7	0.8	0.4
P. palpebra	17.0	0.8	1.1	3.7	4.2	2.9	2.0	2.2	1.5	0.4	3.2	3.1	9.4	7.5	3.0	1.8	4.3	0.4
P. punctulata	1.9	7.0	18.5	16.0	7.4	8.0	11.7	7.0	5.8	8.0	4.9	10.3	8.9	12.7	2.8	6.0	1.9	4.3
Pseudotextularia deformis	6.4	3.9	6.7	1.7	2.6	2.5	1.4	6.8	3.2	3.0	1.2	4.3	2.4	2.4	1.1	1.6	1.2	6.4
P. elegans	5.3	6.1	11.5	8.6	7.7	1.0	1.4	9.2	4.1	2.3	12.1	11.4	13.6	17.5	9.6	8.1	5.1	18.8
Planoglobulina brazoensis	0.5	0.6	0.6	0.2	0.3	1.0	0.9	1.3	0.3	1.5	1.0	0.8	-	0.5	0.9	0.7	0.4	0.4
P. carseyae	-	-	-	-	-	-	0.7	-	-	-	-	1.2	-	0.5	-	-	-	-
Plummerita hantkeninoides	-	-	-	-	-	-	-	-	0.3	-	-	-	-	-	-	1.1	-	-
Racemiguembelina intermedia	0.5	-	-	0.2	0.3	0.0	-	-	-	-	-	0.3	-	-	-	-	-	0.4
R. fructicosa	1.9		0.3	0.5	0.3	0.3	0.2	-	0.3	-	-	0.2	0.3	-	-	-	-	1.8
R. powelli	-	-	-	-	0.3	1.0	-	0.9	0.9	0.4	0.5	0.9	1.3	0.5	0.9	-	-	-
Rosita contusa	-	-	-	-	-	-	-	0.2	-	-	-	-	-	-	-	-	-	1.1
R. hexacamerata	3.2	1.4	0.8	2.4	1.9	4.1	3.2	3.1	2.0	3.4	1.0	0.5	0.5	3.3	1.3	2.5	0.4	1.4
R. macrocephala	1.9	2.2	1.7	2.7	2.2	0.6	0.7	1.1	2.6	1.9	1.7	3.2	1.6	0.9	3.0	1.3	1.2	1.4
R. pennyi	0.3	3.1	0.6	3.0	2.6	5.1	2.5	5.9	5.5	2.3	3.7	3.5	1.8	1.4	2.4	1.6	1.9	4.3
R. rugosa	6.9	7.3	3.9	7.6	4.2	7.6	2.0	5.3	4.7	6.4	5.7	4.3	5.5	2.4	2.2	1.8	0.8	1.1
R. scotti	2.4	2.2	3.7	3.0	6.4	4.5	3.8	5.5	6.1	4.9	0.2	2.9	3.4	2.4	3.5	4.0	3.5	4.3
Total	**377**	**358**	**356**	**593**	**312**	**314**	**443**	**455**	**344**	**264**	**405**	**648**	**382**	**212**	**460**	**451**	**258**	**283**

*Asterisk indicates number of specimens

TABLE 2. RELATIVE ABUNDANCE OF PLANKTIC FORAMINIFERA AT LA LAJILLA, MEXICO. SAMPLES 4.1 TO 12.1 CLASTIC DEPOSIT AND OVERLYING MARL OF MAASTRICHTIAN AGE

Note that sample 4.1 is at the base of ths clastic deposit.

Plummerita hantkeninoides Zone

Sample No.	4.1	4.2B	4.2	4.3	5.1	6.1B	6.1	6.2	6.3	7.1	7.2	7.3	7.4	8.1	9.1	9.2	9.3	10.1	11.1	12.1
Thickness in cm.	83	86	88	97	103	108	111	133	153	157	163	166	168	170	173	183	187	193	197	203
Abathomphalus intermedius	4.5	2.7	-	-	0.7	-	-	-	-	-	-	-	-	-	-	-	0.7	-	-	-
A. mayaroensis	3.4	0.9	-	-	1.5	-	-	-	-	0.6	*2	1.6	*1	*1	-	-	0.7	-	-	-
Archeoglobigerina cretacea	5.9	4.1	-	-	2.2	-	-	-	-	0.6	-	1.6	-	-	*2	-	1.5	-	-	-
Gansserina gansseri	-	0.5	-	-	-	-	-	-	-	-	-	-	-	-	-	-	-	-	-	-
Globigerinelloides aspera	0.6	0.5	-	-	3.7	-	-	-	-	24.2	*2	7.8	*1	*7	*1	*9	2.2	3.4	*1	7.9
G. rosebudensis	0.8	0.9	-	-	-	-	-	-	-	-	-	-	-	-	-	-	-	-	-	-
G. subcarinatus	-	-	-	-	-	-	-	-	-	-	-	-	-	-	-	-	-	8.2	-	2.4
G. volutus	1.1	-	-	-	-	-	-	-	-	-	-	-	-	-	-	-	-	-	-	0.5
G. yaucoensis	1.7	-	-	-	-	*2	-	*2	-	13.7	*2	12.5	-	*12	*3	*8	5.2	12.9	*1	4.3
"G" monmouthensis	-	-	-	-	-	-	-	-	-	-	-	-	-	-	-	-	3.0	3.4	-	0.5
Globotruncanella petaloidea	2.8	1.4	-	-	0.7	-	-	-	-	-	*2	-	-	*3	*1	*2	1.5	2.7	*3	1.9
Guembelitria cretacea	-	-	-	-	-	-	-	-	-	-	-	-	-	-	*1	*2	0.7	-	-	-
Hedbergella holmdelensis	0.8	-	-	-	-	-	-	-	-	3.1	-	0.8	-	-	-	*3	5.2	0.7	-	1.1
H. monmouthensis	4.8	3.6	-	-	3.7	*1	-	-	-	9.3	*4	17.2	*3	*5	*2	*6	6.7	6.1	*1	4.9
Globotruncana aegyptiaca	2.0	1.4	-	*3	2.2	-	*1	*1	-	-	-	-	-	-	*1	-	1.5	-	*1	-
G. arca	2.0	1.4	-	-	4.4	-	-	-	-	1.9	*2	6.3	*1	-	*2	-	0.7	-	*1	-
G. duwi	-	-	-	-	-	-	-	-	-	-	-	-	-	*1	-	-	-	-	-	-
G. insignis	-	0.5	-	*1	2.9	-	-	-	-	-	-	-	-	*1	-	-	-	-	-	-
G. orientalis	2.8	0.9	-	-	2.2	-	-	-	-	-	-	-	-	-	*1	-	3.7	-	-	-
G. rosetta	2.5	3.6	-	-	2.2	-	-	-	-	-	-	-	-	-	*1	-	0.7	-	-	-
Globotruncanita angulata	0.3	-	-	-	-	-	-	-	-	-	-	-	-	-	-	-	-	-	-	-
G conica	2.0	2.3	-	-	3.7	-	-	-	-	-	-	-	-	-	-	-	-	-	-	-
G. stuarti	1.7	2.3	-	-	-	-	-	-	-	0.6	-	6.3	*8	*1	*1	-	1.5	-	-	-
G. stuartiformis	-	-	-	-	2.2	-	-	-	-	-	-	-	-	-	-	-	-	-	-	-
Heterohelix complanata	0.3	0.9	-	-	-	-	-	-	-	-	-	-	-	-	-	-	-	-	-	-
H. dentata	2.5	-	-	-	-	-	-	*2	-	1.9	-	0.8	-	*8	*2	*2	5.9	11.6	*1	14.1
H. globulosa	0.6	1.8	-	-	0.7	*1	-	-	-	6.2	*2	12.5	*4	*2	*1	*3	3.0	9.5	*6	2.4
H. navarroensis	3.7	1.8	-	-	1.5	-	*1	*2	-	1.9	-	-	-	*6	-	-	10.4	6.1	*1	7.3
H. moremani	0.8	-	-	-	-	-	-	-	-	0.6	-	-	-		-	-	1.5	2.0	-	-
H. planata	2.8	0.9	-	-	-	-	-	-	-	1.9	*1	-	*1	*8	-	*3	0.7	3.4	*1	5.7
H. striata	4.5	7.7	-	*3	11.8	*1	-	-	-	-	-	0.8	-		-			2.0	-	3.8
Pseudoguembelina costulata	12.1	5.0	-	*3	7.4	*4	*2	*8	-	32.9	*7	19.5	-	*21	*7	*30	26.7	22.4	*4	33.4
P. kempensis	1.4	2.3	-	-	0.7	-	-	-	-	0.6	-		-	*2	-	-	-	-	-	1.9
P. palpebra	1.7	6.3	-	-	11.8	-	-	-	-	-	*1	4.7	-	-	*2	-	2.2	1.4	*1	1.1
P. punctulata	13.2	-	-	*1	27.9	*2	-	-	-	-	-	1.6	-	-	*1	-	2.2	1.4	*1	1.9
Pseudotextularia deformis	1.1	5.4	-	-	1.5	-	-	-	-	-	-	3.1	*1	-	*1	-	2.2	-	-	0.3
P. elegans	5.1	31.2	-	*2	3.7	-	-	-	-	-	-	-	-	-	*2	-	-	-	-	-
Planoglobulina brazoensis	-	-	-	-	-	-	-	-	-	-	-	2.3	-	*1	*1	-	-	-	-	-
P. carseyae	-	1.4	-	-	-	-	-	-	-	-	-	-	-	-	-	-	-	-	-	0.3
Plummerita hantkeninoides	-	-	-	-	-	-	-	-	-	-	-	-	-	-	-	-	-	2.0	-	-
Racemiguembelina intermedia	0.3	-	-	-	-	-	-	-	-	-	-	-	-	-	*1	-	-	-	-	-
R. fructicosa	-	0.5	-	-	-	-	-	-	-	-	-	-	-	-	-	-	-	-	-	-
R. powelli	0.6	0.9	-	-	-	-	-	-	-	-	-	-	-	*1	*1	-	-	-	-	-
Rosita contusa	-	-	-	-	-	-	-	-	-	-	-	-	-	-	-	-	-	-	-	-
R. hexacamerata	0.3	0.9	-	*1	-	-	-	-	-	-	-	-	-	-	-	-	-	-	-	1.6
R. macrocephala	2.5	0.5	-	-	-	-	-	-	-	-	-	-	-	-	*1	-	2.2	0.7	-	-
R. pennyi	1.4	2.3	-	-	0.7	-	-	-	-	-	-	0.8	-	-	-	-	-	-	-	-
R. rugosa	1.7	1.4	-	-	-	-	-	*1	-	-	-	-	-	-	-	-	5.2	-	*1	2.4
R. scotti	3.7	2.3	-	-	0.7	*1	-	-	-	-	-	-	-	-	-	-	0.7	-		-
Total	**355**	**221**	**0**	**14**	**136**	**11**	**4**	**16**	**0**	**161**	**25**	**128**	**20**	**80**	**36**	**68**	**135**	**147**	**24**	**368**

*Asterisk indicates number of specimens

Similar to the situation at La Lajilla, the thin Maastrichtian marl layer contains most Maastrichtian species present in the Mendez Formation below the siliciclastic deposit and no size sorting is apparent (Adatte et al., 1994 a, b). Thus, at El Mulato the stratigraphic position of this Maastrichtian-age marl layer above the siliciclastic deposit also indicates that siliciclastic deposition preceded the K/T boundary.

The first Tertiary sediments immediately overlying the Maastrichtian marl are characterized by the simultaneous first appearance of nine Tertiary species which in K/T complete sections evolve sequentially in Zones P0 and P1a (*Globoconusa conusa*, *Woodringina hornerstownensis*, *W. claytonensis*, *Parvularugoglobigerina eugubina*, *P. longiapertura*, *Globanomalina taurica*, *Chiloguembelina crinita*, *C. waiparaensis*, and *Igorina spiralis*). Of these taxa, *C. waiparaensis* evolved in the late Maastrichtian and *I. spiralis* in Zone P1a(2) in the southern high latitude ODP Site 738 (Keller, 1993). The simultaneous first occurrence of these nine species strongly indicates that Zone P0 and the lower part of Zone P1a are missing, as at La Lajilla. Moreover, the high relative abundance of these Danian species at the time of their first appearance also suggests a hiatus because evolving species are generally rare in their initial appearance. It is likely that erosion also removed the topmost Maastrichtian sediments, as at La Lajilla.

La Parida. The Parida outcrop differs from Mimbral, Lajilla, and Mulato in that the siliciclastic deposit reaches a maximum thickness of only 80 cm and over a distance of 50 m laterally thins to 10 cm and disappears (Fig. 2; Stinnesbeck et al., 1994). The topmost siliciclastic sediments consist of the bioturbated rippled sandy limestone characteristic of the top of unit 3 in all other outcrops. Where the siliciclastic unit is absent, marls of the Maastrichtian Mendez Formation conformably underlie marls of the Tertiary Velasco Formation. At Parida, as at El Mulato and Lajilla, a thin layer (5- to 10-cm) of Maastrichtian marl overlies the siliciclastic deposit and contains a typical latest Maastrichtian foraminiferal assemblage including the index species *P. hantkeninoides* (López-Oliva, 1996). This situation indicates that deposition of the siliciclastic sediments also preceded the K/T boundary. Tertiary sediments above this marl layer are characterized by the simultaneous first appearance of seven Danian species (*Eoglobigerina fringa, E. edita, Parvularugoglobigerina longiapertura, P. eugubina, Planorotalites compressus, Chiloguembelina waiparaensis, Woodringina hornerstownensis* (López-Oliva, 1996), which in K/T-complete sections evolve sequentially. This indicates that Zones P0 and the lower part of Zone P1a(1) are missing, a situation similar to that in the other sections examined.

DISCUSSION AND CONCLUSIONS

K/T Boundary Hiatus

Biostratigraphic comparison of the northeastern Mexico sections with the K/T boundary stratotype at El Kef indicates that the K/T transition is not continuous at Mimbral, Mulato, Lajilla, or Parida. At Mimbral, a 4-cm clay layer overlying the top of the siliciclastic sediments is devoid of planktic foraminifera but contains elevated Ir concentrations and rare shocked-quartz grains, suggesting that the clay layer may represent the boundary clay (Keller et al., 1994a, b). However, above the clay layer, Ir concentrations reach maximum values and a well-developed early Danian Zone P1a (2) assemblage is present, suggesting that the lower part of the Zone P1a and possibly also Zone P0 are missing. At the Lajilla, Mulato and Parida sections, no clay layer is present and the lower part of Zone P1a and Zone P0 are missing. This K/T boundary hiatus is marked by a sharp contact between the thin Maastrichtian marl layer (above the siliciclastic deposit) and the Tertiary Velasco shales. No sediment mixing is apparent across this contact, as indicated by the simultaneous and abundant first appearances of 5 to 12 Tertiary species. Since these species all evolve sequentially in Zones P0 and P1a, a hiatus is clearly present. This hiatus spans at least Zones P0 and the lower part of P1a (P1a (1)) of the early Danian and probably the uppermost Maastrichtian. The presence of this hiatus may explain why siliciclastic deposits in northeastern Mexico are overlain by only thin layers of Maastrichtian marls, whereas in southern Mexico 1 m of Maastrichtian sediments overlies the siliciclastic deposits (Macias Pérez, 1988; Montanari et al., 1994). Short hiatuses are frequently present at the K/T boundary and in the early Danian as documented at Agost and Caravaca, Nye Klov, Brazos River, Negev, Brazil, and deep-sea Sites 528, 577, 738 (Keller, 1989, 1993; Keller and Benjamini, 1991; Canudo et al., 1991; D'Hondt and Keller, 1991; MacLeod and Keller, 1991; Keller et al., 1993; Stinnesbeck and Keller, this volume).

Investigation of four K/T boundary sections (Mimbral, Lajilla, Mulato, and Parida) with siliciclastic sediments in northeastern Mexico reveals that all, except Mimbral, contain a thin layer of marl with typical Maastrichtian planktic foraminiferal assemblages above the siliciclastic deposits. These marl layers show no grain-size gradation (laser counter analysis by Adatte et al., 1994a, this volume), or foraminiferal size sorting that would suggest suspension settling of sediments through the water column. Moreover, the faunal assemblages are rich and diverse, containing large and small species and adult and juvenile forms, just as in the Mendez marls below the siliciclastic deposit. Furthermore, the presence of the latest Maastrichtian index species *P. hantkeninoides* indicates that deposition must have occurred prior to the K/T boundary but within the last 170 to 200 k.y. of the Maastrichtian. For these reasons, it is difficult to argue that these marl layers represent suspension settling from the muddy waters after the K/T impact-generated tsunami waves passed (Smit et al., 1994a, b). That interpretation is also difficult to reconcile with the fact that at Parida this marl layer can be laterally traced beyond the siliciclastic deposit, where no terrigenous sediment transport influenced sediment deposition.

In all of the sections examined (Fig. 1), the top of the siliciclastic deposit (top of unit 3) is strongly bioturbated, and in some outcrops two or three discrete sandstone layers are biotur-

TABLE 3. RELATIVE ABUNDANCE (%) OF PLANKTIC FORAMINIFERA AT LA LAJILLA, MEXICO. SAMPLES 13.1 TO 16.7 PALEOCENE VELASCO FM. SHALES

Note that sample 13.1 is at the base of Tertiary sediments.

	P1a Zone											P1b
Sample No.	13.1	14.1	14.2	14.3	14.4	14.5	14.6	14.7	14.8	14.9	14.10	14.11
Thickness in cm.	207	213	217	223	227	233	237	243	247	253	257	275
Abathomphalus intermedius	-	-	-	-	-	-	-	-	-	-	-	-
A. mayaroensis	-	-	-	-	-	-	-	-	-	-	-	-
Archeoglobigerina cretacea	-	-	-	-	2.0	0.4	-	-	-	-	-	-
Gansserina gansseri	-	-	-	-	-	-	-	-	-	-	-	-
Globigerinelloides aspera	2.6	7.6	4.6	3.6	0.5	2.2	1.0	0.9	2.1	2.0	1.2	
G. rosebudensis	-	-	-	-	-	-	-	-	-	-	-	-
G. subcarinatus	1.1	1.9	2.3	-	-	0.4	-	0.9	-	-	-	0.4
G. volutus	-	-	-	-	-	-	-	-	-	-	-	-
G. yaucoensis	3.8	1.3		0.9	-	1.1	-	-	-	-	2.4	0.6
"G" monmouthensis	-	-	-	-	-	-	-	-	-	-	-	-
Globotruncanella petaloidea	-	1.9	-	-	0.5	1.5	-	-	-	-	-	-
Guembelitria cretacea	1.5	-	-	-	-	-	-	-	-	0.4	-	-
Hedbergella holmdelensis	0.4	-	-	-	-	-	-	-	-		-	-
H. monmouthensis	1.5	-	1.5	-	-	-	-	-	-	1.6	-	-
Globotruncana aegyptiaca	-	-	-	-	-	-	-	-	-	-	-	-
G. arca	-	-	-	-	-	-	-	-	-	-	-	-
G. duwi	-	-	-	-	-	-	-	-	-	-	-	-
G. insignis	-	-	-	0.9	1.0	-	-	-	-	-	-	-
G. orientalis	-	-	-	-	-	-	-	-	-	-	-	-
G. rosetta	-	-	0.8	-	1.0	0.7	-	-	-	0.4	1.2	-
Globotruncanita angulata	-	-	-	-	-	-	-	-	-	-	-	-
G conica	-	-	-	-	-	-	-	-	-	-	-	-
G. stuarti	-	-	0.8	2.7	-	-	-	-	-	0.4	-	-
G. stuartiformis	-	-	-	-	-	-	-	-	-	-	-	-
Heterohelix complanata	-	-	-	-	-	-	-	-	-	-	-	-
H. dentata	-	2.5	3.1	2.7	1.5	1.5	-	-	1.4	3.3	0.6	-
H. globulosa	4.5	5.7	3.1	0.9	2.9	0.7	-	-	0.7	2.5	1.2	-
H. navarroensis	10.2	12.0	1.5	1.8	1.0	5.1	0.3	2.8	4.3	1.2	1.8	1.0
H. moremani	-	-	-	-	-	2.2	-	-	0.7	-	-	-
H. planata	-	-	4.6	2.7	2.9	0.4	0.3	-	4.3	-	-	1.0
H. striata	-	-	5.4	-	6.4	3.3	0.7	-	1.4	4.5	2.4	-
Pseudoguembelina costulata	4.9	-	3.8	9.8	9.3	5.9	-	-	6.4	4.5	6.0	1.0
P. kempensis	0.4	-	-	-	0.5	0.7	1.0	10.1	-	-	1.2	-
P. palpebra	-	-	3.1	2.7	1.0	2.6	0.7	-	0.7	-	-	0.4
P. punctulata	-	-	2.3	0.9	2.5	-	1.4	-	-	-	1.8	0.8
Pseudotextularia deformis	-	-	-	-	-	-	-	-	-	-	-	-
P. elegans	-	-	4.6	6.3	1.5	0.7	-	-	-	0.8	0.6	-
Planoglobulina brazoensis	-	-	-	-	-	-	-	-	-	-	-	-
P. carseyae	-	-	-	-	-	-	-	-	-	-	-	-
Plummerita hantkeninoides	-	-	-	-	-	-	-	-	-	-	-	-
Racemiguembelina intermedia	-	-	-	0.9	-	-	-	-	-	-	-	-
R. fructicosa	-	-	-	-	-	-	-	-	-	-	-	-
R. powelli	-	-	-	-	-	-	-	-	-	-	-	-
Rosita contusa	-	-	-	-	-	-	-	-	-	-	-	-
R. hexacamerata	-	-	1.5	4.5	-	1.1	-	-	-	-	0.6	-
R. macrocephala	-	-	0.8	-	-	-	-	-	-	-	-	-
R. pennyi	-	-	-	1.8	-	-	-	-	-	-	-	-
R. rugosa	-	0.6	0.8	-	-	0.7	0.3	-	-	-	-	0.4
R. scotti	-	-	-	0.9	0.5	0.4	-	-	-	-	-	-
Chiloguembelina crinita	6.4	-	-	3.6	-	-	0.3	2.8	0.7	2.5	-	1.7
C. midwayensis	3.8	5.1	-	0.9	0.5	1.1	4.4	-	-	2.5	7.2	9.7
C. taurica	-	-	-	-	-	-	0.3	-	-	-	-	7.0
C. waiparaensis	9.8	2.5	30.0	11.6	7.4	14.7	1.4	2.8	6.4	2.5	7.8	4.5
Eoglobigerina edita	3.8	7.0	4.6	6.3	3.9	3.7	2.7	0.9	0.7	6.6	-	-
E. eobulloides	3.0	-	-	-	-	1.1	1.7	3.7	-	2.9	1.2	-
Eoglobigerina fringa	0.4	2.5	2.3	1.8	-	2.9	5.7	2.8	1.4	2.0	3.6	2.7
E. simplicissima	-	-	0.8	-	2.0	-	-	-	-	-	-	-
E. tetragona	-	-	-	-	-	-	-	-	-	-	-	-
E. trivialis	0.8	-	-	2.7	1.5	0.4	-	1.8	-	-	-	2.1
Globanomalina pentagona	-	2.5	0.8	0.9	-	4.8	3.0	3.7	-	2.5	1.8	2.7
G. taurica	-	-	-	-	2.5	-	4.1	-	-		3.0	3.9
Globoconusa conusa	9.8	-	-	-	1.0	-	0.7	1.8	-	1.6	-	0.4
G. daubjergensis	-	-	-	1.8	3.9	2.6	5.7	6.4	1.4	4.1	1.8	7.0
G. extensa	-	-	-	-	-	-	-	-	-	-	-	1.7
Globorotalia archeocompressa	-	-	-	-	-	-	1.0	4.6	2.1	0.8	-	-
G. polycamera	-	-	-	-	0.5	3.7	3.0	1.8	6.4		2.4	0.8
Guembelitria irregularis	0.4	-	-	-	-	-	1.0	1.8	-	0.4	-	0.8
Igorina spiralis	-	-	-	-	-	-	-	-	-	-	-	1.9
Morozovella inconstans	-	-	-	-	-	-	-	-	-	-	-	-
Parvularugoglobigerina eugubina	11.7	7.0	3.1	15.2	13.7	11.0	23.6	24.8	29.3	23.8	15.0	-
P. longiapertura	17.7	2.5	3.8	2.7	17.6	9.2	20.6	16.5	11.4	14.8	22.2	1.7
Planorotalites compressa	-	5.1	3.1	3.6	-	-	12.5	3.7	9.3	2.0	3.6	18.6
P. planocompressa	-	-	-	-	5.4	4.0	1.4	4.6	-	-	v	10.5
Subbotina moskvini	-	-	-	1.8	0.5	-	0.3	-	-	-	1.2	0.2
S. pseudobulloides	1.5	-	-	0.9	-	0.4	-	-	0.7	-	0.6	-
S. triloculinoides	-	-	-	-	-	0.7	-	0.9		-	0.6	-
S. varianta	-	-	-	-	-	-	-	-	-	-	-	-
Woodringina claytonensis	-	-	-	-	-	0.4	-	-	0.7	0.4	-	-
W. hornerstownensis	-	3.8	6.9	1.8	4.4	8.1	0.7	-	5.7	9.0	6.6	9.3
Total	**265**	**158**	**130**	**112**	**204**	**273**	**296**	**109**	**140**	**244**	**167**	**484**

*Asterisk indicates number of specimens

TABLE 3. RELATIVE ABUNDANCE (%) OF PLANKTIC FORAMINIFERA AT LA LAJILLA, MEXICO. SAMPLES 13.1 TO 16.7 PALEOCENE VELASCO FM. SHALES (continued)

			P1b Zone						P1c Zone			
Sample No.	14.12	14.13	14.14	14.15	15.1	16.1	16.2	16.3	16.4	16.5	16.6	16.7
Thickness in cm.	275	285	295	303	305	310	320	330	340	350	360	370
Abathomphalus intermedius	-	-	-	-	-	-	-	-	-	-	-	-
A. mayaroensis	-	-	-	-	-	-	-	-	-	-	-	-
Archeoglobigerina cretacea	-	-	-	-	-	-	-	-	-	-	-	-
Gansserina gansseri	-	-	-	-	-	-	-	-	-	-	-	-
Globigerinelloides aspera	2.2	-	-	3.5	-	0.8		1.3	0.4	0.2	0.7	0.4
G. rosebudensis	-	-	-	-	-	-	-	-	-	-	-	-
G. subcarinatus	-	-	-	-	-	-	-	-	0.6	-	-	0.4
G. volutus	-	-	-	-	-	-	-	-	-	-	-	-
G. yaucoensis	-	-	-	-	-	-	0.7	0.4	0.2	0.6	0.2	0.3
"G" monmouthensis	-	-	-	-	-	-	-	-	-	-	-	-
Globotruncanella petaloidea	-	-	-	-	-	0.5	-	-	-	-	0.2	-
Guembelitria cretacea	2.2	-	1.4	-	-	0.0	-	-	-	-	-	-
Hedbergella holmdelensis	-	-	-	-	-	-	-	-	-	-	-	-
H. monmouthensis	1.1	1.0		0.6	-	-	-	-	-	-	-	-
Globotruncana aegyptiaca	-	-	-	-	-	-	-	-	-	-	-	-
G. arca	-	-	-	-	-	-	-	-	-	-	-	-
G. duwi	-	-	-	-	-	-	-	-	-	-	-	-
G. insignis	-	-	-	-	-	-	-	-	-	-	-	-
G. orientalis	-	-	-	-	-	-	-	-	-	-	-	-
G. rosetta	2.2	-	-	3.5	-	-	-	-	-	-	-	-
Globotruncanita angulata	-	-	-	-	-	-	-	-	-	-	-	-
G conica	-	-	-	-	-	-	-	-	-	-	-	-
G. stuarti	-	-	-	-	-	-	-	-	-	-	-	-
G. stuartiformis	-	-	-	-	-	-	-	-	-	-	-	-
Heterohelix complanata	-	-	-	-	-	-	-	-	-	-	-	-
H. dentata	2.2	-	-	-	-	-	-	-	-	-	-	-
H. globulosa	-	2.0	-	1.2	-	0.8		0.9	0.8	0.6	-	-
H. navarroensis	-	5.0	0.7	-	-	2.2	0.9	1.1	1.5	0.6	0.2	0.5
H. moremani	-	-	-	-	-	-	-	-	-	-	-	-
H. planata	-	-	-	1.2	-	2.2	-	2.7	1.0	0.8	0.9	0.9
H. striata	2.2	2.0	0.7	3.5	-	0.8	0.7	0.2	-	0.4	0.3	0.5
Pseudoguembelina costulata	1.1	8.0	0.7	16.2	-	4.6	0.7	1.5	0.8	0.8	0.2	3.7
P. kempensis	-	1.0	-	-	-	0.8	-	1.8	0.4	0.6	0.3	0.1
P. palpebra	2.2	-	-	-	-	0.8	-	1.8	-	0.4	0.2	2.3
P. punctulata	7.5	-	-	-	-	1.1	0.7	3.7	-	0.8	0.3	1.6
Pseudotextularia deformis	-	-	-	-	-	-	-	-	-	-	-	-
P. elegans	-	-	0.7	4.0	*1	-	-	1.6	-	0.2	0.2	0.1
Planoglobulina brazoensis	-	-	-	-	-	-	-	-	-	-	-	-
P. carseyae	-	-	-	-	-	-	-	-	-	-	-	-
Plummerita hantkeninoides	-	-	-	-	-	-	-	-	-	-	-	-
Racemiguembelina intermedia	-	-	-	-	-	-	-	-	-	-	-	-
R. fructicosa	-	-	-	-	-	-	-	-	-	-	-	-
R. powelli	-	-	-	-	-	-	-	-	-	-	-	-
Rosita contusa	-	-	-	-	-	-	-	-	-	-	-	-
R. hexacamerata	-	-	-	0.6	-	-	-	0.9	-	-	-	-
R. macrocephala	-	-	-	-	-	-	-	-	-	-	-	-
R. pennyi	-	-	-	-	*1	-	-	-	-	-	-	-
R. rugosa	-	-	0.7	-	-	-	-	-	-	-	-	-
R. scotti	-	-	-	-	-	-	-	-	-	-	-	-
Chiloguembelina crinita	2.2	1.0	2.8	2.3		1.4	7.5	0.9	2.3	4.0	0.5	0.8
C. midwayensis	16.1	5.0	6.4	1.7	-	1.9	2.5	-	-	1.8	1.4	1.9
C. taurica	6.5	-	2.1	0.6	-	3.0	3.2	4.2	-	-	-	-
C. waiparaensis	3.2	8.0	-	-	-	5.9	2.7	3.8	27.5	28.3	34.5	39.5
Eoglobigerina edita	3.2	-	0.7	-	-	0.5	-	-	-	-	-	-
E. eobulloides	-	5.0	-	-	-	-	-	-	2.5	4.0		1.2
Eoglobigerina fringa	-	-	-	-	-	-	-	-	-	-	-	-
E. simplicissima	-	-	1.4	-	-	-	1.1	-	-	-	-	-
E. tetragona	-	-	-	1.2	-	6.2	1.1	0.5	10.0	4.2	2.4	0.9
E. trivialis	-	-	4.3	-	-	-	0.2	1.5	1.9	2.4	0.2	0.5
Globanomalina pentagona	12.9	2.0	5.7	6.4	-	25.7	15.0	21.1	5.8	17.9	7.7	7.5
G. taurica	-	2.0	10.6	3.5	-	1.6	4.1	3.8	0.6	1.6	1.7	0.4
Globoconusa conusa	-	-	-	-	-	-	-	1.6	-	-	-	-
G. daubjergensis	4.3	5.0	15.6	-	-	5.7	1.8	0.7	3.1	1.4	1.4	1.6
G. extensa	-	-	5.7	-	-	-	0.2	-	0.6	-	0.3	1.2
Globorotalia archeocompressa	-	-	3.5	-	-	-	-	-	-	-	-	-
G. polycamera	2.2	-	3.5	1.2	-	0.5	0.5	-	-	-	-	-
Guembelitria irregularis	-	-		0.6	-	0.3	1.1	0.2	-	0.2	0.5	-
Igorina spiralis	-	4.0	9.9	5.2	-	3.0	2.5	3.8	0.8	1.8	1.9	2.6
Morozovella inconstans	-	-	-	0.6	-	-	-	0.4	5.8	2.8	4.9	2.5
Parvulorugoglobigerina eugubina	-	-	-	-	-	-	-	-	-	-	-	-
P. longiapertura	-	-	-	-	-	-	-	-	-	-	-	-
Planorotalites compressa	10.8	27.0	7.1	13.9	-	9.2	7.3	9.5	5.2	0.8	1.0	1.3
P. planocompressa		11.0	2.8	2.3	-	1.6	-	-	8.5	7.0	13.3	4.6
Subbotina moskvini	2.2	-	1.4	5.2	-	1.6	1.8	4.8	1.9	1.8	2.0	4.1
S. pseudobulloides	3.2	10.0	2.8	6.4	-	1.4	2.5	2.9	1.7	1.6	3.9	2.4
S. triloculinoides	4.3	1.0	1.4	4.6	*1	4.1	4.3	8.2	2.5	6.8	6.6	12.2
S. varianta	-	-	-	-	-	-	-	0.7	-	-	-	-
Woodringina claytonensis	1.1	-	2.1	1.7	-	0.8	5.9	0.4	6.5	3.0	3.6	1.1
W. hornerstownensis	5.4	-	5.0	8.7	-	10.8	28.9	12.8	6.9	2.8	8.7	2.8
Total	**93**	**100**	**141**	**173**	**3**	**370**	**440**	**546**	**480**	**502**	**588**	**755**

*Asterisk indicates number of specimens

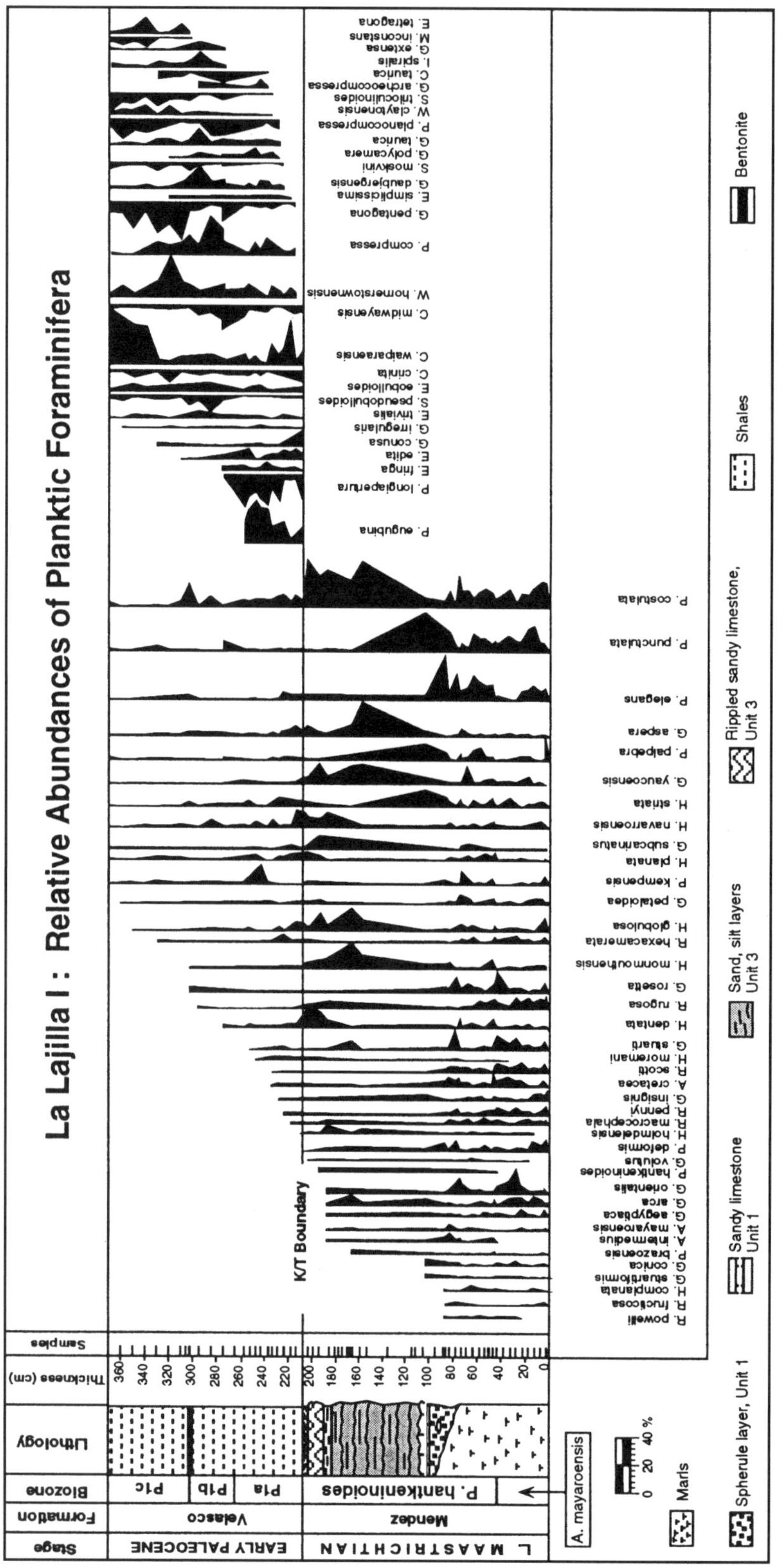

Figure 5. Relative abundance of common planktic foraminifera across the K/T transition at La Lajilla. Note that Maastrichtian foraminifera are generally rare in the siliciclastic deposit but common in the marl layer above and below the siliciclastic deposit. Note also that the simultaneous appearance and high abundance of 12 early Danian species immediately above the marl layer mark a hiatus.

bated. We have examined sediment infilling of burrows in order to examine whether burrowing occurred from overlying Tertiary strata. We found no evidence of Tertiary microfossils within the burrows. Organisms, thus, lived and burrowed down from overlying Maastrichtian strata that were later eroded. The remnants of these Maastrichtian strata appear to be the marl layers overlying the siliciclastic deposits at La Lajilla, El Mulato, and La Parida. Recently, Ekdale and Stinnesbeck (1994, 1996) found evidence of habitation within several layers of unit 3 as well as at the base of unit 2. This evidence indicates the presence of several episodes of colonization by organisms during deposition of the siliciclastic deposit and rules out single-event deposition. Thus, sedimentologic, mineralogic, biostratigraphic, and trace fossil-data all point to multievent deposition of the siliciclastic member, with depositional events separated by scouring and erosion as well as by alternating episodes of terrigenous influx and hemipelagic sedimentation.

The only unusual feature of the siliciclastic deposits in northeastern Mexico sections is the spherule-rich sediment of unit 1, which contains rare glass fragments with compositions similar to those of the glass spherules found at Beloc, Haiti (Smit et al., 1992; Stinnesbeck et al., 1993; Robin et al., 1994; Koeberl, 1994). Whatever the origin of these glass fragments, whether impact or volcanic, the event that produced them preceded the K/T boundary event.

ACKNOWLEDGMENTS

We thank R. Scott, S. Gartner, and W. Sliter for critical reviews and many helpful suggestions and W. Stinnesbeck, T. Adatte, W. Ward, and T. Ekdale for many discussions and for sharing unpublished data. This study was supported by NSF grant INT-9314080, OCE-9021338, The Petroleum Research Fund ACS-PRF grant No. 2670-AC8, and National Geographic Society grant No. 4620-91 and by the National Council for Science and Technology (CONACYT Reg. No. 61183) in Mexico.

REFERENCES CITED

Adatte, T., Stinnesbeck, W., and Keller, G., 1994a, Siliciclastic deposits in NE and Central Mexico predate K/T boundary and are incompatible with tsunami deposition: Geological Society of America Abstracts with Programs, v. 26, p. 334.

Adatte, T., Stinnesbeck, W., and Keller, G., 1994b, Global biotic effects of the K/T boundary event: Mass extinction restricted to low latitudes?, *in* New developments regarding the K/T event and other catastrophes in Earth history [abs.]: Lunar and Planetary Institute, Contribution 825, Houston, Texas, p. 1–2.

Berggren, W. A., and Miller, K. G., 1988, Paleogene tropical planktic foraminiferal biostratigraphy and magnetobiochronology: Micropaleontology, v. 34, p. 362–380.

Bohor, B. F., and Betterton, W. J., 1993, Arroyo El Mimbral, Mexico, K/T unit: Origin as debris flow/turbidite, not a tsunami deposit [abs.], *in* Proceedings, Lunar and Planetary Science Conference 24: Houston, Texas, p. 143–144.

Canudo, J. I., Keller, G., and Molina, E., 1991, Cretaceous/Tertiary boundary extinction pattern and faunal turnover at Agost and Caravaca, S.E. Spain: Marine Micropaleontology, v. 17, p. 319–341.

D'Hondt, S., and Keller, G., 1991, Some patterns of planktic foraminiferal assemblage turnover at the Cretaceous/Tertiary boundary: Marine Micropaleontology, v. 17, p. 77–118.

Ekdale, A. A., and Stinnesbeck, W., 1994, Sedimentologic significance of trace fossils in the KT "Mimbral Beds" of Northeastern Mexico: Geological Society of America Abstacts with Programs, v. 26, p. 395.

Ekdale, A. A., and Stinnesbeck, W., 1996, Ichnology of Cretacious-Tertiary (K/T) boundary beds in northeastern Mexico: Palaios (in press).

Hildebrand, A. R., Penfield, G. T., Kring, D. A., Pilkington, M., Camargo, Z., A., Jacobsen, S. B., and Boynton, W. V., 1991, Chicxulub Crater: A possible Cretaceous/Tertiary boundary impact crater on the Yucatan Peninsula, Mexico: Geology, v. 19, p. 867–871.

Keller, G., 1988, Extinction, survivorship and evolution of planktic Foraminifera across the Cretaceous/Tertiary boundary at El Kef, Tunisia: Marine Micropaleontology, v. 13, p. 239–263.

Keller, G., 1989, Extended Cretaceous/Tertiary boundary extinctions and delayed population change in planktonic Foraminifera from Brazos River, Texas:Paleoceanography, v. 4, p. 287–332.

Keller, G., 1993, The Cretaceous-Tertiary boundary transition in the Antarctic Ocean and its global implications: Marine Micropaleontology, v. 21, p. 1–45.

Keller, G., and Benjamini, C., 1991, Paleoenvironment of the eastern Tethys in the Early Paleocene: Palaios, v. 6, p. 439–464.

Keller, G., and Stinnesbeck, W., 1996, Sea-level changes, siliciclastic deposits and megatsunamis across the Cretaceous-Tertiary boundary, *in* MacLeod, N., and Keller, G., eds., The Cretaceous/Tertiary boundary mass extinction: Biotic and environmental events: New York, W. W. Norton, p. 443–479.

Keller, G., Barrera, E., Schmitz, B., Mattson, E., 1993, Gradual mass extinction, species survivorship, and long term environmental changes across the Cretaceous-Tertiary boundary in high latitudes: Geological Society of America Bulletin v. 105, p. 979–997.

Keller, G., Stinnesbeck, W., Adatte, T., MacLeod, N., and Lowe, D. R., 1994a, Field guide to Cretaceous-Tertiary boundary sections in northeastern Mexico: Houston, Texas, Lunar and Planetary Institute Contribution 827, 110 p.

Keller, G., Stinnesbeck, W., and López-Oliva, J. G., 1994b, Age, deposition and biotic effects of the Cretaceous/Tertiary boundary event at Mimbral, NE Mexico: Palaios, v. 9, p. 144–157.

Koeberl, C., 1994, Stinnesbeck et al., 1993, and Keller, et al., 1993: Comment: Geology, v. 22, p. 957.

Longoria, J. F. and Gamper, M. A., 1992, Planktonic forminiferal biochronology across the KT boundary from the Gulf coastal plain of Mexico: Implications for timing the extraterrestrial bolide impact in the Yucatan: Boletin. Ass. Mexicana Petrol., Geology, v. 425, p. 19–40.

Macias Pérez, F. J., 1988, Estratigrafía detallada del límite Cretácico-Terciario en Xicotepec, Puebla: Una alternativa al catastrofismo [tésis]: México City, Institut Politécnico Nacional, 108 p.

MacLeod, N., and Keller, G., 1991, Hiatus distribution and mass extinctions at the Cretaceous /Tertiary boundary: Geology, v. 19, p. 497–501.

Montanari, A., Claeys, P., Asaro, F., Bermudez, J., and Smit, J., 1994, Preliminary stratigraphy and iridium and other geochemical anomalies across the KT boundary in the Bochil section (Chiapas, southeastern Mexico, *in* New developments regarding the K/T event and other catastrophes in Earth history [abs.]: Houston, Texas, Lunar and Planetary Institute Contribution 825, p. 84–85.

Pardo, A., Ortiz, N. , and Keller, G., 1996, Latest Maastrichtian and K/T boundary foraminiferal turnover at Agost, Spain, *in* MacLeod, N., and Keller, G., eds., The Cretaceous/Tertiary boundary mass extinction: Biotic and environmental events: New York, W. W. Norton Press, p. 157–190.

Pope, K. O., Ocampo, C. A., and Duller, C. E., 1991, Mexican site for K/T impact crater?: Nature, v. 351, p. 105–108.

Robin, E., Rocchia, R., Lyons, J.B., and Officer, C.B., 1994, Deposition of channel deposits near the Cretaceous Tertiary boundary in northeastern Mexico: Catastrophic or normal sedimentary deposits?: by Koeberl, C., on Stinnesbeck et al. (1993) and Keller et al. (1993): Comments and Replies: Geology, v. 22, p. 958.

Sawlowicz, Z., 1993, Iridium and other platinum elements as geochemical markers in sedimentary environments: Paleogeography, Paleoclimatology, Paleoecology, v. 104, p. 253–270.

Schmitz, B., 1988, Origin of microlayering in worldwide distributed Ir-rich marine Cretaceous-Tertiary boundary clays: Geology, v. 16, p 1068–1072.

Sharpton, V. L., Dalrymple, G. B., Marin, L. E., Ryder, G., Schuraytz, B. C., and Urrutia-Fucugauchi, J., 1992, New links between the Chixculub impact structure and the Cretaceous/Tertiary boundary: Science, v. 359, p. 819–821.

Sharpton, V. L., and nine others, 1993, Chicxulub multiring impact basin: Size and other characteristics derived from gravity analysis: Science, v. 261, p. 1564–1567.

Smit, J., 1982, Extinction and evolution of planktic foraminifera after a major impact at the Cretaceous/Tertiary boundary: Geological Society of America Special Paper 190, p. 329–352.

Smit, J., and eight others, 1992, Tektite-bearing, deep-water siliciclastic unit at the Cretaceous-Tertiary boundary in northeastern Mexico: Geology; v. 20, p. 99–103.

Smit, J., Alvarez, W., Claeys, P., Montanari, S., and Roep, Th. B., 1994a, Misunderstandings regarding the K/T boundary deposits in the gulf of Mexico, *in* New developments regarding the KT event and other catastrophes in Earth history [abs]: Houston, Texas, Lunar and Planetary Institute Contribution 825, p. 116–117.

Smit, J., Roep, Th. B., Alvarez, W., Claeys, P., Montanari, S. and Grajales, M., 1994b, Impact-tsunami-generated siliciclastic beds at the K/T boundary of the coastal plain: A synthesis of old and new outcrops, *in* New developments regarding the KT event and other catastrophes in Earth history [abs.]: Houston, Texas, Lunar and Planetary Institute Contribution 825, p. 119–120.

Stinnesbeck, W., and 10 others, 1993, Deposition of channel deposits near the Cretaceous/Tertiary boundary in northeastern Mexico: Catastrophic or "normal" sedimentary deposits?: Geology, v. 21, p. 797–800.

Stinnesbeck, W., Keller, G., and Adatte, T., 1994, K/T boundary deposits in Northeastern Mexico: Bolide impact or sea-level lowstand?, *in* New developments regarding the KT event and other catastrophes in Earth history [abs.]: Houston, Texas, Lunar and Planetary Institute Contribution 825, p. 121–123.

Manuscript Accepted by the Society September 14, 1995

Printed in U.S.A.

Geological Society of America
Special Paper 307
1996

High-resolution biostratigraphy of sea-level low, biotic extinction, and chaotic sedimentation at the Cretaceous-Tertiary boundary in Alabama, north of the Chicxulub crater

Daniel Habib
Department of Geology, Queens College and Graduate School, City University of New York, Flushing, New York 11367
Richard K. Olsson and Chengjie Liu
Department of Geological Sciences, Rutgers University, State University of New Jersey, New Brunswick, New Jersey 08903
Shimon Moshkovitz
Geological Survey of Israel, 30 Malkhei Israel Street, Jerusalem 95-501, Israel

ABSTRACT

Dinoflagellate cysts, planktonic foraminifera, and calcareous nannofossils were used to date and interpret geologic events adjacent to and across the Cretaceous-Tertiary (K/T) boundary in Alabama. High-density sampling and analysis of two sections and their relationship to two previously studied sections reveal the simultaneous occurrence of low sea level, mass extinction of benthic macrofauna and calcareous microplankton, and chaotic sedimentation precisely at the K/T boundary. In the sections at Millers Ferry, Mussel Creek, Moscow Landing, and Braggs, dinoflagellate stratigraphy dating corresponds precisely with that provided by foraminifera and/or nannofossils. Poorly sorted gravels containing reworked pebbles and cobbles of latest Maastrichtian age in the lower part of a low-stand channel fill containing the K/T boundary suggest that this chaotically deposited sediment may have formed from a tsunami generated by the bolide impactor at Chicxulub. Low sea-level stand at this time may have contributed to the extinction of the benthic macrofauna.

INTRODUCTION

Background and purpose of study

Dinoflagellate fossils have been largely neglected in biostratigraphic dating of the Cretaceous-Tertiary (K/T) boundary. Their value in this part of the section is significant, however, when one considers that they did not suffer the extinctions engendered on microfossil groups such as calcareous nannofossils and planktonic foraminifera. Rather, there are now a number of established dinoflagellate species whose first occurrences range across the K/T boundary apparently uninterrupted by the events that caused the terminal Cretaceous mass extinction of life. Although first occurrence datums may still represent time-transgressive migrations or bias in depositional lithofacies, the datums are not subject to the current ambiguities that arise from reworking Cretaceous fossils into the Paleocene (e.g., Pospichal, 1994). The biostratigraphic value of the dinoflagellate fossils can be tested against the evidence of the other microfossil groups and also against the dinocysts' stratigraphic relationship with the major iridium (Ir) spike at the K/T boundary. For example, *Damassadinium californicum* (Drugg) Fensome et al. 1993 (=*Danea californica*) first occurs at 10 cm above the Ir-spike K/T boundary in the Tunisian global stratotype sections (Brinkhuis and Leevreveld, 1988) at the top of the *Guembelitria cretacea* (P0) planktonic foraminiferal zone. On Seymour Island,

Habib, D., Olsson, R. K., Liu, C., and Moshkovitz, S., 1996, High-resolution biostratigraphy of sea-level low, biotic extinction, and chaotic sedimentation at the Cretaceous-Tertiary boundary in Alabama, north of the Chicxulub crater, *in* Ryder, G., Fastovsky, D., and Gartner, S., eds., The Cretaceous-Tertiary Event and Other Catastrophes in Earth History: Boulder, Colorado, Geological Society of America Special Paper 307.

Antarctica, *D. californicum* first occurs approximately 20 cm above the prominent, single Ir peak in the Filo Negro section that was sampled at 10 cm intervals (Askin et al., 1994). The interval across the K/T boundary on Seymour Island is devoid of calcareous nannofossils and planktonic foraminifera (Askin, 1988; Askin et al., 1994), a situation which makes the dinoflagellate evidence particularly relevant. In the Davis-Hopkins well in southeastern Georgia, the first occurrence (FO) of *D. californicum* was used to date the K/T boundary interval in bioclastic calcarenites devoid of nannofossils and foraminifera (Habib and Miller, 1989).

Dinoflagellate evidence of the K/T boundary has been tested against other microfossil groups in sections from south-central and southwestern Alabama. *D. californicum* was used to date the earliest Danian in basal sands of the Clayton Formation (Habib et al., 1992). Splits of the same samples studied for calcareous nannofossils (Moshkovitz and Habib, 1993) and for planktonic foraminifera (Olsson and Liu, 1993) supported the biostratigraphic ages of latest Maastrichtian and early Danian determined by the use of dinoflagellates. In the section at Mussel Creek in south-central Alabama, *D. californicum* first occurs 60 cm above the K/T boundary in cross-laminated sands. This datum is consistent with the corresponding datums reported from Tunisia and Seymour Island, as a result of the higher sedimentation rate in the sands. Significantly, at Mussel Creek and in Tunisia, the species first occurs at the boundary between P0 and Pα foram zones. Both in Alabama and on Seymour Island, (Habib et al. 1992; Askin et al., 1994; Elliot et al., 1994) FO *D. californicum* is associated with the first marine transgression above the K/T boundary, resulting from the first eustatic rise of global sea level in the Cenozoic (Mancini and Tew, 1993, after Haq et al., 1988).

Dinoflagellate species richness is considered to be valuable for interpreting marine transgression and regression. In a sequence stratigraphy–calibrated interval in Alabama, (Donovan et al., 1988), dinoflagellate diversity was found to vary from one individual depositional systems tract to another. The largest number of dinoflagellate species and specimens occurs near the base of the transgressive tract of early Danian age, directly above transgressive lag deposits. The number of species remains large but decreases toward the boundary with the higher high-stand systems tract. In the high-stand tract, the number of *species* decreases rapidly, with a corresponding increase in number of *specimens* of landplant particles, reworked palynomorphs, and amount of detrital inertinite. This palynological alternation of an interval rich in dinoflagellates succeeded by one characterized by phytoclasts and inertinite is interpreted to represent a cycle of marine transgression followed by terrigenous influx of terrestrial and reworked organic matter during shoreline progradation. Gregory and Hart (1992) showed the correspondence between systems tract components and palynological assemblages in their study of various sequence stratigraphy–calibrated sections of Late Cretaceous and Paleogene ages in the Gulf Coastal Plain. Variations in dinoflagellate species richness have been used to distinguish nearshore and offshore assemblages in ancient sediments (Edwards, 1989; Firth, 1993a) and in modern estuarine (McMinn, 1991) and shelf environments (Wall et al., 1977). Organic-walled dinoflagellate fossils (fossil dinocysts) represent the benthic stage of the life cycle in meroplanktonic species. As such, they reflect the number of habitats available to them in the marine neritic environment. In this respect, dinoflagellates follow the general principle in ecology that species richness varies with available area (Krebs, 1985), so that species should increase in number as the marine shelf expands during transgression. Dinocysts also behave as detrital particles that tend to concentrate in deeper and more quiescent offshore environments. In the inshore-offshore direction, the greatest diversity occurs in the offshore area that includes, in addition, the dinocysts of species displaced from nearer shore environments (Traverse and Ginsburg, 1966; Wall et al., 1977).

The purpose of this study is to present dinoflagellate evidence of the precise dating of the K/T boundary in Alabama, especially as it relates to the massive biotic extinction event recorded in the road-cut section near Braggs (Bryan and Jones, 1989) and to the rapid sedimentological event exposed in the Tombigbee River section at Moscow Landing (Fig. 1). The second objective is to interpret these events in the context of a sea level that was falling as it approached the K/T boundary and that began to rise at the very beginning of the Paleocene. The area of study is important because of its geographic position across the Gulf of Mexico from the Chicxulub impact crater (Hildebrand et al., 1991) and its proximity to the sections in Texas that are believed to contain K/T boundary deposits of impact-generated tsunami origin (Jiang and Gartner, 1986; Bourgeois et al., 1988). We submit evidence that dates chaotically deposited gravels at Moscow Landing precisely at the K/T boundary and suggest that they formed directly in response to tsunami-like storm waves generated by the impactor at Chicxulub, presently located approximately 700 km to the south across the Gulf of Mexico. We conclude that this occurred at a time of global low stand of sea level, when the benthic macrofaunal community was stressed by constricted epeiric and marginal seas.

Methods and materials

The first part of this study is to present additional evidence of a precise dinoflagellate stratigraphy at the K/T boundary in Alabama. New data are presented from the latest Maastrichtian–earliest Danian interval in Millers Ferry Core 225 at the Millers Ferry dam site along the Alabama River (Fig. 1) (Olsson and Liu, 1993). Fifty-three samples were studied in the interval 9.1 m thick (Table 1). The core samples represent stratigraphic intervals as small as 7 cm up to 20 cm. The purpose of studying this section is to calibrate the dinoflagellate stratigraphy in a high-resolution sampling interval that is calibrated by planktonic foraminifera in the Danian part and by calcareous nannoplankton in the latest

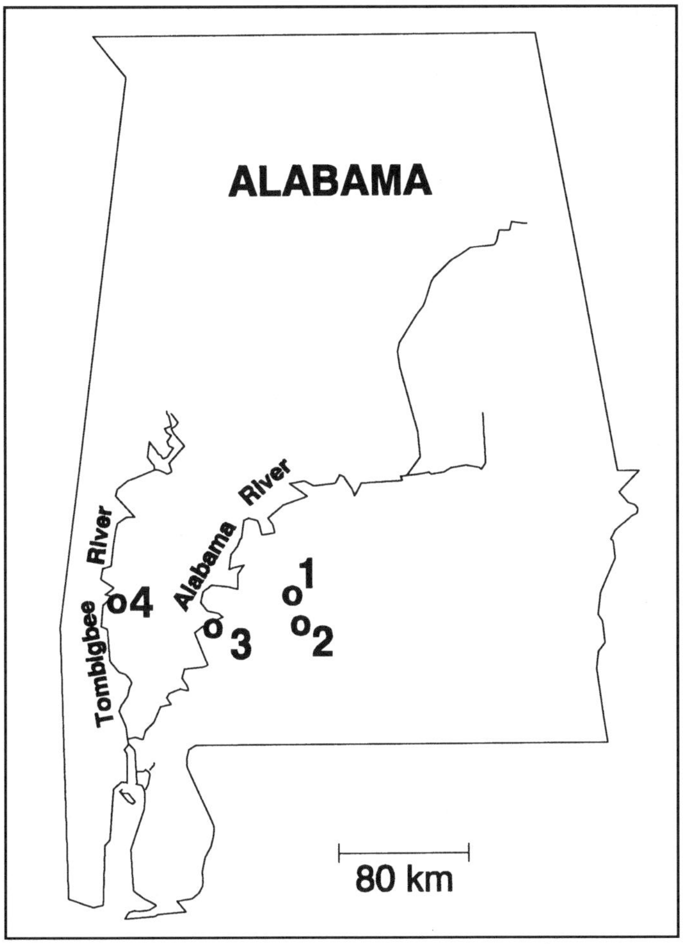

Figure 1. Geographic location of studied sections in Alabama. 1—Braggs roadcut. 2—Mussel Creek. 3—Millers Ferry. 4—Moscow Landing.

Maastrichtian interval and to further test the stratigraphy by precise correlation with the independently dated dinoflagellate stratigraphy in the previously studied Braggs and Mussel Creek sections approximately 55 km to the east (Moshkovitz and Habib, 1993). Based on the sequence stratigraphy–calibrated sections in the Braggs area, dinoflagellate species richness and detrital organic matter are used to interpret marine transgression and shoreline progradation in the investigated section.

The second part of the study is to utilize the biostratigraphy to precisely date the rapid erosional event recorded at Moscow Landing and to relate it to the Chicxulub bolide impactor to the south at a time when global sea level was at low stand. Twenty-six samples were studied in the channel section exposed at Moscow Landing (Table 1). Eleven samples were collected within a sandy gravel section within the buried channel. An additional eight samples were studied in the interval directly above the channel, and seven samples were studied in the interval directly below the channel. The investigated section is 3.02 m thick.

RESULTS OF INVESTIGATION

Millers Ferry Core 225

Figure 2 illustrates the stratigraphy of selected dinoflagellate species calibrated by the other microfossil groups. The ranges are presented in stratigraphic succession, based on first occurrences of species. The interval of latest Maastrichtian age is 1.12 m thick. It is characterized by soft gray calcareous marls of the Prairie Bluff Chalk, which become sandy toward the top. This part of the section lies within the upper part of the *Micula prinsii* calcareous nannoplankton zone, which represents the latest Maastrichtian (Olsson et al., 1992). The early Danian is represented by basal calcareous sands of the Clayton Formation. The basal 44 cm contain the earliest Danian P0 planktonic foraminiferal zone. It is overlain directly by the Pα (*Parvularugoglobigerina eugubina*) zone, which is 1.92 m thick in the Millers Ferry core. The next higher interval is 5.4 m thick and extends to the top of the investigated part of the section. This interval is represented by the lower part of planktonic foraminiferal Subzone P1a (*Parasubbotina pseudobulloides*).

The dinoflagellate assemblages in the investigated section of the Millers Ferry core are richer than those occurring in the same interval at Braggs and Mussel Creek, even though the total number of species is about the same (Moshkovitz and Habib, 1993, table 3). The samples are more highly fossiliferous, especially with respect to the species considered to be important for stratigraphic dating. The latest Maastrichtian interval (Fig. 2) contains *Thalassiphora pelagica* (Eisenack) Eisenack and Gocht 1960, *Palynodinium grallator* Gocht 1970, and *Cyclapophysis monmouthensis* Benson 1976. These species correspond to the latest Maastrichtian *Palynodinium grallator* Zone in the Danish K/T stratotype sections and surrounding area (Hansen, 1979; Hultberg, 1986; Schiøler and Wilson, 1993). *P. grallator* is common to abundant in the Millers Ferry core, where it is represented by many well-preserved, even inflated, specimens. This species has also been reported from Spain (DeConinck and Smit, 1982) but is apparently absent in the Tunisian K/T stratotype sections (Brinkhuis and Zachariasse, 1988). The presence of *T. pelagica* and *C. monmouthensis* in the lowest sample, 1.12 m below the K/T boundary, is significant. These species first occur in the *M. prinsii* nannoplankton zone in Alabama. In the U.S. Geological Survey C. D. Hopkins No. 1 Test Hole in southeastern Georgia (Davis-Hopkins well), *T. pelagica* first occurs in the lowest part of this zone and *C. monmouthensis* first occurs in the uppermost part (Moshkovitz and Habib, 1993, after Habib and Miller, 1989). *C. monmouthensis* was also found to first occur in the latest Maastrichtian in the U.S. Geological Survey Albany core in Georgia (Firth, 1993a) and in the El Kef stratotype section in Tunisia (Brinkhuis and Leereveld, 1988).

The K/T boundary is distinguished within the very short interval from FO *Alisocysta circumtabulata* (Drugg) Stover and Evitt 1978 to FO *Senoniasphaera inornata* (Drugg) Stover and

TABLE 1. LIST OF SAMPLES*

Millers Ferry					
Sample	Cm/KTB	Spp./Spec.	Sample	Cm/KTB	Spp./Spec.
MF314	+272	38/320	MF230	+799	18/302
MF316	+252	55/300	MF232	+792	21/300
MF318	+236	60/292	MF240	+747	24/306
MF320	+226	79/278	MF242	+733	30/300
MF322	+215	41/301	MF246	+707	38/314
MF324	+199	36/310	MF250	+686	20/302
MF326	+185	63/326	MF252	+678	24/297
MF328	+173	28/280	MF258	+646	30/314
MF330	+156	35/280	MF262	+613	33/291
MF332	+141	40/287	MF266	+583	41/310
MF334	+123	15/290	MF274	+539	16/320
MF336	+111	20/300	MF276	+524	21/287
MF338	+ 97	29/300	MF282	+487	27/306
MF340	+ 81	18/326	MF286	+455	39/300
MF342	+ 59	26/292	MF288	+438	46/290
MF344	+ 44	13/278	MF290	+421	25/298
MF346	+ 30	18/301	MF294	+385	30/300
MF348	+ 16	21/320	MF296	+377	40/272
MF350	+ 0	13/298	MF298	+371	29/301
MF352	- 18	19/307	MF300	+364	30/310
MF354	- 30	26/322	MF302	+354	31/296
MF356	- 46	14/280	MF304	+340	36/310
MF358	- 60	21/280	MF306	+331	39/324
MF360	- 74	15/322	MF308	+315	34/306
MF362	- 84	18/297	MF310	+298	43/296
MF364	- 98	33/346	MF312	+287	60/301
MF366	-112	39/313			

Moscow Landing					
Sample	Cm/KTB	Spp./Spec.	Sample	Cm/KTB	Spp./Spec.
ML2B	+67-69	0	ML2A-8*	+247-249	56/310
MF2B-7	+61-63	0	ML2A-7	+219-221	47/300
ML2B-8	+50-52	0	ML2A-6*	+214-216	46/318
ML2B-9*	+35-38	0	ML2A-5*	+169-172	75/294
ML2B-10*	+ 0- 2	0	ML2A-4*	+147-149	69/320
ML2B-11*	- 0- 4	14/316	ML2A-3	+144-146	78/270
ML2B-12	-10-12	10/270	ML2A-2*	+132-135	79/300
ML2B-13*	-15-18	16/340	ML2A-1*	+129-132	85/260
ML2B-14	-24-26	15/325	ML2B-1*	+126-129	7/ 49
ML2B-15	-30-32	25/301	ML2B-2	+102-104	14/150
ML2B-16	-37-40	37/278	ML2B-3*	+ 87- 89	3/ 56
ML2B-17	-45-48	12/298	ML2B-4*	+ 81- 83	9/102
ML2B-18	-50-53	25/320	ML2B-5	+ 77- 79	0

*Millers Ferry samples approximately 2 to 4 cm thick. In Moscow Landing section, samples studied for planktonic foraminifera are marked with an asterisk. Samples representing the uppermost 53 cm of the Prairie Bluff Chalk range from ML2B-18 to ML2B-11. Samples from within the basal gravels of the Clayton Formation range from ML2B-10 to ML2B-1. Samples 2A-1 to 2A-8 represent the transgressive facies of the Clayton Formation immediately above the basal gravels. Number of indigenous species per specimen count (Spp./Spec.) listed for each sample.

Evitt 1978. In the Millers Ferry core, *S. inornata* first occurs in sample no. 348 (Table 1), 16 cm above the K/T boundary defined by foraminifera. Based on the average sedimentation rate for the P0 zone in this core calculated by Olsson and Liu (1993), this species may have appeared as much as 13 k.y. (or less if it is present *within* the 16-cm unsampled interval) after the beginning of the Danian. It first occurs 25 cm above the foraminiferal boundary in the section at Mussel Creek (Moshkovitz and Habib, 1993), where cross-laminated sands were deposited at a higher rate of sedimentation. *S. inornata* was first

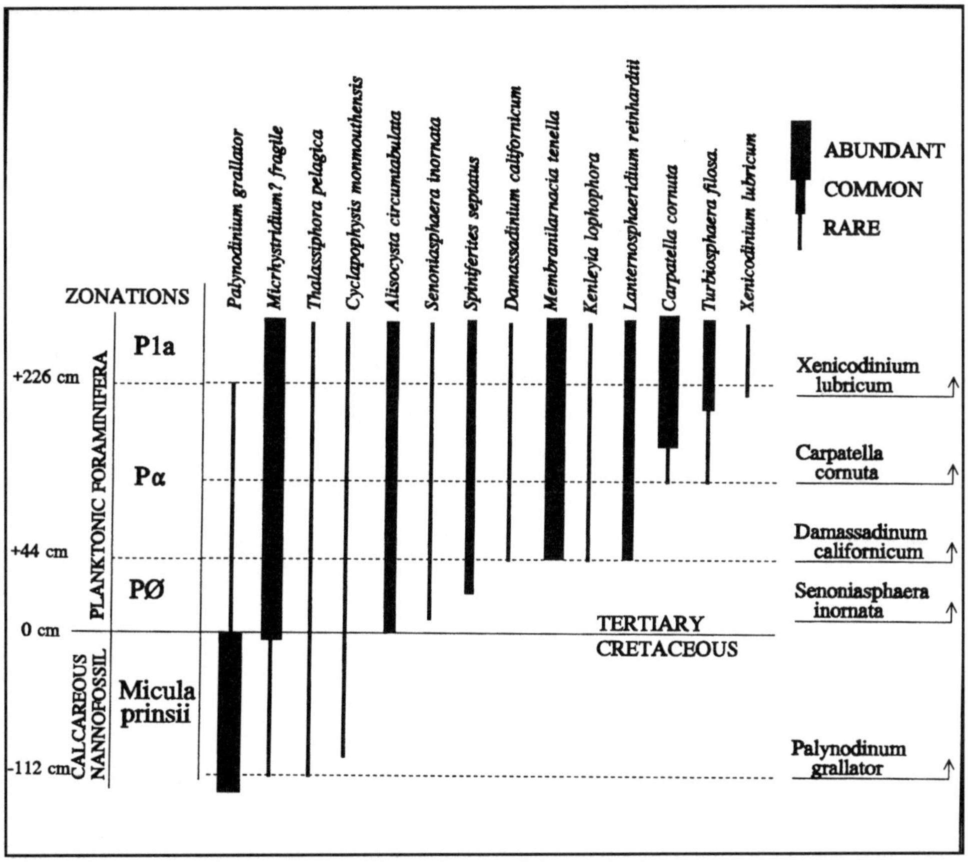

Figure 2. Stratigraphic ranges of selected dinoflagellate species in the Millers Ferry core. Nannofossil-planktonic foraminifera zonation after Olsson and Liu (1993). First occurrence dinoflagellate datums illustrated to the right.

described (as *Chiropteridium inornatum*) in the Clayton Formation of Alabama by Drugg (1970). It first occurs precisely at the K/T boundary in the Danish Basin (Hultberg, 1986, after Hansen, 1979). *A. circumtabulata* first occurs precisely at the K/T boundary in the Millers Ferry core (Fig. 2 and poster presentation at the Lunar and Planetary Institute Snowbird III Conference, Habib, 1994), 16 cm below FO *S. inornata*. *A. circumtabulata* first occurs at the base of the Danian in the El Kef sections, where it defines the lowermost dinoflagellate subzone (Brinkhuis and Zachariasse, 1988). It first occurs also in the early Danian of California, where it was first described (as *Eisenackia circumtabulata*) by Drugg (1967). Elliot et al. (1994) reported *A. circumtabulata* to first occur in the transitional interval across the K/T boundary in Antarctica.

D. californicum first occurs above FO *S. inornata* at the boundary between the P0 and Pα foraminiferal zones at Mussel Creek (Habib et al., 1992; Olsson and Liu, 1993) and in the Millers Ferry core. A number of stratigraphically important species appear at the approximate stratigraphic level of FO *D. californicum*. These include *Membranilarnacia tenella* Morgenroth 1968, *Kenleyia lophophora* Cookson and Eisenack 1965, and *Lanternosphaeridium reinhardtii* Habib 1993. There is a dramatic increase in the total number of dinoflagellate and small acritarch species associated with FO *D. californicum* in the Alabama sections, in samples rich in fecal amorphous debris. Habib et al. (1992) interpreted this increase to have resulted from the first episode of sea-level rise in the Paleocene. This event correlates with the P0/Pα foraminiferal boundary in Alabama. In their study of four samples that cross the K/T boundary in the Barranco del Gredero section in southeastern Spain, DeConinck and Smit (1982) show a very similar succession, from *C. monmouthensis* (plate 1, figs. 4–5, as *Cyclapophysis lemniscata*) and *S. inornata* to *A. circumtabulata* and *M. tenella* (plate 1, fig. 1, as *Adnatosphaeridium* aff. *apenninicum*) to *D. californicum* (as *Danea mutabilis*). *D. californicum* first occurs at the top of the interval of impoverished foraminifera, just below FO *P. eugubina* (basal Pα foraminiferal zone).

Carpatella cornuta Grigorovich 1969 is represented by the thicker-walled, partially tabulated form that is scabrate to distinctly microgranulate (Grigorovich, 1969, p. 74, figs. 2 and 6). This morphotype first occurs in sample 334, 1.23 m above the K/T boundary in the lower half of the Pα zone (Table 1 and Fig. 2). In the Braggs area of Alabama (Fig. 1), the stratigraphic range of this morphotype is restricted to the NP1 calcareous nannofossil zone. It appears in the early Danian of Georgia (Firth, 1993a), the western North Atlantic (Habib and Drugg,

1987), the Danish Basin (Hansen, 1979, figs. 5, a–f), the Carpathians (Grigorovich, 1969), and northern Germany (Stover and Evitt, 1978, after Morgenroth, 1968). *Turbiosphaera filosa* (Wilson) Archangelsky 1969 appears approximately at this level in Alabama. *Xenicodinium lubricum* Morgenroth 1968 represents the highest first occurrence datum in the Pα zone. It appears near the top of this foraminiferal zone, in sample 320, 2.26 m above the K/T boundary (Table 1).

Figure 3 illustrates the dinoflagellate species richness curve in the Millers Ferry core, from 1.12 m below the K/T boundary to 7.99 m above it. The stratigraphic fluctuation of species richness and the nature of the depositional organic matter in the investigated samples are used to interpret marine transgression and regression, according to the evidence provided by sequence stratigraphy and benthic foraminiferal analyses (Donovan et al, 1988; Gregory and Hart, 1992; Olsson and Liu, 1993). Other studies indicate a correlation between dinoflagellate species abundance and sea-level change derived from sedimentologic and lithostratigraphic evidence. These include studies by Edwards (1989), Edwards et al. (1984), Goodman (1979), Firth (1993b), Hultberg and Malmgren (1986), Brinkhuis and Zachariasse (1988), Leckie and Singh (1991), and Li and Habib (1996). In modern environments, dinoflagellate species are fewest in coastal estuarine sediments (McMinn, 1991), despite the abundance of dinocyst specimens, but then increase dramatically offshore in shelf environments (Wall et al., 1977).

The major fluctuations in species richness in the Millers Ferry core (Fig. 3) correlate *in age* with the sequence stratigraphic analysis–calibrated fluctuations in the Braggs and Mussel Creek areas (Moshkovitz and Habib, 1993, text-fig. 3). Species are fewest in the lowstand sands at the base of the Clayton Formation and then increase in number dramatically near the base of the transgressive tract (directly above transgressive lag deposits). The number of species then decreases but remains large until the progradational high-stand tract, where there is a decrease in number of species associated with an increase in the amount of terrigenous organic detritus, inertinite, and degraded dinocysts.

Zigzag fluctuations within the species richness curve are formed by palynological couplets, with an average thickness in the order of 0.5 m to 1.0 m. Each couplet is characterized by a lower part containing an increased number of species associated with amorphous debris and an upper part with fewer species and increased amounts of terrigenous and reworked organic matter. The couplets are interpreted to represent minor cycles of dinoflagellate incursion followed by assemblages of

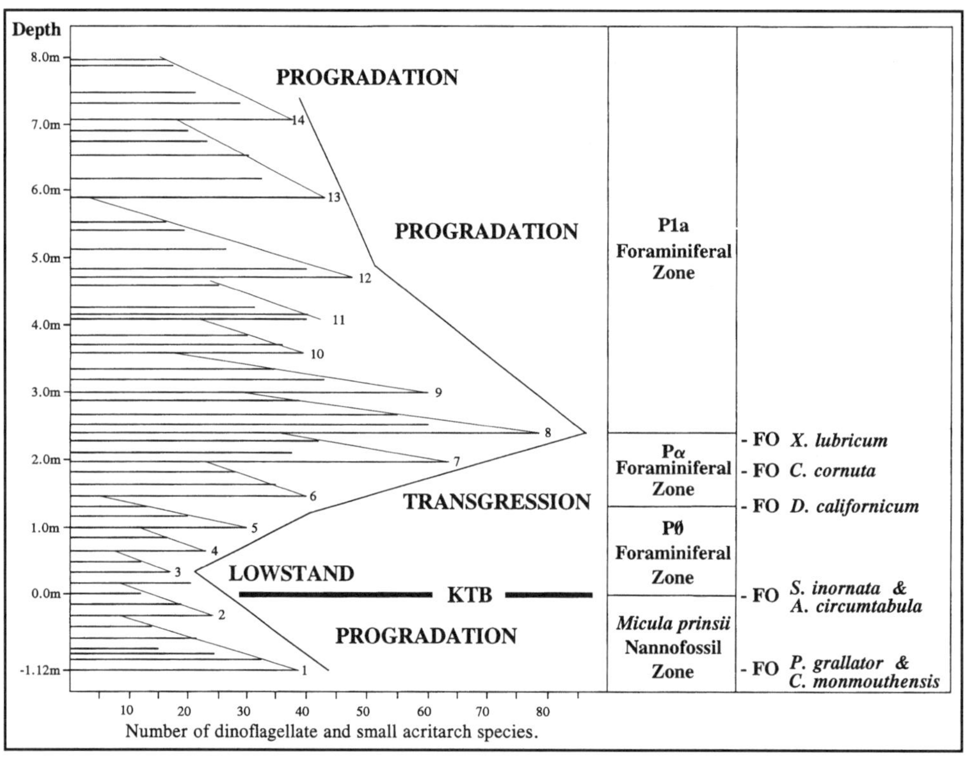

Figure 3. Dinoflagellate–small acritarch species richness curve in Millers Ferry core. Fourteen palynological couplets are distinguished in the 9.1-m interval ranging in age from latest Maastrichtian to early Danian. Each couplet is defined by the largest number of dinoflagellate species at the base and by terrigenous detritus (land plant particles and reworked palynomorphs) in the upper part. Terrigenous detritus is most abundant in the uppermost four samples of each of couplets 12, 13 and 14, where it is associated with brown peridinioid dinocysts. The thickness of these couplets and their terrigenous content imply a higher rate of sedimentation. FO—first occurrence.

prograding organic detritus. The trend of the dinoflagellate incursions in each cycle is used, in turn, to interpret marine transgression and regression that, in the sequence stratigraphy evidence at Braggs, correspond to one third-order depositional sequence. According to this evidence, the trend in the latest Maastrichtian is regressive. Sea level was at low stand at the K/T boundary. Sea level rose in the earliest Danian and transgressed across the Clayton paleoshelf. This transgression was followed by the episode of progradation (Fig. 3).

The interpretations based on dinoflagellate species richness and depositional organic matter correspond to the evidence derived from the comprehensive study of benthic foraminifera in the Millers Ferry core (Olsson et al, this volume) and correlate in age with the palynological interpretations derived from sequence stratigraphy in southwestern Alabama (Donovan et al., 1988; Mancini and Tew, 1993).

Moscow Landing

The outcrop at Moscow Landing is characterized by channel-fill gravels that form the basal lithologic unit (Clayton Basal Sands) of the Clayton Formation. Figure 4 is a photograph of the sections that were sampled and illustrates the channel profile. Figure 5 schematically depicts the two sections (ML2A, ML2B) that were sampled, along with the biostratigraphic datums used to identify the K/T boundary. Section ML2A extends from the upper boundary of the channel fill 120 cm into the overlying shelf deposits of the Clayton Formation. Section ML2B represents a thickness of 182 cm. It extends from the uppermost 53 cm of the underlying Prairie Bluff Chalk through and to the top of the channel deposit.

Figure 4. Photograph of sampled sections (ML2A, ML2B) at Moscow Landing. An erosional unconformity is evident between the channel deposit forming the Clayton Basal Sand unit and the underlying Prarie Bluff Chalk. Gravels are concentrated in the lower half of the channel deposit. Finer-grained sediment in the upper part is heavily burrowed. Geologic pick is shown for scale. It rests against a boulder at the base of the channel deposit.

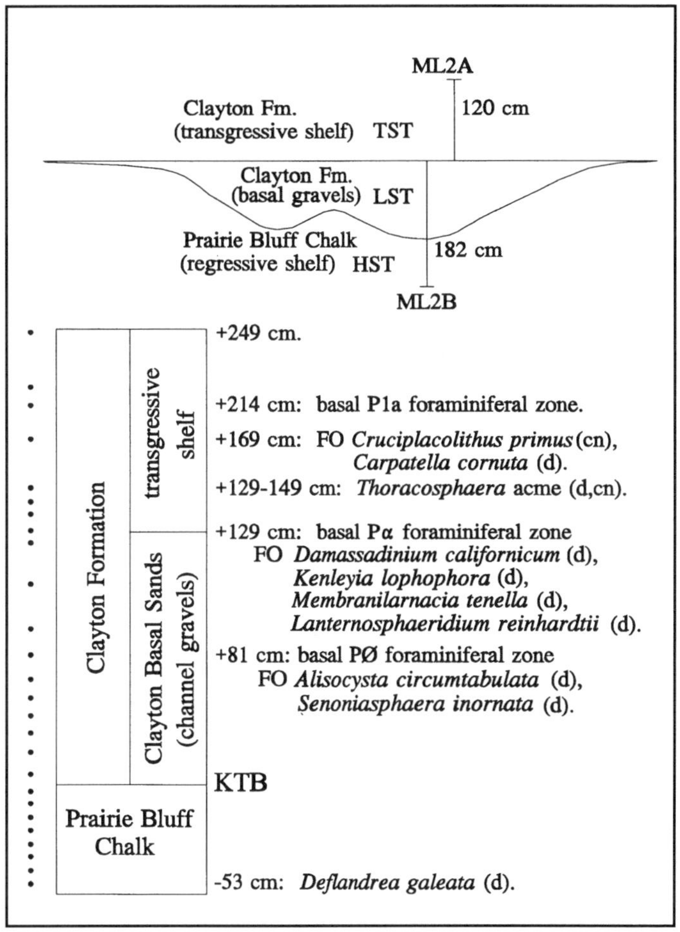

Figure 5. Multiple-fossil biostratigraphic datum across the K/T boundary (KTB) at Moscow Landing. Section sampled in ML2A and ML2B parts of channel profile. Circles to the left of the column indicate sampling levels. d = dinoflagellate, cn = calcareous nannofossil, FO = first occurrence, HST = high-stand systems tract, LST = low-stand systems tract, TST = transgressive systems tract.

At Moscow Landing, the Prairie Bluff Chalk consists of a pale bluish-gray marl. The dinoflagellate *Deflandrea galeata* (Lejeune-Carpentier) Lentin and Williams 1973 occurs in the lowermost sample and –53 cm (Table 1) and ranges discontinuously into the samples of the overlying Clayton Shelf. The uppermost 53 cm of the Prairie Bluff Chalk are characterized by a palynological shelf assemblage containing abundant *Spiniferites* spp., *Areoligera coronata* (Wetzel) Lejeune-Carpentier 1938, *A. medusettiformis* (Wetzel) Lejeune-Carpentier 1938, *Glaphyrocysta retiintexta* (Cookson) Stover and Evitt 1978, and *Cerodinium boloniense* (Riegel) Lentin and Williams 1989. The presence of *D. galeata* indicates an age not older than middle late Maastrichtian (Schiøler and Wilson, 1993). Aurisano (1989) reported FO *D galeata* in the uppermost part of the *Globotruncana gansseri* or lowermost *Abathomphalus mayaroensis* planktonic foraminiferal zones in New Jersey, of late Maastrichtian age. The absence of *Palynodinium grallator*, *Cyclapophysis monmouthensis*, and *Thalassiphora pelagica* in this interval suggests a late Maastrichtian age older than latest Maastrichtian. The presence of *P. grallator* in reworked clasts within the immediately overlying Clayton gravels indicates that sediment of latest Maastrichtian age was eroded.

The lower part of the immediately overlying channel deposit consists of unsorted or very poorly sorted, unoriented pebbles and cobbles packed close together but separated in a matrix of very poorly sorted sand, granules, and pebbles (Fig. 6). The larger clasts are angular to subrounded and consist of the same pale bluish-gray marls that characterize the Prairie Bluff Chalk. Boulders of Prairie Bluff lithology occur near the base of the channel (Fig. 4). Mancini and Tew (1992) reported the occurrence of boulders with diameters up to 1.0 m in channel deposits at Moscow Landing. The pebbles and cobbles in the lower part of the channel contain well-preserved dinoflagellate assemblages of latest Maastrichtian age. In addition to *P. grallator*, the presence of *C. monmouthensis* indicates a latest Maastrichtian age very close to the K/T boundary (Moshkovitz and Habib, 1993). The calcareous nannofossil species *Micula prinsii* occurs in a reworked clast (Sample ML2B-9 in Table 1). which indicates a latest Maastrichtian age. In addition, the clasts contain specimens of *Lithraphidites quadratus* with broader blades and larger specimens of *Arkhangelskiella cymbiformis*, each of which indicates an age high in the late Maastrichtian. Examination of the matrix in the lower part of the channel deposits revealed no additional information of its age. However, both *Senoniasphaera inornata* and *Alisocysta circumtabulata* were found at the level 81 to 83 cm above the base of the deposit, a placement that indicates an earliest Danian age. Foraminifera in the interval from 81 cm to the top of the channel lie within the P0 zone.

The contact between the basal gravels and the overlying shelf facies of the Clayton Formation is sharp. The upper part of the channel deposit is heavily burrowed in finer-grained sediment containing a greater proportion of sand (Fig. 7). The sampled interval directly above the channel contains assemblages rich in dinoflagellate species and specimens. *Damassadinium californicum*, *Membranilarnacia tenella*, *Kenleyia lophophora*, and *Lanternosphaeridium reinhardtii* first occur in the lowest sample (0 to 3 cm) above the channel in an assemblage containing 85 species. This sample also marks the Pα planktonic foraminiferal zone, which correlates precisely with the dinocyst-foraminiferal datum at Millers Ferry. All four samples in the interval from 0 to 20 cm above the channel contain the thoracosphaerid acme and *Biscutum* spp. in otherwise impoverished nannofossil assemblages. *Cruciplacolithus primus* Perch-Neilsen 1977 first occurs at 43 to 45 cm (sample ML2A-5), which indicates a level in the lower part of the NP1 nannofossil zone. The thick-walled morphotype of *Carpatella cornuta* first occurs in this sample, which correlates with the NP1 zone and with the middle part of the Pα foraminiferal zone elsewhere in Alabama (Moshkovitz and Habib, 1993) and at Millers Ferry (Fig. 2).

The K/T boundary is placed at the 81 cm level in the 129-cm-thick channel deposit at Moscow Landing, based on FO *Senoniasphaera inornata* and the foraminiferal P0 zone. However, based on sedimentological evidence, the "boundary" could instead be represented by the 81 cm *interval* between the

Figure 6. Photograph of gravel section in the lower part of the channel deposit at Moscow Landing. Very poor sorting and lack of bedding is indicated by pebbles and cobbles of the Prairie Bluff Chalk in the Basal Sand Unit of the Clayton Formation. Note the random orientation of the larger clasts and the range of grain size in the matrix. After Smit et al., (1994).

Figure 7. Photograph of upper surface of channel deposit. Heavy burrowing of finer-grained sediment is illustrated.

base of the channel (Fig. 5) and the biostratigraphic datums, if these poorly sorted soft marl gravels were deposited rapidly as the result of a catastrophic event. If the formation of the detritus occurred over a span of days as the result of a bolide impact to the south at Chicxulub, then the K/T boundary would be marked by an interval containing no indigenous fossils of Danian age up to the first biostratigraphic evidence of the earliest Danian. Such a deposit could have formed from an impact-generated tsunami.

The finer-grained sediment in the upper 48 cm of the channel and the evidence of bedding suggested by horizontal partings indicate a lower sedimentation rate in the interval from FO *S. inornata* up to FO *D. californicum*. The shelf deposits directly above the channel reflect a continuous stratigraphy of early Danian age, as indicated by the three groups of microfossils.

CONCLUSIONS

The number of datums provided by the three groups of microfossils (dinocysts, foraminifera, nannofossils) presents a high-resolution and precise biostratigraphic dating of the events that occurred across the K/T boundary in Alabama. The accuracy of these datums is successfully tested by their mutual correlation in the Moscow Landing, Millers Ferry, Braggs, and Mussel Creek sections. Based on these correlations, the K/T boundary in Alabama was the time of simultaneous low stand of sea level, mass extinction of benthic macrofaunal and calcareous microplankton marine life, and chaotically deposited coarse detrital sediment. Evidence of sea-level low is taken from benthic foraminiferal assemblages (Olsson et al., this volume), sequence stratigraphy (Donovan et al., 1988), and dinoflagellate species richness. Sea level was already lowering in the latest Maastrichtian as it approached the K/T boundary and did not begin to rise until the Danian. This event correlates in time with the biotic extinctions recorded at Braggs. This correlation suggests that marine regressions at the margins of continents and in the epeiric seaways may have played a contributing factor in the extinction of the shallow-water benthic macrofauna, although not sufficient to be the primary factor. The evidence of a chaotic sedimentological event at the K/T boundary, evident at Moscow Landing and also at Millers Ferry (Olsson et al., this volume), in the Brazos River area of Texas (Bourgeois et al., 1988), and in the Mimbral sections of northern Mexico, suggests a correlation with the Chicxulub bolide impact crater (Hildebrand et al., 1991) in the Yucatan Peninsula of Mexico. These sites form a partial ring in the coastal plain across the Gulf of Mexico and around Chicxulub. Although chaotic sediment may have formed from short-lived catastrophic events such as storms and floods, the position of these sites, their proximity to the Chicxulub crater, and their precise temporal correlation suggest a tsunami event caused by the Chicxulub impactor.

ACKNOWLEDGMENTS

We are grateful to the three reviewers who criticized the manuscript and to Gerald M. Friedman (Graduate School, City University of New York), who criticized the sedimentological interpretation. Leonard J Cinquemani (Queens College, City University of New York) prepared the computer graphics. Supported in part by National Science Foundation Grant EAR 89-03522 and by PSC-CUNY Research Awards Program Grant 664051.

REFERENCES CITED

Askin, R. A., 1988, The palynological record across the Cretaceous/Tertiary transition on Seymour Island, Antarctica: Geological Society of America Memoir 169, p. 155–162.

Askin, R. A., Elliott, D. H., Jacobson, S. R., Kyte, F. T., Li, X., and Zinsmeister, W. J., 1994, Seymour Island: A southern high-latitude record across the K/T boundary, *in* New developments regarding the K/T event and other catastrophes in Earth history: Houston, Texas, Lunar and Planetary Institute Contribution 825, p. 7–8.

Aurisano, R., 1989, Upper Cretaceous dinoflagellate biostratigraphy of the subsurface Atlantic Coastal Plain of New Jersey and Delaware, U.S.A.: Palynology, v. 13, p. 143–179.

Bourgeois, J., Hansen, T. A., Wiberg, L., and Kaufmann, E. G., 1988, A tsunami deposit at the Cretaceous Tertiary boundary in Texas: Science, v. 241, p. 567–568.

Brinkhuis, H., and Leereveld, H., 1988, Dinoflagellate cysts from the Cretaceous/Tertiary boundary sequence of El Kef, northwest Tunisia: Review of Paleobotany and Palynology, v. 56, p. 5–19.

Brinkhuis, H., and Zachariasse, W. J., 1988, Dinoflagellate cysts, sea level changes and planktonic foraminifers across the Cretaceous-Tertiary boundary at El Haria, northwest Tunisia: Marine Micropaleontology, v. 13, p. 153–191.

Bryan, J. R., and Jones, D. S., 1989, Fabric of Cretaceous-Tertiary marine macrofaunal transitions at Braggs, Alabama: Palaeogeography, Palaeoclimatology, Palaeoecology, v. 69, p. 267–277.

DeConinck, J., and Smit, J., 1982, Marine organic-walled microfossils at the Cretaceous-Tertiary boundary in the Barranco del Gredero: Geologie en Mijnbouw, v. 61, p. 173–178.

Donovan, A. D., Baum, G. R., Blechschmidt, G. L., Loutit, T. S., Pflum, C. E., and Vail, P. P., 1988, Sequence stratigraphy setting of the Cretaceous-Tertiary boundary in central Alabama, *in* Sea level changes—an integrated approach: Society of Economic Paleontologists and Mineralogists Special Publication 42, p. 299–307.

Drugg, W. S., 1967, Palynology of the Upper Moreno Formation (Late Cretaceous–Paleocene) Escarpado Canyon, California: Palaeontographica, v. 120, B, p. 1–71.

Drugg, W. S., 1970, Some new genera, species and combinations of phytoplankton from the Lower Tertiary of the Gulf Coast, USA, *in* Ultra microplankton, Proceedings, North American Paleontological Convention, Part G (1969), Chicago: p. 809–843.

Edwards, L. E., 1989, Dinoflagellate cysts from Lower Tertiary formations, Haynesville cores, Richmond County, Virginia: U.S. Geological Survey Professional Paper 1489C, p. C1–C12.

Edwards, L. E., Goodman, D. K., and Witmer, R. J., 1984, Lower Tertiary (Pamunkey Group) dinoflagellate biostratigraphy, Potomac River area, Virginia and Maryland, *in* Frederiksen, N. O., and others, eds., Cretaceous and Tertiary stratigraphy, paleontology, and structure, southwestern Maryland and northeastern Virginia: Reston, Virginia, American Association of Stratigraphic Palynologists Field Trip Volume and Guidebook, p. 137–152.

Elliot, D. H., Askin, R. A., Kyte, F. T., and Zinsmeister, W. J., 1994, Iridium and dinocysts at the Cretaceous-Tertiary boundary on Seymour Island, Antarctica: Implications for the K/T event: Geology, v. 22, p. 675–678.

Firth, J. V., 1993a, Dinoflagellate biostratigraphy of the Maastrichtian to Danian interval in the U.S. Geological Survey Albany core, Georgia, U.S.A.: Palynology, v. 11, p. 199–216.

Firth, J. V., 1993b, Dinoflagellate assemblages and sea-level fluctuations in the Maastrichtian of southwest Georgia: Review of Palaeobotany and Palynology, v. 79, p. 179–204.

Goodman, D., 1979, Dinoflagellate "communities" from the Lower Eocene Nanjemoy Formation of Maryland, U.S.A.: Palynology, v. 3, p. 169.

Gregory, W. A., and Hart, G. F., 1992, Toward a predictive model for the palynologic response to sea-level changes: Palaios, v. 7, p. 3–33.

Grigorovich, A. S., 1969, A new genus *Carpatella* (Dinoflagellata) from Danian-Paleocene deposits of the Carpathians: Paleontologicheskii Sbornik, no. 6, p. 67–72 (in Russian).

Habib, D., 1994, Biostratigraphic evidence of the K/T boundary in the eastern Gulf Coastal Plain, north of the Chicxulub crater, *in* New developments regarding the K/T event and other catastrophes in Earth history: Houston, Texas, Lunar and Planetary Institute Contribution 825, p. 45–46.

Habib, D., and Drugg, W. S., 1987, Palynology of sites 603 and 605. Deep Sea Drilling Project, *in* Initial reports of the Deep Sea Drilling Project, Volume 93: Washington, D.C., National Science Foundation, p. 751–775.

Habib, D., and Miller, J. A., 1989, Dinoflagellate species and organic facies evidence of marine transgression and regression in the Atlantic Coastal Plain: Palaeogeography, Palaeoclimatology, Palaeoecology, v. 74, p. 23–47.

Habib, D., Moshkovitz, S., and Kramer, C., 1992, Dinoflagellate and calcareous nannofossil response to sea level change in K/T boundary sections: Geology, v. 20, p. 165–168.

Hansen, J. M., 1979, A new dinoflagellate zone at the Maastrichtian-Danian boundary in Denmark: Geologiske Undersoegelse Arbog, v. 1978, p. 131–140.

Haq, B. U., Hardenbol, J., and Vail, P. R., 1988, Mesozoic and Cenozoic chronostratigraphy and cycles of sea-level change, *in* Sea level changes: An integrated approach: Society of Economic Paleontologists and Mineralogists Special Publication 12, p. 71–108.

Hildebrand, A. R., Penfield, G. T., Kring, D. A., Pilkington, M., Camargo, A., Jacobsen, S. B., and Boynton, W. V., 1991, Chicxulub crater: A possible Cretaceous/Tertiary boundary impact crater on the Yucatan Peninsula, Mexico: Geology, v. 19, p. 867–871.

Hultberg, S. U., 1986, Danian dinoflagellate zonation, the C-T boundary and the stratigraphical position of the fish clay in southern Scandinavia: Journal of Micropaleontology, v. 5, p. 37–47.

Hultberg, S. U., and Malmgren, B. A., 1986, Dinoflagellate and planktonic foraminiferal paleobathymetrical indices in the boreal uppermost Cretaceous: Micropaleontology, v. 32, p. 316–323.

Jiang, M. J., and Gartner, S., 1986, Calcareous nannofossil succession across the Cretaceous-Tertiary boundary in east-central Texas: Micropaleontology, v. 32, p. 232–255.

Krebs, C. J., 1985, Ecology. The experimental analysis of distribution and abundance (third edition): New York, Harper and Row, 800 p.

Leckie, D. A., and Singh, C., 1991, Estuarine deposits of the Albian Paddy Member (Peace River Formation) and the lowermost Shattesbury Formation. Alberta, Canada: Journal of Sedimentary Petrology, v. 61, p. 835–849.

Li, H., and Habib, D., 1996, Dinoflagellate stratigraphy and its response to sea level change in Cenomanian-Turonian sections of the Western Interior of the United States: Palaios, v. 11, p. 15–30.

Mancini, E. A., and Tew, B. H., 1992, Paleogene unconformity-bounded depositional sequences of southwest Alabama: Lithofacies systems tracts and sequence boundaries: Alabama Geological Society, 29th Annual Field Trip, Guidebook, 72 p.

Mancini, E. A., and Tew, B. H., 1993, Eustasy v. subsidence: Lower Paleocene depositional sequences from southern Alabama, eastern Gulf Coastal Plain: Geological Society of America Bulletin, v. 105, p. 3–17.

McMinn, A., 1991, Recent dinoflagellate cysts from estuaries on the central coast of New South Wales, Australia: Micropaleontology, v. 37, p. 269–287.

Morgenroth, P., 1968, Zur kenntnis der dinoflagellaten und hystrichosphaeridien des Danien: Geologisches Jahrbuch, v. 86, p. 533–578.

Moshkovitz, S., and Habib, D., 1993, Calcareous nannofossil and dinoflagellate stratigraphy of the Cretaceous-Tertiary boundary, Alabama and Georgia: Micropaleontology, v. 39, p. 167–191.

Olsson, R. K., and Liu, C., 1993, Controversies on the placement of Cretaceous-Paleogene boundary and the K/P mass extinction of foraminifera: Palaios, v. 8, p. 127–139.

Olsson, R. K., Liu, C., Aubry, M-P., and Ehrendorfer, T., 1992, A biostratigraphically continuous Cretaceous/Paleogene section at Millers Ferry, Alabama: Support for a bolide impact: Geological Society of America Abstracts with Programs, v. 24, p. A28.

Pospichal, J. J., 1994, Calcareous nannofossils at the K/T boundary, El Kef: No evidence for stepwise, gradual or sequential extinctions: Geology, v. 22, p. 99–102.

Schiøler, P. J., and Wilson, G. J., 1993, Maastrichtian dinoflagellate zonation in the Dan Field, Danish North Sea: Review Palaeobotany and Palynology, v. 78, p. 321–351.

Smit, J., Roep, T. B., Alvarez, W., Montanari, S., and Claeys, P., 1994, Stratigraphy and sedimentology of K/T clastic beds in Moscow Landing (Alabama) outcrop: Evidence for impact-related earthquakes and tsunamis, *in* New developments regarding the K/T event and other catastrophes in Earth history: Houston, Texas, Lunar and Planetary Institute Contribution 825, p. 119–120.

Stover, L. E., and Evitt, W. R., 1978, Analyses of pre-Pleistocene organic walled dinoflagellates: Stanford University Publications, Geological Sciences, v. 15, 300 p.

Traverse, A., and Ginsburg, R. N., 1966, Palynology of the surface sediments of Great Bahama Bank, as related to water movement and sedimentation: Marine Geology, v. 4, p. 417–459.

Wall, D., Dale, B., Lohmann, G. P., and Smith, W. K., 1977, The environmental and climatic distribution of dinoflagellate cysts in modern marine sediments from regions in the North and South Atlantic Oceans and adjacent seas: Marine Micropaleontology, v. 2, p. 121–200.

MANUSCRIPT ACCEPTED BY THE SOCIETY SEPTEMBER 14, 1995

Printed in U.S.A.

Geological Society of America
Special Paper 307
1996

Sulfur and boron isotope study of high-Ca impact glasses from the K/T boundary: Constraints on source rocks

Marc Chaussidon
CRPG-CNRS, BP20, 54501 Vandoeuvre-lès-Nancy, France
Haraldur Sigurdsson
Graduate School of Oceanography, University of Rhode Island, Narragansett, Rhode Island 02882
Nicole Métrich
Laboratoire Pierre Süe, GST CEN-Saclay 91191 Gif-sur-Yvette, France

ABSTRACT

A boron and sulfur isotope study has been conducted by ion microprobe on 20 high-Ca yellow impact glass particles from the K/T boundary deposit in Haiti. The yellow glasses are S-rich with up to 1.24 wt% SO_3 and show a large range of $\delta^{34}S$ values (+1.5 ± 2‰ to +13.2 ± 4‰) and $\delta^{11}B$ values (–9.8 ± 2‰ to +11.0 ± 2‰). In addition to the previously recognized andesite (or continental crust), anhydrite, and carbonate components, continental sedimentary rocks containing clays and diagenetic minerals (e.g., loess or sandstones) are indicated by the high B contents and low $\delta^{11}B$ values of some yellow glasses. This component could correspond to the Todos Santos red-bed formation, recognized by drilling in the Yucatan and likely to be present at Chicxulub. Mass balance mixing calculations (based on SiO_2, CaO, B contents, and $\delta^{11}B$ values) show that all the isotopic characteristics ($\delta^{11}B$, $\delta^{34}S$, and $\delta^{18}O$) of the yellow glasses can be explained by mixing during impact of very variable proportions of andesite (31 to 75%, mean 49%), sulfate (5 to 47%, mean 30%), red sandstone (3 to 26%, mean 14%), and carbonate (0 to 26%, mean 7%). The major element compositions modeled for the yellow glasses, taking into account the large volatilization of SO_2 and CO_2, match closely the natural composition of the yellow glasses. The present observations suggest that the amount of SO_2 released in the atmosphere represents ≈18.0 wt% of the impacted terranes, whereas only ≈2.6 wt% of the impacted terranes is released as CO_2 to the atmosphere.

INTRODUCTION

High-Ca yellow glasses form a volumetrically minor component of the impact spherules in the K/T boundary deposit in Haiti, but these glasses contain valuable information about lithology of the geologic terrane at the impact site. The high-Ca glasses define a mixing trend with abundant high-silica glasses, which led to the hypothesis that they are derived from fusion of Ca-rich sediment in presence of silicate melt (Sigurdsson et al., 1991a). Two different sources of Ca have been proposed for the yellow glasses: either anhydrite-rich evaporitic terrane, as suggested by their high S content and elevated $\delta^{34}S$ (+13.2 ± 4‰; Sigurdsson et al., 1991b), or platform carbonates as suggested by their high $\delta^{18}O$ values (+13.6 ± 0.8‰; Blum and Chamberlain, 1992). These two hypotheses have important and radically different implications for atmospheric and other environmental effects following bolide impact: either global cooling due to sulfate aerosol formation (Sigurdsson et al., 1992) or a greenhouse effect because of increased CO_2 concentration in the atmosphere (O'Keefe and Ahrens, 1989). Because of the correlations observed between S and CaO in the yellow glasses (Sigurdsson et al., 1991b), we conducted a detailed sulfur and boron isotopic

Chaussidon, M., Sigurdsson, H., and Métrich, N., 1996, Sulfur and boron isotope study of high-Ca impact glasses from the K/T boundary: Constraints on source rocks, *in* Ryder, G., Fastovsky, D., and Gartner, S., eds., The Cretaceous-Tertiary Event and Other Catastrophes in Earth History: Boulder, Colorado, Geological Society of America Special Paper 307.

study of 20 high-Ca yellow glass particles in order to constrain their source. We have also carried out isotopic analyses on the analogous synthetic high-Ca yellow glasses, which were produced in the laboratory by 1-atmosphere fusion of andesite with gypsum (Sigurdsson et al., 1991b).

THE K/T YELLOW IMPACT GLASSES

In Haiti the K/T boundary is well exposed in uplifted carbonate sediments that constitute the southern peninsula (Maurrasse et al., 1979; Desreumaux, 1981, 1987; Andreieff and Desreumaux, 1984). In the Beloc region the carbonate sediments contain a microfossil assemblage that suggests a relatively continuous succession of sedimentation across the K/T boundary. A prominent smectite clay layer at the boundary was interpreted initially as "a 50 cm thick volcanogenic turbidite, showing typical subaqueous flow structures" (Maurrasse, 1982, p. 31). On the basis of the discovery of shocked-quartz grains in the upper part of the smectite layer, Hildebrand and Boynton (1990) proposed that it represented an impact ejecta layer and suggested that the smectite clay was the alteration product of impact glass. In 1985, J. Zachos conducted studies of the Beloc formation and sampled the "volcanogenic turbidite layer" of Maurrasse (our sample BE-04), leading to the discovery of impact glass spherules (or micro-tektites) at the K/T boundary (Sigurdsson et al., 1991a; Izett, 1991). The smectite-rich boundary layer varies in thickness from 20 to 120 cm and consists of three units. At the base is a 20- to 30-cm-thick smectite unit, which contains little or no carbonate but contains smectite spherules ranging from 1 to 8 mm, with glassy cores, and this unit represents ballistic ejecta fallout. The second unit consists of 20 to 40 cm of smectite, mixed with carbonate, and also contains abundant smectite spherules. This unit is poorly sorted, is ungraded, and represents a density current deposit. Although dominantly of smectite, there are regions within the unit where glassy cores of spherical to ellipsoidal spherules are well preserved, such as in sample BE-04, studied by Sigurdsson et al., (1991a, b) and analyzed in this work. The topmost unit is a reddish yellow brown smectite, 0.5 to 1 cm thick. This unit contains shocked-quartz particles (Hildebrand and Boynton, 1990) and also the iridium anomaly (Maurasse, 1982; Jéhanno et al., 1992). Some of the black impact glass spherules have an outer coating of yellowish high-calcium glass, which also forms streaks within the black glass spherules. The chemical composition of the dominant black glass from Haiti is similar to that of average continental crust or andesite (Sigurdsson et al., 1991a, b). Glass spherules of nearly identical composition and comparable size have also been found in a similar deposit at the K/T boundary in Mimbral, northeastern Mexico, and in drill cores from the Gulf of Mexico (Smit et al., 1991), indicating that the dispersal of these glass spherules was certainly regional, and perhaps a global event. Rare particles of dark gray high-silica glass are also present in the Haiti deposit (Koeberl and Sigurdsson, 1992). As discussed elsewhere, the chemistry of the impact glasses found in Haiti is consistent with their derivation from the Chicxulub impact structure in the Yucatan, Mexico, where the geologic succession is also in agreement with the nature of the impact target that has been inferred from geochemistry of the Haiti glasses (Sigurdsson et al., 1992).

ANALYTICAL TECHNIQUES

Sulfur contents in natural and synthetic high-Ca glasses were measured with a Camebax electron probe (Camparis service, Paris), operating with 15 kV accelerating voltage, 30 nA beam current, and a spot size of ≈10 μm. The position of sulfur Kα peaks in standards (pure sulfate and sulfide) and yellow glasses were determined by spectral scans. Counting times for peak and for background on either side of the peak were 30 sec.

Boron content and isotopic composition and sulfur isotopic composition were determined with a Cameca ims3f ion microprobe at Centre de Recherches Petrographiques et Geochimiques–Centre National de la Recherche Scientifique (CRPG-CNRS) (Nancy, France). The techniques for B analyses in glasses have been previously described (Chaussidon and Libourel, 1993; Chaussidon and Jambon, 1994). The B contents are determined with an accuracy of ±5% in the range 10 to 30 ppm and the $\delta^{11}B$ values with an accuracy of ±1.5‰ (1 sigma) relative to NBS-951 standard and a precision of about ±1 to ±1.5‰. The techniques developed for the measurement of S isotope composition in sulfides (Chaussidon and Lorand, 1990) were slightly modified for the analysis of glasses. Counting times on ^{32}S and ^{34}S peaks were increased to 4 and 8 sec per cycle, respectively, for typical counting rates on peak ^{32}S of 2,000 to 3,000 counts/sec (with 25 nA O-primary beam and mass resolution of ≈4,000). One $^{34}S/^{32}S$ measurement lasts approximately 90 min.

Synthetic glass standards were prepared by fusion in a vertical quench furnace at 1 atmosphere of mixtures of andesite and gypsum (Sigurdsson et al., 1991b). Glass #1 was synthesized at 1,208 °C and has variable S content (between 0.07 and 1.2 wt% SO_3). Glass #4 was synthesized at 1,306 °C and has a homogeneous S content of 0.95 ± 0.05 wt% SO_3. Conventional sulfur isotope analysis of the gypsum that was used in the melting experiments gives a $\delta^{34}S$ of +24.45 ± 0.04‰ relative to Canyon Diablo Troilite (CDT) standard. Instrumental mass fractionation (α_{inst}) for S isotopes during ion probe analysis was determined from the ion probe $^{34}S/^{32}S$ analysis of these synthetic glasses, which have major and trace element compositions very similar to the yellow glasses (Table 1). The homogeneity, within the analytical uncertainty, of the $^{34}S/^{32}S$ ratio of the synthetic glasses and the lack of relationship between the S contents and the $^{34}S/^{32}S$ ratios (Table 1) show that no significant isotopic fractionation occurred during the synthesis of the glasses. Assuming that the $\delta^{34}S$ value of these glasses is that of the starting gypsum, a value of 0.9546 (±2.8‰) is obtained for α_{inst} (Table 1). This value of α_{inst} is similar to values found for sulfides (Chaussidon and Lorand, 1990).

The $\delta^{34}S$ of the yellow glasses were determined from the average of three to five individual spots on each glass chip, and they are given with 1 sigma of ±2‰ or ±4‰ (Table 2). The

TABLE 1. ION PROBE $^{34}S/^{32}S$ ANALYSIS OF SYNTHETIC STANDARD GLASSES

Sample	$^{34}S/^{32}S$ (x10^{-2})	1 sigma (x10^{-2})	^{32}S intensity (counts/sec)
Synthetic glass #1			
	4.4058	0.0240	1914.4
	4.4050	0.0332	916.2
	4.3908	0.0415	1093.3
	4.3813	0.0293	970.1
Synthetic glass #4			
Grain no. 1	4.3857	0.0360	521.1
	4.3896	0.0332	417.3
	4.4074	0.0430	593.5
	4.4185	0.0400	321.3
Synthetic glass #4			
Grain no. 2	4.4153	0.0082	2642.1
	4.4092	0.0067	2711.7
	4.3980	0.0131	2483.2
Average	4.4006	0.0123	
Mass discrimination:*	0.9546 (± 2.8 ‰)		

*Mass discrimination is given as α_{inst} with: α_{inst} = ($^{34}S/^{32}S$ measured/ $^{34}S/^{32}S$ true).

TABLE 2. B, S CONTENTS AND ISOTOPIC COMPOSITIONS OF HAITI HIGH-CA IMPACT GLASSES

Sample	B (ppm)	SO_3 (Wt. %)	$\delta^{11}B$* (‰)	$\delta^{34}S$ (‰)
BE4-av-1	15.2	0.479	-4.5	1.5 ± 2
BE4-av-2	11.7	0.345	5.4	5.7 ± 4
BE4-av-3	13.3	0.551	-8.4	2.7 ± 2
BE4-av-6	28.6	1.243	0.4	6.3 ± 2
BE4-av-7	11.5	0.138	-3.1	3.2 ± 4
BE4-av-8	18.6	1.095	-7.4	5.4 ± 4
BE4-av-9	16.2	0.568	-9.8	5.4 ± 4
BE4-av-10	23.1	1.245	3.3	6.0 ± 2
BE4-av-13	10.2	0.773	0.9	3.1 ± 4
BE4-av-14	13.8	0.713	-0.3	5.9 ± 4
BE4-av-15	16.8	0.654	-4.6	11.1 ± 2
BE4-av-16	14.4	0.601	-4.6	3.0 ± 4
BE4-av-17	13.1	0.768	-5.9	11.2 ± 4
BE4-av-18	10.4	0.486	11.0	6.6 ± 4
BE4-av-19	11.0	0.371	1.1	2.0 ± 4
BE4-av-22	17.4	0.922	-7.8	1.9 ± 4
BE1-4†	28.0	0.83	-2.1	13.2 ± 4

*$\delta^{11}B$ values are determined at ±2‰ (see text).
†Data from Sigurdsson et al., 1991b.

poorer reproducibility of most of the yellow glasses (1 sigma = ±4‰), in comparison with the synthetic glasses of similar S contents, suggests that a significant part of this variation is due not to analytical techniques but to significant isotopic heterogeneity of the natural glass samples.

RESULTS

B and S contents

The B and S contents of the yellow glasses are given in Table 2. They range between 10.4 and 28.6 ppm for B and between 0.138 and 1.243 wt% for SO_3. These B contents are lower than the range previously reported for Muong-Nong–type tektites, which are thought to originate from a B-rich source rock containing river or deltaic clay-rich sediments, but are similar to the range of the impact glasses from Aouelloul and Darwin areas, which are believed to have derived their B from a sandstone component in the source (Koeberl, 1992). The yellow glasses are S-saturated or close to saturation according to melting experiments of mixtures of andesite and gypsum (Sigurdsson et al., 1991b). However, correlations exist between S and other elements, such as B (Fig. 1), or other major elements (Fig. 2, Tables 2 and 3). This indicates that the S and B contents of the yellow glasses are also likely controlled by the composition of their source material. Sulfur content, however, is likely to be primarily controlled by its temperature-dependent solubility. The spectral position of the sulfur Kα line in the yellow impact glasses shows that sulfur speciation is ≈94.5% sulfate in the high-Ca yellow glasses (sulfate/sulfide = 17.2), confirming their derivation from a highly oxidized source rock.

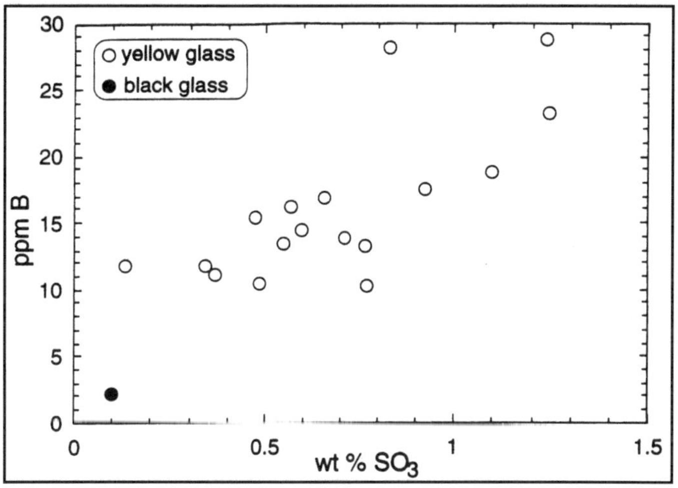

Figure 1. B (ppm) versus S (wt% SO_3) contents of the yellow glasses (open circle). The yellow glasses have high B and S contents—much higher than the black glasses (solid circle; data from Sigurdsson et al., 1991b). B and S contents are positively correlated, suggesting B-rich and S-rich source rocks, likely sedimentary rocks, in the impacted terrane.

B and S isotopic compositions

As shown in Table 2, a very large range of variation is present in the yellow glasses, both for B isotopes ($\delta^{11}B$ values ranging between −9.8 ± 2‰ and +11.0 ± 2‰) and for S isotopes ($\delta^{34}S$ values ranging between +1.5 ± 2‰ and +13.2 ± 4‰). Contrary to the B and S contents, which are similar to some volcanic glasses, the variability of the B and S isotopes in the yellow glasses is a factor of ≈5 times higher than what has been observed up to now in volcanic glasses (Taylor, 1986; Palmer, 1991). This range for B

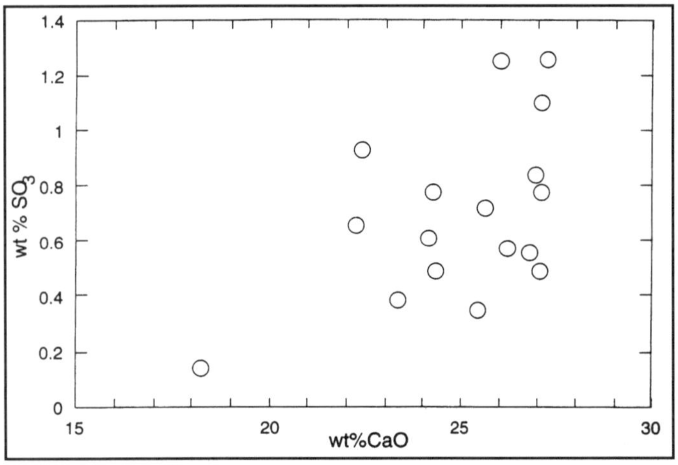

Figure 2. Sulfur contents versus CaO contents for the yellow glasses. The highest S contents are found for the highest CaO contents, which suggests anhydrite as the source of S. However, S contents vary by a factor of ≈3 for a given CaO content, probably because of variable volatilization of S during impact and of the presence of carbonate in the source.

isotopes is also higher than the range of variation in other tektites and impact glasses from the four known strewn fields (the North American, Central European, Ivory Coast, and Australasian strewn fields) that have $\delta^{11}B$ values between –9.3 ± 1.5‰ and +2.7 ± 1.5‰. No data are available at present for S isotopes in other tektites or impact glasses, but the present results are consistent (for both B and S) with our previous measurements made on one yellow glass (Sigurdsson et al., 1991b). Two positive trends are observed between $\delta^{34}S$ and $\delta^{11}B$ values for the yellow glasses, a high $\delta^{11}B$–low $\delta^{34}S$ trend and a low $\delta^{11}B$–high $\delta^{34}S$ trend (Fig. 3). No simple relationship is observed between $\delta^{34}S$ and S contents (Table 2) but a negative correlation is observed between B contents and $\delta^{11}B$ for most of the yellow glasses (Fig. 4), whatever their $\delta^{34}S$ and $\delta^{11}B$ values are.

Isotopic fractionation during melting and quenching

In order to apply the B and S isotopic compositions of the yellow glasses to constrain the nature of the source rocks, the isotopic fractionation that may occur during the loss of B and S by volatilization during impact and/or degassing during cooling and quenching must be evaluated.

Sulfur: The large difference in S content between the yellow glasses (0.14 to 1.24 wt% SO_3) and a pure anhydrite endmember (59 wt%) would imply that most of the sulfur was devolatilized during the impact if anhydrite was present in large amounts in the source rocks. However, it seems unlikely that this process fractionated the S isotopes to any great extent, as shown by the constant $\delta^{34}S$ values of the synthetic yellow glasses produced during melting experiments with anhydrite and silicate melt (Sigurdsson et al., 1992, table 1). In addition, further S degassing from the impact melts can only change the $\delta^{34}S$ of the melts by 1 or 2 permil at most. The change in $\delta^{34}S$ of a melt during S degassing can be described by a simple Rayleigh fractionation process:

$$\delta^{34}S_{yg} = \delta^{34}S_0 + \Delta_{SO_2\text{-melt}} * \ln([SO_3]_{yg} / [SO_3]_0)$$

with $\delta^{34}S_0$ and $[SO_3]_0$ being the sulfur isotope composition and S content of the melt before sulfur degassing. The $\Delta_{SO2\text{-melt}}$ depends on (1) the speciation of S in the melt (94.5% sulfate for one high-Ca yellow glass) and (2) the temperature of the melt. For these conditions we have (Zheng, 1990):

TABLE 3. MAJOR ELEMENT COMPOSITION OF HAITI HIGH-CA IMPACT GLASSES
Electron microprobe analyses; in wt%

Sample	SiO_2	TiO_2	Al_2O_3	FeO	MnO	MgO	CaO	Na_2O	K_2O	P_2O_5
BE4-av-1	46.9	0.59	12.33	4.94	0.06	4.37	27.03	2.38	0.58	0.39
BE4-av-2	49.1	0.61	12.53	5.16	0.15	4.08	25.49	2.25	0.65	0.29
BE4-av-3	48.0	0.64	12.58	4.95	0.13	4.47	26.8	1.97	0.56	0.26
BE4-av-6	46.6	0.65	12.56	4.91	0.18	4.49	27.3	1.87	0.58	0.35
BE4-av-7	53.5	0.58	14.29	5.09	0.08	3.59	18.28	2.84	1.17	0.30
BE4-av-8	46.8	0.64	12.52	5.04	0.16	4.35	27.12	1.93	0.56	0.40
BE4-av-9	47.5	0.64	12.94	4.99	0.11	4.28	26.25	2.24	0.66	0.40
BE4-av-10	48.1	0.60	12.97	5.02	0.05	4.48	26.07	2.03	0.61	0.35
BE4-av-13	47.6	0.63	12.89	4.56	0.11	3.86	27.08	2.18	0.64	0.36
BE4-av-14	48.3	0.63	12.98	5.08	0.06	4.15	25.67	2.43	0.72	0.40
BE4-av-15	51.7	0.61	13.36	4.55	0.13	3.66	22.33	2.73	0.94	0.31
BE4-av-16	49.5	0.69	13.52	5.01	0.10	4.33	24.15	2.15	0.68	0.29
BE4-av-17	48.7	0.66	13.29	5.16	0.11	4.18	24.26	2.28	0.77	0.30
BE4-av-18	48.9	0.67	13.25	5.04	0.14	4.07	24.34	2.48	0.80	0.33
BE4-av-19	49.0	0.68	13.43	5.06	0.13	4.19	23.37	2.36	0.74	0.37
BE4-av-22	48.7	0.71	13.75	4.82	0.13	3.99	22.42	2.7	0.92	0.38
BE1-4*	46.8	0.61	12.50	4.42	n.d.	4.27	27.00	2.36	0.62	n.d.

*Data from Sigurdsson et al., 1991b.
n.d. = not determined.

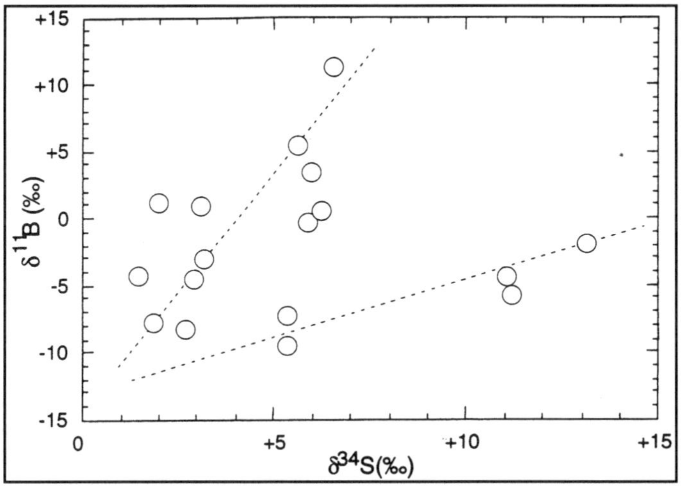

Figure 3. A plot of $\delta^{11}B$ versus $\delta^{34}S$ for the yellow glasses (data from this study and from Sigurdsson et al., 1991b). The yellow glasses present a very large range of variation for both isotopic ratios, and two trends can be seen: a high $\delta^{11}B$–low $\delta^{34}S$ one and a high $\delta^{34}S$–low $\delta^{11}B$ one. These relationships can be well explained from the mixing in different proportions of "andesite" or continental crust, anhydrite (high $\delta^{34}S$ component), carbonate (high $\delta^{11}B$ component) and sandstone (low $\delta^{11}B$ component), in the source rocks (see text).

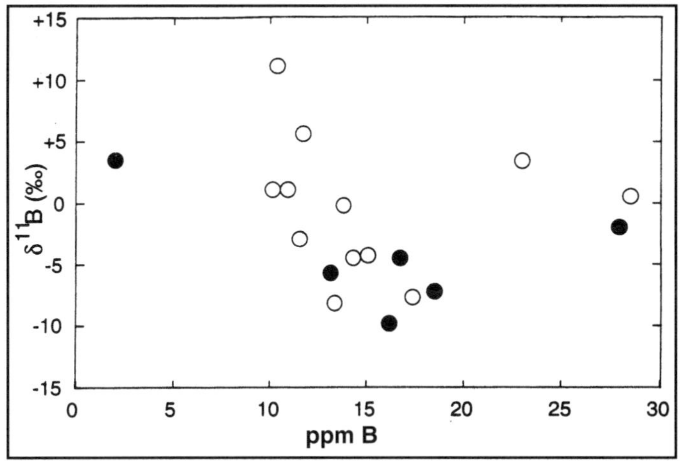

Figure 4. A plot of $\delta^{11}B$ versus B contents for the yellow glasses. Open symbols are for the high $\delta^{11}B$–low $\delta^{34}S$ yellow glasses and shaded symbols for the high $\delta^{34}S$–low $\delta^{11}B$ ones. A negative trend is observed for all except three of the samples. The relationships between $\delta^{11}B$ and B contents are best explained by mixing processes during melting of a heterogeneous source (see text).

$$\Delta_{SO_2\text{-melt}} = -1.43 * 10^6/T^2 - 0.5$$

Assuming that the yellow glasses are S-saturated, their quenching temperatures can be derived from the experimental relationship between the solubility of S and the temperature of the melt, as shown by the fusion of mixtures of andesite and gypsum (Sigurdsson et al., 1992). This indicates that the glasses were quenched in the range 1,250 to 1,360 °C. Therefore, $\Delta_{SO_2\text{-melt}}$ can be bracketed between –1.1‰ and –1.0‰. This implies a maximum isotopic shift of +2.4‰ due to degassing, over the range of S contents (0.14 and 1.24wt% SO_3) for the yellow glasses. This shift is much lower than the total range of $\delta^{34}S$ values (≈10‰) observed for the yellow glasses, and such a process would produce a negative correlation between $\delta^{34}S$ versus wt% SO_3, which is not observed (Table 2). In addition, isotopic fractionation during degassing cannot produce the range of $\delta^{34}S$ values observed in the yellow glasses, if we adopt the hypothesis of evaporitic anhydrite in the source rock. A minimum $\delta^{34}S$ for Cretaceous marine anhydrite is about +12‰ (Claypool et al., 1980), and any degassing fractionation should therefore increase the $\delta^{34}S$ values of the impact melts to above +12‰. Therefore, degassing is not considered the major process that controls the $\delta^{34}S$ of the yellow glasses, and mixing between rocks of different $\delta^{34}S$ values seems more likely the dominant process.

Boron. Some loss of B by devolatilization during the impact and/or by degassing of the impact melts is suggested by the fact that B and S contents of the yellow glasses are correlated and that the S contents are in part controlled by these processes. However, three lines of evidence argue against a strong effect of devolatilization and/or degassing on the $\delta^{11}B$ values of the yellow glasses:

1. Contrary to S, the yellow glasses, which contain between ≈10 and ≈28 ppm B, are strongly undersaturated with respect to B: Silicate melts of similar major element composition are able to dissolve much higher amounts of B (2.8 wt% B, for instance, in synthetic melts produced by Chakraborty et al., 1994).
2. The trends observed for some of the samples between $\delta^{11}B$, ppm B, wt% SO_3, and $\delta^{34}S$ cannot be explained by isotopic fractionation during degassing.
3. Devolatilization and degassing processes have been shown, from a study of tektites and impact glasses from all the major occurrences known, to be able to decrease the B contents by a factor of 2 or 5 without changing significantly the $\delta^{11}B$ values (1 or 2 per mil). From this study the B isotopic fractionation between vapor and melt ($\Delta_{vapor\text{-}melt}$) was estimated to be close to 0‰ (and less than 3‰ in absolute value). In fact, the rim of one flanged australite tektite, which is supposed to have been remelted in the high atmosphere, has $\delta^{11}B$ values similar to those of Muong-Nong–type tektites and thailandites, which represent the range of impact melts produced by the same impact. For the K/T boundary yellow glasses, a completely unrealistic value of ≈–33‰ for $\Delta_{vapor\text{-}melt}$ would be necessary to explain the trend of decreasing $\delta^{11}B$ with decreasing B content (Fig. 4), by a Rayleigh-type isotopic fractionation during degassing.

B AND S ISOTOPIC CONSTRAINTS ON IMPACTED ROCKS

Nature of the source rocks

Previous studies on the geochemistry of the Haitian impact glasses (Sigurdsson et al., 1991a; Blum and Chamberlain, 1992; Koeberl and Sigurdsson, 1992) have shown that the three main types of glasses described (black glasses, yellow glasses, and high Si-K glasses) can be the result of melting of various source rocks (continental crust + sediments, continental crust + carbonate and anhydrite-rich sediments, and shales or argillites, respectively). In addition, the fact that the different types of glasses are found mixed together in the same ejecta deposit and that the yellow glass occurs in some cases as bands or schlieren in the black glass (Sigurdsson et al., 1991a) suggests that several different source rocks were melted on the impact and that the melts produced were only partially mixed.

Because of the lack of evidence of significant isotopic fractionation during the formation of the glasses (see above), the range of variations of $\delta^{34}S$ (from +1.5 to +13.2‰) and $\delta^{11}B$ values (from –9.8 to +11.0‰) found for the yellow glasses implies isotopically variable impacted terranes. The high $\delta^{34}S$ and high $\delta^{11}B$ values, associated with high B (up to ≈30 ppm) and S (up to 1.24 wt% SO_3), require the presence of marine carbonates and marine sulfates. Late Cretaceous sulfates have a $\delta^{34}S$ value of ≈+12‰ (Claypool et al., 1980). No data are available on the $\delta^{11}B$ value of Cretaceous carbonates and sulfates, but no significant change in $\delta^{11}B$ value of seawater is expected in the Cenozoic (Chaussidon and Albarède, 1992; Ishikawa and Nakamura, 1993). Present-day marine carbonates and sulfates have different B isotopic compositions because of the isotopic fractionation taking place during their precipitation from seawater (Vengosh et al., 1991, 1992; Ishikawa and Nakamura, 1993). However, both have similar positive $\delta^{11}B$ values of ≈+10‰ for carbonates and ≈+12.5‰ for sulfates, with B contents of ≈25 ppm for carbonates and ≈5 ppm for sulfates.

The presence of another component with a negative $\delta^{11}B$ value <–10‰ and a B content >10 ppm is required to account for the composition of some of the yellow glasses. This component is likely not a marine sediment, because modern pelagic and neritic sediments have $\delta^{11}B$ values between –6.2 and –1.8‰ and because lower $\delta^{11}B$ values are only found in old marine sediments that have undergone isotopic fractionation during diagenesis (Ishikawa and Nakamura, 1993). Such sediments were likely not present within the source rocks at the time of the impact. On the other hand, this low $\delta^{11}B$ value falls within the range estimated from analyses of granitic tourmalines for the "mean continental crust" (between ≈–13 and ≈–8‰, Chaussidon and Albarède, 1992) but requires slightly higher B contents than those typical of continental magmatic rocks (Harder, 1978). It corresponds well to continental sediments for which, however, very few bulk B isotope analyses are available. Soils, loess and river or delta sediments have $\delta^{11}B$ values ranging between –12.4‰ and +1.9‰ and have high B contents, up to 90 ppm (Spivack et al., 1987; Ishikawa and Nakamura, 1993). Borate minerals from continental evaporites have variable but low $\delta^{11}B$ values, between 0 and –22‰ (Swihart et al., 1986). Moreover, most continental sediments are expected to have low $\delta^{11}B$ values because (1) detrital minerals such as tourmalines retain their magmatic low $\delta^{11}B$ values (Chaussidon et al., 1991), (2) diagenetic minerals have lower $\delta^{11}B$ values than primary minerals or fluids because of B isotopic fractionations occurring at low temperatures (Spivack et al., 1987), and (3) clay minerals have low $\delta^{11}B$ values because they are at equilibrium with low $\delta^{11}B$ continental waters. In addition, the low $\delta^{11}B$ component cannot correspond to the andesitic or continental crustal component, which has been established from the major and trace element composition of the black and yellow glass, because the black glass that is the closest in composition to average continental crust or andesite has a very low B content (2 ppm; i.e., an order of magnitude lower than the yellow glasses) and a $\delta^{11}B$ value of +3.4‰ (Sigurdsson et al 1991b).

The rock components indicated by the B and S isotopes—marine carbonates, marine evaporites, and continental sediments—are in accordance with the derivation of the yellow glasses from the Chicxulub structure, as already suggested by several other geochemical parameters (Sigurdsson et al., 1992). However, unlike carbonate and anhydrite, in previous studies continental sediments have not been considered as an important source component. The continental sediment component observed in isotopic geochemistry of the K/T boundary impact glasses could well correspond to the Todos Santos red-bed formation, which has been found by drilling to underlie the Cretaceous carbonate and evaporite deposits in the Yucatan region, including the Chicxulub area. A high B content of ≈50 ppm for this type of formation can be postulated from the B contents of clays and loess, most of the B probably being present in the cement between the grains or in the diagenetic minerals. Harder (1978) gives a range of variations for sandstone between 5 and 70 ppm B. However, further geochemical studies of these red-bed formations are required to ascertain their composition. The Todos Santos Formation is a red bed sequence, which began to accumulate in the Jurassic and generally overlies the crystalline basement in the Yucatan and southern Mexico (Donnelly et al., 1990). It is commonly a few meters to tens of meters thick in the Yucatan, and limestones and evaporites occur sporadically near the top. The Todos Santos Formation fringed a vast salt basin in Mexico, developed during the separation of North and South America, and in well Y-1 in north Yucatan there are more than 400 m of Jurassic red-beds of this type (Bass and Zartman, 1969; Viniegra, 1971). Viniegra (1981) considers the Todos Santos Formation to be of Middle Jurassic age (~150 Ma), and to consist largely of continental red beds, or Lecho Rojos. The Todos Santos Formation also contains some volcanic material, and the red beds are diachronous and are of Late Jurassic age in southeastern Mexico and Guatemala.

Proportions of "andesite," carbonate, sulfate, and red sandstone in the impacted rocks

Unlike S, which might be largely lost by volatilization during the formation of the yellow glasses if anhydrite is present in the source rock, B is only present as a trace constituent in the yellow glasses and might not have been strongly devolatilized.

The large $\delta^{11}B$ variation can be used to make a first-order estimate of the proportions of the different rocks melted during the impact. Four possible rock types, as suggested by B and S isotope data, are considered in this approach: "andesite" (or continental crust), carbonate, sulfate, and red sandstone. The fractions of these different components in the source (f_{and}, f_{carb}, f_{sulf}, and f_{sand}, respectively) can be obtained from mass balance equations written for parameters that are typical of each component, SiO_2, CaO and B contents and $\delta^{11}B$ values. The mass balance equations are of the simple form:

$$[SiO_2]_{yg} = [SiO_2]_{and} * f_{and} + [SiO_2]_{carb} * f_{carb} + [SiO_2]_{sulf} * f_{sulf} + [SiO_2]_{sand} * f_{sand}$$

and for the isotopic compositions:

$$[B]_{yg} * \delta^{11}B_{yg} = [B]_{and} * \delta^{11}B_{and} * f_{and} + [B]_{carb} * \delta^{11}B_{carb} * f_{carb} + [B]_{sulf} * \delta^{11}B_{sulf} * f_{sulf} + [B]_{sand} * \delta^{11}B_{sand} * f_{sand}$$

where $\Sigma f = 1$.

The composition of andesite is taken as that used in melting experiments of Sigurdsson et al. (1991b), a Mexican andesite with 58% SiO_2 and 6.8% CaO (Cantagrel et al., 1984). Its B content and isotopic composition are assumed to be similar to that of black glass BE-04 (Sigurdsson et al., 1991b). It is also assumed in this first-order estimate that $[SiO_2]_{carb} = [SiO_2]_{sulf} = 0$, that $[CaO]_{carb} = 56$ wt% and that $[CaO]_{sulf} = 41$ wt%. The composition of the red sandstone component is taken as a mean between sandstone in general (Pettijohn et al., 1972) and North American loess, which would represent the clay fraction (Taylor et al., 1983). B contents and isotopic compositions assumed for marine carbonate, marine sulfate and sandstone are:

$$[B]_{carb} = 25,\ [B]_{sulf} = 5,\ [B]_{sand} = 50,\ \delta^{11}B_{carb} = +10‰,\ \delta^{11}B_{sulf} = +12.5‰ \text{ and } \delta^{11}B_{sand} = -10‰.$$

The results of this mass balance calculation are shown in Figure 5. It must be stressed that this aproach does not postulate any "a priori" proportion of the different components in the source rock. The calculation gives consistent results for the majority of the yellow glasses analyzed (11 out of 17), which are, interestingly, all of the glass samples present on the low-$\delta^{34}S$ high-$\delta^{11}B$ trend of Figure 3. This suggests that for some high-$\delta^{34}S$ yellow glasses, that is, those on the other trend in Figure 3, these four components do not adequately describe the source or that some chemical fractionation may have occurred in addition to simple mixing. For four of these samples, consistent results are obtained by mixing only three components: andesite, anhydrite, and sandstone (Figure 5). It can be seen from Figure 5 that the fractions of the different components in the source are very variable but generally decrease from "andesite" (31 to 75%, mean 49%), to sulfate (5 to 47%, mean 30%), to red sandstone (3 to 26%, mean 14%) and to carbonate (0 to 26%, mean 7%). The fractions of "andesite" generally parallel those of red sandstone, and the fractions of carbonates and sulfates are also broadly positively correlated.

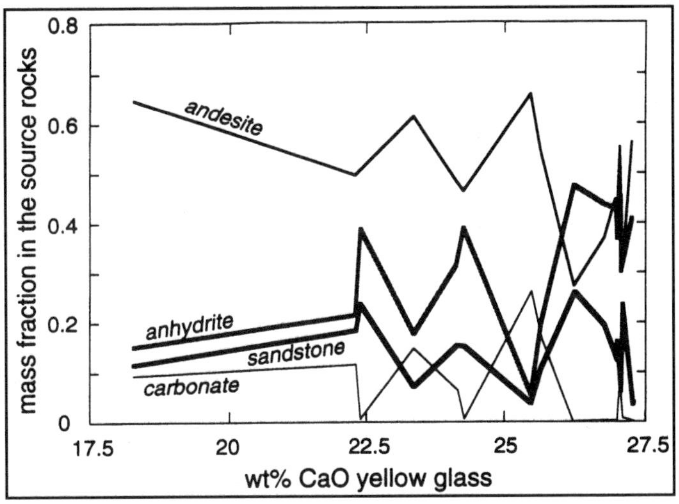

Figure 5. Fractions of the different components in the source rocks of the yellow glasses, given by the mixing calculation based on SiO_2, CaO, and B contents and B isotopes only (see text). The fractions of the different components decrease from "andesite" or continental crust, to anhydrite, carbonate, and sandstone. Note that the fractions of andesite and sandstone show parallel variations, which are anti-correlated to the variations of the fractions of anhydrite and carbonate.

The fraction of carbonate found in this approach is lower than the estimate obtained from $\delta^{18}O$ data by Blum and Chamberlain (1992), who estimated ≈34% platform carbonate for a yellow glass with a CaO content of 27 wt%. Two glasses with about 27 wt% CaO correspond to 5 and 25% carbonate in the model given above. This range of variation in the estimate of source rock composition clearly shows the heterogeneity of the impacted material and the lack of homogenization during melting. The difference between the present estimate and that of Blum and Chamberlain is due to the fact that they did not consider sulfate as a major constituent in the impacted rocks. Their argument against a sulfate source came from the $\delta^{18}O$ values of the yellow glasses, ranging between +13.1 and +14.5‰. After consideration of the decrease in $\delta^{18}O$ of the yellow glass due to mixing with an andesitic or continental crustal source, a postulated evaporite end-member would have a higher $\delta^{18}O$ than the $\delta^{18}O$ value of marine anhydrite (≈+14‰), and in their view this requires a large proportion of marine carbonate that has a high $\delta^{18}O$ value of ≈+29‰ (Blum and Chamberlain, 1992). However, the $\delta^{18}O$ values calculated for the yellow glasses with the present model (between +9.8 and +13.8‰) fall within the range measured by Blum and Chamberlain (1992), if we postulate a high $\delta^{18}O$ value of +22‰ for the red-bed component. We recognize that no O isotope data exist on these red bed formations, but they could have high $\delta^{18}O$ because clay minerals of continental origin have variable but high $\delta^{18}O$ values of ≈+21 ± 7‰ (Savin and Epstein, 1970), and similarly diagenetic overgrowths in sandstones have $\delta^{18}O$ of ≈+23‰ (Lee and Savin, 1985). Therefore, it appears that the $\delta^{18}O$ of the yellow glasses cannot be used in isolation to preclude the presence of a significant sulfate component in the source of Haitian impact glasses.

C and S volatilization during impact melting

Once the fractions of carbonate and sulfate in the source rocks are known, it is possible to estimate the fraction of C and S volatilized during the impact. The theoretical sulfur content of the yellow glasses after melting can be estimated by simple mass balance from the fractions of the different components, assuming 0.2 wt% SO_3 in andesite and 0.5 wt% SO_3 in the red sandstone. A broad positive correlation is obtained between the S contents thus calculated and the S contents measured in the yellow glasses (Fig. 6). This implies that (1) the estimation of the fractions of the different components in the source from B contents and isotopic compositions gives results consistent with the S contents directly measured in the glasses and (2) the fraction of S lost by volatilization is very large and rather constant for all the glasses (between 89 and 98%). The amount of SO_2 released to the atmosphere represents ≈18.0 wt% of the impacted terranes, for a mean value of 30% for the fraction of anhydrite in the source rocks. Using a similar line of reasoning, ≈2.6 wt% of the impacted terranes is released as CO_2 to the atmosphere. With this knowledge from mass balance of the amounts of SO_2 and CO_2 lost by volatilization during the formation of the yellow glasses, it is possible to calculate from the fractions of "andesite," carbonate, sulfate, and sandstone found in the present model, the expected major and trace element content of the impact glasses produced. The SiO_2 contents calculated for the impact melts before volatilization range between 35.0 and 46.4 wt%. This is significantly lower than the SiO_2 contents of the yellow glasses, ranging between 46.6 and 53.5 wt%. However, after correction for the loss of SO_2 and CO_2, SiO_2 contents are increased to between 46.6 and 53.3 wt%, and a good correlation is found between modeled and measured contents (Fig. 7). Similar results are shown for CaO in Figure 6, with even a better correlation between modelled contents and measured contents, which might be due to a good homogeneity of the CaO contents of carbonate and anhydrite and to a larger heterogeneity in SiO_2 contents of the continental sediments (red beds).

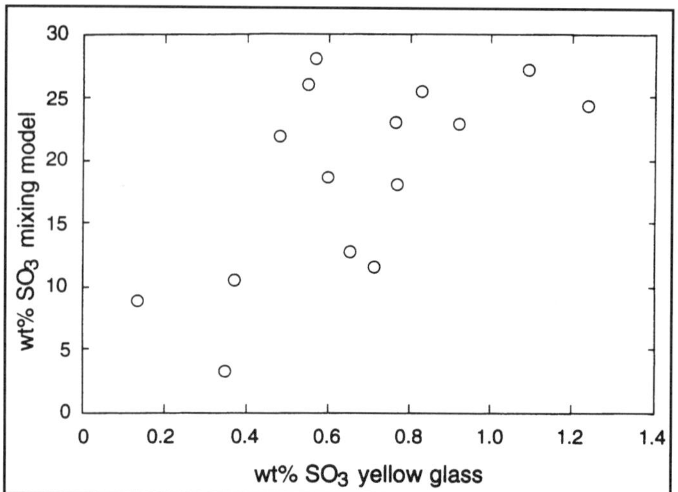

Figure 6. Sulfur contents calculated from the mixing model presented in the text versus S contents of yellow glasses. A broad positive correlation is obtained, showing that the yellow glasses that are the richest in S have, from mass balance arguments based on SiO_2, CaO, and B contents and B isotopes only, the highest proportion of anhydrite in the source. The trend also implies that the fraction of total S lost by volatilization is rather constant.

Behavior of S during volatilization

The $\delta^{34}S$ values of the glasses can be used to constrain the process of volatilization of S. If S was lost after melting and isotopic homogenization of the melts, the $\delta^{34}S$ of the yellow glasses should be very close to +12‰, because of the small isotopic fractionations anticipated during degassing and volatilization and because their composition would be dominated by S

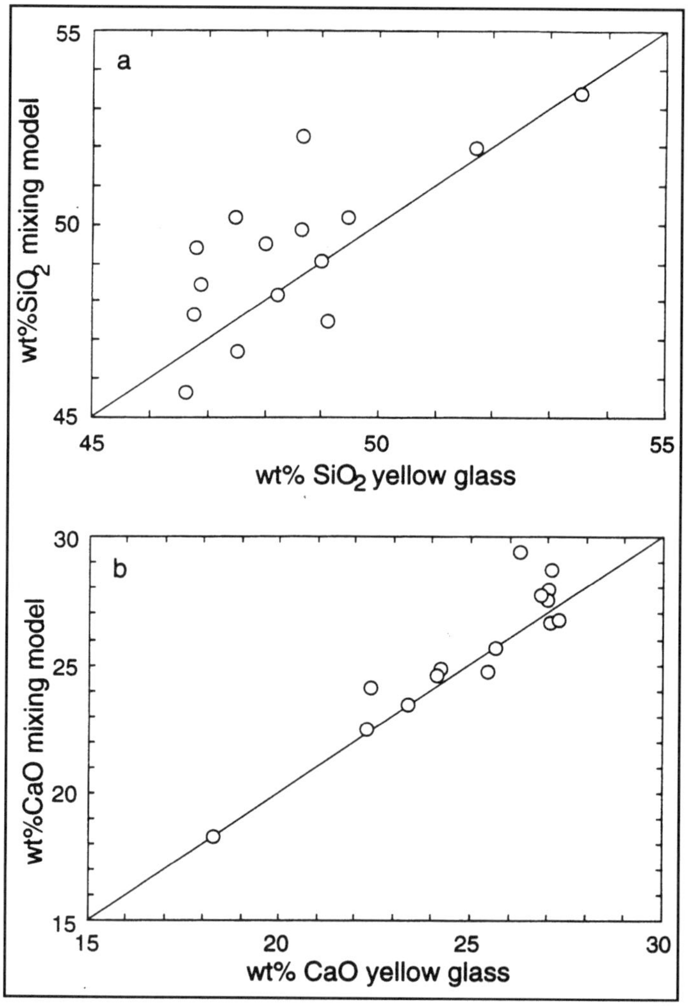

Figure 7. SiO_2 (a) and CaO (b) calculated from the mixing model versus observed CaO and SiO_2 of the yellow glasses. The calculation of the SiO_2 and CaO contents takes into account: (1) the variable proportions of "andesite," anhydrite, carbonate, and sandstone in the source and (2) the loss by volatilization of SO_2 and CO_2 in the proportions given by the model. A better agreement is found for CaO than for SiO_2 contents, suggesting perhaps a higher variability in SiO_2 content of the continental sediments involved in the impact than for the CaO contents of anhydrite and carbonates.

from anhydrite containing ≈59 wt% SO_3 (0.5 wt% SO_3 being a maximum content for sandstone; Pettijohn et al., 1972). The variation of $\delta^{34}S$ observed in the yellow glasses implies that S derived from anhydrite and from other rock components was volatilized to different extent during impact melting. The fraction of S volatilized from anhydrite and from sandstone can be estimated by mass balance. If we define f_{yg}^{sulf} and f_{yg}^{sand} as the fractions of S remaining in the melt that forms the yellow glass after volatilization from the sulfate component and from the sandstone component respectively, we have:

$$f_{yg}^{sulf} * [SO_3]_{sulf} + f_{yg}^{sand} * [SO_3]_{sand} = [SO_3]_{yg}$$

and:

$$f_{yg}^{sulf} * [SO_3]_{sulf} * \delta^{34}S_{sulf} + f_{yg}^{sand} * [SO_3]_{sand} * \delta^{34}S_{sand} = [SO_3]_{yg} * \delta^{34}S_{yg}$$

A $\delta^{34}S$ value of −10‰ has been adopted for the sandstone, corresponding to low $\delta^{34}S$ sandstones (Nielsen, 1978) and a value of +12‰ for anhydrite. Results of this calculation indicate values of f_{yg}^{sulf} and f_{yg}^{sand} in the range 0.001 to 0.012 and 0.1 to 0.85, respectively, as shown in Figure 8. This mass balance implies that, in order to obtain the $\delta^{34}S$ values measured in the yellow glasses, S must be volatilized ≈100 times more efficiently from sulfate than from sandstone. Such a difference in volatilization between anhydrite and silicate (S in sandstone is probably mostly as pyrite in the matrix formed of clays) follows the difference in energy of vaporization between carbonates and/or sulfates (low energy of vaporization) and silicates (high energy of vaporization; O'Keefe and Ahrens, 1989). In addition, sandstones have a low S content and a high SiO_2 content, so that no dominant SO_2 gas phase is expected during their fusion because the silicate melts produced will be S-undersaturated. On the contrary, the amount of S from anhydrite and incorporated in the yellow glass will never exceed the solubility limit in the impact melt and is therefore much less than the initial S content of anhydrite. This implies a strong loss of S by volatilization and degassing. Finally, f_{yg}^{sulf} and f_{yg}^{sand} appear positively correlated for most of the glasses (Fig. 8), which could imply that an external parameter, such as the temperature, for instance, controls the process of volatilization.

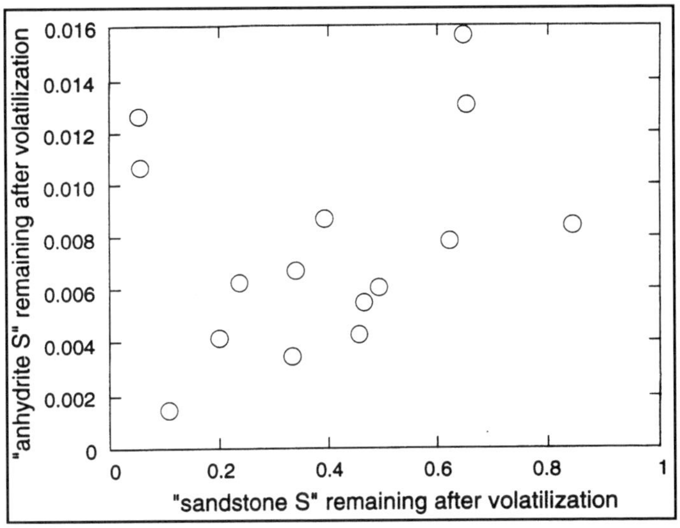

Figure 8. Mass fractions of S remaining after volatilization in anhydrite and in sandstone, calculated from mass balance of S and S isotopes in the yellow glasses (see text). Sandstone appears to have retained its S ≈ 100 times more efficiently than anhydrite, which is consistent with the differences in energy of volatilization between sulfate and silicates (see text). The poor quality of the correlation observed could be due to the uncertainty in the $\delta^{34}S$ measurements and/or to the natural variability of all the process.

CONCLUSIONS

Isotopic analysis of sulfur and boron in the high-Ca yellow impact glasses from the K/T boundary has provided additional constraints on the rock types that were involved in the source region of these melts. Previous studies have proposed that these glasses can be accounted for by a reaction between a silicate melt and either an evaporite- or carbonate-rich source. Our determination of boron and sulfur isotope ratios by ion microprobe on 20 glass particles shows that the yellow glasses have a large range in $\delta^{34}S$ values and $\delta^{11}B$ values. Isotopic and trace element evidence indicates that in addition to the previously recognized impact terrain components of "andesite" (or continental crust), anhydrite, and carbonate, some continental sedimentary rocks containing clays and diagenetic minerals (e.g., loess and/or sandstones) are indicated, especially by the high B contents and low $\delta^{11}B$ values of some of the high-Ca glasses. We propose that this component corresponds to the Todos Santos red-bed formation, a Jurassic deposit known to underlie the Cretaceous carbonate and evaporite sediments in the Yucatan, and also present in drill holes near the Chicxulub impact structure. Mass balance mixing calculations show that all the isotopic characteristics ($\delta^{11}B$, $\delta^{34}S$, and $\delta^{18}O$) of yellow glasses can be explained by mixing during impact of very variable proportions of andesite (mean 49%), sulfate (mean 30%), red sandstone (mean 14%), and carbonate (mean 7%).

The results of this study thus support our earlier proposal that a sulfate-rich source, most likely an evaporite deposit, was an important component in the source region of the yellow impact glasses found at the K/T boundary (Sigurdsson et al., 1991b). They are therefore consistent with a major emission of sulfur dioxide gas during impact, with likely formation of a stratospheric sulfuric acid aerosol. As discussed elsewhere (Sigurdsson et al., 1992), the potentially catastrophic effects of such an aerosol include short-term but severe global surface cooling.

ACKNOWLEDGMENTS

Francis Saupé is thanked for conventional sulfur isotope analyses of the starting gypsum. We thank two anonymous reviewers for very constructive remarks. This work was in part supported by grants for NASA (NA6W3496) and the US National Science Foundation (EAR-9219121).

REFERENCES CITED

Andreieff, P., and Desreumaux, C., 1984, Le passage Maestrichtien-Danien dans le sud d'Haiti (Grandes Antilles): Congres National des Societes Savantes, 109th, Dijon: Sciences Fascia I, p 311–322.

Bass, M. N., and Zartman , R. E., 1969, The basement of Yucatan peninsula: Eos (Transactions, American Geophysical Union), v. 50, p. 313.

Blum, J. D, and Chamberlain, C. P., 1992, Oxygen isotope constraints on the origin of impact glasses from the Cretaceous-Tertiary boundary: Science, v. 257, p. 1104–1107.

Cantagrel, J. M., Gourgaud, A., and Robin, C., 1984, Repetitive mixing events and Holocene pyroclastic activity at Pico de Orizaba and Popocatepetl (Mexico): Bulletin Volcanologique, v. 47, p. 735–748.

Chakraborty, S., Dingwell, D. B., and Chaussidon, M., 1993, Chemical diffusivity of boron in melts of haplogranitic composition: Geochimica et Cosmochimica Acta, v. 57, p. 1741–1751.

Chaussidon, M., and Albarède, F., 1992, Secular boron isotope variations in the continental crust: An ion microprobe study: Earth and Planetary Science Letters, v. 108, p. 229–241.

Chaussidon, M., and Jambon, A., 1994, Boron content and isotopic composition of oceanic basalts: Geochemical and cosmochemical consequences: Earth and Planetary Science Letters, v. 121, p. 277–291.

Chaussidon, M., and Libourel, G., 1993, Boron partitioning in the upper mantle: An experimental and ion probe study: Geochimica et Cosmochimica Acta, v. 57, p. 5053–5062.

Chaussidon, M., and Lorand, J.-P., 1990, Sulphur isotope composition of orogenic spinel lherzolite massifs from Ariège (North Eastern Pyrenees, France): An ion microprobe study: Geochimica et Cosmochimica Acta, v. 54, p. 2835–2846.

Chaussidon, M., France-Lanord, C., and Palmer, M., 1991, Boron isotope variations of tourmalines from the Himalayan leucogranites, their possible source rocks and their alteration products in the bay of bengal: Eos (Transactions, American Geophysical Union), v. 72, p. 524.

Claypool, G. E., Holser, W. T., Sakai, I. R., and Zak, I., 1980, The age curves of sulfur and oxygen in marine sulfate and their mutual interpretation: Chemical Geology, v. 28, p. 199–260.

Desreumaux, C., 1981, Geologie structurale de la presquile sud Haiti, entre la region de Jacmel et al plaine du Cul de Sac: Congres National des Societes Savantes, 106th, Perpignan: Fascia III, p. 389–399.

Desreumaux, C., 1987, Geologie structurale de la presquile sud Haiti [Thesis Doctorat]: Marseille, France, Universite Marseille, 388 p.

Donnelly, T. W., Horne, G. S., Finch, R. C., and Lopez-Ramos, E., 1990, Northern Central America: The Maya and Chortis blocks: Boulder, Colorado, Geological Society of America, Geology of North America, v. H, p. 37–76.

Harder, H., 1978, Boron: Granitic rocks and related effusives, *in* Wedepohl, K. H., ed., Handbook of Geochemistry, Volume II-1, Springer-Verlag, p. 5E5–5E10.

Hildebrand, A. R., and Boynton, W. V., 1990, Proximal Cretaceous-Tertiary boundary impact deposits in the Caribbean: Science, v. 248, p. 843–847.

Ishikawa, T., and Nakamura, E., 1993, Boron isotope systematics of marine sediments: Earth and Planetary Science Letters, v. 117, p. 567–580.

Izett, G. A., 1991, K/T boundary tektites from near Beloc, Haiti [abs.]: Proceedings, 22nd Annual Lunar and Planetary Science Conference: Houston, Texas, Lunar Planetary Institute, p. 625–626.

Jéhanno, C., Boclet, D., Froget, L., Lambert, B., Robin, E., Rocchia, R., and Turpin, L., 1992, The Cretaceous-Tertiary boundary at Beloc, Haiti: No evidence for an impact in the Caribbean area: Earth and Planetary Science Letters v. 109, p. 229–241.

Koeberl, C., 1992, Geochemistry and origin of Muong Nong–type tektites: Geochimica et Cosmochimica Acta, v. 56, p. 1033–1064.

Koeberl, C., and Sigurdsson, H., 1992, Geochemistry of impact glasses from the K/T boundary in Haiti: Relation to smectites, and a new type of glass: Geochimica et Cosmochimica Acta, v. 56, p. 2113–2129.

Lee, M., and Savin, S. M., 1985, Isolation of diagenetic overgrowths on quartz sand grains for oxygen isotopic analyses: Geochimica et Cosmochimica Acta, v. 49, p. 497–501.

Maurrasse, F. J.M. R., 1982, Guide to the field excursions of Haiti: Miami, Miami Geological Society, 66 p.

Maurrasse, F. J.M. R., Pierre-Louis, F., and Rigaud, J. J. G., 1979a. Transactions of the Fourth Latin American Geological Congress, Trinidad, 328 p.

Maurrasse, F. J.M. R., Husler, J., Georges, G., Schmitt, R., and Damond, P., 1979, Upraised Caribbean sea floor below acoustic reflector B″ at the southern peninsula of Haiti: Geologie en Mijnbouw, v. 8, p. 71–83.

Nielsen, H., 1978, Sulfur: Isotopes in nature, *in* Wedepohl, K. H., ed., Handbook of geochemistry, Volume II-2: Berlin, Springer-Verlag, p. 16B1–16B40.

O'Keefe, J. D., and Ahrens, T. J., 1989, Impact production of CO_2 by the Cretaceous/Tertiary extinction bolide and the resultant heating of the Earth: Nature, v. 338, p. 247–249.

Palmer, M., 1991, Boron isotope systematics of Halmahera arc (Indonesia) lavas: Evidence for involvement of the subducted slab: Geology, v. 19, p. 215–217.

Pettijohn, F. J., Potter, P. E., and Siever, R., 1972, Sand and sandstone: Berlin, Springer-Verlag, and New York, Heidelberg, 618 p.

Savin, S. M., and Epstein, S., 1970, The oxygen and hydrogen isotope geochemistry of clay minerals: Geochimica et Cosmochimica Acta, v. 34, p. 25–42.

Sigurdsson, H., D'Hondt, S., Arthur, M. A., Bralower, T., Zachos, J. C., Fossen, M., and Channell, J., 1991a, Glass from the Cretaceous-Tertiary boundary in Haiti: Nature, v. 349, p. 482–487.

Sigurdsson, H., Bonté, P. H., Turpin, L., Chaussidon, M., Métrich, N., Steinberg, M., Pradel, Ph., and D'Hondt, S., 1991b, Geochemical constraints on source region of Cretaceous/Tertiary impact glasses: Nature, v. 353, p. 839–842.

Sigurdsson, H., D'Hondt, D., and Carey, S., 1992, The impact of the Cretaceous/ Tertiary bolide on evaporite terrane and generation of major sulfuric acid aerosol: Earth and Planetary Science Letters, v. 109, p. 543–559.

Smit, J., Alvarez, W., Montanari, A., Swinburne, N., Van Kempen, T. M., Klaver, G. T., and Lustenhouwer, W. J., 1991, Tektites and microkrystites at the Cretaceous/Tertiary boundary: Two strewnfields, one crater: Proceedings, 22nd Annual Lunar and Planetary Science Conference: Houston, Texas, Lunar and Planetary Institute, p. 1277–1278.

Spivack, A. J., Palmer, M., and Edmond, J. M., 1987, The sedimentary cycle of boron isotopes: Geochimica et Cosmochimica Acta, v. 51, p. 1939–1949.

Swihart, G. H., Moore, P. B., and Callis, E. L., 1986, Boron isotopic composition of marine and non-marine evaporite borates: Geochimica et Cosmochimica Acta, v. 53, p. 1297–1301.

Taylor, B., 1986, Magmatic volatiles: Isotopic variations of C, H and S, *in* Valley, J. W., Taylor, H. P., and O'Neil, J. R., eds., Stable isotopes in high temperature geological processes: Reviews in Mineralogy, v. 16, p. 185–225.

Taylor, S. R., McLennan, S. M., and McCulloch, M. T., 1983, Geochemistry of loess, continental crustal composition and crustal model ages: Geochimica et Cosmochimica Acta, v. 47, p. 1897–1905.

Vengosh, A., Kolodny, Y., Starinsky, A., Chivas, A. R., and McCulloch, M. T., 1991, Coprecipitation and isotopic fractionation of boron in modern biogenic carbonates: Geochimica et Cosmochimica Acta, v. 55, p. 2901–2910.

Vengosh, A., Starinsky, A., Kolodny, Y., Chivas, A. R., and Raab, M., 1992, Boron isotope variations during fractional evaporation of sea water: New constraints on the marine vs. nonmarine debate: Geology, v. 20, p. 799–802.

Viniegra, O. F., 1971, Age and evolution of salt basins of southeastern Mexico: American Association of Petroleum Geologists Bulletin, v. 55, p. 478–494.

Viniegra, O. F., 1981, Great carbonate bank of Yucatan, southern Mexico: Journal of Petroleum Geology, v. 3, p. 247–278.

Zheng, Y. F., 1990, The effect of Rayleigh degassing magma on sulphur isotope composition: A quantitative evaluation: Terra Nova, v. 2, p. 74–78.

MANUSCRIPT ACCEPTED BY THE SOCIETY SEPTEMBER 14, 1995

Geological Society of America
Special Paper 307
1996

The Cretaceous-Tertiary catastrophic event at Millers Ferry, Alabama

Richard K. Olsson and Chengjie Liu
Department of Geological Sciences, Rutgers University, New Brunswick, New Jersey 08903
Mickey van Fossen
Department of Geological Sciences, Rutgers University, New Brunswick, New Jersey 08903 and Lamont-Doherty Earth Observatory, Columbia University, Palisades, New York 10964

ABSTRACT

Photographs of the 1967 excavation for the powerhouse at the Millers Ferry dam site on the Alabama River and analysis of two cores taken on the site have provided an opportunity to view the surface of the late Maastrichtian Prairie Bluff Chalk prior to deposition of the early Danian Clayton basal sands. The surface of the chalk is severely scoured with groove markings, striations, and flute markings that suggest erosion, as by a bolide-generated tsunami wave at K/T boundary time. Numerous normal faults that offset the Prairie Bluff Chalk and that are in turn buried by the overlying Clayton basal sands indicate impact-related earthquakes. The Clayton basal sands consist of a chaotic layer of large-sized blocks of the Prairie Bluff Chalk and coarse sand at its contact with the Prairie Bluff Chalk and an overlying Danian inner-shelf micaceous clayey sand. The chaotic layer is interpreted as a tsunami-generated deposit. Analysis of benthic foraminifera in the Prairie Bluff Chalk and in the Danian shelf sands indicates that there was no significant change in sea level at K/T boundary time and that sea level was at a low stand. The abrupt cessation of Prairie Bluff Chalk deposition is evidence of a sudden extinction of marine biota. These data, taken together, point to a geologically instantaneous catastrophic event such as could have been generated by a bolide impact at Chicxulub Crater in Yucatan.

INTRODUCTION

The Cretaceous-Tertiary (K/T) stratigraphy at Millers Ferry, Alabama, was exposed in an excavation in 1966 and 1967 in preparation for the construction of the powerhouse for the Millers Ferry dam on the Alabama River (Fig. 1). The excavation exposed not only the contact between the Cretaceous Prairie Bluff Chalk and the Tertiary Clayton Formation but also a large area of the top of the Prairie Bluff Chalk. Hay and Mohler (1967) included data generated from the Millers Ferry excavation when they established the first calcareous nannofossil zones for the basal Danian. They noted that the two basal Danian zones, the *Markalius astroporus* Zone (= *Markalius inversus* Zone) and the *Cruciplacolithus tenuis* Zone, were present in the Clayton Formation at the Millers Ferry excavation. Planktonic foraminiferal studies on samples from the excavation and two engineering cores also showed a basal Danian age for the Clayton Formation (Olsson, 1970). Smit (1982) noted that his basalmost Danian planktonic foraminiferal Zone P0 was probably present at Millers Ferry because Olsson had reported a basal interval with the P0 marker species *Guembelitria cretacea*. As interest in the scientific community on the causes of the end-Cretaceous extinction grew in the following decade, the Millers Ferry cores were reexamined with a view toward high-resolution biostratigraphy and the phylogeny of the early Danian planktonic foraminifera (Hemleben et al., 1991; Olsson et al., 1992a; Liu and Olsson, 1992, 1994; Olsson and Liu, 1993). These studies established the presence of Zone P0 at Millers Ferry. Furthermore, the unaltered original wall textures observed on planktonic foraminifera at Millers Ferry established a basis on which the phylogeny of early Danian planktonic foraminifera could be constructed (Hemleben et al., 1991; Olsson et al., 1992b).

Olsson, R. K., Liu, C., and van Fossen, M., 1996, The Cretaceous-Tertiary catastrophic event at Millers Ferry, Alabama, *in* Ryder, G., Fastovsky, D., and Gartner, S., eds., The Cretaceous-Tertiary Event and Other Catastrophes in Earth History: Boulder, Colorado, Geological Society of America Special Paper 307.

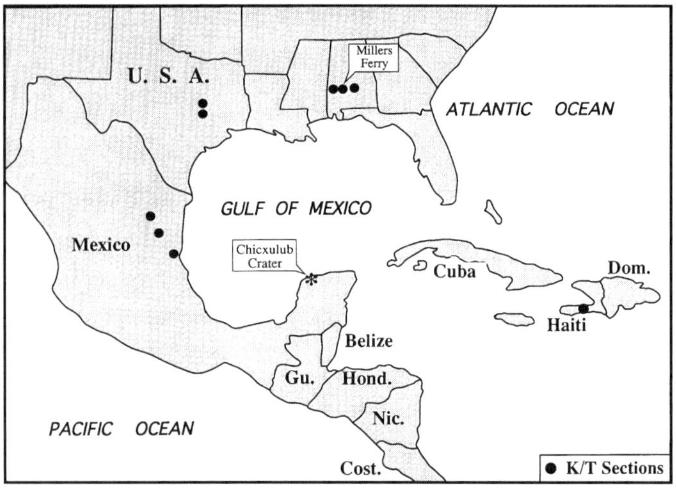

Figure 1. Location map showing Millers Ferry, Alabama, in relation to the Chicxulub Crater and other K/T boundary sections. Gu., Guatemala; Hond., Honduras; Nic., Nicaragua; Cost., Costa Rica.

The stratigraphy of the K/T boundary in Alabama has been interpreted in different ways (Olsson and Liu, 1993). At issue is the extent of a hiatus that separates the Prairie Bluff Chalk from the overlying Clayton basal sands. The Clayton basal sands are discontinuous; where they are absent, the Pine Barren Member directly overlies the Prairie Bluff Chalk. The Clayton basal sands have been interpreted as low-stand deposits overlying a type 1 unconformity that separates the sands from the Prairie Bluff chalk (Donovan et al., 1988; Mancini et al., 1989). The hiatus of the unconformity has been interpreted to include the upper Maastrichtian *Abathomphalus mayaroensis* planktonic foraminiferal zone and part of the basal Danian (Mancini et al., 1989). However, Olsson and Liu (1993) have established that the basal Danian Zones Pα and P0 are present. Although the upper Maastrichtian marker species, *A. mayaroensis*, is absent in the Prairie Bluff Chalk, the uppermost Maastrichtian calcareous nannofossil *Micula prinsii* Zone is identified at Millers Ferry (Olsson et al., 1992a), at Braggs, Alabama (Habib et al., 1992), and at Moscow Landing, Alabama (Habib, personal communication, 1994). Thus, the absence of the foraminifer *A. mayaroensis* is probably due to ecologic exclusion, the species being a deeper-water dweller, rather than to a hiatus. Consequently, the erosional surface on the top of the Prairie Bluff Chalk must have been caused by an event of very short duration.

Since the discovery of the Chicxulub Crater in the Yucatan as the bolide impact site at the K/T boundary (Hildebrand et al., 1991), coarse-grained deposits at the K/T boundary at Mimbral and elsewhere in the Gulf Coastal Plain of Mexico have been attributed to a tsunami generated by the bolide impact (Smit et al., 1992). Previously, similar deposits at the K/T boundary at Brazos River, Texas, were ascribed to a tsunami event (Bourgeois et al., 1988). There appears to be ample evidence that a tsunami did affect the K/T boundary in the Gulf of Mexico. However, few data have so far been presented from Alabama K/T boundary sections, with the exception of Smit et al. (1994a), who reported tsunami deposits at Moscow Landing on the Tombigbee River, Alabama. However, Savdra (1993) also studied the K/T boundary section at Moscow Landing and interpreted the Clayton sands as low-stand infills of incised valleys.

Hay and Mohler (1967), in their study on Danian calcareous nannofossil biostratigraphy, noted the observations of a site geologist at the Millers Ferry excavation who described the surface of the Prairie Bluff Chalk as gently undulating with a maximum relief of 0.9 m and without interconnections between topographic lows. No evidence of karsting was evident. Interestingly, Hay and Mohler interpreted the unconformity at the K/T boundary at Millers Ferry to be the product of submarine erosion. These comments suggested to us that the topography of the surface might contain evidence of a tsunami event at the end of Cretaceous time. In preparation for the construction of the Millers Ferry powerhouse, the Danian Clayton Formation was excavated to the top of the upper Maastrichtian Prairie Bluff Chalk, the top surface of which served as the foundation for the footings of the powerhouse. Photographs of the construction activities taken in 1967 by the U.S Army Corps of Engineers during this phase of construction show that the top of the Prairie Bluff was carefully cleaned in preparation for laying the cement foundation blocks. Close-up views of approximately 3,546 m^2 of the top of the Prairie Bluff Chalk can be observed in these photographs, thus offering an unique opportunity to examine the Cretaceous/Tertiary contact and the surface on which the Danian Clayton basal sands were deposited. The surface of the Prairie Bluff Chalk as it was described in 1967 is seen to be irregular and undulating without any interconnections that would indicate fluvial erosion during a period of subaerial exposure. Furthermore, there is no evidence of weathering in the photographs or in two cores (Olsson, 1970) taken from the site. On the contrary, the surface is marked by scour features, flute markings, groove markings, and striations that have a north-south orientation.

Our interpretation is that (1) these features were eroded by a tsunami that was generated by a bolide impact at Chicxulub Crater at K/T boundary time, (2) chaotic beds lying on the Prairie Bluff Chalk were deposited by this tsunami, (3) faults cutting the uppermost Maastrichtian Prairie Bluff Chalk were caused by earthquakes generated by the Chicxulub impact, and (4) the cessation of Prairie Bluff Chalk deposition was a result of terminal Cretaceous extinctions of biota.

STRATIGRAPHY AT MILLERS FERRY EXCAVATION

In an overview of the site (Figs. 2, 3, 4), the horizontal, thinly bedded Pine Barren Member of the Clayton Formation is visible in the walls of the excavation. The Pine Barren Member, which consists of thin alternating chalky beds and more silty, sandy beds, is separated from the medium- to coarse-grained Clayton basal sands by a transgressive surface. The top of the Prairie Bluff Chalk is at the bottom of the excavation.

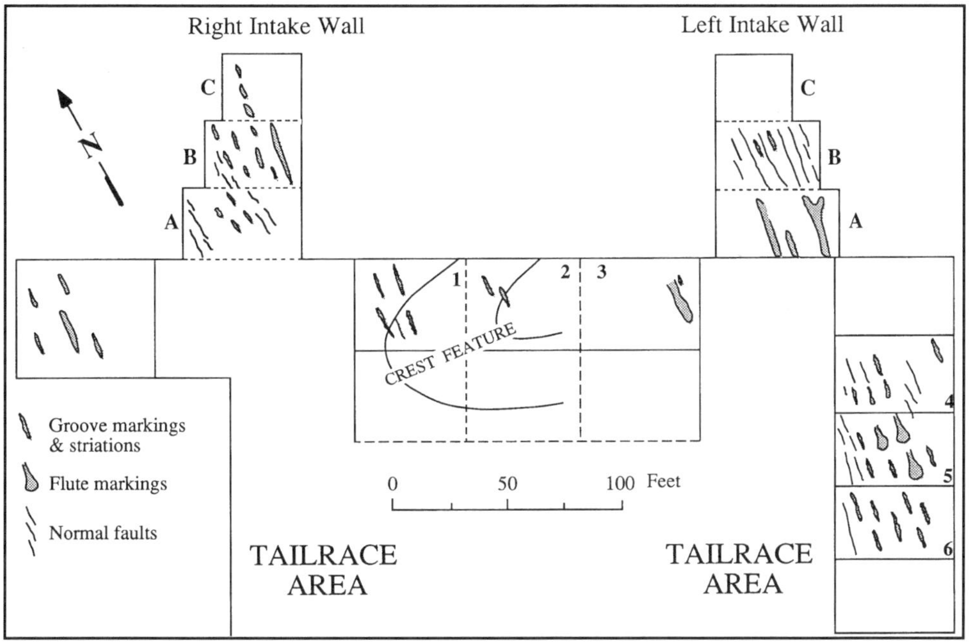

Figure 2. Plan view of Millers Ferry excavation showing location of construction blocks where figures in text are located. Also shown are location and orientation of features identified on surface of the Prairie Bluff Chalk.

The laying of the foundation for the powerhouse proceeded by segments or blocks. In each block the surface of the Prairie Bluff Chalk was carefully cleaned, first by use of a power shovel to remove the bulk of the overburden and then finally by high pressure air hoses, shovels, and brooms to expose the surface of the Prairie Bluff Chalk (Figs. 5, 6).

The stratigraphy of the Clayton basal sands and the contact with the Prairie Bluff Chalk is clearly seen in the northwest section of the excavation (Fig. 5). In these photographs and in two cores, the Clayton basal sands consist of three units. The lowermost unit of the Clayton basal sands at Millers Ferry consists of a thin clayey chalk that rests on the Prairie Bluff Chalk (Figs. 5, 6, 7). The middle unit is a chaotic mixture of coarse sand, clay, and clasts of Prairie Bluff Chalk that vary in size from chips to blocks up to 2 m wide and 0.25 m thick (Figs. 5, 8). Cretaceous foraminifera are abundant in this unit, being derived from the Prairie Bluff Chalk clasts. Scattered and rare occurrences of calcite-replaced spherules, which are comparable to the replaced spherules found at the K/T boundary at Mimbral, Mexico (Smit et al., 1992), are found in this unit. The uppermost unit is a horizontal, thinly bedded, micaceous, clayey sand, the upper surface of which is burrowed. This unit contains primitive Danian planktonic foraminifera (Liu and Olsson, 1992) and is separated from the Pine Barren Member by a transgressive surface, possibly produced by submarine currents. Deposition of the micaceous, clayey sand occurred in an inner-shelf environment. Why this unit was not extensively burrowed is unclear. The lack of an infaunal macrofauna may be related to the K/T boundary extinctions of molluscan assemblages. The analysis of the K/T marine macrofaunal transition at Braggs, Alabama, by Bryan and Jones (1989) indicated that the Prairie Bluff Chalk contains a molluscan assemblage characterized by a high-diversity, mixed trophic nucleus dominated by infaunal suspension-feeders and deposit-feeders. In the

Figure 3. View of excavation in 1967. Horizonal alternating thin limestones and clayey, silty sands observed in walls of excavation are the Pine Barren member of the Clayton Formation. They overlie on a transgressive surface the Clayton basal sands, which is the dark gray interval (arrow) at the base of the excavation. The top of the Prairie Bluff Chalk lies just beneath the floor of the excavation.

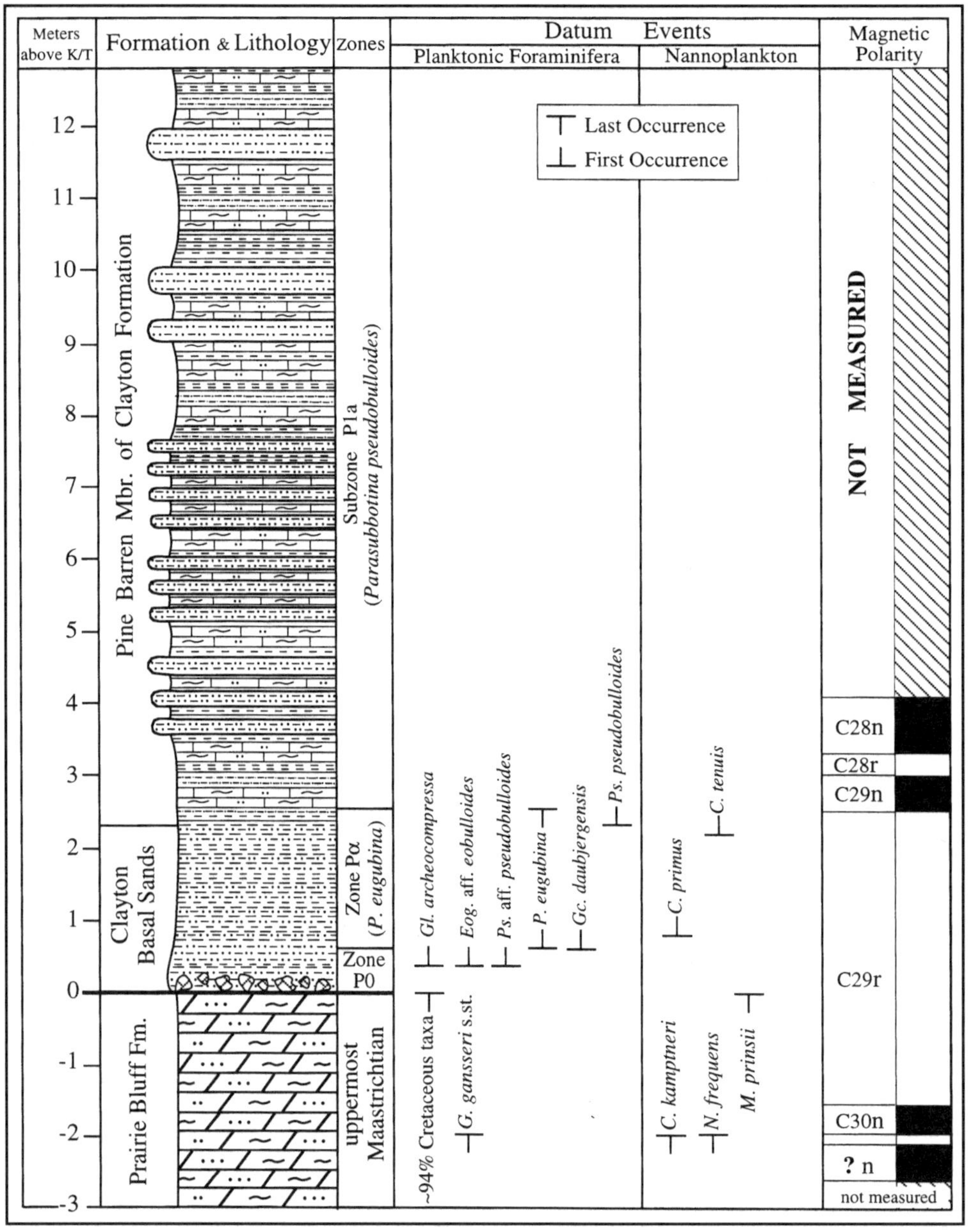

Figure 4. Stratigraphic column of the excavation drawn to scale from the photographs. Paleomagnetic stratigraphy (by M. Van Fossen) and ranges of key calcareous microfossil taxa are shown. Zone P0 (*Guembelitria cretacea* Zone) is defined (Smit, 1982) as the interval between the extinction of Cretaceous planktonic foraminifera and the first occurrence of *Parvularugoglobigerina eugubina*, which defines the base of Zone Pα. The chaotic bed is included in Zone P0 because its deposition follows the K/T terminal event. Foraminifera in the chaotic bed include reworked Cretaceous taxa derived from chalk clasts and from transported coarse sands. The upper part of Zone P0 lies in the basal part of the thinly bedded micaceous sands and contains the survivor species *Guembelitria cretacea*, which is also present in the chalk clasts of the chaotic bed. The species is also present in the underlying Prairie Bluff Chalk. The first Danian taxa (*Globanomalina archeocompressa, Eoglobigerina* aff. *eobulloides,* and *Parasubbotina* aff. *pseudobulloides*) appear in the uppermost part of Zone P0 above the interval where only *G. cretacea* is present. Rare occurrences of two other survivor (and ancestral) species (Liu and Olsson, 1994)—*Hedbergella holmdelensis* and *Hedbergella monmouthensis*—also lie above the chaotic bed. These species are rare in the Prairie Bluff Chalk. The three survivor species make up the 6% of Cretaceous planktonic taxa that survived the K/T terminal event (Olsson and Liu, 1993).

Figure 5. Views of Clayton basal sands and contact with the Prairie Bluff Chalk in the northwest part of the excavation (blocks A,B,C of right intake wall in Fig. 2). A, view to west in block B showing Clayton basal sands contact with Prairie Bluff Chalk. Clayton basal sands (about 1.5 m thick) show onlap to the north and thickening to the south. White layer (small arrow) just above the contact with Prairie Bluff Chalk is the clayey chalk layer. Large chalk clasts within coarse sand unit are the chaotic layer between the clayey chalk layer and the thinly bedded micaceous clayey sands above (separated by large arrow). Large arrowhead points to the burrowed transgressive surface, above which is the Pine Barren Member. Note the undulating surface of the Prairie Bluff Chalk, outlined by its contact with the cement wall to the right (long arrow), and numerous grooves, striations, and scourlike features with a northerly orientation (small arrowheads). Parallel, evenly spaced, deep, sharp, right-angled grooves are from power shovel. B, view to the west in block C showing Clayton basal sands contact with Prairie Bluff Chalk (small arrow). The clayey chalk layer and the chaotic bed thin and onlap the Prairie Bluff Chalk to the north. Note undulating contact with the Prairie Bluff Chalk. Contact between chaotic bed and the micaceous sand is indicated by large arrow. Transgressive surface is pointed out by large arrowhead. Surface of the Prairie Bluff Chalk is largely obscured by power shovel grooves. C, view to the east in block B showing Clayton basal sands contact with Prairie Bluff Chalk and the transgressive surface of the Pine Barren Member (large arrowhead). Note undulating surface of Prairie Bluff Chalk outlined by contact with the construction support walls to the right and left (small arrows). Note numerous grooves, striations, and scourlike features with a northerly trend (small arrowheads). Parallel, evenly spaced, deep, sharp, right-angled grooves are from power shovel. D, view to the west in block A. Clayton basal sands are approximately 2.4 m thick. The clayey chalk layer (small arrow) is about 15.2 cm thick. A 15.2-cm-thick layer of broken chalk clasts (large arrow) marks the top of the chaotic layer. The upper thinly bedded micaceous clayey sand is about 1.5 m thick. The burrowed transgressive surface marks the top of this sand (large arrowhead). Note undulating surface of the Prairie Bluff Chalk outlined by contact with the cement wall to the left and the groove markings and depressions with a northerly orientation (small arrowheads). Deep grooves in lower left corner are from power shovel. See Figure 2 for location of photographs.

A

B

C

Figure 6. View of excavation wall along north wall of excavation. A, view of block 1 showing west front of crestlike feature (large arrowhead). All three units of the Clayton basal sands thin and onlap this feature to the east. Note the undulating surface of the Prairie Bluff Chalk outlined by contact with cement wall on the south and the east (small arrows). Groove markings and striations are visible in foreground (small arrowhead). Parallel, evenly spaced, deep, sharp, right-angled grooves are from power shovel. B, view to the north in block 2 showing the sharply irregular contact between the Clayton basal sands and the Prairie Bluff Chalk (small arrows) and the burrowed transgressive surface separating the Pine Barren Member (large arrowhead). Note large clast in the chaotic bed resting on Prairie Bluff Chalk on right side of photograph (small arrow). The clayey chalk layer and the chaotic bed onlap and disappear on this scour feature. The upper micaceous sands thin dramatically to the east. Large groove markings and possible prod marks are visible in the foreground (small arrowheads). Normal faults with downthrow to the west occur on right side of photograph (large arrow). Note burrowed contact between horizontally bedded Clayton basal sands and the Pine Barren Member (large arrowhead). C, view to the north in block 3 showing Clayton basal sands contact with Prairie Bluff Chalk (small arrow). The micaceous sand has thinned to about 0.6 m (large arrowhead). Note slightly undulating surface of Prairie Bluff Chalk outlined along contact and construction support wall. Large groove markings are visible in the foreground as well as outlines of large megafossils on surface of Prairie Bluff Chalk (small arrowheads). See Figure 2 for location of photographs.

overlying basal Clayton Formation, molluscan diversity is low. A pronounced Lazarus Effect is displayed across the K/T boundary, which Bryan and Jones interpret as the result of local extinction and/or migration. Thus, the preservation of bedding in the micaceous sand at Millers Ferry may be explained by the temporary absence of a burrowing macrofauna following the K/T extinctions.

All three units are present where the Clayton basal sands are thickest. The lower two units onlap on the irregular surface of the Prairie Bluff Chalk and are absent where the Clayton basal sands thin over topographic highs of the chalk surface (Figs. 5, 6). The uppermost micaceous sand thins over these highs but is present throughout the excavation. In total, the thickness of the three units of the Clayton basal sands is as great as 2.5 m in the northwest section of the excavation but thins to the northeast to less than 10 cm. In the southern sections of the excavation the sands are more or less 1 m in thickness.

According to Smit et al. (1994b), the buildup of clastic deposits at the K/T boundary follows a consistent pattern around the Gulf of Mexico, although they explained that this pattern may be modified by local variables such as paleodepth, sea-level change, and distance from shoreline. This pattern con-

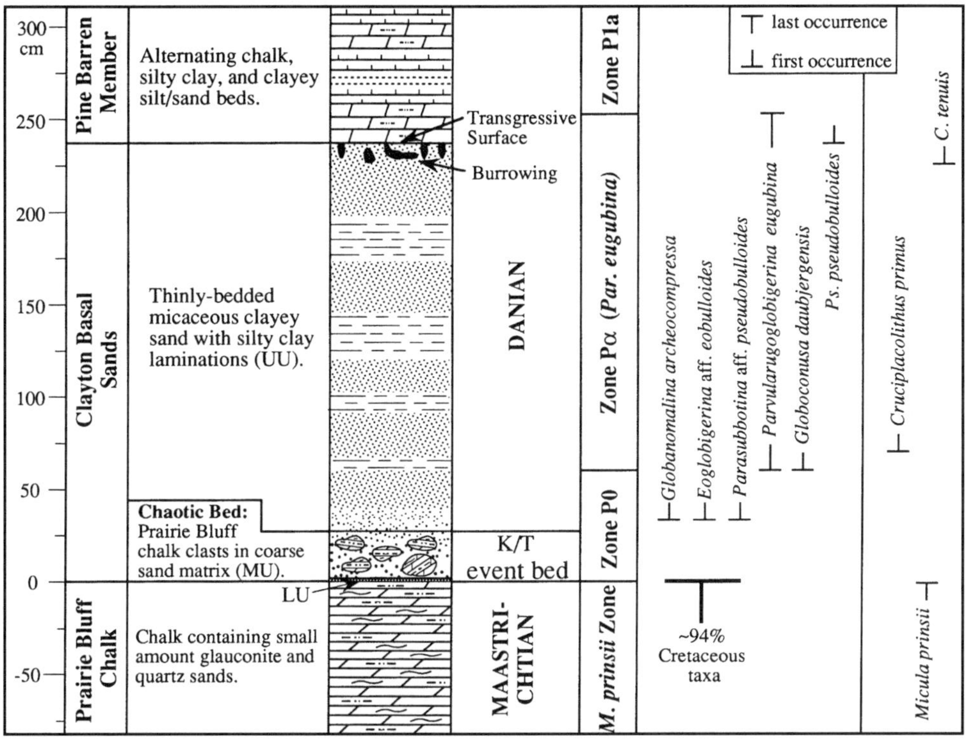

Figure 7. Division of the Clayton basal sands at Millers Ferry. Lower unit (LU), a thin clayey chalk layer, is present only in topographic lows of the Prairie Bluff Chalk. Middle unit (MU) (chaotic bed) is the coarse sand bed with broken blocks of the Prairie Bluff Chalk. The upper (UU) thinly bedded, micaceous clayey sand is Danian in age and was deposited in an inner-shelf environment. It overlaps the chaotic bed and onlaps the Prairie Bluff Chalk on topographic highs. The micaceous sands are separated from the overlying Pine Barren Member by a transgressive surface. See Figure 4 for explanation of planktonic foraminiferal biostratigraphy.

sists of, in ascending order, a basal unit (unit 1) with abundant spherules mixed with locally derived material, a second unit (unit 2) of detrital-rich sands with coarse basal layers that record bidirectional paleocurrents, a third unit (unit 3) of thin intercalated layers of silt and fine sand with a bioturbated upper part, and an uppermost unit (unit 4) of micritic limestone and mudstone.

No distinct spherule-rich layer is present at the base of the the Clayton basal sands at Millers Ferry, but as noted above, rare and scattered calcite-replaced spherules occur in the chaotic middle unit, suggesting reworking and dispersal of a spherule layer. The lower clayey chalk layer may represent liquified and slumped Prairie Bluff Chalk following the interpretation of Smit et al. (1994a) for a similar bed at Moscow Landing, Alabama. According to the classification of Smit et al. (1994b), these beds are characteristic of their unit 1. At Millers Ferry the middle coarse sand unit with abundant clasts derived from the Prairie Bluff Chalk compares well with the backwash deposits of unit 2 of Smit et al. This unit at Millers Ferry may be more complex, however, in that it contains some very large blocks of Prairie Bluff Chalk, which suggests that this unit is a mixture of clasts derived from the initial erosive effects from the passage of a tsunami and coarse sands derived from subsequent backwash. Units 3 and 4 of Smit et al. are not recognizable at Millers Ferry. According to them, unit 3 represents the settling of fine sediment from weakening tsunami or seiche waves, and unit 4 represents the settling out of the suspension cloud that was stirred up by passing megawaves. The upper micaceous sand of the Clayton basal sands was deposited in an inner-shelf environment in early Danian time. It is possible that units 3 and 4 grade imperceptibly into the basal part of these inner-shelf sands.

THE PRAIRIE BLUFF SURFACE

Heretofore it has been difficult to obtain data on the topography of the surface of the Cretaceous deposits in outcrops of the K/T boundary in the Gulf Coastal Plain because of their two-dimensional geometry. Diapirism of Cretaceous shales has been recognized in some outcrops (Smit et al., 1992, 1994b), but there appears to be little information on erosive effects on the K/T boundary seafloor. An unusual opportunity to view the surface of the Prairie Bluff Chalk is provided by photographs taken during the construction of the powerhouse at Millers Ferry. At the time, unfortunately, the possibility of a tsunami event at K/T time had not been presented to the scientific com-

Figure 8. View of chaotic bed along east wall of excavation in blocks 4 and 5. A, view of block 4 showing large blocks of the Prairie Bluff Chalk in the chaotic bed. Possible shingling of blocks (long arrows) would suggest southerly movement due to backwash and seiche wave. The upper micaceous sand is about 0.3 m thick (large arrow) and is overlain by the transgressive Pine Barren Member (large arrowhead). B, view to the north along east wall of excavation of block 5, with block 4 in the background. The chaotic bed rests directly on the Prairie Bluff Chalk (small arrow) and, in turn, is overlain by the micaceous sands (large arrow). Burrowed transgressive surface of the Pine Barren Member caps the Clayton basal sands (large arrowhead). Large flute markings and grooves show erosive effect of tsunami wave on the Prairie Bluff Chalk (small arrowheads). Direction of flow is to the north. See Figure 2 for location of photographs.

munity, and practically no study was made on the stratigraphy of the excavation except for two general biostratigraphic studies (Hay and Mohler, 1967; Olsson, 1970). This scientific slight can be partially corrected by study of the numerous photographs taken during construction and the analysis of two cores taken on the site.

Eight different features can be observed in the photographs of the Millers Ferry excavation: parallel lineations, groove markings, flute casts, prod marks, scour features, disrupted blocks of chalk, shingling orientation, and normal faulting.

The surface of the Prairie Bluff Chalk is clearly irregular and undulating. This is shown in profiles of excavation walls and in the contact of the Prairie Bluff Chalk with construction walls (Figs. 6, 9). The nature of the irregularity is evident in the numerous scour features that cut the surface of the Prairie Bluff Chalk. Most numerous are parallel groove markings and lineations (Figs. 5, 6, 9). Other features include apparent prod marks and flute marks (Figs. 6, 8). The orientation of all these features is north-south. The flute marks are of interest in that their orientation indicates a northerly flow for the eroding current. Possible shingling of Prairie Bluff Chalk blocks is suggestive of a southerly movement (Fig. 8). Severe scouring is evident along the northern wall of the excavation. Here (Fig. 6) the upper micaceous sand thins dramatically over a crestlike feature. Taken together, all these features indicate that the top of the Prairie Bluff Chalk was violently eroded, as by passage of a tsunami that tore up large blocks of the chalk.

EVIDENCE FOR EARTHQUAKES

Numerous normal faults break the surface of the Prairie Bluff Chalk (Figs. 6, 10). They also have a north-south trend. A spectacular series of northerly trending synthetic, slickensided, normal faults with relay ramps between the fault segments was uncovered along the northeast side of the excavation (Fig. 10). Displacement on faults observed in the excavation is a few centimeters to about a half meter. These faults are buried by the overlying Clayton basal sands. Normal faulting of the Prairie Bluff Chalk is also reported at Moscow Landing, Alabama (Smit et al., 1994a). Although some faults cut the overlying Danian strata, these authors report that most faults terminate at the K/T boundary. Smit et al. interpret these faults as listric faults that created seafloor depressions on a south downdropped block. These depressions became the loci for the deposition of tsunami deposits.

At Millers Ferry all faults have a north-south strike. They do not appear to have created downdropped depressions that would have been sites for deposition of clastic deposits. Rather, at Millers Ferry the chaotic beds of the Clayton basal sands appear to have been deposited in broad channels that have a

Figure 9. View of the eroded surface of the Prairie Bluff Chalk. A, view to the east of surface in block A of the right intake wall showing undulating surface of the Prairie Bluff Chalk due to scouring by tsunami (small arrows). B, view of surface in block 6 showing undulating surface outlined by construction walls (small arrows) and groove markings and striations with a north-south orientation (small arrowheads). See Figure 2 for location of photographs.

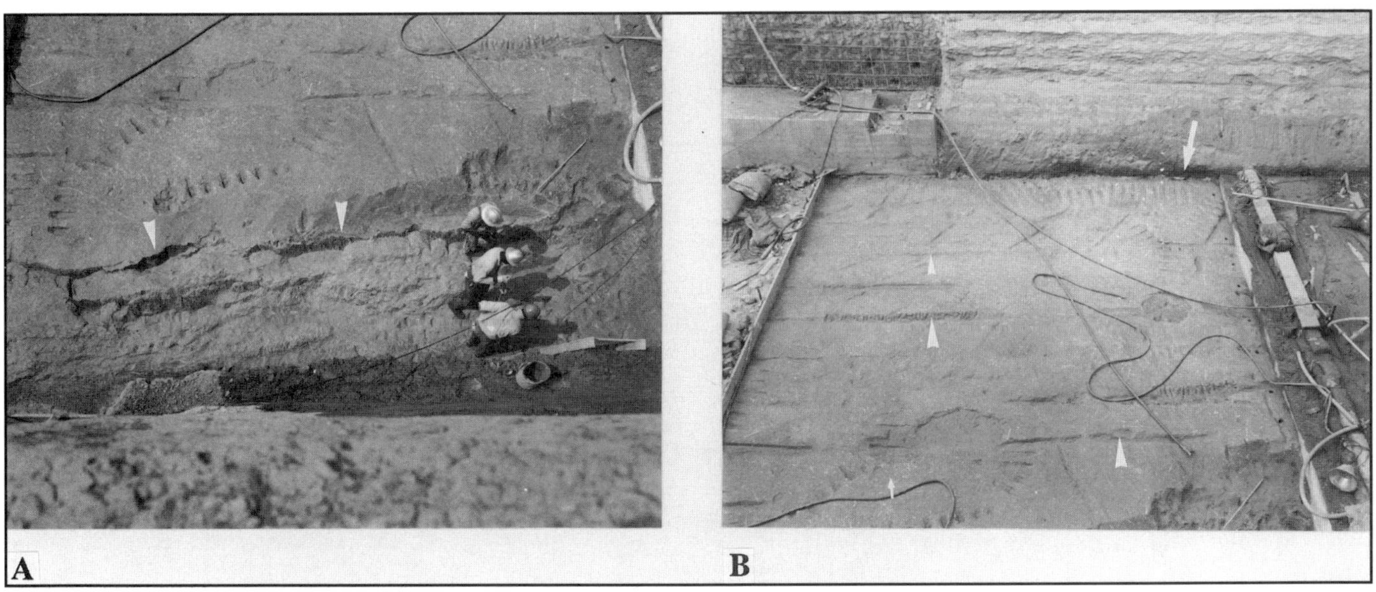

Figure 10. View of faulting of the Prairie Bluff Chalk surface in block B of left intake wall. A, view along eastern side of block showing synthetic slickensided, normal faults that were subsequently buried by the Clayton basal sands (large arrowheads). Relay ramps are visible between the fault segments. Parallel, evenly spaced, sharp, right-angled, deep grooves are from power shovel. Note abundant outlines of megafossils that are visible on the Prairie Bluff Chalk surface. B, view to the west showing surface of Prairie Bluff Chalk with numerous synthetic normal faults (large arrowheads) and groove markings (small arrowhead) with a northerly trend. Abundant outlines of megafossils are visible on the Prairie Bluff Chalk surface (small arrow). The micaceous sands are about 2.3 cm thick in this block (large arrow). Parallel, evenly spaced, sharp, right-angled, deep grooves are from power shovel. See Figure 2 for location of photographs.

northerly strike. They are confined to the channels and onlap the Prairie Bluff Chalk surface. The inner-shelf micaceous sands and silts of the upper part of the Clayton basal sands are more extensive, onlap, and thin over topographic highs. In places in Alabama, as at Braggs, they are absent. We have identified these sands at Moscow Landing in faulted depressions noted by Smit et al. (1994a) beneath the Pine Barren transgressive surface.

EROSION AND THE K/T BOUNDARY HIATUS

The extent of erosion of the surface of the Prairie Bluff Chalk can be estimated in cores 225 and 226 (Fig. 11). Core 225 penetrated all three units of the Clayton basal sands whereas core 226 penetrated only the upper micaceous sand before entering the Prairie Bluff Chalk. Core 225 drilled through about 2.5 m of Clayton basal sands. It contains the lowermost Danian Zones P0, Pα, and P1a. Core 226 intersected a topographic high on the Prairie Bluff Chalk. It penetrated a little over 1 m of the micaceous sand of the Clayton basal sands. The lowermost part of Zone Pα and all of Zone P0 are missing. The Prairie Bluff Chalk in both cores contains the uppermost Maastrichtian *Micula prinsii* calcareous nannofossil zone, and in core 226 the topmost part of the zone is present (Olsson et al., 1992b). Paleomagnetic stratigraphy (Fig. 4) shows that the K/T boundary here lies within Chron 29R. Two datums serve as useful horizons from which to measure the hiatus separating the Danian and the Maastrichtian: the last occurrence (LO) of the planktonic foraminifer *Gansserina gansseri* and a local last occurrence of a benthic foraminifer *Clavulinoides trilaterus*. The last occurrence of *G. gansseri* is found in both cores at the same elevation (Fig. 11). This species has its last occurrence within the *Abathomphalus mayaroensis* Zone. The last occurrence of the benthic species *Clavulinoides trilaterus* occurs just below the LO of *G. gansseri*. These datums suggest that up to a meter of the top of the Prairie Bluff Chalk may have been eroded or peeled off if the seabed was flat. Alternately, the Prairie Bluff seabed may have had some relief that was subsequently modified by earthquakes and by severe current action. In any case, we have the interesting stratigraphic situation whereby the erosional hiatus of the Prairie Bluff Chalk closes on the topographic highs and the nondepositional hiatus of the Clayton basal sands closes in the topographic lows (Fig. 11). This indicates that the hiatus separating the Clayton basal sands and the Prairie Bluff at Millers Ferry must be very short, a situation that would support a brief or even instantaneous erosional event at the K/T boundary.

The paleogeographic setting of the Prairie Bluff Chalk at the end of the Maastrichtian indicates that deposition of chalk occurred south of a siliciclastic environment that is represented by the Providence Sand (Fig. 12). Sandy intervals observed in the cores of the Prairie Bluff Chalk indicate periods of increased influx of siliciclastic sediment from this source into the chalk environment. The coarse sands of the chaotic unit of the Clayton basal sands most probably occurred from backwash of a tsunami at K/T time. The broken blocks of Prairie Bluff Chalk in this unit were probably ripped up by the passage of the wave as it moved northward from its point of origin. Subse-

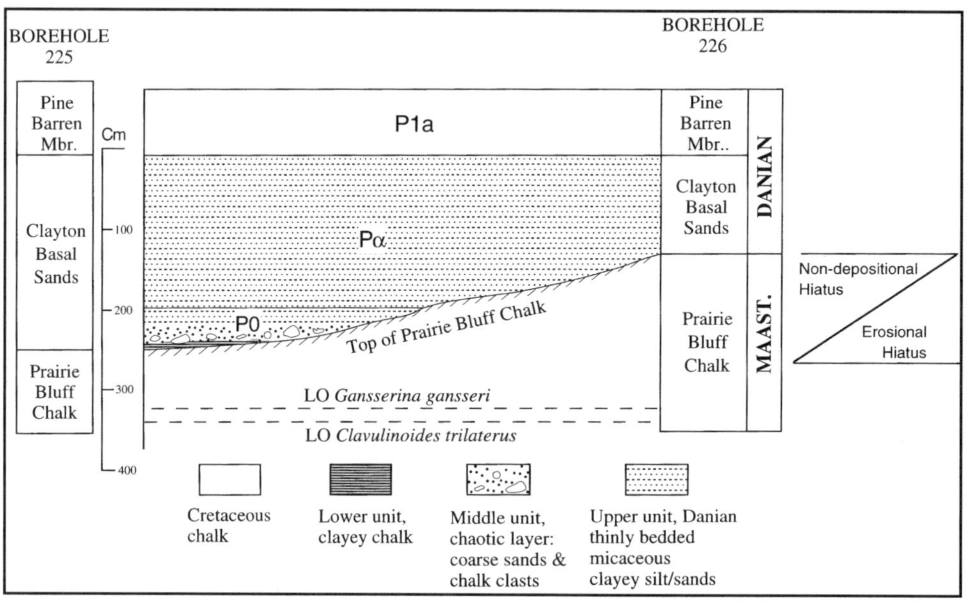

Figure 11. Hiatus at the K/T boundary due to the K/T event as expressed in cores 225 and 226. Datum levels of the last occurrences of *Gansserina gansseri* and *Clavulinoides trilaterus* are used as a measure of erosion of the top of the Prairie Bluff Chalk. This indicates that about 0.9 m of the chalk was removed by the tsunami if the sea floor was level. Following the hiatus from the oldest Danian sediments (Zone P0) to the youngest Prairie Bluff Chalk on topographic highs suggests a very brief, perhaps instantaneous, event at K/T boundary time.

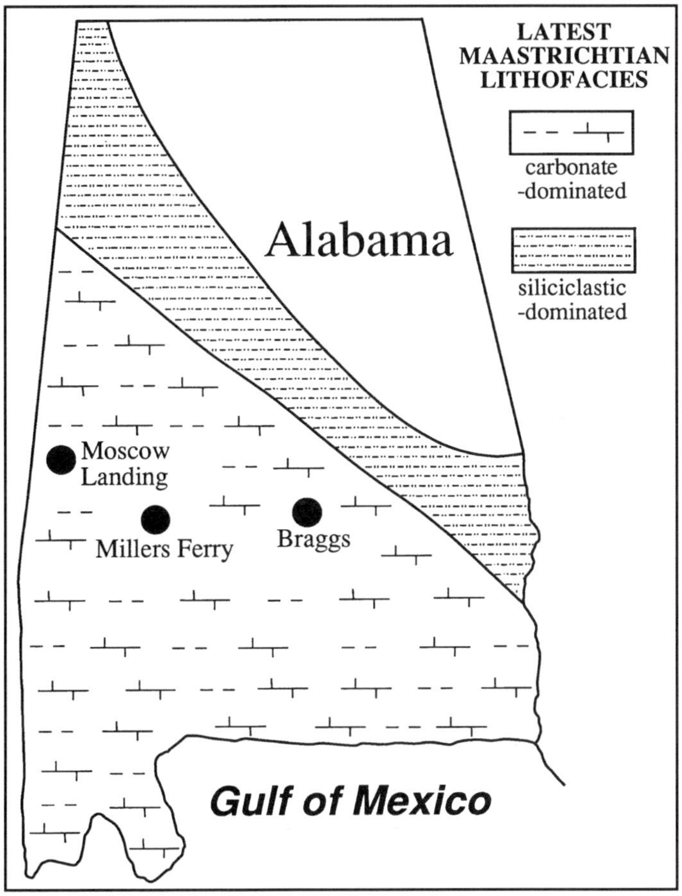

Figure 12. Paleogeography of Alabama during the Maastrichtian showing chalk and siliciclastic environments of deposition. Adapted from figure 5.10 in Sohl and Koch (1986).

quent movement of these blocks may have occurred during the backwash event (Fig. 8). Deposition during earliest Danian time of the finer-grained micaceous clayey sands indicates that chalk deposition had ended. In fact, the stratigraphic evidence indicates that deposition of the Prairie Bluff Chalk ended abruptly (Figs. 6, 10). Chalky environments do not reappear in Alabama until deposition of the transgressive Pine Barren Member. The abrupt end of the Prairie Bluff Chalk implies that the biota that composes the chalk also disappeared abruptly. The abundant molluscs, foraminifera, calcareous nannofossils, and other calcareous fossil groups that compose the Prairie Bluff Chalk contrast sharply with the much sparser assemblages found in the Clayton basal sands (Figs. 6, 10). We suggest not only that the physical evidence at Millers Ferry excavation points to a high-energy catastrophic event at K/T boundary time but that the disappearance of the Prairie Bluff Chalk is also evidence of a massive biotic crisis and extinction.

The upper micaceous sand of the Clayton basal sands contains the early phylogeny of Danian planktonic foraminifera. Our studies (Liu and Olsson, 1992, 1994; Olsson and Liu, 1993) have shown that the derivation of Danian planktonic foraminifera can be traced to three Cretaceous taxa: *Guembelitria cretacea* and two species of *Hedbergella*. *G. cretacea* and its descendants were opportunistic species that flourished in the earliest Danian oceans but ultimately, like all opportunistic species, gave way to better-adapted species once the environment settled down. *Hedbergella* was ancestral to the species that radiated to fill the available niches of the Danian oceans, a radiation which gave rise to the major groups of planktonic foraminifers that characterized Cenozoic oceans to the present day. If other Cretaceous taxa survived into the Danian (Keller, 1988, 1989; Canudo et al., 1991), they did not play an important role in repopulating the Danian oceans.

SEA LEVEL AT K/T BOUNDARY TIME

Olsson and Liu (1993) pointed out that the placement of the K/T boundary at Braggs, Alabama, within the transgressive Pine Barren Member by the Exxon group (Donovan et al., 1988) was in error. Habib et al. (1992) showed that the K/T boundary at Braggs is correctly placed at the top of the Prairie Bluff Chalk rather than within the Pine Barren Member. Our analysis at Braggs supports the conclusion of Habib et al. since we have recovered the marker species for Zone Pα, *Parvularugoglobigerina eugubina*, in the basal beds of the Pine Barren. We use the paleoslope models of Maastrichtian and Paleocene paleodepth distribution of benthic foraminifera developed by Olsson and Nyong (1984) and Olsson and Wise (1987) in analyzing benthic assemblages of foraminifera across the K/T boundary at Millers Ferry (Fig. 13). These models indicate that paleodepths were on the order of 30 m or so at the time of the boundary event. Significant changes in paleodepth across the K/T boundary or changes in species associations that differ from the paleoslope models cannot be demonstrated in these data. A disruption of benthic assemblages following the K/T event may have occurred, but if it did, benthic assemblages recovered rapidly, as can be observed in their stratigraphic distribution (Figs. 13, 14). Benthic foraminiferal changes that do occur are related to substrate changes (i.e., chalk to sand substrate). According to the benthic foraminiferal morphogroup categories of Koutsoukos and Hart (1990), most species in the Prairie Bluff Chalk were infaunal deposit feeders, whereas most species in the Clayton basal sands were epifaunal deposit feeders. Most of these benthic species have geologic ranges in the Cretaceous and in the Tertiary. These data suggest that sea level was probably changing slowly and was in a low-stand phase (possibly early transgressive). A rapid rise in sea level at the transgressive surface separating the Pine Barren Member from the Clayton basal sands is evident in the benthic foraminiferal assemblages (Fig. 14).

Donovan et al. (1988) placed the K/T boundary in the Pine Barren Member within a transgressive systems tract based on their study of the Braggs, Alabama, section. They placed the K/T boundary in cycle TA 1.1 on the cycle chart (Fig. 15). However, the Exxon cycle chart, published in 1987 and in 1988 by Haq et al., placed the K/T boundary within cycle TA 1.2

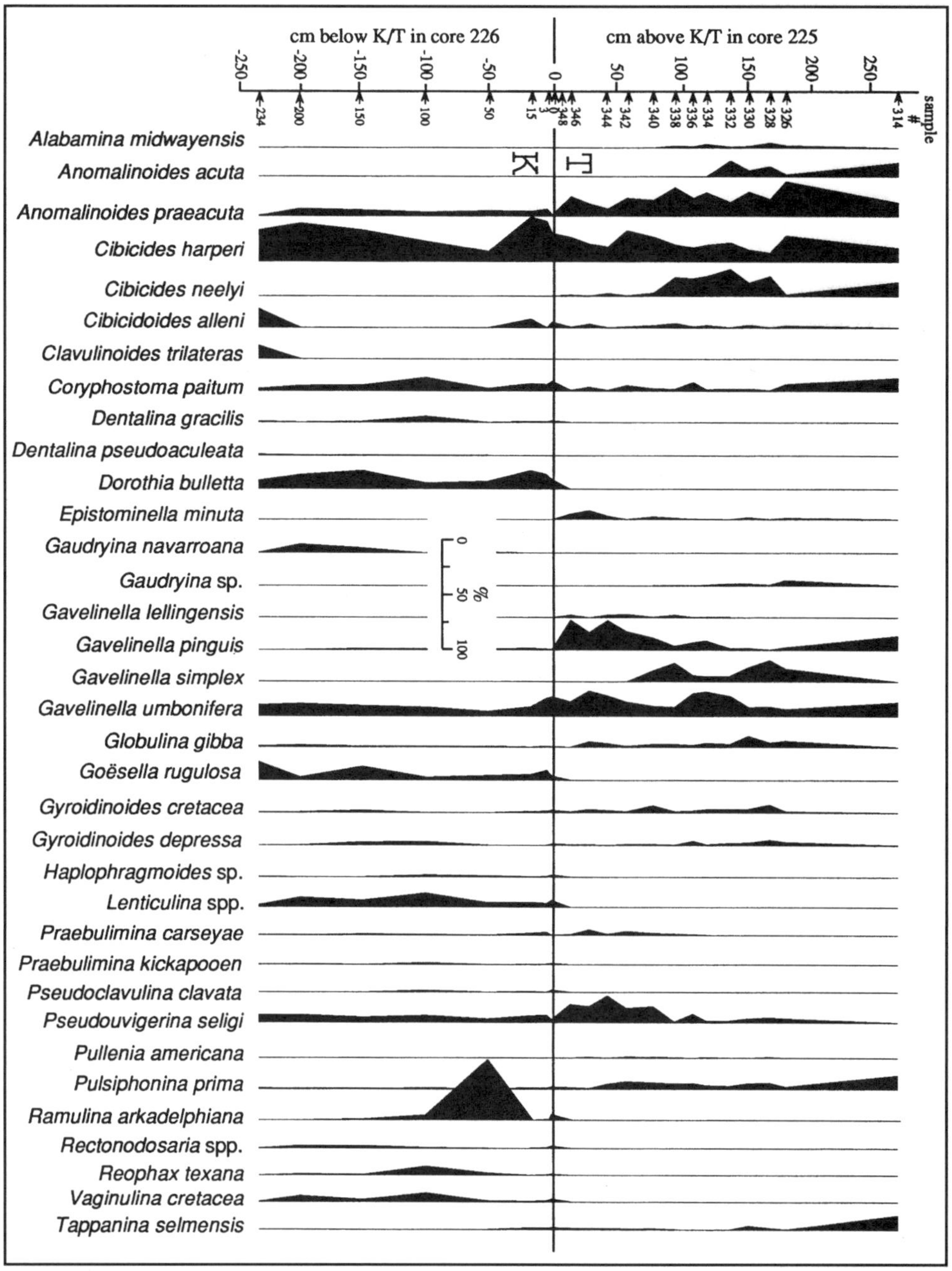

Figure 13. Relative abundance of benthic foraminifera across the K/T boundary at Millers Ferry. Cretaceous benthic abundances are from core 226, and Danian benthic abundances are from core 225. There is relatively little change in the composition of benthic assemblages across the boundary, suggesting that no significant change (up or down) in sea level occurred during the transition from the Cretaceous to the Tertiary. Change in abundance of some species across the boundary is related to substrate change from chalk to sand. See text for explanation. Most species have both Cretaceous and Tertiary ranges. Paleodepth is estimated by using paleoslope models of Olsson and Nyong (1984) and Olsson and Wise (1987).

(Fig. 15). The study by the Exxon group of Donovan et al. (1988) at Braggs revised the K/T data for the cycle chart. In the revision by Donovan et al. the Prairie Bluff Chalk is placed in cycle UZA 4.6 (Fig. 15). But Habib et al. (1992) showed that the K/T boundary at Braggs should be placed at the top of the Prairie Bluff Chalk and not in the Pine Barren. This means that on the cycle chart the K/T boundary would then separate cycles TA 1.1 and UZA 4.6. If the top of the Prairie Bluff Chalk was eroded during a fall in sea level, it would be appropriate to identify these cycles. However, if the top of the chalk was eroded by an impact-generated tsunami, such an event might not coincide with a cycle boundary. Our data, which indicate

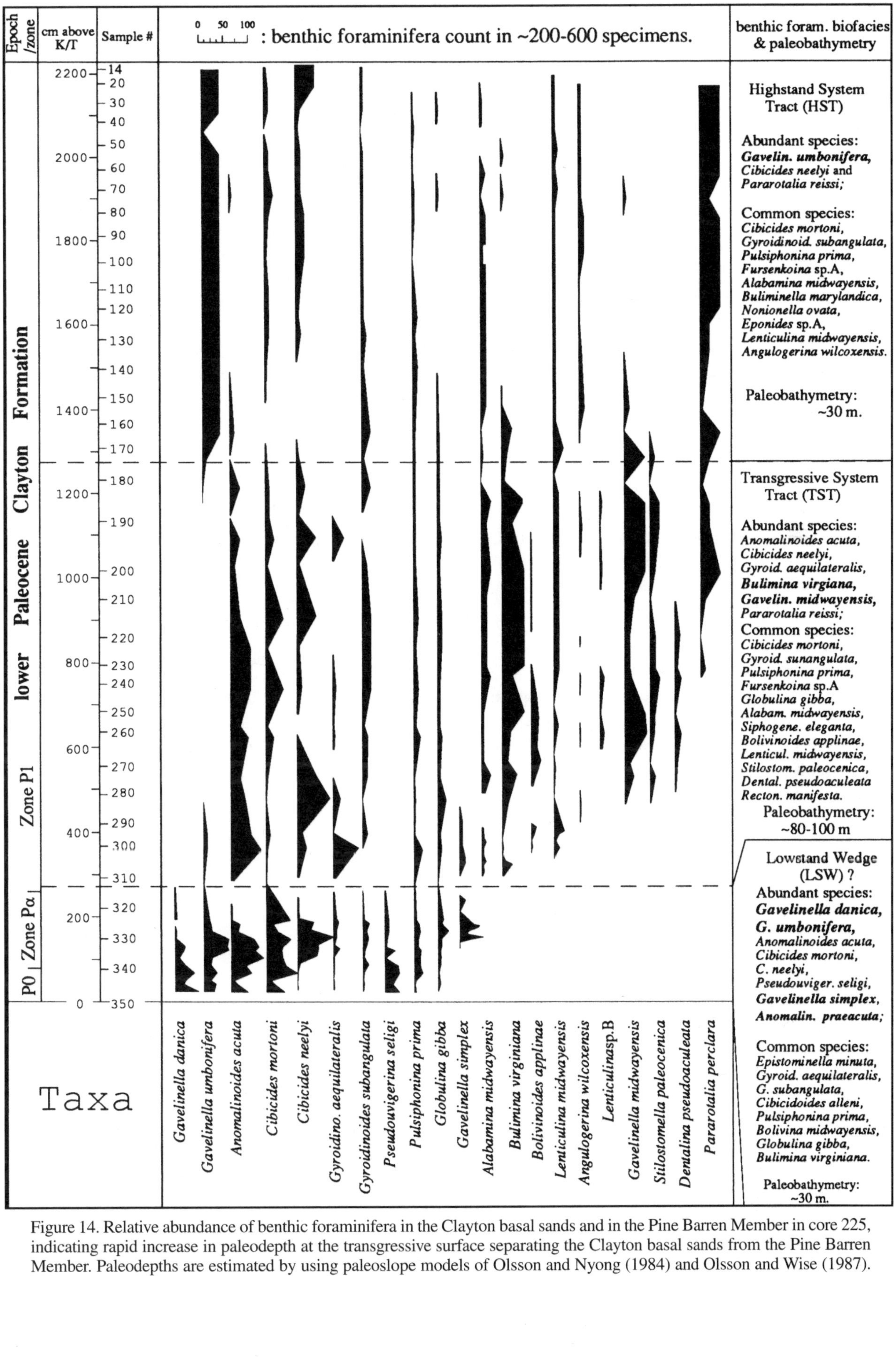

Figure 14. Relative abundance of benthic foraminifera in the Clayton basal sands and in the Pine Barren Member in core 225, indicating rapid increase in paleodepth at the transgressive surface separating the Clayton basal sands from the Pine Barren Member. Paleodepths are estimated by using paleoslope models of Olsson and Nyong (1984) and Olsson and Wise (1987).

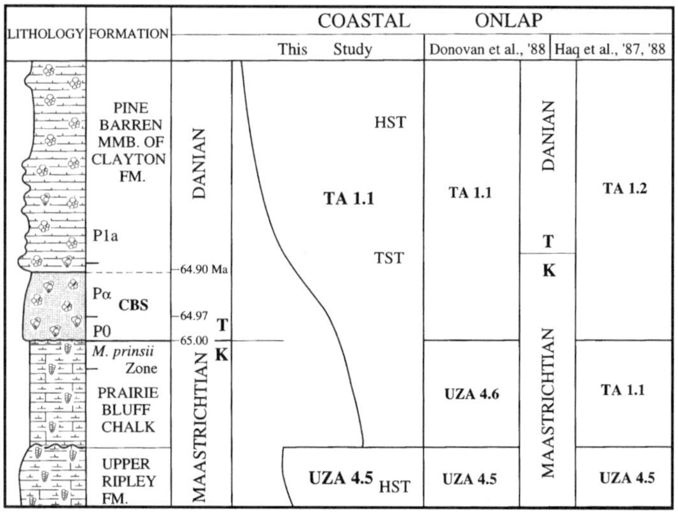

Figure 15. A comparison of the eustatic cycle charts of et al. (1987), Donovan et al. (1988), and this study based on sequence stratigraphy and biostratigraphy in the Alabama Gulf Coast. Adapted from Olsson and Liu (1993).

that there was no abrupt change in sea-level across the K/T boundary, suggest that the boundary lies within sea-level cycle TA 1.1, a placement similar to the proposal of Donovan et al. (1988) but at a different stratigraphic level (Fig. 15).

CONCLUSION

Analysis of photographs of the 1967 excavation at Millers Ferry and two cores taken at the site indicates that the K/T contact at Millers Ferry separates chaotic beds above the contact and a severely scoured, faulted surface below the contact. The erosion and deposition of these features can be related to the bolide impact at Chicxulub, which lay across the Gulf of Mexico directly south of Millers Ferry (Fig. 1). Earthquakes and seismic waves caused normal faulting and liquification of the Prairie Bluff Chalk. A tsunami followed and severely eroded the surface of the Prairie Bluff Chalk, tearing up large blocks of the chalk.

Subsequent backwash from the Providence Sands, which lay to the north, deposited coarse sands that mixed with the chalk clasts, producing a chaotic bed in broad channels. Normal inner-shelf sedimentation resulted in the upper micaceous sand of the Clayton basal sands in which are found primitive Danian planktonic foraminifers. These foraminifers, in turn, led to the radiation that repopulated the early Paleocene seas. The K/T event occurred during a time of sea-level low stand (early transgressive?) within cycle TA 1.1 of Haq and others (1987). The collapse and mass extinctions of marine biota due to the K/T event caused the sudden cessation of chalk deposition. Reestablishment of chalk deposition is recorded in the Pine Barren transgressive systems tract and is related to recovery of the marine biota in late Zone Pα.

ACKNOWLEDGMENTS

We thank Jay Shelley and John Stuckey for their assistance and hospitality in making available photographs of the construction activity at the 1966–1967 Millers Ferry excavation during a visit to the Millers Ferry powerhouse in February 1993. We also thank Bruce McCall, Department of the Army, Mobile District, Corps of Engineers, who assisted in obtaining prints of the 1966 and 1967 activity. We appreciate the advice given to us by Drs. Gail M. Ashley and Roy W. Schlische, Department of Geological Sciences, Rutgers University, on the interpretation of sedimentary and fault structures observed in the photographs of the Millers Ferry excavation. We thank S. Gartner, E. Mancini, and C. Savdra for critical reviews that led to improvement of the manuscript. These reviewers do not necessarily support our conclusions, for which we are solely responsible. This research is partially supported by the Department of Geological Sciences, Rutgers University.

REFERENCES CITED

Bourgeois, J., Hansen, T. A., Wiberg, L., and Kauffman, E. G., 1988, A tsunami deposit at the Cretaceous-Tertiary boundary in Texas: Science, v. 241, p. 567–568.

Bryan, J. R., and Jones, D. S., 1989, Fabric of the Cretaceous-Tertiary marine macrofaunal transition at Braggs, Alabama: Palaeogeography, Palaeoclimatology, Palaeoecology, v. 69, p. 279–301.

Canudo, J. I., Keller, G., and Molina, E., 1991, Cretaceous-Tertiary boundary extinction pattern and faunal turnover at Agost and Caravaca, S.E. Spain: Marine Micropaleontology, v. 17, p. 319–341.

Donovan, A. D., Baum, G. R., Blechschmidt, G. L., Loutit, L. S., Pflum, C. E., and Vail, P. R., 1988, Sequence stratigraphic setting of the Cretaceous-Tertiary boundary in central Alabama, *in* Wilgus, C. K., and five others, eds., Sea-level changes—An integrated approach: Society of Economic Paleontologists and Mineralogists Special Publication 42, p. 299–307.

Habib, D., Moskovitz, S., and Kramer, C., 1992, Dinoflagellate and calcareous nannofossil response to sea-level change in Cretaceous-Tertiary boundary sections: Geology, v. 20, p. 165–168.

Haq, B. U., Hardenbol, J., and Vail, P. R., 1987, Chronology of fluctuating sea levels since the Triassic: Science, v. 235, p. 1156–1167.

Haq, B. U., Hardenbol, J., and Vail, P. R., 1988. Mesozoic and Cenozoic chronostratigraphy and cycles of sea-level change, *in* Wilgus, C. K., and five others, eds., Sea-level changes—An integrated approach: Society of Economic Paleontologists and Mineralogists Special Publication 42, p. 72–108.

Hay, W. W., and Mohler, H. P., 1967, Calcareous nannoplankton for early Tertiary rocks at Pont Labau, France, and Paleocene—Early Eocene correlations: Journal of Paleontology, v. 41, p. 1505–1541.

Hemleben, Ch., Mühlen, D., Olsson, R. K., and Berggren, W. A., 1991, Surface texture and the first occurrence of spines in planktonic foraminifera from the early Tertiary: Geologisches Jahrbuch, v. 128, p. 117–146.

Hildebrand, A. R., Penfield, G. T., Kring, D. A., Pilkington, M., Camargo Z., A., Jacobsen, S. B., and Boynton, W. V., 1991, Chicxulub Crater: A possible Cretaceous/Tertiary boundary impact crater on the Yucatán Peninsula, Mexico: Geology, v. 19, p. 867–871.

Keller, G., 1988, Extinction, survivorship and evolution of planktonic foraminifers across the Cretaceous/Tertiary boundary at El Kef, Tunisia: Marine Micropaleontology, v. 13, p. 239–263.

Keller, G., 1989, Extended Cretaceous/Tertiary boundary extinctions and delayed population change in planktonic foraminifera from Brazos

River, Texas: Paleoceanography, v. 4, p. 287–332.

Koutsoukos, E. A. M., and Hart, M. B., 1990, Cretaceous foraminiferal morphogroup distribution patterns, palaeocommunities and trophic structures: A case study from the Sergipe Basin, Brazil: Transactions of the Royal Society of Edinburgh: Earth Sciences, v. 81, p. 221–246.

Liu, C., and Olsson, R. K., 1992, Evolutionary adaptive radiation of microperforate planktonic foraminifera following the K/T mass extinction event: Journal of Foraminiferal Research, v. 22, p. 328–346.

Liu, C., and Olsson, R. K., 1994, On the origin of Danian normal perforate planktonic foraminifera from *Hedbergella*: Journal of Foraminiferal Research, v. 24, p. 61–74.

Mancini, E. A., Tew, B. H., and Smith, C. C., 1989, Cretaceous-Tertiary contact, Mississippi and Alabama: Journal of Foraminiferal Research, v. 19, p. 93-104.

Olsson, R. K., 1970, Planktonic foraminifera from base of Tertiary, Millers Ferry, Alabama: Journal of Paleontology, v. 44, p. 598–604.

Olsson, R. K., and Liu, C., 1993, Controversies on the placement of Cretaceous-Paleogene boundary and the K/P mass extinction of planktonic foraminifera: Palaios, v. 7, p. 127–139.

Olsson, R. K., and Nyong, E. E., 1984, A paleoslope model for Campanian–lower Maastrichtian foraminifera of New Jersey and Delaware: Journal of Foraminiferal Research, v. 14, p. 50–68.

Olsson, R. K., and Wise, S. W., Jr., 1987, Upper Paleocene to middle Eocene depositional sequences and hiatuses in the New Jersey Atlantic margin, *in* Ross, C.A., and Hyman, D., eds., Timing and depositional history of eustatic sequences: Cushman Foundation Foraminiferal Research Special Publication 24, p. 99–112.

Olsson, R. K., Liu, C., Aubry, M-P., and Ehrendorfer, T., 1992a, A biostratigraphically continuous Cretaceous/Paleogene section at Millers Ferry, Alabama: Support for a bolide impact: Geological Society of America Abstracts with Programs, p. A28.

Olsson, R. K., Hemleben, Ch., Berggren, W. A., and Liu, C., 1992b, Wall texture classification of planktonic foraminifera genera in the lower Danian: Journal of Foraminiferal Research, v. 22, p. 195–213.

Savdra, C. E., 1993, Ichnosedimentologic evidence for a noncatastrophic origin of Cretaceous-Tertiary boundary sands in Alabama: Geology, v. 21, p. 1075–1078.

Smit, J., 1982, Extinction and evolution of planktonic foraminifera after a major impact at the Cretaceous/Tertiary boundary: Geological Society of America Special Publication 190, p. 329–352.

Smit, J., and eight others, 1992, Tektite-bearing, deep-water clastic unit at the Cretaceous-Tertiary boundary in northeastern Mexico: Geology, v. 20, p. 99–103.

Smit, J., Roep, Th. B., Alvarez, W., Montanari, S., and Claeys, P., 1994a, Stratigraphy and sedimentology of KT clastic beds in the Moscow Landing (Alabama) outcrop: Evidence for impact-related earthquakes and tsunamis, *in* New developments regarding the K/T event and other catastrophes in Earth history [abs.]: Houston, Texas, Lunar and Planetary Institute Contribution 825, p. 119–120.

Smit, J., Roep, Th. B., Alvarez, W., Claeys, P., Montanari, S., and Grajales, M., 1994b, Impact-tsunami-generated clastic beds at the KT boundary of the Gulf Coast Plain: A synthesis of old and new outcrops, *in* New developments regarding the K/T event and other catastrophes in Earth history [abs.]: Houston, Texas, Lunar and Planetary Institute Contribution 825, p. 117–119.

Sohl, N., and Koch, C., 1986, Molluscan biostratigraphy and biofacies of the *Haustator bilira* assemblage zone (Maastrichtian) of the East Gulf Coastal Plain, *in* Reinhardt, J., ed., Stratigraphy and sedimentology of continental, nearshore, and marine Cretaceous sediments of the eastern Gulf Coastal Plain: Guidebook No. 3, Georgia Geological Society annual meeting, Atlanta, Georgia, p. 45–56.

Manuscript Accepted by the Society September 14, 1995

Printed in U.S.A.

Geological Society of America
Special Paper 307
1996

Stratigraphic distribution of extraterrestrial markers at the Cretaceous-Tertiary boundary in the Gulf of Mexico area: Implications for the temporal complexity of the event

R. Rocchia, E. Robin, L. Froget, and J. Gayraud
Centre des Faibles Radioactivités, Laboratoire mixte CEA-CNRS, Avenue de la Terrasse, 91198 Gif-sur-Yvette cedex, France

ABSTRACT

In the Gulf of Mexico area, Cretaceous-Tertiary sections are characterized by a rather unusual depositional sequence. The lithological units, which are not all present in all sections, are a bed of spherules containing rare glass particles, succeeded by a sandstone, with a rather thick series of marly ripple beds at the top. Ir and Ni-rich spinels, markers derived from extraterrestrial material, are found only in the uppermost part of the sequence. The clear stratigraphical separation of the units and the late deposition of the cosmic markers are not easily explained by a single collisional event. Tentatively, we propose a multiple collision process resulting from the fragmentation of a single-incident bolide, colliding with the Earth at grazing incidence, with ricochet and rebound of fragments.

INTRODUCTION

In the Gulf of Mexico area, Cretaceous-Tertiary boundary sedimentary sequences are markedly distinct from those found in the rest of the world. In this region, the sequences are rather complex and contain lithological units that are either absent or not clearly identified elsewhere. These units are a basal bed of spherules or coarse globules containing glass particles, overlain by a massive sandstone bed, and capped by fine-grained ripple beds. These units have been observed at several sites, separated by several hundreds of kilometers, from Alabama and Texas to southern Mexico. Divergent explanations have been proposed for this sedimentary sequence. The sandstone bed could represent a series of turbidites, deposited over 10^5 years or so, without a genetic link to the K/T event (Stinnesbeck et al., 1993). Another explanation is that the sandstone was produced by a tsunami or a series of megawaves generated by a proximal impact (Smit et al., 1992). In that last hypothesis, the spherule bed is interpreted as the remains of the ejecta produced by the proposed impact on the Yucatan peninsula (Hildebrand et al., 1991; Sharpton et al., 1992). If so, the entire sequence, including the ripple beds and the marly layers overlying the massive sandstone, should represent a single and brief event. Clearly, the evaluation of the timing of the K/T transition sequence is critical for the understanding of the boundary event. The stratigraphic position of the usual K/T markers (overabundances of Platinum Group Elements, nickel-rich spinels, and shocked quartz) is a potential tool in evaluating the formation of the sequence with a fine time resolution. We report here the distributions of Ni-rich spinels and iridium.

IMPORTANCE OF NI-RICH SPINELS

Ni-rich spinels are minerals formed when a piece of extraterrestrial matter (rich in Ni) is melted and oxidized in the Earth's atmosphere (high O_2 fugacity). Such minerals are systematically found in cosmic spherules (Brownlee, 1981; Robin et al., 1992), in the fusion crust of meteorites (Robin et al., 1992; Doehm and Margolis, 1990), in the ablation products found in a Jurassic horizon (Jéhanno et al., 1988), and in particles associated with oceanic impact debris from late Pliocene sediments (Margolis et al., 1991). They differ from terrestrial

Rocchia, R., Robin, E., Froget, L., and Gayraud, J., 1996, Stratigraphic distribution of extraterrestrial markers at the Cretaceous-Tertiary boundary in the Gulf of Mexico area: Implications for the temporal complexity of the event, *in* Ryder, G., Fastovsky, D., and Gartner, S., eds., The Cretaceous-Tertiary Event and Other Catastrophes in Earth History: Boulder, Colorado, Geological Society of America Special Paper 307.

spinels by their high Ni content (>1 wt%), which indicates an undifferentiated source material (meteoritic or mantle material), and by their high iron oxidation state ($Fe^{3+}/Fe_{tot} > 70$ atom.%) resulting from their formation under relatively high oxygen fugacities (>10^{-4}atm.) (Robin et al., 1991; Gayraud et al., 1994). Terrestrial magmas, poor in Ni, do not evolve under such high oxygen pressures (Haggerty, 1976). This explains why highly oxidized Ni-rich spinels have no counterpart in terrestrial rocks. They are specific markers of meteoritic materials that interacted at high temperature with the Earth's atmosphere, the only understandable mechanism known to produce these minerals.

Ni-rich spinels are also found in most K/T boundary sections. The presence of such minerals in K/T sediments, in close coincidence with the planktonic crisis (Robin et al., 1991), is a major argument supporting an extraterrestrial origin for the K/T boundary biological crisis. In addition, Ni-rich spinels are very useful for timing considerations because they were contained in rather big droplets or partly melted fragments of meteoritic material (Jéhanno et al., 1992; Robin et al., 1993) of up to several hundred microns that settled rapidly. In addition, after sedimentation Ni-rich spinels, like other primary minerals and unlike Ir, cannot move in the compacted sediment even after alteration of their embedding matrix. Therefore, they provide a high resolution record of a cosmic event. This is clearly illustrated by our data from the site of El Kef, the stratotype of the K/T boundary in the mesogean realm. The El Kef section is characterized by a high sedimentation rate and appears to have been little affected by bioturbation. Under such exceptional conditions the K/T event should have been recorded with minimum distortion. This is confirmed by the distribution of Ni-rich spinels: These minerals are found only in a millimeter-thick rusty-brown layer, consisting of goethite and marking the paleontological boundary (Robin et al., 1991) (Fig. 1). The Ir distribution is markedly different and extends over more than 2 m.

Such a difference is partly explained by the mobility of Ir in the sediment (upper Maastrichtian overabundances) and partly by the residence time of this element (Hodge et al., 1986) in sea water (distribution tail extending over the first 2 m of basal Danian). Obviously, the occurrence of maximum concentrations of Ir and spinels in the same horizon points to a common source. However, the most constraining information about the timing of the K/T event is provided by the millimeter-thick distribution of Ni-rich spinels, which indicates an instant event or an event shorter than 100 yr, on the basis of a constant clay deposition rate across the boundary (Robin et al., 1991). These results suggest that some valuable information about the controversial depositional sequence of the Caribbean K/T sections could be derived from the distribution of Ir and Ni-rich spinels.

IR AND SPINEL DISTRIBUTIONS

Analytical methods

Iridium concentrations are measured by Instrumental Neutron Activation Analysis. After irradition of whole rock samples in the neutron flux of the Osiris reactor (Laboratoire Pierre Süe, Saclay), iridium is counted with a γ–γ spectrometer detecting the coincidence of 468 and 316 keV γ-ray lines emitted during the decay of ^{192}Ir.

Ni-rich spinels are strongly magnetic minerals and can therefore be easily extracted from a liquid phase with an electro-

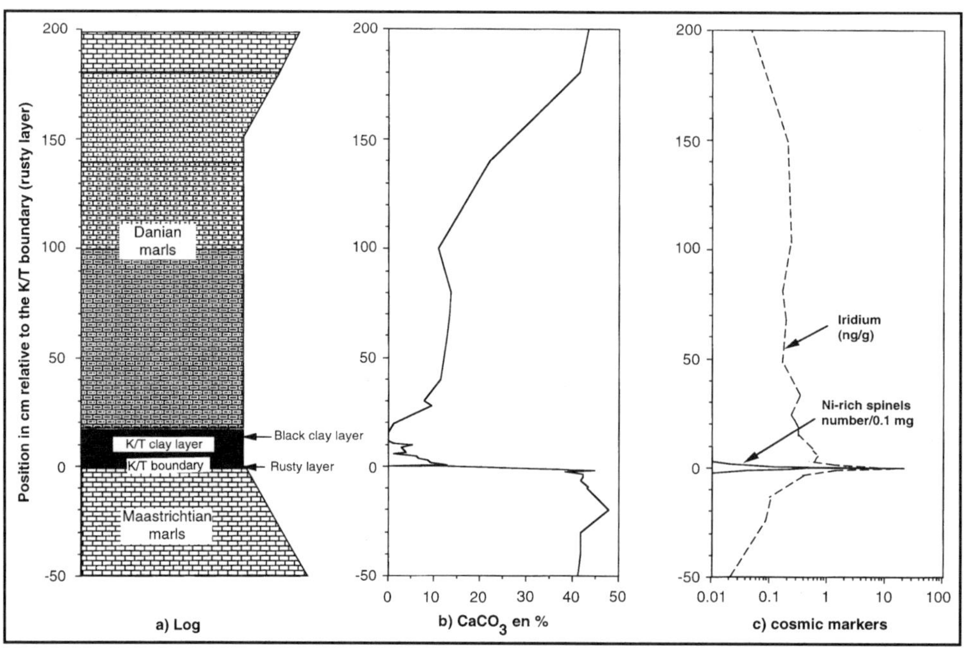

Figure 1. Distribution of Ir and Ni-rich spinels across the K/T boundary of El Kef (Tunisia). Note the pulselike distribution of Ni-rich spinels. The Ir background level is below 30 pg/g.

magnet. Sediment samples were first ground and homogenized. After dissolution of the carbonate fraction with acetic acid, the residue circulates in a closed-loop funnel activated by a peristaltic pump. Magnetic particles are collected in a glass tube placed in the gap of the magnetic separator. The system is run for 20 to 30 minutes to ensure that all magnetic particles are collected. The funnel is then rinsed with water. The magnetic separator is switched off and the particles that stopped in the magnet gap are recovered by washing out the glass tube with alcohol and directly injecting into the filtration system. The collection efficiency of the system for magnetic particles such as Ni-rich spinels is higher than 95%. After filtration on a 0.4-µm Nuclepore filter, spinels larger than 1 µm are counted and sorted according to their size, using an automatic search routine on the scanning electron microscope.

Sites

For this work, we have studied four sites at selected locations. Beloc (Haiti) is a deep-water site located close to, but outside, the Gulf of Mexico. Two sites, Mimbral (Mexico) and Brazos River (Texas), that are presently in the Gulf coastal region were formed under lower water depth (≈500 m and ≈100 m respectively). The fourth is a deep-water site inside the Gulf, Offshore Drilling Project (ODP) Hole 536.

Beloc (Haiti)

Results from this site have already been published (Jéhanno et al., 1992). In this section, formed under about 2,000 m of water, the sandstone is missing. The spherule bed (25 to 30 cm thick) is nicely size graded. Ir and Ni-rich spinels appear significantly higher up section, in a clayey layer about 50 to 60 cm above the base of the spherule bed and separated from the top of this bed by several tens of centimeters of carbonate-rich sediments (Fig. 2).

Comments. At Beloc, the cosmic markers and the spherules are not mixed: They are located in stratigraphically distinct layers. On the other hand, glass from the spherule bed differs significantly from tektites (Jéhanno et al., 1992). On stratigraphic and compositional grounds, there is no evidence for a genetic link between the glass spherules and the K/T cosmic markers. However, the interval of time separating the deposition of spherules, on one hand, and Ir and Ni-rich spinels, on the other hand, is not necessarily long. The size grading observed in the spherule bed, likely due to the size-dependent transit time through 2,000 m of water, suggests that at least this part of the sequence was deposited rapidly. We cannot presently evaluate the deposition rate of the intercalated carbonate-rich layer. However, the fact that spinel-bearing spheroids and the smallest glass spherules with the same size (0.2 mm) are found in well-separated layers implies a minimum delay of about one week, but not necessarily much longer (Jéhanno et al., 1992).

Mimbral (Mexico)

This section (Smit et al., 1992) is typical of the K/T sequence observed at several sites around the Gulf of Mexico (Keller et al., 1994). The basal unit, the spherule layer, has a highly variable thickness (from 10 cm to about 1 m) and a sharp contact with the overlying 2- to 3-meter-thick massive sandstone bed. Here again the cosmic markers, Ir and spinels, are stratigraphically separated from the spherule bed. They appear in the sedimentary sequence on top of the massive sandstone bed in the 15- to 20-cm-thick series of rippled alternating layers of sandy and clayey marl beds. Ni-rich spinels must not be con-

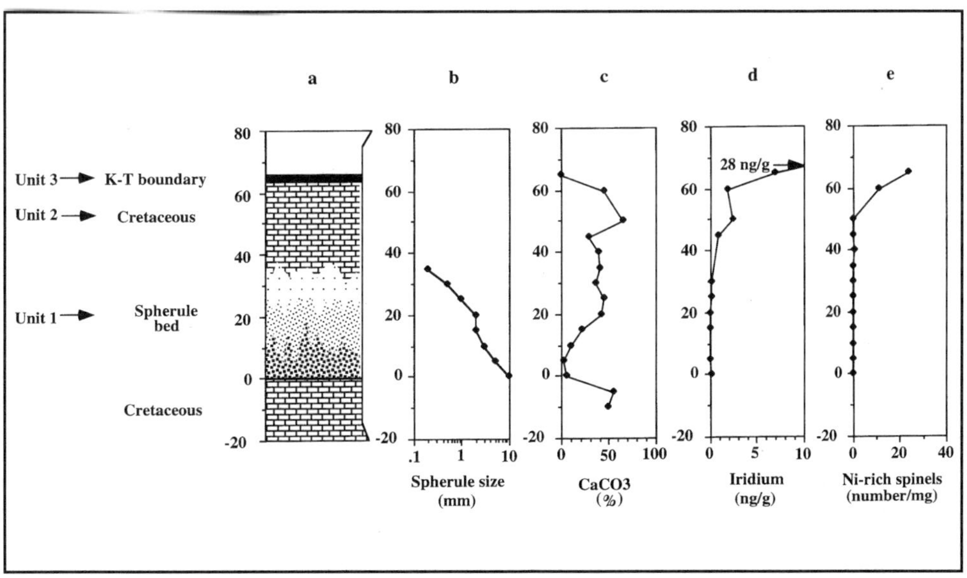

Figure 2. Distribution of Ir and Ni-rich spinels across the K/T boundary of Beloc (Haiti).

fused with the spinels found in the spherule layer by Smit et al. (1992), which are terrestrial alumino- and titano-chromites similar to those found all over the section. Both Ir and Ni-rich spinel distributions are rather extended (Fig. 3). Concentrations are highly varied, with a clear tendency to be higher in the clayey marl layers. Concentrations (maximum values: 0.5 ng/g and 300 spinel/g, respectively) are lower than in distant sections (9 to 28 ng/g and 30,000 spinels/g at Beloc, for instance). The fluences of Ir (15 ng/cm^2) and spinels (3 10^3 spinels/cm^2, compared with 10^5 spinels/cm^2 at Caravaca) are also very low.

Comments. The extent of the spinel distribution—all over the ripple beds—is quite different from the pulselike distribution observed at El Kef. It could be explained by a strong bioturbation. However, the alternate high and low concentrations of Ir and Ni-rich spinels observed in the clayey and sandy marl layers of the ripple beds suggest another explanation: periods of coarse-grain deposition in a rapid flow (sandy marls) followed by periods of more quite sedimentation conditions (clayey marls with accumulation of Ir and Ni-rich spinels). If this interpretation is correct, the 20- to 30-cm-thick ripple beds should have been deposited in an extremely short time, corresponding to the settling time of Ni-rich spinels–bearing spheroids (transit through the atmosphere and the column of water) and therefore not exceeding a few days.

Brazos river (site 1—Texas)

In this section, already described several times (Hansen et al., 1987, for instance), we have not found a single Ni-rich spinel. Iridium overabundances are observed only above the presumed tsunami deposit, with a maximum concentration of 1.4 ng/g in the mudstone 15 cm above the top of this deposit (Fig. 4). Curiously, a sharp minimum Ir concentration is found immediately above the maximum value in a rust-colored thin sandy layer.

Comments. According to chemical data, the K/T boundary should be placed at the level where the maximum Ir concentration is found, consistent with paleontological data (Keller, 1989). The time represented by the 15-cm-thick mudstone layer below the maximum Ir concentration is uncertain: Beeson et al. (1994) claim that there is no evidence of upward fining of sediments or of size sorting of fossils in this interval. Smit et al. (1994a) disagree and present arguments favoring the idea that the basal 15-cm-thick mudstone layer was deposited in a few days at most. The only point derived from Ir data that could suggest a rapid deposition is the existence of the rust-colored sandy layer. The low Ir concentration observed in this layer could result from the dilution produced by a short period of accelerated deposition.

ODP Hole 536 (Gulf of Mexico)

This site, located far from the coast, corresponds to rather deep water sedimentation conditions. In this core, the spherule bed, *sensu stricto,* does not exist. Some rare spherules with glass shards have been found at the bottom of the core (in the core catcher), but these are below Ir and Ni-rich spinels (Alvarez et al, 1992). The sandstone bed is also missing. Ir and Ni-rich spinels (Fig. 5) are found mainly in a calcareous ooze containing Cretaceous fossils (Keller et al., 1993) (section 536-9-5, 75 to

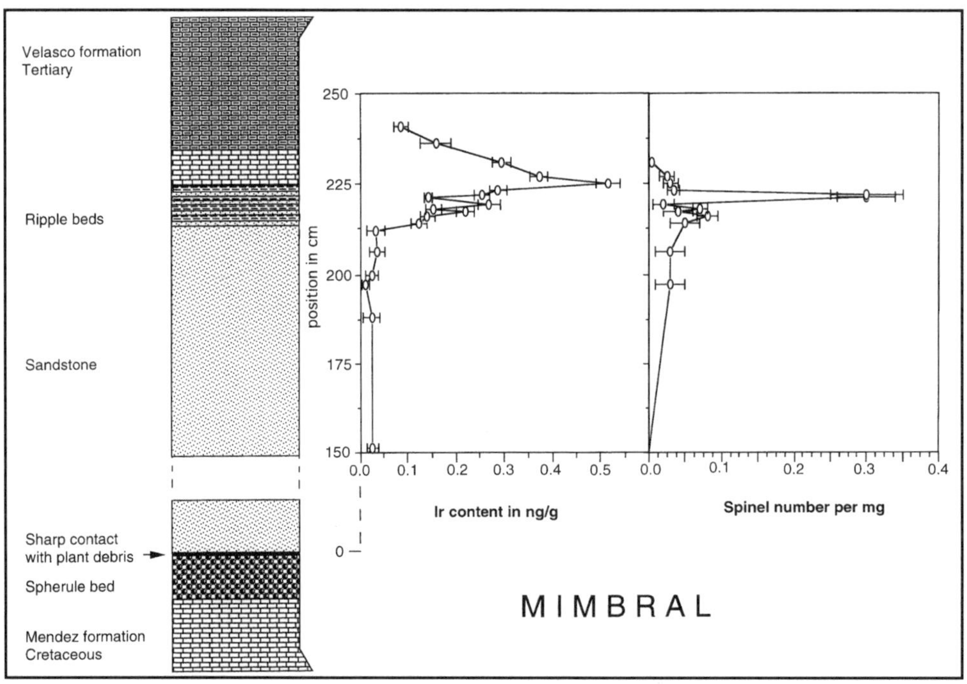

Figure 3. Distribution of Ir and Ni-rich spinels across the K/T boundary of Mimbral (Mexico).

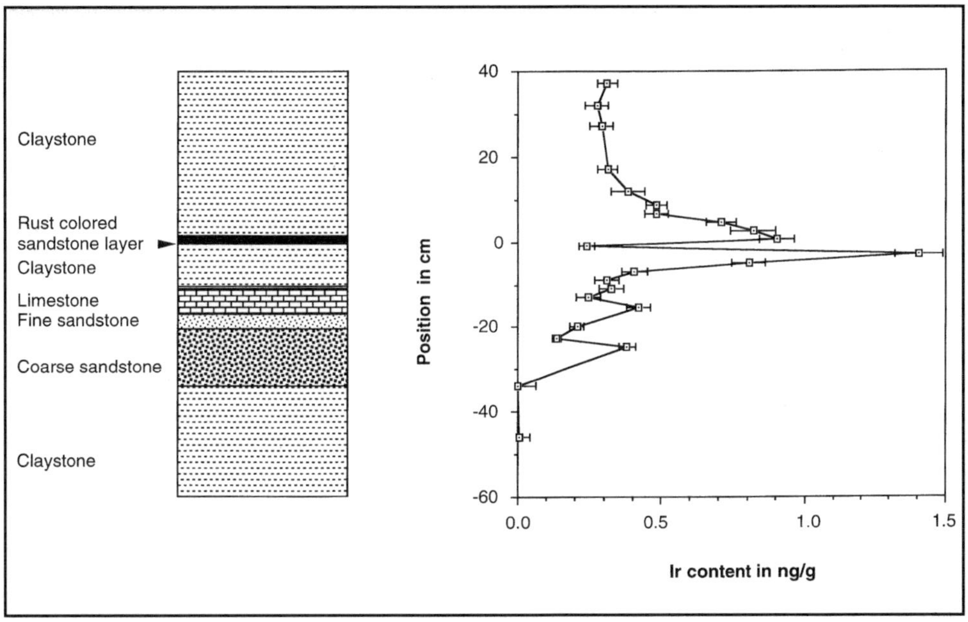

Figure 4. Distribution of Ir across the K/T boundary of Brazos River 1 (Texas). Ni-rich spinels have not been found in this section.

120 cm), the unit 4 of Alvarez et al. (1992). The distributions of Ir and Ni-rich spinels are both extended over more than 50 cm. Ir concentrations are low (maximum 0.7 ng/g), but the fluence is rather elevated, 55 ng/cm^2, and similar to the global average. Spinel concentrations (15,000 spinels/g) and fluence (10^5 spinels/cm^2) are also high.

Comments. The extent of Ir and Ni-rich spinel distributions, when compared with El Kef data, suggests that deposition occurred in a short time. This is also the opinion of Alvarez at al. (1992), who think that this part of the sedimentary sequence represents the finest fraction of Cretaceous sediments stirred up by the K/T impact wave. If so, most of the Ir and Ni-rich spinel-bearing particles were deposited at the very end of this sedimentation phase. However, this deposition mode is questioned by the work of Keller et al. (1993), who claim that both small- and large-sized species are present in the sediment, implying "normal" deposition conditions. Whatever the exact explanation for this sequence, it is remarkable that the record of the cosmic event (Ir and Ni-rich spinels) is apparently complete (fluences close to the global average values) in reworked sediments of upper lower Maastrichtian age (Keller et al., 1993) containing no glass particles.

DISCUSSION

The four sections studied in this chapter, even if they differ on some points, have in common the fact that the different markers of the K/T boundary transition (glass spherules, turbiditic or tsunami-generated deposits, cosmic markers), when present, are found unmixed in very distinct lithological units.

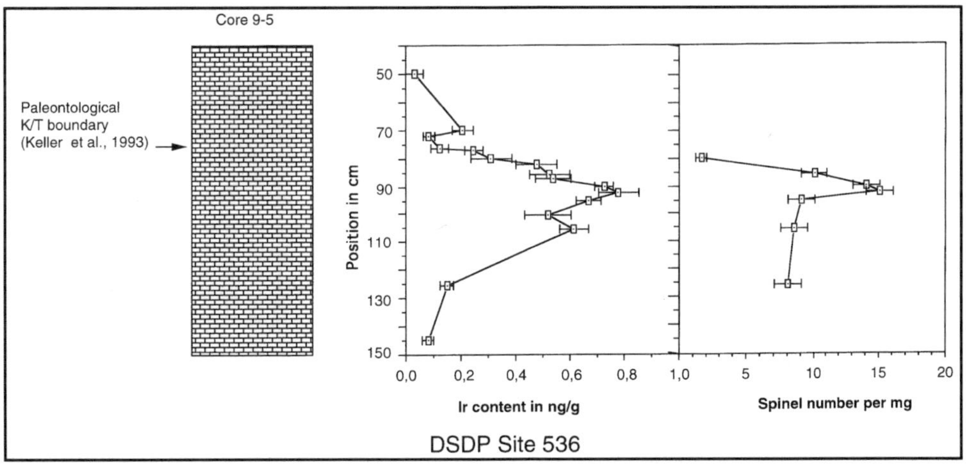

Figure 5. Distribution of Ir and Ni-rich spinels across the K/T boundary in ODP Hole 536 (Leg 77).

In particular, spherules are never found mixed with Ir and Ni-rich spinels. This is rather strange if we hypothesize that these three components are all produced by the same collisional event and if we consider the expected stirring effects of giant tsunami waves generated by an impact on the Yucatan peninsula. This separation is especially surprising at the site of Mimbral, which was formed in rather deep water (500 m). The existence of very distinct lithological units implies that the deposition of spherules was already completed when the formation of the sandstone started. Stinnesbeck et al. (1993) even say that the spherule bed was almost lithified when the first sand grains settled out. Simple considerations about the propagation of a tsunami wave through the Gulf of Mexico show that this is not so easy. With a velocity of about 500 km/s, a solitary wave generated by the Chicxulub impact could reach the paleosite of Mimbral in less than two hours: This is about the time spent by 1-mm spherules in the atmosphere and in the 500-m water column before reaching the seafloor. Obviously, the schedule of the event is rather tight but not totally unlikely.

Now consider the upper part of the sequence, which is far more constraining. Spinels and their host bodies sedimented after the attenuation of turbiditic or tsunami-generated flows, when relatively quiet conditions again prevailed. Stinnesbeck et al. (1993) claim that this delay likely represents a "significant amount of time." They do not give any estimation for this amount, but it is clear that, according to the usual wording of geologists, they mean a time much longer than hours or days. An estimate is given by Smit et al. (1994b), who estimate that the entire sequence, exhibiting the record of at least six successive waves with different directions, has been deposited in one pulsating event, lasting for several days. Jéhanno et al. (1992), considering the Beloc site, deduced similar values. These lower limits are much longer than the timing from model predictions (Melosh et al., 1990).

Two possibilities can be envisaged to account for this apparent inconsistency:

1. Glass spherules are not linked with the rest of the sequence. They are derived from an intra-Maastrichtian terrestrial event unrelated to the K/T boundary cosmic catastrophe. This would explain why these glass remains do not have the chemical characteristics of tektites (see Jéhanno et al., 1992).

2. The glass remains are relics of nontypical tektites (high Fe^{3+}/Fe^{2+} and no lechatelierite). The time interval between the deposition of glass spherules and Ni-rich spinels is still too long for a single impact. However, that can be explained by two distinct bolides colliding with the Earth over a rather short period of time. This two-collision scenario has already been proposed to account for the traces of erosion that are observed at the interface between the boundary clay (kaolinite due to a first impact) layer and the fireball layer (second impact) in continental deposits of North America (Shoemaker and Izett, 1992). Paleobotanical data suggesting an interval of several weeks or months could also be explained in this way (Wolfe, 1991). This leaves ample time for the deposition of the massive sandstone bed. However, the occurrence over a rather short interval of time of a couple of exceptional and *independent* impact events is unlikely on statistical grounds. Simpler explanations should be considered.

So, if we consider the most simple case of a single impactor, the presence of spherules and spinels in distinct layers implies a delay, a gap, or a dead time somewhere in the development of the collisional sequence. This delay does not exist (or is very short) in the case of a normal incidence impact. The situation, however, is completely different if the bolide was not volatilized but only disrupted by the collision. Big fragments could then have been injected for a few days on rather distant ballistic orbits. In that scenario, glass spherules and tsunamis are supposed to have been generated during the first contact, the first splash. Ir and Ni-rich spinels would result from the final accretion phase when fragments recollided with the Earth after damping of the turbiditic flows or tsunami megawaves. This kind of rebound, or ricochet, with delayed effects is expected to result from oblique collisions at grazing incidences (Schultz and Gault, 1991). Laboratory analyses and simulations show that such a fragmentation process should produce a global deluge of hypervelocity debris reentering the atmosphere and striking a variety of geological settings. Schultz and Gault (1989) showed that the ricochet velocity, approaching 0.5 Vi (Vi = the impactor velocity) for an impact angle of 15°, is easily sufficient to place big fragments up to 0.1 × the diameter of the projectile) on very eccentric orbits. It might even exceed the escape velocity. A delay of several days can be obtained for apogees lower than the Moon-Earth distance. The reentry of ejected debris, after several hours or days spent in orbit, will produce the effects of multiple-Tunguska events. This oblique collision-fragmentation model provides a nice solution to our timing problem. In addition, it gives the right conditions for the formation of spinels. It also accounts for the compositional variation of spinels observed from site to site (Kyte and Smit, 1986, Robin et al., 1992, 1994). The various fragments, with a wide distribution of mass, velocity, angle of incidence, and, possibly, different composition might have interacted in different ways with the atmosphere, leading to an uneven geographical distribution and variable compositions of Ni-rich spinels. These regional variations, which few people take into account, are presently unexplained in the case of a vertical impact. On the contrary, in the fragmentation process they are the inescapable consequence of different conditions of interaction with the atmosphere, combined with the possible inhomogeneity of the incident bolide. Finally, we will note that an oblique collision is strongly suggested by the asymmetry of the Chicxulub crater (Schultz, 1994).

CONCLUSION

Although some sedimentary features (size grading of the spherule bed at Beloc, extent of the spinel distribution at Mimbral and Hole 536) suggest that some of the sedimentary sequences in the Gulf of Mexico area formed rapidly, the distinct layers of Ir and spinels separated from the other sedimentary units imply a significant event duration that is hardly explained in the case of a single vertical impact. In the most simple hypothesis, that is, one involving a single impactor, the fragmentation (possibly resulting from an oblique collision) of the incident bolide and the wide dispersion of debris with delayed deposition are consistent with the available observations. We expect that stratigraphical data about other impact products, such as shocked minerals, will provide useful complementary information.

ACKNOWLEDGMENTS

We would like to thank G. Byerly from Louisiana State University, F. Kyte from University of California, Los Angeles, and G. Ryder from Lunar and Planetary Institute, Houston, for their helpful comments and critical reviews. We are also indebted to the Ocean Drilling Project, which provided the DSDP core samples, and to ODP personnel, who helped with the sampling in these cores. This is CFR contribution No. 1683.

REFERENCES CITED

Alvarez, W., Smit, J., Lowrie, W., Asaro, F., Margolis, S. V., Claeys, Ph., Kastner, M., and Hildebrand, A. R., 1992, Proximal impact deposits at the Cretaceous-Tertiary boundary in the Gulf of Mexico: A restudy of DSDP Leg 77 Sites 536 and 540: Geology, v. 20, p. 697–700.

Beeson, D., Gartner, S., Keller, G., MacLeod, N., Médus, J., Rocchia, R., and Robin, E., 1994, The KT boundary along the Brazos river, Falls County, Texas: Multidisciplinary stratigraphy and depositional environment [abs.], *in* New developments regarding the K/T event and other catastrophes in Earth history: Houston, Texas, Lunar and Planetary Institute Contribution 825, p. 9–10.

Brownlee, D. E., 1981, Extraterrestrial components, *in* Emiliani, C., ed., The sea, Volume 7: New York, Wiley, p. 733–762.

Doehm, E., and Margolis, S. V., 1990, Trace-element geochemistry and mineralogy of the Cretaceous/Tertiary boundary; Identification of extraterrestrial components: Geological Society of America Special Paper 247, p. 367–382.

Gayraud, J., Robin, E., Rocchia, R., and Froget, L., 1994, Preliminary results regarding the formation conditions of meteoric spinels, *in* New developments regarding the KT event and other catastrophes in Earth history: Houston, Texas, Lunar and Planetary Institute Contribution 825, p. 41–42.

Haggerty S. E., 1976, Opaque mineral oxides in terrestrial ignous rocks, in Ruble, D., ed., Oxide minerals, Volume 3: Mineralogical Society of America, p. 101–300.

Hansen, T. A., Farrand, R., Montgomery, H., Billman, H., and Blechschmidt, G., 1987, Sedimentology and extinction patterns across the Cretaceous-Tertiary boundary interval in east Texas: Cretaceous Research, v. 8, p. 229–252.

Hildebrand, A. R., Penfield, G. T., Kring, D. A., Pilkington, M., Camargo Z., A., Jacobsen, S. B., and Boynton, W. V., 1991, Chicxulub crater: A possible Cretaceous-Tertiary boundary impact crater on the Yucatan Peninsula, Mexico: Geology, v. 19, p. 867–871.

Hodge, V., Stallard, M., Koide, M., and Goldberg, E. D., 1986, Determination of platinum and iridium in marine waters, sediments and organisms: Analytical Chemistry, v. 58, p. 616–620.

Jéhanno, C., Boclet, D., Bonté, Ph., Castellarin, A., and Rocchia, R., 1988, Identification of two populations of extraterrestrial particles in a Jurassic hardground of the Southern Alps: Proceedings, Lunar and Planetary Science Conference, v. 18, p. 623–630.

Jéhanno, D., Boclet, D., Froget, L., Lambert, B., Robin, E., Rocchia, R., and Turpin, L., 1992, The Cretaceous-Tertiary boundary at Beloc, Haiti: No evidence for an impact in the Caribbean area: Earth and Planetary Science Letters, v. 109, p. 229–241.

Keller, G., 1989, Extended Cretaceous/Tertiary boundary extinctions and delayed population change in planktonic foraminifera from Brazos River, Texas: Paleoceanography, v. 4, p. 282–332.

Keller, G., MacLeod, N., Lyons, J. B., and Officer, C. B., 1993, Is there evidence for Cretaceous-Tertiary boundary-age deep-water deposits in the Caribbean and Gulf of Mexico?: Geology, v. 21, p. 776–780.

Keller, G., Stinnesbeck, B., Adatte, T., MacLeod, N., and Lowe, D. R., 1994, Field guide to Cretaceous-Tertiary boundary sections in Northeastern Mexico: Houston, Texas, Lunar and Planetary Institute Contribution 827, 110 p.

Kyte, F T., and Smit, J., 1986, Regional variations in spinel compositions: An important key to the Cretaceous-Tertiary event: Geology, v. 14, p. 485–487.

Margolis, S. V., Claeys, Ph., and Kyte, F. T., 1991, Microtektites, microkrystites and spinels from a late Pliocene asteroid impact in the southern ocean: Science, v. 251, p. 1594–1597.

Melosh, H. J., Schneider, N. M., Zahnie, K. J., and Latham, D., 1990, Ignition of global wildfires at the Cretaceous/Tertiary boundary: Nature, v. 343, p. 251–254.

Robin, E., Boclet, D., Bonté, Ph., Froget, L., Jéhanno, C., and Rocchia, R., 1991, The stratigraphic distribution of Ni-rich spinels in Cretaceous-Tertiary boundary rocks at El Kef (Tunisia), Caravaca (Spain) and Hole 761C (Leg 122): Earth and Planetary Science Letters, v. 107, p. 715–721.

Robin, E., Bonté, Ph.., Froget, L., Jéhanno, C., and Rocchia, R., 1992, Formation of spinels in cosmic objects during atmospheric entry: A clue to the Cretaceous-Tertiary boundary event: Earth and Planetary Science Letters, v. 108, p. 181–190.

Robin, E., Froget, L., Jéhanno, C., and Rocchia, R., 1993, Evidence for a K/T impact event in the Pacific Ocean: Nature, v. 363, p. 615–617.

Robin, E., Gayraud, J., Froget, L., and Rocchia, R., 1994, On the origin of the regional variations in spinel compositions at the KT boundary [abs.], *in* New developments regarding the KT event and other catastrophes in Earth history: Houston, Texas, Lunar and Planetary Institute Contribution 825, p. 96–97.

Schultz, P. H., 1994, Visualizing the nature and consequences of the Chicxulub impactor: Clues from Venus [abs.], *in* New developments regarding the KT event and other catastrophes in Earth history: Houston, Texas, Lunar and Planetary Institute Contribution 825, p. 104–106.

Schultz, P. H., and Gault, D. E., 1989, Protracted global catastrophes from oblique impacts: Houston, Texas, Lunar and Planetary Science Conference Abstracts, v. 20, p. 970–971.

Schultz, P. H., and Gault, D. E., 1991, Prolonged global catastrophes from oblique impacts: Geological Society of America Special Paper 247, p. 239–261.

Sharpton, V. L., Dalrymple, G. B., Marin, L. E., Ryder, G., Schuraytz, B. C., and Urrutia-Fucugauchi, J., 1992, New links between the Chicxulub impact structure and the Cretaceous/Tertiary boundary: Nature, v. 359, p. 819–821.

Shoemaker, E. M., and Izett, G. A., 1992, Stratigraphic evidence from western North America for multiple impacts at the K/T boundary: Houston, Texas, Lunar and Planetary Science Conference Abstracts, v. 23, p. 1293–1294.

Smit, J., and eight others, 1992, Tektite-bearing, deep-water clastic unit at the Cretaceous-Tertiary boundary in Northeastern Mexico: Geology, v. 20,

p. 99–103.

Smit, J., Alvarez, W., Clayes, Ph., Montanari, S., and Roep, Th. B., 1994a, Misunderstandings regarding the K/T boundary deposits in the Gulf of Mexico [abs.], *in* New developments regarding the KT event and other catastrophes in Earth history: Houston, Texas, Lunar and Planetary Institute Contribution 825, p. 116–117.

Smit, J., Roep, Th. B., Alvarez, W., Claeys, Ph., Montanari, S., and Grajales, M., 1994b, Impact-tsunami-generated clastic beds at the KT boundary of the gulf coastal plain: A synthesis of old and new outcrops [abs.], *in* New developments regarding the KT event and other catastrophes in Earth history: Houston, Texas, Lunar and Planetary Institute Contribution 825, p. 117–119.

Stinnesbeck, W., and 10 others, 1993, Deposition of channel deposits near the Cretaceous-Tertiary boundary in northeastern Mexico: Catastrophic or "normal" sedimentary deposits?: Geology, v. 21, p. 797–800.

Wolfe, J. A., 1991, Paleobotanical evidence for a June "impact winter" at the Cretaceous-Boundary: Nature, v. 352, p. 420–422.

Manuscript Accepted by the Society September 14, 1995

Printed in U.S.A.

Geological Society of America
Special Paper 307
1996

Testing patterns of Cretaceous-Tertiary planktonic foraminiferal extinction at El Kef (Tunisia)

Norman MacLeod
Department of Palaeontology, The Natural History Museum, Cromwell Road, London, SW7 5BD United Kingdom

ABSTRACT

The bolide impact–mass extinction model predicts that last appearances of most upper Maastrichtian taxa should be simultaneous and coincident with impact debris in all complete boundary sections and cores. Progressive (rather than catastrophic) planktonic foraminiferal K/T turnover patterns have been interpreted to result from various sources of bias in the fossil record (e.g., Signor-Lipps Effect, reworking). In addition, some proponents of bolide-driven models have argued that mass-extinction events can be distinguished by their characteristically broader patterns of morphotypic and paleoecologic extinction susceptibility. These hypotheses are tested against a record of K/T planktonic foraminifers faunal turnover at the El Kef (Tunisia) boundary stratotype.

Statistical tests for three different Signor-Lipps Effect models within the El Kef stratotype section fail to provide unambiguous support for the hypothesis that these data have been significantly biased in favor of a progressive, rather than abrupt, turnover pattern. Signor-Lipps–type bias does exist within these data, but it is insufficient to account for the progressive character of the observed turnover pattern. There also appears to be no evidence in these data or in data from other K/T sections or cores supporting the interpretation that all so-called Cretaceous species' occurrences recorded from horizons above the K/T boundary are the result of reworking. Finally, analyses of extinction susceptibility throughout the El Kef K/T transition interval suggests that the boundary event imparted the same morphotypic signature to the surviving fauna as background-level extinction events in Neogene planktonic foraminifera.

These results fail to corroborate predictions of a catastrophic bolide impact–mass extinction model for planktonic foraminiferal extinctions within the El Kef succession. They are consistent, however, with many recent summaries of faunal turnover across the K/T boundary for planktonic foraminifera as well as many other biotic groups. Since the predictions of the catastrophic bolide impact-mass extinction model have not been unambiguously confirmed by paleontological analyses, the question of what factor(s) may have played a role in determining the character of the K/T mass extinction remains open.

INTRODUCTION

Among the debates collectively known as the Cretaceous-Tertiary (K/T) boundary controversy, none has been more hotly contested than interpretations of uppermost Cretaceous–lowermost Tertiary biotic records. Whereas the textbook model of a catastrophic collapse of such diverse clades as dinosaurs, ammonites, rudistid and inoceramid bivalves, calcareous nannoplankton, and planktonic foraminifera at the close of the Mesozoic has been taught to generations of earth scientists (Schindewolf, 1962; Bramlette and Martini, 1964; Bramlette, 1965; Newell, 1967; Percival and Fischer, 1977), several recent

MacLeod, N., 1996, Testing patterns of Cretaceous-Tertiary planktonic foraminiferal extinction at El Kef (Tunisia), *in* Ryder, G., Fastovsky, D., and Gartner, S., eds., The Cretaceous-Tertiary Event and Other Catastrophes in Earth History: Boulder, Colorado, Geological Society of America Special Paper 307.

high-resolution biostratigraphic investigations have shown the pattern of this extinction to be much more complex than previously appreciated. For example, it is now widely accepted that rudistid and inoceramid bivalves disappear well below the K/T boundary, prior to any trace of impact debris (e.g., Johnson and Kauffman, 1993, 1996; Kauffman, 1984, respectively). More controversial are reports of so-called Cretaceous dinosaur, nannoplankton, and planktonic foraminiferal species in Danian sediments, sometimes extending for several meters above the local K/T boundary (e.g., Sloan et al., 1986; Perch-Nielsen et al., 1982; Keller, 1988a, respectively). In addition, an even larger number of studies show that many ecologically diverse clades exhibit no unusual trends in either species richness or relative abundance across the boundary horizon (e.g., Askin, 1990 [dinoflagellates]; Hollis, 1993 [radiolaria]; Keller, 1988b; Thomas, 1990; Widmark and Malmgren, 1992 [benthic foraminifera]; Brouwers and De Deckker, 1993 [ostracodes]; Surlyk, 1990 [brachiopods]; Labandiera, 1992 [insect and insect plant associations]; Archibald and Clemens, 1984; Archibald and Bryant, 1990 [microvertebrates]; Lerbekmo et al., 1987; Sweet and Braman, 1992; Sweet et al., 1990; McIver et al., 1991; Méon, 1990 [pollen]; Knoboch et al., 1993; Golovnea, 1994 [plant body fossils]). Most of these data have appeared since the original proposal of the bolide-impact–mass-extinction scenario (Alvarez et al., 1980), and in many cases these studies were initiated for the expressed purpose of testing the biotic predictions of that scenario.

What predictions does the Alvarez et al. (1980, 1982) bolide-impact–mass-extinction model make with respect to the K/T fossil record? Many proponents of this model have consistently argued that impact-related extinction events should be characterized by large numbers of co-occurring last appearance datums (LADs) that either coincide with direct evidence for concentrations of impact debris in the same section/core or that can be chronostratigraphically correlated to such deposits in other sections/cores (e.g., Alvarez et al., 1980; Alvarez et al., 1982; Alvarez et al., 1984; Hut et al., 1987). Owing to the short time interval over which extinctions directly linked to a bolide impact are thought to occur (1 to 10 yr; see Alvarez et al., 1984), there would be little chance of preserving a biogeographic signature of variations in extinction intensity. Accordingly, an instantaneous mass extinction should not exhibit pronounced biogeographic structure. Lastly, Gould (1984, 1985), Raup (1989, 1991), and Jablonski (1986a, 1986b, 1986c, 1989, 1994), among others, have argued that mass extinctions play a fundamental role in shaping the outcome of evolutionary processes, since "some traits that tend to confer extinction resistance during times of normal (background) levels of extinction are ineffectual during mass extinction" (Jablonski, 1989, p. 117). In this sense, a geologically instantaneous mass extinction might be characterized by different morphologic and/or taxonomic patterns of extinction susceptibility. Thus, the various bolide-impact–mass-extinction models make at least three predictions for the K/T fossil record: (1) Local patterns of extinction should be abrupt and precisely coeval with independent physical evidence of impact, (2) global extinction patterns should show no overt biogeographic structure, and (3) patterns of extinction and survivorship should be morphologically (and perhaps also taxonomically) distinct from episodes of normal or background extinction.

These statements represent alternative counterparts to the following null hypotheses: (1) Local patterns of extinction extend over recognizable stratigraphic intervals that either may or may not be coeval with independent physical evidence of bolide impact, (2) global extinction patterns exhibit a recognizable biogeographic structure, and (3) global patterns of extinction and survivorship sort taxa in a manner commensurate with morphologic and/or taxonomic patterns observed during background events. If these latter statements cannot be rejected on the basis of paleontological observations, or if those observations do not conform to the bolide-impact–mass-extinction model's predictions, that model must either be rejected outright or, if possible, altered to bring it into concordance with available biotic data.

This investigation analyzes the K/T record of planktonic foraminifera at the El Kef section in Tunisia. Any study focusing exclusively on any single clade must be regarded as an incomplete representation of biotic change across this interval. Planktonic foraminifera, nevertheless, represent an excellent place at which to begin the high-resolution systematic study of the fossil record. If predictions of the bolide-impact–mass-extinction model are not corroborated by the planktonic foraminiferal fossil record, they are unlikely to be met by the much larger number of clades whose fossil records across this interval are not, and indeed may never be, known in comparable detail (e.g., Alvarez et al., 1984).

This study also concerns itself with only a single locality: the El Kef section in Tunisia. For many of the same reasons that the study of single organismal groups are necessarily incomplete relative to the K/T biota as a whole, the study of single localities is necessarily incomplete with respect to the global record of any biotic group. Still, detailed analysis of the El Kef section, once again, represents a good place to begin this systematic reassessment of the K/T paleontologic record, for it is here that the K/T boundary stratotype is placed (Cowrie et al., 1989). Consequently, there is no disagreement as to the position of the K/T boundary within the El Kef succession. Comparison of the El Kef stratotype with over 30 K/T boundary sections and cores worldwide (MacLeod and Keller, 1991) also indicates that its upper Maastrichtian–lower Danian interval contains no hiatuses and was deposited at a uniform time-averaged sediment accumulation rate.

MATERIAL AND METHODS

The El Kef boundary stratotype is located ~7 km west of the town of El Kef in northwestern Tunisia. Lithostratigraphically, this sequence consists of soft gray marls making up the El

Haria Formation. The boundary is placed at the base of a 50-cm-thick black clay layer that, in turn, has a thin red clay layer at its base. These basal red and overlying black clay layers are the only beds of their type within the succession. Aside from the observed contrast in sediment type, the base of this clay bed is marked geochemically by an iridium anomaly and marked mineralogically by an unusually high concentration of Ni-rich spinel crystals (Robin et al., 1991).

The planktonic foraminiferal faunas that form the basis of this study were collected by Gerta Keller during the 1992 International Workshop on Cretaceous-Tertiary Transitions at El Kef, organized under the joint sponsorship of the Global Sedimentary Geology Program (Working Group 6), the Association Tunisienne des Etudes Internationales de Géologie, and the Service Géologique National de Tunisie. Sample processing procedures, details of analyzed size fractions, and a discussion of relevant systematic issues are presented in Keller et al. (1995).

Pattern of Cretaceous planktonic foraminiferal extinctions at El Kef

Disagreement currently exists between Jan Smit and Gerta Keller as to the pattern of observed planktonic foraminiferal last appearance datums (LADs) across the El Kef K/T boundary. Smit (1982, 1990; Smit et al., 1988) claimed that in the El Kef stratotype, planktonic foraminifera exhibit a pattern of abrupt and simultaneous disappearance at the boundary horizon for all Cretaceous taxa (Fig. 1A). Smit reported that the single exception to this pattern was the Cretaceous planktonic foraminiferal species *Guembelitria cretacea*, whose stratigraphic range continued on into lowermost Danian sediments. Smit reported no LADs occurring in the interval below the boundary horizon, no foreshadowing of the boundary event by reductions in species-specific relative abundances, and no other Cretaceous taxa occurring above the boundary horizon, though Smit (1990, p. 192) noted the presence of "a few reworked larger Cretaceous species [with] smaller species [being] somewhat more abundant relatively." In contrast, Keller (1988a) reported observing ~10% of Cretaceous species disappearing from the El Kef sequence at least 50 cm below the boundary horizon, ~60% disappearing within a few centimeters above or below the boundary itself, and ~30% of Cretaceous species occurring at least 50 cm above the boundary horizon (with some species persisting several meters above the last sampled horizon) (Fig. 1B).

A literature survey of planktonic foraminiferal biostratigraphic data from El Kef and other K/T boundary sections and cores reveals widespread support for Keller's turnover pattern. These data include her own work that documents similar patterns of Cretaceous species LADs prior to, at, and above local K/T boundaries at Brazos River, Texas (Keller, 1989), and ODP Sites 690 and 738, southern Indian Ocean (Keller, 1993), along with collaborative studies that document the same patterns at Agost and Caravaca, southern Spain (Canudo et al., 1991); Israel (Keller and Benjamini, 1991); Nye Kløv, Denmark (Schmitz et al., 1992), and Mimbral, Mexico (Keller et al., 1994). Moreover, independent analyses of the El Kef stratotype section (Brinkhuis and Zachariasse, 1988; Ben Abdelkader et al., 1992) corroborate Keller's pattern. Liu and Olsson's (1992; see also Olsson and Liu, 1993) analysis of the Millers Ferry K/T boundary section also supports Keller's progressive turnover pattern across the boundary horizon, though these authors differ with Keller as to the interpretation of that pattern (see below). None of the aforementioned studies has reproduced Smit's turnover pattern of a single Cretaceous species extending into lowermost Danian sediments at El Kef or at any other biostratigraphically complete K/T boundary section or core.

Distinct from the simple observation of species' LADs throughout a stratigraphic interval, the interpretation of such patterns must take into consideration other aspects of the local fossil record. Of great importance to the issue of extinction patterns is the question of whether or not the observed LAD for a species in a particular section can be regarded as an adequate estimate of its true LAD. This matter is the subject of literature organized around the topics of the so-called Signor-Lipps Effect, postdepositional reworking, and stratigraphic confidence intervals.

Signor-Lipps Effect

The Signor-Lipps Effect, which was first described by Shaw (1964), states that because of vagaries in sampling, ecologic occurrence, and preservation, it cannot be automatically assumed that a fossil species' last observed occurrence (its apparent LAD) provides an accurate estimate of its true last occurrence (its true LAD). Hence Signor and Lipps's (1982, p. 291) conclusion that "the recorded ranges of fossils, especially of uncommon taxa or taxa in habitats not represented by a continuous record, may be inadequate to test either gradual or catastrophic [extinction] hypotheses." Thus, the Signor-Lipps Effect represents an attempt to set limits on the interpretability of biostratigraphic data that are contingent upon either the assumption or (preferably) the demonstration of significant differences between apparent and true LADs.

The Signor-Lipps Effect makes several predictions of patterns that should be observed in any fauna whose stratigraphic record has been systematically biased to favor a progressive, rather than abrupt, turnover pattern. First, Signor and Lipps (1982) noted that this effect can be generated in faunas whose total abundance or species whose relative abundance is either already low or declining as they approach an arbitrarily defined stratigraphic horizon (Fig. 2A; see also Raup, 1972, 1976; Signor, 1978; Koch and Morgan, 1988). The predicted pattern of decreasing abundance with decreasing distance from the horizon can be statistically tested using a wide variety of descriptive techniques (see Sokal and Rohlf, 1981) or modeled using the method of Koch and Morgan (1988). If such a relationship is confirmed, the biased species richness patterns present in the raw data can be corrected using rarefaction (Maguran, 1988).

Second, a Signor-Lipps Effect may be produced by the

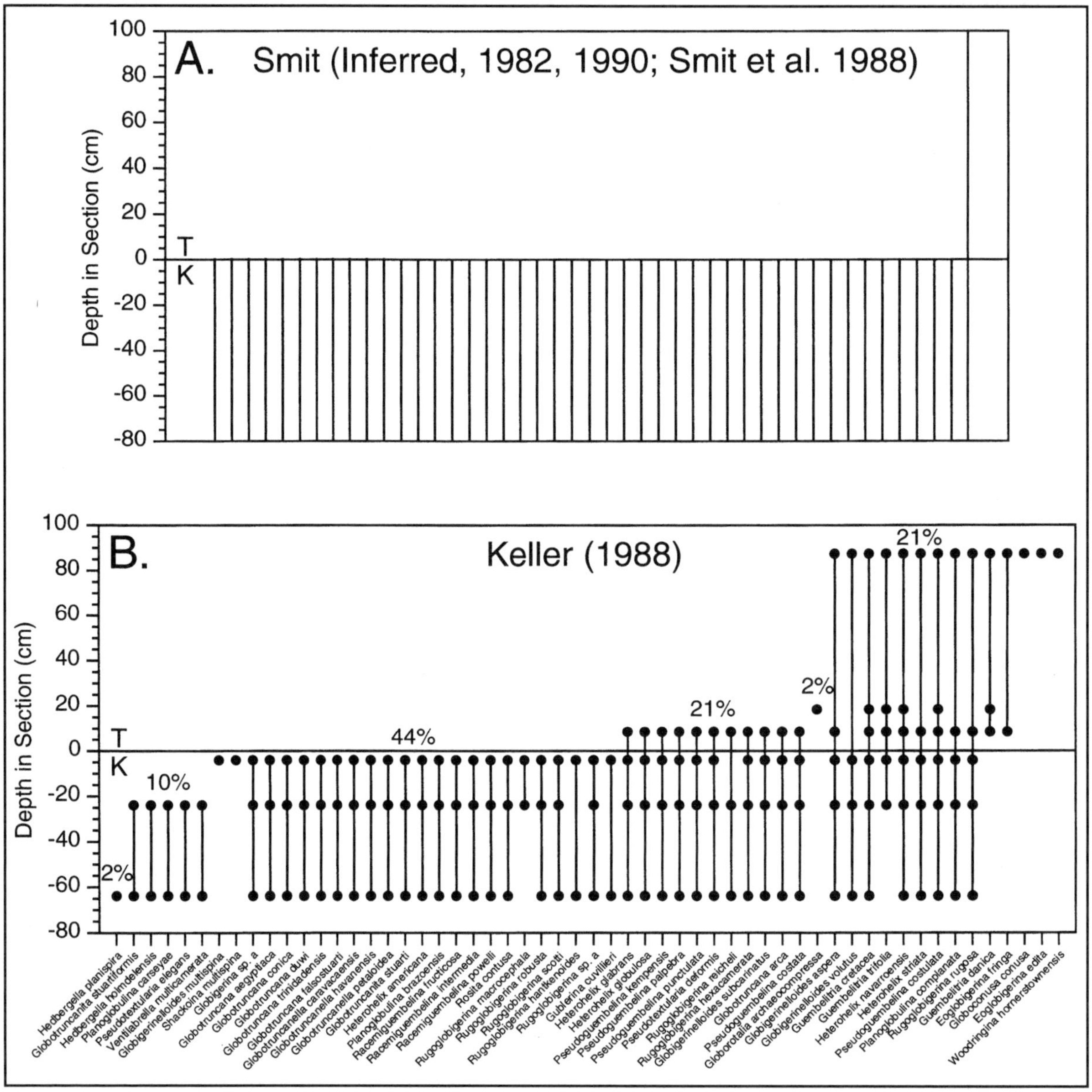

Figure 1. Alternative patterns of K/T planktonic foraminifer faunal turnover for the El Kef stratotype section. A, Pattern inferred from statements in Smit (1982, 1990; Smit et al., 1988). B, a selection of four samples from Keller's (1988a) biostratigraphy.

selective removal of species susceptible to either syndepositional or postdepositional diagenesis. Such effects are known to be problematic in assemblages of calcareous microfossils (such as planktonic foraminifera) in which it is not unusual to encounter intervals of moderate to severe faunal dissolution corresponding to fluctuations in physical oceanographic parameters (e.g., depths of the marine lysocline and carbonate compensation depth (CCD); see Berger 1968, 1970, 1979). This source of bias may be examined by assessing relative dissolution susceptibilities for various species and then determining if there is a systematic bias in the ratio of dissolution-resistant to dissolution-susceptible species as the K/T boundary is approached.

Third, a Signor-Lipps Effect may be produced if the local environment shifts from a set of conditions within the tolerance range of a particular species to a set of conditions outside that range. The existence of such habitat shifts was cited by Signor and Lipps (1982) as responsible for the patchy distribution of many species occurrences throughout a stratigraphic sequence. Such distributions often represent an intrinsic paleoecological attribute of individual species that may play an important role in controlling their extinction susceptibility (see Hanski, 1982; Hanski and Gilpin, 1991).

There are two ways to approach the testing of this potential source of bias. The most direct of these involves the quantitative

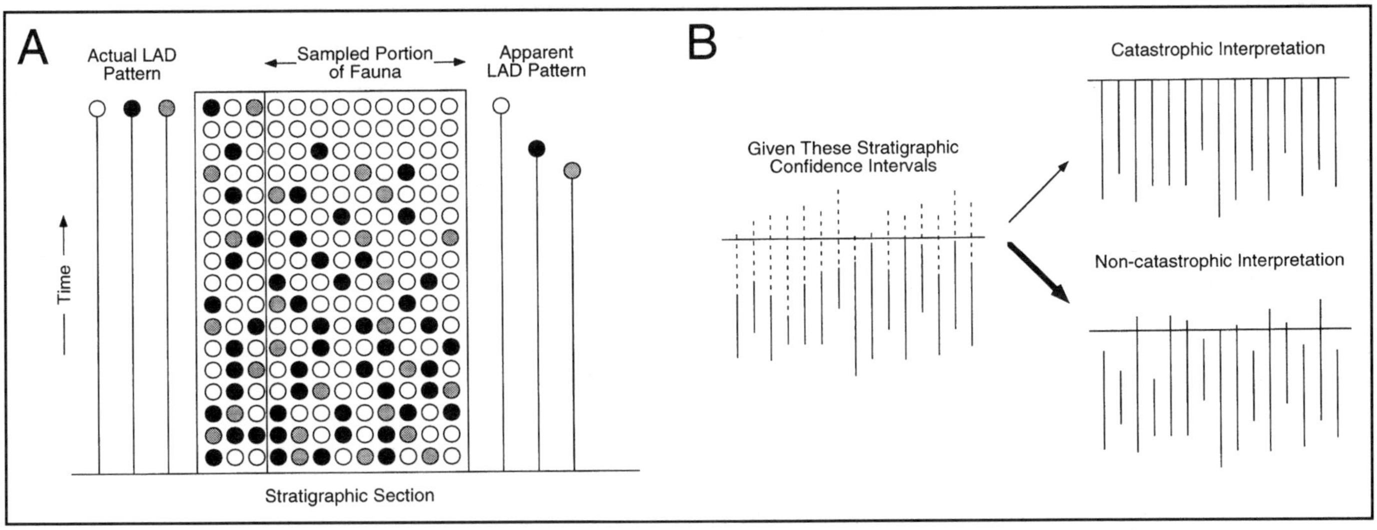

Figure 2. A, conceptual basis for the Signor-Lipps Effect. Given a fossiliferous stratigraphic section (variously shaded circles), sampling effects may result in the erroneous appearance of a gradual faunal turnover across any arbitrarily chosen stratigraphic horizon irrespective of the actual pattern of last appearance datums (LADs). A Signor-Lipps–type pattern can be produced by a wide variety of phenomena, including sampling pattern, sample size, diagenetic effects, and lateral habitat migration. Note that since paleontologists can only examine fossils included in their sample, the actual turnover pattern present in an assemblage that has been biased by one or more of these factors will always be uncertain. Signor and Lipps (1982, p. 291) explicitly acknowledge this inherent ambiguity by categorically stating, "Our arguments do not eliminate the possibilities that the Cretaceous-Tertiary extinction was gradual *or* catastrophic" (emphasis mine). B, alternative interpretations of stratigraphic confidence intervals. Because of the inherent indeterminism of these statistical constructs, a catastrophic interpretation of LAD patterns represents but one of a large number of possible alternatives. In order to determine the probability of any particular pattern of LAD geometries being correct, it must be compared to the total number of all possible geometries compatible with a given set of confidence intervals. Size of arrows indicates the relative probabilities of catastrophic and noncatastrophic turnover geometries that will be consistent with any set of confidence interval data.

comparison of sediment analysis results with the distribution of particular taxa in an effort to determine whether or not the presence of a particular species is correlated with the presence or absence of some physical factor. Alternatively one may use available biotic data to identify assemblages of species that share common occurrence or relative abundance patterns throughout the sampled interval. Once these ecological guilds are identified, it can be determined whether or not they disappear in a structured manner as the boundary is approached.

If any of these tests yield statistically significant patterns consistent with predictions of the Signor-Lipps Effect, this explanation can be advocated to account for the observed data. On the other hand, if such tests fail to yield the predicted patterns, continued suggestion that the Signor-Lipps Effect has nevertheless biased the observed patterns (e.g., Kerr, 1994; Smit, 1994) must be dismissed as lacking credibility. Since the disappearance of very rare taxa is inherently ambiguous with respect to the testing of the bolide-impact–mass-extinction model (see Signor and Lipps, 1982) appeals to a SignorLipps Effect alone cannot provide the information necessary to discriminate between alternative explanatory scenarios for such taxa. In addition, if evidence supporting a Signor-Lipps bias is found, the implied range extensions must be recognized to have potentially affected all species in the fauna, and not just those whose apparent LAD's fall below the K/T boundary (e.g., Kerr, 1994; Smit 1994).

Stratigraphic confidence limits

Calculation of stratigraphic confidence limits has also been proposed as an alternative approach to testing hypotheses of faunal turnover across the K/T boundary (D. Jablonski, 1994, oral communication). Starting with the same assumption as the Signor-Lipps Effect with regard to the relationship between apparent and true LADs (or first appearance datums [FADs]), this method uses observed data to infer the hypothetical limits of species occurrence within a local section or core, subject to various assumptions. Unfortunately, there are both practical and theoretical difficulties involved in using stratigraphic confidence limits to constrain hypotheses that predict alternative faunal turnover patterns.

The techniques currently used for inferring stratigraphic confidence intervals were developed by Strauss and Sadler (1989). These require that sections or cores be sampled continuously for species occurrences and that those occurrences be randomly distributed throughout the sampled interval. Marshall (1994a) developed a variation of the Strauss and Sadler method that enables confidence intervals to be estimated for species with nonrandom occurrence patterns. However, this alternative method still requires continuous sampling. Micropaleontological samples are typically taken over discrete stratigraphic intervals, transported to a laboratory, and processed to separate the fauna from the surrounding inorganic sediment. This results in a time-averaged, discrete sample of species occurrence patterns that violates the continuous-sampling assumption of the Strauss and Sadler (1989) and Marshall (1994a) methods. (Note: It is currently unknown to what degree the results of these confidence interval estimation techniques are robust to deviations from continuous sampling. Many statistical tests [e.g., t-tests, ANOVA] are routinely applied to paleontological data that violate the basic assumptions of these tests [e.g., normality], and it may be possible to estimate reliable stratigraphic confidence intervals for some types of discretely sampled data. Also no test is available for determining whether or not a section that is simply scanned for fossil occurrences has truly been continuously sampled.)

Aside from these practical considerations, several serious problems are inherent in using stratigraphic confidence intervals to identify a particular faunal turnover geometry as more likely than another given a particular pattern of species occurrences. Stratigraphic confidence intervals are measures of uncertainty. They represent intervals within which, to a stated degree of certainty, one expects a taxon's true LAD (or FAD) to reside. Determination of such intervals does not provide for the a posteriori selection of certain horizons that might fall within the confidence interval (e.g., the K/T boundary) as being more likely candidates for the true LAD (or FADs; see Fig. 2B) than any other included horizon (see Smith, 1994). In order to estimate the probabilities associated with particular geometries of LADs (or FADs) within the confidence band, those geometries must be compared with all possible LAD (or FAD) geometries that could be constructed.

RESULTS

Signor-Lipps Effect

Figure 3 shows the distribution of planktonic foraminiferal sample sizes across the El Kef K/T boundary within the Keller et al. (1995) data. These data represent the sizes of counted samples after the processed residue had been randomly split into aliquots for relative abundance analysis. Sample sizes do not decrease sharply across the boundary horizon. There is an abrupt increase in sample size between Cretaceous samples 32 and 31 (500 cm and 450 cm below the K/T boundary respectively) and a similar abrupt drop in sample size between Tertiary samples 10 and 9 (10.5 cm and 17.5 cm above the K/T boundary respectively). Within the broad middle interval that includes the K/T boundary, sample size undergoes a progressive decline. Although the average sample size is relatively high ($\bar{x}$ = 277.97), the distribution is nonuniform $\bar{x}_{Tertiary}$ = 171.38, $\bar{x}_{Cretaceous}$ = 338.22). A two-sample t-test (inhomogeneous variances) reveals that the Tertiary sample sizes are significantly smaller than the underlying Cretaceous sample sizes ($\alpha = 0.99$). Because overall sample size does decline across the boundary, it is necessary to test the observed data for systematic bias that may have led to the production of a Signor-Lipps Effect.

Inspection of the pooled species abundance data for the entire interval (Fig. 4) also suggests that diminishing sample sizes may affect our perception of the El Kef K/T planktonic

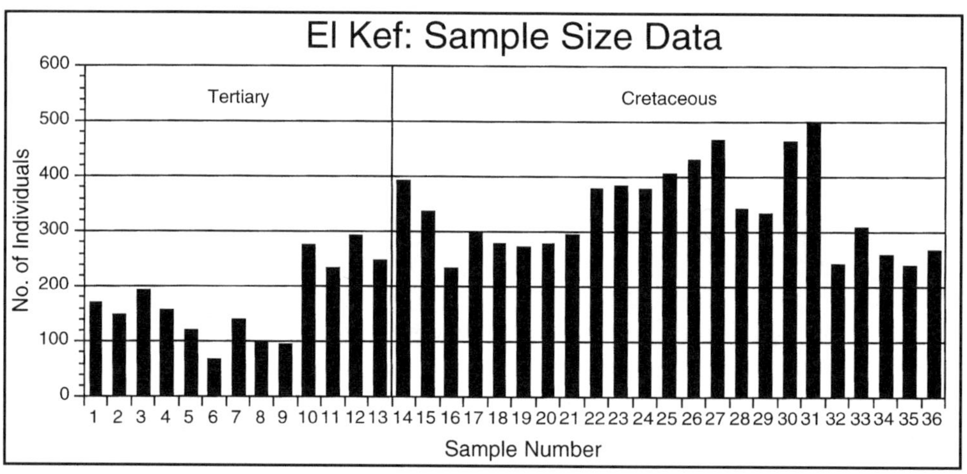

Figure 3. Variation in sample size across the K/T boundary in the Keller et al. (1996) El Kef data. Sample sizes do not reflect total residue counts, but rather the sizes of residue aliquots used to determine specific patterns of relative abundance variation. Systematic changes in the total number of specimens counted may influence species richness counts. Note, however, that sample sizes do not change at the boundary horizon but undergo an abrupt increase between samples 32 and 31 (500 and 450 cm below the K/T boundary) and an abrupt decrease between samples 10 and 9 (10.5 and 17.5 cm above the boundary).

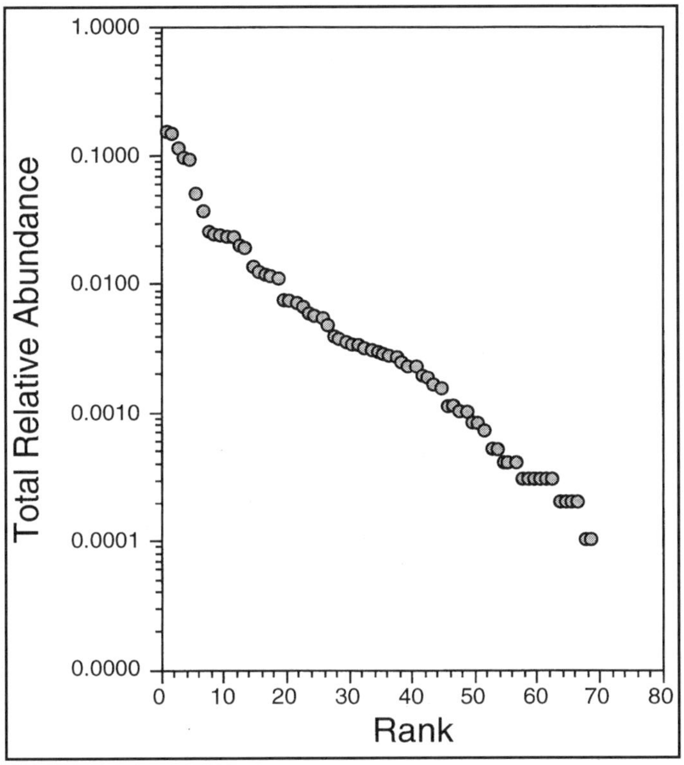

Figure 4. Ranked total abundance plot for the Keller et al. (1995) El Kef planktonic foraminiferal data. This pattern is indicative of a log-normal distribution of taxic abundances that is typical of the majority of large, mature, and varied modem communities studied by both terrestrial and marine ecologists (see Maguran, 1988).

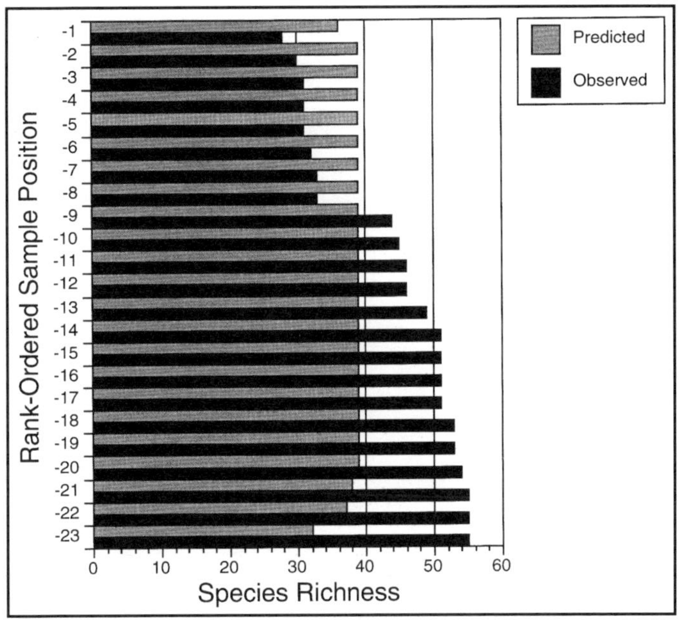

Figure 5. Comparison between observed and predicted species range distributions for the Keller et al. (1995) El Kef planktonic foraminiferal data made on the basis of variations in sample size (Fig. 3) and total rank abundances (Fig. 4). Sample positions below the K/T boundary event horizons (negative numbers along y-axis) are rank-ordered and do not reflect actual depths within the section. Sample –1 represents the highest Maastrichtian sample. Predicted species ranges are based on the Koch and Morgan (1988) model. Note strong discrepancy between the predicted and observed species richness values throughout the entire interval. This discrepancy indicates that these data have not been systematically biased by changing sample sizes leading to production of a Signor-Lipps Effect. See text for discussion.

foraminiferal faunal turnover. Like most natural populations (see Maguran, 1988), these samples are composed of a very large number of comparatively rare species. Only three species in this assemblage exhibit gross relative abundances greater than 10%, and none makes up more than 15% of the sample. As sample size diminishes it is reasonable to suspect that the rarer species will disappear prior to their true extinction level simply as a result of failure to collect adequately sized samples.

The potential effect of varying sample sizes in the pre-K/T interval can be evaluated using the method of Koch and Morgan (1988). This procedure uses sample sizes along with the pooled relative abundance of each species to determine the "average or expected number of species that will range across any two specified [stratigraphic] levels" (Koch and Morgan, 1988, p. 126). Substantial agreement between expected and observed species range distributions constitutes evidence that observed biostratigraphic patterns have been systematically influenced by variations in sample size. Failure to detect significant levels of correspondence between expected and observed range distributions indicates a lack of systematic sample-size bias and supports alternative interpretations of the observed data (e.g., biological-environmental causation).

Results of the Koch and Morgan (1988) test are shown in Figure 5. Whereas observed species richness values exhibit a consistent decline throughout the 23 upper Maastrichtian samples below the K/T boundary, with a pronounced step between horizons –9 and –8 (–65 cm and –45 cm respectively), values predicted on the basis of observed sample sizes and species numbers do not show a cumulative decline in species richness until the uppermost Maastrichtian sample.

Since the pattern of predicted values may strike some readers as odd, a brief explanation of these results is necessary. The predicted species richness values shown in Figure 5 were determined by estimating the number of species expected to compose a sample of the observed size, given the observed abundances of each species (see Koch and Morgan, 1988). Sample sizes over this interval exhibit a peaked distribution (Fig. 6). This suggests that expected species richness values should also peak in the middle of the sampled interval and decline toward either end. The remarkable consistency of the central peak in expected species richness values for the El Kef data results from the fact that most species in any given sample are represented by very low numbers of specimens. Almost half of all species observed in any given sample are represented by fewer than three specimens. Overall, 73% of all upper Maastrichtian species exhibit average relative abundance values of less than 0.02 (Fig. 7). Since the probability of finding such low-abundance species is

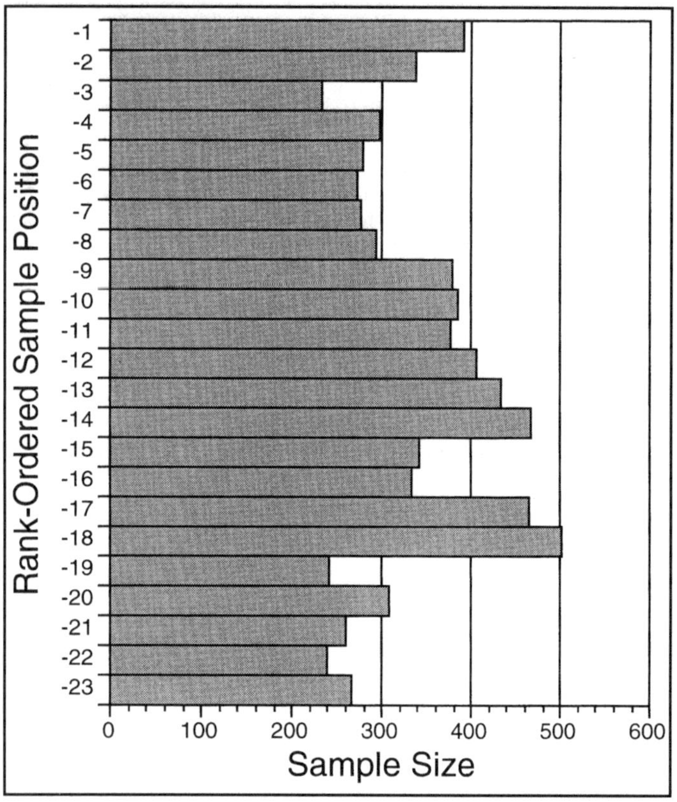

Figure 6. Sample size variation within the 23 uppermost Maastrichtian samples from the Keller et al. (1995) data. Sample positions below the K/T boundary event horizons (negative numbers along y-axis) are rank-ordered and do not reflect actual depths within the section. Sample –1 represents the highest Maastrichtian sample. Note peaked distribution of sample sizes in the middle of this uppermost Maastrichtian interval. In terms of relative abundance–sample size interactions, this distribution would be expected to result in an apparent species richness pattern that was similarly peaked in the middle of the sampled interval and fell off toward the upper (K/T boundary) and lower boundaries regardless of whether or not all species ranged through the entire interval (see Koch and Morgan, 1988).

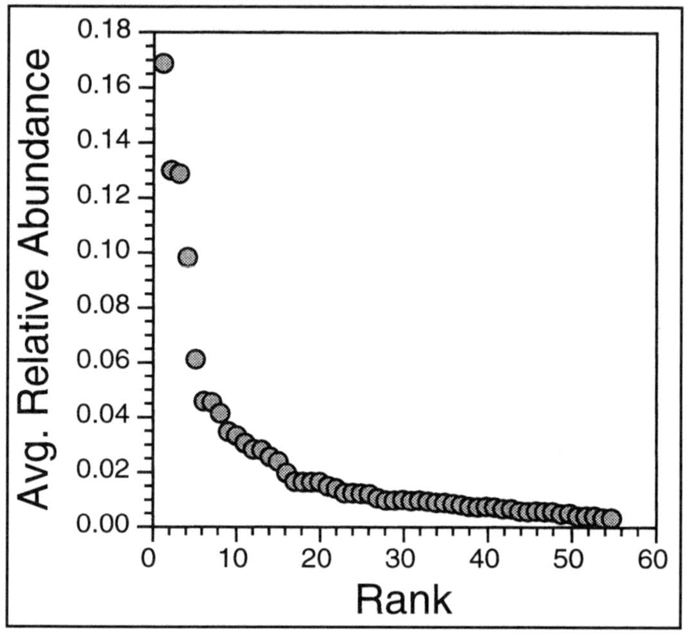

Figure 7. Ranked average relative abundance plot for Cretaceous species in the uppermost 23 Maastrichtian samples of the Keller et al. (1995) El Kef planktonic foraminiferal data. Relative abundances throughout this sampled interval are remarkably consistent for almost all species. Note the large proportion of species with small (>0.05) relative abundances. Most of these species are represented by >5 specimens per sample. Accordingly, the chance of finding a representative of these species in any sample is relatively low. This relative abundance distribution is primarily responsible for the constant species rioiness values predicted by the Koch and Morgan model results in the middle portion of the analyzed interval (see Fig. 5). It is also important to note that (1) all upper Maastrichtian species whose last appearance datum falls below the K/T boundary are rare species (>0.05), (2) all abundant species (<0.05) within this interval are also found in overlying lowermost Danian strata, and (3) many rare species are also found in overlying lowermost Danian strata. There appears to be no rank order correlation between average relative abundance and apparent survivorship for species whose average relative abundance is >0.05.

necessarily low, the predicted values reflect a relatively small core of moderately abundant species that are present in almost every sample, along with a larger group of less abundant species whose occurrence pattern is much more patchy but whose collective influence on estimated species richness is more or less constant. (Note: predicted species richness values have been rounded to the nearest integer.)

The predicted species richness values shown in Figure 5 represent the pattern expected if interactions between sample size and species relative abundances were the only factors influencing the observed data. A clear Signor-Lipps pattern is present in the predicted species richness values as the K/T boundary is approached. Nonetheless, the decline in species observed in these data is much larger than that predicted on the basis of interactions between sample size and species relative abundances. Application of the nonparametric Spearman's rank correlation test to the association between observed and predicted species richness values results in a correlation of –0.313 (n = 23), which is not significant at the 95% confidence level. Accordingly, there seems little reason to suspect that the decline in planktonic foraminiferal species richness in the uppermost Maastrichtian interval at El Kef can be adequately accounted for by a Signor-Lipps Effect due to variations in sample size.

Another potential source of potential Signor-Lipps bias in these data is variation in species-specific resistance to diagenetic dissolution. Presence of the basal red and black clay layers at the El Kef K/T boundary may have served as a barrier to the migration of fluids such that diagenetic effects might be concentrated in the uppermost Maastrichtian interval. Under this model, taxa disappearing prior to the El Kef K/T boundary may represent a suite of species exhibiting relatively low resistance to corrosive fluids moving through the sediment after deposition.

To test this effect, the dissolution susceptibility scales of

Douglas (1971) and Malmgren (1987) for Upper Cretaceous planktonic foraminifers were used to classify the El Kef fauna into dissolution-susceptible, dissolution-resistant, and unaffected categories (see MacLeod and Keller, 1994). The El Kef biostratigraphic data were then further subdivided into species observed to disappear below the K/T boundary (34 species) and species observed to extend at least to the boundary horizon (26 species). If the dissolution-based Signor-Lipps model is correct, the pre–K/T-extinction fauna should be differentially composed of dissolution-susceptible forms relative to the cohort of taxa that reach the K/T boundary. Results are shown in Figure 8.

Contrary to predictions of this Signor-Lipps model, these data indicate that the set of species disappearing prior to the K/T boundary is differentially composed of dissolution-resistant species, whereas the set of species extending at least to the K/T boundary is overwhelmingly composed of either dissolution-susceptible or unaffected species. A nonparametric G-test (Zar, 1974; Sokal and Rohlf, 1981) of these frequency distributions shows that there is less than a 0.001 chance (G = 17.63) that these results are attributable to random sampling from a single underlying distribution. These data also have direct bearing on the question of reworking within the El Kef sequence (see below).

A final test of a potential systematic Signor-Lipps–style bias within the observed El Kef biostratigraphic data deals with the possibility of environmental changes excluding a component of the Cretaceous fauna from this locality prior to their ultimate extinction horizon. Studies of other El Kef and several other K/T boundary sections (e.g., Hsü and McKenzie, 1985, 1990; Zachos and Arthur, 1986; Zachos et al., 1989; Barrera and Keller, 1990) have documented a decrease in $\delta^{13}C$ in the vicinity of the boundary. However, in all biostratigraphically complete sequences this change begins well after many indigenous Cretaceous species disappear and prior to the first occurrence of impact debris (see Fig. 1B). The lack of obvious lithostratigraphic change covarying with the faunal disappearance pattern offers little support for the environmental change scenario. Nevertheless, systematic variation in many types of environmental factors would not be expected to leave a lithostratigraphic mark. A more powerful test of this scenario lies in the examination of species-specific presence-absence and/or relative abundance logs.

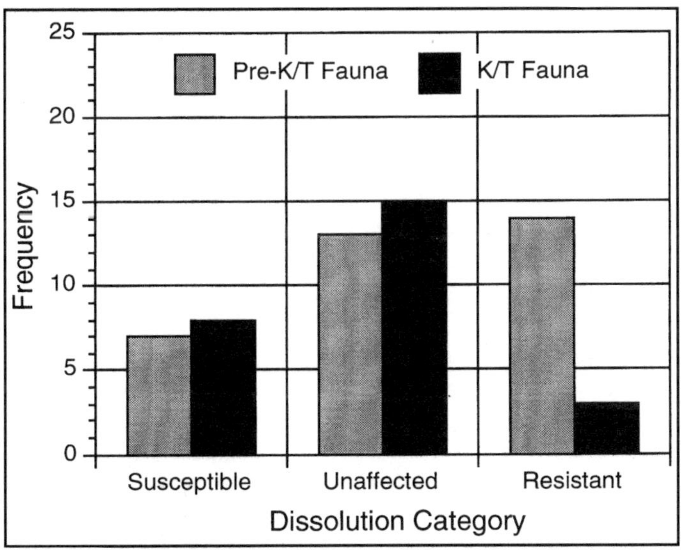

Figure 8. Frequency histogram of numbers of dissolution-susceptible, unaffected, and dissolution-resistant species composing the assemblage of species disappearing prior to the K/T boundary (gray) and the assemblage of species extending to (at least) the K/T boundary (black) in the Keller et al. (1995) data. Note the large proportion of dissolution-resistant taxa included in the K/T fauna and the small proportion of dissolution-resistant taxa included in the K/T fauna. These data suggest that the biostratigraphic ranges of this fauna have not been systematically biased by selective dissolution close to the boundary horizon producing a Signor-Lipps Effect.

If different cohorts of planktonic foraminiferal species with contrasting environmental tolerances exist within the El Kef K/T biostratigraphic data, these cohorts should exhibit similar within-group responses to environmental fluctuations. Accordingly, systematic patterns of covariation and contrast between presence-absence or relative abundance logs can be used to identify groups of species with similar environmental tolerances. In this investigation faunal similarities based on presence-absence observations were used to identify cohorts of species exhibiting similar environmental tolerances. (Note: See MacLeod, 1995a, for a discussion of the advantages of using presence-absence data.) Interspecific similarities were quantified using the Euclidean distance metric with pairwise deletion of missing values (samples above the observed LAD). Patterns within the resulting distance matrix were then summarized using multidimensional scaling (MDS). Once MDS coordinates were obtained, these data were subdivided into species disappearing prior to the K/T boundary and those extending at least to the boundary horizon itself. If this subdivision reflects significant biotic differences between these two groups' response to geologically cryptic environmental limiting factors, the two groups should form distinct clusters within a scatter plot of MDS scores. Such clustering would suggest that species disappearing prior to the K/T boundary may do so as a result of natural pattems of environmental variation that led to the production of apparent, rather than real, LADs. Results of this analysis for a two-dimensional MDS solution are shown in Figure 9.

Both pre-K/T and K/T cohorts exhibit comparable dispersions. The empty space in the middle of the scatter plot in Figure 9 indicates that each species' occurrence pattern is unique. About this empty zone, pre-K/T LAD species and K/T LAD species exhibit mixed distributions with no significant clustering. Discriminant analysis of these data is unable to locate a linear axis that classifies more than 77% of the species correctly. These results suggest that many species disappearing prior to the El Kef K/T boundary exhibit occurrence patterns that are indistinguishable from species extending to the boundary horizon and beyond. Consequently, there appears to be no evidence in these data to support the inference of a Signor-Lipps bias

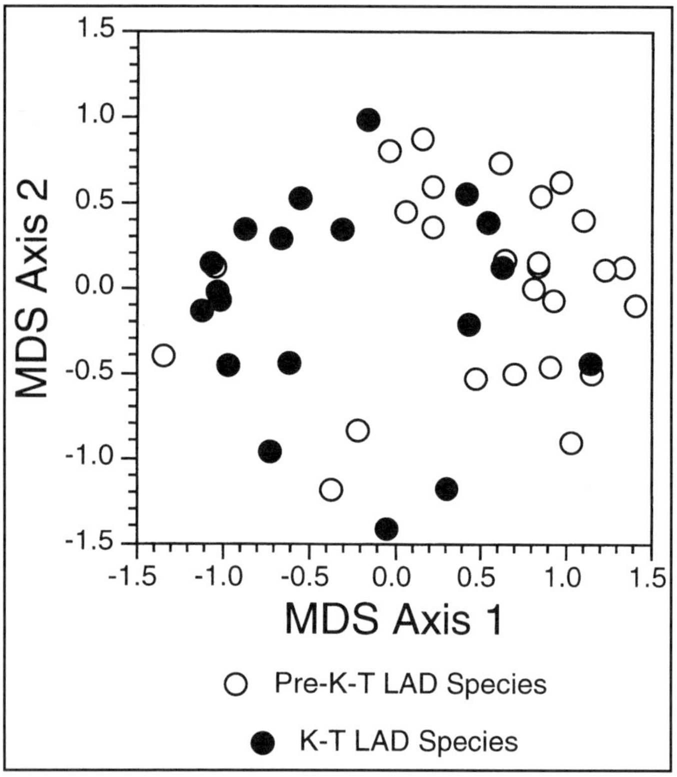

Figure 9. A two-axis multidimensional scaling (MDS) solution for the Euclidean distance matrix of occurrence patterns for Cretaceous taxa within the 23 uppermost Maastrichtian samples of the Keller et al. (1995) data. Species exhibiting similar occurrence patterns exhibit small interpoint distances, whereas those whose occurrence patterns differ exhibit large interpoint distances. The gap in the middle of the plot indicates that no two species occurrence patterns are identical to one another. About this gap, species whose apparent LADs fall below the K/T boundary (open circles) and those whose LADs extend to the K/T boundary (closed circles) are mixed with no overt clustering. This suggests that no distinctive pattern of occurrences (reflecting common responses to local patterns of environmental variation) characterizes either of these groups. See text for discussion.

resulting from coordinated interspecific responses to uppermost Maastrichtian environmental variation.

Stratigraphic confidence intervals

Whereas violation of the assumptions necessary for the calculation of stratigraphic confidence intervals prevent their application to these El Kef planktonic foraminiferal data, a simple experiment can be performed that illustrates the ambiguity of this proposed approach to the analysis of faunal turnover patterns. Let us assume that the 30 species disappearing from the El Kef succession within the first 10 m below the K/T boundary (Fig. 10) all possess 95% stratigraphic confidence intervals that include the boundary horizon. Would we then be able to say anything definite about the true pattern of faunal turnover? We could not reject the possibility that all of these species were synchronously eliminated at the boundary horizon. However, we would also not be able to reject any of a large number of alternative turnover geometries, each of which is consistent with such data. Even if we arbitrarily exclude the Danian portions of each species' confidence interval and only take into consideration the sampled pre-K/T horizons, there exist ~1.09×10^{27} different possible arrangements of LADs for these data. Since stratigraphic confidence intervals are measures of uncertainty, none of these alternative turnover geometries can be excluded from consideration on an a priori basis. All are equally likely to represent the "true" pattern.

Springer (1990) and Marshall (1994b) have developed confidence-interval–based methods for testing the hypothesis of whether data such as these are compatible with a coextinction model and identifying the stratigraphic interval over which such a model may be valid. Although both of these authors applied their methods to the Macellari (1986) upper Maastrichtian ammonite data in an attempt to test the mass extinction model for this group in the Seymour Island section, neither was able to demonstrate a unique association with an inferred coextinction (mass extinction) horizon and the K/T boundary. Moreover, both approaches assume coextinction (which is, in fact, the very hypothesis we would like to test) and fail to take into consideration the large number of independent extinction and combined independent-coextinction models that are also consistent with the Macellari (1986) data and their associated stratigraphic confidence intervals (see also Marshall, 1995).

Reworking of so-called Cretaceous species

In addition to disagreements arising from the interpretation of species disappearing in the uppermost Maastrichtian, the status of so-called Cretaceous species recovered from lowermost Danian sediments has also been a source of controversy. In his investigation of several K/T boundary sections, including El Kef, Smit (1982, see also Smit, 1990) reported observing a mass extinction event that "exterminated all but one species of the planktonic foraminifers" (p. 329). Smit (1977; Smit et al., 1988) did note the presence of Cretaceous planktonic foraminiferal species in lowermost Danian portions of the Agost and Caravaca sections, but he has never explicitly identified these species as K/T survivors. Instead, Smit (1982, 1990; Smit et al., 1988) have consistently maintained that only *Guembelitria cretacea* actually survived the K/T boundary event and that all other Cretaceous species in Danian sediments are reworked from underlying strata.

Liu and Olsson (1992) and Olsson and Liu (1993) have documented the existence of numerous Cretaceous planktonic foraminiferal species in the lowermost Danian sequence at Millers Ferry, Alabama, some of which extend in continuous occurrences many meters above the local K/T boundary. Like Smit, these authors interpret *G. cretacea* as surviving the boundary event, but add *Hedbergella holmdelensis* and *H. monmouthensis* to their list of survivors. Olsson and Liu consider all other so-called Cretaceous taxa observed above the K/T boundary in the Millers Ferry section to be reworked.

In contrast, studies summarized by MacLeod and Keller (1991, 1994) and MacLeod (1995a, 1995b, 1995c, 1996a,

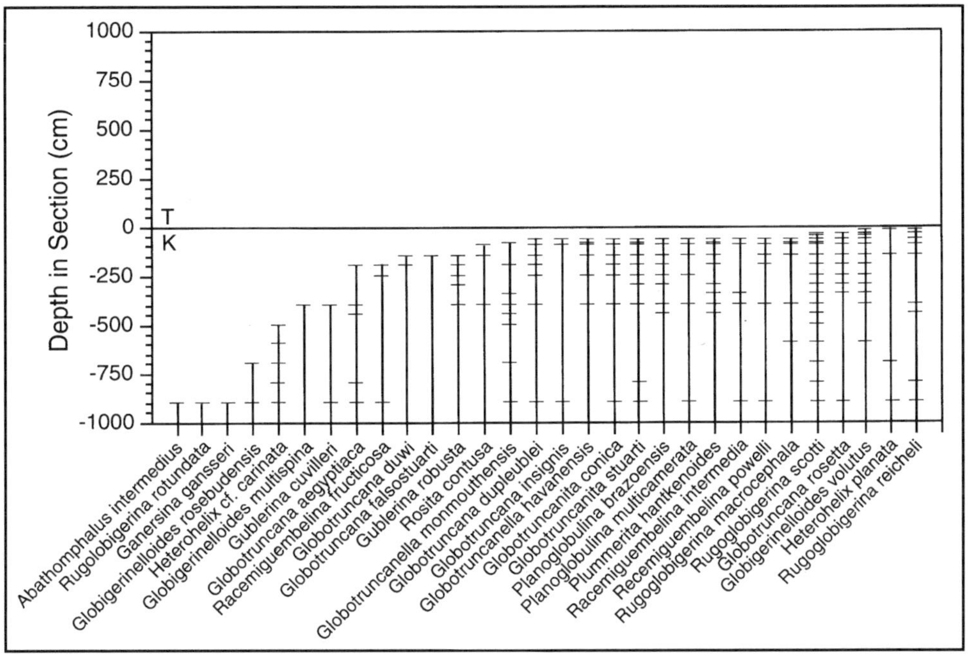

Figure 10. Occurrence pattern of the 30 Cretaceous species whose LAD falls below the K/T boundary horizon in the Keller et al. (1995) data. Although it is not possible to assign confidence intervals to these stratigraphic ranges (see text), if we assume that the confidence intervals of all 30 species included the K/T boundary horizon, the true pattern of LADs would still remain unknown. Even if any post-K/T confidence interval segments are excluded from consideration and only sampled horizons are taken into account, there exist more than 1.09×10^{27} possible LAD geometries for these data. These geometries are consistent with a wide variety of independent extinction, coextinction, and combined independent/coextinction models. Using the techniques of stratigraphic confidence interval estimation, it is impossible to determine which of these alternative geometries represents the "true" LAD pattern.

1996b) have documented the existence of this Cretaceous planktonic foraminiferal fauna in more than 30 different K/T boundary sections and cores. MacLeod and Keller (1994) and MacLeod (1996b) have subjected these data to comparative tests of the reworking and survivorship scenarios via comparisons of stratigraphic continuity, depositional setting, relative abundance/population ratio, stable isotopic signature, quantitative morphology, comparative biogeography, and preservational state (see MacLeod, 1994, for a description of these tests). In all instances, results indicated that patterns exhibited by many species individually, and the entire fauna in aggregate, were consistent with the predictions of K/T survivorship and inconsistent with predictions of hypotheses grounded on the assumption that these occurrences represent reworked Cretaceous faunas.

Whereas data from El Kef faunas were included in the MacLeod and Keller (1994) biogeographic study, none of the other reworking-survivorship tests (MacLeod, 1994) has been applied to the El Kef data by itself. However, Olsson and Liu's (1993) relative abundance–population test can be applied to the Keller et al. (1996) data. Olsson and Liu (1993, p. 136) describe their relative abundance–population ratio test as follows: "If one species survived [the K/T boundary event] its population size in the lower Paleocene would be composed of both the reworked and indigenous surviving fractions while that of an extinct species consists of only the reworked fraction. As a result, the relative abundance of the survivor taxon to that of the extinct species would significantly increase after the extinction of the other species." Since the relative abundance of true survivor species should increase above the extinction horizon, the relative abundance of reworked species should dramatically decline at that same horizon and undergo a more or less consistent decay to its final observed appearance.

Relative abundance–population ratio comparisons among seven abundant so-called Cretaceous taxa occurring in lowermost Danian sediments within the Keller et al. (1996) data (Fig. 11) show that, as is typical, *G. cretacea* undergoes a marked relative abundance increase coincident with the K/T boundary. Given this pattern, there seems little room for any interpretation other than that of survivorship for this species. However, intraspecific and interspecific patterns of relative abundance variation in five of the other six species fail to correspond to predictions of Olsson and Liu's (1993) relative abundance–population ratio test. Only *Hedbergella monmouthensis* (one of Liu and Olsson's survivor taxa) exhibits the pattern expected for reworked species. Patterns of lowermost Danian relative abundance variation for the remaining five species are remarkably similar to the patterns exhibited by these taxa prior to the K/T boundary and markedly different from Olsson and Liu's (1993) predictions for reworked species.

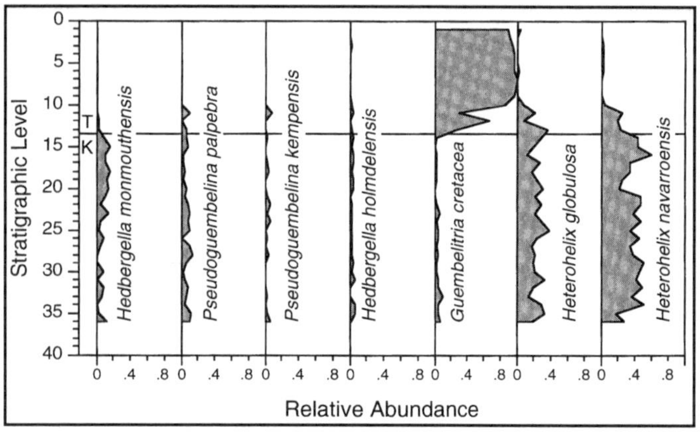

Figure 11. Relative abundance variation in seven abundant upper Maastrichtian planktonic foraminiferal taxa in the Keller et al. (1995) data. This group includes both widely accepted survivor species (*Guembelitria cretacea, Hedbergella holmdelensis, Hedbergella monmouthensis*) and so-called Cretaceous species frequently observed in lowermost Danian sediments (*Pseudoguembelina palpebra, Pseudoguembelina kempensis, Heterohelix globulosa, Heterohelix navarroensis*). Note that, with the exception of *G. cretacea* and *H. monmouthensis,* all species exhibit intraspecific and interspecific patterns of relative abundance variation in lowermost Danian samples consistent with their uppermost Maastrichtian condition. Note also that *H. globulosa* increases its relative abundance in lowermost Danian sediments while the gradual decline in *H. navarroensis* relative abundance appears to be part of a long-term decline initiated prior to the boundary horizon. See text for discussion.

Phenotypic patterns of extinction susceptibility

The most far-reaching implication of the bolide-impact–mass-extinction scenario lies in the realm of evolutionary theory. If, as Gould (1984, 1985), Raup (1986, 1991), Jablonski (1986a, 1986b, 1986c, 1989, 1994), and others claim, the five or so widely recognized mass extinctions represent fundamental discontinuities in the fabric of natural selection, then mass extinction can legitimately be seen as constituting a unique evolutionary process. Jablonski (1986a, 1986b, 1986c) proposed that an ecological distinction between the nature of selection pressures characterizing background and mass extinction events could be recognized in K/T molluscs by comparing observed survivorship levels in different life-history guilds as inferred from morphologic evidence. In particular, Jablonski concluded that since mollusc groups with planktotrophic larvae (large protoconch) were differentially resistant to extinction during background events but suffered the same rate of extinction susceptibility as nonplanktotrophic groups (small protoconch) during the K/T mass extinction, the nature of selection pressures during these two types of biotic crises differed profoundly.

Whereas planktonic foraminiferal life histories cannot presently be inferred from fossil material, the question of whether or not planktonic foraminifera experienced any differential phenotypic selection pressures as a result of the K/T extinction event can be addressed. Stanley et al. (1988) surveyed the Neogene history of the globogeriniform (spherical chambers, globular test) and globorotaliform (angular chambers, flattened test, peripheral keel[s]) planktonic foraminifera and concluded that during this interval of background extinction globigeriniform species were significantly less prone to extinction than globorotaliform species. Phenotypic differences between these two groups are thought to be correlated with ecological distinctions such as habitat, food source, and population size (see Stanley et al., 1988; Hemleben et al., 1989).

Did selection pressures phenotypically sort planktonic foraminiferal species during the K/T extinction event in a manner similar to Neogene background extinction events, or is there a pronounced morphologic distinction between the patterns of extinction and survivorship that characterize these episodes of Earth history? Figure 12 shows the relative frequency of heterohelicid, globigerinid, and globorotalid morphotypes for indigenous Cretaceous species in the uppermost El Kef Maastrichtian strata and putative K/T survivor species from the lowermost El Kef Danian strata. These data illustrate that although the relative percentage of heterohelicids underwent a modest increase across the K/T boundary, the relative percentage of generalized globorotalid morphotypes declined dramatically. Since heterohelicid morphotypes were not considered in the Stanley et al. (1988) study, these data cannot be assessed for conformance to patterns characteristic of background extinction. Nonetheless, a X^2 analysis of the 2 × 2 contingency table for globigerinid and globorotalid morphotypes before (19 and 11 species respectively) and after (8 and 1 species respectively) the event horizon yields a value of 1,217.42, which is highly significant ($\alpha >> 0.99$). Thus, patterns of phenotypic (and presumed ecologic) variation in planktonic foraminiferal extinction susceptibility across the

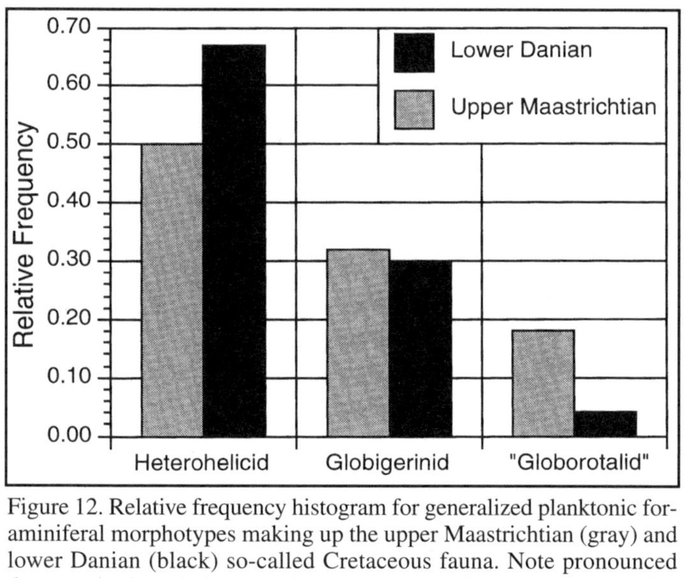

Figure 12. Relative frequency histogram for generalized planktonic foraminiferal morphotypes making up the upper Maastrichtian (gray) and lower Danian (black) so-called Cretaceous fauna. Note pronounced decrease in the relative abundance of globorotaliform taxa across the boundary horizon. These data suggest that generalized globorotalid taxa suffered differentially higher extinction rates in the uppermost Maastrichtian, whereas heterohelicid and globigerinid taxa either increased their proportional representation or remained unchanged across the boundary horizon. See text for discussion.

El Kef K/T boundary are consistent with patterns documented from a Neogene interval characterized by background extinction intensities. This characteristic El Kef extinction susceptibility pattern is also consistent with similar data summaries for the global K/T fauna (MacLeod and Keller, 1994; MacLeod, 1996b).

SUMMARY

The original description of the bolide impact–mass extinction hypothesis proposed that "an asteroid struck the earth, formed an impact crater, and some of the dust-sized material ejected from the crater reached the stratosphere and was spread around the globe. This dust effectively prevented sunlight from reaching the surface for a period of several years. . . . Loss of sunlight suppressed photosynthesis, and as a result most food chains collapsed and the extinctions resulted" (Alvarez et al., 1980, p. 1105). Alvarez et al. (1982) revised their estimate of stratospheric dust residence times to a few months and expanded the range of conceivable killing mechanisms to include acid rain and the release of noxious chemicals. Since 1982, a large number of other earth scientists have also cited the geologically instantaneous disappearance of Cretaceous taxa as either integral to or evidence for the bolide-impact–mass-extinction scenario (e.g., see papers included in Silver and Schultz, 1982; Donovan, 1989; and Sharpton and Ward, 1990, in addition to the single-author treatments of Raup, 1986 and 1991; Stanley, 1987; and Eldredge, 1991). The only conclusion that can be drawn from these statements is that the bolide impact–mass extinction scenario, as generally understood by most scientists (as well as its authors), includes the proposition that extinctions causally linked to a K/T bolide impact should have taken place over such a small interval of time as to appear instantaneous in the stratigraphic record.

This study has focused on the uppermost Cretaceous and lowermost Danian record of planktonic foraminifers in the El Kef boundary stratotype section because of the quality of that group's fossil record as well as the importance geologists traditionally have attached to stratotype sections. However, planktonic foraminiferal data from the El Kef section remains only a single biotic record from a single locality and cannot be assumed to adequately portray the nature of the K/T faunal turnover in all marine (to say nothing of terrestrial) environments.

The raw data for planktonic foraminiferal turnover across the El Kef K/T boundary clearly exhibit a progressive, rather than an instantaneous, pattern (Keller, 1988a; Keller et al., 1995; MacLeod and Keller, 1996), as do comparable data from most biostratigraphically complete K/T boundary successions (MacLeod and Keller, 1991, 1994). This apparent turnover pattern may reflect the operation of various sources of bias in the stratigraphic record (e.g., Signor-Lipps Effect, reworking), or it may represent a reasonably accurate approximation of the actual turnover pattern.

Tests of the Keller et al. (1995) El Kef planktonic foraminiferal data for the Signor-Lipps Effect due to sample size–relative abundance interactions, preservation potential, and coordinated presence-absence response to environmental change all yielded results that are inconsistent with the predictions of a substantial bias. A small Signor-Lipps Effect due to sample size–relative abundance interactions was predicted for these data by the Koch and Morgan (1988) modeling method. This bias results from the peaked nature of sample sizes just below the K/T boundary in the Keller et al. (1995) data. Nevertheless, this source bias is unable to entirely account for the progressive decline in planktonic foraminiferal species richness in the El Kef succession as the boundary horizon is approached.

Owing to the nature of microfossil sample collection procedures, the reliability of the apparent LAD distribution patterns in the uppermost Maastrichtian interval could not be evaluated via the method of stratigraphic confidence intervals. However, since stratigraphic confidence intervals are measures of uncertainty, a very large number of alternative faunal turnover geometries are undoubtedly consistent with these data whatever the lengths of the true confidence intervals may be.

A relative abundance–population ratio test of the survivorship and reworking explanations for occurrences of so-called Cretaceous species in overlying Danian sediments at El Kef shows these data to be consistent with an interpretation of significant planktonic foraminiferal survivorship across the K/T boundary and into the lower portion of the Tertiary. Finally, comparative phenotypic analyses show that patterns of extinction susceptibility across the El Kef K/T boundary are very similar to planktonic foraminiferal morphotypic extinction patterns characteristic of Neogene background extinction events.

These results suggest that the El Kef K/T planktonic foraminiferal faunal turnover must have taken place across an extended time interval—several orders of magnitude longer than the interval alluded to in the original bolide impact–mass extinction model. Results from studies of other organismal groups that have shown local faunal turnover patterns to be consistent with a catastrophic turnover model (e.g., Springer, 1990; Ward et al., 1991; Sheehan et al., 1991; Marshall, 1994b, 1995) are also consistent with a large number of other noncatastrophic turnover models (see Ward et al., 1986; Kennedy, 1989; Ward and Kennedy, 1993). Until such time as the biotic predictions of the bolide impact–mass extinction scenario are made more explicit in terms of the time interval involved, causal processes in operation, and the taxonomic identity of predicted victims/survivors, it is important to keep an open mind with respect to the question of what factor(s) may have been responsible for the K/T mass extinction.

ACKNOWLEDGMENTS

I thank Gerta Keller, David Archibald, Andrew Smith, Stephen Culver, Charles Marshall, Peter Ward, David Fastovsky, and three anonymous reviewers for reading, commenting on, and/or discussing previous versions of the arguments presented herein. This report represents a contribution from The Natural History Museum's Global Change and the Biosphere project.

REFERENCES CITED

Alvarez, L. W., Alvarez, W., Asaro, F., and Michel, H. V., 1980, Extraterrestrial cause for the Cretaceous-Tertiary extinction: Science, v. 208, p. 1095–1108.

Alvarez, W., Alvarez, L. W., Asaro, F., and Michel, H. V., 1982, Current status of the impact theory for the terminal Cretaceous extinction, *in* Silver, L. T., and Schultz, P. H., eds., Geological implications of impacts of large asteroids and comets on Earth: Geological Society of America Special Paper 190, p. 305–315.

Alvarez, W., Kauffman, E. G., Surlyk, F., Alvarez, L. W., Asaro, F., and Michel, H. V., 1984, Impact theory of mass extinctions and the invertebrate fossil record: Science, v. 223, p. 1135–1141.

Archibald, J. D., and Bryant, L., 1990, Differential Cretaceous-Tertiary extinctions of nonmarine vertebrates: Evidence from northeastern Montana, *in* Sharpton, V. L., and Ward, P., eds., Global catastrophes in Earth history: An interdisciplinary conference on impacts, volcanism, and mass mortality: Geological Society of America Special Paper 247, p. 549–562.

Archibald, J. D., and Clemens, W. A., 1984, Mammal evolution near the Cretaceous-Tertiary boundary, *in* Berggren, W. A., and Vancouvering, J. A., eds., Catastrophes and Earth history: The new uniformitarianism: Princeton, New Jersey, Princeton University Press, p. 229–371.

Askin, R. A., 1990, The palynological record across the Cretaceous/Tertiary transition on Seymour Island, Antarctica, *in* Feldmann, R. M., and Woodburne, M. O., eds., Geology and paleontology of Seymour Island, Antarctic Peninsula: Geological Society of America Memoir 169, p. 131–156.

Barrera, E., and Keller, G., 1990, Foraminiferal stable isotope evidence for gradual decrease of marine productivity and Cretaceous species survivorship in the earliest Danian: Paleoceanography, v. 5, p. 867–870.

Ben Abdelkader, O., Haj Ali, N. B., Ben Salem, H., and Razgallah, S., 1992, Proceedings, Intemational Workshop on Cretaceous-Tertiary Transitions (El Kef Section. Part II: Field Trip Guidebook) Tunis: Tunis, IUGS/GSGP/ATEIG/Tunisian Geological Survey, 25 p.

Berger, W. H., 1968, Planktonic foraminifers: Selective solution and paleoclimatic interpretation: Deep-Sea Research, v. 15, p. 31–43.

Berger, W. H., 1970, Planktic foraminifers: Selective solution and the lysocline: Marine Geology, v. 8, p. II 1 13 8.

Berger, W. H., 1979, Preservation of foraminifera, *in* Lipps, J. H., and Berger, W. H., eds., Foraminiferal ecology and paleoecology: Houston, Texas, Society of Economic Paleontologists and Mineralogists Short Course No. 6, p. 105–155.

Bramlette, M. N., 1965, Massive extinctions in biota at the end of Mesozoic time: Science, v. 148, p. 1696–1699.

Bramlette, M. N., and Martini, E., 1964, The great change in calcareous nannoplankton fossils between the Maastrichtian and Danian: Micropaleontology, v. 10, p. 291–322.

Brinkhuis, H., and Zachariasse, W. J., 1988, Dinoflagellate cysts, sea level changes and planktonic foraminifers across the Cretaceous-Tertiary boundary at El Haria, northwest Tunisia: Marine Micropaleontology, v. 13, p. 153–191.

Brouwers, E. M., and De Deckker, P., 1993, Late Maastrichtian and Danian ostracode faunas from northern Alaska: Reconstructions of environment and paleogeography: Palaios, v. 8, p. 140–154.

Canudo, J. J., Keller, G., and Molina, E., 199 1, Cretaceous/Tertiary boundary extinction pattern and faunal turnover at Agost and Caravaca, S.E. Spain: Marine Micropaleoontology, v. 17, p. 319–341.

Cowrie, J. W., Ziegler, W., and Remaine, J., 1989, Stratigraphic Commission accelerates progress, 1984–1989: Episodes, v. 112, p. 79–83.

Donovan, S. K., 1989, Mass extinctions: Processes and evidence: London, Belhaven, 266 p.

Douglas, R. G., 1971, Cretaceous foraminifera from the northwestern Pacific Ocean: Leg 6, Deep Sea Drilling Project, *in* Initial reports of the Deep Sea Drilling Project, Volume 6: Washington, D.C., U.S. Government Printing Office, p. 1027–1053.

Eldredge, N., 1991, The miner's canary: Unraveling the mysteries of extinction: New York, Prentice Hall, 246 p.

Golovnea, L. B., 1994, The flora of the Maastrichtian-Danian deposits of the Koryak Upland, NE Russia: Cretaceous Research, v. 15, p. 89–100.

Gould, S. J., 1984, The cosmic dance of Siva: Natural History, v. 93, p. 8.

Gould, S. J., 1985, The paradox of the first tier: An agenda for paleobiology: Paleobiology, v. 11, p. 2–12.

Hanski, I., 1982, Dynamics of regional distribution: The core and satellite species hypothesis: Okios, v. 31, p. 210–211.

Hanski, I., and Gilpin, M., 1991, Metapopulation dynamics: Brief history and conceptual domain: Biological Journal of the Linnaean Society, v. 42, p. 3–16.

Hemleben, C., Spindler, M., and Anderson, O. R., 1989, Modern planktonic foraminifera: London, Springer-Verlag, 363 p.

Hollis, C. J., 1993, Latest Cretaceous to Late Paleocene radiolarian biostratigraphy: A new zonation from the New Zealand region: Marine Micropaleontology, v. 21, p. 295–327.

Hsü, K. J., and McKenzie, J. A., 1985, A "strangelove" ocean in the earliest Tertiary, *in* Sundquist, E. T., and Broecker, W. S., eds.. Natural variations: Archean to Present: American Geophysical Union Monograph 32, p. 487–492.

Hsü, K. J., and McKenzie, J. A., 1990, Carbon-isotope anomalies at era boundaries; Global catastrophes and their ultimate cause, *in* Sharpton, V. L., and Ward, P. D., eds., Global catastrophes in Earth history: An interdisciplinary conference on impacts, volcanism, and mass mortality: Geological Society of America Special Paper 247, p. 61–70.

Hut, P., Alvarez, W., Elder, W. P., Hansen, T., Kauffman, E. G., Keller, G., Shoemaker, E. M., and Weissman, P. R., 1987, Comet showers as a cause of mass extinctions: Nature, v. 329, p. 118–126.

Jablonski, D., 1986a, Causes and consequences of mass extinctions: Implications for macroevolution, *in* Elliott, D. K., ed., Dynamics of extinction: New York, Wiley, p. 183–229.

Jablonski, D., 1986b, Background and mass extinctions: The alteration of macroevolutionary regimes: Science, v. 231, p. 129–133.

Jablonski, D., 1986c, Evolutionary consequences of mass extinctions, *in* Raup, D. M., and Jablonski, D., eds., Patterns and processes in the history of life: Berlin, Springer-Verlag, p. 313–329.

Jablonski, D., 1989, The biology of mass extinction: A palaeontological view: Philosophical Transactions of the Royal Society, Ser. B, v. 325, p. 357–368.

Jablonski, D., 1994, Mass extinctions: Persistent problems and new directions, *in* New developments regarding the K/T event and other catastrophes in Earth history: Houston, Texas, Lunar and Planetary Institute Contribution 825, p. 56–57.

Johnson, C. C., and Kauffman, E. G., 1993, Rates and patterns of Maastrichtian reef extinction associated with environmental changes in the Caribbean province: Geological Society of America Abstracts with Programs, v. 25, p. A363.

Johnson, C. C., and Kauffman, E. G., 1996, Maastrichtian extinction patterns of Caribbean province rudistids, *in* MacLeod, N., and Keller, G., eds., The Cretaceous-Tertiary mass extinction: Biotic and environmental changes: New York, W.W. Norton (in press).

Kauffman, E. G., 1984, The fabric of Cretaceous extinctions, *in* Berggren, W. A., and Van Couvering, J. A., eds., Catastrophes and Earth history: The new uniformitarianism: Princeton, New Jersey, Princeton University Press, p. 151–246.

Keller, G., 1988a, Extinction, survivorship and evolution of planktic foraminifera across the Cretaceous/Tertiary boundary at El Kef, Tunisia: Marine

Micropaleontology, v. 13, p. 239–263.

Keller, G., 1988b, Biotic turnover in benthic foraminifers across the Cretaceous-Tertiary boundary at El Kef, Tunisia: Palaeogeography, Palaeoclimatology, Palaeoecology, v. 66, p. 153–171.

Keller, G., 1989, Extended Cretaceous/Tertiary boundary extinctions and delayed population change in planktonic foraminiferal faunas from Brazos River, Texas: Paleoceanography, v. 4, p. 287–332.

Keller, G., 1993, The Cretaceous-Tertiary boundary transition in the Antarctic Ocean and its global implications: Marine Micropaleontology, v. 21, p. 1–46.

Keller, G., and Benjamini, C., 1991, Paleoenvironment of the eastern Tethys in the Early Paleocene: Palaios, v. 6, p. 439–464.

Keller, G., Stinnesbeck, W., and Lopez-Oliva, J. G., 1994, Age, deposition, and biotic effects of the Cretaceous/Tertiary boundary event at Mimbral, NE Mexico: Palaios, v. 9, p. 144–157.

Keller, G., Li, L., and MacLeod, N., 1995, The Cretaceous/Tertiary boundary stratotype section at El Kef, Tunisia: How catastrophic was the mass extinction?: Palaeogeography, Palaeoclinatology, Palaeoecology, v. 19, p. 251–273.

Kennedy, W. J., 1989, Thoughts on the evolution and extinction of Cretaceous ammonites: Proceedings of the Geologists' Association, v. 100, p. 251–279.

Kerr, R. A., 1994, Testing an ancient impact's punch: Science, v. 263, p. 1371–1372.

Knoboch, E., Kvacek, Z., Buzek, C., Mai, D. H., and Batten, D. J., 1993, Evolutionary significance of floristic changes in the northern hemisphere during the late Cretaceous and Palaeogene, with particular reference to central Europe: Review of Palaeobotany and Palynology, v. 78, p. 41–54.

Koch, C. F., and Morgan, J. P., 1988, On the expected distribution of species ranges: Paleobiology, v. 14, p. 126–138.

Labandiera, C. C., 1992, Diets, diversity, and disparity: Determining the effect of the terminal Cretaceous extinction on insect evolution: North American Paleontological Convention, 5th, Abstracts with Programs, Special Publication of the Paleontological Society, v. 6, p. 174.

Lerbekmo, J. F., Sweet, A. R., and St. Louis, R. M., 1987, The relationship between the iridium anomaly and palynological floral events at three Cretaceous-Tertiary boundary localities in western Canada: Geological Society of America Bulletin, v. 99, p. 325–330.

Liu, C., and Olsson, R. K., 1992, Evolutionary radiation of microperforate planktonic foraminifera following the K/T mass extinction: Journal of Foraminiferal Research, v. 22, p. 328–346.

Macellari, C. E., 1986, Late Campanian–Maastrichtian ammonites from Seymour Island, Antarctic Peninsula: Journal of Paleontology, v. 60, p. 1–55.

MacLeod, N., 1994, An evaluation of criteria that may be used to identify species surviving a mass extinction, *in* New developments regarding the K/T event and other catastrophes in Earth history: Houston, Texas, Lunar and Planetary Institute Contribution 825, p. 75–77.

MacLeod, N., 1995a, Cretaceous/Tertiary (K/T) biogeography of planktic foraminifera: Historical Biology, v. 10, p. 49–101.

MacLeod, N., 1995b, Graphic correlation of high latitude Cretaceous-Tertiary boundary sequences at Nye Kløv (Denmark), ODP Site 690 (Weddell Sea), and ODP Site 738 (Kerguelen Plateau): Comparison with the El Kef (Tunisia) boundary stratotype: Modern Geology, v. 19, p. 109–147.

MacLeod, N., 1995c, Graphic correlation of new Cretaceous/Tertiary (K/T) boundary sections, *in* Mann, K. O., and Lane, H. R., eds., Graphic correlation: The method and its application: Tulsa, Oklahoma, Society for Sedimentary Geology Special Publication 53, p. 215–233.

MacLeod, N., 1996a, Stratigraphic completeness and planktic foraminiferal survivorship across the Cretaceous-Tertiary (K/T) boundary: A reply to Olsson and Liu, *in* Moguilevsky, A., ed., ODP and the marine biosphere: Aberystwyth, University of Wales (in press).

MacLeod, N., 1996b, Nature of the Cretaceous-Tertiary (K/T) planktonic foraminiferal record: Stratigraphic confidence intervals, Signor-Lipps effect, and patterns of survivorship, *in* MacLeod, N., and Keller, G., eds., The Cretaceous-Tertiary mass extinction: Biotic and environmental changes: New York, W.W. Norton (in press).

MacLeod, N., and Keller, G., 1996, K-T Redux: Paleobiology (in press).

MacLeod, N., and Keller, G., 1991, How complete are Cretaceous/Tertiary boundary sections? A chronostratigraphic estimate based on graphic correlation: Geological Society of America Bulletin, v. 103, p. 1439–1457.

MacLeod, N., and Keller, G., 1994, Comparative biogeographic analysis of planktic foraminiferal survivorship across the Cretaceous/Tertiary (K/T) boundary: Paleobiology, v. 20, p. 143–177.

Maguran, A. E., 1988, Ecological diversity and its measurement: Princeton, New Jersey, Princeton University Press, 179 p.

Malmgren, B. A., 1987, Differential dissolution of Upper Cretaceous planktonic foraminifera from a temperate region of the South Atlantic Ocean: Marine Micropaleontology, v. 11, p. 251–271.

Marshall, C. R., 1994a, Confidence intervals on stratigraphic ranges: Partial relaxation of the assumption of randomly distributed fossil horizons: Paleobiology, v. 20, p. 459–469.

Marshall, C. R., 1994b, Using the ammonite fossil record to predict the position of the K-T boundary Ir anomaly on Seymour Island, Antarctica: Geological Society of America Abstracts with Programs, v. 26, no. 7, p. A394.

Marshall, C. R., 1995, Distinguishing between sudden and gradual extinctions in the fossil record: Predicting the position of the Cretaceous-Tertiary iridium anomaly using the ammonite fossil record on Seymour Island, Antarctica: Geology, v. 23, p. 731–734.

McIver, E. E., Sweet, A. R., and Basinger, J. F., 1991, Sixty-five-million-year-old flowers bearing pollen of the extinct triprojectate complex—a Cretaceous-Tertiary boundary survivor: Review of Paleobotany and Palynology, v. 70, p. 77–88.

Méon, H., 1990, Palynologic studies of the Cretaceous/Tertiary boundary interval at El Kef outcrop, northwestern Tunisia: Review of Paleobotany and Palynology, v. 65, p. 85–94.

Newell, N. D., 1967, Revolutions in the history of life: Geological Society of America Special Paper 89, p. 63–91.

Olsson, R. K., and Liu, C., 1993, Controversies on the placement of Cretaceous-Paleogene boundary and the K/P mass extinction of planktonic foraminifera: Palaios, v. 8, p. 127–139.

Perch-Nielsen, K., McKenzie, J., and He, Q., 1982, Biostratigraphy and isotope stratigraphy and the catastrophic extinction of calcareous nannoplankton at the Cretaceous/Tertiary boundary, *in* Silver, L. T., and Schultz, P. H., eds., Geological implications of impacts of large asteroids and comets on the Earth: Geological Society of America Special Paper 190, p. 353–371.

Percival, S. F., and Fischer, A. G., 1977, Changes in calcareous nannoplankton in the Cretaceous-Tertiary biotic crisis at Zumaya, Spain: Evolutionary Theory, v. 2, p. 1–35.

Raup, D. M., 1972, Taxonomic diversity during the Phanerozoic: Science, v. 177, p. 1065–1071.

Raup, D. M., 1976, Species diversity in the Phanerozoic: An interpretation: Paleobiology, v. 2, p. 289–297.

Raup, D. M., 1986, The nemesis affair: A story of the death of the dinosaurs and the ways of science: New York, W.W. Norton, 220 p.

Raup, D. M., 1989, The case for extraterrestrial causes of extinction: Philosophical Transactions of the Royal Society of London, Ser. B, v. 325, p. 421–435.

Raup, D. M., 199 1, Extinction: Bad genes or bad luck: New York, W.W. Norton, 210 p.

Robin, E., Boclet, D., Bonté, P., Froget, L., Jéhanno, C., and Rocchia, R., 1991, The stratigraphic distribution of Ni-rich spinels in Cretaceous-Tertiary boundary rocks at El Kef (Tunisia), Caravaca (Spain) and Hole 761 C (Leg 122): Earth and Planetary Science Letters, v. 107, p. 715–721.

Schindewolf, O., 1962, Neokatastophismus?: Deutsches Geologische Gesellschaft Zeitschrift Jahrgang, v. 114, p. 430–445.

Schmitz, B., Keller, G., and Stenvall, O., 1992, Stable isotope and foraminiferal changes across the Cretaceous-Tertiary boundary at Stvens Klint, Denmark: Arguments for long-term oceanic instability before and after bolide-impact event: Palaeoceanography, Palaeoclimatology, Palaeo-

geography, v. 96, p. 233–260.
Sharpton, V. L., and Ward, P. D., eds., 1990, Global catastrophes in Earth history: An interdisciplinary conference on impacts, volcanism, and mass mortality: Geological Society of America Special Paper 247, 631 p.
Shaw, A., 1964, Time in stratigraphy: New York, McGraw-Hill, 365 p.
Sheehan, P. M., Fastovsky, D. E., Hoffmann, R. G., Berghaus, C. B., and Gabriel, D., 1991, Sudden extinction of the dinosaurs: Latest Cretaceous, Upper Great Plains, U.S.A.: Science, v. 254, p. 835–839.
Signor, P. W., III, 1978, Species richness in the Phanerozoic: A reflection of labor by systematists?: Paleobiology, v. 7, p. 36–53.
Signor, P. W., III, and Lipps, J. H., 1982, Sampling bias, gradual extinction patterns and catastrophes in the fossil record, *in* Silver, L. T., and Schultz, P. H., eds., Geological implications of impacts of large asteroids and comets on the Earth: Geological Society of America Special Paper 190, p. 291–296.
Silver, L. T., and Schultz, P. H., eds., 1982, Geological implications of impacts of large asteroids and comets on the Earth: Geological Society of America Special Paper 190, 500 p.
Sloan, R. E., Rigby, J. K., Van Valen, L. M., and Gabriel, D., 1986, Gradual dinosaur extinction and simultaneous ungulate radiation in the Hell Creek Formation: Science, v. 232, p. 629–633.
Smit, J., 1977, Discovery of a planktonic foraminifers association between the *Abathomphalus mayaroensis* Zone and the *'Globigerina' eugubina* Zone at the Cretaceous/Tertiary boundary in the Barranco del Gredero (Caravaca, SE Spain): Koninklijke Nederlandese Akademie van Wetenschappen Proceedings, Series B, v. 80, p. 280–301.
Smit, J., 1982, Extinction and evolution of planktonic foraminifera after a major impact at the Cretaceous/Tertiary boundary, *in* Silver, L. T., and Schultz, P. H., eds., Geological implications of impacts of large asteroids and comets on the Earth: Geological Society of America Special Paper 190, p. 329–352.
Smit, J., 1990, Meteorite impact, extinctions and the Cretaceous-Tertiary boundary: Geologie en Mijnbouw, v. 69, p. 187–204.
Smit, J., 1994, Blind tests and muddy waters: Nature, v. 368, p. 809–810.
Smit, J., Groot, H., de Jonge, R., and Smit, P., 1988, Impact and extinction signatures in complete Cretaceous Tertiary (KT) boundary sections [abs.]: Global catastrophes in Earth history: An interdisciplinary conference on impacts, volcanism, and mass mortality: Houston, Texas, Lunar and Planetary Institute, p. 182–183.
Smith, A. B., 1994, Systematics and the fossil record: Documenting evolutionary patterns: London, Blackwell, 223 p.
Sokal, R. R., and Rohlf, J. F., 1981, Biometry: San Francisco, W.H. Freeman, 859 p.
Springer, M. S., 1990, The effect of random range trunctions on patterns of evolution in the fossil record: Paleobiology, v. 16, p. 512–520.
Stanley, S. M., 1987, Extinctions: New York, W.H. Freeman, 242 p.
Stanley, S. M., Wetmore, K. L., and Kennett, J. P., 1988, Macroevolutionary differences between two major clades of Neogene planktonic foraminifera: Paleobiology, v. 14, p. 235–249.
Strauss, D., and Sadler, P. M., 1989, Classical confidence intervals and Bayesian probability estimates for ends of local taxon ranges: Mathematical Geology, v. 21, p. 411–427.
Surlyk, F., 1990, Mass extinction events. Section 2.13.6: Cretaceous-Tertiary (Marine), *in* Briggs, D. E. G., and Crowther, P. R., eds., Palaeobiology: A synthesis: Oxford, Blackwell, p. 198–203.
Sweet, A. R., and Braman, D. R., 1992, The K-T boundary and contiguous strata in western Canada: Interactions between paleoenvironments and palynological assemblages: Cretaceous Research, v. 13, p. 31–79.
Sweet, A. R., Braman, D. R., and Lerbekmo, J. F., 1990, Palynofloral response to K/T boundary events: A transitory interruption within a dynamic system, *in* Sharpton, V. L., and Ward, P. D., eds., Global catastrophes in Earth history: Geological Society of America Special Paper 247, p. 457–469.
Thomas, E., 1990, Late Cretaceous–Early Eocene mass extinction in the deep sea, *in* Sharpton, V. L., and Ward, P. D., eds., Global catastrophes in Earth history: Geological Society of America Special Paper 247, p. 481–495.
Ward, P. D., and Kennedy, W. J., 1993, Maastrichtian ammonites from the Biscay region (France, Spain): Paleontological Society Memoir 34, p. 1–58.
Ward, P., Wiedmann, J., and Mount, J. F., 1986, Maastrichtian molluscan biostratigraphy and extinction patterns in a Cretaceous/Tertiary boundary section exposed at Zumaya, Spain: Geology, v. 14, p. 899–903.
Ward, P. D., Kennedy, W. J., MacLeod, K. G., and Mount, J. F., 1991, Ammonite and inoceramid bivalve extinction patterns in Cretaceous/Tertiary boundary sections of the Biscay region (southwestern France, northern Spain): Geology, v. 19, p. 1181–1184.
Widmark, J. G., and Malmgren, B. A., 1992, Benthic foraminiferal changes across the Cretaceous-Tertiary boundary in the deep sea; DSDP Sites 525, 527, and 465: Journal of Foraminiferal Research, v. 22, p. 81–113.
Zachos, J. C., and Arthur, M. A., 1986, Paleoceanography of the Cretaceous/Tertiary boundary event from stable isotopic and other data: Paleoceanography, v. 1, p. 526.
Zachos, J. C., Arthur, M. A., and Dean, W. E., 1989, Geochemical evidence for suppression of pelagic marine productivity at the Cretaceous/Tertiary boundary: Nature, v. 337, p. 61–64.
Zar, J. H., 1974, Biostatistical analysis: Englewood Cliffs, New Jersey, Prentice Hall, 620 p.

MANUSCRIPT ACCEPTED BY THE SOCIETY SEPTEMBER 14, 1995

Printed in U.S.A.

Geological Society of America
Special Paper 307
1996

Planktic foraminifera, asteroids, and marine production: Death and recovery at the Cretaceous-Tertiary boundary

Steven D'Hondt
Graduate School of Oceanography, University of Rhode Island, Narragansett, Rhode Island 02882
Timothy D. Herbert
Department of Geological Sciences, Brown University, Providence, Rhode Island 02912
John King and Carol Gibson
Graduate School of Oceanography, University of Rhode Island, Narragansett, Rhode Island 02882

ABSTRACT

The Cretaceous/Tertiary (K/T) impact radically changed the long-term state of the open-marine ecosystem. Comparison of range data from multiple sites indicates that more than 90% of planktic foraminiferal species went extinct at or shortly after the time of the K/T impact. Nearly complete replacement of planktic foraminiferal faunas began at the impact horizon. This turnover resulted in the disappearance of onshore-offshore provincialization of low- and mid-latitude planktic foraminiferal faunas. Drastic decreases in calcareous nannofossil production and organic export from surface to deep waters coincided with this destruction of faunal provincialization. Precessional chronometry of South Atlantic sites indicates that the initiation of faunal replacement and the decrease in nannofossil production occurred in an unresolvably short interval of time.

Recovery of planktic foraminiferal assemblages was very dynamic. A rapid burst of first occurrences closely followed the impact. Despite the rapidity of this evolutionary radiation, long-term assemblage stabilization and recovery of organic export did not occur for at least a few hundred k.y. after the impact. Calcareous nannoplankton production remained below preimpact levels for two million years or more. There is some evidence that organic export to deep waters remained below preimpact levels for a similar interval of time.

The coincidence of the K/T impact with planktic foraminiferal mass extinction, faunal turnover, and strong decreases in organic export and nannofossil production is very compatible with an impact model of mass extinction. In contrast, the proximate causes of the geologically long delay in postimpact recovery of organic export and nannofossil production remain unresolved. This presents a major challenge to our understanding of the ecologic and oceanographic consequences of large impacts and mass extinctions.

INTRODUCTION

The marine microfossil record provides a very important key to our understanding of the Cretaceous/Tertiary (K/T) boundary (65 Ma). Recognition of an impact at the marine K/T boundary is historically based on the close stratigraphic association of impact debris with the planktic foraminiferal or nannofossil biostratigraphic boundary (Alvarez et al., 1980). Hence, our understanding of the K/T impact and mass extinctions is intimately tied to the marine microfossil record. Furthermore,

D'Hondt, S., Herbert, T. D., King, J., and Gibson, C., 1996, Planktic foraminifera, asteroids, and marine production: Death and recovery at the Cretaceous-Tertiary boundary, *in* Ryder, G., Fastovsky, D., and Gartner, S., eds., The Cretaceous-Tertiary Event and Other Catastrophes in Earth History: Boulder, Colorado, Geological Society of America Special Paper 307.

analysis of calcareous microfossil chemistry, preservation, and rates of accumulation provides most of the primary evidence for K/T changes in marine production, ocean chemistry, and climate (e.g., Hsü et al., 1982; Zachos et al., 1989).

The marine microfossil record has long served as a flashpoint for disagreement over the timing, rapidity, and possible causes of mass extinction at the K/T boundary (e.g., Thierstein, 1981; Perch-Nielsen et al., 1982; Smit, 1982; Keller, 1989). Much of this disagreement has focused on a relatively narrow set of issues, such as interpretation of pre- and postimpact last occurrence patterns at a small number of localities. For example, differences in published planktic foraminiferal ranges of the El Kef (Tunisia) K/T sequence have been a major focus of argument for the last decade (Smit, 1982, 1990; Keller, 1988; Ginsberg et al., 1994; Kerr, 1994).

This chapter has a broader intent—to demonstrate the widespread and long-lasting effect of the K/T impact on the open-marine ecosystem. Subsidiary goals are to (1) clarify—and, where possible, resolve—the often-contentious issues that have plagued discussion of the K/T impact and its consequences for planktic foraminiferal extinction, (2) review the relationship between the K/T impact and the long-term (100-k.y. to m.y.) state of the early Paleocene open-marine ecosystem, and (3) highlight one of the major remaining challenges to our understanding of the oceanographic and biotic response to the K/T impact. In approaching these goals, we will compare planktic foraminiferal, isotopic, and sedimentologic records from the South Atlantic with records from other regions.

STRATIGRAPHIC CONTROL OF THE SOUTH ATLANTIC K/T BOUNDARY

Biostratigraphy and K/T event stratigraphy

At South Atlantic K/T boundary sequences, Upper Cretaceous and Paleocene strata are composed of alternating chalks and marls or alternating limestones and marls (Barker et al., 1983; Moore et al., 1984). The K/T boundary immediately overlies the carbonate-rich portion (chalk or limestone) of one such carbonate/marl couplet at South Atlantic sites. The K/T boundary of Deep Sea Drilling Project (DSDP) Site 528 is marked by a thin clay layer (slightly less than 1 cm thick). This boundary clay and the uppermost Cretaceous chalks of DSDP Sites 527 and 529 are directly overlain by lower Paleocene marl. Anomalous concentrations of iridium (Ir) have been documented at the K/T boundary of all three sequences (Alvarez et al., 1982; Lindinger, 1988; Huffman et al., 1990). Furthermore, feldspar and glauconite spherules have been discovered at the K/T boundary of Site 527 (Smit and Romein, 1985). These spherules morphologically resemble K/T microtektites from Haiti and Mexico (Sigurdsson et al., 1991; Izett, 1991; Smit et al., 1992) and are generally interpreted to be altered microtektites (Smit and Romein, 1985).

Sites 527, 528, and 529 are biostratigraphically and magnetostratigraphically well characterized. The K/T boundary occurs within the 29R magnetochron at all three sites (Chave, 1984). With the possible exception of the short *Guembelitria cretacea* foraminiferal zone (Zone P0), all latest Cretaceous and earliest Paleocene foraminiferal and nannofossil zones are also present at these sites (Boersma, 1984; Manivit, 1984; D'Hondt and Keller, 1991). At Site 528, Zone P0 is limited to the K/T boundary clay (D'Hondt and Keller, 1991). Zone P0 is an interval zone, spanning the interval from the base of the K/T boundary clay to the first occurrence of *Parvularugoglobigerina eugubina* (Smit, 1982). The biogeographic and isotopic characteristics of *P. eugubina* indicate that it was primarily a low-latitude open-ocean taxon (Boersma and Premoli Silva, 1983; D'Hondt and Keller, 1991; Liu and Olsson, 1992; D'Hondt and Zachos, 1993). As such, its first occurrence in open-ocean sequences may have greatly preceded its first occurrence in shelf sequences where Zone P0 is represented by a thicker stratigraphic column (e.g., El Kef, Tunisia [Keller, 1989]). Whether Zone P0 was diachronous or not, its apparent absence at most open-ocean sites probably simply reflects a short zonal duration, sedimentary mixing on centimeter scales, and/or the lack of biostratigraphic samples from the lowermost 1 to 2 cm of Paleocene sediment.

The lithostratigraphic, biostratigraphic, and event-stratigraphic similarity of Sites 527, 528, and 529 suggests that they are similarly complete on submagnetochron timescales. That stratigraphic similarity, the presence of the latest Cretaceous *Micula prinsii* (nannofossil) zone, and the occurrence of iridium and microspherules from the K/T impact suggest that little geologic time is unrepresented at the K/T boundary of these South Atlantic sites. These sites are of special chronostratigraphic significance because most K/T boundary studies of the last decade followed the Berggren et al. (1985) timescale, which in turn relied on South Atlantic DSDP sequences and Tethyan sequences from Gubbio (Italy) for estimating the age of the K/T boundary and early Paleocene biostratigraphic data.

Milankovitch chronostratigraphy

The Upper Cretaceous and Paleocene records of South Atlantic DSDP sites are characterized by repeated variation in sedimentary color, carbonate content, and magnetic susceptibility (Borella, 1984; Herbert and D'Hondt, 1990; D'Hondt et al., 1992). Multitaper spectral analysis of optical densitometry and magnetic susceptibility records demonstrates that this variation is regularly periodic (Herbert and D'Hondt, 1990; D'Hondt et al., 1992; Park et al., 1993). This periodic variation defines the carbonate/marl alternations visible throughout these sections (Barker et al., 1983; Moore et al., 1984). Comparison to absolute age estimates for paleomagnetic and biostratigraphic data (Berggren et al., 1985) indicates that the mean period of the most consistently prominent cycle in these records was 23.5 ± 4.4 k.y. This is very close to the predicted mean precessional period of the Late Cretaceous (20.8 k.y.: A. L. Berger et al., 1989), especially given timescale uncertainties and the possibility of unconformities. As a result of this close correspondence, we interpreted these cycles as precessionally driven (Herbert and D'Hondt, 1990). This interpretation is consistent with global circulation modeling studies, which predict strong climatic sensitivity to precession in Cretaceous low- and mid-latitude regions (Barron et al., 1985; Oglesby and Park, 1989; Park and Oglesby, 1991).

Precessional chronometry can be used to assess the completeness of magnetochron 29R relative to standard geomagnetic timescales. We have previously identified precessional cycles within South Atlantic records of C29R (Herbert and D'Hondt, 1990). Our primary criteria for precessional identification were (1) the spectral similarity of the uppermost Cretaceous (C29R) records to the underlying Upper Cretaceous records (C33R through C30N) and (2) the spectral similarity of the lowermost Paleocene C29R records to the overlying Paleocene records (C29N through C27R) (Herbert and D'Hondt, 1990). Duration estimates of the C29R interval were not used to identify the precessional cycle. Not doing so allowed us to avoid relying on previous estimates of K/T boundary age for our orbital interpretation.

The Cretaceous interval of C29R at DSDP Site 528 spans 635 ± 25 cm and is characterized by a mean precessional wavelength of 39.7 cm (Figs. 1, 2). This corresponds to 16 ± 0.6 precessional cycles. Given a Late Cretaceous precessional duration of 20.8 k.y. and recent estimates of the variability in mean precessional duration (Herbert and D'Hondt, 1990), 16 ± 0.6 latest Cretaceous precessional cycles are equivalent to 333 ± 20 (1 σ) k.y.

The situation is more complicated at Site 529, where the last 10 m of the uppermost Cretaceous record contain a brief normal-polarity interval and a 150-cm thick interval that appears to have been strongly affected by reduction diagenesis (Fig. 1). If the short normal-polarity interval is a subchron and the 29R/30N reversal is the underlying normal to reversed transition, then there are up to 31.2 ± 1 precessional cycles (649 ± 33 k.y.) in the Cretaceous portion of C29R at this site (with a mean wavelength of 40.2 cm in core 42 and a mean wavelength of 38.9 cm in core 43). If, however, the short normal-polarity interval is the remnant of a greatly truncated C30N, then only 11.7 ± 0.5 precessional cycles are represented in the Cretaceous C29R interval at this site (243 ± 16 k.y.) (Figs. 1, 2). Chronostratigraphic interpretation of this Cretaceous record is further complicated by the absence of obvious erosional contacts in the short normal-polarity interval and a lack of readily identifiable subchrons in the C29R interval of other records.

In the South Atlantic, well-constrained precession-based estimates of Paleocene C29R duration are presently limited to Site 529. Precessional cycles are not well developed in the Paleocene C29R interval of other South Atlantic DSDP sites. This situation may largely reflect the greater effect of bioturbation on the precession-scale records of deeper sites, which were characterized by lower mean rates of accumulation. At Site 529, there are 13 ± 0.5 precessional cycles with a mean wavelength of 18.1 cm in the Paleocene interval of C29R (Figs. 1, 2). This corresponds to 270 ± 17 (1 σ) k.y. A composite of the Cretaceous C29R record from Site 528 and the Paleocene C29R record from Site 529 is equivalent to 603 ± 26 k.y.

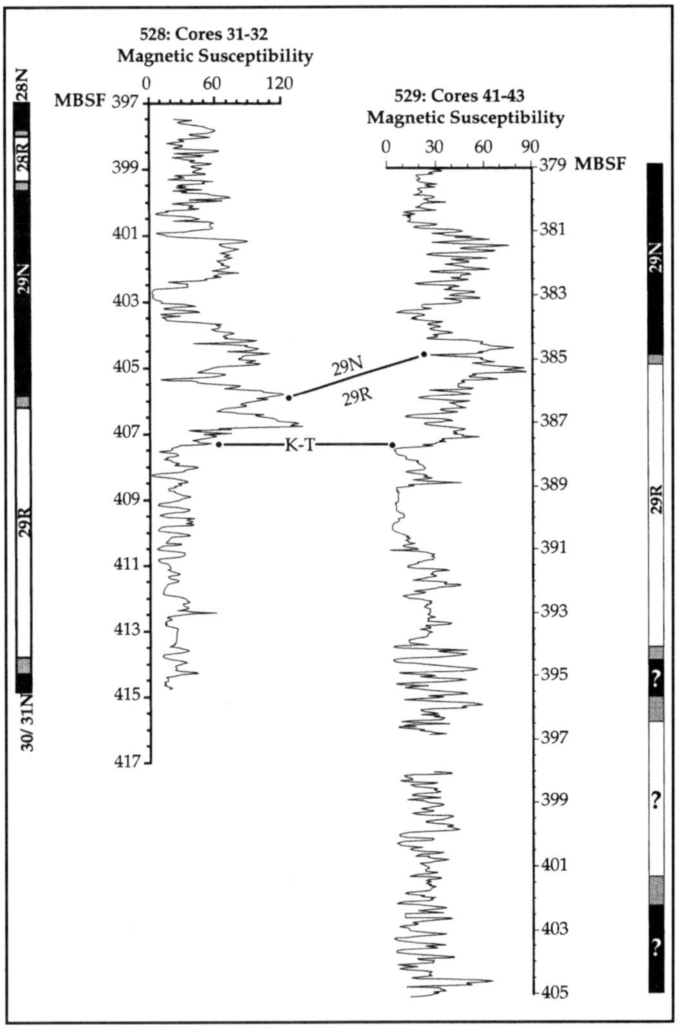

Figure 1. Magnetic susceptibility records across the K/T boundary of DSDP Sites 528 and 529. These were determined by stratigraphically continuous nondestructive analysis of archived core halves. Magnetic susceptibility is a measure of magnetic concentration. At these sites, it is inversely related to carbonate concentration. Paleomagnetic reversal data from Chave (1984). Core depth is in meters below seafloor (mbst).

These precession-based estimates indicate that a composite South Atlantic record of the 29R magnetochron is chronostratigraphically complete relative to the Berggren et al. (1985) geomagnetic timescale. In fact, these orbitally derived estimates suggest that most previous timescales (Berggren et al., 1985; Kent and Gradstein, 1985; Harland et al., 1989) have slightly underestimated the duration of C29R. Those timescales estimated the duration of C29R to have been 570 or 580 k.y. The chronostratigraphic record of these South Atlantic K/T sequences stands in stark contrast to that of many more widely studied records. For example, the magnetostratigraphic and cyclostratigraphic completeness of the K/T marine stratotype at El Kef (Tunisia) remains unknown.

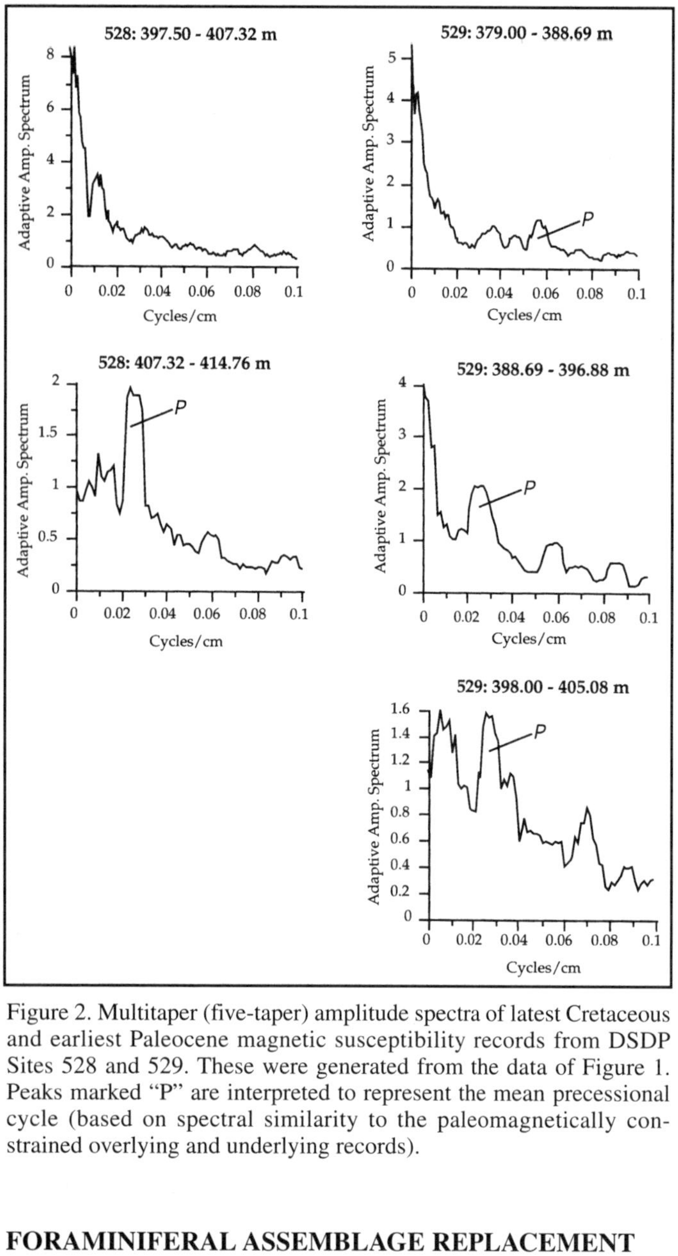

Figure 2. Multitaper (five-taper) amplitude spectra of latest Cretaceous and earliest Paleocene magnetic susceptibility records from DSDP Sites 528 and 529. These were generated from the data of Figure 1. Peaks marked "P" are interpreted to represent the mean precessional cycle (based on spectral similarity to the paleomagnetically constrained overlying and underlying records).

FORAMINIFERAL ASSEMBLAGE REPLACEMENT ACROSS THE K/T BOUNDARY

The composition of low- and mid-latitude planktic foraminiferal assemblages radically changed across the K/T boundary (Olsson, 1960, 1970; Berggren, 1962; Luterbacher and Premoli Silva, 1964; Smit, 1982, 1990; Keller, 1988; D'Hondt and Keller, 1991; Liu and Olsson, 1992). At some sites, Late Cretaceous faunas were completely replaced at the stratigraphic level of the K/T impact (Fig. 3). At other sites, the record of faunal turnover spans a few centimeters immediately above the impact horizon (Fig. 4). There has been considerable argument about the fidelity of these patterns of faunal replacement. Some authors have interpreted the slightly protracted records of assemblage turnover at some sites to indicate "gradual" K/T turnover of foraminiferal assemblages (Keller, 1988, 1989; Barrera and Keller, 1990). In contrast, others have interpreted the same records as the result of catastrophic turnover and subsequent reworking of Late Cretaceous foraminiferal tests into basal Paleocene sediments (Olsson and Liu, 1993).

Despite these interpretative differences, several points are clear about the timing of K/T assemblage turnover. First, Late Cretaceous foraminiferal faunas appear to have been stable prior to the K/T impact (Figs. 3, 4). Second, the drastic turnover started at the impact horizon. Third, the rate of faunal replacement was extraordinarily rapid (whether instantaneous or spread over a very short interval of geologic time). Fourth, open-ocean sites tend to be marked by very sharp faunal turnover (D'Hondt and Keller, 1991), whereas nearshore sequences more commonly exhibit "extended" faunal turnover (Keller, 1988, 1989; Liu and Olsson, 1992). It is very unlikely that this contrast between nearshore and open ocean sites resulted from an unconformity at the K/T boundary in open-ocean sequences, since many open-ocean sites contain geochemical and mineralogic evidence of the K/T impact (also compare MacLeod and Keller, 1991 with D'Hondt and Herbert, 1992). Possible explanations of the small nearshore-offshore discrepancy include (1) higher levels of sedimentary reworking in nearshore environments, (2) lower rates of sediment accumulation in open-ocean sequences, and (3) short-lived differential survivorship between nearshore and open-ocean sequences.

Prior to the K/T impact, open-ocean faunas were typically dominated by biserial heterohelicids *(Heterohelix, Pseudoguembelina,* and *Pseudotextularia)* and planispiral and trochospiral taxa *(Globigerinelloides, Hedbergella, Rugoglobigerina,* and *Globotruncanella)* (Fig. 3). Distinctive large forms were relatively common but were not numerically dominant. The latter taxa included keeled forms (e.g., *Globotruncana, Globotruncanita, and Contusotruncana),* and multiserial ones (e.g., *Racemiguembelina, Planoglobulina,* and *Gublerina).* Throughout the Late Cretaceous, planktic foraminiferal faunas of nearshore environments were generally less diverse than those of the open ocean (Douglas, 1972; Sliter, 1972a, b; Leckie, 1987; D'Hondt and Keller, 1991) (Fig. 4). Nearshore assemblages were dominated by small serial taxa. Larger taxa and trochospiral forms appear to have been much less common in nearshore environments than in the open ocean.

Nearshore-offshore faunal provincialization broke down as a result of the drastic impact-driven assemblage turnover—at least at low and middle latitudes (D'Hondt and Keller, 1991). Prior to the impact, the small triserial taxon, *Guembelitria cretacea,* was rare in open-ocean faunas but relatively common in nearshore and shelf assemblages. Following the impact, *G. cretacea* ecologically radiated to open-ocean environments and dominated both nearshore and open-ocean assemblages of low and middle latitudes (Figs. 3, 4).

After the short-lived interval of assemblage dominance by *G. cretacea,* earliest Paleocene foraminiferal assemblages underwent an extended interval of successive faunal replacement (Figs. 3, 4). Nearshore-offshore faunal provincialization

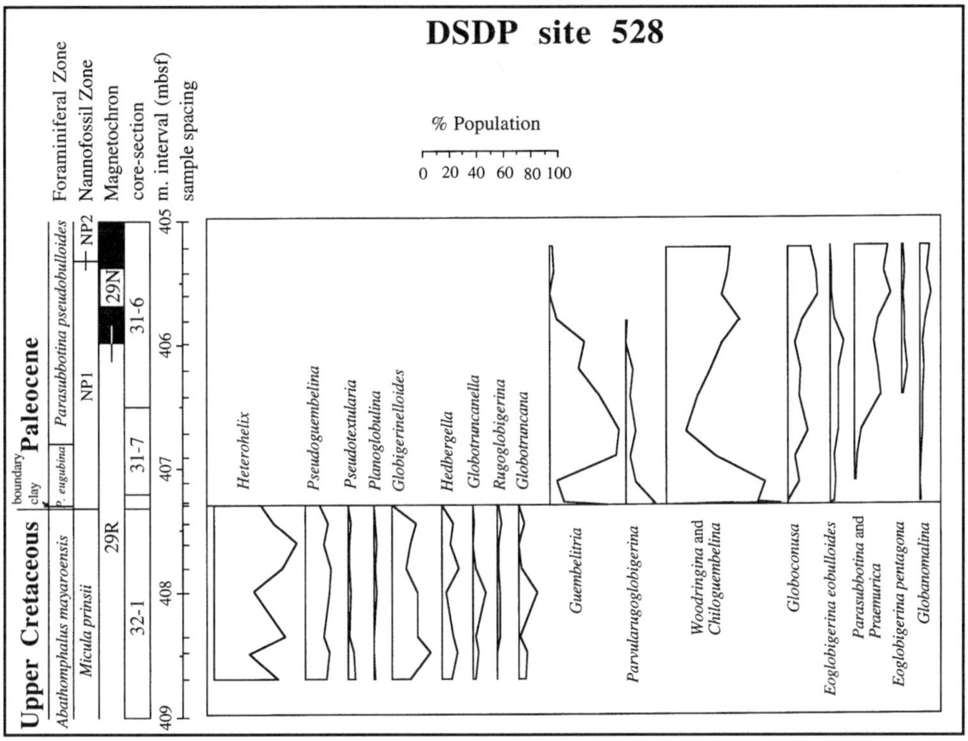

Figure 3. Foraminiferal assemblage turnover across the K/T boundary of DSDP Site 528. Figure modified from D'Hondt and Keller, 1991. The nannofossil zonation is from Manivit (1984) and T. J. Bralower (unpublished). The paleomagnetic zonation is from Chave (1984).

was reinitiated when *Parvularugoglobigerina* began to dominate low-latitude open-ocean assemblages (D'Hondt and Keller, 1991; Liu and Olsson, 1992). After a brief interval when *Parvularugoglobigerina* was the most abundant open-ocean taxon, open-ocean faunas of low and middle latitudes alternated in dominance between biserial taxa *(Woodringina and Chiloguembelina)* and triserial *Guembelitria.* Open-ocean assemblages began to stabilize as "'normal-perforate" *Praemurica* and *Parasubbotina* became the most abundant faunal constituents a few hundred k.y. after the impact event (during magnetochron 29N).

FORAMINIFERAL LAST AND FIRST OCCURRENCES AT THE K/T BOUNDARY

Last occurrences

The relative timing of the K/T impact and extinctions has been hotly argued for over a decade. This debate is exacerbated by the nature of the fossil record. For example, preimpact last occurrences could represent faunal events unrelated to the K/T boundary impact—or they could simply reflect the uncertainty of sampling the true last occurrences of rare taxa that actually went extinct at the boundary (Signor and Lipps, 1982). Postimpact last occurrences are subject to a slightly different array of interpretations. They could be interpreted to indicate that the K/T impact was a transient event superimposed on a longer-term trend of extinctions primarily driven by long-timescale endogenous processes (Keller, 1988, 1989; Canudo et al., 1991). They could also be interpreted to indicate that some taxa survived the impact but did not survive the subsequent reorganization of the Earth's physical and biotic system. Finally, they could be interpreted to result from sedimentary reworking of extinct taxa into lowermost Paleocene sediments (Huber, 1991; Olsson and Liu, 1993; Huber, this volume).

We have investigated whether planktic foraminiferal last occurrences at open-ocean sites closely preceded the K/T impact. Globally significant last occurrences did not precede the impact at the sites we have examined (South Atlantic DSDP Sites 527, 528, and 529) (Figs. 5, 6, 7). In our census of these sites, three taxa last occurred below the impact horizon—*Gublerina acuta, Gublerina cuvillieri, and Schackoina multispina* (all at Site 528). Although *G. acuta* did not range to the K/T boundary in our Site 528 samples, it did range to the impact horizon in earlier studies of El Kef (Tunisia), Brazos River (Texas), and Millers Ferry (Alabama) (Keller, 1988, 1989; Olsson and Liu, 1993). Similarly, *G. cuvillieri* ranges to the impact horizon at El Kef (Keller, 1988), and *S. multispina* ranges to the K/T boundary at Caravaca (Spain) (Canudo et al., 1991). These three taxa are generally very rare in Upper Cretaceous sediments (e.g., the latter two taxa have not been documented at Brazos River, Millers Ferry, or DSDP Sites 527 and 529). These comparisons show that the preboundary last occurrences of these three taxa at Site 528 are either artifacts of taxonomic rarity or local events.

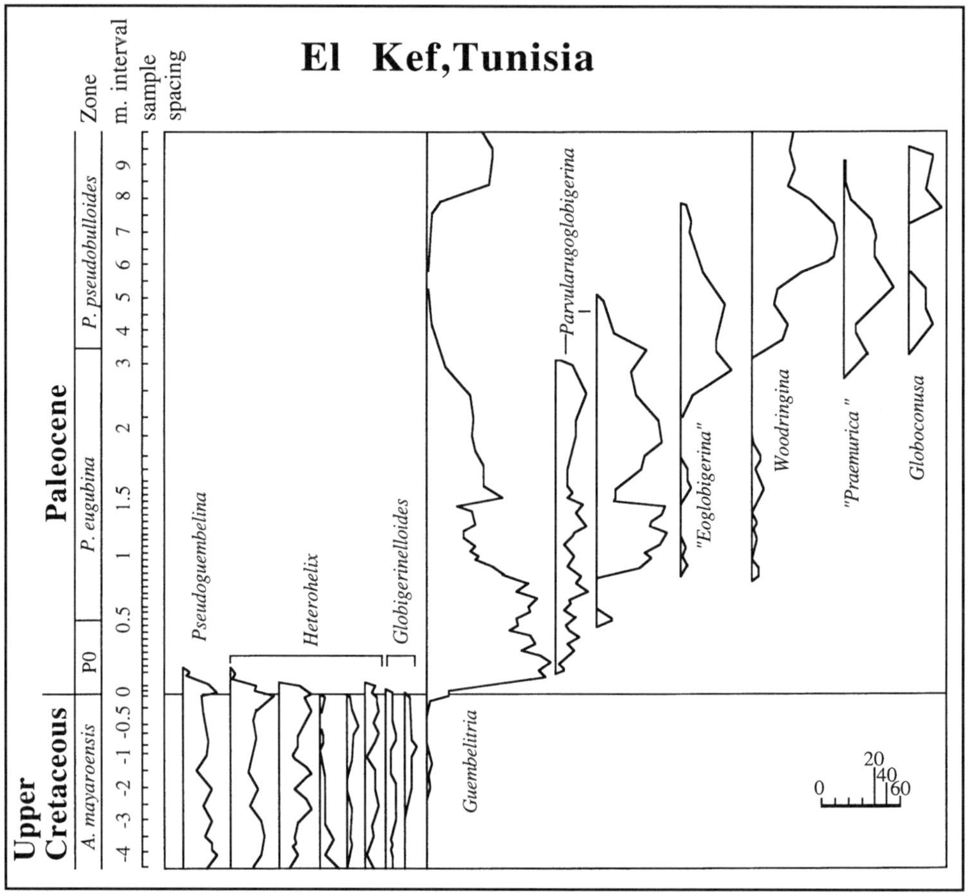

Figure 4. Foraminiferal assemblage turnover across the K/T boundary of El Kef, Tunisia. Figure modified from Keller, 1988.

Previous studies have often found planktic foraminiferal last occurrences spread over a few centimeters or tens of centimeters above and/or below the impact horizon (e.g., Smit, 1977; Keller, 1988; Canudo et al., 1991). The taxa that last occur below the impact horizon are typically rare and may occur in as few as one or two of 10 to 20 latest Cretaceous samples (Keller, 1988; Canudo et al., 1991). Furthermore, the taxa that last occur below the impact horizon at one site generally range to that horizon at another site. For example, Canudo et al. (1991) reported that nine taxa last occur below the impact horizon in their study of a K/T section from Agost (Spain) and that four taxa last occurred below that horizon in their study of the nearby Caravaca section (also in southern Spain). However, only three taxa did not range to the impact horizon in one or both of the two sections. One of these three taxa, *Gublerina cuvillieri,* ranges to the impact horizon at El Kef (Keller, 1988) and Osinaga (Spain) (Canudo, 1994). One of the remaining two taxa (*Globotruncanella pshadae* = *Globotruncanella havanensis* [Pessagno, 1967]) ranges to the K/T boundary at all of the South Atlantic sites (Figs. 5, 6, 7). The last taxon, *Globotruncanita pettersi,* ranges to the impact horizon at Osinaga (Canudo, 1994).

These comparisons underscore two points about K/T planktic foraminiferal extinctions. First, if enough records are compared, latest Cretaceous taxa range to the impact horizon. Hence, global mass extinction of planktic foraminifera did not closely precede the K/T impact but either coincided with the impact or closely followed it. Second, several planktic foraminiferal taxa range a few centimeters to a few tens of centimeters above the impact horizon at one locality or another (Figs. 5, 6, 7). With the exception of *Guembelitria cretacea,* most taxa that range above the impact horizon are abundant in Upper Cretaceous sediments but are not abundant in lowermost Paleocene sediments (compare Fig. 3 to Fig. 5). The drastic K/T changes in foraminiferal assemblages and the limited extent of ranges above the impact horizon suggest that either postimpact last occurrences simply resulted from reworking of extinct taxa into lowermost Paleocene sediments (Olsson and Liu, 1993), or that some Late Cretaceous taxa survived the impact but did not survive the ensuing reorganization of the marine ecosystem.

First occurrences

The earliest Paleocene was characterized by extraordinarily rapid rates of planktic foraminiferal first appearance (Figs. 5, 6, 7). Several major morphotypes quickly appeared in the after-

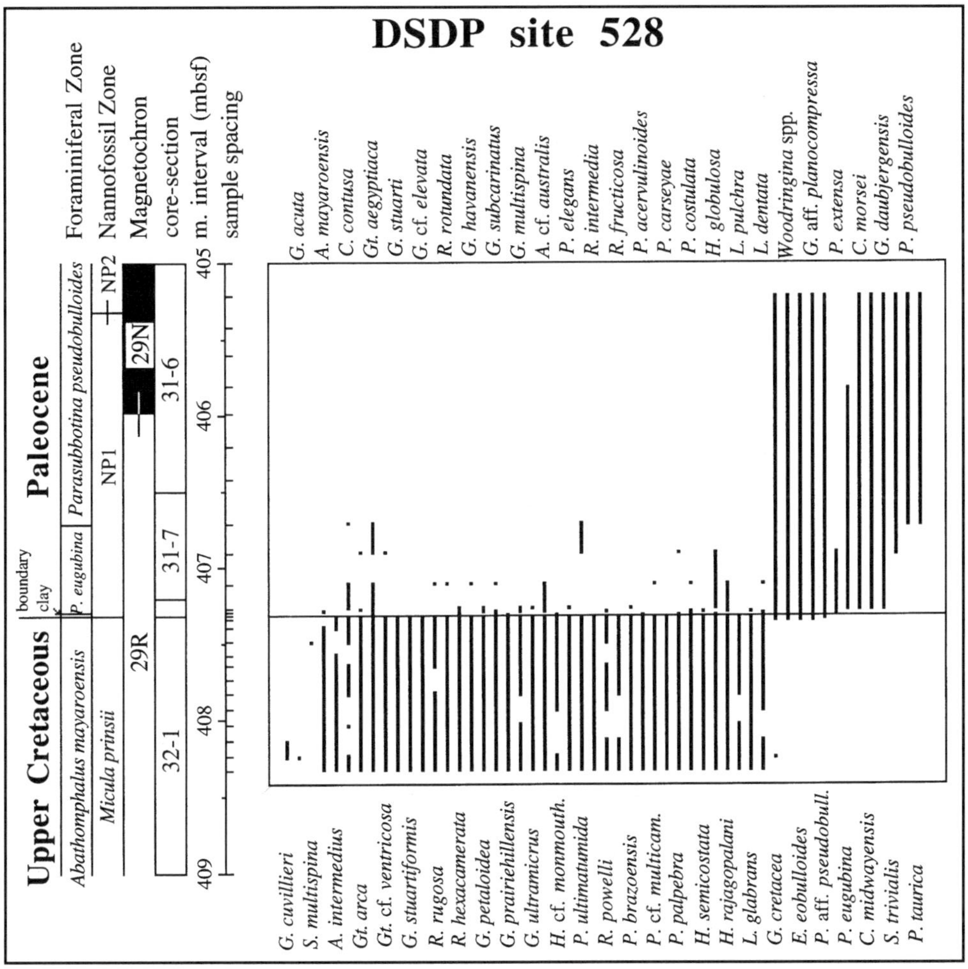

Figure 5. Stratigraphic ranges of planktic foraminiferal taxa in the K/T interval of DSDP Site 528.

math of the K/T impact. In fact, the classic spinose globigerine forms of Cenozoic planktic foraminifers first appeared at that time (Hemleben et al., 1991; Olsson et al., 1992; Liu and Olsson, 1994). The timing of first occurrences was largely decoupled from the sequence of faunal replacement. Microperforate biserial *Woodringina,* microperforate trochospiral *Parvularugoglobigerina,* and "normal-perforate" *Praemurica* and *Parasubbotina* all appeared in the short earliest Paleocene P0 Zone (D'Hondt, 1991; Olsson et al., 1992). However, the microperforate taxa quickly and briefly ascended to faunal dominance whereas the normal-perforate taxa did not become dominant faunal elements for several hundred k.y.

THE EFFECT OF THE K/T IMPACT ON THE STATE OF MARINE PRODUCTION

Carbonate accumulation and nannofossil production across the K/T boundary

Several studies have suggested that marine accumulation of calcium carbonate decreased by 50% to 85% at the K/T boundary (Zachos et al., 1985; Zachos and Arthur, 1986; Arthur et al., 1987; D'Hondt and Keller, 1991). Such studies have relied on standard biogeomagnetic timescales for chronometric control. These studies have demonstrated that early Paleocene carbonate accumulation rates were generally much lower than Late Cretaceous rates (i.e., that mean rates were much lower in earliest Paleocene magnetochron 29N than they were in latest Cretaceous magnetochron 30N). However, on submagnetochron scales (i.e., within C29R), estimates of the magnitude and timing of changes in accumulation rates have been subject to significant uncertainties. This is primarily because standard timescales have estimated the age of biostratigraphic data (i.e., the K/T boundary) by interpolating between paleomagnetic reversals. Such interpolations typically assume linear relationships between time and sedimentary thickness or constant accumulation of some sedimentary component. Without chronostratigraphic control on submagnetochron scales, these assumptions can be difficult to justify—particularly when extrapolating across catastrophic-event horizons, such as the K/T impact horizon.

Orbital chronometry can be used to constrain stratigraphic age-estimates on submagnetochron timescales. For example, the precession-based chronometry of South Atlantic DSDP

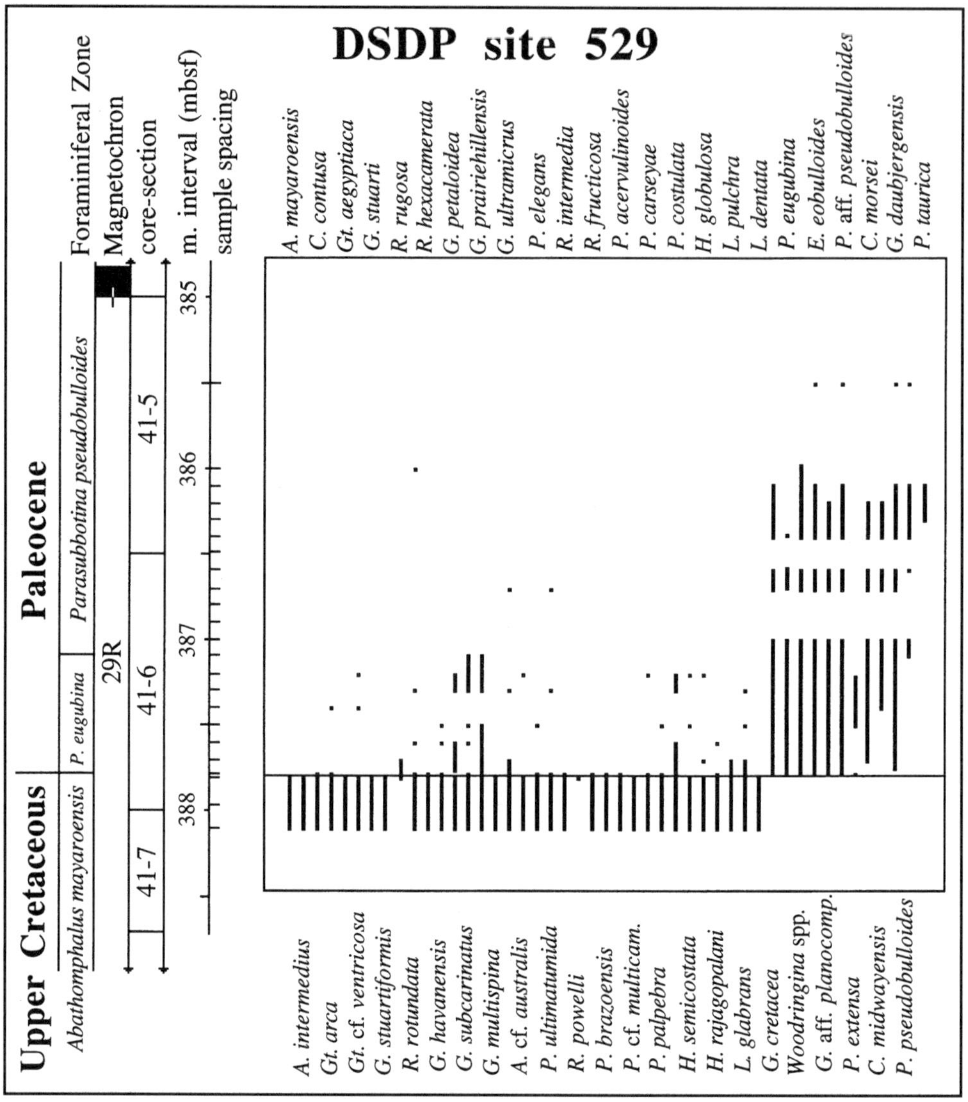

Figure 6. Stratigraphic ranges of planktic foraminiferal taxa in the K/T interval of DSDP Site 529. The paleomagnetic zonation is from Chave (1984).

sites can be used to estimate the duration of time that separated the K/T impact from the older C30N/29R paleomagnetic reversal (~333 k.y.) and the younger C29R/29N reversal (~270 k.y.). Use of these precession-based estimates for subdivision of C29R suggests that an 84% decrease in carbonate accumulation rate occurred at the K/T boundary of South Atlantic DSDP Site 528 (Fig. 8). In turn, this suggests that the K/T decrease in carbonate accumulation was a real event and not merely an artifact of biogeomagnetic timescales.

The K/T decrease in carbonate accumulation was a unique but long-lasting event in the Cretaceous and Paleogene records of the Atlantic and Pacific Oceans. Carbonate accumulation rates decreased radically at the approximate time of the asteroid impact and remained below latest Cretaceous levels for more than two million years after the impact (Fig. 8) (Shackleton et al., 1984; Zachos et al., 1985; D'Hondt, 1989).

The primary cause of the K/T decrease in carbonate accumulation rate was decreased accumulation of calcareous nannofossils (Fig. 8) (Zachos et al., 1985). Improvement in carbonate preservation across the K/T boundary suggests that the decrease in nannofossil accumulation resulted from decreased production of calcareous nannofossils rather than from increased dissolution of carbonate. For example, planktic foraminifers are typically more susceptible than calcareous nannofossils to carbonate dissolution (Kennett, 1982). However, preservation of planktic foraminifers is generally better in lowermost Paleocene sediments than in uppermost Cretaceous sediments (Arthur et al., 1987; Smit and van Kempen, 1987; D'Hondt and Keller, 1991), and the mass-ratio of planktic foraminifers to calcareous nannofossils increased by as much as an order of magnitude across the K/T boundary (Zachos et al., 1985, 1989).

The combination of decreased carbonate accumulation

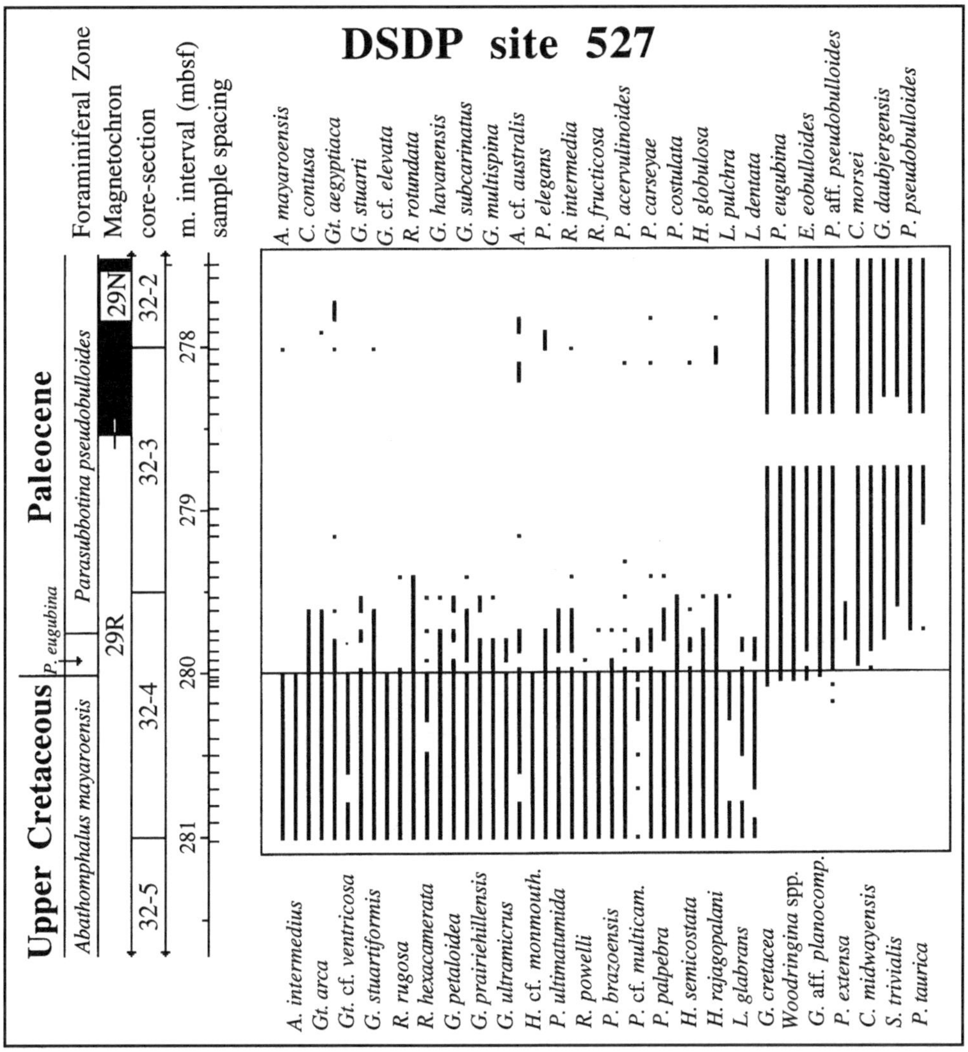

Figure 7. Stratigraphic ranges of planktic foraminiferal taxa in the K/T interval of DSDP Site 527. The paleomagnetic zonation is from Chave (1984).

rates and increased preservation of carbonate sediments also suggests that the corrosiveness of deep water and sedimentary pore water decreased at the K/T boundary. Because carbonic acid derived from decay of organic carbon is the primary source of acid to the deep ocean and sediments, this in turn suggests that export of organic carbon from surface to deep waters decreased. Such a decrease in organic export indicates that either marine primary production decreased or a smaller proportion of primary production was exported to deep waters.

Carbon isotopes and decreased organic export at the K/T boundary

At least two lines of isotopic evidence indicate that the K/T impact resulted in a drastic and long-lasting decrease in export of organic carbon from surface to deep seawaters. The first isotopic measure of this decrease is a pronounced drop in the difference between surface and deep carbon isotopic records. In the best-controlled K/T isotopic studies, the surface isotopic records are derived from planktic foraminiferal skeletons (tests), whereas the deep records are derived from benthic foraminiferal tests (Boersma and Shackleton, 1981; Zachos et al., 1989, 1992; Stott and Kennett, 1990a, b). Surface marine waters are typically marked by $^{13}C/^{12}C$ ratios that are much higher than average for marine waters. This discrepancy results from preferential uptake of the lighter isotope by photosynthetic organisms and from the rain of ^{12}C- and nutrient-enriched organic matter from the photic zone to deeper waters. Upwelling and vertical diffusion result in dissolved nutrients and ^{12}C-enriched carbon returning to near-surface waters within an oceanic mixing cycle (~1,000 yrs) or less. The carbon isotopic discrepancy between surface and deep waters is maintained because, on a global scale, rates of biological stripping and organic export of ^{12}C and nutrients from surface water are at

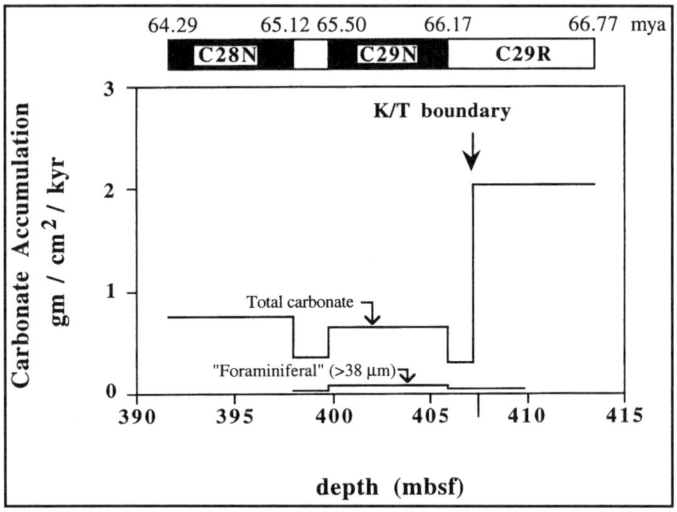

Figure 8. Mean accumulation rates of carbonate and foraminifer-sized (>38 µm diameter) sediment across the K/T boundary of DSDP Site 528. Within magnetochron 29R, accumulation rate estimates are based on sedimentation rates derived from precessional chronometry (see text). Accumulation rate estimates for magnetochrons 29N through 28N are based on sedimentation rates derived from Kent and Gradstein's (1985) estimates of magnetochron duration. Accumulation rates were calculated by using the approach of Zachos et al. (1985) [AR = (% C / 100) * [W – (P / 100) * (1.01 g/cm^2)] * S, where AR is accumulation rate (g/k.y.); % C is % carbonate, noncarbonate, or coarse sediment; W is wet bulk density (g/cm^3); P is % porosity; and S is sedimentation rate (cm/k.y.)]. Estimates of wet bulk density and porosity are from Moore et al. (1984). Note that the rate of accumulation of foraminifer-sized sediment did not appreciably change at the K/T boundary. Hence, the decrease in carbonate accumulation rate resulted from decreased nannofossil (coccolith and calcareous dinoflagellate) accumulation.

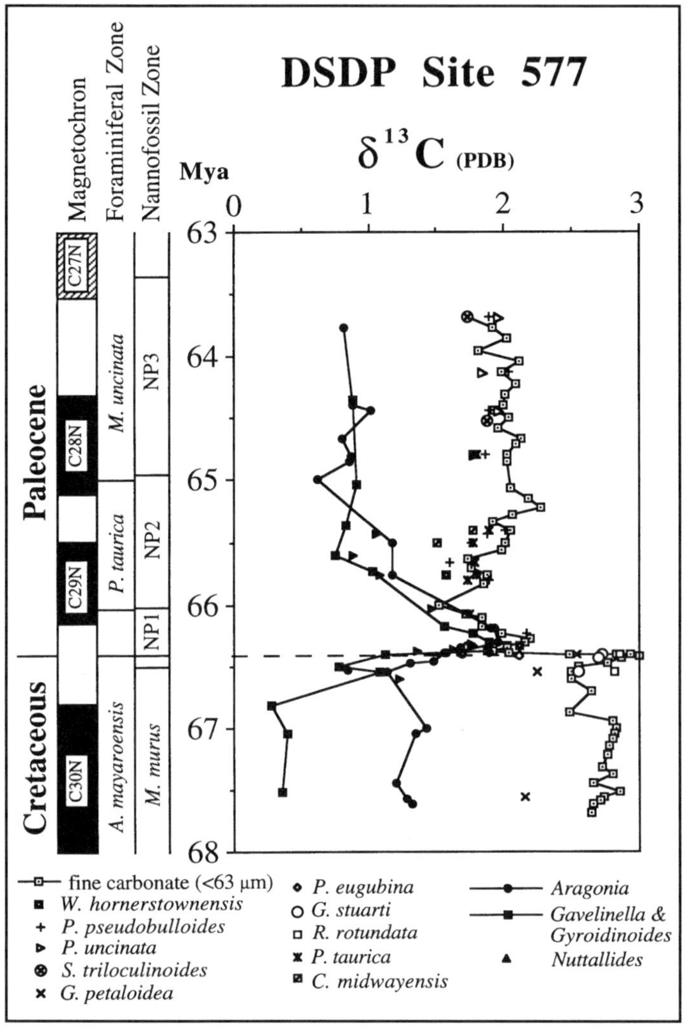

Figure 9. Stable isotopic record of the uppermost Cretaceous and lowermost Paleocene at DSDP Site 577. *Aragonia, Gavelinella, Gyroidinoides,* and *Nuttallides* are benthic foraminifera. The other taxa are planktic foraminifera. Fine carbonate (<63 µm) is composed of calcareous nannofossils, micrite, small foraminifera, and fragments of larger carbonate particles; its isotopic composition is often assumed to approximate that of carbonate precipitated in equilibrium with surface seawater. Figure modified from Zachos et al. (1989).

least two orders of magnitude faster than rates of upwelling and upward diffusion of ^{12}C and nutrients from deep water. Since $\delta^{13}C$ values represent the ratio of the heavier carbon isotope to the photosynthetically preferred light isotope,

$$\delta^{13}C = [(^{13}C/^{12}C_{sample} - {}^{13}C/^{12}C_{std}) * 10^3] / (^{13}C/^{12}C_{std}),$$

the organic carbon pump generally results in a steady-state $\delta^{13}C$ difference between surface and deep waters, with surface waters exhibiting $\delta^{13}C$ values that are almost 2.0‰ more positive than deep waters throughout most of the world ocean. At the K/T boundary, this cycle appears to have been strongly modified, resulting in planktic and benthic $\delta^{13}C$ values that approach each other much more closely (Fig. 9).

The K/T decrease in the planktic to benthic $\delta^{13}C$ gradient was geologically rapid and coincided with the impact (Boersma and Shackleton, 1981; Hsü et al., 1982; Zachos et al., 1985, 1989, 1992; Zachos and Arthur, 1986; D'Hondt and Lindinger, 1988; Stott and Kennett, 1990a, b). Recovery of that gradient was delayed for at least several hundred thousand years (Fig. 9). Comparison of latest Cretaceous and earliest Paleocene data from then-equatorial Pacific DSDP Site 577 suggests that the tropical planktic to benthic $\delta^{13}C$ difference decreased from between 1.2 and 1.6‰ to between 0.3 and 0.4‰ across the K/T boundary (Zachos et al., 1989). By 0.5 m.y. later, the planktic to benthic difference rose to ~0.8 to 1.0‰. Taken at face value, these results suggest a three- to fourfold K/T decrease in the tropical surface to deep $\delta^{13}C$ gradient. However, it remains possible that part of this decrease reflected interspecies differences between Cretaceous and Paleocene planktic foraminifera (Stott and Kennett, 1990a; D'Hondt and Zachos, 1993). This possibility can be avoided by assuming that the magnitude of the K/T $\delta^{13}C$ decrease is better estimated by the differential between the minimum earliest Paleocene planktic to benthic gradient and the postrecovery gradient between the same benthic and planktic taxa (Fig. 9). This conservative estimate still indicates a drastic (twofold) decrease in the tropical planktic to benthic $\delta^{13}C$ gradient. The two to fourfold decrease in the $\delta^{13}C$ gradient in turn indicates a two- to fourfold decrease in the flux of ^{12}C-enriched organic matter from shallow to deep waters.

The second isotopic measure of decreased organic export at the K/T boundary is collapse of the $\delta^{13}C$ gradient between infaunal and epifaunal benthic foraminifera (Zachos et al., 1989) (Fig. 9). The $\delta^{13}C$ values of infaunal benthic tests typically track the $\delta^{13}C$ signature of sedimentary pore waters, whereas the $\delta^{13}C$ values of epifaunal benthic tests track that of the overlying bottom water (McCorkle et al., 1985). Pore-water $\delta^{13}C$ values are typically more negative than bottom-water values as a result of oxidation of ^{12}C-enriched organic carbon within the sediment. Hence, the K/T collapse of the infaunal to epifaunal $\delta^{13}C$ gradient indicates a strong decrease in sedimentary oxidation of organic carbon. This decrease is consistent with a large decrease in the flux of organic carbon to the sediment (Zachos et al., 1989).

The K/T decrease in organic export to deep waters indicates either that a much smaller proportion of marine production was exported to deep waters or that marine production was unusually low in the earliest Paleocene.

SUMMARY AND DISCUSSION

The effect of the K/T impact on the open-ocean biosphere

Diverse evidence indicates that the K/T impact had a tremendous effect on the open-ocean biosphere. The impact coincided with catastrophic reductions in nannofossil production and in the export of organic matter from the photic zone to deep water. Nannofossil production remained below Late Cretaceous levels for more than two million years. Organic export remained below Late Cretaceous levels for at least 500 k.y. and possibly as much as two million years (D'Hondt and Lindinger, 1988). Late Cretaceous planktic foraminiferal taxa range to, or in some cases slightly above, the impact horizon. This strongly suggests that the K/T mass extinction of planktic foraminifers was initiated at the time of impact.

Extension of planktic foraminiferal ranges above the impact horizon allows (but does not require) the possibility that several taxa survived the immediate consequences of the impact. However, only one or two of 40 or 50 Late Cretaceous taxa exhibit stratigraphic ranges that could be interpreted to indicate survival beyond the early stages of post-impact ecosystem reorganization. Regardless of the extent of short-term taxonomic survival, essentially complete replacement of low- and mid-latitude open-ocean planktic foraminiferal assemblages was initiated at the time of impact and finished within an unresolvably short interval of time.

The coincidence of the K/T impact with initiation of the planktic foraminiferal mass extinction, open-ocean faunal replacement, and drastic reductions in nannofossil production and organic export is readily compatible with an impact model of mass extinction. Alvarez et al. (1980) proposed that cessation of photosynthesis by dust-induced global darkness was the primary cause of the K/T mass extinctions. Subsequent studies have noted that impact volatilization of abundant SO_2 from anhydrites of the Chicxulub target may have significantly contributed to global darkness (Brett, 1992; Pope et al., 1994).

Such cessation of photosynthesis would have directly resulted in catastrophic short-term reduction of marine production—as suggested by coincidence of the K/T impact with radical decreases in organic export and nannofossil production. In turn, a collapse of marine production at the K/T boundary could have been the direct cause of marine extinctions. Those extinctions preferentially affected organisms that relied on living phytoplankton and zooplankton for nutrition (Sheehan and Hanson, 1986; Arthur et al., 1987). In contrast, detritus-feeding organisms were much less affected by the K/T extinctions. For example, cessation of phytoplankton production for an oceanographically short interval of time (several months to a few years) would have had catastrophic consequences for planktic foraminifers, which prey on phytoplankton and zooplankton, sometimes directly depend on photosymbiont activity for nutrition, and, at least in the modem ocean, are largely incapable of populating turbid nearshore environments where organic debris is most abundant. In fact, planktic foraminifers may be particularly susceptible to extinction when driven to very low population densities since, unlike many radiolaria, modern planktic foraminifers do not appear to reproduce asexually (compare Hemleben et al., 1989, with Anderson, 1983).

Organic export and marine production after the K/T impact

The drastic K/T changes in planktic foraminiferal assemblages, organic export, and nannofossil production clearly indicate that the K/T impact had a long-lasting effect on the marine ecosystem. However, the proximate causes that linked the impact to the long-term decrease in organic export remain enigmatic. Most researchers interpret the low early Paleocene levels of organic export to indicate a 500-k.y. to 2 m.y. interval of decreased marine production following the K/T boundary events (Hsü et al., 1982; Zachos et al., 1985, 1989; Arthur et al., 1987; D'Hondt and Lindinger, 1988; Keller and Lindinger, 1989; Barrera and Keller, 1990; Stott and Kennett, 1990a, b). This interpretation poses a real challenge to current models of marine production and to our present understanding of the K/T impact.

Clarification of this challenge requires a brief introduction to some definitions and concepts that underpin current models of marine production. Marine production can be defined in several different ways. For example, *primary production* can be defined as biomass production by photosynthetic processes in a defined volume over a defined unit of time, *new production* as the portion of primary production that directly relies on nutrients introduced to a defined volume per unit time, *regenerated production* as the portion of primary production that relies on nutrients recycled in the defined volume per unit time, and *export production* as the total mass of organic matter that leaves a defined volume per unit time (W. A. Berger et al., 1989). Export production is often assumed to be equal to new production under steady-state conditions (W. A. Berger et al., 1989).

For our discussion, the relevant steady-state timescales are the mixing time of the surface ocean (~20 yrs) and longer. If we assume that our paleoceanographic measures of organic export to deep waters and sediment are appropriate proxies for export of organic matter from the Late Cretaceous and early Paleocene photic zone, then under steady-state conditions, they may also be appropriate proxies for new production. They are not, however, necessarily appropriate proxies for primary production. The ratio of new to primary production varies widely in the modern ocean, ranging from ~0.1 in oligotrophic central gyres to between 0.2 and 0.6 in upwelling regions (Eppley, 1989). This variation underscores the possibility that ratios of primary to new production (and export production) were considerably different in Late Cretaceous and early Paleocene oceans, depending on oceanographic conditions and ecosystem structure.

On a global basis, new production in the modern ocean is largely limited by nutrient availability—to a first approximation, dissolved nutrients that enter the photic zone are utilized until they exit the photic zone in organic detritus. The primary exception occurs in the high-latitude southern ocean, where winter darkness and a deep mixed layer limit production and allow export of dissolved nutrients from the photic zone to deeper waters via winter downwelling (Mitchell et al., 1991). Marine production in the equatorial Pacific is often hypothesized to be limited by micronutrient availability or grazing (Barber and Chavez, 1991; Cullen et al., 1992). However, such limitation of equatorial production has no significant effect on global net production, because the underutilized nutrients are horizontally advected to other parts of the surface ocean (e.g., the mid-Pacific gyres), where they eventually exit the photic zone in organic export.

Given our understanding of the modern ocean, a drastic long-lasting decrease in export production at the K/T boundary would have required one or more of the following steady-state conditions: (1) greatly decreased nutrient availability in the world ocean, (2) a radical decrease in the export ratio of carbon to nutrients, (3) enhanced light limitation in critical regions, or (4) an extraordinarily poised ecosystem (i.e., one where grazing consistently prevents phytoplankton from successfully utilizing available nutrients on a global scale). There are possibilities and problems associated with application of all of these explanations to a 500-k.y. to 2 m.y. interval of greatly reduced organic export.

Given nutrient limitation of marine production, decreased concentrations and input rates of biolimiting nutrients could readily limit net and export production for millions of years. However, the rapidity of the K/T decrease in nannofossil and export production is difficult to explain by decreased marine concentrations of primary nutrients. For example, phosphate, a biolimiting nutrient, has a mean oceanic residence time of about 70 k.y. (Broecker and Peng, 1982), and there is no obvious mechanism for reducing oceanic phosphate concentrations by 50 to 75% on the 1,000-yr timescale of oceanographic mixing—much less the timescale of an impact event.

A 50 to 75% decrease in the export ratio of carbon to nutrients could conceivably have occurred at the K/T boundary, but only if preboundary ratios were much higher than those of today. For example, the modern carbon to nitrogen ratio in export production is ~7.6(±1.2):1 (Takahashi et al., 1985). Cultures of extant phytoplankton exhibit C:N ratios that range between 5:1 and 10.5:1 (Parsons et al., 1961), and zooplankton fecal pellets (a major vector of particulate export) exhibit C:N ratios between 5.3:1 and 7.4:1 (Small et al., 1983). C:N ratios of pure proteins are ~4:1, and the other major components of marine plankton have higher C:N ratios. Since marine plankton cannot drive their C:N ratios below those of their constituent molecules, a K/T decrease in the C:N export ratio can explain the long-lasting early Paleocene decrease in organic export only if Late Cretaceous C:N export ratios were much higher than modern ratios. Such a difference between modern and Late Cretaceous export-ratios would have been possible only if Late Cretaceous open-marine ecosystems were much less efficient in their use of biolimiting nutrients than modern open-marine ecosystems.

On timescales of a few years, light limitation of marine production is clearly compatible with models of the K/T impact (e.g., Alvarez et al., 1980). It may also prove to be compatible on timescales of 100 k.y. or more, since one possible consequence of a major impact would be a terrestrial ring centered on the equator (Schultz and Gault, 1990). Given the 22.5° mean angle of the earth's axis relative to its orbit around the sun, such a ring could conceivably cause strongly seasonal light limitation at middle and low latitudes. However, the effect of such a ring should have decayed as the ring degraded and its components rained earthward. Schultz and Gault (1990) proposed that such a ring could have been stable for 100 k.y. It is unclear whether this mechanism could have been sustained as long as two million years.

Could new and export production in the early Paleocene open-ocean ecosystem have been grazing-limited on a global scale? Given the population dynamics of modern plankton, it would have been difficult to sustain such a tightly coupled ecosystem for hundreds of thousands to millions of years—even in a taxonomically impoverished ocean. Phytoplankton typically have populational doubling rates of a couple of hours to a few days (Banse, 1992). Furthermore, high- and mid-latitude regions that are characterized by strong seasonality in mixed-layer depth and light availability are subject to strong seasonal variation in phytoplankton/zooplankton ratios, resulting in phytoplankton doubling readily outstripping zooplankton grazing early in the spring bloom (Banse, 1992). Consequently, in large regions of the world ocean, phytoplankton-zooplankton dynamics may be too unstable on diel to annual timescales to remain tightly coupled on oceanographic and geologic timescales.

The preceding discussion illustrates that it would have been very difficult to sustain extremely low levels of marine biological production on geologic timescales. Alternatively, it is possible that the long-term earliest Paleocene decrease in organic export to deep waters and sediment did not result from decreased production but simply from degradation of exported organic matter at shallower depths (perhaps, in large part, because of the relatively small size of postimpact phytoplankton and zooplankton). Regardless of the ultimate cause of low earliest Paleocene organic export, the catastrophic and long-lasting changes in nannofossil production and organic export indicate that the K/T impact drove the marine ecosystem to a radically different metastable state that lasted more than two million years.

CONCLUSIONS

The K/T impact had an immediate and drastic effect on the long-term state of the marine ecosystem. More than 90% of planktic foraminiferal taxa went extinct at or shortly after the time of the K/T impact. Following the impact, a previously nearshore species, *G. cretacea,* dominated both nearshore and open-ocean environments—resulting in the disappearance of onshore-offshore provincialization of low- and mid-latitude planktic foraminiferal faunas at the K/T boundary. Carbonate accumulation rates and export of organic carbon from surface to deep waters respectively decreased by about 70 to 85% and 50 to 80% at the boundary.

Recovery of marine production and planktic foraminiferal assemblages was very dynamic. Rapid planktic foraminiferal evolution closely followed the impact. In contrast, assemblage stabilization and recovery of organic export did not occur for at least a few hundred k.y. Over those few hundred k.y., faunal turnover was marked by successive replacement of dominant faunas. Unusual assemblages of microperforate taxa dominated the postimpact interval of lowest organic export. "Normal-perforate" taxa began to dominate assemblages as vertical gradients of marine organic export began to recover. Calcareous nannofossil production remained below preimpact levels for more than two million years. Organic export to deep waters may have remained below preimpact levels for a similar interval of time.

The coincidence of the K/T impact with planktic foraminiferal mass extinction, faunal turnover, and strong decreases in indices of marine production is readily compatible with an impact model of mass extinction. In contrast, the geologically long delay in postimpact recovery of marine organic export and calcareous nannofossil production presents a major challenge to current models of marine production and to our understanding of the long-term consequences of large impacts.

ACKNOWLEDGMENTS

This research was supported by the Earth Sciences Division of the National Science Foundation and by the Petroleum Research Fund of the American Chemical Society. T. J. Bralower and R. M. Leckie provided thoughtful and constructive reviews of an earlier draft. We thank J. J. Sepkoski for suggesting the more interesting half of this manuscript's title and drawing our attention to the possible applicability of terrestrial-ring models to postimpact light limitation. We also thank P. Donaghay, M. Pilson, P. Schultz, and the members of a GSO seminar on marine production for engaging and informative discussions.

REFERENCES CITED

Alvarez, L. W., Alvarez, W., Asaro, F., and Michel, H. V., 1980, Extraterrestrial cause for the Cretaceous-Tertiary extinction: Science, v. 208, p. 1095–1108.

Alvarez, W., Alvarez, L. W., Asaro, F., and Michel, H. V., 1982, Current status of the impact theory for the terminal Cretaceous extinctions, *in* Silver, L. T., and Schultz, P. H., eds., Geological implications of impacts of large asteroids and comets on the Earth: Geological Society of America Special Paper 190, p. 305–315.

Anderson, O. R., 1983, Radiolaria: New York, Springer-Verlag, 355 p.

Arthur, M. A., Zachos, J. C., and Jones, D. S., 1987, Primary productivity and the Cretaceous/Tertiary boundary event in the oceans: Cretaceous Research, v. 8, p. 43–54.

Banse, K., 1992, Grazing, temporal changes of phytoplankton concentrations, and the microbial loop in the open sea, *in* Falkowski, P. G., and Woodhead, A. D., eds., Primary productivity and biogeochemical cycles in the sea: New York, Plenum Press, p. 409–440.

Barber, R. T., and Chavez, F. P., 1991, Regulation of primary productivity rate in the equatorial Pacific: Limnology and Oceanography, v. 36, p. 1803–1815.

Barker, P. F., Carlson, R. L., and Johnson, D. A., 1983, Initial reports of the Deep Sea Drilling Project, Volume 72: Washington, D.C., U.S. Government Printing Office, 1023 p.

Barrera, E., and Keller, G., 1990, Foraminiferal stable isotope evidence for gradual decrease of marine productivity and Cretaceous species survivorship in the earliest Danian: Paleoceanography, v. 5, p. 867–890.

Barron, E. J., Arthur, M. A., and Kauffman, E. G., 1985, Cretaceous rhythmic bedding sequences: A plausible link between orbital variations and climate: Earth and Planetary Science Letters, v. 72, p. 327–340.

Berger, A. L., Loutre, M. F., and Dehant, V., 1989, Influence of the changing lunar orbit on the astronomical frequencies if pre-Quaternary insolation patterns: Paleoceanography, v. 4, p. 555–564.

Berger, W. A., Smetacek, V. S., and Wefer, G., 1989, Ocean productivity and paleoproductivity—An overview, *in* Berger, W. A., Smetacek, V. S., and Wefer, G., eds., Productivity of the ocean: present and past: New York, Wiley, p. 1–34.

Berggren, W. A., 1962, Some planktonic foraminifera from the Maastrichtian and type Danian stages of southern Scandinavia: Stockholm Contributions in Geology, v. 9, p. 1–102.

Berggren, W. A., Kent, D. V., and Flynn, J. J., 1985, Jurassic to Paleogene. Part 2: Paleogene geochronology and chronostratigraphy, *in* Snelling, N. J., ed., The chronology of the geological record: Geological Society of London Memoir 10, p. 141–195.

Boersma, A., 1984, Cretaceous-Tertiary planktonic foraminifers from the South-Eastern Atlantic, Walvis Ridge area, Deep Sea Drilling Project Leg 74, *in* Moore, T. C., Jr., Rabinowitz, P. D., and others, eds., Initial reports of the Deep Sea Drilling Project, Volume 74: Washington, D.C., U.S. Government Printing Office, p. 501–524.

Boersma, A., and Premoli Silva, I., 1983, Paleocene planktonic foraminiferal biogeography and paleoceanography of the Atlantic Ocean: Micropaleontology, v. 29, p. 355–381.

Boersma, A., and Shackleton, N. J., 1981, Oxygen- and carbon-isotope variations and planktonic-foraminifer depth habitats, Late Cretaceous to

Paleocene, Central Pacific, Deep Sea Drilling Project Sites 463 and 465, *in* Thiede, J., Vallier, T. L., and others, eds., Initial reports of the Deep Sea Drilling Project, Volume 62: Washington, D.C., U.S. Government Printing Office, p. 513–526.

Borella, P. E., 1984, Sedimentology, petrology and cyclic sedimentation patterns, Walvis Ridge transect, Leg 74, Deep Sea Drilling Project, *in* Moore, T. C., Jr., Rabinowitz, P. D., and others, eds., Initial reports of the Deep Sea Drilling Project, Volume 74: Washington, D.C., U.S. Government Printing Office, p. 645–663.

Brett, R., 1992, The Cretaceous-Tertiary extinction: A lethal mechanism involving anhydrite target rocks: Geochimica et Cosmochimica Acta, v. 56, p. 3603–3606.

Broecker, W. S., and Peng, T.-H., 1982, Tracers in the sea: Palisades, New York, Lamont-Doherty Geological Observatory, Columbia University, 620 p.

Canudo, J. I., 1994, Bioestratigrafia y evolucion de los foraminiferos planctonicos en el transito Cretacico-Terciario en España, *in* Molina, E., ed., Extincion y Registro Fosil: Cuadernos Interdisciplinares 5, Seminario Interdisciplinar de la Universidad de Zaragoza, p. 141–164.

Canudo, J. I., Keller, G., and Molina, E., 1991, Cretaceous/Tertiary boundary extinction pattern and faunal turnover at Agost and Caravaca, S. E. Spain: Marine Micropaleontology, v. 17, p. 319–341.

Chave, A. D., 1984, Lower Paleocene–Upper Cretaceous Magnetostratigraphy, sites 525, 527, 528, and 529, Deep Sea Drilling Project Leg 74, *in* Moore, T. C., Jr., Rabinowitz, P. D., and others, eds., Initial reports of the Deep Sea Drilling Project, Volume 74: Washington, D.C., U.S. Government Printing Office, p. 525–532.

Cullen, J. J., Lewis, M. R., Davis, C. O., and Barber, R. T., 1992, Photosynthetic characteristics and estimated growth rates indicate grazing is the proximate control of primary production in the equatorial Pacific: Journal of Geophysical Research, v. 97 (C1), p. 639–654.

D'Hondt, S., 1989, Environmental change and the ecology and the evolution of latest Cretaceous and earliest Paleocene planktic foraminifera [Ph.D. thesis]: Princeton, New Jersey, Princeton University, 164 p.

D'Hondt, S., 1991, Phylogenetic and stratigraphic analysis of earliest Paleocene triserial and biserial planktonic foraminifera: Journal of Foraminiferal Research, v. 21, p. 168–181.

D'Hondt, S., and Herbert, T. D., 1992, Hiatus distributions and mass extinctions at the Cretaceous/Tertiary boundary: Comment: Geology, v. 20, p. 380–382.

D'Hondt, S., and Keller, G., 1991, Some patterns of planktic foraminiferal assemblage turnover at the Cretaceous-Tertiary boundary: Marine Micropaleontology, v. 17, p. 77–118.

D'Hondt, S., and Lindinger, M., 1988, An extended Cretaceous-Tertiary boundary stable isotope record: Implications for paleoclimate and the nature of the K/T boundary event, *in* Global catastrophes in Earth history: Houston, Texas, Lunar and Planetary Institute Contribution 673, p. 40–41.

D'Hondt, S., and Zachos, J. C., 1993, On stable isotopic variation and earliest Paleocene planktonic foraminifera: Paleoceanography, v. 8, p. 527–547.

D'Hondt, S. L., King, J. W., Gibson, C., and Park, J., 1992, The evolution of Late Cretaceous climatic cycles at South Atlantic DSDP Site 516F: Eos (Transactions, American Geophysical Union), v. 73, p. 271.

Douglas, R. G., 1972, Paleozoogeography of Late Cretaceous Planktonic Foraminifera in North America: Journal of Foraminiferal Research, v. 2, p. 14–34.

Eppley, R. W., 1989, New production: History, methods, problems, *in* Berger, W. A., Smetacek, V. S., and Wefer, E., eds., Productivity of the ocean: Present and past: New York, Wiley, p. 85–98.

Ginsberg, R. N., and 13 others, 1994, Results of blind tests to resolve controversies: Iridium at Gubbio; extinctions at El Kef, *in* New developments regarding the K/T event and other catastrophes in Earth history: Houston, Texas, Lunar and Planetary Contribution 825, p. 44.

Harland, W. B., Armstrong, R. L., Cox, A. V., Craig, L. E., Smith, A. G., and Smith, D. G., 1989, A geological timescale: Cambridge, Cambridge University Press, 263 p.

Hemleben, C., Spindler, M., and Anderson, O. R., 1989, Modern planktonic foraminifera: New York, Springer-Verlag, 363 p.

Hemleben, C., Muhlen, D., Olsson, R. K., and Berggren, W. A., 1991, Surface texture and the first occurrence of spines in planktonic foraminifera from the early Tertiary: Geologisches Jahrbuch, Series A, v. 128, p. 117–146.

Herbert, T. D., and D'Hondt, S. L., 1990, Precessional climate cyclicity in Late Cretaceous–Early Tertiary marine sediments: A high resolution chronometer of Cretaceous-Tertiary boundary events: Earth and Planetary Science Letters, v. 99, p. 263–275.

Hsü, K. J., McKenzie, J. A., and He, Q. X., 1982, Terminal Cretaceous environmental and evolutionary changes, *in* Silver, L. T., and Schultz, P. H., eds., Geological implications of impacts of large asteroids and comets on the Earth: Geological Society of America Special Paper 190, p. 317–328.

Huber, B. T., 1991, Maastrichtian planktonic foraminifer biostratigraphy and the Cretaceous-Tertiary boundary at ODP Hole 738C (Kerguelen Plateau, Southern Indian Ocean), *in* Barron, J., Larsen, B., and others, eds., Proceedings, Ocean Drilling Programs Scientific Results, Volume 119: College Station, Texas, Ocean Drilling Program, p. 451–465.

Huffman, A. Crocket, J., Carter, N., Borella, P., and Officer, C., 1990, Chemistry and mineralogy across the Cretaceous-Tertiary boundary at DSDP site 527, Walvis Ridge, South Atlantic Ocean, *in* Sharpton, V. L., and Ward, P. D., eds., Global catastrophes in Earth history: Geological Society of America Special Paper 247, p. 319–334.

Izett, G. A., 1991, Tektites in Cretaceous-Tertiary boundary rocks on Haiti and their bearing on the Alvarez impact extinction hypothesis: Journal of Geophysical Research, v. 96, p. 20879–20905.

Keller, G., 1988, Extinction, survivorship, and evolution across the Cretaceous/Tertiary boundary at El Kef: Marine Micropaleontology, v. 13, p. 239–263.

Keller, G., 1989, Extended period of extinctions across the K/T boundary in planktonic foraminifera of continental shelf sections: Implications for impact and volcanism theories: Geological Society of America Bulletin, v. 101, p. 1408–1419.

Keller, G., and Lindinger, M., 1989, Stable isotope, TOC and $CaCO_3$ record across the Cretaceous-Tertiary boundary at El Kef, Tunisia: Palaeogeography, Palaeoclimatology, Palaeoecology, v. 73, p. 243–265.

Kennett, J. P., 1982, Marine geology: Englewood Cliffs, New Jersey, Prentice-Hall, 813 p.

Kent, D. V., and Gradstein, F. M., 1985, A Cretaceous and Jurassic geochronology: Geological Society of America Bulletin, v. 96, p. 1419–1427.

Kerr, R. A., 1994, Testing an ancient impact's punch: Science, v. 263, p. 1371.

Leckie, R. M., 1987, Paleoecology of mid-Cretaceous planktonic foraminifera: A comparison of open ocean and epicontinental sea assemblages: Micropaleontology, v. 33, p. 164–176.

Lindinger, M., 1988, The Cretaceous/Tertiary boundaries of El Kef and Caravaca: Sedimentological, geochemical and clay mineralogical aspects [Ph.D. thesis]: Zürich, Swiss Federal Institute of Technology (ETH), 253 p.

Liu, C., and Olsson, R. K., 1992, Evolutionary radiation of microperforate planktonic foraminifera following the K/T mass extinction event: Journal of Foraminiferal Research, v. 22, p. 328–346.

Liu, C., and Olsson, R. K., 1994, On the origin of Danian normal perforate planktonic foraminifera from Hedbergella: Journal of Foraminiferal Research, v. 24, p. 61–74.

Luterbacher, H. P., and Premoli Silva, I., 1964, Biostratigrifia del limite Cretaceo-Terziario nell' Appenino centrale: Rivista Italiana di Paleontologia e Stratigrafia, v. 70, p. 67–128.

MacLeod, N., and Keller, G., 1991, Hiatus distributions and mass extinctions at the Cretaceous/Tertiary boundary: Geology, v. 19, p. 497–501.

Manivit, H., 1984, Paleogene and Upper Cretaceous calcareous nannofossils from Deep Sea Drilling Project Leg 74, *in* Moore, T. C., Jr., Rabinowitz, P. D., and others, eds., Initial reports of the Deep Sea Drilling Project, Volume 74: Washington, D.C., U.S. Government Printing Office, p. 475–500.

McCorkle, D. C., Emerson, S. R., and Quay, P. D., 1985, Stable carbon isotopes in marine porewaters: Earth and Planetary Science Letters, v. 74, p. 13–26.

Mitchell, B. G., Brody, E. A., Holm-Hansen, O., McClain, C., and Bishop, J., 1991, Light limitation of phytoplankton biomass and macronutrient utilization in the Southern Ocean: Limnology and Oceanography, v. 36, p. 1662–1677.

Moore, T. C., Jr., Rabinowitz, P. D., and others, 1984, Chapter 5, Site 528, *in* Moore, T. C., Jr., Rabinowitz, P. D., and others, eds., Initial reports of the Deep Sea Drilling Project, Volume 74: Washington, D.C., U.S. Government Printing Office, p. 307–405.

Oglesby, R. J., and Park, J., 1989, The effect of precessional insolation changes on Cretaceous climate and cyclic sedimentation: Journal of Geophysical Research, v. 94, p. 14693–14816.

Olsson, R. K., 1960, Foraminifera of latest Cretaceous and early Tertiary age in the New Jersey coastal plain: Journal of Paleontology, v. 34, p. 1–58.

Olsson, R. K., 1970, Planktonic foraminifera from base of Tertiary, Millers Ferry, Alabama: Journal of Paleontology, v. 44, p. 598–604.

Olsson, R. K., and Liu, C., 1993, Controversies on the placement of the Cretaceous-Paleogene boundary and the K/P mass extinction of planktonic foraminifera: Palaios, v. 8, p. 127–139.

Olsson, R. K., Hemleben, C., Berggren, W. A., and Liu, C., 1992, Wall texture classification of planktonic foraminifera genera in the Lower Danian: Journal of Foraminiferal Research, v. 22, p. 195–213.

Park, J., and Oglesby, R., 1991, Milankovitch rhythms in the Cretaceous: A GCM modeling study: Palaeogeography, Palaeoclimatology, Palaeoecology, v. 90, p. 329–355.

Park, J., D'Hondt, S., King, J. W., and Gibson, C., 1993, Late Cretaceous precessional cycles in double time: A warm-earth Milankovitch response: Science, v. 261, p. 1431–1434.

Parsons, T. R., Stephens, K., and Strickland, D. H., 1961, On the chemical composition of eleven species of marine phytoplankters: Journal Fisheries Research Board of Canada, v. 18, p. 1001–1016.

Perch-Nielsen, K., McKenzie, J., and He, Q., 1982, Biostratigraphy and isotope stratigraphy and the "catastrophic" extinction of calcareous nannoplankton at the Cretaceous/Tertiary boundary: Geological Society of America Special Paper 190, p. 353–371.

Pessagno, E. A., 1967, Upper Cretaceous planktonic foraminifera from the Western Gulf Coastal Plain: Palaeontographica Americana, v. 5, p. 245–445.

Pope, K. O., Baines, K. H., Ocampo, A. C., and Ivanov, B. A., 1994, Impact winter and the Cretaceous/Tertiary extinctions: Results of a Chicxulub asteroid impact model: Earth and Planetary Science Letters, v. 128, p. 719–725.

Schultz, P. H., and Gault, D. E., 1990, Prolonged global catastrophes from oblique impacts, in Sharpton, V. L., and Ward, P. D., eds., Global catastrophes in Earth history: Geological Society of America Special Paper 247, p. 239–261.

Shackleton, N. J., and members of the shipboard scientific party, 1984, Accumulation rates in Leg 74 sediments, *in* Moore, T. C., Jr., Rabinowitz, P. D., and others, eds., Initial reports of the Deep Sea Drilling Project, Volume 74: Washington, D.C., U.S. Government Printing Office, p. 621–644.

Sheehan, P. M., and Hansen, T. A., 1986, Detritus feeding as a buffer to extinction at the end of the Cretaceous: Geology, v. 14, p. 868–870.

Signor, P., III, and Lipps, J., 1982, Sampling bias, gradual extinction patterns and catastrophes in the fossil record, in Silver, L. T., and Schultz, P. H., eds., Geological implications of impacts of large asteroids and comets on the Earth: Geological Society of America Special Paper 190, p. 291–296.

Sigurdsson, H., D'Hondt, S., Arthur, M. A., Bralower, T. J., Zachos, J. C., van Fossen, M., and Channell, J. E. T., 1991, Glass from the Cretaceous/Tertiary boundary in Haiti: Nature, v. 349, p. 482–487.

Sliter, W. V., 1972a, Upper Cretaceous planktonic foraminiferal zoogeography and ecology—Eastern Pacific margin: Palaeogeography, Palaeoclimatology, Palaeoecology, v. 12, p. 15–31.

Sliter, W. V., 1972b, Cretaceous foraminifers—Depth habits and their origin: Nature, v. 239, p. 514–515.

Small, L. F., Fowler, S. W., Moore, S. A., and LaRosa, J., 1983, Dissolved and fecal pellet carbon and nitrogen release by zooplankton in tropical waters: Deep-Sea Research, v. 30 (12A), p. 1199–1220.

Smit, J., 1977, Discovery of a planktonic foraminiferal association between the Abathomphalus mayaroensis Zone and the Globigerina eugubina Zone at the Cretaceous-Tertiary boundary in the Barranco del Gredera (Caravaca, SE Spain), a preliminary report: Proceedings Koninklijke Nederlandse Akademie, van Wetenschappan, Series B., Physical Sciences, v. 80, p. 280–301.

Smit, J., 1982, Extinction and evolution of planktonic foraminifera at the Cretaceous/Tertiary boundary after a major impact, in Silver, L. T., and Schultz, P. H., eds., Geological implications of impacts of large asteroids and comets on the Earth: Geological Society of America Special Paper 190, p. 329–352.

Smit, J., 1990, Meteorite impact, extinctions and the Cretaceous-Tertiary boundary: Geologie en Mijnbouw, v. 69, p. 187–204.

Smit, J., and Romein, A., 1985, A sequence of events across the Cretaceous-Tertiary boundary: Earth and Planetary Science Letters, v. 75, p. 155–170.

Smit, J., and van Kempen, Th. M. G., 1987, Planktonic foraminifers from the Cretaceous/Tertiary boundary at DSDP Sea Drilling Project Site 605, North Atlantic, *in* van Hinte, J. E., Wise, S. W., Jr., and others, eds., Initial reports of the Deep Sea Drilling Project, Volume 93: Washington, D.C., U.S. Government Printing Office, p. 549–553.

Smit, J., and eight others, 1992, Tektite-bearing deep-water clastic unit at the Cretaceous-Tertiary boundary in northeastern Mexico: Geology, v. 20, p. 99–103.

Stott, L. D., and Kennett, J. P., 1990a, The paleoceanographic and paleoclimatic signature of the Cretaceous/Tertiary boundary in the Antarctic: Stable isotopic results from ODP Leg 113, *in* Proceedings, Ocean Drilling Program, Scientific Results, Volume 113: College Station, Texas, Ocean Drilling Program, p. 829–848.

Stott, L. D., and Kennett, J. P., 1990b, New constraints on early Tertiary palaeoproductivity from carbon isotopes in foraminifera: Nature, v. 342, p. 526–529.

Takahashi, T., Broecker, W. S., and Langer, S., 1985, Redfield ratio based on chemical data from isopycnal surfaces: Journal of Geophysical Research, v. 90, p. 6907–6924.

Thierstein, H. R., 1981, Late Cretaceous nannoplankton and the change at the Cretaceous-Tertiary boundary: Society of Economic Paleontologists and Mineralogists Special Publication 32, p. 355–394.

Zachos, J. C., and Arthur, M. A., 1986, Paleoceanography of the Cretaceous-Tertiary boundary event: Inferences from stable isotopic and other data: Paleoceanography, v. 1, p. 5–26.

Zachos, J. C., Arthur, M. A., Thunell, R. C., Williams, D. F., and Tappa, E. J., 1985, Stable isotope and trace-element geochemistry of carbonate sediments across the Cretaceous/Tertiary boundary at Deep Sea Drilling Project Hole 577, Leg 86, *in* Heath, G. R., Burckle, L. H., and others, eds., Initial reports of the Deep Sea Drilling Project, Volume 86: Washington, D.C., U.S. Government Printing Office, p. 513–532.

Zachos, J. C., Arthur, M. A., and Dean, W. E., 1989, Geochemical evidence for suppression of pelagic marine productivity at the Cretaceous/Tertiary boundary: Nature, v. 337, p. 61–64.

Zachos, J. C., Aubry, M.-P., Berggren, W. A., Ehrendorfer, T., and Heider, F., 1992, Magnetobiochemostratigraphy across the Cretaceous/Paleogene boundary at ODP Site 750A, Southern Kerguelen Plateau, *in* Wise, S. W., Jr., Schlich, R., and others, eds., Proceedings, Ocean Drilling Program, Scientific Results, Volume 120, part 2: College Station, Texas, Ocean Drilling Program, p. 961–977.

Manuscript Accepted by the Society September 14, 1995

Printed in U.S.A.

Geological Society of America
Special Paper 307
1996

Evidence for planktonic foraminifer reworking versus survivorship across the Cretaceous-Tertiary boundary at high latitudes

Brian T. Huber
Department of Paleobiology, NHB-121, National Museum of Natural History, Smithsonian Institution, Washington, DC 20560

ABSTRACT

Comparison of nannofossil and planktonic foraminifer distributions and carbon isotope analyses across the Cretaceous/Tertiary (K/T) boundary at Ocean Drilling Program (ODP) Sites 738, 690, and 750 in the Southern Ocean indicates extensive reworking of Cretaceous specimens into Tertiary sediments.

At Site 738, which yields the most complete K/T boundary sequence recovered from the deep sea, Cretaceous nannofossil and planktonic foraminifer species consistently occur throughout the Danian sequence, and several Cretaceous species have been identified in lower Eocene samples more than 100 m above the K/T boundary. In addition, the planktonic foraminifer *Globigerinelloides impensus* and inoceramid prisms have been identified in several Danian samples at Site 738, although these are known to have become extinct in the uppermost Campanian and in the mid-Maastrichtian, respectively. Strontium isotope analysis of reworked inoceramid prisms yield $^{87}Sr/^{86}Sr$ ratios expected for late Campanian biogenic calcite, whereas the $^{87}Sr/^{86}Sr$ ratios of planktonic foraminifera from near their evolutionary appearance are appropriate for their stratigraphic position.

Previous studies of K/T boundary sections from the low latitudes have determined that surface waters during the earliest Danian were more enriched in ^{12}C than in the late Maastrichtian, and the $\delta^{13}C$ vertical gradient was markedly reduced. Thus, it is possible to use the $\delta^{13}C$ values of postboundary Cretaceous species as a test of whether they lived in the Danian ocean or were reworked.

In the Southern Oceans, carbon isotope measurements of the putative K/T survivor taxon *Globigerinelloides* show no difference between Cretaceous and Tertiary values. Moreover, the $\delta^{13}C$ values of this taxon in Danian sediments are consistently higher than those of co-occurring in situ Danian species, and at Site 738 the $\delta^{13}C$ values of in situ Danian species follow a different trend than that of the *Globigerinelloides* values. This suggests that these specimens, and probably other specimens of *Globigerinelloides* in Danian sediments, were redeposited.

Taken together, these data indicate that reworking at these sites compromises arguments for enhanced postboundary survivorship of Cretaceous species in the southern high-latitude oceans.

INTRODUCTION

The pronounced turnover of planktonic foraminifera at the Cretaceous-Tertiary (K/T) boundary has been well documented for over 30 years (e.g., Olsson, 1960; Berggren, 1962; Luterbacher and Premoli Silva, 1964). These and many other studies have revealed that very large, ornate, and diverse assemblages from Maastrichtian sequences were completely replaced by minute, unornamented, low-diversity assemblages within a small stratigraphic interval. Discovery of iridium at

Huber, B. T., 1996, Evidence for planktonic foraminifer reworking versus survivorship across the Cretaceous-Tertiary boundary at high latitudes, *in* Ryder, G., Fastovsky, D., and Gartner, S., eds., The Cretaceous-Tertiary Event and Other Catastrophes in Earth History: Boulder, Colorado, Geological Society of America Special Paper 307.

the K/T boundary and a proposal that the terminal Cretaceous extinctions were caused by a bolide impact (Alvarez et al., 1980) served to stimulate intensive efforts toward precisely defining the duration of this turnover. Rather than resolving this issue, however, the detailed studies that followed have led to considerable debate as to whether the planktonic foraminiferal extinctions were abrupt, occurring over the course of a few days to hundreds of years, or gradual, with extinctions spanning hundreds of thousands of years below and above the K/T boundary.

This disagreement stems from different opinions on the significance of the last species occurrences below and above the boundary. According to some workers, the gradual pattern of extinction below the K/T boundary may be an artifact of the low probability of identifying the true last occurrence of rare species, and post-K/T occurrences of Maastrichtian planktonic foraminifera may result from sediment reworking by currents or bioturbation (e.g., Smit, 1982; Smit and Romein, 1985; Huber, 1991a; Olsson and Liu, 1993; Huber et al., 1994). The number of planktonic foraminifer species thought to have survived into the Danian by proponents of an abrupt extinction scenario range from one (Smit, 1982) to three (Olsson et al., 1992; Olsson and Liu, 1993) and, most recently, four (Huber and Boersma, 1994). On the other hand, a more literal interpretation of the pattern of planktonic foraminiferal turnover has led other workers to infer that long-term endogenous factors, such as changes in sea level, temperature, and global volcanism, led to prolonged extinctions below and above the K/T boundary (e.g., Brinkhuis and Zachariasse, 1988; Keller, 1988, 1989a, 1989b, 1993; Canudo et al., 1991; MacLeod and Keller, 1994). According to authors ascribing to this view, the number of Maastrichtian species surviving into the Danian ranges from one (Brinkhuis and Zachariasse, 1988) to 21 (Keller, 1989a) or 26 (MacLeod and Keller, 1991).

Determination of whether the post-K/T occurrences of Maastrichtian species represent true survivorship or result from sediment reworking is critical to understanding what factor or factors triggered the planktonic foraminiferal turnover. This is not a simple problem, as criteria used to identify reworked microfossils are often equivocal. For example, differences in shell color and preservation may be used to distinguish reworked from *in situ* specimens, but this can only be applied to moderately to well-preserved assemblages obtained from sections where there is a significant difference between Maastrichtian and Danian lithologies. Consistency at different localities in the last occurrence ordering of Cretaceous taxa above the K/T boundary has been used as evidence of Cretaceous survivorship (Keller, 1989b), but this may be an artifact of the greater probability that the most abundant Cretaceous species will be reworked higher in a section (Liu and Olsson, 1992; Olsson and Liu, 1993). Rare species with a discontinuous stratigraphic record should not be used to characterize patterns of species turnover since their absence above their recorded last occurrence may be an artifact of sampling effects (Signor and Lipps, 1982). A change in the relative abundance of some trans-K/T species has been used as evidence for survivorship (e.g., Smit, 1982; Keller, D'Hondt and Keller, 1991; Liu and Olsson, 1992), but such abundance changes may have been strongly influenced by differential preservation or reworking (Olsson and Liu, 1993).

A less equivocal approach for identifying survivorship and/or reworking is by comparison of the carbon isotope compositions of foraminifera and calcareous nannofossils (from fine-fraction bulk carbonate) in samples below and above the K/T boundary and with those of species that first occur within the Danian (Perch-Nielsen et al., 1982; Barrera and Keller, 1990a). This approach is based on the assumption that $\delta^{13}C$ values of near-surface water carbonates and the vertical carbon isotopic gradient (the difference between the $\delta^{13}C$ values of nannofossil carbonate or planktonic foraminifers and those of benthic foraminifers) decreased across the K/T boundary to relatively low values throughout the world's oceans during the early Danian, as a result of a decline in surface water productivity (Boersma and Shackleton, 1981; Hsü et al., 1982; Zachos and Arthur, 1986; Barrera and Keller, 1990a, 1990b, 1994; Zachos et al., 1989, 1992).

Barrera and Keller (1990a, b) used $\delta^{13}C$ analyses as a test for survivorship of the Cretaceous planktonic species *Heterohelix globulosa* and *Guembelitria cretacea* in a core from Brazos River, Texas. Their study revealed that the vertical $\delta^{13}C$ gradient dropped to negative values and the $\delta^{13}C$ composition of *H. globulosa, G. cretacea,* and the benthic *Lenticulina* sp. gradually decreased by 2.5 to 3.0‰ beginning at the K/T boundary and extending to the P1a Zone. They interpreted this as evidence that *G. cretacea* and *H. globulosa* did survive to inhabit the early Danian ocean for at least 40,000 yr. On the other hand, Zachos et al. (1992) found no difference between Maastrichtian and Danian $\delta^{13}C$ values of *H. globulosa* and *Globigerinelloides* at ODP Site 750 (southern Indian Ocean) and suggested that the Danian occurrences of those species were due to reworking. However, a significant amount of the lowermost Danian record is missing at Site 750 because of a hiatus, so comparison with the record from Brazos River may be irrelevant. Moreover, reworking is common in sediments directly above deep-sea unconformities (J. Pospichal, personal communication, 1994).

Barrera and Keller (1994) have determined that there is little change in $\delta^{13}C$ values at the K/T boundary and no significant decrease in the $\delta^{13}C$ vertical gradient in high latitudes, but there is a decrease followed by an increase in bulk carbonate, and to a lesser extent planktonic foraminiferal, $\delta^{13}C$ values beginning about 200,000 years after the K/T event. The decoupling of the low- versus high-latitude $\delta^{13}C$ records and the bulk carbonate versus planktonic foraminiferal $\delta^{13}C$ records confounds interpretation of the cause for changes in seawater $\delta^{13}C$ during the early Danian.

The purpose of this study is to compare the planktonic foraminiferal and nannofossil distributions and carbon isotope records across the K/T boundaries at ODP sites 738 and 750 in the southern Indian Ocean and ODP Site 690 in the southern South Atlantic to evaluate the possibility of reworking of Cretaceous specimens in Tertiary sediments. This issue is important because it sheds light on whether the relatively high abundance of Cretaceous planktonic foraminifera and nannofossils in Tertiary sediments in these sections represents a low extinction rate or extended survival in high latitudes after the K/T boundary event, as some have suggested (e.g., Keller, 1993; Keller et al., 1993; MacLeod and Keller, 1994), or whether the Cretaceous specimens were simply redeposited. The Site 738 section is considered biostratigraphically complete (Thierstein et al., 1991; Pospichal and Huber, 1992), whereas a hiatus interrupts the lowermost Danian record at Site 750 (Ehrendorfer and Aubry, 1992; Pospichal and Huber, 1992) and bioturbation obscures the K/T depositional record at Site 690 (Pospichal and Wise, 1990). Planktonic and benthic foraminifera that cross the K/T boundary at all these sites are sufficiently preserved for isotopic analysis.

METHODS

Most Site 738 samples discussed in this paper were included in the biostratigraphic study of Huber (1991a, b), who described the foraminiferal preservation, species distributions, relative abundance, and biozonation as well as the sample preparation methods. The abundance, preservation, and size of the trans-K/T species and undoubtedly *in situ* Danian species dictated which samples were analyzed for oxygen and carbon isotopic ratios. *Globigerinelloides multispinus* (=*G. asperus* and *G. messinae* of other authors) was chosen because it is the most common, most consistently occurring, and largest trans-K/T species in the Danian sequence and because this species has been cited as a high-latitude K/T survivor species (Keller, 1993; Keller et al., 1993). Although *G. multispinus* occurs in every sample analyzed from Section 738C-20R-5 and makes up a significant proportion of the lower Danian assemblages, no samples from the laminated interval and few samples above yielded an adequate number of sufficiently preserved specimens larger than 125 μm for stable isotopic analyses. Such is also the case for the *in situ* Danian species *Eoglobigerina eobulloides*. Results of the isotopic analyses of these species are included in Table 1. Sample preparation and techniques for isotopic analysis are in Barrera and Keller (1994), who reported carbon and oxygen isotope values of other planktonic foraminifer taxa and bulk carbonate from the Site 738 samples discussed in this chapter and from additional Site 738 samples.

Because conventional marker species are often absent, the biostratigraphy of calcareous plankton from the southern high latitudes differs from low-latitude zonations. The Antarctic zonations used by Wei and Pospichal (1991) for calcareous nannoplankton and Huber (1991a) for planktonic foraminifera are shown in Figure 2 relative to zonal schemes used at lower latitudes.

HOLE 738C—KERGUELEN PLATEAU

Stratigraphy

The most complete deep-sea K/T boundary sequence in the southern high latitudes was recovered from Hole 738C on the southern Kerguelen Plateau (62°42′S, 82°47′E) at 2,253-m water depth (Fig. 1). The paleodepth of this site was estimated at less than 1,000 m during the early Paleocene (Barrera and Huber, 1991). The boundary is placed at a thin clay layer 96.2 cm from the top of Section 119-738C-20R-5 (377.16 m below seafloor or mbsf), about 2 cm above the base of a clay-rich, laminated interval that is about 15 cm thick (Fig. 2). Presence of the laminated interval eliminates bioturbation as a cause for upward reworking of Cretaceous material into Danian sediments at this site. Immediately underlying the boundary clay is a 6-cm-thick, well-indurated white chalk of latest Maastrichtian age, followed by a 14-cm-thick chert layer and about 2 m of unrecovered core (Barron, Larsen, et al., 1989). Above the laminated interval is a less indurated, dark-colored chalk of early Danian age that lightens and becomes increasingly bioturbated upward. Carbonate content of the bulk sediment changes from 97% in the Maastrichtian chalk to 69% in the boundary clay, then fluctuates between about 77 and 84% up to 43 cm in the K/T core section (Ehrmann, 1991; Thierstein et al., 1991). By Section 119-738C-20R-3, carbonate is about 93%.

Schmitz et al. (1991) recorded 18 parts per billion (ppb) iridium within the boundary clay, which is four times the values obtained from the uppermost Maastrichtian chalk and over 100 times the Danian background values. High concentrations of Ir gradually tail off to low values over a 1.1-m interval in Core 738C-20R. This profile was interpreted by Schmitz et al. (1991) as evidence for a single input of extraterrestrial Ir and subsequent reworking by oceanographic processes. Elevated Ir concentrations in the uppermost Maastrichtian chalk probably resulted from postdepositional transport by microbes or by downward absorption and diffusion processes (Schmitz et al., 1991; Thierstein et al., 1991).

The nannofossil and planktonic foraminiferal biostratigraphies of this sequence indicate that the K/T boundary interval is biostratigraphically complete (Wei and Pospichal,1991; Huber, 1991a; Pospichal and Huber, 1992; Keller, 1993). The calcareous nannofossil biostratigraphy of Wei and Pospichal (1991) was modified by Pospichal and Huber (1992) such that the top of Zone NA1 (based on the first occurrence [FO] of the Tertiary genus *Hornibrookina*) is now placed at 86 cm and the top of Zone NA2 (= top of Zone NP1) is at 16 cm in Section 738C-20R-5. The last high-latitude Mesozoic marker, *Nephrolithus frequens*, is common to abundant throughout the upper Maastrichtian section (Wei and Thierstein, 1991).

The upper Maastrichtian foraminiferal zonal marker *Abathomphalus mayaroensis* was not identified in the limestone immediately below the boundary clay in Section 119-738C-20R-5 (Fig. 2), perhaps because of the strongly indurated

TABLE 1. OXYGEN AND CARBON ISOTOPIC VALUES OBTAINED FOR THIS STUDY FROM FINE FRACTION (<38 μm) CARBONATE AND PLANKTONIC FORAMINIFERA ACROSS THE K/T BOUNDARY AT SITE 738*

Core interval, ODP Hole 738C	Depth (mbsf)	*Globigerinelloides multispinus*		*Eoglobigerina* spp.		Bulk Carbonate	
		$\delta^{18}O$	$\delta^{13}C$	$\delta^{18}O$	$\delta^{13}C$	$\delta^{18}O$	$\delta^{13}C$
20R-3, 139-141	374.49			-0.08	1.52		
20R-4, 17-19	374.77			0.19	1.54	-0.35	2.48
20R-4, 19-21	374.79			0.80	1.54		
20R-4, 59-61	375.19			0.21	1.54		
20R-4, 99-100	375.59			0.20	1.31		
20R-4, 108-110	375.68					0.01	1.79
20R-4, 129-130	375.89	-0.06	2.25				
20R-4, 130-131	375.90			0.18	1.54		
20R-4, 134-135	375.94			0.24	1.32		
20R-5, 4-5	376.14			-0.04	1.42	-0.07	0.55
20R-5, 8-10	376.18			0.11	1.61	-0.45	0.94
20R-5, 15-16	376.25	-0.35	2.13	-0.11	1.69		
20R-5, 16-17	376.26					-0.35	0.78
20R-5, 27-28	376.37	-0.25	2.10	-0.26	1.51	-0.27	0.86
20R-5, 33-34	376.53	-0.49	1.95	-0.21	1.62		
20R-5, 55-56	376.75	-0.42	2.04	-0.17	1.67	-0.38	1.55
20R-5, 65-66	376.85	-0.38	2.06	-0.19	1.91		
20R-5, 66-67	376.86					-0.59	1.54
21R-1, 34-36	380.25	-0.30	2.50				
22R-1, 10-12	389.76	-0.11	2.01			-0.32	2.04

*All stable isotope values are expressed relative to the PDB standard.

nature of the sediment. This species occurs consistently in samples below, from the core-catcher of Core 119-738C-20R through Core 119-738C-23R. The foraminifer Zone P0 was not recognized by Huber (1991a) since the marker species *Parvularugoglobigerina eugubina*, whose first occurrence defines the upper P0 zonal boundary, was not found in any Site 738 sample. A subsequent search for this taxon by Huber et al. (1994) using a larger number of specimens has yielded similar results. Because of the absence of this species, Huber (1991a) used the Antarctic Paleogene zonal scheme of Stott and Kennett (1990a), with a modified definition of Zone AP1 for subdivision of the Paleogene sequence. Zone APα, defined from the FO of *Eoglobigerina fringa* to the FO of *Globoconusa daubjergensis*, extends from the K/T boundary clay to 57 cm in the K/T core section. Although the nominate species of this zone is now considered a junior synonym of *Eoglobigerina eobulloides* (Olsson et al., 1992), the latter species was also reported by Huber (1991a) immediately above the boundary clay. Zone

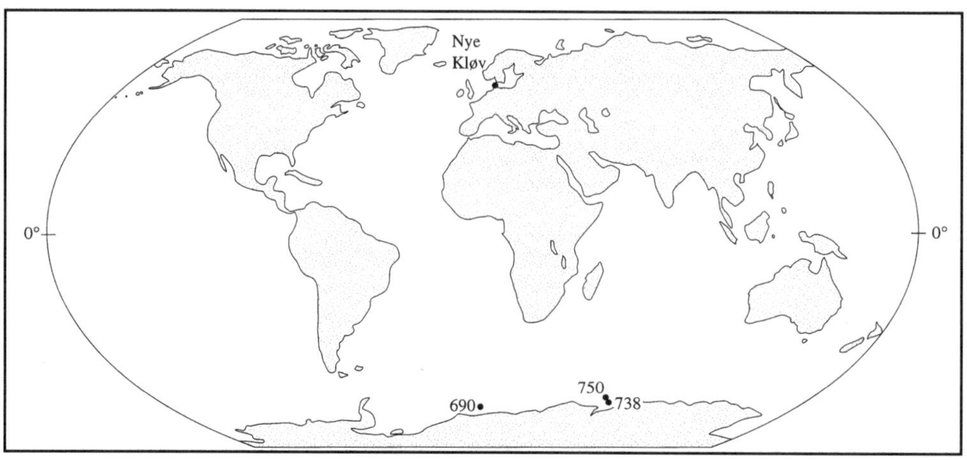

Figure 1. Location of land-based and deep-sea K/T boundary sites discussed in this study.

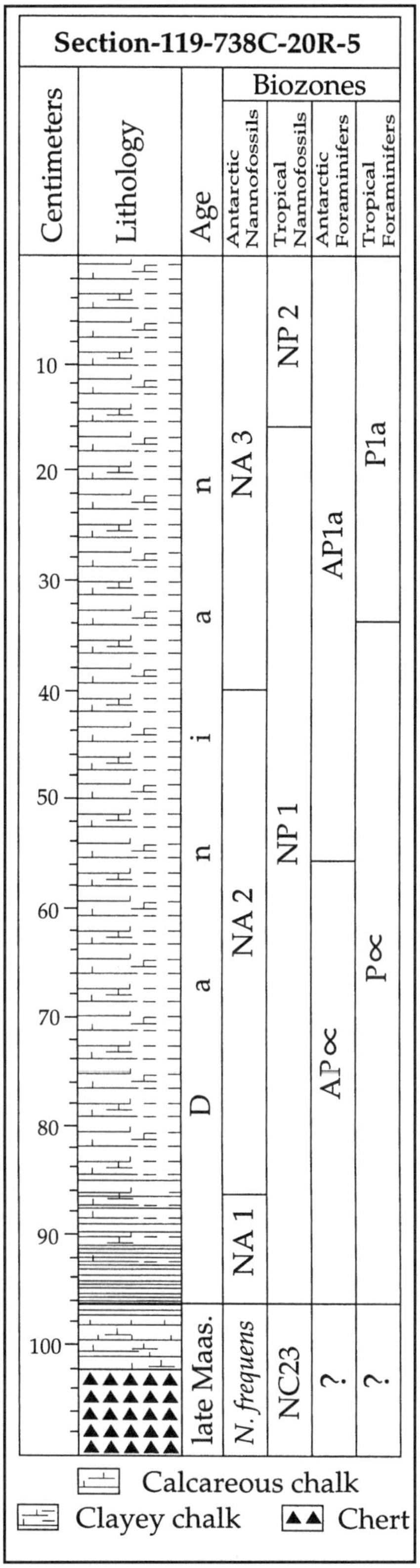

Figure 2. Lithology and biostratigraphy of Section 20R-5 at ODP Site 738. The Antarctic nannofossil zonal scheme of Wei and Pospichal (1991) is shown correlated with the tropical nannofossil zonation of Martini (1971). The Antarctic Paleogene planktonic foraminifer zonal scheme of Huber (1991a), modified from Stott and Kennett (1990a), is shown relative to the tropical planktonic foraminifer scheme of Berggren and Miller (1988).

AP1a comprises the interval from the FO of *G. daubjergensis* to the FO of *Praemurica inconstans*, extending from 376.76 to 372.88 mbsf (Huber, 1991b).

Trans-K/T species occurrences

Relative abundance counts of nannofossils below and above the Site 738 K/T boundary (Wei and Pospichal, 1991) reveal that Cretaceous species decrease from 100% of the total assemblage at the top of the Maastrichtian chalk to less than 5% within 2 m above the boundary (Fig. 3). Wei and Pospichal (1991) recognized a significant reworking component among the Danian nannofossil assemblages they identified; 12 species were considered by these authors as reworked, and some of these were found at sporadic intervals throughout the Danian sequence up to the middle of Zone NA-6. Four species were identified as K/T survivors. These occur sporadically in the lower Danian sequence in common to rare abundance. Incoming Tertiary species gradually increase in abundance and diversity upward from 8 cm above the boundary. After noting a strong correlation between the decreasing abundance of Cretaceous nannofossils and the iridium profile above the K/T at Site 738, Pospichal (1996) suggested that most of the trans-K/T nannofossils were probably redeposited by bottom currents.

Fourteen species of Cretaceous planktonic foraminifera have been identified in sediments above the K/T boundary at Site 738 (Fig. 4). *Globigerinelloides multispinus*, *G. subcarinatus*, *Heterohelix planata*, *H. globulosa*, and *Archaeoglobigerina australis* are the most common and consistently occurring of the trans-K/T species, and three of these species—*G. multispinus*, *G. subcarinatus*, and *H. planata*—occur in samples as high as lower Eocene foraminiferal Zone AP6a. Such anomalously high stratigraphic occurrences of Cretaceous taxa and similar abundances of Cretaceous species above and below the K/T boundary strongly indicate postboundary reworking.

Evidence for reworking of the coarser sediment fraction at Site 738 is also derived from the occurrence in Tertiary sediments of species whose extinction level is below the K/T boundary as well as by comparison of the strontium isotope ratios of suspected reworked specimens with *in situ* specimens. A single specimen of *Globigerinelloides impensus*, which has a well-documented last occurrence datum that correlates with upper polarity Subchron C33N and is used to identify the Campanian/Maastrichtian boundary in the southern high latitudes (Huber, 1992), was identified in one sample from Zone AP1a in Section 119-738C-20R-5 (Huber, 1991a). *Inoceramus* prisms were also identified in Danian sediments, well above their

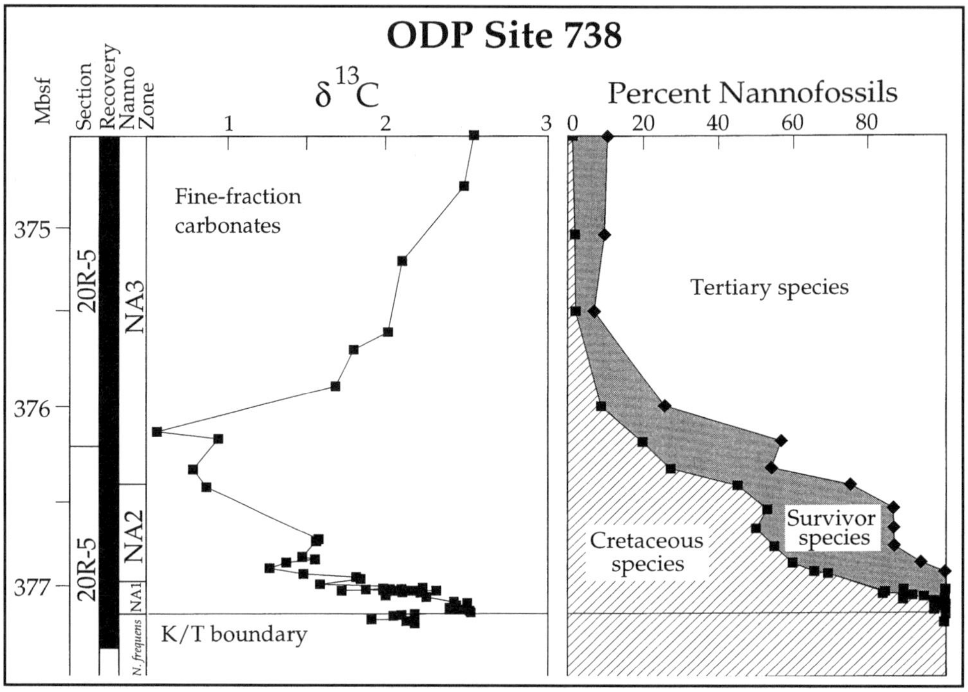

Figure 3. Carbon isotopic composition of fine-fraction carbonate and relative abundance of Cretaceous, survivor, and Tertiary nannofossils in sections 4 and 5 of Core 119-738C-20R. Nannofossil zonation and abundance data from Wei and Pospichal (1991), Pospichal and Huber (1992), and Pospichal (1996). Modified after Barrera and Keller (1994). Solid black represents interval recovered. All carbon isotope values are expressed relative to the PDB (Peedee belemnite) standard. Mbsf = meters below sea floor.

extinction level, which occurred during the mid-Maastrichtian, near the base of the *Abathomphalus mayaroensis* Zone (MacLeod, 1994; MacLeod et al., this volume). MacLeod and Huber (1996) have documented that prisms isolated from upper Maastrichtian and Danian intervals of Site 738 yield strontium isotopic ratios expected for late Campanian fossils, whereas samples of co-occurring planktonic foraminifera near their evolutionary first appearance yield strontium isotopic ratios appropriate for their stratigraphic position. Occurrence of *G. impensus* in Danian sediments at Site 738 and the strontium isotope data provide strong evidence of an upper Campanian source bed for the reworked microfossils. Occurrence of *A. mayaroensis* and *G. subcarinatus* in Danian cores indicates that upper Maastrichtian units also served as a source for the reworked sediments.

Among the trans-K/T planktonic foraminifera, three species, including *Guembelitria cretacea*, *Hedbergella monmouthensis*, and *Zeauvigerina waiparaensis*, have been well documented as survivors of the terminal Cretaceous extinction event (Liu and Olsson, 1992; Keller, 1993; Huber and Boersma, 1994). The sporadic distribution within the Danian of *G. cretacea* and *H. monmouthensis* precludes assumptions as to whether or not their occurrences at Site 738 represent survivorship or reworking. But *Z. waiparaensis* is one of the most abundant and consistently occurring planktonic species in the lower Danian at Site 738 (Huber, 1991a, b; Keller, 1993; Huber and Boersma, 1994), and its distribution within the Danian sediments therefore is not considered an artifact of reworking.

The information presented above clearly indicates that there is a significant reworked component among the Danian microfossil assemblages at Site 738. However, it is still possible that some of the trans-K/T species considered as reworked also include *in situ* survivors within the lower Danian. Further insight to the survivorship versus reworking question can be gained by isolating specimens of the most common "reworked" species and obtaining measurements of their carbon isotopic ratios.

Carbon isotope results

Carbon isotopic data for Site 738 are presented in Table 1 and Figure 5 for bulk carbonate, the Danian planktonic foraminifers *Eoglobigerina* spp. and *Parasubbotina pseudobulloides*, and the putative survivor planktonic foraminifer *Globigerinelloides multispinus*. Most of the bulk carbonate and all of the *Gavelinella beccariiformis* and *P. pseudobulloides* data presented in Figure 5 were reported previously by Barrera and Keller (1994). Too few specimens of trans-K/T and *in situ* Danian foraminifera occur below 376.85 mbsf for stable isotopic analysis.

One of the most significant features of the curves illustrated in Figure 5 is that the fine-fraction bulk carbonate and *in situ*

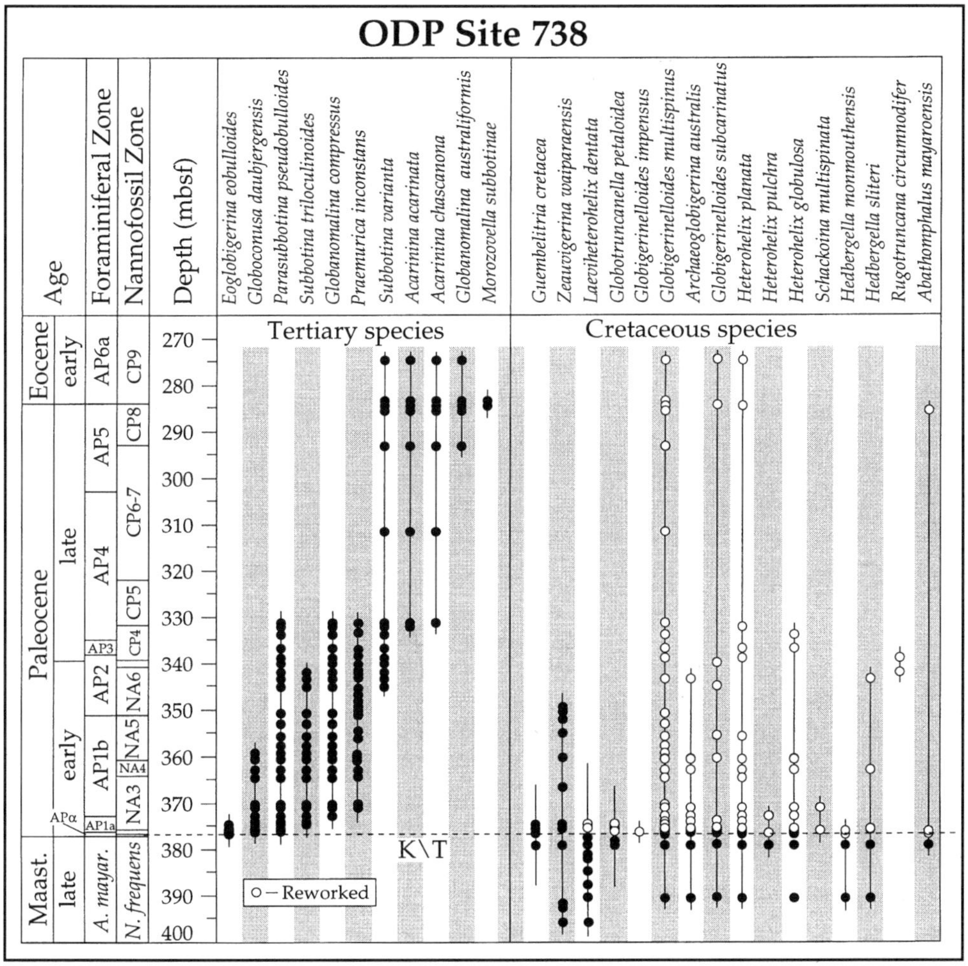

Figure 4. Distribution of Cretaceous and selected Tertiary planktonic foraminifera in Maastrichtian through lower Eocene sediments at Site 738. Planktonic foraminifer biozonation based on Huber (1991a, b) and nannofossil biozonations based on Wei and Pospichal (1991) and Wei and Thierstein (1991).

Danian foraminifer curves are decoupled. The bulk carbonate curve shows a 0.4‰ positive shift across the boundary, with values of 2.41 and 2.52‰ up to 90 cm in Section 119-738C-20R-5. This is followed by a decrease, increase, then marked decrease to a minimum of 0.55‰ at 376.94 mbsf within lower Zone NA3. The bulk carbonate $\delta^{13}C$ values then return to about 2.0‰ by 375.84 mbsf. *Eoglobigerina* spp. shows a less dramatic change in $\delta^{13}C$ values within this stratigraphic interval, with a decrease from 1.69‰ at 376.25 mbsf to between 1.32‰ and 1.54‰ from 375.91 mbsf to 375.59 mbsf. Where they overlap, the $\delta^{13}C$ values of *Eoglobigerina* spp. and *P. pseudobulloides* are within 0.2‰ of each other and are ~1.0‰ more negative than bulk carbonate $\delta^{13}C$ by the middle of Zone NA3. A more detailed discussion of the Site 738 isotopic records and their correlation to those of Sites 690 and 750 is in Barrera and Keller (1994).

Comparison of the percentage of Cretaceous nannofossils and the bulk carbonate $\delta^{13}C$ curve at Site 738 (Fig. 3) reveals parallel trends up to the bottom of Section 119-738C-20R-4, where Cretaceous nannofossils make up less than 5% of the total assemblage. This correspondence may suggest that the Danian bulk carbonate curve is largely controlled by changes in the taxonomic composition of nannofossil species (Thierstein et al., 1991; Barrera and Keller, 1994). Hence bulk carbonate $\delta^{13}C$ may not be a useful proxy for estimation of changes in surface water paleoproductivity without a better understanding of nannofossil vital effects.

The $\delta^{13}C$ analyses of *Globigerinelloides multispinus* reveals no significant change in $\delta^{13}C$ composition across the K/T boundary. Moreover, this species maintains a $\delta^{13}C$ composition between 1.95 and 2.25‰, while the *in situ* Danian species *Eoglobigerina* spp. became more negative in $\delta^{13}C$, decreasing from 1.91‰ at 376.85 mbsf to a minimum of 1.31‰ at 375.89 mbsf. These results suggest that *G. multispinus* was not a survivor species but was likely redeposited from Cretaceous sediments elsewhere on the southern Kerguelen Plateau.

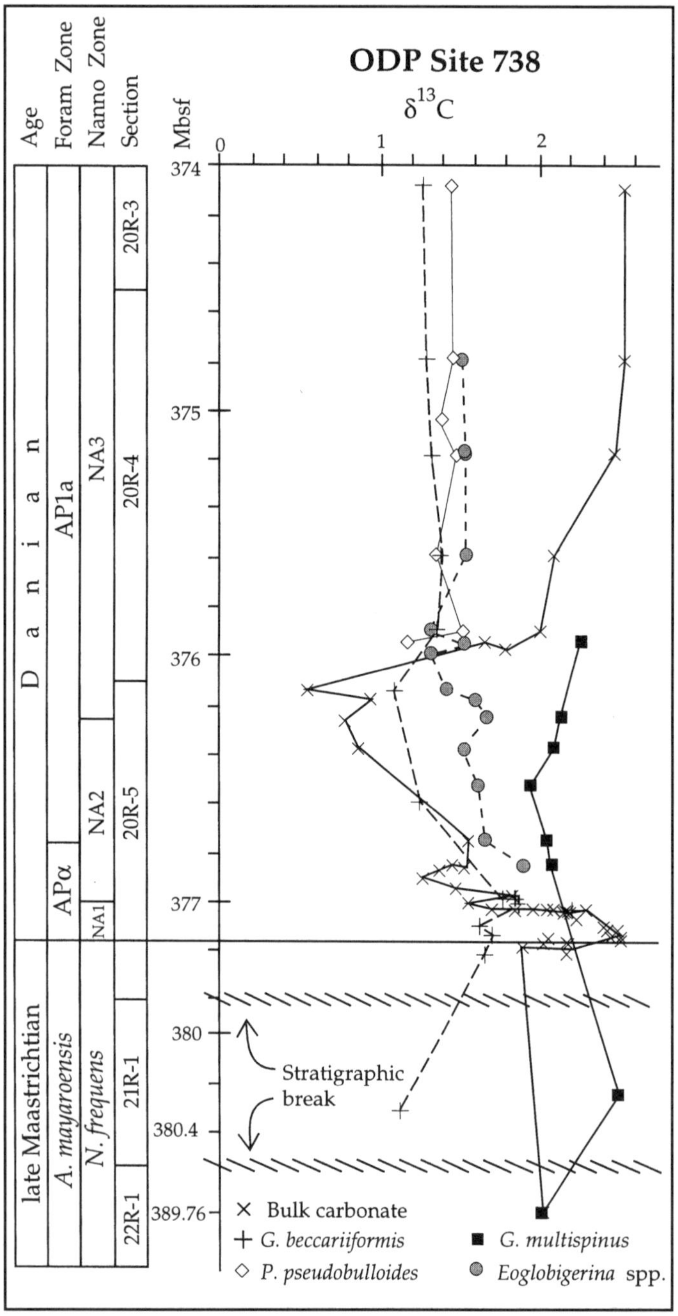

Figure 5. Carbon isotope data from fine-fraction carbonate, the benthic *Gavelinella beccariiformis*, the trans-K/T species *Globigerinelloides multispinus*, and the *in situ* Danian species *Parasubbotina pseudobulloides* and *Eoglobigerina* spp. from cores 119-738C-20R through -22R. Data from this study and Barrera and Keller (1994). All carbon isotope values are expressed relative to the PDB standard.

HOLE 750A—KERGUELEN PLATEAU

Stratigraphy

A discontinuous K/T boundary section was recovered from Hole 750A, which is located at 57°35.5′S and 81°14.4′E on the Kerguelen Plateau (southern Indian Ocean) in 2,031-m water depth. Quilty (1992b) assigned a middle- to upper-bathyal paleodepth to this site for the Maastrichtian, based on the taxonomic composition of the benthic foraminiferal assemblages. The K/T boundary is placed at 91.5 cm in Section 120-738C-15R-3 (349.50 mbsf), where there is a fractured and discontinuous contact between a semilithified, white, nannofossil chalk of late Maastrichtian age and a more lithified dark grayish-green marl of early Danian age (Schlich, Wise et al., 1989). The sharpness of the K/T contact and breaks in the cored sediment indicate that the discontinuous recovery of the boundary interval was at least partly due to drilling disturbance. Unlike those in Site 738, the lowermost Danian sediments are not laminated. Whole-rock carbonate measurements are near 95% in the Maastrichtian chalk, then decrease to 83% just above the boundary level, and rebound to 95% by about 70 cm above the boundary (Zachos et al., 1992). The increasing carbonate content within the lower Danian sediments is reflected by a gradual lightening of the sediment color.

The biostratigraphy of the K/T sequence from Hole 750A was originally reported by Ehrendorfer and Aubry (1992) for calcareous nannofossils and by Quilty (1992a) and Berggren (1992) for planktonic foraminifera. Further discussion of the Site 750 biostratigraphy was presented in Pospichal and Huber (1992). All of these authors reported the presence of uppermost Maastrichtian biozones (*Nephrolithus frequens* Zone for nannofossils and *Abathomphalus mayaroensis* Zone, *Pseudotextularia elegans* Subzone for planktonic foraminifera), suggesting that the upper Maastrichtian is biostratigraphically complete. Absence of calcareous nannoplankton zones NA1 and lowermost NA2 indicates that the lowermost Danian is incomplete (Pospichal and Huber, 1992). Poor preservation and lack of diagnostic species has precluded recognition of planktonic foraminiferal zonal subdivisions of the lowermost Danian based on definitions of the tropical P-zones (Berggren, 1992).

Trans-K/T species occurrences

Relative abundance counts of calcareous nannoplankton across the K/T boundary at Site 750 (Fig. 6) reveal that the Cretaceous assemblages were gradually replaced by Danian assemblages within 3.58 m above the K/T contact (Ehrendorfer and Aubry, 1992). A total of 22 Cretaceous nannofossil species found in Danian sediments at Site 750 were considered by Ehrendorfer and Aubry (1992) to have been reworked. These become increasingly rare and sporadic in occurrence upward, then disappear by 3.42 m above the boundary. These authors identified eight additional taxa that cross the K/T boundary as persistent species (similar to survivor species). The first incoming species of *Hornibrookina* appears immediately above the boundary. This is followed by the gradual appearance and increasing dominance of other incoming taxa.

The distribution of planktonic foraminifera in the upper Maastrichtian–lower Danian sequence from Hole 750A is illustrated in Figure 7, which is modified from Zachos et al. (1992,

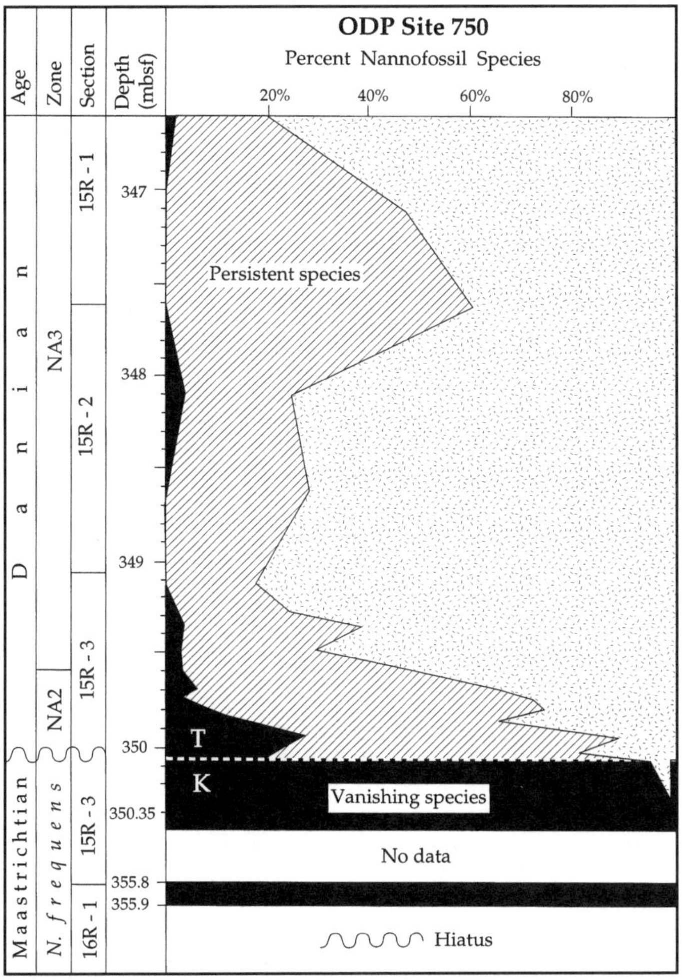

Figure 6. Relative abundance of nannofossils (excluding *Thoracosphaera* spp. and *Prinsius* sp.) from upper Maastrichtian through lower Danian sediments at Site 750 based on the quantitative data presented in Ehrendorfer and Aubry (1992, Table 8) relative to the Antarctic nannofossil zonation defined by Wei and Pospichal (1991). "Vanishing" species include all taxa that became extinct at or below the K/T boundary, "persistent" species include all taxa that lived in the Cretaceous and Tertiary ocean, and "incoming" species include all taxa that first appeared in the Danian.

Fig. 6). Three species (*Rugoglobigerina macrocephala*, *R. pustulata*, and *R. rugosa*) were identified in only the lowermost Danian sample, two species (*Heterohelix* sp. 1 and *Heterohelix* sp. 2) consistently occur up to the top of Zone NA2, and two species (*Globigerinelloides multispinus* and *G. subcarinatus*) have consistent occurrences into Zone NA3, 30 cm above the K/T boundary. Within the samples that were studied, these species were represented by one to several specimens.

No species with an extinction below the K/T boundary was found in the Danian sediments at Site 750, and no discernable color differences were reported for the trans-K/T species. Hence, it is not clear based on the microfossil distributions whether the post-K/T occurrences of Cretaceous species resulted from reworking or survivorship.

Carbon isotope results

A carbon isotopic record across the K/T boundary at Site 750 was obtained by Zachos et al. (1992) for bulk carbonate, the trans-K/T species *H. globulosa* and *Globigerinelloides* spp., the Danian species *Eoglobigerina eobulloides*, and the benthic species *Nuttalides truempyi*. All of the specimens analyzed by Zachos et al. (1992) were selected from the >150-µm fraction. These data are replotted in Figure 8.

Although the Paleocene sedimentary record below upper Zone NA2 is missing at Site 750, comparison of the Site 750 and Site 738 $\delta^{13}C$ plots (Figs. 5, 8) reveals a similar interval of overlap in upper Zone NA2 between the bulk carbonate and benthic foraminifer curves. This interval is followed at both sites by a positive shift in bulk carbonate values of more than 1‰ in lower Zone NA3, then relatively little change in values through middle Zone NA3. It is not clear at Site 750 whether or not the bulk carbonate record is decoupled from the Danian planktonic foraminifer $\delta^{13}C$ record, as there are too few measurements of *E. eobulloides* for comparison of long-term trends.

The trans-K/T planktonic foraminifera show little difference between their Cretaceous and Tertiary carbon isotopic compositions. As at Site 738, the mean $\delta^{13}C$ values of these taxa are significantly heavier than the bulk carbonate and Danian planktonic foraminifer values. On this basis, Zachos et al. (1992) concluded that the trans-K/T species recovered from Danian sediments were reworked.

HOLE 690C—MAUD RISE

Stratigraphy

A biostratigraphically complete, but heavily bioturbated, K/T boundary was recovered in Hole 690C, located on Maud Rise (65°9′S, 1°12′E) in the Weddell Sea sector of Antarctica at 2,925-m water depth. Thomas (1990) assigned a lower bathyal water depth to this site for K/T boundary time based on the composition of benthic foraminiferal assemblages. The boundary occurs within Section 113-690C-15X-4 and is marked by a distinct change in color from a white nannofossil ooze that contains isolated blebs of Danian marl to a pale brown marl containing isolated blebs of Maastrichtian chalk (Barker, Kennett et al., 1988). Whole-rock measurements yield percent carbonate values of 80 to 90% within the white chalk and about 44 to 60% within the pale brown marl (Stott and Kennett, 1990b).

Stott and Kennett (1990a) placed the K/T boundary within a 10-cm interval between 35 and 45 cm in Section 113-690C-15X-4, where Tertiary planktonic foraminifera make up more than 50% of the assemblage. Foraminiferal assemblages from the white nannofossil ooze below were assigned to the *Abathomphalus mayaroensis* Zone, and assemblages from above were assigned to Zone APα, which Stott and Kennett (1990a) defined as the partial range of *Eoglobigerina fringa* from the last occurrence (LO) of *A. mayaroensis* to the initial "common"

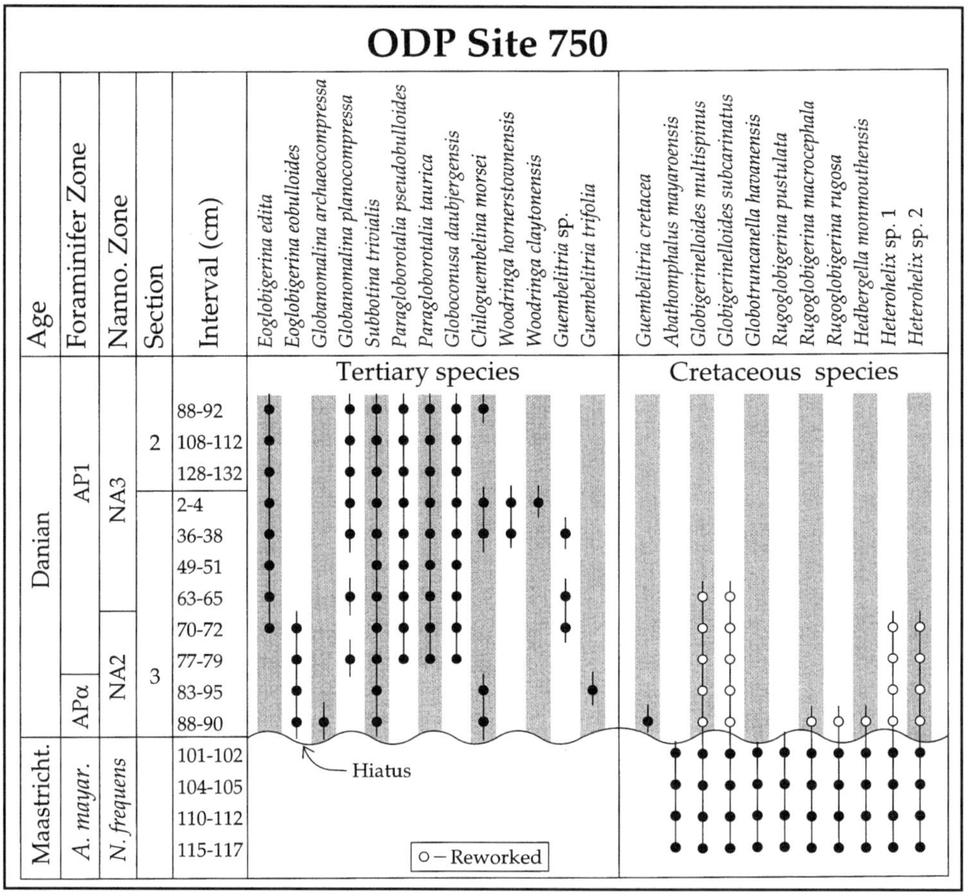

Figure 7. Distribution of Cretaceous and selected Tertiary planktonic foraminifera in Maastrichtian through lower Danian samples from Core 120-750A-15R, modified from Zachos et al. (1992) to include the Antarctic Paleogene biozonation of Huber (1991a) for planktonic foraminifera and Wei and Pospichal (1991) for nannofossils.

occurrence of *P. pseudobulloides*. These authors attributed the absence of *P. eugubina* from the lower Danian at Site 690 to the high-latitude location outside of its biogeographic range.

Pospichal and Wise (1990) more precisely constrained the boundary at 41.5 cm within Section 113-690C-15X-4 (247.815 mbsf), based on the first occurrence of the Tertiary nannofossil *Biantholithus sparsus* in sediments considered as in situ. This level is near the top of a magnetically reversed interval assigned to polarity Subchron C29R (Hamilton, 1990), and it corresponds with an iridium anomaly measuring 1,566 ± 222 parts per trillion (Michel et al., 1990). Nannofossils from the white ooze levels below the K/T boundary were assigned to the *Nephrolithus frequens* Zone, and those from the brown marl above were assigned to Zone CP1a by Pospichal and Wise (1990). These authors did not recognize a hiatus immediately above the boundary, but they did suggest that a hiatus may separate zones CP1a and CP1b between 247.39 and 247.23 mbsf, based on distinct changes in the nannofossil assemblages and a corresponding abrupt shift in the δ13C values reported by Stott and Kennett (1990b).

More recently, Wei and Pospichal (1991) identified their basal Danian Zone NA1 immediately above the boundary level and recognized the NA1/NA2 zonal boundary about 15 cm above the iridium enrichment layer. Lower occurrences of *Hornibrookina*, the zonal marker for the NA1/NA2 boundary, are considered the result of burrowing (Wei and Pospichal, 1991). The most recent interpretation of the nannofossil biostratigraphy suggests that the lower Danian at Site 690 is more condensed than at Site 738 and may contain hiatuses within and between Zones NA1 and NA2 (Pospichal, personal communication, 1994).

Trans-K/T species occurrences

The pattern of species turnover at the K/T boundary in Hole 690C is obscured by the intense bioturbation that has occurred from about 130 cm below to 39 cm above the boundary (Pospichal and Wise, 1990). Relative abundance counts of nannofossils reported in Pospichal and Wise (1990) reveal high proportions of Tertiary and survivor species from dark blebs well within the upper Maastrichtian *N. frequens* Zone and at

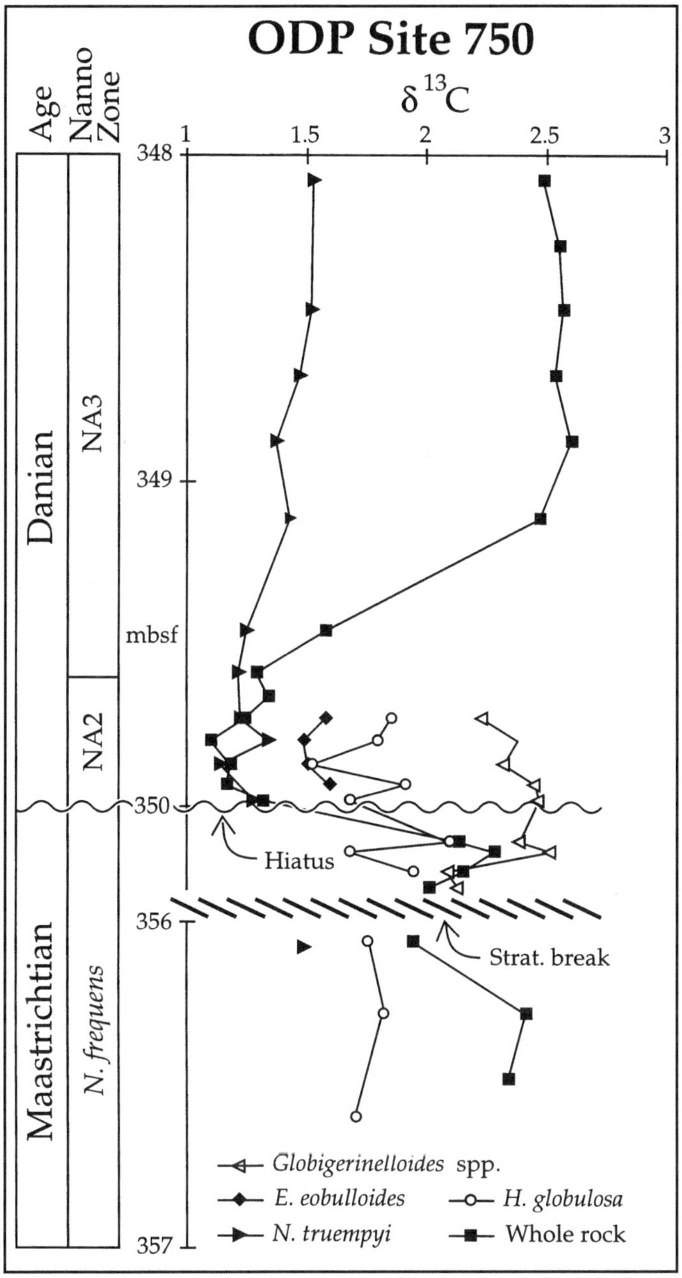

Figure 8. Carbon isotope data for ODP Hole 750A from bulk carbonate, the benthic foraminifer *Nuttalides truempyi*, the trans-K/T taxon *Globigerinelloides* spp., and the Danian species *Eoglobigerina eobulloides*. Modified from Zachos et al. (1992) to include the Antarctic nannofossil biostratigraphy of Wei and Pospichal (1991). All carbon isotope values are expressed relative to the PDB standard.

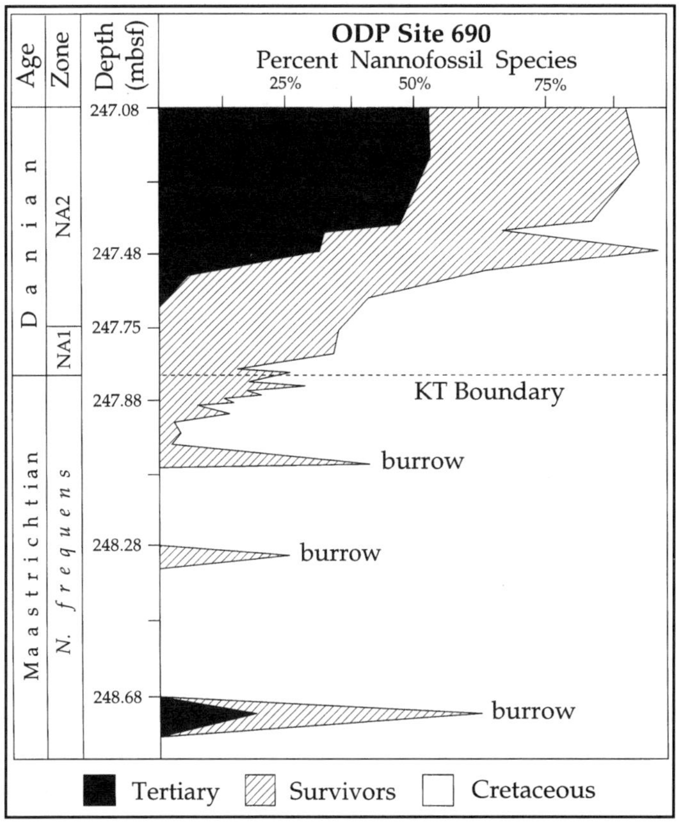

Figure 9. Relative abundance of nannofossils from upper Maastrichtian through lower Danian sediments from ODP Hole 690C modified after Pospichal and Wise (1990). Burrowed intervals were identified by Pospichal and Wise (1990) based on sediment color differences and abrupt incursions of Tertiary and survivor species.

least one sharp increase in Cretaceous species within Zone NA2 (Fig. 9). These authors reported the presence of Cretaceous species as high as 5 m above the boundary level.

Similarly, the planktonic foraminiferal turnover is spread across several tens of centimeters as the result of bioturbation. There is no way to determine, based on distributional analysis, whether some of the trans-K/T species from within the bioturbated interval were survivors of the terminal Cretaceous extinction event. But occurrences of *G. subcarinatus* and *G. multispinus* at sample levels that are 10 m and as high as 32 m above the K/T (Fig. 10) are clearly the result of sediment redeposition.

Carbon isotope results

The stable isotopic stratigraphy of the K/T transition at Site 690 was reported by Stott and Kennett (1990b). Material analyzed includes fine-fraction carbonate, the benthic species *Gavelinella beccariiformis*, the Danian species *Eoglobigerina fringa* and *P. pseudobulloides,* and the trans-K/T species *H. globulosa* and *Globigerinelloides multispinus*. Results of those analyses are plotted in Figure 11.

The $\delta^{13}C$ values of fine-fraction carbonate are similar to the co-occurring *Globigerinelloides multispinus* values and are consistently heavier than the *H. globulosa* and benthic foraminifer values throughout the upper Maastrichtian section. As occurs at sites 738 and 750, the fine-fraction carbonate then becomes very negative in the lower Danian, with the

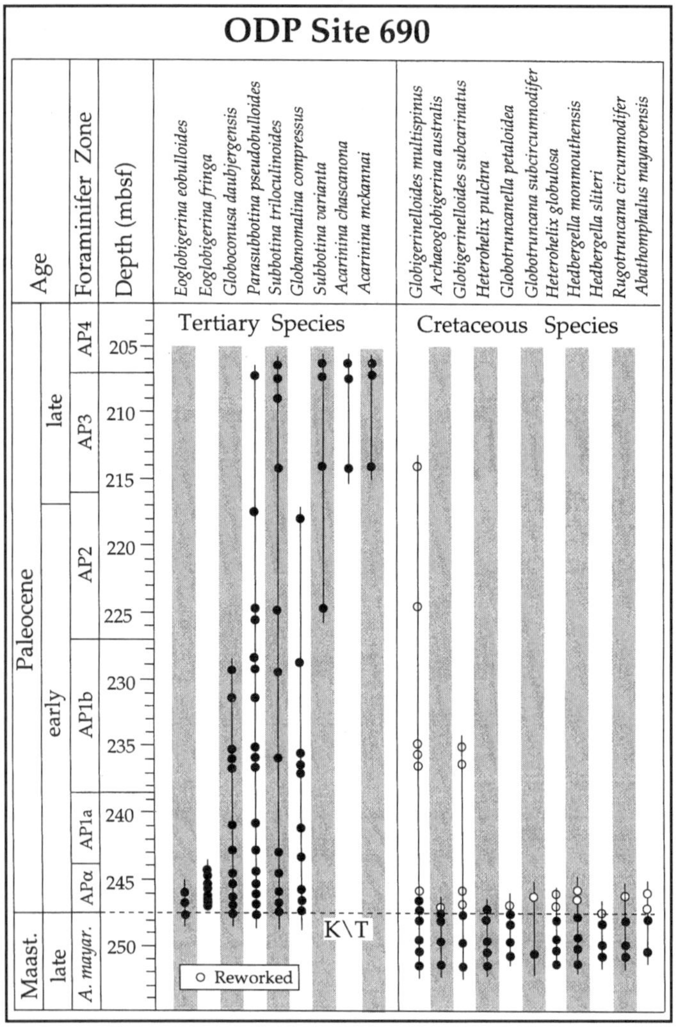

Figure 10. Cretaceous and selected Tertiary planktonic foraminifer distributions at ODP Site 690, based on the biostratigraphy of Stott and Kennett (1990a).

lowest value at 247.24 mbsf in lowermost Zone NA3, where the benthic foraminfer and fine-fraction carbonate values overlap. The fine-fraction $\delta^{13}C$ then returns to more positive values in lower to middle Zone NA3.

Lowermost Danian values of *E. fringa* are similar to the upper Maastrichtian $^{13}C/^{12}C$ ratios of *H. globulosa* but are heavier by about 0.5‰ than the lowermost value of *P. pseudobulloides* from 0.45 m higher in the core. Since the isotopic records of the latter two species do not overlap, the significance of this offset is uncertain. But the 0.64‰ difference between the $\delta^{13}C$ composition of *P. pseudobulloides* and the fine-fraction carbonate and their contrasting isotopic trends in lowermost Zone NA3 suggest that, like Site 738, the Danian planktonic $\delta^{13}C$ record is decoupled from the fine-fraction carbonate record.

The Danian $\delta^{13}C$ record of *G. multispinus* does not overlap sufficiently with co-occurring *in situ* planktonic species for

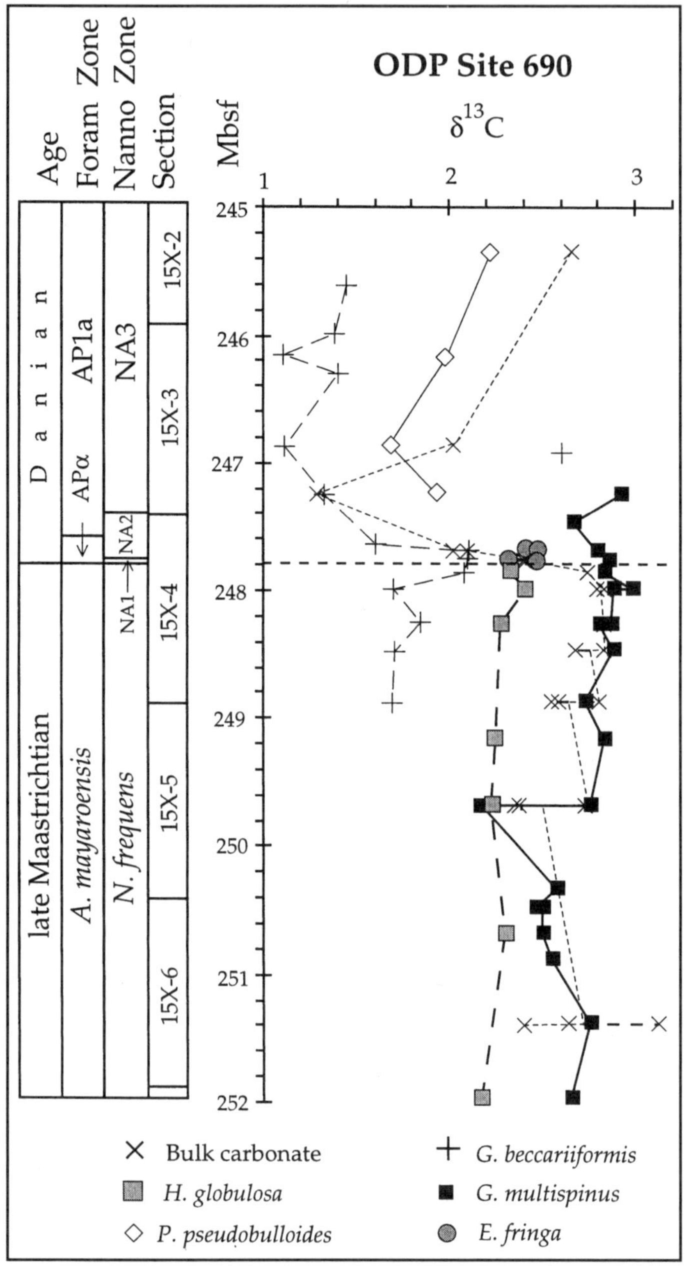

Figure 11. Carbon isotope data for ODP Hole 690C from bulk carbonate, the benthic foraminifer *Gavelinella beccariiformis*, the trans-K/T species *Heterohelix (H.) globulosa* and *Globigerinelloides (G.) multispinus*, and the Danian species *Eoglobigerina (E.) fringa* and *Parasubbotina (P.) pseudobulloides*. Modified from Stott and Kennett (1990b) to include the Antarctic Paleogene foraminifer biozonation of Huber (1991a) and the Antarctic nannofossil biostratigraphy of Wei and Pospichal (1991). All carbon isotope values are relative to the PDB standard.

longer-term comparison of their isotopic trends. Nonetheless, the carbon isotopic composition of *G. multispinus* from above and below the K/T boundary is nearly identical, and lowermost Danian values are about 0.4‰ heavier than *E. fringa* from the same stratigraphic interval. It is therefore possible that the lower

Danian occurrences of *G. multispinus* are due to reworking. As no samples of *H. globulosa* were measured from above the K/T boundary level at Site 690, the cause for post-K/T occurrences of this species remains an open question at this site.

NYE KLØV, DENMARK

The K/T boundary exposed at Nye Kløv, Denmark (~58°N), occurs in a thin clay-rich marl that separates a 6-m section of light-colored Maastrichtian chalk from a 6.5-m section of dark gray, clay-rich marl of Danian age that contains several layers of flint nodules (Keller et al., 1993). An iridium anomaly of 1.6 ppb has been identified within the boundary clay (Nazarov et al., 1983; Hansen et al., 1986). Keller et al. (1993) reported that with the exception of a short hiatus at the base of Zone P0, the K/T transition is biostratigraphically continuous. Although biomarkers for the uppermost Maastrichtian are absent from the section, these authors assumed that the Cretaceous record is complete, based on comparison with the stratigraphic record at Stevns Klint, Denmark.

No detailed calcareous nannofossil distributions have been reported for this sequence.

Trans-K/T species occurrences

Several Cretaceous planktonic foraminifer species, including *Guembelitria cretacea*, *Heterohelix* spp., *Globigerinella* sp., *Hedbergella* sp., and *Rugoglobigerina* sp., were reported by Bang (1979) as reworked within the Danian section at Nye Kløv. However, no criteria for identifying these specimens as reworked were specified. Trans-K/T species identified within the same section by Keller et al. (1993) include *Guembelitria* spp., *Heterohelix complanata*, *Globigerinelloides aspera*, *H. globulosa.*, and *Hedbergella* spp. *Guembelitria* spp. and *H. globulosa* were characterized by these authors as common to abundant within 1 m above the boundary, with last occurrences at about 5.2 and 3.9 m above the K/T respectively. Similarity in the extinction patterns at Nye Kløv and Brazos River, Texas, and a difference in $\delta^{13}C$ values of *H. globulosa* above and below the K/T boundary were cited by Keller et al. (1993) as evidence that the trans-K/T species identified at Nye Kløv were survivors.

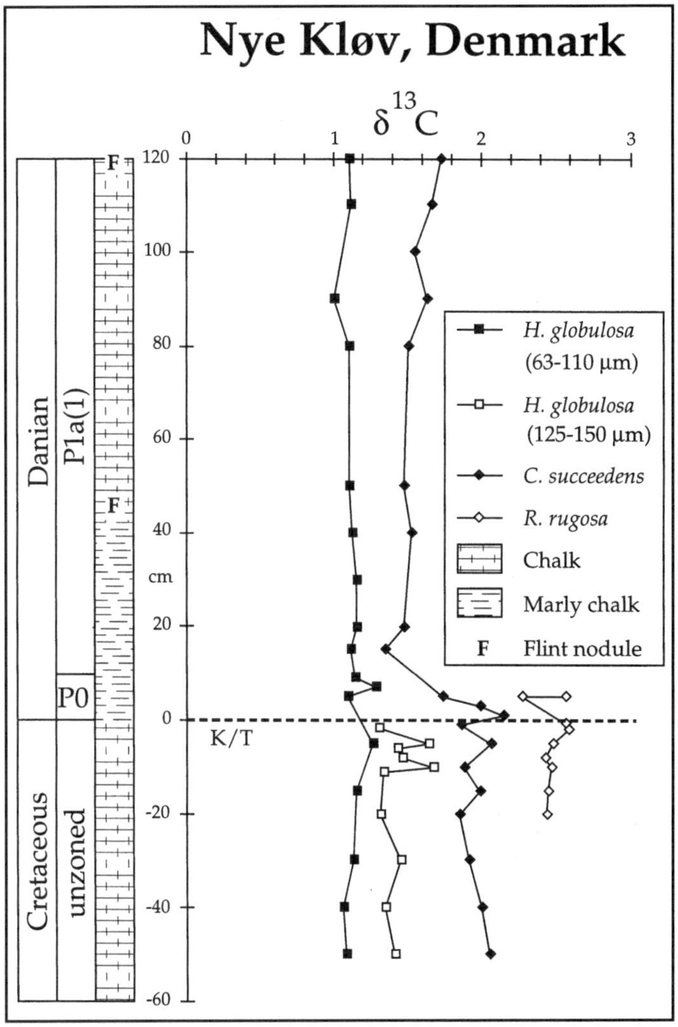

Figure 12. Carbon isotopic values across the K/T boundary at Nye Kløv, Denmark, of the benthic foraminifer *Cibicides (C.) succeedens* and the planktonic foraminifera *Heterohelix (H.) globulosa* and *Rugoglobigerina (R.) rugosa.* Data published in Keller et al. (1993) and Barrera and Keller (1994). Note that there is no significant change in the $\delta^{13}C$ values of *H. globulosa* from the small-size fraction (63 to 110 μm) across the K/T boundary. There is an apparent shift in values when comparing larger-sized Maastrichtian specimens of this species with small-sized specimens from Danian sediments, an approach used by Keller et al. (1993). All carbon isotope values are expressed relative to the PDB standard.

Carbon isotope results

Results of $\delta^{13}C$ measurements of small (63 to 110 μm) *H. globulosa* from below the K/T at Nye Kløv (Fig. 12) and comparison with the isotopic values of small specimens from above the boundary reveal that there is no significant change in $\delta^{13}C$ values (Barrera and Keller, 1994). Measurement of the larger (125 to 150 μm) *H. globulosa* specimens from Cretaceous sediments revealed ^{13}C values that are consistently more positive by about 0.3‰. A number of authors have demonstrated that size-related disequilibrium fractionation of ^{13}C can be significant within planktonic foraminifer species, with an increase of as much as 2.0‰ over a 500-μm range in test diameter in modern species (Berger et al., 1978) and an increase of as much as 1.1‰ over a 130-μm range in lower Danian species (D'Hondt and Zachos, 1993). These results demonstrate that the same size fraction must be used when comparing the $\delta^{13}C$ values of a species from below and above the K/T boundary.

The absence of a $\delta^{13}C$ shift across the K/T boundary at Nye Kløv for *H. globulosa* cannot be used as unequivocal evidence for survivorship or reworking. The extinction pattern in that section may be further elucidated by analyzing the $\delta^{13}C$

composition of other trans-K/T species from below and above the boundary together with at least one co-occurring Danian species.

CONCLUSIONS

Distributional analyses of planktonic foraminifera and nannofossil faunas and determination of the carbon isotope composition of trans-K/T boundary species at ODP sites 738, 750, and 690 in the region of the Southern Oceans suggest that the high occurrences of Cretaceous species in lower Paleocene sediments is likely the result of extensive reworking in these sections.

Evidence for reworking in Hole 738C, which has yielded the most complete K/T boundary sequence drilled in the deep sea, includes (1) Cretaceous nannofossils and foraminifer species occur consistently throughout the Danian section, with some species occurrences documented in lower Eocene sediments, more than 100 m above the K/T boundary (Wei and Pospichal, 1991; Huber, 1991b); (2) *Inoceramus* prisms and the planktonic foraminifer *Globigerinelloides impensus*, which are known to have become extinct in the mid Maastrichtian and uppermost Campanian, respectively, occur in Danian sediments; (3) reworked *Inoceramus* prisms within upper Maastrichtian and lower Danian sediments have strontium isotope ratios expected for late Campanian biogenic calcite, whereas samples of co-occurring planktonic foraminifera near their evolutionary first appearance have strontium isotope ratios appropriate for their stratigraphic position; and (4) carbon isotope ratios of the putative survivor *Globigerinelloides multispinus* show very little difference above and below the boundary, although values of co-occurring Danian planktonic species are consistently lighter and have a different trend.

At Site 690, specimens of Cretaceous taxa are found as high as 10 to 32 m above the boundary in upper Paleocene sediments. Moreover, carbon isotope analyses of *Globigerinelloides* from Sites 690 and 750 (Stott and Kennett, 1990; Zachos et al., 1992) also show no change across the K/T boundary and yield consistently higher values than co-occurring in situ Danian species. More conclusive evidence bearing on survivorship vs. reworking at these sites will require additional carbon isotope analyses of trans-K/T and in situ Danian species to determine whether or not their $\delta^{13}C$ stratigraphy plots independently.

The only trans-K/T species considered ancestral to Cenozoic planktonic foraminifer lineages are *Guembelitria cretacea*, *Zeauvigerina waiparaensis*, *Hedbergella holmdelensis*, and *H. monmouthensis* (Olsson et al., 1992; Liu and Olsson, 1992, 1994; Huber and Boersma, 1994). The possibility remains that some additional Cretaceous species did survive the terminal Cretaceous extinction and grew in the earliest Danian ocean (e.g., *H. globulosa* at Brazos River, Texas; Barrera and Keller, 1990a, b). However, those species are considered inconsequential to the evolutionary radiation that followed the mass extinction at the K/T boundary. We conclude that the magnitude of survivorship after the K/T boundary event was probably minimal and that the K/T boundary event was the cause of the termination of most Cretaceous planktonic foraminifer lineages at this time.

ACKNOWLEDGMENTS

I thank J. Pospichal for the use of unpublished Site 738 nannofossil data, K. C. Lohmann and J. O'Neil for making available the stable isotope facilities at the University of Michigan, E. Barrera for analyzing the samples, J. Zachos and B. Sliter for reviewing the manuscript, and E. Barrera, S. Gartner, and K. MacLeod for their editorial suggestions. Funds for the isotopic analyses were provided by NSF grants HRD91-03312 and EAR-9104784 to E. Barrera.

REFERENCES CITED

Alvarez, L. W., Alvarez, W., Asaro, F., and Michel, H. V., 1980, Extraterrestrial cause for the Cretaceous-Tertiary extinction: Science, v. 208, p. 1095–1108.

Bang, I., 1979, Foraminifera from the type sections of the *eugubina* Zone compared with those from Cretaceous/Tertiary boundary localities in Jylland, Denmark, *in* Birkelund, T., and Bromley, R. G., eds., Cretaceous-Tertiary Boundary Events. I: The Maastrichtian and Danian of Denmark: Copenhagen, University of Copenhagen, p. 127–130.

Barker, P. F., Kennett, J. P., and others, 1988, Proceedings, Ocean Drilling Program, Initial Reports, Volume 113: Ocean Drilling Program, College Station, Texas, v. 113, p. 785 p.

Barrera, E., and Huber, B. T., 1991, Paleogene and early Neogene oceanography of the southern Indian Ocean: Leg 119 foraminifer stable isotope results, *in* Barron, J. A., Larsen, B. L., and others, Proceedings, Ocean Drilling Program, Scientific Results, Volume 119: College Station, Texas, Ocean Drilling Program, p. 693–717.

Barrera, E., and Keller, G., 1990a, Foraminiferal stable isotope evidence for a gradual decrease of marine productivity and Cretaceous species survivorship in the earliest Danian: Paleoceanography, v. 5, p. 867–890.

Barrera, E., and Keller, G., 1990b, Foraminiferal and stable isotope evidence for gradual environmental changes across the K/T boundary near Brazos River, Texas: Geological Society of America Abstracts with Programs, v. 22, p. A279.

Barrera, E., and Keller, G., 1994, Productivity across the Cretaceous-Tertiary boundary in high latitudes: Geological Society of America Bulletin, v. 106, p. 1254–1266.

Barron, J. A., Larsen, B. L., and others, 1989, Proceedings, Ocean Drilling Program, Initial Reports, Volume 119: College Station, Texas, Ocean Drilling Program, p. 942 p.

Berger, W. H., Killingly, J. S., and Vincent, E., 1978, Stable isotopes in deep sea carbonates: Box Core ERDC-92, west equatorial Pacific: Oceanologica Acta, v. 1, p. 203–216.

Berggren, W. A., 1962, Some planktonic foraminifera from the Maestrichtian and type Danian stages of southern Scandinavia: Stockholm Contributions to Geology, v. 9, p. 1–106.

Berggren, W. A., 1992, Paleogene planktonic foraminifer magnetobiostratigraphy of the southern Kerguelen Plateau (sites 747-749), *in* Wise, S. W., Jr., Schlich, R., and others, Proceedings, Ocean Drilling Program, Scientific Results, Volume 120: College Station, Texas, Ocean Drilling Program, p. 551–568.

Berggren, W. A., and Miller, K. G., 1988, Paleogene tropical planktonic foraminiferal biostratigraphy and magnetobiochronology: Micropaleontology, v. 34, p. 362–380.

Boersma, A., and Shackleton, N. J., 1981, Oxygen and carbon isotope variations and planktonic foraminiferal depth habitats: Late Cretaceous to Paleocene, Central Pacific, DSDP Sites 463 and 465, Leg 65, *in* Thiede, J., Vallier, T. L., and others, Initial reports of the Deep Sea Drilling Project, Volume 65: Washington, D.C., U.S. Government Printing Office, p. 513–526.

Brinkhuis, H., and Zachariasse, W. J., 1988, Dinoflagellate cysts, sea level changes and planktonic foraminifers across the Cretaceous-Tertiary Boundary at El Haria, Northwest Tunisia: Marine Micropaleontology, v. 13, p. 153–191.

Canudo, J. I., Keller, G., and Molina, E., 1991, Cretaceous/Tertiary boundary extinction pattern and faunal turnover at Agost and Caravaca, S.E. Spain: Marine Micropaleontology, v. 17, p. 319–341.

D'Hondt, S., and Keller, G., 1991, Some patterns of planktic foraminiferal assemblage turnover at the Cretaceous-Tertiary boundary: Marine Micropaleontology, v. 17, p. 77–118.

D'Hondt, S., and Zachos, J. C., 1993, On stable isotopic variation and earliest Paleocene planktonic foraminifera: Paleoceanography, v. 8, p. 527–547.

Ehrendorfer, T., and Aubry, M. P., 1992, Calcareous nannoplankton changes across the Cretaceous/Paleocene boundary in the southern Indian Ocean, *in* Wise, S. W., Jr., Schlich, R., and others, Proceedings, Ocean Drilling Program, Scientific Results, Volume 120: College Station, Texas, Ocean Drilling Program, p. 451–470.

Ehrmann, W. U., 1991, Implications of sediment composition of the southern Kerguelen Plateau for paleoclimate and depositional environment, *in* Barron, J. A., Larsen, B. L., and others, Proceedings, Ocean Drilling Program, Scientific Results, Volume 119: College Station, Texas, Ocean Drilling Program, p. 185–210.

Hamilton, N., 1990, Mesozoic magnetostratigraphy of Maud Rise, Antarctica, *in* Barker, P. F., Kennett, J. P., and others, Proceedings, Ocean Drilling Program, Scientific Results, Volume 113: College Station, Texas, Ocean Drilling Program, p. 255–260.

Hansen, H. J., Gwozdz, R., Hansen, J. M., Poromley, R. G., and Rasmussen, K. L., 1986, The diachronous C/T plankton extinction in the Danish Basin, *in* Walliser, O., ed., Global bioevents: Amsterdam, Springer-Verlag, p. 381–384.

Hsü, K. J., McKenzie, J. A., and He, Q. X., 1982, Terminal Cretaceous environmental and evolutionary changes, *in* Silver, L. T., and Schultz, P. H., eds., Geological implications of impacts of large asteroids and comets on the Earth: Geological Society of America Special Paper 190, p. 329–352.

Huber, B. T., 1991a, Maestrichtian planktonic foraminifer biostratigraphy and the Cretaceous/Tertiary boundary at ODP Hole 738C (Kerguelen Plateau, southern Indian Ocean), *in* Barron, J. A., Larsen, B. L., and others, Proceedings, Ocean Drilling Program, Scientific Results, Volume 119: College Station, Texas, Ocean Drilling Program, p. 451–465.

Huber, B. T., 1991b, Paleogene and early Neogene planktonic foraminifer biostratigraphy of ODP Leg 119 Sites 738 and 744, Kerguelen Plateau (southern Indian Ocean), *in* Barron, J. A., Larsen, B. L., and others, Proceedings, Ocean Drilling Program, Scientific Results, Volume 119: College Station, Texas, Ocean Drilling Program, p. 427–449.

Huber, B. T., 1992, Upper Cretaceous planktonic foraminiferal biozonation for the Austral Realm: Marine Micropaleontology, v. 20, p. 107–128.

Huber, B. T., and Boersma, A., 1994, Cretaceous origination of *Zeauvigerina* and its relationship to Paleocene biserial planktonic foraminifera: Journal of Foraminiferal Research, v. 24, p. 268–287.

Huber, B. T., Liu, C., Olsson, R. K., and Berggren, W. A., 1994, Comment on "The Cretaceous-Tertiary boundary transition in the Antarctic Ocean and its global implications" by Gerta Keller: Marine Micropaleontology, v. 24, p. 91–99.

Keller, G., 1988, Extinction, survivorship and evolution of planktic foraminifers across the Cretaceous/Tertiary Boundary at El Kef, Tunisia: Marine Micropaleontology, v. 13, p. 239–263.

Keller, G., 1989a, Extended Cretaceous-Tertiary boundary extinctions and delayed population change in planktonic foraminiferal faunas from Brazos River, Texas: Paleoceanography, v. 4, p. 287–332.

Keller, G., 1989b, Extended period of extinctions across the Cretaceous/Tertiary boundary in planktonic foraminifera of continental-shelf sections: Implications for impact and volcanism theories: Geological Society of America Bulletin, v. 101, p. 1408–1419.

Keller, G., 1993, The Cretaceous-Tertiary boundary transition in the Antarctic Ocean and its global implications: Marine Micropaleontology, v. 21, p. 1–45.

Keller, G., Barrera, E., Schmitz, B., and Mattson, E., 1993, Gradual mass extinction, species survivorship, and long-term environmental changes across the Cretaceous-Tertiary boundary in high latitudes: Geological Society of America Bulletin, v. 105, p. 979–997.

Liu, C., and Olsson, R. K., 1992, Evolutionary radiation of microperforate planktonic foraminifera following the K/T mass extinction event: Journal of Foraminiferal Research, v. 22, p. 328-346.

Liu, C., and Olsson, R. K., 1994, On the origin of Danian normal perforate planktonic foraminifera from *Hedbergella*: Journal of Foraminiferal Research, v. 24, p. 61–74.

Luterbacher, H. P., and Premoli Silva, I., 1964, Biostratigrafia del limite Cretaceo-Terziario nell' Appennino centrale: Revista Italiana di Paleontologia e Stratigrafia, v. 70, p. 67-128.

MacLeod, K. G., 1994, Bioturbation, inoceramid extinction, and mid-Maastrichtian ecological change: Geology, v. 22, p. 139–142.

MacLeod, K. G., and Huber, B. T., 1996, Strontium isotopic evidence for significant reworking in sediments spanning the Cretaceous/Tertiary boundary at ODP Site 738: Geology, v. 24, p. 463–466.

MacLeod, N., and Keller, G., 1991, Hiatus distributions and mass extinctions at the Cretaceous/Tertiary boundary: Geology, v. 19, p. 497–501.

MacLeod, N., and Keller, G., 1994, Comparative biogeographic analysis of planktic foraminiferal survivorship across the Cretaceous/Tertiary (K/T) boundary: Paleobiology, v. 20, p. 143–177.

Martini, E., 1971, Standard Tertiary and Quaternary calcareous nannoplankton zonation, *in* Farinacci, A., ed., Proceedings, Second Planktonic Conference, Rome: Rome, Edizioni Technoscienza, v. 2, p. 739–785.

Michel, H. V., Asaro, F., Alvarez, W., and Alvarez, L., 1990, Geochemical studies of the Creaceous/Tertiary boundary in ODP holes 689B and 690C, *in* Barker, P. F., Kennett, J. P., and others, Proceedings, Ocean Drilling Program, Scientific Results, Volume 113: College Station, Texas, Ocean Drilling Program, p. 159–168.

Nazarov, M. A., Barsukov, L. D., Kolesov, G. M., Naydin, D. P., and Alekseyev, A. S., 1983, Origin of the iridium anomaly at the boundary between the Maastrichtian and the Danian stages: Geochemistry, v. 20, p. 142–159.

Olsson, R. K., 1960, Foraminifera of latest Cretaceous and earliest Tertiary age in the New Jersey Coastal Plain: Journal of Paleontology, v. 34, p. 1–58.

Olsson, R. K., Hemleben, C., Berggren, W. A., and Liu, C., 1992, Wall texture classification of planktonic foraminifera genera in the lower Danian: Journal of Foraminiferal Research, v. 22, p. 193–213.

Olsson, R. K., and Liu, C., 1993, Controversies on the placement of Cretaceous-Paleogene boundary and the K/P mass extinction of planktonic foraminifera: Palaios, v. 8, p. 127–139.

Perch-Nielsen, K., McKenzie, J., and He, Q., 1982, Biostratigraphy and isotope stratigraphy and the catastrophic extinction of calcareous nannoplankton at the Cretaceous/Tertiary boundary, *in* Silver, L. T., and Schultz, P. H., eds., Geological implications of impacts of large asteroids and comets on the Earth: Geological Society of America Special Paper 190, p. 353–371.

Pospichal, J. J., 1996, High austral latitude Cretaceous/Tertiary boundary calcareous nannofossils from ODP Site 738, *in* Whatley, R., and Moguilevsky, A., eds., Microfossils and oceanic environments: Aberystwyth, Wales, Aberystwyth Press (in press).

Pospichal, J. J., and Huber, B. T., 1992, The Cretaceous/Tertiary boundary in the southern Indian Ocean: results from the coring operations of the Ocean Drilling Program, *in* Kidd, R. B., and Rea, D. K., eds., Synthesis of results from scientific drilling in the Indian Ocean: Washington, D.C., American Geophysical Union, Geophysical Monograph 70,

p. 275–294.

Pospichal, J. J., and Wise, S. W. Jr., 1990, Calcareous nannofossils across the K/T boundary, ODP Hole 690C, Maud Rise, Weddell Sea, *in* Barker, P. F., Kennett, J. P., and others, Proceedings, Ocean Drilling Program, Scientific Results, Volume 113: College Station, Texas, Ocean Drilling Program, p. 515–532.

Quilty, P. G., 1992a, Upper Cretaceous planktonic foraminifera and biostratigraphy, ODP Leg 120, southern Kerguelen Plateau, *in* Wise, S. W., Jr., Schlich, R., and others, Proceedings, Ocean Drilling Program, Scientific Results, Volume 120: College Station, Texas, Ocean Drilling Program, p. 371–392.

Quilty, P. G., 1992b, Upper Cretaceous benthic foraminifers and paleoenvironments, southern Kerguelen Plateau, Indian Ocean, *in* Wise, S. W., Jr., Schlich, R., and others, Proceedings, Ocean Drilling Program, Scientific Results, Volume 120: College Station, Texas, Ocean Drilling Program, p. 393–443.

Schlich, R., Wise, S. W., Jr., and others, 1989, Proceedings, Ocean Drilling Program, Initial Reports, Volume 120: College Station, Texas, Ocean Drilling Program, 648 p.

Schmitz, B., Asaro, F., Michel, H. V., Thierstein, H. R., and Huber, B. T., 1991, Element stratigraphy across the Cretaceous/ Tertiary boundary in Hole 738C, *in* Barron, J. B., Larsen, B., and others, Proceedings of the Ocean Drilling Program, Scientific Results, Volume 119: College Station, Texas, Ocean Drilling Program, p. 719–730.

Signor, P. W., and Lipps, J. H., 1982, Sampling bias, gradual extinction patterns and catastrophes in the fossil record, *in* Silver, L. T., and Schultz, P. H., eds., Geological implications of impacts of large asteroids and comets on the Earth: Geological Society of America Special Paper 190, p. 291–296.

Smit, J., 1982, Extinction and Evolution of planktonic foraminifera after a major impact at the Cretaceous/Tertiary boundary, *in* Silver, L. T., and Schultz, P. H., eds., Geological implications of impacts of large asteroids and comets on the Earth: Geological Society of America Special Paper 190, p. 329–352.

Smit, J., and Romein, A. J. T., 1985, A sequence of events across the Cretaceous-Tertiary boundary: Earth and Planetary Science Letters, v. 74, p. 155–170.

Stott, L. D., and Kennett, J. P., 1990a, Antarctic Paleogene planktonic foraminifer biostratigraphy: ODP Leg 113 sites 689 and 690, *in* Barker, P. F., Kennett, J. P., and others, Proceedings, Ocean Drilling Program, Scientific Results, Volume 113: College Station, Texas, Ocean Drilling Program, p. 549–569.

Stott, L. D., and Kennett, J. P., 1990b, The paleoceanographic and paleoclimatic signature of the Cretaceous/Paleogene boundary in the Antarctic: stable isotopic results from ODP Leg 113, *in* Barker, P. F., Kennett, J. P., and others, Proceedings, Ocean Drilling Program, Scientific Results, Volume 113: College Station, Texas, Ocean Drilling Program, p. 829–848.

Thierstein, H. R., Asaro, F., Ehrmann, W. U., Huber, B. T., Michel, H., Sakai, H., and Schmitz, B., 1991, The Cretaceous/Tertiary boundary at Site 738, southern Kerguelen Plateau, *in* Barron, J. A., Larsen, B., and others, Proceedings of the Ocean Drilling Program, Scientific Results, Volume 119: College Station, Texas, Ocean Drilling Program, p. 849–867.

Thomas, E., 1990, Late Cretaceous through Neogene deep-sea benthic foraminifers (Maud Rise, Weddell Sea, Antarctica), *in* Barker, P. F., Kennett, J. P., and others, Proceedings, Ocean Drilling Program, Scientific Results, Volume 113: College Station, Texas, Ocean Drilling Program, p. 571–594.

Wei, W., and Pospichal, J., 1991, Danian calcareous nannofossil succession at ODP Site 738 in the southern Indian Ocean, *in* Barron, J. A., Larsen, B. L., and others, Proceedings, Ocean Drilling Program, Scientific Results, Volume 119: College Station, Texas, Ocean Drilling Program, p. 495–512.

Wei, W., and Thierstein, H. R., 1991, Upper Cretaceous and Cenozoic calcareous nannofossils of the Kerguelen Plateau (southern Indian Ocean) and Prydz Bay (East Antarctica), *in* Barron, J. A., Larsen, B. L., and others, Proceedings, Ocean Drilling Program, Scientific Results, Volume 119: College Station, Texas, Ocean Drilling Program, p. 467–494.

Zachos, J. C., and Arthur, M. A., 1986, Paleoceanography of the Cretaceous/Tertiary boundary event: Inferences from stable isotopic and other data: Paleoceanography, v. 1, p. 5–26.

Zachos, J. C., Arthur, M. A., and Dean, W. E., 1989, Geochemical evidence for suppression of pelagic marine productivity at the Cretaceous/Tertiary boundary: Nature, v. 86, p. 513–532.

Zachos, J. C., Berggren, W. A., Aubry, M. P., and Mackensen, A., 1992, Chemostratigraphy of the Cretaceous/Paleocene boundary at Site 750, southern Kerguelen Plateau, *in* Wise, S. W., Jr., Schlich, R., and others, Proceedings, Ocean Drilling Program, Scientific Results, Volume 120: College Station, Texas, Ocean Drilling Program, p. 961–977.

MANUSCRIPT ACCEPTED BY THE SOCIETY SEPTEMBER 14, 1995

Printed in U.S.A.

Geological Society of America
Special Paper 307
1996

Calcareous nannoplankton mass extinction at the Cretaceous/Tertiary boundary: An update

James J. Pospichal
Department of Geology, Florida State University, Tallahassee, Florida 32306

ABSTRACT

Quantitative analyses of calcareous nannofossil assemblages across Cretaceous/ Tertiary (K/T) boundary sections from the Tethys region, the Indian Ocean and the South Atlantic Ocean reveal that nannoplankton extinctions were abrupt and synchronous at the K/T boundary. Extinctions correlate with previously documented shifts in $\delta^{13}C$, abrupt decline in carbonate sedimentation, and anomalous iridium abundance. Nannoplankton extinctions appear to have been as severe in the high austral latitudes as in the tropics.

There is no unequivocal evidence for survivorship into the earliest Danian of numerically important Cretaceous species, those traditionally considered to have become extinct at the K/T boundary. Overall abundance patterns, associations with sea-level lowstands and occasionally iridium reworked from the K/T boundary suggest that Cretaceous specimens in Danian sediments are redeposited by processes of bioturbation and/or current activity.

Although a notable turnover of nannoplankton occurs in the early to mid-Maastrichtian, nannofossil assemblages did not change significantly during the last million years or so of the Maastrichtian, which indicates that stable conditions immediately preceded the K/T boundary event. During the latest Maastrichtian the austral latitudes were characterized by a distinct latitudinal thermal gradient, as indicated by the strong abundance decrease of *Prediscosphaera stoveri* toward the equator. A similar distribution of *Hornibrookina* in the early Danian indicates the persistence of this thermal gradient into the earliest Tertiary.

The lowermost Danian is characterized by sequential abundance peaks of surviving and new species. The individual species that participate in this sequential turnover vary significantly from site to site. This variation may reflect unstable conditions around the globe following the K/T boundary crisis.

INTRODUCTION

A fundamental problem for micropaleontologists in regard to the Cretaceous/Tertiary (K/T) boundary event is to correctly interpret the pattern of extinction across the boundary interval itself. Over the past few decades, K/T faunal and floral analyses have been presented that suggest several basic patterns. These include sequential or stepwise extinctions across the boundary, gradual extinctions occurring above and below the boundary, and abrupt or one mass extinction. The differences are illustrated by Smit's (1982) and Keller's (1988) accounts of the pattern of planktonic foraminifer extinctions across the K/T boundary at El Kef, Tunisia. Smit reported that nearly all Cretaceous species abruptly became extinct at the K/T boundary, with one lone survivor, *Guembelitria cretacea*. Keller concluded that extinctions occurred sequentially, in increments beginning below the boundary, with at least 30% of the Cretaceous species surviving into the Paleocene. The conflicting interpretations may result from

Pospichal, J. J., 1996, Calcareous nannoplankton mass extinction at the Cretaceous/Tertiary boundary: An update, *in* Ryder, G., Fastovsky, D., and Gartner, S., eds., The Cretaceous-Tertiary Event and Other Catastrophes in Earth History: Boulder, Colorado, Geological Society of America Special Paper 307.

different methodology or discrepancies in taxonomic concepts (see Olsson and Liu, 1993).

Central to this problem is how to determine which species became extinct at the K/T boundary and which ones survived. This problem exists because, in nearly all marine K/T sections, Cretaceous foraminifers and calcareous nannofossils, with no obvious change in abundance or preservation, are present in basal Paleocene sediments. Their presence there has been attributed either to reworking or to survival beyond the K/T event, but unequivocal evidence for either interpretation has been elusive. Several investigators have attempted to use independent means, such as stable isotope analysis, to solve this problem (Perch-Nielsen et al., 1982; Barrera and Keller, 1990). Ideally this should work best for planktonic foraminifers because individual specimens can be analyzed. However, isotopic evidence for survivorship of the planktonic foraminifer species *Heterohelix globulosa*, as presented by Barrera and Keller (1990), for example, has not been borne out by further analyses (Zachos et al., 1992; Huber, this volume). Bulk sediment analyses that purportedly measure predominantly the nannofossils in the fine fraction have also suggested survivorship of Cretaceous species (Perch-Nielsen et al., 1982). This too has been questioned in light of the potential problems of diagenesis or with the uncertain origin of the microcarbonate measured in fine fraction or whole-rock analyses (e.g., Jiang and Gartner, 1986).

Another disagreement among paleontologists is whether or not biotic diversity was on the decline well before the K/T event, with the latter being the final stroke (see Alvarez et al., 1984). This argument persists among investigators ranging from students of dinosaurs (Sheehan et al., 1991, versus Sloan et al., 1986) to ammonite specialists (see Ward, 1990). Sampling biases can explain some of these results (Signor and Lipps, 1982), but with marine microfossils, distortions should be less apparent.

A third controversy among paleontologists is the disagreement on the magnitude of the extinctions between low and high latitudes. For macrofossils, some investigators have reported that the effects of the K/T event were global, reaching environments at all latitudes (Raup and Jablonski, 1993). Others have suggested that the bulk of the extinctions occurred in the tropics and that the higher-latitude regions were much less affected (e.g., Zinsmeister et al., 1989; Elliot et al., 1994). The latter interpretation has been suggested for planktonic foraminifers by Keller (1993) and Keller et al. (1993) but has not been demonstrated for the calcareous nannoplankton.

Accurate identification of the extinction pattern is critical to the hypotheses and models of the K/T kill mechanism(s). A sequential or gradual pattern may result from long-term changes in climate and/or sea level (Clemens et al., 1981; Hallam, 1987) or from an extended period of intense volcanism (McLean, 1982, 1985; Officer and Drake, 1983; Officer et al., 1987), whereas an abrupt synchronous pattern of mass extinction may indicate a catastrophic event such as impact of a large asteroid (Alvarez et al., 1980).

For the most part, calcareous nannofossil specialists agree that a geologically abrupt extinction took place among this microfossil group and, particularly, that nothing in the record predicts the K/T extinction event (Percival and Fischer, 1977; Romein, 1977; Romein and Smit, 1981; Thierstein, 1981, 1982; Perch-Nielsen, 1986; Jiang and Gartner, 1986; Alcalá-Herrera et al., 1993; Gartner et al., 1994; Pospichal, 1994a, b; 1996a, b, c).

Calcareous nannofossils are the remains of unicellular, planktonic marine golden-brown algae (nannoplankton) that secrete tiny (2 to 30 μm) calcareous plates (coccoliths). Also included in this fossil group are other calcareous nannoliths of approximately the same size and of an inferred similar biological affinity. The nannoplankton evolved in the Late Triassic and reached chalk-forming abundances by Late Jurassic times. They are abundant in today's oceans, with higher diversity in the low latitudes. The most dramatic event in the more than 210-m.y. history of this group is their near total extinction at the K/T boundary (Bown et al., 1991). Bramlette and Martini (1964) first documented this event. Subsequently, numerous nannopaleontologists have noted this remarkable extinction in outcrop sections and deep-sea cores around the globe (see summaries of Thierstein, 1981, 1982; Perch-Nielsen et al., 1982; Pospichal and Huber, 1992). Over 100 nannoplankton species became extinct across the K/T boundary, while only 15 to 20 survived the event (Perch-Nielsen, 1986). In the wake of the nannoplankton extinctions, several species occur sequentially in bloomlike abundances in the earliest Tertiary seas. Some of these bloom abundances have a near worldwide distribution, whereas others are highly provincial, reflecting particular paleoceanographic conditions or paleoenvironments.

This chapter briefly summarizes nannofossil data accumulated in the last several years from a number of deep-sea cores and outcrop K/T sections from around the globe and provides an update to the K/T nannofossil reviews presented by Perch-Nielsen (1979a, b, c, d), Perch-Nielsen et al. (1982), Smit and Romein (1985), and Pospichal and Huber (1992). The most comprehensive summary to date of nannofossil assemblages across the K/T boundary is provided by Thierstein (1981), who details late Maastrichtian nannofloral biogeography and supplies valuable discussion on the environmental affinities of surviving K/T taxa.

THE SECTIONS

Recent drilling by the Ocean Drilling Program (ODP), particularly in the high austral latitudes, has produced several K/T boundary cores from a number of locations (Fig. 1; Table 1). Studies of these cores have shed much light on the K/T extinction scenario among nannoplankton in the austral realm (Pospichal and Wise, 1990a; Pospichal, 1991; Thierstein et al., 1991; Wei and Thierstein, 1991; Wei and Pospichal, 1991; Ehrendorfer and Aubry, 1992; Pospichal and Bralower, 1992; Pospichal and Huber, 1992; Pospichal and Wise, 1992). A summary of the pertinent lithologic and paleoenvironmental information

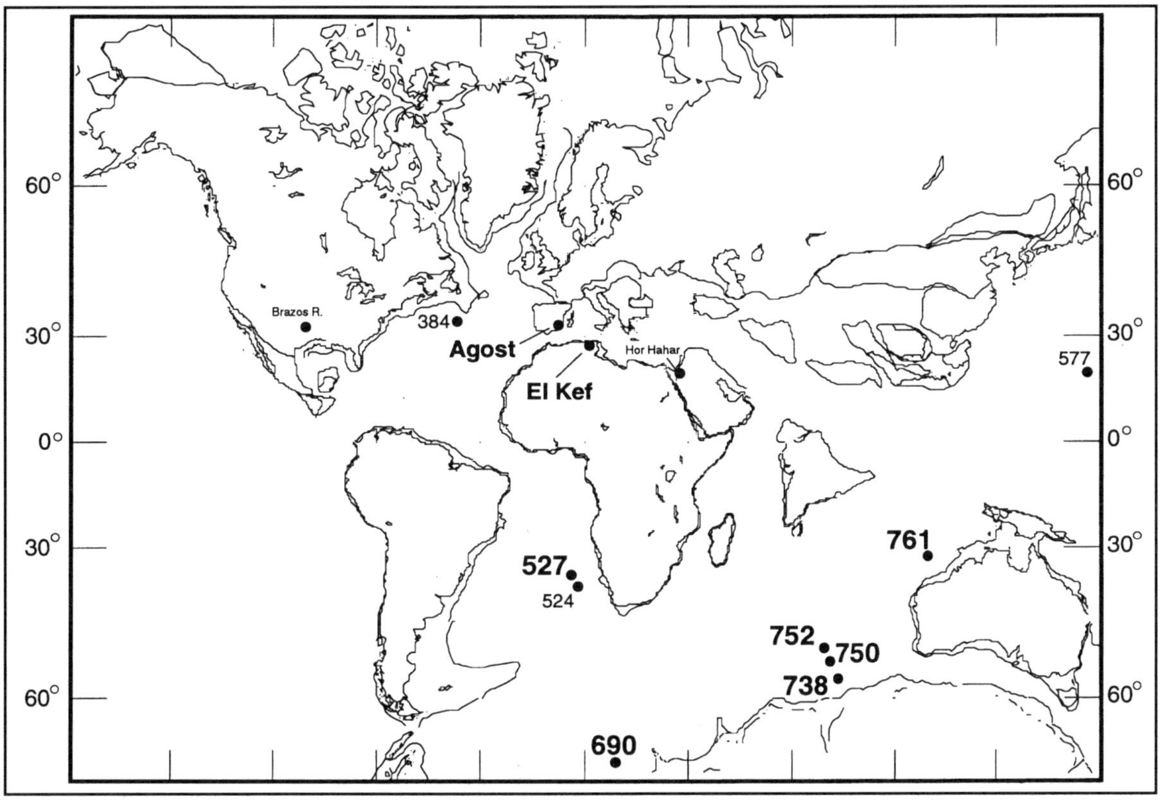

Figure 1. Paleogeographic (66 Ma) locations of K/T sections studied in this chapter (bold) and others mentioned in the text. Reconstructions from Scotese and Denham (1988), Terra Mobilis™.

TABLE 1. LOCATION, PALEOENVIRONMENTAL, AND LITHOLOGIC INFORMATION OF THE K/T SECTIONS STUDIED IN THIS CHAPTER

ODP/ DSDP Sites	Present Day Water Depth (m)	Ocean	Physiographic Feature	K/T Paleoenvironment	K/T Lithology
527	4428.0	S. Atlantic	Walvis Ridge	lower bathyal	bioturbated light orange nannofossil chalk overlain by thin gray clay layer and dark orange clay-rich chalk
690	2914.0	S. Atlantic	Maud Rise	middle to upper bathyal	bioturbated cream-colored nannofossil chalk overlain by orange clay rich chalk
738	2252.5	S. Indian	Southernmost Kerguelen Plateau	middle to upper bathyal	indurated white chalk with chert overlain by 2 mm clay layer and laminated clay-rich chalk
750	2030.5	S. Indian	Northern Kerguelen Plateau	middle to lower bathyal	white nannofossil chalk with chert nodules overlain by gray green marl
752	1086.3	S. Indian	Broken Ridge	middle to lower bathyal	light green nannofossil chalk with dark green ash layers and chert
761	2167.9	E. Indian	Wombat Plateau N.W. Australian Margin	middle to upper bathyal	light green chalk overlain by dark green bioturbated nannofossil chalk
Outcrop Sections					
Agost, Spain		W. Tethys	Southern Iberian Margin	upper bathyal	light gray marl overlain by 2 mm rust-colored layer with 6-7 cm black clayey marl above
El Kef, Tunisia		W. Tethys	North African Margin	outer neritic to upper bathyal	light gray tan marl overlain by 2 mm rust-colored layer with 50 cm black clay and clayey marl above

Notes: DSDP Site 527 (Hsü, K. J., and LaBrecque, J. L., et al., 1984; paleobathymetry, Fütterer, 1984); ODP Site 690 (Barker, P. F., Kennet, J. P., et al., 1988); ODP Site 738 (Barron, J., Larsen, B., et al., 1989); ODP Site 750 (Schlich, R., Wise, S. W., Jr., et al., 1989); ODP Site 752 (Peirce, J. W., Weissel, J. K., et al., 1989); ODP Site 761 (Haq, B. U., von Rad, U., O'Connell, S., et al., 1990); Agost (Canudo et al., 1991); El Kef (Keller, 1988; paleobathymetry, Keller, 1992).

on each of these K/T drill core sites and the outcrop sections discussed in this chapter is given in Table 1.

The South Atlantic is represented by several well-studied K/T boundary cores. Deep Sea Drilling Project (DSDP) Sites 524 through 529 on the Walvis Ridge produced several important K/T papers (e.g., Hsü et al., 1982; Perch-Nielsen et al., 1982; Boersma, 1984; Manivit, 1984; Huffman et al., 1990; Alcalá-Herrera et al., 1993; Gartner et al., 1994).

The first biostratigraphically complete, extreme high southern latitude K/T boundary was cored at Site 690 on Maud Rise in the Weddell Sea, during ODP Leg 113 (Fig. 1; Table 1). The nannofossil record in this heavily bioturbated section is detailed in Pospichal and Wise (1990a) and Pospichal et al. (1990) and briefly summarized here.

In the Indian Ocean, ODP Leg 119 recovered a K/T section on the southern end of the Kerguelen Plateau (Site 738), and Legs 120 and 121 cored sections on the mid- to northern portion of the plateau (Sites 750, 752). All of these sections record the K/T event at high latitudes (see summary of Pospichal and Huber, 1992). The boundary was also cored at the lower-latitude Site 761 on the northwest Australian margin during Leg 122.

Two K/T boundary localities from the region of the ancient Tethys Sea are also reviewed here: the Agost section in southeast Spain and the boundary stratotype at El Kef, Tunisia. The latter is perhaps the most complete and expanded section (MacLeod and Keller, 1991).

METHODS

All samples were collected at centimeter-spaced intervals across the K/T boundary and with careful attention to sedimentary structures and bioturbation. For the drill core samples, photos were taken of sample locations for correlation with these structures and sediment color changes.

Smear slides from each sample were prepared at consistent thickness to ensure reliable abundance comparisons. Slides were analyzed with the light microscope at 1,560× and at least 500 nannofossil specimens per sample were counted along a random traverse on the slide. Slides were generally examined beyond the 500-specimen count to uncover any rare species and to ensure that no biostratigraphic information was lost. For the El Kef and Agost samples, the number of fields of view traversed to achieve the count of 500 was also tallied. From these data, the individual species percent abundance was calculated as the percentage of the total number (~500) counted. For the El Kef and Agost sections, the individual species abundance per field of view was also calculated, which is simply the total specimens counted per number of fields of view (s/fv) traversed. This is a means of measuring individual species and total nannofossil abundance in the sediment independent of the abundance of other species.

Diversity measurements are given as species richness, or the total number of species recorded in each sample within the count of 500. In addition, diversity for Agost, El Kef, Site 527, and Site 738 is also given by the Shannon-Weaver index ($H = -\Sigma p_i \log p_i$), where p_i = number of individual specimens/total number counted (see Shannon, 1949).

Abundances were compiled for three basic assemblages. The Cretaceous assemblage consists of those species that lived during the Maastrichtian; these are traditionally viewed as having become extinct at the K/T boundary. The second assemblage includes those genera that survive or persist across the boundary (see Percival and Fischer, 1977; Thierstein, 1981; Perch-Nielsen et al., 1982; Pospichal, 1991). This assemblage is made up of 15 to 20 species and includes some that are considered environmental generalists or opportunistic "disaster" species. Survivor taxa such as *Thoracosphaera*, a calcareous dinoflagellate, and *Braarudosphaera* may, in the absence of competitive forms, exhibit sequential blooms just above the boundary. The third assemblage, Tertiary species, consists of taxa that evolved in the Danian and includes, for the most part, representatives of new Tertiary genera. A fourth, but statistically insignificant group, include older reworked Cretaceous specimens, usually lower Maastrichtian and Campanian, which are often present in low numbers both below and above the boundary.

Figure 2 is a summary of the cosmopolitan and regional biostratigraphic schemes and associated markers for the late Maastrichtian–Danian interval that have been proposed by a number of authors over the past few decades. For the Cretaceous, the commonly used nannofossil biozonations are the general schemes of Sissingh (1977) and Roth (1978). The *Micula prinsii* Zone was proposed for the uppermost Maastrichtian in low to mid-latitudes (Perch-Nielsen, 1979d). For the Upper Cretaceous of the high austral latitudes, the zonation of Watkins et al. (1996) is employed.

The commonly used schemes for the Danian are those of Martini (1971) and Okada and Bukry (1980). These numbered schemes use essentially the same criteria to define the initial Danian zones (Fig. 2). Also represented on Figure 2 are several regional schemes that employ additional markers to produce a finer subdivision of the Danian (Romein, 1979; Perch-Nielsen, 1979a, 1981; Jiang and Gartner, 1986; van Heck and Prins, 1987; Wei and Pospichal, 1991).

A variety of criteria is used for defining the K/T boundary with calcareous nannofossils. The last occurrence of Cretaceous taxa is commonly used (Martini, 1971; Okada and Bukry, 1980; Sissingh, 1977) or, in some cases, the first appearance of *Biantholithus sparsus* (Perch-Nielsen, 1979a; Pospichal and Wise, 1990a; Wei and Pospichal, 1991). Van Heck and Prins (1987) suggest the sudden change in nannofossil assemblages to delimit the K/T boundary. The use of any of these methods for defining the boundary can present problems, as bioturbation and reworking can lead to a misplacement of the boundary by more than 1 m (Pospichal et al., 1990). Although traditionalists may prefer strict adherence to a proper biostratigraphic definition of any stratigraphic boundary, the color changes, lithologic changes, and geochemical signatures that are present at the K/T boundary facilitate its precise location (see Smit and Romein, 1985).

The K/T boundary stratotype is located at the El Kef,

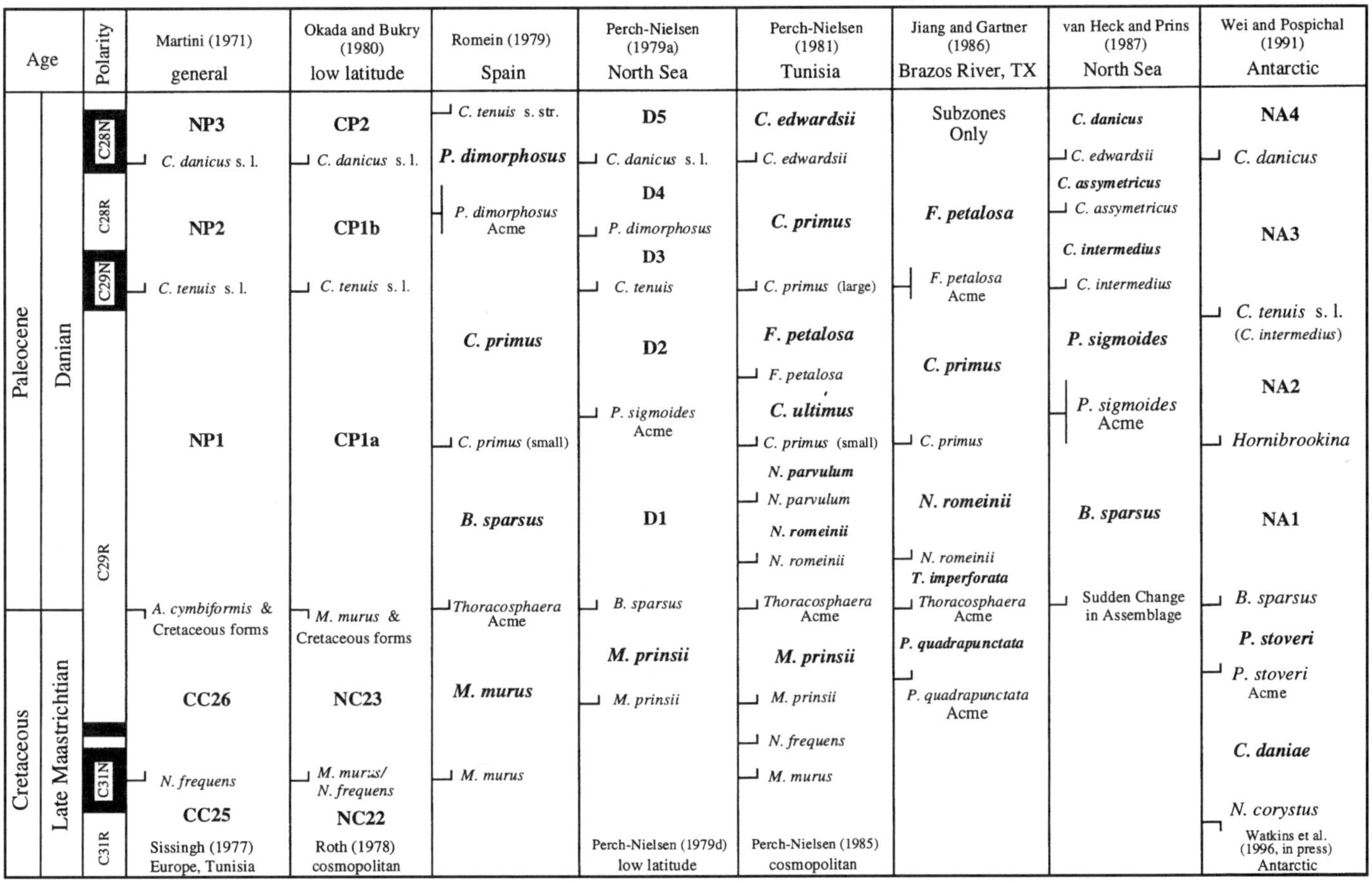

Figure 2. Nannofossil biozonations of the K/T boundary (modified from Perch-Nielsen, 1985). Correlation with the magnetic scale follows Berggren et al. (1985) and is to the CP and NP zones only for the Danian and to the zones of Watkins et al. (1996) for the Cretaceous. Authors of the Cretaceous zonations are listed at the bottom of the diagram. Note: The zonal boundaries, datum levels, and magnetic chrons are depicted here in relative order only. The length of each does not depict relative duration (i.e., the K/T boundary actually occurs in the upper part of C29R, which is expanded here to accommodate Danian markers).

Tunisia, section and is defined at the base of the boundary clay, which includes the millimeter-scale red clay layer containing an Ir anomaly, shocked quartz, and microspherules. The base of the boundary clay is characterized by the abrupt decrease in abundance of both planktonic foraminifera (Smit, 1982) and calcareous nannofossils (Pospichal, 1994a).

Age models and estimations given here follow the time scale of Berggren et al. (1985), who suggest an age of 66.4 Ma for the K/T boundary and 65.9 Ma for the boundary of nannofossil Zones NP1 and NP2. These ages differ on the more recent scale of Cande and Kent (1992), who give an age of 66.0 Ma for the K/T boundary; I estimate the NP1/NP2 boundary at 65.4 Ma using this scale. Swisher et al. (1992) suggest an age of 65.0 Ma for the K/T boundary derived from $^{40}Ar/^{39}Ar$ measurements of Chicxulub melt rock and K/T boundary tektites. Sedimentation rates for the uppermost Maastrichtian at Agost and El Kef were calculated using an estimated 225 k.y. as the duration of the *Micula prinsii* Zone (after Henriksson, 1993).

NANNOFOSSIL ABUNDANCE AND DIVERSITY AT THE K/T BOUNDARY

Figure 3 shows the abundance of nannofossils plotted versus age across the boundary at Agost, Spain, and El Kef, Tunisia. The major change in nannofossil abundance occurs at the base of the boundary clay at each outcrop locality, where at El Kef, for example, Cretaceous nannofossil abundance drops sharply by nearly 99%. For the most part, small fluctuations in abundance below the boundary are due to variations in preservation. At Agost, preservation is poorest just below the K/T boundary, but preservation at El Kef is much better, and thus the data from this section perhaps better reflect the true nannofossil assemblage dynamics.

In the deep-sea cores, in which no distinct boundary clay layer is observed, the drop in Cretaceous specimen abundance is apparent but is not as sharp as in the shallower sections. At Site 738, where a distinct 2-mm boundary clay layer is present,

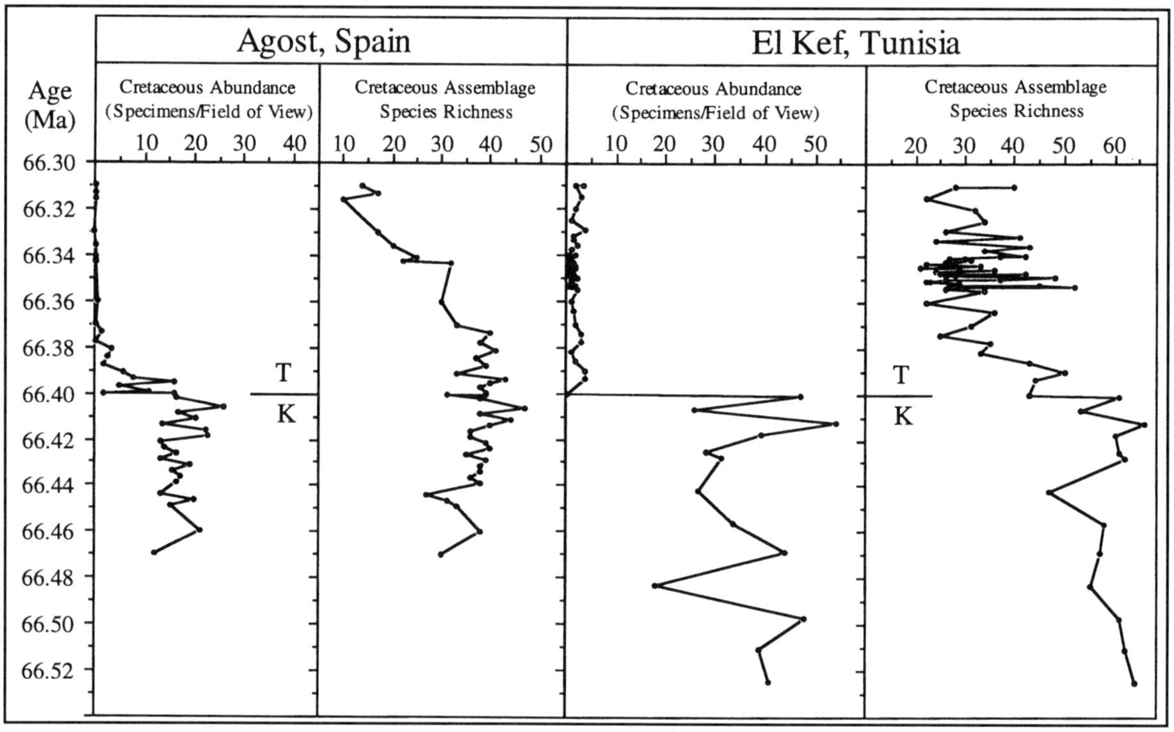

Figure 3. Abundance (specimens/field of view) and diversity (species richness) across the K/T boundary for the Cretaceous nannofossil assemblages from Agost, Spain (Pospichal, 1996a), and El Kef, Tunisia (Pospichal, 1994a). Ages (Ma) after Berggren et al. (1985).

very poor preservation in the well-indurated chalks just below this layer does not permit an accurate evaluation of the abundance changes. In fact, there is an increase in abundance going from the chalk to the clay, which is strictly a function of enhanced preservation in the clay. Moreover, poor core recovery due to a thick chert layer just below the boundary excludes confident assessment of any nannofossil changes preceding the boundary event at this site.

Cretaceous assemblage species richness at El Kef and Agost declines sharply across the K/T boundary (Fig. 3). The section at El Kef shows a more dramatic change, as the number of species drops from an average of about 60 species/sample in the topmost Maastrichtian to about 40 at the base of the boundary clay. Species richness strongly fluctuates between 20 and 50 within the reworked Cretaceous assemblage above the boundary. The Shannon-Weaver index, which incorporates individual species abundance as well as the number species, exhibits a similar pattern at El Kef (Fig. 4). Shannon-Weaver values for the Cretaceous assemblage are fairly stable below the boundary. After a drop at the K/T boundary, values fluctuate above the boundary but never exceed preboundary values. The sequential planktonic foraminifera extinction levels noted by Keller (1988) from the same set of samples are also shown in Figure 4. Unlike the planktonic foraminifera extinctions, there is no systematic reduction across the boundary in Cretaceous nannofossil diversity represented by either the species richness data or the Shannon-Weaver index. The prominent change in nannofossil diversity occurs at the K/T boundary itself, with the fluctuating Cretaceous species diversity values above the boundary attributed to variations in reworking.

The Cretaceous assemblage species richness at middle latitude Site 527 is less than in the Tethyan sections, and changes across the boundary are less distinct (Fig. 5). In the topmost 4 m of Maastrichtian, Cretaceous assemblage species richness and Shannon-Weaver index values actually increase toward the K/T boundary (except for the interval corresponding to poor preservation, Fig. 5), then decline across the boundary. Alcalá-Herrera et al. (1993) performed similar detailed diversity measurements across the boundary of Site 527, and the patterns noted here are comparable to their results.

At high-latitude Site 738 (S. Kerguelen Plateau), Cretaceous assemblage species richness and Shannon-Weaver index values are lower than at Site 527 and in the Tethyan sections. About 20 to 25 Cretaceous species are consistently present in the samples from several meters below the K/T core. Because of very poor preservation in thc well-indurated chalks in the 3-cm interval just below the K/T boundary, however, the number of species drops to about 12 (Fig. 6). On the other hand, Cretaceous assemblage species richness, Shannon-Weaver index values, and overall abundance actually increase across the K/T boundary because of improved preservation in the clay-rich chalks above the boundary. Nonetheless, poor nannofossil preservation and poor core recovery below the K/T boundary

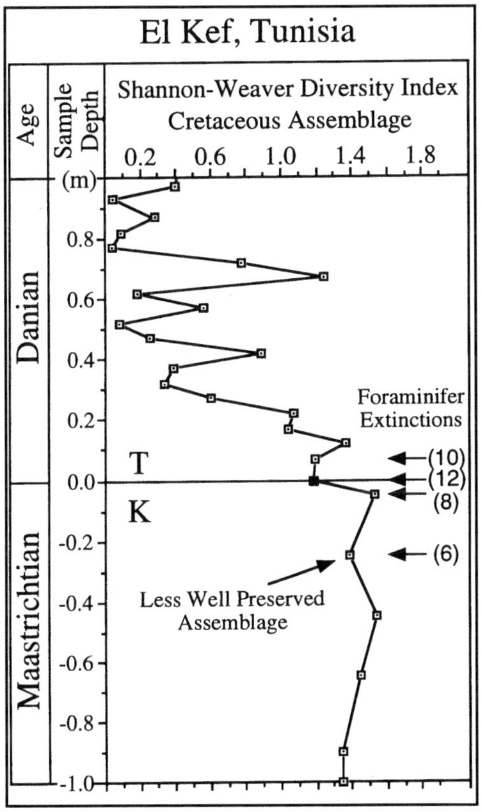

Figure 4. Shannon-Weaver diversity index for the Cretaceous assemblage from a 2-m interval across the K/T boundary, El Kef. Planktonic foraminifer extinctions of Keller (1988) from the same sample set are shown (number of species extinct) (figure from Pospichal, 1994a).

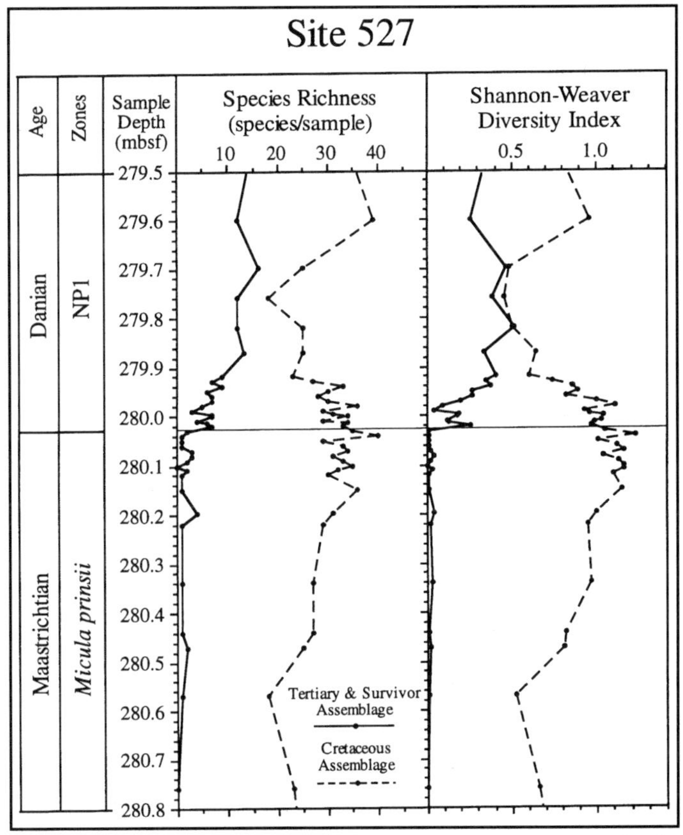

Figure 5. Species richness and Shannon-Weaver diversity index for the Cretaceous assemblage and the Tertiary plus survivor assemblages across the K/T boundary, South Atlantic DSDP Site 527.

limit the assessment of diversity and subboundary changes at this site.

Figures 5 and 6 also show the diversity (species richness, Shannon-Weaver index) of the combined Tertiary plus survivor assemblages. At most locations, the survivor assemblage in the Maastrichtian generally consists of one or two species (usually *Thoracosphaera*) that are of very low abundance. At or just above the boundary, more survivor species appear, along with the incremental addition of new Tertiary species. Substantial fluctuations in abundance and diversity of new species and survivors occur within nannofossil Zones NP1 and NP2.

A synthesis of the nannofossil abundance and diversity data indicates no major changes occur below the K/T boundary within at least the last 100 k.y. or so. As noted by Thierstein (1981), Cretaceous assemblages were stable, which suggests that climate, ocean temperatures, and nutrient cycling were in a steady state for the time period immediately leading up to the terminal Cretaceous. Based on qualitative observations of assemblages beyond

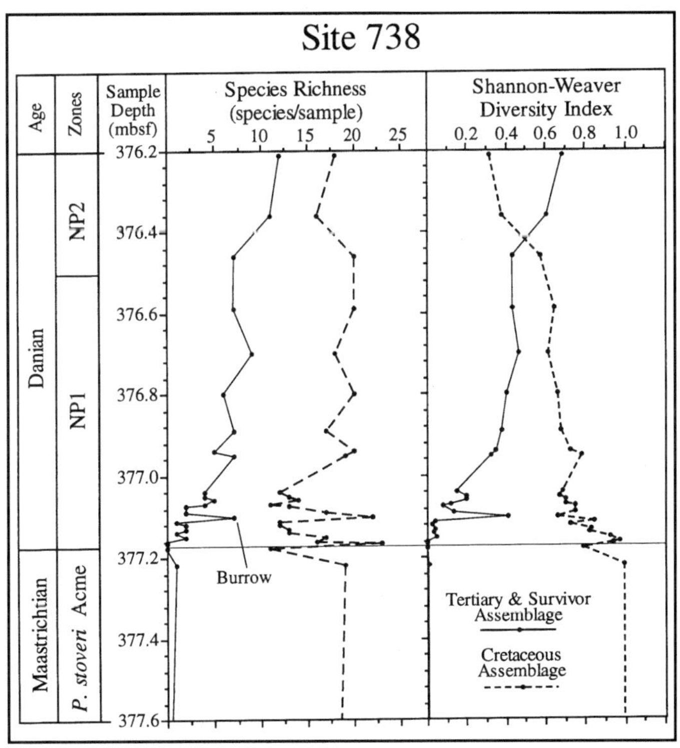

Figure 6. Species richness and Shannon-Weaver diversity index for the Cretaceous assemblage and the Tertiary plus survivor assemblages across the K/T boundary, southern Indian Ocean ODP Site 738. Nannofossil data are given in Pospichal (1996b).

this 100-k.y. time interval, the period of assemblage stability probably extends back several million years to the upper Campanian to mid-Maastrichtian, where notable nannofossil assemblage changes do occur. However, there is no apparent link between the near total annihilation of the nannofossil assemblages at the K/T boundary and these earlier changes.

Above the boundary, the fluctuating diversity and abundance values of the Cretaceous assemblage at each locality reflect variable reworking and redeposition. Preservation generally does not change much within the Danian sediments and is not considered to have a strong influence on the assemblages there.

NANNOFOSSIL ABUNDANCE, ISOTOPES, AND PRODUCTIVITY

Figures 7, 8, and 9 illustrate nannofossil abundance along with carbon and oxygen isotopes and percent carbonate. At the base of the K/T boundary clay, the sharp decline in Cretaceous nannofossil abundance at El Kef correlates precisely with a sharp, approximately 40% decline in $CaCO_3$ and a nearly 2.0 per mil negative shift in $\delta^{13}C$ noted in the fine-fraction by Keller and Lindinger (1989) (Fig. 7). An iridium anomaly (Kuslys and Krähenbühl, 1983) is also present at the base of the boundary clay and correlative with the isotope and nannofossil changes. Cretaceous nannofossil abundance remains low in the 3-m interval above the boundary, but survivors plus Tertiary forms show variably fluctuating peaks. The $\delta^{13}C$ values also fluctuate widely, although the nannofossil and isotope peaks do not correlate. The carbonate content remains low through the lowermost 3 to 4 m of the Danian and rises to subboundary values above that. Concomitant with the rise in $\%CaCO_3$ and $\delta^{13}C$, the Tertiary nannofossil assemblage abundance also increases, although irregularly, starting at about 4 m. The fine-fraction $\delta^{18}O$ values at El Kef decrease across the boundary and also strongly fluctuate in the lowermost 3 m of the Danian. As with

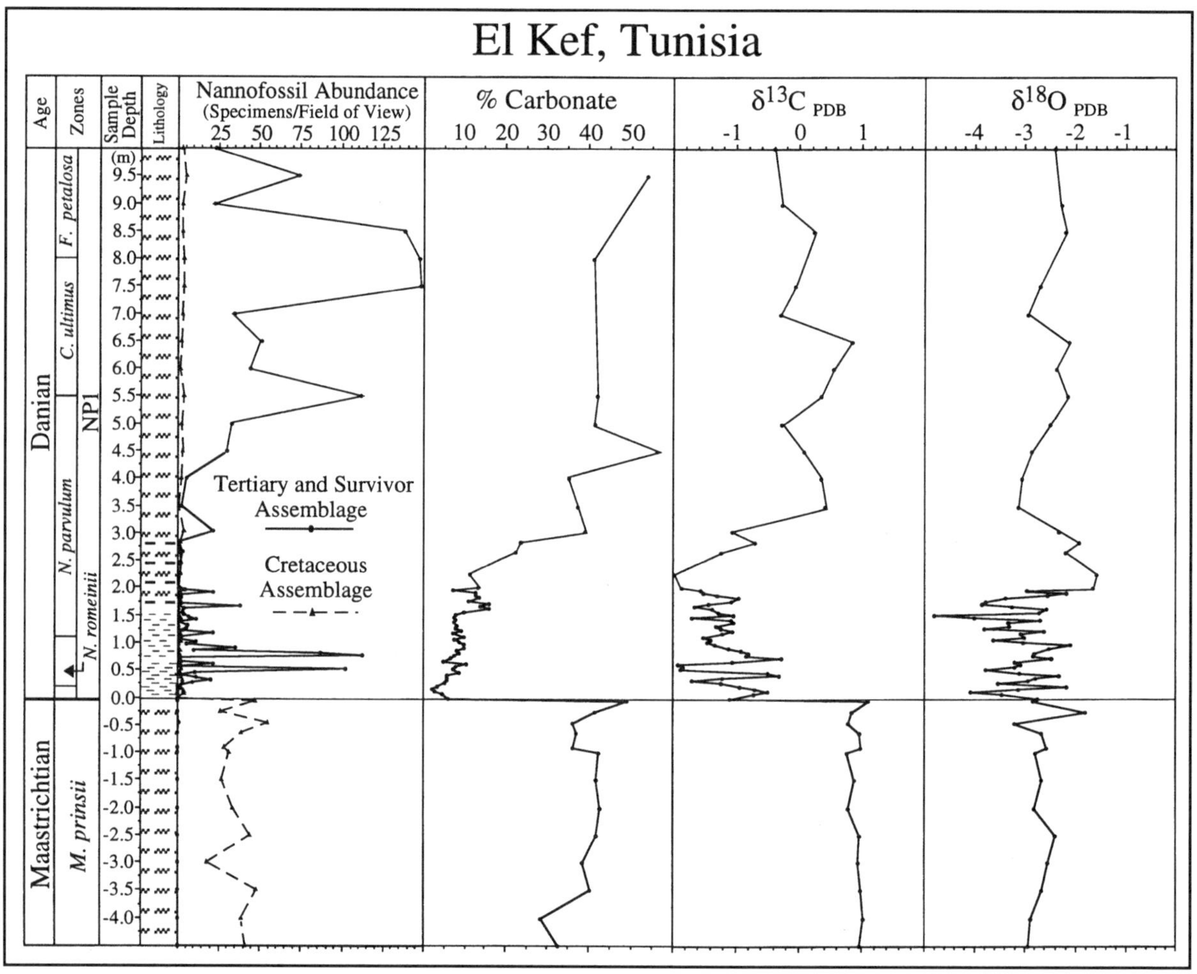

Figure 7. El Kef, Tunisia, nannofossil abundance (specimens/field of view) correlated with % carbonate and the fine-fraction carbon and isotope data of Keller and Lindinger (1989). Interval is from 4.5 m below the K/T boundary to 10.0 m above. Lithology key: wavy lines = marls; dashed lines = boundary clay (modified from Pospichal, 1994a).

the $\delta^{13}C$ data, there does not appear to be a correspondence between the fluctuating $\delta^{18}O$ peaks and nannofossil peaks above the boundary.

At the base of the K/T boundary clay at Agost, the Cretaceous nannofossil abundance shows a similar marked decline along with a drop in $\%CaCO_3$, a 2.0 per mil negative shift in $\delta^{13}C$, 1.0 per mil negative shift in the $\delta^{18}O$ fine-fraction, and an iridium anomaly (Smit, 1990) (Fig. 8). The deep-sea Site 527 (Fig. 9) shows the onset of the decline in the percent abundance of Cretaceous species, which correlates with the iridium anomaly and a 30% drop in carbonate (Huffman et al., 1990). A 1.0 per mil negative shift in $\delta^{13}C$ and a nearly 0.5 per mil positive shift in $\delta^{18}O$ (bulk sediment) are noted at this level by Shackleton and Hall (1984).

The decline in abundance of Cretaceous nannofossils across the K/T boundary corresponds well with the changes in iridium, carbonate, and stable isotope compositions, as shown from these various localities and others not illustrated here (see Romein and Smit, 1981; Perch-Nielsen et al., 1982). Nannofossils, as a component of the sediment, are also generally less abundant in the intervals of lower carbonate and the more negative $\delta^{13}C$. This agrees with the scenario of a collapse in surface-water primary productivity at K/T time (e.g., Hsü et al., 1982; Zachos and Arthur, 1986; Arthur et al., 1987; Zachos et al, 1989). These authors have proposed that the extinction of the marine phyto- and zooplankton at the end of the Cretaceous resulted in the decreased photosynthetic fixation of carbon in surface waters by the phytoplankton and a collapse of the food chain in the marine biosphere. Primary productivity may have remained suppressed for nearly 500 k.y. (Zachos et al., 1989). The return of productivity to more "normal" levels as measured in the isotopic signal generally corresponds to the observed increase and dominance by newly evolved Tertiary nannoplankton. In the deep sea, this usually occurs near the NP1/NP2 boundary, or about 500 k.y. after the K/T event. The recovery time may vary from site to site; for example, it may have been more rapid in the Tethyan region, as indicated by nannofossil abundances and increasing $\delta^{13}C$ values well within nannofossil Zone NP1 at El Kef. Keller (1988) inferred a low productivity interval for El Kef of 300 to 400 k.y. from planktonic foraminifer data.

As discussed further in the following sections, the strongly fluctuating abundances of Tertiary nannofossil species above the boundary at El Kef (Fig. 7) and the variable $\delta^{13}C$ values are perhaps indicators of unstable conditions that existed following the K/T crisis. On the other hand, the $\delta^{18}O$ values may be slightly more difficult to interpret, particularly when correlated with nannofossil data. The Agost and El Kef sections show a negative $\delta^{18}O$ shift across the boundary, which can be interpreted as warming, whereas values at Site 527 show a positive shift, or cooling. In either case, the overall nannofloral extinction pattern is the same.

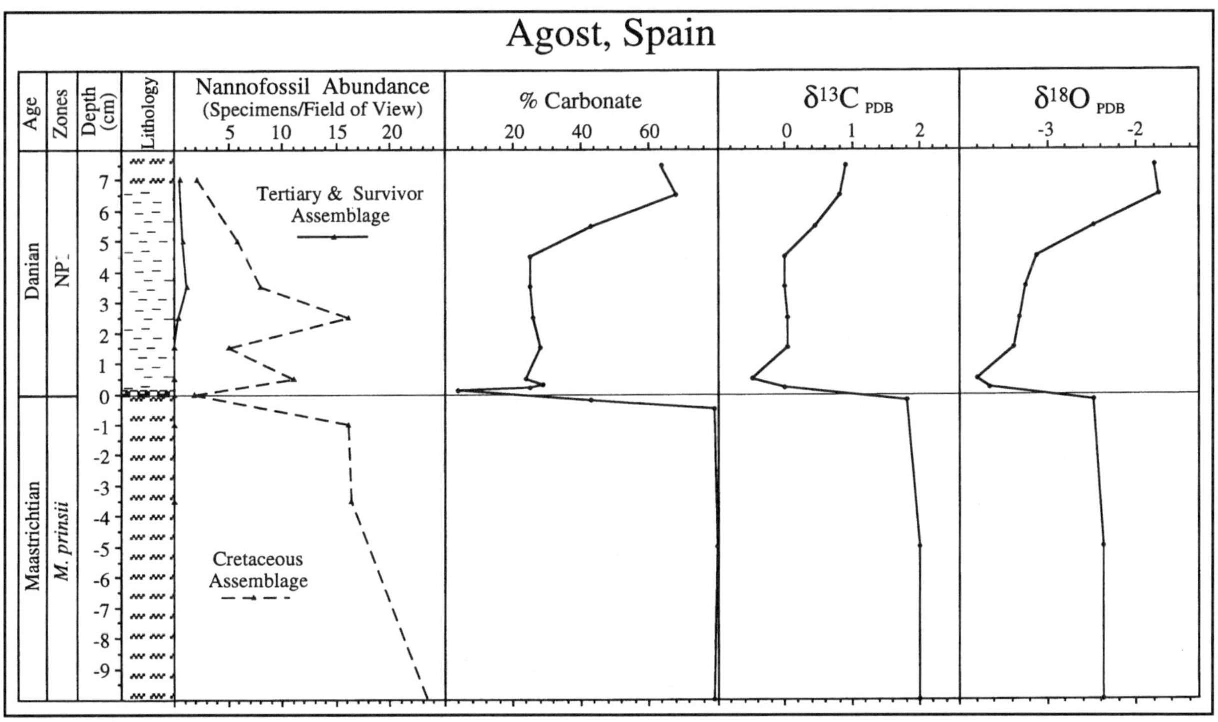

Figure 8. Agost, Spain nannofossil abundance (specimens/field of view) correlated with % carbonate, and the fine-fraction carbon and isotope data of Smit (1990). Interval is from 10 cm below the K/T boundary to 8 cm above. Lithology key: wavy lines = marls; dashed lines = boundary clay; dots = 2-mm basal rusty lamina (modified from Pospichal, 1996a).

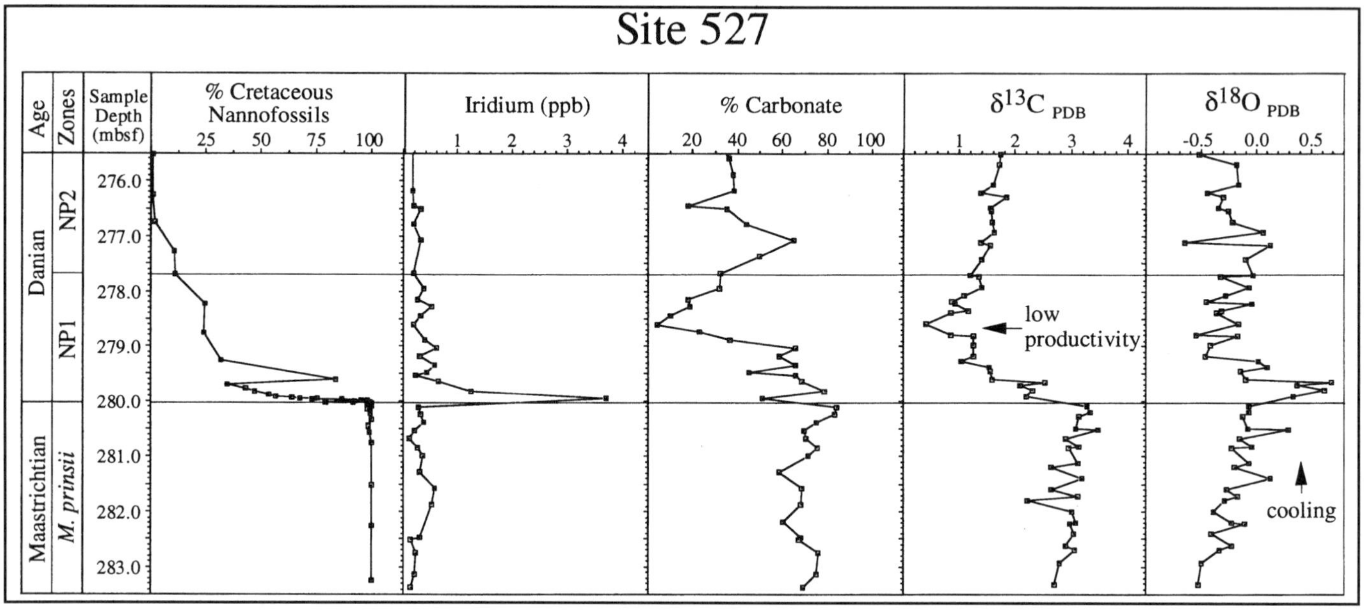

Figure 9. Cretaceous nannofossil percent abundance correlated with Ir (ppb = parts per billion) and % carbonate (Huffman et al., 1990) and with bulk-sample carbon and oxygen isotopes (Shackleton and Hall, 1984). mbsf = meters below seafloor.

NANNOFOSSIL TURNOVER ACROSS THE K/T BOUNDARY

The graphs of Figures 10, 11, and 12 represent the percent abundance changes across the K/T boundary among the three main nannofossil assemblages in the Tethyan, South Atlantic, and Indian Ocean sections. All the graphs show the same overall pattern but may differ in detail. In general, each exhibits a marked decline in the percentage of Cretaceous specimens starting at the boundary, followed by the abrupt replacement by survivors and the eventual dominance by the Tertiary assemblage. This pattern is similar at all latitudes represented in this study, as illustrated by the comparison of the turnovers at Sites 690 and 738 to the lower-latitude sections.

At El Kef the pattern is different, with the appearance and dominance of the new Tertiary species just above the boundary and probably not more than 50 to 100 k.y. after the K/T event. Survivors are present but do not compose a higher proportion of the total assemblage. Other sections have intervals within NP1 in which survivors are the dominant forms. In these sections, Tertiary species do not dominate the assemblage until 200 to 500 k.y. after the K/T event.

Cretaceous assemblage

Figures 13, 14, and 15 represent the individual species percent abundances across the K/T boundary in the different sections. In the Tethyan sections (Fig. 13), the dominant Maastrichtian species are *Micula decussata* and *Watznaueria barnesae.* These species are extremely rare or absent in the high latitude assemblages, which are characterized by abundant *Nephrolithus frequens* and *Kamptnerius magnificus* (Figs. 14, 15). Several species are common to all locations, and the most prevalent of these include *Arkhangelskiella cymbiformis*, *Cribrosphaerella ehrenbergii*, *Prediscosphaera cretacea*, and *P. stoveri*. A comparison of these cosmopolitan species in Figures 13 through 15 indicates that all exhibit a similar general pattern of abundance decline across the boundary at each locality.

The abundance distribution of the species *Prediscosphaera stoveri*, illustrated in Figure 16, may be quite significant. It is extremely abundant just below the boundary in the high austral latitudes and shows a distinct latitudinal abundance gradient in the Indian Ocean. This species is less common in the Tethyan sections (Fig. 13) and is also present in the Brazos River sections (= *P. quadripunctata* of Jiang and Gartner, 1986). This distribution probably reflects the latitudinal surface temperature gradients in late Maastrichtian times.

Survivor assemblage

In the Tethyan sections, the dominant survivor species above the K/T boundary are *Thoracosphaera* and *Braarudosphaera*, the latter being more abundant at Agost (Fig. 13) and the former being more widely distributed among all the localities (including Brazos River; Jiang and Gartner, 1986). The deep-sea localities lack *Braarudosphaera*, and the higher latitudes lack definitive *Thoracosphaera* blooms (Figs. 14, 15) within nannofossil Zone NP1. The mid- and higher-latitude sections are dominated by *Zygodiscus sigmoides* in NP1 with *Markalius inversus* and *Biscutum castrorum* (= *B. constans* of

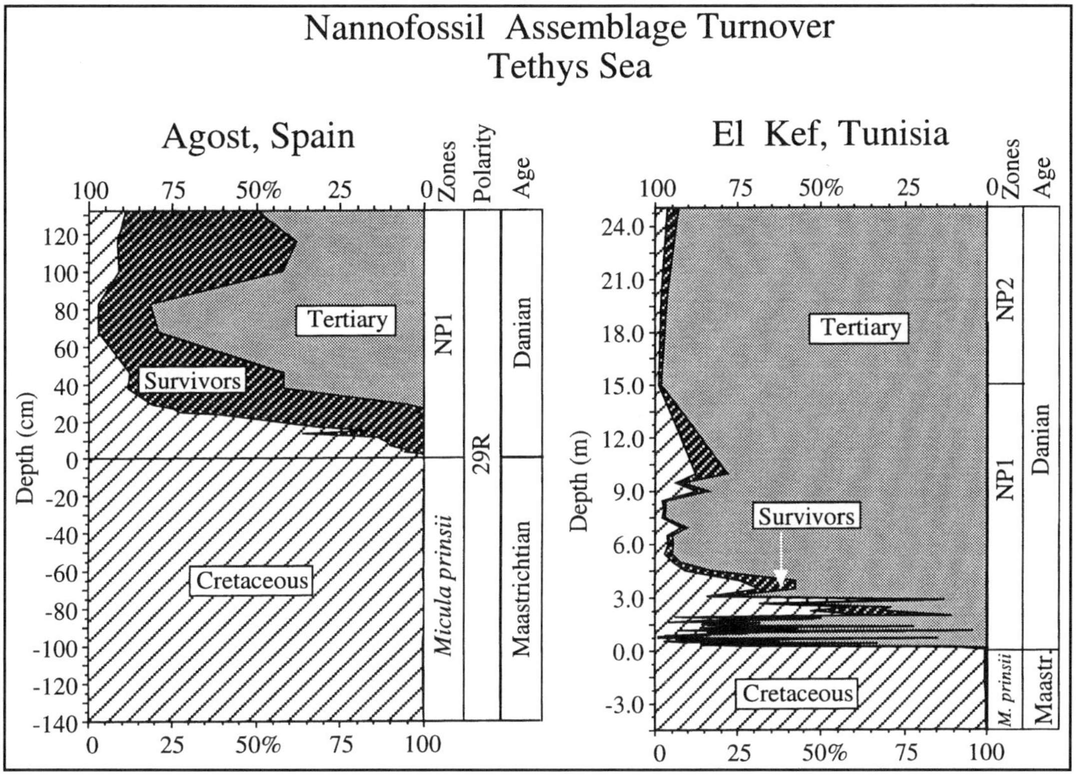

Figure 10. Nannofossil assemblage turnover represented by the percent abundance of Cretaceous, Tertiary, and survivor nannofossil assemblages across the K/T boundary of Agost, Spain, and El Kef, Tunisia (Tethys sites). Agost figure modified from Pospichal (1996a). Magnetostratigraphy for the Agost section is from Groot et al. (1989).

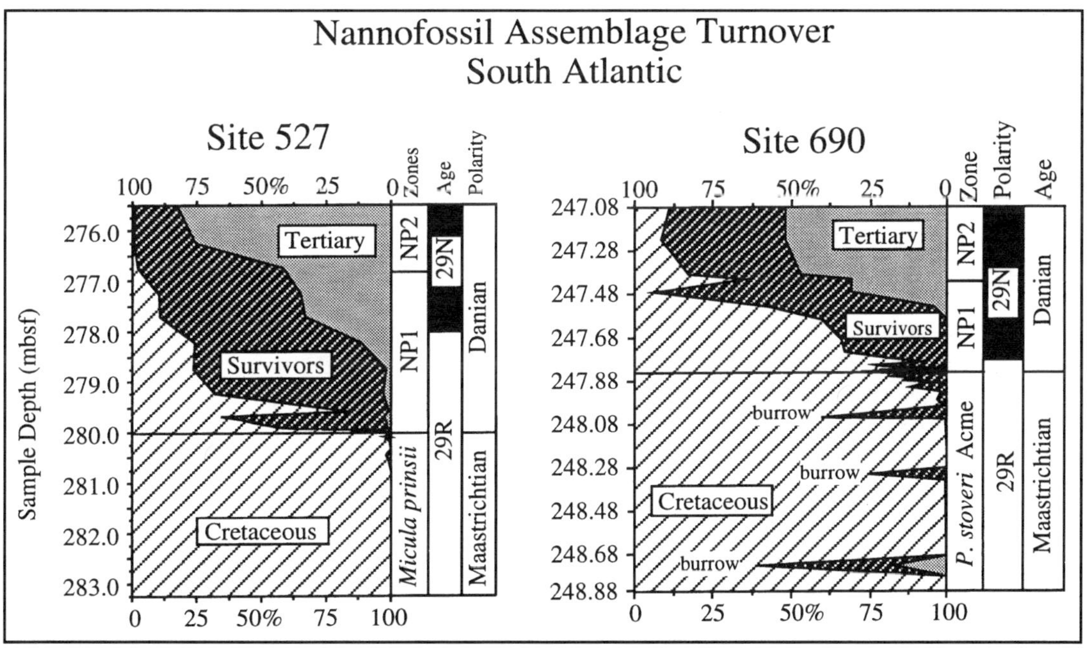

Figure 11. Nannofossil assemblage turnover represented by the percent abundance of Cretaceous, Tertiary, and survivor nannofossil assemblages across the K/T boundary of South Atlantic DSDP Site 527 and Weddell Sea Site 690. Magnetostratigraphy for Site 527 is from Chave (1984); and for Site 690, Hamilton (1990).

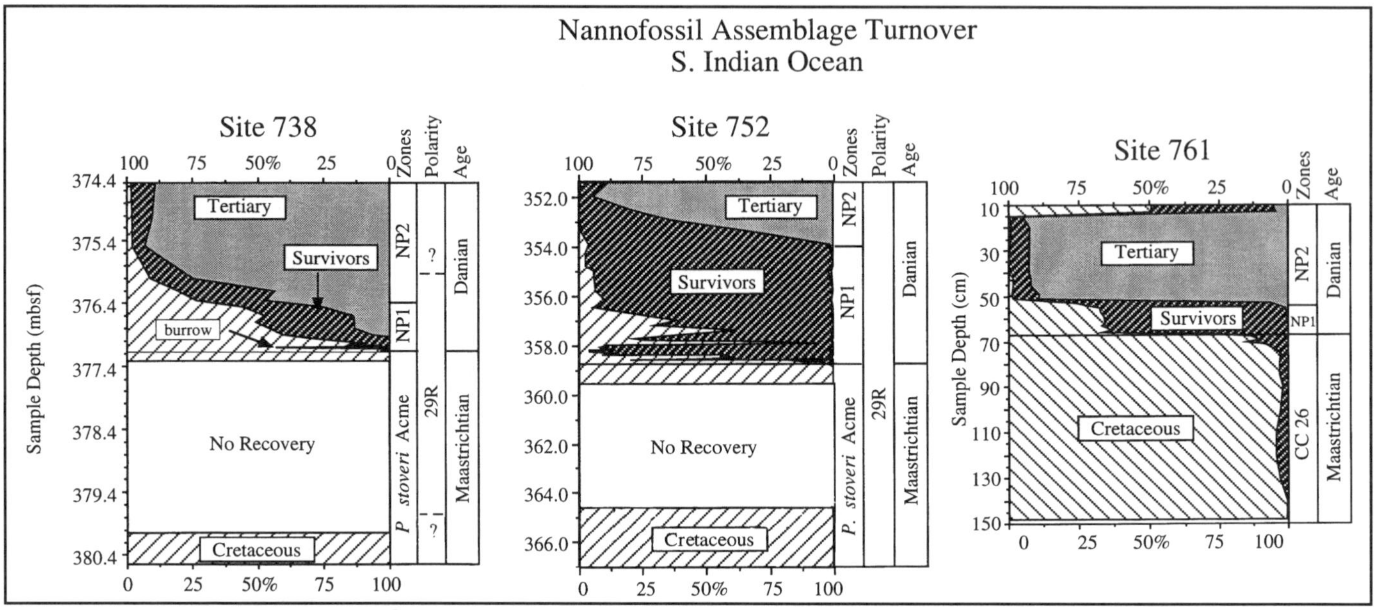

Figure 12. Nannofossil assemblage turnover represented by the percent abundance of Cretaceous, Tertiary, and survivor nannofossil assemblages across the K/T boundary of Indian Ocean ODP Sites 738, 752, and 761. Magnetostratigraphy for Site 738 is from Sakai and Keating (1991); and for Site 752, Gee et al. (1991). Nannofossil data for Site 738 are given in Pospichal (1996b).

some authors). Another species, *Cyclagelosphaera reinhardtii*, is prevalent in the mid-latitude sections of Sites 527 and 761 (Figs. 14, 15).

The distribution of some of the less common survivor species is a little more enigmatic. For example, within NP1 at Agost, *Octolithus multiplus* is an important component but is rare to absent in this zone at El Kef (Fig. 13). This overall variability in the distribution and abundance of the survivor species is perhaps an indication of the heterogeneous environmental conditions that existed around the globe immediately following the K/T boundary event.

Tertiary assemblage

The new Tertiary taxa appearing within NP1 are even more variably distributed than the survivor forms (Figs. 13 through 15). The El Kef section contains one key overwhelmingly abundant Tertiary species, *Neobiscutum romeinii*, which is not present at the other localities. This species appears just above the boundary and rapidly dominates the assemblage (up to 90%) at El Kef. At Agost, this species has not been identified, but its absence or rarity may be due to poorer preservation. *Neobiscutum romeinii* has been noted at Brazos River (Jiang and Gartner, 1986) and the Hor Hahar section in Israel (Eshet et al., 1992), but it is much less abundant than at El Kef. In the deep-sea sites studied here, *N. romeinii* has not been identified although it was noted at Site 577 in the Pacific Ocean (Monechi, 1985). The slightly more robust *Neobiscutum parvulum*, on the other hand, is preserved at Agost but is less abundant than at El Kef (Fig. 13). This species is also present at the localities where *N. romeinii* has been noted but has not been observed in the deep-sea sites of this study.

Another Tertiary species, *Futyania petalosa*, is more broadly distributed, with abundance peaks in the upper part of NP1 and in NP2 at El Kef (Fig. 13). The samples studied from Agost are stratigraphically below the first occurrence of this species, but abundant specimens were noted at nearby Caravaca, Spain (Romein, 1977), which suggests that it may also occur at Agost. Elsewhere, it is common across the NP1/NP2 boundary in the Brazos River section (Jiang and Gartner, 1986) and at the Hor Hahar section in Israel (Eshet et al., 1992). At the deep-sea sites, only very rare specimens were noted at Sites 750 and 752.

In the extreme high austral latitudes, *Hornibrookina* is dominant in the upper part of NP1. The distribution of this genus shows a distinct abundance pattern (Fig. 17), particularly across the Kerguelen Plateau in the southern Indian Ocean. The abundance of this form drops off considerably between Site 750 (up to 30% of the assemblage) and Site 752, despite the fact that these sites were separated by no more than 10° latitude at the K/T transition. At Site 752 only few to rare specimens are found in NP2. In the South Atlantic, *Hornibrookina* achieves its highest abundance near the NP1/NP2 boundary at Site 690, and none were noted at Site 527. As with *Prediscosphaera stoveri* in the Maastrichtian (Fig. 16), the distribution of *Hornibrookina* may reflect continuation of distinct latitudinal temperature gradients into the early Danian.

Prinsius dimorphosus (including *P. tenuiculus*) is perhaps the most nearly global of the Tertiary species that show dominant abundances. Although the actual abundance peak is dia-

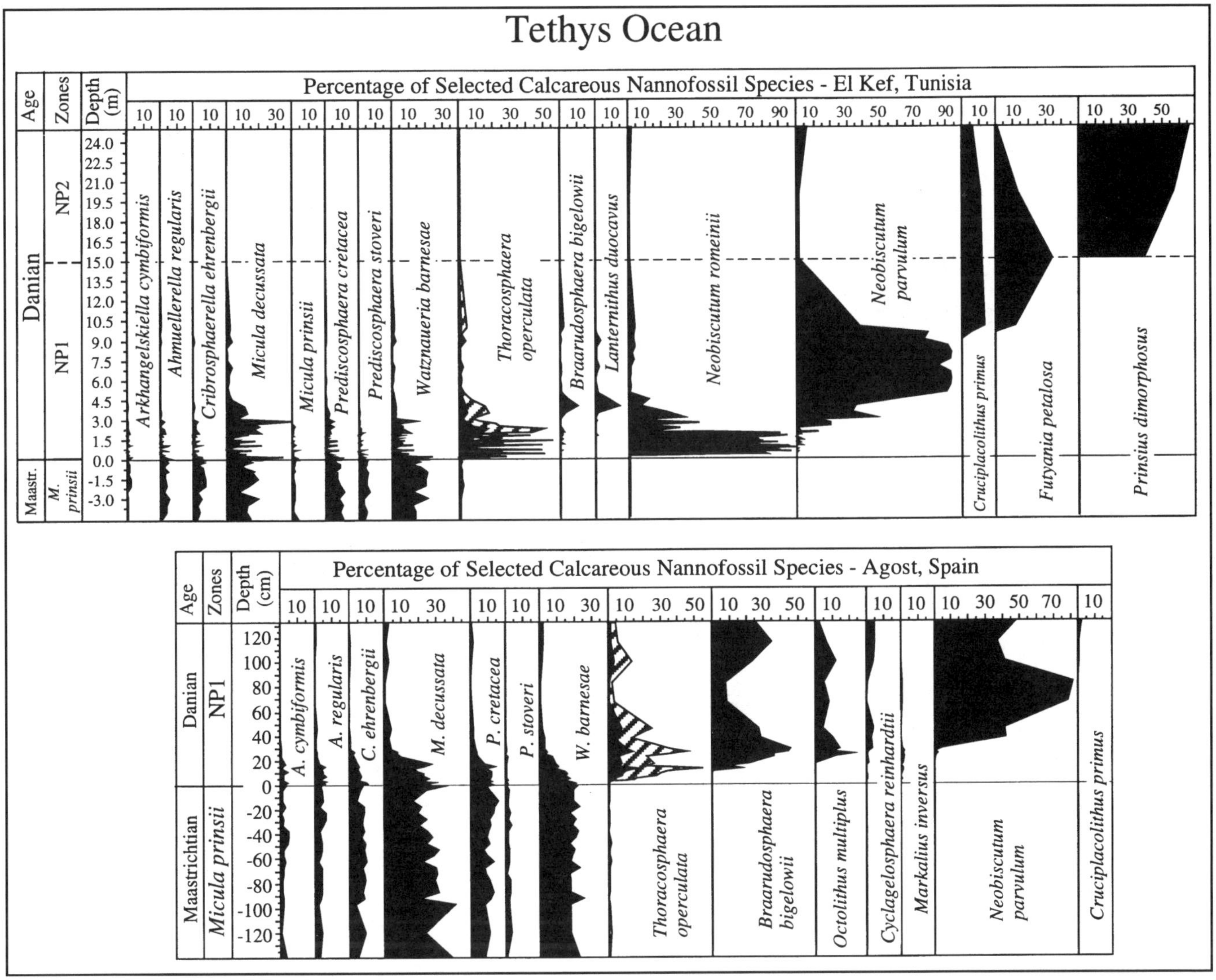

Figure 13. Percent abundance of select individual nannofossil species across the K/T boundary of Agost, Spain, and El Kef, Tunisia. Stippled pattern on *Thoracosphaera* graph represents adjustment if fragments (~1/3 of a whole specimen) are considered as one specimen. Other graphs do not show this adjustment. Agost figure modified from Pospichal (1996a).

chronous from site to site, this species is abundant in NP2 and NP3 sediments in marginal to deep sea sections (compare, e.g., Site 738 with El Kef). It is abundant at Pacific Ocean Site 577 (Monechi, 1985), as well as, North Atlantic Site 384 (Thierstein and Okada, 1979). Samples reported here from Sites 527 and 690 are below this species' peak abundance, which has been noted higher up in NP2 in these sections.

Recovery and repopulation of the Danian Seas

Recovery at each locality, for the purposes of this section, is considered to be the time at which conditions are favorable for the evolution and proliferation (dominance) of new species over surviving and opportunistic species. This point is reached at varying levels from site to site, and in this case does not necessarily correspond to productivity recovery as indicated in the carbon isotope record discussed previously. The initial appearance and dominance of the species *Neobiscutum romeinii* at El Kef marks a comparatively rapid recovery time of perhaps no more than 50 k.y. Recovery at Agost is slightly later if the absence of *N. romeinii* there is attributed to environmental factors and not to preservational differences. For the deep-sea sites, recovery occurs near the NP1/NP2 boundary or about 200 to 500 k.y. after the K/T event. The appearance and abundance peak of *Hornibrookina* in the high austral latitude sites near the top of NP1 indicate a recovery slightly sooner than in the mid-latitude localities.

The early Danian at each site is characterized by the

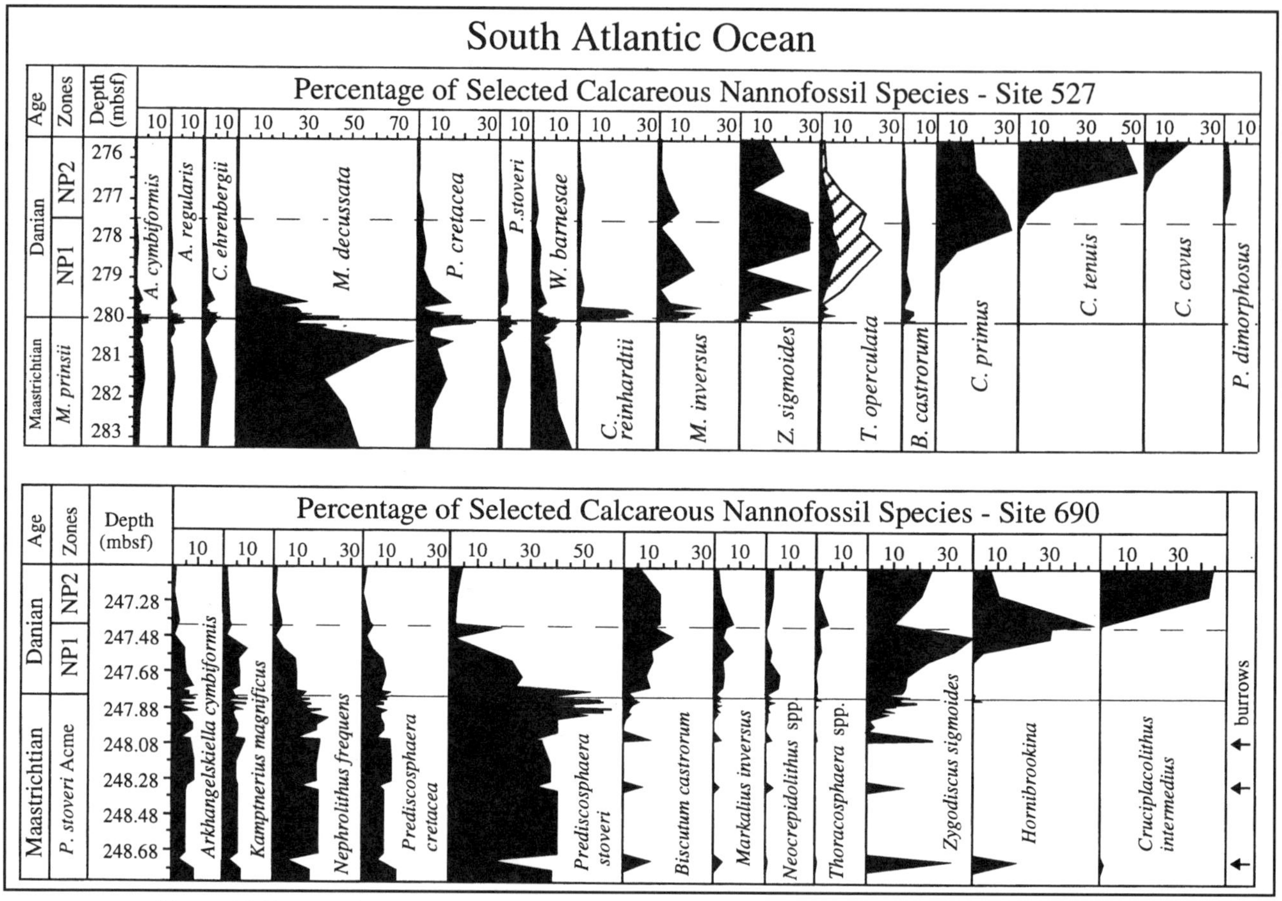

Figure 14. Percent abundance of select individual nannofossil species across the K/T boundary of South Atlantic DSDP Site 527 and Weddell Sea Site 690. Stippled pattern on *Thoracosphaera* graph of Site 527 represents adjustment if fragments (~1/3 of a whole specimen) are considered as one specimen. Other graphs do not show this adjustment.

sequential abundance peaks of surviving and new Tertiary species. However, repopulation of the oceans following the K/T event is not the same at each locality in terms of assemblage composition, although the general pattern of repopulation is similar. This is perhaps an indication of extreme environmental variability immediately following the K/T crisis, with these conditions persisting 0.5 to 1.0 m.y. into the Danian. The first Tertiary species to dominate globally, *P. dimorphosus* (within NP2), marks the onset of the amelioration of these variable conditions as nannofossil assemblages become more homogenous between globally disparate sites. Furthermore, the distribution of the Maastrichtian species *Prediscosphaera stoveri* (Fig. 13) and the Danian genus *Hornibrookina* illustrates the presence of distinct latitudinal temperature gradients, at least in the Southern Hemisphere, during the K/T transition.

It should be further emphasized that, as mentioned above, the overall pattern of disappearance of Cretaceous taxa and sequential abundance peaks of survivors and Tertiary species is similar at all latitudes according to the data shown here. This demonstrates that the K/T event affected calcareous nannoplankton of the high latitudes to the same degree as those in the tropics.

HIGH SOUTHERN LATITUDE LATE MAASTRICHTIAN COOLING?

Figure 18 represents the ranges of important austral high-latitude species for the upper Maastrichtian–lower Danian interval as derived from Sites 690, 750, and 752. The prominent extinctions through the early to late Maastrichtian are staggered, and very few originations are noted in the late Maastrichtian (Pospichal and Wise, 1990b; Watkins, 1992; Huber and Watkins, 1992; Watkins et al., 1996). These early/late Maastrichtian extinctions correlate with oxygen isotopic trends that reflect cooling in the Maastrichtian (Barrera and Huber, 1990). Moreover, the overall long-term cooling of the south polar region is perhaps indicated by strong latitudinal provincialism among planktonic foraminifera and calcareous nannoplankton, which began in the late Campanian and early Maastrichtian (Huber and Watkins, 1992). The equatorward migration of *Nephrolithus frequens* (Worsely, 1974; Pospichal and Wise, 1990b) as well as the abundance peak of *Prediscosphaera stoveri* in the latter part of the Maastrichtian may also reflect this long-term change. Furthermore, at Site 690,

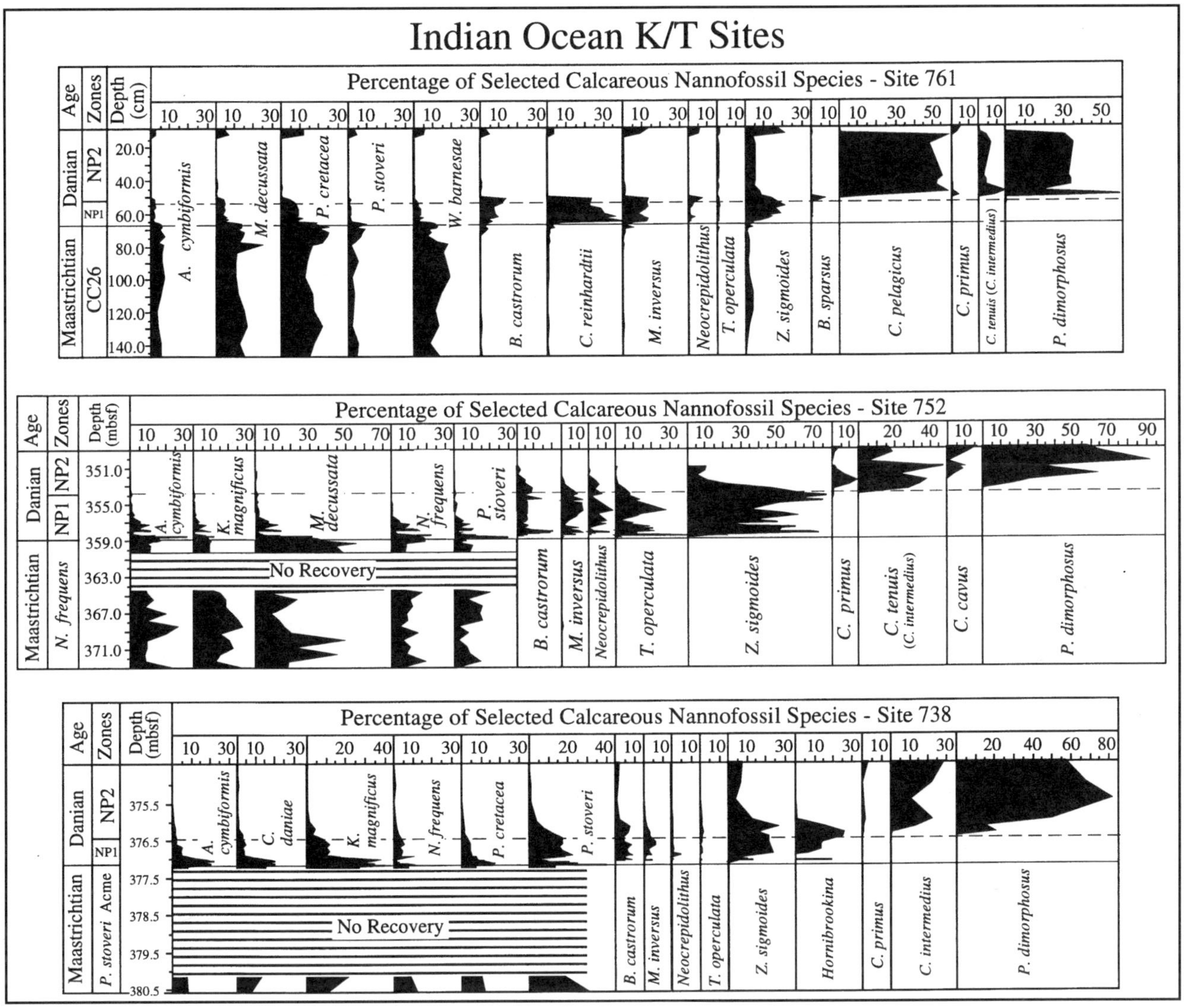

Figure 15. Percent abundance of individual nannofossil species across the K/T boundary of Indian Ocean ODP Sites 738, 752, and 761. Nannofossil data for Site 738 are given in Pospichal (1996b).

cooling is supported by the disappearance from the section of *Watznaueria barnesae* near the lower/upper Maastrichtian boundary (as shown in Fig. 18). This disappearance corresponds with a distinct positive excursion $\delta^{18}O$ values (Barrera and Huber, 1990).

Cooling near the end of the Maastrichtian may also be indicated by the nannofossil record at mid-latitude South Atlantic Site 527. In Figure 19, the oxygen isotope record (Shackleton and Hall, 1984) is shown along with the cool-water Cretaceous nannofossil species *Nephrolithus frequens*, *Kamptnerius magnificus*, and *Cribrosphaerella daniae* and the warm-water form *Watznaueria barnesae*. In the interval represented, the cool-water taxa exhibit very slight abundance increases below the boundary, whereas the warm-water species shows an abundance decrease: the $\delta^{18}O$ values indicate a slight cooling trend. Taken together, these data may suggest cooling temperatures in the latest Maastrichtian. However, although the 10 to 15% drop in the percent abundance of *W. barnesae* may seem significant, the percent abundance changes of the cool-water taxa are very low and well within the limits of random error. In addition, the section must be analyzed below the interval sampled here in order to more clearly determine what may be short-term fluctuations in nannofossil abundances characteristic of the entire upper Maastrichtian as opposed to significant changes specific to the interval represented here.

SURVIVORSHIP VERSUS REWORKING

In all the graphs presented in this chapter thus far, nowhere are the Cretaceous species shown to disappear completely at the K/T boundary. Specimens are always present in variable amounts within nannofossil Zones NP1 and NP2, and there is

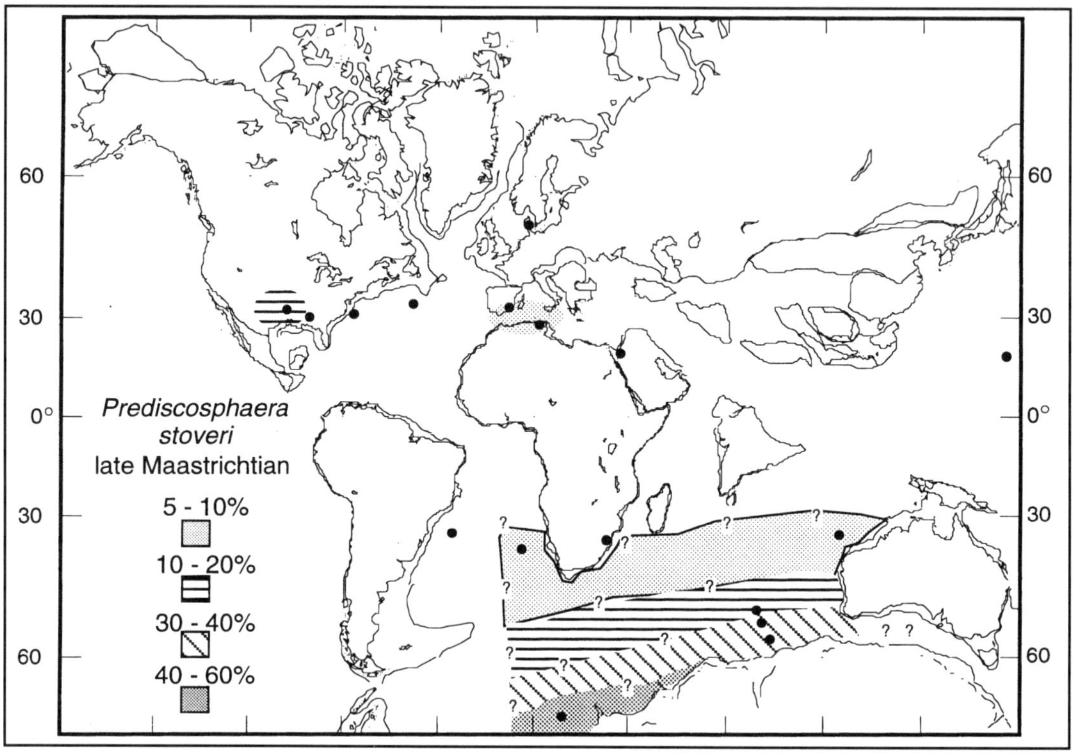

Figure 16. Percent abundance distribution of *Prediscosphaera stoveri* for the topmost Maastrichtian. Data (*P. quadripunctata* = *P. stoveri*) for Brazos River, Texas, are from Jiang and Gartner (1986). See Figure 1 for site designations. Plate (66 Ma) reconstructions are from Scotese and Denham (1988), Terra Mobilis™.

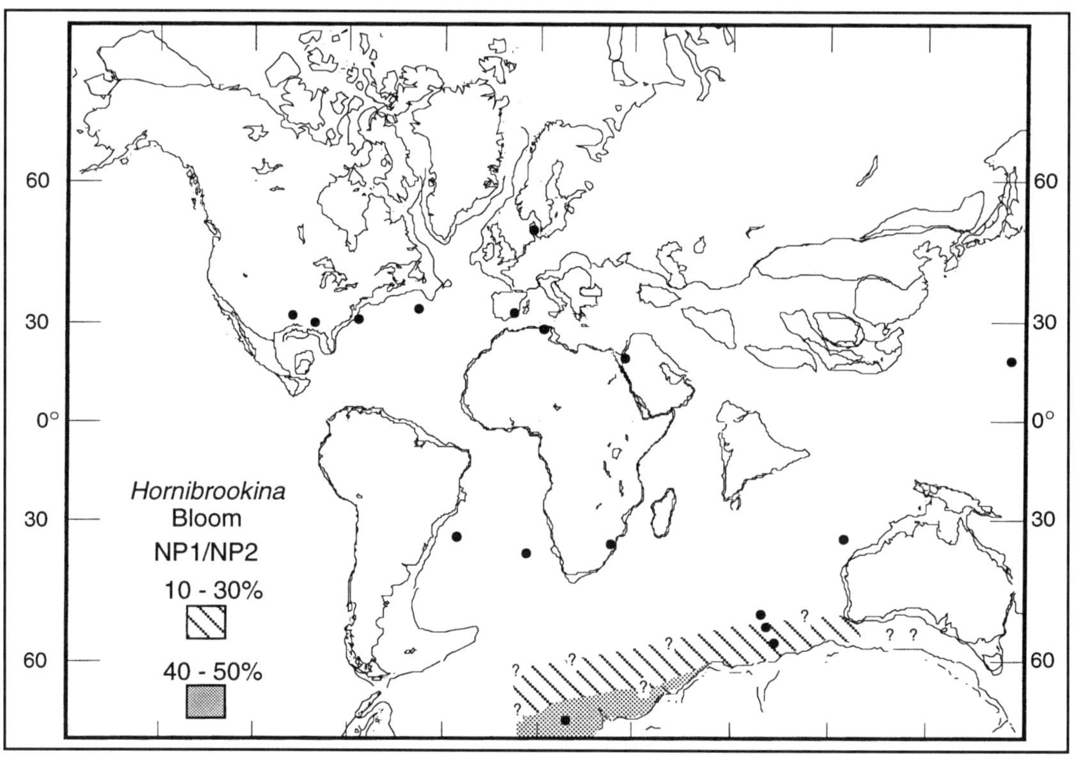

Figure 17. Percent abundance distribution of *Hornibrookina* for near the Danian NP1/NP2 boundary. See Figure 1 for site designations. Plate (66 Ma) reconstructions are from Scotese and Denham (1988), Terra Mobilis™. Note drop in abundance between Sites 750 and 752.

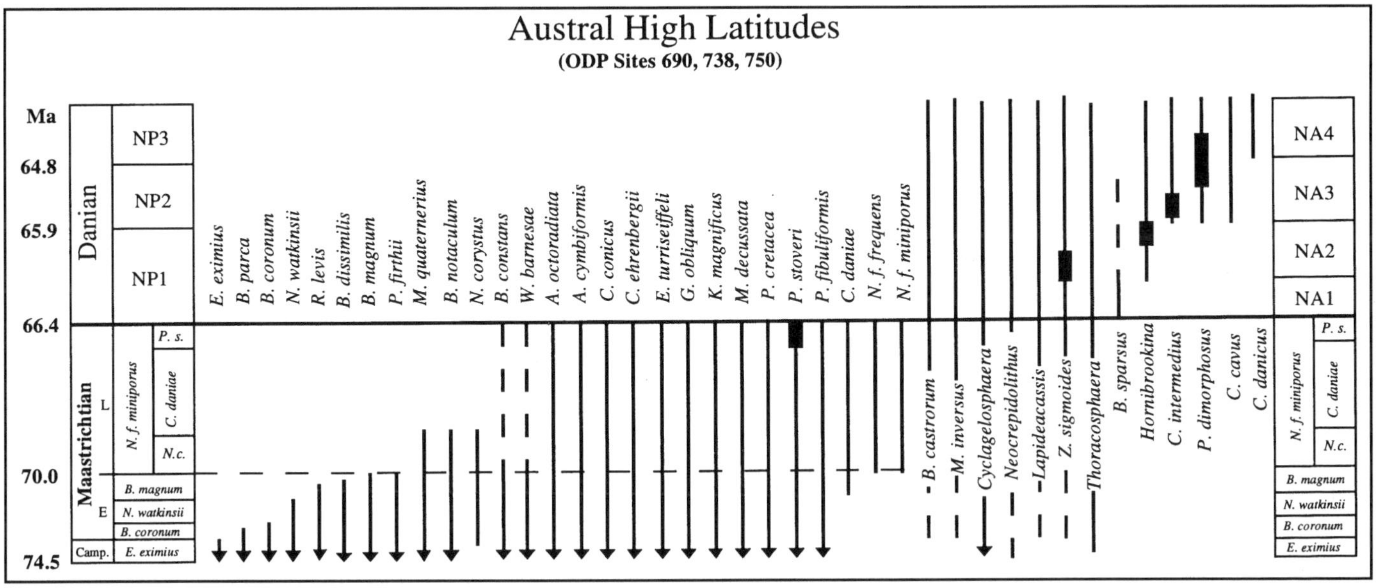

Figure 18. Ranges of selected taxa for the early Maastrichtian and Danian. Note the high number of extinctions near the early/late Maastrichtian boundary. Danian zonation on right is from Wei and Pospichal (1991); Maastrichtian zonation (Watkins et al., 1996). Ages (Ma) are from Berggren et al. (1985). P. s. = *Prediscosphaera stoveri* and N. c. = *Nephrolithus corystus*.

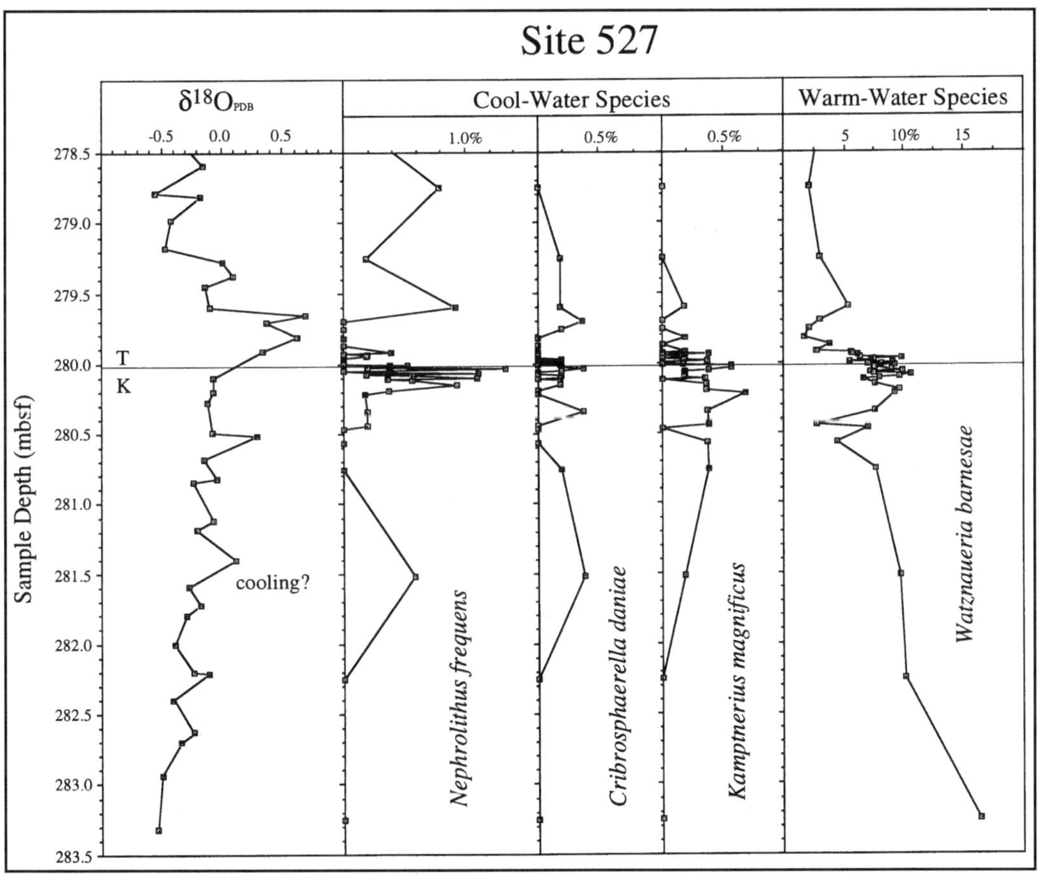

Figure 19. Percent abundance of high-latitude cool-water Cretaceous nannofossil species and the warm-water species correlated with bulk sediment oxygen isotopes (Shackleton and Hall, 1984) across the K/T boundary of South Atlantic DSDP Site 527. Isotopes indicate a slight cooling trend in the latest Maastrichtian. Nannofossil changes among cool-water taxa are within margins of error.

always a gradual decline up through these zones. These occurrences fuel the controversy of whether these specimens are reworked extinct forms or actual survivors. The fact that they are present in the Danian sediments led some observers to conclude that these species did indeed survive the K/T event. However, their overall abundance and individual species abundance patterns argue against survivorship.

Because of the small size of nannofossils this conclusion is difficult to test. Presently, only two basic ways to approach the problem are available. The first is to look at the total and individual abundances of Cretaceous specimens above the boundary and attempt to rationalize population changes with what an extinct population should look like compared to an extant one. The second is to relate these abundance changes to independent evidence such as isotopes or, as presented later, iridium and sea level changes. None of these methods individually has produced unequivocal results or a definitive answer to the problem (see MacLeod, 1994).

Cretaceous nannofossil abundance patterns above the K/T boundary

As shown in Figures 10 through 12, the general pattern of Cretaceous nannofossil reduction above the K/T boundary follows bioturbation mixing models (e.g., Berger and Heath, 1968), a point emphasized by Thierstein and Okada (1979) and Thierstein (1981). However, the pattern is the same in sections such as at Site 738 where a short laminated sequence is preserved with little bioturbation. This has led to the suggestion that perhaps, instead of bioturbation, the cause of this pattern (Fig. 12) is actually the result of the gradual decline of Cretaceous species in the Danian seas (Thierstein et al., 1991). The absence of bioturbation does not suggest the lack of reworking because redeposition by bottom current activity must be considered a strong possibility (Pospichal, 1991; Pospichal and Huber, 1992). It is apparent that just looking at an overall pattern does not reveal a clear solution to the problem.

One way to determine whether Cretaceous specimens in Danian sediments are reworked or were actual survivors is to look at individual species abundance patterns. Figure 20 (from Pospichal, 1994a) shows the two most abundant species at El Kef, *Watznaueria barnesae* and *Micula decussata*. Below the boundary, *W. barnesae* is generally the dominant form, whereas above the boundary *M. decussata* is always more abundant. The latter is a more robust, dissolution-resistant species (see Thierstein, 1980) and may be expected to be more abundant in reworked assemblages. Also, above the boundary, unlike below, the abudnance of these two species changes in concert. This uniformity may be expected for a reworked assemblage if one horizon is being eroded elsewhere to supply the reworked specimens.

Another clue that points to a Cretaceous origin for the assemblages is that at most localities the Cretaceous species abundance changes occur in proportion to one another above the boundary (Fig. 21). There is no variability within the Cretaceous assemblage as with the haracteristic sequential abundance changes amont the Tertiary and survivor species. When the abundance of one species rises or falls, generally the others do likewise and in the same proportions. This also suggests that the specimens have been reworked but again does not provide definitive proof.

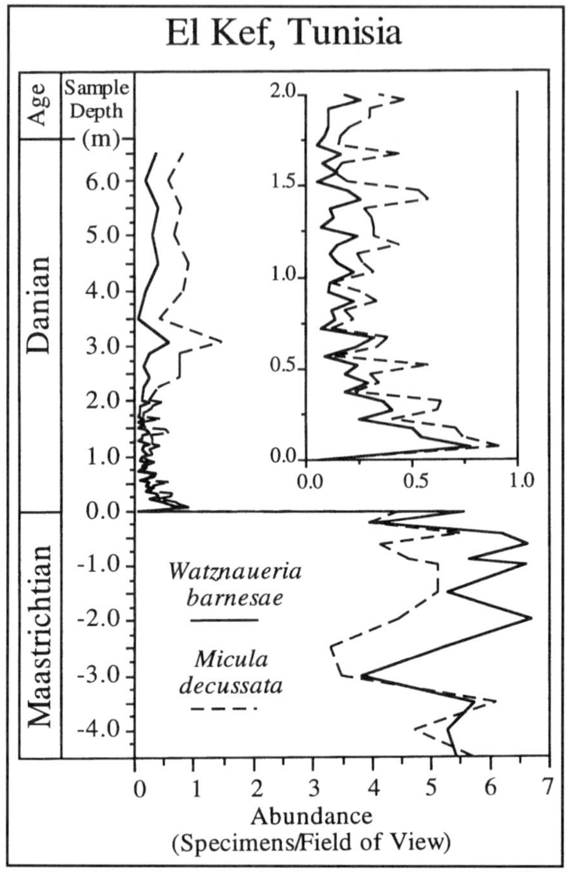

Figure 20. Abundance (specimens/field of view) of the Cretaceous taxa *Watznaueria barnesae* and *Micula decussata* across the K/T boundary at El Kef, Tunisia. Note uniform changes above the boundary and switch in dominance to more the robust species *M. decussata*. (Modified from Pospichal, 1994a.)

Correlating Cretaceous abundance with independent data sets

Isotopes. In arguing for survivorship, Perch-Nielsen et al. (1982) noted a "Tertiary" isotope signal in bulk samples from the lowermost centimeters of the Danian at Site 524 and El Kef. This signal is predicted to be derived from mostly Cretaceous nannofossils, which dominate the fine fraction just above the K/T boundary. Hence, it was concluded that these "Cretaceous" specimens lived in the early Danian seas. However, bulk sediment isotope measurements in this interval must be interpreted with caution because of the uncertain origin of much of the abundant microcarbonate present in these sediments. Sources include foraminifer and thoracosphaerid fragments that may contribute substantially to this "Tertiary" signal.

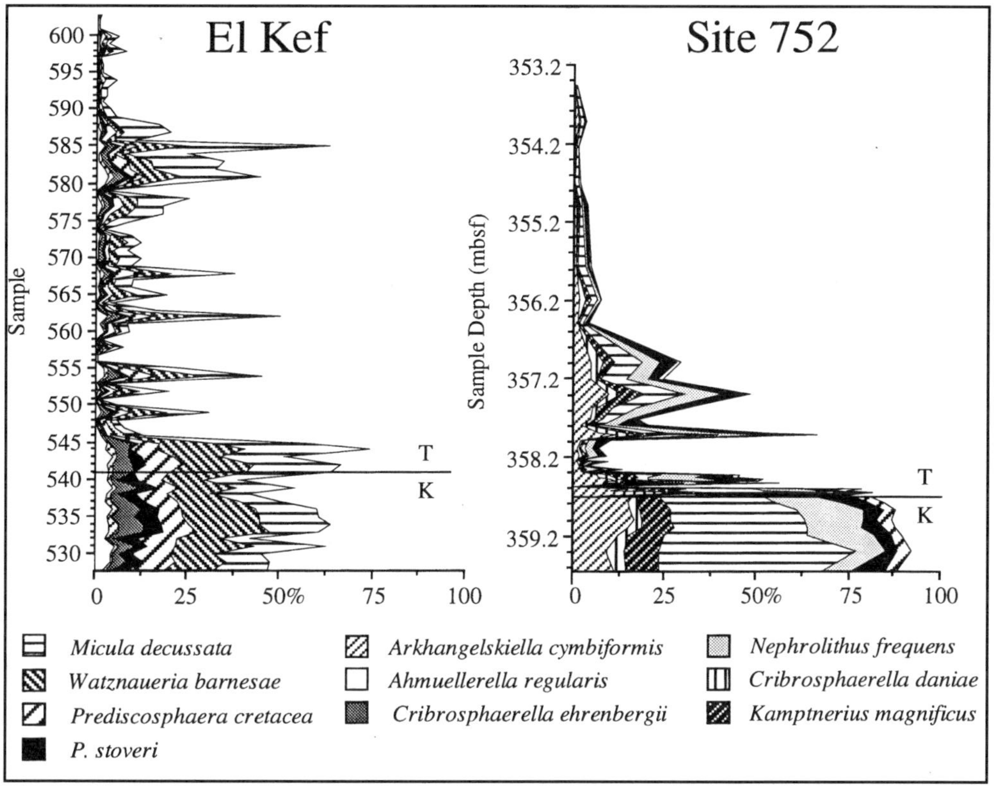

Figure 21. Percent abundance of the most abundant Cretaceous species across the K/T boundary of El Kef, Tunisia, and Indian Ocean ODP Site 752 (Pospichal, 1991). Note uniform and proportional changes in samples above the boundary.

The bulk sample analysis approach was recently challenged by Gartner et al. (1994), who measured $\delta^{13}C$ from K/T boundary samples from Site 527 and Site 529 (also on the Walvis Ridge). Their analysis was on samples processed by repeated steps of settling the sediment through a column of water in order to systematically remove all particles smaller and larger than the coccoliths. Isotopic composition was measured on the resultant residue, which contained significantly more coccolith than non-coccolith debris. Their results indicate that the isotope signal in the sediments just above the boundary is not necessarily derived from the "Cretaceous" component and that these species probably did not live into the early Danian.

It is obvious that with nannofossils, the isotope approach may produce ambiguous results because of factors such as diagenesis or variations in methodology, but the detailed work of Gartner and coworkers is highly suggestive and must be tested on additional sections.

Sea-level lowstands. Figure 22 represents the abundances of near-shore palynomorphs (Brinkhuis and Zachariasse, 1988)

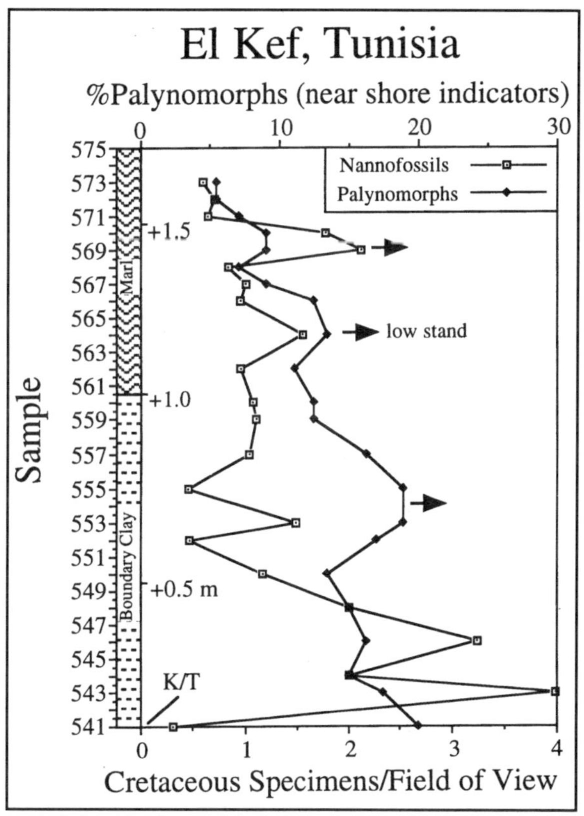

Figure 22. Cretaceous specimen abundance (specimens/field of view) plotted with near-shore indicator palynomorphs (Brinkhuis and Zachariasse, 1988) above the boundary at El Kef, Tunisia. Arrows indicate sea-level low stands. Nannofossil and palynomorph data are from the same sample set (modified from Pospichal, 1994a).

plotted with Cretaceous nannofossil abundance (Pospichal, 1994a) from the same El Kef K/T sample set. The palynomorph peaks indicate near-shore conditions or short-term sea-level low stands in the earliest Danian. A rough correlation is present between the low stands and Cretaceous specimen abundance. Reworking of Cretaceous nannofossils is expected in this scenario if the potential for reworking is considered to be enhanced during shoaling conditions. A similar conclusion was drawn by Eshet et al. (1992) from nannofossil and palynomorph data of the Hor Hahar K/T section in Israel.

Iridium and nannofossils. Reworking of Cretaceous nannofossils upsection may be inferred if a correlation can be found to exist between these specimens and other components of the sediment that may also be redistributed by physical processes on the seafloor. A relationship between the platinum group element, iridium, and calcareous nannofossils was first noted in the heavily bioturbated K/T boundary of Site 690 (Pospichal et al., 1990). At this site, dark-colored burrows sampled from below the K/T boundary were found to contain Tertiary nannofossils. These Tertiary specimens were not present in the surrounding light-colored chalk matrix, and it was estimated that in one burrow, Tertiary specimens were carried a minimum of 1.3 m downward by burrowing organisms. Figures 11 and 14 illustrate the abundances of Tertiary and survivors in three distinct burrows below the K/T boundary at Site 690. Figure 23 shows the iridium and percent abundance of Cretaceous specimens replotted from Michel et al. (1990) and Pospichal and Wise (1990a), respectively. The three Cretaceous low abundances below the K/T boundary are from burrows. The iridium measurements from bulk samples near the burrows show similar peaks above background. It is surmised from this that burrowing organisms moved Danian nannofossils and iridium downsection and that similar distinct secondary iridium peaks that have been found in other sections above and below the K/T boundary may be the result of the same process.

Figures 24 through 26 represent the continued search for a nannofossil-iridium relationship in deep-sea cores from the Indian Ocean (Pospichal and Huber, 1992). At Site 738, the Cretaceous nannofossil abundance declines upsection, and a similar pattern is noted with the iridium abundances by Schmitz et al. (1991), who suggest reworking to explain the high iridium concentrations above the main peak. A small (5-mm) burrow was sampled from within the laminated sediments of Site 738. The burrow contains Danian specimens of *Hornibrookina* moved an estimated 60 cm from upsection. No noticeable change was noted in iridium measurements taken near this burrow, however.

The section at Site 761 is marked by distinct erosional disconformities, illustrated in Figure 25 by the sharp changes in Cretaceous nannofossil percent abundance. Erosional contacts are visible in the core, which also shows moderate bioturbation. The iridium profile of Rocchia et al. (1992) displays a similar pattern of decline above the boundary as the Cretaceous nannofossil percent abundance. As at Site 738, however, the relationship between nannofossil abundance and iridium decline that

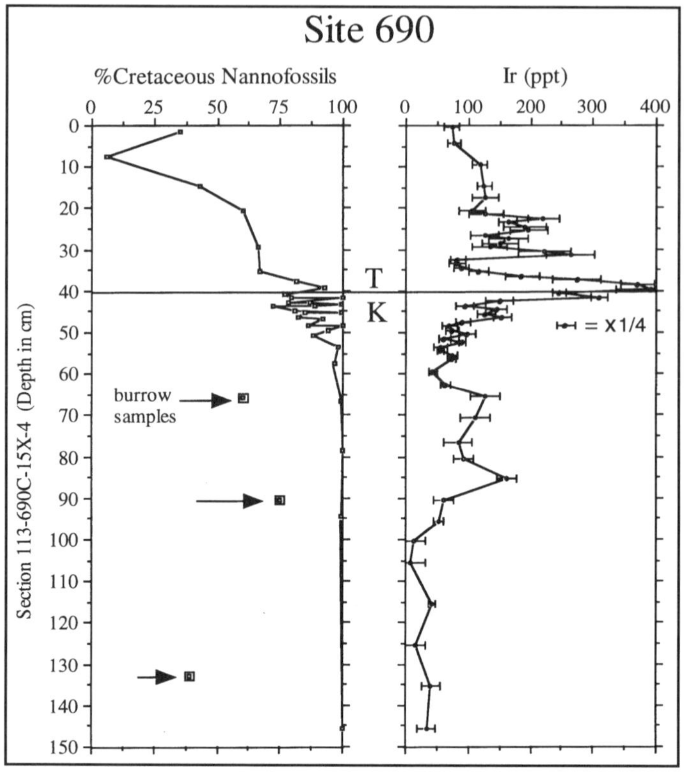

Figure 23. Percent abundance of Cretaceous nannofossils (Pospichal and Wise, 1990a) and iridium abundance (ppt = parts per trillion) from Michel et al. (1990) across the K/T boundary of Weddell Sea ODP Site 690. Note secondary iridium peaks in vicinity of burrows.

suggests reworking upsection of these components is not as convincing as the nannofossil-iridium-burrow relationship at Site 690.

A more convincing case for mechanical reworking and redeposition of Cretaceous nannofossils and iridium is present at Site 752 (Fig. 26). The nannofossil profile and the iridium profile (Michel et al., 1991) above the boundary are similar. Cretaceous nannofossils begin their abundance decline at the peak iridium anomaly and rapidly decline just above the boundary along with the iridium abundance. Interestingly, both rise in abundance again at about 358.5 mbsf (meters below sea floor). The peak at about 357.9 mbsf (80 cm above the K/T boundary), however, is most suggestive of reworking. No other mechanism has been proposed that can account for the near disappearance of both iridium and nannofossils in the section and then their reappearance together at high abundances in the same level. No burrows were noted at this level; thus it is suggested that Cretaceous specimens in Danian sediments are reworked by bottom currents in addition to burrowing.

The K/T iridium profile in all marine sections shows elevated abundances beginning just below and for several centimeters or more above the boundary (Rocchia et al., 1994). This halo of high iridium abundance around the boundary is attributed by these authors to chemical diffusion of this element after deposition rather than to physical redistribution by biotur-

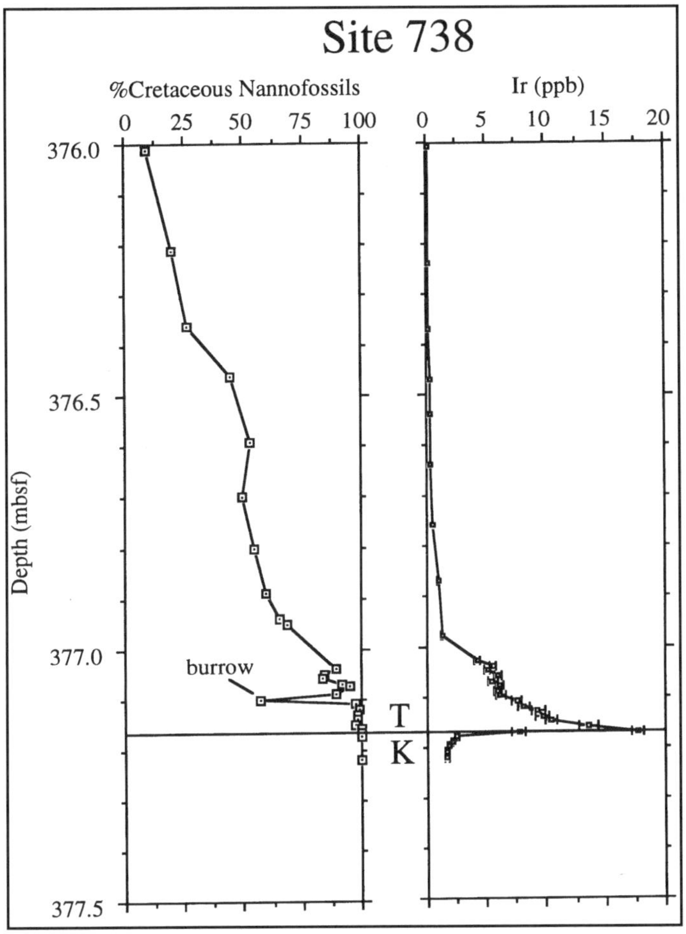

Figure 24. Percent abundance of Cretaceous nannofossils and iridium abundance (ppb = parts per billion) from Schmitz et al. (1991) across the K/T boundary of South Indian Ocean ODP Site 738 (modified from Pospichal and Huber, 1992). Nannofossil data are given in Pospichal (1996b).

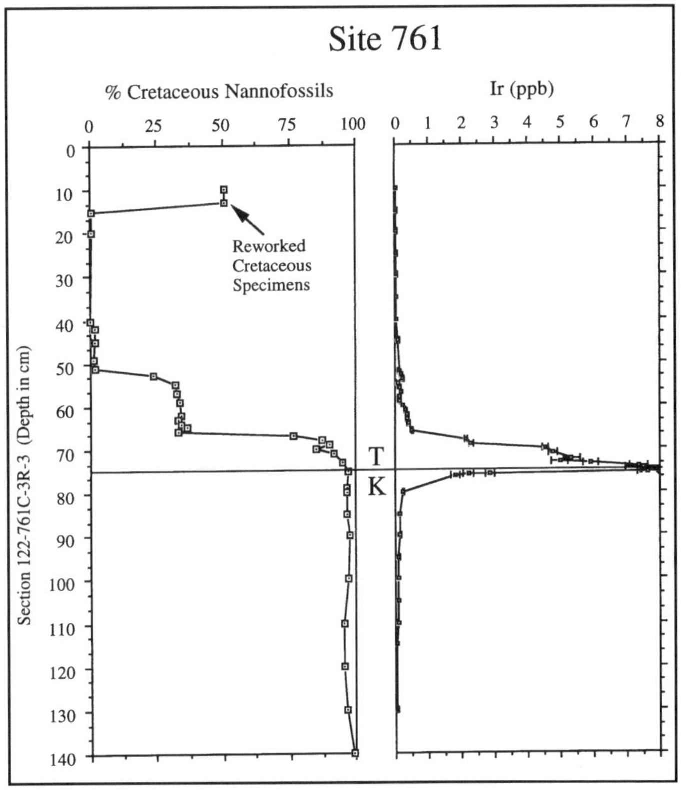

Figure 25. Percent abundance of Cretaceous nannofossils (Pospichal and Bralower, 1992) and iridium abundance (ppb = parts per billion) from Rocchia et al. (1992) across the K/T boundary of Indian Ocean ODP Site 761 (modified from Pospichal and Huber, 1992).

bation (see also Schmitz et al., 1988). This conclusion is supported by the record of Ni-rich meteoric spinels associated with the iridium at the K/T boundary. The spinels do not show a similar broad abundance pattern in the boundary interval but have a distinct narrow peak, which corresponds precisely to the peak iridium abundance at the K/T boundary. If iridium and Cretaceous nannofossils are being physically moved up and down section, then it follows that the spinels should be too. Although physical redistribution of the iridium as proposed here is probable and cannot be ignored, the fact that the spinels do not appear to be mobilized as noted by Rocchia et al. (1994) is an

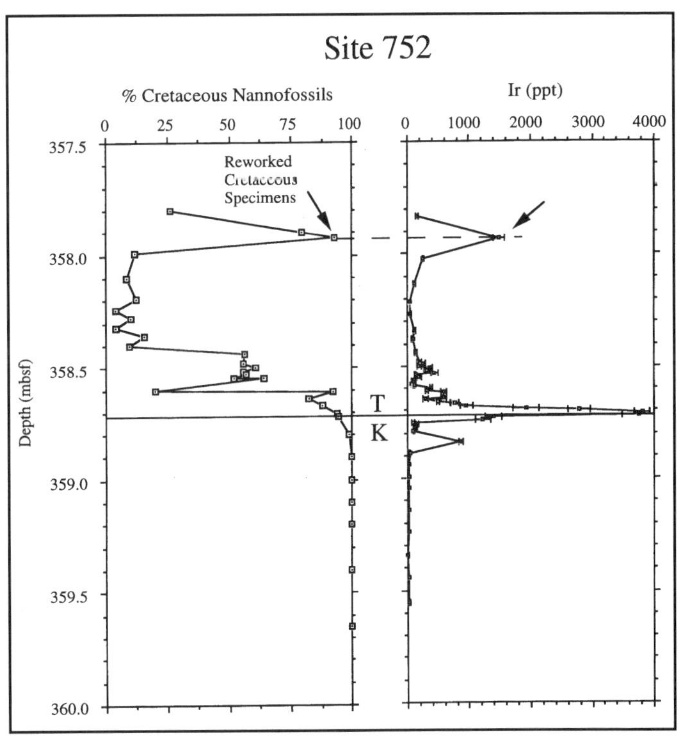

Figure 26. Percent abundance of Cretaceous nannofossils (Pospichal, 1991) and iridium abundance (ppt = parts per trillion) from Michel et al. (1991) across the K/T boundary of South Indian Ocean ODP Site 752. Note secondary iridium peak and Cretaceous nannofossil abundance peak 80 cm above K/T boundary (modified from Pospichal and Huber, 1992).

interesting point. A simple test would be to check for meteoric spinels in the burrows at Site 690 and at the level of high Cretaceous nannofossil abundance and anomalous iridium 80 cm above the boundary at Site 752.

DISCUSSION

Short-term climate changes in the 250 k.y. or so years immediately preceding the K/T event, which may be inferred from planktonic foraminifer record from El Kef, Tunisia, and Agost, Spain (Keller, 1988; Canudo et al., 1991), are not confirmed by the nannofossil data. In regard to long-term changes, detailed quantitative studies of nannofloral and micro- and macrofaunal changes in the Late Cretaceous, which extend beyond the latest Maastrichtian, have only recently begun.

In austral high latitudes, important climatic changes may have occurred 5 to 10 m.y. before the terminal Cretaceous event. This is suggested by the increased extinction rates through the lower to mid-Maastrichtian in the Southern Ocean (Pospichal and Wise, 1990b; Huber and Watkins, 1992). Aside from a brief mention by Perch-Nielsen (1986), a similar phenomenon among nannofossil assemblages from the mid- and low latitudes or for the high boreal latitudes has not yet been documented in detail. However, many scientists have argued that global changes were occurring elsewhere during the Late Cretaceous, notably those that led to the reduction in the diversity of a number of marine organisms, including the ammonites (Ward et al., 1986; Wiedmann, 1986; Hallam, 1987). Furthermore, as with the Southern Ocean nannoflora, the critical interval of extinction cited for the inoceramids is at the early/late Maastrichtian transition (MacLeod and Ward, 1990). The potential relationship between these events has yet to be adequately explored.

Another matter of debate among paleontologists concerns the assertion that extinctions were more severe in the low latitudes and nearly inconsequential in the high latitudes (Zinsmeister et al., 1989; Barrera and Keller, 1992; Keller, 1993; Keller et al., 1993; Elliot et al., 1994). Other investigators have reported that the K/T extinctions were as massive in relative magnitude in the highest latitudes as in the tropics (Raup and Jablonski, 1993). As stated previously, the calcareous nannoplankton extinctions were global in extent, irrespective of latitude. This is true whether one considers the "Cretaceous" specimens in Danian sediments as survivors or as redeposited. The important point is that the pattern and magnitude of the overall nannofossil turnover are identical in low and high latitudes. Cosmopolitan Cretaceous species common to both the Southern Ocean and Tethys Sea suffered the same fate at the K/T boundary, whether it was an abrupt end or a million years of gradual dying.

CONCLUSIONS

The immediate objective of this chapter has been to summarize the record of calcareous nannofossil assemblages across the K/T boundary of several biostratigraphically complete sections in order to determine the pattern of nannoplankton extinctions. This includes noting whether or not any distinct changes occurred among nannofossil assemblages that may have foreshadowed the K/T event. An additional objective was to compare uppermost Maastrichtian and lowermost Danian nannofossil assemblages across latitude in the Indian, Tethys, and South Atlantic Oceans in order to determine any biogeographic gradients indicative of latitudinal thermal gradients of this time interval. The main conclusions are as follows.

1. No pattern of stepwise, sequential, or gradual nannoplankton extinctions is noted in this study. Extinctions are abrupt and synchronous and correlate precisely with shifts in carbon isotopes, the drop in carbonate sedimentation, and anomalous iridium abundance peaks.

2. There is no unequivocal evidence for survivorship into the Danian of species considered members of the Cretaceous assemblage as defined here. Consideration of the associations with iridium, short-term sea-level low stands and the overall pattern of abundance above the boundary suggest that Cretaceous specimens present in Danian sediments are reworked either by processes of bioturbation or by current activity. Survivorship for any measurable time into the Danian as suggested by some isotope analyses (Perch-Nielsen et al., 1982) cannot be confirmed in this study or by results of Gartner et al. (1994).

3. The effects of the K/T event on nannoplankton were as severe in the austral high latitudes as in the tropics. Regardless of whether one views "Cretaceous" specimens in Danian sediments as survivors or as redeposited, the pattern and magnitude of the overall nannofossil turnover are comparable between low and austral high latitudes.

4. There is no evidence of significant changes in nannofossil assemblages in the last 250 k.y. or so of the Maastrichtian. This reflects relatively stable conditions immediately preceding the K/T boundary event.

5. Nannoplankton extinction at the K/T boundary was not total. Depending on taxonomic concepts, about 15 to 20 species survived the K/T crisis (Perch-Nielsen, 1986) to make up the survivor assemblage discussed here. In general, these species are either high-latitude taxa (*Zygodiscus sigmoides*) or those associated with neritic environments (*Braarudosphaera*). Their preference for environments characterized by fluctuating salinities, temperature, and—in the case of the high-latitude species—seasonal light intensity variations may be an indication of variable conditions that existed around the globe in the earliest Danian.

6. At all sites the lowermost Danian is characterized by a pattern of sequential abundance peaks of surviving and new species. No two localities are exactly the same in terms of the species composition for these sequential abundance peaks.

7. Well-developed latitudinal thermal gradients were already well established in the Southern Hemisphere by latest Maastrichtian time and continued into the early Danian, as indicated by the biogeographic provinces defined by *Prediscosphaera stoveri* and *Hornibrookina*. A similar gradient for these

taxa has not been established for the Northern Hemisphere. The austral nannofloral provinces reflect the continuation of a climatic trend and associated nannofossil provincialism, perhaps related to geographic reorganization and subsequent circulation changes, that began in the Campanian (S. W. Wise, personal communication, 1990; Huber and Watkins, 1992). A distinct turnover in nannofossil assemblages in the mid-Maastrichtian (Pospichal and Wise, 1990b; Watkins et al., 1996) characterizes this ongoing deterioration of climates in the region of the Southern Ocean during the Late Cretaceous.

ACKNOWLEDGMENTS

Nannofossil studies were supported by a Smithsonian Graduate Fellowship, Geological Society of America Grant #4772-91, and NSF grants DPP 91-18480 and EAR 93-16960. I thank Dr. G. Keller for providing the samples from Agost and El Kef and for helpful discussions. Special thanks to S. Gartner for a detailed review that greatly improved the manuscript and to S. W. Wise, Jr., D. K. Watkins, and W. Wei for helpful suggestions.

REFERENCES CITED

Alcalá-Herrera, J. A., Grossman, E. L., and Gartner, S., 1993, Nannofossil diversity and equitability and fine-fraction $\delta^{13}C$ across the Cretaceous/Tertiary boundary at Walvis Ridge Leg 74, South Atlantic: Marine Micropaleontology, v. 20, p. 77–88.

Alvarez, L. W., Alvarez, W., Asaro, F., and Michel, H. V., 1980, Extraterrestrial cause for the Cretaceous-Tertiary extinction: Science, v. 208, p. 1095–1108.

Alvarez, W., Alvarez, L. W., Asaro, F., and Michel, H. V., 1984, The end of the Cretaceous: Sharp boundary or gradual transition?: Science, v. 223, p. 1183–1186.

Arthur, M. A., Zachos, J. C., and Jones, D. S., 1987, Primary productivity and the Cretaceous/Tertiary boundary event in the oceans: Cretaceous Research, v. 8, p. 43–45.

Barker, P. F., Kennett, J. P., and others, 1988, Proceedings, Ocean Drilling Program, Initial Reports, Volume 113: College Station, Texas, Ocean Drilling Program, 785 p.

Barrera, E., and Huber, B. T., 1990, Evolution of Antarctic waters during the Maestrichtian: Foraminifer oxygen and carbon isotope ratios, Leg 113, *in* Barker, P. F., Kennett, J. P., and others, Proceedings, Ocean Drilling Program, Scientific Results, Volume 113: College Station, Texas, Ocean Drilling Program, p. 159–168.

Barrera, E., and Keller, G., 1990, Stable isotope evidence for gradual environmental changes and species survivorship across the Cretaceous/Tertiary boundary: Paleoceanography, v. 5, p. 867–890.

Barrera, E., and Keller, G., 1992, Productivity and planktonic foraminiferal changes across the Cretaceous-Tertiary boundary in high latitudes: Geological Society of America Abstracts with Programs, v. 24, p. A332.

Barron, J., Larsen, B., and others, 1989, Proceedings, Ocean Drilling Program, Initial Reports, Volume 119: College Station, Texas, Ocean Drilling Program, 942 p.

Berger, W. H., and Heath, G. R., 1968, Vertical mixing in pelagic sediments: Journal of Marine Research, v. 2, p. 134–143.

Berggren, W. A., Kent, D. V., Flynn, J. J., and van Couvering, J. A., 1985, Cenozoic geochronology: Geological Society of America Bulletin, v. 96, p. 1407–1418.

Boersma, A., 1984, Cretaceous-Tertiary planktonic foraminifers from the south-eastern Atlantic, Walvis Ridge area, Deep Sea Drilling Project Leg 74, *in* Moore, T. C., Jr., Rabinowitz, P. D., and others, Initial reports of the Deep Sea Drilling Project, Volume 74: Washington, D.C., U.S. Government Printing Office, p. 501–524.

Bown, P. R., Burnett, J. A., and Gallagher, L. T., 1991, Critical events in the evolutionary history of the calcareous nannoplankton: Historical Biology, v. 5, p. 279–290.

Bramlette, M. N., and Martini, E., 1964, The great change in calcareous nannoplankton fossils between the Maestrichtian and Danian: Micropaleontology, v. 10, p. 291–322.

Brinkhuis, H., and Zachariasse, W.J., 1988, Dinoflagellate cysts, sea level changes and planktonic foraminifers across the Cretaceous-Tertiary boundary at El Haria, Northwest Tunisia: Marine Micropaleontology, v. 13, p. 153–191.

Cande, S., and Kent, D. V., 1992. A new geomagnetic polarity time scale for the Late Cretaceous and Cenozoic: Journal of Geophysical Research, v. 97, p. 13917–13951.

Canudo, J. I., Keller, G., and Molina, E., 1991, Cretaceous/Tertiary boundary extinction pattern and faunal turnover at Agost and Caravaca, S.E. Spain: Marine Micropaleontology, v. 17, p. 319–341.

Chave, A. D., 1984, Lower Paleocene–Upper Cretaceous magnetostratigraphy, Sites 525, 527, 528, and 529, Deep Sea Drilling Project Leg 74, *in* Moore, T. C., Jr., Rabinowitz, P. D., and others, Initial reports of the Deep Sea Drilling Project, Volume 74: Washington, D.C., U.S. Government Printing Office, p. 525-531.

Clemens, W. A., Archibald, J. D., and Hickey, L. J., 1981, Out with a whimper not a bang: Paleobiology, v. 7, p. 293–298.

Ehrendorfer, T., and Aubry, M.-P., 1992, Calcareous nannoplankton changes across the Cretaceous/Paleocene boundary in the Southern Indian Ocean (ODP Site 750A), *in* Wise, S. W., Jr., Schlich, R., and others, Proceedings, Ocean Drilling Program, Scientific Results, Volume 120: College Station, Texas, Ocean Drilling Program, p. 451–470.

Elliot, D. H., Askin, R. A., Kyte, F. T., and Zinsmeister, W. J., 1994, Iridium and dinocysts at the Cretaceous/Tertiary boundary on Seymour Island, Antarctica: Implications for the K/T event: Geology, v. 22, p. 675–678.

Eshet, Y., Moshkovitz, S., Habib, D., Benjamini, C., and Magaritz, M., 1992, Calcareous nannofossil and dinoflagellate stratigraphy across the Cretaceous/Tertiary boundary at Hor HarHar, Israel: Marine Micropaleontology, v. 18, p. 199–228.

Fütterer, D. K., 1984, Bioturbation and trace fossils in deep sea sediments of the Walvis Ridge, Southeastern Atlantic, Leg 74, *in* Moore, T. C., Jr., Rabinowitz, P. D., and others, Initial reports of the Deep Sea Drilling Project, Volume 74: Washington, D.C., U.S. Government Printing Office, p. 543–556.

Gartner, S., Alcalá, J., and Grossman, E., 1994, Coccolithophore extinction at the KT boundary: Gradual or abrupt [abs.], *in* New developments regarding the K/T event and other catastrophes in Earth history: Houston, Texas, Lunar and Planetary Institute Contribution 825, p. 40–41.

Gee, J., Klootwijk, C. T., and Smith, G. M., 1991, Magnetostratigraphy of Paleogene and Upper Cretaceous sediments from Broken Ridge, eastern Indian Ocean, *in* Weissel, J., Peirce, J., and others, Proceedings, Ocean Drilling Program, Scientific Results, Volume 121: College Station, Texas, Ocean Drilling Program, p. 359–376.

Groot, J. J., de Jonge, R. B. G., Langereis, C. G., ten Kate, W. G., and Smit, J., 1989, Magnetostratigraphy of the Cretaceous-Tertiary boundary at Agost, Spain: Earth and Planetary Science Letters, v. 94, p. 385–397.

Hallam, A., 1987, End-Cretaceous mass extinction event: Argument for terrestrial causation: Science, v. 238, p. 1237–1242.

Hamilton, N., 1990, Mesozoic magnetostratigraphy of Maud Rise, Antarctica, *in* Barker, P. F., Kennett, J. P., and others, Proceedings, Ocean Drilling Program, Scientific Results, Volume 113: College Station, Texas, Ocean Drilling Program, p. 255–260.

Haq, B. U., von Rad, U., and others, 1990, Proceedings, Ocean Drilling Program, Initial reports, Volume 122: College Station, Texas, Ocean Drilling Program, 826 p.

Henriksson, A. S., 1993, Biochronology of the terminal Cretaceous calcareous nannofossil zone of *Micula prinsii*: Cretaceous Research, v. 14, p. 59–68.

Hsü, K. J., Mckenzie, J. A., and He, Q. X., 1982, Terminal Cretaceous environmental and evolutionary changes, *in* Silver, L. T., and Schultz, P. H., eds., Geological implications of impact of large asteroids and comets on earth: Geological Society of America Special Paper 190, p. 317–328.

Hsü, K. J., LaBrecque, J. L., and others, 1984, Initial reports of the Deep Sea Drilling Project, Volume 73: Washington, D.C., U.S. Government Printing Office, 798 p.

Huber, B. T., and Watkins, D. K., 1992, Biogeography of the Campanian-Maastrichtian calcareous plankton in the region of the Southern Ocean: Paleogeographic and paleoclimatic implications, *in* The Antarctic paleoenvironment: A perspective on global change: Washington, D.C., American Geophysical Union, Antarctic Research Series, v. 56, p. 31–60.

Huffman, A. R., Crocket, J. H., Carter, N. L., Borella, P. E., and Officer, C. B., 1990, Chemistry and minerology across the Cretaceous/Tertiary boundary at DSDP Site 527, Walvis Ridge, South Atlantic Ocean, *in* Sharpton, V. L., and Ward, P. D., eds., Global catastrophes in Earth history: An interdisciplinary conference on impacts, volcanism, and mass mortality: Geological Society of America Special Paper 247, p. 319–334.

Jiang, M. J., and Gartner, S., 1986, Calcareous nannofossil succession across the Cretaceous/Tertiary boundary in east-central Texas: Micropaleontology, v. 32, p. 232–255.

Keller, G., 1988, Extinction, survivorship and evolution of planktonic foraminifers across the Cretaceous/Tertiary boundary at El Kef, Tunisia: Marine Micropaleontology, v. 13, p. 239–263.

Keller, G., 1992, Paleoecologic response of Tethyan benthic foraminifera to the Cretaceous-Tertiary boundary transition, *in* Takayanagi, Y., and Saito, T., eds., Studies in benthic foraminifera: Tokyo, Takai University Press, p. 77–91.

Keller, G., 1993, The Cretaceous-Tertiary boundary transition in the Antarctic Ocean and its global implications: Marine Micropaleontology, v. 21, p. 1–45.

Keller, G., and Lindinger, M., 1989, Stable isotope, TOC and $CaCO_3$ record across the Cretaceous/Tertiary boundary at El Kef, Tunisia: Palaeogeography, Palaeoclimatology, Palaeoecology, v. 73, p. 243–265.

Keller, G., Barrera, E., Schmitz, B., and Mattson, E., 1993, Gradual mass extinction, species survivorship, and long-term environmental changes across the Cretaceous-Tertiary boundary in high latitudes: Geological Society of America Bulletin, v. 105, p. 979–997.

Kuslys, M., and Krähenbühl, U., 1983, Noble metals in Cretaceous/Tertiary sediments from El Kef: Radiochimica Acta, v. 34, p. 139–141.

MacLeod, K.G., and Ward, P.D., 1990, Extinction pattern of Inoceramus (Bivalvia) based on shell fragment biostratigraphy, *in* Sharpton, V. L., and Ward, P. D., eds., Global catastrophes in Earth history: An interdisciplinary conference on impacts, volcanism, and mass mortality: Geological Society of America Special Paper 247, p. 509–518.

MacLeod, N., 1994, An evaluation of criteria that may be used to identify species surviving a mass extinction, *in* New developments regarding the K/T event and other catastrophes in Earth history: Houston, Texas, Lunar and Planetary Institute Contribution 825, p. 75–77.

MacLeod, N., and Keller, G., 1991, Hiatus distributions and mass extinctions at the Cretaceous/Tertiary boundary: Geology, v. 19, p. 497–501.

Manivit, H., 1984, Paleogene and Upper Cretaceous calcareous nannofossils from Deep Sea Drilling Project Leg 74, *in* Moore, T. C., Jr., Rabinowitz, P. D., and others, Initial reports of the Deep Sea Drilling Project, Volume 74: Washington, D.C., U.S. Government Printing Office, p. 475–499.

Martini, E., 1971, Standard Tertiary and Quaternary calcareous nannoplankton zonation, *in* Farinacci, A., ed., Proceedings, Second Planktonic Conference, Rome: Rome, Edizioni Tecnoscienza, v. 2, p. 739–785.

McLean, D. M., 1982, Deccan volcanism: The Cretaceous-Tertiary marine boundary timing event: Eos (Transactions, American Geophysical Union), v. 63, p. 562.

McLean, D. M., 1985, Deccan Traps mantle degassing in the terminal Cretaceous marine extinctions: Cretaceous Research, v. 6, p. 235–259.

Michel, H. V., Asaro, F., Alvarez, W., and Alvarez, L. W., 1990, Geochemical studies of the Cretaceous-Tertiary boundary in ODP Holes 689B and 690C, *in* Barker, P. F., Kennett, J. P., and others, Proceedings, Ocean Drilling Program, Scientific Results, Volume 113: College Station, Texas, Ocean Drilling Program, p. 159–168.

Michel, H. V., Asaro, F., and Alvarez, W., 1991, Geochemical study of the Cretaceous-Tertiary boundary region in ODP 121-752B, *in* Weissel, J., Peirce, J., and others, Proceedings, Ocean Drilling Program, Scientific Results, Volume 121: College Station, Texas, Ocean Drilling Program, p. 415–422.

Monechi, S., 1985, Campanian to Pleistocene calcareous nannofossils stratigraphy from the northwest Pacific Ocean, Deep Sea Drilling Project Leg 86, *in* Heath, G. R., Burkle, L. H., and others, Initial reports of the Deep Sea Drilling Project, Volume 86: Washington, D.C., U.S. Government Printing Office, p. 301–336.

Officer, C. B., and Drake, C. L., 1983, The Cretaceous/Tertiary transition: Science, v. 219, p. 1383–1390.

Officer, C. B., Hallam, A., Drake, C. L., and Devine, J. D., 1987, Late Cretaceous and paroxysmal Cretaceous/Tertiary extinctions: Nature, v. 326, p. 143–149.

Okada, H., and Bukry, D., 1980, Supplementary modification and introduction of code numbers to the low-latitude coccolith biostratigraphic zonation: Marine Micropaleontology, v. 5, p. 321–325.

Olsson, R. K., and Liu, C., 1993, Controversies on the placement of the Cretaceous-Paleogene boundary and the K/P mass extinction of planktonic foraminifera: Palaios, v. 8, p. 127–139.

Peirce, J. W., Weissel, J. K., and others, 1989, Proceedings, Ocean Drilling Program, Initial Reports, Volume 121: College Station, Texas, Ocean Drilling Program, 1,000 p.

Perch-Nielsen, K., 1979a, Calcareous nannofossil zonation at the Cretaceous/Tertiary boundary in Denmark, *in* Birkelund, T., and Bromley, R. G., eds., Cretaceous-Tertiary boundary Events, Volume 1: Copenhagen, University of Copenhagen, p. 115–135.

Perch-Nielsen, K., 1979b, Calcareous nannofossils in Cretaceous/Tertiary boundary sections in Denmark, *in* Christensen, W. K., and Birkelund, T., eds., Cretaceous-Tertiary boundary events, Volume 2: Copenhagen, University of Copenhagen, p. 120–126.

Perch-Nielsen, K., 1979c, Calcareous nannofossils at the Cretaceous/Tertiary boundary near Biarritz, France, *in* Christensen, W. K., and Birkelund, T., eds., Cretaceous-Tertiary boundary events, Volume 2: Copenhagen, University of Copenhagen, p. 151–155.

Perch-Nielsen, K., 1979d, Calcareous nannofossils from the Cretaceous between the North Sea and the Mediterranean. International Union of Geological Sciences Series A, v. 6, p. 223–272.

Perch-Nielsen, K., 1981, Nouvelles observations sur les nannofossiles calcaires à la limite Crétacé-Tertiare près de El Kef (Tunisie): Cahiers de Micropaléontologie, v. 3, p. 25–36.

Perch-Nielsen, K., 1985, Mesozoic calcareous nannofossils, *in* Bolli, H. M., Saunders, J. B., and Perch-Nielsen, K., eds., Plankton stratigraphy: Cambridge, Cambridge University Press, p. 329–426.

Perch-Nielsen, K., 1986, Geologic events and the distribution of calcareous nannofossils—some speculations: Bulletin des Centres Recherches Exploration-Production Elf Acquitaine, v. 10, p. 421.

Perch-Nielsen, K., McKenzie, J. A., and He, Q., 1982, Bio- and isotope-stratigraphy and the "catastrophic" extinction of calcareous nannoplankton at the Cretaceous/Tertiary boundary, *in* Silver, L. T., and Schultz, P. H., eds., Geological implications of impacts of large asteroids and comets on the Earth: Geological Society of America Special Paper 190, p. 353–371.

Percival, S. F., and Fischer, A. G., 1977, Changes in calcareous nannoplankton in the Cretaceous-Tertiary biotic crisis at Zumaya, Spain: Evolutionary theory: Chicago, Illinois, University of Chicago, v. 2, p. 1–35.

Pospichal, J. J., 1991, Calcareous nannofossils across the Cretaceous/Tertiary boundary at Site 752, Eastern Indian Ocean, *in* Weissel, J., Peirce, J., and others, Proceedings, Ocean Drilling Program, Scientific Results, Volume 121: College Station, Texas, Ocean Drilling Program, p. 395–414.

Pospichal, J. J., 1994a, Calcareous nannofossils at the K/T boundary, El Kef: No evidence for stepwise, gradual or sequential extinctions: Geology, v. 22, p. 99–102.

Pospichal, J. J., 1994b, Calcareous nannofossils and the KT boundary crisis: An update, *in* New developments regarding the K/T event and other catastrophes in Earth history [abs.]: Houston, Texas, Lunar and Planetary Institute Contribution 825, p. 90.

Pospichal, J. J., 1996a, Cretaceous/Tertiary boundary calcareous nannofossils from Agost, Spain, *in* Proceedings, International Nannoplankton Meeting, 5th: Salamanca, Spain, University of Salamanca Press, p. 185–217.

Pospichal, J. J., 1996b, High austral latitude Cretaceous/Tertiary boundary calcareous nannofossils from ODP Site 738, *in* Whately, R., and Moguilevsky, A., eds., Microfossils and Oceanic environments: Aberystwyth, University of Wales Aberystwyth Press (in press).

Pospichal, J. J., 1996c, Calcareous nannofossils and clastic sediments of the Cretaceous-Tertiary boundary of northeastern Mexico: Geology, v. 24, no. 3, p. 255–258.

Pospichal, J. J., and Bralower, T. J., 1992, Calcareous nannofossils across the Cretaceous/Tertiary boundary, Site 761, Northwest Australian Margin, *in* Haq, B. U., von Rad, U., and others, Proceedings, Ocean Drilling Program, Scientific Results, Volume 122: College Station, Texas, Ocean Drilling Program, p. 735–752.

Pospichal, J. J., and Huber, B. T., 1992, The Cretaceous/Tertiary boundary in the southern Indian Ocean: Results from Ocean Drilling Program coring operations, *in* Duncan, R. A., Rea, D. K., Kidd, R. B., von Rad, U., and Weissel, J. K., eds., The Indian Ocean: A synthesis of results from the Ocean Drilling Program: Washington, D.C., American Geophysical Union, Geophysical Monograph 70, p. 275–294.

Pospichal, J. J., and Wise, S. W., Jr., 1990a. Calcareous nannofossils across the K/T boundary, ODP Hole 690C, Maud Rise, Weddell Sea, *in* Barker, P. F., Kennett, J. P., and others, Proceedings, Ocean Drilling Program, Scientific Results, Volume 113: College Station, Texas, Ocean Drilling Program, p. 515–532.

Pospichal, J. J., and Wise, S. W., 1990b, Maestrichtian calcareous nannofossil biostratigraphy of Maud Rise ODP Leg 113 Sites 689 and 690, Weddell Sea, *in* Barker, P. F., Kennett, J. P., and others, Proceedings, Ocean Drilling Program, Scientific Results, Volume 113: College Station, Texas, Ocean Drilling Program, p. 465–487.

Pospichal, J. J., and Wise, S. W., 1992, Calcareous nannofossils from southern high latitude K/T boundaries, *in* Proceedings, International Nannoplankton Association: Memorie Di Scienze Geologiche, v. 43, p. 133–147.

Pospichal, J. J., Wise, S. W., Asaro, F., and Hamilton, N., 1990, The effects of bioturbation across a biostratigraphically complete, high southern latitude K/T boundary, *in* Sharpton, V. L., and Ward, P. D., eds., Global catastrophes in Earth history: An interdisciplinary conference on impacts, volcanism, and mass mortality: Geological Society of America Special Paper 247, p. 497–508.

Raup, D. M., and Jablonski, D., 1993, Geography of end-Cretaceous marine bivalve extinctions: Science, v. 260, p. 971–973.

Rocchia, R., Boclet, D., Bonte, P., Froget, L., Galbrun, B., Jéhanno, C., and Robin, E., 1992, Iridium and other element distributions, mineralogy and magnetostratigraphy near the Cretaceous-Tertiary boundary in Hole 761C, *in* von Rad, U., Haq, B. U., and others, Proceedings, Ocean Drilling Program, Scientific Results, Volume 122: College Station, Texas, Ocean Drilling Program, p. 753–762.

Rocchia, R., Robin, E., Froget, L., and Gayraud, J., 1994, Ni-rich spinels (meteoric spinels) as indicators of the KT event timing *in* New developments regarding the K/T event and other catastrophes in Earth history [abs.]: Houston, Texas, Lunar and Planetary Institute Contribution 825, p. 99–100.

Romein, A. J. T., 1977, Calcareous nannofossils from the Cretaceous/Tertiary boundary interval in the Barranco del Gredero (Caravaca, Prov. Murcia, S.E. Spain), I and II: Proceedings, Koninklijke Nederlandse Akademie van Wetenschappen, Amsterdam, ser. B., v. 80, p. 256–279.

Romein, A. J. T., 1979, Lineages in early Paleogene calcareous nannoplankton: Utrecht Micropaleontological Bulletin, v. 22, p. 1–231.

Romein, A. J. T., and Smit, J., 1981, The Cretaceous/Tertiary boundary: Calcareous nannofossils and stable isotopes: Proceedings, Koninklijke Nederlandse Akademie van Wetenschappen, Amsterdam, ser. B., v. 83, p. 295–313.

Roth, P. H., 1978, Cretaceous nannoplankton biostratigraphy and oceanography of the Northwestern Atlantic Ocean, *in* Benson, W. E., Sheridon, S. E., and others, Initial reports of the Deep Sea Drilling Project, Volume 44: Washington, D.C., U.S. Government Printing Office, p. 731–759.

Sakai, H., and Keating, B., 1991, Paleomagnetism of Leg 119-Holes 737A, 738C, 742A, 745B, and 746A, *in* Barron, J., Larsen, B., and others, Proceedings, Ocean Drilling Program, Scientific Results, Volume 119: College Station, Texas,Ocean Drilling Program, p. 51–770.

Schlich, R., Wise, S. W., Jr., and 27 others, 1989, Proceedings, Ocean Drilling Program, Initial Reports, v. 120: College Station, Texas, Ocean Drilling Program, 647 p.

Schmitz, B., Andersson, P., and Dahl, J., 1988, Iridium, sulfur isotopes and rare elements in the Cretaceous-Tertiary boundary clay at Stevns Klint, Denmark: Geochemica et Cosmochimica Acta, v. 52, p. 229–236.

Schmitz, B., Asaro, F., Michel, H. V., Thierstein, H. R., and Huber, B. T., 1991, Element stratigraphy across the Cretaceous/Tertiary boundary in ODP Hole 738C, *in* Barron, J., Larsen, B., and others, Proceedings, Ocean Drilling Program, Scientific Results, Volume 119: College Station, Texas,Ocean Drilling Program, p. 719–730.

Scotese, C. R., and Denham, C. R., 1988, Terra Mobilis™: Plate tectonics for the Macintosh: Austin, Texas, Earth in Motion Technologies.

Shackleton, N. J., and Hall, M. A., 1984, Carbon isotope data from Leg 74 sediments, *in* Moore, T. C., Jr., Rabinowitz, P. D., and others, Initial reports of the Deep Sea Drilling Project, Volume 74: Washington, D.C., U.S. Government Printing Office, p. 613–620.

Shannon, C. E., 1949. The mathematical theory of communication, *in* Shannon, C. E., and Weaver, W., eds., The mathematical theory of communication: Chicago, University of Illinois Press, p. 29–125.

Sheehan, P. M., Fastovsky, D. E., Hoffmann, R. G., Berghaus, C. B., and Gabriel, D. L., 1991, Sudden extinction of the dinosaurs: Latest Cretaceous, Upper Great Plains, U.S.A.: Science, v. 254, p. 835–839.

Signor, P. W., and Lipps, J. H., 1982, Sampling bias, gradual extinction patterns and catastrophies in the fossil record, *in* Silver, L. T., and Schultz, P. H., eds., Geological implications of impacts of large asteroids and comets on the Earth: Geological Society of America Special Paper 190, p. 291–296.

Sissingh, W., 1977, Biostratigraphy of Cretaceous calcareous nannoplankton: Geologie en Mijnbouw, v. 56, p. 37–65.

Sloan, R. E., Rigby, J. K., Van Valen, L., and Gabriel, D. L., 1986, Gradual dinosaur extinction and simultaneous ungulate radiation in the Hell Creek Formation: Science, v. 232, p. 629–633.

Smit, J., 1982, Extinction and evolution of planktonic foraminifera after a major imact at the Cretaceous/Tertiary boundary, *in* Silver, L. T., and Schultz, P. H., eds., Geological implications of impacts of large asteroids and comets on the Earth: Geological Society of America Special Paper 190, p. 329–352.

Smit, J., 1990, Meteorite impact, extinctions and the Cretaceous-Tertiary boundary: Geologie en Mijnbouw, v. 69, p. 187–204.

Smit, J., and Romein, A. J. T., 1985, A sequence of events across the Cretaceous-Tertiary boundary: Earth and Planetary Science Letters, v. 74, p. 155–170.

Swisher, C. C., III, and 11 others, 1992, Coeval $^{40}Ar/^{39}Ar$ ages of 65.0 million years ago from Chicxulub crater melt rock and Cretaceous-Tertiary boundary tektites: Science, v. 257, p. 954–958.

Thierstein, H. R., 1980, Selective dissolution of late Cretaceous and earliest Tertiary calcareous nannofossils: Experimental evidence: Cretaceous Research, v. 2, p. 165-176.

Thierstein, H. R., 1981, Late Cretaceous nannoplankton and the change at the

Cretaceous/Tertiary boundary, *in* Warme, J. E., Douglas, R. G., and Winterer, E. L., eds., The Deep Sea Drilling Project: A decade of progress: Society of Economic Paleontologists and Minerologists Special Publication 32, p. 355–394.

Thierstein, H. R., 1982, Terminal Cretaceous plankton extinctions: a critical assessment, *in* Silver, L. T., and Schultz, P. H., eds., Geological implications of impacts of large asteroids and comets on the Earth: Geological Society of America Special Paper 190, p. 385–400.

Thierstein, H. R., and Okada, H., 1979, The Cretaceous/Tertiary boundary event in the North Atlantic, *in* Tucholke, B. E., Vogt, P. R., and others, Initial reports of the Deep Sea Drilling Project, Volume 43: Washington, D.C., U.S. Government Printing Office, p. 601–616.

Thierstein, H. R., Asaro, F., Ehrmann, W. U., Huber, B., Michel, H., Sakai, H., and Schmitz, B., 1991, The Cretaceous-Tertiary boundary at Site 738, south Kerguelen Plateau, *in* Barron, J., Larsen, B., and others, Proceedings, Ocean Drilling Program, Scientific Results, Volume 119: College Station, Texas, Ocean Drilling Program, p. 849–868.

van Heck, S. E., and Prins, B., 1987, A refined nannoplankton zonation for the Danian of the Central North Sea: Abhandlunger der Geologischen Brundesanstalt, v. 39, p. 285–303.

Ward, P., 1990, A review of Maastrichtian ammonite ranges, *in* Sharpton, V. L., and Ward, P. D., eds., Global catastrophes in Earth history: An interdisciplinary conference on impacts, volcanism, and mass mortality: Geological Society of America Special Paper 247, p. 519–530.

Ward, P., Wiedmann, J., and Mount, J., 1986, Maastrichtian molluscan biostratigraphy and extinction patterns in a Cretaceous-Tertiary boundary section exposed at Zumaya, Spain: Geology, v. 14, p. 899–903.

Watkins, D. K., 1992, Upper Cretaceous nannofossils from ODP Leg 120, Kerguelen Plateau, Southern Ocean, *in* Wise, S. W., Jr., Schlich, R., and others, Proceedings, Ocean Drilling Program, Scientific Results, Volume 120: College Station, Texas, Ocean Drilling Program, p. 343–370.

Watkins, D. K., Wise, S. W., Jr., Pospichal, J. J., and Crux, J., 1996, Upper Cretaceous calcareous nannofossil biostratigraphy of the Southern Ocean, *in* Whately, R., and Moguilevsky, A., eds., Microfossils and Oceanic environments: Aberystwyth, University of Wales Aberystwyth Press (in press).

Wei, W., and Pospichal, J. J., 1991, Danian calcareous nannofossil succession at ODP Site 738 in the southern Indian Ocean *in* Barron, J., Larsen, B., and others, Proceedings, Ocean Drilling Program, Scientific Results, Volume 119: College Station, Texas, Ocean Drilling Program, p. 495–512.

Wei, W., and Thierstein, H. R., 1991, Upper Cretaceous and Cenozoic calcareous nannofossils of the Kerguelen Plateau (southern Indian Ocean) and Prydz Bay (East Antarctica), *in* Barron, J., Larsen, B., and others, Proceedings, Ocean Drilling Program, Scientific Results, Volume 119: College Station, Texas, Ocean Drilling Program, p. 467–493.

Wiedmann, J., 1986, Macro-invertebrates and the Cretaceous-Tertiary boundary, *in* Walliser, O., ed., Global bioevents: Berlin, Springer-Verlag, Lecture Notes in Earth Sciences, v. 8, p. 397–409.

Worsley, T. R., 1974, The Cretaceous/Tertiary boundary event in the ocean, *in* Hay, W. W., ed., Studies in paleo-oceanography: Society of Economic Paleontologists and Minerologists Special Publication 20, p. 94–125.

Zachos, J. C., and Arthur, M. A., 1986, Paleoceanography of the Cretaceous/Tertiary boundary event: Inferences from stable isotopic and other data: Paleoceanography, v. 1, p. 5–26.

Zachos, J. C., Arthur, M. A., and Dean, W. E., 1989, Geochemical evidence for suppression of pelagic marine productivity at the Cretaceous-Tertiary boundary: Nature, v. 337, p. 61–64.

Zachos, J. C., Aubry, M.-P., Berggren, W. A., Ehrendorfer, T., Heider, F., and Lohmann, K. C., 1992, Chemobiostratigraphy of the Cretaceous/Paleocene boundary at ODP Site 750, southern Kerguelen Plateau, *in* Wise, S. W., Jr., Schlich, R., and others, Proceedings, Ocean Drilling Program, Scientific Results, Volume 120: College Station, Texas, Ocean Drilling Program, p. 961977.

Zinsmeister, W. J., Feldmann, R. M., Woodburne, M. O., and Elliot, D. H., 1989, Latest Cretaceous/earliest Tertiary transition on Seymore Island, Antarctica: Journal of Paleontology, v. 63, p. 731–738.

Manuscript Accepted by the Society September 14, 1995

Printed in U.S.A.

Geological Society of America
Special Paper 307
1996

The biostratigraphy and paleobiogeography of Maastrichtian inoceramids

Kenneth G. MacLeod and Brian T. Huber
Department of Paleobiology, National Museum of Natural History, Smithsonian Institution, MRC: NHB 121, Washington, DC 20560
Peter D. Ward
Department of Geological Sciences, AJ-20, University of Washington, Seattle, Washington 98195

ABSTRACT

During the mid-Maastrichtian there was a pulse of extinction that affected inoceramids living in all of the world's oceans. We have documented this event at localities from tropical to austral paleolatitudes and from shelfal to abyssal paleodepths. Locally, the decline in inoceramid abundance occurred over a resolvable interval of geologic time, and globally, the extinction occurred at different times in different areas. This event is distinct from the Cretaceous/Tertiary boundary event; however, it is an important part of the transition from the Cretaceous Period to the Tertiary Period. Cooling and enhanced production of deep waters at high latitudes during the Maastrichtian could have caused increased ventilation of the bottom waters in the regions in which many inoceramids were living. As a causal mechanism for the extinction, this change is consistent with constraints provided by the observed distribution of inoceramid remains.

INTRODUCTION

Discussions concerning the Maastrichtian have been dominated by the Cretaceous/Tertiary (K/T) boundary debate, but significant events that occurred during the Maastrichtian Age are increasingly well documented. Among the biological changes documented in the marine realm are extinction among inoceramid bivalves (e.g., Dhondt, 1983a; Kauffman, 1988; MacLeod and Orr, 1993; MacLeod, 1994a), extinction among rudist reef faunas (Kauffman, 1988; Johnson and Kauffman, 1990), and shifts in the latitudinal distribution of planktonic foraminifera and calcareous nannoplankton (Huber, 1992; Huber and Watkins, 1992). The latter, in combination with oxygen isotopic analyses, indicate that surface and deep water temperatures decreased during most of the Maastrichtian (e.g., Douglas and Savin, 1975; Boersma and Shackleton, 1981; Boersma, 1984; Barrera et al., 1987; Barrera and Huber, 1990). Further, pole-to-equator temperature gradients were greater during the Maastrichtian than during earlier ages of the Cretaceous (Spicer and Corfield, 1992; Huber et al., 1995). Other proposed changes in the Maastrichtian oceans include positive carbon isotopic shifts of greater than 1‰ in benthic foraminiferal tests (Barrera and Huber, 1990; Seto et al., 1991), change in the oceanic strontium budget (Nelson et al., 1991), regression (e.g., Haq et al., 1987), and an increase in the intensity of bioturbation in bathyal sediments (MacLeod, 1994b). In the terrestrial realm, both floral (Wolfe and Upchurch, 1987; Johnson and Hickey, 1990; 1992; Spicer and Parrish, 1990; Spicer and Corfield, 1992) and faunal (Clemens, 1986; Lehman, 1987; Clemens and Nelms, 1993) biogeographic shifts have been recognized. Predating the K/T boundary by up to millions of years, these changes could not have been caused by the K/T bolide, but they are an important part of the larger K/T interval.

It has been suggested that there was a global bio-event during the Maastrichtian (e.g., Kauffman, 1988; MacLeod and Ward, 1990; MacLeod, 1994b), but rigorous documentation has been lacking. Although the studies cited in the previous paragraph demonstrate that changes occurred across the globe dur-

MacLeod, K. G., Huber, B. T., and Ward, P. D., 1996, The biostratigraphy and paleobiogeography of Maastrichtian inoceramids, *in* Ryder, G., Fastovsky, D., and Gartner, S., eds., The Cretaceous-Tertiary Event and Other Catastrophes in Earth History: Boulder, Colorado, Geological Society of America Special Paper 307.

ing the Maastrichtian, they do not necessarily demonstrate that there was a coordinated episode of global change in the mid-Maastrichtian. The terrestrial data are limited to North America and are mostly derived from the Great Plains region; samples for stable isotopic analyses were largely collected from either high southern paleolatitudes or from a few paleotropical sites in the central Pacific; detailed study of inoceramids has been concentrated in the Basque region of France and Spain; and rudists and associated faunas are best documented in the Caribbean. In addition to geographic limitations, temporal resolution across these studies varies considerably. Correlation, both regionally and globally, is often crude, and paleoecological conclusions are not always in agreement. For example, climatic changes inferred for the late Maastrichtian based on leaf physiognomy of fossils from the Great Plains (Wolfe and Upchurch, 1987; Johnson and Hickey, 1990) are opposite to those inferred from similar data collected in contemporaneous deposits in Alaska (Spicer and Parrish, 1990), and the stratigraphic distribution of dinosaur fossils collected in Montana has been used to support proposals of both declining (Clemens, 1986) and constant (Sheehan et al., 1991) diversity through the Maastrichtian. Despite the geographic and topical spread of evidence concerning Maastrichtian paleoecology, relationships among various events are speculative and potential cause(s) are not well constrained.

To test whether there was an episode of global change during the mid-Maastrichtian, we have examined the stratigraphic distribution of the remains of inoceramids from 31 localities that collectively provide wide geographic coverage (Fig. 1). The Inoceramidae is an excellent group to focus on in an examination of Maastrichtian events because (1) inoceramids were common and globally distributed during the early Maastrichtian, (2) they did not survive the age (i.e., they undergo change during the interval), and (3) they have left a rich, accessible record in the form of characteristic microfossils. Some inoceramids grew to be very large; however, even the largest often passively disaggregated and are preserved as hundreds of millions of columnar, polygonal prisms of calcite ~100 μm across. This taphonomic process has greatly increased the inoceramid fossil record and provides a means of objectively estimating changes in their standing population (MacLeod and Orr, 1993). In addition, because these prisms commonly occur in Deep Sea Drilling Project (DSDP) and Ocean Drilling Program (ODP) cores, it is relatively easy to generate a truly global data base (MacLeod and Ward, 1990).

MATERIALS AND METHODS

In this chapter the observed abundance of inoceramid prisms is used as a proxy for the local abundance of inoceramid bivalves (MacLeod and Ward, 1990; MacLeod and Orr, 1993). Samples discussed are from upper Campanian, Maastrichtian,

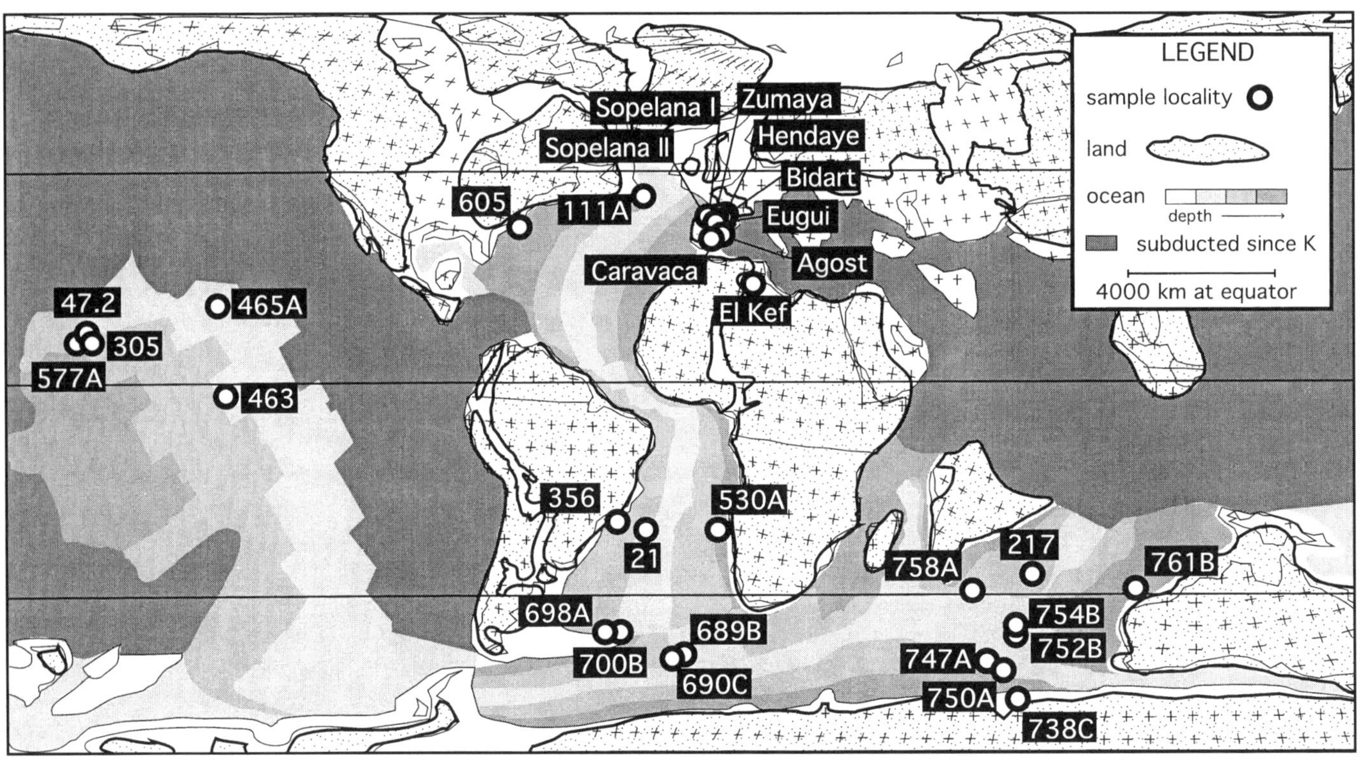

Figure 1. Approximate early Maastrichtian (70-Ma) paleogeographic position of the localities studied. Reconstruction was generated using the PGIS/Mac™ software package. Land/sea distributions modified from Barron (1987) and Huber (1992). Paleobathymetry modified from Barron (1987).

and lower Danian strata. The various sections are correlated using the base of the Maastrichtian, the base of the *Abathomphalus mayaroensis* planktonic foraminiferal zone, and the K/T boundary. In samples from DSDP/ODP cores, the position of these datums was taken from published reports. In other sections, they are based on our observations. Samples were disaggregated using standard micropaleontological techniques and washed on a 90-µm or a 63-µm screen. The abundance of the prisms in these washed samples was scored on a qualitative scale relative to co-occurring microfossils from absent to abundant or as a quantitative estimate of the number of prisms in 50 g of bulk sample (MacLeod and Orr, 1993). Qualitative estimates are based on washed residues of either a >63-µm or a >90-µm fraction; quantitative results are all based on the >90-µm fraction. For the purpose of discussing methodology, the samples for which we report results can be divided into three categories.

One category represents samples for which one or more of us has independently examined the distribution of inoceramid prisms. Samples that fall into this category include all the samples from DSDP holes 21, 111A, and 530A (MacLeod and Ward, 1990) and ODP Hole 689B (Huber, 1990); most of the samples from ODP holes 698A and 700B (Huber, 1991a); about half of the samples from ODP Hole 690C (Huber, 1990); and all of the samples from the Bidart, Hendaye, Zumaya, Sopelana I, and Sopelana II sections (MacLeod and Orr, 1993). Qualitative estimates of the abundance of prisms were recorded for all the DSDP/ODP material, but two different scales were used. MacLeod and Ward (1990) estimated abundance on a relative scale from 0 (prisms absent) to 10 (prisms abundant), whereas Huber (1991a) scored abundance as absent, rare, few, common, or abundant. We integrated these two schemes by equating absent with 0, rare with 2.5, few with 5, common with 7.5, and abundant with 10. Huber (1990) found no prisms, so we report a score of 0/absent for the samples examined in that study. MacLeod and Orr (1993) reported quantitative estimates.

A second category is composed of DSDP/ODP Micropaleontological Reference Center samples, processed samples from other studies not previously examined for prism abundance, or samples reexamined for the purpose of this chapter. This category includes all the material from DSDP holes 217 and 747A, a small number of samples from ODP holes 698A and 700B, and the upper nine samples from ODP Hole 738C. For these samples we used the qualitative scale of MacLeod and Ward (1990).

The final category is composed of samples from sections chosen to expand the available data such that they provide global constraints on the stratigraphic distribution of Maastrichtian inoceramids. Samples in this category include all samples from Eugui, Agost, Caravaca, and El Kef; all samples from DSDP holes 47.2, 305, 356, 463, 465A, 577A, and 605A; all samples from ODP holes 750A, 752B, 754B, 758A, and 761B; about half of the samples from ODP Hole 690C; and the lower 15 samples from ODP Hole 738C. Site selection was based on each section's paleogeographic position relative to other available Maastrichtian sections, on the presence of relatively undisturbed Maastrichtian sediments that could be disaggregated, on the reported occurrence of calcareous microfossils, on the amount of lithologic variability (i.e., we chose sections that seemed to represent a relatively constant depositional setting over those with evidence of high frequency variation; e.g., compare ODP Hole 761B [selected] with 762C [passed over]), and on the absence of evidence for large-scale reworking. Previous reports of the occurrence of inoceramid remains in a section were not considered in the selection process. Samples were collected with the goal of achieving complete coverage with highest sampling density in the region of the first appearance of *A. mayaroensis*. For core samples, we collected at relatively even intervals (at any given sampling density). If the "evenest" level fell in an interval of apparent reworking (e.g., turbidity current deposits, drilling disturbance), where there was a largely void or previously sampled interval, or where lithology was inappropriate (e.g., chert); we sampled at the nearest horizon that was not disqualified based on these criteria. Proximity to visible inoceramid remains on the cut surface of the cores was not considered in sample selection. Sample preparation and counting procedures follow MacLeod and Orr (1993).

RESULTS

Land-based sections, southwestern Europe and northern Africa

These sections preserve Tethyan to Tethyan/warm temperate faunas. Deposited in a shelfal setting (Keller, 1989), the section at El Kef, Tunisia, represents the shallowest environment sampled. Based on paleogeographic reconstructions (Mathey, 1988) and the higher proportion of terrigenous material, the section at Eugui (near Pamplona, Spain) represents a more shoreward, shallower position in the Basque–Cantabrian Basin than the Bidart, Hendaye, Zumaya, and two Sopelana sections. The latter five sections represent bathyal, basin floor to slope settings (Mount and Ward, 1986; Mathey, 1988; Ward and Kennedy, 1993; MacLeod, 1994a). The Agost and Caravaca sections (southern Spain) were deposited in a similar setting; Coccioni and Galeotti (1994) estimated paleodepths for the Caravaca section at 600 to 1000 m. Figure 2 shows the abundance of prisms through these nine sections. Inoceramid prisms are most abundant below the *A. mayaroensis* Zone, they decline in abundance near the base of this zone, and there is an extended stratigraphic interval between last appearance datum (LAD) of prisms and the K/T boundary. At Eugui prisms are less common, and their LAD is lower than in other sections from the region. The base of the *A. mayaroensis* Zone is defined by the first appearance datum (FAD) of the nominative taxon at all sections except El Kef, where *A. mayaroensis* is extremely rare and therefore an unreliable index fossil. For the

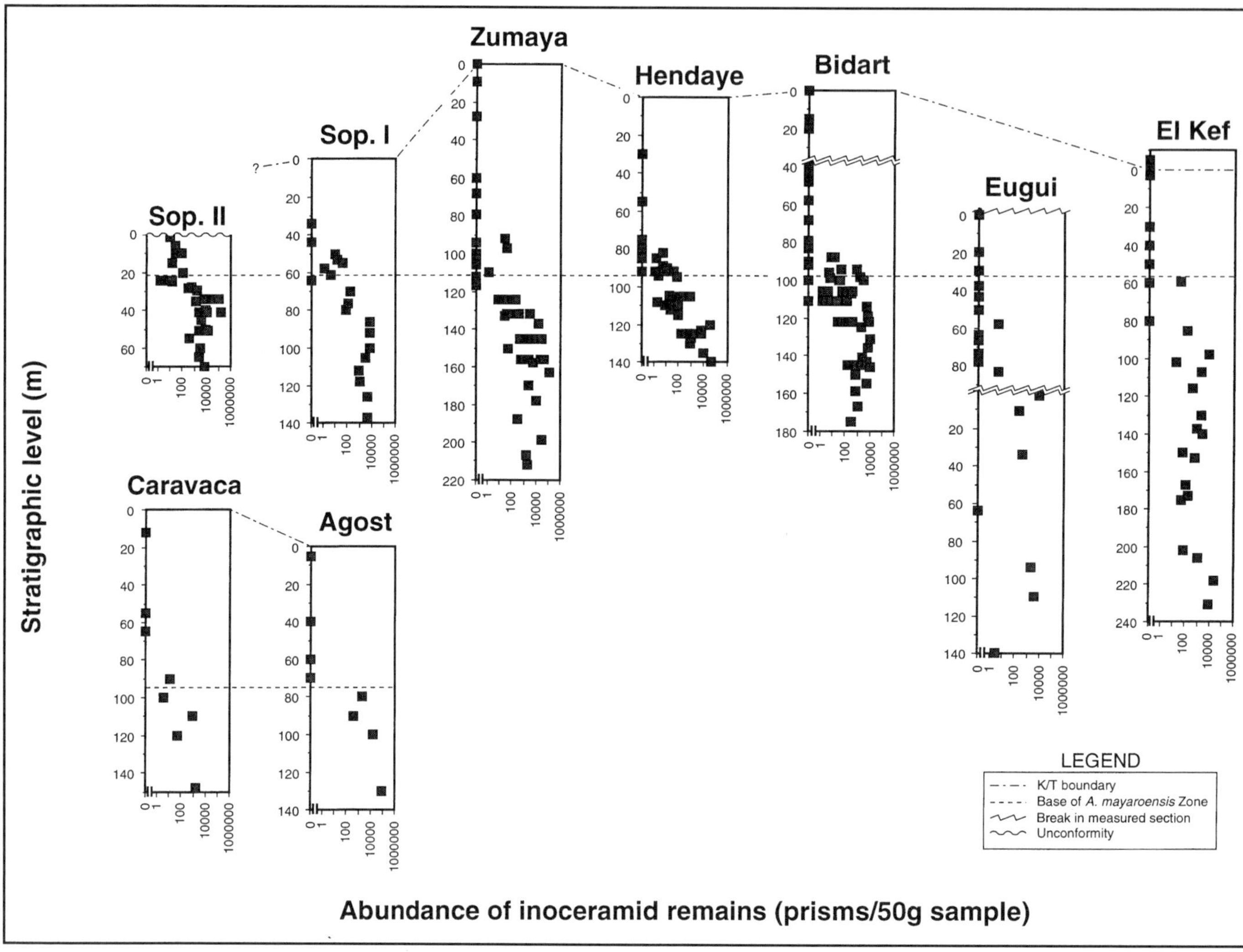

Figure 2. The abundance of prisms plotted on a log scale against stratigraphic level in meters for the nine land-based sections from southwestern Europe and Tunisia. Data for the Sopelana II, Sopelana I, Zumaya, Hendaye, and Bidart sections from MacLeod and Orr (1993).

El Kef section the *A. mayaroensis* Zone is recognized based on associated taxa and is placed slightly above the FAD of *Rugoglobigerina scotti* (W. N. Orr, unpublished data).

Atlantic Ocean deep-sea cores

Sections from the Atlantic represent mid-latitudes from both sides of the equator. Paleobathymetric estimates range from 200 to 400 m for Hole 111A (Ruffman, 1972) to 3,500 to 4,500 m for Hole 530A (Saltzman and Barron, 1982; Barron et al., 1984). Hole 530A is the deepest site in the data set, and washed residues from this hole are small and show considerable evidence of dissolution. Estimated paleodepths for the other three sections are approximately 500 m for Site 21 (Saltzman and Barron, 1982), 2,000 to 2,500 m for Site 356 (Sliter, 1977), and about 2,400 m for Site 605 (Jansen and Kroon, 1987). As with the land-based sections, the LAD of inoceramid prisms occurs near or below the base of the *A. mayaroensis* Zone (Fig. 3). Samples from sites 356 and 605 have relatively low abundances of prisms and a relatively early LAD of prisms (i.e., in the Campanian and before the FAD of *A. mayaroensis*, respectively). In addition, prismatic shell fragments and individual small prisms were common to abundant in samples from the upper Maastrichtian of Site 605 (Fig. 4). These prisms (both individually and in the shell fragments) are smaller than the prisms recorded lower in the section and at other sites in this study; however, they are similar to presumed *Tenuipteria argentea* shell fragments found in the upper Maastrichtian strata of the coastal sections from the Basque region of France and Spain (MacLeod and Orr, 1993).

Indian Ocean deep-sea cores

Site 217 was migrating northward with the Indian Plate during the Late Cretaceous, and by the late Maastrichtian it had entered the Tethyan Realm. Holes 752B, 754B, 758A, and

761B contain subtropical to mid-latitude foraminiferal assemblages, and holes 738C, 747A, and 750A (discussed below) contain austral assemblages. Thus, our Indian Ocean samples form part of a latitudinal transect of a Maastrichtian ocean. Specific paleodepth estimates have not been published for the Indian Ocean sites examined, but benthic foraminifera in the *Initial Reports* volumes (e.g., *Nuttalides truempyi* and *Stensioina beccariiformis*) indicate generally bathyal depositional settings. Figure 5 shows the distribution of prisms through the sections. At all five sites prisms are found in the *A. mayaroensis* Zone. Holes 754B and 758A are truncated by unconformities above the FAD of *A. mayaroensis* but below the LAD of inoceramid prisms. The other three sites show a gradual decline in the abundance of prisms in the mid-Maastrichtian. The LAD of prisms is relatively low in Hole 761B, and sporadic occurrences of prisms occur quite high at Site 217.

Pacific Ocean deep-sea cores

All five sections examined from the Pacific represent low paleolatitudes and contain tropical to subtropical planktonic foraminiferal assemblages. Lower bathyal (about 1,500 m) paleodepths have been estimated for Hole 465A (Boersma, 1981). Paleodepth estimates are not available for the other sites, and we have not attempted to supply such constraints. Prism abundances through the Pacific sections are shown in Figure 6. These data are unusual in the relative scarcity of inoceramid remains. In our samples prisms were found only in material from Hole 465A. Although inoceramid remains are known from holes 47.2 and 48.2 (Heezen et al., 1971a, b; Boersma, 1981; Saltzman and Barron, 1982), we found no prisms in our samples from Hole 47.2 and saw no inoceramid remains on the cut surface of the archive half of the Cretaceous cores from Hole 48.2. Where we did observe prisms in Pacific sediments (Hole 465A), they occurred below the *A. mayaroensis* Zone.

Southern Ocean deep-sea cores

Samples from seven sites representing high southern paleolatitudes (>55°S) were examined. Estimated paleodepths for the Maastrichtian intervals are 1,000 to 1,500 m for Hole 689B (Thomas, 1990), 1,500 to 2,000 m for Hole 690C (Thomas, 1990), approximately 1,000 m for Hole 747A (Quilty, 1992), and 650 to 850 m for Hole 750A (Quilty, 1992). Barrera and Huber (1991) estimated a 1,000-m paleodepth for Hole 738C at K/T boundary time. Estimated paleodepths for the Paleocene of Holes 698A and 700B are 800 to 1,500 m and 2,000 to 2,500 m, respectively (Katz and Miller, 1991); we assume that Maastrichtian depths were slightly shallower but generally comparable. Figure 7 shows the distribution of prisms in these high-latitude sections. A single

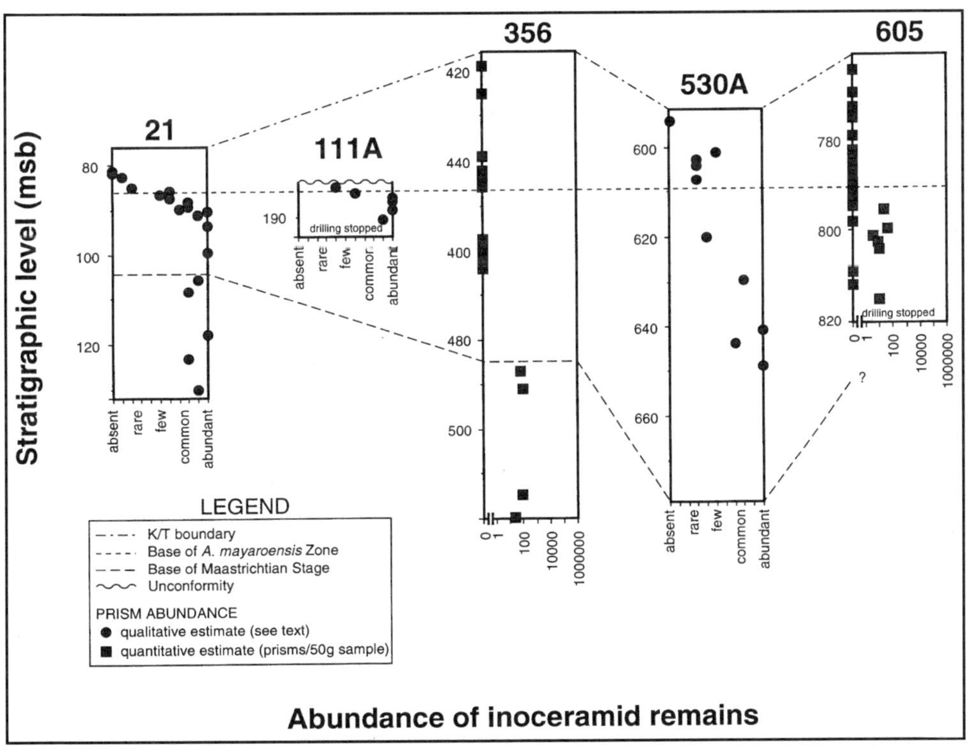

Figure 3. The abundance of prisms plotted against stratigraphic level (meters subbottom [msb]) for the five Atlantic mid-latitude localities. Qualitative estimates are from MacLeod and Ward (1990); their scale is equated with the scale of Huber (1991a) as described in methodology section.

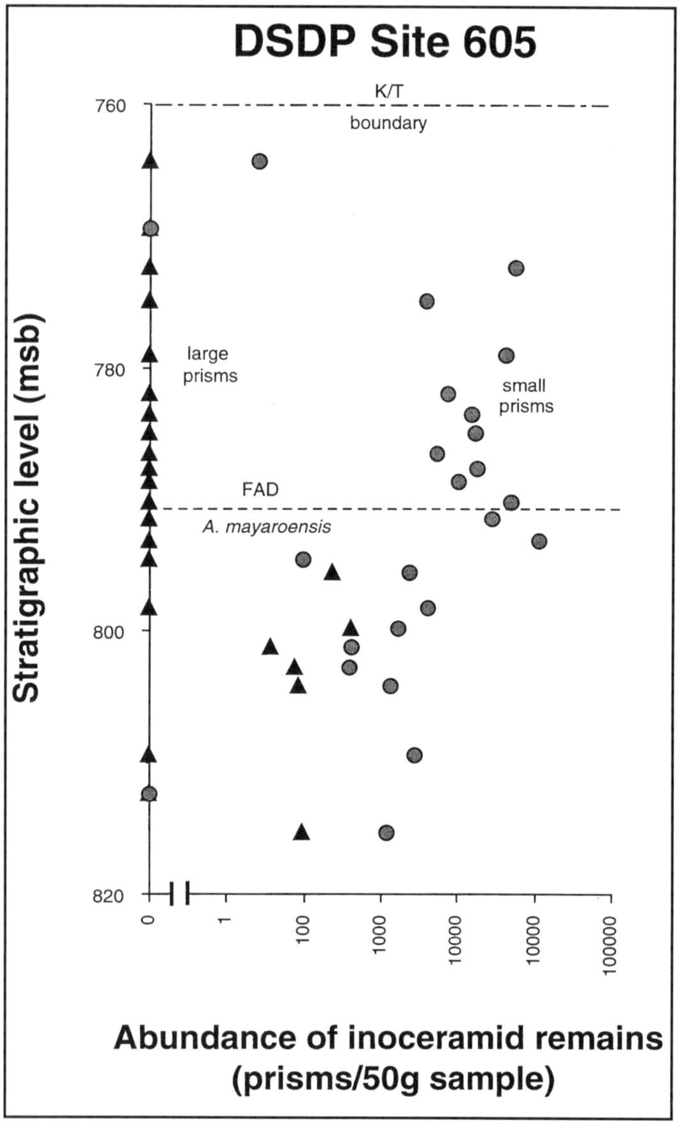

Figure 4. The abundance of two morphologic categories of prisms and prismatic shell fragments (number of pieces per 50-g bulk sample) plotted against stratigraphic level for DSDP Site 605. The category of small prisms and prismatic shell fragments (circles) matches material previously identified as *Tenuipteria* remains, whereas the large prisms (triangles) match *Inoceramus* remains (MacLeod and Orr, 1993). These results thus support a stratigraphic overlap between the two inoceramid taxa (MacLeod, 1994a).

prism was observed in a single sample from Hole 690C, but prisms are absent from all the rest of our samples from holes 689B and 690C. In Hole 738C prisms occur at high, relatively constant abundances throughout the Maastrichtian and into the Danian; prisms at this site occur in sediments as young as the Eocene (Huber, 1991b). At Hole 747A prisms are also common to abundant in every sample analyzed, but all samples come from below the *A. mayaroensis* Zone. With the exception of a single prism in a single sample in Hole 750A, the LAD of prisms occurs below the *A. mayaroensis* Zone in the remaining three sites.

DISCUSSION

First-order patterns

The global distribution of Maastrichtian inoceramids demonstrates that these bivalves were ubiquitous during the early Maastrichtian. Inoceramids are known from all continents, and we found prisms in 25 out of our 30 localities containing strata below the FAD of *A. mayaroensis*. These 25 inoceramid-bearing localities span shelfal (El Kef) to abyssal (DSDP Hole 530A) paleodepths, represent a wide range of paleolatitudes, and include localities in the Atlantic, Tethys, Pacific, Indian, and Southern oceans. In most sections containing inoceramids, prisms are found in every sample collected across tens to hundreds of meters of section. Prism abundances of 10,000 to 100,000 prisms/50 g of bulk sample are typical.

Despite their wide paleogeographic and paleobathymetric distribution, inoceramids largely disappeared in a pulse of extinction that occurred during the mid-Maastrichtian. Inoceramids decline gradually but over a relatively short stratigraphic interval near the base of the *A. mayaroensis* planktonic foraminiferal Zone. The rate at which the extinction progressed falls between the catastrophic (e.g., Kauffman, 1988) and gradual (Dhondt, 1983a) scenarios previously proposed for this event. One unusual inoceramid taxon, *Tenuipteria*, survived after the mid-Maastrichtian event and disappeared at the K/T boundary (e.g., Speden, 1970; Dhondt, 1983b; MacLeod, 1994a). We think it likely that the shell fragments and small prisms that we found in DSDP Site 605 are remains of *Tenuipteria*. If this identification is correct, it represents the first time *Tenuipteria* shell fragments have been reported in DSDP/ODP core material and supports a stratigraphic overlap between *Tenuipteria* and other inoceramids (MacLeod, 1994a).

The distribution of prisms in Hole 738C seems to contradict a proposed mid-Maastrichtian extinction event, but we think the record at that locality is influenced by large-scale reworking. Age diagnostic foraminiferal and nannofossil taxa have been recognized at anomalous positions at this site, and prisms are reported to occur sporadically as high as the lower Eocene (Huber, 1991b; Wei and Thierstein, 1991). The K/T boundary occurs in a laminated interval, a circumstance that initially raised hopes that this stratigraphic interval did not contain reworked fossils (Barron, Larsen, et al., 1989); Huber (this volume) addresses reworking in the immediate region of the boundary in more detail. The amount of mixing necessary to account for uniformly high abundance of prisms into the Danian, though, exceeds all previous estimates. In support of a reworking explanation for the anomalous prisms (i.e., prisms isolated from above the expected LAD in mid-Maastrichtian strata), MacLeod and Huber (1996) reported that these prisms have strontium isotopic ratios expected for Campanian fossils, whereas co-occurring planktonic foraminifera near their FAD have strontium isotopic ratios appropriate for their stratigraphic position.

At a much smaller scale, six other samples from four separate sites may be compromised by reworking or contamination. The uppermost inoceramid-bearing sample for holes 690C, 750A, and 752B and one sample from Site 217 contained a single prism each (Figs. 5 and 7). We think the simplest and most conservative explanation for these isolated occurrences is contamination during collection or processing. The two highest inoceramid-bearing samples from Site 217, on the other hand, contain enough prisms to make this explanation untenable. Pessango and Michael (1974) suggested common reworking in the late Maastrichtian portion of this section, which would explain the anomalous occurrences of prisms, but we cannot cite any independent corroboration of reworking in our samples. Regardless, holes 217, 750A, and 752B all show the dramatic mid-Maastrichtian decline in inoceramid abundance, and Hole 690C is otherwise devoid of prisms.

Second-order patterns

Although the data presented above suggest a single global pulse of extinction among inoceramids during the mid-Maastrichtian, the LAD of prisms occurs earlier in high southern latitudes than elsewhere. In Antarctic localities (ODP holes 698A, 700B, and 750A), prisms disappear at or below the FAD of *A. mayaroensis*. At many low to mid-latitude sites (e.g., Bidart, Sopelana II, ODP Hole 758A), inoceramid remains are found above that datum. The first appearance of *A. mayaroensis* is itself time transgressive, but in such a fashion that it strengthens the evidence for a diachronous LAD of inoceramids. *A. mayaroensis* appears earliest in high southern latitudes (Huber, 1992; Huber and Watkins, 1992; Fig. 8) where inoceramids disappear first. Therefore, there seems to be a general Antarctic-to-equator progression in the timing of the pulse of extinction among inoceramids.

Inoceramids also seem to disappear earlier in onshore sites than they do in offshore sites. In the Basque country of France and Spain, the decline and disappearance of inoceramid macrofossils and microfossils occurs at a lower level (based on lithostratigraphic as well as biostratigraphic correlation) in the shoreward Eugui section than it does in nearby sections (Bidart, Hendaye, Zumaya, Sopelana I, and Sopelana II) representing more offshore environments. ODP Hole 761B, located off the northwest coast of Australia, records the lowest (relative to the FAD of *A. mayaroensis*) LAD of prisms among localities in the Indian Ocean. Along the São Paulo Plateau–Rio Grande Rise in

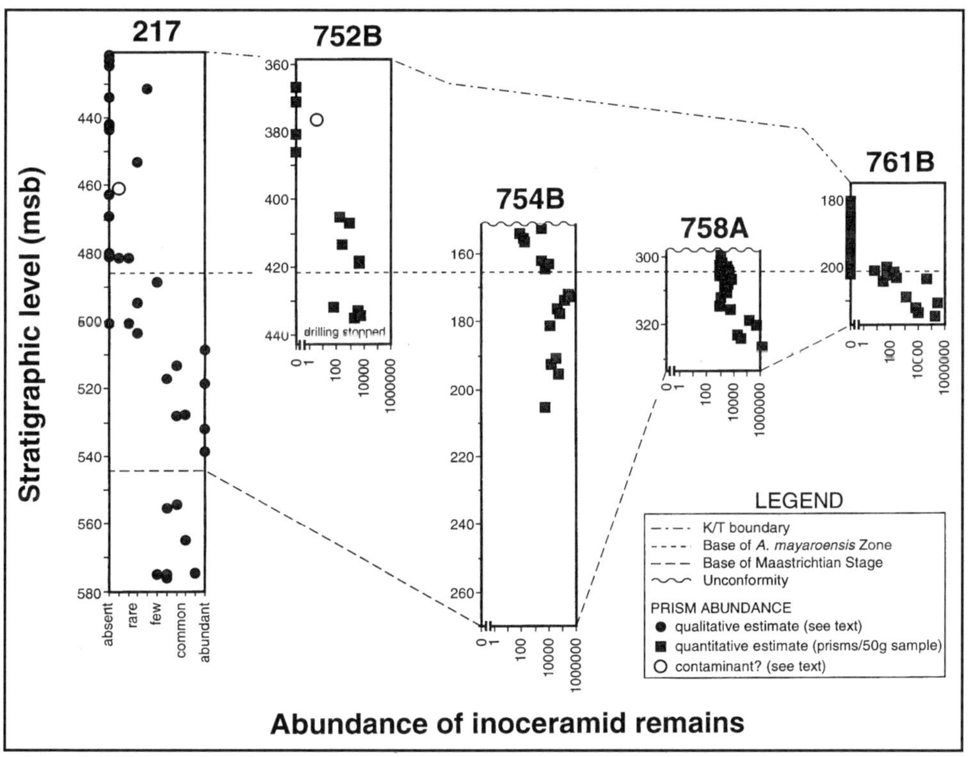

Figure 5. The abundance of prisms plotted against stratigraphic level (meters subbottom [msb]) for the five Indian Ocean localities. Qualitative estimates follow the scale of MacLeod and Ward (1990) and are equated with the scale of Huber (1991a) as described in methodology section. Larger, open circles for holes 217 and 752B represent samples containing a single prism whose occurrence is attributed to contamination.

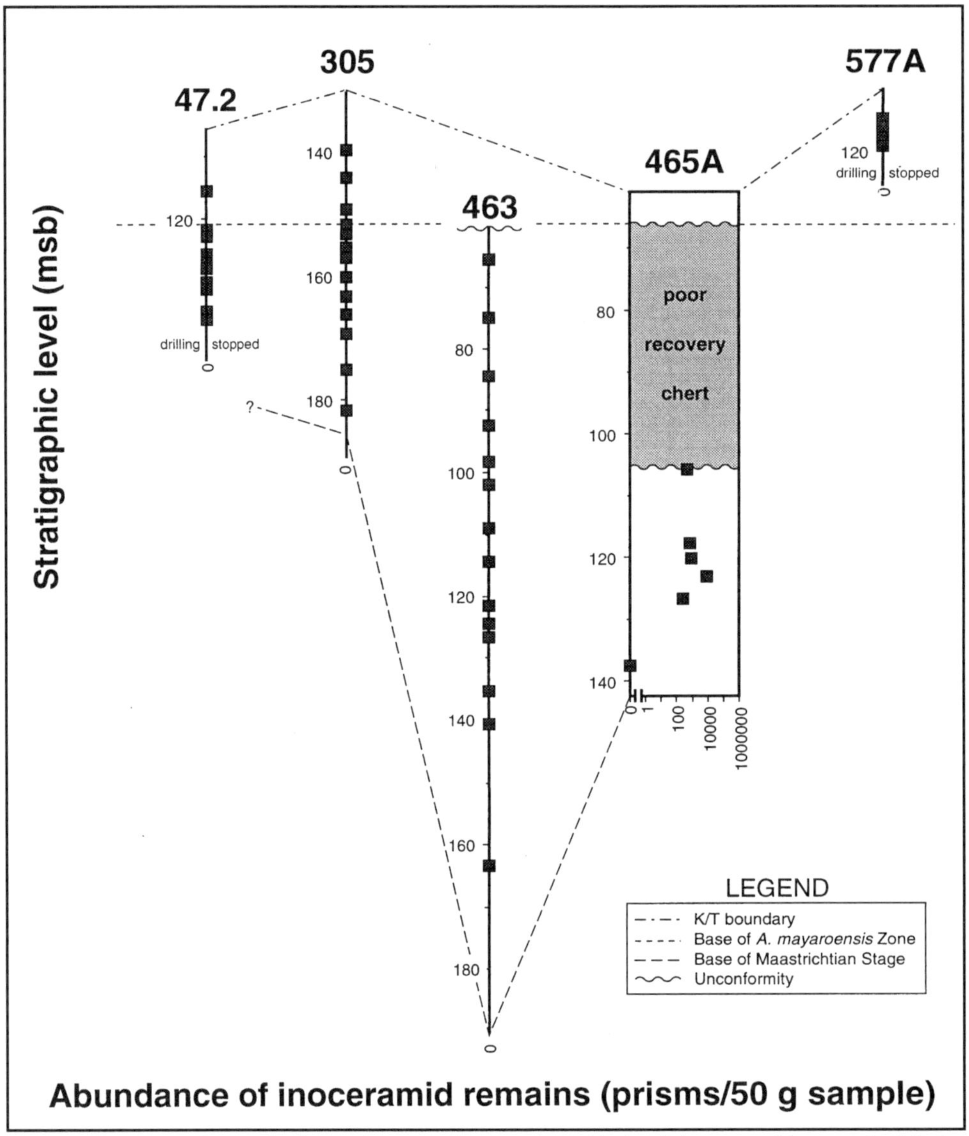

Figure 6. The abundance of prisms plotted against stratigraphic level (meters subbottom [msb]) for the five Pacific Ocean localities. Prisms were absent from most of the samples examined, but where they occur, they are found below the first appearance datum (FAD) of *A. mayaroensis*.

the South Atlantic, inoceramid prisms last occur in Campanian strata in a shoreward site (DSDP Site 356) but continue to be found above the FAD of *A. mayaroensis* in a more offshore site (DSDP Site 21). Finally, inoceramids last occur in shelfal Campanian strata on James Ross Island—neither body-fossils nor prisms have ever been reported from Maastrichtian strata on Seymour Island (Zinsmeister and Macellari, 1988; Huber, 1988; Zinsmeister and Feldmann, 1994)—whereas in offshore ODP sites from high southern latitudes inoceramid prisms range into mid-Maastrichtian strata. Interestingly, the shoreward site in the South Atlantic (DSDP Site 356) is interpreted to have been deposited in deeper water than the corresponding offshore site (DSDP Site 21). Thus, if the observed offshore trend is meaningful, it may not be the result of a simple depth progression of the extinction pulse.

Inoceramid remains are ubiquitous below the FAD of *A. mayaroensis*, but they are absent from five (including Hole 690C) of the 30 sections that include lower Maastrichtian strata. Based on our samples, inoceramids were less common in the central Pacific than they were elsewhere. Perhaps inoceramid occurrences are reflecting oceanographic differences between the relatively old, wide Maastrichtian Pacific Ocean, and the relatively young, narrow Maastrichtian Atlantic and Indian oceans. Conversely, the scarcity of inoceramids in Pacific sites may be an artifact of sampling or dilution. The recovery of Cretaceous calcareous sediments is relatively sparse in

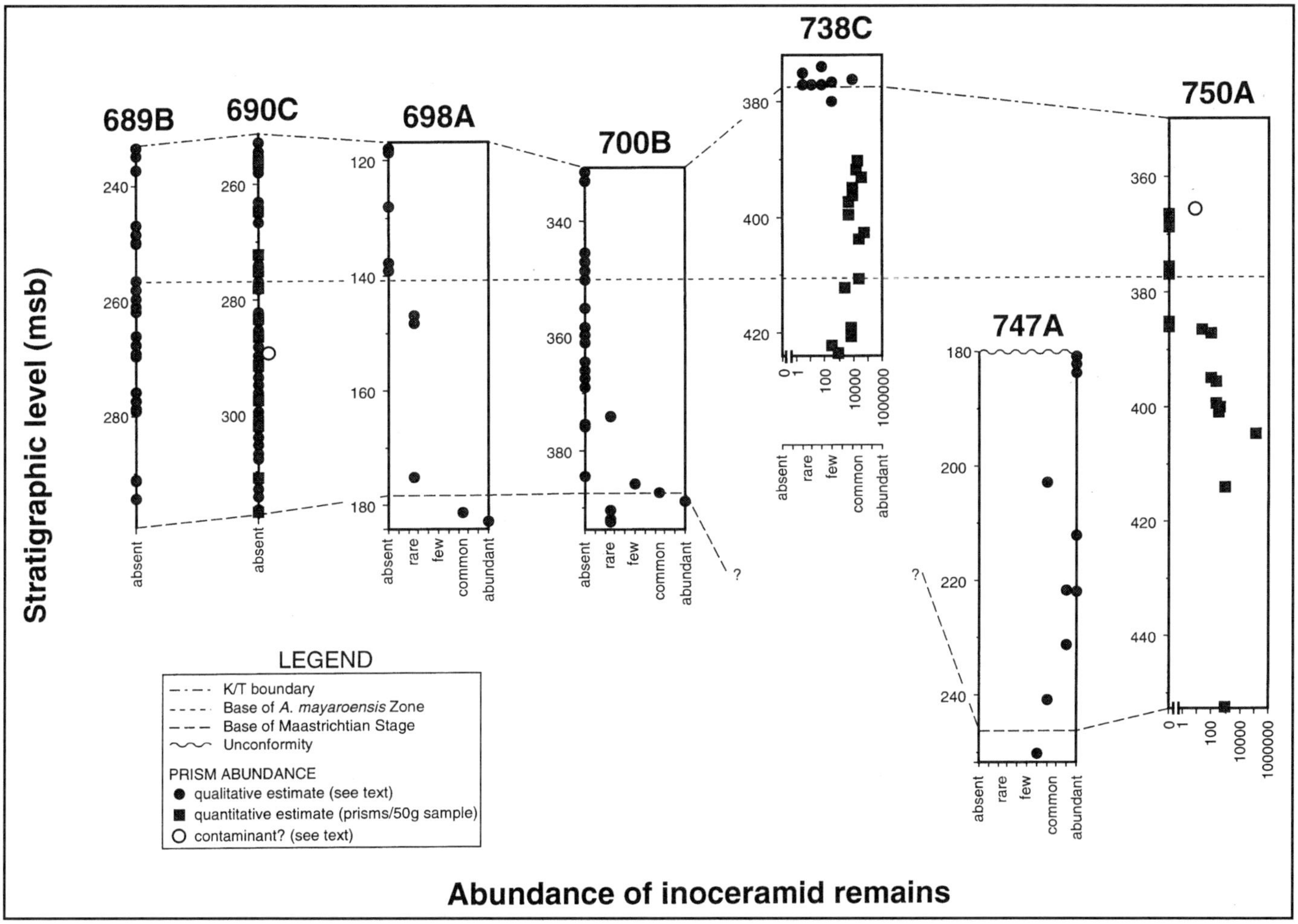

Figure 7. The abundance of prisms plotted against stratigraphic level (meters subbottom [msb]) for the seven Southern Ocean localities. Qualitative estimates follow the scale of MacLeod and Ward (1990) and are equated with the scale of Huber (1991a) as described in methodology section. Both qualitative and quantitative estimates of prism abundance were made on material from Hole 738C, so two axes are shown. Larger, open circles for holes 690C and 750A represent samples containing a single prism whose occurrence is attributed to contamination. Most of the prisms from Hole 738C are also interpreted as reworked. Data for holes 698A and 700B from Huber (1991a), except that LAD of prisms was moved up one sample in Hole 698A based on reexamination of that sample for this study.

the Pacific Ocean and the sites sampled represent areas of high surface productivity. More problematic is the absence of inoceramid remains in two ODP holes from Maud Rise (689B and 690C). Compared to nearby, inoceramid-bearing holes (698A and 700B), the Maud Rise sites were deposited a little farther south. However, planktonic foraminiferal assemblages indicate no environmental differences among the sites (Huber, 1991a), and within the resolution of available estimates, benthic foraminifera indicate broadly similar paleodepths (Thomas, 1990; Katz and Miller, 1991). We do not yet have an explanation for the absence of inoceramids on Maud Rise.

Associated patterns

Data regarding the noninoceramid benthos, although scarce, confirm widespread changes in the deep oceans during the Maastrichtian. Typically inoceramid-bearing strata contain few other benthic macrofossils, but trace-fossils indicate that a significant community of burrowing organisms lived with the inoceramids. Changes in the distribution of inoceramid prisms through the Basque sections suggest that the population of these burrowers increased at the same time that inoceramids declined (MacLeod, 1994b). Among benthic foraminifera in ODP holes 689B and 690C there is a slight increase in diversity during the mid-Maastrichtian (Thomas, 1990) and a nearly correlative 1‰ positive excursion in benthic foraminiferal $\delta^{13}C$ values (Barrera and Huber, 1990). In ODP Hole 752B there is an increase in abundance of benthic foraminifera and a greater than 1‰ positive shift in benthic foraminiferal ^{13}C values across the inoceramid extinction interval (Nomura, 1991; Seto et al., 1991, Fig. 9). Finally, relative abundance of benthic foraminiferal morphotypes

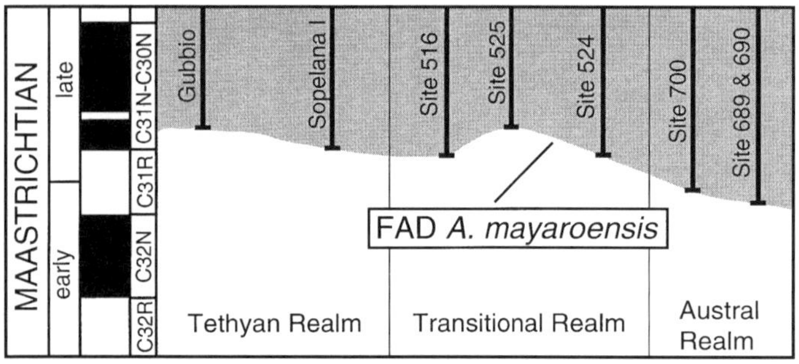

Figure 8. Stratigraphic position of the FAD of *A. mayaroensis* along a latitudinal transect (modified from Huber and Watkins, 1992). This taxon first appears earlier in high southern latitudes than it does in lower-latitude sites. Because the last appearance datum (LAD) of prisms occurs at or below the first appearance datum (FAD) of *A. mayaroensis* in the high-latitude sites but after that datum in mid- to low latitudes, the extinction pulse among inoceramids seems to progress from the Antarctic toward the equator.

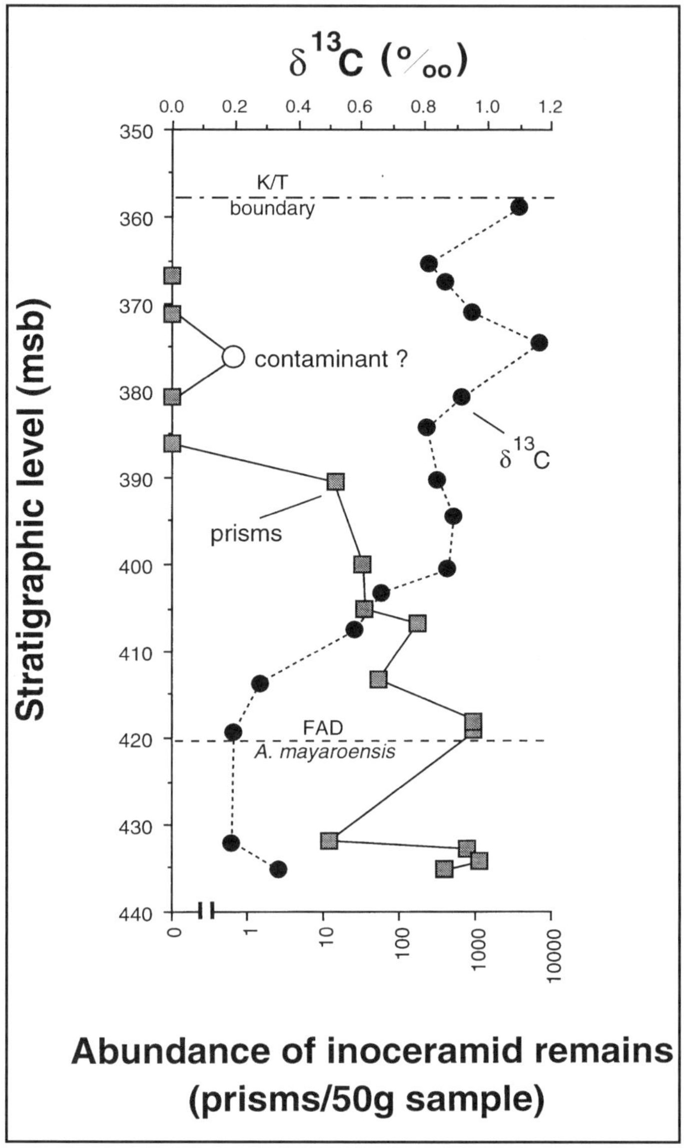

Figure 9. The carbon isotopic value measured for groups of 10 individuals of *Stensioina beccariiformis* (Seto et al., 1991) and the abundance of inoceramid prisms (number of prisms per 50-g bulk sample) plotted against stratigraphic position (meters subbottom) in Hole 752B.

changes across the inoceramid extinction interval in the Basque sections (M. Ducharme, unpublished data).

Except for evidence of cooling, examination of planktonic and nektonic organisms shows that surface waters were not greatly affected by the mid-Maastrichtian changes. Cooling is demonstrated by a ~1‰ increase in planktonic foraminiferal calcite $\delta^{18}O$ values (e.g., Barrera et al., 1987; Barrera and Huber, 1990), an equatorward expansion of the biogeographic range of austral planktonic foraminifera and nannofossils species (Huber, 1992; Huber and Watkins, 1992), and an equatorward contraction of the range of Tethyan species (Huber, 1992). On the other hand, ammonites preserved in the well-studied coastal sections of the Basque region (e.g., Wiedmann, 1988; Ward and Kennedy, 1993) show no apparent changes in abundance or diversity at the time that the inoceramid fauna collapsed. In these sections the planktonic foraminiferal assemblage also does not change significantly through the Maastrichtian (M. Ducharme, unpublished data; W. N. Orr, unpublished data). Similarly, we found no significant difference in the planktonic foraminiferal assemblages before and after the inoceramid extinction in ODP holes 750A and 761B.

Paleoceanographic implications

In addition to demonstrating an interval of global change during the Maastrichtian, inoceramid biostratigraphy and paleobiogeography also place constraints on the nature of that change. Whereas the global stratigraphic distribution of inoceramid remains requires an event that affected all the world's oceans, the effects are time transgressive and highly selective. As discussed above, change is concentrated in the benthos, and among benthic taxa only inoceramids seem to have been adversely affected.

MacLeod (1994b) proposed a model in which a change in ocean circulation during the mid-Maastrichtian led to enhanced ventilation of the deep oceans and to extinction among inoceramids. The temporal progression of the inoceramid extinction is consistent with cool, deep waters sourced in austral latitudes displacing warm, saline deep waters sourced in tropical latitudes (Saltzman and Barron, 1982). The $\delta^{13}C$ shift associated with the decline of inoceramid prisms in Hole 752B (Seto et al., 1991; Fig. 9) could be the signature of a change from one bottom water mass to another. However, the absence of inoceramids at some sites suggests complexity and regional variability in details of the vertical structure of the Maastrichtian oceans not addressed by the model. The circulation model does allow the possibility of refugia (presumably in low latitudes) where locally the seafloor continued to be bathed in warm, saline waters long after these conditions had been eliminated from the rest of the globe. Prisms from the upper Maastrichtian of Site 217 could be evidence of a surviving population of inoceramids. Until the anomalous prisms at this site can be shown to be in place, though, we think it is more conservative to interpret them as reworked.

The circulation model also explains associated changes among the benthic and planktonic organisms. Both increases in the intensity of bioturbation (MacLeod, 1994b) and changes among benthic foraminifera (M. Ducharme, unpublished data) suggest increasing oxygen levels at or near the sediment-water interface in sections in France and Spain. Ultimate causes of a reorganization of ocean circulation (e.g., global cooling) might cause shifts in geographic distributions of planktonic organisms (Huber, 1992; Huber and Watkins, 1992) but might not eliminate any planktonic habitats.

One significant assumption of this paleoceanographic model is that most Maastrichtian inoceramids must have required environmental conditions associated with low concentrations of dissolved oxygen in bottom waters. Many inoceramids are known to have lived under low-oxygen conditions (e.g., Kauffman, 1981; Elder, 1985). MacLeod (1994b) hypothesized that such individuals would have benefited from warm temperatures, substrate stability (due to low populations of burrowing organisms), exclusion of molluscovores, and/or conditions favorable to chemosynthetic symbionts (MacLeod and Hoppe, 1992, 1993; Grossman, 1993). Determining which, if any, of these variables was most important to any given population of inoceramids may be an intractable problem; however, a consistent stratigraphic association between the inoceramid extinction and evidence for increasing oxygen concentrations (e.g., MacLeod, 1994b) would support the low-oxygen assumption. It should be noted that although inoceramids occur in well-aerated nearshore facies during the Late Cretaceous, we know of no such occurrences in the mid- to late Maastrichtian. Ecological restriction of inoceramids toward the end of their range is an intriguing evolutionary prediction implicit in the model, but it falls outside the scope of this chapter.

Even assuming that Maastrichtian inoceramids occupied a narrow ecological niche, the geographic and bathymetric range of Maastrichtian inoceramids poses difficulties for any model proposing to explain their disappearance. Regression during the Maastrichtian (Haq et al., 1987) may ameliorate these difficulties. Falling sea levels could have forced the shoreward margin of a low-oxygen benthic environment over the edge of the continental shelf. At the same time, draining of epicontinental seas in low latitudes would have reduced the surface area of effective evaporative basins and thereby have reduced the production of warm, saline deep water and promoted the expansion of cool deep water sourced in high latitudes. Thus, the habitat of inoceramids may have been encroached upon from above and below. Unfortunately, stratigraphic resolution on a global scale precludes testing this hypothesis. On the other hand, examination of independent signals (e.g., stable isotopic ratios) in conjunction with paleontological changes has considerable potential in future investigations.

CONCLUSIONS

The pattern of disappearance among Maastrichtian inoceramids demonstrates that there was a major ecological event during the mid-Maastrichtian that affected all the world's oceans across a wide range of depths and latitudes. Although there were changes in the temperature of surface waters, planktonic foraminifera and ammonites were not greatly affected by the event. The duration and the diachronous nature of the inoceramid extinctions suggest that the forcing mechanism was gradual change and not a sudden, catastrophic perturbation. If most Maastrichtian inoceramids were adapted to low-oxygen environments, a reorganization of ocean circulation leading to increasing influence of oxygenated, Antarctic bottom waters fits the constraints imposed by inoceramid biostratigraphy and paleobiogeography and also explains associated data (MacLeod, 1994b). In this scenario typical inoceramids would be expected to survive longest in somewhat isolated basins in low latitudes. Documenting such trends is at the limit of stratigraphic resolution, and accurate predictions are compromised by uncertainties regarding details of Maastrichtian paleobathymetry. However, parallel study of other paleoecological indicators can provide independent constraints on the nature of mid-Maastrichtian changes and, thus, on the interplay of ecological variables on a global scale across a geologically resolvable interval of time.

ACKNOWLEDGMENTS

We thank W. N. Orr and M. Ducharme for sharing unpublished data; R. J. Stanton, S. Gartner, and M. Ducharme for reviewing early drafts of this manuscript; and the Ocean Drilling Program for providing samples from DSDP/ODP cores. Funding was provided by NSF EAR 92-05555 and a postdoctoral fellowship at the Smithsonian Institution.

REFERENCES CITED

Barrera, E., and Huber, B. T., 1990, Evolution of Antarctic waters during the Maastrichtian: Foraminifer oxygen and carbon isotope ratios, ODP Leg 113, *in* Barker, P. F., Kennett, J. P., and others, eds., Proceedings, Ocean Drilling Program, Scientific Results, Volume 113: College Station, Texas, Ocean Drilling Program, p. 813–823.

Barrera, E., and Huber, B. T., 1991, Paleogene and early Neogene oceanography of the southern Indian Ocean: Leg 119 foraminiferal stable isotope results, *in* Barron, J. A., Larsen, B. L., and others, eds., Proceedings, Ocean Drilling Program, Scientific Results, Volume 119: College Station, Texas, Ocean Drilling Program, p. 693–717.

Barrera, E., Huber, B. T., Savin, S. M., and Webb, P. N., 1987, Antarctic marine temperatures: Late Campanian through early Paleocene: Paleoceanography, v. 2, p. 21–47.

Barron, E. J., 1987, Global Cretaceous paleogeography—International Geologic Correlation Program Project, Volume 191: Palaeogeography,

Palaeoclimatology, Palaeoecology, v. 59, p. 207–216.

Barron, E. J., Saltzman, E., and Price, D. A., 1984, Occurrence of Inoceramus in the South Atlantic and oxygen isotopic paleotemperatures in Hole 530A, *in* Hay, W. W., and others, eds., Initial reports of the Deep Sea Drilling Project, Volume 75: Washington, D.C., U.S. Government Printing Office, p. 893–904.

Barron, J. A., Larsen, B. L., and others, 1989, Proceedings, Ocean Drilling Program, Initial reports, Volume 119: College Station, Texas, Ocean Drilling Program, 942 p.

Boersma, A., 1981, Cretaceous-Tertiary foraminifers from Deep Sea Drilling Project Leg 62 sites in the Central Pacific, *in* Thiede, J., Vallier, T. L., and others, eds., Initial reports of the Deep Sea Drilling Project, Volume 62: Washington, DC, U.S. Government Printing Office, p. 377–396.

Boersma, A., 1984, Campanian through Paleocene paleotemperature and carbon isotope sequence and the Cretaceous-Tertiary boundary, *in* Berggren, W. A., and Van Couvering, J. A., eds., Catastrophes and Earth history: Princeton, New Jersey, Princeton University Press, p. 247–277.

Boersma, A., and Shackleton, N. J., 1981, Oxygen and carbon isotope variations and planktonic foraminiferal depth habitats: Late Cretaceous to Paleocene, Central Pacific, DSDP Sites 463 and 465, Leg 65, *in* Thiede, J., Vallier, T. L., and others, eds., Initial reports of the Deep Sea Drilling Project, Volume 65: Washington, D.C., U.S. Government Printing Office, p. 513–526.

Clemens, W. A., 1986, Evolution of the terrestrial vertebrate fauna during the Cretaceous-Tertiary transition, *in* Elliot, D. K., ed., Dynamics of extinction: New York, John Wiley and Sons, p. 63–85.

Clemens, W. A., and Nelms, L. G., 1993, Paleoecological implications of Alaskan terrestrial vertebrate fauna in latest Cretaceous time at high paleolatitudes: Geology, v. 21, p. 503–506.

Coccioni, R., and Galeotti, S., 1994, K-T boundary extinction: Geologically instantaneous or gradual event? Evidence from deep-sea benthic foraminifera: Geology, v. 22, p. 779–782.

Dhondt, A. V., 1983a, Campanian and Maastrichtian inoceramids: A review: Zitteliana, v. 10, p. 689–701.

Dhondt, A. V., 1983b, Tegulated inoceramids and Maastrichtian biostratigraphy: Newsletters on Stratigraphy, v. 12, p. 43–53.

Douglas, R. G., and Savin, S. M., 1975, Oxygen and carbon isotope analyses of Tertiary and Cretaceous microfossils from the Shatsky Rise and other sites in the North Pacific Ocean, *in* Larson, R. L., Moberly, R., and others, eds., Initial reports of the Deep Sea Drilling Project, Volume 32: Washington, D.C., U.S. Government Printing Office, p. 509–520.

Elder, W. P., 1985, Biotic patterns across the Cenomanian- Turonian extinction boundary near Pueblo, Colorado, *in* Pratt, L. M., Kauffman, E. G., and Zelt, F. B., eds., Fine-grained deposits and biofacies of the Cretaceous Western Interior Seaway: Evidence of cylic sedimentary processes: SEPM Field Trip Guidebook No. 4, p. 157–169.

Grossman, E. L., 1993, Evidence that inoceramid bivalves were benthic and harbored chemosynthetic symbionts: Comment: Geology, p. 94–95.

Haq, B. V., Hardenbol, J., and Vail, P. R., 1987, Chronology of fluctuating sea levels since the Triassic: Science, v. 235, p. 1156–1167.

Heezen, B. C., et al., 1971a, Site 47, *in* Fischer, A. G., and others, eds., Initial reports of the Deep Sea Drilling Project, Volume 6: Washington, D.C., U.S. Government Printing Office, p. 67–143.

Heezen, B. C., et al., 1971b, Site 48, *in* Fischer, A. G., and others, eds., Initial reports of the Deep Sea Drilling Project, Volume 6: Washington, D.C., U.S. Government Printing Office, p. 145–169.

Huber, B. T., 1988, Upper Campanian–Paleocene foraminifera from the James Ross Island region (Antarctic Peninsula), *in* Feldmann, R. M., and Woodburne, M. O., eds., Geology and paleontology of Seymour Island, Antarctica: Geological Society of America Memoir 169, p. 163–252.

Huber, B. T., 1990, Maestrichtian planktonic foraminifer biostratigraphy of the Maud Rise (Weddell Sea, Antarctica): ODP Leg 113 Holes 689B and 690C, *in* Barker, P. F., Kennett, J. P., and others, eds., Proceedings, Ocean Drilling Program, Scientific Results, Volume 113: College Station, Texas, Ocean Drilling Program, p. 489–513.

Huber, B. T., 1991a, Planktonic foraminifer biostratigraphy of Campanian-Maestrichtian sediments from ODP Leg 114, Sites 698 and 700, southern South Atlantic, *in* Ciesielski, P. F., Kristoffersen, Y., and others, eds., Proceedings, Ocean Drilling Program, Scientific Results, Volume 114: College Station, Texas, Ocean Drilling Program, p. 281–297.

Huber, B. T., 1991b, Maastrichtian planktonic foraminifer biostratigraphy and the Cretaceous/Tertiary boundary at ODP Hole 738C (Kerguelen Plateau, southern Indian Ocean), *in* Barron, J. A., Larsen, B. L., and others, eds., Proceedings, Ocean Drilling Program, Scientific Results, Volume 119: College Station, Texas, Ocean Drilling Program, p. 451–465.

Huber, B. T., 1992, Paleobiogeography of Campanian-Maastrichtian foraminifers in the southern high latitudes: Palaeogeography, Palaeoclimatology, Palaeoecology, v. 92, p. 325–360.

Huber, B. T., Hodell, D. A., and Hamilton, C. P., 1995, Middle–Late Cretaceous climate of the southern high latitudes: Stable isotopic evidence for minimal equator-to-pole thermal gradients: Geological Society of America Bulletin, v. 107, p. 1164–1191.

Huber, B. T., and Watkins, D. K., 1992, Biogeography of Campanian-Maastrichtian calcareous plankton in the region of the Southern Ocean: Paleogeographic and paleoclimatic implications, *in* Kennett, J. P., and Warnke, D. A., eds., The Antarctic paleoenvironment: A perspective on global change: Washington, D.C., American Geophysical Union, Antarctic Research Series, v. 56, p. 31–60.

Jansen, H., and Kroon, D., 1987, Maestrichtian foraminifers from site 605, Deep Sea Drilling Project Leg 93, northwest Atlantic, *in* van Hinte, J. E., Wise, S. W., and others, eds., Initial reports of the Deep Sea Drilling Project, Volume 93: Washington, D.C., U.S. Government Printing Office, p. 555–575.

Johnson, C. C., and Kauffman, E. G., 1990, Originations, radiations and extinctions of Cretaceous rudistid bivalve species in the Caribbean, *in* Kauffman, E. G., and Walliser, O. H., eds., Extinction events in Earth history: Berlin and Heidelberg, Springer-Verlag, p. 305–324.

Johnson, K. R., and Hickey, L. J., 1990, Megafloral change across the Cretaceous/Tertiary boundary in the northern Great Plains and Rocky Mountains, U.S.A., *in* Sharpton, V. L., and Ward, P. D., eds., Global catastrophes in earth history: An interdisciplinary conference on impacts, volcanism, and mass mortality: Geological Society of America Special Paper 247, p. 433–444.

Johnson, K. R., and Hickey, L. J., 1992, Foliar physiognomy of Maastrichtian leaf floras from the northern Great Plains: Implications for paleoclimate: Mesozoic of the western interior: SEPM 1992 Theme Meeting, Mesozoic of the Western Interior, Ft. Collins, Colorado, Abstracts, p. 36.

Katz, M. E., and Miller, K. G., 1991, Early Paleogene benthic foraminiferal assemblages and stable isotopes in the Southern Ocean, *in* Ciesielski, P. F., Kristoffersen, Y., and others, eds., Proceedings, Ocean Drilling Program Scientific Results, Volume 114: College Station, Texas, Ocean Drilling Program, p. 481–512.

Kauffman, E. G., 1981, Ecological reappraisal of the German Posidonienschiefer (Toarcian) and the stagnant basin model, *in* Gray, J., Boucot, A. J., and Berry, W. B. N., eds., Communities of the past: Stroudsburg, Pennsylvania, Hutchinson Ross, p. 311–382.

Kauffman, E. G., 1988, The dynamics of marine stepwise extinction: Revista Española de Paleontología, v. Extraordinario, p. 57–71.

Keller, G., 1989, Extended period of extinctions across the Cretaceous/Tertiary boundary in planktonic foraminifera of continental-shelf sections: Implications for impact and volcanism theories: Geological Society of America Bulletin, v. 101, p. 1408–1419.

Lehman, T. M., 1987, Late Maastrichtian paleoenvironments and dinosaur biogeography in the western interior of North America: Palaeogeography, Palaeoclimatology, Palaeoecology, v. 60, p. 189–217.

MacLeod, K. G., 1994a, The extinction of inoceramid bivalves in Maastrichtian strata of the Bay of Biscay region of France and Spain: Journal of Paleontology, v. 68, p. 1048–1066.

MacLeod, K. G., 1994b, Bioturbation, inoceramid extinction, and mid-

Maastrichtian ecological change: Geology, v. 22, p. 139–142.

MacLeod, K. G., and Hoppe, K. A., 1992, Evidence that inoceramid bivalves were benthic and harbored chemosynthetic symbionts: Geology, v. 20, p. 117–120.

MacLeod, K. G., and Hoppe, K. A., 1993, Evidence that inoceramid bivalves were benthic and harbored chemosynthetic symbionts: Reply: Geology, p. 95–96.

MacLeod, K. G., and Huber, B. T., 1996, Strontium isotopic evidence for extensive reworking in sediments spanning the Cretaceous-Tertiary boundary at ODP Site 738: Geology (in press).

MacLeod, K. G., and Orr, W. N., 1993, The taphonomy of Maastrichtian inoceramids in the Basque region of France and Spain and the pattern of their decline and disappearance: Paleobiology, v. 19, p. 235–250.

MacLeod, K. G., and Ward, P. D., 1990, Extinction pattern of *Inoceramus* (Bivalvia) based on shell fragment biostratigraphy, *in* Sharpton, V. L., and Ward, P. D., eds., Global catastrophes in Earth history: An interdisciplinary conference on impacts, volcanism, and mass mortality: Geological Society of America Special Paper 247, p. 509–518.

Mathey, B., 1988, Paleogeographical evolution of the Basco-Cantabrian domain during the Upper Cretaceous: Revista EspaRevista Española de Paleontología, v. Extraordinario, p. 142–147.

Mount, J. F., and Ward, P. D., 1986, Origin of limestone/marl alternations in the upper Maastrichtian of Zumaya, Spain: Journal of Sedimentary Petrology, v. 56, p. 228–236.

Nelson, B. K., MacLeod, K. G., and Ward, P. D., 1991, Rapid change in strontium isotopic composition of sea water before the Cretaceous/Tertiary boundary: Nature, v. 351, p. 644–647.

Nomura, R., 1991, Paleoceanography of upper Maastrichtian to Eocene benthic foraminiferal assemblages at sites 752, 753, and 754, eastern Indian Ocean, *in* Weissel, J., Peirce, J., Taylor, E., Alt, J., and others, eds., Proceedings, Ocean Drilling Program, Scientific Results, Volume 121: College Station, Texas, Ocean Drilling Program, p. 3–29.

Pessagno, E. A., and Michael, F. Y., 1974, Mesozoic foraminifera, leg 22, site 217, *in* von der Borch, C. C., Sclater, J. G., and others, eds., Initial reports of the Deep Sea Drilling Project, Volume 22: Washington, D.C., U.S. Government Printing Office, p. 629–634.

Quilty, P. G., 1992, Upper Cretaceous benthic foraminifers and paleoenvironments, southern Kerguelen Plateau, Indian Ocean, *in* Wise, S. W., Schlich, R., and others, eds., Proceedings, Ocean Drilling Program, Scientific Results, Volume 120: College Station, Texas, Ocean Drilling Program, p. 393–443.

Ruffman, A. S., 1972, Discussion of site 111, *in* Laughton, A. S., and others, eds., Initial reports of the Deep Sea Drilling Project, Volume 12: Washington, D.C., U.S. Government Printing Office, p. 77–78.

Saltzman, E. S., and Barron, E. J., 1982, Deep circulation in the Late Cretaceous: Oxygen isotope paleotemperatures from *Inoceramus* remains in DSDP cores: Palaeogeography, Palaeoclimatology, Palaeoecology, v. 40, p. 167–181.

Seto, K., Nomura, R., and Niitsuma, N., 1991, Data report: Oxygen and carbon isotope records of the upper Maastrichtian to Lower Eocene benthic foraminifers at site 752 in the eastern Indian Ocean, *in* Weissel, J., Peirce, J., Taylor, E., Alt, J., and others, eds., Proceedings, Ocean Drilling Program, Scientific Results, Volume 121: College Station, Texas, Ocean Drilling Program, p. 885–889.

Sheehan, P. M., Fastovsky, D. E., Hoffman, R. G., Berghaus, C. B., and Gabriel, D. L., 1991, Sudden extinction of the dinosaurs: Latest Cretaceous, upper Great Plains, U.S.A.: Science, v. 254, p. 835–839.

Sliter, W. V., 1977, Cretaceous foraminifers from the southwestern Atlantic Ocean, Leg 36, Deep Sea Drilling Project, *in* Barker, P. F., Dalziel, I. W. D., and others, eds., Initial reports of the Deep Sea Drilling Project, Volume 36: Washington, D.C., U.S. Government Printing Office, p. 519–573.

Speden, I. G., 1970, Generic status of the *Inoceramus*? *tegulatus* species group (Bivalvia) of the latest Cretaceous of North America and Europe: Postilla, v. 145, p. 1–45.

Spicer, R. A., and Corfield, R. M., 1992, A review of terrestrial and marine climates in the Cretaceous with implications for modelling the "Greenhouse Earth": Geological Magazine, v. 129, p. 169–180.

Spicer, R. A., and Parrish, J. T., 1990, Latest Cretaceous woods of the central North Slope, Alaska: Palaeontology, v. 33, p. 225–242.

Thomas, E., 1990, Late Cretaceous through Neogene deep-sea benthic foraminifers (Maud Rise, Weddell Sea, Antarctica), *in* Barker, P. F., Kennett, J. P., and others, eds., Proceedings, Ocean Drilling Program, Scientific Results, Volume 113: College Station, Texas, Ocean Drilling Program, p. 571–594.

Ward, P. D., and Kennedy, W. J., 1993, Maastrichtian ammonites from the Biscay region (France, Spain): Journal of Paleontology, Memoir v. 34, p. 1–58.

Wei, W., and Thierstein, H. R., 1991, Upper Cretaceous and Cenozoic calcareous nannofossils of the Kerguelen Plateau (southern Indian Ocean) and Prydz Bay (East Antarctica), *in* Barron, J. A., Larsen, B., and others, eds., Proceedings, Ocean Drilling Program, Scientific Results, Volume 119: College Station, Texas, Ocean Drilling Program, p. 467–494.

Wiedmann, J., 1988, The Basque coastal sections of the K/T boundary—A key to understanding "Mass Extinction" in the fossil record: Revista Española de Paleontología, v. Extraordinario, p. 127–140.

Wolfe, J. A., and Upchurch, G. R., Jr., 1987, North American nonmarine climates and vegetation during the Late Cretaceous: Palaeogeography, Palaeoclimatology, Palaeoecology, v. 61, p. 33–77.

Zinsmeister, W. J., and Feldmann, R. M., 1994, Antarctica, the forgotten stepchild: A view of KT extinction from the high southern latitudes [abs.]: New developments regarding the KT event and other catastrophes in Earth history: Houston, Texas, Lunar and Planetary Institute Contribution 825, p. 134–135.

Zinsmeister, W. J., and Macellari, C. E., 1988, Bivalvia (Mollusca) from Seymour Island, Antarctic Peninsula, *in* Feldmann, R. M., and Woodburne, M. O., eds., Geology and paleontology of Seymour Island, Antarctica: Geological Society of America Memoir 169, p. 253–284.

Manuscript Accepted by the Society September 14, 1995

Printed in U.S.A.

Geological Society of America
Special Paper 307
1996

Models of vertebrate mass mortality events at the K/T boundary

Alan H. Cutler and Anna K. Behrensmeyer
Evolution of Terrestrial Ecosystems Program, Department of Paleobiology, National Museum of Natural History, Smithsonian Institution, MRC: NHB 121, Washington, DC 20560

ABSTRACT

It has been argued that the absence of mass mortality deposits at the K/T boundary constitutes a weakness in the asteroid impact hypothesis for the extinction of the dinosaurs. We present two simple models to assess the probability that such mortality deposits would be observable in the rock record. Based on the inferred population density of dinosaurs in Late Cretaceous communities and the skeletal content of large animals such as dinosaurs, even 100% mortality would not yield enough bones to produce widespread bone beds. Also, because much of the terrestrial fossil record is time-averaged, any peak in bone abundance resulting from a mortality pulse is likely to be obscured by background attritional mortality when remains accumulated before (or after) the pulse are combined with those resulting from it. No special explanation is needed for the absence of bone beds at the K/T boundary.

INTRODUCTION

Since the Alvarez impact model was first proposed in 1980, increasing evidence has supported the idea that the Mesozoic Era was brought to a close by the impact of one or more extraterrestrial bodies on the Earth (Alvarez et al., 1980; Alvarez et al., 1994). Thus far, the most convincing lines of evidence have been mineralogic, geologic, and geochemical rather than paleontologic (Glen, 1994). Indeed, the fossil record remains ambiguous, and paleontologists disagree as to whether the putative impact was in fact responsible for abrupt and widespread extinction of species (Flessa, 1990; Glen, 1994; Kerr, 1994). The problem is especially acute for the terrestrial record, where complete K/T boundary sections are few and the fossil record tends to be fragmentary and facies dependent (Fastovsky, 1990).

One source of confusion has been the absence of a clear picture of how a global biotic catastrophe might be represented in the fossil record. Signor and Lipps (1982) made an early contribution along these lines by pointing out that vagaries of preservation would tend to truncate the apparent ranges of species below their actual extinction, with the result that abrupt extinction events could appear gradual in the fossil record. In their analysis, Signor and Lipps assumed that taphonomic processes remained constant up to and across the extinction boundary. Other authors (Carpenter, 1988; Zinsmeister and Feldmann, 1994; Williams, 1994) have suggested that the scale of mortality implied by an impact should be reflected at the extinction boundary by widespread bone beds and mass mortality horizons. In this scenario, both an unusual number of vertebrate remains and taphonomic processes favorable to their preservation are required immediately after the meteor impact.

Mass mortality associated with local or regional events such as droughts (Rogers, 1991) and volcanic ashfalls (Voorhies, 1985) have been inferred in the vertebrate fossil record. Such records of small-scale mass death events raise the question of whether mass mortality of vertebrates on a global scale might also leave a recognizable signal in the record. It is obviously too much to expect to find vast horizons of dinosaur carcasses in their final death postures dusted with a layer of iridium, but it is less obvious whether such an event would produce a detectable elevation in the frequency or scale of bone concentrations (a "bone spike") at the extinction horizon. Should we expect to find a bone spike in the record corresponding to the iridium and fern spikes already documented?

Theoretically, an impact event would result in gradients or concentric zones of different types of mortality and potential fossil deposits (Fig. 1). Dispersed carcasses and single species

Cutler, A. H., and Behrensmeyer, A. K., 1996, Models of vertebrate mass mortality events at the K/T boundary, *in* Ryder, G., Fastovsky, D., and Gartner, S., eds., The Cretaceous-Tertiary Event and Other Catastrophes in Earth History: Boulder, Colorado, Geological Society of America Special Paper 307.

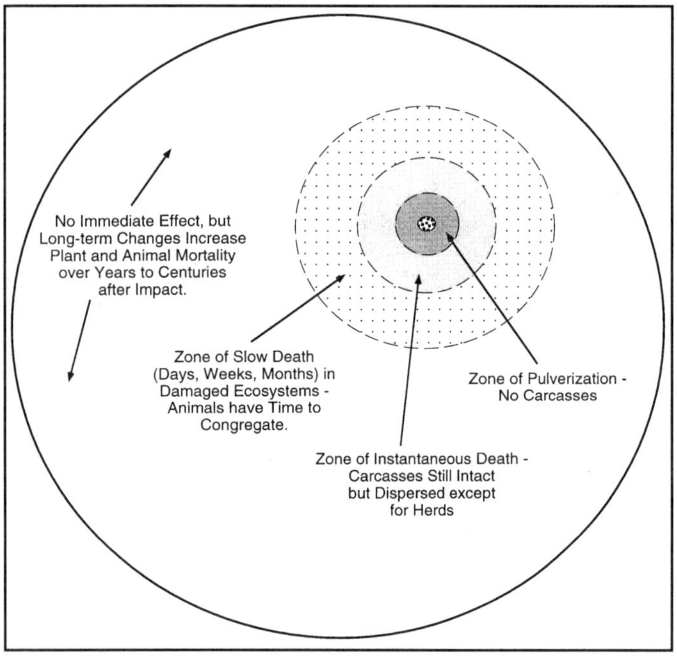

Figure 1. Hypothetical effects of an impact event on land vertebrate taphonomy.

bone beds would occur close to the impact area (but beyond the zone of immediate impact where organic remains would be pulverized), with individuals and herds dropping in their tracks and 100% mortality of standing populations. In more distant areas, dinosaurs could have survived long enough to congregate in areas of remaining resources, possibly resulting in multispecies bone beds associated with the last sources of water or food. In both cases, the probability that a modern-day observer will find evidence for such a mortality event in the fossil record depends on the number of carcasses, postevent preservational circumstances, and present availability of outcrops that span the critical time interval.

We assess here the likely effects of catastrophic mortality on bone abundance in the fossil record by posing two questions: (1) Were overall population densities in dinosaur communities high enough to produce large-scale bone beds? (2) Given the vagaries of time-averaging and preservation of bones, can the terrestrial record be resolved finely enough for a single global mortality event to stand out over "background" fossil bone occurrences? We argue that the answer to both of these questions is no. Local bone beds are possible in the K/T boundary sediments, as they are elsewhere in the rock record, but widespread bone beds generated by impact-related mortality are improbable.

POPULATION DENSITY MODEL

Dinosaurs disappear from the fossil record at the K/T boundary and are proposed as direct or indirect victims of an asteroid impact, but were they present in sufficient abundance to produce a global mortality layer marking the K/T boundary? Béland and Russell (1978) published estimates of dinosaur population densities in the Upper Cretaceous Oldman Formation (now referred to as the Dinosaur Park Formation [Eberth and Hamblin, 1993]). They estimated that the total biomass of the Dinosaur Park dinosaur community would have been 200 metric tons per km^2, equivalent to 100 2-ton animals per km^2. This is an extremely high density of large animals compared to Recent terrestrial ecosystems (Farlow, 1976; Béland and Russell, 1978), but high population density alone is not sufficient to produce a widespread bone bed, even in the event of 100% mortality and 100% preservation.

Consider hadrosaurs, which dominated the Dinosaur Park megafauna (Béland and Russell, 1978) and can serve as a "standard" dinosaur for this analysis. Counting each tooth battery as a single "bone unit," an individual hadrosaur had about 337 bones in its body (M. Brett-Surman, personal communication, 1994). Given a population density of 100 2-ton individuals per km^2, a mass kill of the total population would yield an overall bone density of only .034 bones/m^2, or 1 bone/30 m^2, assuming even dispersion. If the animals congregated prior to death so that a bone bed of 1 bone/m^2 was produced, this bed would occupy an area equal to only 3.4% of the habitat area. A denser bone bed would, of course, be even more areally constrained, which would reduce the odds of its being intersected by outcrop exposures at any given time.

An alternative approach to this analysis would be to consider the bone mass, rather than the number of bones, represented by a dinosaur population. Prange et al. (1979) found that skeletal mass scaled to body mass in birds and mammals according to the following relationship: skeletal mass [kg] = 0.061 (body mass $[kg])^{1.090}$. Applying this to a population of 2-ton individuals at the density given above yields 24,200 kg/km^2, or 24.2 g/m^2, assuming even dispersal of animals across the landscape. The bone versus body mass relationship tends to underestimate the bone mass for larger animals (it is off by a factor of 2 for a 6.6-ton elephant; Prange et al., 1979, Table A2), but even if the above result were doubled, it would only yield 48.4 g/m^2. Mammalian bone density ranges between 0.64 and 2.3 g/cm^3 (Behrensmeyer, 1975). Taking 1.5 g/cm^3 as a representative bone density for vertebrates, 48 g of bone would equal a volume of 32 cm^3, equivalent to a piece of bone approximately 4 cm × 8 cm × 1 cm in size. Thus, even if preservation were 100%, a uniform distribution of bone from the overall population would be sparsely spread over the paleolandscape. If the bone were concentrated in patches, these patches would occupy a very small proportion of the potential preserved deposits.

TIME-RESOLUTION MODEL

For a mass mortality–produced bone spike to be resolvable in the rock record, the abundance of bone material must be substantially enriched over what is produced by normal attritional mortality and time-averaging (Behrensmeyer and Chapman, 1993). Consider the dinosaur population described above, in which the standing crop of bone within the habitat is on the order of 48 g/m^2. The rate at which this bone material is supplied to taphonomic processes and, ultimately, the sedimentary record is determined by the mortality rate of the population. Annual mortality (turnover) rates for large (>100 kg) mammals in East Africa range from 9 to 34%, mainly as a function of body size (Western, 1980). Much of this mortality consists of juveniles, however, which contribute a relatively small proportion of the standing bone mass. Also, if dinosaurs were much longer-lived than modern large mammals (Dodson, 1990), their turnover rates would be lower than the mammalian rates. Taking 5% as an admittedly speculative figure for dinosaur annual turnover, this would yield a bone flux of 2.4 g/m^2 per year. A mass mortality event killing 100% of the population would release the entire standing bone mass, an increase in the bone flux by a factor of 20. In effect, 20 yr worth of mortality would be compressed into a single event.

Whether this bone spike would be resolvable in the fossil record would depend on two other factors: the degree of time-averaging within the sedimentary sequence (the acuity of the sequence; Schindel, 1982) and the preservational rate of bones (i.e., the percent that survive to become fossils). Extensive time-averaging would tend to dilute and attenuate the mass mortality spike; poor preservation would degrade or completely obliterate it.

The issue of time-averaging can be addressed with the following model. We define the enrichment factor *(E)* as:

$$E = B_m / B_a \qquad (1)$$

where B_m is the bone abundance resulting from mass mortality and B_a is the bone abundance resulting from normal attritional mortality. This describes the amplitude of the bone spike over background bone abundance.

Disregarding preservational factors, the enrichment factor produced by a single mass mortality event is given by

$$E = [(M_m / M_a) + T - 1]/T \qquad (2)$$

where M_m is the percent mortality during a mass mortality event and M_a is percent mortality under background (attritional) mortality conditions. T is the degree of time-averaging represented in a deposit, measured in years. The model assumes that burial occurs at intervals of T and that mass mortality occurs at the end of the time-averaged interval. This maximizes the total bone input into the deposit, because if mass mortality (and extinction) occurred earlier in the time-averaging interval, there would be no population to contribute bones for the remainder of this interval and the deposit would be correspondingly depleted. In effect, the model assumes a best case scenario for producing an enriched bone deposit from an impact event.

Figure 2 (solid line) shows a plot of Equation 2 for an attritional mortality rate of 5% and a mass mortality rate of 100%. At $T = 1$, the enrichment factor equals 20, but as time-averaging

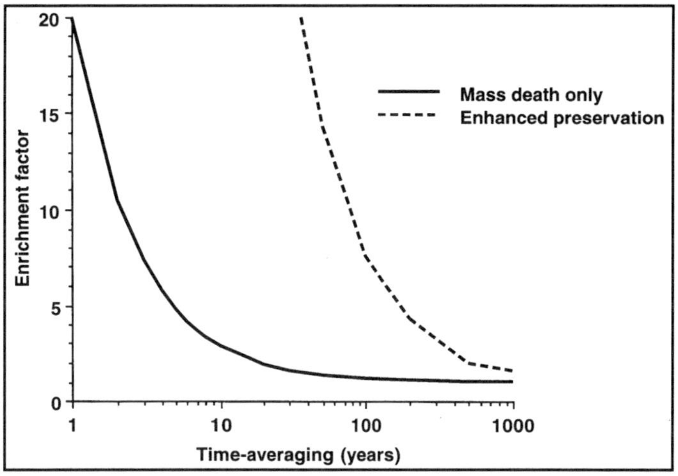

Figure 2. Enrichment of boundary sediments under conditions of mass death with and without enhanced preservation of bones. Enrichment factor is the ratio of the bone abundance resulting from mass mortality to bone abundance resulting from normal attritional mortality. See text for explanation of mortality and preservational rates.

increases, the factor rapidly decreases. At $T = 100$ (i.e., 100 yr of time-averaging), the enrichment factor only slightly exceeds 1 ($E = 1.19$). At higher values E converges to 1. Thus, only at relatively low values of time-averaging will a mass mortality spike be discernable. Although time-averaging can locally be very low (e.g., 10^1 to 10^2 yr; Rogers, 1993), more typical values in various types of fluvial vertebrate assemblages are estimated at 10^2 to 10^4 yr (Behrensmeyer, 1982; Behrensmeyer and Chapman, 1993), which would be too long to preserve a detectable mass mortality spike.

The fraction of bones actually preserved as fossils is another factor in the production of bone beds. In sedimentary facies where the potential for bone preservation is very poor or nonexistent, such as acidic floodplain soils (Behrensmeyer and Hook, 1992), bone beds are unlikely no matter how mortality rates vary, and the search for K/T bone beds would be futile. It is possible, however, that in some circumstances normal processes of bone destruction would be inhibited following a mass mortality event, and bone preservation in a wide range of depositional settings would be enhanced. For example, carcass decomposition might proceed more slowly during a postimpact "winter," local scavengers might be swamped by the sheer volume of carcasses, allowing more time for burial of bones, or loss of vegetation (e.g., due to fire or overgrazing by starving herbivores) could increase local sedimentation rates. This might leave a distinctive set of taphonomic features as a signature of the mortality event (e.g., weathered but still articulated carcasses), and—more important for this analysis—it would tend to increase the difference in bone accumulation rate between normal and catastrophic conditions. Enhanced preservation following mass mortality would therefore augment any bone spike produced by mortality alone.

Equation 2 can be modified to take such enhanced preservation into account:

$$E = [(M_m/M_a)(P_m/P_a) + T - 1]/T \quad (3)$$

where P_m is the rate of bone preservation following mass mortality and P_a is the rate of bone preservation following attritional mortality. Other terms are the same as in Equation 2.

Figure 2 (dashed line) shows a plot of Equation 3. Mortality values are as above, and P_m = 100%, i.e. preservational enhancement is maximized. A value of 3% is used for P_a; this is based on estimates of Recent bone survivorship in Amboseli Park in Kenya, where semiarid climate and alkaline soils are favorable to bone preservation (Behrensmeyer and Dechant Boaz, 1980). As with the plot in Figure 1, the value of E is strongly controlled by the degree of time-averaging, and at high values of time-averaging E converges on 1. Compared with the mass-death-only model, however, there is a greater tendency for the production of a bone spike. At T = 100 yr, for example, a modest spike (E = 6.7) would be present. For a bone bed to occur at the K/T boundary, then, the boundary would have to coincide with an interval of relatively low time-averaging and enhanced bone preservation. This may occur locally at some as yet undiscovered section or sections, but there is no evidence as yet for a global reduction in time-averaging (increased sediment accumulation rates) due to an asteroid or comet impact.

DISCUSSION AND CONCLUSIONS

Obviously, any modeling exercise is sensitive to numbers used, and those presented here are estimates subject to continued deliberation and revision. We feel that the important point is not that mass mortality bone beds will be completely absent at the K/T boundary, because it is possible that such mortality deposits exist at some as yet undiscovered boundary section or sections. Our point is that given what we know about the ecology of large animals, the taphonomy of bone beds, the temporal resolution of the sedimentary record, and the scarcity of sedimentary sections that span the boundary, the absence of bone beds at the K/T boundary is unsurprising and there is no need to invoke any special explanation for this absence.

It may be that there was dissolution of bones by acid rain following the impact. as has been suggested (Retallack et al., 1987), but the apparent absence of bone beds at the K/T boundary does not require any such extensive dissolution. Dinosaurs may or may not already have been in decline before the inferred impact occurred, but the absence of bone beds does not require abnormally low population levels at the boundary nor does it refute any scenario of global catastrophic mortality.

An iridium spike is evident in the rock record because the impact of a 10-km asteroid (assuming it occurred) would have vastly increased the flux of extraterrestrial matter into the Earth's atmosphere. Assuming a background accretion rate of 78×10^9 g per year (Kyte and Wasson, 1986) and an object mass of 11.52×10^{17}, the iridium flux would have been elevated by a factor of 1.47×10^7 by an impact. In spite of this, the high input of iridium is represented in the sedimentary record by attenuated spikes of only $10^2 \times$ background (Orth, 1989), presumably as a result of time-averaging and bioturbational smearing. In contrast, the mortality pulse at the K/T boundary would be on a scale of 10^2 over background at most. This huge difference in scale provides a useful perspective on the apparent absence of a global mortality spike from the record, even if such an event occurred.

Although we have based our arguments on large-bodied terrestrial vertebrates, similar lines of reasoning should be applicable to other organisms that experienced the K/T extinction event. It should be emphasized that our analysis addresses only the expected abundance of bone material at the extinction horizon; we do not exclude the possibility that bone material at the actual K/T boundary, should any be found, might yet possess some geochemical or taphonomic signature that could shed light on the final days of the dinosaurs.

ACKNOWLEDGMENTS

Cutler acknowledges support from the Office of the Director, National Museum of Natural History, through the Smithsonian Institution Office of Fellowships and Grants. We thank Michael Brett-Surman, Peter Dodson, David Fastovsky, Ken MacLeod, and participants in the 1994 Paleobiology Symposium at the University of Pennsylvania for helpful information and encouragement. Thanks also to David Eberth and Raymond Rogers for their suggestions for improving the manuscript. This is contribution Number 18 of the Evolution of Terrestrial Ecosystems Program at the National Museum of Natural History.

REFERENCES CITED

Alvarez, L. W., Alvarez, W., Asaro, F., and Michel, H. V., 1980, Extraterrestrial cause for the Cretaceous-Tertiary extinction: Science, v. 208, p. 1095–1108.

Alvarez, W., Asaro, F., Claeys, P., Grajales-Nishimura, J. M., Montanari, A., and Smit, J., 1994, Developments in the KT impact theory since Snowbird II, *in* New developments regarding the K/T event and other catastrophes in Earth history: Houston, Texas, Lunar and Planetary Institute Contribution 825, p. 3–5.

Behrensmeyer, A. K., 1975, The taphonomy and paleoecology of Plio-Pleistocene vertebrate assemblages east of Lake Rudolf, Kenya: Bulletin of the Museum of Comparative Zoology, v. 146, p. 473–578.

Behrensmeyer, A. K., 1982, Time resolution in fluvial vertebrate assemblages: Paleobiology, v. 8, p. 211–227.

Behrensmeyer, A. K., and Chapman, R. E., 1993, Models and simulations of time-averaging in terrestrial vertebrate assemblages, *in* Kidwell, S. M., and Behrensmeyer, A. K., eds., Taphonomic approaches to time resolution in fossil assemblages: Short Courses in Paleontology 6, The Paleontological Society, p. 125–149.

Behrensmeyer, A. K., and Dechant Boaz, D. E., 1980, The recent bones of Amboseli National Park, Kenya, in relation to East African paleoecology, *in* Behrensmeyer, A. K., and Hill, A., eds., Fossils in the making: Chicago, Illinois, University of Chicago Press, p. 72–93.

Behrensmeyer, A. K., and Hook, R. W., 1992, Paleoenvironmental contexts and taphonomic modes, *in* Behrensmeyer, A. K., Damuth, J. D., DiMichele, W. A., Potts, R., Sues, H.-D., and Wing, S. L., eds., Terrestrial ecosystems through time: Chicago, Illinois, University of Chicago Press, p. 15–136.

Béland, P., and Russell, D. A., 1978, Paleoecology of Dinosaur Provincial Park (Cretaceous), Alberta, interpreted from the distribution of articulated vertebrate remains: Canadian Journal of Earth Sciences, v. 15, p. 1012–1024.

Carpenter, K., 1988, Dinosaur bone beds and mass mortality: Implications for the KT extinction, *in* Global catastrophes in Earth history: An interdisciplinary conference on impacts, volcanism and mass mortality: Houston, Texas, Lunar and Planetary Institute Contribution 673, p. 24–25.

Dodson, P., 1990, Sauropod paleoecology, *in* Weishampel, D. B., Dodson, P., and Osmólska, H., eds., The Dinosauria: Berkeley and Los Angeles, University of California Press, p. 402–407.

Eberth, D. A., and Hamblin, A. P., 1993, Tectonic, stratigraphic, and sedimentologic significance of a regional disconformity in the Upper Judith River Group (Belly River wedge) of southern Alberta, Saskatchewan, and northern Montana: Canadian Journal of Earth Sciences, v. 30, p. 174–200.

Farlow, J., 1976, A consideration of the trophic dynamics of a Late Cretaceous large-dinosaur community (Oldman Formation): Ecology, v. 57, p. 841–857.

Fastovsky, D. E., 1990, Rocks, resolution and the record: A review of depositional constraints on fossil vertebrate assemblages at the terrestrial Cretaceous-Tertiary boundary, eastern Montana and western North Dakota, *in* Sharpton, V. L., and Ward, P. D., eds., Global catastrophes in Earth history: Geological Society of America Special Paper 247, p. 541–548.

Flessa, K. W., 1990, The "facts" of mass extinctions, *in* Sharpton, V. L., and Ward, P. D., eds., Global catastrophes in Earth history: Geological Society of America Special Paper 247, p. 1–7.

Glen, W., 1994, The mass extinction debates: How science works in a crisis: Stanford, California, Stanford University Press, 370 p.

Kerr, R. A., 1994, Testing an ancient impact's punch: Science, v. 263, p. 1371–1372.

Kyte, F. T., and Wasson, J. T., 1986, Accretion rate of extraterrestrial matter: Iridium deposited 33 to 67 million years ago: Science, v. 232, p. 1225–1229.

Orth, C. J., 1989, Geochemistry of the bio-event horizons, *in* Donovan, S. K., ed., Mass extinctions: Processes and evidence: New York, Columbia University Press, p. 37–72.

Prange, H. D., Anderson, J. F., and Rahn, H., 1979, Scaling of skeletal mass to body mass in birds and mammals: American Naturalist, v. 113, p. 103–122.

Retallack, G. J., Leahy, G. D., and Spoon, M. D., 1987, Evidence from paleosols for ecosystem changes across the Cretaceous/Tertiary boundary in eastern Montana: Geology, v. 15, p. 1090–1093.

Rogers, R. R., 1991, Taphonomy of three dinosaur bone beds in the Upper Cretaceous Two Medicine Formation of northwestern Montana: Evidence for drought-related mortality: Palaios, v. 5, p. 394–413.

Rogers, R. R., 1993, Systematic patterns of time-averaging in the terrestrial vertebrate record: a Cretaceous case study, *in* Kidwell, S. M., and Behrensmeyer, A. K., eds., Taphonomic approaches to time resolution in fossil assemblages, Short Courses in Paleontology 6, The Paleontological Society, p. 228–249.

Schindel, D. E., 1982, Resolution analysis: A new approach to gaps in the fossil record: Paleobiology, v. 8, p. 340–353.

Signor, P. W., and Lipps, J. H., 1982, Sampling bias, gradual extinction patterns and catastrophes in the fossil record, *in* Silver, L. T., and Schultz, P. H., eds., Geological implications of impacts of large asteroids and comets on the Earth: Geological Society of America Special Paper 190, p. 291–296.

Voorhies, M. R., 1985, A Miocene rhinoceros herd buried in volcanic ash: National Geographic Research Reports, v. 19, p. 671–688.

Western, D., 1980, Linking the ecology of past and present mammal communities, *in* Behrensmeyer, A. K., and Hill, A. P., eds., Fossils in the making: Chicago, Illinois, University of Chicago Press, p. 41–54.

Williams, M. E., 1994, Catastrophic versus noncatastrophic extinction of the dinosaurs: Testing, falsifiability, and the burden of proof: Journal of Paleontology, v. 68, p. 183–190.

Zinsmeister, W. J., and Feldmann, R. M., 1994, Antarctica, the forgotten stepchild: A view of KT extinction from the high southern latitudes, *in* New developments regarding the K/T event and other catastrophes in Earth history: Houston, Texas, Lunar and Planetary Institute Contribution 825, p. 134–135.

MANUSCRIPT ACCEPTED BY THE SOCIETY SEPTEMBER 14, 1995

Printed in U.S.A.

Geological Society of America
Special Paper 307
1996

The significance of the extinction of the dinosaurs

Dale A. Russell*
Canadian Museum of Nature, P.O. Box 3443, Station "D," Ottawa, Ontario K1P 6P4, Canada

ABSTRACT

The relative "fitness" of dinosaurs (reflected by the biodiversity of their environments, encephalizations, and metabolic rates) is consistent with their association with intermediate "evolutionary faunas." After a period of early to middle Mesozoic stagnation, indicators of dinosaurian "fitness" rapidly increased through to the terminal Cretaceous mass extinction. The indicators, as manifested in surviving vertebrates, then continued apparently without great change into Cenozoic time, suggesting that bolide-induced mass extinctions are of less importance in driving evolution than natural selection and the global physical environment.

INTRODUCTION

Dinosaurs are unquestionably among the more interesting creatures to have inhabited our planet, and the association of their extinction with a catastrophe of extraterrestrial origin has done nothing to diminish this interest. From the perspective of the geologic record, their disappearance was almost certainly the result of an uncommon event (Raup, 1991). However, if dinosaurs resembled modern mammals and birds in their competitive abilities, their demise would represent a "watershed" event in Earth history, not only because one dominant form of life on land was replaced by others but also because a very long period of time elapsed before animals with comparable (e.g., modern) competitive abilities reappeared on land.

The existence of broad trends in metazoan evolution is a matter of some controversy. Whether or not a given trend has resulted from changes in the variability in organisms and is thus free from directional effects imposed by "intrinsic drives" is also debated (for recent discussions, see Gould, 1988; McKinney, 1990). However, a few such trends are here postulated to have occurred, having been driven by competition and natural selection. The "fitness" (considered here as the competitive abilities) of dinosaurs is accordingly examined within the context of these trends in order to assess the place of dinosaurs in metazoan evolution. Fitness is here considered in an absolute rather than relative sense; were it possible to place organisms from different times directly into competition with each other, then Paleozoic tetrapods would usually be outperformed by and therefore replaced by modern tetrapods.

One correlate of fitness—biodiversity (Vermeij, 1991; Lodge, 1993)—has been considered to increase approximately linearly through geologic time as a result of a progressively broadening utilization of ecospace (Sepkoski, 1991, 1992b). Alternatively, fitness is here considered to be an exponential function of time resulting from the effects of biological interactions. The choice is largely one of preference, for a clear separation between linearity and exponentiality has not been made on the basis of existing evidence from the geologic record. It is evident that biodiversity is also modulated in a quasilogistical manner by the physical environment (Sepkoski, 1992b). Both biological and physical variables will be considered in assessing the evolutionary position of dinosaurs and thereby the significance of their extinction.

THE PALEOBIOLOGICAL POSITION OF DINOSAURS

Fitness

Intuitively, it would seem that "the survival of the fittest" is of more fundamental importance in interorganismal competition and natural selection than "the survival of the most com-

*Present address: Department of Marine, Earth, and Atmospheric Sciences, North Carolina State University, Box 8208, Raleigh, North Carolina 27695-8208; and North Carolina State Museum of Natural Sciences, Box 29555, Raleigh, North Carolina 27626.

Russell, D. A., 1996, The significance of the extinction of the dinosaurs, *in* Ryder, G., Fastovsky, D., and Gartner, S., eds., The Cretaceous-Tertiary Event and Other Catastrophes in Earth History: Boulder, Colorado, Geological Society of America Special Paper 307.

plex" (cf. Ruse, 1993; Lewin, 1994). The relative fitness of an organism may be equated with its relative competitive success in extending its distribution. For example, it has long been known that island biota are vulnerable to invasions from larger land masses. Thus, island biota may be considered to be relatively less fit.

Organisms that are known to have extended their distributions are characterized by their relatively superior competitive, defensive, and reproductive performances and by their origination within areas of relatively greater biodiversity. Predators and disease are particularly potent causes of extinction in invaded biota (Aronson, 1990; Jablonski and Bottjer, 1991; Sepkoski, 1991; Vermeij, 1991; Tunnicliffe, 1992; Lodge, 1993). Diversity is in turn positively correlated with primary productivity (Wylie and Currie, 1992).

In oceans, the per unit area productivity of green plants is less than on land by a factor of about seven (Rodin et al., 1975); marine species are one to two orders of magnitude less diverse than terrestrial species (Sepkoski, 1992b), and metabolic rates and the relative size of the central nervous system in marine fishes are generally less than in mammals and birds (Schmidt-Nielsen, 1984; Jerison, 1973). The invasion of the marine realm over the past several hundred million years by predators of terrestrial origin (initially reptiles: placodonts, ichthyosaurs, plesiosaurs, chelonians, mosasaurs; and later mammals: cetaceans, sirenians, carnivorans) is consistent with circumstances associated with organisms extending their distributions. Marine vertebrates have not reinvaded the land. Similarly, the success of Asian placental mammals in invading Australia may be facilitated by the generally lower life-time fecundities, lower metabolic rates, and smaller brains of indigenous Australian marsupials (Vermeij, 1991; Hinds and MacMillen, 1984; Murray, 1991).

A "low-fitness" fauna might thus be characterized by a low diversity of sluggish animals capable of only limited and stereotyped behavior; a "high-fitness" fauna might be characterized by a high diversity of animals, many of which were very active and capable of great variety of behavioral responses. The relative fitness of animal associations in the past can crudely be approximated from their position within a sequence of "evolutionary faunas."

Evolutionary faunas

Metazoan (multicellular) organisms, appearing in late Precambrian time, have been grouped into partly contemporaneous "evolutionary faunas" that have remained statistically coherent through long intervals of geological time. Although the archaic marine "Cambrian fauna" (Sepkoski, 1991, 1992a, 1992b) has vanished, members of the "Paleozoic fauna" persist in relatively unproductive environments in which they can survive, perhaps because of their typically low metabolic rates and low levels of activity. They may prosper in more productive environments but are usually excluded by active predators (e.g., bony fishes) belonging to the "Modern fauna" (Aronson and Sues, 1987; Aronson, 1990; Bambach, 1993). With the passage of geologic time the older "faunas" have successively been displaced from the more productive nearshore waters into less productive environments located at greater depths offshore.

The ranking of areas in biodiversity and productivity and as sources of emigrant organisms is thus from abyssal offshore to pelagic-onshore to rich lowland terrestrial environments (Aronson, 1990; Jablonski and Bottjer, 1991; Sepkoski, 1991; Tunnicliffe, 1992). Because fitness appears to increase along this gradient at present and the displacement of evolutionary faunas suggests that it probably did so in the past, it is likely that fitness has also increased with time. This is supported by observations, based on anatomical and ecological evidence, that locomotor and feeding efficiency as well as activity levels have increased in marine animals through geologic time (Aronson and Sues, 1987; Vermeij, 1987; Bambach, 1993).

Three "evolutionary faunas" of terrestrial vertebrates have also been statistically identified: a late Paleozoid–Triassic "labyrinthodont amphibian, 'anapsid,' mammal-like reptile fauna," a Jurassic-Cretaceous "early diapsid, dinosaur, pterosaur fauna," and a Late Cretaceous–Cenozoid "frog, salamander, lizard, snake, turtle, crocodile, bird, mammal fauna" (Benton, 1989). By analogy with the history of marine evolutionary faunas, it is likely that the general fitness of dinosaurs was intermediate between that of the Paleozoic-Triassic and late Cretaceous–Cenozoic evolutionary faunas.

Quantification of trends

As noted above, it is difficult to isolate simple trends from evidence preserved in the fossil record. Nevertheless, because evolutionary processes are generally held to be driven by competition and natural selection (i.e., autocatalytically) and because the environment of Earth has been generally stable over very long periods of time (cf. the long-sustained presence of liquid water on the planetary surface), increase in fitness is here axiomatically accepted to be autocatalytic, or exponential with respect to time. The position of dinosaurs in Earth history is then assessed using the exponential curve as a measuring device. It must be emphasized that the following relationships are exceedingly crude and require much more careful analysis (cf. McKinney, 1990).

Biodiversity (surrogate for fitness). As has been noted previously, the "fittest" organisms seem to be associated with the areas of highest biodiversity. Prior to 700 Ma a gradual increase in diversity may have been dominated by the progressive oxygenation of the atmosphere, which in turn may have been effectively controlled by tectonic processes (Des Marais et al., 1992). Fitting an exponential curve to the time of appearance of the oldest, hypothetically single metazoan species and to the 10 million metazoan species assumed to be present on Earth today (May, 1992)—700 million years later—yields (T, elapsed time in millions of years):

$$\text{Metazoan diversity (species)} = e^{0.023T} \quad (1)$$

The equation implies that half of the modern number of species was present on our planet during the dinosaurian era (Fig. 1), in general agreement with the paleontologic record (cf. Harland et al., 1990; Sepkoski, 1991, fig. 3A).

In actual plots of marine diversity (Sepkoski, 1992a, 1993), the age of dinosaurs does seem largely to have coincided with a midpoint in diversity. However, the latter curves are quite irregular and often broadly depart from exponentiality (Fig. 2). If the process of the diversification of life is fundamentally exponential, it has nevertheless been strongly affected by other factors (see the section "Planetary Environments and Dinosaurs").

Encephalization (surrogate for biodiversity). In vertebrates, the mass of the central nervous system relative to that of the body has increased exponentially through geologic time, the points for dinosaurs apparently conforming to the general trend (Hopson, 1977; Russell, 1983; Wyles et al., 1983). A relationship linking brain proportions in primitive chordates and average brain proportions in modern mammals through 530 million years (Jerison, 1973; Russell, 1983) may be approximated by:

$$\text{Encephalization (average)} = 0.001e^{0.013T} \quad (2)$$

The small size of the dinosaurian brain, relative to that of modern mammals and birds, is well known (Hopson, 1977).

Metabolism (surrogate for activity). It has been proposed on the basis of anatomic, behavioral, and ecological data that activity levels have increased in animals through geologic time (Vermeij, 1987; Bambach, 1993). That metabolic rate per-unit-rate-body-weight increases with decreasing body mass (Schmidt-Nielsen, 1984) is also well known. It is not so widely appreciated that truly giant foliage-eating dinosaurs inhabiting lowland environments during Late Jurassic time were replaced by much smaller dinosaurs inhabiting similar and probably at least equally productive environments during Late Cretaceous time (mean body weights are, respectively, 7,000 and 2,700 kg; see Coe et al., 1985, tables 9.6 and 9.8). This body-size trend alone would imply that dinosaurian metabolic rates were increasing.

An animal as large as 7,000 kg would almost certainly have a low (reptilian) per kg metabolic rate in order to avoid heat prostration. A typical modern foliage-eating mammal weighs 200 kg (Coe et al., 1985, table 9.5) and is known to possess a high (mammalian, eutherian) metabolic rate. Food requirements reflect metabolism, and the points defining those of a 7,000-kg dinosaur and a modern 200-kg mammal some 150 million years later (Coe et al., 1985, equations 9.6 and 9.7) define the following relationship:

$$\text{Tetrapod metabolism (kcal/kg/day)} = 3.7e^{0.019T} \quad (3)$$

The curve passes through the center of a cluster of points representing the food requirements for a "reptilian," a "marsupialian" (or intermediate, Hinds and MacMillen, 1984), and a "mammalian" metabolism postulated for a typical Late Cretaceous foliage-eating dinosaur (Fig. 3). This coincidence suggests that Late Cretaceous dinosaurs probably had intermediate metabolic rates.

General trend for metazoan fitness. If fitness is correlated with biodiversity, encephalization (behavioral flexibility), and metabolism (activity level), then these correlates may reveal how fitness changes with respect to time within an average global physical environmental framework. The relationship is accordingly described by (averaging the constants in Equations 1, 2, and 3):

$$\text{Fitness} \approx e^{0.02T} \quad (4)$$

Whereas dinosaurs would have been scourges of late Paleozoic terrestrial ecosystems (i.e., they would have been relatively very "fit"), it is likely that they could survive only in protected environments today (i.e., they would be relatively very "unfit").

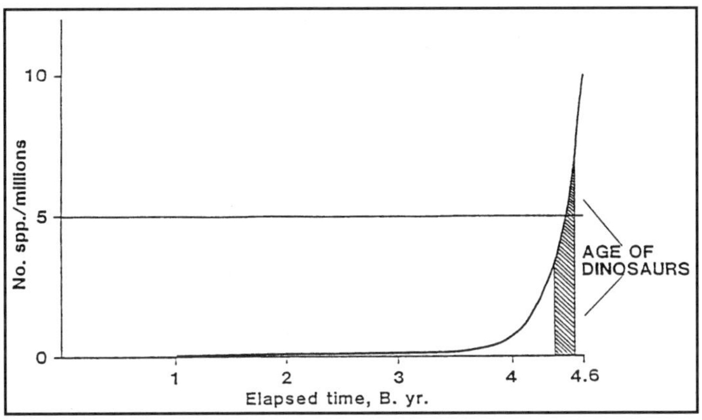

Figure 1. An exponential curve fitted to endpoints of one species and 10 million species on a geological time scale. The age of dinosaurs brackets the midpoint on the biodiversity curve.

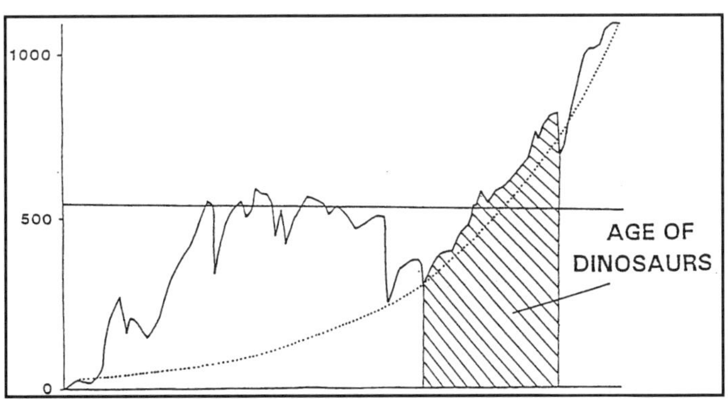

Figure 2. The age of dinosaurs indicated on a plot of the diversity of families of marine animals over the last 600 million years (modified after Sepkoski, 1993). The dotted line indicates an exponential trend in diversity.

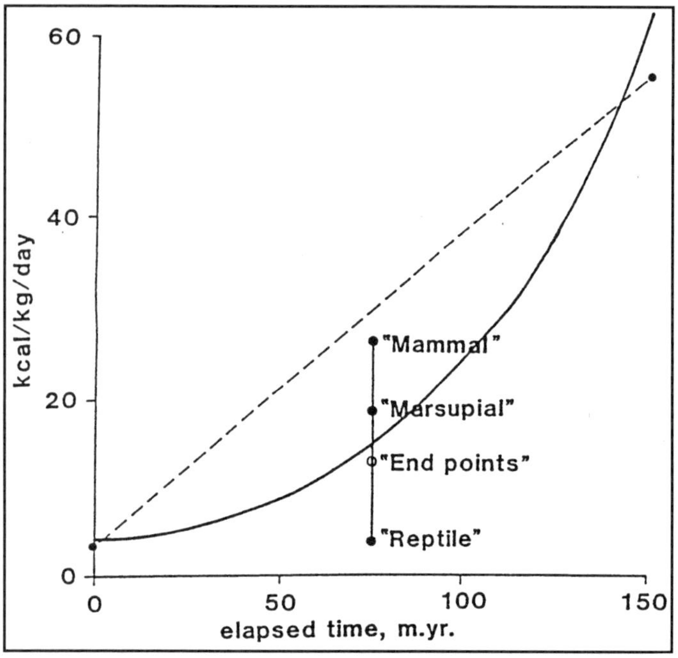

Figure 3. Food requirements of foliage-eating dinosaurs, plotted against elapsing geologic time; the dotted line represents a linear fit between endpoints. For further explanation see text.

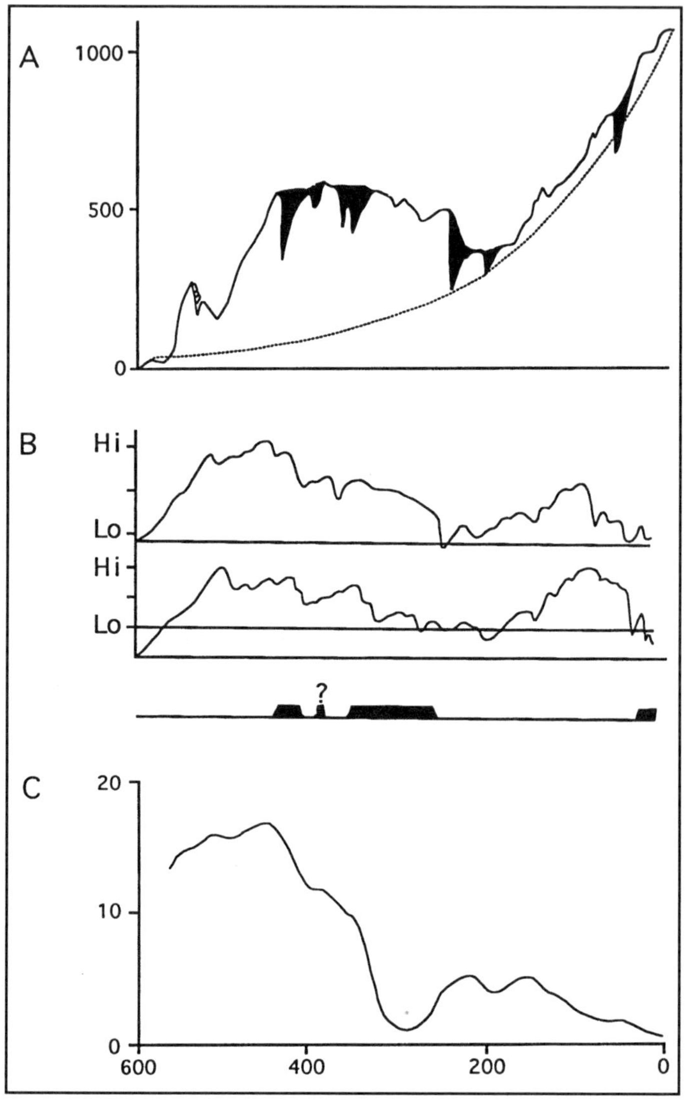

Figure 4. A, As in Figure 2 but with recovery times after mass extinctions in black; B, sea-level curves and black bar representing glacial intervals (after Crowley and North, 1991); C, atmospheric carbon dioxide levels, in multiples of present atmospheric level (after Berner, 1992). The horizontal axis represents time in millions of years before the present.

PLANETARY ENVIRONMENTS AND DINOSAURS

Physical variables modulating biodiversity

Biodiversity is known to be linked to various physical variables over broad areas of the Earth's surface, including area (Brown, 1986; Wylie and Currie, 1992), temperature (Heinrich, 1977; Brock, 1985), solar radiation, atmospheric warmth, availability of water and plant fodder (Wylie and Currie, 1992), and atmospheric carbon dioxide levels (Melillo et al., 1993). Diversity trends should therefore also be affected by significant changes in the global physical environmental framework. Such changes might produce broad departures from exponentiality in biodiversity trends, which under an unchanged physical regime would otherwise be smooth.

When exponential curves are fitted to points at the extremes of family-group diversity trends (e.g., marine animals, Sepkoski [1993], fig. 4; terrestrial plants, Knoll and Niklas [1987], fig. 6B; terrestrial vertebrates, Benton [1989], fig. 9.4B), it is evident that broad departures from exponentiality do exist (they are larger with linear fits). Brief "hyperexponential bursts" of diversity due to biological innovation (Miller and Sepkoski, 1988) and long-term compensatory interactions between a warming sun and declining greenhouse gases (Schwartzman and Volk, 1991) appear to operate on time scales dissimilar to those associated with these departures. Mass extinctions, which produce abrupt, downward "spikes" in diversity and are followed by rapid logistic rebounds (Fig. 4A), reflect profound but transient deteriorations in the physical environment (Miller and Sepkoski, 1988; Sepkoski, 1992b).

Paleozoic departure from exponentiality

In the case of marine animals (Fig. 4A through C), an early interval of approximately exponential diversification coincided with the absence of polar ice, an expansion of shallow seas over the continents (Crowley and North, 1991), and very high atmospheric carbon dioxide levels (Berner, 1992; marine phytoplankton productivity today is limited by the availability of carbon dioxide, Riebesell et al., 1993). The succeeding departure from exponentiality coincided with a reduction in the area of epicontinental seas, return of high-latitude glaciation, and rapid fall in carbon dioxide levels. It is perhaps relevant that marine organisms with calcareous skeletons, which are preserved in abundance in the fossil record, become less diverse in progressively cooler waters (Clarke, 1992). To what extent the departure was due to the retention by expanding terrestrial ecosystems of nutrients formerly available to marine ecosystems is difficult to ascertain. Near the end of the Paleozoic, animals of terrestrial origin were invading the seas, suggesting that biodiversity and productivity had become relatively greater on land, as they are today (Sepkoski, 1992b; Rodin et al., 1975). Thus, there is some evidence not bearing directly on dinosaurian evolution that large departures from fitness trends may be associated with broad changes in the global physical environment.

Mesozoic departure from exponentiality

In terrestrial environments vertebrate diversity stagnated over a long interval during which the diversity of marine invertebrates was increasing exponentially (Fig. 5; Benton, 1989; Jablonski and Bottjer, 1991). This interval coincided with a time when most of the land areas of the Earth were united into a giant supercontinent, and arid environments were extreme and widespread (Crowley and North, 1991). Plant diversity and productivity were evidently concentrated in higher latitudes instead of the broader equatorial region (Tiffney and Niklas, 1990; Ziegler et al., 1987, 1993). The general pattern of diversification in terrestrial vertebrates paralleled that of terrestrial plants, in conformity with the dependence of vertebrates upon plant productivity. A trend in maximum brain size appears to reflect diversity, increasing little through much of the dinosaurian era (Fig. 6; Russell, 1983, fig. 1C; Knoll and Niklas, 1987; Benton 1989, fig. 9.4). Encephalization thus may be a function of the diversity of interorganismal reactions.

Insect diversification (Labandeira and Sepkoski, 1993) does not show the prominent departures from exponentiality present in the three previous cases. One relatively minor negative departure coincides with the arid interval encompassing most of the dinosaurian era and another with the origin of small insectivorous birds (Sereno and Gao, 1992). The most recent phase of maximal diversification, in which dinosaurs participated (Fig. 7; Dodson, 1990), began during the latter part of the dinosaurian era and continues to the present. It coincided with the dispersal of the continents and onset of more equable climates (Crowley and North, 1991).

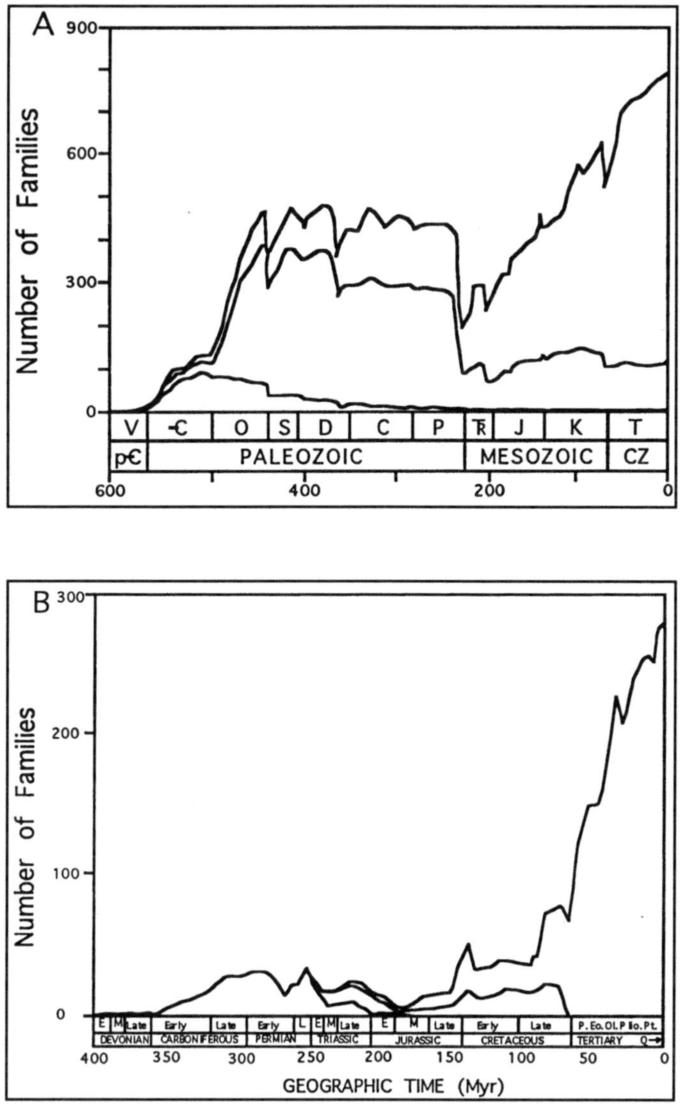

Figure 5. A, Diversity of marine families through the last 600 million years, showing Cambrian, Paleozoic, and Modern "evolutionary faunas" (after Sepkoski, 1991); B, diversity of families of terrestrial backboned animals, showing three "evolutionary faunas" (after Benton, 1989).

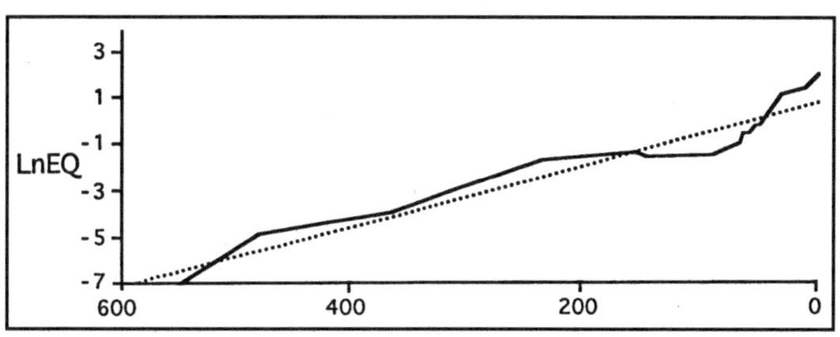

Figure 6. Natural log of maximum brain weight corrected for body weight effects (encephalization, lnEQ) plotted against geologic time. The solid line links the points from which the regression (dotted line) was calculated.

THE EXTINCTION OF THE DINOSAURS

At the time of greatest diversity of dinosaurs (during the final 15 million years of the Cretaceous), area-diversity relationships suggest that, worldwide, more than 400 genera of dinosaurs should have been present (Russell, 1995). Only 111 have so far been described, and only 28 are known from relatively

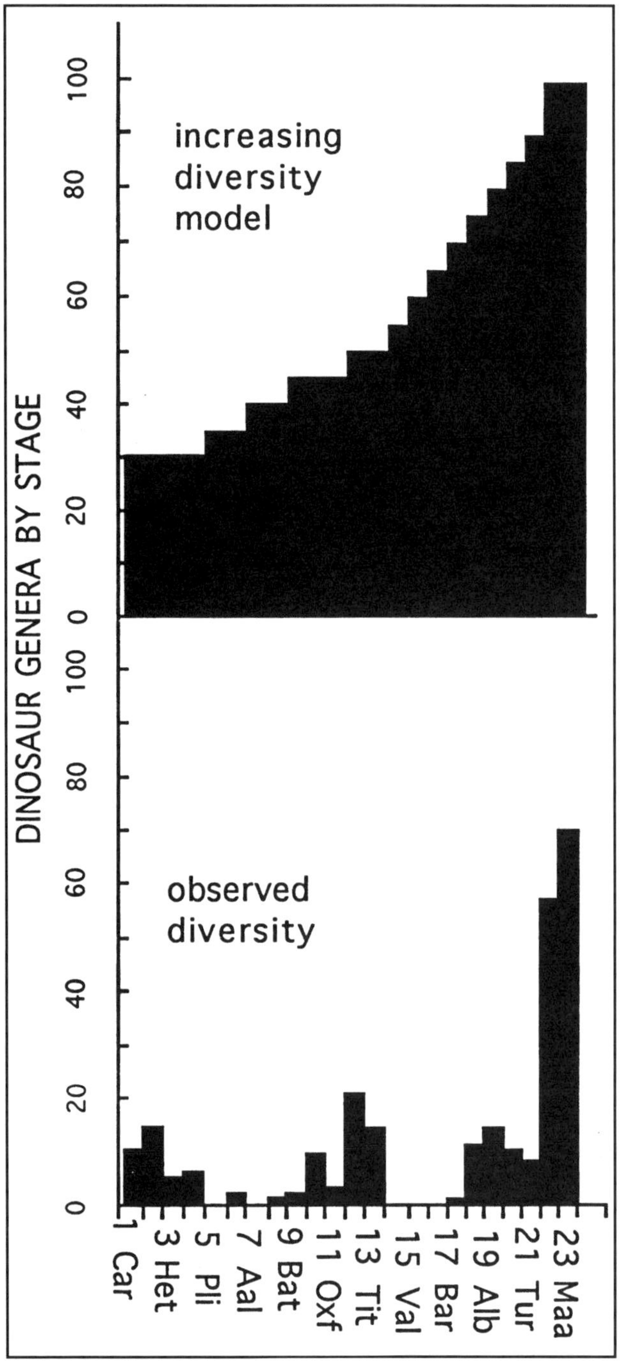

Figure 7. Diversity of dinosaurs through Late Triassic–Late Cretaceous time (modified from Dodson, 1990). Maa—Maastrichtian; Tur—Turonian; Alb—Albian; Bar—Barremian; Val—Valanginian; Tit—Tithonian; Oxf—Oxfordian; Bat—Bathonian; Aal—Aalemian; Pli—Pliensbachian; Het—Hettangian; Car—Carnian.

complete skeletal materials (all from either North America or Asia; Weishampel et al., 1990). Thus, reasonably complete information on the general form of but 7% of dinosaurian genera is available for the richest portion of their record. It should be apparent that their record is too incomplete to resolve time scales comparable to those of a bolide impact. However, that a bolide impact was the cause of their extinction is abundantly implied by trace element, isotopic, mineralogic, and microfossil evidence (Hildebrand, 1993; Nicols et al., 1992; Olsson and Liu, 1993). This evidence suggests that the stresses that annihilated the dinosaurs were not freezing temperatures (D. Archibald, personal communication, 1994) but probably starvation through a global collapse in the productivity of green plants (Sheehan and Fastovsky, 1992). This collapse may have been caused by darkness and acid rain (Hildebrand, 1993).

CONCLUSIONS

It is evident that the foregoing conceptual framework at best represents only the vaguest of outlines of an exceedingly complex and fascinating reality. If it is generally accurate, the trends in metazoan fitness suggest that the appearance of dinosaurlike creatures was expectable in the history of life. They can be regarded as characteristic of a "medieval" or a middle interval, as the name of the Mesozoic Era suggests. Relative to the dinosaurs, modern vertebrates are probably hyperdiverse, hyperactive, and hypercerebral. During early Mesozoic time an increase in the diversity of terrestrial metazoans appears to have been restrained by a reduction in the area of tropical lands and increased tropical desertification. However, dinosaurs responded to the return of more mesic conditions in late Mesozoic time by rapidly diversifying.

Probably because of the brevity of mass extinctions, they appear not to have been as hostile to metazoan fitness as long-sustained suboptimal conditions in the physial environment. After the terminal Cretaceous extinctions, the return to former levels of biodiversity was rapid. Major trends in diversity, behavioral complexity, and activity, which had begun during the Late Cretaceous, continued with little apparent interruption into Cenozoic time. The dinosaurian assemblages that were exterminated resembled their early Cenozoic replacements much more closely than did their Jurassic antecedents. The era of large dinosaurs was ending not because dinosaurs were in decline but rather because they were changing into creatures that were more diverse, more active, and more cerebral and thereby more similar to modern birds (to which they were ancestral) and mammals.

The effects of the Cretaceous-terminating bolide may therefore not have been of watershed importance in the history of life, although the effects of selective pressures that were driving dinosaurs to higher levels of fitness may have been of great import. "Behavioral flexibility" and "activity level" correspond rather well to "thought" and "work" in defining fitness in human terms.

ACKNOWLEDGMENTS

It is with deep gratitude that I acknowledge my intellectual indebtedness to colleagues in invertebrate paleontology who during the past decade have searched for pattern in the history of life with such fruitful results. Warm thanks are extended to Drs. David Jablonski, Michael Benton, and Peter Dodson for their construction comments and candid reservations. Responsibility for errors of interpretation, for which the scope is large, is surely mine alone.

REFERENCES CITED

Aronson, R., 1990, The rise and fall of life at sea: New Scientist, no. 1736, p. 34–37.

Aronson, R., and Sues, H. D., 1987, The paleoecological significance of an anachronistic ophiuroid community, *in* Kerfoot, W. C., and Sih, A., eds., Predation: Direct and indirect impacts on aquatic communities: Hanover, New Hampshire, University Press of New England, p. 355–366.

Bambach, R. K., 1993, Seafood through time: Changes in biomass, energetics, and productivity in the marine ecosystem: Paleobiology, v. 19, p. 372–397.

Benton, M. J., 1989, Patterns of evolution and extinction in vertebrates, *in* Allen, K. C., and Briggs, D. E. G., eds., Evolution and the fossil record: London, Belhaven Press, p. 218–241.

Berner, R. A., 1992, Palaeo-CO_2 and climate: Nature, v. 358, p. 114.

Brock, T. D., 1985, Life at high temperatures: Science, v. 230, p. 132–138.

Brown, J. H., 1986, Two decades of interaction between the MacArthur-Wilson model and the complexities of mammalian distributions: Biological Journal of the Linnean Society, v. 28, p. 231–251.

Clarke, A., 1992, Is there a latitudinal diversity cline in the sea?: Trends in Ecology and Evolution, v. 7, p. 286–287.

Coe, M. J., Dilcher, D. L., Farlow, J. O., Jarzen, D. M., and Russell, D. A., 1985, Dinosaurs and land plants, *in* Friis, E. M., Chaloner, W. G. E., and Crane, P. R., eds., The origins of angiosperms and their biological consequences: Cambridge, Cambridge University Press, p. 225–258.

Crowley, T. J., and North, G. R., 1991, Paleoclimatology: Oxford Monographs on Geology and Geophysics, v. 16, 339 p.

Des Marais, D. J., Strauss, H., Summons, R. E., and Hayes, J. M., 1992, Carbon isotope evidence for the stepwise oxidation of the Proterozoic environment: Nature, v. 359, p. 605–609.

Dodson, P., 1990, Counting dinosaurs: How many kinds were there?: Proceedings of the National Academy of Sciences U.S.A., v. 87, p. 7608–7612.

Gould, S. J., 1988, Trends as changes in variance: New slant on progress and directionality in evolution: Journal of Paleontology, v. 62, p. 319–329.

Harland, W. B., Armstrong, R. L., Cox, A. V., Craig, L. E., Smith, A. G., and Smith, D. G., 1990, A geologic time scale 1989: Cambridge, Cambridge University Press, xvi + 263 p.

Heinrich, B., 1977, Why have some animals evolved to regulate a high body temperature?: The American Naturalist, v. 111, p. 623–640.

Hildebrand, A. R., 1993, The Cretaceous/Tertiary boundary impact (or the dinosaurs didn't have a chance): Journal of the Royal Astronomical Society of Canada, v. 87, p. 77–118.

Hinds, D. S., and MacMillen, R. E., 1984, Energy scaling in marsupials and eutherians: Science, v. 225, p. 335–337.

Hopson, J. A., 1977, Relative brain size and behavior in archosaurian reptiles: Annual Review of Ecology and Systematics, v. 8, p. 429–448.

Jablonski, D., and Bottjer, D. J., 1991, Environmental patterns in the origins of higher taxa: The post-Paleozoic record: Science, v. 252, p. 1831–1833.

Jerison, H. J., 1973, Evolution of the brain and intelligence: New York, Academic Press, xiv + 482 p.

Knoll, A. H., and Niklas, K. J., 1987, Adaptation, plant evolution, and the fossil record: Review of Palaeobotany and Palynology, v. 50, p. 127–149.

Labandeira, C. C., and Sepkoski, J. J., 1993, Insect diversity in the fossil record: Myth and reality: Science, v. 261, p. 310–315.

Lewin, R., 1994, A simple matter of complexity: New Scientist, no. 1911, p. 37–40.

Lodge, D. M., 1993, Biological invasions: Lessons for ecology: Trends in Ecology and Evolution, v. 8, p. 133–137.

May, R. M., 1992, How many species inhabit the Earth?: Scientific American, v. 267, p. 42–48.

McKinney, M. L., 1990, Classifying and analysing evolutionary trends, *in* McNamara, K. J., ed., Evolutionary trends: London, Belhaven Books, p. 28–58.

Melillo, J. M., McGuide, A. D., Kicklighter, D. W., Moore, B., Vorosmarty, C. J., and Schloss, A. L., 1993, Global climate change and terrestrial net primary production: Nature, v. 363, p. 234–240.

Miller, A. I., and Sepkoski, J. J., 1988, Modeling bivalve diversification: The effect of interaction on a macroevolutionary system: Paleobiology, v. 14, p. 364–369.

Murray, P., 1991, The Pleistocene megafauna of Australia, *in* Vickers-Rich, P., Monaghan, J. M., Baird, R. F., and Rich, T. H., eds., Vertebrate palaeontology of Australasia: Victoria, Pioneer Design Studio, p. 1071–1164.

Nichols, D., Brown, J. L., Attrep, M., and Orth, C. J., 1992, A new Cretaceous-Tertiary boundary locality in the western Powder River basin, Wyoming: Biological and geological implications: Cretaceous Research, v. 13, p. 3–30.

Olsson, R. K., and Liu, C. J., 1993, Controversies on the placement of the Cretaceous-Paleogene boundary and the K/P mass extinction of planktonic foraminifera: Palaios, v. 8, p. 127–139.

Raup, D., 1991, Extinction: Bad genes or bad luck?: New Scientist, no. 1786, p. 46–49.

Riebesell, U., Wolf-Gladrow, D. A., and Smetacek, V., 1993, Carbon dioxide limitation of marine phytoplankton growth rates: Nature, v. 361, p. 249–251,

Rodin, L. E., Bazilevich, N. I., and Rozov, N. N., 1975, Productivity of the world's main ecosystems, *in* Reichle, D. E., Franklin, J. F., and Goodall, D. W., eds., Productivity of world's ecosystems: Washington, D.C., U.S. National Academy of Sciences, p. 13–26.

Ruse, M., 1993, Evolution and progress: Trends in Ecology and Evolution, v. 8, p. 55–59.

Russell, D. A., 1983, Exponential evolution: Implications for intelligent extraterrestrial life: Advances in Space Research, v. 3, p. 95–103.

Russell, D. A., 1995, China and the lost worlds of the dinosaurian era; Historical Biology, v. 10, p. 3–12.

Schmidt-Nielsen, K., 1984, Scaling; why is animal size so important?: Cambridge, Cambridge University Press, ix + 241 p.

Schwartzman, D. W., and Volk, T., 1991, Biotic enhancement of weathering and surface temperatures on earth since the origin of life: Palaeogeography, Palaeoclimatology, Palaeoecology (Global and Planetary Change Section), v. 90, p. 357–371.

Sepkoski, J. J., 1991, Diversity in Phanerozoic oceans: A partisan review, *in* Dudley, E. C., ed., The unity of evolutionary biology: Proceedings, International Congress of Systematic and Evolutionary Biology, 4th, Portland, Oregon: Portland, Dioscorides Press, p. 210–236.

Sepkoski, J. J., 1992a, Proterozoic–early Cambrian diversification of metazoans and metaphytes, *in* Schopf, J. W., and Klein, C., eds., The Proterozoic biosphere: A multi-disciplinary study: Cambridge, Cambridge University Press, p. 553–561.

Sepkoski, J. J., 1992b, Phylogenetic and ecologic patterns in the Phanerozoic history of marine biodiversity, *in* Eldredge, N., ed., Systematics, ecology and the biodiversity crisis: New York, Columbia University Press, p. 77–100.

Sepkoski, J. J., 1993, Ten years in the library: New data confirm paleontological patterns: Paleobiology, v. 19, p. 43–51.

Sereno, P. C., and Gao, C. G., 1992, Early evolution of avian flight and perching: New evidence from the Lower Cretaceous of China: Science, v. 255, p. 845–848.

Sheehan, P. M., and Fastovsky, D. E., 1992, Major extinctions of land-dwelling vertebrates at the Cretaceous-Tertiary boundary, eastern Montana: Geology, v. 20, p. 556–560.

Tiffney, B. H., and Niklas, K. J., 1990, Continental area, dispersion, latitudinal distribution, and topographic variety: A test of correlation with terrestrial plant diversity, *in* Ross, R. M., and Allmon, W. D., eds., Causes of evolution: A paleontological perspective: Chicago, University of Chicago Press, p. 76–102.

Tunnicliffe, V., 1992, The nature and origin of the modern hydrothermal vent fauna: Palaios, v. 7, p. 338–350.

Vermeij, G. J., 1987, Evolution and escalation: An ecological history of life: Princeton, New Jersey, Princeton University Press, 527 p.

Vermeij, G. J., 1991, When biotas meet: Understanding biotic interchange: Science, v. 253, p. 1099–1104.

Weishampel, D. B., Dodson, P., and Osmolska, H., 1990, The Dinosauria: Berkeley, University of California Press, xvi + 733 p.

Wyles, J. S., Kunkel, J. G., and Wilson, A. C., 1983, Birds, behavior, and anatomical evolution: Proceedings of the National Academy of Sciences U.S.A., v. 80, p. 4394–4397.

Wylie, J. L., and Currie, D. J., 1992, Species-energy theory and patterns of species richness. I: Patterns of bird, angiosperm, and mammal species richness on islands: Biological Conservation, v. 63, p. 137–144.

Ziegler, A. M., Raymond, A. L., Gierlowski, T. C., Horrell, M. A., Rowley, D. B., and Lottes, A. L., 1987, Coal, climate and terrestrial productivity: The Present and early Cretaceous compared, *in* Scott, A. C., ed., Coal and coal-bearing strata: Recent advances; Geological Society of London Special Publication 32, p. 25–49.

Ziegler, A. M., and eight others, 1993, Early Mesozoic phytogeography and climate: Philosophical Transactions of the Royal Society of London, ser. B, v. 341, p. 297–305.

Manuscript Accepted by the Society September 14, 1995

Printed in U.S.A.

Geological Society of America
Special Paper 307
1996

The Cretaceous-Tertiary boundary on the Pacific plate: Composition and distribution of impact debris

Frank T. Kyte, Jennifer A. Bostwick, and Lei Zhou
Institute of Geophysics and Planetary Physics, University of California, Los Angeles, California 90024

ABSTRACT

The Cretaceous-Tertiary (K/T) boundary has been identified at seven localities on the Pacific plate: LL44-GPC3; DSDP (Deep Sea Drilling Project) Sites 465, 576, 577, and 596; and ODP (Ocean Drilling Project) Sites 803 and 886. At 66 Ma these sites were in the center of the paleo-Pacific basin and at distances of ~5800 to 11,000 km from the Chicxulub structure, Yucatán. All seven sites have large Ir anomalies at the boundary, with Ir fluences ranging from 52 to 320 ng/cm^2. Six of the K/T boundaries are bioturbated, indicating oxic deep and bottom waters in the central Pacific at 66 Ma. At DSDP Site 465A, anoxic conditions are indicated by preservation of a ~3-mm-thick, pyrite-rich boundary clay.

Abundances of quartz grains (>30 μm) with single and multiple planar deformation features (i.e., shocked) range from ~1,000 to 2,000 grains/cm^2 at six of the seven sites; a notable exception is ODP Site 886, with an estimated 13,800 shocked grains/cm^2. The largest shocked-quartz grains (up to 320 μm) are found at sites nearest Chicxulub: LL44-GPC3, DSDP 465, and ODP Site 886. Large, hollow spherules that may represent distal deposits of altered impact melt glasses occur at several sites. Magnesioferrite spinel, recovered at six of the seven sites, is compositionally and texturally distinct from spinel at other K/T boundary sites and from meteoroid ablation debris.

Our results suggest an asymmetric distribution of impact ejecta related to the Chicxulub structure. If Chicxulub is the source of the K/T boundary debris, then available data suggest that melt glass and larger shocked minerals were ejected farther toward the west. Also, it appears that the first-formed, hottest condensates from the impact vapor cloud must have been deposited in the North Pacific.

INTRODUCTION

Since the initial discovery of anomalous Ir (Alvarez et al., 1980) and other siderophile elements (Smit and Hertogen, 1980; Ganapathy, 1980; Kyte et al., 1980) in Cretaceous/Tertiary (K/T) boundary sediments, it has become increasingly evident that the Mesozoic Era was terminated by a major large-body impact event. Reports of abundant spheroidal debris (Smit and Klaver, 1981; Montanari et al., 1983) and shocked minerals (Bohor et al., 1984) in these sediments, as well as anomalous Ir in nonmarine boundary sections (Orth et al., 1981), quickly demonstrated that the impact was a truly global phenomenon with a mineralogical as well as a chemical record of large-body accretion. The succeeding decade of research led to the discovery of numerous K/T boundary sites around the world (e.g., Alvarez et al., 1984; Smit and Romein, 1985) and the identification of high-energy deposits around the Gulf of Mexico (Bourgeois et al., 1988; Smit et al., 1992b) and Caribbean (Hildebrand and Boynton, 1990) and culminated in considerable interest in a prime candidate for the K/T impact site: the Chicxulub struc-

Kyte, F. T., Bostwick, J. A., and Zhou, L., 1996, The Cretaceous-Tertiary boundary on the Pacific plate: Composition and distribution of impact debris, *in* Ryder, G., Fastovsky, D., and Gartner, S., eds., The Cretaceous-Tertiary Event and Other Catastrophes in Earth History: Boulder, Colorado, Geological Society of America Special Paper 307.

ture (Hildebrand et al., 1991; Sharpton et al., 1992) on the Yucatán Peninsula.

With a possible K/T impact site now identified, it is important to interpret the boundary sediments within a geographic framework. Quantitative analyses of the abundance and composition of chemical and mineralogical components of K/T boundary sediments can be used to constrain the chemical and physical evolution of ejecta following the impact event. In this chapter, we describe the K/T boundary as it occurs on the Pacific plate. We summarize results of many studies of the K/T boundary in the Pacific, but we concentrate on our own work, including some previously unpublished data that emphasize the chemical and mineralogical record of large-body accretion.

K/T BOUNDARY SITES IN THE PACIFIC

At 66 Ma, the Pacific basin was ~40% of the Earth's surface. The only remnant of this once-vast region is now the Mesozoic-age crust on the western portion of the Pacific plate (Fig. 1). At 66 Ma, this portion of crust was located roughly in the center of the paleo-Pacific basin. We have now identified the K/T boundary at seven sites on the Pacific plate (Table 1). At 66 Ma these sites were at distances of ~5,800 to 11,000 km from the Yucatán. This is similar to the distance of sites in western Europe (7,000 to 8,000 km) and the Walvis Ridge in the South Atlantic (9500 km).

DSDP Site 465. The K/T boundary at this site is a calcareous sequence from the Hess Rise, North Pacific (33°49.23′N, 178°55.14′E). It is present in sections 3 and 4 of core 3 from Hole 465A. The boundary was recognized by the shipboard scientific party (Thiede et al., 1981), based on foraminiferal biostratigraphy (Boersma, 1981). Unfortunately, this boundary was cored by rotary drilling and was severely deformed by drilling disturbance (Fig. 2). White Cretaceous oozes are swirled and mixed with light brown Tertiary oozes. Thiede et al. (1981) place the boundary near the bottom of section 3 at ~465A-3-3, 144 cm (62.44 mbsf [meters below seafloor]). Portions of Cretaceous and Tertiary sediments are displaced over intervals of >40 cm above and below this datum. Throughout this interval are numerous small (<1 cm) patches of black, pyrite-rich sediment that are displaced fragments of the K/T boundary clay. Near the top of section 4, (465A-3-4, 8 to 12 cm) we identified an intact fragment of the boundary interval that includes at least 1.3 cm of stratigraphy around the boundary. A piece of this fragment (Fig. 3) contains a 4-mm layer of white Cretaceous ooze and a ~3-mm, pyrite-rich boundary clay, most of which is dark black but which grades upward into 6 mm of light brown Tertiary ooze.

Although DSDP Hole 465A was one of the first localities reported to contain an Ir anomaly (Kyte et al., 1980), it has remained poorly studied and understood because of the severe drilling disturbance. Alvarez et al. (1982) reported that the net Ir fluence (integrated anomalous Ir above background) at this site is ~320 ng/cm^2. Our estimate of the Ir fluence, based on Ir analyses and interpretation of the distribution of sediments in the disturbed interval, is ~100 ng/cm^2. We suggest that the

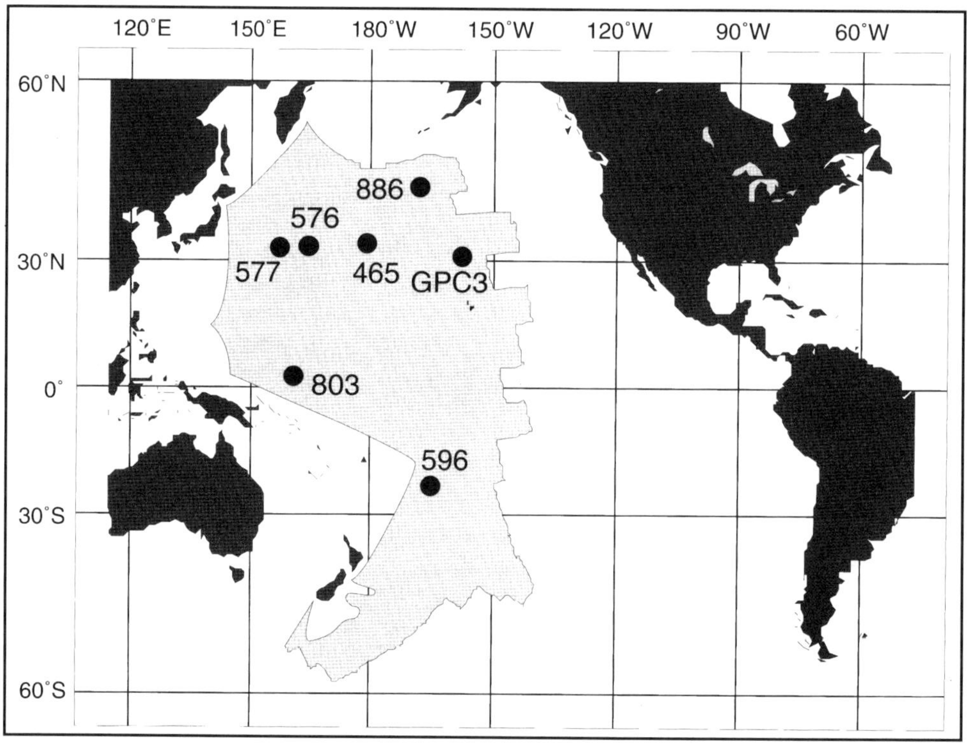

Figure 1. Location map for the seven localities on the Pacific plate where the K/T boundary has been recovered. Highlighted area is the portion of the Pacific plate with Mesozoic-age crust.

TABLE 1. DESCRIPTIONS OF K/T BOUNDARY LOCALITIES ON THE PACIFIC PLATE

Pacific Locality	Distance from Chicxulub at 66 Ma (km)	K/T Depth (mbsf)	Biostratigraphy	Lithology	Iridium (ng/cm^2)
LL44-GPC3	5,800	20.5	Ichthyoliths	Pelagic clay	70
ODP 886	6,450	65.2	None	Pelagic clay	52
DSDP 465	7,900	62.5	Forams/nannos	Calc. ooze	~100
DSDP 576	9,100	54.5	Ichthyoliths	Pelagic clay	110*
DSDP 577	9,600	109.6	Forams/nannos	Calc. ooze	61
DSDP 596	9,700	20.1	Ichthyoliths	Pelagic clay	320
ODP 803	11,000	622.4	Nannnos	Claystone	140

*Average from Holes 576 and 576B.

320 ng/cm^2 report is an overestimate and may reflect a tendency to sample darker (more Ir-rich) patches of sediment in the disturbed interval.

DSDP Site 577. The K/T boundary at this site is a calcareous sequence from the Shatsky Rise, western North Pacific (32°26.51′N, 157°43.40′E). It was recovered in three holes (577, 577A, and 577B) at a depth of ~109.6 mbsf. The boundary was recognized by the DSDP Leg 86 shipboard scientific party, based on nannofossil biostratigraphy (Heath et al., 1985). Michel et al. (1985) reported a broad Ir anomaly with a net Ir fluence of 61 ng/cm^2 in Hole 577B. The K/T boundary at Site 577 has been the subject of numerous studies of its paleontologic, mineralogic, isotopic, and geochemical composition (e.g., Wright et al., 1985; Smit and Romein, 1985; Gerstel et al., 1986; Zachos and Arthur, 1986). This boundary sequence is bioturbated and contains abundant spherules. Many spherules contain high-temperature mineral phases of probable impact origin, including magnesioferrite spinel (Smit and Kyte, 1984; Kyte and Smit, 1986) and hi-Ca clinopyroxene (Smit et al., 1992a).

DSDP Site 576. Sediments at this site are all pelagic clays collected from the deep basin ~500 km east of DSDP Site 577 in the western North Pacific (32°21,37′N, 164°16.52′E). Iridium anomalies occur in Holes 576 and 576B (Kyte et al., 1995). The anomalies are ~1 m above calcareous turbidites containing Cretaceous foraminifera (Heath et al., 1985) and within a ~5-m interval that should contain the K/T boundary, based on ichthyolith biostratigraphy (Doyle and Riedel, 1985). The Ir fluences and the depths of the Ir anomaly peaks are somewhat different in the two holes: 130 ng/cm^2 at 54.5 mbsf in 576 and 90 ng/cm^2 at 54.0 mbsf in 576B. It is not known whether these differences reflect real variability of the boundary at this site or distortions caused by hydraulic piston coring of these sediments. The upper tail of the Ir anomaly in Hole 576 is interrupted by a gap between cores 7 and 8, so we prefer the estimates for Hole 576B. The Ir anomaly is confined mostly to a sharp peak within ~20 cm, although in Hole 576 anomalous Ir at levels <2 ng/g may persist upsection for ~100 cm. Sediments in the boundary are a homogenous, dark-brown clay with no signs of layering. We assume that the boundary has been smeared by bioturbation.

LL44-GPC3. This is a long, large-diameter piston core of pelagic clay from the central North Pacific (30°19′N, 157°49.9′W). Ichthyolith biostratigraphy (Doyle and Riedel, 1979; Doyle, 1980) narrowed the location of the K/T boundary down to a 20-cm interval in which Kyte and Wasson (1986a) located the K/T Ir anomaly with a peak at 20.55 mbsf and a net Ir fluence of 70 ng/cm^2. The K/T Ir anomaly occurs near the base of a homogenous unit of clay extending from 20.67 to 20.10 mbsf (Corliss and Hollister, 1982). Both above and below this interval clays are laminated. Mottled sediments between 20.60 and 20.67 mbsf mark a gradational unit between the laminated Cretaceous clays and the homogenous K/T unit. We have found variable Ir within this mottled sediment, with one sample as high as 16 ng/g Ir. The K/T boundary in LL44-GPC3 is marked by abundant spherules and shocked quartz as well as anomalous Ir.

DSDP Site 596. The sediments at this site are pelagic clays from the western South Pacific (23°51.2′S, 165°39.27′W). The K/T boundary was not recognized by the DSDP Leg 91 shipboard scientific party, but ichthyolith biostratigraphy indicated that it should be present at ~20 mbsf (Winfrey et al., 1987). Zhou et al. (1991) discovered an Ir anomaly with a peak at 20.1 mbsf, coincident with abundant shocked quartz, spherules, and magnesioferrite spinel. The Ir anomaly at this site is extraordinarily thick (~2.0 m), and the net Ir fluence—320 ng/cm^2—is among the highest values anywhere. Zhou et al. (1991) attributed the breadth of this anomaly to bioturbation, but distortion of sediment during hydraulic piston coring cannot be excluded. A K/T boundary was probably also recovered at nearby DSDP Site 595 (only 15 km away), but sediments here were cored by rotary drilling and are heavily disturbed. To date, no attempt has been made to locate this boundary.

ODP Site 803. The K/T boundary at ODP Site 803, on the Ontong Java Plateau (2°25.98′N, 160°32.46′E), is more deeply buried than at other Pacific sites. Poorly preserved nannofossils indicated that the boundary should exist at ~622.3 mbsf in Hole 803D (Kroenke et al., 1991). Kyte et al. (1995) found a large Ir anomaly at 622.41 m (803D-68R-1, 61 to 62 cm) with an estimated net Ir fluence of 140 ng/cm^2. Sediments here are com-

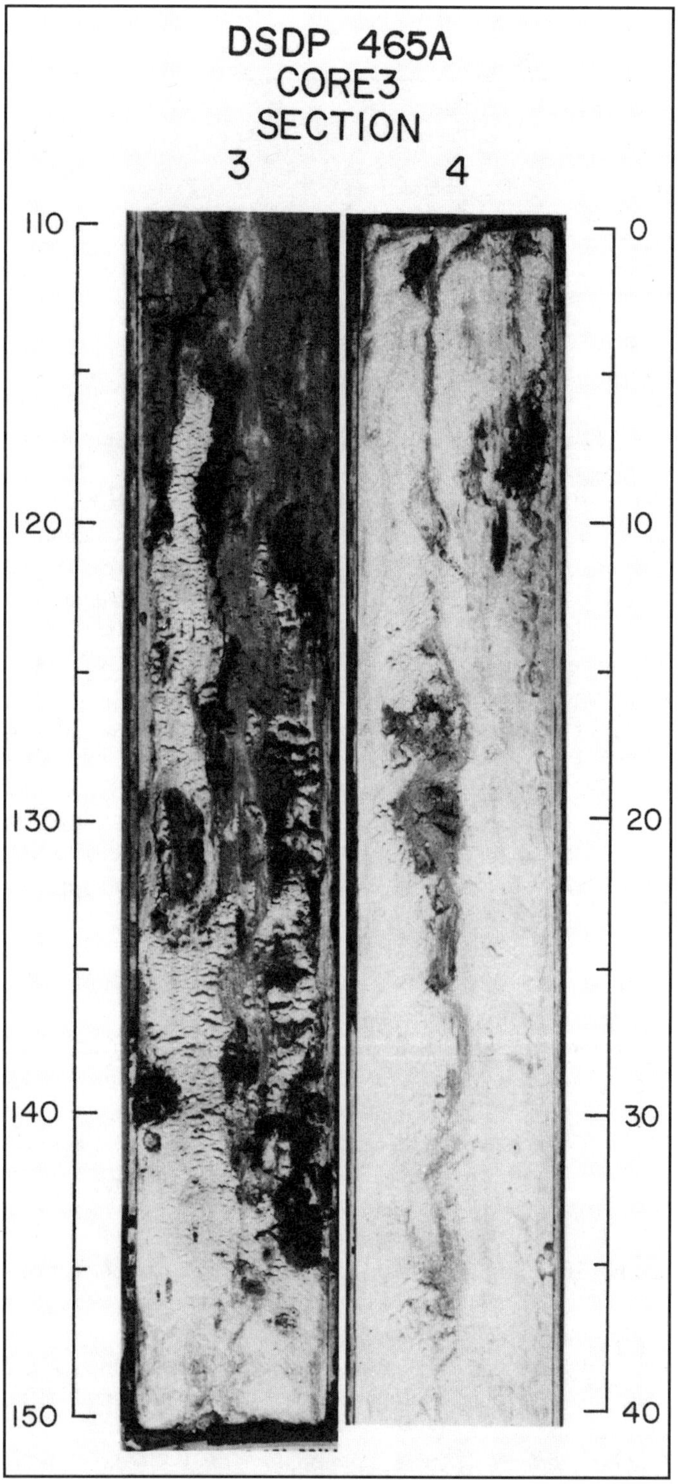

Figure 2. The K/T boundary in DSDP Hole 465A is severely disturbed by rotary drilling. The boundary is placed at 144 cm depth in core 3 section 3. However, drilling disturbance has mixed white Cretaceous oozes with light brown Tertiary oozes across >80 cm of core.

Figure 3. Portions of the K/T boundary clay can be recovered from DSDP Site 465. This clast is from a fragment of the boundary that was disturbed to 465A-3-4, 8 to 12 cm (see text). Only the uppermost 4 mm of the white Cretaceous ooze is part of the boundary fragment. The remainder of the lower white ooze is from surrounding Cretaceous sediments of unknown stratigraphic position.

pacted and semi-lithified. Most of the Ir peak is confined to ~15 cm, although anomolous Ir is found across at least 25 cm. We assume that the peak has been smeared by bioturbation. Kyte et al. (1995) reported numerous millimeter-sized flat lenses of white sediment at the K/T Ir anomaly that could be flattened spherules. Magnesioferrite spinel and quartz with planar deformation features are abundant in trace mineral extracts.

ODP Site 886. This is the most recently recovered K/T boundary in the Pacific basin, from a site near the Chinook Trough (44°41.38′N, 168°14.40′W) in the North Pacific. It occurs within a 18.1-m-thick unit of pelagic clay of Late Cretceous to late Miocene age (Unit III of Rea et al., 1993). No biostratigraphic data exist for pre-Miocene sediments at this site, and the K/T boundary was located with a survey of Ir down the 72.4-m length of cores from ODP Hole 886C (Kyte et al., 1995). The peak in Ir was found at 65.24 mbsf (886C-8H-1, 144 to 145 cm) with a net Ir fluence of 52 ng Ir/cm^2. This is a rather broad Ir anomaly, extending across at least 50 cm of section. Spherules, magnesioferrite spinel, and quartz with planar deformation features are abundant within the Ir anomaly.

BIOTURBATION AND OXIC PACIFIC DEEP WATER

Of the seven K/T boundary sites in the Pacific, all appear to be bioturbated except for the boundary in DSDP Hole 465A. Ironically, although 465A has the only intact boundary clay on the Pacific plate, it is the most poorly preserved because of drilling disturbance. Clearly, most of the deep central Pacific appears to have been well oxygenated at 66 Ma, unlike many shallow-marine sites throughout western Europe and Tethys and in New Zealand that commonly record anoxic conditions at K/T time (e.g., Schmitz, 1988). Only the pyrite-rich boundary clays in Hole 465A are highly enriched in chalcophile elements such as As and Sb (Kyte et al., 1980; Kyte and Wasson, 1982). These elements are not enriched at other localities, such as DSDP Sites

577 (Michel et al., 1985) and 596 (Zhou et al., 1991). Clearly, chalcophile element enrichment is mostly related to local chemical environments, and anoxic conditions are unnecessary for the formation of the K/T boundary Ir anomaly.

Bioturbation of K/T boundary sediments at the LL44-GPC3 site is an interesting anomaly in itself. At 66 Ma, this site was just north (within 10°) of the zone of high biological productivity at the Equator (Leinen and Heath, 1981; Kyte et al., 1993). In the uppermost Cretaceous sediments, suboxic conditions are indicated by sediment lamination and alternating bands of brown and grayish-green sediment up to 20 cm thick. Such color differences have been associated with differences in the organic carbon flux to the sediment and with differences in the extent of redox diagenesis in the sediment column (Lyle, 1983). The homogenous clays of the K/T unit containing the Ir anomaly indicate that for a period of time the flux of organic matter to the ocean floor decreased, allowing for oxic conditions in the sediment column and active bioturbation by benthic fauna. This observation would be consistent with an extended period of reduced biological productivity in the surface waters of the Equatorial Pacific. Laminated sediments return in the Paleocene, from 20.1 to 19 mbsf, although there is no evidence of Fe reduction (greenish sediment) in this part of the core, and no geochemical evidence of suboxic sediments.

IRIDIUM ANOMALIES AND IRIDIUM IN THE CENOZOIC

Other than the large, thick, Ir anomaly at DSDP Site 596, the Ir fluence of the K/T boundary is relatively uniform in the Pacific, averaging ~90 ng/cm^2 at the other six sites. Such large anomalies are easily detected by neutron activation analysis. At the four pelagic clay sites (GPC3, 576, 596, and 886), discovery of the anomalies was the first strong evidence that the K/T boundary was preserved. Over the past several years we have searched for similar anomalies by systematically measuring Ir in Late Cretaceous through Recent sediments at these four sites. Iridium was measured in continuous strips down the entire 24.3-m length of LL44-GPC3 (Kyte et al., 1993), at closely spaced (10 to 20 cm) intervals across 25 m of Hole 596 (Zhou and Kyte, 1992), and at somewhat larger intervals across the Cenozoic of Hole 886C (Kyte et al., 1995) and a composite section of DSDP Site 576 (Kyte, unpublished data).

These four Cenozoic Ir profiles are illustrated in Figure 4. At each locality the highest Ir concentrations measured in any sediment deposited over the last 70 to 80 m.y. are found at the K/T boundary. Although small bumps are observed in some profiles, none could be described as anomalies. There are points of interest, such as 1.1 ng/g Ir at 60.6 mbsf in ODP Site 886 sediments, that have been replicated but not investigated further. Generally, iridium concentrations in this type of sediment are a function of sediment accumulation rates, with highest concentrations found in the most slowly accumulating sediments (e.g., in the Eocene and Oligocene from ~18 to 9 mbsf in LL44-GPC3). This inverse relationship between Ir concentrations and sedimentation rates was first noted by Barker and Anders (1968). It was the intention of Alvarez et al. (1980) to use this relationship to measure the accumulation rate of sediment when they first measured Ir in the K/T boundary at Gubbio.

The absence of additional large and correlated Ir anomalies in our Cenozoic Ir profiles effectively excludes the possibility that global Ir anomalies of a magnitude similar to that at the K/T boundary formed at any time during the Cenozoic. This is compelling evidence against the possibility of periodic comet showers invading the inner solar system during this interval (Kyte and Wasson, 1986a). Further, these data clearly illustrate the truly anomalous nature of the K/T boundary Ir concentrations, as recognized by Alvarez et al. (1980).

THE BOUNDARY CLAY AT DSDP SITE 465A

The only intact K/T boundary clay recovered from the Pacific basin was in the clast from DSDP Hole 465A (Fig. 3). Our identification of this black layer as the K/T boundary was confirmed with paleontological analysis by H. R. Thierstein (personal communication, 1985), who kindly analyzed three smear-slides of toothpick samples from the clast. A sample from the white ooze adjacent to the base of the black layer contained only Cretaceous nannofossils. An adjacent sample from the base of the black layer was barren, containing only a few highly etched shields. The brown ooze was found to contain a mixture of Cretaceous and Tertiary nannofossils as well as Tertiary foraminiferans. The persistence of Cretaceous nannofossils into the Tertiary is well documented (Thierstein, 1981, 1982), and we are confident that this clast contains the actual biostratigraphic K/T boundary.

We separated this clast into four samples for analysis, including the white Cretaceous ooze (4 mm), upper and lower halves of the ~3-mm black boundary clay (1.5 mm each), and the light-brown Tertiary ooze (6 mm). Additionally we analyzed a light-brown sample of sediment that we believe was from the lowermost Tertiary, within ~10 cm of the boundary clay prior to disturbance (and probably much closer; 465A-3-3, 120 cm), and white oozes from the Cretaceous (465A-3-4, 50 cm) and Tertiary (465A-3-3, 25 cm) that were selected as backgrounds. DePaolo and Kyte (1984) reported initial results of leaching experiments and isotopic analyses of these samples. We have also performed detailed geochemical analyses (using methods of Kyte et al., 1992) on splits of these samples. The concentrations of $CaCO_3$, Cr, and the siderophile elements Fe, Co, Ni, Pd, Re, Os, Ir, Pt, and Au in these samples are illustrated in Figure 5.

Carbonate concentrations in the background calcareous oozes at this site are typically nearly 100%, with insoluble residues <1%. In the lower 1.5 mm of the boundary clay we estimate only ~5% calcite (calculated assuming that all Ca is in carbonate), and the upper 1.5 mm contains only ~25% calcite. It is reasonable to assume that this clay layer is a nearly pure

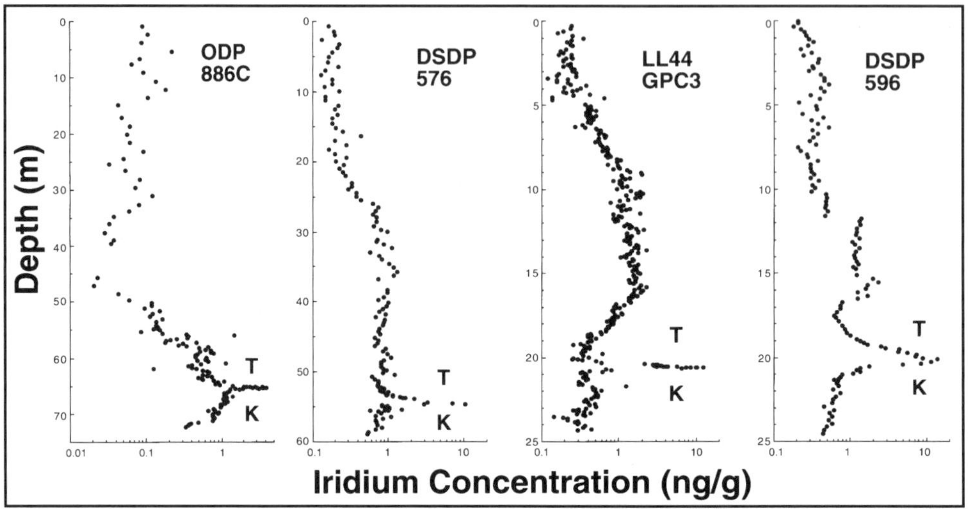

Figure 4. Iridium concentrations have been measured in sediments across the Cenozoic in four pelagic clay sequences from the Pacific. At every site the K/T boundary Ir anomaly coincides with spherules, magnesioferrite spinel, and shocked quartz. The sediments with the highest Ir concentrations at every site occur at the K/T boundary.

sample of the fallout from the K/T boundary impact event. Interestingly, the ~4-mm sample of white ooze from just below the black clay layer contains only ~85% calcite and thus has a clay content considerably higher than underlying carbonates. We suspect that this sample may also contain some impact fallout in sediments slightly below the level of pyrite formation.

All of the siderophile elements measured are enriched in the clay-rich sediments of the clast. They are most enriched in the base of the black boundary clay; the one notable exception is Ir, which was found to increase slightly in concentration toward the top of the clast. The siderophile interelement abundances are highly variable from one layer to another, with abundances varying significantly from those found in chondritic meteorites. This is not surprising, as previous studies of noble metals in K/T boundary sediments (Kastner et al., 1984; Kyte et al., 1985; Evans et al., 1993) and Precambrian ejecta deposits from the Acraman impact (Wallace et al., 1990) have shown that diagenesis can dramatically enhance the mobility of these elements. Kastner et al. (1984) and Kyte et al. (1985) both showed that interelement abundances of noble metals can be quite varied among individual samples but will most closely approximate chondritic abundances when the entire boundary deposit is integrated.

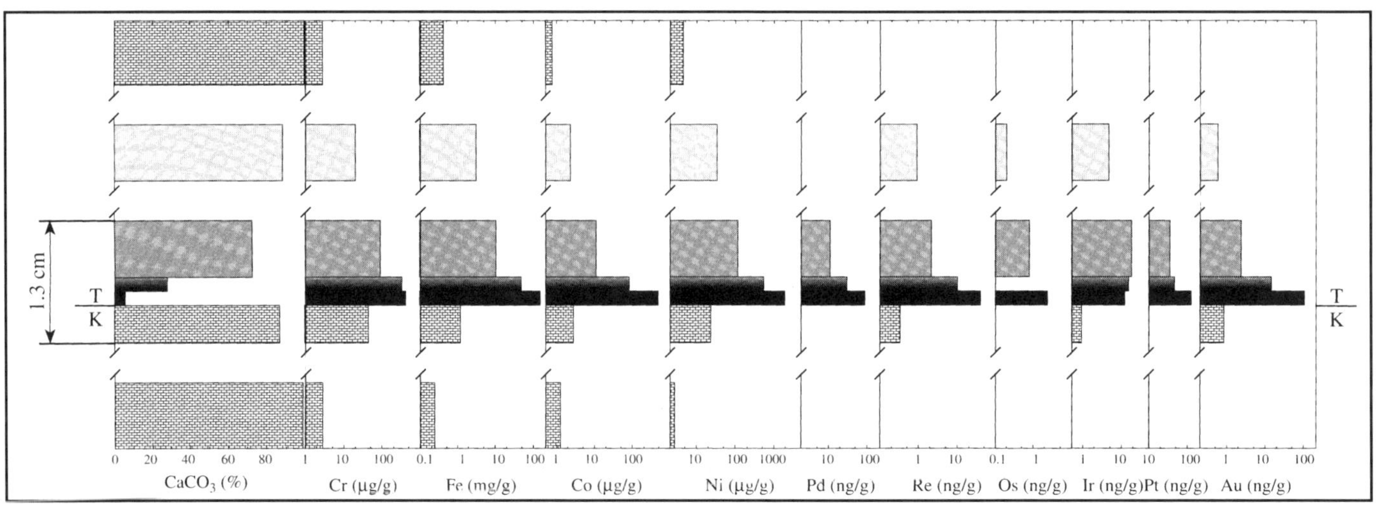

Figure 5. Calcite, Cr, and siderophile element concentrations (bulk sediment) in the K/T boundary from DSDP Hole 465A. Siderophile elements are all most concentrated right at the biostratigraphic boundary marked by the pyrite-rich black clay layer. Four samples are from the clast containing the boundary clay (Fig. 3). The exact stratigraphic positions of the other three samples are unknown (see text). Where elemental concentrations are not plotted, they were not measured.

The most extreme variation in siderophile interelement ratios at DSDP Site 465 is that for Ir/Au, which varies by a factor of 65 from the base of the black layer to the top of the clast. However, abundances for the total noble metal content of the clast (integrated over the black layer and 6 mm of brown ooze) are generally closer to chondritic than is any single sample. For the entire clast, Ni/Ir and Pd/Ir ratios are ~1.2× chondritic and Pt/Ir is ~1.5×. On the other hand, Au/Ir is ~4× chondritic, and we estimate Os/Ir at 0.08×. The observed trends in the noble metal profiles all suggest that if we had more section to analyze and integrate, Ni/Ir, Pd/Ir, Pt/Ir, and Au/Ir ratios would decrease to levels closer to ratios measured in chondrites. Os/Ir ratios have been notably lower than chondritic in K/T boundary sediments from Caravaca, Spain (Kyte et al., 1985) and in magnetic spherules from the K/T boundary in Italy (Smit and Kyte, 1984), although in neither case were Os/Ir ratios as low as we observe at DSDP Site 465. Kyte et al. (1985) suggested that low Os/Ir ratios might reflect fractionation of the volatile OsO_4 during evolution of the impact vapor cloud. Clearly, if the siderophile element enrichment is largely a result of fallout of meteoritic components in ejecta from a major impact event, the siderophiles must be significantly fractionated by processes related to the impact itself or subsequent diagenetic alteration of the fallout.

SPHERULES

Spheroidal debris is nearly ubiquitous in K/T boundary sediments, occurring in a variety of sizes, textures, and compositions (e.g., Smit and Klaver, 1981; Montanari et al., 1983; Smit and Kyte, 1984). The K/T boundary in the Pacific is no exception. We have recovered spherules from K/T boundary sediments at LL44-GPC3, DSDP Site 577, and ODP Site 886. Spherules have been observed in bulk specimens of sediments from DSDP Sites 465, 576, and 596 but not recovered except for a total of three spherules from a sample of DSDP 596. In these sediments, spherules did not survive sample disaggregation. With the Site 576 and 596 clays, our failure to recover spherules may have been because samples were allowed to dry out and thus contract to <50% of their initial volume. Spherules can probably be extracted from fresh, wet samples at these sites, and future sampling will attempt to accomplish this.

Our microscopic and mineralogical analyses of K/T boundary spherules in the Pacific are generally consistent with interpretations by other workers (Bohor, 1990; Bohor and Betterton, 1990, 1992; Smit et al. 1991, 1992a) who have argued that at least two types of spheroidal debris occur in two distinct strewnfields. One type of spherule, which has been reported only from regions near North America (e.g., Smit et al., 1992a), is larger than the other. These contain no magnesioferrite spinel and are interpreted as originally composed of target-rock melt glass, although they are now mostly hollow. These spherules (Type 1 of Bohor and Betterton, 1990; K/T "tektites" of Smit et al., 1992a) are confined to the lower of the two boundary layers in the Western Interior (the kaolinitic layer) that has a restricted radial distribution around the putative impact crater in the Yucatán (Hildebrand et al., 1991; Sharpton et al., 1992). These larger spherules are interpreted to be derived from the tektite-like glass that has been found in large spherules from Haiti (Sigurdsson et al., 1991) and northeastern Mexico (Smit et al., 1992b).

The other spherules (Type 2 of Bohor and Betterton, 1990; microkrystites of Smit et al., 1992a) are restricted to the upper layer in the Western Interior (based on the presence of magnesioferrite spinel; Bohor and Foord, 1987) and are globally distributed. This type of spherule displays textures interpreted as pseudomorphic after melt-crystallized phases (Smit et al., 1992a) and has often been desribed as containing magnesioferrite spinel.

Both types of spherule have been identified in the Pacific. A decade ago we found large spherules in LL44-GPC3 that have very smooth surfaces; many of these are hollow with mammillary growths on interior surfaces; however, we did not recognize their significance and concentrated our work on smaller, spinel-bearing "microkrystites." Bohor and Betterton (1991) have suggested that these hollow spherules and interior surfaces reflect dissolution of interior microtektite-like glass. Smit had access to our coarse separates from LL44-GPC3, and we presume that it was the presence of these hollow spherules that led Smit et al. (1992a) to report this site as containing "tektites." Most recently, similar large spherules, up to 1.5 mm in diameter, have been recovered from ODP Site 886 (Kyte et al., 1995). Ingram (1995) recovered numerous large spherules in separates from a large-volume sample of the ODP Site 886 K/T boundary and has described dumbbell, as well as spheroidal, morphologies.

These large hollow spherules also occur at sites more distant from Chicxulub. Smit et al. (1992a) describe a few smaller (500-μm), hollow spherules from DSDP Site 577. Zhou et al. (1991) reported observing hollow spherules in samples from Site 596 during disaggregation and dissolution of samples processed for shocked quartz by B. F. Bohor. We have recovered three of these spherules from a DSDP 596 sample, the largest with a diameter of 600 μm. DSDP Sites 577 and 596, which were 9,600 and 9,700 km from Chicxulub, respectively, at 66 Ma (Table 1), were farther than most European K/T sites and at a distance similar to that of K/T boundary sites on the Walvis Ridge, South Atlantic.

The smaller, crystallized spherules are very abundant throughout the Pacific. Although they have not specifically been recovered from Sites 576, 596, and 803, magnesioferrite spinel, which is known to crystallize in spherules (e.g., Smit and Kyte, 1984), has been recovered from each of these sites. The crystallized spherules that we have recovered from LL44-GPC3 and Sites 577 and 886 occur in a wide range of colors and sizes. We have concentrated our work on spinel-bearing spherules, which are magnetic and can be extracted from coarse fractions with a slightly magnetized iron needle. These are commonly smaller than most other spherules (usually <200 μm) and often have a rusty, brown color (analyses of magnesioferrite are discussed below).

Most of the larger crystallized spherules are white. In DSDP Site 577, these often contain long green crystals that Smit et al. (1992a) found to be a Ca-rich clinopyroxene. We have sectioned several of these and confirmed these observations. Clinopyroxenes have a feathery, dendritic habit, and microprobe analyses provided elemental concentrations essentially identical to those reported by Smit et al. (1992a). Interestingly, these clinopyroxenes have detectable concentrations of NiO (0.5 wt%) and significant Al_2O_3 (5 wt%). However, in the six clinopyroxene-bearing spherules we examined, we found no traces of spinel as reported by Smit et al. (1992a). We infer that their sample may have been a rare occurrence and that spinel-bearing and clinopyroxene-bearing spherules are different types of impact debris that crystallized from distinctly different liquids.

MAGNESIOFERRITE SPINEL

Magnesioferrite spinel was first reported in K/T spherules from Petriccio, Italy, by Montanari et al. (1983). Smit and Kyte (1984) recognized that the spinels represented unaltered high-temperature phases that crystallized in molten silicate droplets from a large-body impact. K/T boundary spinels were characterized by magnesioferrite compositions with high ratios of Fe_2O_3/FeO, high concentrations of Ni, and low Ti and Cr (Kyte and Smit, 1986). All these chemical characteristics make K/T spinel distinct from virtually all known igneous or metamophic sources. Kyte and Smit (1986) showed that the compositions of North Pacific spinel from LL44-GPC3 and DSDP Site 577 were distinct from those recovered from Italian (Furlo) and South Atlantic (DSDP 524) localities. Pacific spinels were found to have almost no FeO and much higher MgO and Al_2O_3 contents than non-Pacific sites.

Kyte and Smit (1986) inferred that spinel-bearing spherules must form as either impact-melt droplets, vapor condensates, or projectile ablation debris. They favored the latter two, as they might be most likely to experience high oxygen fugacities. Robin et al. (1992) have shown that spinels in chondritic cosmic spherules and meteorite fusion crusts have somewhat similar compositions, with high Ni and a significant magnesioferrite component, and they proposed that K/T spherules formed under conditions somewhat analogous to meteoroid ablation debris. Recently, Robin et al. (1993) reported that spinel-bearing spherules from DSDP Site 577 had a broad range of Ir concentrations, a few with Ir contents at meteoritic levels. Some of the high-Ir spherules were described as irregular in shape, with spinel concentrated at the rim. These were interpreted to be only partially melted and oxidized meteoritic material. Robin et al. (1993) concluded that DSDP Site 577 contained locally derived meteoroid ablation debris and that the highly oxidized spinel of the North Pacific reflected derivation from local oceanic impact events that were independent of the event that formed the Chicxulub structure.

We have now recovered magnesioferrite spinel, both in spherules and as grains in trace mineral concentrates from six of the seven Pacific Sites. Spinel was not recovered in our sample from DSDP Site 465. This may have been because of analytical problems related to our attempts to dissolve the abundant pyrite or to diagentic alteration of the spinel under the extreme reducing conditions that must have occurred at this site. Details of our recent analyses are given by Kyte and Bohor (1995) and Kyte and Bostwick (1995).

Kyte and Bohor (1995) reported the discovery of a new high-temperature mineral phase found intergrown with magnesioferrite grains from DSDP Site 596. This is a Ni-rich magnesiowüstite that is preserved in the cores of some larger spinel crystals (Fig. 6A). The most MgO-rich specimen analyzed had a chemical formula approximated as $(Mg_{.85}Ni_{.11}Fe_{.04})O$. Magnesiowüstite does not occur in terrestrial igneous systems because it can crystallize only from liquids with extremely ultramafic, MgO-rich liquids (e.g., with molar MgO/SiO_2 ratios greater than 2; Muan and Osborne, 1954). Kyte and Bohor (1995) argued that liquids with such compositions cannot evolve as a result of meteoroid ablation, and they infer that such liquids only evolve in the vapor cloud of a major impact event. They suggest that magnesiowüstite-magnesioferrite intergrowths occur in droplets that equilibrate with the vapor at sufficently high temperatures (>2,300 °C) that refractory MgO is enriched in the liquid relative to more volatile SiO_2. Kyte and Bostwick (1995) have found additional samples of spinel with magnesiowüstite in two spherules from DSDP Site 577.

Our results on analyses of magnesioferrite compositions (Kyte and Bostwick, 1995) are summarized in Figure 7. Spinels from Pacific localities are clearly distinct from those from the Indian Ocean (DSDP 761), South Atlantic (DSDP 524), Furlo (Italy), and Caravaca (Spain). Pacific spinels have significantly higher MgO and Al_2O_3 contents and higher Fe_2O_3/FeO ratios than those at other sites around the world. Further, there appears to be a trend across the Pacific, with the most extreme compositions in the northwest at Sites 576 and 577.

In addition to the regional variation in composition, magnesioferrite spinel in the Pacific is also distinct texturally from that at other sites or in meteoroid ablation debris. Kyte and Bostwick (1995) observed that Pacific spinel commonly occurs as

Figure 6. A, Magnesioferrite spinel grain from the K/T boundary from DSDP Site 596. In the elliptical region in the center of the grain (arrow), magnesioferrite (bright) is intergrown with Ni-rich magnesiowüstite (dark). B, Spinel-bearing spherule from the K/T boundary in DSDP Site 577 with large anhedral grains of magnesioferrite (bright grains) with numerous voids in a clay groundmass (gray). C, Porous, anhedral magnesioferrite grain from the K/T boundary in ODP Site 803. D, Cosmic spherule from Recent Pacific clay sediments that has skeletal, dendritic spinels that are common in meteoroid ablation debris. Large dark grains are olivine. Voids between olivine grains formerly contained an interstitial glass that has been etched by seawater. E, SEM (scanning electron microscope) image of shocked quartz from the K/T boundary in DSDP Site 465. F, Optical photomicrograph of a quartz grain from the K/T boundary in LL44-GPC3 with planar deformation features.

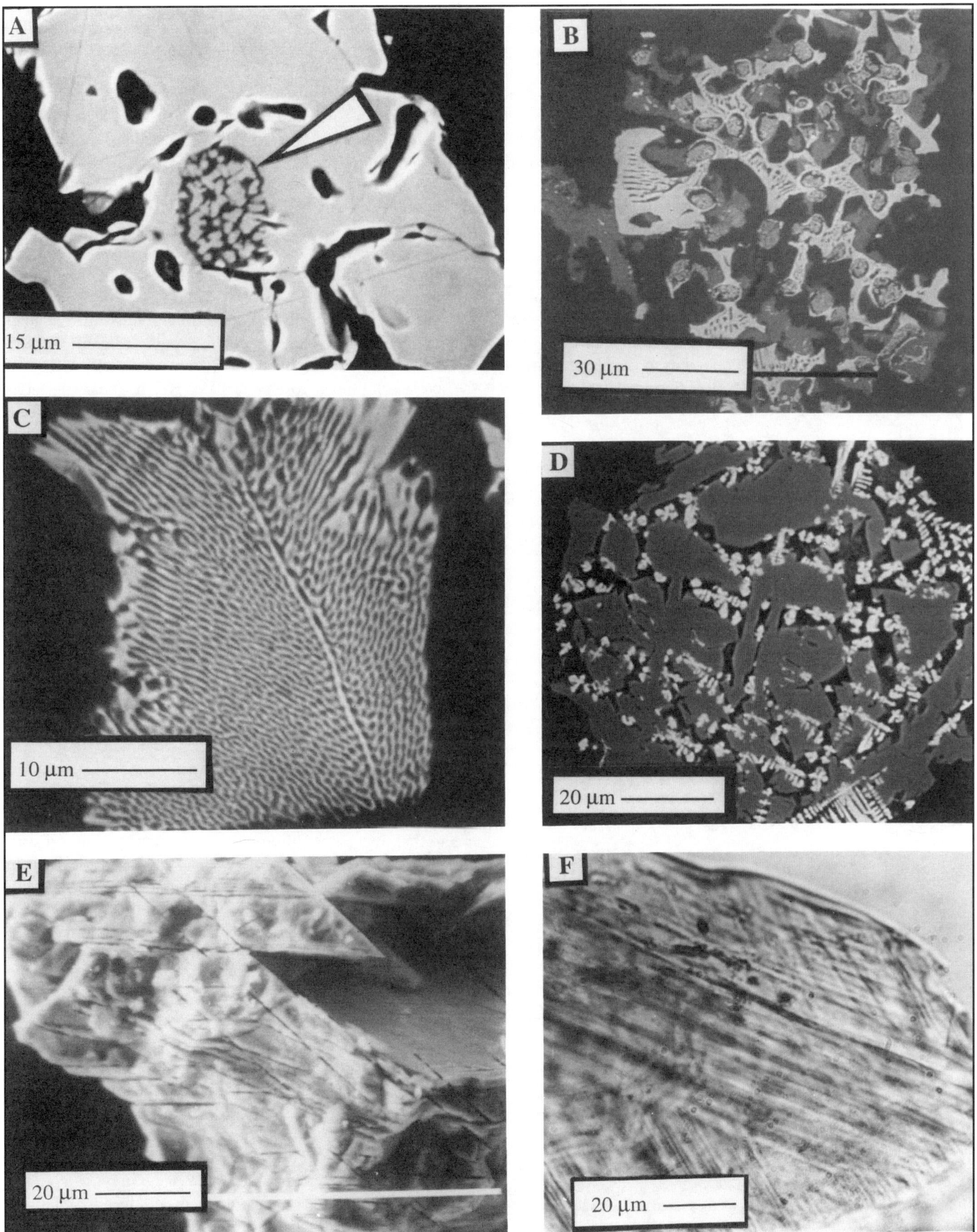
A
15 μm
B
30 μm
C
10 μm
D
20 μm
E
20 μm
F
20 μm

porous, anhedral masses with voids that are sometimes elongated and tube-shaped (Fig. 6B, C). These textures are suggestive of a myrmekitic intergrowth of spinel with some preexisting phase that they infer was most likely magnesiowüstite. Although these anhedral growths are common at all Pacific sites, spinels at other K/T sites and in cosmic spherules are invariably subhedral to euhedral grains that are commonly skeletal and can be dendritic (Fig 6D). Kyte and Bostwick (1995) failed to find convincing evidence that any of the spinel-bearing spherules at DSDP Site 577 or elsewhere are chemically or texturally similar to cosmic spherules or meteorite fusion crusts, as inferred by Robin et al. (1993). Kyte and Bostwick (1995) conclude that Pacific spinel compositions, textures, and coexisting magnesiowüstite cannot be explained by meteoroid ablation processes, and they infer that the spinel-bearing spheroids must be forming within an impact vapor cloud, possibly as the hottest, earliest materials condensing from a portion of the cloud that is enriched in projectile material.

SHOCKED QUARTZ

Shocked quartz (i.e., quartz with planar deformation features) was first identified in the Pacific by Bohor and Izett (1986) in a single sample from LL44-GPC3, but their analysis

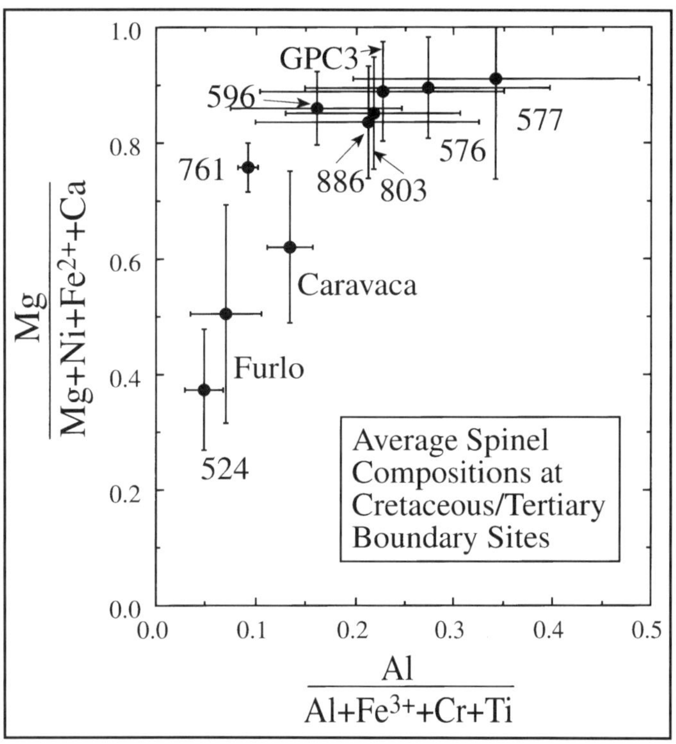

Figure 7. Magnesioferrite spinels from the K/T boundary in the Pacific are compositionally distinct from those in non-Pacific sites. Non-Pacific sites plotted are ODP Site 761, eastern Indian Ocean; DSDP Site 524, Walvis Ridge, South Atlantic; Caravaca, Spain; Furlo, Italy. Points plotted are mean compositions measured at each site. Bars represent the 1 σ variance in the range of compositions at each site.

provided no quantitative description of the abundance of this phase in the Pacific. The first quantitative analysis of shocked quartz in the Pacific was at DSDP Site 596 (Zhou et al., 1991) and showed that this phase was abundant (~1800 grains cm^{-2}) in this Southern Hemisphere site. Bostwick and Kyte (this volume) have presented data on the size distribution of shocked quartz and abundance of grains >30 μm at all seven Pacific sites. Most of their results were based on identification of planar features by scanning electron microscope (SEM) and microprobe analyses of grain mounts, but they included some optical examination of grains using universal stage microscopy, including measurement of the orientation of planar features in quartz from ODP Site 886 and LL44-GPC3.

Shocked quartz with planar deformation features is abundant at every K/T boundary site in the Pacific (Fig. 6E, F). Estimates of the abundance of shocked grains are imprecise, due to the sparse sampling across bioturbated K/T boundary horizons. Within uncertainties, the abundance shocked quartz is relatively uniform across the Pacific, with the number of shocked grains >30 μm in diameter in the range of ~1,000 to 2,000 grains/cm^2 (Bostwick and Kyte, this volume). One notable exception is at ODP Site 886, where the estimated fluence of shocked grains is 13,800 grains/cm^2.

There appear to be signifcant trends in the size distribution of shocked quartz in the Pacific, with the largest grains at the sites nearest the Yucatán (886, GPC3, and 465). At each of these sites, shocked grains ≥200 μm were recovered; the largest (320 μm) was from LL44-GPC3. These are significantly larger than the maximum size previously reported at European K/T boundary sites (190 μm; Bohor and Izett, 1986; Izett, 1990), despite the fact the shocked-mineral studies in the Pacific used samples with orders-of-magnitude less fallout material.

CONCLUSIONS

The K/T boundary on the Pacific plate has many similarities to, but also some differences from, other globally distributed sites. Relative to the putative impact site in the Yucatán, our results suggest a significant asymmetry in the global abundance and composition of some ejecta components.

Six of the seven Pacific sites have relatively similar Ir fluences, averaging ~90 ng/cm^2. Site 596, western South Pacific, is distinct in having an Ir anomaly ~2 m thick, with 320 ng/cm^2. We now suspect that the K/T boundary at this site may have been distorted by hydraulic piston coring and that high fluences of Ir and shocked quartz should be used with caution, as they might be overestimated. The Ir fluence across the Pacific is comparable to that estimated from other localities around the world (e.g., Alvarez et al., 1984; Kyte et al., 1985; Strong et al., 1987). Further, our analyses of Ir across the Cenozoic at four sites show that the K/T Ir anomaly is a global sedimentary event that was unique over at least the last 70 m.y.

All K/T boundaries in the Pacific except that at DSDP Site 465 have been smeared by bioturbation. This shows that at most Pacific sites deep and bottom waters were well oxidized at K/T time and that anoxic sedimentary environments are not a prerequisite for deposition of large Ir anomalies. Unfortunately, although a thin clay-rich fallout layer was preserved at Site 465, this section was severely disturbed by rotary drilling. However, the ~3-mm-thick clay layer at Site 465 occurs at the biostratigraphic boundary and is enriched in several siderophile elements and contains spherules and shocked quartz. This layer is comparable in thickness (~3 mm) to the upper "fireball" layer in the Western Interior of North America (e.g., Bohor, 1990) and the fallout layer preserved in a few other marine localities, such as Caravaca, Spain (Smit and Romein, 1985). This suggests that the intital K/T fallout was rather uniform in thickness over large regions of the Earth.

Although all the preceding observations suggest a rather uniform K/T deposit, detailed examination of the spherule and mineral data suggests otherwise. Large, hollow spherules (up to 1.5 mm) that may be the remnants of impact-melt glasses are present at LL44-GPC3 and ODP Site 886, extending the strewnfield for this component to distances of ~6,500 km from the Yucatán. This is comparable to the distance of western European sites (e.g., Caravaca), but such components have never been reported that far east. Smaller hollow spherules at DSDP Sites 577 and 596 may extend this strewnfield to a distance of ~10,000 km. If these spherules are related to the lower, kaolinitic layer of the K/T couplet in the Western Interior of North America (Pollastro and Bohor, 1993), then these data are consistent with decreasing concentrations and sizes of tektitic glasses with increased distance from Chicxulub but with more debris deposited toward the west.

Shocked-quartz abundances provide a similar picture. The Pacific sites nearest Chicxulub—GPC3, 465 and 886—have the largest shocked-quartz grains, significantly larger the the maximum sizes reported at European localities. There are no quantitative data to compare shocked-quartz fluences in the Pacific to those elsewhere. However, in contrast to analyses of quartz in European (e.g., Izett, 1990), South Atlantic (Huffman et al., 1990), and Indian Ocean (Thierstein et al., 1991) localities, shocked quartz constitutes the major fraction of nonopaque mineral grains >30 μm in trace mineral residues.

The most striking asymmetry in impact components is the clear regional variation in magnesioferrite spinel compositions, with the highest MgO, Al_2O_3 and Fe_2O_3/FeO in the Pacific, particularly in the western North Pacific at Sites 576 and 577. In addition to this compositional variation, Pacific spinel has distincly different textures, and several samples have been found with intergrowths of magnesiowüstite, a phase that can only crystallize from liquids with extremely MgO-rich and SiO_2-poor liquids. The regional variations in spinel do not easily fit a model with Chicxulub as a source, as the most extreme compositions are in the western North Pacific—Sites 576 and 577—some 4,000 km farther west than GPC3. Robin et al. (1993) have proposed that spinel variations reflect numerous, local impact events and that Pacific spinels at each site are locally derived. This scenario cannot be completely discounted, but we have difficulty imagining how multiple small impacts could produce what appear to be regional trends over the entire Pacific basin and possibly a much larger area. Further, their model fails to account for the distribution of shocked quartz and hollow spherules. We believe that the significant difference between Pacific K/T spinel and that at other sites is that it formed in liquid silicate droplets at much higher temperatures and that it was not quenched to form skeletal, dendritic crystals.

If these observations are to be explained by a single impact event at the Chicxulub structure, they require an impact that ejected larger shocked minerals and more tektitic impact melt farther to the west than to the east. Our data do not strongly constrain ejecta distribution in a north-south direction, but our preliminary data from the new ODP Site 886 suggest that this most-northerly locality has the most ejecta of any Pacific site. The asymmetry is further complicated by magnesioferrite compositions that are interpreted to indicate that the hottest vapor condensates were deposited in the western North Pacific.

Factors that could cause the asymmetry could include a heterogeneous target or a low-angle impact (Schultz and Gault, 1990) or could simply be related to expansion of an impact vapor cloud and fallout over a rotating Earth (e.g., Alvarez, 1994). Unfortunately, our relatively detailed database is inadequate to constrain any of these scenarios. We note that despite the numerous investigations of K/T boundary sediments over the last 15 years, uniform quantitative analyses of the distribution of impact debris have not been performed over most of the Earth's surface. Only further detailed analyses of other regions can confirm and elucidate the asymmetry that our data seem to support.

Finally, we note that the K/T boundary appears to be very well preserved on the Pacific plate. Within the pelagic clay provinces, it is buried at depths of only ~20 to 70 m. These depths are trivial for the current Ocean Drilling Program and with very little effort, a large number of new K/T boundary sections could be recovered.

ACKNOWLEDGMENTS

We thank H. R. Thierstein for paleontological analyses of samples from DSDP Site 465. A. R. Huffman, G. R. Byerly, and G. Ryder provided helpful reviews of this paper. Sediment cores are curated with support by the National Science Foundation. This research was supported by NSF grant EAR-91-18701.

REFERENCES CITED

Alvarez, L. W., Alvarez, W., Asaro, F., and Michel, H. V., 1980, Extraterrestrial cause for the Cretaceous–Tertiary extinction: Science, v. 208, p. 1095–1108.

Alvarez, L. W., Alvarez, W., Asaro, F., and Michel, H. V., 1982, Current status of the impact theory for the terminal Cretaceous extinction, *in* Silver, L T., and Schultz, P. H., eds., Geological implications of impacts of large asteroids and comets on the Earth: Geological Society of America Special Paper 190, p. 305–315.

Alvarez, W., 1994, Trajectories of ballistic impact ejecta on a rotating Earth [abs], *in* New developments regarding the KT event and other catastrophes in Earth history: Houston, Texas, Lunar and Planetary Institute Contribution 825, p. 2.

Alvarez, W., Alvarez, L. W., Asaro, F., and Michel, H. V., 1984, The end of the Cretaceous: Sharp boundary or gradual transition? Science, v. 223, p. 1183–1186.

Barker, J. L., Jr., and Anders, E., 1968, Accretion rate of cosmic matter from iridium and osmium contents of deep-sea sediments: Geochimica et Cosmochimica Acta, v. 32, p. 627–645.

Boersma, A., 1981, Cretaceous and early Tertiary foraminifers from deep sea drilling project Leg 62 sites in the central Pacific, *in* Initial reports of the Deep Sea Drilling Program, Volume 62: Washington, D.C., U.S. Government Printing Office, p. 377–396.

Bohor, B. F., 1990, Shocked quartz and more: Impact signatures in Cretaceous/Tertiary boundary clays, *in* Sharpton, V. L., and Ward, P. D., eds., Global catastrophes in Earth history: Geological Society of America Special Paper 247, p. 335–342.

Bohor, B. F., and Betterton, W. J., 1990, K/T boundary spherules: Clarifying the concept [abs]: Lunar and Planetary Science 21: Houston, Texas, Lunar and Planetary Institute, p. 108–109.

Bohor, B. F., and Betterton, W. J., 1991, K/T spherules are altered microtektites [abs]: Meteoritics, v. 26, p. 320.

Bohor, B. F., and Betterton, W. J., 1992, Ejection and dispersal mechanisms of the K/T impact [abs]: Lunar and Planetary Science 23: Houston, Texas, Lunar and Planetary Institute, p. 135–136.

Bohor, B. F., and Foord, E. E., 1987, Magnesioferrite from a nonmarine K-T boundary clay in Wyoming [abs]: Lunar and Planetary Science 18: Houston, Texas, Lunar and Planetary Institute, p. 101–102.

Bohor, B. F., and Izett, G. A., 1986, Worldwide size distribution of shocked quartz at the K-T boundary [abs]: Lunar and Planetary Science 17: Houston, Texas, Lunar and Planetary Institute, p. 68–69.

Bohor, B. F., Foord, E. E., Modreski, P. J., and Tripplehorn, D. M., 1984, Mineralogical evidence for an impact event at the Cretaceous-Tertiary boundary: Science, v. 224, p. 867–869.

Bourgeois, J., Hansen, T. A., Wiberg, P. L. and Kauffman, E. G., 1988, A tsunami deposit at the Cretaceous-Tertiary boundary in Texas: Science, v. 241, p. 567–570.

Corliss, B. H., and Hollister, C. D., 1982, A paleoenvironmental model for Cenozoic sedimentation in the central North Pacific, *in* Scrutton, R. A., and Talwani, M., eds., The Ocean Floor: New York, J. Wiley & Sons, p. 277–304.

DePaolo, D. J., and Kyte, F. T., 1984, Isotopic and chemical characteristics of Cretaceous-Tertiary boundary sediments from DSDP Site 465A [abs]: Lunar and Planetary Science 15: Houston, Texas, Lunar and Planetary Institute, p. 220–221.

Doyle, P. S., 1980, Seabed disposal program: Improvement of ichthyolith stratigraphy for giant piston core 3: Progress Report: Albuquerque, New Mexico, Sandia Laboratories, 14 p.

Doyle, P. S., and Riedel, W. R., 1979, Cretaceous to Neogene ichthyoliths in a giant piston core from the central North Pacific: Micropaleontology, v. 25, p. 337–364.

Doyle, P. S., and Riedel, W. R., 1985, Ichthyolith biostratigraphy of western North Pacific pelagic clays, Deep Sea Drilling Project Leg 86, *in* Heath, G. R., Burckle, L. H., and others, Initial reports of the Deep Sea Drilling Project, Volume 86: Washington, D.C., U.S. Government Printing Office, p. 349–366.

Evans, N. J., Gregoire, D. C., Goodfellow, W. D., McInnes, B. I., Miles, N., and Veizer, J., 1993, Ru/Ir ratios at the Cretaceous-Tertiary boundary; implications for PGE source and fractionation within the ejecta cloud: Geochimica et Cosmochimica Acta, v. 57, p. 3149–3158.

Ganapathy, R., 1980, A major meteorite impact on the Earth 65 million years ago: Evidence from the Cretaceous-Tertiary boundary: Science, v. 209, p. 921–923.

Gerstel, J., Thunell, R. C., Zachos, J. C., and Arthur, M. A., 1986, The Cretaceous/Tertiary boundary event in the North Pacific: Planktonic foraminiferal results from Deep Sea Drilling Project Site 577, Shatsky Rise: Paleoceanography, v. 1, p. 97–117.

Heath, G. R., Burckle, L. H., and others, 1985, Initial reports of the Deep Sea Drilling Project, Volume 86: Washington, D.C., U.S. Government Printing Office, 804 p.

Hildebrand, A. R., and Boynton, W. V., 1990, Proximal Cretaceous-Tertiary impact deposits in the Caribbean: Science, v. 248, p. 843–847.

Hildebrand, A. R., Penfield, G. T., Kring, D. A., Pilkington, M., Camargo Z., A., Jacobsen, S. B., and Boynton, W. V., 1991, Chicxulub crater: A possible Cretaceous/Tertiary boundary impact crater on the Yucatán Peninsula, Mexico: Geology, v. 19, p. 867–871.

Huffman, A. R., Crocket, J. H., Carter, N. L., Borella, P. E., and Officer, C. B., 1990, Chemistry and mineralogy across the Cretaceous/Tertiary boundary at DSDP Site 527, Walvis Ridge, South Atlantic Ocean, *in* Sharpton, V. L., and Ward, P. D., eds., Global catastrophes in Earth history: Geological Society of America Special Paper 247, p. 319–334.

Ingram, B. L., 1995, Ichthyolith strontium-isotopic stratigraphy of deep sea clays: ODP Sites 885 and 886, Leg 145 (North Pacific Transect), *in* Rea, D. K., Basov, I. A., and others, Proceedings, Ocean Drilling Program, Scientific Results, Volume 145: College Station, Texas, Ocean Drilling Program, p. 399–412.

Izett, G. A., 1990, The Cretaceous-Tertiary (K/T) boundary interval, Raton Basin, Colorado and New Mexico, and its content of minerals—Evidence relevant to the Cretaceous-Tertiary impact-extinction theory: Geological Society of America Special Paper 249, 100 p.

Kastner, M., Asaro, F., Michel, H. V., Alvarez, W., and Alvarez, L. W., 1984, The precursor of the Cretaceous-Tertiary boundary clays at Stevns Klint, Denmark, and DSDP Hole 465A: Science, v. 226, p. 137–143.

Kroenke, L. W., Berger, W. H., and others, 1991, Proceedings, Ocean Drilling Program, Volume 130: College Station, Texas, Ocean Drilling Program, 1240 p.

Kyte, F. T., and Bohor, B. H., 1995, Ni-rich magnesiowüstite in Cretaceous/Tertiary boundary spherules crystallized from ultramafic, refractory silicate liquid: Geochimica et Cosmochimica Acta, v. 59, p. 4967–4974.

Kyte, F. T., and Bostwick, J. A., 1995, Magnesioferrite spinel in Cretaceous/Tertiary boundary sediments of the Pacific basin: Remnants of hot, early ejecta from the Chicxulub impact?: Earth and Planetary Science Letters, v. 132, p. 113–127.

Kyte, F. T., and Smit, J., 1986, Regional variation in spinel compositions: An important key to the Cretaceous-Tertiary event: Geology, v. 14, p. 485–487.

Kyte, F. T., and Wasson, J. T., 1982, Geochemical constraints on the nature of large accretionary events *in* Silver, L. T., and Schultz, P. H., eds., Geological Society of America Special Paper 190, p. 235-242.

Kyte, F. T., and Wasson, J. T., 1986a, Accretion rate of extraterrestrial matter: Iridium deposited 33 to 67 million years ago: Science, v. 232, p. 1225–1229.

Kyte, F. T., and Wasson, J. T., 1986b, Comet showers, periodic extinctions, and iridium: Science, v. 234, p. 1485–1486.

Kyte, F. T., Zhou, Z., and Wasson, J. T., 1980, Siderophile-enriched sediments from the Cretaceous-Tertiary boundary: Nature, v. 288, p. 651–656.

Kyte, F. T., Smit, J., and Wasson, J. T., 1985, Siderophile interelement variations in the Cretaceous-Tertiary boundary sediments from Caravaca, Spain: Earth and Planetary Science Letters, v. 73, p. 183–195.

Kyte, F. T., Zhou, L., and Lowe, D. R., 1992, Noble metal abundances in an Early Archean impact deposit: Geochimica et Cosmochimica Acta, v. 56, p. 1365–1372.

Kyte, F. T., Leinen, M., Heath, G. R., and Zhou, L., 1993, Cenozoic sedimentation history of the central North Pacific: Inferences from the elemental geochemistry of core LL44-GPC3: Geochimica et Cosmochimica Acta, v. 57, p. 1719–1740.

Kyte F. T., Bostwick, J. A., and Zhou, L., 1995, Identification and characterization of the Cretaceous-Tertiary boundary at ODP Sites 886 and 803 and DSDP Site 576, *in* Rea, D. K., Basov, I. A., and others, Proceedings, Ocean Drilling Program, Scientific Results, Volume 145: College Station, Texas, Ocean Drilling Program, p. 427–434.

Leinen, M., and Heath, G. R., 1981, Sedimentary indicators of atmospheric activity in the northern hemisphere during the Cenozoic: Paleogeography, Paleoclimatology, Paleoecology, v. 36, p. 1–21.

Lyle, M., 1983, The brown-green color transition in marine sediments: A marker of the Fe(III)-Fe(II) redox boundary: Limnology and Oceanography, v. 28, p. 1026–1033.

Michel, H. V., Asaro, F., and others, 1985, Elemental profile of iridium and other elements near the Cretaceous/Tertiary boundary in Hole 577B, *in* Heath, G. R., Burckle, L. H., and others, Initial reports of the Deep Sea Drilling Project, Volume 86: Washington, D.C., U.S. Government Printing Office, p. 533–538.

Montanari, A., Hay, R. L., Alvarez, W., Asaro, F., Michel, H. V., Alvarez, L. W., and Smit, J., 1983, Spheroids at the Cretaceous-Tertiary boundary are altered impact droplets of basaltic composition: Geology, v. 11, p. 668–671.

Muan, A., and Osborn, E. F., 1954, Phase equilibria at liquidus temperatures in the system $MgO-FeO-Fe_2O_3-SiO_2$: Journal of the American Ceramic Society, v. 39, p. 121–140.

Orth, C. J., Gilmore, J. S., Knight, J. D., Pillmore, C. L., Tschudy, R. H., and Fassett, J. E., 1981, An iridium abundance anomaly at the palynological Cretaceous-Tertiary boundary in northern New Mexico: Science, v. 214, p. 1341–1343.

Pollastro, R. M., and Bohor, B. F., 1993, Origin and clay-mineral genesis of the Cretaceous/Tertiary boundary unit, Western Interior of North America: Clays and Clay Minerals, v. 41, p. 7–25.

Rea, D. K., Basov, I. A., and others, 1993, Proceedings, Ocean Drilling Program, Initial reports, Volume 145: College Station, Texas, Ocean Drilling Program, 1040 p.

Robin, E., Bonté, Ph., Froget, L., Jéhanno, C., and Rocchia, R., 1992, Formation of spinels in cosmic objects during atmospheric entry: a clue to the Cretaceous-Tertiary boundary event: Earth and Planetary Science Letters, v. 108, p. 181–190.

Robin, E., Froget, L., Jéhanno, C., and Rocchia, R., 1993, Evidence for a K/T impact event in the Pacific Ocean: Nature, v. 363, p. 615–617.

Schmitz, B., 1988, Origin of microlayering in world wide distributed Ir-rich marine Cretaceous-Tertiary boundary clays: Geology, v. 16, p. 1068–1072.

Schultz, P. H., and Gault, D. E., 1990, Prolonged global catastrophes from oblique impacts, *in* Sharpton, V. L., and Ward, P. D., eds., Global catastrophes in Earth history: Geological Society of America Special Paper 247, p. 239–261.

Sharpton, V. L., Dalrymple, G. R., Marin, L. E., Ryder, G., Schuraytz, B. C., and Urrutia-Fucugauchi, J., 1992, New links between the Chicxulub impact structure and the Cretaceous/Tertiary boundary: Nature, v. 359, p. 819–821.

Sigurdsson, H., D'Hondt, S., Arthur, M. A., Bralower, T. J., Zachos, J. C., van Fossen, M., and Channell, J. E. T., 1991, Glass from the Cretaceous/Tertiary boundary in Haiti: Nature, v. 349, p. 482–487.

Smit, J., and Hertogen, J., 1980, An extraterrestrial event at the Cretaceous-Tertiary boundary: Nature, v. 285, p. 198–200.

Smit, J., and Klaver, G., 1981, Sanidine spherules at the Cretaceous-Tertiary boundary indicate a large impact event: Nature, v. 292, p. 47–49.

Smit, J., and Kyte, F. T., 1984, Siderophile-rich magnetic spheroids from the Cretaceous-Tertiary boundary in Umbria, Italy: Nature, v. 310, p. 403–405.

Smit, J., and Romein, A. J. T., 1985, A sequence of events across the Cretaceous-Tertiary boundary: Earth and Planetary Science Letters, v. 74, p. 155–170.

Smit, J., Montanari, A., and Alvarez, W., 1991, Microkrystites and microtektites at the K/T boundary: Two different sources or one? [abs]: Lunar and Planetary Science 22: Houston, Texas, Lunar and Planetary Institute, p. 1277–1278.

Smit, J., Alvarez, W., Montanari, A., Swinburne, N., van Kempen, T. M., Klaver, G. T., and Lustenhouwer, W. J., 1992a, "Tektites" and microkrystites at the Cretaceous Tertiary boundary: Two strewn fields, one crater?: Lunar and Planetary Science 22: Houston, Texas, Lunar and Planetary Institute, p. 87–100.

Smit, J., and eight others, 1992b, Tektite-bearing, deep-water clastic unit at the Cretaceous-Tertiary boundary in northeastern Mexico: Geology, v. 20, p. 99–103.

Strong, C. P., Brooks, R. R., Wilson, S. M., Reeves, R. D., Orth, C. J., Mao, X., Quintana, L. R., and Anders, E., 1987, A new Cretaceous-Tertiary boundary site at Flaxbourne River, New Zealand: Biostratigraphy and geochemistry: Geochimica et Cosmochimica Acta, v. 51, p. 2769–2777.

Thiede, J., and others, 1981, Initial reports of the Deep Sea Drilling Project, Volume 62: Washington, D.C., U.S. Government Printing Office, 1120 p.

Thierstein, H. R., 1981, Late Cretaceous nannoplankton and the change at the Cretaceous-Tertiary boundary: *in* Warme, J. E., et al., eds., The Deep Sea Drilling Project: A decade of progress: Society of Economic Paleontologists and Mineralogists Special Publication 32, p. 355–394.

Thierstein, H. R., 1982, Terminal Cretaceous plankton extinctions: A critical assessment, *in* Silver, L. T., and Schultz, P. H., eds., Geological implications of impacts of large asteroids and comets on the Earth: Geological Society of America Special Paper 190, p. 385–400.

Thierstein, H. R., Asaro, F., and others, 1991, The Cretaceous/Tertiary boundary at Site 738, southern Kerguelen Plateau: *in* Barron, J., Larsen, B., and others, Proceedings, Ocean Drilling Program, Scientific Results, Volume 119: College Station, Texas, Ocean Drilling Program, p. 719–730.

Wallace, M. W., Gostin, V. A., and Keays, R. R., 1990, Acraman impact ejecta and host shales: Evidence for low-temperature mobilization of iridium and other platinoids: Geology, v. 18, p. 132–135.

Winfrey, C. E., Doyle, P. S., and Riedel, W. R., 1987, Preliminary ichthyolith biostratigraphy, southwest Pacific, Deep Sea Drilling Project Leg 91, *in* Menard, H. W., Natland, J., and others, Initial reports of the Deep Sea Drilling Project, Volume 91: Washington, D.C., U.S. Government Printing Office, p. 447–473.

Wright, A. A., Bleil, U., and others, 1985, Summary of Cretaceous/Tertiary boundary studies, Deep Sea Drilling Project Site 577, Shatsky Rise, *in* Heath, G. R., Burckle, L. H., and others, Initial reports of the Deep Sea Drilling Project, Volume 86: Washington, D.C., U. S. Government Printing Office, p. 799–804.

Zachos, J. C., and Arthur, M. A., 1986, Paleoceanography of the Cretaceous/Tertiary boundary event: Inferences from stable isotopic and other data: Paleoceanography, v. 1, p. 5–26.

Zhou, L., and Kyte, F. T., 1992, Sedimentation history of the South Pacific pelagic clay province over the last 80 million years inferred from the geochemistry of DSDP Hole 596: Paleoceanography, v. 7, p. 441–465.

Zhou, L., Kyte, F. T., and Bohor, B. F., 1991, The Cretaceous/Tertiary boundary of DSDP Site 596, South Pacific: Geology, v. 19, p. 694–697.

Manuscript Accepted by the Society September 14, 1995

Printed in U.S.A.

Geological Society of America
Special Paper 307
1996

The size and abundance of shocked quartz in Cretaceous-Tertiary boundary sediments from the Pacific basin

Jennifer A. Bostwick and Frank T. Kyte
Institute of Geophysics and Planetary Physics, University of California, Los Angeles, California 90024

ABSTRACT

Quantitative analyses of the size distribution and abundance of shocked quartz (quartz with planar deformation features [PDFs]) were performed on Cretaceous-Tertiary (K/T) boundary sediments from seven localites in the Pacific basin: LL44-GPC3; DSDP (Deep Sea Drilling Project) Sites 465, 576, 577, and 596; and ODP (Ocean Drilling Project) Sites 803 and 886. Quartz with PDFs was found at each site. On average, 63% of the quartz grains in Pacific K/T sediments with diameters >30 µm have PDFs. The fluences of quartz with PDFs are relatively uniform across the Pacific; estimates for the abundances of shocked grains >30 µm in diameter range from ~1,000 to 2,000 grains/cm^2. One exception is ODP site 886, where we estimate a fluence of 13,800 grains/cm^2. Ten PDF-bearing quartz grains larger than 200 µm were recovered, including one with a maximum diameter of 320 µm. These largest grains were all from the sites nearest North America: DSDP 465, ODP 886 and LL44-GPC3. This may reflect proximity of these sites to Chicxulub.

Comparison to other globally distributed sites is difficult. We conclude that shocked quartz in the Pacific is on average significantly smaller and about an order of magnitude less abundant (in mass/cm^2) than that deposited in North America. Maximum grain sizes of shocked quartz in the North Pacific appear to be larger than at sites in Europe, although these sites were similar distances from Chicxulub at 66 Ma.

INTRODUCTION

A decade ago shocked quartz was first described in Cretaceous-Tertiary (K/T) boundary sediments from Montana (Bohor et al., 1984). This was submitted as compelling evidence in support of the asteroid impact event hypothesized to have occurred at the end of the Cretaceous (Alvarez et al., 1980). Quartz with planar deformation features (PDFs) has been recognized by many researchers as an important component of many K/T deposits (Bohor et al., 1984, 1987; Grieve and Alexopoulos, 1988; Izett and Pillmore, 1985; Izett, 1990; Bohor, 1990; Hildebrand and Boynton, 1990; Sharpton et al., 1990).

Several studies have shown that the distribution of quartz with PDFs is widespread—possibly global—and several maps illustrating the distribution of the largest grains at each site have been published (Bohor and Izett, 1986; Izett, 1989; 1990; Bohor, 1990; Hildebrand, 1993). These maps show that the largest grains are found in the vicinity of North America, the Gulf of Mexico, and the Caribbean (i.e., Haiti), with maximum diameters ranging from 500 to 1,250 µm. Quartz with PDFs was reported to be much smaller (<190 µm maximum diameter) and much less abundant at European sites and in New Zealand (e.g., Bohor, 1990). Using the available data, Izett (1990) concluded that shocked minerals are several orders of magnitude more abundant at the Western Interior K/T boundary sites than at European localities or at other sites (e.g., New Zealand) around the globe.

These results were generally supported by analyses of

Bostwick, J. A., and Kyte, F. T., 1996, The size and abundance of shocked quartz in Cretaceous-Tertiary boundary sediments from the Pacific basin, *in* Ryder, G., Fastovsky, D., and Gartner, S., eds., The Cretaceous-Tertiary Event and Other Catastrophes in Earth History: Boulder, Colorado, Geological Society of America Special Paper 307.

deep-ocean K/T sediments. Bohor and Betterton (1989) recovered shocked-quartz grains as large as 300 μm from marine sediment in samples from DSDP Site 603B off the coast of North Carolina, where large grains might be expected. In a study of quartz from the K/T boundary at DSDP Site 527, Walvis Ridge, South Atlantic, Huffman et al. (1990) found that only 2% of the quartz in K/T boundary sediments contained PDFs and that nearly all of these were <63 μm in diameter. No shocked quartz was found in the K/T boundary clay at ODP Site 738, southern Indian Ocean (Thierstein et al., 1991). Traces of shocked quartz were found in a single sample from North Pacific site LL44-GPC3, but Bohor and Izett (1986) provided no quantitative analysis of its abundance. The only study that reported shocked quartz at a site distant from North America with abundances that might be similar to those at sites in North America was that of Zhou et al. (1991), who reported abundances of ~1,800 shocked grains/cm^2 in the K/T boundary at DSDP Site 596, western South Pacific.

With the discovery of the Chicxulub structure on the Yucatán Peninsula, Mexico (Hildebrand et al., 1991; Sharpton et al., 1992), we now have a potential candidate for the K/T impact site as well as a source of the shocked quartz. The global distribution of the size and abundance of shocked quartz may be used to constrain the evolution of ejecta following the impact event and help to confirm whether Chicxulub is indeed the source. However, the currently available data may not actually be representative of the K/T impact fallout and may be biased by variable analytical and sampling procedures, limitations on sample availability, or variability in preservation of the K/T fallout at different localities.

The *absolute abundance* of shocked quartz at any locality will be a function of the size-fractions sampled. Most samples are processed with the intention of recovering large grains (>100 μm). A sample processed for this purpose cannot provide an adequate mineral concentrate to use in describing the abundance or the size distribution of shocked quartz because it is likely that the size fraction smaller than 100 μm, a significant portion of the total, is not completely retained. The study by Zhou et al. (1991) gives a very good estimate of the absolute abundance of shocked quartz, as they analyzed all nonopaques at DSDP Site 596 down to a size of 30 μm. The *absolute size* of the largest grain found at any site will largely be a function of sample availability. At North American sites, kilogram-sized samples of a pure, 3-mm-thick fallout layer can be processed (Izett, 1990). In such samples the maximum reported size is more likely to be representative of the fallout than at oceanic sites such as DSDP Site 603 (Bohor and Betterton, 1989), DSDP Site 527 (Huffman et al., 1990) and LL44-GPC3. In the latter case, Bohor and Izett (1986) had only ~1 g of sample of GPC3 clay (provided by F. T. Kyte) from a smeared, bioturbated, and thus diluted boundary sequence. The probablility that their sample contained a grain approaching the maximum size deposited in this region is extremely remote. The *absolute concentration* of shocked quartz relative to total quartz at any locality will largely be a function of preservation of the fallout layer. At K/T sites in North America where the fallout was preserved in quiescent coal swamps, a pure fallout layer with shocked-quartz concentrations representative of the ejecta may be found. At these sites, clastic quartz typically only constitutes ~0.6 wt.% of the fallout layer (Izett, 1990). At shallow marine localities such as in western Europe and New Zealand there is a much greater probability of sediment reworking to either destroy fragile shocked quartz or to dilute it with local detritus. Dilution by local detritus can be a serious problem that is not often addressed. At some deep-ocean sites such as DSDP 596 and LL44-GPC3, which are distant from terrigenous sediment sources (Leinen and Heath, 1981; Zhou and Kyte, 1992; Kyte et al., 1993), dilution by detrital quartz is likely a minor or nonexistent problem.

In this chapter we attempt to quantify the size and abundance of shocked quartz in K/T boundary sediments in the Pacific basin. At 66 Ma the Pacific Ocean covered ~40% of the Earth's surface. The only remnant of this enormous region is now the portion of the Pacific plate with Mesozoic-age crust in the western Pacific (Fig. 1). K/T boundary sediments at these Pacific sites are particularly amenable to trace mineral analyses, as biogenic and authigenic components are easily dissolved and eolian terrigenous components are dominantly very fine silts and clays. For example, the mean size of quartz in lower Cenozoic sediments from LL44-GPC3 is typically ~3 μm (Rea et al., 1985). We have found shocked quartz at all seven known sites in the Pacific that have anomalous Ir. This study greatly extends the region of the Earth known to contain this impact component.

Samples and procedures

Samples are typically trace mineral extracts from ~1 cm^3 (typically >0.4 g) of pelagic sediment from K/T localities with reported Ir anomalies. The seven localities in the Pacific (Fig. 1; Table 1) are described in detail by Kyte et al. (this volume). Of these sites, all are bioturbated except for DSDP 465, which was disturbed by rotary drilling. Typically, bioturbation has resulted in mixing of the fallout layer over intervals from ~15 to >100 cm.

We had only one sample each from calcareous sequences in the North Pacific at DSDP Sites 465 and 577. The sample from Site 577 was from near the base of the spherule- and Ir-rich, bioturbated boundary in Hole 577B (Heath and Burckle, 1985; Michel et al., 1985; Smit and Romein, 1985). Sampling the K/T boundary at DSDP Site 465 is much more problematic. Although it contains a significant Ir anomaly (Kyte et al., 1980) it is poorly characterized because of the drilling disturbance. However, fragments of a thin, black, pyrite-rich boundary clay have been recognized (fig. 3 in Kyte et al. [this volume]). We

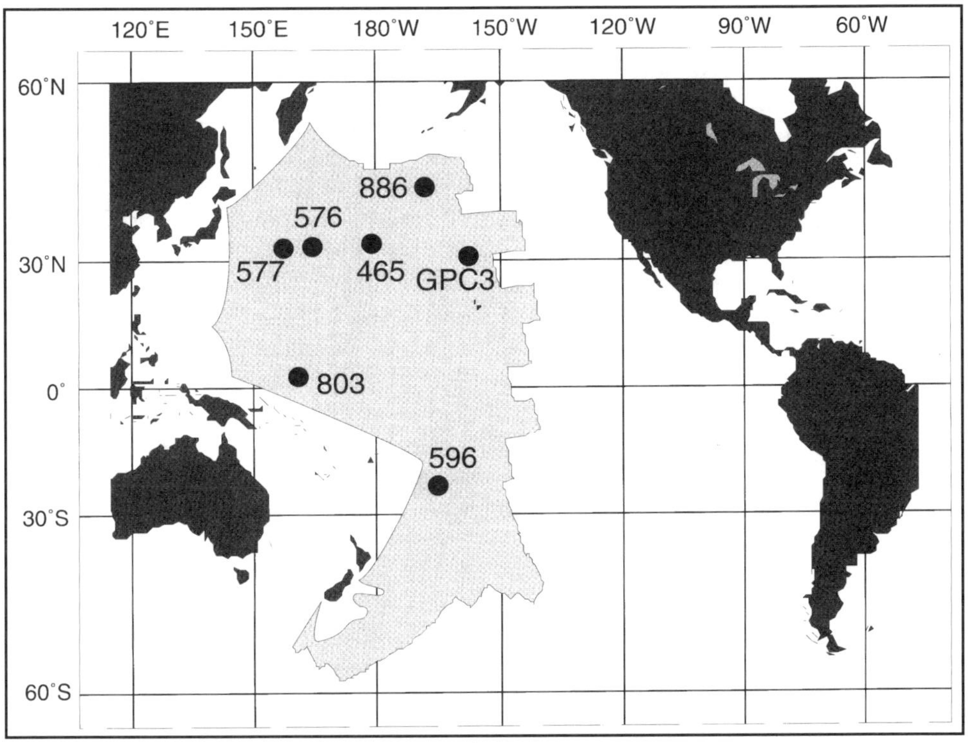

Figure 1. Location map of the seven localities where the K/T boundary has been recovered in the Pacific Basin. Shaded area is the portion of the Pacific plate with Mesozoic-age crust.

analyzed a sample containing a portion of this clay layer; we estimate that it contains ~20% black clay layer and equal amounts of carbonate-rich sediments from immediately above and below.

Shocked quartz had been analyzed previously in samples of pelagic clay from DSDP Site 596 (Zhou et al., 1991) and LL44-GPC3 (Bohor and Izett, 1986). We performed quantitative analyses on only one sample from DSDP 596 in order to compare our method to previous quantitative analyses (Zhou et al., 1991) and to obtain grain-size distributions that were not recorded in that study. At LL44-GPC3 we quantitatively analyzed three samples from within the Ir anomaly (Kyte and Wasson, 1986). The other three K/T boundaries—DSDP 576 and

TABLE 1. K/T BOUNDARY SITES ON THE PACIFIC PLATE

Locality	Latitude	Longitude	Lithology
LL44-GPC3	30°19.00'N	157°49.9'W	Pelagic clay
DSDP Site 465	33°49.23'N	178°55.14'E	Calcareous ooze/clay
DSDP Site 576	32°21.37'N	164°16.52'E	Pelagic clay
DSDP Site 577	32°26.51'N	157°43.40'E	Calcareous ooze
DSDP Site 596	23°51.20'S	165°39.27'W	Pelagic clay
ODP Site 803	2°25.98'N	160°32.46'E	Claystone
ODP Site 886	44°41.38'N	168°14.40'W	Pelagic clay

ODP Sites 803 and 886—are all recently discovered (Kyte et al., 1995) and are also clay-rich sections. A detailed profile of 12 samples across the K/T section of DSDP Site 576 was studied, but only preliminary surveys of one sample from ODP 803 and two from ODP 886 were analyzed quantitatively.

Samples were gently dispersed in a 10% sodium hexametaphosphate solution and clays were pipetted off. Biogenic (calcite, opal, phosphate) and authigenic (Fe-Mn oxyhydroxide, zeolite) phases were removed by sequential leaching with 1N oxalic acid, 0.5N HCl, 25% HF, and 12N H_2SO_4. At all times samples were allowed to settle so that larger grains were not washed away. We attempted to quantitatively recover all trace minerals larger than 15 μm. Residues were typically dominated by quartz and magnesioferrite spinel.

Nonopaque grains were picked, mounted, and examined in either an ISI Alpha-9 scanning electron micrscope (SEM) or a CAMEBAX model Cameca electron microprobe. Qualitative elemental compositions were determined by energy dispersive (EDS) X-ray analysis. Backscatter or secondary electron imaging was used to measure grain sizes; maximum diameters for each grain were recorded. We interpret multiple straight, sharp, linear or planar features observed on grain surfaces to be PDFs. Some cover the entire grain, but others do not. We found that our methods were not completely reliable for very small grains.

Our ability to mount and identify PDFs on very small grains (<20 μm) is low. In order to assure statistical reliability, all grain counts and calculations thereof were only for those grains larger than 30 μm.

In addition to the samples described above, several samples (including at least one from each site) have been examined for a variety of purposes, including analyses of spinel and spherules (e.g., Kyte et al., this volume). In each case, we attempted to find the largest quartz grains and determine whether they had PDFs. A few of these samples were used to prepare separate mineral concentrates for examination in thin section and on a spindle stage. A preliminary optical survey of nonopaque minerals was performed using universal-stage microscopic techniques consistent with those of Engelhardt and Bertsch (1969). We understand that SEM identification of PDFs in not a conventional method for characterizing shock deformation, and we use this optical survey to corroborate the general results obtained in the SEM.

SEM analyses of trace mineral separates

DSDP Site 576. Our initial and most detailed analysis of trace minerals was in sediments from the K/T boundary sequence at DSDP Site 576. Initial analyses were conducted on sediments with anomalous Ir found at depths of ~54.5 mbsf (meters below seafloor; based on the composite section of Heath et al., 1985) in Hole 576 and ~54 mbsf in Hole 576B (Kyte et al., 1995). Several samples were processed as we developed our procedure. Quartz grains with apparent PDFs were abundant in samples known to have high concentrations of Ir. Additionally, we found abundant grains of magnesioferrite spinel. These results were taken as confirmation that we had indeed found the K/T boundary at this locality (Kyte et al., 1995).

Our first quantitative analyses were performed on a suite of 11 samples across the K/T boundary in core 6 section 4 of Hole 576B (Table 2). Nonopaque grains in these samples were mostly quartz and feldspar. Very few other nonopaque grains were recovered, including biotite grains (45 to 60 μm) in the sample at 80 to 81 cm. However, most grains listed as "other nonopaques" for these samples in Table 2 had EDS spectra indicating compositions of Al or Ca silicates. We are uncertain as to whether these are natural minerals or laboratory contaminants. The sample with the peak Ir concentration (13 ng/g) at 54.00 mbsf (576B-6-4, 40 to 42 cm) was found to contain the largest grains and the greatest concentration of both total quartz (>30 μm) and quartz with PDFs. This sample had a concentration of 276 PDF-bearing quartz grains per gram, representing 70% of the total quartz. Feldspar constituted only 2% of the nonopaques in this sample. Although three quartz grains larger than 100 μm were found in the background samples below the boundary, quartz with PDFs was found in only three samples: in the sample at the Ir peak and in samples immediately above and below, which also had anomalous Ir concentrations (Table 2).

TABLE 2. TRACE MINERAL CONTENT OF K/T BOUNDARY SEDIMENTS ON THE PACIFIC PLATE

Sample		Depth (mbsf)	Dry Mass (g)	Ir (ng/g)	Nonopaque Grains	Total Quartz Grains	Quartz Grains With PDFs	Total Felsdpar Grains	Other Nonopaque Grains
576B-6-4,	10-11 cm	53.70	0.51	1.2	5	4	0	0	1
576B-6-4,	15-16 cm	53.75	0.63	1.4	21	4	0	12	5
576B-6-4,	30-31 cm	53.90	0.40	2.9	13	11	0	2	0
576B-6-4,	35-37 cm	53.95	0.31	6.8	36	18	5	17	1
576B-6-4,	39-41 cm	53.99	0.52	....	264	259	167	3	2
576B-6-4,	40-42 cm	54.40	0.63	13.4	254	250	174	4	0
576B-6-4,	45-47 cm	54.05	0.41	1.5	20	15	4	5	0
576B-6-4,	50-52 cm	54.10	0.51	1.4	6	5	0	1	0
576B-6-4,	56-57 cm	54.16	0.50	0.8	9	5	0	3	1
576B-6-4,	60-61 cm	54.20	0.71	0.8	47	35	0	10	2
576B-6-4,	80-81 cm	55.26	0.71	1.2	56	42	0	7	7
576B-6-4,	100-101 cm	55.46	0.41	0.4	30	15	0	6	9
LL44-GPC3,	2,053-54 cm	20.53	0.47	5.2	69	55	43	10	4
LL44-GPC3,	2,056-57 cm	20.56	0.55	12.0	245	234	194	4	7
LL44-GPC3,	2,060-62 cm	20.60	0.55	10.9	345	330	192	12	3
886C-8-1,	141-142 cm	65.21	0.25	3.6	319	301	223	4	14
886C-8-1,	147-148 cm	65.27	0.27	2.7	393	368	207	13	12
465A-3-4,	10-11 cm	62.60	1.71	....	479	474	299	5	0
577B-1-4,	70-71 cm	109.60	0.60	....	178	168	72	3	7
596-3-4,	50-51 cm	20.10	0.49	....	118	108	69	5	5
803D-63R-1,	63-64 cm	622.43	0.40	10.8	166	112	38	48	6

One additional sample from 53.99 mbsf (576B-6-4, 39 to 41 cm; no Ir data are available for this sample) was analyzed in order to confirm our results and improve our statistics. In this sample we found even higher concentrations of quartz (324 grains with PDFs per gram; Table 2). The results on samples from 39 to 41 cm and from 40 to 42 cm depths are very similar and confirmed our expectations that our methods could provide reproducible results.

Other sites in the Pacific. We analyzed at least one sample from each of the other six K/T boundaries in this region (Fig. 1; Table 2). Iridium concentrations have been measured on splits of most of these samples. Quartz grains with PDFs were found in every sample, as well as abundant magnesioferrite spinel. As at DSDP Site 576, quartz constitutes more than 90% of the non-opaque minerals at every site except ODP Site 803. Feldspars were quite abundant in the sample from Site 803, and we believe that these reflect contribution of local detritus from the volcanics of the Ontong–Java Plateau. Also, nonopaques other than quartz and feldspar were a minor component at all sites. These included pyroxene, zircon, mica, and other unidentified phases.

Three trace mineral separates from LL44-GPC3 were analyzed quantitatively (Table 2). These were from a recently collected suite of samples that have also been analyzed for Ir. The actual peak in Ir (22 ng/g) was found between two of these samples at 2,059 to 2,060 cm. This is a somewhat higher concentration than the 12 to 15 ng/g Ir found in earlier studies (e.g., Kyte and Wasson, 1986), and this peak is 4 cm lower in the core than previously reported (2,055 cm). We are uncertain as to the cause of this discrepancy, but we find that the Ir fluence estimated from this sample suite is not significantly different from the 70 ng/cm^2 estimated by Kyte and Wasson (1986).

Examples of shocked-quartz grains observed by SEM analyses are illustrated in Figures 2 and 3. The largest quartz grain studied by SEM was 300 μm long (Fig. 2A) but was not classified as containing PDFs. Commonly, grains were observed to break along PDFs during mounting (Fig. 2E, F). On some grains, PDFs extend entirely across grains (Fig. 2C, D, E; 3B). In others, PDFs are more distinct on one portion of the grain surface or can be discontinuous (Fig. 3A, B, C). In some cases, this is almost certainly due to differential etching by HF (Fig. 3A), but may also result from secondary overgrowths of silica (Zhou et al., 1991) or the presence of PDFs within domains of multidomain quartz grains. In some cases, we distinquish both thin and thick PDFs on the same grain, possibly reflecting heterogeneous spacing of submicroscopic planes (Fig. 3C, D; Huffman, 1993).

Optical analyses of trace mineral separates

Optical analyses by universal stage microscopy were performed on quartz separated from two K/T samples from ODP Site 886 (886C-8-1, 138 to 139 cm and 144 to 145 cm) and one K/T sample from LL44-GPC3 (2,059 to 2,060 cm). The 52 largest nonopaque grains (≥100 μm) were mounted in thin section. Because some grains were highly shocked or had very fine-grained textures, it was not always possible to distinguish quartz from feldspar optically. In these cases compositions were confirmed by SEM-EDS analyses.

Five different types of grains were observed (Table 3). (1) The first type of quartz grains has one to five or more sets of PDFs per grain (Fig. 4A, B, C). These grains typically have irregular shapes. (2) Quartz with mosaic texture having large (>10 μm) subgrain domains that can contain PDFs might be polycrystalline. However, considering the irregularity of the subgrain domain boundaries, we cannot distinguish whether this mosaic texture is due to recovery of a single grain or an aggregate (Fig. 4D, E). (3) Quartz with mosaic texture having very small (<10 μm) subgrain domains has migrating domain boundaries due to recrystallization (Fig. 4F). (4) Some quartz has neither mosaic texture or PDFs. This apparently unmetamorphosed quartz is milky and free of PDFs. (5) A single feldspar grain was twinned and lacked PDFs.

The results of these analyses are generally consistent with the results of our SEM analyses. Clear quartz with PDFs constitutes 55% of all quartz observed optically. Additionally, at least three mosaic grains with large domains were found to contain PDFs in subdomains. Thus, 63% of the quartz grains observed optically (32 out of 51) have PDFs, results comparable to our SEM analyses that identified 69 and 64% of all quartz >30 μm as shocked in GPC3 and 886, respectively.

The angles of poles of PDFs relative to the c-axis were measured in 21 quartz grains. A total of 45 sets of PDFs were measured. The frequency distribution of the PDF orientations (Fig. 5) peaks between 20 and 30 degrees (ω) and lacks basal PDFs, but the orientations are otherwise broadly distributed. These results are similar to those of a study of K/T ejecta by Grieve and Alexopoulos (1988; Fig. 5), who measured 62 sets of PDFs in 35 grains from the K/T boundary at Scollard Canyon, Alberta. Their study also reported the absence of basal PDFs. Our results differ somewhat, however, from those of Sharpton et al. (1992; Fig. 5) who reported 7% of 71 sets to be basal PDFs in 30 grains from a Chicxulub breccia. This may reflect a significant difference in the shock history or preservation of PDFs in distal K/T ejecta relative to Chicxulub crater breccias.

Size and abundances of shocked quartz in the Pacific

Histograms of the size distributions of quartz (quartz with PDFs and total quartz; Fig. 6) show that most of the grains recovered are in the size fractions <100 μm. In most samples, there appears to be a modal peak in the size distribution in the range of 40 to 60 μm. We believe that this is real and not an artifact of sample processing (i.e., loss of small grains), as we went to great trouble to be reproducible down to at least 30 μm. There

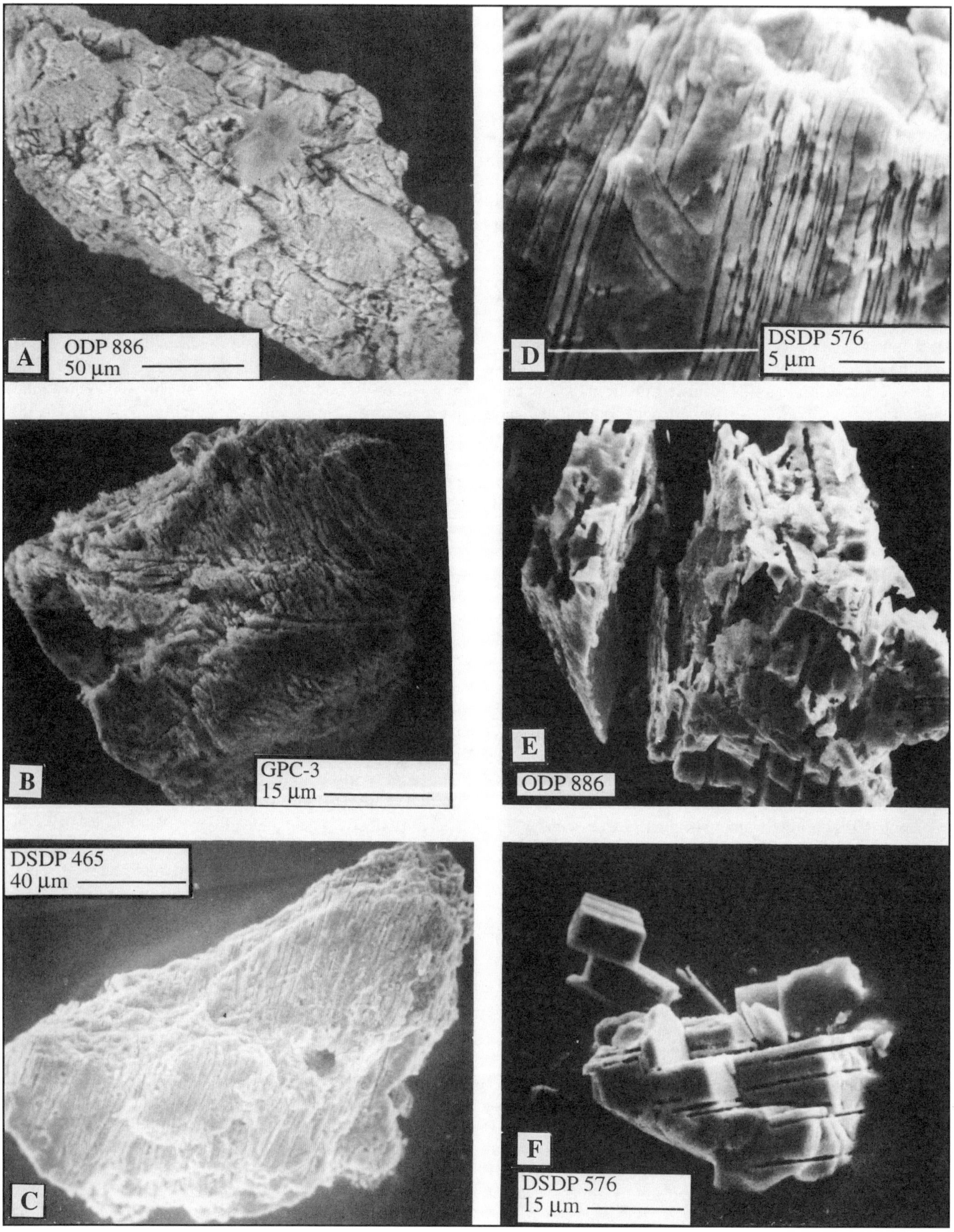
A
ODP 886
50 µm
D
DSDP 576
5 µm
B
GPC-3
15 µm
E
ODP 886
DSDP 465
40 µm
C
F
DSDP 576
15 µm

Figure 2. SEM photomicrographs of six quartz grains (30 to 300 μm) from K/T boundary sediments in the Pacific Basin. A, A 300-μm quartz grain from ODP 886. Faint linear features are observed in a few places, but this grain was not classified as shocked. In contrast, some grains are entirely composed of planar features, as in B, a 60-μm grain from GPC3, and C, a 200-μm grain from 465. D, This 30-μm grain from 576 has variable spacing between different sets of PDFs. E, A ~35-μm grain from 886 that appears to be breaking off along PDFs. F, A 45-μm quartz grain from 576 that has broken along three distinct sets of PDFs. Note the discontinuous character of some planar features, which may reflect differential etching or secondary overgrowth of silica.

is a clear tendency for the largest proportion of unshocked grains to occur in the smallest size bins. Several possible processes might be responsible for this: (1) the finer fraction of the impact ejecta may be less shocked, (2) there may be significant amounts of non-impact-derived quartz in the smaller sizes, or (3) this tendency may reflect our analytical methods and a decreased efficiency at identifying PDFs on the smallest grains. We suspect the latter.

A summary of the results of the analyses is given in Table 4, including the maximum diameter of the largest grain with PDFs from each site, the percent of total quartz that has PDFs, and our best estimate of the total abundance of quartz with PDFs at each locality. For comparison, distances from Chicxulub at 66 Ma and net Ir fluences for each site are included. The maximum diameter data include grains from samples that were not included in the quantitative SEM work. In particular, a sample from LL44-GPC3 that was used for spindle-stage work (not reported here) had two grains >300 μm. One of these had PDFs and a maximum diameter of 320 μm, and the other was a grain with large mosaic domains and a diameter of 350 μm. Another 300-μm grain found in ODP Site 886 sediments (Fig. 3A) had a few faint linear features observed in the SEM analysis, but it has not been included in the category with PDFs.

The estimated total shocked-quartz fluences (grains with PDFs/cm^2) are subject to considerable uncertainty and cannot be determined with a high degree of precision. At each site except DSDP 465, the Ir and ejecta components are smeared across fairly broad intervals by bioturbation. The 1,800 grains/cm^2 for DSDP Site 596 is from Zhou et al. (1991), who analyzed 10 samples and found shocked quartz distributed across at least 57 cm of core within the very broad Ir peak at this site. In estimating the shocked-quartz fluence, Zhou et al. (1991) normalized the concentrations of grains with PDFs to the Ir concentration and Ir fluence using the general formula

$$F_{PDF} = \left(\frac{C_{PDF}}{C_{Ir}}\right)_{average} \times F_{Ir} \qquad (1)$$

where *F* is the fluence and *C* the concentration of Ir or quartz with PDFs. We have applied a similar method to Sites 576, 803, 886, and GPC3, where we have Ir data, with the exception that at Site 576 we used only the sample at the Ir peak (54.4 mbsf). This method is probably accurate within a factor of 2, even though it may overestimate the actual quartz fluence if some Ir has diffused out of the peak following mixing by bioturbation. Lacking Ir data for the samples from DSDP Sites 465 and 577, we have estimated these using the method of the "best-educated-guess-possible" of a lower limit on the fluence.

Despite the uncertainties inherent in these estimates, it appears that the fluence of quartz grains with PDFs that have maximum diameters >30 μm is in the range of ~1,000 to 2,000 grains/cm^2 over most of the Pacific plate. One notable exception is ODP Site 886 where we obtain an estimate of 13,800 grains/cm^2. We consider this to be a reliable estimate based on two samples with >800 shocked grains/gram as well as qualitative observations of abundant quartz in three other samples. These analyses confirm that shocked quartz is abundant across much of the Ir anomaly, at least from 65.10 to 65.28 mbsf in site 886.

There appears to be a discernable trend in the distribution of maximum grain sizes of shocked quartz. The three sites nearest to North America and Chicxulub were the localities in which we found the largest grains with PDFs. Altogether we found 10 grains with PDFs that had maximum diameters ≥200 μm: three at DSDP Site 465, two at ODP Site 886, and five at LL44-GPC3. Two other grains ≥300 μm that did not contain PDFs were also recovered from 886 and GPC3. This is not just an artifact of larger sample sizes or more grains at these sites. Both sites 576 and 596 have been extensively sampled in addition to the data reported in Table 2. Zhou et al. (1991) analyzed 10 samples from Site 596, and we have qualitatively analyzed another three from this core; the largest grain with PDFs is only 150 μm (only two are larger than 120 μm). At DSDP Site 576, we have qualitatively picked through a large number of trace mineral separates as we developed our procedures, and 190 μm is the maximum found. Altogether, we have not examined an extraordinary number of grains, probably only two or three times the 1,687 total grains with PDFs reported in Table 2. Certainly, larger grains can probably be found at all seven sites, but our observation is that large grains appear to be most abundant in samples in the northwestern part of the study area.

Comparison to K/T boundary sites outside the Pacific

Direct comparisons between this study and previous work outside the Pacific are hindered by an absence of common goals and analytical procedures employed by various researchers as well as sample variability and availability. In our study, samples are typically from bioturbated and presumably well-homogenized boundary intervals. Many other pelagic sequences (e.g., DSDP 527, Walvis Ridge) are similarly dis-

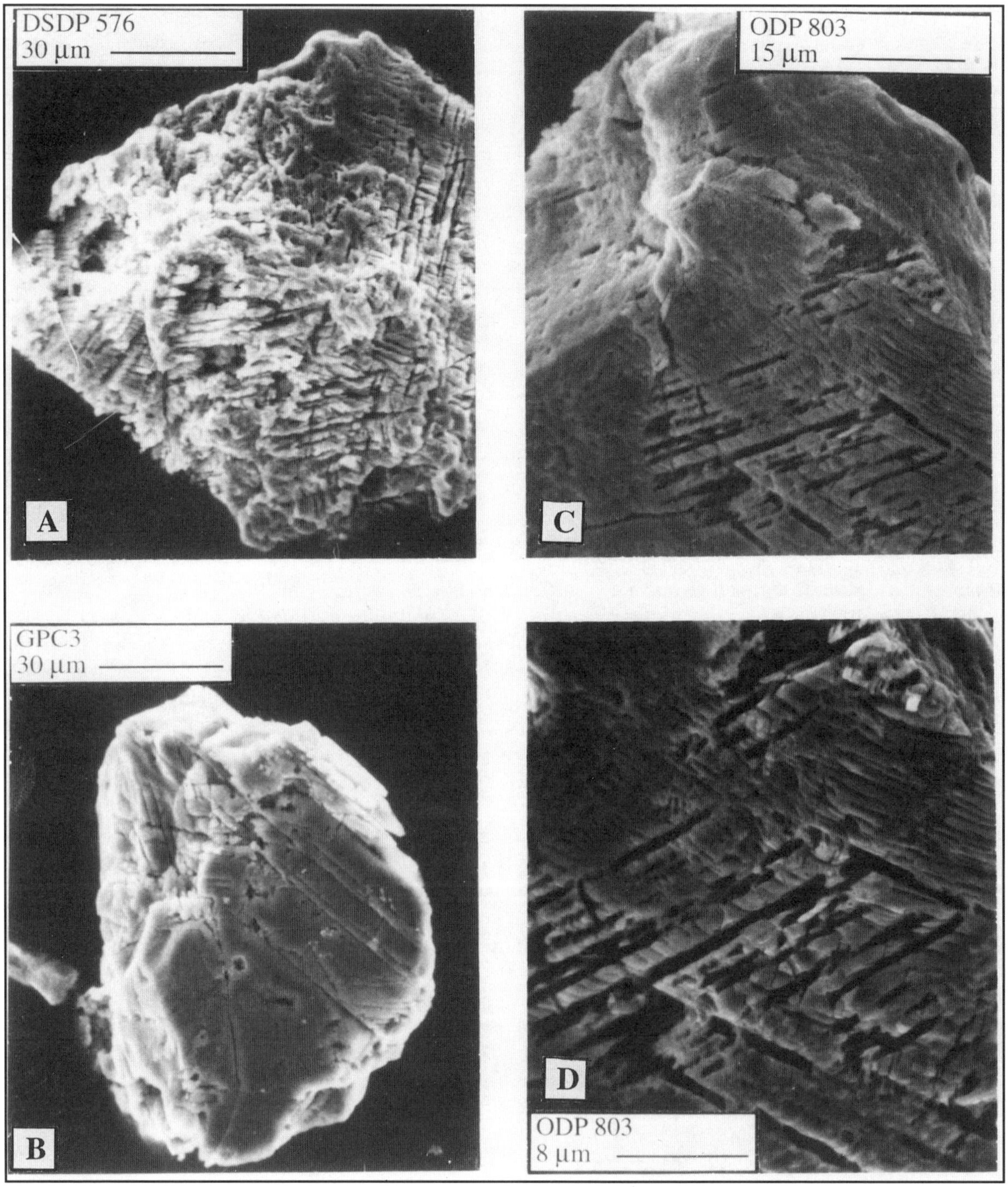

Figure 3. SEM photomicrographs of three quartz grains from KT boundary sediments in the Pacific Basin that illustrate discontinuous exposure of PDFs. A, Grain from DSDP Site 576 with at least three sets of PDFs and a heavily, yet differentially etched surface texture. B, Grain from the North Pacific with one clear set of PDFs that extends across the entire grain (from lower right to upper left) and two other sets that are discontinuous. C, Grain from ODP Site 803, western Equatorial Pacific (same as D), with two clear sets of thick PDFs on two parts of the grain and hints of a third set of thinner PDFs. D, Same grain as in C but at a higher magnification. Along with the two thick PDFs evident at lower magnification, we can now see two sets of very thin PDFs.

TABLE 3. OPTICAL ANALYSES OF K/T QUARTZ AND FELDSPAR

K/T Boundary Samples	Quartz				Feldspar
	PDFs Present	Milky	Mosaic Small	Mosaic Large	
ODP Site 886 (65.24 mbsf)	11	1	1	5	1
ODP Site 886 (65.18 mbsf)	5	0	3	1	0
LL44-GPC3 (20.59 mbsf)	13	5	4	2	0
Total number of grains	29	6	8	8	1
Percent of total	55%	11%	15%	15%	2%

turbed, and only in a few localities such as DSDP 465, Caravaca, and Agost, does a pure fallout layer appear to be preserved. Throughout North America, the ~3-mm "fallout" layer is well preserved and easy to sample in large quantities. Izett (1990) reports analyses of kilogram-sized samples from North America, and Bohor (personal communication, 1995) typically analyzes ~50-g samples, including his analyses of Caravaca and New Zealand. By comparison, our total sample in this study (the cumulative sample from all seven sites) contains orders of magnitude smaller amounts of fallout (in cm^2 of fallout) than individual samples in most preceding work.

At 66 Ma the Pacific sites were at distances of ~5,800 to 11,000 km from Chicxulub, the putative impact site for the KT event (Table 4). This is ~two to four times farther than most North American localities, but similar to the distances of European and South Atlantic (i.e., Walvis Ridge) sites. The Pacific K/T boundaries contain a trace mineral component that is very similar to that which occurs in the upper layer of the K/T unit in North America (e.g., Bohor, 1990; Izett, 1990). Clastic silicates are mainly quartz, and about half of these contain single or multiple sets of PDFs. When combined with the North American data, ours implies that this component was distributed over at least a quarter of the Earth's surface. In the Pacific basin this component is not diluted by detrital quartz, as a result of an absence of local sources. At other sites such as New Zealand, Europe, and the Walvis Ridge (Bohor and Izett, 1986; Izett, 1990; Huffman et al., 1990), most clastic quartz does not contain PDFs. For example, the peak concentration of quartz with PDFs at DSDP Site 527 is ~2% relative to total quartz (Huffman et al., 1990). We consider it unlikely that these differences reflect variability in the ejecta fallout, and we infer that these low concentrations of shocked quartz reflect contamination and dilution of the impact signature by local detrital sources.

The shocked quartz in the Pacific appears to be smaller than that deposited in the region around North America. Izett (1990) reports mean grain sizes ~200 μm in North America, whereas in the Pacific these are in the range of 50 to 70 μm (Fig. 5). Maximum sizes are also considerably larger in the North American region, although we infer that the maximum size in the Pacific basin is poorly constrained because of the small sample sizes we have worked with.

We suggest that one significant result of our study is that we have found several shocked-quartz grains in the North Pacific that are considerably larger than those reported for sites outside the North American–Caribbean region. Previously, the maximum size reported was 190 μm (Bohor and Izett, 1986; Izett, 1990) in Spain (Caravaca) and Italy (Pontedazzo). Considering that we have found 10 grains with PDFs ranging in size from 200 to 320 μm in small samples from three sites (465, 886, and GPC3), this is unlikely to be a statistical fluke. Kyte et al. (this volume) show that spinel compositions and spherule abundances are quite different in the Pacific than at European or South Atlantic K/T sites. These differences appear to include maximum sizes of shocked quartz as well.

Comparison of fluences of shocked quartz between Pacific and non-Pacific sites is also problematic. Based on a fluence of ~1,800 shocked grains/cm^2 at DSDP Site 596, Zhou et al. (1991) estimated that the number of grains deposited in the South Pacific was comparable to those in North America, although with mean sizes >two times lower. They concluded that the mass of shocked quartz deposited at Site 596 is only ~four times lower than in North America. Our results corroborate this estimate for the entire Pacific basin. We estimate that the amount of shocked quartz deposited there has fluences within an order of magnitude of that deposited in North America. We have no means for a direct quantitative comparison to K/T ejecta at other globally distributed sites. However, if Izett's (1990) estimate that shocked minerals are "several orders of magnitude more abundant" in North America than in Europe is correct, then this would also imply that considerably more shocked quartz was deposited in the Pacific than in Europe.

CONCLUSIONS

We have found that shocked quartz is abundant at all seven K/T boundary sites on the Pacific plate. Dilution by detrital quartz and other windblown components is a minor, or nonexistent, problem, and impact-derived quartz is easily separated from other pelagic sediment components. We find that in most samples the fraction of the total quartz (>30 μm) with PDFs is

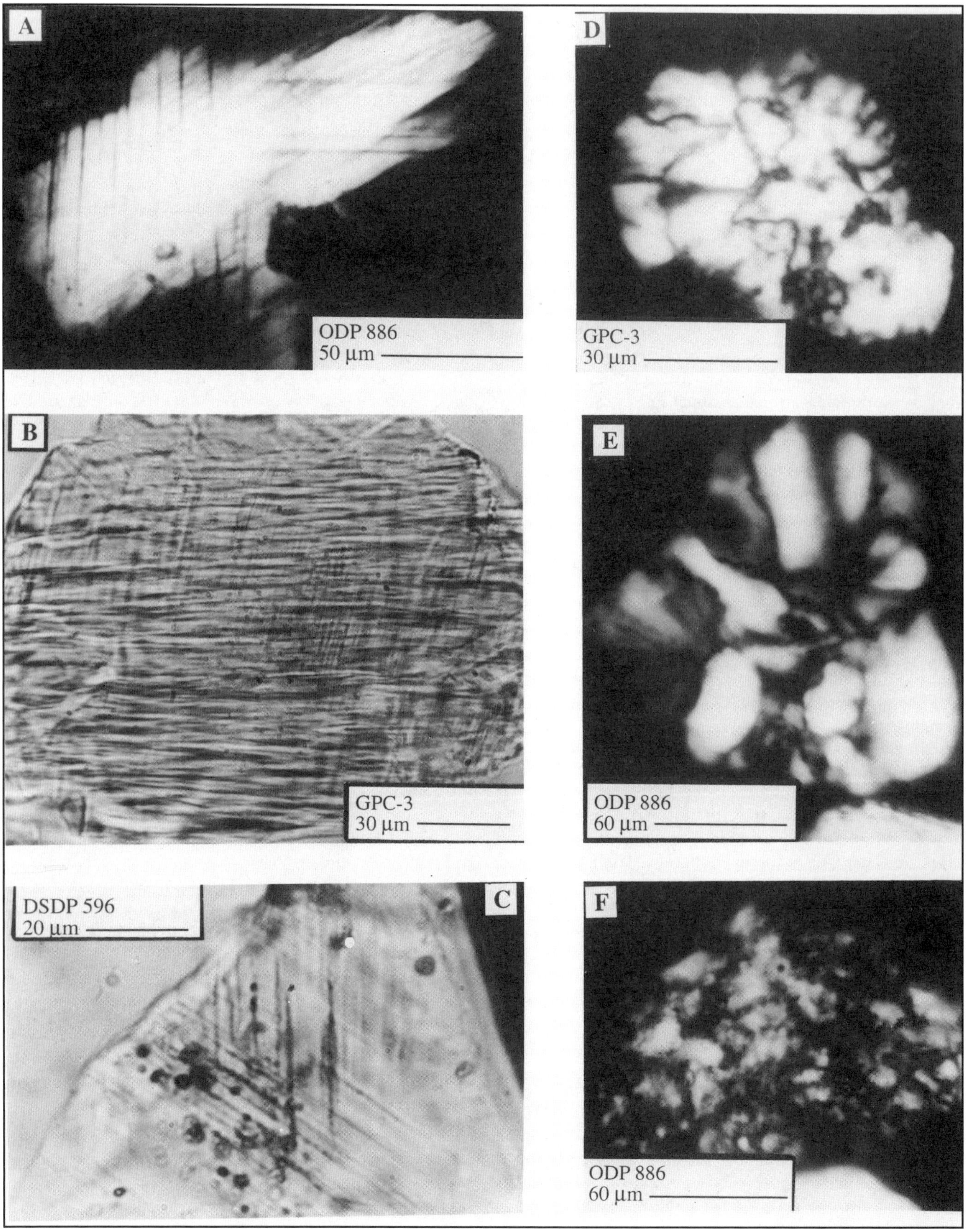
A
ODP 886
50 μm
D
GPC-3
30 μm
B
GPC-3
30 μm
E
ODP 886
60 μm
DSDP 596
20 μm
C
F
ODP 886
60 μm

Figure 4. Optical photomicrographs of six quartz grains from ODP Site 886. All but B and C were taken on the U-stage with crossed nichols; B and C were on spindle stage mounts. The highly variable character of PDFs can be seen in photos A, B, and C, which show at least two sets of PDFs each. Photos D, E, and F show mosaic structure (possibly as a result of shock) and may also be partly annealed.

more than 60%. In all probability, virtually all of the larger quartz grains (including mosaic-textured quartz, which was not classified as shocked in this study) are derived from the K/T impact event. Quantitative analysis of the abundance of shocked quartz (grains/cm^2) is difficult and imprecise. Given these uncertainties, there is little difference in abundance from site to site: typically ~1,000 to 2,000 shocked grains/cm^2. The notable exception is ODP Site 886 with an estimated shocked quartz fluence of 13,800 grains/cm^2. The largest shocked quartz grains were found at sites DSDP 465, ODP 886, and LL44-GPC3, the three sites nearest to North America. This may result from their proximity to Chicxulub.

Comparison of our results to previous analyses of K/T quartz at other sites around the world is difficult at best because of differing sampling strategies and procedures employed by various researchers. Shocked quartz generally appears to be less abundant (in grains/cm^2) and smaller in the Pacific than in the North American region, a trend consistent with a source at Chicxulub, Mexico. Prior studies that reported extremely low concentrations of shocked quartz (relative to total quartz) in Europe and the South Atlantic were probably hindered by contamination and dilution of this trace component by quartz from local detrital sources.

Several quartz grains with PDFs >200 μm, including one with a maximum diameter of 320 μm, have been recovered. These are significantly larger than previously reported maximum quartz sizes for K/T sites in western Europe (<190 μm), localities that were similar distances from Chicxulub at 66 Ma. If Chicxulub was the source of the globally distributed shocked quartz, then this may reflect a nonuniform distribution of this component of the ejecta.

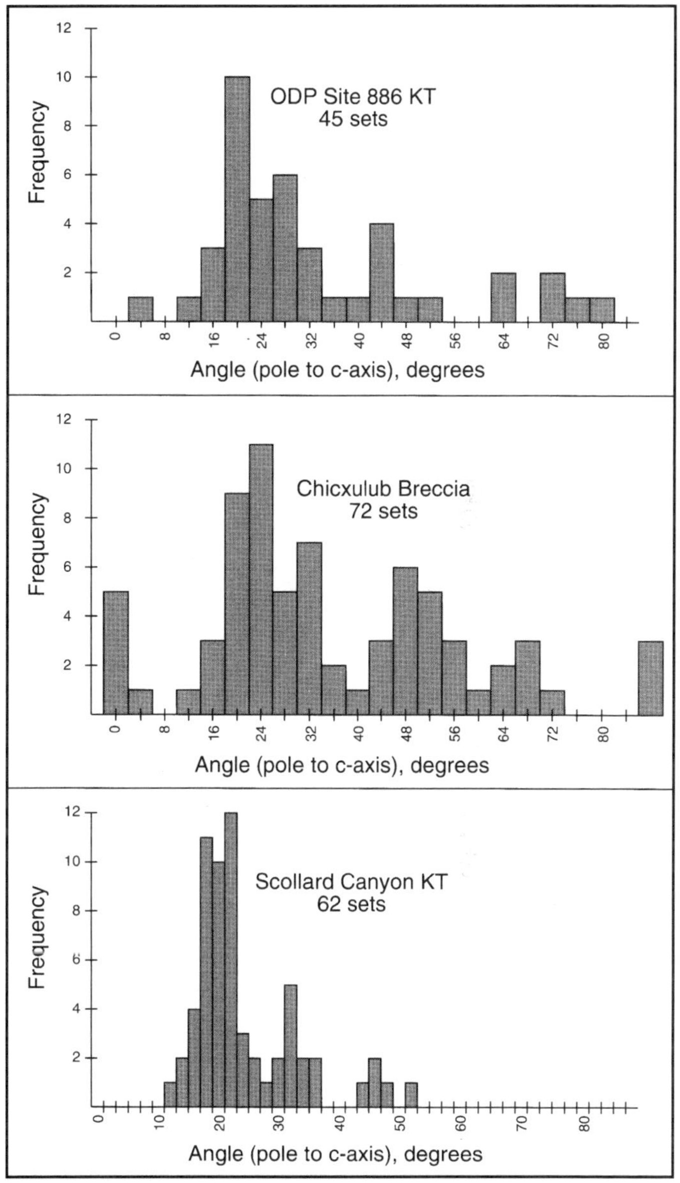

Figure 5. Histograms of the frequency distribution of the angle between the pole to the PDF and the c-axis in shocked-quartz grains. Results for grains from the K/T boundary in ODP Site 886 are similar to results from the K/T boundary in Scollard Canyon, Canada (after Grieve and Alexopoulos, 1988), and Chicxulub (after Sharpton et al., 1992).

ACKNOWLEDGMENTS

We thank B. F. Bohor and W. J. Betterton for assistance in developing our trace mineral extraction procedures. J. M. Christie generously instructed J. A. B in the use of the universal stage. G. A. Izett provided instruction for spindle-stage analyses. We thank A. R. Huffman and V. R. Sharpton for thoughtful reviews of an earlier version of this paper. Sediment cores are curated with support by the National Science Foundation. This research was supported by NSF grant EAR-91-18701.

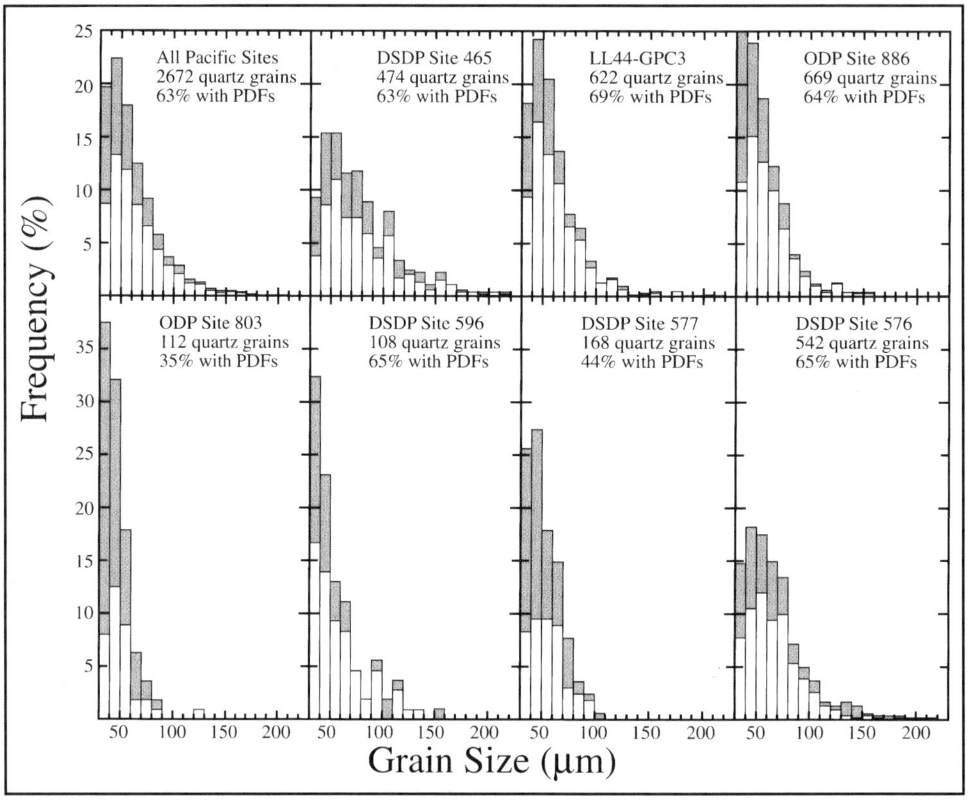

Figure 6. Size frequency distribution histograms for quartz grains >30 μm recovered from seven sites in the Pacific basin. Frequencies of shocked grains (white) and unshocked grains (dark gray) are plotted as a percent of total quartz.

TABLE 4. K/T BOUNDARY SITES ON THE PACIFIC PLATE - SUMMARY TABLE

Distance to Chicxulub at 66 Ma (km)	K/T Boundary Site	Largest Quartz Grain with PDFs (mm)	Percent Quartz with PDFs	Quatrz Grains with PDFs/cm^2	Iridium* (ng/cm^2)
5,800	LL44-GPC3	320	69	1,840	70
6,450	ODP 886	225	64	13,820	52
7,900	DSDP 465	212	63	>650	~100
9,100	DSDP 576†	190	65	2,270	110
9,600	DSDP 577	90	43	>1,200	61
9,700	DSDP 596	150	65	1,800§	320
11,000	ODP 803	125	35	1,230	140

*Iridium fluences from the literature: GPC3, Kyte and Wasson, 1986; DSDP 465, Kyte et al., this volume; DSDP 596, Zhou et al., 1991; DSDP 577, Michel et al., 1985; DSDP 576 and ODP 803 and 886, Kyte et al., 1995.
†Samples from Ir anomaly,
§From Zhou et al., 1991.

REFERENCES CITED

Alvarez, L. W., Alvarez, W., Asaro, F., and Michel, H. V., 1980, Extraterrestrial cause for the Cretaceous–Tertiary extinction: Science, v. 208, p. 1095–1108.

Bohor, B. F., 1990, Shocked quartz and more; Impact signatures in Cretaceous/Tertiary boundary clays: Geological Society of America Special Paper 247, p. 335–342.

Bohor, B. F., and Betterton, W. J., 1989, Glauconite spherules and shocked quartz at the K-T boundary in DSDP Site 603 B [abs.], *in* Proceedings, Lunar and Planetary Science 20: Houston, Texas, Lunar and Planetary Institute, p. 92–93.

Bohor, B. F., Foord, E. E., Modreski, P. J., and Triplehorn, D. M., 1984, Mineralogic evidence for an impact event at the Cretaceous-Tertiary boundary: Science, v. 224, p. 867–869.

Bohor, B. F., and Izett, G. A., 1986, World-wide size distribution of shocked quartz at the K/T boundary: Evidence for a North American impact site, *in* Proceedings, Lunar and Planetary Science 17: Houston, Texas, Lunar and Planetary Institute, p. 68–69.

Bohor, B. F., Modreski, P. J., and Foord, E. E., 1987, Shocked Quartz in the Cretaceous-Tertiary boundary clays: Evidence for a global distribution, Science, v. 236, p. 705–709.

Engelhardt, W. v., and Bertsch, W., 1969, Shock induced planar deformation structures in quartz from the Ries Crater, Germany: Contributions to Mineralogy and Petrology, v. 20, p. 203–234.

Grieve, R. A. F., and Alexopoulos, J., 1988, Microscopic planar features in quartz from Scollard Canyon, Alberta, and the Cretaceous-Tertiary boundary event: Canadian Journal of Earth Sciences, v. 25, p. 1530–1534.

Heath, G. R., Kovar, R. B., and others, 1985, Elemental composition of Cenozoic pelagic clays from Deep Sea Drilling Project Sites 576 and 578, Western North Pacific, *in* Heath, G. R., and Burckle, L. H., and others, 1985, Initial reports of the Deep Sea Drilling Project, Volume 86: Washington D.C., U.S. Government Printing Office, p. 605–646.

Heath, G. R., Burckle, L. H., and others, 1985, Initial reports of the Deep Sea Drilling Project, Volume 86: Washington D.C., U.S. Government Printing Office, 804 p.

Hildebrand, A. R., 1993, The Cretaceous/Tertiary boundary impact (or the dinosaurs didn't have a chance): Journal of the Royal Astronomical Society of Canada, v. 87, p. 77–118.

Hildebrand, A. R., and Boynton, W. V., 1990, Proximal Cretaceous-Tertiary boundary impact deposits in the Caribbean: Science, v. 248, p. 843–847.

Hildebrand, A. R., Penfield, G. T., Kring, D. A., Pilkington, M., Carmargo Z., A., Jacobsen, S. B., and Boynton, W. V., 1991, Chicxulub Crater: A possible Cretaceous/Tertiary boundary impact crater on the Yucatan Penisula, Mexico: Geology, v. 19, p. 867–871.

Huffman, A. R., 1993, The microstructural response of quartz and feldspar under shock loading at variable temperatures: Journal of Geophysical Research, v. 98, p. 22171–22197.

Huffman, A. R., Crocket, J. H., Carter, N. L., Borella, P. E., and Officer, C. B., 1990, Chemistry and mineralogy across the Cretaceous/Tertiary boundary at DSDP Site 527, Walvis Ridge, South Atlantic Ocean, *in* Sharpton, V. L., and Ward, P. D., eds., Global catastrophes in Earth history: Geological Society of America Special Paper 247, p. 319–334.

Izett, G. A., 1989, Cretaceous-Tertiary (K-T) boundary interval, Western North America, and its implication for impact-extinction theory [abs.]: International Geological Congress, 28th, Washington D.C., Abstracts, v. 2, p. 2-106–2-107.

Izett, G. A., 1990, The Cretaceous-Tertiary (K-T) boundary interval, Raton Basin, Colorado and New Mexico, and its content of shock-metamorphosed minerals—Evidence relevant to the Cretaceous-Tertiary impact-extinction theory: Geological Society of America Special Paper 249, 100 p.

Izett, G. A., and Pillmore, C. L., 1985, Abrupt appearance of shocked quartz at the Cretaceous Tertiary Boundary, Raton Basin, Colorado and New Mexico: Geological Society of America Abstracts with Programs, v. 17, p. 617.

Kyte, F. T., and Wasson, J. T., 1986, Accretion rate of extraterrestrial matter: Iridium deposited 33 to 67 million years ago: Science, v. 232, p. 1225–1229.

Kyte, F. T., Zhou, Z., and Wasson, J. T., 1980, Siderophile-enriched sediments from the Cretaceous-Tertiary boundary: Nature, v. 288, p. 651–656.

Kyte, F. T., Leinen M., Heath, G. R., and Zhou, L., 1993, Cenozoic sedimentation history of the central North Pacific: Inferences from the elemental geochemistry of core LL44-GPC3: Geochimica et Cosmochimica Acta, v. 57, p. 1719–1740.

Kyte, F. T., Bostwick, J. A., and Zhou, L., 1995, Identification of the Cretaceous-Tertiary boundary at ODP Site 886, ODP Site 803, and DSDP Site 576, *in* Proceedings, Ocean Drilling Program, Scientific Results, Volume 145: College Station, Texas, Ocean Drilling Program, p. 427–434.

Leinen, M., and Heath, G. R., 1981, Sedimentary indicators of atmospheric activity in the northern hemisphere during the Cenozoic: Paleogeography, Paleoclimatology, Palaeoecology, v. 36, p. 1–21.

Michel, H. V., Asaro, F., and others, 1985, Elemental profile of iridium and other elements near the Cretaceous/Tertiary boundary in Hole 577B, *in* Heath, G. R., Burckle, L. H., and others, 1985, Initial reports of the Deep Sea Drilling Project, Volume 86: Washington D.C., U.S. Government Printing Office, p. 533–538.

Rea, D. K., Leinen, M., and Janecek, T.R., 1985. Geologic approach to the long-term history of atmospheric circulation: Science, v. 227, p. 721–725.

Sharpton, V. L., Schuraytz, B. C., Burke, K., Murali, A. V., and Ryder, G., 1990, Detritus in K/T boundary clays of western north America; Evidence against a single oceanic impact, *in* Sharpton, V. L., and Ward, P. D., eds., Global catastrophes in Earth history: Geological Society of America Special Paper 247, p. 349–357.

Sharpton, V. L., Dalrymple, G. B., Marin, L. E., Ryder, G., Schuraytz, B. C., and Urrutia-Fucugauchi, J., 1992, New links between the Chicxulub impact structure and the Cretaceous/Tertiary boundary: Nature, v. 359, p. 819–821.

Smit, J., and Romein, A. J. T., 1985, A sequence of events across the Cretaceous-Tertiary boundary: Earth and Planetary Science Letters, v. 74, p. 155–170.

Thierstein, H. R., Asaro, F., and others, 1991, The Cretaceous/Tertiary boundary at Site 738, southern Kerguelen Plateau, *in* Hare, C., and Mazzullo, E. K., Proceedings, Ocean Drilling Program, Scientific Results, Volume 119: College Station, Texas, Ocean Drilling Program, p. 719–730.

Zhou, L., and Kyte, F. T., 1992, Sedimentation history of the South Pacific pelagic clay province over the last 80 million years inferred from the geochemistry of DSDP Hole 596: Paleoceanography, v. 7, p. 441–465.

Zhou, L., Kyte F. T., and Bohor, B. F., 1991, The Cretaceous/Tertiary boundary of DSDP Site 596: South Pacific Geology, v. 19, p. 694–697.

Manuscript Accepted by the Society September 14, 1995

Geological Society of America
Special Paper 307
1996

K/T boundary layer in Deccan Intertrappeans at Anjar, Kutch

N. Bhandari and P. N. Shukla
Physical Research Laboratory, Navrangpura, Ahmedabad 380009, India
Z. G. Ghevariya and S. M. Sundaram
Geological Survey of India, Gandhi Nagar 382043, India

ABSTRACT

We have identified the iridium-rich Cretaceous-Tertiary boundary (KTB) layer within the third intertrappean sediment bed at Anjar, located in Kutch near the western periphery of the Deccan volcanic province. This chocolate-colored limonitic layer (<1 cm thick) is characterized by high concentration of Ir (1,271 pg/g) and Os (1,414 pg/g), about 15 times higher than in the adjacent sediments and more than 2 orders of magnitude higher compared to the underlying basalt. High concentration of Ir and Os and other siderophiles is accompanied by enrichment of such chalcophiles as Se, Sb, Ag, As, and Zn and depletion of such lithophiles as Sc, Hf, and Al compared to the underlying basalt. Some of these characteristics are similar to those found in other continental and marine KTB sections. The Os/Ir ratio of ~1.1, close to the meteoritic value and other chemical and stratigraphic criteria, indicates that the limonite layer is the fallout ejecta layer, resulting from the bolide impact at the KTB. Absence of enrichment of Ir and other platinum group elements in many other intertrappean sediments indicates that the volcanic contribution of iridium, during Deccan eruption, was small and cannot account for the Ir enhancement at the KTB.

INTRODUCTION

The discovery of anomalously high iridium at the Cretaceous-Tertiary boundary (KTB) (Alvarez et al., 1980) and its global occurrence in many marine and continental sections have resulted in an intense debate about the origin of the iridium. Whereas the low iridium concentration of crustal rocks and the high abundance of iridium in meteoritic matter have been taken as evidence for a large (~10 km) bolide impact at KTB, occurrence of large-scale Deccan volcanism in central India at about the same time has given rise to alternative scenarios. For example, McLean (1985), Officer and Drake (1985), and Courtillot et al. (1986) have proposed that mantle material rich in Ir can emanate during volcanic episodes, resulting in an iridium-rich KTB layer. Discovery of iridium-rich particulates in Hawaiian volcanic emissions by Zoller et al. (1983) has further strengthened this hypothesis. On the other hand, the presence of certain features that are easy to generate in extraterrestrial impact, such as shocked quartz (Bohor et al., 1987), silicate spherules (Smit and Klaver, 1981), spinels (Robin et al., 1991), and tektites (Izett et al., 1991), found at the KTB, and the identification of the Chicxulub impact crater (Hildebrand et al., 1991; Krog et al., 1993) have strengthened the impact hypothesis. To explain all these features, Alt et al. (1988) have proposed a hybrid model involving impact-induced volcanism, similar to that which occurred on the moon and generated the lunar maria. Rampino and Caldeira (1993) suggest impact-induced tectonic cycles on the Earth, whereas Sutherland (1994) believes that the impact and volcanism are fortuitously coincidental but independent events, occurring by chance at the KTB.

The important role of the chemical characteristics of the KTB sediments within Deccan in resolving some of these questions was recognized, and a search for an iridium-rich layer has been made in the past within the Takli intertrappean beds at Nagpur (Rocchia et al., 1988; Bhandari et al., 1993a) and at several other locations (Bhandari et al., 1993b). Now we have

Bhandari, N., Shukla, P. N., Ghevariya, Z. G., and Sundaram, S. M., 1996, K/T boundary layer in Deccan Intertrappeans at Anjar, Kutch, *in* Ryder, G., Fastovsky, D., and Gartner, S., eds., The Cretaceous-Tertiary Event and Other Catastrophes in Earth History: Boulder, Colorado, Geological Society of America Special Paper 307.

identified such an iridium-rich layer at Anjar and determined the abundance of some diagnostic elements in this layer. The chemical characteristics of this layer in terms of its origin and its relation to Deccan volcanism are discussed here.

SITE DESCRIPTION

Deccan traps are extensively exposed in the Kutch region of western India. The Anjar trap sequence, shown in Figure 1, consists of seven gently warped basalt flows with four well-developed intertrappean sediment beds (between flows I, II, III, and IV and between VI and VII) and two red bole horizons interfacing flows IV, V, and VI. This section gently dips and becomes compressed toward the west and has been briefly described earlier (Ghevariya, 1988; Bhandari et al., 1995). The third intertrappean sediment bed (IT-III) consists of a sequence of gypseous clays, splintery black-gray shales, cherty limestone, and bedded chert, a maximum of 9 m thick, which is exposed about 1.5 km southeast of the village of Viri (23°4′50″N, 70°30′E). The gray and black shale in the third intertrappean bed contains three chocolate-colored layers (Br 1, 2, and 3) separated by about 32 and 25 cm respectively. These layers are generally patchy and less than 1 cm thick, the second layer being the most prominent. This layer (Br 2) sometimes bifurcates into two layers and is traceable along the bedding surface. The shale above this layer contains charophytes and fragments of ornithoid eggshells and fish plates. Ghevariya (1988) has reported the presence of dinosaur bones, eggshell fragments, and other Upper Cretaceous fossils in the second and third (IT-II and IT-III) intertrappeans. Bajpai et al. (1993) have studied the dinosaur eggshells from the third intertrappean. The IT-III contains the uppermost horizon having *in situ* dinosaur fossils, as shown in Figure 1. Two of the traps, FIII and FIV, have been dated in our laboratory by $^{39}Ar/^{40}Ar$ method to be 65.1 ± 0.6 and 64.7 ± 1.5 Ma, coincident with the 65 Ma KTB event (K. Pande and T. R. Venkatesan, 1993, personal communication). All these observations indicate that the trap sequence was formed close to the terminal Cretaceous period.

Some chocolate-colored limonitic layers, similar in appearance, are present in the nearby Bhachao section, which has at least five traps of the same sequence exposed on the surface. Five samples of the limonitic layers were collected between the Bhuj sandstone and the lowermost flow, which is equivalent of Anjar FII, to compare their chemical patterns with the limonitic layers of the Anjar section.

EXPERIMENTAL PROCEDURES

Several samples from different basalt flows and intertrappean sediments were collected. A 1.5-m-thick section in IT-III was sampled at closely spaced intervals (1 to 10 cm). The relative locations of these samples are shown in Figure 1 and their depths from the black clay horizon are given in Table 1. Three samples (23, 24, 27) contain a thin chocolate-brownish band. Since this color is typical of K/T boundary clay, these bands were separated, labeled as fraction B, and separately processed. Fraction A represents the rest of the sample, stratigraphically located within a couple of centimeters above and below fraction B. The samples were analyzed for Ir, Os, and about 25 other elements (including 9 REE) by radiochemical and instrumental neutron activation analyses using standard procedures described by Bhandari et al. (1994). Columbia River Basalt BCR-1, Allende meteorite, IAEA (International Atomic Energy Agency) soil SL-1, or IAEA reference soil S-5 were used as standards, which together with the samples were irradiated to a fluence of about 10^{18}n/cm^2 in the Cirus reactor of Bhabha Atomic Research Center, Bombay. The irradiated samples were counted at predetermined intervals of time on suitable germanium detectors and then processed radiochemically for Ag, Os, and Ir.

The concentration of most of the elements (Tables 1 through 3) are calculated with respect to BCR-1, except for Cr, Ag, Ir, and Os, for which Allende standard was used. Arsenic was

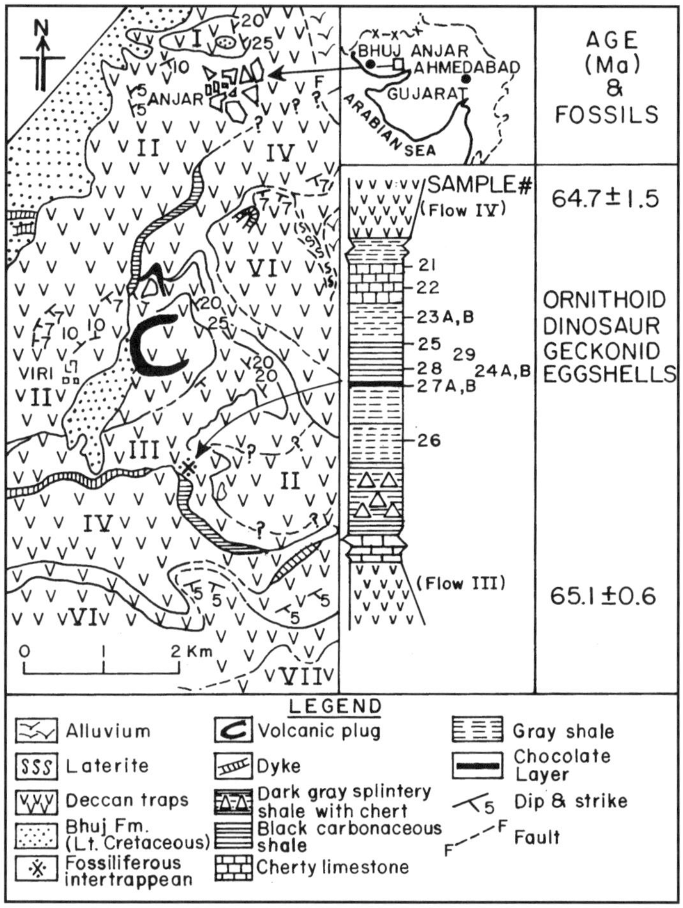

Figure 1. Geological map showing the location of the Anjar trap sequence (after Ghevariya, 1988). Lithostratigraphy, ages, fossils, and sample horizons are shown. The relative depths of various samples are given in Table 1. Doubtful faults are indicated by ?.

TABLE 1. CONCENTRATIONS OF SIDEROPHILE ELEMENTS IN ANJAR III INTERTRAPPEAN SECTION, KUTCH

Sample	Depth* (cm)	Description	Ir (pg/g)	Os (pg/g)	Cr (μg/g)	Co (μg/g)	Fe (%)	Mn (μg/g)	Ni (μg/g)	Os/Ir
#21	2	Black clay	97		25.76	20.86	5.24			
#22	52	L.S.	94		1.82	4.20	1.08			
#23-A	62	White chert	104		2.47	5.09	0.62			
#23-B	63	Black chert	22		1.15	4.34	0.70			
#25	70	Black mud	119		10.94	17.30	4.71			
#29	73	Brownish shale	95		12.16	27.40	4.50			
#28	75	Blackish shale	95	299 ± 76	8.40	13.60	2.73	594	23.8	3.1 ± 1
#24-A	80	Choc. white shale	186	<171	5.54	14.50	3.11			<0.91
#24-B	81	Brown shale	168	400 ± 131	8.54	23.09	4.83			2.4 ± 0.8
#27-A	82	Gray shale	733	1123 ± 155	4.19	61.00	31.42	330	73.5	1.53 ± 0.2
#27-B	83	Chocolate layer	1271	1414 ± 125	4.97	121.2	37.43	427	99.8	1.11 ± 0.1
#26	103	White shale	78	<80	5.15	17.2	2.38	473	41.6	<1
F-III-32	650	Basalt	<10	-	8.57	45.7	9.10	1643	54	
S-5					29.76	14.8	4.68			
S-5(R)†					28.9±2.8	14.8±0.76	4.45±0.19			
%Errors (1σ)			10	-	4	0.2	1	5	5	
EF§			>127	-	0.58	2.65	4.11	0.26	1.85	

*The depth is measured from the black clay horizon.
†Recommended values.
§EF is enhanced factor of sample 27-B with respect to basalt FIII.

calculated with respect to soil S-5. Concentration of some elements such as Al, Mn, Cu, and so on were measured by atomic absorption spectrometer using standard W-1 as a reference. The values of various elements for reference standard S-5 calculated with respect to BCR agree well with the recommended values within error limits as indicated in Tables 1, 2 and 3.

CHEMICAL STRATIGRAPHY

The concentrations of various elements measured are given in Tables 1, 2, and 3. Typical errors based on counting statistics (1σ) are also listed in these tables. Figure 2 shows the depth profiles of some of these elements around the chocolate-colored layer (27B). The analyses indicate that the concentration of iridium in sediments is generally around 100 pg/g, similar to the maximum values found in other intertrappeans in the Deccan (Bhandari et al., 1993b). The one basalt (FIII) analyzed has Ir below our detection limit of ~10 pg/g, consistent with low iridium content (6 to 26 pg/g) of Deccan basalts (Orth et al., 1990). However, just below the uppermost horizon containing dinosaur fossils, the concentration of iridium increases abruptly, reaching a peak value of 1,271 pg/g within the thin, compacted chocolate-colored limonitic layer, located about 5.8 m above the basalt flow FIII and 3.18 m below FIV. The high Ir is accompanied by unusually high Os (1,414 pg/g), Fe, Co, Sb, Zn, Se, As, and Ag (Tables 1 and 2). The concentrations of some elements in the chocolate-colored layer are very high compared to those in basalt FIII. The enhancement factors are Sb(795), Se(14.5), Fe(4.1), Co(2.6), Cu(1.8), and Ni(1.8). Sc, Hf, Th, Mg, K, and Na are relatively depleted by factors of 5 or more, and so are Al, Cr, Mn, and rare earth elements, by factors of 1.5 to 22. These enhancement factors are also given in Tables 1 through 3 and will be discussed later in terms of their source characteristics and postdepositional alterations. Here we mention that the depth profiles of Co, Zn, Se, and Sb closely follow the Fe profile. Maximum Ag, Sb, and Zn concentrations are found in sample 27A and not in 27B. The Cr profile, on the other hand, is not similar to that of Fe or Co. Analysis of other samples, including the other two chocolate layers (Br 1 and Br 3), is in progress and will be reported in due course.

The REE patterns of some typical samples normalized to chondrites and North American Shale Composite (NASC) are shown in Figure 3 (a and b). The Ce anomaly, relative to its neighbors La and Nd, has been calculated following Liu et al. (1988) and is given in Table 3. It may be noted that sample 27B shows a peculiar REE pattern and the highest Ce anomaly.

DISCUSSION

We first discuss the source of the limonitic layer, which shows anomalously high concentration of Ir, Os, and other siderophile and chalcophile elements described above. Its composition may represent the original source or may have been modified by the environmental conditions under which it was deposited. On the other hand, it is possible that some elements were enriched or depleted by postdepositional diagenesis. The characterization of the source composition is difficult if depositional or postdepositional processes have severely altered its

TABLE 2. CONCENTRATION OF LITHOPHILE AND CHALCOPHILE ELEMENTS IN ANJAR III INTERTRAPPEAN SECTION, KUTCH

Sample	Ca (%)	Ba (μg/g)	Na (%)	Sb (μg/g)	Zn (μg/g)	Sc (μg/g)	Hf (μg/g)	Th (μg/g)	Se (μg/g)	Ag (ng/g)	Cu (μg/g)	As (μg/g)	Al (%)	Mg (%)	K (%)
#21	....	134.7	1.13	0.13	48.7	12.74	2.96	3.10	1.97	17.0		<2			
#22	40.6	43.9	0.28	0.04	15.7	2.51	0.62	0.60	0.42	5.3		BDL*			
#23-A	....	53.7	0.25	0.06	18.1	2.46	0.46	0.52	0.45	107.7					
#23-B	....	59.2	0.33	0.12	12.7	3.91	0.44	0.44	0.01	386.0					
#25	....	95.5	0.70	0.86	37.6	9.94	2.67	2.55	1.78	39.5					
#29	....	86.3	0.63	0.40	28.0	9.06	2.58	2.39	1.51	40.6					
#28	6.26	86.7	0.42	0.27	25.1	9.19	2.32	2.44	1.24	....	64.9		2.97	4.22	0.34
#24-A	3.4	122.7	0.30	2.14	58.5	4.18	1.35	1.29	1.25	304.0	....		....	....	....
#24-B	....	118.0	0.34	2.75	55.0	6.01	1.71	2.12	1.11	26.2	....		....	....	
#27-A	1.2	240.3	0.23	17.3	97.9	3.77	0.48	0.51	7.78	4300.0	202	494	0.38	0.27	0.029
#27-B	0.43	166.7	0.13	15.9	84.6	2.39	0.35	0.43	37.17	38.2	271	750	0.32	0.52	0.024
#26	20.3	40.9	0.37	0.25	24.1	5.17	1.35	1.34	1.00	....	34.6		1.63	2.36	0.17
F-III-32	6.7	296.0	1.07	0.02	84.7	23.90	3.76	2.72	2.56	72.4	151.3	BDL*	7.29	3.03	0.83
S-5		480.2	2.31	19.16	....	14.7	4.62	10.2	2.93						
S-5(R)†		562±53	1.92±0.11	14.3±2.2	....	14.8±0.66	6.3±0.3	11.3±0.73	....						
%Errors (1σ)	5	11	1	13	10	0.1	8	9	18	3	5	3	5	5	5
EF§	0.06	0.56	0.12	795	1	0.1	0.09	0.16	14.5	59.4	1.79		0.444	0.17	0.02

*BDL = below detection limit.
†R = recommended values.
§EF for Ag is for sample 27-A.

TABLE 3. CONCENTRATION OF RARE EARTH ELEMENTS IN ANJAR III INTERTRAPPEAN SECTION, KUTCH

Sample	La (μg/g)	Ce (μg/g)	Nd (μg/g)	Sm (μg/g)	Eu (μg/g)	Gd (μg/g)	Tb (μg/g)	Yb (μg/g)	Lu (μg/g)	Ce Anomaly*
#21	20.4	38.30	23.11	3.81	1.18	2.67	0.50	1.51	0.24	0.40
#22	7.02	11.30	5.10	1.01	0.32	0.98	0.14	0.60	0.13	0.39
#23-A	7.77	9.20	4.90	1.29	0.29	0.92	0.15	0.67	0.12	0.29
#23-B	1.84	2.90	1.60	0.40	0.14	0.45	0.06	0.37	0.07	0.36
#25	12.97	26.70	14.30	2.25	0.82	1.55	0.27	0.70	0.14	0.44
#29	11.00	29.40	14.80	2.04	0.70	1.36	0.24	0.52	0.12	0.53
#28	11.38	26.70	13.80	2.33	0.72	1.53	0.26	0.66	0.13	0.48
#24-A	9.04	15.80	9.20	2.05	0.49	1.47	0.23	0.76	0.13	0.38
#24-B	9.32	23.50	8.80	1.91	0.58	2.06	0.27	0.60	0.12	0.56
#27-A	5.44	14.63	7.80	1.39	0.32	0.33	0.15	2.14	0.29	0.52
#27-B	5.25	17.14	9.60	1.12	0.32	0.39	0.15	0.91	0.17	0.57
#26	10.55	16.95	9.20	1.93	0.52	1.71	0.25	0.68	0.12	0.37
F-III-32	23.30	47.40	33.50	4.90	1.99	3.47	0.78	1.71	0.35	0.39
S-5	26.9	54.72	26.4	5.32	1.19	5.12	0.64	1.96	0.35	....
S-5(R)†	28.1±1.5	59.7±3	29.9±1.6	5.42±0.4	1.18±0.08	....	0.66±0.07	2.24±0.2	0.37±0.04	
%Error (1σ)	2	0.8	8	2	2	13	7	4	4	....
EF	0.22	0.36	0.29	0.23	0.16	0.11	0.19	0.62	0.48	....

$^*A_{Ce} = (Ce/[1.44La + 0.66\ Nd])_{NASC}$.
†R = recommended values.

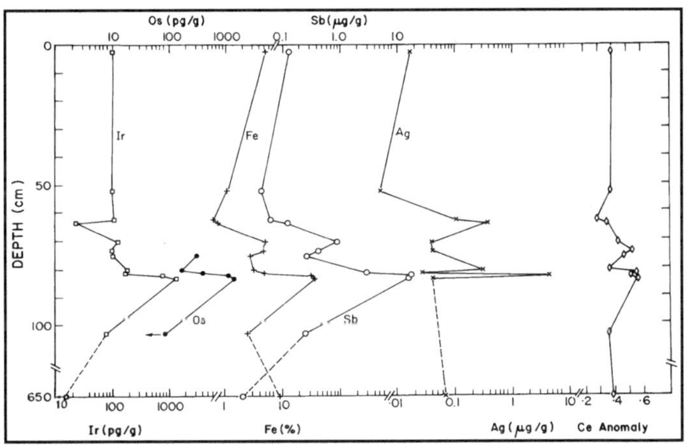

Figure 2. Depth profiles of Ir, Os, Fe, Sb, Ag, and Ce anomaly (normalized to NASC) in the intertrappean IT-III. The sample depths are slightly shifted for the sake of clarity.

chemical composition. Therefore we use some diagnostic elements that are not much affected by such processes and discuss the implications of their observed pattern.

The concentration of iridium in the limonitic layer (27B) in Anjar IT-III is not only the highest among Anjar intertrappeans but also 10 to 20 times higher compared to several intertrappean sediments studied so far. Iron is one of the most abundant (>30%) elements in this layer. We have analyzed other sediments and limonitic layers from the nearby Bhachao site as well as faraway sections from the eastern periphery of the Deccan plateau—from Nagpur, Jabalpur, Bargi, and Padwar (Bhandari et al., 1993a, b)—but none of them shows Ir and Os concentration above 120 pg/g. Concentrations of some elements in the limonitic layers from the Bhachao section are given in Table 4. Ir and Os in all of them were below our detection limit, and Sb was also low, in spite of the fact that Fe is high (9 to 25.2%). It therefore appears that high Fe is not always accompanied by high Ir or Os. Therefore, the processes that are responsible for enrichment of Fe in the Bhachao limonitic layers did not lead to the enrichment of Ir, Os, and chalcophiles. Os/Ir in the sample 27B is about 1, close to the meteoritic value but different from the value of ≥2 found in the other intertrappeans. Considering high Ir, Os, and nearly meteoritic Os/Ir ratio to be characteristic of the KTB layer found all over the Earth, it is likely that this limonitic layer is the sediment deposited at the KTB. The geochronological framework is consistent with our inference, as the ages of Flows III and IV are close to KTB age of 65 Ma. However, the $^{39}Ar/^{40}Ar$ dates do not have sufficient precision to show that the layer was deposited at KTB. So far, the KTB in this section has not been defined based on palynological or paleontological data. The paleomagnetic stratigraphy of the Anjar section has also not been constructed so far. All these studies, some of which are in progress, should enable us to decide whether the limonitic layer represents the KTB. Meanwhile, we take the enhanced Ir and Os concentration and $^{39}Ar/^{40}Ar$ dates as a reasonable indication of the KTB and discuss the implications of these results.

Apart from high Ir and other siderophiles, sample 27B

shows high concentration of such chalcophiles as Se and As (Tables 1 and 2 and Fig. 2). Ag and Sb, instead, have their maximum concentration in sample 27A. High concentration of Ir and chalcophiles has been seen in Hawaiian volcanic particulate emissions (Olmez et al., 1986) and in ashes from Kamchatkan volcanoes (Felitsyn and Vaganov, 1988), but we have analyzed many sediments from a large number of sites in the Deccan and have not found such an enhancement of chalcophiles in any of them (Bhandari et al., 1993a, b). If Deccan volcanism was contributing these elements, then they ought to be frequently enriched in many intertrappean sediments. Our observations thus indicate that the Deccan is not a likely source of these elements. Gilmour and Anders (1989) favored ocean water as a likely source for some of these elements, but as the Anjar intertrappeans are shallow coastal deposits, significant marine contribution of these elements can be ruled out. One obvious process that can enrich these elements is leaching of the surface sediments and rocks by the acid rain associated with the KTB bolide impact (Prinn and Fegley, 1987; Sigurdsson et al. 1992) and their subsequent precipitation during suitable redox conditions. Schmitz (1992) found similar enrichment in several continental KTB sections and attributed them to weathering processes.

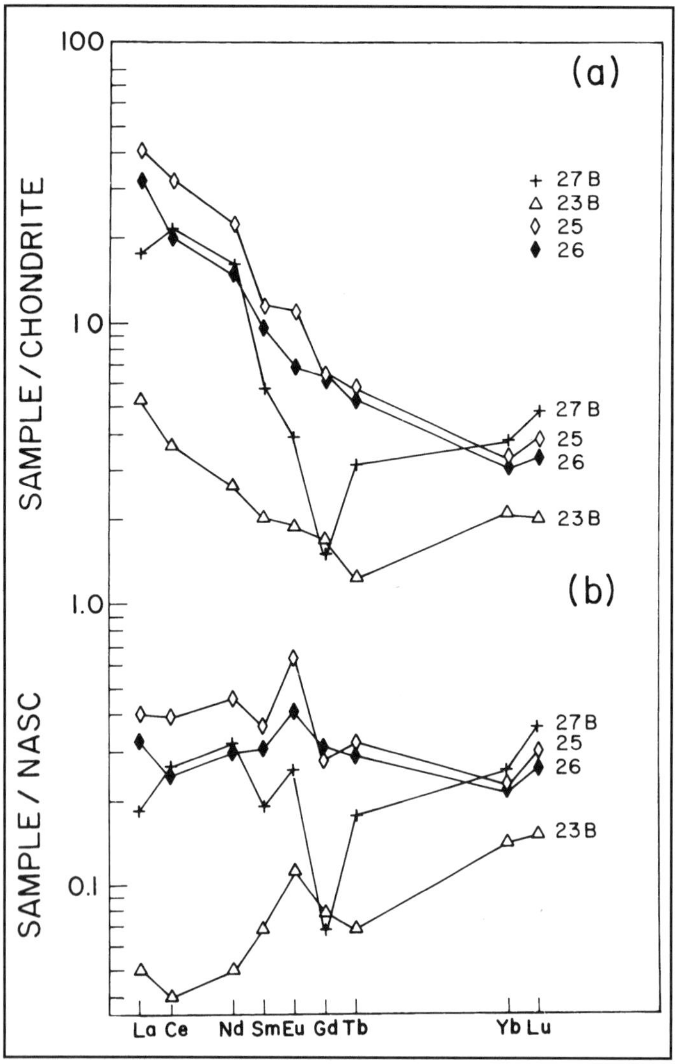

Figure 3. REE patterns of some typical intertrappean samples normalized to (a) chondrites and (b) North American Shale Composite (NASC).

TABLE 4. CONCENTRATIONS OF SOME ELEMENTS IN BHACHAO LIMONITIC LAYERS

Sample	Fe (%)	Co (μg/g)	Ir (pg/g)	Os (pg/g)	Sb (μg/g)
BH1	9.22	25.7	<156	BDL*	....
BH2	22.6	24.8	<48	BDL*	0.26
BH3	25.2	24.7	....	....	0.53
BH4	17.8	20.1	....	....	0.21
BH5	20.8	23.4	....	....	0.21

*BDL = below detection limit (≤ 200 pg/g).

The REE are generally depleted in this limonitic layer compared to the underlying basalt (Table 3). The Ce anomaly for most of the IT-III samples is about 0.44 (Fig. 2). However, within a depth interval of about 20 cm (between samples 27 and 23), there are fluctuations in the cerium anomaly. Starting from a value of 0.37 in the bottommost intertrappean analyzed here (sample 26), the anomaly rises to the highest value of 0.57 in the boundary sample 27B and remains in the same range up to about 10 cm above it (with the exception of fraction 24A), thereafter decreasing to its lowest value of 0.29 in sample 23A. The change in cerium anomaly can arise either as a result of change in the provenance of the depositing material or of change in redox conditions, as has been found to occur in marine sediments (Liu et al., 1988). If it is assumed that the provenance is similar around KTB, then the variations of cerium anomaly would imply a change in redox conditions. Apart from a high Ce anomaly, the boundary sample (27B) shows a peculiar REE pattern with high concentration of HREE, as is clear from Figure 3, where REE patterns of some typical samples are shown. There appears to be a negative Gd anomaly for both fractions of the boundary sample, 27A and 27B, which can be seen in both chondrite and shale normalized patterns (Fig. 3 [a, and b]), but the uncertainities in measurement of the neighboring Tb require that this anomaly should be confirmed by more precise measurements. The reason for the negative Gd anomaly is not clearly understood, although positive Gd anomalies have been observed for waters of the Atlantic and Pacific Oceans by De Baar et al. (1985).

A small positive Eu anomaly in the shale normalized REE pattern is mainly because the NASC is depleted in Eu. Small Eu anomalies have been ascribed to diagenetic remobilization under reducing conditions (MacRae et al., 1992). Such a possibility of diagenetic remobilization may be responsible for positive Eu anomaly observed, for example, in sample 26.

We have described above the main chemical characteristics of the various intertrappean sediments from IT-III and specifically the chocolate-colored limonitic layer, which shows enrichment of Ir, Os, and other siderophiles and chalcophiles and depletion of rare earths and some lithophiles. These characteristics must be discussed in relation to field observations and lithostratigraphy to understand their origin.

First, we may mention the field observation that the Ir-rich layer is located in the third intertrappean sediments at Anjar, overlying three lava flows. Quick succession of lower flows indicates that the Deccan volcanism was already active when this KTB layer was deposited. Based on these observations and the fact that the KTB layer is located not as an infratrappean but as an intertrappean, it has already been concluded (Bhandari et al., 1995) that the K/T event, if it was an extraterrestrial impact, did not trigger Deccan volcanism. Recent dates of Deccan flows at different sites, which range between 63 and 67 Ma, indicate that the peak Deccan activity predated the KTB by at least 2 m.y. (Venkatesan et al., 1993). This chronological framework is consistent with our observation, but the location of the KTB layer within the Anjar intertrappeans provides a high-resolution time frame, free of the uncertainties in absolute chronology, and enables us to accurately establish the relative time sequence of Deccan episodes and the K/T event. From the Anjar sequence it is clear that the dinosaurs survived the initiation of Deccan volcanism and that the volcanism itself was not triggered by the impact of the K/T bolide.

Second, the iridium-rich layer is not associated with any of the flows at Anjar. It is located in the sedimentary sequence, after 5.8 m of sediments had accumulated over Flow III. Taking the sedimentation rate as being in the range of 1 to 10 cm/ka in the intertrappeans, the time lapse between FIII and KTB layer turns out to be approximately 5×10^4 to 5×10^5 years. Flow IV is also separated from this limonitic layer by 3.18 m of sediments corresponding to a similar interval of time (3×10^4 to 3×10^5 years). Such a long lapse of time indicates that high Ir deposition was not accompanied by any of the Anjar flows. Furthermore, if Anjar or any of the Deccan flows were the source of Ir, higher Ir concentration would be expected, because of the proximity of Anjar to Deccan. Comparison with iridium concentration measured in other continental sections shows that the concentration of Ir in this layer is similar to and certainly not higher than that found elsewhere as far away as the western United States (Gilmour and Anders, 1989; Schmitz, 1992). This observation makes Deccan volcanic episodes an unlikely source of Ir and is consistent with the low iridium content (≤120 pg/g) of several intertrappean sediments, at Anjar and elsewhere, analyzed by us (Bhandari et al., 1993a, b). Our observations thus exclude volcanic origin of Ir and favor an extraterrestrial source, such as a comet or an asteroid, that impacted on the Earth, resulting in this KTB fallout layer. The high time resolution chemical profiles possible in the rapidly depositing Anjar intertrappean sequence, now in progress, should enable us to distinguish between cometary and asteroidal impact, as different depth profiles are expected in the two cases (Bhandari et al., 1994).

ACKNOWLEDGMENTS

We thank Drs. K. Pande and T. R. Venkatesan for their unpublished dates of Anjar flows. Our appreciation is due to R. Rengarajan for helping with the Atomic Absorption Spectrometer, to N. R. Manchanda for looking after the gamma ray spectrometers, and to S. K. Shah for computer programming. K. M. Suthar provided valuable help in both the field and the laboratory.

NOTE ADDED IN PROOF

Subsequent to the work reported here, the intertrappean (IT-III) was completely exposed at the sampling site and its thickness was found to be about 5.8 m, less than 9 m reported here. The iridium-rich layer Br2 (Sample 27) is located about 0.8 m above flow III and 5 m below flow IV. The deposition chronology reported here should therefore be revised, but this will not change the conclusions drawn in this chapter. The other two layers Br1 and Br3 also show enhancement of iridium (691 and 1,333 pg/g, respectively) and osmium (662 and 1,165 pg/g, respectively) compared to the adjacent horizons.

REFERENCES CITED

Alt, D., Sears, J. W., and Hyndman, D. W., 1988, Terrestrial maria: The origins of large basalt plateaus, hotspot tracks and spreading ridges: Journal of Geology, v. 96, p. 647–662.

Alvarez, L. W., Alvarez, W., Asaro, F., and Michel, H. V., 1980, Extraterrestrial causes of the Cretaceous-Tertiary extinction: Science, v. 208, p. 1095–1108.

Bajpai, S., Sahni, A., and Srinivasan, A., 1993, Ornithoid eggshells from Deccan intertrappean beds near Anjar (Kachchh), Western India: Current Science, v. 64, p. 42–45.

Bhandari, N., Gupta, M., and Shukla, P. N., 1993a, Deccan volcanic contribution of Ir and other trace elements near the K/T boundary, India: Chemical Geology v. 103, p. 129-139.

Bhandari, N., Shukla, P. N., and Castagnoli, G. C., 1993b, Geochemistry of some K/T sections in India: Paleogeography, Paleoclimatology, and Paleoecology, v. 104, p. 199–211.

Bhandari, N., Gupta, M., Pandey, J., and Shukla, P. N., 1994, Chemical profiles in K/T boundary section of Meghalaya, India: Cometary, asteroidal or volcanic: Chemical Geology, v. 113, p. 45–60.

Bhandari, N., Shukla, P. N., Ghevariya, Z. G., and Sundaram, S. M., 1995, Impact did not trigger Deccan volcanism: Evidence from Anjar K/T boundary intertrappean sediments: Geophysical Research Letters, v. 22, p. 433–436.

Bohor, B. F., Modreski, P. J., and Foord, E. E., 1987, Shocked quartz in the Cretaceous-Tertiary boundary clays: Evidence for a global distribution: Science, v. 236, p. 705–709.

Courtillot, V., Besse, J., Vandemme, D., Montigny, R., Jaeger J. J., and Cappetta, H., 1986, Deccan flood basalts at the Cretaceous-Tertiary boundary: Earth and Planetary Science Letters, v. 80, p. 361–374.

De Baar, H. J. W., Brewer, P. G., and Bacon, M. P., 1985, Anomalies in rare earth distribution in sea water: Gd and Tb: Geochimica et Cosmochimica Acta, v. 49, p. 1961–1969.

Felitsyn, S. B., and Vaganov, P. A., 1988, Iridium in the ash of Kamchatkan volcanoes: International Geological Reviews, v. 30, p. 1288–1291.

Ghevariya, Z. G., 1988, Intertrappean dinosaurian fossils from Anjar area,

Kachchh district, Gujarat: Current Science, v. 57, p. 248–251.

Gilmour, I., and Anders, E., 1989, Cretaceous-Tertiary boundary event: Evidence for a short time scale: Geochimica et Cosmochimica Acta, v. 53, p. 503–511.

Hildebrand, A. R., Penfield, G. T., Kring, D. A., Pilkington, M., Carmargo, Z. A., Jacobson, S. B., and Boynton, W. V., 1991, Chicxulub crater: A possible Cretaceous/Tertiary boundary impact on the Yucatan peninsula, Mexico: Geology, v. 19, p. 867–871.

Izett, G. A., Dalrymple, G. B., and Snee, L. W., 1991. $^{40}Ar/^{39}Ar$ age of Cretaceous-Tertiary Boundary tektites from Haiti: Science, v. 252, p. 1539–1542.

Krog, T. E., Kamo, S. L., Sharpton, V. L., Marin, L. E., and Hildebrand, A. R., 1993, U-Pb ages of single shocked zircons linking distal K/T ejecta to the Chicxulub crater: Nature, v. 366, p. 731–734.

Liu, Y. G., Miah, M. R. U., and Schmitt, R. A., 1988, Cerium: A chemical tracer for paleooceanic redox conditions: Geochimica et Cosmochimica Acta, v. 52, p. 1361–1371.

MacRae, N. D., Nesbitt, H. W., and Kronberg, B. I., 1992, Development of a positive Eu anomaly during diagenesis: Earth and Planetary Science Letters, v. 109, p. 585–591.

Mclean, D. M., 1985, Deccan Traps mantle degassing in the terminal Cretaceous marine extinction: Cretaceous Research, v. 6, p. 235–259.

Officer, C. B., and Drake, C. L., 1985, The terminal Cretaceous environmental events: Science, v. 227, p. 1161–1167.

Olmez, I., Finnegan, D. L., and Zoller, W. H., 1986, Iridium emissions from Kilauea volcano: Journal of Geophysical Research, v. 91, p. 653–663.

Orth, C. J., Attrep, M., Jr., and Quintana, L. R., 1990, Iridium abundance patterns across bioevent horizons in the fossil record: Geological Society of America Special Paper 247, p. 45–59.

Prinn, R. G., and Fegley, B., 1987, Bolide impacts, acid rain and biospheric traumas at the Cretaceous-Tertiary boundary: Earth and Planetary Science Letters, v. 83, p. 1–15.

Rampino, M. R., and Caldeira, K., 1993, Major episodes of geological change: Correlations, time structure and possible causes: Earth and Planetary Science Letters, v. 114, p. 215–227.

Robin, E., Boclet, D., Bonté, Ph., Froget, L., Jéhanno, C., and Rocchia, R., 1991, The stratigraphic distribution of Ni-rich spinels in the Cretaceous-Tertiary boundary at El Kef (Tunisia), Caravaca (Spain) and Hole 761C (Leg 122): Earth and Planetary Science Letters, v. 107, p. 715–721.

Rocchia, R., Boclet, D., Courtillot, V., and Jaeger, J. J., 1988, A search for iridium concentration in the deccan traps and intertraps: Geophysical Research Letters, v. 15, p. 812–815.

Schmitz, B., 1992, Chalcophile element and iridium in continental Cretaceous-Tertiary boundary clays from the western interior of the U.S.A.: Geochimica et Cosmochimica Acta, v. 56, p. 1695–1703.

Sigurdsson, H., D'Hondt, S., and Carey, S., 1992. The impact of the Cretaceous/Tertiary bolide on evaporite terraine and generation of major sulphuric acid aerosol: Earth and Planetary Science Letters, v. 109, p. 543–559.

Smit, J., and Klaver, G., 1981, Sanidine spherules at the C/T boundary indicate a large impact event: Nature, v. 292, p. 47–49.

Sutherland, F. L., 1994, Volcanism around K/T boundary time—its role in an impact scenario for the K/T extinction events: Earth Science Reviews, v. 36, p. 1–26.

Venkatesan, T. R., Pande, K., and Gopalan, K., 1993, Did Deccan volcanism predate the K/T transition?: Earth and Planetary Science Letters, v. 119, p. 181–189.

Zoller, W. H, Parrington, J. R., and Phelan Kotra, J. M., 1983, Iridium enrichment in air-borne particulates from Kilauea volcano: January, 1983: Science, v. 222, p. 1118–1121.

Manuscript Accepted by the Society September 14, 1995

Printed in U.S.A.

Geological Society of America
Special Paper 307
1996

Formation conditions of oxidized Ni-rich spinel and their relevance to the K/T boundary event

J. Gayraud, E. Robin, R. Rocchia, and L. Froget
Centre des Faibles Radioactivités, Laboratoire mixte CEA/CNRS, Avenue de la Terrasse, 91198 Gif-sur-Yvette cedex, France

ABSTRACT

Highly oxidized Ni-rich spinel crystals have been found throughout the world at the Cretaceous-Tertiary (K/T) boundary. Although these crystals are believed to derive from the K/T impactor, their formation conditions remain obscure. In this chapter, we present experimental results concerning the crystallization of oxidized Ni-rich spinel from melts of chondritic composition. The effects of varying oxygen fugacity, temperature, duration of the heat exposure, and cooling rate on the morphology, abundance, and composition of spinel have been studied for two types of chondrite, a CI-type carbonaceous chondrite (Orgueil) and an LL6-type ordinary chondrite (Saint Severin). This study supports the view that the spinel crystals reported worldwide at the K/T boundary are derived from meteoritic material that equilibrated at $f_{O_2} \geq 10^{-2}$ bar and T ≤ 1520 °C. Such formation conditions rule out derivation of K/T spinel from known volcanic or impact processes. In particular, K/T spinel cannot be formed in terrestrial magmas that evolved under extremely low oxygen fugacities, near the fayalite-magnetite-quartz (FMQ) buffer. For the same reason, direct crystallization from impact melt rocks or direct condensation from the cloud of vaporized target and projectile material can be excluded. We propose that spinel crystals from the K/T boundary have crystallized in liquid ablation droplets produced during the interaction with the atmosphere of large meteoroids decelerated at low altitudes, under high oxygen fugacities. Our scenario implies multiple meteoroid impacts all over the Earth, resulting either from the fragmentation of the K/T impactor upon an oblique impact or from the encounter with a fragmented comet.

INTRODUCTION

The most popular theory accounting for the high iridium enrichment (and other siderophile element enrichments) observed worldwide at the Cretaceous-Tertiary (K/T) boundary is the collision with an asteroid or a comet (Alvarez et al., 1980; Ganapathy, 1980; Kyte et al., 1980; Smit and Hertogen, 1980; Hsü, 1980). This theory was subsequently and masterfully confirmed by the discovery in the K/T clay layer of abundant quartz grains exhibiting multiple sets of planar deformation features (Bohor et al., 1984; Bohor et al., 1987), which are unique and specific features in quartz from terrestrial impact craters (see on this topic the recent work of Goltrant et al., 1991, 1992). More recently, field investigations all around the Gulf of Mexico (Smit et al., 1992; Stinnesbeck et al., 1993) and in the Caribbean area (Hildebrand and Boynton, 1990; Jéhanno et al., 1992) reported the existence of a massive spherule bed underlying the K/T clay layer. This spherule bed contains glass particles that are interpreted as impact glasses (Izett, 1991; Sigurdsson et al., 1991; Kring and Boynton, 1991; Blum and Chamberlain, 1992; Koeberl, 1994), though this interpretation is still controversial (Lyons and Officer, 1992; Jéhanno et al., 1992; Robin et al., 1994b). Several lines of evidence (Hildebrand et al., 1991; Kring and Boynton, 1992; Swisher et al.,

Gayraud, J. Robin, E., Rocchia, R., and Froget, L., 1996, Formation conditions of oxidized Ni-rich spinel and their relevance to the K/T boundary event, *in* Ryder, G., Fastovsky, D., and Gartner, S., eds., The Cretaceous-Tertiary Event and Other Catastrophes in Earth History: Boulder, Colorado, Geological Society of America Special Paper 307.

1992; Sharpton et al., 1992; Blum et al., 1993; Krogh et al., 1993; Koeberl et al., 1994) suggest that the buried circular structure near Chicxulub on the north coast of Yucatan (Penfield and Camargo, 1981; Pope et al., 1991; Hildebrand et al., 1991) is the impact site.

In addition to Ir, shocked quartz, and glass particles, spinel crystals with unusual, if not unique, compositions in the spinel group (Smit and Kyte, 1984; Kyte and Smit, 1986), have been found throughout the world at the K/T boundary: in marine sections in Spain (Bohor et al., 1986; Jéhanno et al., 1987), in Italy (Montanari et al., 1983; Smit and Kyte, 1984; Kyte and Smit, 1986; Jéhanno et al., 1987; Robin et al., 1994a), in France (Robin et al., 1994a), in Austria (Robin et al., 1994a), in Tunisia (Robin et al., 1991), in Haiti (Jéhanno et al., 1992), in Mexico (Rocchia et al., 1994), and in New Zealand (Bohor et al., 1986); in continental sections in North America in Wyoming (Bohor and Foord, 1987); and in deep sea cores in the Gulf of Mexico (Rocchia et al., 1994) and in the Pacific (Kyte and Smit, 1986; Robin et al., 1993; Kyte et al., 1994), Atlantic (Kyte and Smit, 1986), and Indian Oceans (Rocchia et al., 1992). These minerals are, therefore, an important marker of the K/T boundary, as important as Ir, shocked quartz, and glass particles. However, the question of their origin is still a subject of debate.

K/T spinel differs from spinel present in terrestrial and extraterrestrial rocks (Haggerty, 1976; Rumble, 1976; El Goresy, 1976a; El Goresy, 1976b) by its high Ni content and high Fe^{3+}/Fe_{total} ratio (Smit and Kyte, 1984; Kyte and Smit, 1986; Bohor et al., 1986; Robin et al., 1992). As such chemical characteristics are unknown in volcanic spinel, the K/T spinel was interpreted as relict mineral from the K/T impact event. The high Ni content argues for derivation of the K/T spinel from predominantly projectile material (Bohor et al., 1986; Robin et al., 1992), and potential sources include vapor condensate (Kyte and Smit, 1986; Bohor et al., 1986; Kyte et al., 1994) and projectile ablation debris (Kyte and Smit, 1986; Robin et al., 1992). Bohor et al. (1986), Kyte and Smit (1986), and Kyte et al. (1994) favor derivation of K/T spinel directly from a vapor phase composed either of pure projectile material or of mixed projectile and target material, whereas Robin et al. (1992, 1993) favor derivation from ablation in the Earth's atmosphere of projectile material. Although there is no consensus about the mode of formation of the K/T spinel, the authors cited above consider oxidizing conditions, and in particular high oxygen fugacities, as an important factor for spinel crystallization. In this chapter we present experimental results concerning the equilibrium crystallization of oxidized Ni-rich spinel that shed light on the formation and, hence, the source of the K/T spinel.

EXPERIMENTAL TECHNIQUES

Spinels were synthetized by melting a few milligrams of meteorite in a platinum crucible enclosed in a quartz cell and placed in the coil of a high-frequency generator (Fig. 1). This technique permits peak temperatures as high as 2500 °C and cooling within a few seconds. The principal advantage of this technique is that it permits the study of the kinetic of oxidation and mineral crystallization by running short (a few seconds) experiments. Short experimental run times are required in order to simulate as closely as possible the short pulse heating (a few seconds; see Hills and Goda, 1993) of a meteoroid decelerating in the Earth's atmosphere. By contrast, experimental runs of longer (a few hours) duration provided information about equilibrium conditions for spinel crystallization. Another advantage of this technique is that only the crucible (and therefore the charge therein contained) is heated, whereas the quartz cell remains at ambient temperature. This allows one to cool the samples within a few tens of seconds (cooling rate ≈50 °C/s) at the pressure of the experiment (from 10^{-6} to $10^{0.7}$ bar, see below), simply by switching the generator off. We believe this closely reproduces the cooling conditions for meteoroid ablation droplets falling in the Earth's atmosphere. Undercooled melts were obtained by quenching the samples either in water or liquid nitrogen. This was done in air at different temperatures for determining the temperature of first appearance, or liquidus, of the different mineral phases. In such experiments, the platinum crucible was suspended on a platinum wire in the coil of the high frequency generator, and at the end of the heat exposure period, the crucible was dropped into a container filled with water or liquid nitrogen.

Crucible material

We chose a platinum crucible for several reasons. Platinum has a high melting temperature (1773 °C) and conductivity and does not react readily with oxygen. The major problem with platinum is the iron loss from the charges (Grove, 1981). However, this problem is important only at low oxygen fugacity ($f_{O_2} < 10^{-6}$ bar) and for long experiments (several tens of hours) (for further detailed information on this problem see Grove, 1981). At $f_{O_2} > 10^{-6}$ bar and for experimental run times of less than a few hours, iron loss is insignificant, as shown by the constant Fe/Si ratio between melted and unmelted samples (Table 1). The crucible was kept open: first, to enable the charge to react with the gas flow through the quartz cell and, second, to permit temperature measurements on the surface of the charge.

Temperature measurements

The temperature was measured with an optical pyrometer calibrated at the melting points of gold (1063 °C), diopside (1391.5 °C), palladium (1555 °C), anorthite (1557 °C), platinum (1773 °C), and forsterite (1910 °C). The range of temperatures studied in the experiments varies from 1335 °C to 1670 °C. The absolute temperature measurements are accurate to ±10 °C in the range 1335 to 1400 °C and ±20 °C in the range 1400 to 1670 °C. Most experimental techniques employ Pt/Rh thermocouples, which are more accurate (usual precision is ±4 °C; see Hill and Roeder, 1974). However, this technique

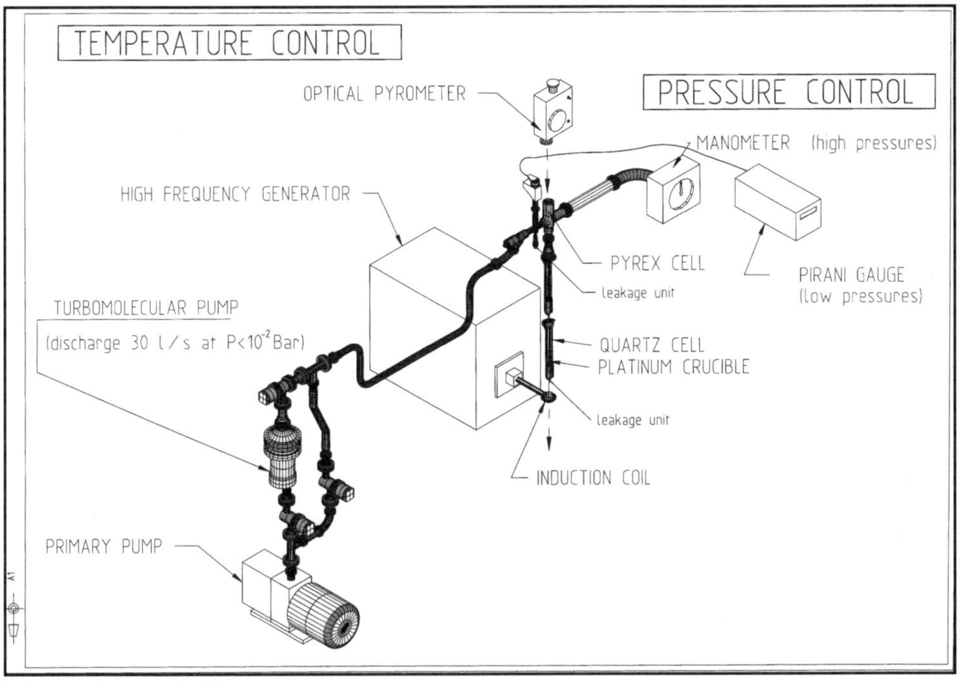

Figure 1. Sketch of apparatus for short heating experiments (a few seconds) under controlled temperatures (from 1000 to 1700 °C) and oxygen fugacities (from 10^{-7} bar up to 10^{1} bar).

TABLE 1. AVERAGE COMPOSITIONS OF SAMPLES OF SAINT SEVERIN LL6 CHONDRITE AND ORGUEIL C1 CHONDRITE*

	Saint Severin					Orgueil				
	Meteorite composition†	Powder§ n**=15		Melt§ n** =15		Meteorite composition†	Powder§ n** = 8		Melt§ n** = 8	
		< >	(σ)‡	< >	(σ)‡		< >	(σ)‡	< >	(σ)‡
SiO_2	41.5	40.6	(1.3)	39.4	(1.6)	32.2	34.6	(1.4)	33.5	(2.5)
TiO_2	0.1	0.1	(0.1)	0.1	(0.1)		0.1	(0.1)	0.1	(0.1)
Al_2O_3	2.5	2.9	(0.2)	2.0	(0.5)	2.3	2.5	(0.2)	2.1	(0.3)
Cr_2O_3	0.6	0.7	(0.1)	0.7	(0.2)	0.6	0.6	(0.1)	0.6	(0.1)
Fe_2O_3§§				28.2	(1.2)				36.9	(3.0)
FeO§§	18.9	20.9	(1.0)			17.0	16.4	(1.1)		
FeS§§	5.9	8.4	(2.5)			20.4	18.6	(1.4)		
MnO	0.3	0.3	(0.1)	0.4	(0.1)	0.4	0.3	(0.1)	0.3	(0.1)
MgO	25.8	22.7	(0.9)	25.7	(1.5)	22.3	22.6	(1.1)	22.6	(0.7)
CaO	1.9	2.2	(0.3)	1.8	(0.3)	1.8	1.6	(0.2)	1.6	(0.2)
Na_2O	1.0	0.9	(0.5)	0.6	(0.2)	1.0	0.5	(0.3)	0.4	(0.3)
K_2O	0.1	0.1	(0.1)	0.1	(0.1)		0.1	(0.1)	0.1	(0.1)
NiO	1.4	0.2	(0.1)	1.0	(0.3)	2.0	2.1	(0.1)	1.8	(0.3)
Total	100.0	100.0		100.0		100.0	100.0		100.0	
Fe/Si§§	484.0	578.0	(33.0)	543.0	(32.0)	881.0	771.0	(50.0)	837.0	(92.0)

*All data in wt. %, except Fe/Si in Fe atoms/10^3 Si atoms.
†Meteorite composition normalized to 100%. Original compositions are from Jarosewich, 1990, for Saint Severin and from Anders and Ebihara, 1982, for Orgueil.
§Compositions were obtained from defocused EDS microprobe analyses (see text).
** n is the number of analyzed samples.
‡< > is the average value and σ is the standard deviation.
§§Fe_2O_3, FeO, and FeS are calculated assuming that all sulfur is in the form of FeS and that the remaining iron is in the ferrous state in the powder and in the ferric state in the melt.

does not permit direct temperature measurements of the charge, as thermocouples are induced in the coil of the high frequency generator. On the other hand, optical pyrometry offers a reasonably high accuracy for the range of temperatures studied. In addition, an uncertainty of ±20 °C of the absolute temperature is small in comparison with the wide range of temperature studied (1335 to 1670 °C).

Pressure control

The oxygen fugacity was controlled by varying the total pressure of an oxygen-rich gas flow through the cell. The gas phase used was air for total pressures from 10^{-6} bar up to 10^{0} bar. Over this range, the total pressure in the cell was fixed as follows: In the first step, the total pressure was lowered well below 10^{-6} bar by evacuating the cell with a turbomolecular pump (discharge ≈30 liters per second at pressures below 10^{-2} bar). We used a turbomolecular pump instead of a classical vacuum pump to minimize contamination of the cell by oil. The fact that the quartz cell was at ambient temperature also prevents contamination of the gas phase by possible outgassing from the quartz, which could occur at higher temperature. In the second step, the total pressure in the cell was then adjusted and stabilized at the pressure of the experiment (≥10^{-6} bar) using two leakage-controlled systems. The oxygen fugacity was assumed to be equivalent to the oxygen partial pressure, assuming that air contains 21 wt% of oxygen. The maximum oxygen fugacity that we can obtain is $f_{O_2} = 10^{-0.68}$ bar. To obtain higher oxygen fugacities, we used a gas cylinder of pure oxygen at 200 bars connected to a pressure-reducer. Total pressures between 10^{-3} bar up to $10^{0.7}$ bar were measured with a manometer, and for lower pressures (down to 10^{-7} bar), a Pirani gauge was used. Total pressure measurements were accurate to ±10% of the reported value.

Starting materials

The starting materials for the runs were charges of 2 to 3 mg of powdered meteorites. Two meteorite samples were used: the first from Saint Severin, an LL6-type chondrite, and the second from Orgueil, a CI-type chondrite. Fifty g and 300 mg of the Saint Severin and Orgueil meteorite were ground in an agate mortar, respectively. No particular control on the grain size was exercised during grinding. In our opinion, it is not a problem regarding Orgueil, as this meteorite is an aggregate of very fine mineral grains (grain size <<1 µm), with only very few large crystals (1 to 10 µm) of olivine, serpentine, and pure magnetite (Fe_3O_4). However, a grain size artifact can be expected with Saint Severin, as this meteorite is mainly composed of large crystals (>100 µm) of olivine, hypersthene, and FeNi metal alloys. In Table 1, we show the bulk chemical composition for 15 charges prepared from Saint Severin powder and for eight charges prepared from Orgueil powder. Bulk chemical compositions were obtained by defocused EDS (electron dispersive spectrometer) microprobe analyses of several large surfaces (as large as 1 × 1 mm; see below for the description of the analytical technique). Only minor heterogeneities were observed from one charge to the other, except for Ni in the case of Saint Severin, which likely reflects the heterogeneous distribution of FeNi metal alloys in the charge. The average composition for Saint Severin powders differs from the composition in the literature, on the basis of a distinctly lower Ni content (0.2 wt% in the powder compared with 1.4 wt% in the meteorite; see Table 1). This difference can be explained, considering that Ni is essentially concentrated in metal particles that are heterogeneously distributed in the powder. This effect is not observed on the fused samples, the metal particles being dissolved in the liquid, and the measured Ni concentrations (≈1 wt%) are consistent with those given in the literature. Our measured composition of Orgueil, for both the powder and fused samples, is consistent with that found in the literature (Table 1).

Analytical techniques

The phase assemblages, that is, the different mineral and amorphous phases present in the fused and recrystallized samples, were determined by X-ray powder diffraction (XRD), using a Philips goniometer equipped with a copper anode tube. Samples were mounted in epoxy and polished for scanning electron microscopy and electron microprobe analysis.

The spinel crystals were analyzed in the same way as reported by Hill and Roeder (1974) for spinels synthetized from basaltic liquids. We used a LINK-AN10000 X-ray EDS attached to a JEOL JSM 840 scanning electron microscope (SEM). Multielement spectra were obtained using a lithium-drifted silicon detector with a resolution of 148 eV for the manganese Kαline. The spectra were compared to a series of pure reference spectra, and X-ray absorption and fluorescence effects were corrected using the ZAF-4 program supplied by Link Analytical System. This enabled us to carry out quantitative analyses for the elements Na, Mg, Al, Si, S, K, Ca, Ti, Cr, Mn, Fe, and Ni simultaneously. The sensitivity of this technique is 0.1 wt% for most of the elements analyzed, except for Na, which is 0.3 wt% because of the preferential absorption of low-energy X-rays in the window of the detector.

We used an automated search routine for the counting and the chemical analysis of spinel crystals. The routine selects features of the proper density (spinel density is higher than matrix density) and size (>2 µm). The mass abundance of spinel was considered as proportional to the fraction of surface occupied by the spinel, assuming that the total amount of spinel observed in a given section reflects the total amount of spinel present in the volume of the sample. We used a coefficient of proportionality of 5/3, corresponding to the density difference between spinel and matrix, assuming a density of 5 for the spinel (magnetite density: 4.9 to 5.2; Parfenoff et al., 1970) and a density of 3 for the matrix (the matrix contains forsterite crystals, density of 3.2, and SiO_2-rich glass with a density around 2.7, Parfenoff et al., 1970).

The oxide compositions of the spinel crystals were determined as follows:

1. We considered only the divalent cations Mg^{2+}, Mn^{2+}, Fe^{2+}, and Ni^{2+}; the trivalent cations Al^{3+}, Cr^{3+}, and Fe^{3+}; and the tetravalent cation Ti^{4+} as being in the spinel structure. We systematically discarded data on spinels that contained a significant contribution from the silicate matrix. This is the reason why spinel crystals <2 μm were not analyzed, as there was a major contribution from the surrounding matrix to the spinel analysis because of the limited resolution of the probe (1 to 2 μm). For spinels >2 μm, this contribution remains low and was corrected from the Ca and Si contents of the matrix, assuming that these elements are absent in the spinel.

2. We assumed that spinel crystals have a stoichiometric composition, and we considered the following spinel end members: spinel *sensu stricto* ($Mg^{2+}[Al^{3+}]_2O_4$), hercynite ($Fe^{2+}[Al^{3+}]_2O_4$), picrochromite ($Mg^{2+}[Cr^{3+}]_2O_4$), chromite ($Fe^{2+}[Cr^{3+}]_2O_4$), magnesioferrite ($Fe^{3+}[Mg^{2+}Fe^{3+}]O_4$), magnetite ($Fe^{3+}[Fe^{2+}Fe^{3+}]O_4$), trevorite ($Fe^{3+}[Ni^{2+}Fe^{3+}]O_4$), jacobsite ($Fe^{3+}[Mn^{2+}Fe^{3+}]O_4$), and ulvöspinel ($Fe^{2+}[Fe^{2+}Ti^{4+}]O_4$).

3. The concentrations of the cations Mg^{2+}, Mn^{2+}, Ni^{2+}, Al^{3+}, Cr^{3+}, and Ti^{4+} and of total iron are known from the EDS analyses, and the concentrations of oxygen and ferrous and ferric iron were determined solving the following system of equations (all elements are expressed in atom%):

$$Fe^{2+} + Fe^{3+} = Fe_{total}$$
$$2(Mg^{2+} + Mn^{2+} + Fe^{2+} + Ni^{2+}) = Al^{3+} + Cr^{3+} + Fe^{3+} + 4Ti^{4+}$$
$$O = Mg^{2+} + Mn^{2+} + Fe^{2+} + Ni^{2+} + 1.5(Al^{3+} + Cr^{3+} + Fe^{3+}) + 2Ti^{4+}$$

4. The oxides MgO, Al_2O_3, TiO_2, Cr_2O_3, MnO, FeO, Fe_2O_3, and NiO were then expressed in weight % and normalized to 100%. The entire analytical procedure was checked with spinel crystals of known compositions (magnetite, titano-magnetite, chromite, and spinel *sensu stricto*), and we found that the recalculated oxide concentrations are correct to better than 5% (1 sigma) of the known values. Such a precision is sufficient, as it is within the statistical variation in spinel composition observed from crystal to crystal for the same run.

EXPERIMENTAL RESULTS

Experimental data for 41 runs of Saint Severin and for eight runs of Orgueil are presented in Table 2. These data cover a wide range of oxygen fugacities (from $10^{-6.52}$ to $10^{0.70}$ bar), temperatures (from 1335 °C to 1670 °C), and experimental run times (from 10 seconds to 4 hours). Spinel crystals, with high nickel contents, together with olivine and glass were observed in all samples (Table 2). Pyroxene and FeNi metal alloys were not observed. The morphology, abundance, and chemical composition of spinel crystals vary with oxygen fugacity, temperature, cooling rate, and initial composition of the crystallizing material. In the following sections, we also present results on the composition and morphology of olivine crystals coexisting with spinel in the liquid.

Rate of oxidation

Most of the results reported below are for experimental run times (heating, heat exposure, and cooling) of ≈2 minutes at $f_{O_2} \geq 10^{-4}$ bar, ≈20 minutes at $f_{O_2} = 10^{-4.7}$ bar, and ≈60 minutes at $f_{O_2} = 10^{-6.52}$ bar. At $f_{O_2} = 10^{-0.68}$ bar and T = 1335 °C, no change either in the abundance, composition, and Fe^{3+}/Fe_{total} ratio of spinel or in the composition and fayalite content of olivine (Table 3 and Fig. 2) is observed at any experimental run times, even for run times as short as a few seconds. At $f_{O_2} = 10^{-4.7}$ bar, and T = 1335 °C, a change in the Fe^{3+}/Fe_{total} ratio of spinel and in the fayalite content of olivine is observed (Fig. 2) for experimental run times shorter than ≈20 minutes, and this time increases with decreasing oxygen fugacity, reaching ≈60 minutes at $f_{O_2} = 10^{-6.52}$. In no case does this reflect a change in the kinetic of oxidation, though the oxidation rate is probably lower at low oxygen fugacity than at high oxygen fugacity. In fact, the change in the Fe^{3+}/Fe_{total} ratio of spinel and in the fayalite content of olivine is likely a consequence of the limited flow of oxygen through the cell at low pressures. Indeed, in order to have complete oxidation of the charge, the mass of oxygen must be in excess compared with the mass of the charge (at least 10 times higher in order to maintain a constant pressure). Considering a charge of ≈2 mg and a flux of oxygen flowing through the cell of ≈5 g/mn at 10^{-2} bar total pressure (pump discharge ≈30 liters per second), oxidation is completed in less than 1 minute at total pressures higher than 10^{-4} bar, about 4 minutes at 10^{-5} bar, and at least 40 minutes at 10^{-6} bar. In a preliminary study, we reported a marked change in the abundance, chemical composition, and Fe^{3+}/Fe_{total} ratio of spinel formed at total pressures below 10^{-4} bar (Gayraud et al., 1994). This change was probably due to incomplete oxidation of the charge, resulting from too short experimental run times (≈1 to 2 minutes).

Spinel and olivine morphology

The morphology of spinel crystals essentially depends on the temperature and cooling rate. At relatively low cooling rates (≈50 °C/s for samples cooled in air at the pressure of the experiment) and T = 1335 °C, massive and relatively large octahedral spinel crystals (up to 20 μm), with only a few dendrites (crystals with multiple cross-arms; see Figs. 3b, 3c, and 3d), and large euhedral olivine crystals (up to 30 μm) are observed (Figs. 3a and 3b), even for experimental run times as short as a few seconds. At higher temperatures, up to 1500 °C, the abundance of octahedral spinel crystals decreases, whereas that of dendrites increases (Figs. 3b and 3c). At T ≥ 1520 °C, very fine and long (entire length up to 100 μm) dendritic spinel and "barred" olivine crystals are observed (Fig. 3d).

At high cooling rates (samples quenched in water and liquid nitrogen) and at $f_{O_2} = 10^{-0.68}$ bar, the same assemblage of

TABLE 2. EXPERIMENTAL RUNS WITH SAINT SEVERIN AND ORGUEIL CHONDRITES IN ORDER OF INCREASING TEMPERATURES

METEORITE	Run No.	T°C	$\log_{10}P_{tot}$ f_{O_2} (in bar)	$\log_{10}f_{O_2}$	Run time (minute)	Cooling	Phase present *
SAINT SEVERIN	SEV-6A	1335	-5.82	-6.52	60	in air	ol (p) + sp (o) + liq
	SEV-118A	1335	-4.00	-4.70	2	in air	ol (p) + sp (o) + liq
	SEV-119A	1335	-4.00	-4.70	15	in air	ol (p) + sp (o) + liq
	SEV-120A	1335	-4.00	-4.70	30	in air	ol (p) + sp (o) + liq
	SEV-121A	1335	-4.00	-4.70	60	in air	ol (p) + sp (o) + liq
	SEV-122A	1335	-4.00	-4.70	180	in air	ol (p) + sp (o) + liq
	SEV-84A	1335	-3.70	-4.40	2	in air	ol (p) + sp (o) + liq
	SEV-85A	1335	-3.52	-4.22	2	in air	ol (p) + sp (o) + liq
	SEV-12A	1335	-3.30	-4.00	2	in air	ol (p) + sp (o) + liq
	SEV-7A	1335	-3.00	-3.70	2	in air	ol (p) + sp (o) + liq
	SEV-13A	1335	-2.30	-3.00	2	in air	ol (p) + sp (o) + liq
	SEV-15A	1335	-1.30	-2.00	2	in air	ol (p) + sp (o) + liq
	SEV-16A	1335	-1.00	-1.70	2	in air	ol (p) + sp (o) + liq
	SEV-17A	1335	-0.30	-1.00	2	in air	ol (p) + sp (o) + liq
	SEV-161A	1335	0.00	-0.68	0.16	in air	ol (p) + sp (o) + liq
	SEV-3A	1335	0.00	-0.68	2	in air	ol (p) + sp (o) + liq
	SEV-152A	1335	0.00	-0.68	15	in air	ol (p) + sp (o) + liq
	SEV-151A	1335	0.00	-0.68	30	in air	ol (p) + sp (o) + liq
	SEV-150A	1335	0.00	-0.68	60	in air	ol (p) + sp (o) + liq
	SEV-144A	1335	0.00	-0.68	180	in air	ol (p) + sp (o) + liq
	SEV-66A	1335	+0.48	-0.22	2	in air	ol (p) + sp (o) + liq
	SEV-67A	1335	+0.70	0.00	2	in air	ol (p) + sp (o) + liq
	SEV-65A	1335	+1.18	+0.48	2	in air	ol (p) + sp (o) + liq
	SEV-68A	1335	+1.40	+0.70	2	in air	ol (p) + sp (o) + liq
	SEV-153A	1380	0.00	-0.68	10	in liquid N_2	ol (p) + sp (o) + liq
	SEV-23A	1390	-3.00	-3.70	2	in air	ol (p) + sp (o) + liq
	SEV-154A	1390	0.00	-0.68	10	in water	ol (p) + sp (o) + liq
	SEV-29A	1390	0.00	-0.68	2	in air	ol (p) + sp (o) + liq
	SEV-35A	1445	-3.00	-3.70	2	in air	ol (p) + sp (o) + liq
	SEV-41A	1445	0.00	-0.68	2	in air	ol (p) + sp (o) + liq
	SEV-158A	1470	0.00	-0.68	10	in water	ol (p+b) + sp (o+d) + liq
	SEV-159A	1490	0.00	-0.68	10	in water	ol (p+b) + sp (o+d) + liq
	SEV-46A	1500	-3.00	-3.70	2	in air	ol (p) + sp (o+d) + liq
	SEV-157A	1500	0.00	-0.68	0.16	in liquid N_2	ol (p+b) + sp (o+d) + liq
	SEV-52A	1520	0.00	-0.68	2	in air	ol (b) + sp (d) + liq
	SEV-155A	1520	0.00	-0.68	10	in water	liq
	SEV-160A	1520	0.00	-0.68	10	in water	liq
	SEV-58A	1560	-3.00	-3.70	2	in air	ol (p) + sp (o+d) + liq
	SEV-64A	1560	0.00	-0.68	2	in air	ol (b) + sp (d) + liq
	SEV-156A	1630	0.00	-0.68	10	in water	liq
	SEV-65Abis	1670	0.00	-0.68	2	in air	ol (b) + sp (d) + liq
ORGUEIL	ORG-125A	1335	-3.30	-4.00	2	in air	ol (p) + sp (o) + liq
	ORG-126A	1335	-3.00	-3.70	2	in air	ol (p) + sp (o) + liq
	ORG-127A	1335	-3.00	-3.00	2	in air	ol (p) + sp (o) + liq
	ORG-129A	1335	-2.00	-2.00	2	in air	ol (p) + sp (o) + liq
	ORG-130A	1335	-1.00	-1.70	2	in air	ol (p) + sp (o) + liq
	ORG-131A	1335	-0.30	-1.00	2	in air	ol (p) + sp (o) + liq
	ORG-132A	1335	0.00	-0.68	2	in air	ol (p) + sp (o) + liq
	ORG-134A	1335	+0.70	0.00	2	in air	ol (p) + sp (o) + liq

* Dominant phase: ol = olivine; sp = spinel; liq = liquid.
Structure: p = porphyric; o = octahedral; b = barred; d = dendritic.

TABLE 3. AVERAGE COMPOSITIONS OF SPINEL AND OLIVINE CRYSTALS SYNTHETIZED FROM SAINT SEVERIN LL6 CHONDRITE AT TWO DIFFERENT EXPERIMENTAL RUN TIMES, AT T =1335°C AND AT f_{O_2} = $10^{-0.68}$ BAR*

	Spinel				Olivine			
t in minute	2		180		2		180	
	n† = 106		n† = 90		n† = 5		n† = 5	
	< >	(σ)§	< >	(σ)§	< >	(σ)§	< >	(σ)§
TiO_2	0.1	(0.1)	0.1	(0.1)	0.1	(0.1)	0.1	(0.1)
SiO_2					40.8	(0.7)	40.4	(0.2)
Al_2O_3	0.9	(0.3)	1.5	(0.4)	0.1	(0.1)	0.1	(0.1)
Cr_2O_3	3.1	(4.6)	2.0	(0.5)	0.1	(0.1)	0.1	(0.1)
Fe_2O_3**	70.9	(5.4)	70.8	(1.3)				
FeO**	7.5	(5.4)	9.4	(2.6)	6.6	(0.8)	5.5	(0.3)
MnO	0.4	(0.1)	0.5	(0.1)	0.3	(0.1)	0.3	(0.1)
MgO	12.1	(3.7)	11.1	(1.7)	50.4	(0.5)	50.1	(0.5)
NiO	5.0	(1.1)	4.6	(0.6)	1.6	(0.1)	3.4	(0.5)
Total	100.0		100.0		100.0		100.0	
Fe^{3+}/Fe_{total}	89.7	(8.1)	87.1	(3.3)				
Fayalite					8.4	(0.7)	9.0	(0.3)
Abundance	20.0		20.0		50.0		50.0	
Crystal type	euhedral		euhedral		euhedral		euhedral	

*All data in wt. %, except Fe^{3+}/Fe_{total} in atom % and fayalite in mole %.
†n is the number of crystals analyzed.
§< > is the average value and σ is the standard deviation
**Fe_2O_3 and FeO are calculated assuming a stoechiometric composition for spinel and olivine (see text).

euhedral spinel and olivine crystals is observed over the range of temperatures 1335 to 1500 °C. The abundance of octahedral crystals also decreases and that of dendritic crystals increases as the temperature of 1500 °C is approached. At T > 1520 °C, we did not observe any spinel or olivine crystals. This defines the onset of crystallization of olivine and spinel, or liquidus, between 1500 °C and 1520 °C at f_{O_2} = $10^{-0.68}$ bar. It was impossible to define the liquidus of a separate phase, olivine or spinel, at f_{O_2} = $10^{-0.68}$ bar, as both minerals seem to appear in a very narrow temperature range of 1500 to 1520 °C. As the oxygen fugacity is lowered, the liquidus of spinel and olivine increases and is above 1560 °C at f_{O_2} = $10^{-3.7}$ bar.

Spinel and olivine abundance

The abundance of spinel crystals depends on oxygen fugacity, temperature, cooling rate, and initial composition of the crystallizing material. At low cooling rates (samples cooled in air at the pressure of the experiment), the abundance of spinel only depends on oxygen fugacity and initial iron content of the crystallizing material. The lower the oxygen fugacity and the initial iron content of the crystallizing material, the lower the spinel abundance. For instance, the abundance of spinel synthetized from Saint Severin, at T = 1335 °C, varies from ≈1 wt% at f_{O_2} = $10^{-6.52}$ bar to ≈20 wt% at f_{O_2} = 10^{-2} bar and then remains nearly constant up to f_{O_2} = $10^{0.7}$ bar (Fig. 4). A similar trend is observed with Orgueil (Fig. 4), but the abundance of spinel is higher, even for lower oxygen fugacities, as a result of a higher initial iron content in this meteorite (see Table 1). At high cooling rates (samples quenched in water and liquid nitrogen), the spinel abundance also decreases with decreasing oxygen fugacity and initial iron content of the crystallizing material, but all the more rapidly as the temperature approaches the spinel liquidus (≈1500 to 1520 °C at f_{O_2} = $10^{-0.68}$ bar). Above liquidus, the abundance of spinel crystals drops to zero.

Olivine crystals are abundant in all samples (≈50 wt%), except in the samples heated above liquidus and quenched in water and liquid nitrogen, which are exclusively glassy.

Spinel and olivine composition

The composition of spinel depends essentially on oxygen fugacity and initial composition of the crystallizing material. The composition of spinel varies continuously as a function of oxygen fugacity (Fig. 5, Tables 4 and 5): the higher the oxygen fugacity, the higher the MgO, Fe_2O_3, and NiO contents and the lower the Al_2O_3, Cr_2O_3, and FeO contents. Consequently, the Fe^{3+}/Fe_{total} ratio of spinel also varies continuously as a function of oxygen fugacity, from ≈50 atom% at f_{O_2} = $10^{-6.52}$ bar to ≈90 % at f_{O_2} = $10^{0.7}$ bar (Fig. 6). At a fixed oxygen fugacity, the spinel crystals from Orgueil and Saint Severin runs have roughly the same composition (Tables 4 and 5). Some minor differences are observed that can be attributed to differences in the initial compositions of the meteorites. For instance, the higher nickel content of spinel from Orgueil is likely due to the higher concentration of this element in this meteorite (Table 1).

The difference in the spinel chromium content from Saint Severin and Orgueil cannot be explained in the same way, as both meteorites have the same chromium content ($Cr_2O_3 \approx 0.6$ to 0.7 wt%, Table 1). As chromium is almost entirely contained in spinel (the Cr_2O_3 content in the whole rock sample is 0.6 to 0.7 wt% and 0.5 to 0.6 wt% derived from the spinel contribution), the higher chromium content of spinel from Saint Severin (×2) is accounted for by the two times lower abundance of spinel crystals in the melt sample.

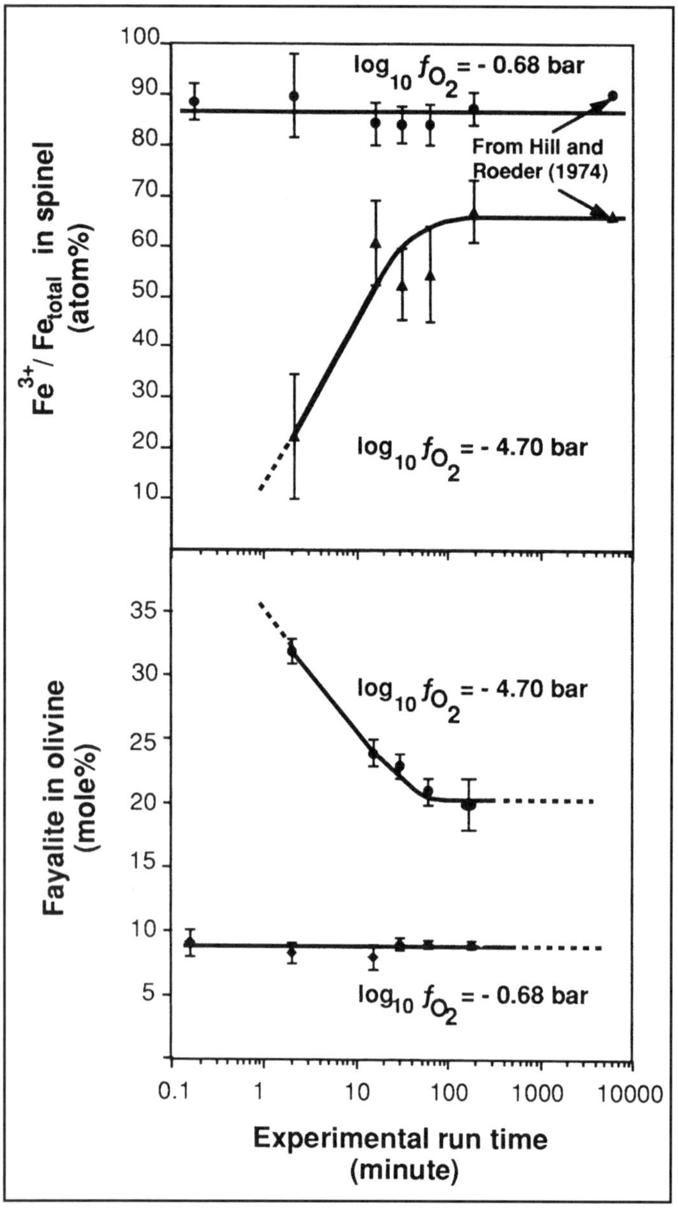

Figure 2. Fe^{3+}/Fe_{total} ratio of spinel (top) and fayalite content of olivine (bottom) in Saint Severin melts as a function of experimental run time at 1335 °C and at two different oxygen fugacities. The vertical bars represent one standard deviation, including the variation in composition from crystal to crystal and the error on the measurement. At high oxygen fugacity, equilibrium is obtained in less than a few seconds and no change in the Fe^{3+}/Fe_{total} ratio of spinel and in the fayalite content of olivine is observed for longer experiments. At low oxygen fugacity, it takes more time to obtain equilibrium, as indicated by the variation with time of the Fe^{3+}/Fe_{total} ratio of spinel and of the fayalite content of olivine, because of the limited oxygen flow through the cell (see text).

No significant variation in the composition and Fe^{3+}/Fe_{total} ratio of spinel is observed with temperature, at $f_{O_2} = 10^{-3.7}$ bar and $f_{O_2} = 10^{-0.68}$ bar (Fig. 7, Tables 6 and 7).

The olivine composition determined at T = 1335 °C also varies as a function of oxygen fugacity (Fig. 8), from $Fa_{7.5}$ at $f_{O_2} = 10^{0.7}$ bar (= 7.5 mole% of fayalite in olivine) to $Fa_{22.5}$ at $f_{O_2} = 10^{-4.7}$ bar and then decreasing at lower oxygen fugacities. Olivine composition also varies as a function of temperature (Fig. 9). At $f_{O_2} = 10^{-0.68}$ bar, the fayalite content in olivine increases from $Fa_{8.5}$ at T = 1335 °C to $Fa_{13.5}$ at T = 1670 °C, but at $f_{O_2} = 10^{-3.7}$ bar the fayalite content in olivine decreases from Fa_{19} at T = 1335 °C, to $Fa_{13.5}$ at T = 1550 °C.

DISCUSSION

The main parameters controlling the morphology, abundance, and chemical composition of spinel in our experiments are oxygen fugacity, temperature, and cooling rates. Let us first consider the effect of changing oxygen fugacity on spinel crystallization.

As several authors have extensively demonstrated (e.g., Fudali, 1965; Hill and Roeder, 1974), the principal effect of changing oxygen fugacity is to increase or decrease the ferric/ferrous ratio of the system. At a fixed temperature, the higher the oxygen fugacity, the higher the ferric iron and the lower the ferrous iron content. As the total ferric iron content of the liquid increases, Fe^{3+} progressively replaces Cr^{3+} and Al^{3+} in the spinel structure, and Ni^{2+} and Mg^{2+} replace Fe^{2+}. The spinel becomes richer in magnetite ($Fe^{3+}[Fe^{2+}Fe^{3+}]O_4$), magnesioferrite ($Fe^{3+}[Mg^{2+}Fe^{3+}]O_4$), and trevorite ($Fe^{3+}[Ni^{2+}Fe^{3+}]O_4$) and poorer in chromite ($Fe^{2+}[Cr^{3+}]_2O_4$) and hercynite ($Fe^{2+}[Al^{3+}]_2O_4$). Consequently, the $Fe^{3+}/(Cr^{3+} + Al^{3+})$ and Fe^{3+}/Fe_{total} ratios of spinel increase continuously with increasing oxygen fugacities. A similar variation of the $Fe^{3+}/(Cr^{3+} + Al^{3+})$ and Fe^{3+}/Fe_{total} ratios of spinel was reported by Hill and Roeder (1974) on spinel crystallizing from basaltic liquid. The composition of spinel crystals formed from this later material is distinguishable from that of spinel crystals reported here. Particularly, the basaltic spinel crystals do not contain Ni because of the low concentration of this element in a basalt (<0.1 wt%; Taylor and McLennan, 1985) and have higher Al and Ti contents (≈10 times higher) as a result of the higher abundance of these elements in basaltic material compared with their abundance in chondritic material (≈10 times higher in a basalt than in a chondrite; Taylor and McLennan, 1985). Despite their chemical differences, the spinel crystals formed from basaltic and chondritic liquids have the same Fe^{3+}/Fe_{total} ratio (see Fig. 6) at a fixed oxygen fugacity, indicating that this ratio is relatively independent of the composition of the starting material. It is not the case for the $Fe^{3+}/(Cr^{3+} + Al^{3+})$ ratio of spinel which also depends on the Fe, Al, and Cr contents of the crystallizing material.

In addition to a change in the spinel composition with oxygen fugacity, there is a change in the number of spinel crystals formed (Fig. 4). The formation of magnetite/magnesioferrite-type spinel crystals is essentially driven by the total amount of ferric iron present in the liquid. As the oxygen fugacity is lowered, the ferric iron content in the liquid, and thus the spinel abundance, decreases. As discussed below, the oxygen fugacity is not the sole parameter controlling the abundance of spinel; the temperature and cooling rates also play an important role.

The change in olivine composition is explained in the same way as the change in spinel composition: the lower the oxygen fugacity, the higher the ferrous iron content of the liquid and the more fayalitic the olivine. However, at low oxygen fugacity, the fayalitic content of olivine reaches a saturation level (Fig. 8) and then decreases, as a result of reduction of ferrous iron to metal iron (Hill and Roeder, 1974).

Let us now consider the effect of changing the temperature and cooling rates on spinel crystallization. As the ferric/ferrous ratio of the liquid varies by changing oxygen fugacity, so this ratio varies by changing the temperature. At a fixed oxygen fugacity, the higher the temperature, the lower the ferric iron and the higher the ferrous iron contents of the liquid. As the temperature increases, the ferric iron content of the system is lowered, and we could expect Fe^{3+} to be progressively replaced by Cr^{3+} and Al^{3+} in the spinel, leading to a decrease of the $Fe^{3+}/(Cr^{3+} + Al^{3+})$ ratio. This effect of the temperature on spinel

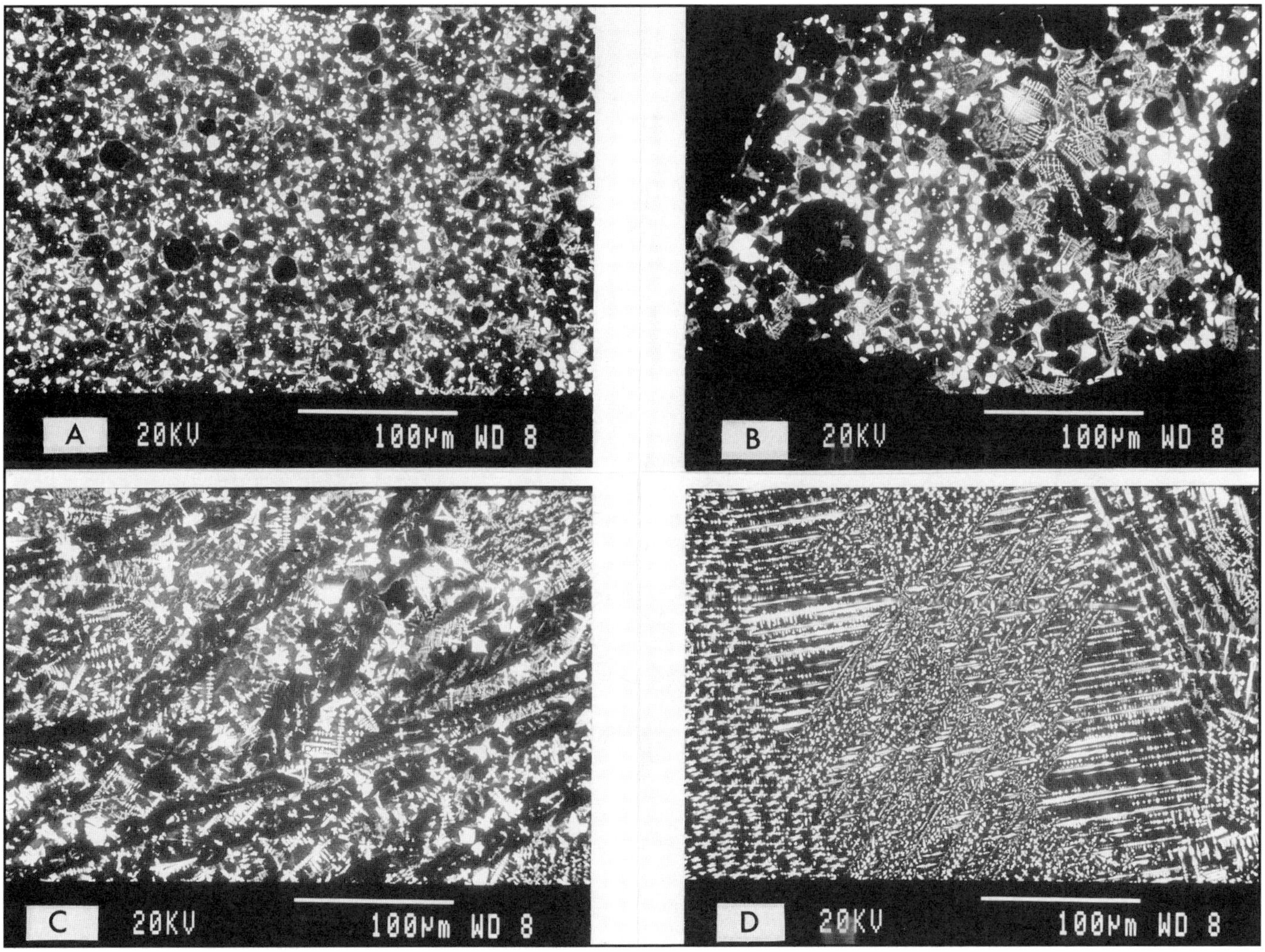

Figure 3. Backscattered electron microscope images of polished sections of Saint Severin melts showing the effect of the temperature on crystal morphology. All samples were fused and cooled in air ($f_{O_2} = 10^{-0.68}$ bar). A, T = 1390 °C. Numerous euhedral crystals of spinel (white) and olivine (dark) are present within glass. B, T = 1470 °C. Two types of spinel crystals, octahedral and dendritic, are visible. C, T = 1500 °C (near liquidus temperature). Spinel crystals are mostly dendritic, and some olivine crystals are barred. D, T = 1670 °C. Only very fine dendritic and barred crystals of spinel and olivine, respectively, are present. Crystals are oriented following parallel directions.

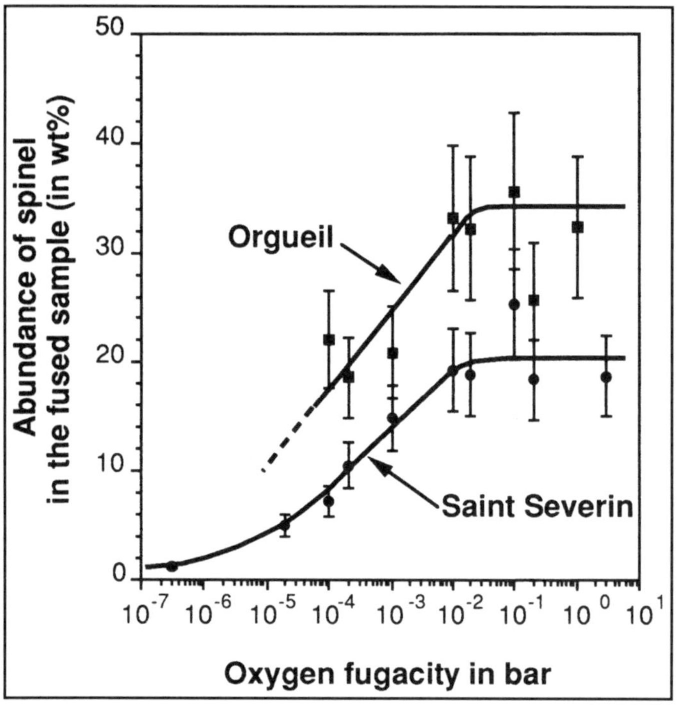

Figure 4. Abundance of spinel crystals versus oxygen fugacity in Saint Severin and Orgueil melts at 1335 °C. The vertical bars represent one standard deviation, including the variation in composition from crystal to crystal and the error on the measurement.

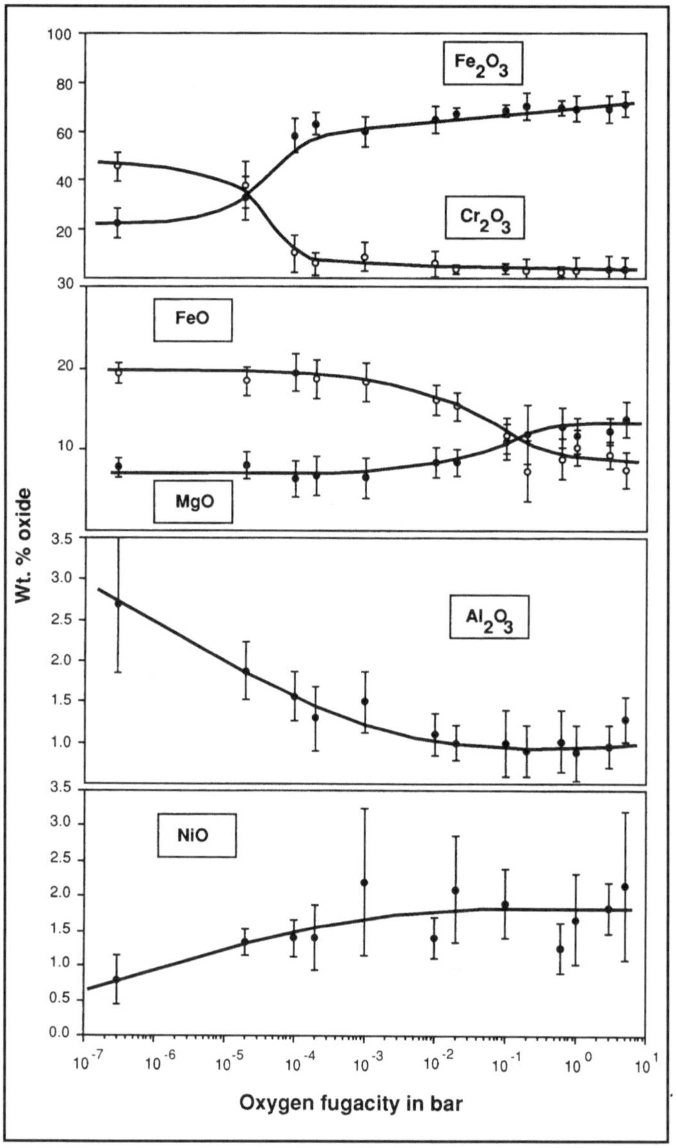

Figure 5. Variation of the composition of spinel crystals with oxygen fugacity in Saint Severin melts at 1335 °C. The vertical bars represent one standard deviation, including the variation in composition from crystal to crystal and the error on the measurement.

composition is indeed reported by Hill and Roeder (1974) for spinel crystallizing from basaltic liquid at low oxygen fugacity (f_{O_2} = $10^{-7.0}$ bar) but is not observed in our experiments. We believe it is because this lowering of the ferric iron is effective only at low oxygen fugacities ($f_{O_2} < 10^{-4.0}$ bar). Indeed, at high oxygen fugacities, the liquid is saturated with ferric iron, even at high temperatures. The amount of ferric iron incorporated in the spinel remains high, and the $Fe^{3+}/(Cr^{3+}+Al^{3+})$ ratio of spinel is virtually unchanged.

As the ferric iron content in the liquid is lowered, the ferrous iron content increases, and one could expect an increase of the ferrous iron content in both spinel and olivine. For instance, at f_{O_2} = $10^{-0.68}$ bar, an increase of the fayalite content in olivine is indeed observed with increasing temperatures (Fig. 9), whereas we did not observe any significant change in the Fe^{3+}/Fe_{total} ratio of spinel (Fig. 7). At lower oxygen fugacity (f_{O_2} = $10^{-3.7}$ bar), we observed a decrease with temperature of the fayalite content of olivine (Fig. 9), due to reduction of ferrous iron to metal iron, which becomes effective at low oxygen fugacity and high temperature (Hill and Roeder, 1974). However, even at such a low oxygen fugacity (f_{O_2} = $10^{-3.7}$ bar) there is still no change with temperature in the Fe^{3+}/Fe_{total} ratio of spinel (Fig. 7). Hill and Roeder (1974) reported a decrease with temperature in the ferrous iron content of spinel crystallized from basaltic liquids at f_{O_2} = $10^{-7.0}$ bar. However, this variation parallels the decrease in the ferric iron content of spinel, leading again to a constant Fe^{3+}/Fe_{total} ratio of spinel at any temperature. It appears that the Fe^{3+}/Fe_{total} ratio of spinel does not vary as a function of temperature, though it is clear that the ferric/ferrous ratio of the liquid changes with temperature as indicated by the change with temperature in the olivine composition. We have no definitive explanation to date, but this absence of variation might be due to the higher liquid solubility of ferrous relative to ferric iron at high temperature (Hill and Roeder, 1974).

The principal effect of changing the temperature and cooling rates is to change the morphology and, to a lesser extent, the abundance of spinel and olivine. The change in crystal morphology with temperature can be interpreted as follows: Below liquidus (<1520 °C), spinel and olivine coexist in equilibrium with the liquid, and numerous large euhedral crystals can rap-

TABLE 4. AVERAGE COMPOSITIONS OF SPINEL CRYSTALS SYNTHETIZED FROM SAINT SEVERIN CHONDRITE AT T = 1335° C AND AT DIFFERENT OXYGEN FUGACITIES*

$\log_{10} f_{O_2}$ (f_{O_2} in bar)	0.70 n† = 254		0.48 n† = 143		0.00 n† = 209		-0.22 n† = 182		-0.68 n† = 106		-1.00 n† = 137		-1.70 n† = 87		-2.00 n† = 201	
	< >	(σ)§	< >	(σ)§	< >	(σ)§	< >	(σ)§	< >	(σ)§	< >	(σ)§	< >	(σ)§	< >	(σ)§
TiO_2	0.1	(0.1)	0.1	(0.1)	0.1	(0.1)	0.1	(0.1)	0.1	(0.1)	0.1	(0.1)	0.1	(0.1)	0.1	(0.1)
Al_2O_3	1.3	(0.3)	1.1	(0.3)	0.9	(0.3)	1.0	(0.4)	0.9	(0.3)	1.0	(0.4)	1.0	(0.2)	1.1	(0.3)
Cr_2O_3	3.7	(4.9)	4.0	(5.1)	3.5	(5.3)	2.8	(2.3)	3.1	(4.6)	4.3	(2.2)	3.7	(1.8)	5.9	(5.4)
Fe_2O_3**	70.0	(5.2)	70.3	(5.4)	70.7	(5.4)	71.9	(3.0)	70.9	(5.4)	69.2	(2.7)	68.5	(2.0)	66.0	(5.5)
FeO**	7.4	(3.2)	9.6	(2.4)	10.6	(3.4)	9.0	(4.3)	7.5	(5.4)	11.9	(3.2)	15.6	(2.6)	16.4	(2.9)
MnO	0.6	(0.1)	0.5	(0.1)	0.5	(0.1)	0.6	(0.1)	0.4	(0.1)	0.5	(0.1)	0.4	(0.1)	0.5	(0.1)
MgO	13.8	(2.2)	12.6	(1.6)	12.0	(2.2)	13.3	(2.7)	12.1	(3.7)	11.1	(2.2)	8.6	(1.7)	8.6	(1.9)
NiO	2.2	(1.1)	1.8	(0.4)	1.7	(0.7)	1.3	(0.4)	5.0	(1.1)	1.9	(0.5)	2.1	(0.8)	1.4	(0.3)
Total	100.0		100.0		100.0		100.0		100.0		100.0		100.0		100.0	
Fe^{3+}/Fe_{total}	89.6	(4.3)	86.8	(3.5)	85.8	(4.3)	87.9	(5.4)	89.7	(8.1)	84.1	(4.1)	79.9	(3.2)	78.3	(3.8)

$\log_{10} f_{O_2}$ (f_{O_2} in bar)	-3.00 n† = 130		-3.70 n† = 100		-4.00 n† = 85		-4.22 n† = 123		-4.40 n† = 77		-4.70 n† = 7		-6.52 n† = 15	
	< >	(σ)§	< >	(σ)§	< >	(σ)§	< >	(σ)§	< >	(σ)§	< >	(σ)§	< >	(σ)§
TiO_2	0.2	(0.1)	0.1	(0.1)	0.3	(0.1)	0.3	(0.1)	0.2	(0.1)	0.1	(0.1)	0.5	(0.5)
Al_2O_3	1.5	(0.4)	1.3	(0.4)	1.6	(0.3)	2.4	(0.7)	2.2	(0.3)	1.9	(0.4)	2.7	(0.8)
Cr_2O_3	8.9	(5.8)	6.1	(4.6)	10.3	(7.7)	14.2	(6.2)	13.3	(4.7)	37.6	(8.9)	45.7	(5.9)
Fe_2O_3**	61.4	(6.0)	64.7	(4.8)	59.5	(7.1)	54.3	(6.6)	55.8	(5.0)	32.4	(8.7)	22.6	(5.9)
FeO**	18.8	(4.1)	19.1	(3.9)	19.9	(3.4)	21.3	(3.1)	20.3	(3.7)	18.3	(2.6)	19.5	(2.0)
MnO	0.3	(0.1)	0.3	(0.1)	0.3	(0.1)	0.3	(0.2)	0.3	(0.1)	0.3	(0.1)	0.4	(0.1)
MgO	6.7	(2.4)	7.0	(2.4)	6.6	(2.3)	5.7	(1.7)	6.4	(2.3)	8.0	(1.7)	7.8	(1.2)
NiO	2.2	(1.1)	1.4	(0.5)	1.5	(0.3)	1.5	(0.8)	1.5	(0.3)	1.4	(0.2)	0.8	(0.4)
Total	100.0		100.0		100.0		100.0		100.0		100.0		100.0	
Fe^{3+}/Fe_{total}	75.1	(5.1)	75.4	(4.9)	73.3	(3.6)	69.5	(4.9)	71.3	(5.2)	60.6	(8.4)	50.3	(6.5)

*All data in wt. %, except Fe^{3+}/Fe_{total} in atom %.
†n is the number of crystals analyzed.
§< > is the average value and σ is the standard deviation.
**Fe_2O_3 and FeO are calculated assuming a stoechiometric composition for spinel (see text).

TABLE 5. AVERAGE COMPOSITIONS OF SPINEL CRYSTALS SYNTHETIZED FROM ORGUEIL C1 CHONDRITE AT T = 1335° C AND AT DIFFERENT OXYGEN FUGACITIES*

$log_{10} f_{O_2}$ (f_{O_2} in bar)	0.00		-0.68		-1.00		-1.70		-2.00		-3.00		-3.70		-4.00	
	n† = 78		n† = 63		n† = 119		n† = 115		n† = 126		n† = 137		n† = 126		n† = 137	
	< >	(σ)§	< >	(σ)§	< >	(σ)§	< >	(σ)§	< >	(σ)§	< >	(σ)§	< >	(σ)§	< >	(σ)§
TiO_2	0.1	(0.1)	0.1	(0.1)	0.1	(0.1)	0.1	(0.1)	0.1	(0.1)	0.1	(0.1)	0.2	(0.1)	0.2	(0.1)
Al_2O_3	1.5	(0.4)	2.5	(0.9)	1.1	(0.2)	1.4	(0.3)	1.6	(0.3)	2.1	(0.5)	2.8	(1.4)	3.2	(1.4)
Cr_2O_3	1.2	(1.3)	1.2	(0.2)	1.9	(0.3)	1.5	(0.5)	1.9	(0.4)	2.0	(0.9)	2.3	(1.1)	3.1	(1.9)
Fe_2O_3**	72.7	(1.2)	71.8	(1.6)	70.9	(1.0)	69.9	(1.1)	69.3	(1.5)	68.1	(1.8)	66.7	(2.5)	64.7	(2.4)
FeO**	6.3	(2.7)	6.2	(4.0)	11.5	(2.4)	14.9	(2.4)	15.2	(3.6)	17.3	(4.2)	18.2	(4.6)	20.4	(4.0)
MnO	0.4	(0.1)	0.5	(0.1)	0.4	(0.1)	0.3	(0.1)	0.3	(0.1)	0.3	(0.1)	0.3	(0.1)	0.2	(0.1)
MgO	12.9	(1.8)	13.6	(2.6)	10.2	(1.5)	8.2	(1.6)	8.1	(2.3)	7.3	(2.8)	6.9	(3.0)	5.8	(2.5)
NiO	4.9	(0.9)	4.1	(0.7)	3.9	(0.4)	3.7	(0.4)	3.5	(0.5)	2.8	(0.4)	2.6	(0.4)	2.4	(1.0)
Total	100.0		100.0		100.0		100.0		100.0		100.0		100.0		100.0	
Fe^{3+}/Fe_{total}	91.3	(3.6)	91.5	(5.3)	84.8	(3.0)	80.9	(2.8)	80.6	(4.3)	78.2	(4.9)	77.0	(5.4)	74.3	(4.5)

*All data in wt. %, except Fe^{3+}/Fe_{total} in atom %.
†n is the number of crystals analyzed.
§< > is the average value and σ is the standard deviation.
**Fe_2O_3 and FeO are calculated assuming a stoechiometric composition for spinel (see text).

idly grow, as shown by our short (less than a few seconds) experiments. When the sample is removed from the high temperature condition, spinel and olivine can still be formed during cooling because of the rapid growth of dendritic and barred crystals. Rapid crystallization of dendritic spinel and barred olivine crystals during cooling is demonstrated by the occurrence of such crystals in samples melted above liquidus and cooled in air at the pressure of the experiment. Indeed, as no crystal can be formed above liquidus, it is clear that they were formed during the cooling stage at subliquidus temperatures. As a general rule, dendritic spinel crystals are all the more numerous as the liquidus is approached. Indeed, as the liquidus is approached the amount of spinel crystals formed in equilibrium in the liquid decreases, and the concentration of ferric iron in the liquid and, hence, the amount of dendrites formed during cooling increase. At high cooling rates (samples quenched in water or liquid nitrogen), there is not sufficient time for dendritic crystal growth during cooling, and the abundance of spinel crystals drops dramatically near the liquidus.

In summary, it appears that the morphology of spinel essentially depends on the temperature and cooling rates and can be used as an indicator of the temperature and cooling rates of the material from which spinel crystallized. The Fe^{3+}/Fe_{total} ratio of spinel depends essentially on oxygen fugacity. In particular, this ratio is relatively insensitive to variations in liquid composition (at least for basaltic, chondritic, and all intermediate liquid compositions) and in temperature (at least in the temperature range 1100 to 1700 °C, for oxygen fugacities ranging from 10^{-7} to $10^{0.7}$ bar). So, this ratio can be used as an indicator of oxygen fugacity of the environment in which spinel crystallized. The composition and in particular the $Fe^{3+}/(Cr^{3+} + Al^{3+})$ ratio of spinel depends on the oxygen fugacity and on the initial composition of the crystallizing material and not on the temperature. However, it is expected that at low oxygen fugacity ($f_{O_2} < 10^{-3.7}$ bar), the composition and $Fe^{3+}/(Cr^{3+} + Al^{3+})$ ratio of spinel might also depend on the temperature. So, the oxygen fugacity and temperature being known, a change in this ratio could reflect a change in the initial composition of the starting material. The abundance of spinel crystals depends on the temperature, cooling rates, oxygen fugacity, and initial composition of the crystallizing material and can be used for the determination of one of these parameters, the others being known.

IMPLICATIONS

As mentioned in the introduction, spinel crystals with a wide variety of morphology and composition have been found throughout the world at the K/T boundary. These crystals are commonly octahedral, which is a typical morphology of crystals formed below liquidus temperature. Few spinel crystals with dendritic morphology have been reported (Smit and Kyte, 1984; Kyte and Smit, 1986; Bohor et al., 1986; Robin et al.,

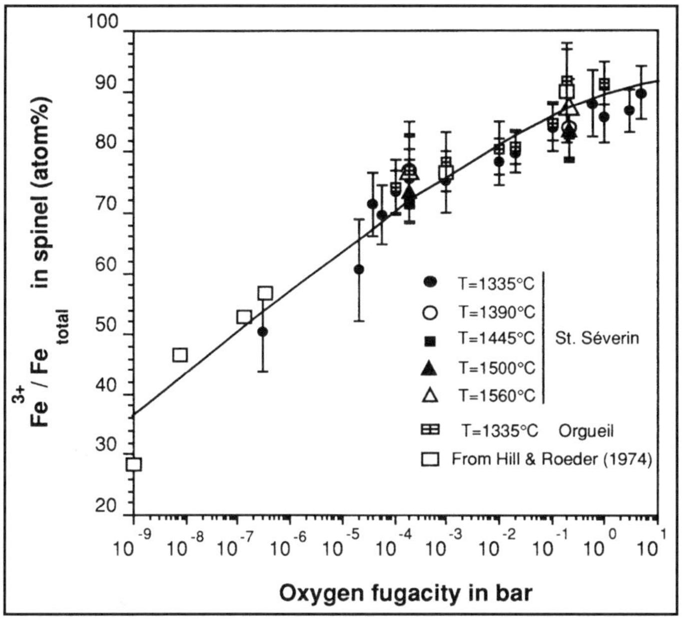

Figure 6. Fe^{3+}/Fe_{total} ratio of spinel crystals versus oxygen fugacity in Saint Severin and Orgueil melts at different temperatures. The vertical bars represent one standard deviation, including the variation in composition from crystal to crystal and the error on the measurement.

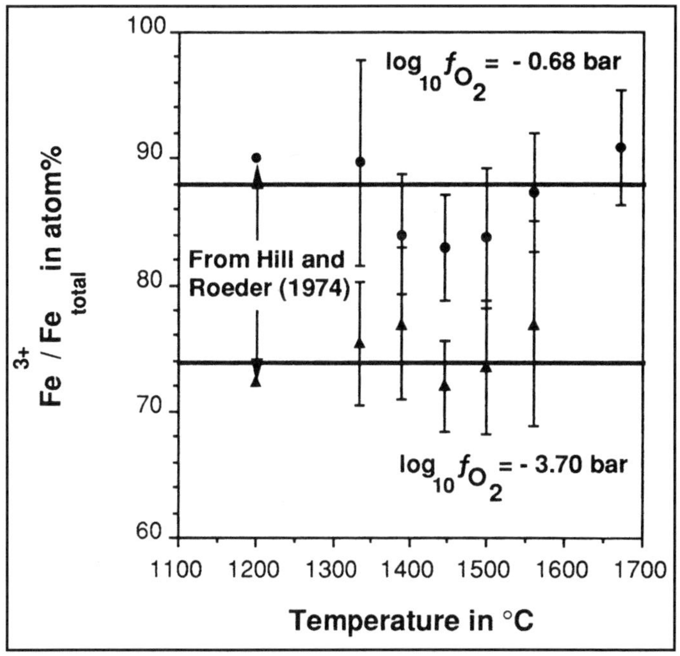

Figure 7. Fe^{3+}/Fe_{total} ratio of spinel crystals versus temperature in Saint Severin melts at two different oxygen fugacities. The vertical bars represent one standard deviation, including the variation in composition from crystal to crystal and the error on the measurement.

1992). This result indicates that most of the spinel crystals found at the K/T boundary are formed below liquidus temperatures, that is, below 1300 °C for a basalt (Hill and Roeder, 1974) and below 1520 °C for a chondrite (this chapter).

The K/T spinel has systematically high Ni content and high Fe^{3+}/Fe_{total} ratio (Table 8), consistent with formation from Ni-rich material in a very oxidizing environment. As already discussed by other authors (Bohor et al., 1986; Robin et al., 1992), the high Ni content of K/T spinel excludes derivation from crustal material, which is strongly depleted in Ni (<0.1 wt%; Taylor and McLennan, 1985), but points to a meteoritic or mantle source. Several authors (Kyte and Smit, 1986; Bohor et al., 1986; Robin et al., 1992) have claimed that the high ferric/ferrous ratio of K/T spinel indicates formation under extremely oxidizing conditions. According to the present study, the mini-

TABLE 6. AVERAGE COMPOSITIONS OF SPINEL CRYSTALS SYNTHETIZED FROM SAINT SEVERIN LL6 CHONDRITE AT DIFFERENT TEMPERATURES AND AT $f_{O_2} = 10^{-3.70}$ BAR*

	T = 1335° C		T = 1390° C		T = 1445° C		T = 1500° C		T = 1560° C	
	n† = 100		n† = 51		n† = 84		n† = 59		n† = 101	
	< >	(σ)§	< >	(σ)§	< >	(σ)§	< >	(σ)§	< >	(σ)§
TiO_2	0.1	(0.1)	0.1	(0.1)	0.2	(0.1)	0.1	(0.1)	0.2	(0.1)
Al_2O_3	1.3	(1.4)	1.5	(0.4)	2.2	(0.5)	2.5	(1.4)	2.9	(1.6)
Cr_2O_3	6.1	(4.6)	13.7	(5.7)	8.3	(7.6)	13.0	(9.0)	11.7	(9.5)
Fe_2O_3**	64.7	(4.8)	58.2	(5.5)	60.7	(6.9)	56.4	(7.8)	58.2	(9.0)
FeO**	19.1	(3.9)	15.8	(4.0)	21.3	(3.7)	18.5	(4.4)	16.0	(5.9)
MnO	0.3	(0.1)	0.4	(0.1)	0.4	(0.1)	0.4	(0.1)	0.5	(0.1)
MgO	7.0	(2.4)	9.6	(2.5)	5.8	(2.3)	7.6	(2.7)	9.4	(3.6)
NiO	1.4	(0.5)	0.7	(0.3)	1.1	(0.4)	1.5	(0.5)	1.1	(0.5)
Total	100.0		100.0		100.0		100.0		100.0	
Fe^{3+}/Fe_{total}	75.4	(4.9)	76.9	(6.0)	72.0	(3.6)	73.5	(5.3)	76.9	(8.1)

*All data in wt. %, except Fe^{3+}/Fe_{total} in atom %.
†n is the number of crystals analyzed.
§< > is the average value and σ is the standard deviation.
**Fe_2O_3 and FeO are calculated assuming a stoechiometric composition for spinel (see text).

TABLE 7. AVERAGE COMPOSITIONS OF SPINEL CRYSTALS SYNTHETIZED FROM SAINT SEVERIN LL6 CHONDRITE AT DIFFERENT TEMPERATURES AND AT f_{O_2} = $10^{-0.68}$ BAR*

	T = 1335° C n† = 106		T = 1390° C n† = 88		T = 1445° C n† = 72		T = 1500° C n† = 114		T = 1560° C n† = 80		T = 1670° C n† = 163	
	< >	(σ)§	< >	(σ)§	< >	(σ)§	< >	(σ)§	< >	(σ)§	< >	(σ)§
TiO_2	0.1	(0.1)	0.1	(0.1)	0.1	(0.1)	0.1	(0.1)	0.1	(0.1)	0.1	(0.1)
Al_2O_3	0.9	(0.3)	1.3	(0.5)	1.6	(0.5)	2.1	(2.2)	1.5	(0.5)	1.1	(0.4)
Cr_2O_3	3.1	(4.6)	2.4	(1.4)	5.2	(6.2)	4.3	(7.0)	1.9	(1.5)	2.3	(0.3)
Fe_2O_3**	70.9	(5.4)	70.4	(2.1)	67.3	(6.1)	68.0	(7.4)	69.9	(3.1)	72.0	(2.2)
FeO**	7.5	(5.4)	12.2	(3.9)	12.4	(3.1)	11.9	(3.8)	9.5	(3.5)	6.7	(3.1)
MnO	0.4	(0.1)	0.5	(0.1)	0.5	(0.1)	0.5	(0.1)	0.4	(0.1)	0.6	(0.1)
MgO	12.1	(3.7)	10.6	(2.5)	10.5	(2.0)	11.4	(2.5)	11.1	(2.1)	13.2	(1.7)
NiO	5.0	(1.1)	2.5	(0.7)	2.4	(0.4)	1.7	(0.4)	4.6	(1.4)	4.0	(0.3)
Total	100.0		100.0		100.0		100.0		100.0		100.0	
Fe^{3+}/Fe_{total}	89.7	(8.1)	84.0	(4.8)	83.0	(4.2)	83.7	(5.5)	87.3	(4.7)	90.9	(4.5)

*All data in wt. %, except Fe^{3+}/Fe_{total} in atom %.
†n is the number of crystals analyzed.
§< > is the average value and σ is the standard deviation.
**Fe_2O_3 and FeO are calculated assuming a stoechiometric composition for spinel (see text)

mum oxygen fugacity required for the formation of K/T spinel is ≈10^{-2} bar (considering a minimum Fe^{3+}/Fe_{total} ratio for spinel of ≈78 atom%, Table 8). Some spinel crystals were formed at oxygen fugacity higher than $10^{-0.68}$ bar (Fe^{3+}/Fe_{total} >90 atom%, Table 8) or higher than the oxygen partial pressure existing at sea level. In other terms, the K/T spinel derive from meteoritic or mantle material that equilibrated at $f_{O_2} \geq 10^{-2}$ bar and T ≤ 1520 °C. These formation conditions are extremely constraining and allow us to rule out some of the potential sources that have been proposed for the K/T spinel.

The K/T spinel cannot be derived from igneous rocks, even from ultramafic rocks that contain a small amount of Ni

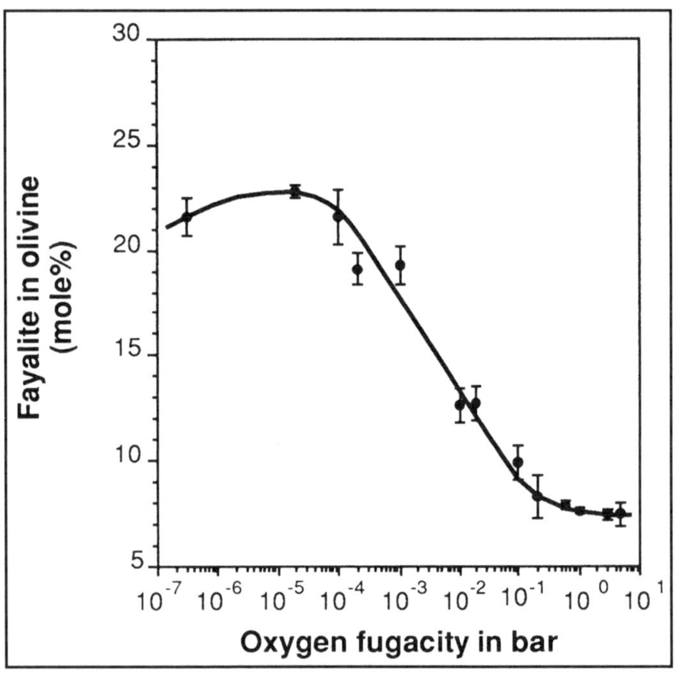

Figure 8. The fayalite content of olivine versus oxygen fugacity in Saint Severin melts at 1335 °C. The vertical bars represent one standard deviation, including the variation in composition from crystal to crystal and the error on the measurement.

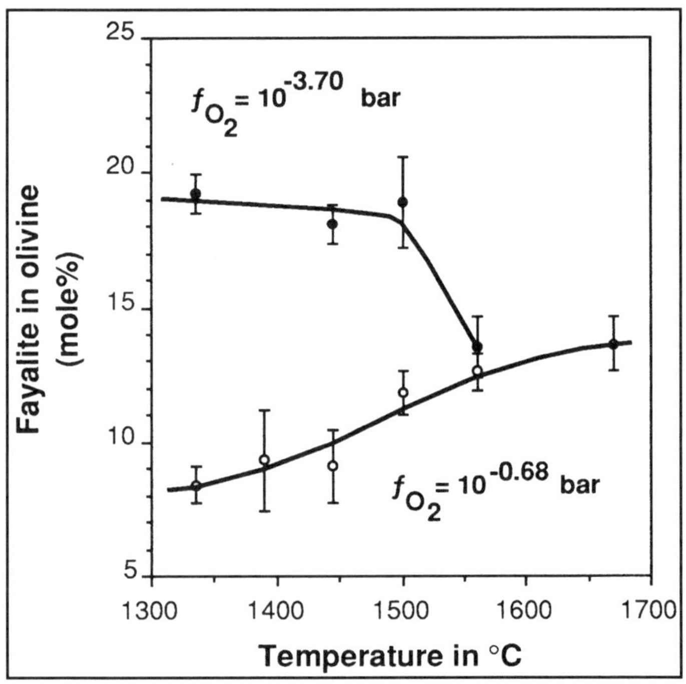

Figure 9. The fayalite content of olivine versus temperature in Saint Severin melts at two different oxygen fugacities. The vertical bars represent one standard deviation, including the variation in composition from crystal to crystal and the error on the measurement.

TABLE 8. AVERAGE COMPOSITIONS OF NI-RICH SPINEL CRYSTALS FROM DIFFERENT K/T BOUNDARY SITES*

	El Kef, Tunisia n† = 151		Sopelana, Spain n† = 224		Caravaca, Spain n† = 138		Bidart, France n† 97		Gosau, Austria n† = 157	
	< >	(σ)§	< >	(σ)§	< >	(σ)§	< >	(σ)§	< >	(σ)§
TiO_2	0.2	(0.2)	0.5	(0.5)	0.4	(0.2)	0.3	(0.4)	0.5	(0.3)
Al_2O_3	5.8	(1.8)	5.2	(1.8)	5.2	(1.7)	3.8	(1.7)	4.5	(1.5)
Cr_2O_3	17.2	(5.0)	8.1	(4.6)	9.1	(3.9)	5.4	(5.1)	6.3	(3.4)
Fe_2O_3**	49.4	(5.8)	57.9	(6.3)	57.7	(5.2)	63.9	(6.4)	61.4	(4.4)
FeO**	12.2	(3.8)	15.0	(5.7)	12.8	(3.9)	11.1	(4.8)	12.1	(4.3)
MnO	0.5	(0.5)	0.6	(0.6)	0.5	(0.3)	0.6	(0.5)	0.5	(0.4)
MgO	10.0	(2.5)	7.8	(3.3)	9.1	(2.4)	10.5	(3.2)	9.4	(2.9)
NiO	4.7	(1.5)	4.9	(1.6)	5.2	(1.4)	4.4	(1.9)	5.3	(1.6)
Total	100.0		100.0		100.0		100.0		100.0	
Fe^{3+}/Fe_{total}	78.6	(5.8)	77.9	(7.6)	80.4	(5.4)	83.9	(6.6)	82.2	(5.8)

	Petriccio, Italy n† = 213		Mimbral, Mexico n† = 45		Beloc, Haiti n† = 318		DSDP Leg 77, Hole 536‡ n† = 31		ODP Leg 122, Hole 761C§§ n† = 80	
	< >	(σ)§	< >	(σ)§	< >	(σ)§	< >	(σ)§	< >	(σ)§
TiO_2	0.2	(0.2)	0.2	(0.3)	0.4	(0.4)	0.4	(0.2)	0.1	(0.2)
Al_2O_3	3.5	(2.4)	4.7	(2.4)	4.7	(2.3)	5.9	(3.4)	4.4	(2.2)
Cr_2O_3	2.8	(3.2)	7.2	(5.9)	5.7	(5.0)	1.7	(2.2)	1.8	(1.4)
Fe_2O_3**	66.6	(4.6)	61.5	(7.1)	63.5	(6.6)	67.5	(5.7)	69.7	(2.9)
FeO**	11.5	(6.4)	12.2	(7.6)	7.5	(5.0)	5.8	(6.2)	3.5	(3.4)
MnO	1.0	(0.8)	0.1	(0.2)	0.5	(0.4)	0.6	(0.6)	0.5	(0.5)
MgO	9.4	(4.7)	10.5	(4.8)	12.7	(3.7)	15.5	(5.5)	15.6	(3.2)
NiO	5.0	(2.1)	3.6	(2.4)	5.0	(2.5)	2.6	(2.0)	4.4	(1.9)
Total	100.0		100.0		100.0		100.0		100.0	
Fe^{3+}/Fe_{total}	84.5	(7.7)	82.7	(10.0)	88.7	(7.6)	91.7	(8.6)	94.9	(4.8)

*All data in wt. %, except Fe^{3+}/Fe_{total} in atom %.
†n is the number of crystals analyzed.
§< > is the average value and σ is the standard deviation.
**Fe_2O_3 and FeO are calculated assuming a stoechiometric composition for spinel (see text).
‡Deep Sea Drilling Program Leg 77, Hole 536.
§§Ocean Drilling Program Leg 122, Hole 761C.

(≈0.2 wt%, Taylor and McLennan, 1985). Indeed, most of these rocks have equilibrated within a few log units of oxygen fugacity of the fayalite-magnetite-quartz (FMQ) buffer (Haggerty, 1976) or several log units below the minimum oxygen fugacity required for the formation of K/T spinel (Fig. 10). These equilibration conditions account for the absence of oxidized Ni-rich spinel in igneous rocks (Haggerty, 1976), even in Ni-rich ultramafic rocks; these rocks have crystallized at extremely low oxygen fugacities. For the same reason, oxidized Ni-rich spinel has no counterpart in meteorites (El Goresy, 1976b), as most meteorites have equilibrated at oxygen fugacities close to those of the iron-wustite (IW) buffer (Fig. 10), as indicated by the occurrence of FeNi metal.

As a consequence, direct crystallization of K/T spinel from impact melt rocks or direct condensation from the cloud of vaporized target and projectile material (see Bohor et al., 1986; Kyte et al., 1994) can be excluded. Indeed, the oxygen fugacity in the melt rocks or in the vapor cloud is imposed by the FMQ and/or IW buffers, as both target and projectile material have equilibrated near these oxygen buffers (Fig. 10), and is therefore low compared with the oxygen fugacity required for the formation of K/T spinel. In addition, the very high temperatures involved in such events would prevent the oxidation of ferrous iron and would even reduce ferric iron initially present in the target rocks. This effect of the temperature on the ferric/ferrous ratio was reported by Friedman et al. (1960) for oxidized rocks fused in air at high temperatures (>2000 °C). Evidence of high temperature and low oxygen fugacity in impact events is also demonstrated by the systematically reduced state of projectile- and target-derived products (Schultz et al., 1994, and Fudali et al., 1987, respectively).

It therefore appears that the formation of K/T spinel requires

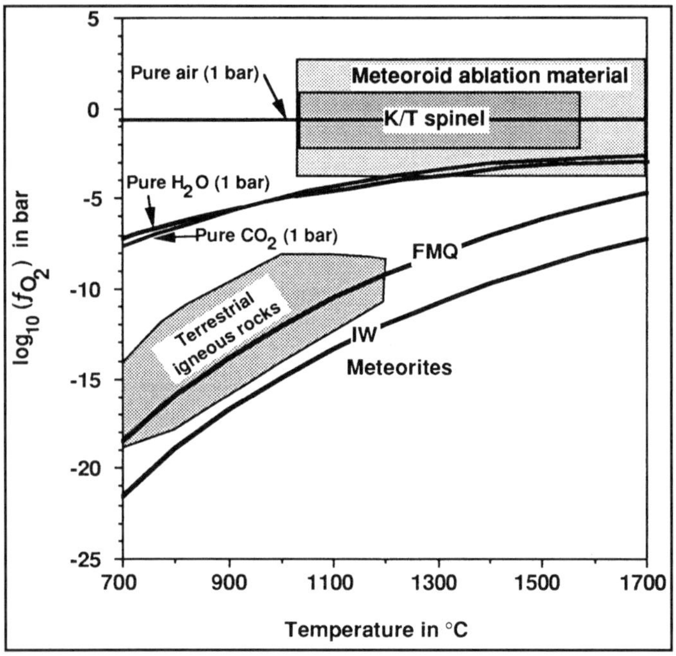

Figure 10. Log $_{10}f_{O_2}$ versus temperature, illustrating the different conditions of formation of the K/T spinel, meteoroid ablation material, terrestrial igneous rocks, and meteorites (see text). The domain of crystallization of terrestrial rocks is from Haggerty (1976). FMQ is the fayalite-magnetite-quartz buffer curve, and IW is the iron-wustite buffer curve (see Frost, 1990). H_2O and CO_2 curves are determined from thermodynamic data of the pure elements from Robie et al. (1978).

an external source for oxygen. It may be the dissociation of H_2O, CO_2, or SO_2 or of any other gaseous species containing oxygen. Dissociation of H_2O molecules can be envisaged if one considers large asteroid impacts in the ocean. At 1 bar total pressure, the oxygen fugacity resulting from the dissociation of pure H_2O remains below the minimum oxygen fugacity required for the formation of K/T spinel (Fig. 10). Higher pressures (as high as a few Mbar) and, hence, higher oxygen fugacities can be expected in the cloud of shock-vaporized water during the early stage of an oceanic impact (Melosh, 1982). However, at this time, the temperature of the highly compressed water is around 100,000 °K (Melosh, 1982), preventing spinel crystallization. As the shocked vapor cloud expands, the pressure declines and the temperature as well. However, there is no evidence that a range of P-T conditions exists that could provide the kind of environment needed for the formation of spinel.

Dissociation of CO_2 and SO_2 may be due to devolatilization of carbonate and sulfate-rich evaporite. Interestingly, the impact supposed to have occurred on the Yucatan peninsula penetrated a carbonate and evaporite sequence (Sigurdsson et al., 1991; Kring and Boynton, 1991). Moreover, oxygen isotope data about yellow glasses (Blum and Chamberlain, 1992) recovered in K/T boundary sediments from Haiti and interpreted as impact glasses (Sigurdsson et al., 1991) indicate that they are essentially a mixture of devolatilized carbonate and silicate rocks. However, the study of shocked carbonate-rich fragments from the Haughton impact crater (Martinez et al., 1994) indicates that the amount of CO_2 derived from the carbonates is not important. Moreover, we have no evidence that dissociation of CO_2 at high pressure and temperature could provide the kind of environment needed for the formation of spinel.

Direct oxidation of projectile material in the Earth's atmosphere appears as the most simple mechanism for producing the K/T spinel. The Earth's atmosphere is an important source of free-oxygen, and the high-velocity interaction with the atmosphere for meteoritic material is known as a source for oxidized Ni-rich spinel (Robin et al., 1992). In addition, a meteoroid experienced a wide range of aerodynamic pressures (from less than 10^{-3} bar up to more than 10^8 bars; see Whipple, 1950; Baldwin and Sheaffer, 1971; Bronshten, 1983; Hills and Goda, 1993) and thus of oxygen fugacities (Fig. 10), depending on its size, density, initial velocity, and entry angle. The oxygen partial pressure (P_{O2}) at the stagnation point in front of the meteoroid can be considered as roughly proportional to the ram pressure (P_{ram}):

$P_{O2} = \alpha P_{ram} = \alpha \rho_{air} V^2$ where α is the weight fraction of O_2 in air at a given altitude ($\approx 1/5$ at sea level); ρ is the local air density, and V is the meteoroid velocity.

The small dust particles ($\approx$100 μm) are decelerated from their cosmic velocity (>11.2 km/s, the escape velocity) in the upper atmosphere (>80 km; see Whipple, 1950; Bronshten, 1983) ($\alpha \approx 1/5$ and $\rho_{air} \leq 10^{-5}$ kg m^{-3}; Weast et al., 1989). The oxygen partial pressure (P_{O2}) at the stagnation point for these particles is $\approx 10^{-3}$ bar, corresponding to $Fe^{3+}/Fe_{total} \approx 75$ atom% in the spinel (see Fig. 6). This value is consistent with the maximum Fe^{3+}/Fe_{total} ratio of spinel measured in micrometeorites (Robin et al., 1992). Large meteoroids are decelerated lower (<40 km) in the atmosphere (Baldwin and Sheaffer, 1971; Bronshten, 1983; Hills and Goda, 1993) and therefore experience higher oxygen pressures. For instance, a stony meteorite 100 m in diameter (density of 3 g cm^{-3}) is decelerated from its cosmic velocity at an altitude of $\approx$10 km (Hills and Goda, 1993) ($\alpha \approx 1/5$ and $\rho_{air} \approx 0.4$ kg m^{-3}; Weast et al., 1989) and experiences oxygen partial pressure as high as 10^2 bars, corresponding to $Fe^{3+}/Fe_{total} > 90$ atom%.

We propose that the K/T spinel is derived from the interaction with the atmosphere of large meteoroids. We exclude the interaction of small dust particles that are decelerated in the upper atmophere and consequently have lower Fe^{3+}/Fe_{total} ratio (<75 atom%) than spinel crystals from the K/T boundary (Fe^{3+}/Fe_{total} >75 atom%). In the proposed scenario, spinel crystallized in liquid ablation droplets mainly produced at low altitudes along the meteoroid trajectory. The major problem with that source is that ablation is important only for stony meteoroids with diameters (D) < 200 m and comets with D < 1,600 m (Hills and Goda, 1993). As the estimated K/T impactor diameter is D $\approx$ 10 km, our scenario supposes the fragmentation of the bolide either before or upon the impact. Two mechanisms can be envisaged: oblique impact (Schultz and Gault,

1982, 1990) and encounter with a fragmented comet (Schmitt, 1989; Hut et al., 1991).

Oblique trajectories (<45°) represent 50% of the impacts. The model developed by Schultz and Gault (1982; 1990) predicts that, for this type of impact, the projectile and part of the target are fragmented. The fragments, more and more numerous as the incidence angle decreases, can be widely dispersed over the Earth. At impact angles <30°, representing 25% of the impacts, the projectile fragments dominate the downrange debris (Schultz and Gault, 1990), leading to a global meteor deluge and, hence, to the deposition all over the Earth of numerous ablation droplets containing oxidized Ni-rich spinel crystals. Interestingly, geophysical data for Chicxulub reveal several intriguing asymmetries in the ring structure that could result from an object impacting at ≈30° from the horizontal (Schultz, 1994).

The encounter with a fragmented comet of a diameter equivalent to the one of the presumed K/T impactor (≈10 km; Alvarez et al., 1980) may also account for a global meteor deluge. Such a phenomena was recently observed on Jupiter with the crash of P/Shoemaker-Levy 9 comet (Schenk and Moses, 1994). Breakup occurs through tidal disruption, when a comet passes within the Roche lobe of a large planet, or by disintegration during volatile loss (Schmitt, 1989; Hut et al., 1991). As the Earth is smaller than Jupiter, the breakup must have occurred far from the Earth—for example near the Sun (Hut et al., 1991)—in order to separate the fragments over a distance comparable with one Earth diameter. In that case, several impacts can occur but only within one hemisphere (Hut et al., 1991). However, 25% of them will have impact angles <30°, leading here again to a global meteor deluge and to the worldwide deposition of spinel crystals.

Both scenarios can account for the regional variations in spinel compositions (Kyte and Smit, 1986; Robin et al., 1994a). Indeed, our study, combined with that of Hill and Roeder, demonstrates that the composition of spinel varies as a function of oxygen fugacity, temperature, and initial composition of the crystallizing material. Variable spinel compositions would thus reflect variations in the physical and chemical environment in which they crystallized, depending on the size, density, velocity, entry angle, and initial composition of projectile fragments reentering the Earth's atmosphere.

CONCLUSIONS

The study of the formation conditions of oxidized Ni-rich spinel leads to the following conclusions:

1. Oxidized Ni-rich spinel can be obtained by melting meteoritic material at high temperatures (>1300 °C) and high oxygen fugacities ($f_{O_2} > 10^{-6}$ bar), even for heating exposure periods as short as a few seconds.

2. The morphology, abundance, and composition of spinel crystals depend on oxygen fugacity, temperature, cooling rates, and initial composition of the crystallizing material. The morphology of spinel crystals is essentially controlled by temperature and cooling rate and can be used as an indicator of these parameters. The Fe^{3+}/Fe_{total} ratio of spinel is essentially controlled by oxygen fugacity and is relatively insensitive to liquid composition (at least for basaltic, chondritic, and all intermediate liquid compositions) and temperature (at least in the temperature range 1100 to 1700 °C, for oxygen fugacities ranging from 10^{-7} to $10^{0.7}$ bar). This ratio can therefore be used as an indicator of the oxygen fugacity of the environment in which spinel crystallized. The composition and in particular the $Fe^{3+}/(Cr^{3+}+Al^{3+})$ ratio of spinel depend on the oxygen fugacity and on the initial composition of the crystallizing material. A change in the composition and $Fe^{3+}/(Cr^{3+}+Al^{3+})$ ratio of spinel may indicate a change either in the oxygen fugacity of the environment in which spinel crystallized or in the composition of the starting material from which it crystallized. It is expected that at low oxygen fugacity ($f_{O_2} < 10^{-3.7}$ bar), the composition and $Fe^{3+}/(Cr^{3+}+Al^{3+})$ ratio of spinel also depend on the temperature. The abundance of spinel crystals depends on oxygen fugacity, temperature, cooling rates and initial composition of the crystallizing material.

3. The spinel crystals found worldwide at the K/T boundary are derived from meteoritic material that equilibrated at $f_{O_2} \geq 10^{-2}$ bar and T ≤ 1520 °C, considering their morphology, composition, and Fe^{3+}/Fe_{total} ratio.

4. Such formation conditions rule out derivation of K/T spinel from known volcanic or impact processes. In particular, K/T spinel cannot be formed from terrestrial magmas, which are known to evolve under extremely low oxygen fugacities, near the FMQ buffer. For the same reason, direct crystallization from impact melt rocks or direct condensation from the cloud of vaporized target and projectile material can be excluded.

5. We propose that spinel crystals from the K/T boundary crystallized in liquid ablation droplets produced during the interaction with the atmosphere of large meteoroids decelerated at low altitudes. Our scenario supposes multiple meteoroid impacts all over the Earth, resulting either from the fragmentation of the K/T impactor upon an oblique impact or from the encounter with a fragmented comet.

ACKNOWLEDGMENTS

J. Gayraud is indebted to M. Guiraud from the Laboratoire de Minéralogie du Museum National d'Histoire Naturelle de Paris for helpful discussions and valuable suggestions. Meteorite samples were kindly provided by C. Perron, B. Zanda, and M. Ghelis from the Museum National d'Histoire Naturelle de Paris. We thank S. Haggerty, C. Koeberl, and G. Ryder for the critical review of the manuscript. This is CFR contribution no. 1674.

REFERENCES CITED

Alvarez, L. W., Alvarez, W., Asaro, F., and Michel, H. V., 1980, Extraterrestrial cause for the Cretaceous-Tertiary extinction: Science, v. 208, p. 1095–1108.

Anders, E., and Ebihara, M., 1982, Solar-system abundances of the elements: Geochimica et Cosmochimica Acta, v. 46, p. 2363–2380.

Baldwin, B., and Sheaffer, Y., 1971, Ablation and breakup of large meteoroids during atmospheric entry: Journal of Geophysical Research, v. 76, p. 4653–4668.

Blum, J. D., and Chamberlain, C. P., 1992, Oxygen isotope constraints on the origin of impact glasses from the Cretaceous-Tertiary boundary: Science, v. 257, p. 1104–1107.

Blum, J. D., Chamberlain, C. P., Hingston, M. P., Koeberl, C., Marin, L. E., Schuraytz, B. C., and Sharpton, V. L., 1993, Isotopic comparison of K-T boundary impact glass with melt rock from the Chicxulub and Manson impact structures: Nature, v. 364, p. 325–327.

Bohor, B. F., and Foord, E. E., 1987, Magnesioferrite from a nonmarine K-T boundary clay in Wyoming [abs.]: Lunar and Planetary Science, v. 18, p. 101–102.

Bohor, B. F., Foord, E. E., Modreski, P. J., and Triplehorn, D. M., 1984, Mineralogical evidence for an impact event at the Cretaceous-Tertiary boundary: Science, v. 224, p. 867–869.

Bohor, B. F., Foord, E. E., and Ganapathy, R., 1986, Magnesioferrite from the Cretaceous-Tertiary boundary, Caravaca, Spain: Earth and Planetary Science Letters, v. 81, p. 57–66.

Bohor, B. F., Modreski, P. J., and Foord, E. E., 1987, Shocked quartz in the Cretaceous-Tertiary boundary clays: Evidence for a global distribution: Science, v. 236, p. 705–709.

Bronshten, V. A., 1983, Physics of meteoric phenomena: Dordrecht, D. Reidel Publishing, 218 p.

El Goresy, A., 1976a, Oxide minerals in lunar rocks, *in* Rumble, D., ed., Reviews in mineralogy, Volume 3: Washington, D.C., Mineralogical Society of America, p. EG1–EG46.

El Goresy, A., 1976b, Opaque oxide minerals in meteorites, *in* Rumble, D., ed., Reviews in mineralogy, Volume 3: Washington, D.C., Mineralogical Society of America, p. EG47–EG72.

Friedman, I., Thorpe, A., and Senftle, F., 1960, Comparison of the chemical composition and magnetic properties of tektites and glasses formed by fusion of terrestrial rocks: Nature, v. 187, p. 1089–1092.

Frost, B. R., 1990, Introduction to oxygen fugacity and its petrologic importance, *in* Frost, B. R., ed., Reviews in mineralogy, Volume 12: Washington, D.C., Mineralogical Society of America, p. 1–9.

Fudali, R. F., 1965, Oxygen fugacities of basaltic and andesitic magmas: Geochimica et Cosmochimica Acta, v. 29, p. 1063–1075.

Fudali, R. F., Dyar, M. D., Griscom, D. L., and Schreiber, H. D., 1987, The oxidation state of iron in tektite glass: Geochimica et Cosmochimica Acta, v. 51, p. 2749–2756.

Ganapathy, R., 1980, A major meteorite impact on the Earth 65 million years ago: Evidence from the Cretaceous-Tertiary boundary clay: Science, v. 209, p. 921–923.

Gayraud, J., Robin, E., Rocchia, R., and Froget, L., 1994, Preliminary results regarding the formation conditions of meteoric spinels [abs]: New developments regarding the K/T event and other catastrophes in Earth history: Houston, Texas, Lunar and Planetary Institute Contribution 825, p. 41–42.

Goltrant, O., Cordier, P., and Doukhan, J. C., 1991, Planar deformation features in shocked quartz; a transmission electron microscopy investigation: Earth and Planetary Science Letters, v. 106, p. 103–115.

Goltrant, O., Leroux, H., Doukhan, J. C., and Cordier, P., 1992, Formation mechanisms of planar deformation features in naturally shocked quartz: Physics of the Earth and Planetary Interiors, v. 74, p. 219–240.

Grove, T. L., 1981, Use of FePt alloys to eliminate the iron loss problem in 1 atmosphere gas mixing experiments: Theoretical and practical considerations: Contributions to Mineralogy and Petrology, v. 78, p. 298–304.

Haggerty, S. E., 1976, Opaque mineral oxides in terrestrial igneous rocks, *in* Rumble, D., ed., Reviews in mineralogy, Volume 3: Washington, D.C., Mineralogical Society of America, p. Hg101–Hg300.

Hildebrand, A. R., and Boynton, W. V., 1990, Proximal Cretaceous-Tertiary boundary impact deposits in the Caribbean: Science, v. 248, p. 843–846.

Hildebrand, A. R., Penfield, G.T., Kring, D. A., Pilkington, M., Camargo, Z. A., Jacobsen, S. B., and Boynton, W. V., 1991, Chicxulub crater: A possible Cretaceous/Tertiary boundary impact crater on the Yucatan Peninsula, Mexico: Geology, v. 19, p. 867–871.

Hill, R., and Roeder, P., 1974, The crystallization of spinel from basaltic liquid as a function of oxygen fugacity: Journal of Geology, v. 82, p. 709–729.

Hills, J. G., and Goda, M. P., 1993, The fragmentation of small asteroids in the atmosphere: The Astronomical Journal, v. 105, p. 1114–1144.

Hsü, K. J., 1980, Terrestrial catastrophe caused by cometary impact at the end of Cretaceous: Nature, v. 285, p. 201–203.

Hut, P., Shoemaker, E. M., Alvarez, W., and Montanari, A., 1991, Astronomical mechanisms and geologic evidence for multiple impacts on Earth [abs]: Lunar and Planetary Science, v. 22, p. 603–604.

Izett, G. A., 1991, Tektites in Cretaceous-Tertiary boundary rocks on Haiti and their bearing on the Alvarez impact extinction hypothesis: Journal of Geophysical Research, v. 96, p. 20879–20905.

Jarosewich, E., 1990, Chemical analyses of meteorites: A compilation of stony and iron meteorite analyses: Meteoritics, v. 25, p. 323–337.

Jéhanno, C., Boclet, D., Bonté, Ph., Devineau, J., and Rocchia, R., 1987, L'iridium dans les minéraux à la limite Crétacé-Tertiaire de plusieurs sites européens et africains: Mémoire de la Société Géologique de France, v. 150, p. 81–94.

Jéhanno, C., Boclet, D., Froget, L., Lambert, B., Robin, E., Rocchia, R., and Turpin, L., 1992, The Cretaceous-Tertiary boundary at Beloc, Haiti: No evidence for an impact in the Caribbean area: Earth and Planetary Science Letters, v. 109, p. 229–241.

Koeberl, C., 1994, Deposition of channel deposits near the Cretaceous-Tertiary boundary in northeastern Mexico: Catastrophic or "normal" sedimentary deposits?: Comment: Geology, v. 22, p. 957.

Koeberl, C., Sharpton, V. L., Schuraytz, B. C., Shirey, S. B., Blum, J. D., and Marin, L. E., 1994, Evidence for a meteoritic component in impact melt rock from the Chicxulub structure: Geochimica et Cosmochimica Acta, v. 58, p. 1679–1684.

Kring, D. A., and Boynton, W. V., 1991, Altered spherules of impact melt and associated relic glass from the K/T boundary sediments in Haiti: Geochimica et Cosmochimica Acta, v. 55, p. 1737–1742.

Kring, D. A., and Boynton, W. V., 1992, Petrogenesis of an augite-bearing melt rock in the Chicxulub structure and its relationship to K/T impact spherules in Haiti: Nature, v. 358, p. 141–144.

Krogh, T. E., Kamo, S. L., Sharpton, V. L., Marin, L. E., and Hildebrand, A. R., 1993, U-Pb ages of single shocked zircons linking distal K/T ejecta to the Chicxulub crater: Nature, v. 366, p. 731–734.

Kyte, F. T., and Smit, J., 1986, Regional variations in spinel compositions: An important key to the Cretaceous/Tertiary event: Geology, v. 14, p. 485–487.

Kyte, F. T., Zhou, Z., and Wasson, J. T., 1980, Siderophile-enriched sediments from the Cretaceous-Tertiary boundary: Nature, v. 288, p. 651–656.

Kyte, F. T., Bostwick, J. A., and Zhou, L., 1994, The KT boundary on the Pacific Plate [abs.], New developments regarding the K/T event and other catastrophes in Earth history: Houston, Texas, Lunar and Planetary Institute Contribution 825, p. 64–65.

Lyons, J. B., and Officer, C. B., 1992, Mineralogy and petrology of the Haiti Cretaceous/Tertiary section: Earth and Planetary Science Letters, v. 109, p. 205-224.

Martinez, I., Agrinier, P., Schärer, U., and Javoy, M., 1994, A SEM-ATEM and stable isotope study of carbonates from the Haughton impact crater, Canada: Earth and Planetary Science Letters, v. 121, p. 559–574.

Melosh, H. J., 1982, The mechanics of large meteoroid impacts in the Earth's oceans, *in* Silver, L. T., and Schultz, P. H., eds., Geological implications of impacts of large asteroids and comets on the Earth: Geological Society of America Special Paper 190, p. 121–127.

Montanari, A., Hay, R. L., Alvarez, W., Asaro, F., Michel, H. V., Alvarez, L. W., and Smit, J., 1983, Spheroids at the Cretaceous-Tertiary boundary are altered impact droplets of basaltic composition: Geology, v. 11, p. 668–671.

Parfenoff, A., Pomerol, C., and Tourenq, J., 1970, Les minéraux en grains: Paris, Masson et Cie, 578 p.

Penfield, G. T., and Camargo, Z. A., 1981, Definition of a major igneous zone in the central Yucatan platform with aeromagnetics and gravity [abs.]: Society of Exploration Geophysicists Technical Program, Abstracts, and Biographies, v. 51, p. 37.

Pope, K. O., Ocampo, A. C., and Duller, C. E., 1991, Mexican site for K/T impact crater: Nature, v. 351, p. 105.

Robie, R. A., Hemingway, B. S., and Fisher, J. R., 1978, Thermodynamic properties of minerals and related substances at 298.15K and 1 bar (10^5 pascals) pressure and at higher temperatures: U.S. Geological Survey Bulletin 1452, 301 p.

Robin, E., Boclet, D., Bonté, Ph., Froget, L., Jéhanno, C., and Rocchia, R., 1991, The stratigraphic distribution of Ni-rich spinels in Cretaceous-Tertiary boundary rocks at El Kef (Tunisia), Caravaca (Spain) and Hole 761C (Leg 122): Earth and Planetary Science Letters, v. 107, p. 715–721.

Robin, E., Bonté, Ph., Froget, L., Jéhanno, C., and Rocchia, R., 1992, Formation of spinels in cosmic objects during atmospheric entry: A clue to the Cretaceous-Tertiary boundary event: Earth and Planetary Science Letters, v. 108, p. 181–190.

Robin, E., Froget, L., Jéhanno, C., and Rocchia, R., 1993, Evidence for a K/T impact event in the Pacific Ocean: Nature, v. 363, p. 615–617.

Robin, E., Gayraud, J., Froget, L., and Rocchia, R., 1994a, On the origin of the regional variations in spinel compositions at the KT boundary [abs.]: New developments regarding the K/T event and other catastrophes in Earth history: Houston, Texas, Lunar and Planetary Institute Contribution 825, p. 96–97.

Robin, E., Rocchia, R., Lyons, J. B., and Officer, C. B., 1994b, Deposition of channel deposits near the Cretaceous-Tertiary boundary in northeastern Mexico: Catastrophic or "normal" sedimentary deposits?: Reply: Geology, v. 22, p. 958.

Rocchia, R., Boclet, D., Bonté, Ph., Froget, L., Galbrun, B., Jéhanno, C., and Robin, R., 1992, Iridium and other element distributions, mineralogy, and magnetostratigraphy near the Cretaceous/Tertiary boundary in Hole 761C, *in* von Rad, U., Haq, B. U., O'Connell, S., and others, Proceedings, Ocean Drilling Program, Science Results, Volume 122: College Station, Texas, Ocean Drilling Program, p. 753–762.

Rocchia, R., Froget, L., Robin, E., and Gayraud, J., 1994, Stratigraphical distributions of cosmic markers at the KT boundary in the Carribean area [abs.]: New developments regarding the K/T event and other catastrophes in Earth history, Houston, Texas, Lunar and Planetary Institute Contribution 825, p. 97–99.

Rumble, D., 1976, Oxide minerals in metamorphic rocks, *in* Rumble, D., ed., Reviews in mineralogy, Volume 3: Washington, D.C., Mineralogical Society of America, p. R1-R24.

Schenk, P., and Moses, J., 1994, The horrendous space kablooey at Jupiter (the fate of comet Shoemaker-Levy 9): Lunar and Planetary Information Bulletin, v. 72, p. 2.

Schmitt, R. A., 1989, Cometesimal explosions in Earth's atmosphere (CEEA) and steep angle cometesimal cratering (SACC) events cause Cretaceous/Tertiary (K/T) extinction phenomena [abs]: Lunar and Planetary Science, v. 20, p. 962–963.

Schultz, P. H., 1994, Visualizing the nature and consequences of the Chicxulub impactor: Clues from Venus [abs]: New developments regarding the K/T event and other catastrophes in Earth history: Houston, Texas, Lunar and Planetary Institute Contribution 825, p. 104–106.

Schultz, P. H., and Gault, D. E., 1982, Impact ejecta dynamics in an atmosphere: Experimental results and extrapolations, *in* Silver, L. T., and Schultz, P. H., eds., Geological implications of impacts of large asteroids and comets on the Earth: Geological Society of America Special Paper 190, p. 153–174.

Schultz, P. H., and Gault, D. E., 1990, Prolonged global catastrophes from oblique impacts, *in* Sharpton, V. L., and Ward, P. D., eds., Global catastrophes in Earth history: Geological Society of America Special Paper 247, p. 239–261.

Schultz, P. H., Koeberl, C., Bunch, T., Grant, J., and Collins, W., 1994, Ground truth for oblique impact processes: New insight from the Rio Cuarto, Argentina, crater field: Geology, v. 22, p. 889–892.

Sharpton, V. L., Dalrymple, G. B., Marin, L. E., Ryder, G., Schuraytz, B. C., and Urrutia-Fucugauchi, J., 1992, New links between the Chicxulub impact structure and the Cretaceous/Tertiary boundary: Nature, v. 359, p. 819–821.

Sigurdsson, H., D'Hondt, S., Arthur, M. A., Bralower, T. J., Zachos, J. C., Van Fossen, M., and Channell, J. E. T., 1991, Glass from the Cretaceous-Tertiary boundary in Haiti: Nature, v. 349, p. 482–487.

Smit, J., and Hertogen, J., 1980, An extraterrestrial event at the Cretaceous-Tertiary boundary: Nature, v. 285, p. 198–200.

Smit, J., and Kyte, F. T., 1984, Siderophile-rich magnetic spheroids from the Cretaceous-Tertiary boundary in Umbria, Italy: Nature, v. 310, p. 403–405.

Smit, J., and 8 others, 1992, Tektite-bearing, deep-water clastic unit at the Cretaceous-Tertiary boundary in northeastern Mexico: Geology, v. 20, p. 99–103.

Stinnesbeck, W., and 10 others, 1993, Deposition of channel deposits near the Cretaceous-Tertiary boundary in northeastern Mexico: Catastrophic or "normal" sedimentary deposits?: Geology, v. 21, p. 797–800.

Swisher, C. C., III, and 11 others, 1992, Coeval $^{40}Ar/^{39}Ar$ ages of 65 million years ago from Chicxulub crater melt rock and Cretaceous-Tertiary boundary tektites: Science, v. 257, p. 954–958.

Taylor, S. R., and McLennan, S. M., 1985, The continental crust: Its composition and evolution: Oxford, Blackwell, 233 p.

Weast, R. C., Astle, M. J., and Beyer, W. H., 1989, Handbook of chemistry and physics: Boca Raton, Florida, CRC Press, 370 p.

Whipple, F. L., 1950, The theory of micro-meteorites. I: In an isothermal atmosphere: Proceedings of National Academy of Science of America, v. 36, p. 687–695.

Manuscript Accepted by the Society September 14, 1995

Printed in U.S.A.

Geological Society of America
Special Paper 307
1996

Oxidation state of iron in tektite glasses from the Cretaceous/Tertiary boundary

N. Oskarsson
Nordic Volcanological Institute, University of Iceland, IS-101 Reykjavík, Iceland
O. Helgason
Science Institute, University of Iceland, IS-107 Reykjavík, Iceland
H. Sigurdsson
Graduate School of Oceanography, University of Rhode Island, Narragansett, Rhode Island 02882

ABSTRACT

Tektite glasses from Beloc, Haiti, formed by impact melting, show bimodal composition of dominantly silicic melt of crustal origin (black glass) and a minor CaO-rich melt of sedimentary origin (yellow glass). Low-velocity Mössbauer spectra of the black glass show the typical broad line doublet of amorphous silicates, and high velocity spectra at 78°K confirm the absence of crystalline iron oxides larger than about 10 nm. High wt%Fe(III)/Σwt%Fe ratios are confirmed by Mössbauer spectroscopy and chemical analysis. Thus ferric iron represents 40% and 74% of total iron in the black and yellow glass, respectively. The corresponding calculated oxygen fugacities are the highest reported for tektite glasses, or log fO_2 about –5 to –6 for the yellow glass and log fO_2 about –7 for the black glass, as compared to log fO_2 = –10.5 in typical australite glass.

The composition of the trace of volatiles in the black glass, which is degassed at 1,000 to 1,250 °C, is 69 mol% H_2O and 31 mol% CO_2, indicating that the source material contained water and carbonate. Isotopic composition and anomalously high sulfate and alkaline earths of the glasses indicate material input from marine sediments. The observed high oxidation state of iron in the K/T boundary glasses suggests mixing between sources of different intrinsic oxygen fugacities: a crustal source close to the quartz-magnetite-fayalite (FMQ) buffer and a volatile-bearing sedimentary source of high intrinsic oxygen fugacity. Incipient shock melting of volatile-bearing phases produced a congruent melt. The redox state of iron in the melted material adjusted to the overall oxygen fugacity of the sources during incipient melting. At still higher pressures, the volatiles escaped, leaving a dry melt. The two glass types represent melting of a stratified target, and they preserve the different overall intrinsic oxygen fugacities of their sources. The black glass represents crustal rocks containing few percent of volatile-bearing phases, whereas the yellow glass represents strata richer in marine sediments. During ejection the physically separated melts were only partially mixed. The glasses thus represent impact melts from a stratified sediment-bearing target in contrast to other tektites, which represent total melting of dry rocks with ferric/ferrous ratios controlled by the FMQ oxygen buffer at 870 to 1,120 °C.

Oskarsson, N., Helgason, O., and Sigurdsson, H., 1996, Oxidation state of iron in tektite glasses from the Cretaceous/Tertiary boundary, *in* Ryder, G., Fastovsky, D., and Gartner, S., eds., The Cretaceous-Tertiary Event and Other Catastrophes in Earth History: Boulder, Colorado, Geological Society of America Special Paper 307.

INTRODUCTION

Tektite glasses in the Beloc section of the K/T boundary in Haiti display a range in composition between a dominant black siliceous glass and a rare yellow lime-glass (up to 30 wt% CaO). Sigurdsson et al. (1991a, b) have proposed a derivation of the two glass types by fusion of a terrestrial target rock during bolide impact. The melted materials are likely continental crust of broadly andesitic composition and evaporite- and carbonate-rich sediments. Analysis of $^{18}O/^{16}O$ in the two glass types supports this hypothesis (Sigurdsson et al., 1991b), giving $\delta^{18}O$ relative to Standard Mean Ocean Water (SMOW) of about +7‰ and +13‰ in the black and yellow glasses, respectively. This range, which is wider than observed for all other tektites combined, may reflect a bimodal crustal and marine origin of the glasses; the low $\delta^{18}O$ in the andesitic black glass is similar to that of pristine igneous rocks of a relatively unaltered crust. The $\delta^{18}O$ data thus reveal the terrestrial sources of the glasses and put geological constraints on the strata suffering fusion.

Three tektite types were studied in the present work: the more common black glass and the rare yellow glass variety collected at the Beloc outcrop of the K/T boundary in Haiti (Sigurdsson et al., 1991a), and an australite from Cooper Creek, Australia. In Table 1 the chemical compositions of the black glass and the yellow glass from the K/T boundary are compared with that of typical australite. The black glass has a composition broadly similar to continental crust or average andesite but is distinctly high in CaO and MgO when compared to volcanic glasses at similar silica content (Fig. 1). The outstanding chemical signature of the yellow glass is high sulfur, CaO, Ba, and Sr compared with igneous rock suites (Sigurdsson et al., 1992). There is strong evidence that the enrichment of the alkaline earths in this glass was brought about by the melting of sulfate/carbonate-bearing strata (Sigurdsson et al., 1991b, 1992). It can therefore be assumed that the breakdown and melting of anhydrite/gypsum and calcite/dolomite were important processes during the formation of the K/T boundary impact glass. Degassing of marine sedimentary strata yielding an oxidizing gas phase that is dominated by H_2O, CO_2, and SO_2 is thus among the processes that will be evaluated here to account for the redox state of iron in the K/T boundary impact glasses.

Breakdown and melting of volatile-bearing minerals in high-temperature silicate systems are essentially of three types: (1) incongruent melting of hydrous silicate minerals that produces anhydrous silicate minerals and/or oxides and a volatile-saturated liquid; (2) incongruent melting of hydrous silicate minerals that produces anhydrous silicate minerals and/or oxides, a volatile-saturated liquid, and a gas/fluid phase; and (3) incongruent breakdown of carbonates and/or sulfates associated with the formation of a gas/fluid phase and oxide.

In any case, the new solid phases can be assumed to control the composition of the new liquid, and iron-bearing phases can be assumed to control the oxygen fugacity of the volatiles dissolved in the liquid and in the evolved gas phase. In terms of oxygen fugacity, a continued equilibrium melting of most terrestrial rocks will proceed along the quartz-magnetite-fayalite (FMQ) oxygen buffer as long as the buffering mineral assemblage is present. In volcanic systems the evolved volatiles may escape and alter the overall oxygen budget of the system during their long-term evolution.

In contrast, rapid total melting of a crustal domain, such as during bolide impact, is likely to produce a melt that approaches the intrinsic oxygen fugacity of the material being melted. At temperatures above the liquidus of the crustal material in a closed system, the intrinsic oxygen fugacity of the source is thus preserved, since all iron-containing minerals are exhausted and a nonbuffered system is established. Ferric/ferrous ratios of most tektite glasses reported so far resemble terrestrial rocks at the FMQ oxygen buffer at temperatures between 870 and 1,120 °C (Fudali et al., 1987).

Experimental shock melting of hydrous minerals illustrates the role of volatiles in the formation of tektite glasses (Lambert and Lange, 1984). At pressures above 25 to 30 GPa, congruent melting starts to take place. The congruent shock melting behavior involves the complete dissolution of the volatiles during the formation of the melt and the pervasive infiltration of the melt along grain boundaries of unmelted crystals. The homogenization of the volatile/silicate system is thus an integral part

TABLE 1. CHEMICAL COMPOSITION OF TEKTITES

	Black glass*	Yellow glass†	Australite*
(Wt. %)			
SiO_2	63.33	49.31	72.40
TiO_2	0.63	0.65	0.75
Al_2O_3	14.11	13.11	12.93
FeO	6.07	4.57	4.93
MnO	0.14	....	0.10
CaO	7.90	23.23	3.92
MgO	2.93	4.09	1.97
Na_2O	3.16	2.85	0.71
K_2O	1.34	0.71	2.08
P_2O_5	0.12	....	0.06
SO_3	....	0.83	....
(ppm)			
Ba	1,876	550	440
Cr	24	25	36
Cu	24	...	34
Ni	17	13	22
Sc	18	17	11
Sr	396	1,620	181
Rb	34	17	...
V	104	...	328
Y	27	...	34
Zn	59	50	74
Zr	123	95	291

*New analysis of the black glass and the australite were made by ICP-OES spectrometry. Sample was fluxed with lithium tetraborate and the glass bead dissolved in hydrochloric/nitric acid mixture. Total iron is expressed as FeO.

†Microprobe average-1 of yellow glass is from Sigurdsson et al., 1992.

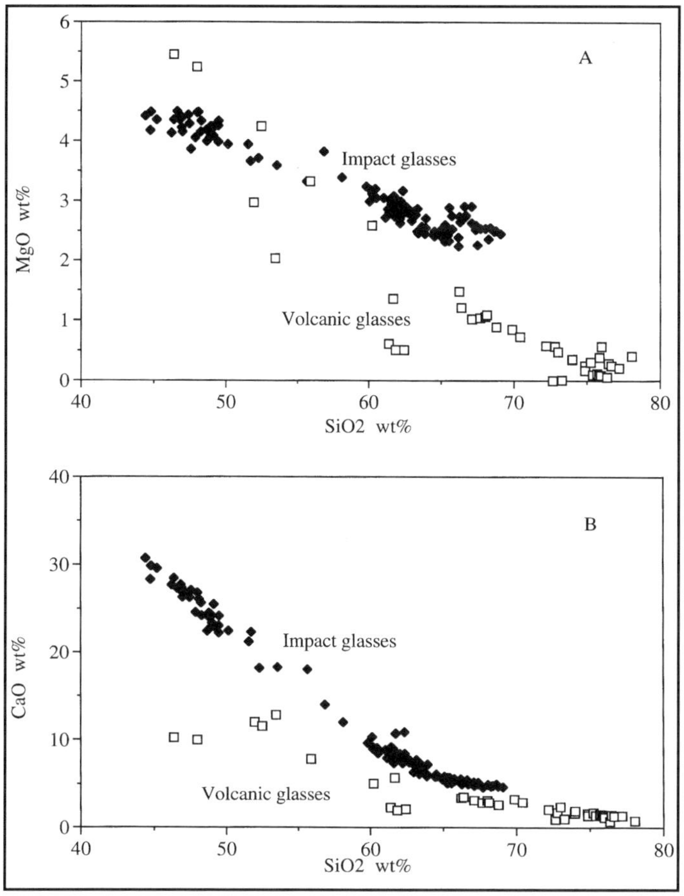

Figure 1. Comparison of the chemical composition of K/T boundary black impact glasses and volcanic glasses. The black impact glasses are systematically high in CaO and MgO as compared to volcanic glasses. Volcanic glass electron microprobe data base from Palais and Sigurdsson (1989) and unpublished data.

of the liquid formation. At still higher pressure total melting is reached, and in the interval 60 to 70 GPa all the volatiles are expelled out of the silicate melt before quenching, leaving the typically dry tektite-type glass. The important experimental evidence for the present discussion is the homogenization of the shock melted system: volatiles are dissolved during congruent breakdown and degassed from solution by the formation of microbubbles and their coalescence and degassing. There is little doubt that the pervasive equilibration of oxygen fugacity, including the redox processes of iron, is inevitable on the scale of melt segregation within a shock-melted rock layer. A time constraint on redox processes in magmatic systems is set by the diffusion speed of oxygen through large magma bodies (Helgason and Steinthorsson, 1992). The diffusion speed criteria are less important in shock melting of volatile-containing phases since dispersion of the melt is a part of the melting process.

A target rock consisting of evaporite- and carbonate-rich sediments and continental crustal or silicic igneous rocks is likely to produce melts of different oxygen fugacities. The high intrinsic oxygen fugacity of the sedimentary component and the presumably lower intrinsic oxygen fugacity of the siliceous rock will approach each other at their melted junctions. The resulting oxidation state of iron in the melt thus largely depends on the magnitude of the different material sources and their mixing. The oxidation state of iron is the single variable that preserves constraints on oxygen fugacities during a bolide impact. It is to be expected that oxygen fugacities observed for tektites, based on ferric/ferrous analysis (Fudali et al., 1987; Helgason and Steinthorsson, 1989; Wivel and Mörup, 1981), will in most cases reflect intrinsic fugacities of dry crustal rocks. An account for the high oxidation state of iron in the K/T boundary glasses, however, involves the evaluation of their relative content of crystalline crust and volatile-bearing marine sediments as well as the placement of constraints on intrinsic oxygen fugacities of the sources.

ANALYTICAL METHODS

A portion of sample from the K/T boundary deposit in Haiti was disaggregated ultrasonically in a dilute HCl solution. The black glass spherules and yellow glass particles were hand picked from the >1-mm sieve fraction. They were washed in 0.1N HNO_3 in an ultrasonic bath for 1 hr to remove minor smectite fragments and coating, leached with deionized water until the reaction was neutral, and finally dried at 110 °C for 2 hr. Microthermometric analysis of the black glass was made on a Linkam heating stage. Chemical analysis of the black tektite glass and the australite (Table 1) was made by induction-coupled plasma optical emission spectrometry (ICP-OES) after fusion with lithium metaborate and dissolution in acid.

Samples for Mössbauer analysis and chemical analysis of ferrous iron were powdered in an agate mortar under propanol. Mössbauer spectra were recorded with a conventional constant accelerator spectrometer. The ^{57}Co source was approximately 10 mCi in a rhodium matrix. Line width of the calibration spectrum (full-width half maximum) is 0.24 mm/s, indicating resolving power near the theoretical limit set by Heisenberg's principle of uncertainty. Velocities are given relative to the centroid of the spectrum for metallic iron at room temperature. Low-velocity scans at room temperature were used to determine the ferric/ferrous ratio. These spectra show the unresolved superposition of doublets of each of the oxidation states characteristic for amorphous silicates (Wivel and Mörup, 1981; Dyar et al., 1987). As shown elsewhere (Helgason and Steinthorsson, 1989), fitting the spectra with two broad line doublets for ferrous iron and one doublet for ferric iron is sufficient to determine the Fe(III)/ΣFe within 3%. As indicated in Figure 2, this procedure was used to fit the tektite spectra. For further testing of the amorphous character of the tektites, low-temperature scans were run at 78°K. At 78°K, superparamagnetic particles larger than 8 to 10 nanometers would show typical six-lines magnetic splitting. No sign of crystalline iron oxides was, however, detected in the tektites at that temperature.

The powder that was used for Mössbauer analysis of the black glass was split into two batches for duplicate chemical

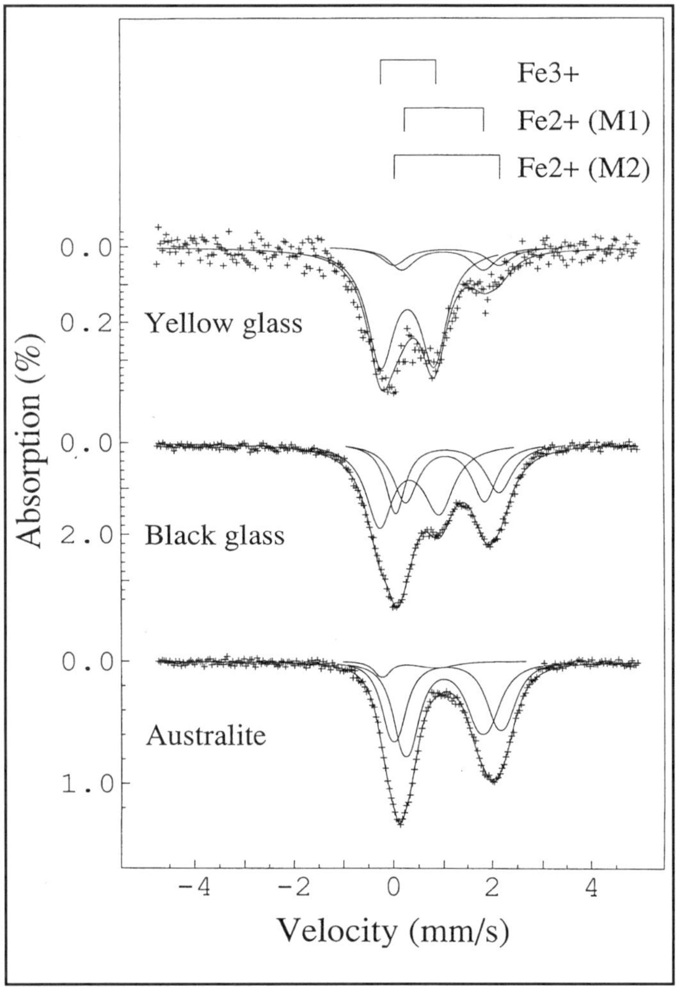

Figure 2. Mössbauer spectra of black and yellow impact glasses from the K/T boundary (upper), compared with australite (lower) The doublets for Fe(II) and Fe(III) and M1 and M2 sites for Fe(II) are marked in the upper right of the figure. Note the significant resolution of the Fe(III) doublet for the yellow glass, as compared with the extremely reduced australite. The different absorption intensity reflects the sample size and iron content. The spectrum for the yellow glass was recorded on only 11.2 mg of sample.

microanalysis of ferrous iron by Wilson's method (Jeffery, 1975). Samples were dissolved in basic vanadate/fluoride solution. Titration of excess vanadate was made in sulfuric acid/boric acid mixture, with ammonium iron sulfate standardized against primary grade potassium dichromate. Barium diphenylamine sulphonate was used as redox indicator in the titration. The results from the Mössbauer analysis and the chemical analysis of ferric/ferrous iron are listed in Table 2.

Analysis of volatiles dissolved in the glasses was performed on a quadropole mass analyzer after vacuum fusion to 1,250 °C in a silica-glass oven (Table 2). The instrumentation and calibration of the mass analyzer was described by Jakobsson and Oskarsson (1990). Prior to the gas analysis, a vacuum pyrolysis experiment was run to 1,250 °C on the black glass in order to determine a degassing profile of the tektite glass.

ANALYTICAL RESULTS

High-velocity Mössbauer spectra of the tektite glasses confirm that the analyzed samples contain no iron-bearing crystals in sizes exceeding about 10 nm. This amorphous state, in itself, suggests high temperatures of formation and rapid cooling of the material. Figure 2 shows Mössbauer spectra of the glasses in terms of absorption versus scan velocity (mm/s), for both the yellow and black glasses from the K/T boundary (upper half) and for the australite. The spectrum for the yellow glass is exceptional. Although this iron-poor sample was only 11.2 mg, as reflected in the low absorption values, the convolution of the spectrum was successful and shows its high oxidation state. A clear resolution of the high ferric iron is also shown in the spectrum for the black glass. The widely different spectrum for the australite is characteristic for a highly reduced iron.

Ferric iron, expressed as weight percentage of total iron (%Fe(III) = 100 • wt%Fe(III)/wt%Fe • ΣFe), in the tektite glasses is listed in Table 2. The value %Fe(III) from chemical analysis and Mössbauer spectra compares very well, as shown in the table. The determined value of %Fe(III) can be used to calculate the fugacity of oxygen in the melt that prevailed during the quenching of the tektite. In Table 3, the calculated log fO_2 for the tektites is given, with the analyzed %Fe(III), based on calculation using the data from Fudali (1965), Hill and Roeder (1974), and Helgason and Steinthorsson (1989) for the relationship between fugacity and %Fe(III). Some discrepancy of calculated fugacities for oxidized liquid is seen for the nonlinear data fit of Helgason and Steinthorsson, which gives systematically higher fO_2 as compared with the earlier studies. We emphasize that 1 to 2 log units in fugacity at the oxidized extreme have little effect upon the subsequent discussion. The important finding is that the data indicate an extreme oxidation of the K/T boundary glasses, corresponding to a minimum of log fO_2 –5 to –6 for the yellow glass and a minimum of log fO_2 –7 for the black glass. The australite sample is, on the other hand, highly reduced with log fO_2 –10.5, which corresponds well with earlier work on other tektites (Fudali et al., 1987).

TABLE 2. IRON AND VOLATILES IN TEKTITE GLASSES

Sample	Fe(III)* (%)	Fe(III)† (%)	Gas§ (wt. %)	CO_2§ (mol %)	H_2O§ (mol %)
Black glass-A	43.7	43.0	0.06	64	36
Black glass-B		40.0			
Black glass (78° K)		39.0			
Yellow glass		74.0	0.01	69	31
Australite		6.0			

*Fe(III) in glasses as determined by chemical analysis. Analysis of total iron was made by ICP-OES spectrometry (Table 1).
†Fe(III) in glasses, as determined by Mössbauer analysis.
§Total volatiles, carbon dioxide, and water mol% in glasses, as determined by pyrolysis and Quadropol mass spectrometry.

TABLE 3. FERRIC/FERROUS RATIO AND OXYGEN FUGACITY IN IMPACT GLASSES*

Sample	Fe(III) (%)	A Log fO_2	B Log fO_2	C Lof fO_2
Black glass-A	43.0	-6.81	-6.92	-4.81
Black glass-B	40.0	-6.95	-7.08	-5.01
Black glass (78° K)	39.0	-7.01	-7.13	-5.08
Yellow glass	74.0	-5.69	-5.72	-3.29
Australite	6.0	-10.87	-11.27	-10.31
Fud-87 minimum	12.0	-9.44	-9.74	-8.37
Fud-87 maximum	2.0	-13.13	-13.69	-13.37

*Table shows percentage of Fe(III) and calculated oxygen fugacities for the K/T boundary black and yellow glasses, the australite glass, and the extreme compositions of tektites reported by Fudali et al., 1987. The oxygen fugacity values in columns A, B, and C are based on calculations using constants from the following sources: A = Fudali, 1965; B = Hill and Roeder, 1974; and C = Helgason and Steinthorsson, 1989.

The water content of the black and yellow glasses was determined by vacuum fusion/mass spectrometry to be 0.013 and 0.021 wt%, respectively, which is typical for tektites and impact glasses (Koeberl and Beran, 1988). The composition of residual volatiles in the glass (Table 2) is 64 mol% H_2O and 36 mol% CO_2 for the black glass and 69 mol% H_2O and 31 mol% CO_2 for the yellow glass. The slightly higher CO_2/H_2O ratio for the black glass may reflect its higher content of alkalis (Table 2). Data from Hampton and Bailey (1984) show that residual CO_2 in volcanic glasses generally increases with increasing alkalis. A vacuum pyrolysis experiment run on the black glass to 1,250 °C reveals a degassing profile in which both carbon dioxide and water are released in the temperature interval 1,000 to 1,250 °C. This is different from most silicic or intermediate volcanic glasses, in which release of volatiles starts above about 800 °C but shows a maximum at the melting point of the glass (Hampton and Bailey, 1984). Although no pyrolysis experiment was made on the yellow glass, it was observed during fusion that maximum pressure and volatile yield occurs at about 1,050 °C.

Microthermometric experiments were also performed on the black glass in nitrogen atmosphere. During slow heating, an opaque phase is formed in the temperature interval 900 to 950 °C which is close to the softening point of the glass. A clear liquid is formed at temperatures above 1,050 °C. During slow cooling, an opaque phase become visible at about 1,050 °C. For the purpose of the following discussion we summarize the results as follows:

1. Compared to other tektites, the K/T boundary impact glasses are exceptionally oxidized, as inferred from ferric/ferrous ratios and the high content of oxidized (sulfate) sulfur (Sigurdsson et al., 1991b).

2. The K/T boundary glasses are low in volatiles, as are all other tektites analyzed so far (Koeberl, 1986; Koeberl and Beran, 1988). The residual gases are released at a higher temperature (1,000 to 1,250 °C) than from volcanic glasses.

3. Microthermometric analysis indicates that the glasses were quenched at temperatures above 1,050 °C and sulfur solubility experiments, simulating the genesis of the yellow glass, indicate temperature of about 1,300 °C for their quenching (Sigurdsson et al., 1991b).

4. The gas phase associated with the ejection of the K/T impact glass contained CO_2 and H_2O in addition to SO_3 and SO_2 suggested by Sigurdsson et al. (1991b).

DISCUSSION

Constraints on the oxygen fugacity/temperature relations during the formation of impact glasses or tektites are set by the intrinsic fugacities of the major systems involved in their formation. The intrinsic oxygen fugacity of the melted silicate material (continental crust or rock with broadly andesitic composition) was likely close to the FMQ oxygen buffer, which is typical for crustal rocks. Only the surface layers of the crust can be assumed to have intrinsic fugacities close to the hematite-magnetite oxygen buffer (HM), and extensive rock suites at the iron-wüstite oxygen buffer (IW) are not known. These oxygen fugacity/temperature relations (FMQ, HM, and IW) are indicated in Figure 3, based on data from Robie et al. (1978). The redox values for iron in other tektites are shown in the lower left of the figure (Fudali et al., 1987). It is evident from the figure that tektites other than the K/T boundary glasses fall within the fugacity range of the FMQ buffer at 870 to 1,120 °C. Total melting of rocks in that redox state would inevitably produce

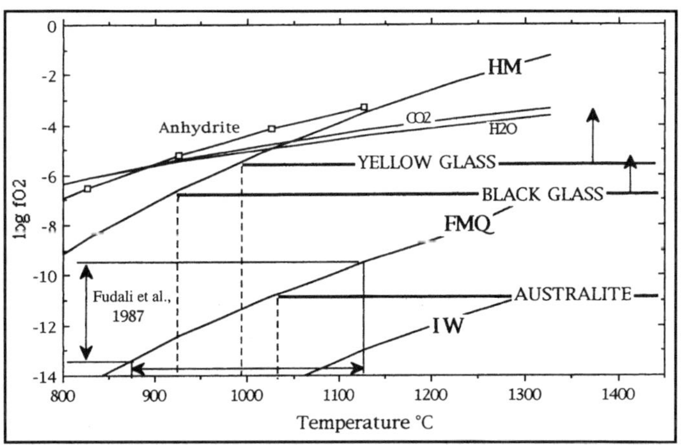

Figure 3. Oxygen fugacity versus temperature for the hematite-magnite (HM), quartz-magnetite-fayalite (FMQ), and iron-wüstite (IW) mineral buffers. The oxygen fugacity–temperature relations for the gas phase derived from breakdown of anhydrite are indicated in the upper left by a curve drawn through open squares. The oxygen fugacity/temperature relations indicated by a pair of curves in the upper middle of the figure are for for steam (lower) and carbon dioxide (upper). Calculations of the fugacity relations (see text) are based on free energy data from Robie et al. (1978). The oxygen fugacity of yellow and the black glasses from the K/T boundary is indicated by horizontal lines. Note the vertical arrows, which indicate their fugacity values, as calculated from Helgason and Steinthorsson (1989). Fugacity ranges of other tektites as reported by Fudali et al. (1987) are shown in the lower left.

glasses that are apparently "reduced" at temperatures about 1,300 °C, as they are no longer buffered by mineral buffers. The australite glass analyzed for reference by us is thus within that range of "normal" tektites (Fudali et al., 1987). Calculated fugacities for the K/T boundary black and yellow glasses are indicated in Figure 3. It may be argued, along similar lines, that the K/T boundary glasses represent sources that initially were on the HM buffer at temperatures in the 900 to 1,000 °C range, as indicated by dashed vertical lines in Figure 3. That hypothesis can, however, be excluded on a number of grounds. First, the K/T boundary glasses are aphyric, and second, their quench temperature was near or above 1,300 °C according to their sulfur solubility (Sigurdsson et al., 1991b). Third, based on their isotopic and chemical composition, the glasses are evidently nonigneous, as discussed above.

In the following paragraphs we examine the effect of total melting of a heterogeneous source, consisting of rocks at the FMQ buffer and marine sediments containing water, carbon dioxide, and sulfates. A complicated system of derived oxygen fugacities can be expected to arise when the oxidizing volatiles of marine sediments are involved, in addition to the crustal mineral buffers. Considering first the release of steam and carbon dioxide, the limiting fugacity relations can be assumed to be close to the fugacity relations of the pure gases during their thermal decomposition:

$$H_2O = H_2 + 1/2O_2,\ K = pH_2 \bullet (pO_2)^{1/2}/pH_2O \quad (1)$$

$$CO_2 = CO + 1/2O_2,\ K = pCO \bullet (pO_2)^{1/2}/pCO_2 \quad (2)$$

At 1 atm total pressure, the mass balance relations of these reactions, when written in the form A = B + 1/2C, require that: pB = 2pC and pA = 1 – pB – pC = 1 – 3C, and the expression for the equilibrium constant becomes:

$$K = 2C \bullet (C)^{1/2}/(1 - 3C) \quad (3)$$

This leads to a third-order expression for C (pO_2 in eqs. 1 and 2):

$$C^3 - 9/4K^2C^2 + 3/2K^2C = K^2/4 \quad (4)$$

These relations for oxygen fugacity of H_2O (lower) and CO_2 (upper) at 1 atm total pressure are plotted in Figure 3 as a function of temperature, based on data from Robie et al. (1978). These relations are practically identical with those for systems in which the partial pressure of the dissociating gases is held at 1 atm. It is clear that both reactions exert oxygen fugacities that are considerably lower than that of the HM buffer at temperatures at about 1,300 °C but significantly higher than the FMQ buffer. It is emphasized here that gas mixtures are not buffer systems but can be looked upon as oxidizing or reducing media.

A different approach is taken for modeling oxygen fugacity relations during the breakdown of anhydrite, according to:

$$CaSO_4(s) = CaO(s) + SO_3(g)^* \quad (5)$$

The reaction produces lime and sulfur trioxide, for which partial pressure is a function of temperature. The partial pressure of SO_3* (marked with asterisk to indicate equilibrium with anhydrite) can be taken as absolute maximum during melting of ahydrite or gypsum-rich marine sediments and can also be taken to prevail as long as anhydrite is present. As long as anhydrite yields a temperature-controlled partial pressure of SO_3*, the rapid breakdown of the trioxide into oxygen and the dominating sulfur dioxide is therefore an indirect oxygen buffer reaction, according to:

$$SO_3^* = SO_2 + 1/2O_2,\ K = pSO_2^*(pO_2)^{1/2}/pSO_3 \quad (6)$$

The mass balance of the reaction requires that pSO_2 = $2pO_2$, which—after substituting into equation 6—gives:

$$pO_2 = (pSO_3^*K^2/4)^{1/3} \quad (7)$$

The oxygen fugacity according to equation 7 is shown in the upper left of Figure 3 (open squares) as a function of temperature. This is the most oxidizing gas phase that can be expected during the melting of marine sediments. Thermodynamic data for the breakdown of anhydrite only allow estimation of this buffer reaction up to 1,127 °C. It is, however, evident that the continued reaction of the SO_2-dominated gas phase toward higher temperatures will reach partial pressures of oxygen close to the HM buffer.

We are aware that equilibrium thermodynamics of ideal gases only offer guidelines for the reconstruction of the redox history of tektite glasses, because of the extremely short duration of the high-energy impact event. However, the oxygen fugacity of the confining gas phase during quenching and dispersion of the tektite glass droplets sets limiting values for their oxygen fugacity. The interpretation of the ferric/ferrous ratio of tektite glasses is therefore less rigid than interpretation of this ratio in igneous rocks, due to the probably short reaction times and transient pressure changes involved. However, reaction rates between the impact melt and associated gas phase are likely to be accelerated, because of the very high or superliquidus temperatures involved. No attempt is made to estimate the effects of very high pressures on the fugacity of oxygen, but we note that increased pressure favors pervasive diffusion of volatiles within the system.

Based on thermodynamic criteria, the intrinsic fugacity of evaporite-containing sediments is very high at the liquidus temperature of its silicate component. It is therefore important to quantify the effect of volatile-bearing phases on the oxygen fugacity of a melt in which these phases are totally dissolved and degassed after equilibration with the iron in the system. Such simulation of shock melting of a volatile-bearing mineral assemblage involves the estimate of oxygen consumption of the ferrous iron during melting in a nonbuffered system. We have emphasized that both the iron system and the available volatiles will approach the same oxygen fugacity, that is, the intrinsic fugacity of the source. It is also underlined that the assumed target layers composed of sediments and igneous rocks are in redox disequilibrium before the impact. Therefore the onset of redox processes is spontaneous when the melting occurs.

A small addition of carbon dioxide is sufficient to oxidize most silicic rocks, originally at the FMQ buffer, to the level of the K/T boundary glasses. In Figure 3 the high oxygen fugacity of the carbon dioxide gas illustrates this property. The gas is reacted toward the CO/CO_2 ratio of the FMQ buffer according to equation 2, and the iron is oxidized toward the pure CO_2 system. Only a fraction of the carbon dioxide is, however, available for oxidation. The CO/CO_2 ratio according to equation 2 at the FMQ buffer at 1,200 °C is about 0.15. This implies that about 15% of added carbon dioxide dissociates while the FMQ buffer is active. In the shock-melting scenario the mineral buffers are exhausted and the final unbuffered CO/CO_2 of the system will reflect the intrinsic fugacity of the source. An important aspect of this discussion is the effect of carbon dioxide release from a carbonate mineral and its contribution to the molten unbuffered system.

A calculation of this effect is illustrated in Figure 4, in which the %Fe(III) of a source, initially with 6wt% FeO and %Fe(III) = 10, is plotted against wt% of incorporated $CaCO_3$. Three cases are illustrated: with 6, 8, and 12% dissociation of carbon dioxide. It is emphasized here that the higher percentages could as well apply to mixtures of carbon dioxide and steam and of sulfur dioxide. The composition of the strata assumed to be the source of the K/T boundary glasses indicates the presence of all three volatiles. The initial %Fe(III) 10 is

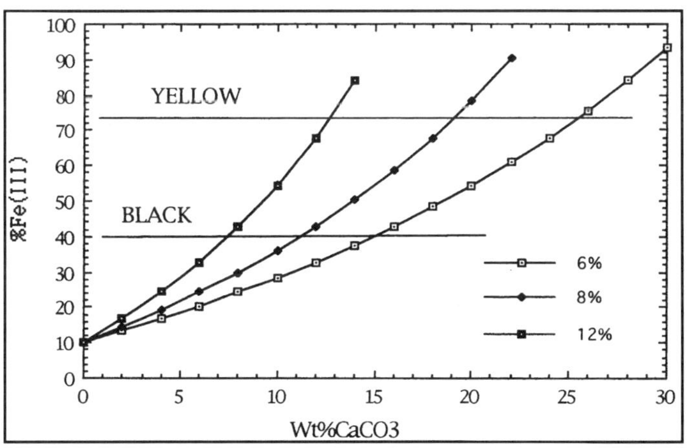

Figure 4. Percentage of oxidized iron in rock/carbonate mixture, relative to total iron, plotted against wt% of calcium carbonate in the mixture. The calculation is based on 6% total iron (as FeO) by weight in the rock fraction. Oxidation of FeO is calculated according to 2FeO + 1/2 O_2 = Fe_2O_3. It is noted that 1 mole of dissociated carbon dioxide could oxidize 2 mole of FeO. The three curves in the figure represent different yields of free oxygen from dissociation of carbon dioxide. The observed percentages of Fe(III) in the K/T boundary impact glasses are indicated by horizontal lines.

selected to be within the normal continental crustal range. This example is conservative in the assumed oxidation effects of the volatiles and in the selection of initial %Fe(III) but serves its purpose of illustrating how only a small amount of volatiles is needed to consume the ferrous iron, provided the reaction is allowed to take place. It is clear from Figure 4 that the black glass can be produced from few percent evaporite. At 12% dissociation of the gas phase, only about 8% calcium carbonate is needed. It may be inferred from Figure 1 and Sigurdsson et al. (1992) that about 2 wt% of CaO and about 2 wt% MgO were added to the continental or igneous source during formation of the melt that quenched to form the black glass. This amount of oxides corresponds to about 8 to 10% calcium carbonate, and that amount is about what is needed to oxidize the igneous iron at %Fe(III) = 10 to the observed value of 40%. Similarly the yellow glass, which is over 20 wt% CaO, can be fully oxidized by the carbon dioxide contained in the corresponding 40% calcium carbonate.

The K/T boundary glasses bear imprints of sedimentary components, and there is a great deal more of this sedimentary component in the yellow glass. The above hypothesis on instantaneous redox equilibration of iron during melting, and the preservation of that equilibriun state also require that the melts have different oxygen fugacities. Different amounts of the volatile-bearing components in the yellow and the black glasses would also result in different oxygen fugacities. This is, indeed, the case and suggests that the glasses represent impact melts from a stratified source. The sediment-rich (upper) source of the yellow glasses produced more-oxidized liquids that did not, in terms of oxygen fugacity, equilibrate with the (lower) sediment-poor melt that formed the black glass. This raises implications for the depth of impact melting in general, since the dominating part of the Beloc glass (>90%) is derived from the strata underlying some 3 km of sedimentary rocks.

CONCLUSIONS

K/T boundary impact glasses are uniquely high in oxygen fugacity, with log fO_2 about –5 to –6 for the yellow glass and log fO_2 about –7 for the black glass. In comparison, an australite glass is highly reduced, with log fO_2 –10.5, which is typical of other tektites. A preferred model for the high oxidation state of iron in the black and yellow glasses from the K/T boundary is that they formed at the intrinsic oxygen fugacity of melted crustal strata that contained different amounts of marine chemical sediments, such as evaporite and carbonate. The oxidation state of an impact melt represents the overall intrinsic fugacity of the crust being melted. The oxygen donors of the marine sediments are carbon dioxide, water, and sulfate. Different oxygen fugacities of the yellow and black glasses suggest melting of a stratified target in which an upper sediment-rich zone yielded the yellow glass and a deeper rock-dominated zone yielded the black glass.

ACKNOWLEDGMENTS

This work was in part supported by grant EAR-9219121 from the National Science Foundation and grant NAGW3496 from NASA. We thank Frank Senftle for supplying a sample of Cooper Creek australite glass for Mössbauer and chemical analysis. We thank Paul Warren and an anonymous reviewer for their comments on the manuscript.

REFERENCES CITED

Dyar, M. D., Naney, M. T., and Swanson, S. E., 1897, Effects of quench methods on Fe^{3+}/Fe^{2+} ratios: A Mössbauer and wet-chemical study: American Mineralogist, v. 72, p. 792–800.

Fudali, R.F., 1965. Oxygen fugacities of basaltic and andesitic magmas: Geochimica et Cosmochimica Acta, v. 29, 1063–1075.

Fudali, R. F., Dyar, M. D., Griscom, D. L., and Schreiber, H. D., 1987, The oxidation state of iron in tektite glass: Geochimica et Cosmochimica Acta, v. 51, p. 2749–2756.

Hampton, C. M., and Bailey, D. K., 1984, Gas extraction experiments on volcanic glasses: Journal of Non-Crystalline Solids, v. 67, p. 147–168.

Helgason, Ö., Steinthorsson, S., and Mörup, S., 1992, Rates of redox reactions in basaltic melts determined by Mössbauer spectroscopy: Hyperfine Interactions, v. 70, p. 985–988.

Helgason, Ö., and Steinthorsson, S., 1989. The ferric/ferrous ratio in basalt melts at different oxygen pressures: Hyperfine Interactions, v. 45, p. 287–294.

Hill, R., and Roeder, P., 1974, The crystallization of spinel from basaltic liquid as a function of oxygen fugacity: Journal of Geology, v. 82, p. 709–729.

Jakobsson, S., and Oskarsson, N., 1990, Experimental determination of fluid composition in the system C-O-H at high P and T and low fO_2: Geochimica et Cosmochimica Acta, v. 54, p. 355–362.

Jeffery, P. G., 1975, Chemical methods of rock analysis: Oxford, Pergamon Press, 525 p.

Koeberl, C., 1986, Geochemistry of tektites and impact glasses: Annual Review of Earth and Planetary Science, v. 14, p. 323–350.

Koeberl, C., and Beran, A., 1988, Water content of tektites and impact glasses and related chemical studies. Proceedings, 18th Annual Lunar and Planetry Science Conference, p. 403–408.

Lambert, P., and Lange, M. A., 1984, Glasses produced by shock melting and devolatilization of hydrous minerals: Journal of Non-Crystalline Solids, v. 67, p. 521–542.

Palais, J. M., and Sigurdsson, H., 1989, Petrologic evidence of volatile emissions from major historic and pre-historic volcanic eruptions, *in* Berger, A., Dickinson, R., and Kidson, J., eds., Understanding climate change: American Geophysical Union Geophysical Monograph 52, p. 31–53.

Robie, R. A., Hemingway, B. S., and Fisher, J. R., 1978. Thermodynamic properties of minerals and related substances at 298,15 K and 1 Bar (10^5 Pascals) pressure and at higher temperatures: U.S. Geological Survey Bulletin 1452, 456 p.

Sigurdsson, H., D'Hondt, S., Arthur, M. A., Bralower, T., Zachos, J. C., Fossen, M., and Channel, J., 1991a, Glass from the Cretaceous-Tertiary boundary in Haiti: Nature, v. 349, p. 482–486.

Sigurdsson, H., Bonté, Ph., Turpin, L., Chaussidon, M., Metrich, N., Steinberg, M., Pradel, Ph., and D'Hondt, S., 1991b, Geochemical constraints on source region of Cretaceous/Tertiary impact glasses: Nature, v. 353, p. 839–842.

Sigurdsson, H., D'Hondt, S., and Carey, S., 1992, The impact of the Cretaceous/Tertiary bolide on evaporite terrane and generation of major sulfuric acid aerosol: Earth and Planetary Science Letters, v. 109, p. 543–560.

Wivel, C., and Mörup, S., 1981, Improved computational procedure for evaluation of overlapping hyperfine parameter distributions in Mössbauer spectra: Journal of Physics E: Scientific Instrumentation, v. 14, p. 605.

Manuscript Accepted by the Society September 14, 1995

Printed in U.S.A.

Geological Society of America
Special Paper 307
1996

Fullerenes of possible wildfire origin in Cretaceous-Tertiary boundary sediments

Dieter Heymann
Department of Geology and Geophysics, Rice University, Houston, Texas 77251-1892
L. P. Felipe Chibante
Rice Quantum Institute and Departments of Chemistry and Physics, Rice University, Houston, Texas 77251-1892
Robert R. Brooks
Department of Soil Science, Massey University, Palmerston North, New Zealand
Wendy S. Wolbach
Department of Chemistry, Illinois Wesleyan University, Bloomington, Illinois 61702-2900
J. Smit
Department of Sedimentary Geology, Faculty Earth Sciences, The Free University of Amsterdam, 1083 HV Amsterdam, Netherlands
A. Korochantsev and M. A. Nazarov
Vernadsky Institute of Geochemistry and Analytical Chemistry, Academy of Sciences of Russia, Kosygin St. 19, 117975 Moscow, Russian Federation
Richard E. Smalley
Rice Quantum Institute and Departments of Chemistry and Physics, Rice University, Houston, Texas 77251-1892

ABSTRACT

A search for fullerenes C_{60} and C_{70} in clays from the Cretaceous-Tertiary boundary (KTB) was started because these fullerenes might have been formed by wildfires after the Chicxulub impact and might have become distributed globally along with iridium, charcoal, and soot. The first sites from which clays were studied by extraction with toluene followed by analysis by High Pressure Liquid Chromatography were Woodside Creek and Flaxbourne River in New Zealand. C_{60} was found in the clays at levels in the range 0.06 to 12.2 ppb. C_{70} was also present. Additional results are reported here for samples from Caravaca (Spain), Stevns Klint (Denmark), Elendgraben (Austria), Sumbar and Malyi Balkhan (Turkmenistan), Khoshak (Khazakhstan), and Tetrichkaro (Georgia). The toluene extracts of several of these clays contained no fullerenes, presumably because these were lost during demineralization or for other causes. C_{60} and C_{70} fullerenes were, however, found at Caravaca, Sumbar, and Malyi Balkhan, which have comparatively undisturbed K/T boundaries. The C_{70}/C_{60} ratios of the extracted fullerenes ranged from 0.21 to 0.36, with the larger values occurring in the more highly weathered and oxidized clays. A special set of fullerenes extracted from nondemineralized Woodside Creek samples had ratios of less than 0.002. These low C_{70} contents cannot be explained by preferential oxidation of C_{70} relative to C_{60} (exactly the oposite happens in actual oxidation) but might be expalined by chromatographic separation on KTB clays near that site. The estimated mean global C_{60} concentration at the KTB today is 1.4 ng/cm^2, which corresponds to a mean C_{60} content of the soot at the boundary of 0.41 ppm. These values are much more consistent with the formation of fullerenes by wildfires than by carbon chemistry in the hot atmospheric plume of the Chicxulub impact.

Heymann, D., Chibante, L. P. F., Brooks, R. R., Wolbach, W. S., Smit, J., Korochantsev, A., Nazarov, M. A., and Smalley, R. E., 1996, Fullerenes of possible wildfire origin in Cretaceous-Tertiary boundary sediments, *in* Ryder, G., Fastovsky, D., and Gartner, S., eds., The Cretaceous-Tertiary Event and Other Catastrophes in Earth History: Boulder, Colorado, Geological Society of America Special Paper 307.

INTRODUCTION

When fullerenes were first discovered to form spontaneously in condensing carbon vapors (Kroto et al., 1985), it was suggested that they might be widely distributed in the Universe. Subsequent searches for fullerenes in interstellar media and meteorites (see Böhme, 1992; Devries et al., 1993) have thus far been unsuccessful, but C_{60} and C_{70} have been reported to occur in the ejecta associated with an impact crater on the Long Duration Exposure Facility Spacecraft (Radicati di Brozolo et al., 1994). C_{60} and C_{70} have been reported to occur on the Earth in certain samples of shungite, a metaanthracite coal from a deposit near Shunga, Russia (Buseck et al., 1992), and in "fulgurite," a substance formed when lightning strikes certain soils or rocks (Daly et al., 1993). The occurrence of fullerenes in shungite was particularly surprising since the only routes to fullerenes discovered thus far in the laboratory have involved gas-phase chemistry at temperatures above 1,000 °C. Such conditions may be attained during lightning strikes, in flames, or in hot atmospheric plumes from crater-forming impacts, but shungite is believed to have formed from carbonaceous material creeping into fissures of a Precambrian rock that metamorphosed under extreme pressures. If the original carbonaceous material did not already contain fullerenes, they must have formed during the metamorphism by as yet unknown solid- or liquid-phase mechanisms. A recent search for fullerenes in shungite, thucolite, and anthraxolite was unsuccessful, but the results could only prove that fullerenes extractable with toluene were not contained in these materials (Heymann, 1995).

The discovery of C_{60} fullerene at the Cretaceous-Tertiary Boundary (KTB) in New Zealand was submitted for publication in December 1993, but the actual paper appeared in print much later, during the review stage of the manuscript of the present chapter (Heymann et al., 1994a). Additional information about discoveries at the KTB in New Zealand, Spain, Denmark, Turkmenistan, Khazakhstan, and Georgia were presented at the Snowbird III Conference and have meanwhile appeared in print or have been submitted for publication during this review stage (Heymann et al., 1994b, c; Cibante et al., 1994; Heymann et al., 1996).

Also during the review stage, Becker et al. (1994) reported the discovery of C_{60} and C_{70} fullerenes in rocks from the Sudbury impact structure. They suggested that the Sudbury fullerenes may have formed as a result of the pyrolysis of organic matter such as polycyclic aromatic hydrocarbons and kerogen contained in the impactor. Although the Ir- and soot-rich deposits at the KTB are due to fallout from the Chicxulub event, we think that the "Sudbury scenario" is less likely to account for the occurrence of fullerenes at the KTB than the "wildfire scenario," which will be discussed in a later section. For one thing, the formation of fullerenes by pyrolysis has only been achieved in an inert atmosphere (Taylor, et al., 1993); for another, the "Chicxulub atmosphere" contained huge amounts of strongly oxidizing substances that, given the very high plume temperatures, could have destroyed all fullerene molecules formed (Brett, 1992a, b; Blum and Chamberlain, 1994; D'Hondt et al., 1994; Sigurdsson, 1994).

Our search for fullerenes in KTB samples was prompted by the discovery that fullerenes can form in the free-burning flames of toluene and of common decorative candles in air and that the soots from these flames contain from 1 to 50 ppm extractable C_{60} and C_{70} (Heymann et al., 1994c). This result suggested that fullerenes could be produced at detectable levels in naturally occurring fires and might therefore still be found in ancient deposits. Perhaps the most extensive combustion to have occurred in Earth's history was the result of the putative meteor impact near Chicxulub in the Yucatan Peninsula of Mexico roughly 65 million years ago, as evidenced by iridium- and soot-rich seams at KTB sites worldwide (Alvarez et al., 1980; Anders et al., 1986a, 1991; Gilmour et al., 1990; Wolbach and Anders, 1989; Wolbach et al., 1985, 1986, 1988, 1990a, b). Therefore we have collected and analyzed samples from the KTB "fallout" layer from two well-characterized, carbon-rich sites in New Zealand: Woodside Creek and Flaxbourne River. The carbon contents of their KTB seams of 4.8 and 3.5 mg/cm^2 after dichromate etch are among the highest anywhere, and the soot contents of 69 and 55% of the carbon are also quite high (Anders et al., 1991). In this chapter we shall briefly review the geology and geochemistry of the KTB sites of New Zealand, the theory of wildfires, and all results obtained from the New Zealand sites to date and present a discussion of their implications in the context of pertinent events and conditions.

During the review stage of this chapter, new results were obtained on samples from Caravaca (Spain), Sumbar and Malyi Balkhan (Turkmenistan), Khoshak (Khazakhstan), Tetrichkaro (Georgia), Stevns Klint (Denmark), and Elendgraben (Austria). The results are presented in the "Post–Snowbird III Results" and are included in the general discussion.

GEOLOGY AND GEOCHEMISTRY OF THE SITES

Reconstructions of land masses and oceans about 80 m.y. ago place southern New Zealand (South Island), where our samples were collected, on the large Campbell plateau, approximately 1,500 km north of Marie Byrd Land, Antarctica, and of Australia, 2,000 km to the west. At the end of the Cretaceous, South Island was probably fully separated from Antarctica and Australia and was located at about 30°S, still on a substantial and shallow marine plateau (Lawver et al., 1985). A new phase of sedimentation began either in the Late Cretaceous or early Tertiary, when a marine transgression covered what appears to have been a peneplain covering much of South Island. It may have resulted either from epeirogenic sinking of a large, but local, region or from a worldwide eustatic rise of sea level (Suggate et al., 1978). The more recent major regression is related to the Kaikoura Orogeny, which began in Miocene time, reached its greatest movements in the Quaternary, and is continuing today (Suggate et al., 1978). It was during this orogeny that rocks in the region of interest were uplifted, faulted, and folded.

The outcrops of KTB rocks that we sampled were located in Marlborough province of the northeastern region of South Island. Five known locations (Woodside Creek, Flaxbourne River, Chancet Rocks, Needles Point, and Wharanui) occur here within 10 km of each other about the village of Ward; a sixth (Waipara) occurs about 200 km to the south. The Woodside Creek site was described in the paper by Alvarez et al. (1980); Needles Point, Chancet Rocks, and Waipara by Brooks et al. (1986). The Needles Point, Chancet Rocks, and Wharanui sites are on the beach of the Pacific Ocean and are flooded at high tide, and the KTB is covered here with thick beach deposits that render sampling very difficult. Any sample is likely to be contaminated with much recent organic remains. In fact, only very small amounts of boundary rock have ever been collected at Chancet Rocks, and samples by that name are more likely to have come from the nearby Flaxbourne River site, which is an outcrop in a limestone quarry along the road from Ward to Ward Beach.

The KTB at Woodside Creek is marked by an approximately 8-mm-thick layer of very strongly weathered ferriginous clay sandwiched between a massive Maastrichtian argillaceous limestone and a Teurian argillaceous limestone, whose beds of about 10 cm are parted by thin seams of shale. These rocks probably represent an upper bathyal environment with water depths of about 600 to 800 m (Brooks et al., 1984, 1986). Brooks et al. (1985) reported that no unweathered material was found in the boundary shale.

The Flaxbourne River boundary clay at the quarry outcrop, which was deposited at a shallower depth than the Woodside Creek sediment, is relatively unweathered, undisturbed by plant roots, and free of small-scale tectonic effects such as shearing and is an average of 20 mm thick (Strong et al., 1987). One clear-cut fault that has displaced the transition zone by about its own width is readily visible in the outcrop close to our sampling location, a testimony to periods of tectonic movement of the Kaikoura Orogeny, which in Marlborough province began in the earliest middle Miocene and has led to the formation of the nearby Kaikoura Ranges (Suggate et al., 1978).

IRIDIUM, CARBON, AND SOOT

Iridium contents at Woodside Creek were reported by Alvarez et al. (1980), by Brooks et al. (1986), and by Wolbach et al. (1988), who presented the most detailed account of iridium, total carbon, elemental carbon, and soot at 22 locations straddling the boundary, from about –40 cm below to about +250 cm above the boundary clay. Iridium and elemental carbon concentrations rise sharply just below the basal 0.3 cm of the boundary layer by factors of 1,500 and 210 compared with the average values of the even lower Cretaceous. The decrease of total carbon, elemental carbon, and iridium is less steep above the boundary, but soot contents drop swiftly within about 5 cm above the boundary. The maximum iridium content is 114 ppb, and the maximum content of elemental carbon is 0.462%. Soot is a major component of the elemental carbon in the basal layer, where its content rises to 71%. The iridium abundance at Woodside Creek is also reported as 197 ng/cm^2 (Brooks et al., 1984), which implies a surface density of the clay of 1.7 g/cm^2.

At Flaxbourne River, Ir was measured at 14 spots straddling the boundary from about –55 to +72 cm. Iridium, Cr, Co, and Ni show a *double maximum* at the boundary in the intervals –2 to –1 cm and 0 to +1 cm, with Ir contents of 20 and 21 ppb respectively. The notch at –1 to 0 cm contains only 7.4 ppb (Strong et al., 1987). It may be due to slight brecciation of the boundary layer with limestone lenses up to 1 cm thickness, presumably in response to the slight faulting at the site. Both seams were present at the location where we sampled in 1993. The lower seam was gray, the upper seam dark gray, appearing black. The integrated specific surface concentration of Ir is 134 ng/cm^2, equivalent to approximately 6.5 g/cm^2. However, Kyte et al. (1985) have suggested that only the Ir in the lower (gray) layer of the boundary clay is due to primary fallout, with Ir in the higher layer mainly due to secondary lateral transport (rivers, creeks, streams), which reduces the "primary" Ir value to 83 ng/cm^2, corresponding to a primary clay thickness of only 3.9 g/cm^2. The carbon abundance at Flaxbourne River is reported as less than 1.3 mg/cm^2 (includes kerogen) and soot as 55% of carbon (Anders et al., 1991).

THEORIES OF WILDFIRES

Wolbach et al. (1985) found that clay samples from three KTB sites (Stevns Klint, Denmark; Caravaca, Spain; and Woodside Creek) contained 1 to 2% of carbonaceous matter that, in turn contained 0.36 to 0.58% elemental carbon. They pointed out that this is from four to five times more than in the Cretaceous and Tertiary marls below and above at the Danish boundary. In a later paper (Wolbach et al., 1988) it was shown that the carbon content peaks quite sharply in the boundary layer of Woodside Creek also. The deduced mean global carbon abundance of 0.021 g/cm^2 in the boundary layers corresponds to a worldwide flux at KTB time 10^4 times the flux recorded in recent marine sediments. At Woodside Creek, the elemental carbon content of the clay is 5,700 ppm, or 0.015 g/cm^2 based on estimates of the specific surface mass of the clay. Most carbon particles are submicrometer sized here and show the morphology of carbon formed in flames, but larger, micrometer-sized particles occur also. Wolbach et al. (1985) considered three possible sources for the spheroid carbon clusters. They pointed out that typical primitive meteorites contain almost half of the total carbon needed to account for the observed global mean value of carbon abundance at the KTB but that the fireball from the impact would have quantitatively oxidized this carbon and that this would be also true if the impactor had been a comet. The same concern was raised about a second possibility, namely the formation from fossil carbon out of the impact crater itself (coal, oil, etc.) vaporized into the atmosphere. The third possibility considered was wildfires triggered by the impact of a 10 km diameter impactor, whose fireball was thought to have ignited wildfires on continents more than 10^3 km from the point of impact with the ensuing winds dispersing the soot worldwide. Wolbach et al. (1985) noted the staggering scale required of the putative wildfires as 10% of the present total biomass carbon on the Earth represented by the carbon at the KTB.

During the next several years the theory became increasingly more specific, partly as a response to constructive comments (see Argyle, 1986), but Anders et al. (1986b) maintained that the fires must have started soon after the impact, because the soot and iridium peaks at Woodside Creek appear simultaneously in the 3-mm-basal layer. However, since this is the only site for which the perfect coincidence of Ir and soot in a basal layer has been demonstrated, there remains some uncertainty as to the exact timing of impact and ignition (Anders et al., 1991).

Wolbach et al. (1988) estimated that the deposition time of the KTB layer at Woodside Creek could have been only weeks or months, perhaps six month at most, and that the fires appear to have started well before all the ejecta had fully settled. Anders et al., (1991) reiterated that Ir from the meteorite (instantaneous), soot from wildfires (the next few weeks), and As, Sb, and Zn absorbed from seawater are correlated in the *primary* fallout of the basal layer at Woodside Creek. They suggested that soot and Ir-bearing ejecta coagulated in the stratosphere, then swept out As, Sb, and Zn from seawater which had become anoxic. The so-called *secondary* fallout, material causing elevated Ir contents in Tertiary rocks above the boundary, is thought to have been eroded from elevated seafloor sites and redeposited in topographic lows. With regard to the fuel that burned, the authors pointed out that the presence of retene (1-methyl-7-isopropyl phenantrene) and other polynuclear aromatic hydrocarbons (PAH) in the boundary layer, the particular isotopic composition of the carbon, and the absence of cenospheres argued in favor of the burning of biomass rather than of fossil fuel.

THE NEW ZEALAND SAMPLES AND THEIR PREPARATION

For this study we obtained two sets of samples each from two KTB sites in New Zealand: Woodside Creek and Flaxbourne River. The first set had been collected by one of us (RRB) in 1986. The second set was collected (DH and RRB) during an expedition to the sites in 1993. At Woodside Creek at the level of the creek, where the strata are inclined by about 45°, the boundary layer had been previously so heavily "mined" that we only collected Cretaceous limestone immediately below and shale from partings in the Tertiary limestone at +22.5 and + 57.5 cm above the boundary. The boundary clay proper was collected on a nearby hillside where the strata were horizontal.

At Flaxbourne River the strata were also inclined by about 45°, and at the spot of collection there were actually two clay seams, one light gray, the other dark gray (black), separated by approximately 1 cm of white limestone, which is consistent with the observations of Strong et al. (1987).

All samples were crushed and then powdered by hand or in a plate grinder prior to extraction with toluene. Some samples were extracted without any further preparation. Others were demineralized with HCl/HF prior to extraction (see Wolbach and Anders, 1989, for details). For one sample, the clay was first extracted, then demineralized and the residue extracted a second time. The digestion with HCl/HF removes most inorganic matter, and the resultant residue consists mostly of kerogen and soot (see Wolbach and Anders, 1989) The kerogen can be removed by oxidation with a dichromate solution at 50 or 85 °C; however, since we found from a test run with synthetic fullerenes that such treatments would also oxidize the fullerenes (see Table 1), kerogen was not removed. Interestingly, as seen from the table, C_{60} was oxidized faster than C_{70}.

EXTRACTION, FILTRATION, EVAPORATION, AND ANALYSIS

All samples were digested with High Pressure Liquid Chromatography (HPLC) toluene, but the extraction procedures varied somewhat. Samples weighing 20 g or less were extracted in centrifuge tubes with from 10 to 50 ml toluene by sonication from 8 to 10 hrs at room temperature. Heavier samples were either sonicated in larger beakers or in erlenmeyers or were extracted with the Soxhlet method in which the dissolved fullerenes become exposed to toluene at its normal boiling point.

Solids and extracts of the small samples were directly separated by centrifuging, whereas the slurries from the large clay samples (100g or heavier) were filtered through standard (Whatman 2V) paper filters and were washed with up to 1l of toluene. The volumes of the extracts were too large at this stage for analysis, hence they were reduced to a few milliliters by evaporation under reduced pressure at temperatures near 40 °C. Additional volume reduction, when needed, was achieved by blowing dry nitrogen into the storage bottles. When needed, a final filtration through 45-μm Millipore was done. For quanti-

TABLE 1. RESULTS OF THE TREATMENT OF *SOLID* FULLERENES C_{60} AND C_{70} WITH A SOLUTION OF $K_2Cr_2O_7$ IN 2 *M* H_2SO_4*

Temp.	Time	C_{60} (Remaining)		C_{70} (Remaining)	
(°C)	(hr.)	(μg)	(%)	(μg)	(%)
50	76.5	29.3	29	27.3	26
50	620.5	1.0	1.0	1.8	1.6
85	76.5	0	0	0	0

*Before the treatment the amounts of C_{60} and C_{70} were 99 μg and 107 μg respectively.

tative results, the extract volumes before analysis were determined by weighing and by assuming that their densities were still that of pure toluene, 0.867 g/cm³.

The analysis for C_{60} and C_{70} was done by HPLC, which has the virtues of being both a low-temperature and a good quantitative method. The HPLC instruments used consisted of two Waters 510 pumps, an U6K injector, a 3.9 × 300 mm Nova-Pak™ C-18 column, and a 996 photodiode array (PDA) taking UV/Vs absorption spectra of the fluid emerging from the column every 2 seconds. The PDA can take spectra in the range 210 to 640 nm, but spectra were taken only in the range 310 to 430 nm because the toluene in the mobile phase absorbs too strongly below 310 nm and because the absorption by the small fullerene samples of this study above 430 nm was negligibly small. The chromatograms shown in this chapter were always graphed from absorption data at 330 nm where C_{60} has a local maximum.

The mobile phase was usually 40:60 toluene:methanol mix flowing at 2.00 ml/min. For these conditions, the retention times of C_{60} and C_{70} were around 6.50 and 11.0 min respectively. Other proportions of methanol and toluene were also used with the result that the retention times became shorter, the peaks narrower and higher, and the separation between hydrocarbon and fullerene peaks less for larger proportions of toluene. Our choice of mobile phase composition was always dictated by the most favorable compromise among these variables.

The volumes of extract injected were always 25 µl, as larger volumes greatly disturbed the efficacy of the column. Using a standard solution of synthetic C_{60} and C_{70}, the limits of detection for these fullerenes were determined to be 0.2 and 0.3 ng injected, As shown in Figure 1, the response of the instrument was linear up to 5,000 ng injected.

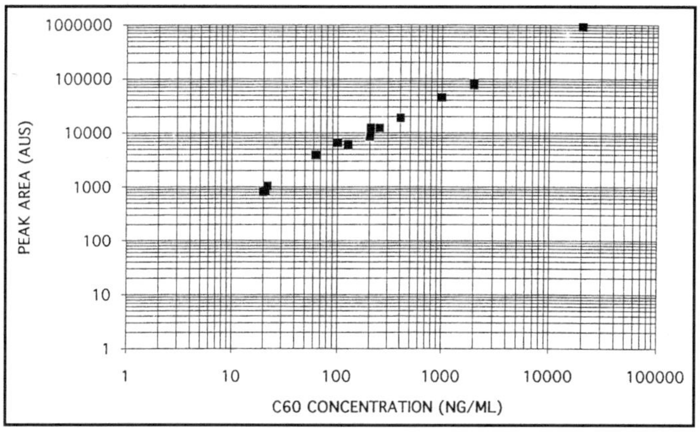

Figure 1. Response of the PDA detector to the injection of samples of 25 µl containing C_{60} in the concentration range 20 to 20,000 ng per ml. At least duplicates were run for every concentration. AUS stands for Absorption Units Seconds. A graph for C_{70} would be very similar, except that the extinction coefficient of C_{70} at 330 nm is only about 0.6 that of C_{60}.

PRECAUTIONS

From the onset of this study we were keenly aware of risks of contaminating samples or extracts with man-made fullerenes and also of the possibility of cross-contamination between samples. Extraordinary precautions were taken to prevent these occurrences.

The jaw crusher and plate grinder were cleaned by standard procedures between any two samples, which were stored in a room free of synthetic fullerenes. The demineralization was done by one of us (WSW) in the Chemistry Department of Illinois Wesleyan University under identical conditions of precaution. All glassware was new, was cleaned with toluene, and was heated to 500 °C in air for at least 12 hr. Extractions and other operations prior to analysis were carried out only in glassware proven "clean" by a blank run, which was a complete run-through without the use of a sample. Extractions, filtrations, centrifugations, and evaporations were carried out in a room free of synthetic fullerenes.

It turned out that the only source of potential contamination was the injector of the HPLC instrument. To erase any memory from it, the injector was always flushed with large amounts of clean toluene, and instrumental blanks—that is to say, injections of clean toluene—were run until at least three such successive blanks showed only baseline noise at the retention times of C_{60} and C_{70}. Only then was a sample from an extract injected. The syringes used for injection were new and dedicated for this work only.

RESULTS

Standards

Figure 2 shows the chromatogram from the injection of 39.6 ng C_{60} and 46.2 ng C_{70} dissolved in toluene. The horizontal coordinate is time; the vertical is absorption. The peak at

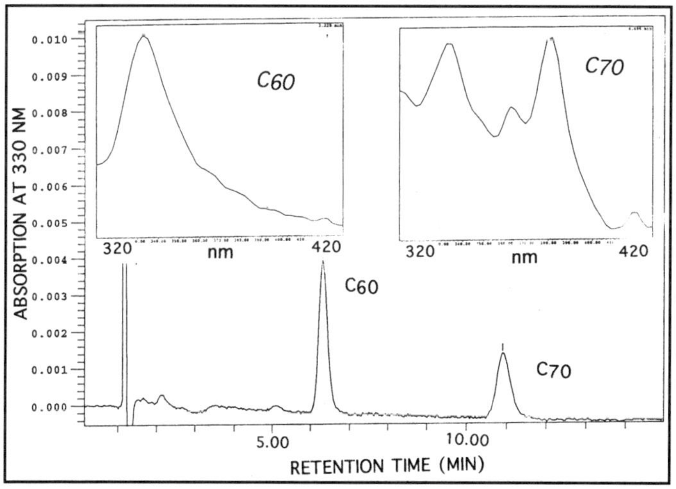

Figure 2. Chromatogram from the injection of 39.6 ng C_{60} and 46.2 ng C_{70} (both synthetic). The narrow but large peak at 1.2 min is due to the injection of pure toluene into the methanol/toluene mix of 60/40. The absorption spectrum of the upper left-hand insert is that of C_{60}; the spectrum of the upper right-hand insert is that of C_{70}.

1.2 min is due to the injection of pure toluene into the mixed mobile phase (methanol:toluene = 60:40; 2 ml/min); the peak at 6.31 min is due to C_{60} and the peak at 10.91 min to C_{70}. For quantitative evaluation, a computation program constructed baselines below peaks and calculated their surface areas.

The inserts of Figure 2 show the absorption spectra at the retention times of C_{60} and C_{70} in the range 310 to 430 nm. The C_{60} spectrum has a strong maximum at 332 nm, with a small peak at 420 nm. The C_{70} spectrum is a characteristic "trident" with maxima at 327, 357, and 374 nm. Figure 2 contains the information used for the identification of fullerenes in extracts, namely retention times and absorption spectra.

Samples of Set 1

The first hints that the boundary clays of Woodside Creek and Flaxbourne River might contain at least some C_{60} were obtained with what turned out to have been samples of marginal mass. Figure 3 presents the chromatogram from an injection of the extract from sample WC2. The peak at 4.44 min juts out from the tail of a very strong hydrocarbon peak (hydrocarbons, including PAHs, have shorter retention times than the fullerenes on this column), but it occurs at exactly the predetermined retention time of C_{60}. No signal above the background is present at 6.5 min, the retention time of C_{70}. If the peak at 4.44 min. were due to C_{60}, its surface area would correspond to 0.75 ng injected.

Confirmation for the presence of C_{60} would normally have come from the absorption spectrum at 4.44 min, but the actual spectrum obtained was a superposition of that of the putative C_{60} and of organics in the "tail" and was too different from that of pure C_{60} to be acceptable as evidence. Attempts were made to obtain a more convincing spectrum by eventually combining all the extracts from the Woodside Creek and Flaxbourne River samples listed in Table 2, separating C_{60} from hydrocarbons with the column itself, evaporating the combined effluents in the range 4 to 5 min to dryness, redissolving in 100 μl toluene, and reanalyzing this fluid. It was learned (and later confirmed by the results of an experiment with synthetic fullerenes) that one may not evaporate nanogram amounts of fullerenes to dryness at elevated temperatures; apparently such small amounts oxidize very swiftly under such conditions.

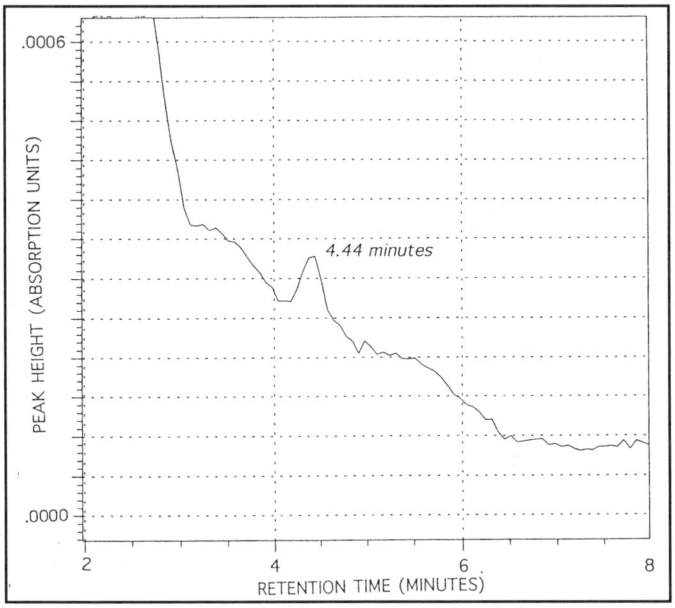

Figure 3. Chromatogram from extract of sample WC2. Methanol/Toluene = 55/45. Only the retention time range 2 to 8 minutes is shown. The "peak" at 4.44 min occurs at exactly the same retention time as that of synthetic C_{60}. It is riding on the tail of a very large hydrocarbon peak that crested at 1.5 min. The amount of the putative C_{60} was too small to yield an acceptable absorption spectrum.

Figure 4 presents the evidence that prompted us to organize an expedition to New Zealand for the collection of larger samples. This bar-graph summarizes all HPLC runs, save standard runs, in true sequence. The height of a bar is equal to the integrated peak area at the retention time of C_{60}. In the case of a chromatogram such as shown in Figure 3, the peak is fairly well defined. In the case of blanks, when all one obtains is noise, the area under a "noise-excursion" at or near that retention time was entered. Injections of extracts from K/T boundary samples are labeled KT. All other bars correspond to injections

TABLE 2. C_{60} CONTENTS IN THE FIRST SET OF BOUNDARY CLAY SAMPLES FROM WOODSIDE CREEK (WC) AND FLAXBOURNE RIVER (FR) AND FROM A TERTIARY SHALE PARTING (TE) +32 Cm ABOVE THE K/T BOUNDARY AT WOODSIDE CREEK

Sample	Weight (g)	Residue (mg)	Clay (wt. %)	C_{60} (ng)	C_{60} in Clay (ppb)	C_{60} in Residue (ppm)	C_{60} (ng/cm^{-2})
WC1*	4.21	37.23	1.1	7.6	2.2	0.17	2.4
WC2	3.32	53.10	1.6	9.1	2.7	0.17	3.6
TE1	0.720	1.41	0.20	<0.8			
TE2	0.842	1.66	0.20	<0.8			
FR1	1.98	12.81	0.60	1.1	0.56	0.10	2.5

*In this case, the untreated clay was first extracted to yield 5.7 ng C_{60}. The demineralization of 3.50 g of the extracted clay yielded 37.23 mg residue, which was extracted to yield 1.58 ng C_{60} (1.9 ng corrected for the original sample weight of 4.21 g).

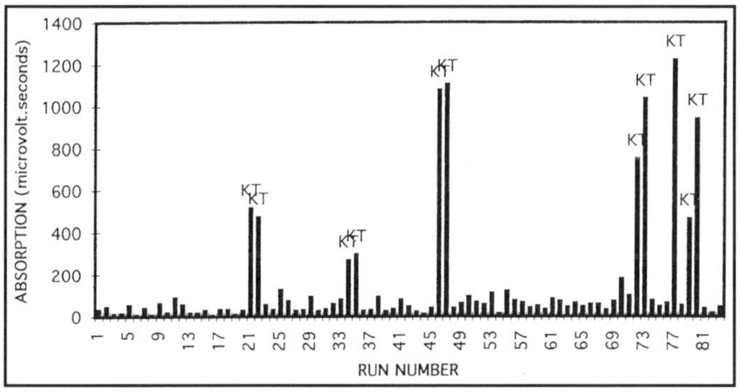

Figure 4. Bar graph of all runs for sample set 1. The height of a bar is equal to the integrated peak area at the retention time of C_{60}. Bars labeled KT were injections from boundary clay extracts of both sites. All other bars represent procedural blanks, instrumental blanks, and extracts from the Tertiary shale sample. This graph led to the tentative conclusion that C_{60} might occur at the K/T boundaries of the two sites.

from the extract of the two Tertiary shales of Table 2 and to procedural blanks and instrumental blanks. The results in Figure 5 suggest very strongly that C_{60} might be present in the K/T clays of Woodside Creek and Flaxbourne River and that contamination of the samples had not occurred.

Although these results do not conclusively demonstrate the presence of fullerenes, it is assumed now, on the basis of the results from the set to be presented next, that the data shown in column 5 of Table 2 are amounts of the fullerene C_{60}. WC1 was the only sample for which the untreated clay was first extracted, then demineralized for a second extraction of the residue. The amount of C_{60} of the first extraction was 5.7 ng; the corrected amount of the second extraction was 1.9 ng. Apparently the presence of the inorganic minerals somehow inhibits the efficient extraction of fullerenes. The percentages of carbon-rich residues agree well with those found earlier by Wolbach et al. (1985), but the C_{60} contents of the residues is considerably less

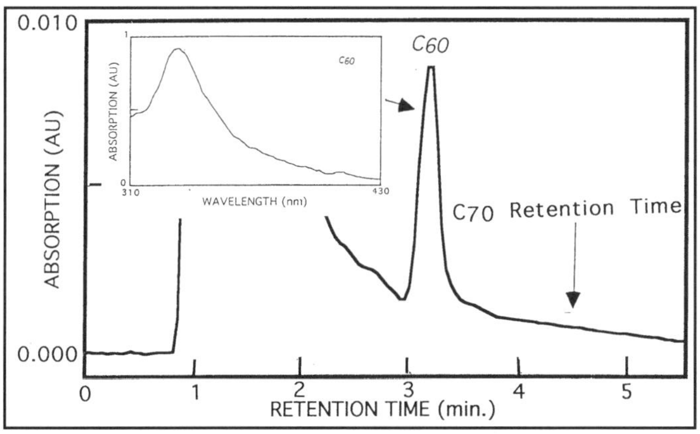

Figure 5. Chromatogram from the extract of sample WC3. Methanol/Toluene = 50/50. Once again, the large, broad early peak is due to hydrocarbons. The area of this peak corresponds to the injection of about 45 ng C_{60}. Note that there is no peak at the retention time of C_{70}, and the noise at this time yields an upper limit of $C_{70}/C_{60} = 0.002$.

than the amounts obtained from soot from the burning of toluene (Heymann et al., 1994c). The data of column 8 of Table 2 offer a fairer evaluation of how much C_{60} is present than the C_{60} content of clay or residue, owing to the fact that the thickness of boundary layers, hence g/cm^2, can be quite variable from site to site. For the Woodside Creek site, a layer thickness of 1.3 g/cm^2; both thicknesses were taken from earlier studies by Anders et al., 1991.

Samples of Set 2

WC3, the very first clay sample of the second set that was extracted, yielded not only the more convincing chromatogram shown in Figure 5 but also a convincing C_{60} absorption spectrum with maximum at 332.6 nm, shown in the insert. All four of the larger WC samples (WC3 through WC6) listed in Table 3 yielded convincing chromatograms and C_{60} absorption spectra. The chromatogram demonstrates the absence of C_{70} from this and all WC extracts, which show only noise at the predicted retention time of that fullerene. The deduced upper limit of the C_{70}/C_{60} ratio in the WC4 extract is 0.002. Such a small ratio is unexpected because both fullerenes are present in every synthesis done thus far in many laboratories, albeit in greatly variable proportions. As a further check on the C_{70} issue, extracts WC3 and WC4 were combined, evaporated down to approximately 100 μl, and analyzed, with the same result: The proportion of C_{70} in the Woodside Creek clay is extremely low.

All of the untreated samples from Flaxbourne River yielded chromatograms similar to the one shown in Figure 6. This chromatogram has a clear-cut C_{60} peak with the attendant C_{60} spectrum (not shown) but also a small peak at the retention time of C_{70}—too small, however, to yield an acceptable C_{70} absorption spectrum. The demineralization of hundreds of grams of rock was difficult and very time-consuming: hence, only two such samples, both from Flaxbourne River, became available in time for this study. The confirmation that C_{70} is present in significant amounts in the clay at Flaxbourne River came from the demineralized samples FR6 and FR7. Figure 7 and its inserts show the chromatogram from extract FR6 and the absorption spectra at the retention times of C_{60} and C_{70}. Both spectra are identical to the spectra of synthetic C_{60} and C_{70} respectively, and the C_{70}/C_{60} weight ratio in the extract is 0.29. The small peak at 5.4 min occurs at the retention time of $C_{60}O$, which may well have formed either in the geologic formation prior to sampling or during the laboratory procedures.

The results from the second set of samples are presented in Table 3. These results have now firmly established that C_{60} occurs in the boundary clays of Woodside Creek and Flaxbourne River, C_{70} occurs in the boundary clay of Flaxbourne River, and the extraction of fullerenes from demineralized samples is much more efficient than it is from the untreated clays.

POST–"SNOWBIRD III" RESULTS

Our search for fullerenes continued during the review stage of this chapter. J. Smit made available a sample from Caravaca

TABLE 3. C_{60} CONTENTS IN THE SECOND SET OF BOUNDARY CLAY SAMPLES FROM WOODSIDE CREEK (WC) AND FLAXBOURNE RIVER (FR), FROM A CRETACEOUS LIMESTONE 1 Cm BELOW AND A TERTIARY SHALE 57.5 Cm ABOVE THE K/T BOUNDARY AT WOODSIDE CREEK*

Sample	Weight (g)	Residue (mg)	Clay (wt. %)	C_{60} (ng)	C_{60} in Clay (ppb)	C_{60} in Residue (ppm)	C_{60} (ng/cm^{-2})
WC3	112.3	....	....	609	5.4	....	7.1
WC4	201.0	....	....	994	4.7	....	6.2
WC5	123.9	....	....	309	2.5	....	3.3
WC6	68.0	....	....	22.1	0.32	....	1.4
CL1	127.7	....	....	<0.8			
TS2	135.6	....	....	<0.8			
FR2(g)	57.2	....	....	65.5	1.14	....	5.0
FR3(g)	200.0	....	....	11.6	0.06	....	0.26
FR4(b)	68.0	....	....	160	2.3	....	10.3
FR5(b)	200.0	....	....	369	1.8	....	8.1
FR6(g)	250.0	1,387	0.6	2,974	11.9	2.1	53.5
FR7(b)	200.0	499	0.25	2,447	12.2	4.9	54.9

*The C_{70} contents of FR6 and FR7 are 1,074 ng and 720 ng respectively; the C_{70}/C_{60} ratios in these samples are 0.36 and 0.29 respectively. About 63% of extract FR7 was spilled after evaporation. Since the weights before and after the spill were known, a correction could be made.

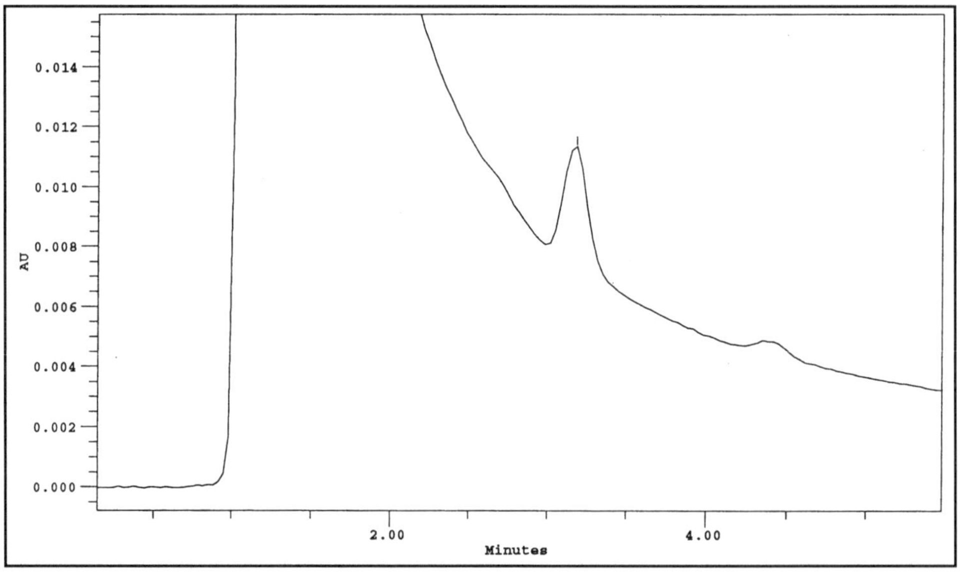

Figure 6. Chromatogram from extract of FR3. Methanol/Toluene = 50/50. Note the C_{60} peak that occurs at the expected retention time and also the small peak at 4.45 min, the retention time of C_{70}.

(Spain) and Stevns Klint (Denmark) that he had collected. M. A. Nazarov made available demineralized samples from Sumbar and Malyi Balkhan (both in Turkmenistan), Khoshak (Khazhakhstan), and Tetrichkaro (Georgia) and nondemineralized samples from Sumbar, Malyi Balkhan, Stevns Klint and Elendgraben (Austria). The extractions and analyses of these samples were done along similar lines to those from New Zealand. Table 4 lists the results together with all the results from the New Zealand sites. This table represents the state of our knowledge as of October 1994.

In addition to the studies of these KTB samples, we have restudied the oxidation of C_{60} and C_{70} by simultaneous heating in air at 245 °C. Both fullerenes were fully decomposed after about 5 hrs, but C_{60} was oxidized faster than C_{70}.

In order to test the hypothesis that the C_{70}-free component at Woodside Creek might be due to chromatographic separation of the fullerenes on natural clay, C_{60} and C_{70} were injected into an HPLC column (4.6 × 250 mm) filled with sieved clay in the 10- to 20-μm size range from Woodside Creek. We used n-Hexane as

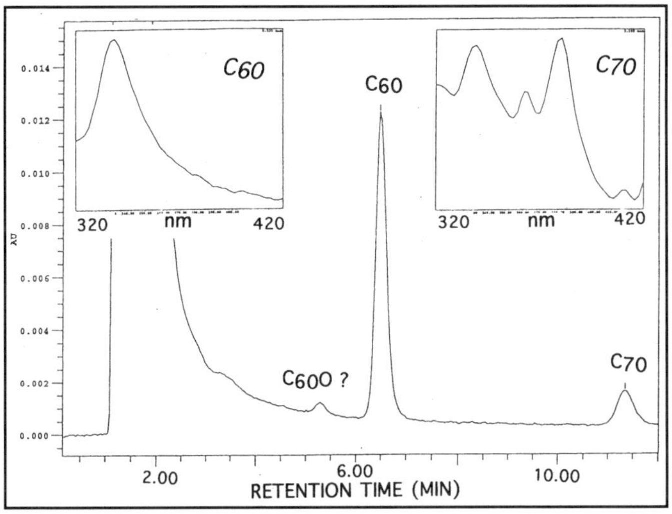

Figure 7. Chromatogram from extract FR6. Methanol/Toluene = 60/40. In addition to the hydrocarbon, C_{60}, and C_{70} peaks, there is a small peak at the expected retention time of $C_{60}O$, an epoxide that can form by the reaction of C_{60} with O_2, or O_3 in the atmosphere (or hydrosphere). The inserts show the absorption spectra at the retention times of C_{60} and C_{70}; these are identical to the spectra of Figure 2. The C_{70}/C_{60} weight ratio in this extract is 0.29.

the mobile phase. The two fullerenes were fairly well separated on this column.

DISCUSSION

Let us begin by discussing what is probably the least controversial issue, namely the increased efficiency of fullerene extraction from the demineralized, carbon-rich residues of some samples. Since it will be argued that the fullerenes were formed in sooting fires, it seems reasonable to assume that they are intimately associated with one, several, or all forms of carbon discovered in the boundary clays by Wolbach et al. (1985). For fullerenes to become extracted, the carbon particles must become wetted by toluene, which may be difficult if these particles are strongly encrusted by silicates and other inorganic substances. Given that the clays formed and resided in geological environments in which the formation of carbonate and silica cements is a very common process, such encrustation seems very likely to have happened. The hypothesis of encrustation is somewhat supported by the observation that black sootlike particles were never observed to settle more slowly that the clay particles from the slurries produced by sonication and stirring, even of samples such as WC4 or FR3(g), which ought to contain at least 1 g of carbon-rich residue. Variable encrustation of the only modestly well homogenized powders is perhaps also responsible for the variations seen in C_{60} contents of clays such as WC3 to WC6 or FR2(g) and FR3(g).

These considerations seem to be contradicted by the fact that extracts from the demineralized samples of Sumbar, Malyi Balkhan, Khoshak, and Tetrichkaro, contained no detectable

TABLE 4. SUMMARY OF RESULTS, INCLUDING POST–SNOWBIRD III DATA*

Sample	C_{60} in Clay (ppb)	C_{70}/C_{60}	Demineralized?	Reference†
WOODSIDE CREEK				
WC1	2.2	n.d.	No/yes	1
WC2	2.7	n.d.	Yes	1
WC3	5.4	<0.002	No	1
WC4	4.7	<0.002	No	1
WC5	2.5	<0.002	No	1
WC6	0.32	<0.002	No	1
WC7	2.48	0.34	Yes	4
CR1	<0.08	n.d.	No	1
TE1	<0.8	n.d.	Yes	2
TE2	<0.8	n.d.	Yes	2
TE3	<0.08	n.d.	No	1
FLAXBOURNE RIVER				
FR1(g)§	0.56	n.d.	No/yes	1
FR2(g)	1.1	n.d.	Yes	1
FR3(g)	0.06	n.d.	No	1
FR4(b)	2.3	*0.28*	No	1
FR5(b)	1.8	*0.28*	No	1
FR6(g)	11.9	0.36	Yes	2
FR7(b)	12.2	0.29	Yes	3
CARAVACA				
CA1	11.9	0.22	No	4
CA2§	2.7	0.26	Yes	4
SUMBAR				
SU1	0.0	n.d.	Yes	4
SU2	0.00	n.d.	Yes	4
SU3	4.0	0.21	No	4
TE	0.0	n.d.	Yes	4
CR	0.0	n.d.	Yes	4
KOSHAK				
KO1	0.0	n.d.	Yes	4
TETRICHKARO				
Te1	0.0	n.d.	Yes	4
MALYI BALKHAN				
MB1	0.00	n.d.	Yes	4
MB2	1.80	0.36	No	4
ELENDGRABEN				
EG1	0.0	n.d.	No	4
STEVNS KLINT				
SK1 (Nazarov)	0.0	n.d.	No	4
SK2 (Smit)	0.0	n.d.	No	3
SK3 (Smit)	0.0	n.d.	No	3

*TE samples are from the Tertiary. CR samples are from the Cretaceous. FR(g) = Flaxbourne River gray seam; FR(b) = Flaxbourne River black seam. C_{70}/C_{60} ratios printed in italics: absorption spectra not available for C_{70}.

†1 = Heymann et al., 1994c; 2 = Heymann et al., 1994a; 3 = Published only in this paper; 4 = Heymann, et al., 1996.

§Sample first extracted then demineralized and extracted.

fullerenes, especially so because fullerenes were extracted from distinct nondemineralized samples from Sumbar and Malyi Balkhan. W. S. Wolbach, who demineralized the samples from New Zealand but not those from the former USSR, pointed out that great care must be taken to recover not only the coarse fraction of the acid-resistant residues but also the very fine, almost colloidal fraction that might have contained the fullerenes of these anomalous demineralized samples (personal communication, 1994).

The second issue is the absence of fullerenes in the extracts from the nondemineralized samples of Elendgraben and Stevns Klint. The case of Elendgraben is the least puzzling, as the clay there has very low carbon and soot abundances (Anders et al., 1991), and the weight of our Elendgraben sample was rather small, too. The amounts in Elendgraben may well be below our limit of detection. The case of Stevns Klint is more puzzling, as the clay there has a rather robust carbon content, and the soot content of 27% is actually higher than that of Caravaca clay (Anders et al., 1991). There are three fair explanations, none of which can currently be tested with the information at hand. First, one posits that the fullerenes at Stevns Klint were quantitatively oxidized just before or else after sedimentation. Second, one posits that these fullerenes were quantitatively mobilized away by solution in organic fluids. And third, one posits that Stevns Klint clay does contain fullerenes but that these cannot be extracted from the nondemineralized samples.

The third issue is whether the burning of wood produces fullerenes, and if so, what range of proportions of C_{70}/C_{60} would result. Three samples from wood burning became available in the course of this study: soot from the chimney of a cedar-burning pizza restaurant and two samples from the Malibu, California, 1993 wildfires. The extract from the cedar burn was so enormously rich in hydrocarbons that these masked any fullerenes that might have been present. Also, this soot had been exposed to hot gases in the chimney for a long period of time. The Malibu samples weighed only 5.69 and 4.25 g; each contained an upper limit of 0.2 and 0.3 ppb C_{60}, that is to say, essentially no fullerenes. Much remains to be done to confirm that wildfires can produce soot with fullerenes in detectable amounts. Nevertheless, the formation of fullerenes in the wildfires as proposed by Wolbach et al. (1985), with the fullerenes remaining bonded to the soot during each of the successive stages of formation, dispersal, settling, and sedimentation, must also be considered a major hypothesis for the presence of fullerenes at the New Zealand sites.

The formation of fullerenes by the free burning of a hydrocarbon has been firmly established by Heymann et al. (1994c), who collected the soot from the burning of toluene and a decorative candle, which contained C_{60} and C_{70} in the range 1 to 50 ppm. Hence the formation of fullerenes by the burning of fossil fuel, particularly crude oil, cannot be completely ruled out if one assumes that only a comparatively small fraction of the soot at the K/T boundaries is from such a source. One problem may be the high required fullerene content of such soot of about 500 ppm, if soot from crude-oil burning would be 10% of the soot in samples FR6(g) or FR7(b). Whereas much higher contents have been achieved in laboratory syntheses of fullerenes, it remains to be seen whether C_{60} contents of 500 ppm or more can be achieved by natural processes.

The fourth issue concerns the C_{70}/C_{60} ratio. C_{70} was formed alongside C_{60} when fullerenes were discovered (Kroto et al., 1985) and when fullerenes were synthesized in macroscopic amounts (Kraetschmer et al., 1990; Haufler et al., 1990). The extracts of the synthetic soots typcally had molar C_{70}/C_{60} ratios from 0.02 to 0.18. Howard et al. (1991), who extracted fullerenes from soot produced by the burning of benzene with oxygen and argon as a diluent at low pressures, found ratios in the range 0.27 to 5.7, depending on the flame conditions. The formation of C_{70} always seems to attend that of C_{60}, albeit in greatly variable proportions. Even if there are minute amounts of C_{70} in the clay at Woodside Creek, the puzzle remains why C_{70} is so abundantly present only 10 km away at Flaxbourne River. One may consider two classes of hypotheses: Either the fullerene at Woodside Creek came from a process that formed little, if any C_{70}, or the fullerenes at Woodside Creek started out with abundant C_{70}, which was quantitatively destroyed or otherwise removed since 65 Ma.

Because of the ease with which C_{70} forms alongside C_{60}, the second class is considered much more probable. Within its context one may consider two subclasses: chemical processes that destroyed C_{70} faster than C_{60} or molecular exchange processes that separated the two fullerenes. The most immediately obvious chemical process is oxidation, either by O_2, O_3, iron-oxides, or clay-catalyzed reactions, all of which are more likely to have occurred in the strongly weathered Woodside Creek than in the less-weathered Flaxbourne River clay. Chibante et al. (1993) have studied the rate of decomposition of solid C_{60} and C_{70} heated in air and have concluded that a mix of $C_{70}/C_{60} = 0.2$ decomposed nearly five times faster than a near-pure sample of C_{60}. This result, however, did not prove that C_{70} proper decomposes faster than C_{60}, because the two fullerenes could not be detected separately. The treatment of C_{60} and C_{70} dissolved in toluene with ozone resulted in 40% faster oxidation of C_{60} than C_{70} (Chibante and Heymann, 1993; Heymann and Chibante, 1993), and the oxidation with chromic acid (Table 1 of this chapter) also attacked C_{60} faster than C_{70}. The most recent result of the heating of fullerenes in air at 245 °C clearly demonstrates that apparently all oxidation processes increase the C_{70}/C_{60} ratio, hence that essentially C_{70} free fullerene cannot have formed by selective oxidation. These results show, however, that the measured C_{70}/C_{60} ratios could be significantly larger than the ratios extant at the time of sedimentation 65 m.y. ago. Perhaps this means that the samples with the smallest ratios (Caravaca, Sumbar) were less oxidized than those with the higher ratios (Woodside Creek, Flaxbourne River, and Malyi Balkhan).

Although little is known about the solubility of fullerenes in naturally occurring mixes of hydrocarbons and other organic compounds ("oils"), it is likely that solubilities are sufficiently high that fullerenes can become mobilized when oil migrates through fullerene-bearing strata (we found that the solubility of C_{60} in Brent crude oil from the North Sea is comparable to that in toluene). It is not suggested here that C_{70} is preferentially removed but that nature has somehow provided clay-based liquid chromatography columns on which the fullerenes can occasionally become quantitatively separated, given great lengths of formations and geologic timescales. Such mobilization may have been significant in the inclined strata of the boundary clays in Marlborough province. The recent result of separating C_{60} and C_{70} on clay from Woodside Creek suggests that this hypothesis is the correct explanation.

The occurrence of fullerenes at five KTB sites representing three distinct and geographically greatly separated areas suggest that much of the Earth's surface became covered not only with Ir- and soot-rich matter but also with C_{60} and C_{70}. The average C_{60} content of the 17 samples of Table 4 that contain fullerene is 4.2 ppb. Unfortunately, we do not know the specific surface mass (g/cm²) for most of our samples, but an average value of 3 g/cm² for these thin sediments seems reasonable. This would imply a specific C_{60} surface concentration of 1.4 ng/cm². The average carbon abundance has been estimated as 11 mg/cm², or 1.1×10^7 ng/cm² (Anders et al., 1991). With a mean soot content of 31%, this corresponds to 3.4×10^6 ng/cm² of soot (Anders et al., 1991). Hence the estimated mean C_{60} content of the KTB soot of 0.41 ppm is fully consistent with the burning of C-bearing matter in air (Heymann et al., 1994c)!

If one tries to account for the C_{60} at the KTB with the "Sudbury scenario" (Becker et al., 1994), one runs into the following problem. The C_{60} content of sample CSF-66-43 has been reported as 10 ppm (Becker et al, 1994). Even if one assumes that CSF-66-43 had the *largest* reported carbon content of the Onaping formation of 1%, then the C_{60} content of CSF-66-43 carbon was 1,000 ppm, or 2,500 times larger than the mean measured at the KTB. Was the Chicxulub event three orders of magnitude less efficient in producing fullerenes than the Sudbury event by the "Sudbury scenario"? Or was 99.9% of the Chicxulub cum Sudbury C_{60} destroyed by oxidation? Such ad hoc extra features make the "Sudbury scenario" much less attractive for explaining the presence of fullerenes at the KTB than the "wildfire scenario."

If the globe had become covered uniformly with C_{60} at K/T time, then at least 7×10^9 g C_{60} were synthesized and distributed by the wildfires. Assuming that X gram of biomass were burned, then the amount of C_{60} formed was:

$$7 \times 10^9 = X \, \varepsilon \, f_c \, \gamma \, ,$$

where ε and f_c are the fractions of burned biomass carbon converted to "smoke" and the fraction of elemental carbon in "smoke." Anders et al., (1991) have used the values $\varepsilon = 0.02$ and $f_c = 0.2$. A reasonable upper limit for γ is 5×10^{-5} (Heymann et al., 1994c). $X = 3.5 \times 10^{16}$ gram, or only 2.2% of the estimated biomass at K/T time. For a γ value of 4.1×10^{-7}, the required burned biomass is about twice the estimated amount present at K/T time, which means that some of the carbon and soot at the KTB may be due to the burning of fossil fuel (see also Anders et al., 1991).

ACKNOWLEDGMENTS

We thank U. A. Th. Brinkman, R. R. Vreuls, and W. H. Voogt of the Department of Analytical Chemistry of the Free University of Amsterdam for their hospitality offered to one of us (DH) and their indispensable assistance with the two experiments done at the Free University: the heating of fullerenes in air and the chromatographic separation of fullerenes on Woodside Creek clay. We thank E. Verdurmen of the Department of Geology of the Free University for his assistance with the sieving of the clay. Funded in part by the NSF and the R. F. Welch Foundation.

REFERENCES CITED

Alvarez, L., Alvarez, W., Asaro, F., and Michel, H. V., 1980, Extraterrestrial cause for the Cretaceous/Tertiary extinction: Science, v. 208, p. 1095–1108.

Anders, E., Wolbach, W. S., and Lewis, R. S., 1986a, Cretaceous extinctions and wildfires: Science, v. 234, p. 261–264.

Anders, C., Wolbach, W. S., and Lewis, R. S., 1986b, Reply to Argyle, E., Cisowski, S. M., Fuller, M., Officer, C. B., and Ekdale, A. A.: Science, v. 234, p. 264–266.

Anders, E., Wolbach, W. S., and Gilmour, I., 1991, Major wildfires at the Cretaceous/Tertiary boundary, *in* Levine, J. S., ed., Global biomass burning: MIT Press, p. 485–492.

Argyle, E., 1986, Cretaceous extinctions and wildfires: Science, v. 234, p. 261–264.

Becker, L., Bada, J. L., Winans, R. E., Hunt, J. E., Bunch, T. E., and French, B. M., 1994, Fullerenes in the 1.85-billion-year-old Sudbury impact structure: Science, v. 265, p. 642–644.

Blum, J. D., and Chamberlain, C. P., 1994, Geochemical constraints on the composition of volatiles released during the formation of KT impact melts: Implications for extinction mechanisms, *in* New developments regarding the KT event and other catastrophes in Earth history: Houston, Texas, Lunar and Planetary Institute Contribution 825, p. 11–13.

Böhme, D. K., 1992, PAH and fullerene ions and ion/molecule reactions in interstellar and circumstellar chemistry: Chemical Reviews, v. 92, p. 1487–1508.

Brett, R., 1992a, Anhydrite: A lethal target rock at the Chicxulub impact site, *in* Lunar and Planetary Science 23: Houston, Texas, Lunar and Planetary Institute, p. 157–158.

Brett, R., 1992b, The Cretaceous-Tertiary extinction: A lethal mechanism involving anhydrite target rocks: Geochimica et Cosmochimica Acta, v. 56, p. 3603–3606.

Brooks, R. R., Reeves, R. D., Yang, X.-H., Ryan, D. E., Holzbecher, J., Collen, J. H., Neall, V. E., and Lee, J., 1984, Elemental anomalies at the Cretaceous-Tertiary boundary, Woodside Creek, New Zealand: Science, v. 226, p. 539–542.

Brooks, R. R., Hoek, P. L., Reeves, R. D., Wallace, R. C., Johnston, J. H., Ryan, D. E., Holzbecher, J., and Collen, J. D., 1985, Weathered spheroids in a Cretaceous/Tertiary boundary shale at Woodside Creek, New Zealand: Geology, v. 13, p. 738–740.

Brooks, R. R., Strong, C. P., Lee, J., Orth, C. J., Gilmore, J. S., Ryan, D. E. and Holzbecher, J., 1986, Stratigraphic occurrence of iridium anomalies at four Cretaceous-Tertiary boundary sites in New Zealand: Geology, v. 14, p. 727–729.

Buseck, P. R., Tsipursky, S. J. and Hettich, R., 1992, Fullerenes from the geological environment: Science, v. 257, p. 215–217.

Chibante, L. P. F., and Heymann, D., 1993, On the geochemistry of fullerenes: Stability of C_{60} in ambient air and the role of ozone: Geochimica et Cosmochimica Acta, v. 57, p. 1879–1881.

Chibante, L. P. F., Pan, C., Pierson, M. L., Haufler, R. E., and Heymann, D., 1993, Rate of decomposition of C_{60} and C_{70} heated in air and the attempted characterization of the products: Carbon, v. 31, p. 185–193.

Chibante, L. P. F., Wolbach, W. S., and Heymann, D., 1994, Determination of nanogram amounts of C_{60} by high pressure liquid chromatography, *in* Lunar and Planetary Science 25: Houston, Texas, Lunar and Planetary Institute, p. 251–252.

Daly, T. K., Buseck, P. R., Williams, P., and Lewis, C. F., 1993, Fullerenes from a fulgurite: Science, v. 259, p. 1599–1601.

Devries, M. S., Reihs, K., Wendt, H. R., Golden, W. G., Hunziker, H. E., Fleming, R., Peterson, E., and Chang, S., 1993, A search for C_{60} in carbonaceous chondrites: Geochimica et Cosmochimica Acta, v. 57, p. 933–935.

D'Hondt, S., Sigurdsson, H., Hanson, A., Carey, S., and Pilson, M., 1994, Sulfate volatilization, surface-water acidification, and extinction at the KT boundary, *in* New developments regarding the KT event and other catastrophes in Earth history: Houston, Texas, Lunar and Planetary Institute Contribution 825, p. 29–30.

Gilmour, I., Wolbach, W. S., and Anders, E., 1990, Major wildfires at the Cretaceous/Tertiary boundary, *in* Clube, S. V. M., ed., Catastrophes and Evolution: Astronomical Foundations: Cambridge, Cambridge University Press, p. 195–213.

Haufler, R. E., and 16 others, 1990, Efficient production of C_{60} (Buckminsterfullerene), $C_{60}H_{36}$, and the solvated buckide ion: Journal of Physical Chemistry, v. 94, p. 8634–8636.

Heymann, D., 1995, Search for extractable fullerenes in anthraxolite, shungite, and thucholite, and formation of fullerenes on earth: Carbon, v. 33, p. 237–239.

Heymann, D., and Chibante, L. P. F., 1993, Reaction of C_{60} and C_{70} with ozone at different temperatures: Recueil des Travaux Chimiques des Pays Bas, v. 112, p. 531–534.

Heymann, D., Wolbach, W. S., Chibante, L. P. F., Brooks, R. R., and Smalley, R. E., 1994a, Search for extractable fullerenes from the Cretaceous-Tertiary boundary of Woodside Creek and Flaxbourne River sites, New Zealand: Geochimica et Cosmochimica Acta, v. 58, p. 3531–3534.

Heymann, D., Wolbach, W. S., Chibante, L. P. F., and Smalley, R. E., 1994b, Search for extractable fullerenes in clays from the KT boundary of the Woodside Creek and Flaxbourne River sites, New Zealand, *in* New developments regarding the KT event and other catastrophes in Earth history: Houston, Texas, Lunar and Planetary Institute Contribution 825, p. 47–48.

Heymann, D., Chibante, L. P. F, Brooks, R. R., Wolbach, W. S., and Smalley, R. E., 1994c, Fullerenes in the K/T boundary layer: Science, v. 265, p. 645–647.

Heymann, D., Korochantsev, A., and Nazarov, M. A., 1996, Search for fullerenes C_{60} and C_{70} at Cretaceous-Tertiary boundary sites in Turkmenistan, Khazakhstan, and Georgia: Cretaceous Geology (in press).

Howard, J. B., McKinnon, J. T., Makarovsky, Y., Lafleur, A. L., and Johnson, M. E., 1991, Fullerenes C_{60} and C_{70} in flames: Nature, v. 352, p. 139–141.

Kraetschmer, W., Lamb, L. D., Fostiropoulos, K., and Huffman, D. R., 1990, Solid C_{60}: A new form of carbon: Nature, v. 347, p. 354–358.

Kroto, H. W., Heath, J. R., O'Brien, S. C., Curl, R. F., and Smalley, R. E., 1985, Buckminsterfullerene: Nature, v. 318, p. 162–163.

Kyte, F. T., Smit, J., and Wasson, J. T., 1985, Siderophile interelement variations in the Cretaceous-Tertiary boundary sediments from Caravaca, Spain: Earth and Planetary Science Letters, v. 73, p. 183–195.

Lawver, L. A., Sclater, J. G., and Meinke, L., 1985, Mezozoic and Cenozoic reconstruction of the South Atlantic: Tectonophysics: v. 114, p. 133–254.

Radicati di Brozolo, F., Bunch, T. E., Fleming, R. H., and Macklin, J., 1994, Observation of fullerenes in an LDEF impact crater: Nature, v. 369, p. 37–40.

Sigurdsson, H., 1994, Environmental consequences of volcanic eruptions and meteorite impacts, *in* New developments regarding the KT event and other catastrophes in Earth history: Houston, Texas, Lunar and Planetary Institute Contribution 825, p. 114–115.

Strong, C. P., Brooks, R. R., Wilson, S. M., Reeves, R. D., Orth, C. J., Mao, X.-Y., Quintana, L. R., and Anders, E. A., 1987, A new Cretaceous-Tertiary boundary site at Flaxbourne River, New Zealand: Biostratigraphy and geochemistry: Geochimica et Cosmochimica Acta, v. 51, p. 2769–2777.

Suggate, R. P., Stevens, G. R., and Tepunga, M. T., 1978, The geology of New Zealand, Volume 2: Wellington, New Zealand, New Zealand Geological Survey, p. 346–624.

Taylor, R., Langley, G. J., Kroto, H. W., and Walton, D. R. M., 1993, Formation of C_{60} by pyrolysis of naphtalene: Nature, p. 366, p. 728–731.

Wolbach, W. S., and Anders, E., 1989, Elemental carbon in sediments: Determination and isotopic analysis in the presence of kerogen: Geochimica et Cosmochimica Acta, v. 53, p. 1637–1647.

Wolbach, W. S., Lewis, R. S., and Anders, E., 1985, Cretaceous extinctions: Evidence for wildfires and search for meteoritic material: Science, v. 230, p. 167–170.

Wolbach, W. S., Lewis, R. S., Anders, E., Grady, M. M., Pillinger, C. T., Brooks, R. R., Orth, C. J., and Gilmore, J. S., 1986, Carbon isotopes and iridium at two Cretaceous/Tertiary (K/T) boundary sites in New Zealand: Meteoritics, v. 21, p. 541–542.

Wolbach, W. S., Gilmour, I., Anders, E., Orth, C. J., and Brooks, R. R., 1988, Global fire at the Cretaceous/Tertiary boundary: Nature, v. 334, p. 665–669.

Wolbach, W. S., Anders, E., and Nazarov, M. A., 1990a, Fires at the K/T boundary: Carbon at the Sumbar, Turkmenia, site: Geochimica et Cosmochimica Acta, v. 54, p. 1133–1146.

Wolbach, W. S., Gilmour, I., and Anders, E., 1990b, Major wildfires at the Cretaceous/Tertiary boundary, *in* Sharpton, V. L., and Ward, P. D., eds., Global catastrophes in Earth history; an interdisciplinary conference on impacts, volcanism, and mass mortality: Geological Society of America Special Paper 247, p. 391–400.

MANUSCRIPT ACCEPTED BY THE SOCIETY SEPTEMBER 14, 1995

Printed in U.S.A.

Geological Society of America
Special Paper 307
1996

A normal polarity subchron that embraces the K/T boundary: A measure of sedimentary continuity across the boundary and synchroneity of boundary events

J. F. Lerbekmo
Department of Earth and Atmospheric Sciences, University of Alberta, Edmonton, Alberta T6G 2E3, Canada
A. R. Sweet
Geological Survey of Canada, 3303 33rd Street., N.W., Calgary, Alberta T2L 2A7, Canada
M. J. M. Duke
SLOWPOKE Reactor Facility, University of Alberta, Edmonton, Alberta T6G 2N8, Canada

ABSTRACT

In order to assess the timing and degree of synchroneity of extinctions near the K/T boundary, global chronologic horizons are required with time spans of a few thousand years or less. Geomagnetic polarity reversal boundaries fulfill these constraints. Detailed magnetostratigraphy has been carried out across the K/T boundary in seven sections from central Alberta, southern Saskatchewan, and northern Montana. The K/T boundary has been identified in these sections on the basis of an Ir anomaly and/or diagnostic palynomorph extinctions. Three of the sections also have bentonites deposited within a few thousand years after the Ir concentration and have been dated by U-Pb, Rb-Sr, K-Ar, and Ar-Ar methods, giving mean ages of 64.4 ± 1.2 and 64.70 ± 0.09 Ma.

The magnetostratigraphic data indicate that there is a short normal subchron that embraces the K/T boundary. The subzone begins about 1 m below the boundary and extends for one to several meters above the boundary. On the basis of mean rate of sedimentation for the 29r magnetozone, the subchron is believed to have lasted at least 50 k.y.

Palynofloral changes across the K/T boundary show range truncations of a diverse latest Cretaceous palynoflora within a few decimeters of the Ir anomaly, followed by an interval of similar thickness of earliest Paleocene opportunistic ferns and/or angiosperms before the reestablishment of the gymnosperm-dominated swamps.

The boundaries of the short normal subchron in the upper part of chron 29r, coupled with the Ir anomaly and rapid changes in palynoflora, provide a temporally unusually well constrained interval for judging the degree of abruptness of the K/T extinction event(s).

INTRODUCTION

A large Ir anomaly has provided a generally accepted global marker for the K/T boundary. However, arguments continue as to the exact time relationships between many Cretaceous extinctions and this horizon. Sea-level and climatic changes can have profound stress effects on life and may produce sedimentary hiatuses that make extinctions appear more abrupt then they actually were.

Such changes were going on near the end of the Creta-

Lerbekmo, J. F., Sweet, A. R., and Duke, M. J. M., 1996, A normal polarity subchron that embraces the K/T boundary: A measure of sedimentary continuity across the boundary and synchroneity of boundary events, *in* Ryder, G., Fastovsky, D., and Gartner, S., eds., The Cretaceous-Tertiary Event and Other Catastrophes in Earth History: Boulder, Colorado, Geological Society of America Special Paper 307.

ceous, and it remains to be sorted out which extinctions were due to climatic or eustatic changes and which may have been due to a bolide impact or possibly strong mantle plume activity. Choices of cause require exceptional precision in time correlation and continuity of record above and below the K/T boundary. Extinctions due to a large impact will be geologically instantaneous within a framework of days or months. Extinctions due to environmental factors can probably occur within hundreds to a few thousand years. Hiatuses in the sedimentary record of a few thousand years are often not detected except by apparent extinctions. Thus, time resolution of the order of a few thousand years near the K/T boundary is necessary to resolve the question of effects due to natural environmental changes versus abnormal terrestrial or extraterrestrial influences. To this end, detailed magnetostratigraphic sampling was undertaken at a number of K/T boundary sections in Alberta, Saskatchewan, and Montana after initial sampling had shown that the level nearest to the boundary consistently showed normal polarity, despite the boundary's being in the upper part of reversed magnetozone 29r. The results of more detailed sampling reported here support the original interpretation and indicate a normal polarity subzone embracing the K/T boundary.

SECTIONS STUDIED

Results will be presented from seven sections studied in Alberta, Saskatchewan, and Montana (Fig. 1). Two sections are in the Red Deer Valley in central Alberta. The Knudsen Coulee section (NW ¼ S.10, T.34, R.22W4) is in a tributary coulee on the Kent and Marion Knudsen farm and is the site of C. M. Sternberg's 1946 discoveries of fragments of a ceratopsian and a tyrannosaurid skeleton (Sternberg, 1947). The tyrannosaurid was not collected at the time but was located again by Lerbekmo in 1977 during magnetostratigraphic study of the K/T boundary interval in the Red Deer Valley (Lerbekmo et al., 1979b). The tyrannosaurid skeleton was 10.5 m below the base of the Nevis (K/T boundary) seam (equivalent to the Ferris seam of Saskatchewan and the lower Z coal of Montana). The skeleton was excavated in 1982 by the Royal Tyrrell Museum of Palaeontology, Drumheller, Alberta, and mounted in the museum.

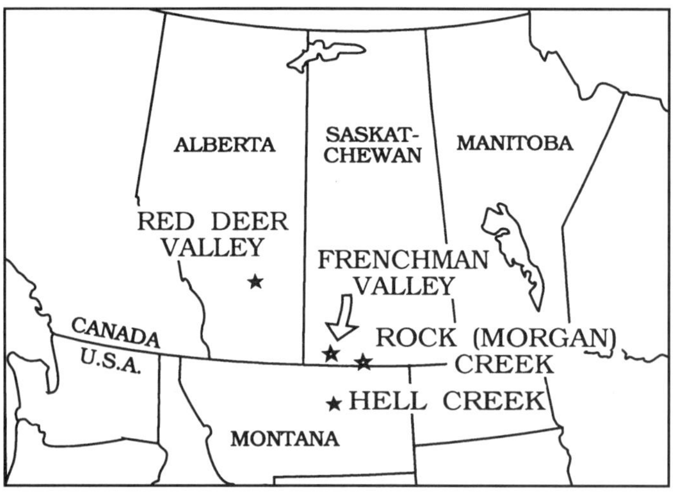

Figure 1. Locations of studied sections.

The West Scollard section is on the west side of the Red Deer Valley in the Scollard Canyon area (NE ¼ S.11, T.34, R.22W4) and is the upper part of the section reported on by Lerbekmo et al. (1979a) and Lerbekmo and Coulter (1985).

Three sections are in the Frenchman River valley near Eastend in southwestern Saskatchewan. The Ravenscrag Butte section (NW ¼ S.27, T.6, R.23W3) is part of the section described by Lerbekmo (1985). The Frenchman River section (SE ¼ S.31, T.4, R.18W3) is a Highway 37 road cut reported on by Lerbekmo (1987), Lerbekmo et al. (1987), and Sweet and Braman (1992). The Eastend section (SW ¼ S.30, T.6, R.21W3) is a road cut on Highway 13.

The Rock (Morgan) Creek East section (S.19, T.1, R.4W3) is on the east side of Rock Creek, across the creek from the section published on by Nichols et al. (1986) and Sweet and Braman (1992).

The Hell Creek section (NW ¼ S.1, T.20N, R.37E) is about 0.8 km west of the Hell Creek Park access road north of Jordan, Montana. The magnetostratigraphy of this section was done in 1979 but not published until now. However, the geochronology and some biostratigraphy were published by Baadsgaard and Lerbekmo (1980) and Baadsgaard et al. (1988).

METHODS

Sampling sites were prepared by digging a horizontal bench deep enough to uncover undisturbed bedrock. Samples were taken by driving a 2.5-cm diameter core-barrel vertically 5 cm into the rock and extruding the core into a 2.5-cm diameter, 5-cm-deep plastic vial (Lerbekmo, 1990). Three to five independently oriented samples were taken at each site (sampling level).

Samples were measured in a Molspin magnetometer. Cleaning to remove overprinting was routinely done by alternating field (AF) demagnetization in steps of 0.5 mT until a stable end point was reached or the sample became too weak to provide a reliable measurement. One sample from most sites was also measured during thermal cleaning in steps of 50°C. The remanent direction of magnetization after demagnetization at 20 mT or 250°C (or the final measurement if at a lesser value) was taken. The site VGP (Virtual Geomagnetic Pole) latitude plotted in Figures 2 through 8 is the latitude of the magnetic pole indicated by the mean direction of magnetization of the samples from one site (elevation). If a sample did not provide a reasonable measuring error it was omitted by the computer program from the computation. For sites for which both AF and thermal data were available, the site VGP latitude join-lines used thermal data points, but the thermal data are not necessarily more reliable than the AF data. If *all* samples at a site yield paleopoles (VGP's) within 70° of that predicted by the normal paleopole for 65 Ma, taken to be at 78°N, 186°E (Marquis and Globerman, 1988), the site is classified as normal by computer program. If one or more samples at a site give a direction within 70° of the reversed palepole, the site is classified as reversed. All other sites are classified as indeterminate, although some, especially those near polarity changes, may in fact be intermediate. Seventy degrees was chosen because it divides the earth's surface into three approximately equal areas (see also Lerbekmo and Coulter, 1985). The bias toward a reversed classification is to compensate for overprinting under the present normal field during weathering (Hillhouse et al., 1977).

Sampling for palynological analysis at intervals of 0.5 to 5 cm was also undertaken near the K/T boundary to determine palynofloral changes near the boundary clay.

Splits of the palynological samples were analyzed for Ir by neutron activation analysis (NAA), to confirm the position of the boundary, especially where the boundary clay is missing. The NAA follows a nickel sulfide fire-assay concentration. The methodology follows closely that described by Lerbekmo and St. Louis (1986), but improvements in gamma ray detectors now allow a detection limit of ≈20 ppt for a starting sample of 20 gm. Background Ir values for mudstones below the K/T boundary and coal and mudrocks above the boundary usually range from 0.02 to 0.06 ppb (Lerbekmo and St. Louis, 1986, and this chapter). Values above 0.2 ppb are considered to be anomalous.

RESULTS

Figures 2 through 8 incorporate the results of paleomagnetic and geochemical analyses across the K/T boundary in the seven sections reported on here. The lithology of all the sections is fairly similar, encompassing greenish fluviatile mudstones below the boundary, with a 1- to 3-cm tan to pinkish clay often present just below a boundary coal. The coal is of variable thickness up to about 1 m but is usually 0.3 to 0.5 m. Overlying the coal are lacustrine brownish gray shales or siltstones. A bentonite 10 to 15 cm thick is preserved in the boundary coal in a number of areas. All of the sections studied are relatively fresh exposures on steep valley slopes or in road cuts.

Figure 2a shows the lithology and detailed magnetostratigraphy near the K/T boundary in the West Scollard section. There is no boundary clay at this location, which was chosen for magnetostratigraphy because of freshness of outcrop and minimal sandstone near the boundary. However, a boundary clay occurs at the original Red Deer Valley discovery location, ≈2.5 km to the south, where Lerbekmo and St. Louis (1986) obtained an Ir peak of 3.36 ± 0.04 ppb (1 σ), and Grieve and Alexopoulos (1988) and Bohor (1990) later reported shocked quartz. The bentonite at the top of the Nevis seam has been dated by Baadsgaard et al. (1988) by the U-Pb, Rb-Sr, and K-Ar methods, which produced a mean age of 64.3 Ma. A sanidine split dated by McWilliams (McWilliams and Baadsgaard, 1991) by the Ar-Ar methods (plateau and step heating) gave a mean age of 64.75 Ma. Palynomorph range truncations in this section have been described by Srivastava (1970), Sweet and Hills (1984), and Sweet and Braman (1992). The continuation of magnetozone 29r above the K/T boundary has been reduced by one or more hiatuses representing a few meters of section. The base of 29r is 25 m below the K/T boundary (Lerbekmo and Coulter, 1985).

Magnetic overprinting at West Scollard and the other sections is not severe. As examples, Fig. 2b shows the behavior of some samples during magnetic cleaning from nine sites in this section. A viscous normal overprint of weak coercivity is usually recognizable as a result of the last 0.7 million years of normal polarity, but most of it is usually removed at 10mT of AF demagnetization or 200°C of thermal cleaning. A normal polarity subzone embraces the K/T boundary with sharply defined limits both below and above. The sample at –1.35 m apparently occurs within the transition from reversed to normal polarity that takes place between –1.5 and –1.2 m. Samples within transition zones are usually very weak magnetically, suggesting that the dipole component of the field is weaker during transition, resulting in less consistency of orientation of magnetic grains. The polarity change at the top of the normal subzone takes place within approximately 0.2 m.

Figure 3a illustrates the paleomagnetic results obtained to date at the Knudsen Coulee location. The normal polarity subzone is well defined below the K/T boundary, but its upward extent has not yet been investigated in the sandstones above the Nevis coal. The distribution of Ir near the boundary is shown in Figure 3b. The peak value of 12.74 ± 0.10 (1 σ) occurs 1 to 2 cm above the base of the coal. The secondary peak in the carbonaceous mudstone just below the 3 cm boundary clay (Fig. 3b) suggests some mobility of the original Ir and reprecipitation in reducing environments of high C concentrations.

The magnetostratigraphy of the Ravenscrag Butte section is shown in Fig. 4. The normal subzone is well constrained at the bottom and top, although some of the inclinations are relatively low. As at West Scollard, the thickness of magnetozone 29r above the boundary is relatively small, suggesting either a slow accumulation rate for the Ferris coal seam or a minor hiatus. There is no boundary clay at this locality, but a small Ir peak of 1.8 ± 0.03 ppb 5 to 10 cm up in the coal suggests the preservation of boundary clay nearby, from which some Ir migrated laterally.

Figure 5 shows the magnetostratigraphy near the K/T boundary in the Highway 13 road cut near Eastend. The base of the normal polarity interval is closely determined, but the subzone is terminated 3 m above the boundary by a disconformity at the base of a major coarse sandstone. A thin discontinuous boundary clay is present but was not itself analyzed for Ir. The lowest 2 cm of coal, however, gave a value of 6.72 ± 0.10 (1 σ) ppb, whereas the uppermost 2 cm of the underlying mudstone yielded 1.97 ± 0.02 ppb.

The results of the paleomagnetic study of the Highway 37 road cut at the Frenchman River are shown in Fig. 6a. This is the second best of the sections for defining the K/T normal subzone, as both bottom and top are well constrained, although the Ferris coal and very carbonaceous shale immediately above the bentonite provide a gap in the magnetic data. Fig. 6b illustrates the magnetic behavior of samples from five sites. The polarity change at the bottom takes place between sites at –1.0 and –0.6 m. The top of the zone is between 0.85 and 1.25 m. These samples, like those at West Scollard, indicate only light overprinting. For instance, the samples at –1.0 m before cleaning show an approximate present pole direction but move toward a reversed direction with minimum AF demagnetization. The samples at 1.25 m have a slightly negative natural remanent magnetization (NRM) even before cleaning. The base of magnetozone 29r in this section is at 19.5 m below the K/T boundary (Lerbekmo, 1987), and the top is at 11.5 m above the boundary in a nearby section (unpublished data).

An Ir peak of more than 7 ppb occurs in the lowest 2 cm of the Ferris seam (Fig. 6c). The boundary clay is up to 2 cm thick but is discontinuous along the road cut. The boundary clay itself has yielded Ir values as high as 4.91 ± 0.04 ppb (unpublished data).

The Ferris coal bentonite mean age from the U-Pb, Rb-Sr, and K-Ar dating methods was 64.9 Ma (Baadsgaard et al., 1988). McWilliam's and Baadsgaard's (1991) mean age on the sanidine split by Ar-Ar was 64.65.

Figure 7a shows the magnetostratigraphy near the K/T boundary in the east Rock (Morgan) Creek area. The normal polarity subzone is fairly well constrained here, although sites below the boundary are somewhat inconsistent in magnetic direction. The reversed site just above the thin Ferris coal may represent a very short reversal or excursion within the K/T normal, because there is some evidence of its presence in one or two other locations where the boundary coal is thin enough to permit sampling at this level. However, confirmation will require further documentation.

A large Ir anomaly is present in western Rock Creek sections where the boundary clay and the Ferris coal seam are thicker. The Rock Creek East section shown here was chosen for magnetostratigraphy because of the thinness of the coal, but the boundary clay is not as well preserved. Figure 7b shows the Ir profile from one of the sections (Rock Creek West "E") on the west side of the creek. Nichols et al. (1986) reported an Ir peak of only 3.0 ppb from another section approximately 0.4 km north of this one.

The magnetostratigraphy of the Hell Creek section is shown in Figures 8a and b. Rates of sedimentation in this area in the late Maastrichtian and early Paleocene were relatively high, as magnetozone 29r is in excess of 50 m thick (Fig. 8a).

The top of 29r in the Hell Creek area is shown by Swisher et al. (1993, Fig. 2) to approximately correspond with the Y coal about 30 m above the Ir horizon. Magnetozone 29r extends downward below the base of the outcrop in our section, which is about 30 m below the base of the boundary Z coal.

The upper limit of the K/T normal subzone is below the first available sample at 1.7 m and has arbitrarily been shown to coincide with the top of the Z coal. This small thickness of the

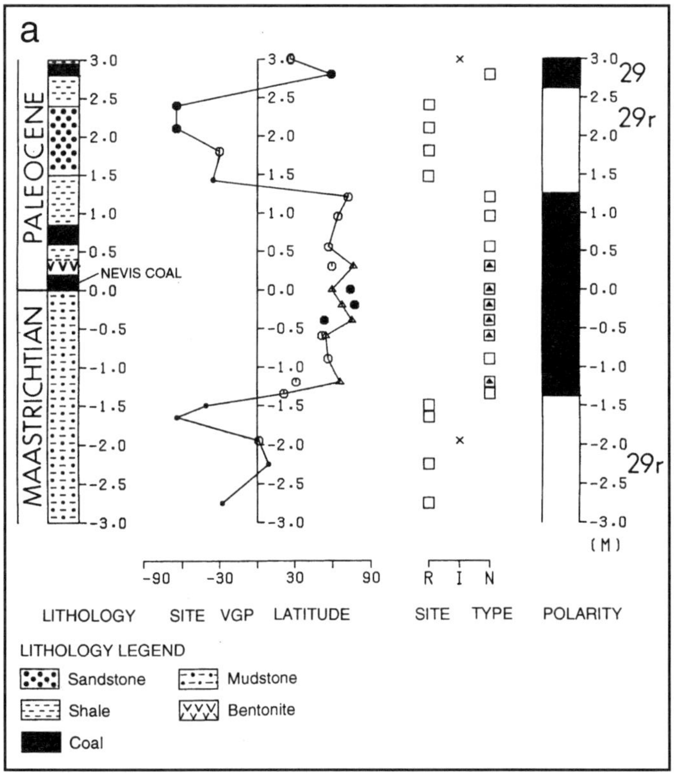

Figure 2a. Magnetostratigraphy and lithostratigraphy of the K/T boundary interval of the West Scollard section. Large solid circles in Site VGP Latitude plot are sites with three or more AF measurements satisfying the Watson (1956) test for nonrandomness. Open circles represent sites with fewer than three samples. Small dots represent sites not satisfying the Watson test. Triangles represent thermal demagnetization. R, I, N = reversed, indeterminate, normal polarities; boxes and x's are after AF demagnetization, triangles after thermal. In the Polarity column, black is normal polarity, white is reversed; numbers to the right are magnetozone assignments.

normal subzone relative to the thickness of 29r suggests a long time span for the Z coal, or a hiatus. The Z coal bentonite has yielded a mean age of 64.1 Ma from the U-Pb, Rb-Sr, and K-Ar methods (Baadsgaard et al., 1988) and 64.8 Ma by Ar-Ar on the sanidine split (McWilliams and Baadsgaard, 1991). There is no boundary clay at this location, but a maximum Ir value of 0.65 ± 0.03 ppb occurs approximately 2 cm above the base of the coal (Baadsgaard et al., 1988, and unpublished data).

PALYNOLOGICAL EVENTS WITHIN AND CONTIGUOUS TO THE NORMAL POLARITY SUBZONE IN WESTERN CANADA AND MONTANA

In all the referenced sections, the base of the short normal subzone occurs in association with a latest Maastrichtian palynoflora (*Wodehouseia spinata* zone of Srivastava [1970]) characterized by a diversity of morphologically complex angiosperm

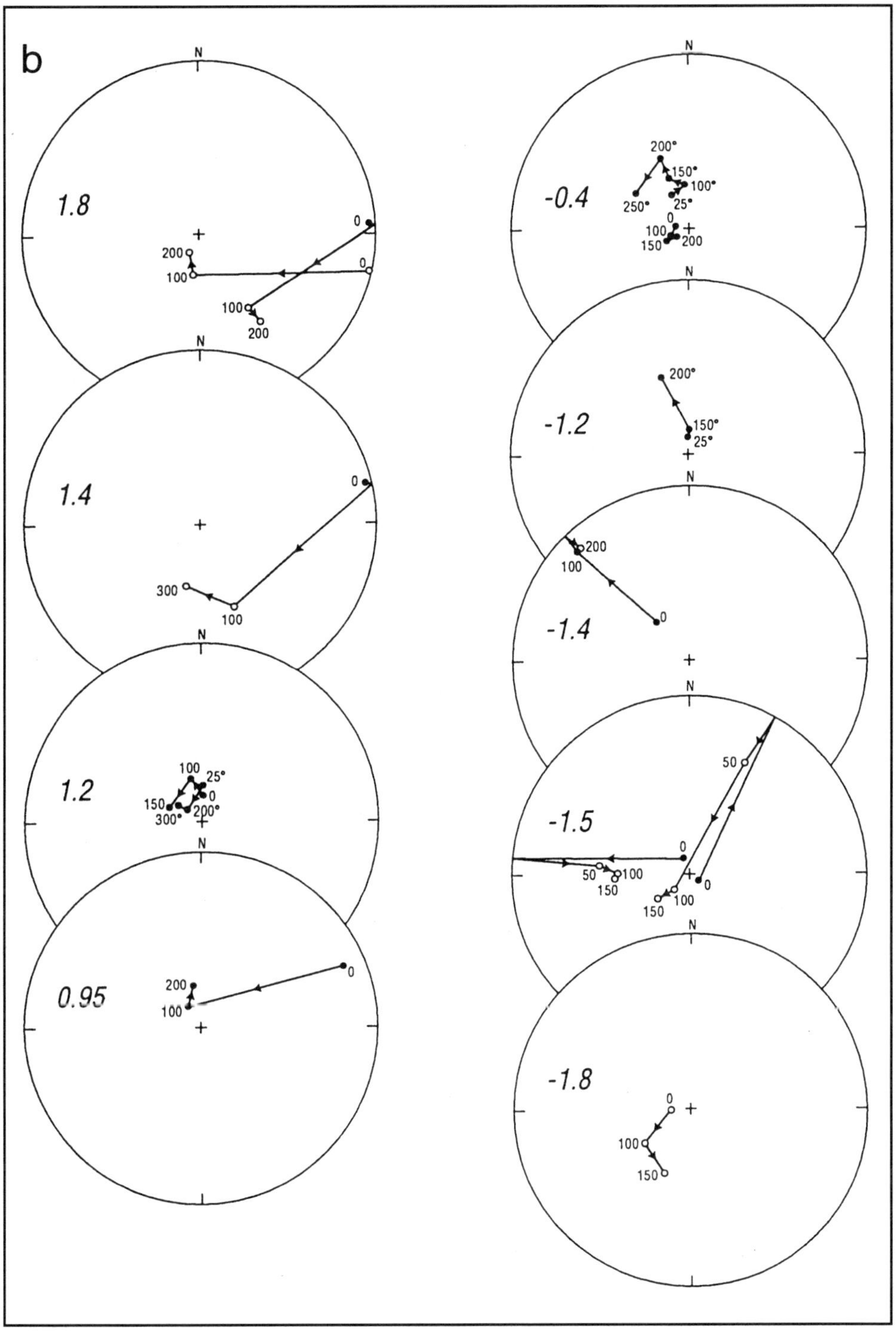

Figure 2b. Examples of directional behavior during magnetic cleaning of samples from nine sites in the West Scollard section (Fig. 2a). Polar-projection equal-area stereographic plots; dots on lower hemisphere, open circles on upper hemisphere. Numbers followed by degree symbol are centrigrade; unlabeled numbers are mT. Large number on left is elevation (m) of site (see Fig. 2a). Note response, especially of reversed samples, to low-level demagnetization, indicating a relatively soft normal overprint.

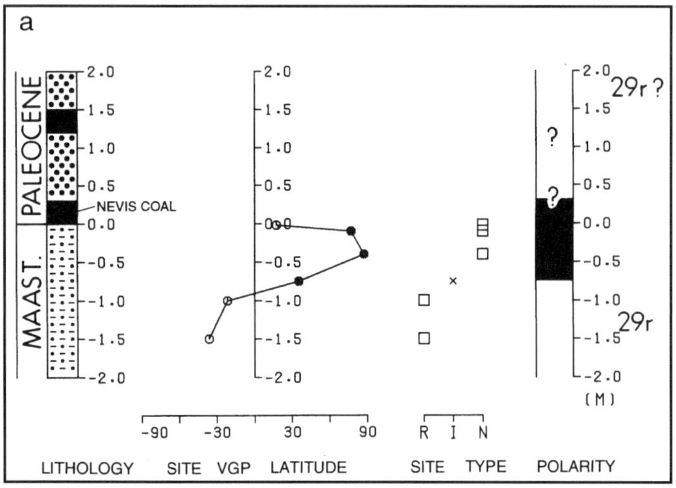

Figure 3a. Magnetostratigraphy and lithostratigraphy of the Knudsen Coulee section. See Figure 2a for explanation.

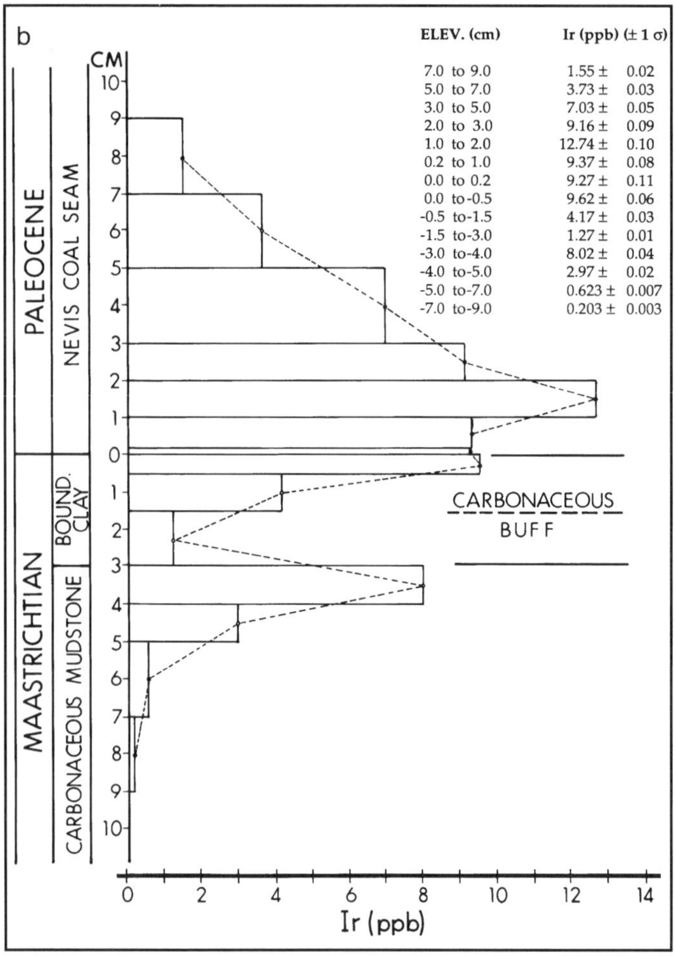

Figure 3b. Iridium profile of the Knudsen Coulee K/T boundary interval.

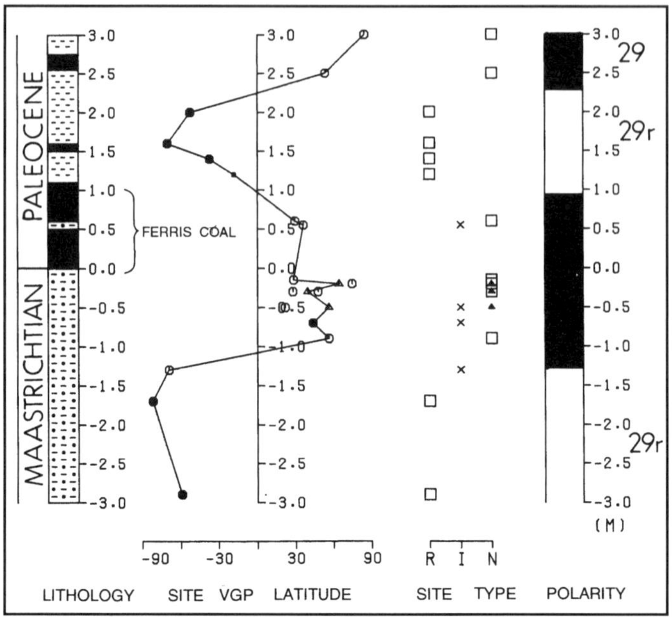

Figure 4. Magnetostratigraphy and lithostratigraphy of the K/T boundary interval of the Ravenscrag Butte section. See Figure 2a for explanation.

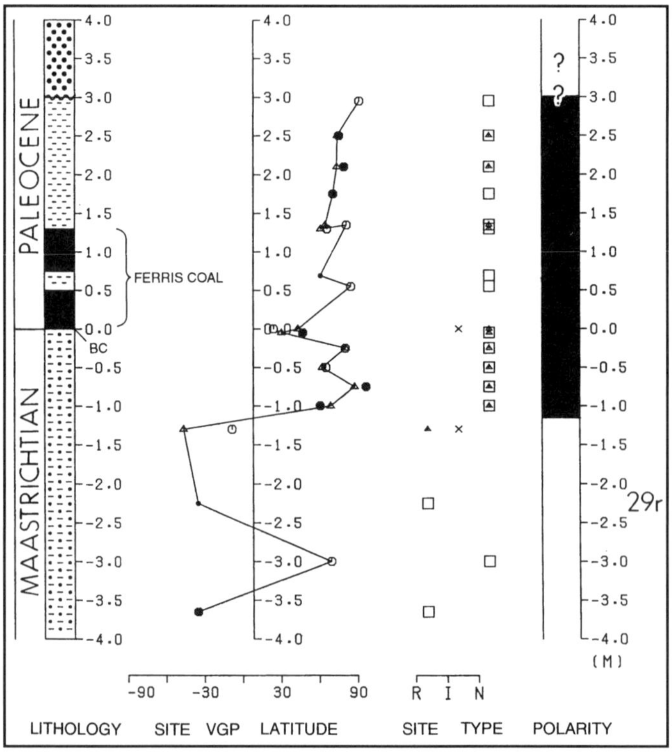

Figure 5. Magnetostratigraphy and lithostratigraphy of the Eastend Highway section. BC = Boundary Clay. See Figure 2a for further explanation.

pollen. Species regionally characteristic of this interval include *W. spinata* in association with *Aquilapollenites conatus, A. delicatus* var. *collaris, A. quadricretaeus, A. quadrilobus, A. reductus, Cranwellia rumseyensis, Gunnera microreticulata, Leptopecopites pocockii, Marsypiletes cretacea, Orbiculapollis lucidus,* and *Striatellipollis striatellus.*

Excluding *W. spinata*, the upward ranges of the above species are usually truncated within the interval of the short normal subzone below the boundary. In the Knudsen Coulee, Frenchman River, and Rock (Morgan) Creek West "E" sections, the truncations occur just before or at the top of the pink to tan K/T boundary claystone (but below the upper "impact layer" of Izett [1990]). Because of the association between these truncations and the physically defined K/T boundary, they are taken to represent the K/T boundary extinction event, and hence, in sections without a boundary claystone, the palynological range truncations can be used to mark the K/T boundary. Associated with the range truncations is an abrupt decrease in the relative abundance of gymnosperm (*Taxodiaceae/ Cupressaceae/Taxaceae*—T/C/T) pollen in sections located in western Canada that represents the loss of the swamp forest canopy.

Above the level of concentrated range truncations is a succession of earliest Paleocene palynofloras representing opportunistic infilling of vacant niches by ferns (trilete and monolete spores) and/or angiosperms prior to the reestablishment of gymnosperm-dominated swamps (Sweet et al., 1991; Sweet and Cameron, 1991; Sweet and Braman, 1992) over an interval of up to about 0.2 m in the sections under discussion. This interval sometimes contains an abundance of angiosperm pollen more typical of the Cretaceous (*W. spinata* and *A. reticulatus*).

In the Knudsen Coulee, Eastend, Frenchman River, and Rock (Morgan) Creek sections there is a coincidence between the above shift from a latest Maastrichtian to an earliest Paleocene successional flora, the base of a regionally persistent coal-bearing interval, a geochemical anomaly typified by a high abundance of iridium (Figs. 3, 5, 6, and 7) and a boundary claystone. In the West Scollard locality there is a 1.04-m fining-upward interval of fine sandstone, siltstone, and rooted mudstone immediately preceding the base of the Nevis coal. The upper 0.24 m of this interval contains a relatively well preserved palynological assemblage that lacks species most typical of the latest Maastrichtian. Starting 0.18 m below the base of the coal, the palynoflora is dominated first by trilete, then monolete fern spores, a succession characteristic of the earliest Paleocene. Hence, in this section the K/T boundary would seem to be some 0.18 m below the base of the coal. This may in part explain the relatively thick normal interval below the coal at the West Scollard locality. At two other localities, Ravenscrag Butte and Hell Creek Road, the palynologically defined position of the K/T boundary also appears to be as much as 0.1 m below the base of the coal. In all three of these localities the boundary clay is missing.

The part of the short normal subzone above the K/T boundary and the upper reversed part of 29r span a palynologically "active" zone that includes the successional fern/angiosperm floras, increases in the abundance of T/C/T pollen, an abundance peak of *W. spinata,* and the first Paleocene occurrence of an unnamed precursor species to *W. fimbriata*, an index species for the early but not earliest Paleocene. The first occurrence of *Wodehouseia* specimens more similar to typical *W. fimbriata* more or less coincides with the 29r-29 contact at West Scollard, at Ravenscrag Butte, and at localities within the vicinity of Rock (Morgan) West "E" and the Hell Creek Road sections. Accompanying the entrance of *W. fimbriata* (s.l.) is the establishment of a low-diversity T/C/T dominated flora that typifies the early Paleocene in western Canada.

DISCUSSION

It has been recognized for some time that the K/T boundary occurs in magnetozone 29r (e.g., Harland et al., 1982). This recognition was based largely on the magnetostratigraphy of deep-sea marine sections near Gubbio, Italy (Premoli Silva et al., 1974; Lowrie and Alvarez, 1977; Roggenthen and Napoleone, 1977), and has been confirmed by more recent studies in other parts of Europe (Mörner, 1982; Mary et al., 1991; Praesinger et al., 1993; Golbrun et al., 1993). None of these studies revealed a normal subzone at the K/T boundary, and so the question naturally arises: If there is such a normal subzone at the K/T boundary, why did none of these studies encounter it?

MacLeod and Keller (1991a, b) and Keller et al. (1993) have shown that most deep sea marine sections have one or more hiatuses at the K/T boundary or in the earliest Paleocene that have removed the temporally rather short PO foraminiferal zone and usually part of the overlying Pla Zone. A hiatus just prior to the K/T boundary is more difficult to identify biostratigraphically because faunal zones of the latest Cretaceous were relatively long lasting. It might be postulated that if the K/T boundary clay is present, a record of latest Maastrichtian time will also be preserved. If the interpretation of Delacotte et al. (1985) on the polarity chronology of the Bidart (Biarritz) K/T boundary section of southwestern France is correct, this may not always be the case. Delacotte et al. (1985) show a 15-cm-thick boundary clay with an Ir anomaly of 6.3 ± 1.1 ppb (Bonte et al., 1984). A normal magnetozone occurs immediately below the boundary claystone, which Delacotte et al. correlate with magnetozone 30. However, strata below the boundary still fall within the latest Cretaceous foraminiferal zone of *Abathomphalus mayaroensis* and the nannofossil zone of *Micula prinsii*, hence there is no biological indication of the inferred hiatus corresponding to that part of 29r below the K/T boundary (some 200 to 300 k.y.), notwithstanding the presence of the boundary Ir anomaly. Alternatively, if sedimentation rates in this section were very high in the latest Cretaceous, the normal polarity interval embracing some 8 m of section below the K/T boundary might belong to the K/T boundary normal subzone.

According to MacLeod and Keller (1991a, b) and Keller et al. (1993), the most complete marine boundary sections are

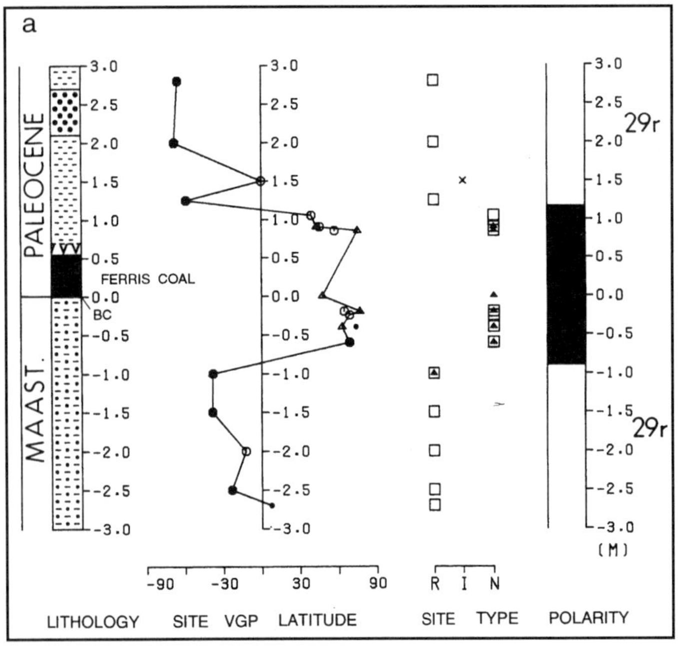

Figure 6a. Magnetostratigraphy and lithostratigraphy of the Frenchman River section. BC = Boundary Clay. See Figure 2a for further explanation.

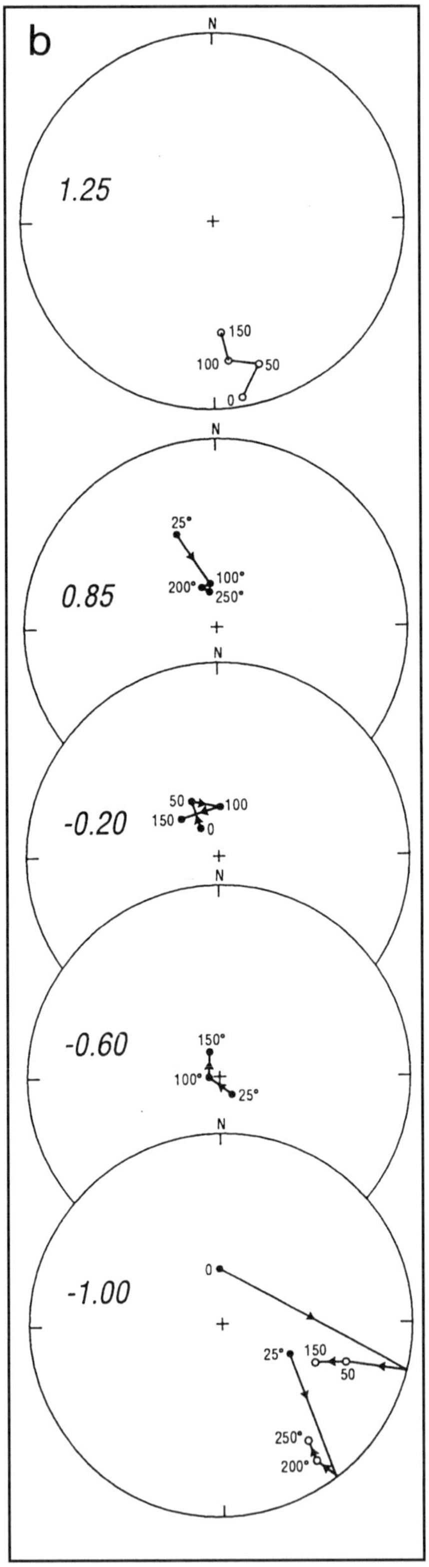

Figure 6b. Examples of directional behavior during magnetic cleaning of samples from five sites in the Frenchman River section. See Figure 2b for explanation. Note that the normal polarity overprinting shown by the reversed samples is mostly removed at relatively low levels of AF or thermal demagnetization.

middle neritic to upper slope, as exemplified by boundary sections at El Kef, Tunisia; Agost and Caravaca, Spain; and the Brazos River area of Texas and in northeastern Mexico. Of these sites, the magnetostratigraphy has been reported upon at Agost, Spain (Groot et al., 1989) and Brazos River, Texas (Keller, 1989). At Agost, the K/T boundary was placed at 11.47 m, and two paleomagnetic samples are shown to have been taken between 11 and 12 m. Both sites showed reversed inclination, but the exact positions with respect to the K/T boundary are difficult to assess from the figure of Groot et al. (1989). The magnetostratigraphy of the Brazos River core was done by W. Gose (Keller, 1989, Fig. 4). No normal subzone is indicated at the K/T boundary, but sample positions are not shown. Praesinger et al. (1993) have presented the magnetostratigraphy of a K/T boundary section on the Black Sea that contains an iridium anomaly peak of about 7 ppb and appears to be complete. No normal polarity subzone was indicated, but no samples appear to have been taken within about 1 m of the boundary. However, the two samples taken closest to the boundary have suspiciously low inclination, suggesting the possibility that they have transitional polarities.

Two Deep Sea Drilling Project (DSDP) sites in the Atlantic have shown indications of a short normal zone at the K/T boundary. At site 384 on the J-Anomaly Ridge in the North Atlantic, Larson and Opdyke (1979) found a normal polarity zone extending from ~18 cm below to at least a meter and a half above the K/T boundary. They interpreted this to be anomaly 29 with a hiatus at the boundary corresponding to at least that part of 29r that extends above the boundary. However,

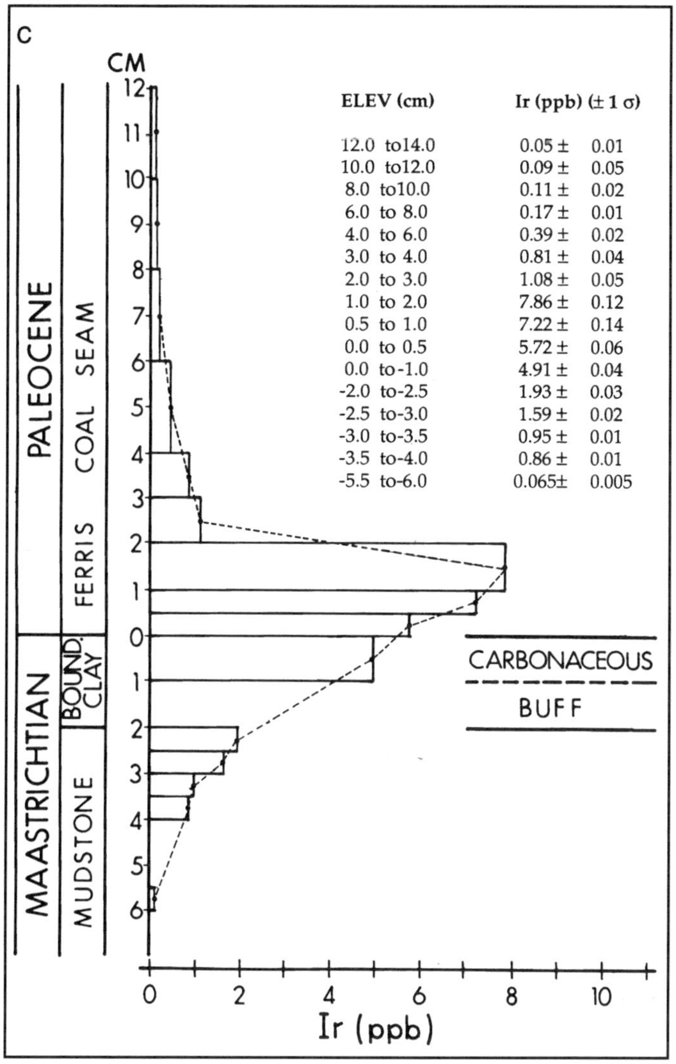

Figure 6c. Iridium profile of the Frenchman River K/T boundary interval.

Thierstein and Okada (1979) disagreed with this interpretation of a hiatus, based upon the nannofossil continuum across the boundary, and considered the normal subzone crossing the K/T boundary to be real but not recognized at Gubbio. Their calculations also indicate that sedimentation rates in the early Paleocene portion of 29r were relatively high compared to the Cretaceous portion (the opposite of the situation at Gubbio). This difference would at least partially account for the relatively large thickness of the K/T normal subzone above the boundary, compared to below. At DSDP site 516 (Hole 516F) on the Rio Grande Rise in the south Atlantic, Hamilton and Suzyumov (1983) and Hamilton (1983) indicate that a short normal subzone occurs just above the K/T boundary. The normal interval is represented by two samples at 4 and 19 cm above the boundary (N. Hamilton, personal communication, 1994), as represented by a 0.5-cm-thick dark clay with an iridium peak of 0.95 ± 0.18 ppb (Michel et al., 1983). Lower, but anomalous, values of iridium extend upward for about 20 cm above the peak. Latest Cretaceous planktonic foraminifers (*Globotruncana* spp.) extend a few centimeters above the peak iridium level, where they occur together with nannofossils of the lowest Paleocene NPl Zone. The Paleocene planktonic foraminiferal zone Pla was only tentatively identified in this hole. Thus, although the K/T boundary interval is essentially intact, the biostratigraphy apparently would permit a small hiatus. Polarity zone 29r is half again as thick in this hole as in the Gubbio section.

It would appear that from the magnetostratigraphy that has been published to date on the K/T boundary in marine sections, both on land and under the sea, the evidence for or against a short normal polarity zone crossing the K/T boundary is inconclusive. The problem is that it is difficult to prove the continuity of sedimentation over periods as short as 50,000 years or so on biological grounds. Yet, this is a significant period of time when assessing whether or not extinctions were geologically instantaneous, as perceived to occur from a large meteorite impact.

It seems at present that, although marine sections have been favored for biostratigraphic studies because of the variety of fossils preserved, swamp and swamp-margin continental K/T strata may be just as rewarding in the search for sedimentary continuity, and palynofloral changes can rival marine organic changes for stratigraphic precision.

SUMMARY AND CONCLUSIONS

Detailed magnetostratigraphy, palynology, and iridium analysis were carried out on seven K/T boundary sections in Alberta, Saskatchewan, and Montana. The K/T boundary is distinguished by a 1- to 4-cm thick, usually bipartite, boundary clay (buff to tan below, dark brown above) in five of the sections. The boundary clay is always immediately overlain by a coal seam 0.2 to 1.3 m thick. Where the boundary clay is present, an Ir peak concentration of 6.7 to 34.7 ppb has been preserved. Where the boundary clay is missing, maximum Ir concentration is 0.65 to 1.85 ppb. The magnetostratigraphy consistently shows a normal polarity subzone, within zone 29r, extending from about 1 m below the boundary to 1 m or more above the boundary. Based on the thickness of magnetozone 29r and the length of chron 29r, the subchron is estimated to have lasted from about 25 k.y. before the boundary to 25 k.y. or more after the boundary, as defined by the boundary clay or Ir anomaly. The palynology shows range truncations of a number of typical latest Maastrichtian palynomorphs in the last ≈1 m preceding the boundary, followed in the earliest Paleocene by ≈10 cm of strata containing an opportunistic succession fern and/or angiosperm flora before reestablishment of the gymnosperm swamp flora.

A normal subzone at the K/T boundary has to date not been authenticated in marine sections on land, although most of those studied have suspected hiatuses at the boundary. Two

DSDP sites are indicated to have normal polarity at the boundary, but the evidence is inconclusive. Swampy continental environments at the end of the Cretaceous may provide as good a record of sedimentary continuity as marine sections.

The definition of a short normal subzone in the upper part of 29r provides the opportunity to develop a sequence of closely spaced, physically defined time horizons (29r/short normal subzone, K/T boundary or contact, short normal subzone/29r, 29r/29) across one of the most important times in the history of life. Development of this sequence in turn provides the exceptional opportunity to examine biological events across this interval regionally in well-constrained, time-bounded units of geologically short duration and should aid in determining the timing and degree of synchroneity of extinctions associated with the K/T boundary event(s).

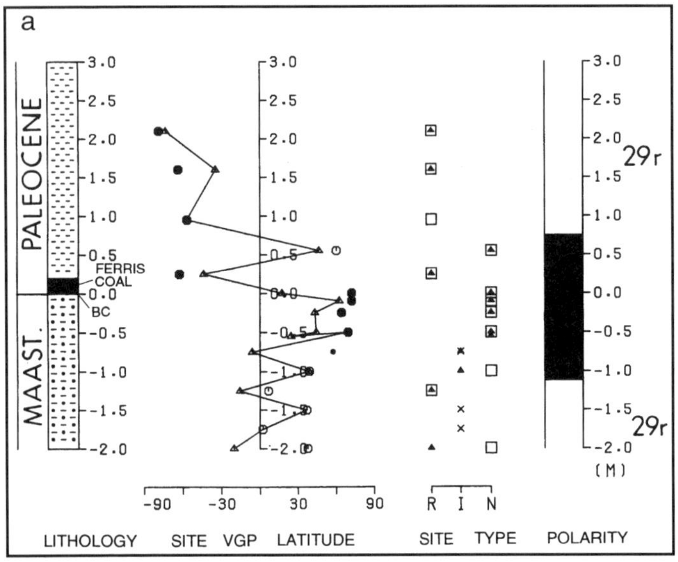

Figure 7a. Magnetostratigraphy and lithostratigraphy of the K/T boundary interval of the Rock (Morgan) Creek East section. BC = Boundary Clay. See Figure 2a for further explanation.

ACKNOWLEDGMENTS

The authors wish to thank Kent and Marion Knudsen, Carl and Jim Anderson, Floyd Dawson, John Trumbo, Ruth Hauso, and IXL Industries for their kindness in allowing repeated access to sections on their respective properties. Particular thanks also to D. R. Braman for his advice and help in collecting samples and to M. E. Evans for continued support and use of facilities in the Paleomagnetism Laboratory of the Department of Physics, University of Alberta. The geochemical preparation for iridum analysis was done by A. Stelmach; and drafting was by F. Dimitrov. The authors also wish to thank W. Gose and D. R. Braman for reading an earlier version of the chapter and for their valuable suggestions that led to addition and improvements to the manuscript.

This study has been financially supported by a Research Grant from the Natural Sciences and Engineering Research Council of Canada to J. F. Lerbekmo. It is also GSC Contribution 35193.

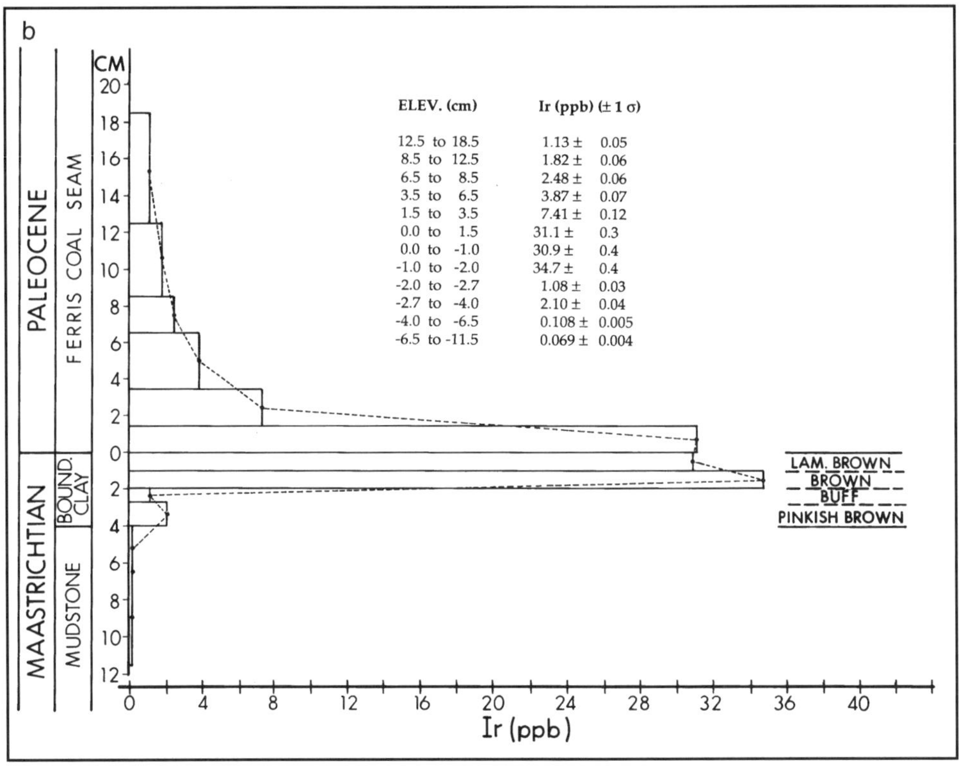

Figure 7b. Iridium profile of the K/T boundary interval of the Rock (Morgan) Creek West "E" section.

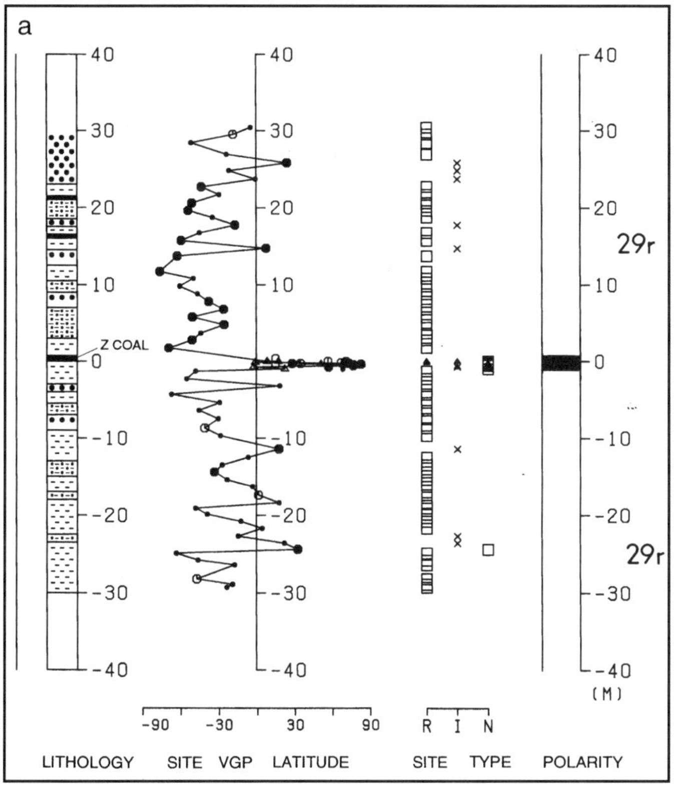

Figure 8a. Magnetostratigraphy and lithostratigraphy of the Hell Creek Park Road section. See Figure 2a for explanation.

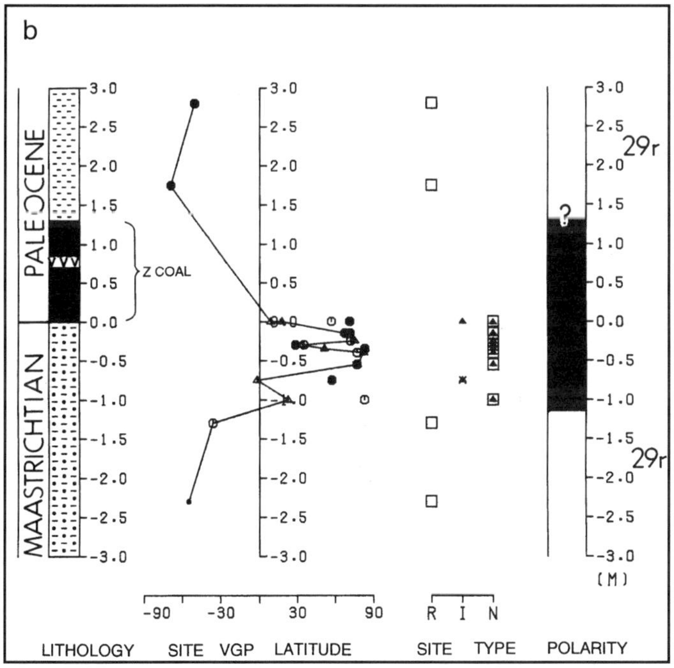

Figure 8b. Detailed magnetostratigraphy and lithostratigraphy of the K/T boundary interval of the Hell Creek Park Road section. See Figure 2a for explanation.

REFERENCES CITED

Baadsgaard, H., and Lerbekmo, J. F., 1980, A Rb/Sr age for the Cretaceous-Tertiary boundary (Z coal), Hell Creek, Montana: Canadian Journal of Earth Sciences, v. 17, p. 671–673.

Baadsgaard, H., Lerbekmo, J. F., and McDougall, I., 1988, A radiometric age for the Cretaceous-Tertiary boundary based upon K-Ar, Rb-Sr, and U-Pb ages of bentonites from Alberta, Saskatchewan, and Montana: Canadian Journal of Earth Sciences, v. 25, p. 1088-1097.

Bohor, B.F., 1990, Shocked quartz and more; impact signatures in Cretaceous/Tertiary boundary clays, in Sharpton, V. L., and Ward, P. D., eds., Global catastrophes in Earth history: An interdis-ciplinary conference on impacts, volcanism, and mass mortality, Snowbird, Utah, 1988: Geological Society of America Special Paper 247, p. 335–342.

Bonte, Ph., Delacotte, O., Renard, M., Laj, C., Boclet, D., Jéhanno, C. and Rocchia, R., 1984, An iridium rich layer at the C/T boundary in the Bidart section (S.W. France): Geophysical Research Letters, v. 11, p. 473–476.

Delacotte, O., Renard, M., Laj, C., Perch-Nielsen, K., Premoli-Silva, I., and Clauser, S., 1985, Magnetostratigraphie et biostratigraphie due passage Crétacé-tertiare de la coupe de Bidart (Pyrénées Atlantiques): Géologie de la France, no. 3, p. 243–254.

Golbrun, B., Feist, M., Colombo, F., Rocchia, R., and Tambareau, Y., 1993, Magnetostratigraphy and biostratigraphy of Cretaceous-Tertiary continental deposits, Ager Basin, Province of Lerida, Spain: Palaeogeography, Palaeoclimatology, Palaeoecology, v. 102, p. 41–52.

Grieve, R. A., and Alexopoulos, J., 1988, Microscopic planar features in quartz from Scollard Canyon, Alberta, and the Cretaceous-Tertiary boundary event: Canadian Journal of Earth Sciences, v. 25, p. 1530–1534.

Groot, J. J., de Jonge, R. B. G., Langereis, C. G., ten Kate, W. G. H. Z. and Smit, J., 1989, Magnetostratigraphy of the Cretaceous-Tertiary boundary at Agost (Spain): Earth and Planetary Science Letters, v. 94, p. 385–397.

Hamilton, N., 1983, Cretaceous/Tertiary boundary studies at Deep Sea Drilling Project site 516, Rio Grande Rise, south Atlantic: A synthesis: Initial Reports of the Deep Sea Drilling Project, v. 72, p. 949–952.

Hamilton, N., and Suzymov, P. P., 1983, Late Cretaceous magnetostratigraphy of site 516, Rio Grande Rise, southwestern Atlantic Ocean, Deep Sea Drilling Project, Leg 72: Initial Reports of the Deep Sea Drilling Project, v. 72, p. 723–730.

Harland, W. B., Cox, A. V., Llewellyn, P. G., Pickton, C. A. G., Smith, A. G., and Walters, R., 1982, A geologic time scale: Cambridge, Cambridge University Press, 131 p.

Hillhouse, J. W., Ndombi, J. W. M., Cox, A., and Brock, A., 1977, Additional results on paleomagnetic stratigraphy of the Koobi Fora Formation, east of Lake Turkana (Lake Rudolph), Kenya: Nature, v. 265, p. 411–415.

Izett, G. A., 1990, The Cretaceous-Tertiary boundary interval, Raton Basin, Colorado and New Mexico, and its content of shock-metamorphosed minerals; Evidence relevant to the K/T boundary impact-extinction theory: Geological Society of America Special Paper 249, 100 p.

Keller, G., 1989, Extended period of extinctions across the Cretaceous/Tertiary boundary in planktonic foraminifera of continental- shelf sections: Implications for impact and volcanism theories: Geological Society of America Bulletin, v. 101, p. 1408–1419.

Keller, G., MacLeod, N., Lyons, J. B., and Officer, C. B. 1993, Is there evidence for Cretaceous-Tertiary boundary-age deep-water deposits in the Caribbean and Gulf of Mexico?: Geology, v. 21, p. 776–780.

Larson, P. A., and Opdyke, N. D., 1979, Paleomagnetic results from early Tertiary/late Cretaceous sediments of site 384: Initial Reports of the Deep Sea Drilling Project, v. 32, p. 785–791.

Lerbekmo, J. F., 1985, Magnetostratigraphic and biostratigraphic correlations of Maastrichtian to early Paleocene strata between south-central Alberta and southwestern Saskatchewan: Bulletin of Canadian Petroleum Geology, v. 33, p. 213–226.

Lerbekmo, J. F., 1987, Magnetostratigraphic restrictions on the age of the

Frenchman Formation and the magnitude of the sub-Frenchman disconformity in southwest Saskatchewan: Bulletin of Canadian Petroleum Geology, v. 35, p. 454–459.

Lerbekmo, J. F., 1990, A tool for obtaining oriented samples of weakly to moderately indurated sedimentary rocks for paleomagnetic measurements: Sedimentary Geology, v. 66, p. 295–299.

Lerbekmo, J. F., and Coulter, K. C., 1985, Late Cretaceous to early Tertiary magnetostratigraphy of a continental sequence: Red Deer Valley, Alberta, Canada: Canadian Journal of Earth Sciences, v. 22, p. 567–583.

Lerbekmo, J. F., and St. Louis, R. M., 1986, The terminal Cretaceous iridium anomaly in the Red Deer Valley, Alberta, Canada: Canadian Journal of Earth Sciences, v. 23, p. 120–124.

Lerbekmo, J. F., Evans, M. E., and Baadsgaard, H., 1979a, Magneto-stratigraphy, biostratigraphy and geochronology of Cretaceous-Tertiary boundary sediments, Red Deer Valley: Nature, v. 279, p. 26–30.

Lerbekmo, J. F., Singh, C., Jarzen, D. M., and Russell, D. A., 1979b, The Cretaceous-Tertiary boundary in south-central Alberta—A revision based on additional dinosaurian and microfloral evidence: Canadian Journal of Earth Sciences, v. 16, p. 1866–1869.

Lerbekmo, J. F., Sweet, A. R., and St. Louis, R. M., 1987, The relationship between the iridium anomaly and palynological floral events at three Cretaceous-Tertiary boundary localities in western Canada: Geological Society of America Bulletin, v. 99, p. 325–330.

Lowrie, W. and Alvarez, W., 1977, Late Cretaceous geomagnetic polarity sequence: Detailed rock and paleomagnetic studies of the Scaglia Rossa limestone at Gubbio, Italy: Geophysical Journal of the Royal Astronomical Society, v. 51, p. 561–582.

MacLeod, N., and Keller, G., 1991a, Hiatus distributions and mass extinctions at the Cretaceous-Tertiary boundary: Geology, v. 19, p. 497–501.

MacLeod, N., and Keller, G., 1991b, How complete are Cretaceous/ Tertiary boundary sections? A chronostratigraphic estimate based on graphic correlation: Geological Society of America Bulletin, v. 103, p. 1439–1457.

Marquis, G., and Globerman, B. R., 1988, Northward motion of the Whitehorse Trough: Paleomagnetic evidence from the Upper Cretaceous Cormacks Group: Canadian Journal of Earth Sciences, v. 25, p. 2005–2016.

Mary, C., Moreau, M., Orue-Etxebarria, X., Apellaniz, E., and Courtellot, V., 1991, Biostratigraphy and magnetostratigraphy of the Cretaceous/Tertiary Sopelana section (Basque country): Earth and Planetary Science Letters, v. 106, p. 133–150.

McWilliams, M., and Baadsgaard, H., 1991. High-resolution 40Ar/39Ar ages from Cretaceous-Tertiary boundary bentonites in western North America [abs.], in Program and Abstracts, Event Markers in Earth History; A Joint Meeting International Geological Correlation Program Projects 261, 293, 303: Calgary, Alberta, p. 53.

Michel, H. V., Asaro, F., Alvarez, W., and Alvarez, L. W., 1983, Abundance profiles of iridium and other elements near the Cretaceous/Tertiary boundary in hole 516F of Deep Sea Drilling Project Leg 72: Initial Reports of the Deep Sea Drilling Proj-ect, v. 72, p. 931–936.

Mörner, N., 1982, The Cretaceous-Tertiary boundary: Chronostratigraphic position and sequence of events: Journal of Geology, v. 90, p. 564–573.

Nichols, D. J., Jarzen, D. M., Orth, C. J., and Oliver, P. Q., 1986, Palynological and iridium anomalies at Cretaceous-Tertiary boundary, south-central Saskatchewan: Science, v. 231, p. 714–717.

Praesinger, A., Aslanian, S., Stoykova, K., Grass, F., Mauritsch, H. J., and Scholger, R., 1993, Cretaceous/Tertiary boundary sections on the coast of the Black Sea near Bjala (Bulgaria): Palaeogeo-graphy, Palaeoclimatology, Palaeoecology, v. 104, p. 219–228.

Premoli Silva, I., Napoleone, G., and Fischer, A. G., 1974, Risultati preliminari sulla stratigrafia paleomagnetica della Scaglia Cretaceo-Paleocenica della sigione di Gubbio (Appennino Centrale): Bollettino Societa Geologica Italiana, v. 92, p. 647–659.

Roggenthen, W. M., and Napoleone, G., 1977, Upper Cretaceous– Paleocene magnetic stratigraphy at Gubbio, Italy. IV: Upper Maastrichtian–Paleocene magnetic stratigraphy: Geological Society of America Bulletin, v. 88, p. 378–388.

Srivastava, S. K., 1970, Pollen biostratigraphy and paleoecology of the Edmonton Formation (Maastrichtian), Alberta, Canada: Palaeogeography, Palaeoclimatology, Palaeoecology, v. 7, p. 221–276.

Sternberg, C. M., 1947, The upper part of the Edmonton Formation of Red Deer Valley, Alberta: Geological Survey of Canada Paper 47-1, 11 p.

Sweet, A. R., and Braman, D. R., 1992, The K/T boundary and contiguous strata in western Canada: Interactions between paleoenvironments and palynological assemblages: Cretaceous Research, v. 13, p. 31–79.

Sweet, A. R., and Cameron, A. R., 1991, Palynofacies, coal petrographic facies, and depositional environments: Amphitheater Formation (Eocene-Oligocene) and Ravenscrag Formation (Maastrichtian-Paleocene), Canada: International Journal of Coal Geology, v. 19, p. 121–144.

Sweet, A. R., and Hills, L. V., 1984, A palynological and sedimentological analysis of the Cretaceous-Tertiary boundary, Red Deer River Valley, Alberta, Canada, in Abstracts, International Palynological Conference, 6th, Calgary: Calgary, Alberta, p. 160.

Sweet, A. R., Braman, D. R., and Lerbekmo, J. F., 1991, Palynofloral response to K/T boundary events—A transitory interruption within a dynamic system: Geological Society of America Special Paper 227, p. 457–469.

Swisher, C. C., Dingus, L., and Butler, R. F., 1993, 40Ar/39Ar dating and magnetostratigraphic correlation of the terrestrial Cretaceous-Paleogene boundary and Puercan Mammal Age, Hell Creek– Tullock formations, eastern Montana: Canadian Journal of Earth Sciences, v. 30, p. 1981–1996.

Thiersten, H. R. and Okada, H., 1979, The Cretaceous-Tertiary boundary event in the north Atlantic: Initial Reports of the Deep Sea Drilling Project, v. 43, p. 601–616.

Watson, G. S., 1956, A test for randomness of direction: Monthly Notices of the Royal Astronomical Society, Geophysical Supplement, v. 7, p. 160–161.

Manuscript Accepted by the Society September 14, 1995

Printed in U.S. A.

Geological Society of America
Special Paper 307
1996

Biotic selectivity during the K/T and Late Ordovician extinction events

Peter M. Sheehan and Patricia J. Coorough
Department of Geology, Milwaukee Public Museum, 800 W. Wells Street, Milwaukee, Wisconsin 53233
David E. Fastovsky
Department of Geology, University of Rhode Island, Kingston, Rhode Island 02881

ABSTRACT

The differing environmental effects of a glaciation and an impact event are reflected in significantly differing ecologic responses during the Late Ordovician and Cretaceous/Tertiary (K/T) mass extinctions. During the Late Ordovician glaciation most biotic changes were associated with an intensifying climatic gradient and glacio-eustatic sea-level changes. During the K/T event most biotic changes were related to a short-term loss of primary productivity, consistent with a loss of sunlight and other ancillary effects of an impact. Animals in food chains that required a supply of food derived from photosynthesis suffered the greatest extinction during the K/T event. Many K/T survivors may have been buffered from the loss of primary productivity by feeding on detritus or being starvation resistant. The marked differences in ecological patterns between the two events affirms a fundamental difference in the mechanisms of extinction.

INTRODUCTION

Major mass extinction events mark ecologic crises that disrupted communities and zoogeographic patterns around the world, followed by massive restructuring of synecologic associations. Extinction events are commonly described in terms of taxonomic extinction, with the focus on the number and percentage of clades that became extinct, permitting the magnitude of the extinction events to be compared. In this chapter we take another approach and compare the ecologic changes experienced by the biota during two extinctions of similar scale. Distinctly different biotic responses during the K/T and Late Ordovician events reflect the different causes.

At least two of the five major mass extinctions were associated with impact events, and it has even been suggested that all major Phanerozoic extinction events were caused by impacts (e.g., McLaren and Goodfellow, 1990). The K/T extinction event is closely associated with an impact, and ecologic changes during the extinction are compatible with the effects of the impact. However, some extinction events, especially the one in the Late Ordovician, lack any evidence of an associated impact. The Ordovician event occurred during a glacial interval, and ecologic changes during the event are compatible with the effects of a glaciation rather than an impact. The K/T event does not have the signature of a climatically induced extinction, as has been suggested by some workers, thus weakening the case for one alternative to the impact scenario.

IMPORTANCE OF EXTINCTION EVENTS

The K/T extinction event has been a contentious issue in paleontology in part because it has required paleontologists to restructure the paradigm they use to understand the history of life. The course of evolution can no longer be viewed as a long-term and continuous process during which new groups originate, diversify, and gain prominence by gradually becoming better adapted to their environment than their competitors. The

Sheehan, P. M., Coorough, P. J., and Fastovsky, D. E., 1996, Biotic selectivity during the K/T and Late Ordovician extinction events, *in* Ryder, G., Fastovsky, D., and Gartner, S., eds., The Cretaceous-Tertiary Event and Other Catastrophes in Earth History: Boulder, Colorado, Geological Society of America Special Paper 307.

postextinction Tertiary biota was derived from clades that happened to survive the extinction, and survival had little to do with adaptations that had promoted survival during the long interval of time preceding the event. Under the new paradigm, groups such as mammals would not have risen to prominence if dinosaurs had not been eliminated. Resistance to nontraditional explanations of biotic transitions has fueled the debate about the extinction of dinosaurs and other animals at the end of the Mesozoic.

Extinction events are important in evolutionary biology because they resulted in unique and massive changes in the global ecosystem. Postextinction evolution involved diversification of clades, but more importantly the recovery intervals involved evolution of new synecologic associations. Some members of surviving clades radiated with little change of habit, but others went through evolutionary radiations that moved them into completely new lifestyles. New synecologic associations appeared, but they were not the result of gradual evolutionary change.

The events we compare here had similar levels of phyletic extinction (Sepkoski, 1986), but the events had different causes. We will explore similarities and differences in the ecologic selectivity during these two events. Comparison of the ecology of the K/T and Ordovician events is hampered by differing information that has been collected for the events. This is partially because workers have concentrated on the unique patterns of survival and extinction that characterize each event.

In making these comparisons it is important to realize that in many cases not all members of a particular ecologic group became extinct. For example, in the K/T event, although detritus feeders preferentially survived while animals requiring green plants for food tended to become extinct, many detritus feeders became extinct while many herbivores survived. Also, in surviving groups, the number of individuals were drastically reduced during the extinction.

LATE ORDOVICIAN EXTINCTION EVENT

While many details of the Late Ordovician extinction are still being investigated, a strong consensus on the cause of the event has developed among scientists who have examined this event. The extinction resulted from the effects of a glaciation, including climatic deterioration, glacio-eustatic sea-level decline, and changes in oxygenation of the oceans resulting from increased oceanic circulation that developed as the climatic gradient steepened (Sheehan, 1973, 1975, 1988; Sheehan and Coorough, 1990; Berry and Boucot, 1973; Lenz, 1976; Brenchley, 1984, 1988; Brenchley and Newall, 1984, Brenchley et al., 1991, 1994; Rong, 1984; Rong and Harper, 1988; Cocks and Rickards, 1988; Wilde et al., 1990; Marshall and Middleton, 1990; Koren, 1991; Melchin and Mitchell, 1991; Middleton et al., 1991; Owen et al., 1991; Goodfellow et al., 1992; Wang et al., 1992, 1993; Long, 1993). No evidence of an associated impact has been found (Orth et al., 1986; Wilde et al., 1986; Robertson et al., 1991; Wang et al., 1992).

Late Ordovician paleogeography is shown on Figure 1. Shallow, epicontinental seas covered most continents, especially in tropical regions (Algeo and Wilkinson, 1991). Tropical areas had highly diverse faunas. A weak latitudinal temperature gradient was present. Prior to the glaciation, warm, saline, anoxic water dominated the deep oceans (Railsback et al., 1990).

Zoogeographic provincialism was strong, with distinct faunas occupying each of the isolated epicontinental seas (Fig. 1). The open oceans, including the margins of the continents, had more cosmopolitan faunas (Jaanusson, 1979; Sweet and Bergström, 1984; Anstey, 1985; Sheehan and Coorough, 1990).

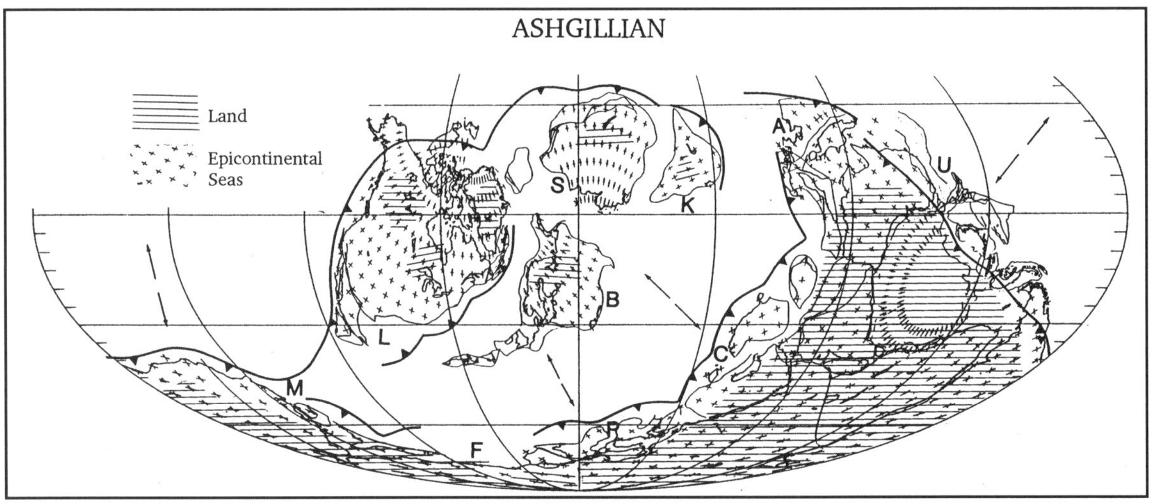

Figure 1. Late Ordovician Geography. The distribution of continental plates and the extent of epicontinental seas is after Scotese and McKerrow, 1990. Geographic regions are indicated by letters. The regions were used in an investigation of brachiopod distribution (see Fig. 2). A = Manchuria, B = Baltica, C = Southern China, F = Africa, K = Kazakhstan, L = Laurentia, M = South America, P = southern Europe, S = Siberia, U = Australia.

A Southern Hemisphere glaciation at the end of the Ordovician has been well documented in North Africa, where glacial features are common (Beuf et al., 1971). Drop-stones from icebergs have been recorded from high latitude oceanic regions (Robardet and Doré, 1988; Brenchley et al., 1991). During the Hirnantian Stage, which corresponds to the glacial interval, there was a marked positive excursion of oxygen and carbon isotopes in marine carbonates (Marshall and Middleton, 1990; Wang et al., 1993; Long, 1993; Brenchley et al., 1994). Similar isotopic excursions occurred in the Cenozoic when glaciation developed.

As water was retained in the developing glaciers, sea level fell, draining the epicontinental seas (Berry and Boucot, 1973; Sheehan, 1973, 1979, 1988; Brenchley et al., 1991, 1994). Temperature gradients increased markedly, and carbonate deposition declined in middle and high latitudes. Oceanic circulation increased as the glaciation intensified (Sheehan, 1979, 1987; Wilde et al., 1990). The glaciation ended suddenly, and sea level rose rapidly.

The most recent estimate of the duration of the Ordovician glaciation in the Hirnantian Stage is 0.5 to 1.0 m.y. (Brenchley et al., 1994). There were pulses of extinction at the beginning and end of the Hirnantian.

Significant extinction occurred in most of the then dominant groups of marine organisms, including brachiopods (Sheehan, 1973; Sheehan and Coorough, 1990; Berry and Boucot, 1973; Cocks, 1988), bryozoans (Anstey, 1985; Tuckey and Anstey, 1992), trilobites (Jaanusson, 1979; Dean, 1985; Chatterton and Speyer, 1989; Fortey, 1989; Briggs et al., 1988; Robertson et al., 1991), graptolites (Skevington, 1978; Berry, 1979; Berry et al., 1991; Berry and Wilde, 1990), conodonts (Sweet, 1990), corals (Kaljo and Klaamann, 1973; Leleshus and Permyakov, 1981; Elias, 1989), ostracods (Copeland, 1977), bivalves (Bretsky, 1973; Hallam and Miller, 1988), cephalopods (Jaanusson, 1979; Crick, 1990), and possibly trace fossils (McCann, 1989). The limited record of phytoplankton also reveals an extinction event (Achab and Duffield, 1982; Duffield and Legault, 1982; Colbath, 1986). Reef communities were decimated (Boucot, 1983; Sheehan, 1985). The radiation onto land was in the very early stages, and there is no faunal record of the extinction on land.

The extent of the event can be illustrated with data on brachiopod distribution across this event. Of 240 brachiopod genera present in the Ashgillian Stage, prior to the Hirnantian glacial interval, 52% did not survive into the Hirnantian. Of 142 genera in the Hirnantian (26 genera originated during the Hirnantian), 53% did not survive into the Silurian. Sixty-seven genera survived into the Silurian, and by the end of the middle Llandoverian, diversity nearly doubled, to 123 genera, but was still well below preextinction diversity.

The initial pulse of extinction was at the beginning of the glaciation when sea level was depressed as much as 100 m because water was retained in the glaciers. The habitat of endemic faunas in the epicontinental seas was eliminated (Sheehan, 1988; Brenchley et al., 1994; Tuckey and Anstey, 1992). Johnson (1974) has shown that these "perched faunas" in Paleozoic epicontinental seas were especially vulnerable to eustatic changes in sea level.

The Ordovician glaciation did not draw down sea level as far as the Pleistocene glaciation did. However, the effect on the biota was much greater because epicontinental seas were much more prevalent in the Ordovician than in the modern world. Today and during the last glaciation, features resembling Ordovician epicontinental seas, such as Hudson Bay, are small compared to the extent of Paleozoic epicontinental seas (Fig. 1). Intensification of the climatic gradient was caused by a combination of the developing glaciation itself, loss of the ameliorating effects of the epicontinental seaways, and increased albedo of the exposed continents (Gildner and Cisne, 1987).

Two primary zoogeographic patterns were present during the extinction. First, narrowly distributed genera that were endemic to isolated epicontinental seas tended to suffer the greatest extinction. The pattern has been documented for brachiopods (Sheehan and Coorough, 1990 and Figs. 2 and 3), bryozoans (Tuckey and Anstey, 1992), trilobites (Robertson et al., 1991), and bivalves (Bretsky, 1973). Second, widely distributed genera preferentially survived, as has been found in many groups during other extinction events (Jablonski, 1986a, b, c, 1989, 1991).

An example of the extent of the extinction of endemic taxa is shown in a study of the distribution of brachiopods. The geographic distribution of the 240 Ashgillian genera is shown in Figure 2, where genera are recorded according to the number of geographic regions in which they have been reported. The 11

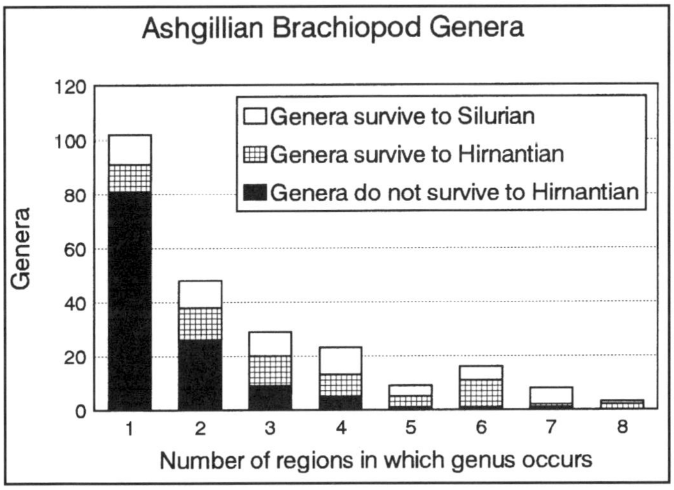

Figure 2. Geographic distribution and survival of Ashgillian brachiopods. Black = do not survive initial pulse of extinction at beginning of the glaciation. Cross-hatched lines = do not survive the final pulse of extinction at the end of the glaciation. Blank = survive to Silurian. The stratigraphic distribution and number of geographic regions from which each brachiopod genus has been recorded were determined by a literature search. Widespread genera were more likely to survive than narrowly distributed genera (see text for discussion).

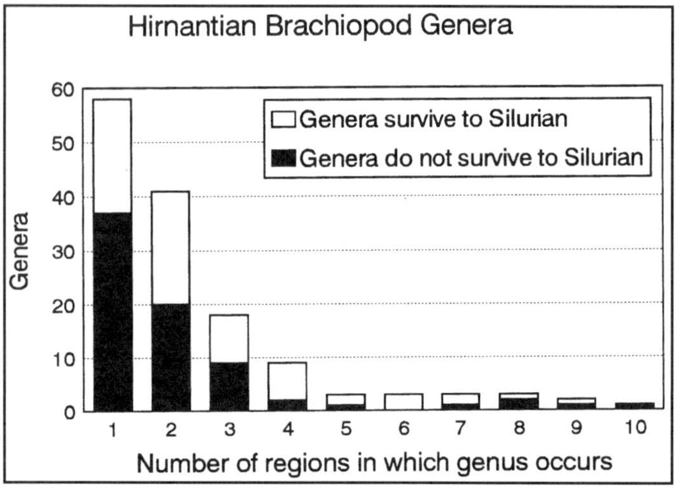

Figure 3. Geographic distribution and survival of Hirnantian brachiopods. Black = do not survive final pulse of extinction at end of glaciation. Blank = survive to Silurian. The stratigraphic distribution and number of geographic regions from which each brachiopod genus has been recorded were determined by a literature search. The regions are indicated by letters in Figure 1. Widespread genera were more likely to survive than narrowly distributed genera, but many widespread genera belonging to the *Hirnantia* fauna succumbed when their habitat disappeared following the glacial interval (see text for discussion).

regions from which brachiopods have been reported are identified on Figure 1. Brachiopods recorded in only one region suffered by far the greatest extinction, with 81 of 102 genera not surviving into the Hirnantian interval. This pattern reflects the elimination of faunas endemic to the isolated epicontinental seas. The second pattern is that genera recorded in progressively more regions were progressively more likely to survive (Fig. 2).

Open-ocean graptolites suffered near extinction as a result of the oxygenation event that eliminated much of their preferred habitat (Berry and Wilde, 1990). Open-ocean groups generally became restricted to the tropics as the glaciation intensified (Skevington, 1978). Planktonic trilobites and trilobites with planktonic larval stages suffered greater extinction than benthic forms (Chatterton and Speyer, 1989). Reefs, which throve during the Middle and Late Ordovician, were very small and were limited to a narrow tropical zone during the glacial interval (Boucot, 1983; Webby, 1984; Sheehan, 1985).

Graptolite shale deposition gave way to bioturbated mudstone deposition in many regions. The transition to bioturbated mudstones is thought to reflect the appearance of oxygenated water as oceanic circulation increased as a result of the increased climatic gradient and the generation of cold, deep-water currents containing oxygenated water masses derived from shallow water in high latitudes (Sheehan, 1979, 1987, 1988; Berry and Wilde, 1990; Berry et al., 1991; Wilde et al., 1990; Brenchley et al., 1991, 1994). A similar aeration of the deep oceans has been documented at the initiation of Tertiary glaciation. A short-lived group of communities dominated by small, thin-shelled brachiopods, such as *Foliomena*, with origins in high latitudes invaded this setting (Sheehan, 1979). The fate of this fauna is unclear, but it did not survive into the Silurian.

Faunal patterns during the glaciation were complex (Rong and Harper, 1988; Brenchley and Cullen, 1984; Sheehan, 1979). Migrating faunas and local extinctions make correlation difficult during the extinction interval. Without an isochron such as the K/T iridium layer, precise correlation of sections around the world is not possible.

During the glaciation, deposition continued around the margins of the continents, and areas in contact with the open oceans suffered less drastic extinctions than did the perched faunas (conodonts are an exception; Bergström; 1990). Faunas became adapted to the new glacial regime. In many places cool-water, siliciclastic sediments replaced warm-water carbonates. The siliciclastic sediments contain a unique cool-water fauna called the *Hirnantia* fauna (Sheehan, 1979, 1988; Rong and Harper, 1988; Brenchley and Cullen, 1984; Owen et al., 1991). Many members of this fauna had origins in high latitudes, et al. were derived from widely distributed groups that had been common in open-ocean environments.

A second wave of extinction occurred when the glaciation ended abruptly, sea level returned to preglaciation levels, and epicontinental seas covered the continents once more. At this time many genera became extinct, even though they had attained wide geographic distribution (Fig. 3 exemplifies this with brachiopods.) Most of these extinct, widespread genera were members of the cool-water *Hirnantia* fauna. Warming at the end of the glaciation eliminated the preferred habitat of the *Hirnantia* fauna, producing an unusual event—the extinction of widely distributed genera (Sheehan and Coorough, 1990).

The extinction was not selective for particular feeding strategies. Detritus-feeders such as trilobites suffered substantial extinction, but trilobites that lived in the water column during some life stage had greater extinction than those that lived their entire life on the bottom (Chatterton and Speyer, 1989). Starvation-resistant groups such as brachiopods suffered substantial extinction. Among bivalves detritus feeders were less likely to become extinct than suspension feeders (Kriz, 1984).

A final pattern is that survivors of many groups tended to have "simple," probably little-evolved morphologies. This has not been examined quantitatively but is a recurrent observation by specialists in many of the groups that suffered extinction (Sheehan, 1988). This pattern may partly reflect another poorly documented trend in which large species, which tend to be relatively derived, may suffer greater extinction than small species.

K/T EXTINCTION EVENT

Unlike the Late Ordovician extinction, the K/T event was strongly selective for trophic ecology. Both on land and in the oceans, animals tied to food chains that were dependent on primary productivity (photosynthetic organisms) were most likely to become extinct. Animals in detritus-based food chains were most likely to survive, as were animals with resistance to starvation over periods of several months.

These patterns are compatible with an impact event that, for a short time, eliminated primary producers through darkness and environmental perturbations such as acid rain. Fossil evidence is strong for a temporary loss of primary productivity, both on land and in the oceans.

Following the impact sunlight was blocked, eliminating photosynthesis on earth for an interval of two to six months (Alvarez et al., 1980; Toon et al., 1982; Pollack et al., 1983). Soot in the impact clay provides evidence of continent-wide forest fires generated by thermal radiation from ballistic reentry of impact ejecta (Wolbach et al., 1985, 1990; Venkatesan and Dahl, 1989; Ivany, 1991; Ivany and Salawitch, 1993; Melosh et al., 1990). The soot contributed to the loss of sunlight, as did acid compounds in the upper atmosphere, as is discussed below.

Secondary effects of the impact also contributed to the extinction. The climate cooled briefly, and continental interiors may have cooled below freezing during this interval (Wolfe, 1991; but see Covey et al., 1990). Nitric acid rain formed by oxidation of atmospheric nitrogen during reentry of dispersed ejecta was another potential environmental perturbation (Lewis et al., 1982; Crutzen, 1987; Prinn and Fegley, 1987; Zahnle, 1990). Release of SO_2 into the stratosphere from the anhydrite-rich impact site may also have been a contributing factor by producing acid rain and blocking sunlight (Sigurdsson et al., 1992; Kring, 1993; Blum and Chamberlain, 1994; D'Hondt et al., 1994). Forest fires, at least on continents adjacent to the impact structure (Hildebrand, 1993), would also have been a killing mechanism. Thus, in addition to the elimination of sunlight, such factors as intense cold, forest fires on some land areas, and acid rain wreaked havoc on the environment. Areas farthest removed from the Chicxulub crater were potential refugia from the effects of the impact (Hildebrand, 1993).

Following the brief interval of darkness, greenhouse warming caused by shock devolatization of CO_2 from the target rock carbonates may have lasted 10^3 to 10^5 years (O'Keefe and Ahrens, 1989). Although the warm climate may have caused some additional extinction, the effect was primarily a control on the evolutionary radiation of surviving clades.

Marine K/T record

Immediately after the K/T impact phytoplankton underwent a sudden extinction, the oceans experienced a crisis in primary productivity, the carbon isotope gradient disappeared, and a "Strangelove" ocean developed (Hsü et al., 1982; Hsü and McKenzie, 1985; Thierstein, 1981, 1982; Arthur et al., 1987; D'Hondt and Lindinger, 1988; Zachos et al., 1989; Stott and Kennett, 1989; Knoll, 1989; Griffis and Chapman, 1988, 1990; Magaritz, 1989; Gilmour et al., 1990; Herbert and D'Hondt, 1990; Kump, 1991; Hollander et al., 1993; Smit, 1994; Pospichal, 1994a, b). Describing the period following the extinction event, Pospichal (1994a, p. 90) notes that "sequential blooms of surviving and new nannoplankton species . . . reflect the unstable conditions around the globe following the KT boundary crisis."

In the marine environment animals living in the water column typically are in food chains that rely directly on phytoplankton (Fig. 4). Most of these animals in phytoplankton-based food chains, from planktonic foraminifera to ammonites, were devastated (Russell, 1979; Thierstein, 1982; Sheehan and Hansen, 1986; Arthur et al., 1987; Roy et al., 1990; Ward et al., 1991).

Some survivors that lived in the water column may have been buffered from extinction because they had starvation-resistant habits. For example, nautiloids lived in the water column but survived with minor extinction. Nautiloids have lecithotrophic eggs with a large yolk that provides nutrition, which may have allowed early ontogenetic stages to survive during the loss of primary production. The ammonites became extinct, but they had planktotrophic larval stages, which required feeding on plankton (Gallagher, 1991).

In the marine benthos, deposit feeders preferentially survived (Arthur et al., 1987; Sheehan and Hansen, 1986; Hansen et al., 1987, 1993; Roy et al., 1990; Kaiho, 1994), as did starvation-resistant suspension feeders such as brachiopods and molluscs with inactive lifestyles and low-energy budgets (Rhodes and Thayer, 1991), and animals with developmental stages that did not require planktonic food (Gallagher, 1991). Molluscs with planktotrophic larval stages or high-energy budgets tended to become extinct (Gallagher, 1991; Rhodes and Thayer, 1991).

Many organisms have the ability to switch food sources (Cadée, 1984), and this ability may have allowed some groups to feed on detritus during the K/T event. For example, many benthic foraminifera (Lipps, 1983), especially those living in soft sediments, are detritus feeders, but many other foraminifera may have been able to switch to detritus feeding when other food sources were eliminated. Foraminifera can even take up dissolved organic matter directly.

Some groups were simply fortunate (or unfortunate) to have adaptations that helped (or hindered) them during the impact event. Most brachiopods were favored by their resistance to starvation. But articulate brachiopods that were specialized for life on the seafloor of the late Cretaceous chalk in northwest Europe suffered severely (Johansen, 1988, 1989; Surlyk and Johansen, 1984). These authors note that the chalk was composed primarily of the skeletons of coccolithophores and pelagic foraminifera. The mass extinction of the chalk-producing groups led to total cessation of chalk production. Brachiopods that were specialized for life on the soft, chalk seafloor became extinct because the extinction eliminated the unusual substrate to which they had become adapted. The specialized chalk brachiopods thus became extinct in spite of the fact that most brachiopods were buffered from extinction (Rhodes and Thayer, 1991).

Among phytoplankton, Kitchell et al. (1986) found that diatoms with nonplanktonic stages in the life cycle had higher survival rates than other phytoplankton. They viewed the nonplanktonic stages as an adaptation to stress in surface waters and suggested that during the impact event the strategy pro-

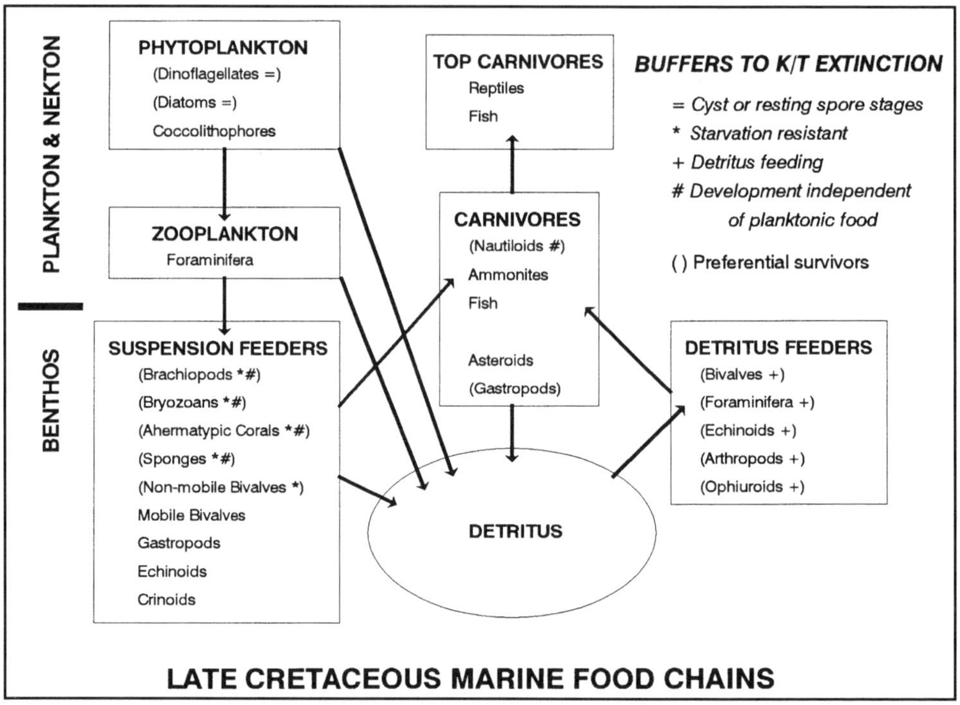

Figure 4. Late Cretaceous marine food webs. Groups that preferentially survived K/T extinction event are enclosed in parentheses. The figure summarizes patterns of survival discussed in Sheehan and Hansen (1986), Arthur et al. (1987), Kitchell et al. (1988), Hansen (1988), Hansen et al. (1993), Knoll (1989), Roy et al. (1990), Rhodes and Thayer (1991), and Gallagher (1991). Buffers to extinction cited in the literature above are indicated by symbols. Development that does not require feeding in the water column includes asexual budding and lecithotrophic eggs or larvae (Gallagher, 1991). Some groups have members in more than one trophic box; for example, suspension-feeding echinoids include irregular echinoids such as sand dollars, and detritus-feeding echinoids include other irregular echinoids such as many atlostomes. Small carnivorous gastropods that preferentially survived the extinction in Texas (Hansen et al., 1987), may have switched to scavenging.

moted survival relative to phytoplankton that were confined to the plankton throughout their life cycle.

Two patterns found in the Late Ordovician are present also. Widely distributed genera preferentially survived (Jablonski, 1986a, b, c, 1989, 1991), and many authors anecdotally cite preferential survival of clades with "simple" morphology.

The record of reefs across the boundary is difficult to evaluate, because nonphototrophic rudist bivalves rather than phototrophic corals dominated reefs in the Late Cretaceous. Mesotrophic conditions may have excluded scleractinian corals and favored the asymbiotic suspension-feeding rudists (Wood, 1993). Reef-building rudists disappeared well before the K/T boundary, possibly because of changing nutrient conditions caused by changing conditions of oceanic stratification (Wood, 1993). The lack of phototrophic scleractinian coral reefs in the latest Cretaceous makes the effect of the impact event on this ecosystem a moot point.

Terrestrial K/T record

As in the oceans, primary productivity on land experienced a crisis at the K/T boundary. An extinction event is recognized both in palynomorphs (Nichols et al., 1990; Tschudy and Tschudy, 1986; Saito et al., 1986; Lerbekmo et al., 1987; Orth et al., 1981) and in the fossil leaf record (Johnson et al., 1989; Johnson and Hickey, 1990; Wolfe and Upchurch, 1986).

Tschudy et al. (1984) recognized a disaster flora immediately above the boundary clay and proposed that recolonization following the impact event began with regrowth dominated by pioneer species (Wolfe and Upchurch, 1986; Nichols et al., 1986; Fleming and Nichols, 1990; Saito et al., 1986). In many regions the disaster flora is indicated by an abundance of fern spores, the so-called fern spike (Wolfe and Upchurch, 1986). Elsewhere—for example, in parts of western Canada—"weedy" angiosperms dominated the postextinction flora (Lerbekmo et al., 1987; Fleming and Nichols, 1990). In the Netherlands, Brinkhuis and Visscher (1994) interpreted an enrichment of bryophyte spores immediately above the K/T boundary to be a brief, local domination of the terrestrial ecosystem by bryophytes following impact. Eventually many plants recovered as seeds germinated and surviving root systems regenerated.

In the terrestrial environment extinction on land was much greater than in streams (Sheehan and Fastovsky, 1992; Sheehan and Hansen, 1986). Stream communities commonly have little primary production and largely rely on food being washed in from land (Mann, 1988; Hildrew et al., 1985; Winemiller, 1990; Closs and Lake, 1994). In addition, many stream-dwelling animals that feed in food chains with living plants at the base can switch to detritus-based food chains. Thus, when primary productivity crashed, animals in streams were buffered from extinction by their ability to feed in detritus-based food chains.

On land (Figs. 5–7), extinction was concentrated in food chains that relied on primary productivity (Sheehan and Fastovsky, 1992; Sheehan and Hansen, 1986; Sheehan and Russell, 1994), including both herbivorous and carnivorous dinosaurs. Other land animals may have preferentially survived because they lived in detritus-based food chains such as forest-soil food chains based on leaf litter, decaying wood, roots, branches, and fungi. Detritus food chains probably included protozoans, annelids, molluscs, many insects, amphibians, lizards, and small mammals. Animals in this food chain were not tied directly to primary productivity. These animals preferentially survived the interval when primary productivity was halted and then radiated in the early Tertiary (Sheehan and Hansen, 1986; Sheehan and Fastovsky, 1992).

As in the marine realm, the extinction event on land was abrupt to within the resolution of the Signor-Lipps effect. Palynomorphs provide the best resolution, and their extinction is coincident with the iridium layer (Orth et al., 1981; Fleming and Nichols, 1990; Nichols et al., 1990). Other groups provide less resolution, as predicted by the Signor-Lipps effect, but even the record of dinosaurs is consistent with abrupt, not gradual, extinction (Sheehan et al., 1991; LeLoeuff et al., 1994).

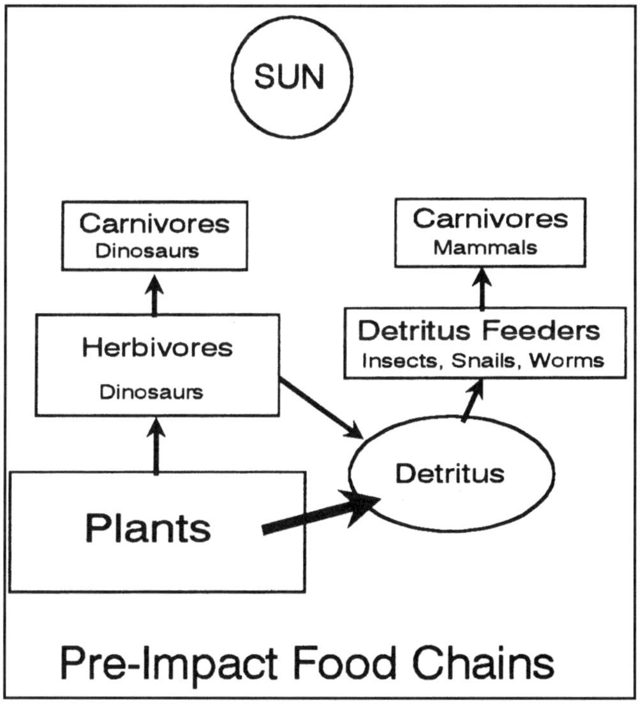

Figure 5. Terrestrial Late Cretaceous food chains.

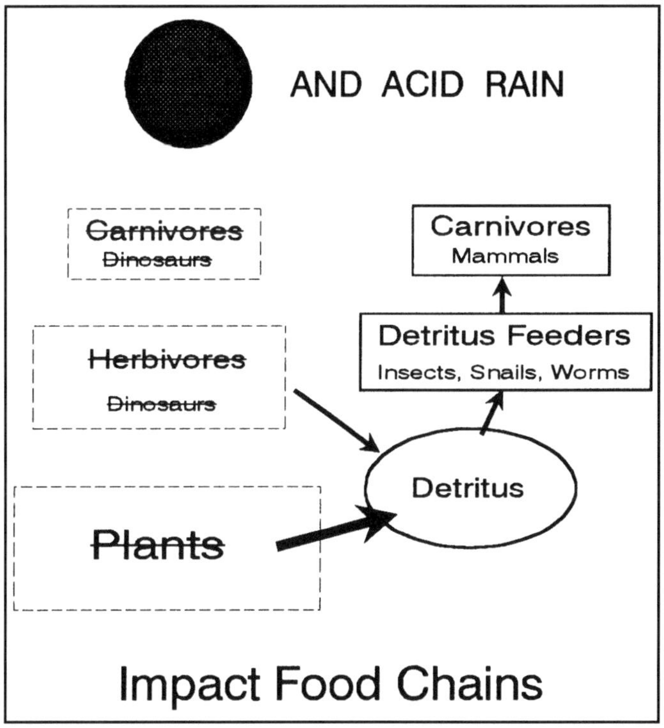

Figure 6. Food chains during the impact crisis at the end of the Cretaceous. Loss of sunlight, acid rain, and other effects discussed in the text killed plants and disrupted animals in food chains based on primary production.

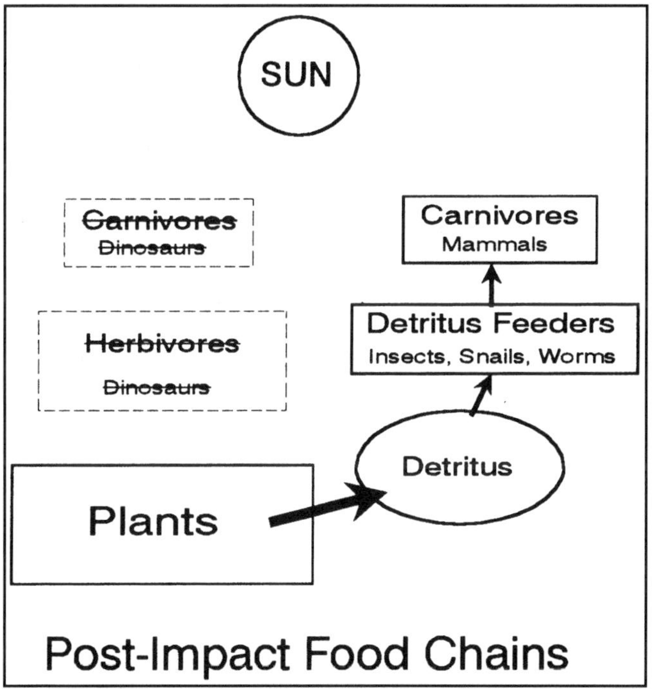

Figure 7. Postimpact, earliest Tertiary food chains. Following the impact, primary productivity resumed. However, many animals previously dependent on primary productivity had become extinct. Note the absence of large-bodied herbivores and carnivores on land. It took several million years before large-bodied mammals and birds evolved to replace the dinosaurs.

COMPARISON OF EXTINCTIONS WITH IMPACT AND TERRESTRIAL CAUSES

Both extinction events share some patterns that appear to characterize all extinction events. However, profound differences in patterns of extinction between the events do exist (Fig. 8), and these differences reflect the distinctive environmental perturbations that produced the events.

Survival of widespread genera

Geographically widespread genera preferentially survived both the K/T and Ordovician events. Of 88 Ashgillian brachiopod genera recorded in three or more of 11 geographic areas, 71 genera (81%) survived (Fig. 2), reflecting the common trend of preferential survival of widely distributed genera. The final pulse of Ordovician extinction ran counter to the general trend, since of 38 Hirnantian genera recorded in three or more areas, only 21 (55%) survived to the Silurian. The low level of survival reflects the elimination of the habitat of the wide-spread, cool-water *Hirnantia* fauna at the end of the glaciation.

	K/T Impact	Late Ord. Glaciation
Abrupt extinction	yes	no
Two pulses of extinction	no	yes
Postextinction disaster flora	yes	?
Preferential survival of:		
1. detritus feeders	yes	no
2. starvation-resistant groups	yes	no
3. cyst- and spore-forming phytoplankton	yes	?
4. widespread genera	yes	yes
Preferential extinction of:		
1. food chains based on living plants	yes	no
2. high-latitude faunas	no	yes
3. epicontinental-sea provinces	no	yes
4. reef biota	- -	yes

Figure 8. Comparison of K/T and Late Ordovician extinction events. Reefs had disappeared early in the Maastrichtian, so their disappearance is not related to the K/T impact event.

Elimination of zoogeographic provinces

The preferential survival of widespread genera that characterizes all extinction events (Jablonski, 1989) was augmented during the Ordovician event by the selective elimination of the highly endemic epicontinental-sea biogeographic provinces. Not only did widespread genera preferentially survive, members of epicontinetal sea provinces were selectively eliminated.

Of 103 brachiopod genera confined to a single geographic region, 81 genera (79%) became extinct, reflecting glacio-eustatic draining of the epicontinental seas. Elimination of the habitat of the epicontinental-sea provinces (Sheehan, 1988) resulted in the loss of provinciality that characterizes the Ordovician event. The highly endemic epicontinental-sea provinces gave way to an interval of extremely cosmopolitan faunas during the Silurian. Local community diversity recovered within a few million years (Sheehan, 1988; Watkins, 1994). However, tens of million of years of evolution were required before endemic provinces were reestablished (Sheehan and Coorough, 1990).

Although widespread genera preferentially survived the K/T event, no loss of provinciality comparable to that in the Ordovician occurred. Extinction of perched faunas did not occur. Although a high sea-level stand in the Late Cretaceous was interrupted by a sea-level decline in the late Maastrichtian, the regression began more than 2 m.y. before the K/T impact and was not associated with the primary extinction event. In fact, sea-level rise of the next sedimentary sequence may have been well underway by the time of the impact (Fastovsky and Sheehan, 1994).

Duration

The K/T extinction was instantaneous in geologic time. Extinctions of inoceramids and rudist reef formers occurred before the extinction event and were unrelated to the event. Extinctions occurring before the event may be related to sea level and climatic changes and were part of normal background extinction. The Ordovician event took place over an interval of 500,000 yr to 1 m.y.

Preferential survival of detritus feeders

During the K/T event, both on land and in the oceans, the crisis in primary productivity selectively eliminated animals in food chains tied directly to living plants. Animals in detritus-feeding food chains were buffered from extinction.

A primary imprint of preferential survival in detritus-feeding food chains was not present during the Ordovician extinction, although some detritus-feeding groups did preferentially survive. For example, byssate, epifaunal suspension-feeding bivalves suffered only marginally greater extinction (64%) than detritus-feeding bivalves (53%) (Kriz, 1984; Hallam and Miller, 1988). Similarly, trilobites with some life stage that fed in the water column underwent slightly greater extinction than trilobites that remained on the bottom throughout life (Chatterton and Speyer, 1989). But again, even detritus feeders that were bottom dwellers throughout their life cycle suffered substantial extinction (Briggs et al., 1988). The pattern is attributed to cooling climate rather than to a global crisis in primary productivity, because in modern oceans cool-water faunas have disproportionate numbers of animals with no life stage during which the animal lives in the water column (Chatterton and Speyer, 1989).

Preferential survival of starvation resistant groups

During the K/T extinction, starvation-resistant plants and animals (Rhodes and Thayer, 1991; Gallagher, 1991) preferentially survived the crisis in primary production. Starvation resistance in some animals, such as brachiopods and inactive bivalves, resulted from low-energy budgets. Others, such as nautiloid cephalopods, which had lecithotrophic eggs and larvae, survived because the yolk provided food during the temporary absence of primary productivity. Differential survival of brachiopods allowed them to become important faunal elements immediately after the K/T event (Johansen, 1988, 1989; Gallagher, 1991; Rhodes and Thayer, 1991).

During the Ordovician, brachiopods were among the hardest-hit groups. Starvation resistance was ineffectual during the Ordovician extinction because the crisis lasted far longer than the life span of individual organisms.

Preferential survival of cyst-forming phytoplankton

Kitchell et al. (1986) and Knoll (1989) found that phytoplankton with cyst-forming and resting-spore stages preferentially survived the K/T extinction. They attributed survival of these phytoplankton to their ability to survive the interval of darkness and environmental stress as cysts or spores. This pattern has not been studied in Ordovician phytoplankton.

Postextinction disaster flora

During the K/T event, primary production was eliminated for a period of a few months on both land and in the oceans. The plant record immediately above the iridium layer permits study of the short-term recovery. In the oceans, recovery of considerable primary productivity probably occurred within a few months because phytoplankton reproduce and grow rapidly. On land, the recovery of primary production probably took more than a year because seeds had to germinate, and brush and trees must have required several years to reattain prior levels of biomass. Recolonization by ferns and other "disaster species" on land is paralleled by the erratic blooms of nannoplankton in the oceans. K/T faunal changes both on land and in the oceans are compatible with a sudden, temporary loss of primary productivity. Similar changes cannot be studied during the Ordovician extinction because the absence of an iridium layer makes the requisite refined correlation impossible.

Reef biota

Comparison of the effects on the reef faunas is difficult because typical reefs with phototrophic frame-builders were not present in the latest Cretaceous. However, during the recovery, phototrophic scleractinian reefs became common by the late Paleocene (Rosen and Turnsek, 1989). The reef biota was severely impaired during the Ordovician extinction. Early Silurian reefs were few in number and very small, and recovery of the reef fauna took several million years.

High-latitude extinction

High-latitude faunas suffered severely during the Ordovician glaciation, and faunas retreated toward the equator. The effects of the glaciation were particularly severe because faunas had adapted to a long interval of greenhouse climate (Brenchley et al., 1994). During the K/T event there was no comparable intensification of extinction in high latitudes. If anything, high-latitude Southern Hemisphere faunas may have escaped the most severe effects of the extinction, possibly because of their distance from the impact site (Hildebrand, 1993).

CONCLUSIONS

The K/T extinction has a clear ecologic signature of the short-term effects of an impact event. Similar patterns are not present in the Late Ordovician extinction. Rather, the Ordovician event is characterized by two episodes of extinction associated with environmental changes at the onset and end of a glaciation. Some similarities between the two extinctions exist, such as preferential survival of both widely distributed clades and small, relatively underived species, but these patterns are inherent to most extinction events.

ACKNOWLEDGMENTS

This work was supported by the U.S. National Science Foundation Grants EAR-90-04589 (Sheehan) and EAR 93-04513 (Fastovsky and Sheehan).

REFERENCES CITED

Achab, A., and Duffield, S. L., 1982, Palynological changes at the Ordovician-Silurian boundary on Anticosti Island, Quebec, *in* Bruton, D. L., and Williams, H. S., eds., Fourth International Symposium on the Ordovician System: Annual Meeting, Oslo, Norway, Abstracts, p. 3.

Algeo, T. J., and Wilkinson, B. H., 1991, Modern and ancient continental hypsometries: Journal of the Geological Society, London, v. 148, p. 643–653.

Alvarez, L. W., Alvarez, W., Asaro, F., and Michel, H. V., 1980, Extraterrestrial cause for the Cretaceous/Tertiary extinctions: Science, v. 208, p. 1095–1108.

Anstey, R. L., 1985, Bryozoan provinces and patterns of generic evolution and extinction in the Late Ordovician of North America: Lethaia, v. 19, p. 33–51.

Arthur, M. A., Zachos, J. C., and Jones, D. S., 1987, Primary productivity and the Cretaceous/Tertiary boundary event in the oceans: Cretaceous Research, v. 8, p. 43–54.

Bergström, S. M., 1990, Relations between conodont provincialism and the changing palaeogeography during the Early Palaeozoic, *in* McKerrow, W. S., and Scotese, C. R., eds., Palaeozoic palaeogeography and biogeography: Geological Society, London, Memoir 12, p. 105–121.

Berry, W. B. N., 1979, Graptolite biogeography: A biogeography of some lowest Paleozoic plankton, *in* Gray, J., and Boucot, A. J., eds., Historical biogeography, plate tectonics, and the changing environment: Corvallis, Oregon State University Press, p. 105–115.

Berry, W. B. N., and Boucot, A. J., 1973, Glacio-eustatic control of Late Ordovician-Early Silurian platform sedimentation and faunal changes: Geological Society of America Bulletin, v. 84, p. 275–284.

Berry, W. B. N., and Wilde, P., 1990, Graptolite biogeography: Implications for palaeogeography and palaeoceanography, *in* McKerrow, W. S., and Scotese, C. R., eds., Palaeozoic Palaeogeography and Biogeography: London, Geological Society Memoir no. 12, p. 129–137.

Berry, W. B. N., Wilde, P., and Quinby-Hunt, M. S., 1991, Late Ordovician graptolite mass mortality and subsequent early Silurian re-radiation, *in* Kauffman, E. G., and Walliser, O. H., eds., Extinction events in Earth history: Berlin, Springer-Verlag, p. 115–123.

Beuf, S. B., Biju-Duval, B., de Charpal, O., Rognon, P., Gariel, O., and Bennacef, A., 1971, Les grès du Paléozoïque inférieur au Sahara: Paris, Editions Technip, 464 p.

Blum, J. D., and Chamberlain, C. P, 1994, Geochemical constraints on the composition of volatiles released during the formation of KT impact melts: Implications for extinction mechanisms, *in* New developments regarding the KT event and other catastrophes in Earth history: Houston, Texas, Lunar and Planetary Institute Contribution 825, p. 11–13.

Boucot, A. J., 1983, Does evolution occur in an evolutionary vacuum? II: Journal of Paleontology, v. 111, p. 545–655.

Brenchley, P. J., 1984, Late Ordovician extinctions and their relationship to the Gondwana glaciation, *in* Brenchley, P. J., ed., Fossils and climate: Chichester, John Wiley & Sons, p. 291–327.

Brenchley, P. J., 1988, Environmental changes close to the Ordovician-Silurian boundary: Bulletin of the British Museum (Natural History), Geology Series, v. 43, p. 377–385.

Brenchley, P. J., and Cullen, B., 1984, The environmental distribution of associations belonging to the *Hirnantia* fauna—Evidence from North Wales and Norway, *in* Bruton, D. L., ed., Aspects of the Ordovician system: Palaeontological Contributions from the University of Oslo 295, p. 113–126.

Brenchley, P. J., and Newall, G., 1984, Late Ordovician environmental changes and their effects on faunas, *in* Bruton, D. L., ed., Aspects of the Ordovician system: Palaeontological Contributions from the University of Oslo 295, p. 65–79.

Brenchley, P. J., Romano, M., Young, T. P., and Stoch, P., 1991, Hirnantian glaciomarine diamictites—Evidence for the spread of glaciation and its effect on Ordovician faunas, *in* Barnes, C. R., and Williams, S. H., eds., Advances in Ordovician geology: Geological Survey of Canada Paper 90-9, p. 325–336.

Brenchley, P. J., Marshall, J. D., Carden, G. A. F., Robertson, D. B. R., Long, D. G. F., Meidla, T., Hints, L., and Anderson, T. F., 1994, Bathymetric and isotopic evidence for a short-lived Late Ordovician glaciation in a greenhouse period: Geology, v. 22, p. 295–298.

Bretsky, P. W., 1973, Evolutionary patterns in the Paleozoic Bivalva: Documentation and some theoretical considerations: Geological Society of America Bulletin, v. 84, p. 2079–2096.

Briggs, D. E. G., Fortey, R. A., and Clarkson, E. N. K., 1988, Extinction and the fossil record of arthropods, *in* Larwood, G. P., ed., Extinction and survival in the fossil record: Systematics Association Special Volume 34, p. 171–209.

Brinkhuis, H., and Visscher, H., 1994, New evidence for terrestrial ecosystem collapse at the KT and Permian/Triassic boundaries, *in* New developments regarding the KT event and other catastrophes in Earth history: Houston, Texas, Lunar and Planetary Institute Contribution 825, p. 17.

Cadée, G. C., 1984, "Opportunistic feeding," a serious pitfall in trophic structure analysis of (paleo)faunas: Lethaia, v. 17, p. 289–292.

Chatterton, B. D. E., and Speyer, S. E., 1989, Larval ecology, life history strategies, and patterns of extinction and survivorship among Ordovician trilobites: Paleobiology, v. 15, p. 118–132.

Closs, G. P., and Lake, P. S., 1994, Spatial and temporal variation in the structure of an intermittent-stream food web: Ecological Monographs, v. 64, p. 1–21.

Cocks, L. R. M., 1988, Brachiopods across the Ordovician-Silurian boundary, *in* Cocks, L. R. M., and Rickards, R. B., eds., 1988, A global analysis of the Ordovician-Silurian boundary: Bulletin of the British Museum (Natural History), Geology Series, v. 43, p. 311–315.

Cocks, L. R. M., and Rickards, R. B., eds., 1988, A global analysis of the Ordovician-Silurian boundary: Bulletin of the British Museum (Natural History), Geology Series, v. 43, p. 1–394.

Colbath, G. K., 1986, Abrupt terminal Ordovician extinction in phytoplankton associations, southern Appalachians: Geology, v. 14, p. 943–946.

Copeland, M. J., 1977, Early Paleozoic Ostracoda of eastern Canada, *in* Swain, F. M., ed., Stratigraphic micropaleontology of Atlantic Basin and borderlands: Amsterdam, Elsevier, p. 1–15.

Covey, C., Ghan, S. J., Walton, J. J., and Weissman, P. R., 1990, Global environmental effects of impact-generated aerosols; Results from a general circulation model, *in* Sharpton, V. L., and Ward, P. D., eds., Global catastrophes in Earth history; An interdisciplinary conference on impact, volcanism, and mass mortality: Geological Society of America Special Paper 247, p. 263–270.

Crick, R. E., 1990, Cambro-Devonian biogeography of nautiloid cephalopods, *in* McKerrow, W. S., and Scotese, C. R., eds., Palaeozoic Palaeogeography and Biogeography: Geological Society, London, Memoir 12, p. 147–161.

Crutzen, P. J., 1987, Acid rain at the K/T boundary: Nature, v. 330, p. 108–109.

Dean, W. T., 1985, Relationships of Cambrian-Ordovician faunas in the Caledonide-Appalachian region, with particular reference to trilobites, *in* Gayer, R. A., ed., The tectonic evolution of the Caledonide-Appalachian Orogen: Braunschweig/Wiesbaden, Friedreich Vieweg & Sohn, p. 17–47.

D'Hondt, S., and Lindinger, M., 1988, An extended Cretaceous-Tertiary boundary stable isotope record: implications for paleoclimate and the nature of the K/T boundary event, *in* New developments regarding the KT event and other catastrophes in Earth history: Houston, Texas, Lunar and Planetary Institute Contribution 825, p. 40–41.

D'Hondt, S., Pilson, M., Sigurdsson, H., Hanson, A. K., and Carey, S., 1994, Surface-water acidification and extinction at the Cretaceous-Tertiary boundary: Geology, v. 22, p. 983–986.

Duffield, S. L., and Legault, J. A., 1982, Acritarch biostratigraphy of Upper

Ordovician–Lower Silurian rocks, Anticosti Island, Quebec: Preliminary results, *in* Lespérance, P. J., ed., Subcommmission on Silurian stratigraphy, Ordovician-Silurian boundary working group, field meeting, Anticosti-Gaspé, Quebec, 1981. Volume 2: Stratigraphy and paleontology: Montréal, Université de Montréal, p. 91–99.

Elias, R. J., 1989, Extinctions and origins of solitary rugose corals, latest Ordovician to earliest Silurian in North America: Association of Australian Paleontologists, Memoir 8, p. 319–326.

Fastovsky, D. E., and Sheehan, P. M., 1994, Habitat vs. asteroid fragmentation in vertebrate extinctions at the KT boundary: The good, the bad, and the untested, *in* New developments regarding the KT event and other catastrophes in Earth history: Houston, Texas, Lunar and Planetary Institute Contribution 825, p. 36–37.

Fleming, R. R., and Nichols, D. J., 1990, The fern-spore abundance anomaly at the Cretaceous-Tertiary boundary: A regional bioevent in western North America, *in* Kauffman, E. G., and Walliser, O. H., eds., Extinction events in Earth history: Berlin, Springer-Verlag, p. 347–349.

Fortey, R. A., 1989, There are extinctions and extinctions: examples from the Lower Palaeozoic: Philosophical Transactions of the Royal Society of London, v. B325, p. 327–355.

Gallagher, W. B., 1991, Selective extinction and survival across the Cretaceous/Tertiary boundary in the northern Atlantic Costal Plain: Geology, v. 19, p. 967–970.

Gildner, R. F., and Cisne, J. L., 1987, Comparison of albedo changes due to shelf exposure and ice during the Late Ordovician glacioeustatic regression: Paleoceanography, v. 2, p. 177–183.

Gilmour, I., Wolbach, W. S., and Anders, E., 1990, Early environmental effects of the terminal Cretaceous impact, *in* Sharpton, V. L., and Ward, P. D., eds., Global catastrophes in Earth history: An interdisciplinary conference on impact, volcanism, and mass mortality: Geological Society of America Special Paper 247, p. 383–390.

Goodfellow, W. D., Nowlan, G. S., McCracken, A. D., Lenz, A. C., and Gregoire, D. C., 1992, Geochemical anomalies near the Ordovician-Silurian boundary, Northern Yukon Territory, Canada: Historical Biology, v. 6, p. 1–23.

Griffis, K., and Chapman, D. J., 1988, Survival of phytoplankton under prolonged darkness: Implications for the Cretaceous-Tertiary boundary darkness hypothesis: Palaeogeography, Palaeoclimatology, Palaeoecology, v. 67. p. 305–314.

Griffis, K., and Chapman, D. J., 1990, Modeling Cretaceous-Tertiary boundary events with extant photosynthetic plankton: Effects of impact-related acid rain: Lethaia, v. 23, p. 379–383.

Hallam, A., and Miller, A. I., 1988, Extinction and survival in the Bivalvia, *in* Larwood, G. P., ed., Extinction and survival in the fossil record, Systematics Association Special Volume 34, p. 121–138.

Hansen, T. A., 1988, Early Tertiary radiation of marine molluscs and the long-term effects of the Cretaceous-Tertiary extinction: Paleobiology, v. 14, p. 37–51.

Hansen, T., Farrand, R. B., Montgomery, H. A., Billman, H. G., and Blechschmidt, G., 1987, Sedimentology and extinction patterns across the Cretaceous-Tertiary boundary interval in east Texas: Cretaceous Research, v. 8, p. 229–252.

Hansen, T. A., Farrell, B. R., and Upshaw, B., III, 1993, The first 2 million years after the Cretaceous-Tertiary boundary in east Texas: Rate and paleoecology of the molluscan recovery: Paleobiology, v. 19, p. 251–265.

Herbert, T. D., and D'Hondt, S., 1990, Precessional climate cyclicity in Late Cretaceous–Early Tertiary marine sediments: A high resolution chronometer of Cretaceous-Tertiary boundary events: Earth and Planetary Science Letters, v. 99, p. 263–275.

Hildebrand, A. R., 1993, The Cretaceous/Tertiary Boundary impact (or the dinosaurs didn't have a chance): Journal of the Royal Astronomical Society, v. 87, p. 77–118.

Hildrew, A. G., Townsend, C. R., and Hasham, A., 1985, The predatory chironomidae of an iron-rich stream: Feeding ecology and food-web structure: Ecological Entomology, v. 10, p. 403–413.

Hollander, D. J., McKenzie, J. A., and Hsü, K. J., 1993, Carbon isotope evidence for unusual plankton blooms and fluctuations of surface water CO_2 in "Strangelove Ocean" after terminal Cretaceous event: Palaeogeography, Palaeoclimatology, Palaeoecology, v. 104, p. 229–237.

Hsü, K. J., and McKenzie, J. A., 1985, A "Strangelove" ocean in the earliest Tertiary, *in* Sundquist, E. T., and Broecker, W. S., eds., The carbon cycle and atmospheric CO_2: Natural variations Archean to Present: Washington, D.C., American Geophysical Union Geophysical Monograph 32, p. 487–492.

Hsü, K. J., and 19 others, 1982, Mass mortality and its environmental and evolutionary consequences: Science, v. 216, p. 249–256.

Ivany, L. C., 1991, Carbon isotope records across the K/T boundary: Implications for biomass burning: Geological Society of America Abstracts with Programs, v. 23, p. A420.

Ivany, L. C., and Salawitch, R. J., 1993, Carbon isotopic evidence for biomass burning at the K-T boundary: Geology, v. 21, p. 487–490.

Jaanusson, V., 1979, Ordovician, *in* Robison, R. A., and Teichert, C., eds., Treatise on invertebrate paleontology. Part A: Introduction: Lawrence, University Press of Kansas, p. A136–A166.

Jablonski, D., 1986a, Background and mass extinctions: The alternation of macroevolutionary regimes: Science, v. 231, p. 129–133.

Jablonski, D., 1986b, Causes and consequences of mass extinctions: A comparative approach, *in* Elliott, D. K., ed., Dynamics of extinction: New York, John Wiley and Sons, p. 183–229.

Jablonski, D., 1986c, Evolutionary consequences of mass extinction, *in* Raup, D. M., and Jablonski, D., eds., Patterns and processes in the history of life: Berlin, Springer-Verlag, p. 313–329.

Jablonski, D., 1989, The biology of mass extinction: A paleontological view: Philosophical Transactions of the Royal Society of London, v. B325, p. 357–368.

Jablonski, D., 1991, Extinctions: A paleontological perspective: Science, v. 253, p. 754–757.

Johansen, M. B., 1988, Brachiopod extinctions in the upper Cretaceous to lowermost Tertiary chalk of northwest Europe: Revista Espadon, v.cts with Programs°. Extraordinario, 1988, Extinction Events, p. 41–56.

Johansen, M. B., 1989, Background extinction and mass extinction of the brachiopods from the chalk of northwest Europe: Palaios, v. 4, p. 243–250.

Johnson, J. G., 1974, Extinction of perched faunas: Geology, v. 2, p. 479–482.

Johnson, K. R., and Hickey, L. J., 1990, Megafloral change across the Cretaceous/Tertiary boundary in the northern Great Plains and Rocky Mountains, U.S.A., *in* Sharpton, V. L., and Ward, P. D., eds., Global catastrophes in Earth history: An interdisciplinary conference on impact, volcanism, and mass mortality: Geological Society of America Special Paper 247, p. 433–444.

Johnson, K. R., Nichols, D. J., Attrep, M., Jr., and Orth, C. J., 1989, High-resolution leaf-fossil record spanning the Cretaceous/Tertiary boundary: Nature, v. 340, p. 708–711.

Kaiho, K., 1994, Planktonic and benthic foraminiferal extinction events during the last 100 m.y.: Palaeogeography, Palaeoclimatology, Palaeoecology, v. 111, p. 45–71.

Kaljo, D., and Klaamann, E., 1973, Ordovician and Silurian corals, *in* Hallam, A., ed., Atlas of Palaeobiogeography: Amsterdam, Elsevier, p. 37–46.

Kitchell, J. A., Clark, D. L, and Gombos, A. M., 1986, Biological selectivity of extinction: A link between background and mass extinction: Palaios, v. 1, p. 504–511.

Knoll, A.H., 1989, Evolution and extinction in the marine realm: Some constraints imposed by phytoplankton: Philosophical Transactions of the Royal Society of London, v. B325, p. 279–290.

Koren, T.J., 1991, Evolutionary crisis of the Ashgill graptolites, *in* Barnes, C. R., and Williams, S. H., eds., Advances in Ordovician geology: Geological Survey of Canada Paper 90-9, p. 157–164.

Kring, D. A., 1993, The Chicxulub impact event and possible causes of K/T boundary extinction, *in* Boaz, D., and Dornan, M., eds., Proceedings, First Annual Symposium on Fossils of Arizona: Mesa, Arizona, Mesa Southwest Museum and Southwest Paleontological Society, p. 63–79.

Kriz, J., 1984, Autecology and ecogeny of Silurian Bivalvia: Special Papers in Palaeontology, v. 32, p. 183–195.

Kump, L. R., 1991, Interpreting carbon-isotope excursions: Strangelove oceans: Geology, v. 19, p. 299-302.

Leleshus, V. L., and Permyakov, V. V., 1981, A characteristic feature of the evolution of corals: Paleontological Zhurnal, v. 1981, p. 3–14.

LeLoeuff, J., Buffetaut, E., and Martin, M., 1994, The last stages of dinosaur faunal history in Europe: A succession of Maastrichtian dinosaur assemblages from the Corbieres (southern France): Geological Magazine, v. 131, p. 625–630.

Lenz, A. C., 1976, Late Ordovician–Early Silurian glaciation and the Ordovician-Silurian boundary in the northern Canadian Cordillera: Geology, v. 4, p. 313–317.

Lerbekmo, J. F., Sweet, A. R., and St. Louis, R. M., 1987, The relationship between the iridium anomaly and palynological floral events at three Cretaceous-Tertiary boundary localities in western Canada: Geological Society of America Bulletin, v. 99. p. 325–330.

Lewis, J. S., Watkins, G. H., and Hartman, H., 1982, Chemical consequences of major impact events on Earth, *in* Silver, L. T., and Schultz, P. H., eds., Geological implications of impacts of large asteroids and comets on the Earth: Geological Society of America Special Paper 190, p. 215–221.

Lipps, J. H., 1983, Biotic interactions in benthic Foraminifera, *in* Tevesz, M. J. S., and McCall, P. L., eds., Biotic interactions in recent and fossil benthic communities: New York, Plenum Press, p. 331–376.

Long, D. G. F., 1993, Oxygen and carbon isotopes and event stratigraphy near the Ordovician-Silurian boundary, Anticosti Island, Quebec: Palaeogeography, Palaeoclimatology, Palaeoecology, v. 104, p. 49–59.

Magaritz, M., 1989, ^{13}C minima follow extinction events: A clue to faunal radiation: Geology, v. 17, p. 337–340.

Mann, K. H., 1988, Production and use of detritus in various freshwater, estuarine, and costal marine ecosystems: Limnological Oceanography, v. 33, p. 910–930.

Marshall, J. D., and Middleton, P. D., 1990, Changes in marine isotopic composition and the Late Ordovician glaciation: Journal of the Geological Society, London, v. 147, p. 1–4.

McCann, T., 1989, Distribution of Ordovician-Silurian ichnofossil assemblages in Wales—Implications for Phanerozoic ichnofaunas: Lethaia, v. 23, p. 243–255.

McLaren, D. J., and Goodfellow, W. D., 1990, Geological and biological consequences of giant impacts: Annual Review of Earth and Planetary Sciences, v. 18, p. 123–171.

Melchin, M. J., and Mitchell, C. E., 1991, Late Ordovician extinction in the Graptoloidea, *in* Barnes, C. R., and Williams, S. H., eds., Advances in Ordovician geology: Geological Survey of Canada Paper 90-9, p. 143–156.

Melosh, H. J., Schneider, N. M., Zahnle, K. J., and Latham, D., 1990, Ignition of global wildfires at the Cretaceous/Tertiary boundary: Nature, v. 343, p. 251–254.

Middleton, P. D., Marshall, J. D., and Brenchley, P. J., 1991, Evidence for isotopic change associated with Late Ordovician glaciation, from brachiopods and marine cements of central Sweden, *in* Barnes, C. R., and Williams, S. H., eds., Advances in Ordovician geology: Geological Survey of Canada Paper 90-9, p. 313–323.

Nichols, D. J., Jarzen, D. M., Orth, C. J., and Oliver, P. Q., 1986, Palynological and iridium anomalies at Cretaceous-Tertiary boundary, south-central Saskatchewan: Science, v. 231, p. 714–717.

Nichols, D. J., Fleming, R. F., and Frederiksen, N. O., 1990, Palynological evidence of effects of the terminal Cretaceous event on terrestrial floras in western North America, *in* Kauffman, E. G., and Walliser, O. H., eds., Extinction events in Earth history: Berlin, Springer-Verlag, p. 351–364.

O'Keefe, J. D., and Ahrens, T. J., 1989, Impact production of CO_2 by the Cretaceous/Tertiary bolide and the resultant heating of the earth: Nature, v. 338, p. 247–249.

Orth, C. J., Gilmore, J. S., Knight, J. D., Pillmore, C. L., Tschudy, R. H., and Fassett, J. E., 1981, An iridium abundance anomaly at the palynological Cretaceous-Tertiary boundary in northern New Mexico: Science, v. 214, p. 1341–1343.

Orth, C. J., Gilmore, J. S., Quintana, L. R., and Sheehan, P. M., 1986, The terminal Ordovician extinction: Geochemical analysis of the Ordovician/Silurian boundary, Anticosti Island, Quebec: Geology, v. 14, p. 433–436.

Owen, A. W., Harper, D. A. T., and Rong, J.-Y., 1991, Hirnantian trilobites and brachiopods in space and time, *in* Barnes, C. R., and Williams, S. H., eds., Advances in Ordovician geology: Geological Survey of Canada Paper 90-9, p. 179–190.

Pollack, J. B., Toon, O. B., Ackerman, T. P., McKay, C. P., and Turco, R. P., 1983, Environmental effects of an impact-generated dust cloud: Implications for the Cretaceous-Tertiary extinctions: Science, v. 219, p. 287–289.

Pospichal, J. J., 1994a, Calcareous nannofossils and the KT boundary crisis: An update, *in* New developments regarding the KT event and other catastrophes in Earth history: Houston, Texas, Lunar and Planetary Institute Contribution 825, p. 90.

Pospichal, J. J., 1994b, Calcareous nannofossils at the K-T boundary, El Kef: No evidence for stepwise, gradual, or sequential extinctions: Geology, v. 22, p. 99–102.

Prinn, R. G., and Fegley, B., 1987, Bolide impacts, acid rain, and biospheric traumas at the Cretaceous-Tertiary boundary: Earth and Planetary Sciences Letters, v. 83, p. 1–15.

Railsback, L. B., Ackerley, S. C., Anderson, T. F., and Cisne, J. L., 1990, Paleontological and isotope evidence for warm saline deep waters in Ordovician oceans: Nature, v. 343, p. 156–159.

Rhodes, M. C., and Thayer, C. W., 1991, Mass extinctions: Ecological selectivity and primary production: Geology, v. 19, p. 877–880.

Robardet, M., and Doré, R., 1988, The Late Ordovician diamictic formations from southwestern Europe: North-Gondwana glaciomarine deposits: Palaeogeography, Palaeoclimatology, Palaeoecology, v. 66, p. 19–31.

Robertson, D. B. R., Brenchley, P. J., and Owen, A. W., 1991, Ecological disruption close to the Ordovician-Silurian boundary: Historical Biology, v. 5, p. 131–144.

Rong, Jia-yu, 1984, Distribution of the *Hirnantia* fauna and its meaning, *in* Bruton, D. L., ed., Aspects of the Ordovician system: Palaeontological Contributions from the University of Oslo 295, p. 101–112.

Rong, Jia-yu, and Harper, D. A. T., 1988, A global synthesis of the latest Ordovician Hirnantian brachiopod faunas: Transactions of the Royal Society of Edinburgh: Earth Sciences, v. 79, p. 383–402.

Rosen, B. R., and Turnsek, D., 1989, Extinction patterns and the biogeography of scleractinian corals across the Cretaceous/Tertiary boundary: Fossil Cnidaria 5, Memoir of the Association of Australian Palaeontologists, v. 8, p. 355–370.

Roy, J. M., McMenamin, M. A. S., and Alderman, S. E., 1990, Trophic differences, originations and extinctions during the Cenomanian and Maastrichtian stages of the Cretaceous, *in* Kauffman, E. G., and Walliser, O. H., eds., Extinction events in Earth history: Berlin, Springer-Verlag, p. 299–303.

Russell, D. A., 1979, The enigma of the extinction of the dinosaurs: Annual Review of Earth and Planetary Sciences, v. 7, p. 163–182.

Saito, T., Yamanoi, T., and Kaiho, K., 1986, End-Cretaceous devastation of terrestrial flora in the boreal Far East: Nature, v. 323, p. 253–255.

Scotese, C. R., and McKerrow, W. S., 1990, Revised world maps and introduction, *in* McKerrow, W. S., and Scotese, C. R., eds., Palaeozoic Palaeogeography and Biogeography: Geological Society of London, Memoir 12, p. 1–21.

Sepkoski, J. J., 1986, Phanerozoic overview of mass extinction, *in* Raup, D. M., and Jablonski, D., eds., Patterns and processes in the history of life: Berlin, Springer-Verlag, p. 277–296.

Sheehan, P. M., 1973, The relation of Late Ordovician glaciation to the Ordovician-Silurian changeover in North American brachiopod faunas: Lethaia, v. 6, p. 147–154.

Sheehan, P. M., 1975, Brachiopod synecology in a time of crisis (Late Ordovician-Early Silurian): Paleobiology, v. 1, p. 205–212.

Sheehan, P. M., 1979, Swedish Late Ordovician marine benthic assemblages and their bearing on brachiopod zoogeography, *in* Gray, J., and Boucot,

A. J., eds., Historical biogeography, plate tectonics, and the changing environment: Corvallis, Oregon State University Press, p. 61–73.

Sheehan, P. M., 1985, Reefs are not so different—They follow the evolutionary pattern of level-bottom communities: Geology, v. 13, p. 46–49.

Sheehan, P. M., 1987, The *Foliomena* community—Life below the thermocline in the Late Ordovician: Geological Society of America Abstracts with Programs, v. 19, no. 4, p. 245.

Sheehan, P. M., 1988, Late Ordovician events and the terminal Ordovician extinction: New Mexico Bureau of Mines and Mineral Resources Memoir 44, p. 405–415.

Sheehan, P. M., 1992, Patterns of synecology during the Phanerozoic, *in* Dudley, E. C., ed., The unity of evolutionary viology, Proceedings, Fourth International Congress of Systematic and Evolutionary Biology: Portland, Oregon, Discorides Press, v. 1, p. 103–118.

Sheehan, P. M., and Coorough, P. J., 1990, Brachiopod zoogeography across the Ordovician-Silurian extinction event, *in* McKerrow, W. S., and Scotese, C. R., eds., Palaeozoic palaeogeography and biogeography: Geological Society, London, Memoir 12, p. 181–187.

Sheehan, P. M., and Fastovsky, D. E., 1992, Major extinctions of land-dwelling vertebrates at the Cretaceous-Tertiary boundary, eastern Montana: Geology, v. 20, p. 556–560.

Sheehan, P. M., and Hansen, T. A., 1986, Detritus feeding as a buffer to extinction at the end of the Cretaceous: Geology, v. 14, p. 868–870.

Sheehan, P. M., and Russell, D. A., 1994, Faunal change following the Cretaceous-Tertiary impact: Using paleontological data to assess the hazards of impacts, *in* Gehrels, T., ed., Hazards due to comets and asteroids: Tucson, University of Arizona Press, p. 879–993.

Sheehan, P. M., Fastovsky, D. E., Hoffmann, R. G., Berghaus, C. B., and Gabriel, D. L., 1991, Sudden extinction of the dinosaurs: Latest Cretaceous, Upper Great Plains, U.S.A.: Science, v. 254, p. 835–839.

Sigurdsson, H., D'Hondt, S., and Carey, S., 1992, The impact of the Cretaceous/Tertiary bolide on evaporite terrane and generation of major sulfuric acid aerosol: Earth and Planetary Science Letters, v. 109, p. 453–559.

Skevington, D., 1978, Latitudinal surface water temperature gradients and Ordovician faunal provinces: Alcheringa, v. 2, p. 21–26.

Smit, J., 1994, Extinctions at the Cretaceous-Tertiary boundary: The link to the Chicxulub impact, *in* Gehrels, T., ed., Hazards due to comets and asteroids: Tucson, University of Arizona Press, p. 859–878.

Stott, L. D., and Kennett, J. P., 1989, New constraints on early Tertiary palaeoproductivity from carbon isotopes in foraminifera: Nature, v. 342, p. 526–528.

Surlyk, F., and Johansen, M. B., 1984, End-Cretaceous brachiopod extinctions in the chalk of Denmark: Science, v. 223, p. 1174–1177.

Sweet, W. C., 1990, The Conodonta. Morphology, taxonomy, paleoecology, and evolutionary history of a long-extinct animal phylum: Oxford, Clarendon Press, 212 p.

Sweet, W. C., and Bergström, S. M., 1984, Conodont provinces and biofacies of the Late Ordovician, *in* Clark, D. L., ed., Conodont biofacies and provincialism: Geological Society of America Special Paper 196, p. 69–87.

Thierstein, H. R., 1981 Late Cretaceous nannoplankton and the change at the Cretaceous-Tertiary boundary, *in* Warme, J., ed., The Deep Sea Drilling Project: A decade of progress: Society of Economic Paleontologists and Mineralogists Special Publication 32, p. 355–394.

Thierstein, H.R., 1982, Terminal Cretaceous plankton extinctions: A critical assessment, *in* Silver, L. T., and Schultz, P. H., eds., Geological implications of impacts of large asteroids and comets on the Earth: Geological Socicty of America Special Paper 190, p. 385–399.

Toon, O. B., Pollack, J. B., Ackerman, T. P., Turco, R. P., McKay, C. P., and Liu, M. S., 1982, Evolution of impact-generated dust cloud and its effects on the atmosphere, *in* Silver, L. T., and Schultz, P. H., eds., Geological Implications of Impacts of Large Asteroids and Comets on the Earth: Geological Society of America Special Paper 190, p. 187–200.

Tschudy, R.H., and Tschudy, B.D., 1986, Extinction and survival of plant life following the Cretaceous/Tertiary boundary event, Western Interior, North America: Geology, v. 14, p. 667-670.

Tschudy, R. H., Pillmore, C. L., Orth, C. J., Gilmore, J. S., and Knight, J. D., 1984, Disruption of the terrestrial plant ecosystem at the Cretaceous-Tertiary boundary, Western Interior: Science, v. 225, p. 1030–1032.

Tuckey, M. E., and Anstey, R. L., 1992, Late Ordovician extinctions of bryozoans: Lethaia, v. 25, p. 111–117.

Venkatesan, M. I., and Dahl, J., 1989, Organic geochemical evidence for global fires at the Cretaceous/Tertiary boundary: Nature, v. 338, p. 57–60.

Wang, K., Chatterton, B. D. E., Attrep, M., and Orth, C. J., 1992, Iridium abundance maxima at the latest Ordovician mass extinction horizon, Yangtze Basin, China: Terrestrial or extraterrestrial?: Geology, v. 20, p. 39–42.

Wang, K., Orth, C. J., Attrep, M., Jr., Chatterton, B. D. E.,Wang, X., and Li, J.-j., 1993, The great latest Ordovician extinction on the south China Plate: Chemostratigraphic studies of the Ordovician-Silurian boundary interval on the Yangtze Platform: Palaeogeography, Palaeoclimatology, Palaeoecology, v. 104, p. 61–79.

Ward, P. D., Kennedy, W. J., and MacLeod, K. G., 1991, Ammonite and inoceramid bivalve extinction patterns in Cretaceous/Tertiary boundary sections of the Biscay region (southwestern France, northern Spain): Geology, v. 19, p. 1181–1183.

Watkins, R., 1994, Evolution of Silurian pentamerid communities in Wisconsin: Palaios, v. 9, p. 488–499.

Webby, B. D., 1984, Ordovician reefs and climate: A review, *in* Bruton, D. L., ed., Aspects of the Ordovician System: Oslo, Universitetsforlaget, p. 89–100.

Wilde, P., Berry, W. B. N., Quinby-Hunt, M. S., Orth, C. J., Quintana, L. R., and Gilmore, J. S., 1986, Iridium abundances across the Ordovician-Silurian stratotype: Science, v. 233, p. 339–341.

Wilde, P., Quinby-Hunt, M. A., and Berry, W. B. N., 1990, Vertical advections from oxic or anoxic water from the main pycnocline as a cause of rapid extinctions or rapid radiations, *in* Kauffman, E. G., and Walliser, O. H., eds., Extinction events in Earth history: Berlin, Springer-Verlag, v. 30, p. 85–98.

Winemiller, K.O., 1990, Spatial and temporal variation in tropical fish trophic networks: Ecological Monographs, v. 60, p. 331–367.

Wolbach, W. S., Lewis, R. S., and Anders, E., 1985, Cretaceous extinctions: Evidence for wildfires and search for meteoritic material: Science, v. 230, p. 167–170.

Wolbach, W. S., Gilmour, I., and Anders, E., 1990, Major wildfires at the Cretaceous/Tertiary boundary, *in* Sharpton, V. L., and Ward, P. D., eds., Global catastrophes in Earth history; An interdisciplinary crence on impact, volcanism, and mass mortality: Geological Society of America Special Paper 247, p. 391–400.

Wolfe, J. A., 1991, Palaeobotanical evidence for a June "impact winter" at the Cretaceous/Tertiary boundary: Nature, v. 352, p. 420–423.

Wolfe, J. A., and Upchurch, G. R., 1986, Vegetation, climatic and floral changes at the Cretaceous-Tertiary boundary: Nature, v. 324, p. 148–152.

Wood, R., 1993, Nutrients, predation and the history of reef-building: Palaios, v. 8, p. 526–543.

Zachos, J. C., Arthur, M. A., and Dean, W. E., 1989, Geochemical evidence for suppression of pelagic marine productivity at the Cretaceous/Tertiary boundary: Nature, v. 337, p. 61–64.

Zahnle, K. J., 1990, Atmospheric chemistry by large impacts, *in* Sharpton, V. L., and Ward, P. D., eds., Global catastrophes in Earth history; An interdisciplinary conference on impact, volcanism, and mass mortality: Geological Society of America Special Paper 247, p. 271–288.

MANUSCRIPT ACCEPTED BY THE SOCIETY SEPTEMBER 14, 1995

Printed in U.S.A.

Geological Society of America
Special Paper 307
1996

Geochemistry of the Frasnian-Famennian boundary in Belgium: Mass extinction, anoxic oceans and microtektite layer, but not much iridium?

P. Claeys*
Department of Geology and Geophysics, University of California, Berkeley, California 94720-4767
F. T. Kyte
Institute of Geophysics and Planetary Physics, University of California, Los Angeles, California 90024
A. Herbosch*
Department of Geology and Geophysics, University of California, Berkeley, California 94720-4767
J.-G. Casier
Paleontology Department, Belgian Royal Institute of Natural Sciences, 29 Rue Vautier, B-1040 Brussels, Belgium

ABSTRACT

The Late Devonian, and in particular the Frasnian-Famennian (F/F) boundary, records one of the five largest mass extinctions in the fossil record. Glassy spherules believed to be of impact origin are associated with the F/F boundary in two Belgian sections (Senzeille and Hony). They have also been reported in sediments deposited approximately 1.5 to 2 m.y. above the boundary in south China, and in the Canning Basin (Australia) this event coincides with a 300 pg/g Ir anomaly. In this study, the F/F boundary in the Hony section was analyzed for trace and major elements to test the possibility of an Ir anomaly associated with the spherule layer. No significant positive Ir anomaly was detected in the 2 m of section investigated. Nevertheless, chalcophile elements show an increase within the dark shale bed marking the F/F boundary. This increase is interpreted to represent a reduction in oxygen concentrations in the depositional environment. This level must be equivalent to the upper part of the Kellwasser anoxic event recognized throughout the paleo-Tethys in what is now western Europe. The F/F boundary seems to be marked by a succession of major events, including impact, oxygen-depleted water on the shelf, and worldwide extinction of organisms.

INTRODUCTION

The Cretaceous-Tertiary (K/T) boundary debate has triggered intense searches for evidence of possible impact-related mass extinctions in the fossil record. So far, only at the K/T boundary is there significant evidence to support an impact-extinction linkage. The Late Devonian, and more particularly the Frasnian-Famennian (F/F) boundary, corresponds to one of the five largest mass extinction periods in the fossil record (Sepkoski, 1982) and is characterized by the disappearance of pelagic and shallow benthic organisms and reef builders (McLaren, 1982). Possible impact products in the form of glassy microtektite-like spherules have been documented recently at or near the F/F boundary in Late Devonian sections in south China and Belgium (Wang, 1992; Claeys et al., 1992; Claeys and Casier, 1994). In China the glass spherules are found 1.5 to 2 m.y. above the F/F

*Present addresses: Claeys, Institut für Mineralogie, Museum für Naturkunde, Invalidenstrasse 43, D-10015 Berlin, Germany; Herbosch, Earth and Environmental Sciences Department, CP 160, Free University of Brussels, 50 Avenue F. D. Roosevelt, B-1050 Brussels, Belgium.

Claeys, P., Kyte, F. T., Herbosch, A., and Casier, J-G., 1996, Geochemistry of the Frasnian-Famennian boundary in Belgium: Mass extinction, anoxic oceans and microtektite layer, but not much iridium?, *in* Ryder, G., Fastovsky, D., and Gartner, S., eds., The Cretaceous-Tertiary Event and Other Catastrophes in Earth History: Boulder, Colorado, Geological Society of America Special Paper 307.

boundary and are stratigraphically equivalent to a small Ir anomaly detected in the Canning Basin in Australia (Playford et al., 1984; Nicoll and Playford, 1990; Wang, 1992). In the Hony section in Belgium, the glass spherules occur a few centimeters above where Sandberg et al. (1988) placed the boundary (Claeys and Casier, 1994). In the Senzeille section in Belgium, 95 km southwest of Hony, the spherule layer is found in basal Famennian strata (Claeys et al., 1992), but the exact stratigraphic position remains unclear because of the lack of reliable conodont biostratigraphy in this section.

Previous studies had failed to show an Ir anomaly associated with the F/F boundary in European and American Late Devonian sections (Orth et al., 1990). The F/F boundary in the Sinsin section in Belgium was unsuccessfully investigated for Ir (McGhee et al., 1984), but that section has a small (40- to 50-cm) gap in sedimentation precisely at the F/F boundary (Streel and Vanguestaine, 1989; Casier and Devleeschouwer, 1994). This gap seems to correspond to the fine dark shale bed containing the spherule layer in Hony. Only in the basal Famennian of the Canning Basin has a small (300 pg/g) Ir anomaly been found in what was then believed to be the Upper *triangularis* conodont Zone (Playford et al., 1984). The increase in Ir was attributed to Ir extraction and concentration from seawater by *Frutexites* algae and bacteria (Playford et al., 1984). Recently, Nicoll and Playford (1990) reevaluated the stratigraphic position of the Ir anomaly and moved it to the lower *crepida* Zone, which is precisely the same stratigraphic level as the Chinese glass spherules.

This chapter presents a multielement geochemical profile across the F/F boundary in the Hony section. The goal of this investigation was to search for a positive Ir anomaly or any other geochemical anomaly associated with the dark shales containing the microtektite-like layer just above the F/F boundary.

THE F/F BOUNDARY MASS EXTINCTION AND THE HONY SECTION

The Hony section is located in the northeastern part of the Dinant Basin in Belgium (Fig. 1). The section provides a continuous exposure of upper Frasnian to lower Famennian strata consisting of olive-green to gray shales, frequently interrupted by beds of coarse bioclastic limestone a few centimeters to a few decimeters thick (Fig. 2). Locally, carbonate nodules of diagenetic origin occur in the shale. The carbonate bioclastic beds are graded, exhibit sharp or erosional bases, and are mainly formed by fragments of brachiopods, crinoids, and, locally, ostracodes. These bioclastic beds are typical storm deposits or tempestites (Flügel, 1982). The depositional setting appears to have been the outer shelf—well under fair-weather wave base but within the zone frequently reworked by storms.

The F/F boundary was recently defined by the International Sub-Commission on Devonian Stratigraphy as the base of the Lower *triangularis* conodont Zone, and the type section

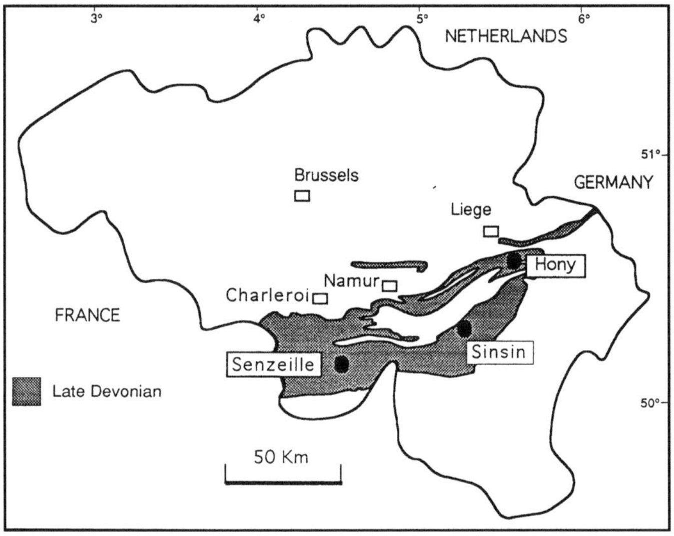

Figure 1. Map of Belgium showing Late Devonian strata and the location of the Hony (Long: E5°34′5″; Lat: N50°32′30″), Senzeille (Long: E4°27′30″; Lat: N50°09′42″), and Sinsin sections (Long: 5°14′00″; Lat: 50°16′40″).

was chosen at Coumiac in southern France. In Hony, the Lower *triangularis* conodont Zone is separated from the uppermost Frasnian *linguiformis* conodont Zone by approximately 150 cm of shale containing no identifiable conodont taxa (Sandberg et al., 1988). This interval has been studied and sampled in detail for the present study and is subdivided into two units (Fig. 2). At the base is 110 cm of massive gray shales containing thin (millimeter-scale) layers of storm-transported coquinoid accumulation. The intensity and frequency of the storm beds appears to decrease upward across the 110 cm. This is inferred from thickening of the massive shales, thinning of the interlayered coquina, and reduction in the size of the bioclasts. The upper unit consists of 40 cm of finer and more fissile dark-gray shales and probably represents a deepening of the depositional environment. In thin section these shales show well-preserved fine quartz silt laminae with some larger basal bioclasts, interpreted as characteristic of very distal tempestites. The shale lacks bioturbation, and its faunal content is low, except for a few ostracod valves. This unit is lithologically different from the overlying olive-green basal Famennian shales, which are more massive and contain abundant crinoids, brachiopods, and some ostracodes.

Sandberg et al., (1988) placed the F/F boundary at the base of this 40 cm dark-gray shale (Fig. 2), based on stratigraphic correlation with the Steinbruch Schmidt F/F boundary section in Germany. At Steinbruch Schmidt, the uppermost Frasnian *linguiformis* conodont Zone is marked by anoxic black bituminous limestones and shales of the upper Kellwasser horizon. This Kellwasser horizon is also recognized in the type section at Coumiac in the form of dark dolomitic mudstones with a positive δC^{13} (≈3‰) anomaly indicating

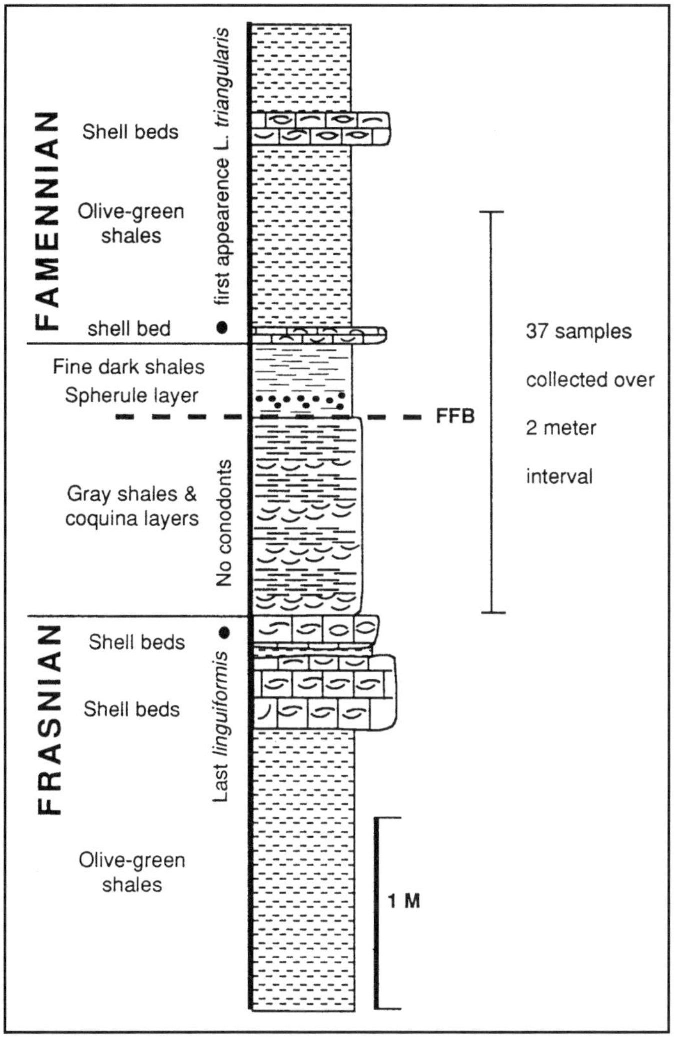

Figure 2. Stratigraphic and lithologic column of the Hony F/F boundary section showing the spherule layer, the changes in lithology, the conodont zonation, and the interval sampled for this study. Location of F/F boundary after Sandberg et al., 1988.

anoxic conditions (Joachimski and Buggisch, 1993). In Coumiac the F/F boundary is located directly above this anoxic horizon (Becker et al., 1989).

In the Hony section, the late Frasnian mass extinction occurred at the base of, or within, the 110-cm-thick bed of gray shales containing storm-generated coquinoid layers (Sandberg et al., 1988). Streel and Vanguestaine (1989) have demonstrated a strong decrease in the abundance of acritarchs, miospores, and other palynomorphs in the dark shale bed above the boundary. They attribute this reduction to a sudden increase in sedimentation rate. In Coumiac, extinction of organisms appears to begin 1 or 2 m below the boundary but seems to culminate in the upper Kellwasser horizon for goniatites, conodonts, and ostracodes and a few centimeters below this horizon for trilobites (Fig. 3 *in* Becker et al., 1989; Lethiers and Casier, 1995).

SAMPLING AND ANALYTICAL TECHNIQUES

This study of major and trace elements is probably one of the most extensive undertaken at the F/F boundary anywhere. The data presented in this study also contribute to the database on the composition of Paleozoic shales. The Hony section was selected for this study because (1) it contains glasslike microtektites just above the F/F boundary, (2) it is well constrained by conodont biostratigraphy, and (3) it probably represents continuous deposition across the F/F boundary.

The F/F boundary was sampled in great detail for this study; 37 samples were taken from a 2-m interval of shale beds above and below the boundary (Fig. 2). In the dark shale bed containing the spherule layer, the sampling interval was less than 5 cm. Farther away, samples were collected on 10- to 20-cm intervals. Five background samples were also taken several meters above and below the boundary. Split samples were analyzed for trace elements by neutron activation at UCLA and by X-ray fluorescence (XRF) for major and minor elements at the Free University of Brussels. The data were then subjected to multivariate statistical analyses.

X-ray fluorescence

Major and minor element concentrations (Si, Al, Fe, Mg, Ca, Na, K, P, S) were determined by X-ray wavelength dispersive spectrometry on pure rock powder pellets using a 3 Kw Philipps PW 1404 with Rh-side window tube. Calibrations were performed with international shales and soils standards (SB0-1, PRI-1, AWI-1, SO-1, SO-3, TB, TS, SGR-1, MAG-1, SDC-1, BCSS-1; Govindaraju, 1989). The standard error of estimate (root mean square on regression) for this method ranges from 2 to 5% for SiO_2, Al_2O_3, Fe_2O_3 (total), and K_2O to about 10% for MgO, Na_2O, P_2O_5, and S at the average concentration of these elements in the Hony shales. CaO measurements must be considered as semiquantitative because of their low value (0.4 wt% on average) in the Hony shales.

Neutron activation

The samples analyzed by neutron activation were split into two subsamples. The first split sample batch was irradiated at the reactor facility of the University of Missouri (25 hrs at 7.6×10^{13} n cm^{-2} s^{-1}). Ir was measured following Ir purification using a radiochemical neutron activation procedure similar to that of Kyte et al., (1992). After counting, the Ir carrier recovery was determined by ICP-emission spectroscopy. Ir yields were all above 26%. The second batch of samples was irradiated for 5 hrs in the University of California, Riverside, reactor, which uses a neutron flux of 1.8×10^{-12} n cm^{-2} s^{-1}. Samples were then analyzed for trace elements (Sc, Cr, Co, Ni, Zn, As, Rb, Sb, Ce, Ba, La, Ce, Nd, Sm, Eu, Tb, Yb, Lu, Hf, Ta, Th, U) by instrumental neutron activation. The samples were counted on coaxial high-purity intrinsic Ge crystal gamma-ray detectors. Samples were counted twice; the first time the week following irradiation and then 3 weeks after irradiation. U.S.G.S. standard rock powders SCO-1 and BHVO-1 and a powder of Allende meteorite were analyzed in parallel.

INTERPRETATION OF RESULTS

Ir analyses

The analytical results are presented in Table 1, Fig. 3 and in the appendix. No significant positive Ir anomaly was detected in the 2 m of section investigated (Fig. 3). The interpretation of the one point precisely at the spherule layer, where Ir increases up to 83 pg/g over a background of about 40 pg/g, is uncertain. This increase may correspond to a small Ir anomaly of unknown origin, but considering analytical uncertainties, it is probably not significant at the two sigma level. Two similar spot increases occur, at 18 cm above the F/F boundary in the dark shale unit and in the basal Famennian shales (Fig. 3). The Hony Ir results demonstrate that there is no Ir anomaly comparable in magnitude to that observed at the K/T boundary. This conclusion is in agreement with previous Ir measurements at the F/F boundary in the nearby Sinsin section (McGhee et al., 1984; Orth et al., 1990). The absence of an anomaly invalidates the hypothesis that an Ir anomaly existed at the F/F boundary but had not been detected in Sinsin because of the missing interval.

Major element analyses

Table 1 shows that on average the Hony shale is comparable in composition to several reference shales. Nevertheless some differences are worth mentioning. The major oxide composition, depleted in SiO_2 and enriched in Al_2O_3, K_2O, and Fe_2O_3, demonstrates that the Hony shales are enriched in a clay-mineral component compared to the reference shales. This is in agreement with the sedimentological interpretation that the depositional setting of the shales was the outer shelf, well below wave base and distal to major sediment sources. Calcium carbonate content is much lower than in reference shales and is probably related to low oxygen conditions that restricted the abundance of carbonate-producing organisms in the depositional environment. Diagenetic remobilization of $CaCO_3$ from bioclasts and its reprecipitation as limestone nodules, found locally in the shales, may also account for the low CaO content of the Hony shales. On average, the S also appears low compared to the value for shelf shales (Wedepohl, 1969). However, S is rarely measured in shales, and the literature value may not accurately represent the true range of S abundance. In the Hony samples, S concentrations vary greatly from 47 ppm to 536 ppm.

Trace element analyses

Trace element compositions are in the range of reference shales from the literature and for most elements have low standard deviations (Table 1). Zinc and As are the only elements that show a significant enrichment compared to reference shales; they also have high standard deviations and larger range when compared to the other trace elements analyzed (Table 1, columns 3, 4). The olive-green shales are homogeneous in trace element composition, whereas there are some significant variations for the darker shale layers across the F/F boundary. When elemental composition is plotted against the stratigraphic column (Fig. 4), a distinct enrichment in S, Zn, As, and Sb appears localized over a 60- to 80-cm interval across the F/F boundary.

This increase in chalcophile elements begins in the gray shale unit, 20 to 50 cm below the base of the F/F boundary, culminates in the finer and darker shales above the F/F boundary, and abruptly terminates just under the limestone beds containing the first Famennian conodonts. The enrichment appears to correlate with a marked variation in shale color from olive-green in the Frasnian and Famennian to dark-gray near the boundary. The four elements seem to be correlated despite their sawtooth pattern, but the rise in As, Zn, and Sb predates that of the S by ≈30 cm.

Rare earth element (REE) abundances in samples of the Hony shales are comparable (Fig. 5) to those of the North American Shale Composite (Taylor and McLennan, 1985). The absence of Ce fractionation probably indicates (Fig. 5) that shelf waters were too reduced for Mn coatings to form or that they were redissolved by deeper low-oxygen water. Under ambient oceanic conditions, dissolved Ce is usually oxidated and incorporated in Mn and Fe coatings (Sholkovitz et al., 1994).

Factor analysis

The data were analyzed by factor analysis with the SPSS/PC$^+$ program (Norusis, 1991) to clarify the nature of the complex interactions between variables and to confirm the qualitative trend observed for chalcophile elements. The aim of factor analysis is to reduce observed relationships among several variables to more fundamental relationships among fewer, newly established variables, called factors. Theoretical and practical aspects of factor analyses in geology are described in details in Jörskog et al. (1976).

The results of the factor analysis are summarized in Figure 6. This figure presents the six factors that account for 80% of the total variance of major and trace elements in the Hony shales and the importance, or loading, of each element in each factor. This schematic representation identifies groups of coherent elements and provides clues to their genetic significance. Factor analysis demonstrates that the distribution of the elements is not subject to important random variations in the Hony shales. Ir is the only exception: Only 25% of its total vari-

TABLE 1. HONY SHALE COMPOSITION COMPARED TO REFERENCE SHALES*

	Average	Hony Shale Std. Dev.	Range	Terrigenous Shale†	Mean Shale§	NASC**	Shelf Shale‡
Oxides in wt. %							
SiO_2	50.0	2.9	43 to 54	59	55.4	58.3	50.7
Al_2O_3	20.1	1.1	18 to 22	17.7	16.5	15.2	15.1
Fe_2O_{3t}	8.0	1.1	6.2 to 11.4	6.6	7.2	5.5	6.7
MgO	2.0	0.2	1.6 to 2.7	2.1	2.4	2.6	3.3
CaO	0.4	0.03	0.35 to 0.45	1.2	3.8	3.2	7.2
Na_2O	0.6	0.07	0.46 to 0.74	1.1	1	1	0.8
K_2O	5.0	0.4	4.1 to 5.7	3.5	3.1	3.6	3.5
P_2O_5	0.12	0.01	0.11 to 0.17	0.15	0.2	0.1	0.1
S (ppm)	163	134	47 to 536	n.a.	n.a.	n.a.	2400
LOI	n.a.			6	≈10	≈10	≈12
Elements in ppm							
Sc	20.1	1	18 to 22	16	13	14.9	
Cr	117	6	109 to 128	110	90	124.5	
Co	20	9	8 to 57	23	19	25.7	
Ni	64	18	28 to 99	55	68	58	
Zn	164	103	74 to 459	85	95	n.a.	
As	39	15	14 to 66	n.a.	13	28.4	
Rb	185	14	150 to 206	160	140	125	
Sb	1.3	0.4	0.69 to 2.02	n.a.	1.5	2.1	
Cs	8.3	1.4	6.3 to 11.0	15	5	5.2	
Ba	448	41	345 to 532	650	580	636	
La	46	3	38 to 53	38	92	31.1	
Ce	82	6	72 to 95	80	59	66.7	
Nd	38	5	30 to 55	32	24	27.4	
Sm	5.9	0.9	4.5 to 8.6	5.6	6.4	5.6	
Eu	1.3	0.4	0.75 to 2.7	1.1	1	1.18	
Tb	0.71	0.19	0.42 to 1.35	0.77	1	0.85	
Yb	2.51	0.33	1.9 to 3.2	2.8	2.6	3.06	
Lu	0.38	0.04	0.30 to 0.45	0.43	0.7	0.46	
Hf	4.1	0.3	3.5 to 4.6	5	2.8	6.3	
Ta	1.24	0.07	1.1 to 1.4	n.a.	0.8	1.12	
Th	14.1	0.7	12.6 to 15.7	14.6	12	12.3	
U	2.4	0.2	2.8 to 2.0	3.1	3.7	2.66	
Ir (pg/g)	50	14	19 to 83.5	20	n.a.	n.a.	

*Average of 37 samples of Hony shales.
†Taylor and McLennan, 1985.
§Turekian and Wedepohl, 1961.
**Gromet et al., 1984.
‡Wedepohl, 1969.
For Turekian and Wedepohl, Gromet et al., and Wedepohl, major element recalculated assuming a loss on ignition (LOI) of 10 wt% (Turekian and Wedepohl and Gromet et al.) and 12 wt% for Wedepohl. n.a. = data not available.

ance is explained by the six factors identified in Fig. 6. The remaining 75% of Ir variance is specific to that element.

It is not the goal of this chapter to discuss in detail the geological significance of the six different factors indentified. This will be the topic of another paper presently in preparation. The discussion here will focus on factor 2 (Fig. 6), which confirms and reinforces the chalcophile trend already recognized in the dark-gray shale beds across the F/F boundary.

Factor 2 is highly loaded by a coherent chalcophile group formed by As, Sb, and Zn. It is less loaded by P_2O_5 and only weakly so by S, U, and Fe_2O_{3t}. Factor 2 explains all the variance of As and Sb, and most of that of Zn and P. Except for P_2O_5 and U, all the elements influencing factor 2 are chalcophiles and concentrate in reducing environments in the form of sulfides and/or are adsorbed on organic compounds (Vine and Tourtelot, 1970; Berner, 1971; Calvert, 1976). Uranium is commonly enriched in anoxic sediments (Calvert, 1976). This enrichment is mainly due to reduction of soluble U^{6+} to rela-

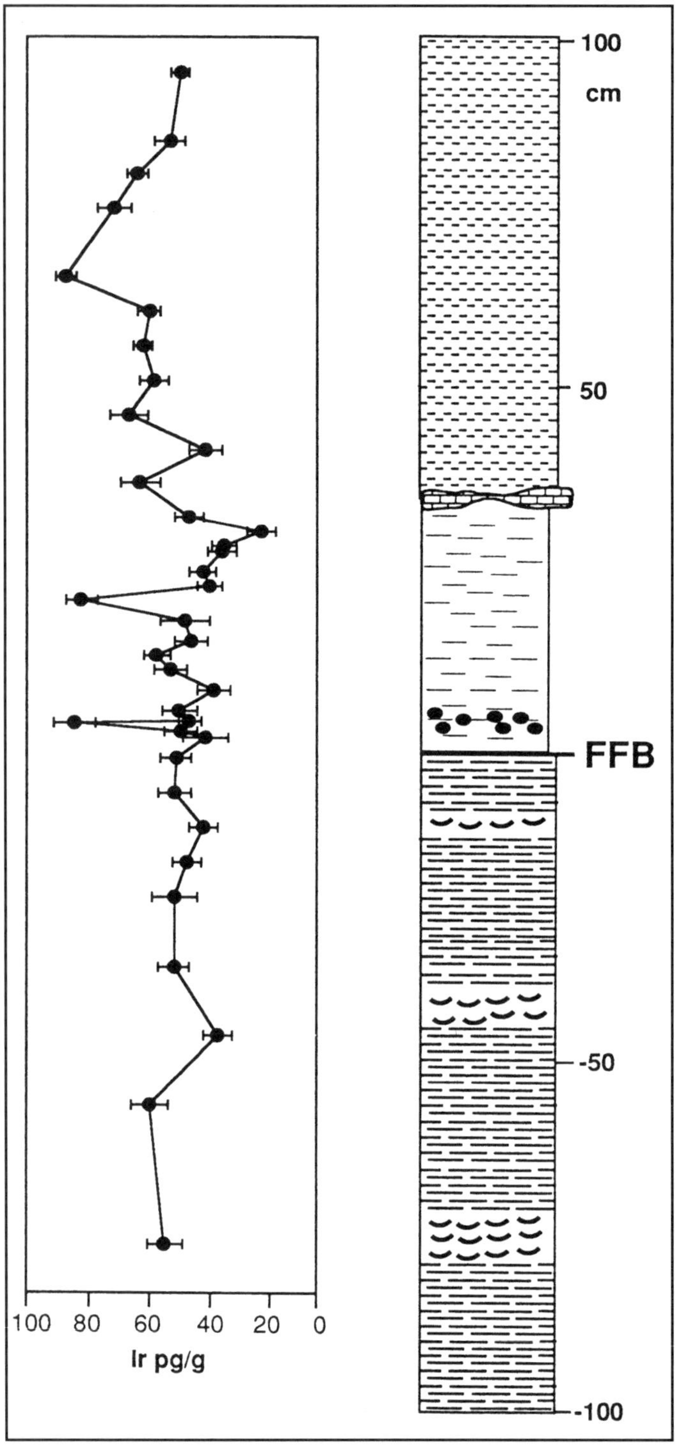

Figure 3. Ir profile across the studied interval. Error bar given at the 1 σ level. No significant Ir anomaly is associated with the spherule layer.

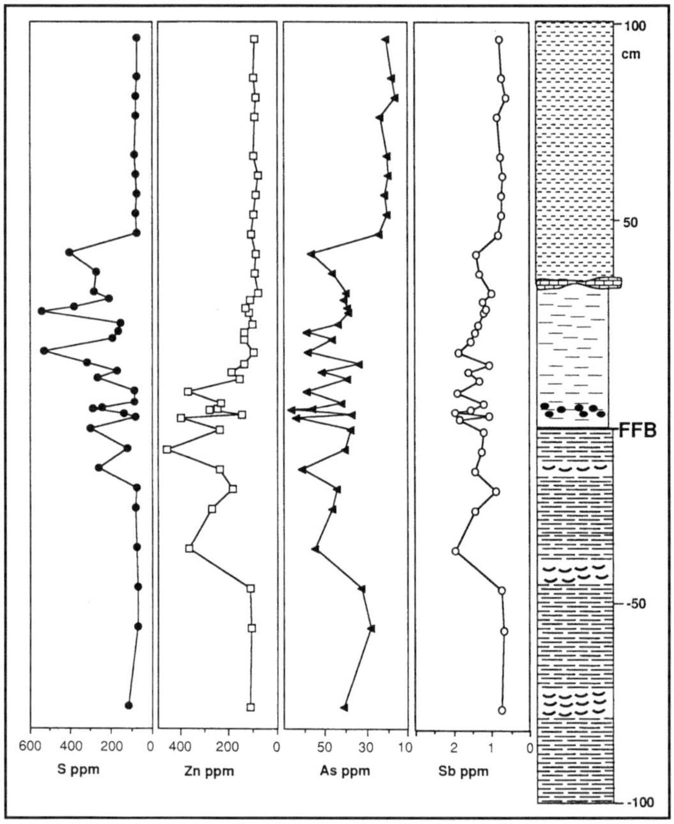

Figure 4. Profile of chalcophile elements (S, Zn, As, Sb) across the F/F boundary. The increase in chalcophiles begins in the gray shale bed, 20 to 50 cm below the boundary and culiminates in the dark shale bed just above the boundary. There is an abrupt return to normal shale content in chalcophile in basal Famennian shales. The increase in S, Zn, As, and Sb is interpreted to represent a significant decrease in oxygen conditions in the depositional environment.

tively insoluble U^{4+} (Klinkhammer and Palmer, 1991). The low U concentration (Table 1) and the weak saturation of factor 2 by U (Fig. 6) show that the depositional environment was dysoxic to suboxic (Tyson and Pearson, 1991) or that the sedimentation rate was too high for significant fixation of U to take place (Jones and Manning, 1994). The variance of U is mainly controlled by factor 1, which is interpreted to represent a global terrigenous-absorption clay factor. Phosphorus loading is probably due to P association with organic matter preserved in suboxic environments. The weak loading by Fe_2O_{3t} and S may indicate the sporadic occurrence of early diagenetic pyrite. Factor 2 can thus be interpreted as representing paleoredox conditions in the depositional environment.

The weak negative loading of factor 2 by Ba, Ir, Rb, and Yb (Fig. 6) may be due to a limited negative correlation between factor 2 and factor 1. These negative loadings must be considered an artifice induced by the use of an orthogonal instead of an oblique model for the factor rotation. They have no effect on the interpretation of factor 2.

The 2 m of F/F boundary section investigated in Hony appears to be characterized by generally limited oxygen conditions, as illustrated by the olive-green color of the shales and the local presence of rare pyrite crystals. Figure 7 shows the variation of factor 2 scores, representing a quantitative estimation of the reducing character of the depositional environment

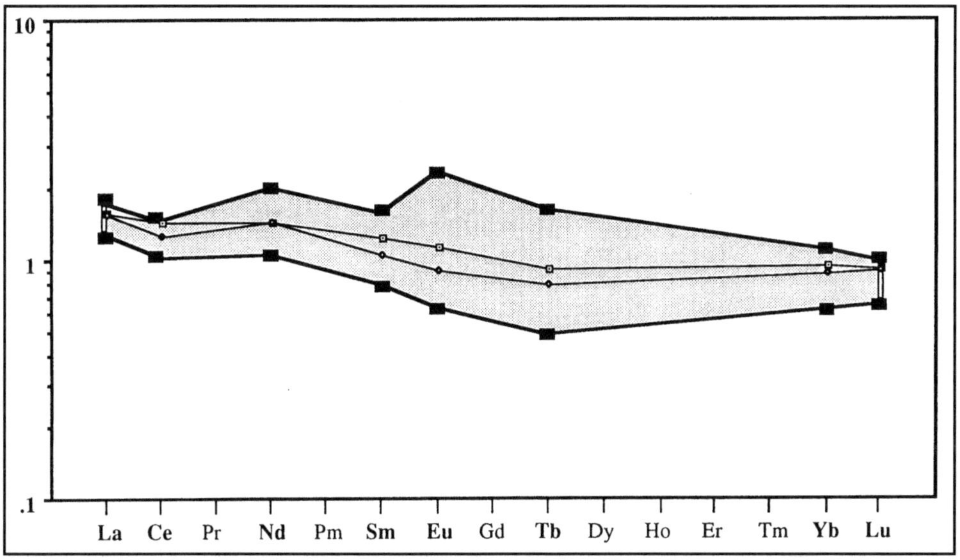

Figure 5. Range of rare earth elements normalized to the North American Shale Composite (NASC) (Gromet et al., 1984). The pattern is flat, indicating the similarity between the Hony shales and NASC. No Ce anomaly is present. Only the eight elements in boldface type were analyzed. The open squares represent a sample from the dark shales at the boundary, the diamonds a sample from the olive-green Famennian shales.

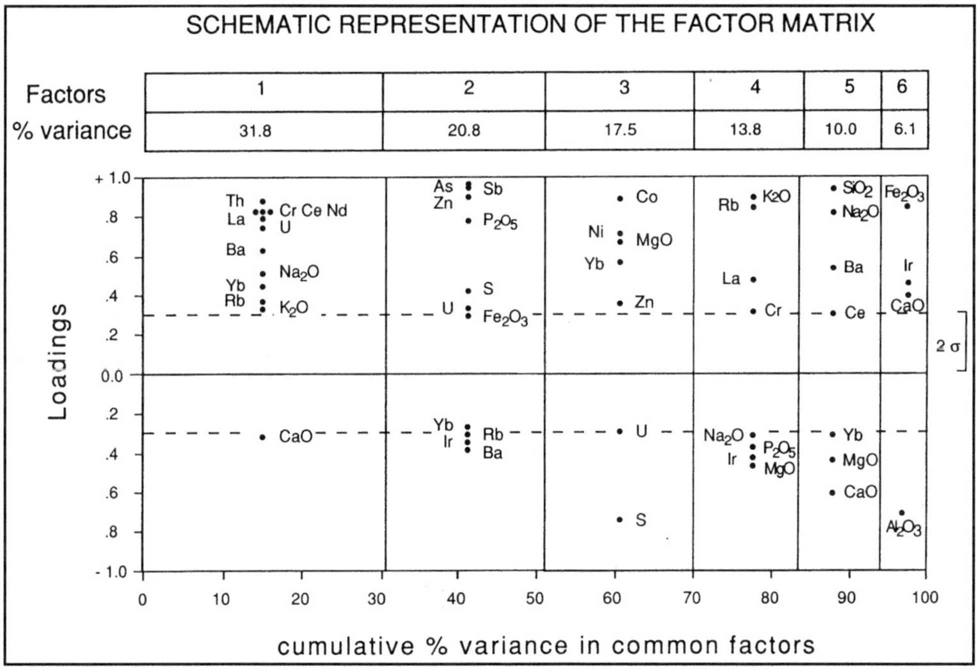

Figure 6. Schematic representation of the varimax factor matrix. Number of elements (variables): 24; number of samples: 37. Variables were $\log_{10}$ transformed, and this figure presents the simple structure obtained by orthogonal rotation using the varimax criteria (Norusis, 1991). The width of the rectangles is proportional to the variance of each factor. The dots represent the load of each element in each factor. Loadings below 0.3 are insignificant and not shown here. Factor 2 is heavily loaded with chalcophile elements and is interpreted to indicate reducing conditions in the depositional environment.

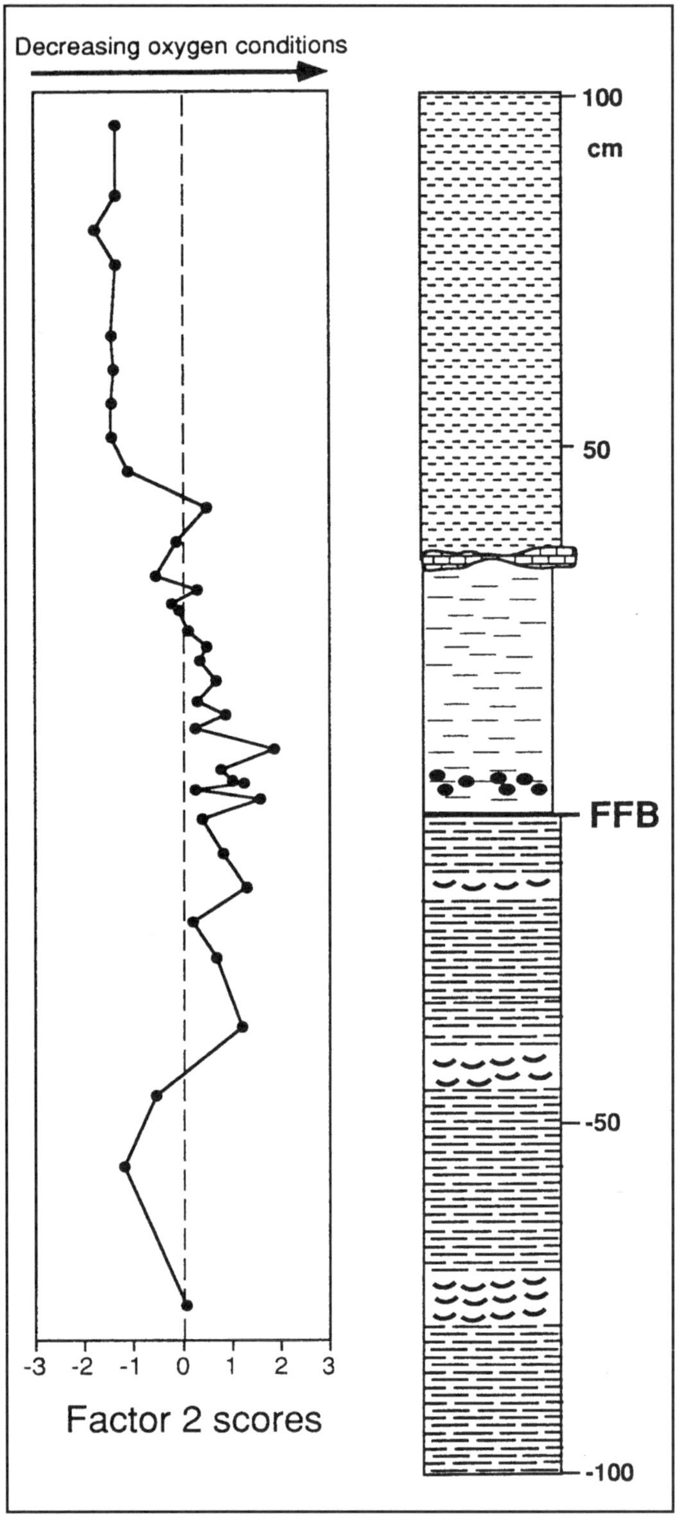

Figure 7. Factor 2 scores profile across the F/F boundary. The scores are presented in standardized values and represent the influence and intensity of factor 2 on each sample. Factor 2 is interpreted to represent reducing conditions in the depositional environment. Factor 2 scores quantatively confirm the reducing trend visible in Figure 4; more intense reducing conditions are in the dark shale bed above the F/F boundary.

for each sample across the F/F boundary. A 70 cm interval right at the boundary is marked by a net increase in low oxygen conditions (Figs. 4 and 7). This increase begins 10 to 20 cm below the F/F boundary and is most intense in the 40-cm, fine, dark, shale bed overlying the boundary. The overlying Famennian is marked by a rapid return to more oxygenated conditions.

DISCUSSION

No Ir anomaly comparable in magnitude to that of the K/T boundary was detected in the Hony section associated with the F/F boundary and microtektite-like spherule layer. The interpretation of the very minor Ir increase found in the spherule remains unclear and inconclusive. As in the Cenozoic, microtektite events do not seem to be associated with major Ir anomalies (Glass, 1990). The rare occurence of Ir anomalies associated with impact events in the sedimentary record was reviewed by Kyte (1988). A small peak in Ir (190 pg/g or ≈2 × the continuum level) has recently been identified in Ocean Drilling Project (ODP) core 758 B from the Ninetyeast Ridge, eastern Indian Ocean, associated with the 0.77-Ma Australasian microtektite strewnfield (Schmidt et al., 1993). In the Late Devonian, the Chinese microtektites apparently coincide with a 300 pg/g Ir anomaly in the Canning Basin in Australia (Wang, 1992). This Ir anomaly had been tentatively interpreted to have been induced by Ir concentration from seawater by *Frutexites* algae (Playford et al., 1984). This Ir peak is higher than that corresponding to the Australasian tektite event and suggests that the Famennian Canning Basin Ir anomaly is related to the impact that produced the Chinese microtektites. These Ir anomalies are more than an order of magnitude below the Ir level detected at the K/T boundary (Alvarez et al., 1980; Kyte et al., 1980) and in sediments recording a late Pliocene impact event in the southern Ocean (Kyte et al., 1988).

Trace element analyses combined with factor analysis in the Hony section indicate a change from oxygen-limited to dysaerobic or perhaps even anoxic conditions at the F/F boundary. Such a change can be caused by a rise of the oxygen minimum zone in the paleo-Tethys ocean and subsequent overflow of oxygen-depleted waters onto the shallow shelf environment. During the late Paleozoic, the paleo-Tethys ocean was probably strongly oxygen-stratified. Low oxygen conditions appear widespread in the paleo-Tethys at the F/F boundary (Buggisch, 1991). There is evidence of low-oxygen conditions at this stratigraphic level in the Canadian Rocky Mountains (Geldsetzer et al., 1987) and perhaps in south China (Wang et al., 1991). Joachimski and Buggisch (1993) described positive $\delta^{13}C$ anomalies indicative of anoxic depositional conditions just below the F/F boundary in the Coumiac (France) and Steinbruch Schmidt (Germany) sections. These anomalies coincide with the deposition of black limestones and shales in the upper Frasnian, known in Germany as the upper Kellwasser horizons. The anoxic Kellewasser horizons so far have not been identified in the Hony section. This study demonstrates that in Hony the F/F boundary is also characterized by an abrupt input of poorly oxygenated waters into shallow marine environments. This event seems less developed or perhaps less intense than in the Coumiac or Schmidt sections, as no truly anoxic or organic-rich bituminous facies formed in Hony. The development of anoxic conditions in the paleo-Tethys appears to vary according to local paleogeography, paleodepths, and paleoceanography.

The exact location of minimum oxygen conditions located in the bed of fine dark shales just above where Sandberg et al., (1988) have placed the boundary diverge between the Hony section and other F/F boundary sections in Western Europe. In other European sections low oxygen conditions are restricted to the upper Frasnian (Joachimski and Buggisch, 1993). This divergence may be due to the 1.5-m interval (Fig. 2) which lacks the detailed conodont biostratigraphy necessary to exactly locate the F/F boundary. The first appearence of Famennian conodonts is 35 cm above the F/F boundary in the first limestone bed topping this oxygen-depleted, dark-gray shale. The microtektite-like spherule layer coincides with the lowest oxygen conditions and occurs at the very base of this dark-gray shale bed (Claeys and Casier, 1994).

Anoxic conditions in the upper Frasnian have been hypothesized to be responsible for the extinction of organisms at the F/F boundary (Becker et al., 1991). This oxygen reduction may be related to rapid changes in sea level occurring at the boundary (Johnson et al., 1985; Sandberg et al., 1988; Becker et al., 1989). Both or either of these factors can potentially stress shallow marine organisms, in particular the large coral-stromatoporoid reefs, and lead to their disappearance. However, these factors are less likely to affect pelagic and deeper-water organisms.

The mechanisms causing these worldwide rapid sea-level fluctuations and changes in ocean chemistry have not yet been documented in detail. It is also unclear at this point how stratigraphically localized anoxic events like the Kellewasser horizons can explain the apparently stepwise extinction pattern over a time frame significantly longer than the brief anoxic periods. According to McGhee (1989), the extinction occurs over six conodont zones across the boundary.

The role of impact event(s) in the mass extinction is also unclear. Two impacts appear to have taken place within a few m.y. of the F/F boundary. Impact products in the form of microtektite-like glass particles have been identified in south China (Wang, 1992) and at two sites in Belgium (Claeys et al., 1992; Claeys and Casier, 1994). The Chinese spherules occur roughly 1.5 to 2 m.y. after the F/F boundary and coincide with a small Ir anomaly in the Canning Basin in Australia (Wang, 1992). At this point, neither of the two impact events appears to have a global distribution. Claeys and Casier (1994) have suggested that the Belgian glass spherules originated from the Siljan Ring impact structure in Sweden. This 54-km impact crater is unlikely to have had an effect comparable in magnitude to that of the much larger K/T boundary Chicxulub crater in Yucatan. On the other hand, the existence of two impacts within 1 to 2 m.y. perhaps suggests that at F/F time the Earth was bombarded by a comet shower (Hut et al., 1987). A hypothetical large oceanic impact could have taken place in the uppermost Frasnian, triggering mass extinctions, altering oceanic circulation, and raising the oxygen minimum zone in the ocean. If this was the case, tsunami deposits might be expected in the uppermost Frasnian. However, no strong evidence for tsunami deposits has so far been reported. Another hypothesis is that a succession of small impacts could induce local or global extinction of species if it coincided with a time when the shallow seas were already under severe anoxic stress. At this point, these scenarios remain purely speculative.

The relationships between possible impact event(s) taking place at or near the F/F boundary, the suboxic events, and the mass extinction need to be further clarified by paleontological, sedimentological, and geochemical studies. This may be a difficult task. The biostratigraphic resolution available for the Late Devonian (one conodont zone ≈0.5 Ma according to Ziegler and Sandberg, 1990), difficulties in worldwide correlations, and the restricted number of F/F boundary sections strongly limit the stratigraphic and temporal resolution that can be attained in the Late Devonian.

CONCLUSIONS

1. No Ir anomaly is associated with the F/F boundary or microtektite-like glass in the Hony section in Belgium. The lack of a positive Ir anomaly in Hony indicates that this impact differed, perhaps in magnitude or type of bolide, from the K/T boundary event and appears more similar to the four tektite-producing impacts that took place in the Cenozoic.

2. There is evidence in Hony of dysaerobic conditions at the F/F boundary. The Hony dark gray shales correlate with the upper Kellwasser anoxic event and further support the general anoxic trend identified at the F/F boundary at several other sites in the paleo-Tethys.

ACKNOWLEDGMENTS

This study was mainly funded by National Science Foundation grant EAR-9118701 to F. T. Kyte at UCLA. P. Claeys also acknowledges support from NASA grant NAGW-3008 and National Science Foundation grant EAR-91-05297 to W. Alvarez at UC Berkeley. A. Herbosch thanks the Belgian Fonds National de la Recherche Scientifique for a travel grant and J. L. Schlachmulder for help with the factor analyses. J-G. Casier and A. Herbosch are supported by a FRFC grant from the Fonds National de la Recherche Scientifique. The authors thank F. Asaro and B. Bohor for careful review and comments, which greatly improved this manuscript.

APPENDIX. NEUTRON ACTIVATION DATA

Sample No	Distance glass	SiO2 wt%	Al2O3 wt%	Fe2O3 wt%	MgO wt%	CaO wt%	Na2O wt%	K2O wt%	P2O5 wt%	S ug/g	Sc ug/g
92-Ho-25	75 cm below	52.4	19.6	7.7	2.1	0.4	0.7	4.9	0.2	92.0	20.8
92-Ho-25'	55 cm below	52.6	19.7	7.5	2.1	0.4	0.7	4.8	0.1	47.0	20.1
92-Ho-25"	45 cm below	53.3	19.3	7.8	2.2	0.4	0.7	4.7	0.1	49.0	20.6
92-Ho-25"'	35 cm below	51.7	19.3	8.1	2.2	0.4	0.7	4.7	0.1	51.0	20.7
92-Ho-26	25 cm below	52.8	19.1	7.8	2.1	0.4	0.7	4.5	0.1	61.0	20.7
92-Ho-26'	20 cm below	53.0	19.6	9.1	2.1	0.4	0.7	4.7	0.1	54.0	20.6
92-Ho-26"	15 cm below	51.6	21.3	6.6	1.9	0.4	0.7	4.4	0.1	248.0	18.5
92-Ho-26"'	10 cm below	50.7	19.9	7.7	1.8	0.4	0.7	5.5	0.1	102.0	20.1
93-Ho-01	5 cm below	51.3	21.3	7.9	2.1	0.4	0.7	4.8	0.1	290.0	19.6
92-Ho-26""	2 cm below	46.2	21.9	7.8	2.0	0.4	0.5	5.6	0.1	58.0	19.3
92-Ho-27	1 cm below	43.4	18.4	9.4	2.1	0.4	0.6	5.6	0.1	118.0	21.0
93-Ho-02	0 cm	52.4	19.5	8.5	1.8	0.4	0.7	4.3	0.1	275.0	18.6
93-Ho-03	0 cm	50.8	21.2	7.0	1.7	0.4	0.7	4.9	0.1	230.0	20.6
93-Ho-04	2 cm above	50.1	20.8	7.7	1.9	0.4	0.6	5.4	0.1	66.0	20.7
93-Ho-05	5 cm above	48.7	18.7	11.4	1.9	0.4	0.5	5.4	0.1	67.0	20.8
92-Ho-27'	8 cm above	43.5	19.7	8.4	2.7	0.5	0.5	4.1	0.1	253.0	20.5
93-Ho-06	10 cm above	49.5	21.2	7.6	2.0	0.4	0.6	4.6	0.1	152.0	18.9
93-Ho-07	12 cm above	47.2	18.2	10.3	2.0	0.4	0.6	5.4	0.1	307.0	20.3
93-Ho-08	15 cm above	48.8	19.3	10.0	1.9	0.5	0.5	4.5	0.1	523.0	19.0
92-Ho-27"	18 cm above	46.3	20.0	9.5	2.4	0.5	0.5	4.5	0.1	179.0	21.3
93-Ho-09	20 cm above	46.1	20.7	7.5	2.2	0.5	0.6	4.8	0.1	148.0	19.3
93-Ho-10	22 cm above	50.3	21.4	8.2	1.8	0.4	0.6	5.2	0.1	136.0	20.3
93-Ho-11	25 cm above	44.7	21.0	6.2	2.1	0.4	0.5	5.5	0.1	536.0	21.6
93-Ho-12	26 cm above	53.6	20.0	7.0	1.7	0.4	0.6	5.3	0.1	373.0	18.8
92-Ho-27"'	28 cm above	53.5	21.8	6.6	1.8	0.4	0.5	5.1	0.1	193.0	19.9
93-Ho-13	30 cm above	50.5	21.0	7.9	1.6	0.4	0.6	5.7	0.1	270.0	21.6
93-Ho-14	35 cm above	49.1	20.1	8.5	1.6	0.4	0.6	5.4	0.1	257.0	21.4
92-Ho-28	40 cm above	50.1	21.3	8.0	1.7	0.4	0.6	5.1	0.1	392.0	20.7
93-Ho-15	45 cm above	45.2	20.5	7.4	2.1	0.4	0.6	5.4	0.1	51.0	20.3
93-Ho-16	50 cm above	51.4	20.3	6.9	2.0	0.4	0.6	5.1	0.1	59.0	20.4
92-Ho-28'	55 cm above	54.3	20.3	7.7	1.9	0.4	0.6	4.9	0.1	55.0	17.6
93-Ho-17	60 cm above	50.4	20.4	7.4	2.0	0.5	0.6	5.1	0.1	60.0	19.1
92-Ho-28"	65 cm above	51.7	20.0	8.4	2.0	0.4	0.6	5.1	0.1	66.0	20.2
93-Ho-18	75 cm above	49.7	20.6	7.7	2.1	0.4	0.6	4.9	0.1	56.0	19.4
92-Ho-28"'	80 cm above	50.7	19.5	8.2	2.0	0.4	0.6	5.1	0.1	56.0	20.2
92-Ho-29	85 cm above	51.8	20.1	8.0	2.1	0.4	0.6	4.8	0.1	54.0	20.2
93-H0-19	95 cm above	51.1	18.2	6.8	2.1	0.4	0.6	4.9	0.1	52.0	20.4
92-Ho-02 BG	6 M below	nd	nd	nd	nd	nd	nd	nd	nd	nd	17.2
92-Ho-14 BG	3 M below	nd	nd	nd	nd	nd	nd	nd	nd	nd	18.8
92-Ho-37 BG	4 M above	nd	nd	nd	nd	nd	nd	nd	nd	nd	4.2
92-Ho-40 BG	6 M above	nd	nd	nd	nd	nd	nd	nd	nd	nd	19.9
93-Ho-20 BG	15 M above	nd	nd	nd	nd	nd	nd	nd	nd	nd	19.2
SCO-1		nd	nd	nd	nd	nd	nd	nd	nd	nd	11.5
BHVO		nd	nd	nd	nd	nd	nd	nd	nd	nd	29.9
Allende		nd	nd	nd	nd	nd	nd	nd	nd	nd	nd

APPENDIX. NEUTRON ACTIVATION DATA

		Cr	Co	Ni	Zn	As	Rb	Sr	Sb	Cs	Ba
Sample No	Distance glass	ug/g	ug/g	ug/g	ug/g	ug/g	ug/g	ug/g	ug/g	ug/g	ug/g
92-Ho-25	75 cm below	120.3	23.3	68.7	108.0	39.6	185.6	91.1	0.8	10.5	483.4
92-Ho-25’	55 cm below	112.6	12.6	75.1	101.4	26.0	180.3	86.3	0.8	9.6	532.1
92-Ho-25”	45 cm below		16.7	62.2	107.8	30.4	180.9	121.2	0.8	9.7	498.2
92-Ho-25”’	35 cm below	117.6	57.0	99.3	366.7	54.1	177.3	109.7	2.0	10.8	513.8
92-Ho-26	25 cm below	116.0	35.3	54.8	270.3	45.6	178.2	<82.5	1.5	8.6	493.1
92-Ho-26’	20 cm below	119.2	27.3	75.7	182.6	43.1	181.8	98.0	1.0	8.7	486.0
92-Ho-26”	15 cm below	111.5	16.5	67.7	237.0	60.5	163.8	86.2	1.5	6.5	414.9
92-Ho-26”’	10 cm below	123.1	18.9	54.0	459.4	39.0	202.0	170.9	1.3	7.5	450.0
93-Ho-01	5 cm below	118.8	10.2	34.8	236.5	35.7	173.6	<111.8	1.3	7.8	483.1
92-Ho-26””	2 cm below	112.4	20.4	52.2	403.7	63.7	189.2	<101.6	1.9	9.7	381.3
92-Ho-27	1 cm below	127.7	18.6	43.9	144.3	35.2	202.5	127.1	1.1	7.5	444.7
93-Ho-02	0 cm	108.5	13.2	76.8	279.9	65.7	154.3	<124.0	2.0	6.3	434.8
93-Ho-03	0 cm	120.0	13.4	74.5	243.4	55.2	179.4	<82.6	1.6	7.3	447.3
93-Ho-04	2 cm above	126.6	24.7	74.3	231.7	40.7	205.8	154.8	1.3	8.1	469.3
93-Ho-05	5 cm above	120.3	25.3	91.2	373.0	58.0	200.9	98.7	2.0	11.0	409.7
92-Ho-27’	8 cm above	112.9	20.3	100.3	151.2	37.5	150.3	<88.6	1.4	6.7	344.8
93-Ho-06	10 cm above	109.5	26.2	79.6	186.6	50.5	172.0	<82.7	1.7	7.8	374.2
93-Ho-07	12 cm above	124.8	12.3	55.6	131.0	31.9	194.9	<80.3	1.1	7.1	428.8
93-Ho-08	15 cm above	109.7	12.1	27.9	94.1	57.6	159.3	161.2	1.9	6.3	413.6
92-Ho-27”	18 cm above	118.1	24.8	62.0	132.3	45.3	166.8	123.9	1.6	7.5	392.7
93-Ho-09	20 cm above	109.0	19.4	83.5	130.1	58.0	174.3	<65.3	1.5	7.3	378.2
93-Ho-10	22 cm above	117.9	16.9	55.9	98.4	42.1	194.5	95.4	1.4	7.8	442.5
93-Ho-11	25 cm above	126.5	12.5	46.4	112.7	37.0	196.2	77.4	1.3	9.3	467.5
93-Ho-12	26 cm above	108.9	8.9	<35.5	129.1	37.8	183.2	103.1	1.2	6.6	416.0
92-Ho-27”’	28 cm above	109.0	14.2	76.0	106.7	39.4	189.0	<67.2	1.3	7.2	422.8
93-Ho-13	30 cm above	120.2	15.6	56.1	73.4	38.1	198.9	93.5	1.1	7.0	482.6
93-Ho-14	35 cm above	126.0	7.6	34.2	86.1	45.1	197.5	83.0	1.4	6.5	481.2
92-Ho-28	40 cm above	117.3	14.0	54.8	81.2	56.1	178.5	<56.6	1.5	6.7	430.1
93-Ho-15	45 cm above	119.4	21.9	42.5	101.5	21.7	194.6	74.6	0.9	10.3	481.9
93-Ho-16	50 cm above	115.8	19.3	66.4	91.5	17.8	199.7	113.5	0.8	8.2	451.6
92-Ho-28’	55 cm above	109.9	16.5	56.8	84.9	18.7	190.9	86.5	0.8	8.6	467.5
93-Ho-17	60 cm above	109.5	26.7	87.1	74.5	17.1	184.9	<53.9	0.8	10.3	443.0
92-Ho-28”	65 cm above	118.4	17.3	65.1	92.5	17.9	197.8	75.5	0.8	8.2	453.7
93-Ho-18	75 cm above	109.5	28.6	45.6	88.8	21.1	180.6	85.5	0.9	10.0	467.8
92-Ho-28”’	80 cm above	115.8	15.6	48.0	84.4	13.6	194.2	99.5	0.7	9.1	466.7
92-Ho-29	85 cm above	114.9	20.0	71.3	91.2	15.1	192.3	77.2	0.8	10.0	435.6
93-H0-19	95 cm above	116.8	21.6	73.9	90.3	18.3	191.8	72.0	0.9	10.4	487.9
92-Ho-02 BG	6 M below	110.3	22.1	64.4	74.4	12.4	165.9	113.0	0.8	8.3	787.6
92-Ho-14 BG	3 M below	106.4	22.7	49.0	108.5	6.9	182.0	114.4	1.1	10.4	526.3
92-Ho-37 BG	4 M above	8.0	11.2	24.1	25.1	9.7	14.1	449.4	0.6	3.0	76.1
92-Ho-40 BG	6 M above	113.8	21.1	71.5	82.3	11.9	205.6	105.1	1.0	13.3	429.0
93-Ho-20 BG	15 M above	110.9	17.6	78.9	71.0	12.8	198.3	113.8	0.9	11.5	398.3
SCO-1		69.0	11.0	26.1	94.1	10.9	104.9	181.3	2.4	7.4	527.8
BHVO		275.4	43.2	114.1	102.8	<1.1	13.5	365.8	0.2	<0.2	122.1
Allende		nd	nd	nd	nd	nd	nd	nd	nd	nd	nd

APPENDIX. NEUTRON ACTIVATION DATA

		La	**Ce**	**Nd**	**Sm**	**Eu**	**Tb**	**Yb**	**Lu**	**Hf**	**Ta**
Sample No	Distance glass	ug/g	ug/g	ug/g	ug/g	ug/g	ug/g	ug/g	ug/g	ug/g	ug/g
92-Ho-25	75 cm below	47.9	88.9	41.9	6.9	1.4	0.9	2.9	0.4	4.3	1.3
92-Ho-25'	55 cm below	47.5	85.5	55.1	7.6	1.6	0.9	2.8	0.4	4.1	1.2
92-Ho-25"	45 cm below	49.6	90.0	40.8	6.7	1.4	0.9	2.8	0.4	4.6	1.2
92-Ho-25'"	35 cm below	48.5	95.6	39.2	6.9	1.3	0.8	2.9	0.4	4.2	1.2
92-Ho-26	25 cm below	46.2	85.9	40.4	6.7	1.3	0.9	3.0	0.5	4.2	1.2
92-Ho-26'	20 cm below	48.1	87.2	43.0	6.0	1.1	0.8	2.8	0.4	4.2	1.3
92-Ho-26"	15 cm below	41.6	75.3	35.9	4.7	0.9	0.5	2.0	0.3	4.0	1.1
92-Ho-26'"	10 cm below	47.2	83.8	39.3	7.2	1.7	0.8	2.6	0.4	3.8	1.2
93-Ho-01	5 cm below	49.5	85.9	40.3	5.8	1.1	0.7	2.6	0.4	4.5	1.3
92-Ho-26""	2 cm below	48.3	82.1	38.7	6.1	1.4	0.7	2.4	0.4	3.7	1.2
92-Ho-27	1 cm below	50.9	88.9	46.2	6.7	1.5	0.7	2.4	0.4	4.8	1.4
93-Ho-02	0 cm	42.3	76.3	31.1	4.8	0.8	0.4	2.0	0.3	4.5	1.1
93-Ho-03	0 cm	43.8	77.4	35.7	5.1	1.0	0.5	2.1	0.3	4.1	1.2
93-Ho-04	2 cm above	46.5	82.5	37.4	5.5	1.0	0.5	2.1	0.3	3.8	1.3
93-Ho-05	5 cm above	48.7	84.0	38.1	6.2	1.3	0.6	2.4	0.3	3.8	1.2
92-Ho-27'	8 cm above	37.6	67.1	31.4	7.4	2.2	1.2	3.1	0.4	3.7	1.2
93-Ho-06	10 cm above	43.6	76.6	32.7	5.3	1.0	0.6	2.3	0.4	4.1	1.2
93-Ho-07	12 cm above	49.2	84.1	36.9	6.0	1.3	0.6	2.5	0.4	4.8	1.4
93-Ho-08	15 cm above	47.4	77.2	41.0	5.5	1.3	0.6	2.2	0.3	4.2	1.1
92-Ho-27"	18 cm above	43.3	72.3	29.8	8.6	2.7	1.3	3.2	0.4	4.1	1.2
93-Ho-09	20 cm above	44.0	72.4	35.4	7.1	2.3	1.1	2.4	0.4	3.8	1.2
93-Ho-10	22 cm above	46.1	76.8	32.4	5.9	1.6	0.7	2.2	0.3	3.8	1.3
93-Ho-11	25 cm above	48.2	75.7	40.3	5.3	1.3	0.6	2.3	0.4	4.2	1.3
93-Ho-12	26 cm above	44.7	74.0	29.5	4.5	1.0	0.4	1.9	0.3	3.7	1.2
92-Ho-27'"	28 cm above	44.3	73.0	30.4	5.1	1.3	0.6	2.1	0.3	3.5	1.1
93-Ho-13	30 cm above	48.2	86.6	41.1	5.1	0.9	0.6	2.6	0.4	4.2	1.4
93-Ho-14	35 cm above	53.2	95.0	48.3	5.5	0.8	0.6	2.5	0.4	4.5	1.4
92-Ho-28	40 cm above	48.1	86.1	39.1	4.8	0.8	0.7	2.6	0.4	4.2	1.3
93-Ho-15	45 cm above	47.0	83.4	37.3	6.0	1.2	0.8	2.8	0.4	4.4	1.3
93-Ho-16	50 cm above	47.2	82.8	39.5	5.4	1.0	0.7	2.7	0.4	4.4	1.2
92-Ho-28'	55 cm above	42.4	75.3	35.2	5.0	0.9	0.5	2.1	0.3	4.2	1.2
93-Ho-17	60 cm above	44.2	81.3	33.9	5.7	1.1	0.8	2.5	0.4	3.8	1.2
92-Ho-28"	65 cm above	47.9	84.0	39.5	5.8	1.1	0.7	2.7	0.4	4.1	1.3
93-Ho-18	75 cm above	44.5	78.6	36.9	5.3	1.0	0.7	2.7	0.4	3.9	1.2
92-Ho-28'"	80 cm above	46.9	82.9	37.6	5.3	1.1	0.7	2.6	0.4	4.2	1.3
92-Ho-29	85 cm above	46.2	82.4	36.3	5.6	1.1	0.8	2.7	0.4	4.2	1.3
93-H0-19	95 cm above	46.0	80.9	34.6	5.4	1.0	0.7	2.6	0.4	4.3	1.3
92-Ho-02 BG	6 M below	44.8	79.7	32.6	5.4	1.1	0.7	2.5	0.4	4.5	1.0
92-Ho-14 BG	3 M below	42.7	78.1	30.8	6.1	1.3	0.7	2.5	0.4	3.9	1.1
92-Ho-37 BG	4 M above	9.4	25.8	19.9	6.0	1.9	0.8	1.2	0.2	0.3	0.1
92-Ho-40 BG	6 M above	44.1	76.9	34.5	5.1	1.0	0.7	2.3	0.4	3.4	1.1
93-Ho-20 BG	15 M above	44.7	77.4	35.1	5.4	1.0	0.7	2.4	0.4	3.8	1.0
SCO-1		29.5	49.7	28.5	4.6	1.1	0.7	2.2	0.3	4.4	0.7
BHVO		14.9	32.0	23.1	5.4	2.0	0.9	1.9	0.3	4.0	1.0
Allende		nd	nd	nd	nd	nd	nd	nd	nd	nd	nd

APPENDIX. NEUTRON ACTIVATION DATA

Sample No	Distance glass	Th ug/g	U ug/g	Ir pg/g
92-Ho-25	75 cm below	14.7	2.6	52.6
92-Ho-25'	55 cm below	13.8	2.4	57.9
92-Ho-25"	45 cm below	14.5	2.4	34.5
92-Ho-25'"	35 cm below	14.2	2.3	49.6
92-Ho-26	25 cm below	14.3	2.4	49.4
92-Ho-26'	20 cm below	14.2	2.4	45.0
92-Ho-26"	15 cm below	13.9	2.5	39.2
92-Ho-26'"	10 cm below	14.1	2.4	49.4
93-Ho-01	5 cm below	14.5	2.7	48.9
92-Ho-26""	2 cm below	13.3	2.3	38.5
92-Ho-27	1 cm below	15.5	2.4	47.2
93-Ho-02	0 cm	13.7	2.6	83.5
93-Ho-03	0 cm	14.4	2.6	44.1
93-Ho-04	2 cm above	14.0	2.5	47.7
93-Ho-05	5 cm above	13.7	2.4	36.0
92-Ho-27'	8 cm above	12.8	2.2	50.6
93-Ho-06	10 cm above	14.1	2.2	55.4
93-Ho-07	12 cm above	15.4	2.5	43.8
93-Ho-08	15 cm above	13.5	2.5	45.8
92-Ho-27"	18 cm above	13.8	2.3	81.4
93-Ho-09	20 cm above	13.0	2.0	37.1
93-Ho-10	22 cm above	13.9	2.2	39.6
93-Ho-11	25 cm above	14.7	2.6	33.2
93-Ho-12	26 cm above	13.0	2.0	32.3
92-Ho-27'"	28 cm above	12.6	2.1	19.3
93-Ho-13	30 cm above	14.9	2.5	44.2
93-Ho-14	35 cm above	15.7	2.8	61.2
92-Ho-28	40 cm above	15.1	2.6	38.5
93-Ho-15	45 cm above	14.5	2.3	65.0
93-Ho-16	50 cm above	14.0	2.4	56.3
92-Ho-28'	55 cm above	13.7	2.2	60.3
93-Ho-17	60 cm above	13.3	2.1	58.2
92-Ho-28"	65 cm above	14.3	2.2	86.5
93-Ho-18	75 cm above	13.3	2.2	69.9
92-Ho-28'"	80 cm above	14.0	2.3	62.2
92-Ho-29	85 cm above	13.9	2.3	51.1
93-H0-19	95 cm above	14.0	2.3	47.4
92-Ho-02 BG	6 M below	12.1	2.4	59.5
92-Ho-14 BG	3 M below	13.6	2.0	62.7
92-Ho-37 BG	4 M above	1.9	0.6	15.7
92-Ho-40 BG	6 M above	12.9	2.0	59.4
93-Ho-20 BG	15 M above	12.7	1.9	63.7
SCO-1		8.8	2.7	nd
BHVO		1.1	0.5	nd
Allende		nd	nd	189.5

REFERENCES CITED

Alvarez, L. W., Alvarez, W., Asaro, F., and Michel, H. V., 1980, Extraterrestrial causes for the Cretaceous-Tertiary extinction: Science, v. 208, p. 1095–1108.

Becker, T. R., Feist, R., Flajs, G., House, M. R., and Klapper G., 1989, Frasnian-Famennian extinction events in the Devonian at Coumiac France: Comptes Rendus Académie des Sciences de Paris, t. 309, ser. 2, p. 259–266.

Becker, T. R., House, M. R., Kirchgasser, W. T., and Playford, P. E., 1991, Sedimentary and faunal changes across the Frasnian-Famennian boundary in the Canning Basin of Western Australia: Historical biology, v. 5, p. 183–196.

Berner, R. A., 1971, Principles of chemical sedimentology: New York, McGraw Hill, 240 p.

Buggisch, W., 1991, The global Frasnian-Famennian Kellwasser event: Geologische Rundschau, v. 80, p. 49–72.

Calvert, S. E., 1976, The mineralogy and geochemistry of near-shore sediments. *in* Riley, J. R., and Chester, R., eds., Chemical Oceanography, Volume 6: New York, Academic Press, 187 p.

Casier, J.-G., and Devleeschouwer, X., 1994, Ostracods and the Upper Devonian mass extinction in the southern border of the Dinant Basin (Belgium): International Symposium on Ostracoda, 12th, Prague, Abstracts, p. 21.

Claeys, P., and Casier, J.-G., 1994, Microtektite-like impact glass associated with the Frasnian-Famennian boundary mass extinction: Earth and Planetary Science Letters, v. 122, p. 303–315.

Claeys, P., Casier, J. G., and Margolis, S. V., 1992, Microtektite glass at the Frasnian/Famennian boundary in Belgium: Evidence for an asteroid impact?: Science, v. 257, p. 1102–1104.

Flügel, E., 1982, Microfacies analysis of limestones: Berlin, Springer-Verlag, 633 p.

Geldsetzer, H. H. J., Goodfellow, W. D., McLaren, D. J., and Orchard, M. J., 1987, Sulfur-isotope anomaly associated with the Frasnian-Famennian extinction in Medicine Lake, Alberta, Canada: Geology, v. 15, p. 393–396.

Glass, B. P., 1990, Tektites and microtektites: Key facts and interferences: Tectonophysics, v. 171, p. 393–404.

Govindaraju, K., 1989, Compilation of working values and sample description for 272 geostandards: Geostandards Newsletter, v. 13, special issue, 30 p.

Gromet, P. L., Dymek, R. F., Haskin, L. A., and Korotev, R. L., 1984, The "North American shale composite": Its compilation of major and trace element characteristics: Geochimica et Cosmochimica Acta, v. 48, p. 2469–2482.

Hut, P., Alvarez, W., Elder, W. P., Hansen, T., Kauffman, E. G., Keller, G., Shoemaker, E. M., and Weissman, P. R., 1987, Comet as a cause of mass extinction: Nature, v. 328, p. 118–126.

Joachimski, M. M., and Buggisch, W., 1993, Anoxic events in the Late Frasnian—Causes of the Frasnian-Famennian faunal crisis?: Geology, v. 21, p. 675–679.

Johnson, J. G., Klapper, G., and Sandberg, C. A., 1985, Devonian eustatic fluctuations in Euramerica: Geological Society of America Bulletin, v. 96, p. 567–587.

Jones, B., and Manning, D., 1994, Comparison of geochemical indices used for the interpretation of paleoredox conditions in ancient mudstones: Chemical Geology, v. 11, p. 111–129.

Jörskog, K. G., Klovan, J. E., and Reyment, R. A., 1976, Geological factor analysis. Methods in geomathematics: Amsterdam, Elsevier, 178 p.

Klinkhammer, G. P., and Palmer, M. R., 1991, Uranium in the oceans: Where it goes and why: Geochimica et Cosmochimica Acta, v. 55, p. 1799–1806.

Kyte, F. T., 1988, The extraterrestrial component in marine sediments: Description and interpretation: Paleoceanography, v. 3, p. 235–247.

Kyte, F. T., Zhou, L., and Wasson, J. T., 1980, Siderophile-enriched sediments from the Cretaceous-Tertiary boundary: Nature, v. 288, p. 651–656.

Kyte, F. T., Zhou, L., and Wasson, J. T., 1988, New evidence on the size and possible effects of a Late Pliocene oceanic asteroid impact: Science, v. 241, p. 63–65.

Kyte, F. T., Leinen, M., Heath, G. R., and Zhou, L., 1992, Cenozoic sedimentation history of the central North Pacific: Inferences from the elemental geochemistry of core LL44-GPC3: Geochimica et Cosmochimica Acta, v. 57, p. 1719–1740.

Lethier, S. F., and Casier, J.-G., 1995, The uppermost Frasnian (Upper Kellwasser) ostracodes from Coumiac (Montagne Noire, France): Revue de Micropaléontologie, v. 38, p. 65–79.

McGhee, G. R., 1989, The Frasnian/Famennian extinction event, *in* Donovan, S. K., ed., Mass extinction: Processes and evidences: London, Belhaven Press, p. 133–151.

McGhee, G. R., Orth, C. J., Gilmore, J. S., and Olsen, E. J., 1984, No geochemical evidence for an asteroidal impact at Late Devonian mass extinction horizon: Nature, v. 308, p. 629–631.

McLaren, D. J., 1982, Frasnian-Famennian extinctions, *in* Silver, L. T., and Schultz, P. H., eds., Geological implications of impacts of large asteroids and comets on the Earth: Geological Society of America Special Paper 190, p. 477–483.

Nicoll, R. S., and Playford, P. E., 1990, Upper Devonian iridium anomaly and the Frasnian-Famennian boundary in the Canning Basin, Western Australia: Geological Society of Australia Abstracts, v. 21, p. 296.

Norusis, M. J., 1991, The SPSS guide to data analysis for SPSS/PC+: Chicago, SPSS Inc., 87 p.

Orth, C. J., Attrep, M., and Quitana, L. R., 1990, Iridium abundance patterns across bioevent horizons in the fossil record, *in* Sharpton, V. L., and Ward, P. D., eds., Global catastrophes in Earth history; an interdisciplinary conference on impact, volcanism and mass mortality: Geological Society of America Special Paper 247, p. 45–59.

Playford, P. E., McLaren, D. J., Orth, C. J., Gilmore, J. S., and Goodfellow, W. D., 1984, Iridium anomaly in the Upper Devonian of the Canning Basin, Western Australia: Science, v. 226, p. 437–439.

Sandberg, C. A., Ziegler, W., Dreesen, R., and Butler, J. L., 1988, Late Frasnian mass extinction: Conodont event stratigraphy global changes and possible causes: Courier Forschungsinstitut Senckenberg, v. 102, p. 267–307.

Schmidt, G. L., Zhou, L., and Wasson, J. T., 1993, Iridium anomaly associated with the Australasian tektite-producing impact: Masses of the impactor and of the Australasian tektites: Geochimica et Cosmochimica Acta, v. 57, p. 4851-4859.

Sepkoski, J. J., Jr., 1982, Mass extinction in the Phanerozoic oceans: A review, *in* Silver, L. T., and Schultz, P. H., eds., Geological implications of impacts of large asteroids and comets on the Earth: Geological Society of America Special Paper 190, p. 283–289.

Sholkovitz, E. R., Landing, W. M., and Lewis, B. L., 1994, Ocean particles chemistry: The fractionation of rare earth elements between suspended particles and seawater: Geochimica et Cosmochimica Acta, v. 58, p. 1567–1579.

Streel, M., and Vanguestaine, M., 1989, Palynomorph distribution in a siliciclastic layer near the Frasnian-Famennian boundary at two shelf facies localities in Belgium: Bulletin Societe Belge Geologie, v. 98, p. 109–114.

Taylor, S. R., and McLennan, S. M., 1985, The continental crust: Its composition and evolution: Oxford, Blackwell, 312 p.

Tyson, R. V., and Pearson, T. H., 1991, Modern and ancient continental shelf anoxia: An overview, *in* Tyson, R. V., and Pearson, T. H., eds. Modern and ancient continental shelf anoxia: Geological Society of London Special Publication 58, p. 1–24.

Turekian, K. K., and Wedepohl, K. H., 1961, Distribution of the elements in some major units of the Earth crust: Geological Society of America Bulletin, v. 72, p. 175–192.

Vine, J. D., and Tourtelot, E. B., 1970, Geochemistry of black shale deposits: A summary report: Economic Geology, v. 65, p. 253–275.

Wang, K., 1992, Glassy microspherules (microtektites) from an Upper Devonian Limestone: Science, v. 256, p. 1546–1549.

Wang, K., Orth, C. J., Attrep, M., Jr., Chatterton, B. D. E., Hou, H., and Geldstzer, H. H. J., 1991, Geochemical evidence for a catastrophic biotic event at the Frasnian/Famennian boundary in south China: Geology, v. 19, p. 776–779.

Wedepohl, K. H., 1969, Composition and abundance of common sedimentary rocks, *in* Wedepohl, K. H., ed., Handbook of geochemistry, Volume 1: Berlin, Springer-Verlag, p. 250–269.

Ziegler, W., and Sandberg, C. A., 1990, The Late Devonian standard conodont zonation: Courier Forschunginstitut Senckenberg, v. 121, 115 p.

MANUSCRIPT ACCEPTED BY THE SOCIETY SEPTEMBER 14, 1995

Printed in U.S. A.

Geological Society of America
Special Paper 307
1996

The Pliensbachian-Toarcian (Lower Jurassic) extinction event

Crispin T. S. Little
Department of Geology, University of Bristol, Bristol BS8 1RJ, United Kingdom

ABSTRACT

The Pliensbachian-Toarcian boundary (187 Ma) has been proposed as one of eight periodically recurring extinction events in the past 250 m.y. Detailed biostratigraphic sampling of invertebrate macrofaunas through complete, expanded sequences in North Yorkshire, England and in southwest Germany shows that a species-level event occurs in the Early Toarcian, some 1 m.y. later. This event is intimately associated with an oceanic anoxic event in northwest Europe. This regional event is part of a global phase of elevated extinctions spanning c. 7.5 m.y. of the late Pliensbachian to early Toarcian.

INTRODUCTION

In the fossil record of marine families and genera during the last 250 m.y., Raup and Sepkoski (1986) have identified eight events in which extinction rates were significantly higher than background rates. The authors postulate that these extinctions were spaced with a mean 26 m.y. periodicity. The periodicity itself suggests a single recurring causative mechanism, such as bolide impacts, or terrestrial processes such as volcanism or sea-level change. Much multidisciplinary research effort has been expended during the past decade on the largest of these extinctions at the Permo-Triassic and Cretaceous-Tertiary (K/T) boundaries. The other five smaller-scale events have been relatively ignored. However, their validity is essential for the 26 m.y. hypothesis (Raup and Sepkoski, 1986; Sepkoski,1990, 1993). One of these smaller extinction events is at the Pliensbachian-Toarcian boundary (Lower Jurassic, 187 Ma), a peak implied by the 26 m.y. periodicity hypothesis. Hallam (1986) has suggested that this event is not a "mass" extinction at all, having had a regional rather than a global effect. In a higher-resolution reanalysis of all Early Jurassic marine families, Little and Benton (1995) have found high rates of extinction in a five-zone extinction phase lasting c. 7.5 m.y. across the late Pliensbachian–early Toarcian, rather than there being a single event at the Pliensbachian-Toarcian boundary.

*Present address: Palaeontology Department, The Natural History Museum, London SW7 5BD, United Kingdom.

The aim of this study is to gain a better understanding of the timing and magnitude of the Pliensbachian-Toarcian event. To this end, 10 Pliensbachian-Toarcian sections in northwest Europe have been sampled in detail. The paleontological and sedimentological evidence from the best of these sections spanning the boundary, in northern England and southwestern Germany, is given. The extinction patterns are compared against evidence for a major anoxic event in northwest Europe during the early Toarcian. Sections further afield are then briefly reviewed for a global perspective.

DATING

One of the most important considerations when undertaking a study of this kind is correlation between sections in different regions. Unfortunately, isotopic records (K-Ar, Rb-Sr, etc.) for the Pliensbachian and Toarcian stages are extremely poor, the Pliensbachian-Toarcian boundary age of 187 Ma having been interpolated (Harland et al., 1990). The fact that all Lower Jurassic oceanic crust has been subducted, coupled with the absence of extrusive igneous rocks in the marine Pliensbachian-Toarcian sections, makes magnetostratigraphic correlation impossible. At present, correlation can best be achieved using the excellent ammonite biostratigraphic scheme, first developed by Quenstedt and others over 130 years ago, for the European Jurassic (Dean et al., 1961; Haq et al., 1988; Harland et al., 1990). Across this area, the resolution is down to the subzonal

Little, C. T. S., 1996, The Pliensbachian-Toarcian (Lower Jurassic) extinction event, *in* Ryder, G., Fastovsky, D., and Gartner, S., eds., The Cretaceous-Tertiary Event and Other Catastrophes in Earth History: Boulder, Colorado, Geological Society of America Special Paper 307.

level of timescales of perhaps 500,000 years, which is considerably better than can be obtained using planktonic microfossil zones (Harland et al., 1990).

As with all biostratigraphic schemes, there are problems of paleobiogeographic provinciality. During the Pliensbachian and Toarcian stages ammonite faunas were roughly separated into the northern Boreal Realm and the southern Tethyan Realm (Hallam, 1975; Smith and Tipper, 1986). Shared and immigrant genera allow correlation between realms, but only to the zonal level of timescales of perhaps 1 m.y.

Figure 1 shows the biostratigraphic and lithostratigraphic schemes for the sections in northern England and southwest Germany. The *tenuicostatum* Subzone is missing in Germany; a parasequence boundary at this level in the Yorkshire section suggests that this absence may be due to omission in the area. The extra subzones in the German *falciferum* Zone can also be recognized in the Yorkshire area but have not been formally described.

RECORD OF PLIENSBACHIAN-TOARCIAN SECTIONS

Northern England

The Pliensbachian-Toarcian Cleveland Basin sections on the Yorkshire coast in northern England (Fig. 2) are some of the thickest and most complete in Europe (Howarth, 1955, 1962, 1973; Powell, 1984; Howard, 1985). The studied section consists

	AMMONITE BIOSTRATIGRAPHY			LITHOSTRATIGRAPHY		
	ZONES	SUBZONES Britain	SUBZONES Germany	N. YORKSHIRE Formations	N. YORKSHIRE Members	S-W GERMANY
TOARCIAN (PART)	*Hildoceras bifrons*	*Catacoeloceras crassum*	*Catacoeloceras crassum*	Whitby Mudstone 71.38 m	Alum Shale	Posidonienschiefer 6 - 14.8 m
		Peronoceras fibulatum	*Peronoceras fibulatum*			
		Dactylioceras commune	*Dactylioceras commune*			
	Harpoceras falciferum	*Harpoceras falciferum*	*Harpoceras falciferum*		Jet Rock	
		Cleviceras exaratum	*Cleviceras elegans*			
			Cleviceras exaratum			
			Eleganticeras elegantulum			
	Dactylioceras (Orthodactylites) tenuicostatum	*D. (O.) semicelatum*	*D. (O.) semicelatum*		Grey Shale	
		D. (O.) tenuicostatum	Absent			
		D. (O.) clevelandicum	*D. (O.) clevelandicum*			
		Protogrammoceras paltum	*Protogrammoceras paltum*			
U. PLIENSBACHIAN	*Pleuroceras spinatum*	*Pleuroceras hawskerense*	*Pleuroceras hawskerense*	Cleveland Ironstone 27.4 m	Kettleness	Costatenkalk 2 - 3 m
		Pleuroceras apyrenum	*Pleuroceras apyrenum*			
	Amaltheus margaritatus	*Amaltheus gibbosus*	*Amaltheus gibbosus*		Penny Nab	Amaltheenton 20 m
		Amaltheus subnodosus	*Amaltheus subnodosus*			
		Amaltheus stokesi	*Amaltheus stokesi*	Staithes Sandstone (part)		Numismalismergel (part)

Figure 1. Biostratigraphic and lithostratigraphic schemes for the North Yorkshire and southwest German Pliensbachian-Toarcian sections. After Riegraf et al. (1980) for southwest Germany and Howard (1985), Howarth (1955, 1962, 1973) and Powell (1984) for North Yorkshire.

of 110 m of marine sediments of late Pliensbachian *margaritatus* Zone to mid-Toarcian *bifrons* Zone age. The lithologies vary from mixed siliciclastic and iron-rich oolitic carbonate rocks of the Pliensbachian to finer-grained organic-poor and organic-rich shales of the Toarcian.

The *falciferum* Zone–age Jet Rock Member contains particularly high Total Organic Carbon (TOC), authigenic uranium, and Degree of Pyritization (DOP) values (Myers and Wignall, 1987; Raiswell et al., 1988), all geochemical indicators of sediment anoxia. This thick unit is the local representative of globally distributed *falciferum* Zone organic-rich facies (see below), interpreted as an Oceanic Anoxic Event (OAE) by Jenkyns (1988).

The North Yorkshire section was sampled during two field seasons. Diversity and abundance counts of all macrofaunal invertebrate species were recorded at 1 m intervals or less from this section.

Figure 3 shows the results from part of the North Yorkshire section. During the late Pliensbachian there was a diverse range of benthic species, including infaunal and epifaunal bivalves, brachiopods, and crinoids. The equally diverse nekton included belemnites and ammonites. These faunas cross the Pliensbachian-Toarcian boundary without noticeable diversity decrease. There was a 22% reduction in species diversity 0.5 m above the boundary beneath a decimeter-scale layer of laminated organic-rich shale, a lithology indicative of sediment anoxia. A diverse fauna, containing many typical Pliensbachian-type species, continued through the Toarcian *tenuicostatum* Zone until near the top of the *semicelatum* Subzone. Here there was a significant species extinction event with a 77% (17 from 22 species) reduction in diversity. The benthos was the most severely affected, with an 87% species drop, all infauna disappearing. Among the nekton,

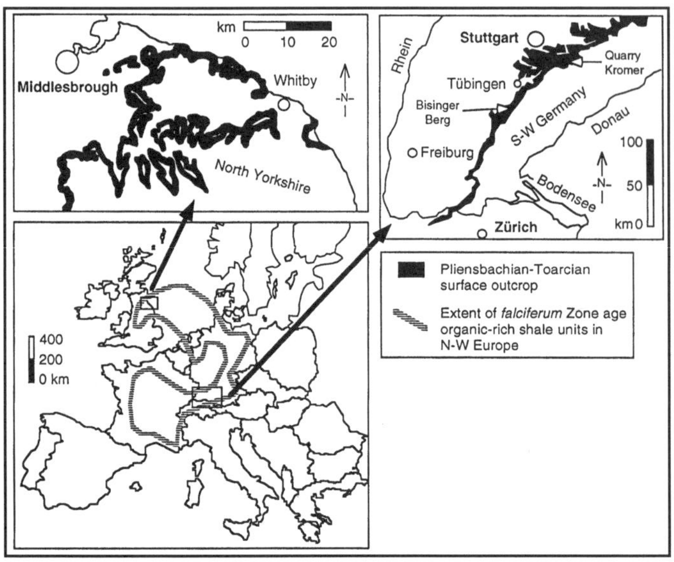

Figure 2. Location maps for the North Yorkshire and southwest Germany sections and the distribution of Lower Toarcian organic-rich shales in northwest Europe.

the common Pliensbachian genus *Passaloteuthis* became extinct, with two species disappearing. This event is intimately associated with the onset of the Jet Rock Member laminated organic-rich shale deposition in the Cleveland Basin.

Survivorship across the extinction event shows patterns of selectivity by lifestyle. Two species (*Seirocrinus subangularis* and *Pentacrinites dichotomous*) were obligate pseudoplanktonic crinoids living attached to floating logs (Simms, 1986), and three (*Serpula sp. A*, *Oxytoma inequivalve,* and *Liostrea hisingeri*) were facultative pseudoplankton, attached to logs and/or ammonite conches. For a comprehensive discussion on the recognition of pseudoplankton, see Wignall and Simms (1990). The rapid turnover of ammonite species, characteristic of the group, continued unchanged during the event, showing patterns of anagenesis and pseudoextinction in the *Dactylioceras* and *Tiltoniceras-Eleganticeras-Cleviceras* lineages (Howarth, 1973, 1992). Thus there was a gradient of extinction from infaunal benthos to epifaunal benthos to nektic belemnites, ammonites, and pseudoplanktonic crinoids. This pattern is consistent with that expected from a long period of sediment anoxia.

Figure 3 clearly shows how further sampling at the top of the Grey Shales Member during the second field season has made the extinction appear more catastrophic by pulling the ranges of more species up to the event, essentially filling in the gaps. This is a good example of the effects of improved sampling on artificial range truncations (Signor and Lipps, 1982; Springer, 1990).

The Jet Rock Member contains an impoverished benthic fauna of two species of epifaunal bivalves (*Pseudomytiloides dubius* and *Meleagrinella substriata*), occurring abundantly in scattered shell beds. The autecology of *Pseudomytiloides*, an inoceramid, has been the subject of much debate (Kauffman, 1981; Seilacher, 1982; Wignall and Simms, 1990; Savrda et al., 1991; Bottjer and Savrda, 1993), with most evidence pointing to its being an epibenthic opportunistic species. It has been suggested, based on the stable isotopic composition of shells of Late Cretaceous species, that the inoceramids were able to form symbiotic relationships with chemosynthetic bacteria, enabling the group to live in conditions of reduced oxygen levels inimical to other benthos (MacLeod and Hoppe, 1992). However, the inoceramids became extinct during the mid-Maastrichtian, and there are no close living analogs. In a review of modern chemosymbiotic bivalves, Oschmann (1994) showed that taxa with this adaptation do not have an inoceramid shell morphology. They live in areas with a sharp O_2/H_2S boundary, but where O_2 levels are high. Therefore the evidence for this feeding strategy for the Toarcian *Pseudomytiloides* appears to be weak. Nektic elements within the Jet Rock Member include scattered fish debris and, especially in the *falciferum* Subzone, an increasing diversity of Tethyan-derived immigrant ammonite and belemnite species.

Benthic diversity in the Cleveland Basin does not increase until the mid-Toarcian *bifrons* Zone–age Alum Shale Member. This unit, formed by bioturbated gray shales, is lithologically similar to the Grey Shales Member and has low TOC and DOP values (Raiswell et al., 1988), indicative of normal sediment oxygen concentrations. Although the species returning in the *bifrons* Zone are different from those of the lowest Toarcian, at the generic level over 90% are Lazarus taxa, shared in sediments below and above the Jet Rock Member. Analyzed at family level, there are no extinctions at all (Little and Benton, 1995). The extinction event recorded in the early Toarcian section is therefore only significant at species level and lasts slightly longer than one ammonite biozone, a time interval of perhaps 1 m.y.

Germany

The Pliensbachian-Toarcian sections in Baden-Württemburg, southwest Germany (Fig. 2), are considerably thinner and more calcareous than those on the North Yorkshire coast (Riegraf et al., 1980). Figure 4 shows a composite of species ranges from two outcrops at Bisinger Berg and Quarry Kromer, supplemented by published work (Riegraf et al., 1980). Although the Baden-Württemburg and North Yorkshire sections are over 1,000 km apart, they share remarkably similar faunal and facies trends. Diverse faunas pass relatively unchanged across the Pliensbachian-Toarcian boundary into the Toarcian *tenuicostatum* Zone. As in Yorkshire, there are a few extinctions at a decimeter-scale organic-rich shale layer, known locally as the Tafelfleins (Savrda et al., 1991), in the *paltum* Subzone. The main extinction event occurs near the top of the *semicelatum* Subzone, coincident with the most organic-rich part of the Posidonienschiefer, with a 61% (8 from 13) species reduction. The organic-rich marls of this unit, equivalent to the Jet Rock Member, contain a very low-diversity, high-abundance benthic fauna of *Pseudomytiloides dubius* and *Meleagrinella substriata* and a diverse nektic fauna, including an increased percentage of Tethyan ammonite species compared to the Yorkshire section. Large ammonite conches of the genera *Lytoceras* and *Harpoceras* supported a particularly diverse pseudoplanktonic assemblage (Seilacher, 1982; cf. Kauffman, 1981, for different interpretation), including *Gervillia lanceolata*, a bivalve species not found in the Jet Rock Member. Benthic diversity similarly does not recover until the top of the *bifrons* Zone.

As in the Yorkshire section, the extinctions recorded in the early Toarcian in southwest Germany are significant only to species level.

Other areas

This pattern of species extinction, linked to *falciferum* Zone–age organic-rich shales, is found in other Pliensbachian-Toarcian sections across northwest Europe (Fig. 2). For example, the sections in the Truc de Balduc basin, southern France (Riegraf, 1982) and northern Switzerland (Kuhn and Etter, 1994), show faunal trends very similar to those in North Yorkshire and southwest Germany. Contemporaneous facies are also developed in many of the carbonate Tethyan sections of south-

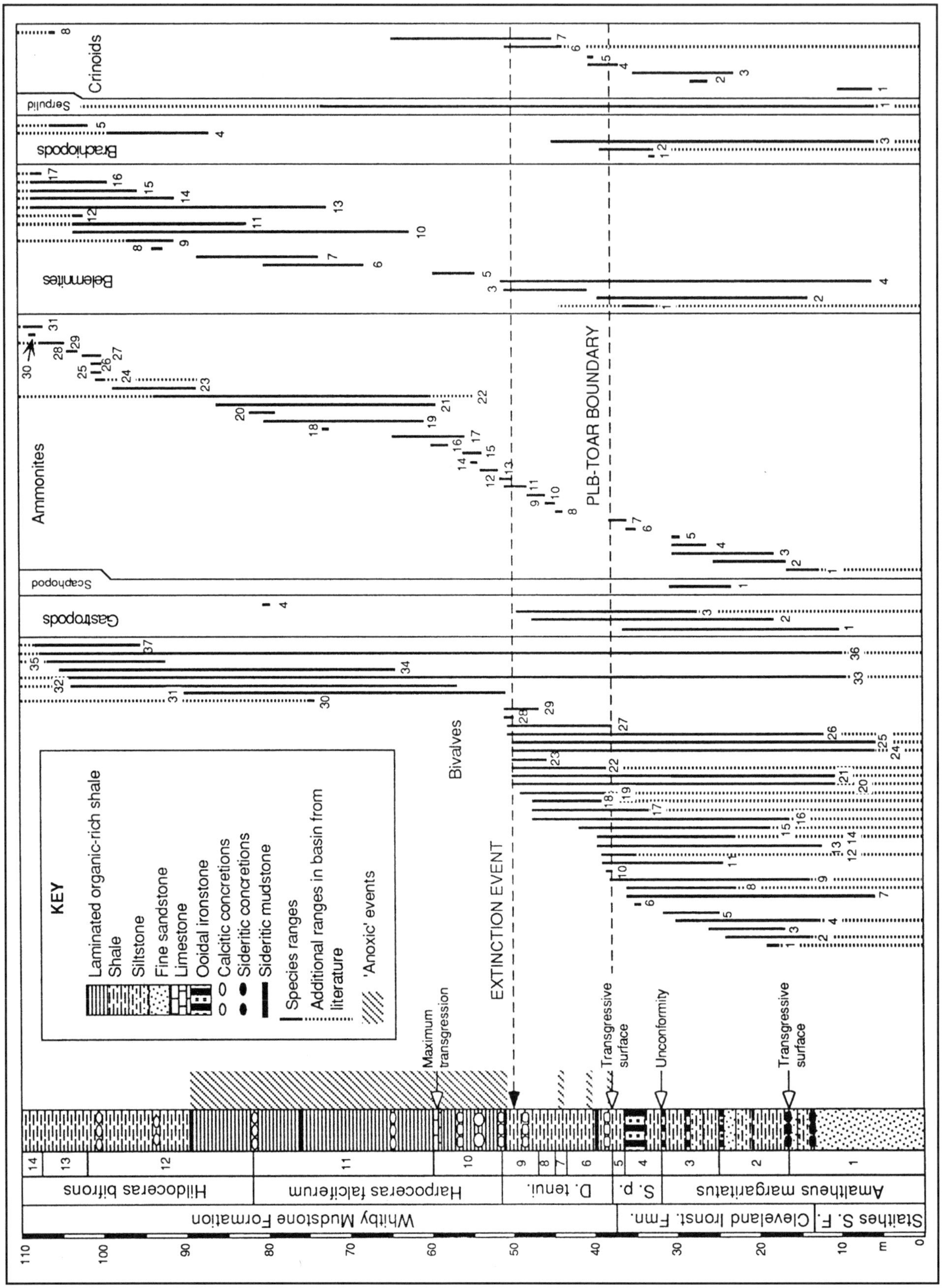

KEY
Laminated organic-rich shale
Shale
Siltstone
Fine sandstone
Limestone
Ooidal ironstone
Calcitic concretions
Sideritic concretions
Sideritic mudstone
Species ranges
Additional ranges in basin from literature
'Anoxic' events
Crinoids
Serpulid
Brachiopods
Belemnites
Ammonites
Scaphopod
Gastropods
Bivalves
PLB-TOAR BOUNDARY
EXTINCTION EVENT
Maximum transgression
Transgressive surface
Unconformity
Transgressive surface
Whitby Mudstone Formation
Cleveland Ironst. Fmn.
Staithes S. F.
Hildoceras bifrons
Harpoceras falciferum
D. tenui.
S. p.
Amaltheus margaritatus
m

ern and eastern Europe. These occur as organic-rich shales and manganese carbonates in Switzerland, Italy, Austria, Hungary, and Greece (Walzebuck, 1982; Jenkyns and Clayton, 1986; Jenkyns, 1988; Jenkyns et al., 1991). The facies distribution is much more patchy than in the northwest European area, and the TOC values tend to be less. It is probable that faunal trends in these sections are similar to those in northwest Europe as a result of the presence of the organic-rich shale units.

Figure 3. Macrofaunal invertebrate species ranges for the late Pliensbachian to early Toarcian sampled sections of the North Yorkshire coast. Ammonite subzone codes: 1. Amaltheus stokesi; 2. Amaltheus subnodosus; 3. Amaltheus gibbosus; 4. Pleuoceras apyrenum; 5. Pleuoceras hawskerense; 6. Protogrammoceras paltum; 7. Dactylioceras clevelandicum; 8. Dactylioceras tenuicostatum; 9. Dactylioceras semicelatum; 10. Cleviceras exaratum; 11. Harpoceras falciferum; 12. Dactylioceras commune; 13. Peronoceras fibulatum; 14. Catacoeloceras crassum. Species codes: **bivalves:** 1. *Tutcheria submulticostata*, 2. *Ryderia graphica*, 3. *Camptonectes subulatus*, 4. *Goniomya hybrida*, 5. *Cardinia laevis*, 6. *Antiquilima* indeterminate species, 7. *Palmoxytoma cygnipes*, 8. *Gryphaea gigantea*, 9. *Mesosaccella galatea*, 10. *Gryphaea sportella*, 11. *Mactromya cardiodes*, 12. *Pholadomya ambigua*, 13. *Gresslya intermedia*, 14. *Plicatula spinosa*, 15. *Pinna folium*, 16. *Modiolus scalprum*, 17. *Entolium lunare*, 18. *Palaeonucula navis*, 19. *Eotrapezium cucculatum*, 20. *Pleuromya costata*, 21. *Grammatodon insons*, 22. *Nicaniella striato-sulcata*, 23. *Nuculana* species A, 24. *Pseudopecten equivalvis*, 25. *Protocardia truncata*, 26. *Pseudolimea acuticostata*, 27. *Parainoceramus* species A, 28. *Nucinella* species A, 29. *Bositra radiata*, 30. *Camptonectes auritis*, 31. *Pseudomytiloides dubius*, 32. *Meleagrinella substriata*, 33. *Liostrea hisingeri*, 34. *Bositra buchii*, 35. *Dacryomya ovum*, 36. *Oxytoma inequivalve*, 37. *Gresslya donaciformis*; **gastropods:** 1. *Ptychomphalus expansus*, 2. *Oolitica clevelandica*, 3. *Katosira blainvillei*, 4. *Procerithium* species A; **scaphopod:** 1. *Dentalium* species A; **ammonites:** 1. *Amaltheus stokesi*, 2. *Amaltheus subnodosus*, 3. *Amaltheus maragaritatus*, 4. *Amaltheus gibbosus*, 5. *Amaltheus laevigatus*, 6. *Pleuroceras apyrenum*, 7. *Pleuroceras hawskerense*, 8. *Dactylioceras crosbeyi*, 9. *Dactylioceras clevelandicum*, 10. *Dactylioceras tenuicostatum*, 11. *Dactylioceras semicelatum*, 12. *Tiltoniceras antiquum*, 13. *Eleganticeras elegantulum*, 14. *Lytoceras crenatum*, 15. *Cleviceras exaratum*, 16. *Cleviceras elegans*, 17. *Hildaites murleyi*, 18. *Nodicoeloceras incrassatum*, 19. *Harpoceras falciferum*, 20. *Ovaticeras ovatum*, 21. *Dactylioceras* indeterminate species, 22. *Phylloceras heterophyllum*, 23. *Dactylioceras commune*, 24. *Hildoceras lusitanicum*, 25. *Dactylioceras crassescens*, 26. *Dactylioceras praepositum*, 27. *Dactylioceras athleticum*, 28. *Peronoceras fibulatum*, 29. *Hildoceras bifrons*, 30. *Porpoceras vortex*, 31. *Catacoeloceras crassum;* **belemnites:** 1. *Pseudohastites longiformis*, 2. *Parapassaloteuthis zeiteni*, 3. *Passaloteuthis milleri*, 4. *Passaloteuthis bisulcata*, 5. *Acrocoelites trisulculosis*, 6. *Youngibelus tubularis*, 7. *Parapassaloteuthis polita*, 8. *Acrocoelites riegrafi*, 9. *Acrocoelites subtriscissus*, 10. *Acrocoelites subtenuis*, 11. *Acrocoelites vulgaris*, 12. *Simpsonibelus lentus*, 13. *Simpsonibelus dorsalis*, 14. *Acrocoelites inaequistriatus*, 15. *Acrocoelites triscissus*, 16. *Simpsonibelus expansus*, 17. *Acrocoelites levidensis;* **brachiopods:** 1. *Homoeorhynchia* species A, 2. *Lobothyris punctata*, 3. *Tetrarhynchia tetrahedra*, 4. *Lingula longovicensis*, 5. *Discinisca papyracea;* **serpulid:** 1. *Serpula* species A; **crinoids:** 1. *Balanocrinus gracilis*, 2. *Balanocrinus solenotis*, 3. *Hispidocrinus schlumbergeri*, 4. *Balanocrinus donovani*, 5. *Isocrinus basaltiformis*, 6. *Seirocrinus subangularis*, 7. *Pentacrinites dichotomus*, 8. *Chariocrinus wuerttembergicus*.

Outside Europe, early Toarcian organic-rich shales are found in offshore western Australia in boreholes, Madagascar (Jenkyns, 1988 and references therein), and Arctic and western Canada (Thomson et al., 1986; Jenkyns, 1988; Jakobs, 1989). Unfortunately there is little published work on faunal trends in these areas. The early Toarcian part of the Japanese Nishinakayama Formation consists of laminated organic-rich shales with nektonic, epibenthic, and pseudoplanktonic faunas very similar to those of northwest Europe (Tanabe et al., 1984).

There are many areas where Toarcian-age organic-rich shales are not developed. In west-central Portugal the Toarcian sections consist of interbedded limestone-marl facies, and significantly, some species that became extinct during the early Toarcian event in northwest Europe persisted into the *falciferum* Zone (Hallam, 1971) in this area. Chilean and Argentinian Pliensbachian-Toarcian sections show mixed clastic-carbonate lithologies. There is no extinction event at the Pliensbachian-Toarcian boundary or in the early Toarcian, and the faunas contain species that are quite different from those found in northwest European sections (Damborenea, 1987; Riccardi et al., 1990; Aberhan, 1993).

Records from other areas known to have been transgressed during the early Toarcian are very poor.

DISCUSSION AND SUMMARY

The early Toarcian species-level extinctions seen in the sampled sections seem to confirm Hallam's (1986) model of a regional event, seen in northwest Europe and possibly Canada and Japan, linked to the Toarcian OAE (Jenkyns, 1988). The sections in Portugal and South America show that where there was no anoxia, diverse faunas persisted during the early Toarcian and served as evolutionary sinks for later repopulation (Hallam, 1986). Thus the early Toarcian species extinction event did not have a global extent, nor was it facies independent. The causes for this period of environmental disturbance are more obscure. One model suggests that thick units of organic-rich shales on epicontinental margins are characteristic of sediments deposited during the early stages of rapid transgressions (Wignall, 1991; Wignall and Hallam, 1991). Indeed, the Toarcian is generally agreed to be a time of global marine transgression (Hallam, 1967; Haq et al., 1988).

Oceanic anoxic events have been linked to two of the other eight extinction events identified by Raup and Sepkoski (1986): those at the Permo-Triassic (Wignall and Hallam, 1993) and the Cenomanian-Turonian boundaries (Kauffman, 1984; Arthur et al., 1985; Elder, 1991). However, OAEs cannot be used to explain all events, as several other Mesozoic OAEs are not linked to mass extinctions.

It is important not to extrapolate the patterns shown by detailed studies of a few sections to explain or dismiss patterns seen in large databases, such as Sepkoski's (1992), that contain global information. Indeed it is difficult to link the early Toarcian species extinction event to the family-level and genus-level event

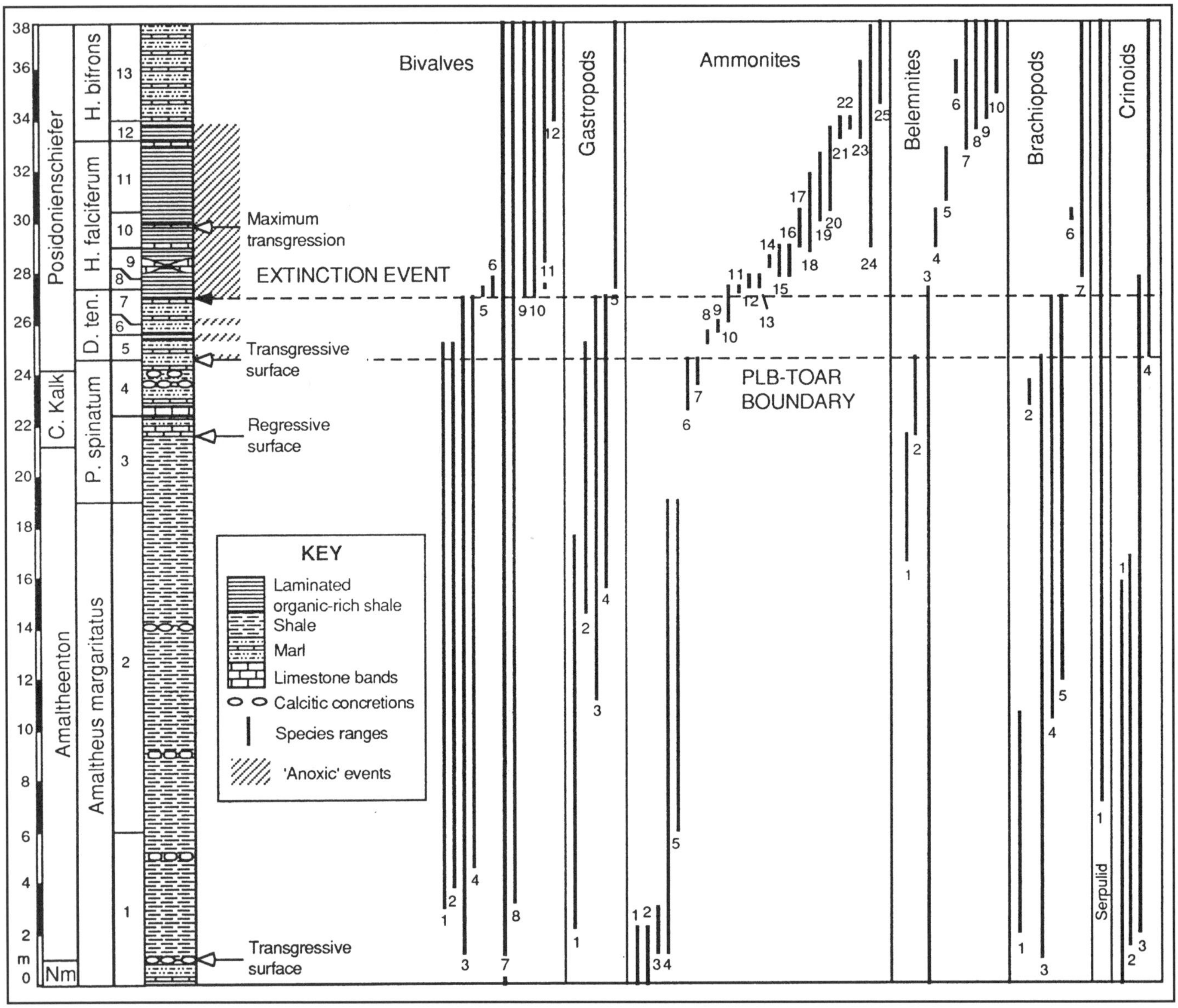

Figure 4. Macrofaunal invertebrate species ranges for the late Pliensbachian to early Toarcian sampled sections of southwest Germany at Bisinger Berg and Quarry Kromer. Ammonite subzone codes: 1. Amaltheus subnodosus; 2. Amaltheus gibbosus; 3. Pleuroceras apyrenum; 4. Pleuroceras hawskerense; 5. Protogrammoceras paltum; 6. Dactylioceras clevelandicum; 7. Dactylioceras semicelatum; 8. Eleganticeras elegantulum; 9. Cleviceras exaratum; 10. Cleviceras elegans; 11. Harpoceras falciferum; 12. Dactylioceras commune; 13. Peronoceras fibulatum. Species codes: **bivalves:** 1. *Pseudopecten equivalvis*, 2. *Pseudolimea acuticostata*, 3. *Entolium lunare*, 4. *Plicatula spinosa*, 5. *Bositra radiata*, 6. *Gervillia lanceolata*, 7. *Liostrea hisingeri*, 8. *Oxytoma inequivalve*, 9. *Pseudomytiloides dubius*, 10. *Meleagrinella substriata*, 11. *Propeamussium pumilum*, 12. *Bositra buchii;* **gastropods:** 1. *Ptychomphalus expansus*, 2. *Katosira blainvillei*, 3. *Aberleya* species A, 4. *Pleurotomaria anglica*, 5. *Coelodiscus minutus;* **ammonites:** 1. *Derolytoceras tortum*, 2. *Lytoceras fimbriatum*, 3. *Amaltheus subnodosus*, 4. *Amaltheus margaritatus*, 5. *Amaltheus gibbosus*, 6. *Pleuroceras hawskerense*, 7. *Pleuroceras spinatum*, 8. *Protogrammoceras paltum*, 9. *Dactylioceras crosbeyi*, 10. *Dactylioceras semicelatum*, 11. *Tiltoniceras antiquum*, 12. *Eleganticeras elegantulum*, 13. *Lytoceras siemense*, 14. *Hildaites forte*, 15. *Lytoceras ceratophagum*, 16. *Cleviceras exaratum*, 17. *Cleviceras elegans*, 18. *Dactylioceras semiannulatum*, 19. *Phylloceras pompeckji*, 20. *Harpoceras falciferum*, 21. *Dactylioceras commune*, 22. *Hildoceras lusitanicum*, 23. *Lytoceras cornucopia*, 24. *Phylloceras heterophyllum*, 25. *Hildoceras bifons*; **belemnites:** 1. *Pleurobelus compressus*, 2. *Parapassaloteuthis zeiteni*, 3. *Passaloteuthis bisulcata*, 4. *Acrocoelites trisulculosus*, 5. *Youngibelus tubularis*, 6. *Simpsonibelus dorsalis*, 7. *Youngibelus gigas*, 8. *Acrocoelites ilminsterensis*, 9. *Acrocoelites oxyconus*, 10. *Acrocoelites pyramidalis*; **brachiopods:** 1. *Zeilleria subdigona*, 2. *Orthotoma quenstedti*, 3. *Lobothyris punctata*, 4. *Gibbirhynchia amalthei*, 5. *Spiriferina villosa*, 6. *Lingula longovicensis*, 7. *Discinisca papyracea*; **serpulid:** 1. *Serpula* species A; **crinoids:** 1. *Isocrinus basaltiformis*, 2. *Balanocrinus subteroides*, 3. *Seirocrinus subangularis*, 4. *Pentacrinites dichotomus*.

identified by Raup and Sepkoski (1986) at the Pliensbachian-Toarcian boundary (Little and Benton, 1995). Certainly the former cannot explain the latter. The regional early Toarcian event appears to be only part of a global five-zone phase of extinction from the late Pliensbachian to early Toarcian (Little and Benton, 1995).

Given its long duration (c. 7.5 m.y.), the Early Jurassic extinction phase is quite unlike the short-term, perhaps catastrophic K/T extinction event, and the forcing mechanisms were probably very different. As yet, no signatures of bolide impact (iridium anomalies, tektites, shocked quartz, etc.) have been found in Pliensbachian-Toarcian sections, although this may be a reflection of a lack of research effort. The Early Jurassic extinction phase thus seems to be a poor candidate for an impact-related event at present. The suggested 26 m.y. periodicity of extinctions (Raup and Sepkoski, 1986) is not disproven by this study, but the hypothesis that all extinctions are caused by impacts is further weakened (Raup, 1987).

ACKNOWLEDGMENTS

The author would like to thank Mike Benton for general help and David Jablonski and Chris Maples for reviewing an earlier draft of the paper. Funding from the Natural Environment Research Council is gratefully acknowledged.

REFERENCES CITED

Aberhan, M., 1993, Faunal replacement in the Early Jurassic of northern Chile: Implications for the evolution in Mesozoic benthic shelf ecosystems: Palaeogeography, Palaeoclimatology, Palaeoecology, v. 103, p. 155–177.

Arthur, M. A., Schlanger, S. O., and Jenkyns, H. C., 1985, The Cenomanian-Turonian oceanic anoxic event. 2: Paleoceanographic controls on organic matter production and preservation, *in* Brooks, J., and Fleet, A., eds., Marine petroleum source rocks: Geological Society of London Special Publication 315, p. 216–218.

Bottjer, D. J., and Savrda, C. E., 1993, Oxygen-related mudrock biofacies, *in* Wright, V. P., ed., Sedimentology Review, v. 1, p. 92–102.

Damborenea, S. E., 1987, Early Jurassic Bivalvia of Argentina. Part 1: Stratigraphical introduction and superfamilies Nuculanacea, Arcacea, Mytilacea and Pinnacea: Palaeontographica Abteilung A, v. 199, p. 23–111.

Dean, W. T., Donovan, D. T., and Howarth, M. K., 1961, The Liassic ammonite zones and subzones of the North-West European Province: Bulletin of The British Museum (Natural History) Geology, v. 4, p. 438–505.

Elder, W. P., 1991, Molluscan paleoecology and sedimentation patterns of the Cenomanian-Turonian extinction interval in the southern Colorado Plateau region, *in* Nations, J. D., and Eaton, J. G., eds., Geological Society of America Special Paper 260, p. 113–137.

Hallam, A., 1967, An environmental study of the Upper Domerian and Lower Toarcian in Great Britain: Philosophical Transactions of the Royal Society of London, v. 252B, p. 393–445.

Hallam, A., 1971, Facies analysis of the Lias in West Central Portugal: Neues Jahrbuch für Geologie und Paläontologie Abhandlungen, v. 139, p. 226–265.

Hallam, A., 1975, Jurassic environments: Cambridge, Cambridge University Press, 269 p.

Hallam, A., 1986, The Pliensbachian and Tithonian extinction events: Nature, v. 319, p. 765–768.

Haq, B. U., Hardenbol, J., and Vail, P. R., 1988, Mesozoic and Cenozoic chronostratigraphy and cycles of sea-level change, *in* Wilgus, C. K., Posamentier, H., Ross, C. A., and Kendall, C. G. St.C., eds., Sea-level changes: An integrated approach: Society of Economic Paleontologists and Mineralogists Special Publication 42, p. 71–108.

Harland, W. B., Armstrong, R. L., Cox, A. V., Craig, L. E., Smith, A. G., and Smith, D. G., 1990, A geologic time scale: Cambridge, Cambridge University Press, 263 p.

Howard, A. S., 1985, Lithostratigraphy of the Staithes Sandstone and Cleveland Ironstone formations (Lower Jurassic) of north-east Yorkshire: Proceedings of the Yorkshire Geological Society, v. 45, p. 261–275.

Howarth, M. K., 1955, Domerian of the Yorkshire coast: Proceedings of the Yorkshire Geological Society, v. 30, p. 147–175.

Howarth, M. K., 1962, The Jet Rock Series and the Alum Shale Series of the Yorkshire coast: Proceedings of the Yorkshire Geological Society, v. 33, p. 381–422.

Howarth, M. K., 1973, The stratigraphy and ammonite fauna of the Upper Liassic Grey Shales of the Yorkshire coast: Bulletin of the British Museum (Natural History) Geology, v. 24, p. 237–277.

Howarth, M. K., 1992, The Ammonite Family Hildoceratidae in the Lower Jurassic of Britain: Monograph of the Palaeontographical Society, London, v. 586 part 1, p. 1–106; v. 590 part 2, p. 107–200.

Jakobs, G. K., 1989, Toarcian (Lower Jurassic) biostratigraphy of the Queen Charlotte Islands, British Columbia: Current Research, Part H, Geological Survey of Canada Paper 89-1H, p. 35–37.

Jenkyns, H. C., 1988, The early Toarcian (Jurassic) anoxic event: Stratigraphic, sedimentary, and geochemical evidence: American Journal of Science, v. 288, p. 101–151.

Jenkyns, H. C., and Clayton, C. J., 1986, Black shales and carbon isotopes in pelagic sediments from the Tethyan Lower Jurassic: Sedimentology, v. 33, p. 87–106.

Jenkyns, H. C., Geczy, B., and Marshall, J. D., 1991, Jurassic manganese carbonates of Central Europe and the early Toarcian anoxic event: Journal of Geology, v. 99, p. 137–149.

Kauffman, E. G., 1981, Ecological Reappraisal of the German Posidonienschiefer (Toarcian) and the stagnant basin model, *in* Gray, J., Boucot, A. J., and Berry, W. B. N., eds., Communities of the past: Stroudsberg, Pennsylvania, Hutchinson Ross Publishing Company, p. 311–381.

Kauffman, E. G., 1984, Paleobiogeography and evolutionary response dynamic in the Cretaceous Western Interior Seaway of North America, *in* Westermann, G. E. G., ed., Jurassic-Cretaceous biochronology and paleobiogeography of North America: Geological Association of Canada Special Paper 27, p. 273–306.

Kuhn, O., and Etter, W., 1994, Der Posidonienschiefer der Nordschweiz: Lithostratigraphie, Biostratigraphie, Fazies: Eclogae geolicae Helvetiae, v. 87, p. 113–138.

Little, C. T. S., and Benton, M. J., 1995, Early Jurassic mass extinction: A global long-term event: Geology, v. 23, p. 495–498.

MacLeod, K. G., and Hoppe, K. A., 1992, Evidence that inoceramid bivalves were benthic and harbored chemosynthetic symbionts: Geology, v. 20, p. 117–120.

Myers, K. J., and Wignall, P. B., 1987, Understanding Jurassic organic-rich mudrocks—new concepts using gamma ray spectrometry and palaeoecology: Examples from the Kimmeridge Clay of Dorset and the Jet Rock of Yorkshire, *in* Leggett, J. K., and Zuffa, G. G., eds., Marine clastic sedimentology: London, Graham and Trotman, p. 175–192.

Oschmann, W., 1994, Adaptive pathways of benthic organisms in marine oxygen-controlled environments: Neues Jahrbuch für Geologie und Paläontologie Abhandlungen, v. 191, p. 393–444.

Powell, J. H., 1984, Lithostratigraphical nomenclature of the Lias Group in the Yorkshire Basin: Proceedings of the Yorkshire Geological Society, v. 45, p. 51–57.

Raiswell, R., Buckley, F., Berner, R. A., and Anderson, T. F., 1988, Degree of pyritization of iron as a paleoenvironmental indicator of bottom-water oxygenation: Journal of Sedimentary Petrology, v. 58, p. 812–819.

Raup, D. M., 1987, Mass extinction: A commentary: Palaeontology, v. 30,

p. 1–13.

Raup, D. M., and Sepkoski, J. J., Jr., 1986, Periodic extinction of families and genera: Science, v. 231, p. 833–836.

Riccardi, A. C., Damborenea, S. E., and Manceñido, M. O., 1990, South America and Antarctic Peninsula: Newsletters on Stratigraphy, v. 21, p. 75–103.

Riegraf, W., 1982, The bituminous Lower Toarcian at the Truc de Balduc near Mende (Départment de la Lozère, S-France), *in* Einsele, G., and Seilacher, A., eds., Cyclic and event stratification: Berlin, Springer-Verlag, p. 506–511.

Riegraf, W., Werner, G., and Lörcher, F., 1980, Der Posidonienschiefer. Biostratigraphie, Fauna und Fazies des südwestdeutschen Untertoarciums (Lias e): Stuttgart, Enke, 195 p.

Savrda, C. E., Bottjer, D. J., and Seilacher, A., 1991, Redox-related benthic events, *in* Einsele, G.,and Seilacher, A., eds., Cycles and events in stratigraphy: Berlin, Springer-Verlag, p. 524–541.

Seilacher, A., 1982, Ammonite shells as habitats in the Posidonia Shales of Holzmaden—floats or benthic islands?: Neues Jahrbuch für Geologie und Paläontologie Monatshefte, p. 98–114.

Sepkoski, J. J., Jr., 1990, The taxonomic structure of periodic extinction: Geological Society of America Special Paper 247, p. 33–44.

Sepkoski, J. J., Jr., 1992, A compendium of fossil marine animal families (second edition): Milwaukee Public Museum Contributions in Biology and Geology 83, p. 1–156.

Sepkoski, J. J., Jr., 1993, Ten years in the library: New data confirm paleontological patterns: Paleobiology, v. 19, p. 43–51.

Signor, P. W., and Lipps, J. H., 1982, Sampling bias, gradual extinction patterns and catastrophes in the fossil record, *in* Silver, L. T., and Shultz, P. H., eds., Geological implications of large asteroids and comets on the Earth: Geological Society of America Special Paper 190, p. 291–296.

Simms, M. J., 1986, Contrasting lifestyles in Lower Jurassic crinoids: A comparison of benthic and pseudopelagic Isocrinida: Palaeontology, v. 29, p. 475-493.

Smith, P. L., and Tipper, H. W., 1986, Plate Tectonics and paleobiogeography: Early Jurassic (Pliensbachian) endemism and diversity: Palaios, v. 1, p. 399–412.

Springer, M. S., 1990, The effect of random range truncations on patterns of evolution in the fossil record: Paleobiology, v. 16, p. 512–520.

Tanabe, K., Inazumi, A.,Tamahama, K., and Katsuta, T., 1984, Taphonomy of half compressed ammonites from the lower Jurassic black shales of the Toyora area, west Japan: Palaeogeography, Palaeoclimatology, Palaeoecology, v. 47, p. 329–346.

Thomson, R. C., Smith, P. L., and Tipper, H. W., 1986, Lower to Middle Jurassic (Pliensbachian to Bajocian) stratigraphy of the northern Spatsizi area, north-central British Columbia: Canadian Journal of Earth Sciences, v. 23, p. 1963–1973.

Walzebuck, J. P., 1982, Bedding types of the Toarcian black shales in NW-Greece, *in* Einsele, G., and Seilacher, A., eds., Cyclic and event stratification: Berlin, Springer-Verlag, p. 512–525.

Wignall, P. B., 1991, Model for transgressive black shales?: Geology, v. 19, p. 167–170.

Wignall, P. B., and Hallam, A., 1991, Biofacies, stratigraphic distribution and depositional models of British onshore Jurassic black shales, *in* Tyson, R. V., and Pearson, T. H., eds., Modern and ancient Continental Shelf anoxia: Geological Society of London Special Publication 58, p. 291–309.

Wignall, P. B., and Hallam, A., 1993, Griesbachian (Earliest Triassic) palaeoenvironmental changes in the Salt Range, Pakistan and southeast China and their bearing on the Permo-Triassic mass extinction: Palaeogeography, Palaeoclimatology, Palaeoecology, v. 102, p. 215–237.

Wignall, P. B., and Simms, M. J., 1990, Pseudoplankton: Palaeontology, v. 33, p. 359–378.

Manuscript Accepted by the Society September 14, 1995

Printed in U.S.A.

Geological Society of America
Special Paper 307
1996

The influence of continental flood basalts on mass extinctions: Where do we stand?

V. Courtillot
Institut de Physique du Globe, 75252 Paris Cedex 05, France
J.-J. Jaeger
Institut des Sciences de l'Evolution, 34095 Montpellier, France
Z. Yang
Institut de Physique du Globe, 75252 Paris Cedex 05, France
G. Féraud and C. Hofmann
Laboratoire de Géodynamique, 06108 Nice Cedex 02, France

ABSTRACT

Recent confirmation of the uniqueness of the Cretaceous-Tertiary (K/T) boundary iridium anomaly and estimates of the possibly associated Chixculub crater diameter make the K/T impact a unique event in the Phanerozoic, leaving other mass extinctions to be explained by other mechanisms. We briefly discuss whether available data can resolve the duration of the peculiar deposits in northeastern Mexico K/T sections, claimed to be directly linked to the impact. Our preliminary paleomagnetic data for samples from these sections show primary reversed magnetizations apparently of K/T (chron 29R) age but fail to discriminate between a short ("100,000 s") and a long ("100,000 yr") duration for the deposition of these sediments. We next review recent age determination of continental flood basalt (CFB) provinces younger than 300 Ma. We restate our earlier conclusion that most of the Deccan traps were emplaced in less than 1 m.y., although the uncertainty in the absolute age (about 2 Ma) remains larger than often quoted in published work; this age corresponds within uncertainties to that of the K/T. The duration of Siberian trap volcanism was also less than 1 m.y., and its absolute age cannot be distinguished from that of the Permo-Triassic boundary. Altogether, the ages of 12 CFBs have been plotted against the ages of 10 mass extinction events. Seven of those events are contemporaneous with a flood basalt, and at least nine flood basalts are contemporaneous with a mass extinction event. The correlation happens to be clearest for the two largest events. There is no doubt that several large impacts have occurred on Earth fairly recently, but apart from Chixculub, they do not coincide with extinctions. An impact did apparently coincide with the K/T and cause the extinction of one-third to one-half of the species that disappeared in the crisis. But the extinction event was already under way as a result of Deccan volcanism. The potential environmental impact of flood basalt eruptions is a function of size, rate of emplacement, location, more or less detrimental chemical composition, and submarine or subaerial emplacement, all variable parameters. In conclusion, CFB volcanism is the main candidate for causing most extinction events. Other internal processes, such as regression and continental breakup, can be associated with volcanism without having to resort to multiple causes.

Courtillot, V., Jaeger, J.-J., Yang, Z., Féraud, G., and Hofmann, C., 1996, The influence of continental flood basalts on mass extinctions: Where do we stand?, *in* Ryder, G., Fastovsky, D., and Gartner, S., eds., The Cretaceous-Tertiary Event and Other Catastrophes in Earth History: Boulder, Colorado, Geological Society of America Special Paper 307.

INTRODUCTION

In the five years between the Snowbird II conference (Sharpton and Ward, 1990) and the Houston conference (Lunar and Planetary Institute, 1994), significant progress has taken place in many fields relevant to the study of global catastrophes in Earth history. Further data have been acquired from a number of boundary sections (not only, if mostly, the Cretaceous-Tertiary or K/T), some in northeastern Mexico (Smit et al., 1992; Stinnesbeck et al., 1993). A K/T impact crater appears to have been found in Yucatan (Penfield and Camargo, 1981; Hildebrand et al., 1991) and dated (Swisher et al., 1992); the most recent geophysical data would make it one of the largest such features among the inner planets since the early large-scale bombardment (Sharpton et al., 1993). At the same time, dating of continental flood basalts (CFBs) has confirmed that they too were extraordinary events, with volumes often in excess of 1,000,000 km^3, yet emplacement times shorter than 1 m.y. (e.g., Richards et al., 1989). The time coincidence (within available resolution) between the main phases of eruption of the Deccan traps and the K/T boundary has been reasonably established (Courtillot et al., 1986; Vandamme et al., 1991) and is further addressed in this chapter. Proposals (Morgan, 1988: Courtillot, 1990) that improved age determinations would show a coincidence between the Siberian traps and the Permo-Triassic boundary (P/T) have been confirmed (Renne and Basu, 1991). Therefore, neither the proponents of an impact scenario for the K/T boundary and possibly other mass extinction events nor those who advocate a volcanic scenario for most mass extinctions remain unchallenged: Both have to account for the new findings and indicate how those findings fit into their respective (possibly revised) theories. Here, we evaluate some of the ever-expanding literature and attempt to summarize where we stand. We first discuss evidence for the impact in Yucatan and then some problems concerning the age and duration of K/T sections and events and finally review the ages of flood basalts and mass extinctions, confirming that many are contemporaneous.

EVIDENCE FOR AN IMPACT IN YUCATAN

High-resolution physical and chemical stratigraphy have confirmed the ubiquitous nature of the iridium anomaly in all K/T sections that appear to be complete enough. This iridium anomaly was declared by Frank Asaro to be unique to the Phanerozoic record (oral communication, Houston conference, 1994), all other anomalies at other boundaries being far smaller (when they exist) and explainable by terrestrial (geologic or biologic) concentration processes. This confirms the earlier views of Orth et al. (1990; but for a different view see Rampino, 1994). Other major observations at the K/T boundary, also unique to the geological record, include shocked minerals. Jean-Claude Doukhan and his students (e.g. Goltrant et al., 1991, 1992; Leroux, 1994) have carefully investigated planar deformation features in shocked quartz by transmission electron microscopy (TEM); they were able to show that these features consist of very thin lamellae (on the order of .1 μm) of pure amorphous silica glass and mechanical twins, seen properly only at the TEM (as opposed to optical) scale. So far these have been found only in samples from nuclear explosion sites, meteorite impacts, some experimentally shocked material, and of course K/T sections. No sample with a volcanic origin has yet revealed such features. To the list of diagnostic shocked minerals at K/T, Leroux (1994) has added diopside and Krogh et al. (1993) zircon. Krogh et al. (1993) find that zircons from K/T sections in Colorado and Haiti and also from breccia in the Chixculub structure, have preserved primary source ages of 545 ± 5 Ma and 418 ± 6 Ma, with variable degrees of isotopic resetting consistent with lead loss at 57 ± 4 Ma. Although this does not fit with the K/T age of 65 Ma, Krogh et al. (1993) feel that this difference may be due to possible mixed-age populations or continued lead loss. Other characteristic observations of K/T sediments include spherules and spinels. Robin et al. (1992) have studied these nickel-rich magnetites, which have preserved some of their original features. They point out that their high oxidation level and nickel content resemble those found in the fine crust of large meteorites, formed by fusion and oxidation during atmospheric entry. Such nickel-rich spinels are apparently unknown in terrestrial rocks. Moreover, the stratigraphic distribution of the spinels appears to be very narrow, with a peak more or less coinciding with the peak of the iridium distribution. Robin et al. (1991) attribute the larger spread of the iridium to chemical mobility in the sediment and longer residence time in seawater; the distribution of magnetites would then be more representative of the actual, very short duration of the event.

All of these data are consistent with an external origin for the K/T extinctions. Melt rock from the Chixculub structure has been dated with the $^{40}Ar/^{39}Ar$ technique at 65 Ma by Swisher et al. (1992), with an uncertainty less than 0.1 m.y. This is virtually indistinguishable from the age of what is considered to be remains of tektite glass from Haiti and northern Mexico K/T sections (see also Izett et al., 1991), and also from the K/T age, although uncertainties in absolute ages from different laboratories are likely to be at least 1%, if only because of the uncertainty on the neutron flux during irradiation. Altogether, there is a strong case that the Chixculub structure marks the impact crater of a giant meteorite or comet, which, because of the unique iridium anomaly and exceptional size, may be the largest known on Earth since at least the beginning of the Phanerozoic. An event able to produce a 300-km multiring basin according to the interpretation of Sharpton et al. (1993) is not a once-in-100-m.y. event as originally proposed by Alvarez et al. (1980), but rather a once-in-1,000-m.y. one. This is in keeping with the fact that no other such sequence of observations has been found for other extinction events at other geological boundaries; it thus implies that the impact hypothesis only accounts for a single mass extinction. Other causes should then be sought for other extinctions.

SOME PROBLEMS OF AGE AND DURATION IN K/T SECTIONS

Some remaining questions are under discussion regarding the impact explanation of the K/T event. The interpretation of logs and samples of cores from the Yucatan Pemex wells is disputed: Meyerhoff et al. (1994) mention interbedded limestone with Campanian microfossils at depths of 1,200 m, Late Cretaceous microfossils in core N21 (near 1,590-m depth) and Turonian limestone in core N22 (near 1,650-m depth). Marin et al. (1994) argue for a lower Tertiary age of fossils in core N12 (near 1,000-m depth), previously considered to be Upper Cretaceous. But this does not address the problem of the deeper cores N21 and N22. An impact of the size advocated for Chixculub would not be likely to have left any unaltered and recognizable layers of preexisting fossil-bearing sediments at such shallow depths, unless they correspond to backfall breccia.

There is disagreement in interpretations of the stratigraphic records of nearby sedimentary sections, both offshore (Alvarez et al., 1992; Keller et al., 1993a) and onshore (Smit et al., 1992; Stinnesbeck et al., 1993). Keller et al. (1993a) find that a hiatus between the lower upper Maastrichtian and lowermost Tertiary is present over most sites in the Gulf of Mexico and Caribbean, whereas Alvarez et al. (1992) find evidence for a giant wave at K/T time. The most complete records seem to occur onshore in Haiti and northeastern Mexico. Some of these Mexican sections were visited during a field trip that preceded the Houston conference. In the field guide (Keller et al., 1994a) and during the field trip, the argument was not settled: Following Smit et al. (1992; this volume), the sections are considered as evidence of proximal impact ejecta and are linked to a tsunami followed by seiches, the whole process lasting less than a week. More precisely, three main subunits (a basal spherule bed, then massive laminated sandstone beds topped by ripple beds) are interpreted as the marks of altered tektites, megawave backwash, and deposits of oscillating currents, respectively. A contrasting interpretation is given by Stinnesbeck et al. (1993), who interpret the three subunits as being distinct in origin and time of deposition, concluding that deposition occurred as discrete events over an extended period of time, preceding deposition of the K/T boundary clay. The corresponding lengths of time could then be on the order of thousands to tens of thousands of years. This controversy was discussed in the field and was summarized by the following question: Were the clastic deposits laid down in 100,000 s or in 100,000 yr? It is our view that the arguments from paleontologists, stratigraphers and sedimentologists show that none of these disciplines has, at present, with the available data, the power to distinguish conclusively between the two hypotheses.

Smit et al. (1992) reported that samples from the Upper Cretaceous Mendez Formation marls and the Paleocene Velasco Formation marls, respectively underlying and overlying the K/T clastics, were unsuitable for determining magnetostratigraphy. They uncovered a present dipole field low-temperature component and an intermediate component, possibly Pliocene in age, and high-temperature components with random directions but in general positive inclination. In a poster at the conference, Urrutia-Fucugauchi et al. (1994) reported some preliminary results from the clastic units. They found some reversed directions and in one sample suggested a normal direction. Detection of a reversal within the clastic sequence clearly would have favored the long-deposition hypothesis and not the tsunami one.

In order to try to contribute to this debate, one of us (V.C.), together with Bill Glen, collected four oriented hand samples. References for the stratigraphic positioning of the samples follow the Keller et al. (1994a) sections in the field guide. The first sample was collected in the uppermost unit (3) of Keller et al. in the Mimbral I section, near the center of a channel. The second came from a similar position in the La Lajilla I section. It was understood only after these two sections had been visited that the polarity change possibly observed by Urrutia-Fucugauchi et al. (1994) would have occurred between the base of unit 2 and the top of unit 3. Therefore, two samples were collected at the third outcrop, El Peñon I, in the lower half of unit 2 and upper half of unit 3. The blocks were oriented with a magnetic compass, and bedding planes were measured in the same way. Measurements of 11 cylindrical samples from the blocks drilled in the laboratory were performed in the Institut de Physique du Globe de Paris (IPGP) paleomagnetic laboratory (by Y.Z.), using a CTF three-component cryogenic magnetometer, a laboratory-built quasi-zero field furnace, and a Schonstedt af demagnetizer, all inside a mu metal and steel–shielded room. Magnetic susceptibility gave no indication for significant alteration or mineralogical transformations during heating. Natural Remanent Magnetization (NRM) intensity was on the order of 2 to 5 $10^{-4}Am^{-1}$ for the three blocks from unit 3 in the three sections and was 100 times stronger for the sample from unit 2 in the Peñon section.

A first component, roughly in the direction of the present field, was generally removed by 300 °C or 30 mT in the samples from unit 3. A higher unblocking temperature component was then removed between 450 and 580 °C. The carrier for this component is most likely a magnetite, as confirmed by isothermal saturation experiments. A comparison of the relative stabilities of Isothermal Remanent Magnetization (IRM) (saturating at 0.4 T) and Anhysteretic Remanent Magnetization (ARM) (acquired in a 0.5 mT dc field) against af demagnetization indicates that this magnetite is in the Pseudo Single Domain (PSD) range (about 2 to 3 μm). Two specimens from the strongly magnetized block from unit 2 do not show a present-day overprint and display quasi-univectorial, yet rather different, reversed magnetization. Although the corresponding directions are not inconsistent with those of other samples, remagnetization, for instance by lightning, cannot be excluded. The higher unblocking-temperature components are all of reversed polarity. The mean direction is D = 155.7°, I = –38.2° (α_{95} = 11.6°), when all 11 samples measured are included. This direction becomes D = 161.1°, I = –36.6° (α_{95} = 8.9°) when the two samples from the block of unit 2 are removed. These two directions are not statistically distinct. Although these data are clearly not numerous enough, the results are consistent. Moreover, the paleopole cannot be distinguished statistically from the Upper Cretaceous pole of Bohnel and Negendank (1988) from Mexico or from the poles predicted for the period 60 to 70 Ma by the North American Apparent Polar Wander Path (APWP) of Besse and Courtillot (1991). The results of this otherwise unpublished research by two of us (V .C. and Y. Z.) are therefore compatible with clastic units 2 and 3 having been deposited within chron 29R (of total duration on the order of .5 m.y.), and no reversal appears to be recorded in the limited interval sampled, although further work on unit 2 is clearly necessary. Therefore, these paleomagnetic results cannot as yet constrain the duration of deposition of the clastics, which could equally have taken 100,000 s or 100,000 yr.

Another point of debate lies with the duration of K/T extinctions as estimated by various paleontologists. For instance, Keller and Barrera (1990) find that a third of species of planktonic foraminifera disappear some 300,000 yr prior to the Cretaceous-Tertiary boundary (KTB) in the El Kef and Brazos River sections, with Cretaceous species surviving into the Tertiary, consistent with stable isotopic data. This finding has been confirmed in a review of K/T sections by MacLeod and Keller (1991, 1996; see also Keller et al., 1994b). They find that the mass extinction at El Kef is real but somewhat less in intensity and longer in duration than often stated and that the killing mechanism was selective across and within groups and restricted to lower latitudes (e.g., Zinsmeister et al., 1989; Keller et al., 1993b; Zinsmeister and Feldmann, 1994).

These views have been challenged, most prominently by Smit (1982). A blind test was organized following the Snowbird II conference: Six unlabeled samples, three from above and three from below the KTB, were collected from the El Kef section in 1992 and distributed to four paleontologists under supervision of R. N. Ginsburg (1994). Ginsburg was unfortunately unable to deliver the results himself at the Houston conference, and both Jan Smit and Gerta Keller felt that what was available vindicated their respective yet contradictory views. All investigators noted that more than one-third of Cretaceous species were found into the Tertiary. However, they differed greatly on the percentage of species dissappearing below the K/T: Between 20 and 2%! Pending full publication of the blind test results, we believe that the bulk of the paleontological literature does argue in favor of a duration on the order of .5 m.y. for the K/T biotic events responsible for unusual extinction levels. The KTB itself does seem to mark a rather short and intense event, at least at some latitudes, but "only" a third to half of the Cretaceous planktonic species become extinct at this point.

A REVIEW OF THE AGES OF FLOOD BASALTS AND MASS EXTINCTIONS

Recent work on large igneous provinces (e.g. Rampino and Stothers, 1988; Richards et al., 1989; Hill et al., 1992; Coffin and Eldholm, 1994) has emphasized their major geodynamic role and potential environmental impact. They are defined as massive crustal emplacements of predominantly mafic extrusive and intrusive rock that originate via processes other than "normal" ocean spreading and are commonly attributed to mantle plumes (Coffin and Eldholm, 1994). Increasing numbers of precise (but not always accurate) ages have become available and strengthen the view that among those emplacements, continental flood basalt (CFB) provinces, most with volumes in excess of 1,000,000 km^3, had major eruptive phases that lasted less than 1 m.y. and may be correlated with times of extinction events.

Compilations by Rampino and Stothers (1988), Richards et al. (1989), Stothers and Rampino (1990), Stothers (1993), Heimann et al. (1994), and Rampino (1994) list from 10 to 15 CFB with an age less than 250 Ma. In a recent review, Coffin and Eldholm (1994) list 14 CFB, 28 oceanic plateaus, and only three ocean basin flood basalts. The CFB status of the Emeishan basalts of South China (e.g., Chung and Jahn, 1993) and of the Wrangellia basalts (Richards et al., 1991) remains uncertain.

A causal link between some flood basalts and some mass extinctions was apparently first suggested by Vogt (1972) and argued for with increasing strength and data by McLean (1982), Officer and Drake (1985), Morgan (1986, 1988), Courtillot et al. (1986), and Officer et al. (1987). A chronology of initiation dates of major continental flood basalts from the last 250 m.y. was compiled by Rampino and Stothers (1988), and these authors indeed found the dates to be close to the estimated dates of mass extinctions of marine organisms (see also Stothers and Rampino, 1990). A recent update was proposed by Stothers (1993), who concluded that correlation of these initiation dates with major extinctions was indeed significant to a high level of confidence, with no measurable lag between the two time series. Sepkoski (1990) has constructed a composite time series of extinction intensity by averaging standardized time series for seven major taxonomic groups (his figure 4). Three outstanding (first-order) events—the end-Permian, end-Cretaceous, and end-Triassic extinctions—stand out clearly. Five smaller (second-order) events occur at the end of the Pliensbachian, Jurassic, Aptian, Cenomanian, and Eocene. Two even smaller (third-order) events are the end-Callovian and end-middle Miocene extinctions (the end-Jurassic could belong to this group). There is no visible end-Paleocene event in Sepkoski's (1990) diagram. Altogether, there appear to be eight prominent events and two or three less clear ones in the last 250 m.y. J. Sepkoski (personal communication, 1994) notes that several of these could be of only regional importance (the middle Miocene being an oceanic and the end-Paleocene a deep-water phenomenon [Thomas, 1990]; the end-Pliensbachian being mostly found in Europe) and that the very existence of an end-Bajocian event, which is included in some earlier compilations, is questionable.

We propose here a new summary of recently published CFB ages in which some new $^{40}Ar/^{39}Ar$ ages are included. Specifically, we briefly review recent data and debates regarding the precise age and duration of the Deccan traps, the Siberian traps, and other continental flood basalts and their relation to mass extinctions (Table 1 and Fig. 1). CFB ages are then compared to the ages of extinction events, as reviewed by Stothers (1993) and more recently by Gradstein et al. (1994) and Odin (1994).

The Deccan traps and the Cretaceous-Tertiary extinctions

Courtillot et al. (1986, 1988, 1990) proposed that then-available paleomagnetic, geochronologic, and paleontological data were together compatible with short eruption of the traps (less than 1 m.y. for the bulk of the flood basalts) near the time of the KTB mostly within reversed chron 29R. In a more recent review, Vandamme et al. (1991) summarized available $^{40}Ar/^{39}Ar$ ages, which they found to range between 63 and 68 Ma, with peaks at 63.5, 65, and 67 Ma. They stressed that the uncertainty in the overall resulting mean (65.5 ± 2.5 Ma) was likely to reflect remaining problems with the argon technique and not the duration of the eruptions, mostly because of the magnetostratigraphic data, which implied eruption within at most three chrons, and paleontological data, particularly from the Narsapur well (off the Bombay coast), which implied that the K/T boundary was indeed to be found within the lava pile. Vandamme et al. therefore confirmed that eruption was likely to have occurred in less than 1 m.y. near (and straddling) K/T time. A number of new $^{40}Ar/^{39}Ar$ plateau and isochron ages have become available recently (Basu et al., 1993; Venkatesan et al., 1993; Baksi, 1994; Radhakrishna et al., 1994) that deserve discussion.

Venkatesan et al. (1993) reported new $^{40}Ar/^{39}Ar$ plateau ages of nine basalt samples collected over more than 2.5 km of section in the western margin of the flood basalt province (western Ghats) and compared them with data obtained previously by Duncan and Pyle (1988) on the same formations. Both series of data were interpreted as reflecting (1) very clustered ages of 67.0 ± 0.3 Ma (plateau ages) or 67.6 ± 1.2 Ma (isochron ages) for the lower 1,800 m of the composite section and (2) significantly younger ages, reaching 62 Ma, in the upper 900 m of the section. Two of the main conclusions of Venkatesan et al. (1993) were that the basal section is distinctly older than the K/T boundary and that the emplacement of the whole section lasted at least 3 m.y. Féraud and Courtillot (1994) challenged this view. As suggested by Venkatesan et al. (1993), some data are probably affected by ^{39}Ar recoil effects, as indicated by higher ages at lower temperatures. This phenomenon is frequent in fine-grained whole rocks and is due to ^{39}Ar displacement from the fine-grain K-rich phases. Although only four samples were interpreted by Venkatesan et al. (1993) as having been affected by ^{39}Ar recoil, these authors considered their other plateau ages as reliable. However, Féraud and Courtillot (1994) pointed out several other samples that could have been affected by the same

TABLE 1. A LIST OF CONTINENTAL FLOOD BASALT AGES AND CORRESPONDING EXTINCTION BOUNDARIES*

Flood Basalt	Age	Unc.	Ref†	Boundary	Age 1§	Unc.	Ref†	Age 2§	Unc.	Age 3§	Unc.
Columbia	17.0	1.0	1	Middle Miocene	10.5	1.0	1				
Ethiopian	35.0	2.0	2	Upper Eocene	35.0	1.0	1				
North Atlantic	59.0	2.0	1	(Upper Paleocene)	57.0	2.0	7				
Deccan	65.5	2.5	3, 8	Maastrichtian	65.0	1.0	1, 10	65.0	0.1	65	0.5
Madagascar	90.0	5.0	1	Cenomanian	91.0	1.0	1	93.5	0.2	92	2
Rajmahal	116.0	2.0	1	Aptian	119.0	6.0	7	112.2	1.1	108	3
Serra Geral	133.0	1.0	1	Late Jurassic	140.0	5.0	1, 6	144.2	2.6	135	5
Etendeka	135.0	5.0	2	Late Jurassic	140.0	5.0	1, 6	144.2	2.6	135	5
Antarctic	176.6	1.8	4	Callovian?	158.0	5.0	7	159.4	3.6	154	5
				(Bajocian?)	171.0	5.0	1, 4	169.2	4.0	164	2
Karoo	182.0	2.0	5	(Bajocian?)	171.0	5.0	1, 4	169.2	4.0	164	2
				Pliensbachian?	190.0	4.0	1	189.6	4.0	184	-
Eastern N. America	201.0	1.0	1	Late Triassic	204.0	4.0	1	205.7	4.0	203	3
Siberian	247.0	2.0	1	Late Permian	250.0	3.0	1	248.2	4.8	250	5

*Ages and uncertainties (unc.) are in Ma.
†References include: 1 = Stothers, 1993; 2 = Rampino and Stothers, 1988; 3 = Vandamme et al., 1991; 4 = Heimann et al., 1994; 5 = Hooper et al., 1993; 6 = Renne et al., 1992; 7 = Kent et al., 1995, and Harland et al., 1990; 8 = this chapter.
§For the ages of boundaries, we give three estimates: Age 1 with references used in this chapter; Age 2 from Gradstein et al., 1994; and Age 3 from Odin, 1994, illustrate remaining differences between very recent numerical scales.

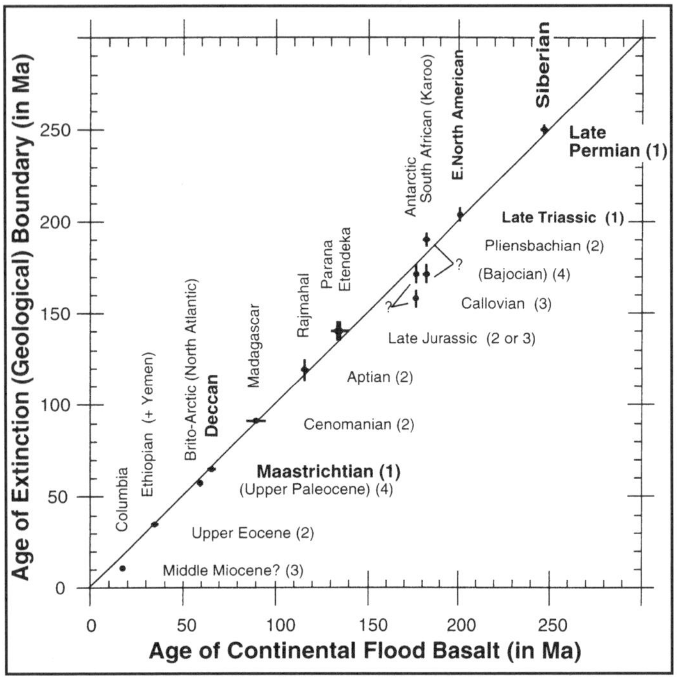

Figure 1. Updated correlation of the ages of post–300 Ma continental flood basalts (CFB) with ages of extinction (geological) boundaries with which they could be associated. Data and uncertainties from Table 1. Names of CFBs are labeled vertically next to data points (geological boundaries, labeled horizontally). The number after the boundary name refers to the order (size) of the extinction following Sepkoski (1990) and this chapter. Boundaries between parentheses correspond to extinctions not in Sepkoski's (1990) list of major events (i.e., "fourth order"). Question marks are indicated next to two CFBs for which either one of two correlations could be proposed. Radius of symbols is about 1 m.y. (uncertainties equal to or smaller than this do not show).

phenomenon. Moreover, two of the four age spectra given by Duncan and Pyle (1988) could have been affected by ^{39}Ar recoil as well. In many of the above-mentioned age spectra, the decreasing apparent ages versus temperature reach a nearly constant value, close to 65 Ma. The pattern continues with a sharp decrease of ages when the very last fractions of argon are released. This age close to 65 Ma also approximately corresponds to the minimum age of the saddle-shaped age spectrum displayed by one sample, which is probably affected by excess argon and/or recoil. Altogether, the true ages of some eight samples out of 17 could be younger than reported by Venkatesan et al. (1993): The ^{40}Ar /^{39}Ar data obtained on the main basal section and reported in their figure 2 may not be significantly older than about 65 Ma, corresponding to several intermediate to high temperature ages measured by both Venkatesan et al. and Duncan and Pyle (1988).

It is of interest to compare more precisely the ranges of ages obtained for the whole western Ghats composite section in the above two studies and in Baksi (1994). Plateau ages range from 62.1 to 67.5 Ma (respectively 61 to 68 Ma when uncertainties are included) for Venkatesan et al. (1993), from 62.7 to 68.5 Ma (respectively about 60 to 70 with uncertainties) for Duncan and Pyle (1988), and 64.9 to 65.9 Ma (respectively about 64.4 to 66.3 with uncertainties) for Baksi (1994). Corresponding ranges for isochron ages are respectively 60.0 to 66.9 (respectively 59 to 70), 66.6 to 68.5 (respectively 65.7 to 69.0) and 66.0 to 66.9 (respectively 65 to 68) Ma. When one dismisses the upper flows (above 800 m) that Venkatesan et al. (1993) claim to be significantly younger, the ranges decrease to 66.1 to 67.5 (respectively 65 to 68) Ma for plateaus and 64.1 to 68.9 (respectively 63-70) Ma for isochrons. It is interesting that for the same lava sequence the ranges of mean values of plateau ages (without uncertainties) are 2 m.y for Venkatesan et al. (1993) and 6 m.y for Duncan and Pyle (1988), whereas for isochron ages they are respectively 5 m.y. and 2 m.y. (that is, the reverse). For Baksi (1994) these ranges decrease to 1 m.y. (respectively 2 m.y.). If one now focuses on the top flows from the Mahabaleshwar formation, plateau ages are about 62 Ma for Venkatesan et al. (samples AM 83-7 and MB 81-24), 65 Ma for Baksi (sample MAP 052), and 68 Ma for Duncan and Pyle (sample MAP 057), with no intersection of the uncertainties for what must be coeval flows based on field relationships! The same applies to the isochron ages, which are individually displaced by 1 to 2 m.y. (either way) with respect to their companion plateau ages. This displacement can be due to undetected alteration in some samples or to the fact that interlaboratory uncertainties are (at least in some cases) much larger than quoted values. It is clear that very few laboratories are able to produce argon ages that have reliable uncertainties less than 0.1 m.y. and that these uncertainties should only be used when comparing the relative ages of samples from the same run. Not only undetected alteration and heterogeneities but also differences in laboratory procedures, uncertainties in neutron fluxes and remaining differences and uncertainties in the ages of monitors render the comparison of absolute ages given in separate runs or by different laboratories at the 1% uncertainty level still often unrealistic.

Basu et al. (1993) have dated three alkalic igneous complexes, two that lie to the north of the traps in the Cambay graben and one (Phena Mata) close to the Narmada valley and intruding some of the trap flows. The first two complexes are found to be 68.5 ± 0.2 and 68.6 ± 0.1 Ma respectively and the third 65.0 ± 0.1 Ma. Basu et al. (1993) interpret the former as evidence for an early phase of magmatism and the latter as a late phase. This last age is, for instance, indistinguishable from the main trap ages of Baksi (1994). Radakrishna et al. (1994) report a single ^{40}Ar/^{39}Ar age from a series of northwest-southeast– trending, mostly reversely magnetized dolerite dykes in central Kerala, south of the present outcrops of the traps. The plateau age is 71.9 ± 1.5 Ma, to which the authors prefer the isotope correlation age of 69.4 ± 1.1 Ma. Radakrishna et al. (1994) interpret these as part of the system of dykes that fed the Deccan flows and think (based in part on paleomagnetic results) that these dykes are correlative with the tholeitic units of the upper Deccan basalts. Radakrishna et al.

propose that the dykes correspond to chron 31R and extend that correlation to the bulk of the traps, hence making the same (unwarranted) placement of the KTB within 31R as Venkatesan et al. (1993, 1994). *To do so runs contrary to all detailed magnetostratigraphic results of the last 15 years, which have clearly positioned the K/T boundary within chron 29R* (e.g., Alvarez et al., 1977, and virtually all magnetostratigraphic time scales since). The dykes could also correlate with the older alkalic complexes of Basu et al. (1993). In any case these older data are very few and correspond to a volumetrically very minor magmatism whose correlation with the main body of the traps is not well established; also the older data all fall in the range of the overall age histogram.

If the new data are added to the compiled histogram of Vandamme et al. (1991, their fig. 9), none of the features of this histogram is changed and the bulk of the data concentrates between 64 and 69 Ma. But this spread clearly does not (entirely) reflect the duration of volcanism: Problems with sample alteration and measurement and major discrepancies between laboratories remain and absolute values cannot be trusted to the nearest m.y. (see, for instance, Féraud and Courtillot, 1994). Comparisons within an individual laboratory are far easier: Based on the internal comparison of their own data, both Duncan and Pyle (1988) and Baksi (1994) supported the views of Courtillot et al. (1986, 1988) that most of the volcanics were erupted in less than 1 m.y. For the same reasons it cannot be said (as do Venkatesan et al., 1993) that the bulk of volcanism started significantly (meaning several million as opposed to several hundred thousand years) prior to the K/T. Khajuria et al. (1994) have reviewed paleontological constraints on the age of the traps, and also conclude that the first flows erupted during the *Abatomphalus mayaroensis* zone, that is, within chron 30N. However, they propose that volcanism lasted longer, into the early Paleocene.

An important clue could be the report by McWilliams et al. (1992) that Deccan volcanism began significantly before the KTB, as characterized by the Haiti tektites and bentonite beds in western North America: "Step heating . . . of the Deccan Trap sequence yields less well-constrained ages. It is clear that Deccan volcanism began before the K/T boundary as recorded by the Haiti tektites and associated Ir anomalies in North America." However, these findings are in an abstract in which numerical ages are not quoted and do not seem to have led to a published paper in which the data can be analyzed; also, this result is not quoted by one of its coauthors in later contradictory work (Baksi, 1994). A particularly important result, which appears to invalidate the proposal of McWilliams et al. (1992), is the finding by Bhandari et al. (1995) of the K/T iridium anomaly layer within intertrappean sediments near Anjar in the province of Kutch. The layer lies between two basalt flows of unknown magnetic polarity, respectively dated at 65.4 ± 0.7 and 65.5 ± 0.7 Ma (Pande et al., 1988), with some paleontological constraints. This finding apparently establishes that Deccan volcanism was active before and during the K/T boundary, as defined by the iridium anomaly.

Altogether, the contemporaneity of the Deccan traps and K/T seems established. Although the total duration of the major phases of volcanism may have been less than 1 m.y. (within chron 29R), the absolute ages of neither the Deccan nor K/T are yet known to that accuracy, given remaining interlaboratory uncertainties. The range of Deccan ages given by the pooled $^{40}Ar/^{39}Ar$ data is about 65.5 ± 2.5 Ma (Vandamme et al., 1991, hardly changed by inclusion of the new data mentioned in this chapter), whereas the range of published K/T ages is 65 ± 1 Ma (Lerbekmo et al., 1979; Baadsgaard et al., 1988; Gillot et al., 1991; Odin, 1989; Izett et al., 1991; Swisher et al., 1992) (see Table 1 and Fig. 1). The uncertainty would be reduced to 0.5 m.y. following Odin (1994) and even 0.1 m.y. following Gradstein et al. (1994)!

The Siberian traps and the Permo-Triassic extinctions

The K/T mass extinction is actually not the largest in the last 300 m.y., but only ranks second after the Late Permian (or Permo-Triassic, or P/T) extinction or possibly third after the Late Triassic extinction (e.g., Sepkoski, 1990, comparing his figs. 1 and 4). That the ages of the Siberian traps and the P/T extinctions would eventually be found to coincide was proposed by several authors (e.g., Morgan, 1988; Courtillot, 1990). Paleontological data for the P/T extinction have recently been reviewed by Erwin (1994a, b), who does not find support for earlier claims that it occurred over as much as 8 m.y. (Teichert, 1990), but rather proposes a duration of 2 m.y. or less. There is as yet no evidence for an extraterrestrial impact at or close to the P/T boundary (but see Rampino, 1994). There is on the other hand significant evidence for volcanism (and on a longer term, as is the case around the K/T, a possibly related sea-level regression). Erwin and Vogel (1992) find pyroclastic volcanism in South China (the Emeishan basalts mentioned above) too small to have had a significant effect on the environment. On the other hand, recent geochronological and paleomagnetic data (the latter must be confirmed) support the fact that the massive Siberian traps did apparently erupt in less than 1 m.y. and at P/T time (Renne and Basu, 1991; Dalrymple et al., 1991; Campbell et al., 1992).

Although Erwin (1994a, b; see also Holser and Magaritz, 1992) stresses that the end-Permian extinction differed from the end-Cretaceous in several respects, we can only point out the similarity of the Deccan and Siberian flood basalts seen from a magnetostratigraphic and geochronologic perspective (the pre-erosional volumes were also similar, in excess of 2,000,000 km^3). Available dates are (depending on how one sees it) both good and bad enough that coincidence between the ages and durations of extinctions and flood basalt volcanism can in neither case be excluded (Table 1 and Fig. 1). Also geochemical anomalies extend in both cases over intervals on the order of 100,000 years. Campbell et al. (1992) argue that the Siberian eruptions were both explosive and very rich in sulfur and would have induced global cooling and expansion of the icecaps, in turn resulting in the large but short-lived regression. A single (if not instantaneous) cause could then have been the main agent of the final P/T extinction, not requiring the complex, multiple "tangled web" invoked by Erwin (1994a).

Other traps and extinctions

Other ages, corresponding to 10 CFBs in addition to the two above, are listed in Table 1. Dates for the Ethiopian (and associated Yemen) traps are rather uncertain (Rampino and Stothers, 1988); we are currently undertaking new fieldwork to provide a remedy for this situation. Coffin and Eldholm (1994) have reviewed data from the North Atlantic (or Brito-Arctic) volcanic province, and they find that temporal correspondence with the end of the Paleocene (at about 57 Ma) is "compelling," despite the fact that this is not generally recognized as a major extinction event (Sepkoski, 1990; but see Thomas, 1992). Earlier suggestions (Courtillot and Cisowski, 1987; Rampino and Stothers, 1988) were that the North Atlantic province might have correlated with the K/T. This is clearly not the case. This province is another example of a situation in which magnetostratigraphy and paleontology provide good clues as to the short duration and actual biostratigraphic placement of the traps (Mussett, 1986; Dickin, 1988; Mussett et al., 1988; Noble et al., 1988). Most of the volcanism probably occurred within less than 2 m.y. around 60 Ma.

Recent work has confirmed that the Parana-Serra Geral traps from South America and the Etendeka traps from Namibia were originally part of the same short-lived province. Renne et al. (1992) propose that, based on new paleomagnetic and $^{40}Ar/^{39}Ar$ data, volcanism started at 133 ± 1 Ma and lasted less than 1 m.y., before initiation of seafloor spreading in the South Atlantic. Mantovani et al. (1993), on the other hand, claim that volcanism lasted 10 m.y., from 137 to 127 Ma, providing a much smaller average eruption rate (hence possibly climatic impact). Renne et al. (1993) see the coast-parallel dykes dated at 125 to 120 Ma as a distinct later phase. It is clear in any case that magmatism preceded rifting. Renne et al. (1992) note that most estimates of the age of the Jurassic-Cretaceous boundary (J/K, a second- or even third-order extinction by Sepkoski's measure) are in the range of 135 to 145 Ma and conclude that temporal correlation is unlikely. However, they also observe that an age near 130 Ma has been suggested (Odin et al., 1992); remaining uncertainties in the absolute age of the J/K do not allow one so far to exclude contemporaneity (see Table 1 and Fig. 1). There is no correlation if one follows the scale of Gradstein et al. (1994) but good correlation following Odin (1994).

Heimann et al. (1994) have recently reported $^{40}Ar/^{39}Ar$ isochron and plateau ages for the Ferrar Group, part of a CFB province in Antarctica. They find that eruptive activity occurred within less than 1 m.y. at 176.8 ± 1.8 Ma; as for the Deccan, for instance (for recent measurements from a single laboratory), ages from different stratigraphic levels or locations could not be distinguished. This event may also be found in South Africa, Tasmania, and South Australia, although dates there do not allow a precise correlation. This CFB is much thinner but has a larger geographical extent than other ones. Its linear shape and other geological arguments (Heimann et al., 1994) point to a control by rifting of Gondwana. Rampino and Stothers (1988) pointed to a possible relationship with the end of the Bajocian (not a major extinction), for which the spread of proposed ages is from 166 to 176 Ma. The correspondence would then be acceptable. We should recall that only end-Callovian and end-Pliensbachian events (about 158 ± 5 Ma and 190 ± 4 Ma respectively) show in Sepkoski's list.

The Karoo basalts of southern Africa are also associated with the breakup of eastern from western Gondwana. They comprise large (now separated) outcrops in Lesotho, Swaziland, Mozambique, and Zimbabwe (Fitch and Miller, 1984; Cox, 1992). Compilation of K-Ar ages by Rampino and Stothers (1988) indicates a start date at 190 ± 5 Ma. Fitch and Miller (1984) find peaks of magmatism at 193 Ma and 178 Ma, the latter being possibly coeval with the Antarctic magmatism (Heimann et al., 1994). Hooper et al. (1993) have recently produced new $^{40}Ar/^{39}Ar$ ages for the basalts of Lesotho. They find only one reversal in the lower quarter of the 1,500-m-thick lava pile, and preliminary dates fall within 182 ± 2 Ma. Hence, we have another indication of a fast-produced CFB. However, it is not yet known whether this extends to other Karoo outcrops farther to the southeast and northeast. It is also not clear whether the new Lesotho and Antarctic dates are significantly different, as the published uncertainties would seem to imply (but is the lack of intersection of uncertainties by only 1 m.y. really significant?). Major volcanism obviously took place over a vast area at about 180 Ma, close to the Early to Middle Jurassic boundary (end-Toarcian). But this is not when the nearest major extinction is located, according to Sepkoski (1990), who places it at the Pliensbachian, dated about 190 ± 4 Ma. Hence, with the data presently available, the Karoo or Antarctic volcanism do not seem to match well with major extinctions. Note, however, that Odin (1994) proposes an age of 184 Ma for the end of the Pliensbachian, which would then fit the Karoo ages. It is expected that further detailed work will settle this matter.

Finally, Rampino and Stothers discuss the Upper Triassic and Lower Jurassic basalts of eastern North America, related to the rifting of Africa and North America and the opening of the Central Atlantic, which they argue have the dimensions of a CFB province. They estimate the start age of magmatism at 200 ± 5 Ma. This age was revised to 201 ± 1 Ma by Stothers (1993) following Dunning and Hodych (1990). The entire Newark extrusive zone is believed to represent a short-lived (≅1 Ma) igneous episode (Olsen and Fedosh, 1988; Fowell and Olsen, 1993). Stothers associates minor volcanism on the west African side at 200 ± 3 Ma with this (Sebai et al., 1991), in a manner reminiscent of the rifted Parana-Etendeka pair. The eastern North America province does not appear in Coffin and Eldholm's (1994) compilation. However, they do list volcanic passive margins along the Baltimore Canyon and Carolina troughs. These are Mesozoic features linked to breakup and opening of the Atlantic and possibly affected by the Cape Verde hot spot (Austin et al., 1990). Sheridan et al. (1993) suggest that the volcanic wedge, whose volume "rivals the Deccan traps," stood above sea level, which may have led to more severe environmental consequences (see below). This feature remains unsufficiently studied and dated. Its importance, among other things, stems from the fact that it could have been associated with the end-Triassic extinction (204 ± 4 Ma), the third largest after the end-Paleozoic and the end-Mesozoic ones. The Triassic-Jurassic boundary is indeed estimated to have an age of ≅202 Ma and to be in close synchroneity with the Newark rift basins (Hodych and Dunning, 1992; D. V. Kent, personal communication, 1994).

In summary, the ages of 12 CFBs are listed in Table 1 and in Figure 1 are plotted against the ages of extinction events suggested to have been associated with them. Although the coincidence appears to be impressive, not all data are equally reliable and some age uncertainties may have been underestimated (see discussion above on $^{40}Ar/^{39}Ar$ ages). Also, because of the type of display chosen, not all the space in Figure 1 is available, and data points are restricted to a steplike band about the diagonal. Yet, it is interesting that in almost all cases when modern $^{40}Ar/^{39}Ar$ ages have been obtained, the age range has significantly been reduced over previous estimates and the correlation is improved. Eight provinces out of 12 where modern $^{40}Ar/^{39}Ar$ techniques (and in most cases magnetostratigraphy) have now been used have fairly well-constrained ages: the Columbia, North Atlantic, Deccan, Serra Geral–Parana and Etendeka, Antarctic, part of Karoo (Lesotho), and Siberian traps. Three of these show a very good correlation with an extinction event, namely, the two most important ones (K/T and P/T) and a minor one (end-Paleocene). Correlation of Parana-Etendeka with the J/K is marginal, but we still consider it as a possibility. Two do not correlate well or at all with extinctions: the Columbia and Antarctic/Karoo (although the Karoo would fit with the Odin [1994] scale). In the case of the Columbia, the correlation would be good if it corresponded to the end of the Lower rather than Middle Miocene. But this is not where Sepkoski (1990) places an extinction event. Note that the Columbia flood basalts constitute one of the smallest CFB provinces. The four CFBs that do not have very well-constrained ages (Ethiopian/Yemen, Madagascar, Rajmahal, and eastern North America) happen to all coincide rather well with four of Sepkoski's first- and second-order events. They are all clearly primary targets for future more detailed work, already in progress for the Ethiopian traps. The eastern North America province is also a major target because of the importance of the end-Triassic extinction. Paleontological work is essential to tie the CFBs with extinctions, independently from the absolute ages. In this way, $^{40}Ar/^{39}Ar$ ages for the CFBs would become key constraints for improving the accuracy of absolute ages of major geological boundaries (as is done for the P/T by Gradstein et al., 1994). Such work has only been performed extensively for the Deccan and largely remains to be done elsewhere.

CONCLUSION

We believe that the correlation shown in Figure 1 and discussed above is sufficiently striking that it cannot be discounted lightly. Of the 10 extinction events listed by Sepkoski (1990), seven are associated with a flood basalt. Of 12 flood basalts observed in the last 250 m.y., at least nine can be associated with an extinction event. The correlation is clearest for the two largest events. The only CFBs in that age range that have been mentioned by others but are not listed in Table 1 are the small or ill-defined West African, Wrangellia, Caribbean, and Emeishan. It seems hardly necessary to evaluate quantitatively the degree of confidence of the correlation, which is found to be higher than 95% in the version of Stothers (1993).

Several authors have claimed some amount of correlation between impacts and extinctions other than K/T. Most recently, Rampino and Haggerty (1994) claimed that six extinction events from the last 550 m.y. were associated with evidence considered to be diagnostic of large impacts. Curiously, they do not mention flood basalts. In the age range of interest to us here, they mention three craters. Two (Manicouagan at 212 ± 2 Ma and Puzech at 220 ± 10 Ma) have ages that do not overlap one of Sepkoski's major extinctions (although this may depend on the time scale used). One, the 100-km-diameter Popigai crater, is dated at 36 ± 1 Ma and correlates with the late Eocene event. This event apparently has a smaller size than would be predicted from Raup's (1991) "species kill curve." Small multiple iridium anomalies have been found in the last 1 m.y. of the Eocene at an Italian and an Antarctic section (Montanari et al., 1993). Shocked quartz has also been found close to the Triassic-Jurassic boundary in Italy (Bice et al., 1992). But the significance of these reports is still, in our view, in need of confirmation, and the number of clear cases is still limited to one (the K/T).

Claims that impacts could generate flood basalts (e.g., Rampino and Stothers, 1988), however interesting, do not seem founded (White, 1989). Observations such as time constraints, lack of iridium for boundaries other than K/T, and probable involvement of the lower mantle (e.g., Richards et al., 1989; Coffin and Eldholm, 1994) seem to argue against such a view. For instance, should the Deccan traps have been triggered by impact, this would place the timing of impact within chron 30N, several hundred thousand years before the Ir anomaly (see also Bhandari et al., 1995).

The potential environmental impact of large and rapid flood basalt eruptions has recently received attention (e.g. McLean, 1985; Stothers et al., 1986; Officer et al., 1987; Rampino and Stothers, 1988; Courtillot, 1990; Sigurdsson, 1990; Caldeira and

Rampino, 1990a, b; Campbell et al., 1992). The damage produced by flood basalts may result from a combination of large size, rapid emplacement, detrimental chemical composition, and subaerial emplacement, all variable parameters. Latitude may also be a factor. Hence CFBs are strong candidates as causes of most mass extinction events, at least since the end of the Carboniferous. Single flows in the Deccan traps may exceed 1,000 km^3 in volume and may have been erupted over a matter of weeks (Claude Jaupart, 1993, personal communication). Such instantaneous rates in excess of 1,000 km^3/yr should be viewed against rates for the 1783 Laki eruption in Iceland (Sigurdsson, 1982), which are 100 times less (a single 12 km^3 flow in eight months). Many such flows may have been emplaced over a comparatively short period. SO_2 concentration and subaerial emplacement seem to have been key parameters. The Deccan and Siberian traps were erupted entirely on land and sent their gaseous discharges directly into the atmosphere. In contrast, the Ontong-Java event (e.g., Tarduno et al., 1991), possibly the largest igneous province on Earth (Coffin and Eldholm, 1994), dated between 124 and 121 Ma, erupted under water and does not correspond to a mass extinction. The North Atlantic CFB is intermediate; it was partly emplaced under water, which is possibly a reason why the associated extinction is relatively minor. Most of the other CFBs mentioned in Table 1 were subaerial; it is particularly important in this respect to know how much of the eastern North America province was erupted above sea level (Sheridan et al., 1993).

On the other hand, the K/T iridium anomaly remains a unique feature in the geological record, and the shocked minerals are strong evidence for an impact, with no unequivocal explanation in terms of volcanism. Results discussed at the Houston conference (iridium anomaly, large size of the crater; see Sharpton et al., 1993) would seem to indicate that the Chixculub K/T impact was a unique event in the entire Phanerozoic. This leaves all other mass extinctions unexplained.

There is no doubt that large impacts have occurred on Earth fairly recently, but (except for Chixculub) with no major biological consequences. Large (40- to 80-km diameter) craters such as Montagnais or Kara and Ust Kara are dated from some 10 m.y. after and before the K/T, respectively, and do not seem to have had a significant effect on extinction rates.

It does appear that an unusual impact did coincide with the K/T and was responsible for the extinction of a third of the species that disappeared in the crisis: Recent data on the Chixculub structure point that way. But the extinction event was already under way due to Deccan volcanism. CFB volcanism, the role of which seems to have been underestimated in the last decade, remains the main candidate for most other extinction events. Other internal processes, such as regression and continental breakup, may have been associated. But for all these events, the primary causes are the same, and they are internal to the Earth. The K/T is the only one in which two events, one internal and the other external, occurred in contemporaneity (see also Sutherland, 1994).

ACKNOWLEDGMENTS

This chapter was written in 1994 while the senior author enjoyed a Sherman Fairchild Distinguished Scholarship at the Division of Geological and Planetary Sciences at the California Institute of Technology, for which he is most thankful. IPGP contribution NS 1359.

REFERENCES CITED

Alvarez, L. W., Alvarez, W., Asaro, F., and Michel, H. V., 1980, Extraterrestrial cause for the Cretaceous-Tertiary extinction: Science, v. 208, p. 1095–1108.

Alvarez, W., Arthur, M. A., Fisher, A. G., Lowrie, W., Napoleone, G., Premoli Silva, I., and Roggenthen, W. M., 1977, Upper Cretaceous-Paleocene magnetic stratigraphy at Gubbio, Italy. V: Type section for the late Cretaceous-Paleocene geomagnetic reversal time scale: Geological Society of America Bulletin, v. 88, p. 383–389.

Alvarez, W., Smit, J., Lowrie, W., Asaro, F., Margolis, S. V., Claeys, P., Kastner, M., and Hildebrand, A. R., 1992, Proximal impact deposits at the Cretaceous-Tertiary boundary in the Gulf of Mexico: A restudy of DSDP Leg 77 Sites 536 and 540: Geology, v. 20, p. 697–700.

Austin, J. A., Stoffa, P. L., Phillips, J. D., Oh, J., Sawyer, D. S., Purdy, G. M., Reiter, E., and Makris, J., 1990, Crustal structure of the southeast Georgia embayment–Carolina trough: Preliminary results of a composite seismic image of a continental suture (?) and a volcanic passive margin: Geology, v. 18, p. 1023–1027.

Baadsgaard, H., Lerbekmo, J. F., and McDougall, I., 1988, A radiometric age for the Cretaceous-Tertiary boundary based upon K-Ar, Rb-Sr and U-Pb ages of bentonites from Alberta, Saskatchewan and Montana: Canadian Journal of Earth Sciences, v. 25, p. 1088–1097.

Baksi, A. K., 1994, Geochronological studies on whole-rock basalts, Deccan Traps, India: evaluation of the timing of volcanism relative to the K-T boundary: Earth and Planetary Science Letters, v. 121, p. 43–56.

Basu, A. R., Renne, P. R., DasGupta, D. K., Teichmann, F., and Poreda, R. J., 1993, Early and late alkali igneous pulses and a high-^3He plume origin for the Deccan flood basalts: Science, v. 261, p. 902–906.

Besse, J., and Courtillot, V., 1991, Revised and synthetic polar wander paths of the African, Eurasian, North American and Indian plates, and true polar wander since 200 Ma: Journal of Geophysical Research, v. 96, p. 4029–4050.

Bhandari, N., Shukla, P. N., Ghevaria, Z. G., and Sundaram, S. M., 1995, Impact did not trigger Deccan volcanism: Evidence from Anjar K/T boundary intertrappean sediments: Geophysical Research Letters, v. 22, p. 433–436.

Bice, D., Newton, C., McCauley, S., Reiners, P., and McRoberts, C., 1992, Shocked quartz at the Triassic-Jurassic boundary in Italy: Science, v. 255, p. 443–446.

Bohnel, H., and Negendank, J. F. W., 1988, Palaeomagnetism of the Puerto Vallarta intrusive complex and the accretion of the Guerrero terrain, Mexico: Physics of the Earth and Planetary Interiors, v. 52, p. 330–338.

Caldeira, K. G., and Rampino, M. R., 1990a, Carbon dioxide emissions from Deccan volcanism and a K/T boundary greenhouse effect: Geophysical Research Letters, v. 17, p 1299–1302.

Caldeira, K. G., and Rampino, M. R., 1990b, Deccan volcanism, greenhouse warming, and the Cretaceous/Tertiary boundary, *in* Sharpton, V. L., and Ward, P. D., eds., Global catastrophes in Earth history: An interdisciplinary conference on impacts, volcanism, and mass mortality: Geological Society of America Special Paper 247, p. 117–123.

Campbell, I. H., Czamanske, G. K., Fedorenko, V. A., Hill, R. I., and Stepanov, V., 1992, Synchronism of the Siberian traps and the Permian-Triassic boundary: Science, v. 258, p. 1760–1763.

Chung, S. L., and Jahn, B., 1993, The permo-triassic Emeishan flood basalt in SW China: Terra Nova, v. 5, abstr. suppl. 1, p. 424.

Coffin, M. F., and Eldholm, O., 1994, Large igneous provinces: Crustal structure, dimensions, and external consequences: Reviews of Geophysics, v. 32, p. 1–36.

Courtillot, V., 1990, A volcanic eruption?: Scientific American, v. 263, p. 85–92.

Courtillot, V., and Cisowski, S., 1987, The Cretaceous-Tertiary boundary events: External or Internal Causes?: Eos (Transactions, American Geophysical Union), v. 68, p. 193–200.

Courtillot, V., Besse, J., Vandamme, D., Montigny, R., Jaeger, J. J., and Cappetta, H., 1986, Deccan flood basalts at the Cretaceous/Tertiary boundary?: Earth and Planetary Science Letters, v. 80, p. 361–374.

Courtillot, V., Féraud, G., Maluski, H., Vandamme, D., Moreau, M. G., and Besse, J., 1988, Deccan flood basalts and the Cretaceous/Tertiary boundary: Nature, v. 333, p. 843–846.

Courtillot, V., Vandamme, D., Besse, J., Jaeger, J. J., and Javoy, M., 1990, Deccan volcanism at the Cretaceous/Tertiary boundary; data and inferences, *in* Sharpton, V. L., and Ward, P. D., eds., Global catastrophes in Earth history: An interdisciplinary conference on impacts, volcanism, and mass mortality: Geological Society of America Special Paper 247, p. 401–409.

Cox, K. G., 1992, Karoo igneous activity and the early stages of the break-up of Gondwanaland, *in* Storey, B. C., Alabaster, T., and Pankhurst, R. J., eds., Magmatism and the causes of continental break-up: Geological Society of London Special Publication 68, p. 137–148.

Dalrymple, G. B., Czamanske, G. K., and Lanphere, M. A., 1991, $^{40}Ar/^{39}Ar$ ages of samples from the Noril'sk-Talnakh ore-bearing intrusions and the Siberian flood basalts, Siberia: Eos (Transactions, American Geophysical Union), v. 72, p. 570.

Dickin, A. P., 1988, The North Atlantic Tertiary Province, *in* MacDougall, J. D., ed., Continental flood basalts: Norwell, Massachusetts, Kluwer Academic, p. 111–149.

Duncan, R. A., and Pyle, D. G., 1988, Rapid eruption of the Deccan flood basalts at the Cretaceous-Tertiary boundary: Nature, v. 333, p. 841–843.

Dunning, G. R., and Hodych, J. P., 1990, U/Pb zircon and baddeleyite ages for the Palisades and Gettysburg sills of the northeastern United States: Implications for the age of the Triassic/Jurassic boundary: Geology, v. 18, p. 795–798.

Erwin, D. H., 1994a, The Permo-Triassic extinction: Nature, v. 367, p. 231–236.

Erwin, D. H., 1994b, The end-Permian mass extinction: a complex, multicausal extinction, *in* New developments regarding the K/T event and other catastrophes in Earth history: Houston, Texas, Lunar and Planetary Institute Contribution 825, p. 33.

Erwin, D. H., and Vogel, T. A., 1992, Testing for causal relationships between large pyroclastic volcanic eruptions and mass extinctions: Geophysical Research Letters, v. 19, p. 893–896.

Féraud, G., and Courtillot, V., 1994, Did Deccan volcanism pre-date the Cretaceous/Tertiary transition?: Comment: Earth and Planetary Science Letters, v. 122, p. 259–262.

Fitch, F. J., and Miller, J. A., 1984, Dating Karoo igneous rocks by the conventional K-Ar and $^{40}Ar/^{39}Ar$ age spectrum methods, *in* Erlank, A. J., ed., Petrogenesis of the volcanic rocks of the Karoo Province: Geological Society of South Africa Special Publication 13, p. 247–266.

Fowell, S. J., and Olsen, P. E., 1993, Time calibration of Triassic/Jurassic microfloral turnover, eastern North America: Tectonophysics, v. 222, p. 361–369.

Gillot, P. Y., Jéhanno, C., Rocchia, R., and Sigurdsson, H., 1991, Datation par la méthode potassium-argon de la limite Crétacé-Paléogène en milieu marin: Âge des verres de Béloc (Haïti): Comptes Rendus de l'Académie des Sciences, Paris, v. 313, p. 193–199.

Ginsburg, R. N., 1994, Results of blind tests to resolve controversies: Iridium at Gubbio, extinctions at El Kef, *in* New developments regarding the K/T event and other catastrophes in Earth history: Houston, Texas, Lunar and Planetary Institute Contribution 825, p. 44.

Goltrant, O., Cordier, P., and Doukhan, J.-C., 1991, Planar deformation features in shocked quartz; a transmission electron microscopy investigation: Earth and Planetary Science Letters, v. 106, p. 103–115.

Goltrant, O., Doukhan, J.-C., Cordier, P., and Courtillot, V., 1992, An investigation by transmission electron microscopy of planar deformation features in naturally shocked quartz: Terra Nova, v. 4, p. 405–412.

Gradstein, F. M., Agterberg, F. P., Ogg, J. G., Hardenbol, J., van Veen, P., Thierry, J., and Huang, Z., 1994, A Mesozoic time scale: Journal of Geophysical Research, v. 99, p. 24051–24074.

Harland, W., Armstrong, R., Cox, A., Craig, L., Smith, A., and Smith, D., 1990, A geologic time scale: Cambridge, Cambridge University Press, 263 p.

Heimann, A., Fleming, T. H., Elliot, D. H., and Fowland, K. A., 1994, A short interval of Jurassic continental flood basalt volcanism in Antarctica as demonstrated by $^{40}Ar/^{39}Ar$ geochronology: Earth and Planetary Science Letters, v. 121, p. 19–41.

Hildebrand, A. R., Penfield, G. T., Kring, D. A., Pilkington, M., Camargo Z., A., S Jacobsen,. B., and Boynton, W. V., 1991, Chixculub crater: A possible Cretaceous-Tertiary boundary impact crater on the Yucatan Peninsula, Mexico: Geology, v. 19, p. 867–871.

Hill, R. I., Campbell, H., Davies, G. F., and Griffiths, R. W., 1992, Mantle plumes and Continental tectonics: Science, v. 256, p. 186–193.

Hodych, J. P., and Dunning, G. R., 1992, Did the Manicouagan impact trigger end-of-Triassic mass extinction?: Geology, v. 20, p. 51–54.

Holser, W.T., and Magaritz, M., 1992, Cretaceous/Tertiary and Permian/Triassic boundary events compared: Geochimica et Cosmochimica Acta, v. 56, p. 3297–3309.

Hooper, P. R., Rehacek, J., Duncan, R. A., Marsh, J. S., and Duncan, A. R., 1993, The basalts of Lesotho, Karoo province, Southern Africa: Eos (Transactions, American Geophysical Union), 1993 Fall meeting abstract volume, p. 553.

Izett, G. A., Dalrymple, G. B., and Snee, L. W., 1991, $^{40}Ar/^{39}Ar$ age of Cretaceous-Tertiary boundary tektites from Haiti: Science, v. 252, p. 1539–1542.

Keller, G., and Barrera, E., 1990, The Cretaceous/Tertiary boundary impact hypothesis and the paleontological record, *in* Sharpton, V. L., and Ward, P. D., eds., Global catastrophes in earth history: An interdisciplinary conference on impacts, volcanism, and mass mortality: Geological Society of America Special Paper 247, p. 563–575.

Keller, G., MacLeod, N., Lyons, J. B., and Officer, C. B., 1993a, Is there evidence for Cretaceous-Tertiary boundary-age deep-water deposits in the Caribbean and Gulf of Mexico?: Geology, v. 21, p. 776–780.

Keller, G., Barrera, E., Schmitz, B., and Mattson, E., 1993b, Gradual mass extinction, species survivorship, and long-term environmental changes across the Cretaceous-Tertiary boundary in high latitudes: Geological Society of America Bulletin, v. 105, p. 979–997.

Keller, G., Stinnesbeck, W., Adatte, T., MacLeod, N., and Lowe, D. R., 1994a, Field guide to Cretaceous Tertiary boundary sections in northeastern Mexico: Houston, Texas, Lunar and Planetary Institute Contribution 827, 110 p.

Keller, G., Li, L., and MacLeod, N., 1994b, The KT boundary stratotype section at El Kef, Tunisia: How catastrophic was the mass extinction?, *in* New developments regarding the K/T event and other catastrophes in Earth history: Houston, Texas, Lunar and Planetary Institute Contribution 825, p. 59.

Kent, D. V., Witte, W. K., and Olsen, P.E., 1995, High resolution late Triassic–early Jurassic geomagnetic polarity sequence from drill-cores in the Neward rift basin, eastern North America: Journal of Geophysical Research, v. 100, p. 14965–14998.

Khajuria, C. K., Prasad, G. V. R., and Manhas, B. K., 1994, Paleontological constraints on the age of Deccan Traps, peninsular India: Newsletters in Stratigraphy, v. 31, p. 21–32.

Krogh, T. E., Kamo, S. L., Sharpton, V. L., Marin, L. E., and Hildebrand, A. R., 1993, U-Pb ages of single shocked zircons linking distal K/T ejecta to the Chixculub crater: Nature, v. 366, p. 731–734.

Lerbekmo, J. F., Evans, M. I., and Baadsgaard, H., 1979, Magnetostratigraphy, biostratigraphy and geochronology of Cretaceous-Tertiary boundary sediments, Red Deer Valley: Nature, v. 79, p. 26–30.

Leroux, H., 1994, Etude par microscopie électronique en transmission du métamorphisme de choc dans le quartz et le diopside: Applications aux impacts, à la limite K/T et aux météorites [Ph.D. thesis]: University of Lille (France), 129 p.

Lunar and Planetary Institute, 1994, New developments regarding the KT event and other catastrophes in Earth history: Houston, Texas, Lunar and Planetary Institute Contribution 825, 138 p.

MacLeod, N., and Keller, G., 1991, Hiatus distributions and mass extinctions at the Cretaceous/Tertiary boundary: Geology, v. 19, p. 497–501.

MacLeod, N., and Keller, G., 1996, Stratigraphic observations, interpretations and paradigms at the Cretaceous-Tertiary boundary, 59 p., 16 fig.: Palaios (in press).

Mantovani, M. S. M., Turner, S., Regelous, M., Kelley, S., and Hawkesworth, C., 1993, Magmatism and continental break-up in the South Atlantic: High precision ^{40}Ar-^{39}Ar geochronology: Eos (Transactions, American Geophysical Union), 1993 Fall meeting abstract volume, p. 553.

Marin, L. E., Sharpton, V. L., Urrutia-Fucugauchi, J., Sikora, P., and Carney, C., 1994, The "upper-Cretaceous unit" in the Chixculub multiring basin: New age based on planktic foraminiferal assemblage, *in* New developments regarding the KT event and other catastrophes in Earth history: Houston, Texas, Lunar and Planetary Institute Contribution 825, p. 77.

McLean, D. M., 1982, Flood basalt volcanism and global extinctions at the Cretaceous-Tertiary transition: American Association for the Advancement of Science, Meeting, Washington, D.C., p. 47.

McLean, D. M., 1985, Deccan traps mantle degassing in the terminal Cretaceous marine extinctions: Cretaceous Research, v. 6, p. 235–259.

McWilliams, M. O., Baksi, A. K., Bohor, B. F., and Murali, A. V., 1992, High-precision relative ages of K/T boundary events in North America and Deccan trap volcanism in India: Eos (Transactions, American Geophysical Union), 1992 Spring Meeting, p. 363.

Meyerhoff, A. A., Lyons, J. B., and Officer, C. B., 1994, Chicxulub structure: A volcanic sequence of late Cretaceous age: Geology, v. 22 , p. 3–4.

Montanari, A., Asaro, F., Michel, H. V., and Kennett, J., 1993, Iridium anomalies of Late Eocene age at Massignano (Italy), and ODP site 689B (Maud Rise, Antarctica): Palaios, v. 8, p. 420–443.

Morgan, W. J., 1986, Flood basalts and mass extinctions: Eos (Transactions, American Geophysical Union), v. 67, p. 391.

Morgan, W. J., 1988, Flood basalts and mass extinctions [abs.]: Houston, Texas, Lunar and Planetary Institute, Conference on global catastrophes in Earth history, p. 126.

Mussett, A. E., 1986, ^{40}Ar/^{39}Ar step-heating ages of the Tertiary igneous rocks of Mull, Scotland: Journal of the Geological Society of London, v. 143, p. 887–896.

Mussett, A. E., Dagley, P., and Skelhorn, R. R., 1988, Time and duration of activity in the British Tertiary igneous province, *in* Morton, A. C., and Parson, L. M., eds., Early Tertiary volcanism and the opening of the NE Atlantic: Geological Society of London Special Publication 39, p. 337–348.

Noble, R.H., Macintyre, R. M., and Brown, P. E., 1988, Age constraints on Atlantic evolution: timing of magmatic activity along the E Greenland continental margin, *in* Morton, A. C., and Parson, L. M., eds., Early Tertiary volcanism and the opening of the NE Atlantic: Geological Society of London Special Publication 39, p. 201–214.

Odin, G. S., 1989, Ages radiométriques récemment obtenus dans la séquence stratigraphique paléogène: Bulletin de la Société Géologique de France, v. 8, p. 145–152.

Odin, G. S., 1994, Geological time scale (1994): Comptes Rendus de l'Académie des Sciences de Paris, v. 318, p. 59–71.

Odin, G. S., Galbrun, B., and Renard, M., 1992, Phanerozoic time scale: Bulletin de Liaison, I.U.G.S. Subcommission on Geochronology, v. 10, p. 39.

Officer, C. B., and Drake, C. L., 1985, Terminal Cretaceous environmental events: Science, v. 227, p. 1161–1167.

Officer, C.B., Hallam, A., Drake, C. L., and Devine, J. D., 1987, Late Cretaceous and paroxysmal Cretaceous/Tertiary extinctions: Nature, v. 326, p. 143–149.

Olsen, P. E., and Fedosh, M. S., 1988, Duration of the early Mesozoic igneous episode in eastern North America determined by use of Milankovitch-type lake cycles: Geological Society of America Abstracts with Programs, v. 20, p. 59.

Orth, C.J., Attrep, M., Jr., and Quintana, L. R., 1990, Iridium abundance patterns across bio-event horizons in the fossil record, *in* Sharpton, V. L., and Ward, P. D., eds., Global catastrophes in earth history: An interdisciplinary conference on impacts, volcanism, and mass mortality: Geological Society of America Special Paper 247, p. 45–59.

Pande, K., Venkatesan, T. R., Gopalan, K., Krishnamurthy, P., and MacDougall, J. D., 1988, ^{40}Ar/^{39}Ar ages of alkali basalts from Kutch, Deccan volcanic province, India, *in* Workshop on Deccan flood basalts, Memoirs of the Geological Society of India, v. 10, p. 145–150.

Penfield, G. T., and Camargo Z., A., 1981, Definition of a major igneous zone in the central Yucatan platform with aeromagnetics and gravity [abs.]: Society of Exploration Geophysics, p. 51.

Radhakrishna, T., Dallmeyer, R. D., and Joseph, M., 1994, Paleomagnetism and ^{36}Ar/^{40}Ar vs. ^{39}Ar/^{40}Ar isotope correlation ages of dyke swarms in central Kerala, India: Tectonic implications: Earth and Planetary Science Letters, v. 121, p. 213–226.

Rampino, M. R., 1994, A general theory of impacts and mass extinctions, and the consequences of large body impact on the Earth, *in* New developments regarding the KT event and other catastrophes in Earth history: Houston, Texas, Lunar and Planetary Institute Contribution 825, p. 91.

Rampino, M. R., and Haggerty, B. M., 1994, Extraterrestrial impacts and mass extinctions of life, *in* Gehrels, T., ed., Hazards due to comets and asteroids: Tucson, University of Arizona Press, p. 827–857.

Rampino, M. R., and Stothers, R. B., 1988, Flood basalt volcanism during the past 250 million years: Science, v. 241, p. 663–668.

Raup, D. M., 1991, A kill curve for Phanerozoic marine species: Paleobiology, v. 17, p. 37–48.

Renne, P. R., and Basu, A. R., 1991, Rapid eruption of the Siberian traps flood basalts at the Permo-Triassic boundary: Science, v. 253, p. 176–179.

Renne, P. R., Ernesto, M., Pacca, I. G., Coe, R. S., Glen, J. M., Prévot, M., and Perrin, M., 1992, The age of the Parana flood volcanism, rifting of Gondwanaland, and the Jurassic-Cretaceous boundary: Science, v. 258, p. 975–979.

Renne, P. R., Mertz, D. F., Ernesto, M., Marques, L., Teixeira, W., Ens, H. H., and Richards, M. A., 1993, Geochronologic constraints on magmatic and tectonic evolution of the Parana province: Eos (Transactions, American Geophysical Union), 1993 Fall meeting abstract volume, p. 553.

Richards, M. A., Duncan, R. A., and Courtillot, V. E., 1989, Flood basalts and hot-spot tracks: Plume heads and tails: Science, v. 246, p. 103-107.

Richards, M. A., Jones, D. L., Duncan, R. A., and DePaolo, D. J., 1991, A mantle plume initiation model for the Wrangellia flood basalt and other oceanic plateaus: Science, v. 254, p. 263–267.

Robin, E., Boclet, D., Bonté, Ph., Froget, L., Jéhanno, C., and Rocchia, R., 1991, The stratigraphic distribution of Ni-rich spinels in Cretaceous-Tertiary boundary rocks at El Kef (Tunisia), Caravaca (Spain) and Hole 761C (Leg 122): Earth and Planetary Science Letters, v. 107, p. 715–721.

Robin, E., Bonté, Ph., Froget, L., Jéhanno, C., and Rocchia, R., 1992, Formation of spinels in cosmic objects during atmospheric entry: A clue to the Cretaceous-Tertiary boundary event: Earth and Planetary Science Letters, v. 108, p. 181–190.

Sebai, A., Féraud, G., Bertrand, H., and Hanes, J., 1991, ^{40}Ar/^{39}Ar dating and geochemistry of tholeiitic magmatism related to the early opening of the Central Atlantic rift: Earth and Planetary Science Letters, v. 104, p. 455–472.

Sepkoski, J. J., Jr., 1990, The taxonomic structure of periodic extinction, *in* Sharpton, V. L., and Ward, P. D., eds., Global catastrophes in earth history: An interdisciplinary conference on impacts, volcanism, and mass mortality: Geological Society of America Special Paper 247, p. 33–44.

Sharpton, V. L., and Ward, P. D., eds, 1990, Global catastrophes in Earth history: An interdisciplinary conference on impacts, volcanism, and mass mortality: Geological Society of America Special Paper 247, 631 p.

Sharpton, V. L., and nine others, 1993, Chicxulub multiring impact basin: Size and other characteristics derived from gravity analysis: Science, v. 261, p. 1564–1567.

Sheridan, D. E., and eight others, 1993, Deep seismic reflection data of EDGE U.S. mid-Atlantic continental-margin experiment: Implications for Appalachian sutures and Mesozoic rifting and magmatic underplating: Geology, v. 21, p. 563–567.

Sigurdsson, H., 1982, Volcanic pollution and climate: The 1783 Laki eruption: Eos (Transactions, American Geophysical Union), p. 63.

Sigurdsson, H., 1990, Assessment of the atmospheric impact of volcanic eruptions, *in* Sharpton, V. L., and Ward, P. D., eds., Global catastrophes in earth history: An interdisciplinary conference on impacts, volcanism, and mass mortality: Geological Society of America Special Paper 247, p. 99–110.

Smit, J., 1982, Extinction and evolution of planktonic foraminifera after a major impact at the Cretaceous/Tertiary boundary, in Silver, L. T., and Schulz, P. H., eds., Geological implications of impacts of large asteroids and comets on Earth: Geological Society of America Special Paper 190, p. 329–352.

Smit, J., and eight others, 1992, Tektite-bearing, deep-water clastic unit at the Cretaceous-Tertiary boundary in northeastern Mexico: Geology, v. 20, p. 99–103.

Stanley, S. M., and Yang, X., 1994, A double mass extinction at the end of the Paleozoic era: Science, v. 266, p. 1340–1344.

Stinnesbeck, W., and 10 others, 1993, Deposition of channel deposits near the Cretaceous-Tertiary boundary in northeastern Mexico: Catastrophic or "normal" sedimentary deposits?: Geology, v. 21, p. 797–800.

Stothers, R. B., 1993, Flood basalts and extinction events: Geophysical Research Letters, v. 20, p. 1399–1402.

Stothers, R. B., and Rampino, M. R., 1990, Periodicity in flood basalts, mass extinctions, and impacts; A statistical view and a model, *in* Sharpton, V. L., and Ward, P. D., eds., Global catastrophes in earth history: An interdisciplinary conference on impacts, volcanism, and mass mortality: Geological Society of America Special Paper 247, p. 9–18.

Stothers, R. B., Wolff, J. A., Self, S., and Rampino, M. R., 1986, Basaltic fissure eruptions, plume heights, and atmospheric aerosols: Geophysical Research Letters, v. 13, p. 725–728.

Sutherland, F. L., 1994, Northern meteoritic impacts and southern volcanic mantle plumes: their roles in generating some differential K/T boundary extinction environments, *in* New developments regarding the KT event and other catastrophes in Earth history: Houston, Texas, Lunar and Planetary Institute Contribution 825, p. 123–125.

Swisher, C. C., and 12 others, 1992, Coeval $^{40}Ar/^{39}Ar$ ages of 65.0 million years ago from Chicxulub crater melt rock and Cretaceous-Tertiary boundary tektites: Science, v. 257, p. 954–958.

Tarduno, J. A., and eight others, 1991, Rapid formation of Ontong Java plateau by Aptian mantle plume volcanism: Science, v. 254, p. 399–403.

Teichert, C., 1990, The Permian-Triassic boundary revisited, *in* Extinction events in Earth history, A contribution to Project Global Bioevents, IUGS-UNESCO-IGCP: Berlin, Springer-Verlag, p. 199–237.

Thomas, E., 1990, Late Cretaceous-Early Eocene mass extinctions in the deep sea, *in* Sharpton, V. L., and Ward, P. D., eds., Global catastrophes in earth history: An interdisciplinary conference on impacts, volcanism, and mass mortality: Geological Society of America Special Paper 247, p. 481–495.

Thomas, E., 1992, Cenozoic deep-sea circulation: evidence from deep-sea foraminifera, *in* Kennett, J. P., and Warnke, D., eds., The Antarctic paleoenvironment: A perspective on global change: Washington, D.C., Antarct. Res. Ser., American Geophysical Union, v. 56, p. 141–165.

Urrutia-Fucugauchi, J., et al., 1994, Preliminary paleomagnetic results on some K-T sections from northeastern Mexico: poster presented at conference, "New developments regarding the K/T event and other catastrophes in Earth history," Houston, Texas, February 9 to 12 (no abstract).

Vandamme, D., Courtillot, V., Besse, J., and Montigny, R., 1991, Paleomagnetism and age determinations of Deccan traps: Results of a Nagpur-Bombay traverse and review of earlier work: Reviews of Geophysics and Space Physics, v. 29, p. 159–190.

Venkatesan, T. R., Pande, K., and Gopalan, K., 1993, Did Deccan volcanism pre-date the Cretaceous/Tertiary transition?: Earth and Planetary Science Letters, v. 119, p. 181–189.

Venkatesan, T. R., Pande, K., and Gopalan, K., 1994, Did Deccan volcanism pre-date the Cretaceous/Tertiary transition?: Reply: Earth and Planetary Science Letters, v. 122, p. 263–265.

Vogt, P. R., 1972, Evidence for global synchronism in mantle plume convection and possible significance for geology: Nature, v. 240, p. 338–342.

White, R. S., 1989, Igneous outbursts and mass extinctions: Eos (Transactions, American Geophysical Union), v. 70, p. 1490–1491.

Zinsmeister, W. J., and Feldmann, R. M., 1994, Antarctica, the forgotten stepchild: A view of KT extinction from the high southern latitudes, *in* New developments regarding the KT event and other catastrophes in Earth history: Houston, Texas, Lunar and Planetary Institute Contribution 825, p. 134.

Zinsmeister, W. J., Feldmann, R. M., Woodburne, M. O., and Elliot, D. H., 1989, Latest Cretaceous/Earliest Tertiary transition on Seymour Island, Antarctica: Journal of Paleontology, v. 63, p. 731–738.

Manuscript Accepted by the Society September 14, 1995

Geological Society of America
Special Paper 307
1996

Impact materials recovered by research core drilling in the Manson Impact Structure, Iowa

Raymond R. Anderson and Brian J. Witzke
Iowa Department of Natural Resources, Geological Survey Bureau, 109 Trowbridge Hall, Iowa City, Iowa 52242-1319
Jack B. Hartung
Iowa Center for Impact Studies, P.O. Box 181, Oakdale, Iowa 52319-0181

ABSTRACT

Soon after Alvarez et al. (1980) proposed that a large hypervelocity impact triggered a sequence of events that culminated in the mass extinction of flora and fauna at the Cretaceous-Tertiary (K/T) boundary, the Manson Impact Structure (MIS) became a leading candidate for the impact site. Several $^{40}Ar/^{39}Ar$ age determinations indicated that the age of the Manson impact was identical to the K/T boundary—ca. 65 Ma. To further investigate the MIS and its relationship to the K/T extinctions, in 1991 and 1992 the Iowa Department of Natural Resources Geological Survey Bureau and U.S. Geological Survey Branch of Astrogeology drilled 12 research core holes in the impact structure. More than 1,200 m of drill core was recovered, almost all impact breccias. These materials include (1) megablocks of crystalline basement rocks uplifted during central peak formation, (2) suevite breccia composed primarily of crystalline rock clasts, (3) a clast-rich impact melt breccia, (4) an enigmatic mixture of brecciated Middle Proterozoic black shales and melt rock that caps the central peak, (5) a late-stage crater ejecta flap of overturned sedimentary strata, and (6) a breccia composed almost exclusively of unshocked sedimentary rocks, dominated by the shallowest target units, and deposited on all terranes of the MIS during the latest stage of crater formation. Recognition of the location and character of these materials has advanced our understanding of this well-preserved complex impact structure and those processes that controlled its formation.

Other investigations of the Manson cores have led to the identification of an older age for the Manson impact (ca. 74 Ma), and recently, probable Manson-derived ejecta was discovered in the Crow Creek Member of the Pierre Shale in South Dakota and Nebraska. These constraints allow an evaluation of any biotic effects in the North American Western Interior that may have resulted from the Manson impact. Review of published paleontologic data suggests that the Manson impact affected portions of the Western Interior biota, primarily certain large vertebrate animals, both terrestrial and marine. However, most lineages of marine and terrestrial organisms were seemingly unaffected by the impact, and the event did not produce global extinctions.

Anderson, R. R., Witzke, B. J., and Hartung, J. B., 1996, Impact materials recovered by research core drilling in the Manson Impact Structure, Iowa, *in* Ryder, G., Fastovsky, D., and Gartner, S., eds., The Cretaceous-Tertiary Event and Other Catastrophes in Earth History: Boulder, Colorado, Geological Society of America Special Paper 307.

INTRODUCTION

The Manson Impact Structure (MIS), located in north-central Iowa, lies at the bedrock surface but is completely buried by 30 to 70 m of glacial drift. The feature was identified as an impact structure by the discovery of multiple sets of planar deformation features (PDFs) in quartz grains recovered from the center of the structure (Short, 1966). A series of $^{40}Ar/^{39}Ar$ age determinations from central peak feldspars (Hartung et al., 1986; Kunk et al., 1987, 1989) indicated that the MIS was formed ca. 65 Ma, "an age indistinguishable from recent estimates of the age of the K-T boundary" (Kunk et al. 1989) and coincident with one of the Earth's great mass extinctions of fauna and flora, possibly related to a large bolide impact (Alvarez et al., 1980).

Responding to the probable relationship between the MIS and the Cretaceous-Tertiary (K-T) boundary, the Iowa Department of Natural Resources Geological Survey Bureau (GSB) and the U.S. Geological Survey (USGS) joined in the drilling of 12 research core holes into the MIS, resulting in the recovery of over 1,200 m of core. Most of the cores were drilled along a line of reflection seismic data (collected by Amoco Production Co.) that approximates a radius of the MIS. Two cores were drilled in the MIS terrace terrane, four cores in the crater moat region, and six cores into the central peak (Fig. 1). Preliminary logs of the cores were produced by the GSB; several cores were split, with half-splits stored at the USGS core repository in Denver.

AGE OF THE MANSON IMPACT STRUCTURE

Prior to study of materials obtained in the recent core-drilling program, the age of the MIS was stratigraphically bracketed between the Upper Cretaceous Dakota Formation (95 Ma), recognized at the time as the youngest rocks to be disturbed by the formation of the MIS, and Pleistocene glacial drift (ca. 2 Ma), which overlies the structure. Four age determinations were reported in the late 1980s on granitic rocks recovered in a 1953 core-drilling program. These ages, one fission track date on apatite and three using $^{40}Ar/^{39}Ar$ techniques on shocked microcline (see Table 1), were increasingly precise and converged on an age of ca. 65.7 ± 1.0 Ma (Kunk et al., 1989). Rocks recovered from the M-1 core were subsequently investigated by three researchers (Table 1). The first two analyzed shocked microcline and melt rock and reported a range of ages from ca. 90 to 65 Ma. They concluded that the older ages resulted from incomplete degassing of preimpact argon or equilibration with nonatmospheric argon.

A more recent age estimate was published by Izett et al. (1994) who used laser degassing $^{40}Ar/^{39}Ar$ techniques to analyze sanidine from the M-1 core that presumably crystallized soon after the impact. They reported an MIS age ca. 9 m.y. older than previous values, 73.8 ± 0.3 Ma (relative to 513.9 Ma for MMhb-1 hornblende standard) or 74.9 ± 0.3 Ma (relative to 520.4 Ma for MMhb-1 hornblende standard). In addition, they reported the presence of shocked-quartz grains in the similar-aged Crow Creek Member of the Cretaceous Pierre Shale Formation in central and eastern South Dakota ~500 km from the MIS. They interpreted these shocked grains as MIS ejecta, possibly emplaced by a tsunami produced by the MIS impact in the Cretaceous seaway. Most recently, Dalrymple and Izett (1994) reported that "fourteen total-fusion for the sanidine from 114.3 m (in the M-1 core) have a mean age of 74.4 ± 1.2 Ma and a weighted mean age of 74.4 ± 0.2 Ma." They concluded that "these results confirm the results of Izett (1993) who found a $^{40}Ar/^{39}Ar$ age of 73.8 ± 0.3

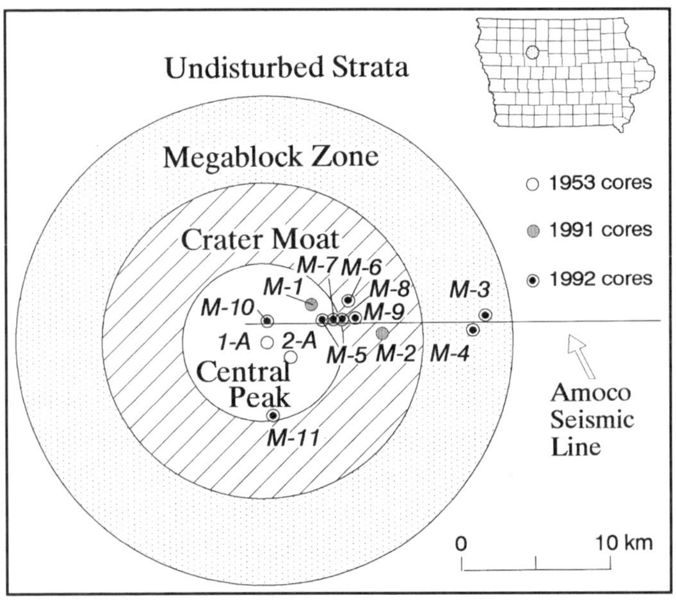

Figure 1. Location of drill holes and three primary terranes in the Manson Impact Structure.

TABLE 1. PUBLISHED AGE DATES FOR MANSON IMPACT STRUCTURE (1986–1994)

Author	Sample	Technique	Age (Ma)
Naser*	Shocked apatite	Fission track	61 ± 9
Kunk*	Shocked K-spar	$^{40}Ar/^{39}Ar$	<70
Kunk et al., 1987	Shocked K-spar	$^{40}Ar/^{39}Ar$	~66
Kunk et al., 1989	Shocked K-spar	$^{40}Ar/^{39}Ar$	65.7 ± 1
Hartung et al., 1990	Shocked K-spar	$^{40}Ar/^{39}Ar$	65.7 ± 1†
Zeitler and Kunk, 1993	Shocked K-spar	$^{40}Ar/^{39}Ar$	65.3 ± 0.5§
Kunk et al., 1993	K-spar from melt	$^{40}Ar/^{39}Ar$**	65.4 ± 0.4‡
Izett et al., 1994	K-spar from melt (sanidine)	$^{40}Ar/^{39}Ar$**	73.8 ± 0.3§§

*Published in Hartung et al., 1986.
†$^{40}Ar/^{39}Ar$ spectra yielded apparent ages of 65.00 to 412.71 Ma, one age of 73.86 ± 0.22 Ma.
§Strong overprint in age spectra; minimum age spectra between 67 and 72 Ma.
**Laser probe.
‡Bulk sample spectrum ~74 Ma: sample range 64.4 ± 0.4 to 97.5 ± 2.6 Ma.
§§Age calculated relative to 513.9 Ma for MMhb-1 hornblende standard; not a plateau age.

for sanidine at 114.3 m," and that "an earlier age assignment of 64–65 Ma for the Manson impact structure (Kunk et al., 1989, 1993), which was based on ambiguous $^{40}Ar/^{39}Ar$ data is incorrect." Kunk and Nord (1994) concurred, evoking a low argon closure temperature (between 100 °C and 50 °C) for the microcline at Manson and deep, postimpact burial (1.5 km) of the structure for keeping the microcline in the impact breccias above its argon closure temperature for ca. 8 Ma after the impact. Zeitler (1994) also agreed with the ca. 74 Ma age for the MIS, noting that Manson basement feldspars contained a significant fraction of small nonretentive domains with slow-cooling closure temperatures (some <100 °C) that would be highly susceptible to alteration and could produce too-young ages. Most impact workers currently accept 73.8 Ma as the age of the Manson impact.

PREIMPACT GEOLOGY OF THE MANSON REGION

Numerous previous investigations of the geology of the MIS and surrounding region (see Hartung and Anderson, 1988, for summary) and estimates of materials eroded since impact have yielded an accurate picture of the geology at the time of the Manson impact. At ca. 74 Ma, the MIS region was covered by ~0.2 km of Upper Cretaceous marine and terrestrial shales, mudstones, and sandstones, and by 65 Ma (Fig. 2) an additional 0.2 km of marine shale–dominated Cretaceous strata were probably deposited (Anderson and Witzke, 1994). Below the Cretaceous rocks lie ~0.75 km of carbonate-dominated Paleozoic strata. Beneath these rocks, a southeast-thickening wedge of Keweenawan (Middle Proterozoic) sandstone–dominated clastic rocks ranged from 0 to 4,200 m thick over the area of the crater, overlying a crystalline basement complex dominated by Early Proterozoic granitic gneiss and Middle Proterozoic granite.

PREDRILLING MODEL OF MIS

Study of rock chip samples collected during the drilling of water wells in the area of the MIS, interpretation of reflection seismic data from a line approximating a radius of the structure collected by Amoco Production Company, and comparison of these data with information from other impact structures led to the synthesis of a model of the structure (Hartung and Anderson, 1988; Hartung et al., 1990; Anderson and Hartung, 1992). The MIS was interpreted as a complex impact structure with an apparent crater diameter of 35 km, a large central peak ~10 km in diameter, a terrace terrane (megablock zone) ~7 km wide, and an intervening crater moat zone (Figs. 1 and 3). Crystalline basement rocks lifted up to the bedrock surface lie at the southeast edge of central peak, and down-dropped blocks within the terrace terrane have structurally preserved Cretaceous marine strata from postimpact erosion. The crater moat and much of the terrace terrane and central peak are covered by a shale matrix–dominated breccia that contains clasts of Cretaceous and Paleozoic strata. Preliminary interpretations regarded this breccia as a postimpact lake deposit mixed with debris flows of ejecta material that covered the terrace terrane and margins of the impact structure. However, study of the most recent drill core has shown that this material is not a lake deposit, and if lake sediments are preserved in the structure, they are much less widespread than initially thought (and were not encountered in the core drilling). Most of the other features of this early interpretation were verified by the analyses of the samples that were collected during the drilling.

1991–1992 RESEARCH CORE DRILLING

The Iowa Geological Survey Bureau and U.S. Geological Survey began an MIS research core-drilling program in the summer of 1991. Major drilling goals were to recover rocks that would (1) increase our understanding of the geology of the feature, one of the largest known intact impact structures in the United States, (2) try to date the structure more accurately by obtaining impact melt rocks for radiometric age determinations and by obtaining postimpact lake sediments to recover palynomorphs and other fossils for paleontologic age determinations, (3) attempt to determine the composition of the impacting body by trace-element analysis of impact melt material, and (4) collect a representative set of Manson rocks to compare with exotic grains recovered from K/T boundary sites in the western United States and other localities. Over the

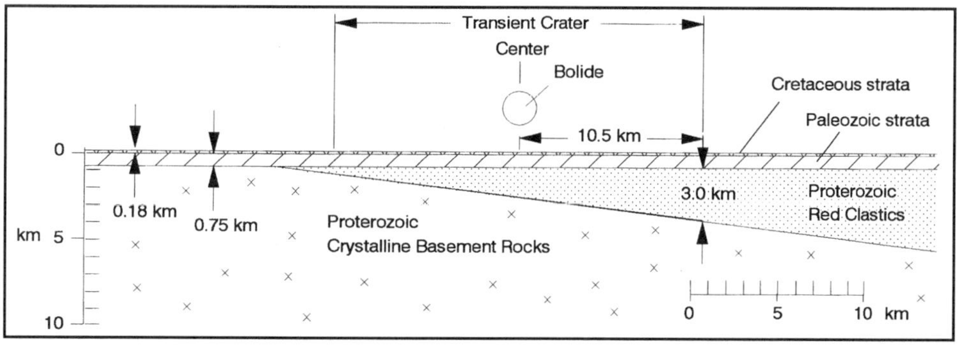

Figure 2. Cross-section showing preimpact geology in the area of the Manson Impact Structure, assuming an impact at 73.8 Ma (from Anderson and Hartung, 1992).

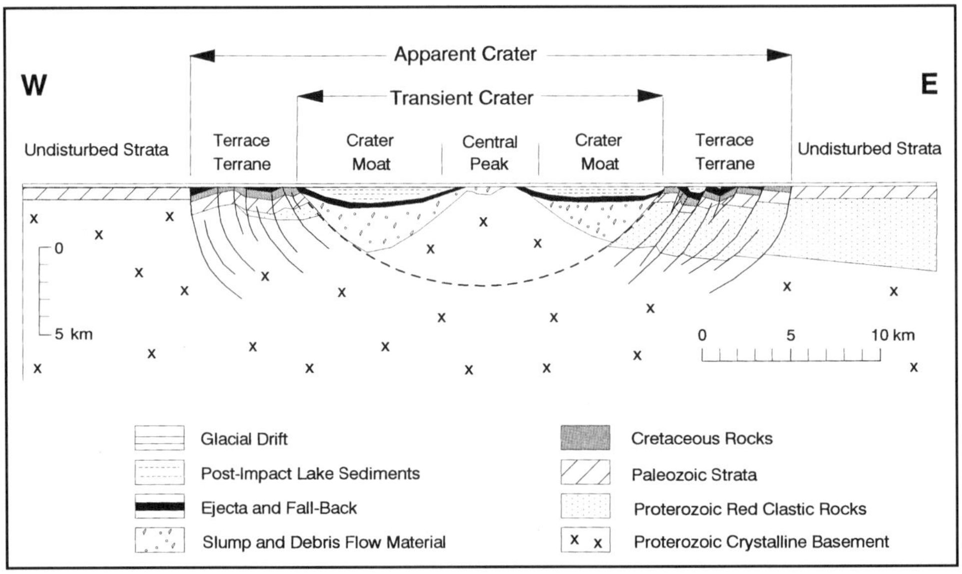

Figure 3. Predrilling model cross section of the Manson Impact Structure constructed using data available prior to initiation of the 1991–1992 GSB-USGS core-drilling project (from Anderson and Hartung, 1992). Subsequent core drilling failed to encounter the lake sediments illustrated in this cross section, but most other features were verified.

course of the drilling program, a total of 1,217 m of Manson impact rock was recovered. Numerous investigations utilizing these cores are complete or in progress. The results of many of these investigations were reported in Koeberl and Anderson (1996a); the results of other investigations will be reported in other journals.

The first core (M-1) was drilled on the flank of the central peak (Fig. 1) near an existing water well from which cuttings had been collected (Anderson et al., 1996). These samples had been interpreted as postimpact lake sediments (Hartung and Anderson, 1996) overlying melt-bearing crystalline clast breccia, which rested on the crystalline rock blocks of the central peak. The M-1 was drilled to sample as many of these materials as possible and as a test of the potential problems in drilling these brecciated materials. The M-2 and M-2A cores were drilled along the Amoco seismic profile, in an area initially interpreted as thick postimpact lake sediments. Both the M-1 and M-2 core holes encountered extremely difficult and time-consuming drilling conditions. No lake sediments were encountered in either core (Table 2). In the spring of 1992 the M-3 and M-4 cores were drilled in the MIS terrace terrane in search of the land surface at the time of impact, presumed to be structurally preserved on down-dropped blocks. M-4 encountered an interval of overturned sedimentary strata. The M-5 core was located on the eastern edge of the central peak in an attempt to duplicate the rocks encountered in M-1 but penetrated directly into crystalline basement rocks below glacial drift. M-6 was drilled northeast of M-5 in an unsuccessful search for lake sediments. M-7 was located ~1 km east of M-5, again in an attempt to replicate the M-1 rocks, and M-8 was located ~200 m east of M-7 in order to drill a complete sequence of melt matrix breccia. M-9 was located ~0.8 km east of M-8 in a last futile attempt to core lake sediments. M-10 was drilled along the Amoco seismic line at a point ~1.6 km due north of the inferred center of the impact structure, and M-11 was drilled on the southern edge of the central peak to examine the symmetry of the central peak. Winter conditions and funding cuts forced the abandonment of drilling in early December 1992.

IMPACT ROCKS CORED

The cores recovered from the Manson Impact Structure encountered six primary classes of impact materials (rocks directly emplaced by the Manson impact). These rocks initially were described by nongenetic terms that conveyed their composition, instead of by more standard, currently used impact lithology terms (e.g., suevite, melt layer, etc.). They include (1) large blocks of crystalline basement rocks (CR), crystalline clast breccias with (2) sandy matrix (SM) and (3) melt-rock matrix (MM), (4) Keweenawan shale breccia (KSB), (5) Phanerozoic clast breccia (PCB), and (6) overturned sedimentary rock strata (OS). The CR, SM, MM, and KSB were encountered only on the MIS central peak (Fig. 4), the OS was drilled on the terrace terrane, and the PCB was cored in all terranes of the MIS. Although additional research is required to definitively interpret the origin and history of the

TABLE 2. MATERIALS ENCOUNTERED IN THE GSB/USGS MANSON CORES

Core	Date Drilled	Materials Drilled*	Depth Interval (m)	Notes
M-1	1991	PCB	53–104	Glassy clasts near base.
		KSB		
		CCB-M	104–159	
		CCB-S	159–211	Few glassy clasts.
M-2	1991	PCB	56–247	Included 105-m-thick block of Cretaceous Carlile Formation.
M-2A	1991	Mudstone	44–60	Possible postimpact lake sediments.
M-3	1992	Cretaceous Strata	39–88	Carlile, Greenhorn, and Graneros formations.
M-4	1992	PCB	39–176	Many Red Clastics clasts.
		Layered Strata	176–375	Inverted stratigraphic sequence, Red Clastics, and Cambrian through Devonian.
M-5	1992	Gneiss	40–42	Shock-deformed quartz.
M-6	1992	KSB	51–60	
M-7	1992	CCB-M	52–132	Abundant black shale clasts.
		Gneiss	132–221	Short intervals of CCB-S.
M-8	1992	PCB	56–83	
		KSB		
		CCB-M	83–124	Abundant black shale clasts.
		Gneiss	124–168	Short intervals of CCB-S.
M-9	1992	PCB	59–64	
M-10	1992	PCB	54–131	Glassy clasts 90–131 m.
		KSB		
		CCB-M	131–149	Calcite druse in fractures.
		Gneiss	149–188	Short intervals of CCB-S and veins of black pseudotachylyte.
M-11	1992	PCB	51–71	Small glassy clasts throughout.
		CCB-S	71–136	Abundant glassy clasts.

*CCB-M = crystalline clast breccia with melt matrix; CCB-S = crystalline clast breccia with sandy matrix; KSB = Keweenawan shale-clast breccia; PCB = Paleozoic-clast breccia.

MIS impact rocks, preliminary investigation provided valuable information.

Crystalline basement rocks (CR)

Large blocks of CR, uplifted with the MIS central peak, comprise the stratigraphically lowest preimpact unit encountered in four research cores on the central peak (M-5, M-7, M-8, M-10). These rocks (Fig. 5) are dominated by biotite and hornblende granite gneisses (ca. 1860 Ma) and granite intrusives (ca. 1450 Ma) but also include minor mafic volcanic and plutonic rocks and metasediments. Mineral grains within all lithology classes display multiple impact deformation features including PDFs (Short and Gold, 1996; Leroux and Doukhan, 1996), kink bands, distorted extinctions, and other features. Veins of melt rock and finely brecciated material (pseudotachylite) are common (Fig. 5).

The CR is interpreted as crystalline basement material uplifted from below the transient crater as a part of the central peak. The large areal distribution of this lithology (up to 5.3 km

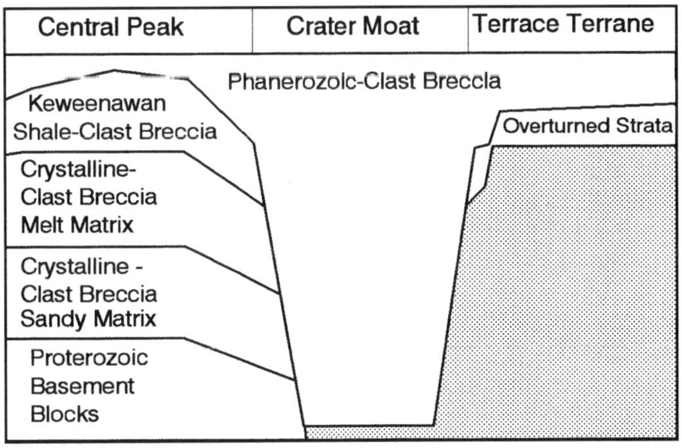

Figure 4. Idealized radial cross section of the Manson Impact Structure displaying the general location of impact rocks encountered in drill cores (not drawn to scale).

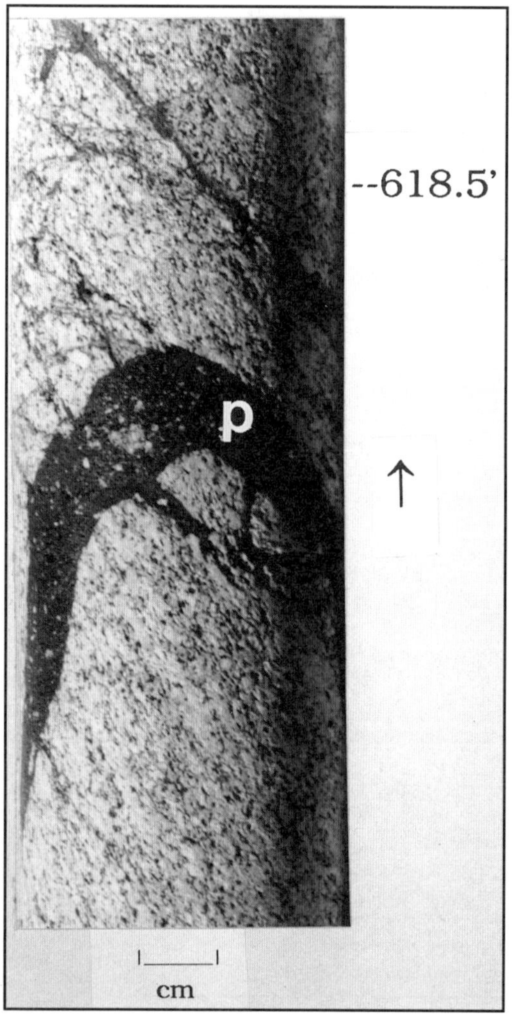

Figure 5. Photograph of pseudotachylyte (p) dike in gneiss of crystalline basement block from M-10, 218.5 ft.

Figure 6. Photograph of crystalling clast breccia with sand matrix from M-7, 290.5 ft (89.4 m); gn = gneiss clast, gr = granite clast, m = sandy matrix.

radius from the center of the MIS) suggests a central peak ~10 km in diameter, very large for a 35-km-diameter crater.

Crystalline clast breccia with sandy matrix (SM)

The SM overlies the CR on the central peak, and was encountered in the M-1, M-7, and M-8 cores. The SM (Fig. 6) is characterized by meter- to centimeter-size angular clasts of crystalline basement rocks (biotite and hornblende gneiss, and granite) in a matrix of sand- to silt-size grains of crystalline rock fragments and mineral grains from disaggregated crystalline rocks. Sand-size grains of Keweenawan clastic rocks and melt rock constitute <1% of matrix materials. Almost every clast and matrix mineral grain displays PDFs or other impact deformation features.

The crystalline-clast breccia with sandy matrix could probably be called suevite breccia because rare melt-rock clasts are present. This unit apparently represents brecciated transient crater floor, uplifted during formation of the central peak. The presence of rare grains of Keweenawan clastics and melt rocks demonstrates minor mixing with overlying lithologies. The sandy matrix of this unit becomes increasingly rich in melt rocks and finely powdered (clay-size particles) rocks upward in the core, as it grades into the overlying crystalline clast breccia with melt matrix.

Crystalline clast breccia with melt matrix (MM)

The SM matrix becomes increasingly rich in melt rock and powder (clay-size particles) upward as the SM grades into the overlying MM. This lithology was encountered in the M-1, M-7, and M-8 cores and is a matrix-dominated breccia. Clasts in the breccia include crystalline rock and melt-rock fragments (Fig. 7). The matrix includes sand- to silt-size grains of crystalline rocks and mineral grains (Bell et al., 1996), finely powdered rocks (Fig. 8) that appear petrographically isotropic, and vesicular and flow-banded light to dark brown melt rock that has been devitri-

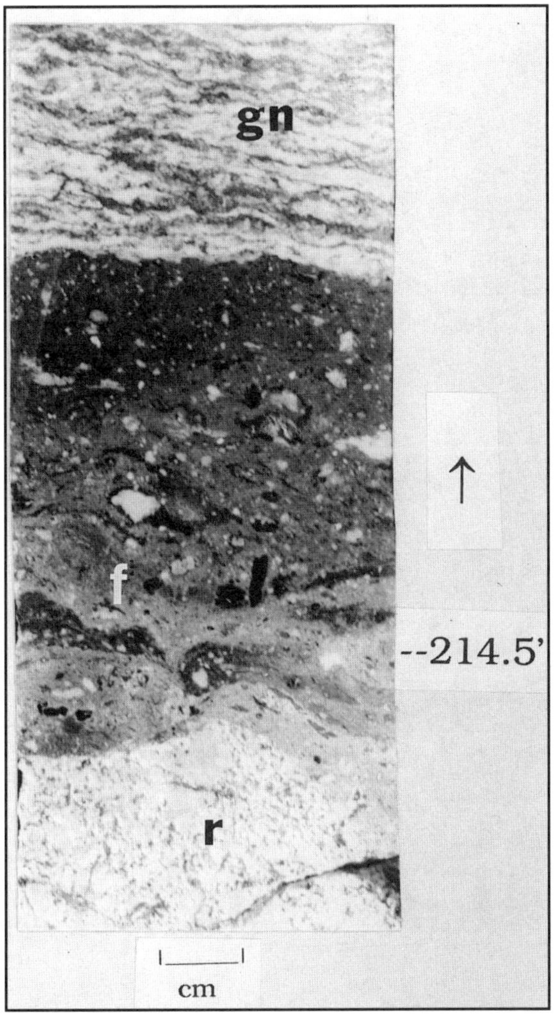

Figure 7. Photograph of crystalline clast breccia with melt matrix from M-7, 214.5 ft (66.0 m); f = flow-banded melt-rock, gn = gneiss clast, r = recrystallized granophyre.

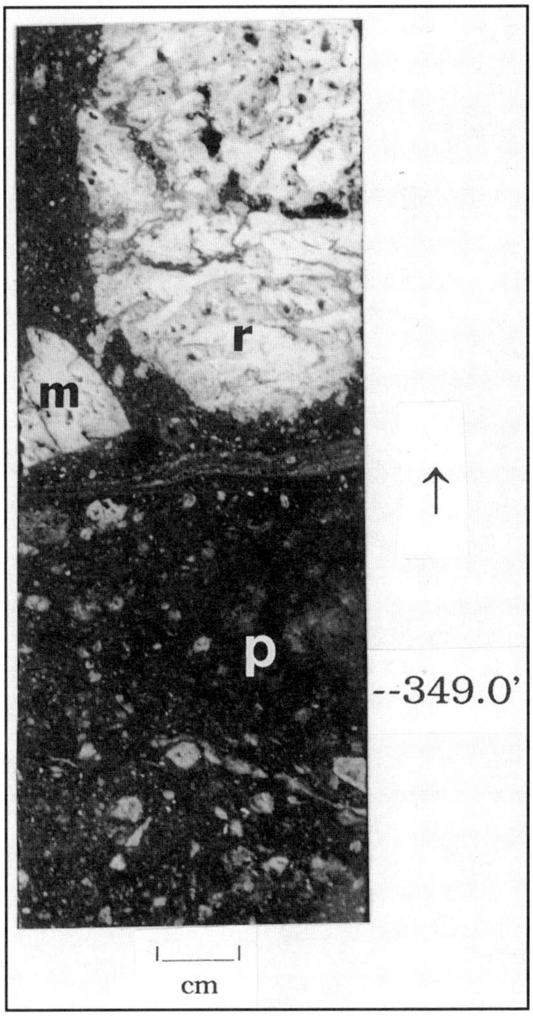

Figure 8. Photograph of crystalline clast breccia with melt matrix from M-1, 349.0 ft (107.4 m); p = finely powdered rock material, r = recrystallized granophyre, m = melt-rock clast.

fied to varying degrees. The melt-rock matrix dominates the upper regions of this unit, grading downward into the finely powdered material and into the sandy-silty material (although some areas display mixing of all matrix lithologies). This unit then grades into the underlying SM.

The central regions of the MM display features representative of the hottest conditions observed in the impact rocks, displaying the maximum melt/clast ratio, sanidine, and zones of solid-state diffusion (Reagan et al., 1993). The zones of solid-state diffusion are dominated by an isotropic powdered-rock matrix. All quartz grains in these zones contain multiple PDFs, fractures, shock isotropism, or partial to total melting and recrystallization. Some grains show evidence of solid-state diffusion with pyroxene, epidote, and biotite grains growing from the quartz margins into isotropic powdered matrix (Reagan et al., 1996). In thin-section, regions of melting appear as tan to dark-brown melt rock, displaying flow banding and varying degrees of recrystallization and devitrification to quartz and sanidine or alteration to clay minerals. The MM contains rare siltstone clasts and crystalline rock clasts with melt-rock mantles (probable airborne accretion) suggesting minor mixing of overlying materials. Angular clasts of melt rock are also present in upper and lower portions of the MM, attesting to its complex evolution. In the M-10 and M-11 cores the MM is present, mixed with blocks of SM and KSB.

The origin of the MM has not yet been determined. It may represent a suevite that fell back to the floor of the transient crater from the cloud of hot gasses and debris that formed as it grew (Bell et al., 1993). If this interpretation is correct, the MM is much richer in melt material than type-suevite described from the Ries Crater (Newsom et al., 1990). Alternatively, the MM may be the crater melt layer (Shoemaker et al., 1993), rock material melted but not ejected during crater growth. If MM is the melt layer, it is richer in clasts that most melt rocks described in the impact literature.

Keweenawan shale breccia (KSB)

The Middle Proterozoic KSB, like the crystalline clast breccias, was encountered only on the MIS central peak, in M-1, M-6, M-7, M-8, M-10, and M-11, and its contact with the underlying MM is abrupt. The KSB clasts are dominated by meter- to centimeter-size blocks of Keweenawan dark gray shale (Witzke and Anderson, 1996) mixed with gray-green devitrified melt rock and minor siltstone and sandstone (Fig. 9). These clasts were likely derived from the lower portion of the Keweenawan sedimentary sequence. The variably calcareous, clay-rich matrix is medium gray and encloses sand- to silt-size grains of quartz, shale, and melt rock. The larger shale clasts appear unmetamorphosed in the core, but many small shale clasts observed in thin section display partial melting. Shock deformation features in quartz grains are observed within the partially melted clasts, but such features are rare within the matrix-grains.

The KSB is the most enigmatic impact lithology encountered. Modeling of MIS crater growth suggests that the transient crater should have excavated below the depth of the Keweenawan shales and well into the crystalline basement near its center. If some of these clastic rocks survived near the center of the transient crater, they would surely have experienced extreme pressure and temperature. The intact nature of the larger KSB suggests possible transport from the sides of the transient crater by an unknown mechanism, probably prior to uplift of the central peak.

Phanerozoic clast breccia (PCB)

The PCB is present in cores from all major terranes of the MIS. It was encountered on the terrace terrane (M-4), in the crater moat (M-2 and M-9), and on the central peak (M-1, M-8, M-10, and M-11). Additionally, the PCB was encountered in an earlier core (1-A) recovered from the center of the MIS, and it is identified in samples from the majority of water wells that penetrate to bedrock within the structure (Witzke and Anderson, 1996). The maximum thickness of PCB penetrated was 189 m, in the M-2 core, including a single block of Upper Cretaceous Carlile Formation over 100 m thick. Wherever it is encountered, the PCB lies at the bedrock surface. The PCB is dominated by clasts of Cretaceous shale, with lesser numbers of Paleozoic carbonate and other sedimentary clasts and rare Keweenawan "Red Clastics" clasts (Fig. 10). The matrix of this unit is light gray and calcareous, with sand- to silt-size grains of Cretaceous shales, mudstones, and sandstones, Paleozoic sedimentary rocks, and "Red Clastics" in approximately the same proportions as the larger clasts. The matrix also contains very rare crystalline rock and melt-rock fragments. Impact shock–altered grains are rare, with only scattered quartz grains displaying PDFs. The PCB locally contains clasts and grains of underlying lithologies near its lower contact.

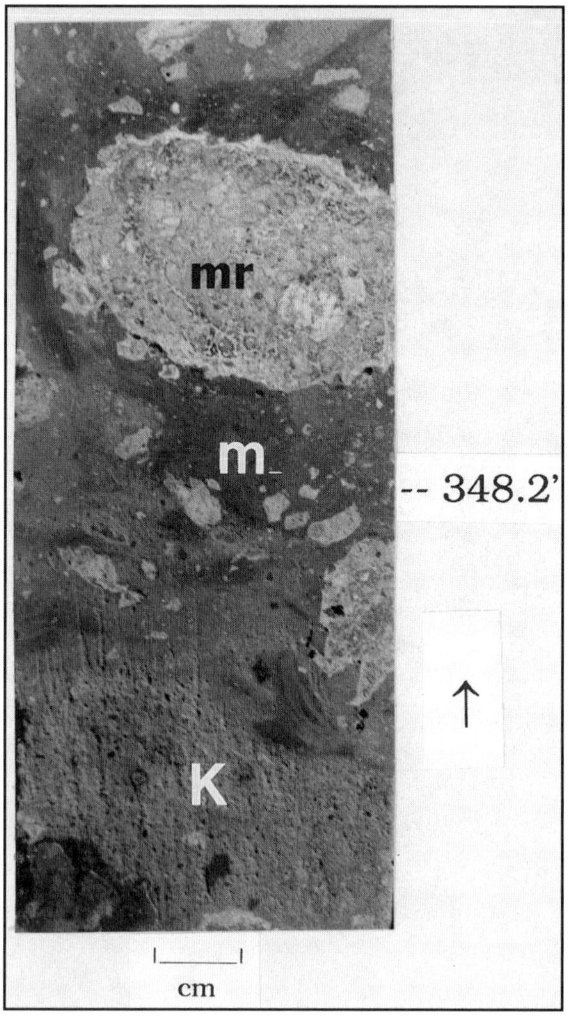

Figure 9. Photograph of Keweenawan shale-clast breccia from M-1, 348.2 ft (107.1 m); K = Keweenawan shale clast, mr = melt-rock clast, m = matrix.

The PCB has a composition closely similar to Bunte breccia described at the Ries Crater (Hörz, 1982; Hörz et al., 1983), a polymictic breccia with clasts dominated by the shallowest rock units (proportionally fewer clasts from deeper preimpact depth) and the matrix derived from the shallowest preimpact units. The Bunte breccia was interpreted by Hörz (1982) and Hörz et al. (1983) as proximal ejecta and secondary crater debris mixed and deposited by a ground surge moving away from the crater (Witzke and Anderson, 1996). This mode of origin requires that the Bunte breccia be deposited outside of the transient crater, which is where the material is presently observed at Ries. However, PCB at Manson is found in all regions of the crater, including the central peak pit in the exact center of the structure (Fig. 4). We interpret the PCB as Bunte breccia material originally deposited outside the crater but carried back into the crater very shortly after crater formation, possibly by water surging back into the area following an impact into a marine environment. However, no structures indicative of

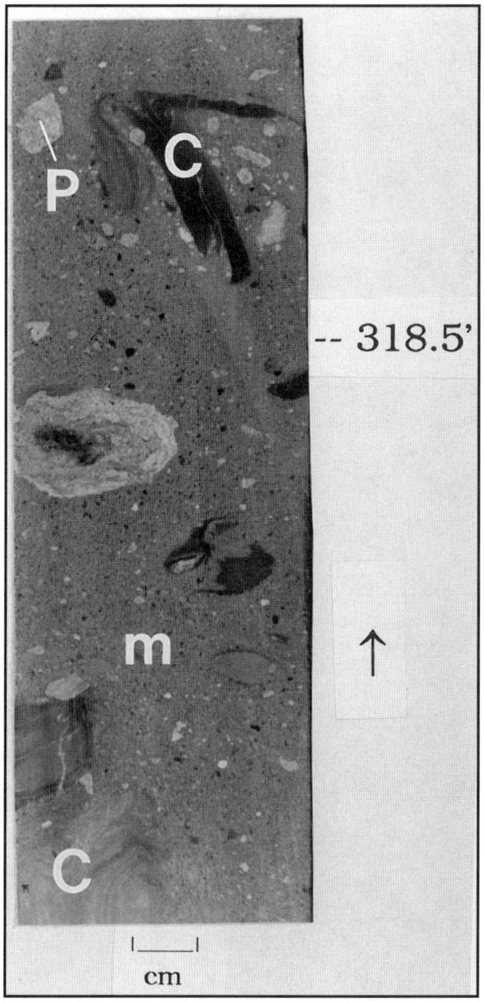

Figure 10. Photograph of Phanerozoic-clast breccia from M-1, 318.5 ft; C = Cretaceous shale clast, P = Paleozoic carbonate clast, m = matrix.

deposition by water have yet been observed in PCB cores. Graded bedding, water escape structures, and similar features usually observed in turbidite-like deposits may have been lost as a result of postimpact regional erosion, estimated to be between 330 m (if the MIS formed at 65 Ma) and 180 m (if it formed at 74 Ma). The interpretation of this material as a series of debris flows emplaced just subsequent to crater formation provides insight into the erosional level of the MIS. The PCB is found in the large majority of holes drilled into the structure (Fig. 11). This implies that the underlying crater is virtually uneroded (Koeberl and Anderson, 1996b), making the MIS one of the best preserved complex craters known on Earth.

Overturned sedimentary strata (OS)

In the M-4 core, beneath PCB, an overturned and thinned brecciated sequence of Proterozoic "Red Clastics" (Fig. 12) and Paleozoic strata was encountered. All preimpact stratigraphic units within the interval cored (upper portion of the Middle Proterozoic "Red Clastics" and Paleozoic Cambrian, Ordovician, and partial penetration of the Devonian) appear to be present but are thinned to an average of ~17% of their preimpact thicknesses. The sandstones are fractured and deformed, as are the carbonate units that display cataclastic textures. Drilling of the M-4 continued to 375 m, the maximum depth capability of the drill rig.

The OS encountered in M-4 must represent the latest stage of crater excavation and growth. An overturned ejecta flap (Fig. 13) fell onto the megablock zone and was structurally preserved on subsiding blocks (Keiswetter et al., 1996). Modeling suggests that these units should be thinned as they are ejected, and this thinning is observed. If M-4 drilling could have been continued for another 100 to 200 m, the preimpact Cretaceous surface may have been encountered. Coring of these materials would likely have provided valuable information about the environment of the impact (marine or terrestrial), the age of the impact, and cratering mechanics, and this region remains a prime target for future research core drilling.

ANSWERS AND NEW QUESTIONS ABOUT THE MANSON IMPACT STRUCTURE

Investigation of the rocks recovered during the GSB–USGS research core drilling confirms that the Manson Impact Structure is an excellently preserved complex impact structure. The MIS displays many features characteristic of complex impact structures, including a well-developed central peak and megablock zone. However, although the cores recovered during the research drilling project have provided a wealth of information, they also have left a number of perplexing questions. Seeking the answers to these enigmas will challenge future researchers.

The Manson Structure central peak, with a diameter of ~10 km, is quite large for a 35-km-diameter impact structure. Most complex impact structures display central peaks that are 0.22 ± 0.03 times the diameter of the apparent crater (Melosh, 1989). Applying this equation to the Manson Structure (diameter of 35 km) suggests a central peak diameter of 7.7 ± 1 km, between 13% and 33% smaller than the diameter of the Manson central peak. The anomalous size of the Manson central peak may be the result of the composition of the impacting body, target lithology, or perhaps angle of impact.

The failure to encounter a well-defined impact melt sheet in the core drilling program is also enigmatic. Cores that were drilled from near the center of the structure, down the eastern flank of the central peak, and into the crater moat (Fig. 1) encountered only a thin melt zone. One explanation for the lack of a thick melt sheet is the sedimentary target rock. Kieffer and Simonds (1980) noted that the amount of melt at craters in crystalline rock was about 2 orders of magnitude greater than at craters in sedimentary rocks. They reported that the volumes of material shocked to pressures sufficient for

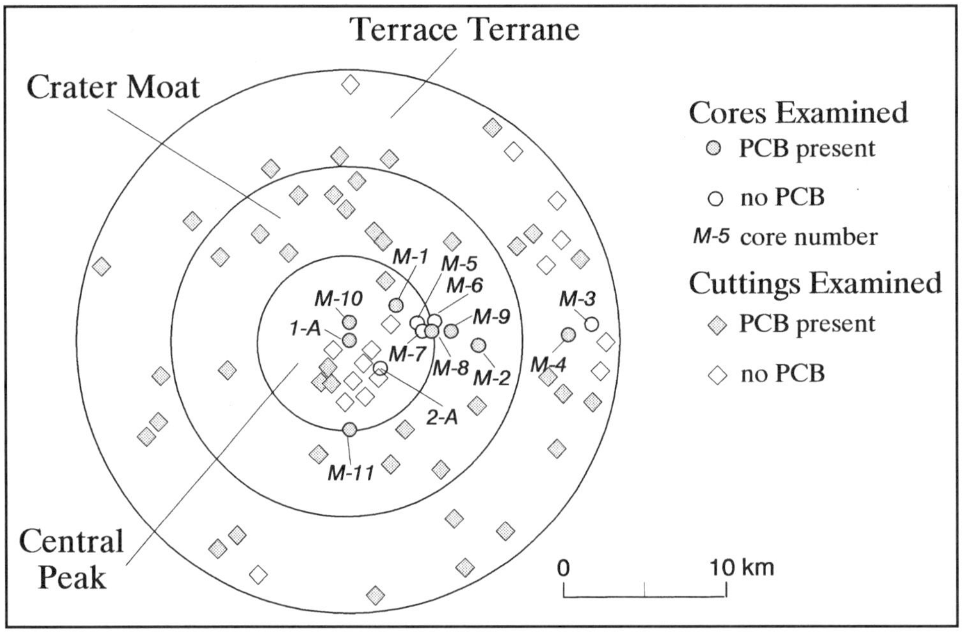

Figure 11. Distribution of Phanerozoic clast breccia in the Manson Impact Structure.

melting were not significantly different in sedimentary and crystalline targets, although the sedimentary rock melt was finely disseminated and dispersed out of the crater by expansion of larger volumes of volatiles released during melting. A second possible explanation for the failure to encounter a significant melt sheet in the drilling was proposed by Schultz and Anderson (1996), who suggested that an oblique impact (from the southeast) could have concentrated the melt sheet in the northwest quadrant of the Manson Structure. This area was not cored in the research drilling program.

One of the most surprising results of the research drilling in the Manson Impact Structure was the absence of postimpact lake sediments. Predrilling interpretations of samples of well cuttings and seismic data had suggested the existence of as much as ~300 m of postimpact lake sediments in the crater moat region. Instead, Phanerozoic-clast breccia (PCB) was encountered in numerous cores as the uppermost unit, suggesting that it was the last impact unit to be deposited within the crater. As previously discussed, the extremely limited number of clasts from deep sources and the very limited occurrence of shock metamorphic features in this last impact unit strongly suggest that it originated as late-stage ejecta and was subsequently transported back into the crater. Although the mechanism for this transport is not known, the PCB filled the crater moat, in the position where the post-impact lake sediments had been expected. If a postimpact lake did form above the PCB, its sediments have been erosionally removed

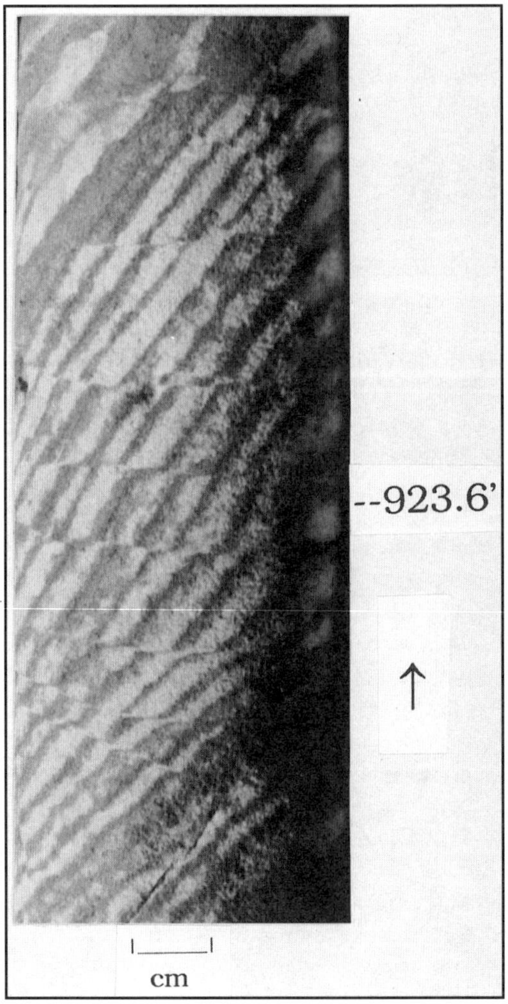

Figure 12. Photograph of overturned sedimentary strata from M-4, 923.6 ft (284.2 m), showing Liesegang banding in Keweenawan Supergroup Red Clastics sandstone offset by small-displacement, impact-related faults.

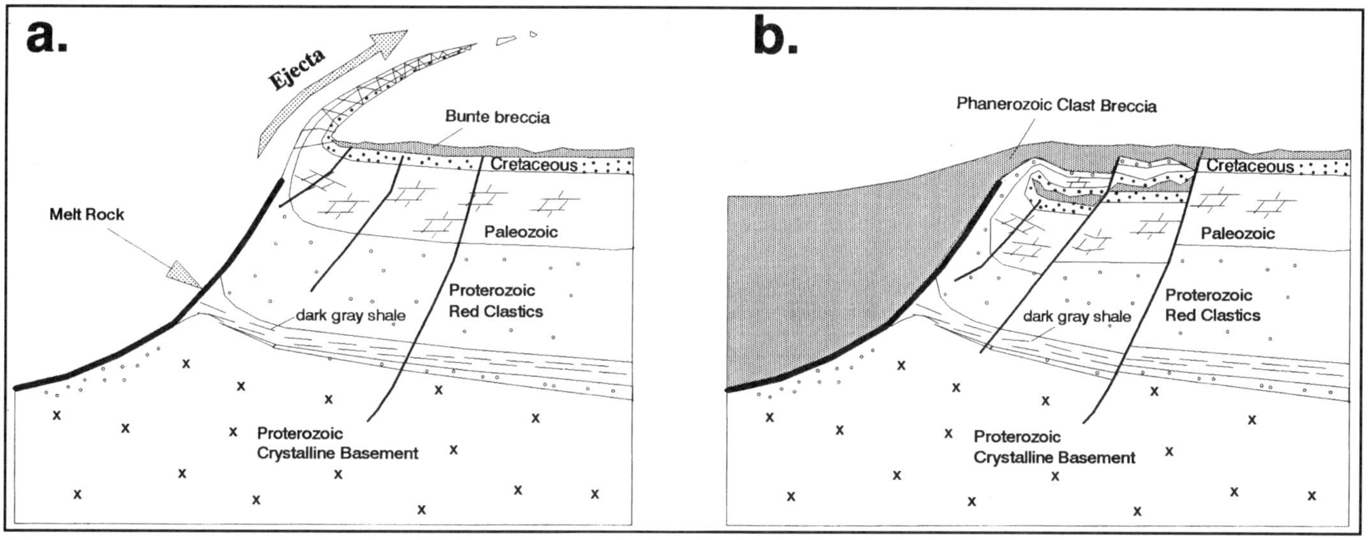

Figure 13. Schematic illustration of late-stage ejecta flap. a, Ejecta flap during latest stage of excavation crater formation. b, Ejecta flap on down-dropped blocks of the megablock zone covered by Phanerozoic clast breccia shortly after crater formation.

in the areas of the research cores. Acquisition and interpretation of shallow seismic data over other areas of the crater moat would be an appropriate tool with which to search for any remnants of these lake sediments.

Perhaps the most enigmatic impact lithology encountered during the Manson research core drilling program was the Keweenawan shale-clast breccia. Modeling (Anderson and Hartung, 1992) suggested that the transient crater reached a maximum depth of ~6 km, with the excavation crater reaching a maximum depth of 4.5 km. Because the Proterozoic crystalline basement was only ~3.5 km below the land surface at impact ground zero, the excavation stage of crater growth should have ejected all Proterozoic Keweenawan clastic rocks from the center of the structure. However, the drilling recovered abundant Keweenawan black shales (KSB) and mudstones mixed with gray-green, devitrified melt rock above crystalline clast breccia in the M-1, M-8, M-10, and M-11 cores, and mixtures of crystalline clast breccia and Keweenawan clastics in M-7, M-8, M-10, and M-11. Many of the Keweenawan shale and mudstone clasts appear relatively unaltered, and matrix quartz grains do not characteristically show PDFs. It therefore seems that all Keweenawan sediments within the Keweenawan shale-clast breccia were not subjected to the same degree of heat and pressure that shocked and melted the underlying crystalline rocks. The KSB displays only minor mixing with the overlying PCB, and it seems unlikely that it was transported to its central peak locations along with the PCB. The Keweenawan clastic rocks must have somehow survived at the bottom or margins of the transient crater, to be lifted up and mixed with crystalline breccias on the central peak.

The resolution of these and other questions that have arisen during the investigation of these cores will provide valuable insight into the history of the Manson Impact Structure and impact processes in general. The Manson drilling program demonstrates the essential role of research core drilling in the investigation of terrestrial impact structures.

BIOTIC EFFECTS OF THE MANSON IMPACT

Although the new age of ca. 74 Ma places the Manson impact ca. 9 m.y. prior to the extinctions at the Cretaceous-Tertiary boundary, the impact may have affected the biota of midcontinent North America. Anderson and Hartung (1992) estimated that ~2.2 ˘ 10^{21} Joules of energy was released by the Manson impact, ~1,600 times the total energy released by the 1989 eruption of Mount St. Helens, and that the impact generated ~600 to 1200 km^3 of ejecta (Anderson and Hartung, 1992; Roddy and Shoemaker, 1993). Such an event would have produced a significant blast zone, igniting combustible material within ~200 km of the impact, leveling all standing vegetation within ~600 km, and killing most exposed terrestrial animals within ~1,000 km (Witzke and Anderson, 1993). Approximately 15% of the ejecta would have been injected above the troposphere (Roddy and Shoemaker, 1993), potentially shrouding the Earth and reducing solar radiation to the surface. Did an event of such magnitude cause any recognizable biotic stresses, including extinctions, at a regional or global scale?

The new age date for the Manson Impact Structure, and the discovery of ejecta (probably from the Manson impact) within the contemporaneous Crow Creek Member of the Pierre Shale in South Dakota and Nebraska, constrain the Manson impact event at ca. 74 Ma, likely within the late Campanian ammonite zones of *Exiteloceras jenneyi* or

Didymoceras stevensoni in the North American Western Interior Basin (Izett et al., 1993; Izett and Cobban, 1994). Any biotic crises or extinction events that occurred during the range of these zones may be evident from the extensive published paleontologic database for the Western Interior (both marine and nonmarine taxa). A preliminary evaluation of various paleontologic compilations was undertaken, but many details remain to be investigated, particularly interregional comparisons.

The Manson impact event occurred near the eastern margin of the Cretaceous Western Interior Seaway, and any effects of the impact on the marine biota would presumably be most apparent in the Western Interior faunal succession. However, there is little evidence for significant extinctions or ecologic reorganization among marine invertebrates at the time of the Manson impact. Western Interior ammonites from the *D. stevensoni* and *E. jenneyi* zones are dominated by evolving lineages of persistent endemic stocks, and no lines became extinct within the interval (Cobban, 1993). Likewise, no significant extinction of benthic bivalves is noted within the interval, and static and evolving lines persisted through the interval (see ranges in Kauffman, 1975; Kauffman et al., 1993).

A foraminiferal zonal boundary is drawn in the Western Interior roughly coincident with the *D. stevensoni* ammonite Zone (Caldwell et al., 1993), but the "foraminiferal spectrum changes very gradually through the Campanian sequences" (McNeil and Caldwell, 1981, p. 109), and no abrupt extinction event occurs coincident with the Manson impact. Similarly, the calcareous nannofossil zone recognized in the Crow Creek Member (Watkins, 1989) does not mark a significant nannoplankton extinction event, and the evolving lineages persisted.

However, the Manson impact event has been implicated in the extinction of an assemblage of marine vertebrates in the Western Interior Seaway by Russell (1993a, b). Characteristic Niobraran (marine vertebrate "age") assemblages "were likely disrupted as a result of the Manson impact event" (Russell, 1993a, p. 670), and "it seems likely that local vertebrate faunas were badly disturbed by the Manson impact and that taxa which were exterminated were replaced by immigrants from beyond North America" (Russell, 1993a, p. 1956). In particular, certain lineages of mosasaurs, marine turtles, and fish vanish from the Western Interior at a time that approximates the Manson impact event.

Interpreting influences of the Manson impact event on the marine biota of the Western Interior is confounded by the general coincidence of the impact and early transgression of the Bearpaw marine cycle. As emphasized by Kauffman (1984), a "patterned biotic response" accompanied regressions and transgressions of the seaway. In general, regression was marked by the southward retreat of many subtropical elements, followed by rapid evolution during transgression. Faunal changes that may have resulted from the Manson impact event, therefore, would have been influenced further by any biotic reorganization during regression of the preceding Claggett marine cycle.

The terrestrial biota that inhabited the western region of the Western Interior prior to the Manson impact is characterized by the extensive fauna and flora of the Judithian vertebrate "age," typified by fossils from the Judith River Group of Montana and Alberta. Although Russell (1993a, b) equated the Manson impact event with the boundary separating the Aquilan and Judithian vertebrate "ages," it is clear from the definitions of Lillegraven and Ostresh (1990) that the Manson impact event roughly coincides with the boundary between the Judithian and Edmontonian vertebrate "ages" (*E. jenneyi* Zone as defined, correlative with upper Judith River strata).

No large-scale terrestrial extinctions are noted at the end of the Judithian vertebrate "age," and many lines seemingly continue uninterrupted into the Edmontonian. Freshwater fish, amphibian, turtle, lizard, and crocodilian faunas show little change from the Judithian into the Edmontonian and Lancian. Mammalian lineages, likewise, show little if any disruption or abrupt extinction coincident with Judithian-Edmontonian boundary (Lillegraven and McKenna, 1986). In addition, most North American dinosaur lineages survived the end of the Judithian with no apparent damage. However, some dinosaur lines apparently did not survive into the Edmontonian, including genera of hadrosaurs, pachycephalosaurs, and centrosaurine ceratopsids. In particular, the diverse assemblage of centrosaurine ceratopsid genera that characterizes Judithian faunas was apparently decimated by the close of the Judithian, and only a single centrosaurine genus survived into the Edmontonian. Giant deinosuchid crocodilians also did not survive the end of the Judithian in the Western Interior.

No extinction event within the terrestrial flora is evident at the close of the Judithian. Most palynologic taxa span the interval in question, ranging through upper Judith River strata into overlying units (Tschudy, 1973). In fact, the Manson impact event occurred within the span of a single long-ranging palynologic zone in the Western Interior region (*Aquiapollenites quadrilobus* interval Zone; Nichols et al., 1982), and no significant extinction event is documented within this zone.

In summary, the Manson impact event produced few recognizable effects on the terrestrial or marine biota in the Western Interior of North America. The evolving lineages in the region, for the most part, continued unimpeded by any catastrophic ecologic changes that may have been brought about by the Manson impact event. Nevertheless, some evidence suggests that portions of the biota in the Western Interior were significantly affected by the impact event, primarily certain large vertebrate animals, both terrestrial and marine. Higher trophic levels may have been more vulnerable to ecologic disruptions created by a regional catastrophic event, but such speculation requires further testing. Although impact events have been of considerable interest in understanding global extinction events, it is likely that the consequences of a Manson-sized impact would largely be evident at a regional, not global, scale.

REFERENCES CITED

Alvarez, L. W., Alvarez, W., Asaro, F., and Michel, H. V., 1980, Extraterrestrial cause for the Cretaceous-Tertiary extinction: Science, v. 208, p. 1095–1108.

Anderson, R. R., and Hartung, J. B., 1992, The Manson impact structure: Its contribution to impact materials observed at the Cretaceous/Tertiary boundary: Houston, Texas, Proceedings, Lunar and Planetary Science Conference 22, p. 101–110.

Anderson, R. R., and Witzke, B. J., 1994, The terminal Cretaceous Manson Impact Structure in North-Central Iowa: A window into the Late Cretaceous history of the eastern margin of the Western Cretaceous Seaway, *in* Shurr, G. W., Ludvigson, G. A., and Hammond, R. H., eds., Perspectives on the eastern margin of the Cretaceous Western Interior Basin: Geological Society of America Special Paper 287, p. 197–210.

Anderson, R. R., Witzke, B. J., and Roddy, D. J., 1996, The drilling of the 1991–1992 Geological Survey Bureau and U.S. Geological Survey Manson Impact Structure research cores, *in* Koeberl, C., and Anderson, R. R., eds., The Manson Impact Structure, Iowa: Anatomy of an impact crater: Geological Society of America Special Paper 302, p. 45–88.

Bell, M. S., Reagan, M. K., Anderson, R. R., and Foster, C. T., Jr., 1993, Petrography of the crystalline breccias from the Manson M-1 Core: Compositional correlation of target lithologies [abs.]: Geological Society of America Abstracts with Program, v. 25, p. A-222.

Bell, M. S. , Reagan, M. K., Anderson, R. R., and Foster, C. T., Jr., 1996, Petrography of crystalline clast breccias from the Manson M-1 core, *in* Koeberl, C., and Anderson, R. R., eds., The Manson Impact Structure, Iowa: Anatomy of an impact crater: Geological Society of America Special Paper 302, p. 221–234.

Caldwell, W. G. E., Diner, R., Eicher, D. L., Fowler, S. P., North, B. R., Stelck, C. R., and von Holdt W. L., 1993, Foraminiferal biostratigraphy of Cretaceous marine cyclothems, *in* Caldwell, W. G. E., and Kauffman, E. G., eds., Evolution of the Western Interior Basin: Geological Association of Canada Special Paper 39, p. 477–520.

Cobban, W. A., 1993, Diversity and distribution of Late Cretaceous ammonites, Western Interior, United States, *in* Caldwell, W. G. E., and Kauffman, E. G., eds., Evolution of the Western Interior Basin: Geological Association of Canada Special Paper 39, p. 435–451.

Dalrymple, G. B., and Izett, G. A., 1994, ^{40}Ar/^{39}Ar Age of Manson Impact Structure, Iowa: Geological Society of America Abstracts with Programs, v. 26, no. 7, p. A337.

Hartung, J. B., and Anderson, R. R., 1988, A compilation of information and data on the Manson Impact Structure: Houston, Texas, Lunar and Planetary Institute, Technical Report 88-08, 32 p.

Hartung, J. B., and Anderson, R. R., 1996, A brief history on investigations of the Manson Impact Structure, *in* Koeberl, C., and Anderson, R. R., eds., The Manson Impact Structure, Iowa: Anatomy of an impact crater: Geological Society of America Special Paper 302, p. 31–44.

Hartung, J. B., Izett, G. B., Naeser, C. W., Kunk, M. J., and Sutter, J. F., 1986, The Manson, Iowa, impact structure and the Cretaceous-Tertiary boundary event [abs.]: Houston, Texas, Lunar and Planetary Institute, Lunar and Planetary Science, v. 27, p. 313–314.

Hartung, J. B, Kunk, M. J., and Anderson, R. R., 1990, Geology and geochronology of the Manson Impact Structure, *in* Sharpton, V. L., and Ward, P. D., eds., Global catastrophes in Earth history: An interdisciplinary conference on impacts, volcanism, and mass mortality: Geological Society of America Special Paper 247, p. 207–222.

Hörz, F., 1982, Ejecta of the Ries Crater, Germany, *in* Silver, L.T., and Schultz, P. H. eds., Geologic implications of impacts of large asteroids and comets on the Earth: Geological Society of America Special Paper 190, p. 39–56.

Hörz, F., Ostertag, R., and Rainey, D. A., 1983, Bunte breccia of the Ries; Continuous deposits of large impact craters: Reviews of Geophysics and Space Physics, v. 21, p. 1667–1725.

Izett G. A., Cobban, W. A., Obradovich, J. D., and Kunk M. J., 1994, The Manson Impact Structure: ^{40}Ar/^{39}Ar Age and its distal impact ejecta in the Pierre Shale in southeastern South Dakota: Science, v. 262, p. 729–732.

Kauffman, E. G., 1975, Dispersal and biostratigraphic potential of Cretaceous benthonic Bivalvia in the Western Interior, *in* Caldwell, W. G. E., ed., The Cretaceous System in the Western Interior of North America: Geological Association of Canada Special Paper 13, p. 163–194.

Kauffman, E. G., 1984, Paleobiogeography and evolutionary response dynamic in the Cretaceous Western Interior Seaway of North America, *in* Westermann, G. E. G., ed., Jurassic-Cretaceous biochronology and paleogeography of North America: Geological Association of Canada Special Paper 27, p. 273–306.

Kauffman, E. G., Sageman, B. B., Kirkland, J. I., Elder, W. P., Harries, P. J., and Vallamil, T., 1993, Molluscan biostratigraphy of the Cretaceous Western Interior Basin, North America, *in* Caldwell, W. G. E., and Kauffman, E. G., eds., Evolution of the Western Interior Basin: Geological Association of Canada Special Paper 39, p. 397–434.

Kieffer, S. W., and Simonds, C. H., 1980, The role of volatiles and lithology in the impact cratering process. Reviews of Geophysics and Space Physics, v. 18, p. 143–181.

Keiswetter, D., Black, R., and Steeples, D., 1996, Structure of the terrace terrane, Manson Impact Structure, Iowa, interpreted from high-resolution seisimc reflection data, *in* Koeberl, C., and Anderson, R. R., eds., The Manson Impact Structure, Iowa: Anatomy of an impact crater: Geological Society of America Special Paper 302, p. 105–114.

Koeberl, C., and Anderson, R. R., eds., 1996a, The Manson Impact Structure, Iowa: Anatomy of an impact crater: Geological Society of America Special Paper 302, 468 p.

Koeberl, C., and Anderson, R. R., 1996b, Manson and company: Impact structures in the United States, *in* Koeberl, C., and Anderson, R. R., eds., The Manson Impact Structure, Iowa: Anatomy of an impact crater: Geological Society of America Special Paper 302, p. 1–30.

Kunk, M. J., and Nord, G. L. Jr., 1994, The Manson Impact Structure is not a K-T Boundary Event: Geological Society of America Abstracts with Programs, v. 26, no. 7, p. A337.

Kunk, M. J., Izett, G. A., and Sutter, J. F., 1987, ^{40}Ar/^{39}Ar age spectra of shocked K-feldspar suggests K-T boundary age for Manson, Iowa, Impact Structure [abs.]: Eos (Transactions, American Geophysical Union), v. 68, p. 1514.

Kunk, M. J., Izett, G. A., Haugerud, R. A., and Sutter, J. F., 1989, ^{40}Ar/^{39}Ar dating of the Manson Impact Structure: A Cretaceous-Tertiary boundary crater candidate: Science, v. 244, p. 1565–1568.

Kunk, M. J., Snee, L. W., French, B. M., and McGee, J. J., 1993, Preliminary ^{40}Ar/^{39}Ar spectrum and laser probe dating of the M1 core of the Manson Impact Structure, Iowa: A K-T boundary crater candidate [abs.]: Houston, Texas, Lunar and Planetary Institute, Lunar and Planetary Science, v. 24, p. 835–836.

Leroux, H., and Doukhan, J., 1996, A transmission electron microscope study of shocked quartz from the Manson Impact Structure, *in* Koeberl, C., and Anderson, R. R., eds., The Manson Impact Structure, Iowa: Anatomy of an impact crater: Geological Society of America Special Paper 302, p. 267–274.

Lillegraven, J. A., and McKenna, M. C., 1986, Fossil mammals from the "Mesaverde" Formation (Late Cretaceous, Judithian) of the Bighorn and Wind River Basins, Wyoming, with definitions of Late Cretaceous North American land-mammal "ages": American Museum Novitates, no. 2840, p. 1–68.

Lillegraven, J. A., and Ostresh, L. M., Jr., 1990, Late Cretaceous (earliest Campanian/Maastrichtian) evolution of western shorelines of the North American Western Interior Seaway in relation to known mammalian faunas, *in* Brown, T. M., and Rose, K. D., eds., Dawn of the age of mammals in the northern part of the Rocky Mountain Interior: Geological Society of America, Special Paper 243, p. 1–30.

McNeil, D. H., and Caldwell, W. G. E., 1981, Cretaceous rocks and their foraminifera in the Manitoba Escarpment: Geological Association of Canada, Special Paper 21, 439 p.

Melosh, H. J., 1989, Impact cratering: New York, Oxford University Press, 245 p.

Newsom, H. E., Graup, G., Iseri, D. A., Geissman, J. W., and Keil, K., 1990, The formation of the Ries Crater, West Germany; Evidence of atmospheric interactions during a larger cratering event, *in* Sharpton, V. L., and Ward, P. D., eds., Global catastrophes in Earth history: An interdisciplinary conference on impacts, volcanism, and mass mortality: Geological Society of America Special Paper 247, p. 195–206.

Nichols, D. J., Jacobson, S. R., and Tschudy, R. H., 1982, Cretaceous palynomorph biozones for the central and northern Rocky Mountain region of the United States, in Powers, R. B., ed., Geologic studies of the Cordilleran Thrust Belt: Rocky Mountain Association of Geologists, Denver, v. 2, p. 721–733.

Reagan, M. K., Foster, C. T., Jr., and Bell, M. S., 1993, Origins of the Upper Crystalline-Clast Bearing Suevite in the Manson M-1 Core [abs.]: Eos (Transactions, American Geophysical Union), v. 74, p. 386–387.

Reagan, M. K., Foster, C. T., Jr., Bell, M. S., and Anderson, R. R., 1996, Preferential feldspar comminution in impact melt breccias from the M-1 core of the Manson Impact Structure, *in* Koeberl, C., and Anderson, R. R., eds., The Manson Impact Structure, Iowa: Anatomy of an impact crater: Geological Society of America Special Paper 302, p. 235–244.

Roddy, D. J., and Shoemaker, E. M., 1993, The Manson impact crater: Estimation of the energy of formation, possible size of the impacting asteroid or comet, and ejecta volume and mass: Houston, Texas, Proceedings, Lunar and Planetary Science Conference 24, p. 1211–1212.

Russell, D. A., 1993a, Manson crater extinctions: Science, v. 262, p. 1956–1957.

Russell, D. A., 1993b, Vertebrates of the Cretaceous Western Interior Sea, *in* Caldwell, W. G. E., and Kauffman, E. G., eds., Evolution of the Western Interior Basin: Geological Association of Canada Special Paper 39, p. 665–680.

Schultz, P. H., and Anderson, R. R., 1996, Asymmetery of the Manson Impact Structure: Evidence for impact angle and direction, *in* Koeberl, C., and Anderson, R. R., eds., The Manson Impact Structure, Iowa: Anatomy of an impact crater, Geological Society of America Special Paper 302, p. 397–418.

Shoemaker, E. M., Roddy, D. J., and Anderson, R. R., 1993, Research program on the Manson Impact Structure, Iowa [abs.]: Houston, Texas, Lunar and Planetary Institute, Lunar and Planetary Science, v. 24, p. 1301–1302.

Short, N, 1966, Shock processes in geology: Journal of Geological Education, v. 14, p. 149–166.

Short, N. M., and Gold, D. P., 1996, Petrography of shocked rocks from the central peak of the Manson Impact Structure, *in* Koeberl, C., and Anderson, R. R., eds., The Manson Impact Structure, Iowa: Anatomy of an impact crater: Geological Society of America Special Paper 302, p. 245–266.

Tschudy, B. D., 1973, Palynology of the upper Campanian (Cretaceous) Judith River Formation, north-central Montana: U.S. Geological Survey, Professional Paper 778, 42 p., 11 plates.

Watkins, D. K., 1989, Biostratigraphy and paleoceanographic implications of nannofossils in marls of the Pierre Shale, Western Interior Basin: Geological Society of America, Abstracts with Programs, v. 21, p. A337.

Witzke, B. J., and Anderson, R. R., 1993, The terminal Cretaceous Manson Impact Structure of Iowa and its possible influence on K/T events in the Western Interior of North America, *in* The Marshall Lambert Symposium: Pioneer Trails Museum, Bowman, North Dakota, published by North Dakota Geological Survey, Bismarck, p. 13–16.

Witzke, B. J., and Anderson, R. R., 1996, Sedimentary-clast breccias of the Manson Impact Structure, *in* Koeberl, C., and Anderson, R. R., eds., The Manson Impact Structure, Iowa: Anatomy of an impact crater: Geological Society of America Special Paper 302, p. 115–144.

Zeitler, P. K., 1994, Dating Shocked Feldspars, Manson Impact Structure: Geological Society of America, Abstracts with Programs, v. 26, no. 7, p. A402.

Zeitler, P. K., and Kunk, M. J., 1993, Age and thermochronology of K-feldspars from the Manson Impact Structure [abs.]: Houston, Texas, Lunar and Planetary Institute, Lunar and Planetary Science, v. 24, p. 1587–1588.

Manuscript Accepted by the Society September 14, 1995

Printed in U.S.A.

Geological Society of America
Special Paper 307
1996

Axial focusing of impact energy in the Earth's interior: A possible link to flood basalts and hotspots

M. B. Boslough, E. P. Chael, T. G. Trucano, D. A. Crawford, and D. L. Campbell
Sandia National Laboratories, Albuquerque, New Mexico 87123-0820

ABSTRACT

Seismological simulations show how energy from a large impact can be coupled to the interior of the Earth. The radially diverging shock wave generated by the impact decays to linearly elastic seismic waves. These waves reconverge (minus attenuation) along the axis of symmetry between the impact and its antipode. The locations that experience the most strain cycles with the largest amplitudes will dissipate the most energy and have the largest increases in temperature at a given depth (for a given attenuation efficiency). We have shown that the locus of maximum energy deposition in the mantle lies along the impact-antipodal axis. Moreover, the most intense focusing is within the asthenosphere at the antipode, over the range of depths where mechanical energy is most readily converted to heat. We suggest that if large impacts on the Earth leave geological evidence anywhere other than the impact site itself, it will be at the antipode. Strong axial focusing also occurs directly beneath the impact site and should be considered in models of impact-induced volcanism.

INTRODUCTION

The most obvious evidence for hypervelocity impacts in the geological record is the craters that impacts leave on the solid surfaces of planets and satellites. The cratered face of the moon led G. K. Gilbert (1893) to the first hypothesis for the process of impact that was given serious consideration, but it was many years before a scientific consensus was reached (e.g., Shoemaker, 1962). Understanding of the effects of impacts has focused on cratering (reviewed by Melosh, 1989), but hypervelocity impact is now known to have had many roles beyond making craters. Examples include control of the interior and surface thermal states of the Earth as it grew by impact-accretion (e.g. Kaula, 1979; Ahrens, 1994), the origin of the moon by the collision of a Mars-sized object with the early Earth (reviewed by Taylor, 1994), and the mass extinction at the Cretaceous-Tertiary (K/T) boundary (Alvarez et al., 1980). Evidence for other impact-induced extinctions has been cited (McLaren and Goodfellow, 1990), and there has been increasing speculation that energetic collisions have been responsible for geological processes as varied as continental flood basalt eruptions, mantle plumes, continental rifting, and geomagnetic pole reversals.

The cratering record has largely been erased on the Earth as a result of resurfacing by plate tectonics, volcanism, glaciation, and weathering (cf. Grieve, 1987). However, comparison to the cratered lunar surface implies that the Earth has experienced many impacts leaving craters the size of Chicxulub or larger (diameter estimates range from 180 to 300 km). The purpose of our computational simulations was to enable us to develop a physical model for how a very large impact could have affected the internal state of the Earth. Such an event may have left evidence that remains long after the impact structure on the surface has disappeared.

Most planetary impact modeling has been aimed at explaining phenomena that have already been observed, and for that reason it has not been a predictive science. A notable exception is the recent impact of comet Shoemaker-Levy 9 on Jupiter, which provided an unprecedented opportunity for modelers to develop hypotheses and use them to make predictions that could be directly tested by observation. Models for features already pres-

Boslough, M. B., Chael, E. P., Trucano, T. G., Crawford, D. A., and Campbell, D. L., 1996, Axial focusing of impact energy in the Earth's interior: A possible link to flood basalts and hotspots, *in* Ryder, G., Fastovsky, D., and Gartner, S., eds., The Cretaceous-Tertiary Event and Other Catastrophes in Earth History: Boulder, Colorado, Geological Society of America Special Paper 307.

ent in the geological record are more difficult to validate because they are not predictive. By contrast, the model described in this chapter can be used to propose a testable hypothesis (subject to either validation or falsification), because it specifically predicts both age and location of as-yet-undiscovered impact structures.

BACKGROUND

A cause-and-effect connection between impacts and various internal geophysical processes has been the subject of extensive discussion and speculation. The earliest suggestions came from Ronca (1966), who presented a hypothesis for impact-triggered volcanism; Glass and Heezen (1967), who invoked impacts as a common cause for both tektites and geomagnetic field reversals; and Urey (1973), who conjectured that (among other things) impacts could lead to extensive seismically initiated lava flows. Support for connections between impacts and the internal dynamics of the Earth came from the association of a geomagnetic reversal with sediments deposited shortly after the Ries impact (Pohl, 1977) and from further correlations of the ages of tektites from strewn fields with the timing of geomagnetic field reversals (Glass et al., 1979). To explain the breakup of tectonic plates, Seyfert and Sirkin (1979) proposed that impact-induced mantle plumes could be responsible and suggested a correlation between major tectonic events and possible impact episodes since the Mesozoic Era. Burek and Wänke (1988) listed possible correlations between impacts and geomagnetic field reversals, unconformity ages, shifts in paleotemperatures, and tectonic episodes. They suggested that shock-induced phase transitions in the upper mantle can generate an instability that has an effect down to the core-mantle boundary. Rampino (1987) proposed that the Deccan flood basalts in India, which were rapidly erupted near the K/T boundary, could have directly resulted from a large impact. Rampino and Stothers (1988) postulated that episodic showers of impacting comets could explain the apparent quasi-periodic correlation between mass extinctions and major continental flood basalt volcanism since the Permian. Glikson (1995) has recently argued for the triggering of major episodes of crustal evolution by large impacts on the basis of the observed episodicity of Precambrian igneous and rifting events.

For major impact events to have an effect on global solid-Earth geophysical processes would probably require that the coupling of impact energy causes a significant change in the thermal state of the Earth's interior. One mechanism that is capable of transporting energy from the Earth's surface to its deep interior is focusing due to axial symmetry (Fig. 1). The convergence of surface and body waves at the antipode of earthquakes is a well-known phenomenon (Gutenberg and Richter, 1934) and has been used by seismologists to provide information about global average properties of the Earth's interior (Rial, 1979; Chael, 1983). Antipodal focusing of seismic waves from major impacts on the Moon, Mercury, and icy satellites has also been invoked by planetary scientists to explain unusual surface features opposite some of the large impact structures on these bodies (Schultz and Gault, 1975; Hughes et al., 1977; Watts et al., 1991). For example, disrupted terrains have been observed antipodal to the Caloris impact basin on Mercury and the Imbrium and Orientale basins on the Moon. A causal relationship was first proposed by Schultz (1974), who suggested the possibility of mass wasting due to the large antipodal ground motion from focused seismic waves (see Melosh, 1989, for a review).

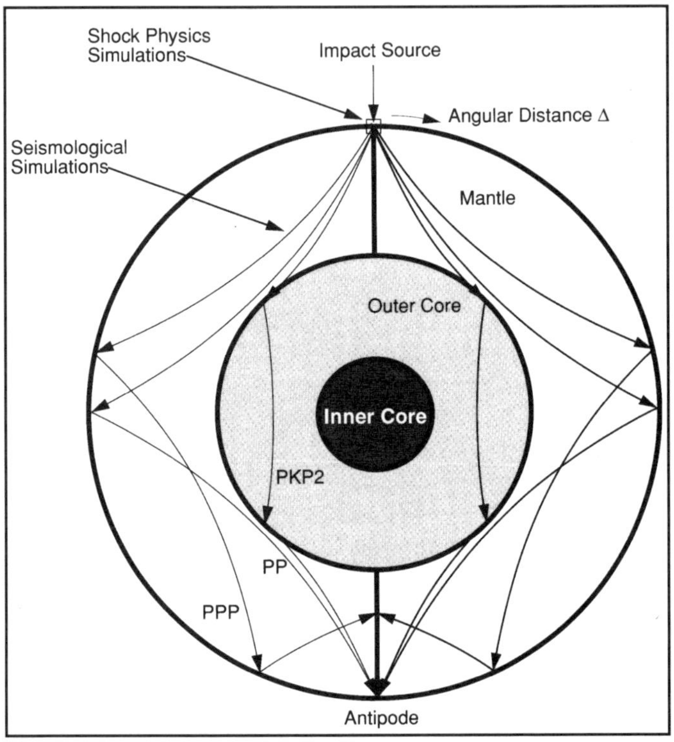

Figure 1. Cross section of the Earth, showing why seismic energy focuses on the axis. The shock physics simulations are limited to the impact source region, and the seismological simulations are over the entire interior. Various families of longitudinal seismic rays are drawn.

Impact-induced antipodal volcanism has also been suggested on the Moon and Mars. Schultz and Gault (1975) proposed that converging tension waves along an axis interior to the Imbrium antipode within the moon generated a shattered zone that could have provided the conduits for mare flooding. An antipodal relationship between the Hellas impact basin and the Alba Patera volcanism on Mars was pointed out by Peterson (1978), who suggested cause and effect. Williams and Greeley (1994) have performed computational modeling to determine the degree of focusing on Mars and have suggested that the convergence of seismic waves was intense enough to fracture the Martian crust and lead to basaltic eruptions by a mechanism similar to that previously proposed for the lunar mare. There have also been recent speculations that antipodal focusing of impact energy could have led to flood basalt and hotspot activity on Earth (Hagstrum and Turrin, 1991; Rampino and Caldeira, 1992).

Several authors have pointed to evidence in the geological

record that suggests the possibility of a causal relationship between large impact events and episodes of basaltic volcanism. This observation has led to the hypothesis that impact-induced melting and excavation can lead to volcanism at the impact site by shock heating, fracturing of the crust, and/or decompressional melting of rocks in the upper mantle (e.g. Ronca, 1966; Green, 1972; Rampino, 1987; Alt et al., 1988; Oberbeck et al., 1992; Chatterjee, 1992; Negi. et al., 1993; Rampino and Caldeira, 1993). Hypotheses such as these have been met by varying degrees of skepticism. White (1989), Loper and McCartney (1990), and Loper (1991) have based much of their criticism on the fact that decompression due to upward motion in an adiabatic fluid leaves a parcel in the same thermal state as its surroundings and cannot cause additional melting. However, this criticism neglects the fact that, for a sufficiently large impact, the upper mantle would experience significant shock heating, generating a localized thermal anomaly in the upper mantle in addition to excavation and fracturing of the crust.

Loper and McCartney (1990) and Loper (1991) also pointed out that the shallow magmatism resulting from an impact is not capable of producing the hotspot tracks associated with many flood basalt eruptions in the geologic record. A reasonable hypothesis that addresses this problem is that seismic energy is coupled deeply into the Earth's interior by focusing on the axis defined by the point of impact resulting from the first-order symmetry of both the energy source and the structure of the Earth's interior. We believe that this is a viable hypothesis because it (1) invokes only processes that are independently known to take place (impacts and axial focusing), (2) it is testable by rigorous computational modeling, and (3) it is potentially predictive, that is, it provides a basis for predicting the existence of currently unknown impact structures at or antipodal to known flood basalt eruptions. The computational simulations presented in this chapter provide the initial proof-of-principle test of the hypothesis and give guidance for future work.

COMPUTATIONAL SIMULATIONS

We use two different types of simulations to model the asteroid impact and resulting seismic disturbance in the Earth's interior (Fig. 1). A single computational method cannot accurately be applied to this problem because of the wide variation in spatial and temporal scales and in peak stresses. During the impact phase of the event, physical processes are taking place at scales on the order of the 10-km-diameter asteroid. Onc-tenth of a second after contact the peak shock pressures exceed 6 Mbar. Thousands of seconds later (when the energy is being dissipated within the Earth) the peak stresses are in the 10 bar range, and the wave motion takes place within the 12,000 km diameter of the Earth. It is difficult to represent both extremes computationally, so we have divided the problem into two parts. First, we investigate the source region where the stresses and strain rates are high, using the multidimensional multimaterial elastic-plastic Eulerian code CTH (McGlaun et al., 1990). Second, we perform a seismological simulation that yields synthetic seismograms for various locations within the Earth by summing normal modes. The seismological simulations are performed in the frequency domain and can avoid some of the problems associated with shock physics codes (such as spatial resolution due to mesh size and attenuation due to artificial viscosity) when applied to linear elastic waves propagating long distances.

We used the strong shock-hydrodynamics code CTH to investigate the source region of a 10-km-diameter asteroid impacting at 20 km/s. Preliminary two-dimensional axisymmetric simulations were performed (Boslough et al., 1995) to determine the influence of an asteroid's shape upon its interaction with the Earth, which was modeled in the near field (to a depth of 50 km) using an analytic equation of state (Thompson, 1989). Within a few seconds after impact, major shape differences become difficult to distinguish, so that (to first order) the impact can be simulated as a point source when modeling far-field effects. The results of such near-field shock physics simulations can be used to generate source functions for the seismological simulations that would be more accurate than point source approximations, which would in turn yield more accurate far-field strain histories. We expect that the deviations from point source functions will become more important as larger impacts and various oblique angles are investigated.

In Figure 2, the computed wave profile from the spherical asteroid impact simulation is plotted as a function of depth. At the top of the asthenosphere (about 75 km depth), the peak shock pressure will have decayed to about 20 Kbar, resulting in significant heating. This heating, combined with uplift due to the removal of overburden by excavation of the crater, might result in generation of much larger quantities of melt than

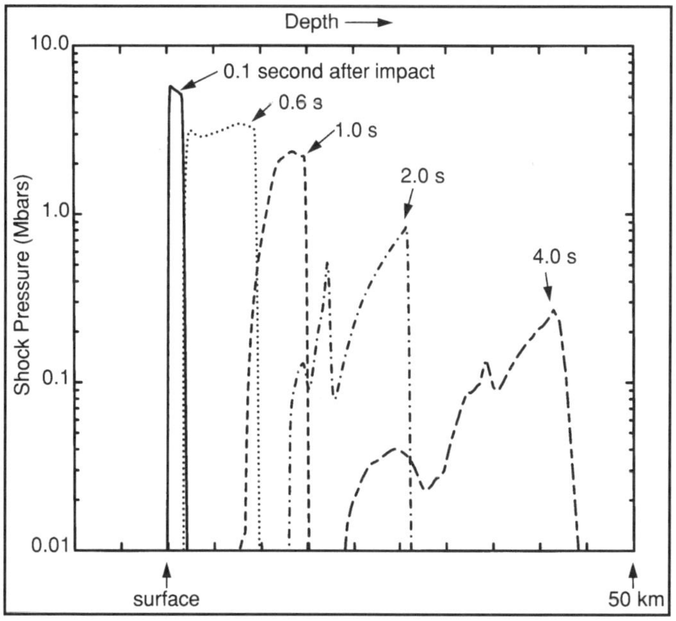

Figure 2. Shock pressure profiles as a function of depth for the impact of a 10-km-diameter sphere at 20 km/s, showing a rapid decay with depth.

would result from uplift alone. This effect is due to the fact that melt volume is extremely sensitive to small variations in the initial temperature of the mantle (White, 1989). Clearly, this is an area for further investigation by computational simulation.

The present seismological calculations were carried out using a point source function. Future work will use CTH to generate more accurate source functions as well as to investigate the effect of oblique impacts. The coupling of energy from major impacts to the mantle by axial focusing of seismic waves is very different than from a great earthquake that, in addition to having less energy, has an asymmetric focal mechanism and a larger area. However, impacts, like explosion sources, are less efficient than earthquakes in coupling their energy to seismic waves. Whereas the fraction of strain energy released in a great earthquake that is radiated as seismic energy can approach unity (Kanamori, 1977), a commonly accepted figure for seismic efficiency due to impact is 10^{-4} (Schultz and Gault, 1975), derived from data based on various laboratory and missile impact tests.

We modeled time histories of displacement, stress and strain through the use of normal mode synthetics. Each of the Earth's normal modes contributes a decaying sinusoid to the particle motion at a point in the Earth. The contribution of an individual mode to the observed motion at some point is a complex function of the source mechanism and the locations of the source and observation point. For our synthetics, we used the elastic, spherically symmetric Earth model 1066A of Gilbert and Dziewonski (1975), for which there are existing compilations of the eigenfrequencies and radial eigenfunctions. The attenuation profile of the Preliminary Reference Earth Model (PREM) (Dziewonski and Anderson, 1981) was used to calculate the Q value for each of the 1066A modes.

For the proof-of-principle simulations, the source was a vertical point force applied at the Earth's surface as a delta function in time. We estimated the impulse to be 3×10^{24} dyne-s, assuming a 20 km/s impact with the Earth of an object with a diameter of 10 km and an average density of 3 gm/cm^3. However, since the seismic simulations are linear, the degree of focusing is independent of the magnitude of the source. The Earth's toroidal modes are not excited by a vertical point force (because of the symmetry), so only spheroidal modes were included in the calculations. The synthetics represent the sum of 3,382 modes, all of 1066A's spheroidal modes with frequencies less than 0.022 Hz (or periods greater than 45 s).

We determined peak amplitudes from synthetic displacement, stress, and strain signals at several locations in and on the Earth. Figure 3 demonstrates the effect of antipodal focusing by comparing the displacement histories at six different angular distances from the source. The amplitudes of the signals decrease as a function of distance from the source. However, this trend reverses as the angular distance approaches 180°, and there is a dramatic increase in amplitude as energy radiated in all azimuthal directions from the source converges. Vertical displacements at the antipode are more than an order of magnitude larger than over most of the Earth's surface. Figure 4 shows how sharply the energy is focused at the antipode by plotting the peak strain registered on the surface as a function of distance from the source. The strain at the antipode reaches a level comparable to that only a few degrees away from the impact. The vertical displacement histories over the entire surface of the Earth are

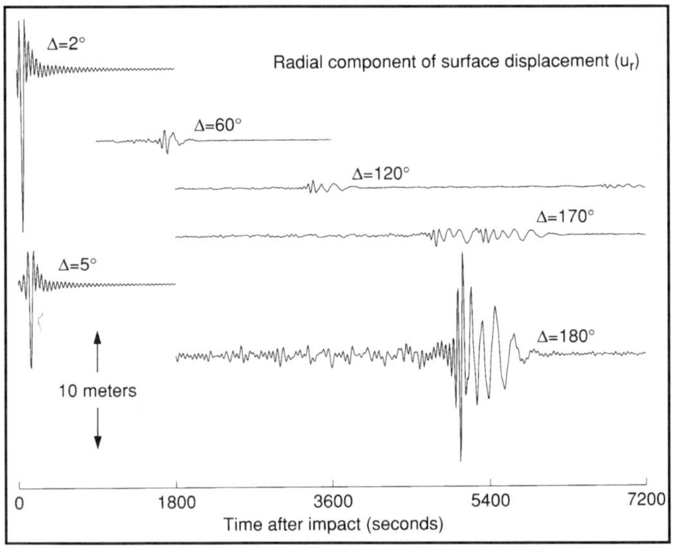

Figure 3. Radial displacement (u_r) histories on the Earth's surface at six different angular distances from the impact. The peak amplitude at the antipode approaches 10 m, similar to what it is only 2° from the source.

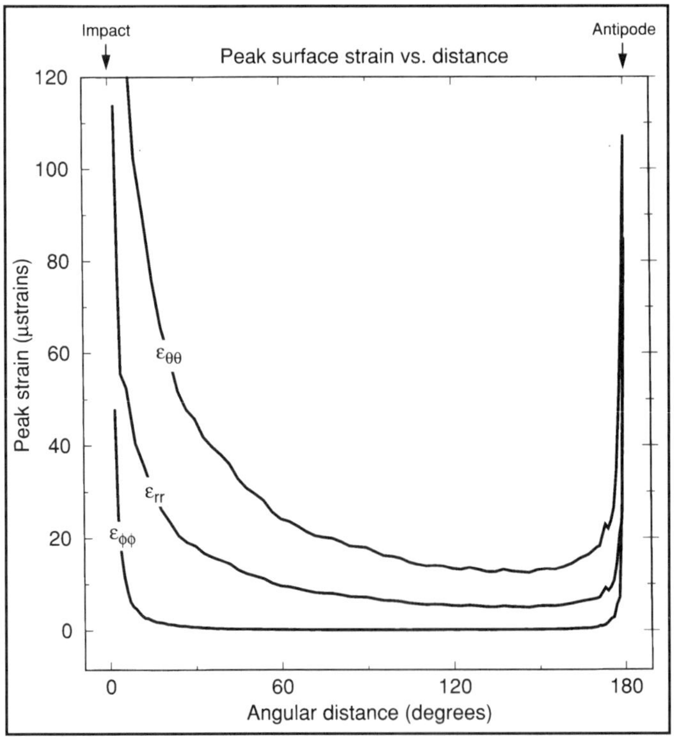

Figure 4. Peak strain amplitudes (three components: $\varepsilon_{\phi\phi}$, $\varepsilon_{\theta\theta}$, and ε_{rr}) as a function of radial distance from the impact source.

imaged in Figure 5 as a function of time and distance from the point of impact. The various seismic phases, which travel at different speeds and have different arrival times, appear as diagonal streaks on this diagram. The same information can be presented in a somewhat more intuitive manner by using it to generate a movie of the surface displacements on a globe. We have generated such a movie, in which surface waves and body wave arrivals can be seen propagating away from the impact site and focusing at the antipode. Selected frames from this movie are shown in Figure 6.

Figure 7 depicts peak displacement amplitude as a function of depth beneath the antipode from the surface to the core-mantle boundary (CMB). Figures 8 and 9 show peak strain and peak stress, respectively, as a function of depth. These plots demonstrate that, within the mantle, the largest displacements, stresses, and strains take place within the asthenosphere, where seismic energy is most strongly attenuated. These large amplitudes are due to the fundamental-mode Rayleigh surface waves. At greater depths the Rayleigh-wave contribution diminishes, and the body arrivals become relatively more prominent. The spatial periodicity in amplitude within the lower mantle may be an artifact of neglecting modes with periods shorter than 45 s. This will be investigated by adding more modes in future work.

The peak amplitudes decrease substantially with depth beneath the antipode. However, the seismic energy at any depth along the impact-antipodal axis is still represented by focused arrivals with amplitudes much larger than elsewhere in the Earth's interior at the same depth. Figure 10 shows the displacement and strain records for three different locations just above the core-mantle boundary. The upper waveforms display the motions at the CMB directly beneath the impact. The direct arrival, with a spherically spreading wavefront, is notably weaker than later focused arrivals. At an angular distance of 90° on the CMB (center waveforms), there is no focusing and the amplitudes remain small throughout the signal. The lower waveforms show the motions on the CMB beneath the impact's antipode. Comparing the 0° and 180° waveforms, one sees that the motions at the CMB are actually greater beneath the antipode than beneath the source location. Future seismic modeling will make use of an expanded set of included modes to extend the synthetics to higher frequencies. The body wave arrivals in particular will be better modeled by using a larger bandwidth. Shock physics simulations will be used to constrain the input source model.

POSSIBLE IMPLICATIONS

The strong focusing of seismic energy along an axis of the Earth between an impact and its antipode, and in particular into the asthenosphere, provides a potential mechanism for the hypothesis that impacts can trigger rapid basaltic eruptions. Moreover, this focusing would produce a thermal anomaly in the mantle to much greater depths than that due to direct shock heating from the impact. This deeper emplacement of impact energy due to seismic focusing suggests the possibility that it could be responsible for the generation of a long-lived hotspot that remains fixed to the mantle as the lithospheric plates move over it. For a sufficiently large impact, this thermal anomaly might mimic the mantle plumes currently favored to be the cause of rapid flood basalt eruptions followed by hotspot tracks.

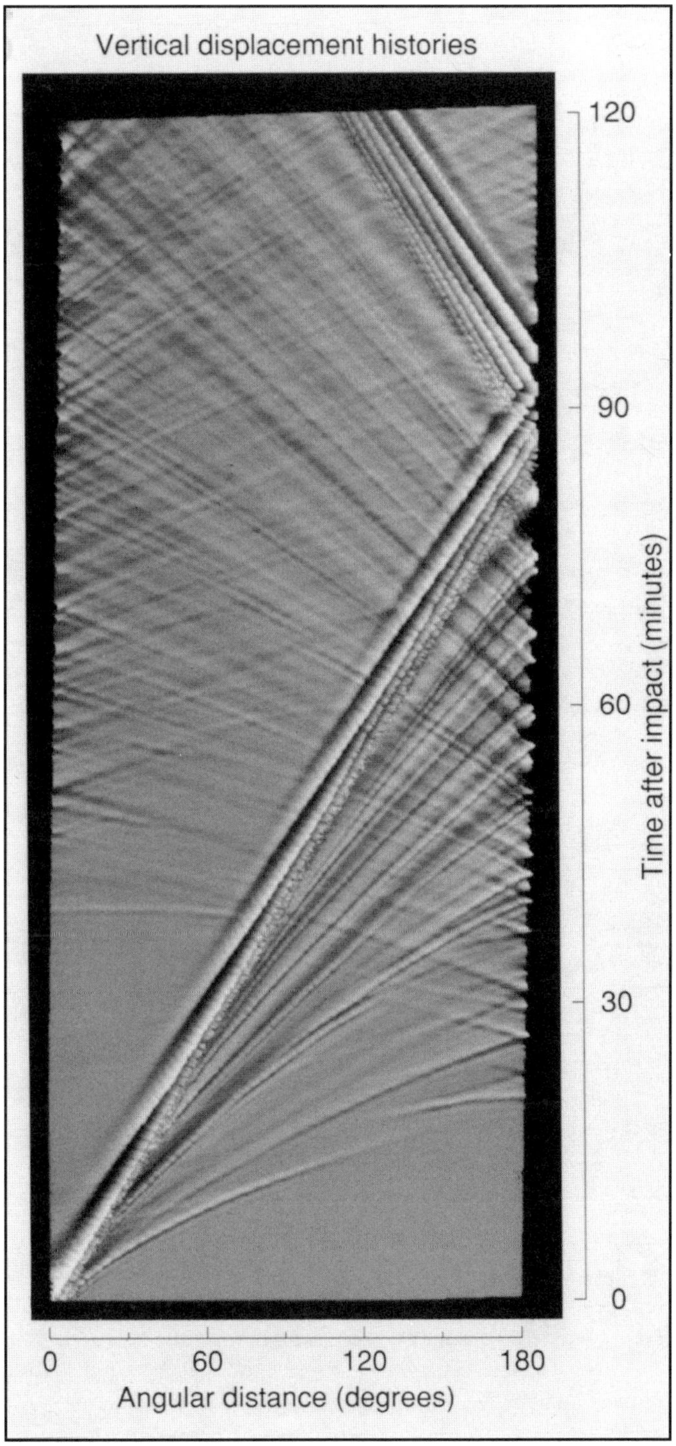

Figure 5. Image of vertical displacement (u_r) at the Earth's surface as a function of time and distance from impact.

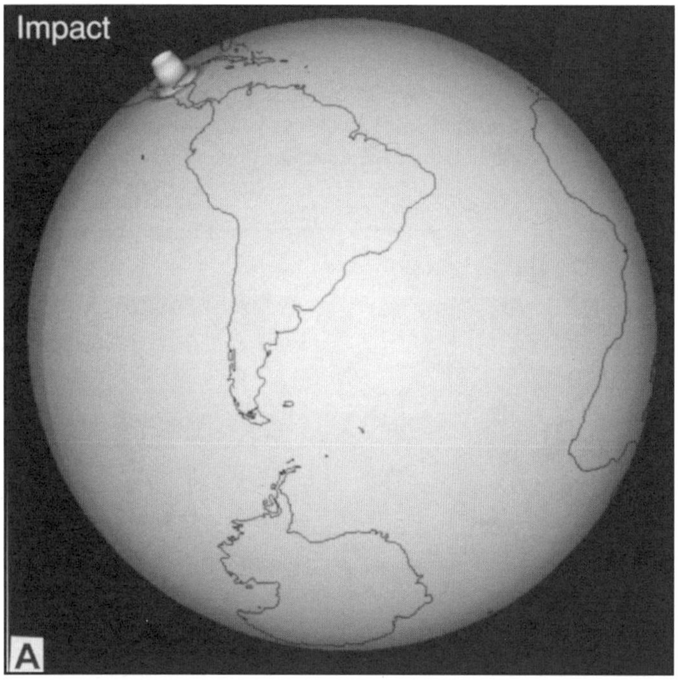

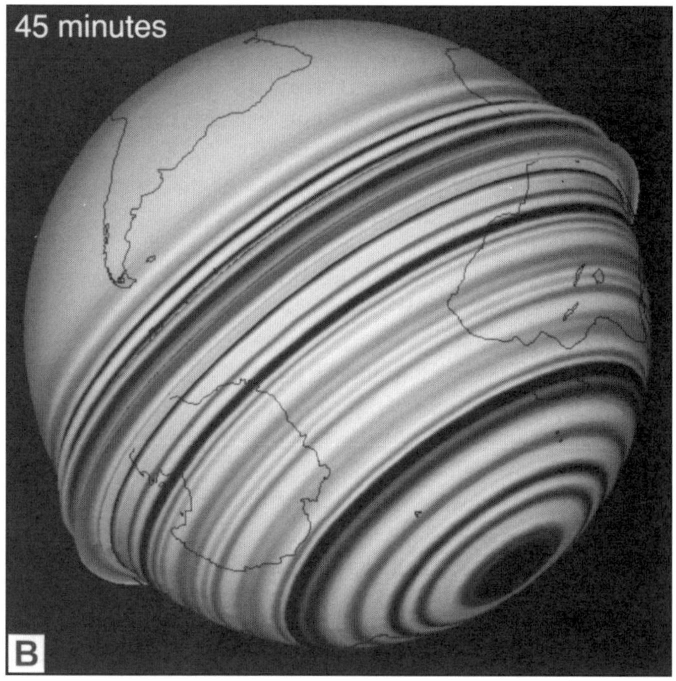

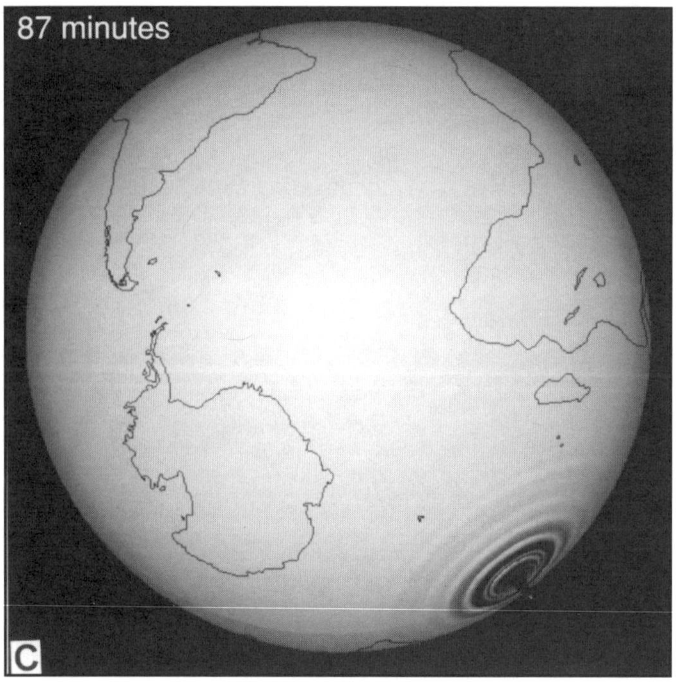

Figure 6. Selected frames from movie of seismic wave propagation away from impact site and focusing on the antipode: A, shortly after impact; B, 45 min after impact; and C, 87 min after impact. Vertical displacements (u_r) are shown as height above globe, relative to maximum displacement at the time step shown. Continents are in their present locations, not at their positions at the time of impact.

The impact-produced flood basalt hypothesis is attractive because it is potentially testable on the basis of predictions of features in the geological record that have not yet been discovered. One can propose that there is more than one type of trigger for flood basalt eruptions—some are triggered by mantle plumes and others by impact—but such a dual explanation is not scientifically satisfying. An economy of assumptions (Occam's razor) would suggest that all episodes of flood basalt volcanism have the same cause. This logically must lead to the prediction of impact structures in coaxial association (either antipodal or collocated) with all hotspots and flood basalt provinces at the time of their formation.

Unlike current plume models for flood basalts and hotspots (e.g., Richards et al., 1989), the impact hypothesis invokes only processes that are independently known to take place; specifically, large impacts and axial focusing of seismic energy. The plume models require the assumption of an instability leading to their formation, but a detailed mechanism that can be subjected to rigorous modeling has not been proposed. In addition, the impact hypothesis is potentially testable as suggested above. The plume hypothesis, by contrast, is inherently untestable for reasons listed by Anderson et al. (1992).

The impact hypothesis also provides a possible link between the K/T impact event and Deccan volcanism in India. The apparent coincidence between the timing of the largest known impact event and the largest episode of volcanic activity since the end of the Paleozoic era has been noted by many. The precise timing and duration of the eruptions and their relationship to the K/T boundary has been the subject of some debate (Vandamme et al., 1991; Féraud and Courtillot, 1994), but the weight of evidence implies that the Deccan basalts were erupting prior to the Chicxulub impact (e.g. Venkatesan et al., 1993; Bhandari et al., 1995).

The position of India relative to the Chicxulub impact structure at the time of impact precludes the possibility of a

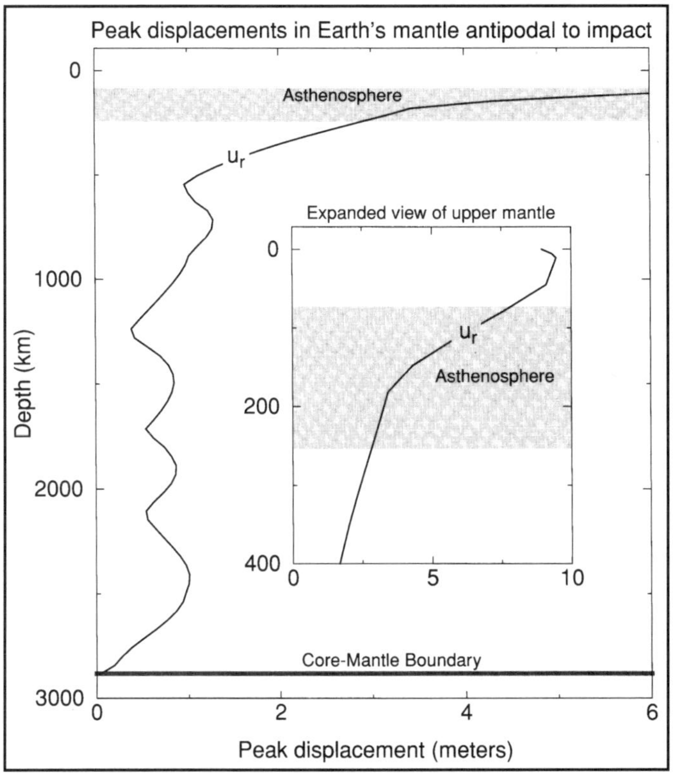

Figure 7. Peak radial displacement (u_r) as a function of depth beneath the antipode down to the core-mantle boundary (CMB).

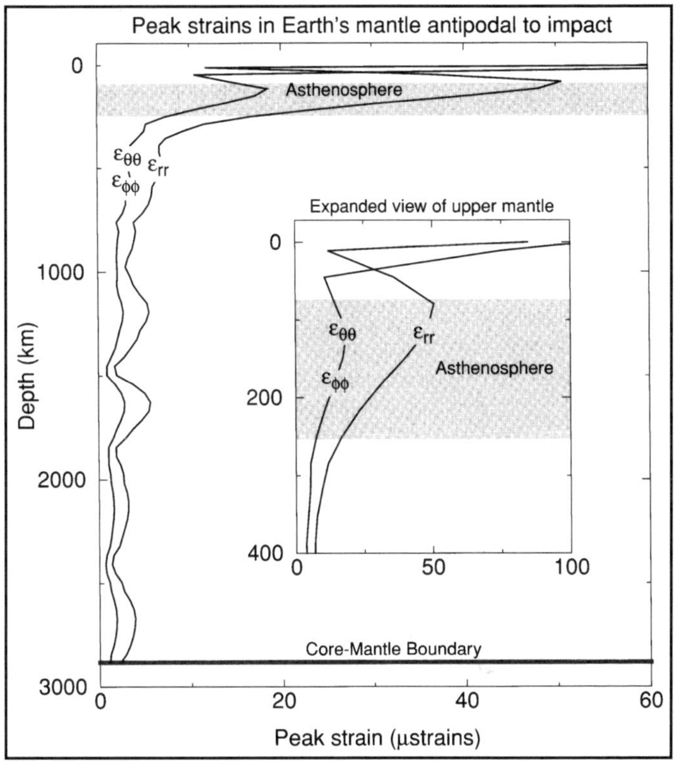

Figure 8. Peak strain (three components: $\varepsilon_{\phi\phi}$, $\varepsilon_{\theta\theta}$, and ε_{rr}) as a function of depth beneath the antipode down to the CMB.

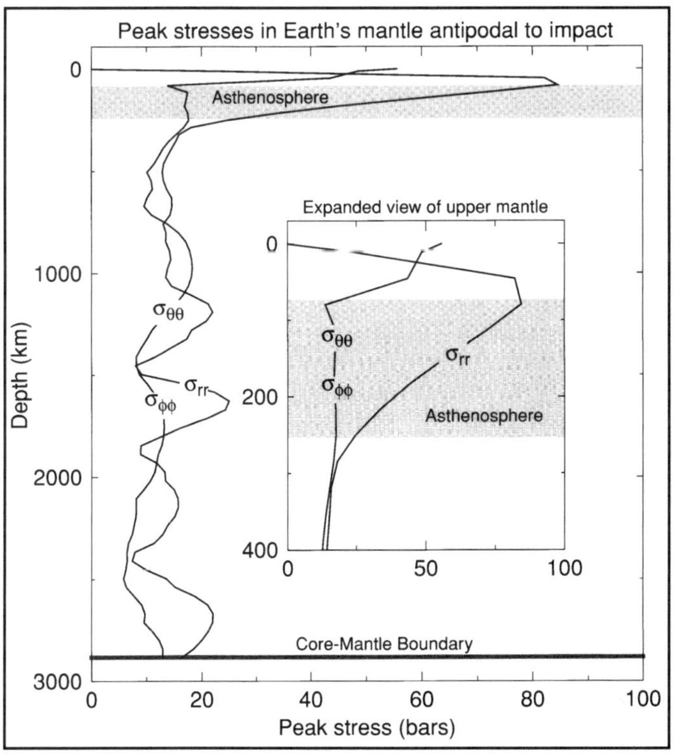

Figure 9. Peak stress (three components: $\sigma_{\phi\phi}$, $\sigma_{\theta\theta}$, and σ_{rr}) as a function of depth beneath the antipode down to the CMB.

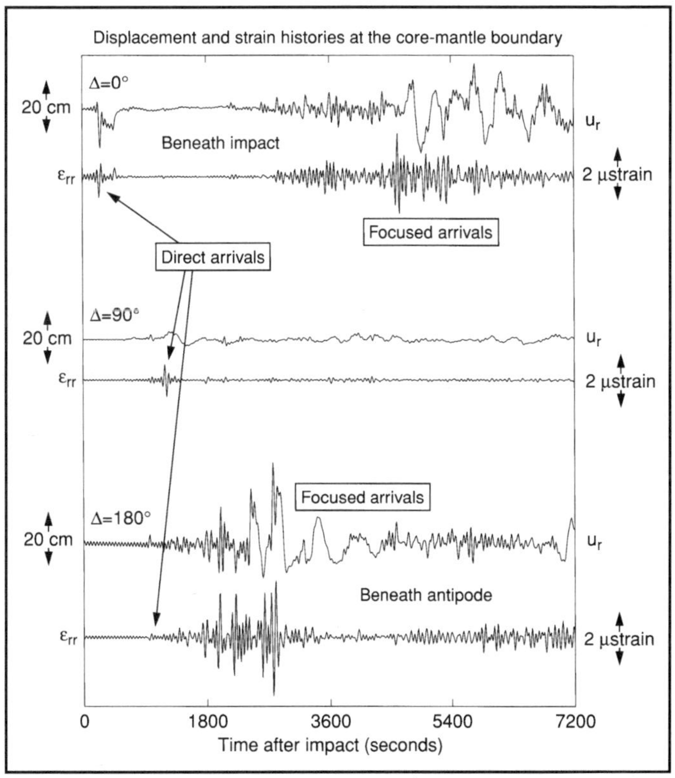

Figure 10. Radial displacement and strain histories at three locations just above the CMB: beneath the impact source ($\Delta = 0°$), equidistant between the impact and antipode ($\Delta = 90°$); and beneath the antipode ($\Delta = 180°$).

direct link; India was thousands of kilometers west of the antipode. The evidence of any basaltic eruption at the antipode of Chicxulub has been subducted and there is no associated hotspot; however, a contemporaneous impact in the Pacific Ocean is plausible. Multiple impacts are known to occur and have been proposed to explain the stratigraphy of the K/T boundary in western North America (Shoemaker et al., 1993). Until the Manson impact structure in Iowa was recently demonstrated to predate the Chicxulub impact by 8 m.y., it was considered as the best candidate for a second K/T impact structure. Some of the proposed mechanisms, such as the splitting of a comet shortly before impact with the Earth (Shoemaker et al., 1993) or impact by a binary asteroid (Grieve, 1993), would be expected to lead to nearly simultaneous impacts on the same hemisphere on Earth. A nearly simultaneous impact would, however, still fail to provide the mechanism for the Deccan eruptions if the earliest basalts indeed predate the boundary.

Another possible mechanism that has been suggested for generating closely spaced impact events (on a geologic time scale) is the perturbation of the Oort cloud by a passing massive object, leading to a comet shower in the inner solar system. Such a mechanism has been used to explain the purported stepwise character of some mass extinctions (Hut et al., 1987) and is supported by geological evidence for multiple impacts at the end of the Eocene (Montanari, 1990). In the case of comet showers, impact events could be spaced over a period of about a million years (Shoemaker et al, 1993). Such widely spaced impacts would not be constrained to the same hemisphere and would not provide the explanation for the closely associated North American boundary layers. Nevertheless, a second K/T impact site in the Pacific has been independently suggested on the basis of spinel spherule populations in K/T boundary deposits in a core from a drill site in the Pacific Ocean (Robin et al., 1993). Clustered impacts over a period on the order of a million years would increase the plausibility of hypotheses invoking a direct impact trigger for the Deccan Traps, because no spatial or temporal association with the Chicxulub impact would be required.

Because of the strong dependence of melt volume on mantle temperature, any mechanism for impact-produced volcanism would strongly depend on the impact location. An impact into, or antipodal to, thin lithosphere and above-average mantle temperatures could conceivably produce an effect where there would otherwise be none. This dependence would be consistent with the observed spatial distribution of flood basalt provinces that can be traced back to regions of the mantle that have anomalously low seismic velocities (Anderson, 1994). A possible antipodal correlation between pairs of hotspots noted by Rampino and Caldeira (1992) would provide support for the axial focusing mechanism unless demonstrated to be a statistical artifact, as suggested by Stothers (1993).

Most currently known impact structures are at least partially on exposed land or continental margins. Oceanic impact structures are difficult to locate because of the lack of detailed topographic and geologic data on the ocean floor (Grieve, 1987). A possible antipodal link between the locations of flood basalts and oceanic impact sites provides a clue for where to begin to search. If such a link were to be confirmed by the existence of antipodal impact structures, it would also provide a strong constraint on plate-tectonic reconstructions independent of paleomagnetic data or a presumed fixed hotspot reference frame. Moreover, such a link would provide a direct probe of mantle convection by providing the initial conditions and locations of hotspots, so their drift could be independently determined.

CONCLUSIONS

Seismological simulations demonstrate that displacement and strain amplitudes at the surface of the Earth near the antipode of a large impact are an order of magnitude larger than those over most of the rest of the Earth's surface and that the seismic energy remains sharply focused down to the core-mantle boundary. These simulations provide the proof-of-principle basis supporting the hypothesis that large impacts on Earth can generate effects deep in the mantle that might lead to observable consequences. For a K/T-sized impact, the peak displacement at the antipode approaches 10 m, exceeding the ground motion only a few hundred kilometers from the source. Peak strains fall off rapidly with depth beneath the antipode, but the region in the mantle that experiences the strongest focusing is the asthenosphere, where seismic energy is most easily converted to heat. Strong axial focusing also takes place directly beneath the impact site. Future work should investigate the possibility of volcanism induced at the impact site by a combination of shock heating, decompressional melting, and axial focusing. Shock physics simulations should be used to provide source functions for the seismological modeling to determine the extent and possible geological consequences of the axial focusing of seismic energy.

ACKNOWLEDGMENTS

This work was conducted at Sandia National Laboratories under the auspices of the Department of Energy under contract DE-AC04-94AL85000 under the LDRD program. We gratefully acknowledge reviews by P. H. Schultz and D. L. Anderson.

REFERENCES CITED

Ahrens, T. J., 1994, The origin of the Earth: Physics Today, August, p. 38–45.

Alt, D., Sears, J. M., and Hyndman, D. W., 1988, Terrestrial maria: The origins of large basalt plateaus, hotspot tracks and spreading ridges: Journal of Geology, v. 96, p. 647–662.

Alvarez, L. W., Alvarez, W., Asaro, F., and Michel, H. V., 1980, Extraterrestrial cause of the Cretaceous/Tertiary extinction: Science, v. 208, p. 1095–1108.

Anderson, D. L., 1994, Superplumes or supercontinents?: Geology, v. 22, p. 39–42.

Anderson, D. L., Zhang, Y., and Tanimoto, T., 1992, Plume heads, continental lithosphere, flood basalts and tomography, *in* Storey, B. C., Alabaster, T., and Pankhurst, R. J., eds., Magmatism and the causes of continental break-up: Geological Society Special Publication 68, p. 99–124.

Bhandari, N., Shukla, P. N., Ghevariya, Z. G., and Sundarum, S. M., 1995, Impact did not trigger Deccan volcanism: Evidence from Anjar K/T boundary intertrappen sediments: Geophysical Research Letters, v. 22, p. 433–436.

Boslough, M. B., Chael, E. P., Trucano, T. G., and Crawford, D. A., 1995, Axial focusing of energy from a hypervelocity impact on Earth: International Journal of Impact Engineering, v. 17, p. 99–108.

Burek, P. J., and Wänke, H., 1988, Impacts and glacio-eustasy, plate-tectonic episodes, geomagnetic reversals: A concept to facilitate detection of impact events: Physics of the Earth and Planetary Interiors, v. 50, p. 183–194.

Chael, E. P., 1983, Constraints on the Earth's anelastic and aspherical structure from antipodal surface waves [Ph.D. thesis]: Pasadena, California Institute of Technology, 139 pp.

Chatterjee, S., 1992, A kinematic model for the evolution of the Indian plate since the late Jurassic, *in* Chatterjee, S., and Hotton, N., III., New concepts in global tectonics: Lubbock, Texas Tech University Press, 450 p.

Dziewonski, A. M., and Anderson, D. L., 1981, Preliminary reference Earth model: Physics of the Earth and Planetary Interiors, v. 25, p. 297–356

Féraud, G., and Courtillot, V., 1994, Did Deccan volcanism pre-date the Cretaceous-Tertiary transition?: Comment: Earth and Planetary Science Letters, v. 122, p. 259–262.

Gilbert, F., and Dziewonski, A. M., 1975, An application of normal mode theory to the retrieval of structural parameters and source mechanisms from seismic spectra: Philosophical Transactions of the Royal Society of London, Series A, Mathematical and Physical Sciences, v. 278, p. 187–269.

Gilbert, G. K., 1893, The moon's face: A study of the origin of its features: Bulletin, Philosophical Society of Washington (District of Columbia), v. 12, p. 241–292.

Glass, B. P., and Heezen, B. C., 1967, Tektites and geomagnetic reversals: Scientific American, v. 217, p. 32–38.

Glass, B. P., Swincki, M. B., and Zwart, P. A., 1979, Australasian, Ivory Coast and North American tektite strewnfields: Size, mass and correlation with geomagnetic field reversals and other earth events, *in* Proceedings, Lunar and Planetary Science Conference, 10th: New York, Pergamon, p. 2535–2545.

Glikson, A. Y., 1995, Asteroid/comet mega-impacts may have triggered major episodes of crustal evolution: Eos (Transactions, American Geophysical Union), v. 76, p. 49–55.

Green, D. H., 1972, Archaen greenstone belts may include terrestrial equivalents of lunar maria?: Earth and Planetary Science Letters, v. 15, p. 263–270.

Grieve, R. A. F., 1987, Terrestrial impact structures: Annual Review of Earth and Planetary Sciences, v. 15, p. 245–270.

Grieve, R. A. F., 1993, When will enough be enough?: Nature, v. 363, p. 670–671.

Gutenberg, B., and Richter, C., 1934, Gerlands Beitraege zur Geophysik, v. 43, p. 56.

Hagstrum, J. T., and Turrin, B. D., 1991, Is flood basalt volcanism a seismically-induced response to large antipodal bolide impacts? [abs.]: Eos (Transactions, American Geophysical Union), v. 72, p. 516.

Hughes, H. G., App, F. N., and McGetchin, T. N., 1977, Global seismic effects of basin-forming impacts: Physics of the Earth and Planetary Interiors, v. 15, p. 251–263.

Hut, P., Alvarez, W., Elder, W. P., Hansen, T., Kauffman, E. G., Keller, G., Shoemaker, E. M., and Weissman, P. R., 1987, Comet showers as a cause of mass extinctions: Nature, v. 329, p. 118–126.

Kanamori, H., 1977, The energy release of great earthquakes: Journal of Geophysical Research, v. 82, p. 2981–2987.

Kaula, W. M., 1979, Thermal evolution of Earth and Moon growing by planetesimal impacts: Journal of Geophysical Research, v. 84, p. 999–1008.

Loper, D. E., 1991, Mantle plumes: Tectonophys., v. 187, p. 373–384.

Loper, D. E., and McCartney, K., 1990, On impacts as a cause of geomagnetic field reversals or flood basalts, *in* Sharpton, V. L., and Ward, P. D., eds., Global catastrophes in Earth history: An interdisciplinary conference on impacts, volcanism, and mass mortality: Geological Society of America Special Paper 247, p. 19–25.

McGlaun, J. M., Thompson, S. L., and Elrick, M. G., 1990, CTH: A three dimensional shock-wave physics code: International Journal of Impact Engineering, v. 10, p. 351.

McLaren, D. J., and Goodfellow, W. D., 1990, Geological and biological consequences of giant impacts: Annual Review of Earth and Planetary Sciences, v. 18, p. 123–171.

Melosh, H. J., 1989, Impact cratering: A geologic process: New York, Oxford University Press, 245 p.

Montanari, A., 1990, Geochronology of the terminal Eocene impacts; an update, *in* Sharpton, V. L., and Ward, P. D., eds., Global catastrophes in Earth history: An interdisciplinary conference on impacts, volcanism, and mass mortality: Geological Society of America Special Paper 247, p. 607–616.

Negi, J. G., Agrawal, P. K., Pandey, O. P. and Singh, A. P., 1993, A possible K-T boundary bolide impact site offshore near Bombay and triggering of rapid Deccan volcanism: Physics of the Earth and Planetary Interiors, v. 76, p. 189–197.

Oberbeck, V. R., Marshall, J. R., and Aggarwal, H., 1992, Impacts, tillites, and the breakup of Gondwanaland, Journal of Geology, v. 101, p. 1–19.

Peterson, J. E., 1978, Antipodal effects of major basin-forming impacts on Mars [abs.]: Lunar and Planetary Science, v. 9, p. 885–886.

Pohl, J., 1977, Paleomagnetische und Gesteinsmagnetische Untersuchungen an den Kermen der Forschungsbohrung Noerdingen: Geologica Bavarica, v. 75, p. 329–348.

Rampino, M. R., 1987, Impact cratering and flood basalt volcanism: Nature, v. 327, p. 468.

Rampino, M. R., and Caldeira, K., 1992, Antipodal hotspot pairs on the Earth: Geophysical Research Letters, v. 19, p. 2011–2014.

Rampino, M. R., and Caldeira, K., 1993, Major episodes of geologic change: Correlations, time structure and possible causes: Earth and Planetary Science Letters, v. 114, p. 215–227.

Rampino, M. R., and Stothers, R. B., 1988, Flood basalt volcanism during the past 250 million years: Science, v. 241, p. 663–668.

Rial, J. A., 1979 Seismic waves at the epicenter's antipode, Part II [Ph.D. thesis]: Pasadena, California Institute of Technology, 117 pp.

Richards, M. A., Duncan, R. A., and Courtillot, V. E., 1989, Flood basalts and hot-spot tracks: Plume heads and tails: Science, v. 246, p. 103–107.

Robin, E., Froget, L., Jéhanno, C., and Rocchia, R., 1993, Evidence for a K/T impact event in the Pacific Ocean: Nature, v. 363, p. 615–617.

Ronca, L. B., 1966, Meteoritic impact and volcanism: Icarus, v. 5, p. 515–520.

Ryder, B., 1994, Coincidence in time of the Imbrium basin impact and Apollo 15 KREEP volcanic flows: The case for impact-induced melting, *in* Dressler, B. O., Grieve, R. A. F., and Sharpton, V. L., eds., Large meteorite impacts and planetary evolution: Geological Society of America Special Paper 293, p. 11–17.

Schultz, P. H., 1974, Moon morphology: Austin, University of Texas Press, 626 p.

Schultz, P. H., and Gault, D. E., 1975, Seismic effects from major basin formation of the Moon and Mercury: The Moon, v. 12, p. 159–177.

Seyfert, C. K., and Sirkin, L. A., 1979, Earth history and plate tectonics: New York, Harper & Row, p. 96–97.

Shoemaker, E. M., 1962, Interpretation of lunar craters, *in* Kopal, Z., ed., Physics and astronomy of the Moon: New York, Academic, p. 283–359.

Shoemaker, E. M., Roddy, D. J., and Anderson, R. R., 1993, Research program on the Manson impact structure, Iowa [abs.]: Lunar and Planetary Science, v. 24, p. 1301–1302.

Stothers, R. B., 1993, Hotspots and sunspots: surface tracers of deep mantle convection in the Earth and sun: Earth and Planetary Science Letters, v. 116, p. 1–8.

Taylor, G. J., 1994, The scientific legacy of Apollo: Scientific American, July, p. 40–47.

Thompson, S. L., 1989, ANEOS analytic equations of state for shock physics codes input manual: Albuquerque, New Mexico, Sandia National Laboratories, SAND89-2951, p. 1–8.

Urey, H. C., 1973, Cometary collisions and geological periods: Nature, v. 242,

p. 32–33.
Vandamme, D., Courtillot, V., Besse, J., and Montigny, R., 1991, Paleomagnetism and age determinations of the Deccan Traps (India): Results of a Nagpur-Bombay traverse and review of earlier work: Reviews of Geophysics, v. 29, p. 159–190.
Venkatesan, T. R., Pande, K., and Gopalan, K., 1993, Did Deccan volcanism pre-date the Cretaceous/Tertiary transition?: Earth and Planetary Science Letters, v. 119, p. 181–189.
Watts, A. W., Greeley, R., and Melosh, H. J., 1991, The formation of terrains antipodal to major impacts: Icarus, v. 93, p. 159–168.
White, R. S., 1989, Igneous outbursts and mass extinctions: Eos (Transactions, American Geophysical Union), v. 70, p. 1480–1490.
Williams, D. A., and Greeley, R., 1994, Assessment of antipodal-impact terrains on Mars: Icarus, v. 110, p. 196–202.

Manuscript Accepted by the Society September 14, 1995

Printed in U.S.A.

Index

[Italic page numbers indicate major references]

C

D

H

M

Q

R

T

U

V

W

Typeset and printed in U.S.A. by Johnson Printing, Boulder, Colorado